Three Risky Decades: A Time for Econophysics?

Three Risky Decades: A Time for Econophysics?

Editors

Ryszard Kutner
Christophe Schinckus
H. Eugene Stanley

MDPI • Basel • Beijing • Wuhan • Barcelona • Belgrade • Manchester • Tokyo • Cluj • Tianjin

Editors

Ryszard Kutner
Faculty of Physics, University of Warsaw, Pasteur Str. 5, PL-02093 Warsaw, Poland

Christophe Schinckus
School of Business, University of the Fraser Valley, 33844 King Road, Abbotsford, BC V2S 7M8, Canada

H. Eugene Stanley
Department of Physics, Boston University, 590 Commonwealth Ave, Boston, MA 02215, USA

Editorial Office
MDPI
St. Alban-Anlage 66
4052 Basel, Switzerland

This is a reprint of articles from the Special Issue published online in the open access journal *Entropy* (ISSN 1099-4300) (available at: https://www.mdpi.com/journal/entropy/special_issues/entropy_econophysics).

For citation purposes, cite each article independently as indicated on the article page online and as indicated below:

LastName, A.A.; LastName, B.B.; LastName, C.C. Article Title. *Journal Name* **Year**, *Volume Number*, Page Range.

ISBN 978-3-0365-4741-1 (Hbk)
ISBN 978-3-0365-4742-8 (PDF)

Contents

About the Editors

Ryszard Kutner

Ryszard Kutner has been working at the Faculty of Physics University of Warsaw, for over fifty years, although he retired for ten years. He has been studying complex systems, especially econophysics, using nonlinear statistical physics methods (see the List of Publications at: https://sciprofiles.com/profile/934121). He is particularly interested in non-equilibrium phase transitions, particularly in critical phenomena in various markets (e.g., financial or company markets). He and his colleagues are Editors of three Topical/Special Issues:
(i) The Continuous Time Random Walk Still Trendy: Fifty-year History, Current State, and Outlook https://link.springer.com/journal/10051/topicalCollection/AC_94c92ef44597467b3e0fc68759b076f1
(ii) Econophysics and sociophysics in turbulent world https://www.sciencedirect.com/journal/physica-a-statistical-mechanics-and-its-applications/special-issue/10ZXGBDQBD0
(iii) Three Risky Decades: A Time for Econophysics? https://www.mdpi.com/journal/entropy/special_issues/entropy_econophysics and several conference proceedings on econophysics, e.g., 2020 vol. 138(1) and 2018 vol. 133(6)
http://przyrbwn.icm.edu.pl/APP/apphome.html
or 2005 vol. 36(8) https://www.actaphys.uj.edu.pl/

Christophe Schinckus

Christophe Schinckus is the Dean of the Faculty of Professional Studies (including the School of Business and the School of Computing) at the University of the Fraser Valley in Canada. He is also an Honorary Professor at the University of Leicester (UK). Chris has a pluridisciplinary background including an MSc in Engineering & Management, a PhD in Economics (University of Paris I – Pantheon Sorbonne) and a PhD in Philosophy of Science (University of Cambridge). Chris's research deals with Econophysics, Physics applied to Finance, FinTech & Cryptocurrencies, Energy and Economics, Social Finance, Anthropocene and Finance, Epistemology of Finance, and more generally with all aspects of epistemic post-modernism. He published more than 130 papers in peer-reviewed journals and a book on Econophysics (Oxford University Press). Chris has been listed in the Stanford University database listing the world's top 2% scientists for the year 2020 (based on standardized citation metrics across all scientists and scientific disciplines).

H. Eugene Stanley

Harry Eugene Stanley is an American physicist and University Professor at Boston University. He has made seminal contributions to statistical physics and is one of the pioneers of interdisciplinary science. His current research focuses on understanding the anomalous behavior of liquid water, but he has made fundamental contributions to complex systems, such as quantifying correlations among the constituents of the Alzheimer brain, and quantifying fluctuations in noncoding and coding DNA sequences, interbeat intervals of the healthy and diseased heart. He is one of the founding fathers of econophysics.

Preface to "Three Risky Decades: A Time for Econophysics?"

Our Special Issue we publish at a turning point, which we have not dealt with since WWII. The interconnected long-term global shocks such as the coronavirus pandemic, the war in Ukraine, and catastrophic climate change have imposed significant humanitary, socio-economic, political, and environmental restrictions on the globalization process and all aspects of economic and social life including the existence of individual people. The planet is trapped—the current situation seems to be the prelude to an apocalypse whose long-term effects we will have for decades. Therefore, it urgently requires a concept of the planet's survival to be built—only on this basis can the conditions for its development be created. The Special Issue gives evidence of the state of econophysics before the current situation. Therefore, it can provide excellent econophysics or an inter-and cross-disciplinary starting point of a rational approach to a new era. This requires the ability to study various coexisting critical phenomena and processes. It seems to us that the combination of physics and economics makes this possible.

Our current Special Issue is divided into nine topic sections (see Sec. 5. Content in Editorial). The topics of the sections show the research diversity of econophysics, even though it contains only a fragment of the topics that are of interest to econophysics (a subjective choice made by the Guest Editors).

Sections i and ix contain review/holistic articles on econophysics through the "glasses" of science of complexity in its historical context, including its present state and perspectives.

Section ii contains works devoted to time series analysis, i.e., analyzing the fundamental empirical data on which econophysics is based. Modern econophysics began its life by analyzing empirical data, such as time series.

The time series analysis has led to identifying an independent research stream correlation, memory, dependence, and relatedness. This subject is the content of section iii.

One of the youngest but already established trends in econophysics is the analysis of cryptocurrency markets, especially their similarities and differences concerning traditional financial markets. We present this trend in section iv.

Perhaps the oldest and, at the same time, the most developed part of the research areas of modern econophysics are financial markets and stock exchanges are included in this. We have included articles on this topic in section v.

Research on the company market has only recently become established in econophysics. Moreover, it is difficult to overestimate the importance of this market for economic and social life. For example, it is a base without which the existence of stock exchanges would not make sense. Section vi contains articles on this topic.

The concept of the relationship between thermodynamic formalism and economics, originating from Paul Ehrenfest, and especially the idea of entropy, has already been absorbed by economics. We have included the works devoted to this direction in section vii.

Ubiquitous financial risk is one of the thematic pillars of econophysics. This subject could not be missing from our Special Issue. We included it in section viii.

An excellent supplement to this Special Issue is our earlier Topical Issue (last update 30 June 2021) entitled: "Econophysics and Sociophysics in Turbulent World", which can be accessed via the following link: https://www.sciencedirect.com/journal/physica-a-statistical-mechanics-and-its-applications/special-issue/10ZXGBDQBD0.

We believe that we have provided extensive inspiration on the path of Special Issues, especially to the young generation.

Ryszard Kutner, Christophe Schinckus, and H. Eugene Stanley
Editors

entropy

MDPI

Editorial

Three Risky Decades: A Time for Econophysics? †

Ryszard Kutner [1,*], Christophe Schinckus [2] and Harry Eugene Stanley [3]

1 Faculty of Physics, University of Warsaw, Pasteur Str. 5, PL-02093 Warsaw, Poland
2 School of Business, University of the Fraser Valley, 33844 King Road, Abbotsford, BC V2S 7M8, Canada; chris.schinckus@ufv.ca
3 Department of Physics, Boston University, 590 Commonwealth Ave, Boston, MA 02215, USA; hes@bu.edu
* Correspondence: ryszard.kutner@fuw.edu.pl

† This Special Issue is dedicated to the founder of econophysics, Professor Harry Eugene Stanley on the occasion of his 80th birthday anniversary.

1. Motivation

The Special Issue comes out in the increasing accumulation of negative global tensions in many areas. The year 2022 seems the most unpredictable of the post-second world war years—a true sanitary/humanitary, climatic, and socio-economic thriller. Among these global challenges, two far-from-stationary (or unstable) phenomena and processes (operating at various spatio-temporal scales) need to be mentioned: (1) the pandemic shock and its economic effects [1], as well as the enormous social frustrations it generates, and (2) the climatic change that is progressing at an alarming pace, resulting in rapidly increasing migrations on a global scale. These aspects overlap, of course, with the tension between different cultures, religions, political systems, and the rivalry of superpowers. Finally, we must bear in mind the impact of local phenomena and processes (such as those caused by Brexit or the attack on the Capitol—both due to solid social polarization). To a greater or lesser extent, all of these are burdened by media information that is playing an increasing role in our society through its growing social impact. In such a context, characterized by extreme/rare and super-extreme events combined with tremendous volatility and giant fluctuations as well as the extraordinary ease of the spread of information and epidemics, our everyday life became more and more uncertain and even more turbulent—all aspects of our society are impacted by these global and local factors. All these, combined with the surprising helplessness of central banks and international financial institutions, result in the decline of the level of investment, an increase in unemployment, extraordinary involvement of states in the economy, inflation, and stagnation, and consequently, a recession.

Given the long-term correlations, multiscale/multifractality, criticality, and complexity of the challenges mentioned above, an interdisciplinary science is essential to structure, understand and eventually predict the way our societies can evolve. By combining scientific methods that is, utilizing physics to study socio-economic realities, econophysics and sociophysics offer necessary interdisciplinary approaches.

Increasing sanitary, climatic, and socio-economic uncertainty can take several forms and impact markets through various market bubbles and collapses (increasing all forms of risk). In this challenging time overwhelmed by information and data, an interdisciplinary approach such econophysics and sociophysics can be particularly useful in rationalizing and reducing the aforementioned risks. This Special Issue illustrates how the combination of scientific fields can provide fruitful conceptual frameworks to understand the current unprecedented transformation of our society.

In conjunction with a previous issue [2,3], this current Special Issue shows the multi-branch nature of econophysics and sociophysics topics and the diversity of econophysicists' and sociophysicists' interests, reflecting the diversity of the world around us. It pushes for not only a qualitative but, above all, quantitative description of reality from very different, complementary points of view. As a society, we are ready to acquire, collect, develop and

Citation: Kutner, R.; Schinckus, C.; Stanley, H.E. Three Risky Decades: A Time for Econophysics?. *Entropy* **2022**, *24*, 627. https://doi.org/10.3390/e24050627

Received: 20 April 2022
Accepted: 21 April 2022
Published: 29 April 2022

publish empirical data, cause analysis, analysis of effects, analysis of mechanisms, and statistical forecasting and proposed actions. The condition is transparency/widespread availability of unadulterated empirical data collected by various independent institutions and portals. We have presented the content of our Special Issue in Sec. V below.

This Special Issue is published under the extraordinary situation—it is a testimony of the pre-current war world, bearing witness and summarizing the era that is just passing. Now we face the challenge of understanding and describing the world to come—the world in which globalization ties with the reevaluation of pre-current war paradigms.

2. Remarks on Prehistory

Econophysics does not come from nowhere and it can be related to some early works developed by some: Louis Bachelier (LB) and especially Jan Tinbergen (JT). Although the former was an expert in mathematical physics, the latter was an active physicist.

Louis Bachelier defended his doctoral dissertation in 1900 [4] under the supervision of Henri Poincaré—his research introduced the hypothesis on the stochastic nature of financial markets. It has been just 100 years since Jan Tinbergen began studying mathematics and physics at the University of Leiden (the Netherlands) with Paul Ehrenfest who appeared on the photo below (see Figure 1).

Figure 1. Group of Paul Ehrenfest students and friends (Leiden 1924). From the left to the right: Gerhard Dieke, Samuel Goudsmit, Jan Tinbergen, Paul Ehrenfest, Ralf Kronig, and Enrico Fermi. *Public photo was taken from the Internet.*

In 1926 Jan Tinbergen graduated from university. In 1929 he defended his doctoral thesis entitled "Minimumproblemen in de natuurkunde en de economie" under the supervision of Paul Ehrenfest. This thesis is the first attempt in the intellectual history to combine natural and economic sciences through a strictly quantitative approach by using physics as theoretical reference. Jan Tinbergen's work was directly influenced by his supervisor's (Paul Ehrenfest) research interest including, among other things, the analogy between thermodynamic formalism and economic processes. Generally speaking, Tinbergen initiated the idea of using physics in economics. Jan Tinbergen was the first Nobel Prize laureate (which he received it with Radgar Frisch) in economics in 1969 and he is nowadays seen as the father of econometrics.

Bachelier and Tinbergen laid down the epistemological foundations for a more quantitative approach of the socio-economic reality. This path became gradually inspiring and generated a constant increase in interest, as illustrated below.

Figure 2 shows a histogram for annual publications in the area of science that we call econophysics today. The plot was built on publications extracted using over 70 characteristic vital names and phrases from nearly 45 journals registered with Web of Science (WofS) database.

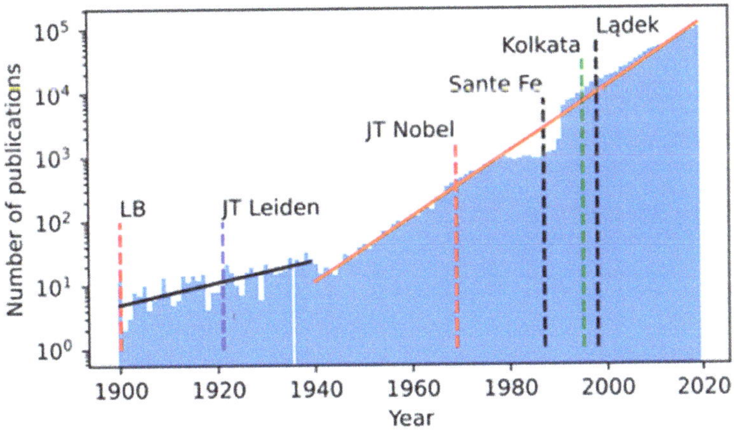

Figure 2. Depending on the time, the annual number of publications (NP) related to the use of physics methods in economics and finance. There are two different regions characterized by different values of the growth factor $1/\tau$. Namely, NP~exp (t/τ), where τ = 25.23 [Year] for the range 1900–1938, and τ = 8.58 [Year] for the range 1939–2019. We have marked the following events in the plot. "LB" (in 1900) marks the appearance of the doctoral dissertation (mentioned in the text) by Louis Bachelier, "JT Leiden" (year 1921) dates Jan Tinberger joining the University of Leiden, "JT Nobel" (year 1969) means the receipt of the Nobel Prize by Jan Tinbereger, "Santa Fe" marks the ground-breaking conference of the Santa Fe Institute (1987) mentioned in the text, "Kolkata" means the historic conference in Kolkata (India, 1995) and "Lądek" the conference in Lądek Zdrój (Poland, 1998)—both related significantly to econophysics. The last two conferences were the precursors of cyclical econophysics conferences held to this day: Econophysics Symposium (FENS in Poland since 2004) and Econophysics Colloquium (organized by Tiziana Di Matteo in various countries since 2005) as well as conferences organized in this century by Wei-Xing Zhou (East China University of Science and Technology) and Hideki Takayasu (Nikkei Institute, Sony, Tokyo, Japan). *The plot was made by Jarosław Klamut, the PhD-student of one of us (RK). The plot was published with his consent.*

The histogram begins in 1900, the year of the publication of the above-mentioned doctoral dissertation by Louis Bachelier [4]. The exponential growth of the histogram is divided into two time periods. The first period was from 1900 to 1938 and the second from the outbreak of World War II in 1939 to 2019. The growth rate of $1/\tau$ for the first period is about three times lower than for the second period. Moreover, an approximately ten-year publication "gap" in the latter half of this later period is clearly visible.

In 1987, in the very center of above-mentioned gap, a conference was held by the Santa Fe Institute, chaired by two Nobel Prize laureates: economist Kenneth Joseph Arrow—Nobel Prize in Economic Sciences (1972) together with John Hicks for their pioneering contributions to general economic equilibrium theory and welfare theory, and physicist Philip Warren Anderson—Nobel Prize in Physics (1977) together with Nevill Francis Mott and John Hasbrouck Van Vleck for their fundamental theoretical investigations of the electronic structure of magnetic and disordered systems. This pioneering conference aimed to answer the question: how economics can benefit from physics, computer science, and biology. This conference initiated a cascade of publications that continues to this day. We can formally treat the right slope of the publication gap shown in Figure 2, as the beginning of modern econophysics [5].

3. Remarks on History

There is a good reason for this Special Issue: the year that has just passed marked the third decade of a new way of dealing with economics through the lens of a physics-based approach on a large scale. Since then, there has been an increasing number of publications (included in the WofS database) devoted to what is now called econophysics. The origins of this movement are complex and manifold. A possible catalyst for this increase is the famous conference at the Santa Fe Institute in 1987, organized indeed by Kenneth Arrow and Philip Anderson. The latter was a co-founder of the Institute, which had been brought into being three years earlier. The mission of the Institute has been defined as „Searching for Order in the Complexity of Evolving Worlds"—the above-mentioned event fits perfectly into it.

The purpose of this event was to see how economics could benefit from physics, computer science, and biology. Econophysics may be related to the ground-breaking work ("Lévy walks and enhanced diffusion in Milan stock exchange") written by the physicist Rosario N. Mantegna in Physica A (1991)—this article, considered by many to be the beginning of modern econophysics, showed that we had entered in an era of extreme and rare events as we experience it almost every day. In addition to these potential origins, other important works also contribute to the development of research related to econophysics: among others, one can quote, "Statistical properties of deterministic threshold elements—the case of market price" by H. Takayasu, H. Miura, T. Hirabayashi, K. Hamada in Physica A (1992), or "The Black-Scholes option pricing problem in mathematical finance: Generalization and extensions for a large class of stochastic processes" by J-P. Bouchaud and D. Sornette in J. Phys. I France (1994). We have just cited some of these works here, realizing that this is a subjective selection that reflects our point of view. In this Special Issue, all perspectives on econophysics are welcome, even though they might generate controversial discussions or opposite viewpoints. The authors will have the opportunity to put forth their way of presenting and working with econophysics.

The new era evoked above cannot be characterized through the classical Brownian and Gaussian behavior (Wiener process) originally discovered by Louis Bachelier in his dissertation [4]; instead, the statistical characterization of our contemporary world is more in line with a Lévy flight process over multiple timescales identified by Mantegna in his article on the Milan Index mentioned above. In this context, the central limit theorem has been replaced by the Lévy–Khintchine generalized central limit theorem. These findings have been confirmed by later works—see Mantegna-Stanley in Nature (Vol. 376(6), 1995).

In a short period of time, an avalanche of publications created an apparently impossible bridge between physics and socio-economic sciences (especially financial markets). In this Special Issue, eminent scholars have been invited, all of whom have significantly contributed to econophysics. We hope their writings will illustrate and exemplify the history of econophysics, the current trends in the field, as well as its future perspectives. We voluntarily keep open the scope of this Issue leaving to the authors' decision what they consider to be the milestones of econophysics and how they see its future. We want econophysics to be presented from different points of view, even though these views might be contradictory or sources of internal scientific tensions. Our work "Econophysics and sociophysics: Their milestones & challenges" in Physica A (2019) can be used as a source of inspiration for the celebration of the development of econophysics. As Guest Editors, we believe that the Special Issue will be scientifically attractive and inspiring. The 30th anniversary is in opportunity to show econophysics as a living and developing field of science related to many other fields. This Special Issue does not aim to be a museum but instead an inspiring collection of writings opening up prospects for the future of the field.

This Special Issue is also a way to present econophysics to the general public and to scholars who are external to the field: its achievements, its challenges, and even the controversial opinions/internal tensions and sometimes contradictions that might have emerged in the field. As Guest Editors, we are keen to show that econophysics is alive and inspiring—especially in the context of the global challenges with which we are faced.

4. Conclusions

We conclude this Editorial with an illustration (shown in Figure 3) characterizing the relationship between econophysics/sociophysics and the fields related to complexity.

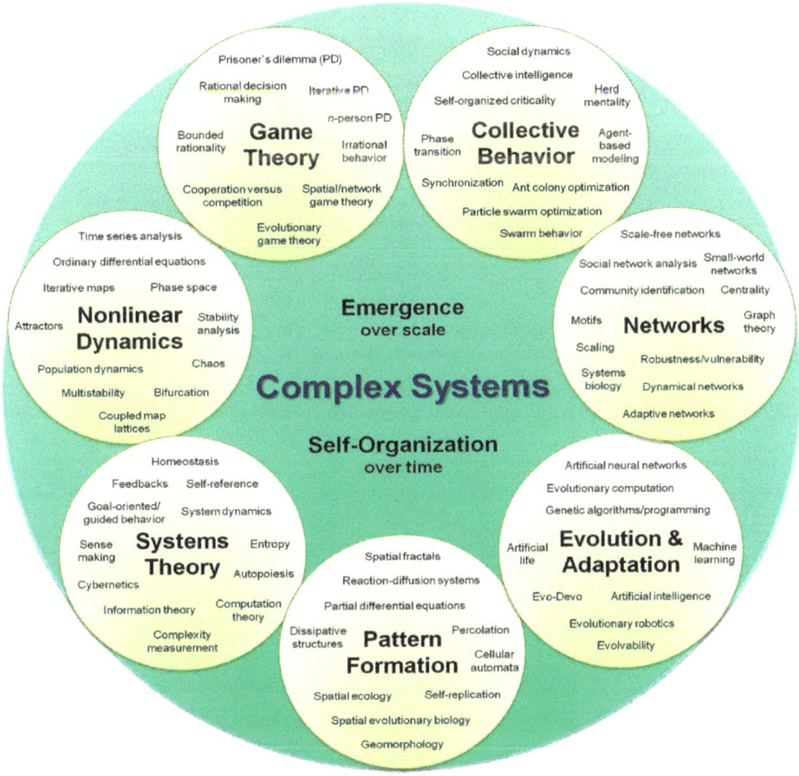

Figure 3. Schematic indication of the wealth of areas of the complex science with which econophysics and sociophysics are related. *Public drawing was taken from the Internet.*

This Special Issue provides 33 articles, we have arranged them, for convenience, in 9 sections exemplifying these epistemic interactions. This arrangement is ambiguous because many works cover several research directions.

However, it is impossible to frame the entire wealth of contemporary econophysics and sociophysics in a single Special Issue. Nevertheless, we hope that we present to the readers work containing new inspiring concepts and an overview of the crucial achievements of econophysics and sociophysics so far.

5. Content

As evoked above, econophysics has many connections with several subfields and this Special Issue aimed at capturing this intellectual richness. With this purpose, the content of this issue can be summarised as follows:

i. **Econophysics as a Complex System: History, Economic Freedom, State of the art, and Econophysics Perspectives**

Economic freedom: The Top, the Bottom, and the Reality. I. 1997–2007 by *Marcel Ausloos and Philipe Broniet*

Plotting the Words of Econophysics *by Gianfranco Tusset*

Development of Econophysics: A Biased Account and Perspective from Kolkata *by Bikas K. Chakrabarti and Antika Sinha*

Radical Complexity *by Jean-Philippe Bouchaud*

Three Decades in Econophysics—From Microscopic Modelling to Macroscopic Complexity and Back *by Alex Smolyak and Shlomo Havlin*

Valuing the Future and Discounting in Random Environments: A Review *by Jaume Masoliver, Miquel Montero, Joseph Perello, J. Doyne Farmer, and John Geanakop*

ii. Time Series Analysis

Relationship between Continuum of Hurst Exponents of Noise-like Time Series and the Cantor Set *by Maria C. Mariani, William Kubin, Peter K. Asante, Joe A. Guthrie, and Osei K. Tweneboah*

Financial Return Distributions: Past, Present, and COVID-19 *by Marcin Watorek, Jarosłw Kwapień, and Stanisław Drożdż*

iii. Correlation, Memory, Dependence and Relatedness

Continuous Time Random Walk with Correlated Waiting Times. The Crucial Role of Inter-trade Times in Volatility Clustering *by Jarosław Klamut and Tomasz Gubiec*

Understanding Changes in the Topology and Geometry of Financial Market Correlations during a Market Crash *by Peter Tsung-Wen Yen, Kelin Xia, and Siew Ann Cheong*

Understanting the Nature of the Long-Range Memory Phenomenon in Socioeconomic Systems by *Rytis Kazakevičius, Aleksejus Kononovicius, Bronislavas Kaulakys, and Vygintas Gontis*

Are Mobility and COVID-19 Related? A Dynamic Analysis for Portuguese Districts *by Antonio Casa Nova, Paulo Ferreira, Dora Almeida, Andreia Dionisio, and Derick Quintino*

Evolving Network Analysis of S&P500 Components: COVID-19 Influence of Cross-Correlation Network Structure *by Janusz Miśkiewicz and Dorota Bonarska-Kujawska*

Effects of Vaccination Efficacy on Wealth Distribution in Kinetic Epidemic Models *by Emanuele Bernardi, Lorenzo Pareschi, Giuseppe Toscani, and Mattia Zanella*

Asymmetric Relatedness from Partial Correlation *by Carlos Saenz de Pipaon Perez, Andrea Zaccaria, and Tiziana Di Matteo*

iv. Currency and Cryptocurrency Markets

Network Analysis of Cross-Correlations on Forex Market during Crises. Globalisation on Forex Market *by Janusz Miśkiewicz*

Neural Networks for Estimating Speculative Attacks Models *by David Alaminios, Fernando Aguilar-Vijande, and José Ramón Sánchez-Serraino*

What Drives Bitcoin? An Approach from Continuous Local Transfer Entropy and Deep Learning Classification Models *by Andrés Garcia-Medina and Toan Luu Duc Huynh*

Cryptocurrency Market Consolidation in 2020–2021 *by Jarosław Kwapień, Marcin Wątorek, and Stanisław Drożdż*

v. Stock Market

Analysis of Individual High-Frequency Traders' Buy–Sell Order Strategy Based on Multivariate Hawkes Process *by Hiroki Watari, Hideki Takyeasu, and Misako Takayasu*

The Stock Market Model with Delayed Information Impact from a Socioeconomic View *by Zhiting Wang, Guiyuan Shi, Mingsheng Shang, and Yuxia Zhang*

A Maximum Entropy Model of Bounded Rational Decision-Making with Prior Beliefs and Market Feedback *by Benjamin Patrick Evans and Mikhail Ptokopenko*

Heterogeneous Criticality in High Frequency Finance: A Phase Transition in Flash Crashes *by Jeremy D. Turiel and Tomasso Aste*

A New Look at Calendar Anomalies: Multifractality and Day-of-the-Week Effect *by Darko Stosic, Dusan Stosic, Irena Vodenska, H. Eugene Stanley, and Tatjana Stosic*

vi. Company Market

Learning Your Options: Option-Based Model of Export Readiness and Optimal Export *by Kirill Ilinski*

Multifractal Company Market: An Application to the Stock Market Indices *by Michał Chorowski and Ryszard Kutner*

On the Mortality of Companies *by Peter Richmond and Bertrand M. Roehner*

vii. Economics vs. Thermodynamics

Econophysics and the Entropic Foundations of Economics *by J. Barkley Rosser, Jr.*

Energy, Entropy, Constraints, and Creativity in Economic Growth and Crises *by Reiner Kümel and Dietmar Lindenberger*

viii. Financial Risk

Aspects of a Phase Transition in High-Dimensional Random Geometry *by Axel Prüser, Imre Kondor, and Andreas Engel*

Optimizing Expected Shortfall under an l_1 Constraint—An Analytic Approach *by Gábor Papp, Imre Konndor, and Fabio Caccioli*

ix. Holistic View

Victory Tax: A Holistic Income Tax System *by Donald J. Jacobs*

Highway Freight Transportation Diversity of Cities Based on Radiation Models *by Li Wang, Jun-Chao Ma, Zhi-Qiang Jiang, Wanfeng Yan, and Wei-Xing Zhou*

Funding: This research received no external funding.

Conflicts of Interest: The authors declare no conflict of interest.

References

1. Ausloos, M.; Grech, D.; Di Matteo, T.; Kutner, R.; Schinckus, C.; Stanley, H.E. Manifesto for a post-pandemic modeling. *Phys. A Stat. Mech. Appl.* **2020**, *559*, 125086. [CrossRef] [PubMed]
2. Ausloos, M.; Grech, D.; Di Matteo, T.; Kutner, R.; Schinckus, C.; Stanley, H.E. Econophysics and Sociophysics in Turbulent World, VSI, Last Update 30 June 2021. Available online: https://www.sciencedirect.com/journal/physica-a-statistical-mechanics-and-its-applications/special-issue/10ZXGBDQBD0 (accessed on 20 April 2022).
3. Kutner, R.; Ausloos, M.; Grech, D.; Di Matteo, T.; Schinckus, C.; Stanley, H.E. Econophysics and sociophysics: Their milestones & challenges. *Phys. A Stat. Mech. Appl.* **2019**, *516*, 240–253. [CrossRef]
4. Bachelier, L. Théorie de la spéculation. *Ann. Sci. l'École Norm. Supérieure* **1900**, *17*, 21–86. [CrossRef]
5. Schinckus, C. The Santa Fe Institute and econophysics: A possible genealogy? *Found. Sci.* **2021**, *26*, 925–945. [CrossRef]

entropy

MDPI

Article

Economic Freedom: The Top, the Bottom, and the Reality. I. 1997–2007

Marcel Ausloos [1,2,3,*] and Philippe Bronlet [3,4]

1 School of Business, University of Leicester, Brookfield, Leicester LE2 1RQ, UK
2 Department of Statistics and Econometrics, Bucharest University of Economic Studies, 6 Piata Romana, 1st District, 010374 Bucharest, Romania
3 Group of Researchers for Applications of Physics in Economy and Sociology (GRAPES), Sart Tilman, B-4031 Liege, Belgium; p.bronlet@gmail.com
4 Advanced Mechanical and Optical Systems (AMOS), Rue des Chasseurs Ardennais 2, B-4031 Liege, Belgium
* Correspondence: marcel.ausloos@ulg.ac.be

Abstract: We recall the historically admitted prerequisites of Economic Freedom (EF). We have examined 908 data points for the Economic Freedom of the World (EFW) index and 1884 points for the Index of Economic Freedom (IEF); the studied periods are 2000–2006 and 1997–2007, respectively, thereby following the Berlin wall collapse, and including 11 September 2001. After discussing EFW index and IEF, in order to compare the indices, one needs to study their overlap in time and space. That leaves 138 countries to be examined over a period extending from 2000 to 2006, thus 2 sets of 862 data points. The data analysis pertains to the rank-size law technique. It is examined whether the distributions obey an exponential or a power law. A correlation with the country's Gross Domestic Product (GDP), an admittedly major determinant of EF, follows, distinguishing regional aspects, i.e., defining 6 continents. Semi-log plots show that the EFW-rank relationship is exponential for countries of high rank ($\geq$20); overall the log–log plots point to a behaviour close to a power law. In contrast, for the IEF, the overall ranking has an exponential behaviour; but the log–log plots point to the existence of a transitional point between two different power laws, i.e., near rank 10. Moreover, log–log plots of the EFW index relationship to country GDP are characterised by a power law, with a rather stable exponent ($\gamma \simeq 0.674$) as a function of time. In contrast, log–log plots of the IEF relationship with the country's gross domestic product point to a downward evolutive power law as a function of time. Markedly the two studied indices provide different aspects of EF.

Keywords: Economic Freedom of the World index; Index of Economic Freedom; rank-size law technique; power law behaviour; exponential behaviour

Citation: Ausloos, M.; Bronlet, P. Economic Freedom: The Top, the Bottom, and the Reality. I. 1997–2007. *Entropy* **2022**, *24*, 38. https://doi.org/10.3390/e24010038

Academic Editors: Ryszard Kutner, H. Eugene Stanley and Christophe Schinckus

Received: 28 November 2021
Accepted: 20 December 2021
Published: 25 December 2021

1. Introduction

Numerous empirical studies [1] pretend to show that Economic Freedom (EF) favours economic growth, prosperity, poverty reduction, and has many other beneficial effects, beside being also a necessary condition for the development of democracy. However, before proposing modern theories of Economic Freedom, it seems that one should first wonder about the EF definition, and have proofs that Economic Freedom exists. The goal of this paper is to study the world EF situation before the recent (21st century) economic crisis. A second paper is intended for later years as explained below. In brief, this is due to different definitions and changes in geo-political economic conditions. It is expected that the paper can be useful for econo-physicists and other researchers, due to the somewhat original approach, more numerical, i.e., along the lines of econophysics thought.

The oldest of these publications, *The Wealth of Nations* by Adam Smith in 1776, shows that the preservation of individual freedom to pursue their own interests is due to the necessity of creating a social and more prosperous civilisation [2]. On the other hand, protectionism and trade performed under a monopoly (like that of the British empire at the time of Adam Smith) serves the purpose of preserving the status quo and privileging a handful of elites. Frederic Bastiat shows, in *Economics Harmonies* [3], that all human actions lead to care and harmony if these actions are motivated by private considerations. Thus, Bastiat recommends, or even advocates, "liberty" [4], in our own words, EF contains so much creativity that it leads to many opportunities for bettering human life.

But what is "Economic Freedom" (EF)? A simple definition among many similarly proposed by others may be as follows: The freedom of the economy is the freedom to produce, exchange and consume any goods and services acquired without use force, fraud, or theft.

In order to have a more complete appraisal of EF, one might consider James Gwartney and Robert Lawson's article [5]. Gwartney and Lawson do not give a proper term for economic freedom, but claim to provide all the conditions to be met in order to obtain "economic freedom": in brief, the foundations of any "economic freedom" is respect for the "rule of law", of property and privacy, i.e., "right to own", and demands freedom for agents wishing to enter into contracts, i.e., "freedom to contract". Thus, before, measuring EF and discussing such measures, let us briefly examine the framework in the following three subsections.

1.1. Rule of Law

Many theoreticians of economic liberalism maintain that the aim of the prerequisites for EF is the establishment of a rule of law; e.g., [6]. A "rule of law" (*"Etat de droit"*) is an institutional system in which the government and the individuals are subject to the law. This right shall apply in an identical way to each individual and to all economic agents.

This principle of equality of individuals before the law is the guarantee that the fundamental rights of citizens will not be violated by those in power. It also excludes any form of privilege, i.e., the application of the law with the purpose of favouring one group of people over another. It restricts also any arbitrary application of the law. Otherwise, one of these "misactions" would lead to a restriction of economic freedom.

1.2. The Right to Own

The second prerequisite for EF is the respect of the individual rights to own property. To achieve this, a system must be established which ensures the right to use (*usus*) and to profit ($fructus$) from this property. The system shall also ensure the right to transfer this property to another person as long as they are both consenting.

These fundamental rights are the guarantees that individuals will be able to be autonomous and will have the opportunity to seek to achieve their own goals. Many economists, such as Milton Friedman [7,8] or Murray Rothbard [9,10], consider the right of ownership as the most fundamental of the rights, of all other rights. It guarantees individuals to have individual freedom and allows for better personal development than otherwise, under a regime of coercion. It also reduces uncertainty and encourages investment by creating favourable conditions for economic development.

Empirical studies [11] show that countries with a right to own have an economic growth rate almost twice larger than countries where this right is not respected. According to (the Peruvian economist) Hernando de Soto [12], a large part of the poverty in third world countries is caused by the system's lack of favouring some equality and by the absence of a right of ownership.

1.3. Freedom to Contract

A contract is an agreement between two or more parties, having the purpose of establishing obligations at the expense of each of those parts. The freedom to contract contains therefore the right to choose the parties with which the contract is formed and to agree on the content of this contract (what to give, to do, or not to do). The parties have the right to choose the subject of the contract, but once the contract has been made, they are obliged to fulfil the terms of the contract.

The main economic function of contracts is to transfer rights of one individual's property to another person.

1.4. Other Definitions of Economic Freedom

The Gwartney and Lawson definition [5] is an ideal one, but accepted by classic liberal economists. It is intimately linked to a respect for the law which in so doing protects individuals against external aggression that would aim to take ownership of their property. This definition is valid only in a "non-negative legal context".

There are many other definitions of EF but none is unanimously accepted. Examples of "economic freedom" in a "positive law context" are given by Amartya Sen [13]; Amartya Sen argues for an understanding of freedom in terms of capacity of an individual to achieve his/her own goals. Notice that in a similar line of thought, Goodin, Rice, Parpo, and Eriksson [14] propose to measure "freedom", even outside financial or economic considerations, from the available time that people have in participating in an activity so chosen by them.

1.5. Paper Content

However, before a theory of economic freedom is proposed, should one not first have proofs of where and when economic freedom exists? In fact, these questions demand a study of other highly fundamental research questions, in particular about the measurement(s) of economic freedom(s?) themselves, and on the meaning of the measures (so called "indices"). Immediately tied to the former and the latter, the correlations with other socio-economic measures should be considered in order to provide stylised data for some preparation of modelling, later on with determinants or/and components. These are huge challenges having led to a vast literature.

Thus, even though the literature is enormous, on many aspects, we have only considered some, in our opinion, very elementary but fundamental, ground level basis, accepting two types of measures, explicitly defined in Section 2: the Economic Freedom of the World (EFW) index and the Index of Economic Freedom (IEF). We have examined 908 data points for the EFW index and 1884 points for the IEF; the studied periods cover 2000–2006 and 1997–2007, respectively, thereby following the 9 November 1989 Berlin wall collapse and including 11 September 2001. Notice that we presently exclude the 2008 financial crisis, and the following years, due to recent economic, geopolitical, changes, and because a new definition of the IEF was recently implemented. Some further work is intended over the more recent period (to be paper II.) in order to provide a complementary analysis. Paper II will also contrast the findings, whence prompting any dynamic aspect.

In order to compare the indices, one needs to study their overlap in time and space. That leaves 138 countries to be examined over a period extending from 2000 to 2006, thus 2 sets of 862 data points. Since each country presents a combination of freedoms, and restrictions to freedoms, it is of interest to observe whether the country ranking contains or hides such a variety of dimensions. Due to the aimed scope of this paper, we will only consider the most often admitted primary determinant of a country's economic growth (EG), i.e., the country's Gross Domestic Product (GDP).

Thus, our data analysis pertains to the rank-size law technique. It is going to be examined whether the measures of EF have a statistical distribution which follows either an exponential or a power law. This is a sort of research question not considered in the classical realms of economics, but should be of interest in econophysics. A correlation with the country's gross domestic product (GDP) follows, distinguishing regional aspects, i.e., defining 6 continents.

The table of contents of this paper may be as follows:

In Section 2, we recall the definition and content of the Economic Freedom of the World (EFW) Index and the Index of Economic Freedom (IEF), respectively.

In Section 3, we present the extracted data, i.e., 908 data points for the EFW index and 1883 for the IEF on the studied periods, 2000–2006 and 1997–2007, respectively.

In Section 4, we provide the empirical laws, on one hand, the rank-size laws for both indices, plus, on the other hand, the (regression) relationship between such indices and the gross domestic product of the countries of interest. We also provide a study of regional aspects through a grouping of countries according to their geographic positions.

In Section 5, we provide conclusions pointing to the weak evolution of indices over the considered time interval. We suggest lines for further research.

2. Economic Freedom Indices

We position our paper within the scholarly contributions having investigated, on one hand "measures of Economic Freedom" in modern times, and the link between EF and EG. Our article explores this possibility by means of a regional analysis, which we conduct on two indicators. Let us summarise the literature from such points of view.

2.1. *Economic Measures*

2.1.1. Economic Freedom of the World (EFW) Index

The Economic Freedom of the World (EFW) Index, published by the Fraser Institute [15], is the result of a project spanning 20 years. It was developed after a set of conferences given by Milton Friedman and Michael Walker between 1986 and 1994, in a project gathering more than 60 of the greatest economists of the time [16]. The aim was to create a ("strong") base with quantifiable and objective data following a transparent procedure. Thus, anyone could use the index, whatever their goals and political ideals.

The EFW index measures the degree of economic freedom in 5 major "areas":

- The size of the government, i.e., public expenditure, taxes, influence on the economy
- The legal structure which guarantees the right to own
- The access to a healthy currency
- Freedom in international trade
- Regulation of costs, work and economy

For each of these 5 domains, several variables are measured, resulting in a set of 21 components included in the index. Each component is placed on a scale going from 0 to 10. The value 0 refers to zero freedom while the value 10 represents total freedom. Once these components are quantified, they are averaged in order to obtain the index value.

Several methods have been studied for doing such an average: without being exhaustive, one considers the weight equivalent to each component; another gives an inversely proportional weight to the standard error of the distribution of the component in the various studied countries. A third method calls upon a panel of economists who estimate the weight that each component must have; the final weight being the average weight obtained from the panel members' appraisals. A fourth method uses the primary component analysis technique to determine each weight. This latter method has the advantage of reducing the importance of anomalies (outliers) in estimating correlations between the components.

Since none of these methods is really satisfactory (from our investigations, the index does not seem to be very sensitive to changes in weight), the weight choice is not further discussed, and taken as the most simple one. Thus, an equal weight for each component is chosen in the forthcoming analysis here below. The index, so constructed, provides a value between 0 and 10 for each country. A country with an index value close to 10 is a country where "economic freedom" is "very large". A country with a value close to 0 is a country where EF is "non-existent".

Of course, it is expected that each country presents a "combination of freedoms". Recently, Lawson et al. [17] have reviewed the determinants of EF, with a time dependent point of view. Some of the most consistent findings are that current levels of EF are strongly correlated with past levels. Lawson et al. deduce that freer countries have more difficulty continuing to improve their economic freedom.

2.1.2. Index of Economic Freedom (IEF)

Another measure of economic freedom, published by the Heritage Foundation [18] and the Wall Street Journal [19], is the Index of Economic Freedom (IEF), which was initiated in 1995 [20].

The index was built on a set of 10 specific components [21]:

- Tax freedom: measures the importance of fiscal fees imposed by the government on the income of individuals and businesses.
- Government spending: measures the total government spending.
- Free trade: it measures the absence of commercial barriers, affecting the import and export of goods or services.
- Investment freedom: measures the freedom of capital flows.
- Financial freedom: measurement of the independence from the government of credit and banking systems.
- Property rights: they are measures of the ease with which individuals acquire a property of their own.
- Corruption: measures the importance of corruption in the economic world.
- Business undertaking freedom: measures the ease with which it is possible to create, develop and close a business.
- Monetary freedom: measures price stability in relation to a price control.
- Labor Code (This item has been added in 2007. Moreover, in 2017, the Heritage Foundation made some methodological changes; the IEF has 12 components nowadays. The new components are "Judicial Effectiveness" belonging to the Rule of Law pillar and "Fiscal Health" as the new factor of the Government size pillar.): it measures the ease with which workers and companies interact without restriction from the state government.

Some of these components are the results of an assembly of additional measures. Each of these components is measured on a scale of 0 to 100. The value 100 represents the maximum freedom. The index was obtained in averaging these 10 components (with an equal weight for each of them).

Notice that more recently, Dialga and Vallée [22] dealt with "methodological issues in the Index of Economic Freedom", indicating that two components, "1. Tax Freedom" and "Government Spending", which define the "2. Government Size" pillar, are negatively correlated to the other "pillars", whence making the index very unstable and thus impairing the country ranking.

2.2. Economic Growth

Most empirical studies, e.g., [23–27] provide evidence that economic freedom, as measured by the Economic Freedom of the World Index, is related to economic growth, income, standard of living, low corruption, etc. Much evidence shows that economic freedom leads to economic growth even where countries have limited political freedom [28–30]. The reverse is not true. The case of IEF is less studied [31]. In most cases, the question turns upon the level of importance of the various independent variables.

One of the first papers that explored the relationship between EF and growth was by Islam [32]. The first study concerning the analysis of the link between different components of EF and economic growth seems due to Ayal and Karras [25]. However identifying which aspects of EF are more conducive to growth has proven difficult, due to multicollinearity among the index areas [33]. Due to the more basic aim of our paper, we will not discuss any further regression models nor (Granger) causality in the freedom–growth relationship, here, whence reducing to a somewhat limited literature review. Nevertheless, for some completeness, let us point out a few papers, either considering EF–EG from the EFW [34,35] or the IEF [31] point of view.

2.3. Criticism/Limitations

These types of indices are often criticised for their methodology. Some "economists" criticise the economic basis on which such indices are based. They consider the measures to be too restrictive and demand that they should include a broader range of freedom concepts. Others, such as John Miller [36], argue that the relationship found for example between a high life level and such indices is the biased result of choices made in the construction of some index. Others, like Heckelman and Stroup [37], criticise the method used in order to average components, which they consider to be arbitrary. See also the previous mention of Dialga and Vallée's recent finding [22].

3. Data

In order to study the spread of EF around the world, its evolution during this past decade, and subsequently its impact on the richness of the world, it is necessary to obtain the values of the EFW index and of the IEF together with the gross domestic product for the studied countries

The EFW index values, obtained from the portal www.freetheworld.com, accessed on 30 October 2006 [38], are provided for 140 countries in the 2000–2006 period, i.e., over 7 years. The values of the IEF can be found on the site of the "Heritage Foundation" [39]. The indices are given for 157 countries in the (12 years) period 1997–2007. The values of the Gross Domestic Product per capita (GDP) of countries for corresponding periods may be downloaded from the IMF website [40]. All values are annual data.

We point out that it was unfortunately necessary to exclude certain countries for which the data was unavailable for various reasons. This is, for example, the case of Iraq. Iraq's second war has made the measurement of economic indicators quite dubious: the values obtained for the IEF and EFW indices or for GDP could not be considered to be significant. That being said, there are still 908 data points for the EFW index and 1784 for the IEF for the studied periods.

3.1. Statistical Characteristics of Indices Distribution

The first step in the study of the indices concerns the distribution of their values. The histograms and cumulated probability densities of the EFW and of the IEF are reproduced in Figures 1 and 2, respectively. The main statistical characteristics (mean, standard deviation, variance, coefficient of variation, skewness and kurtosis) of these distributions are included in Table 1.

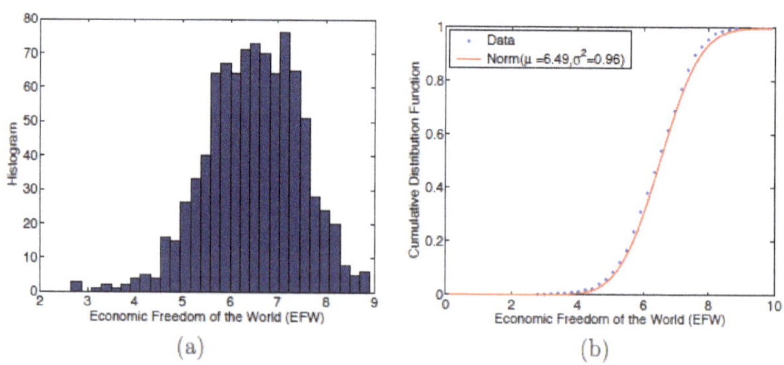

Figure 1. (**a**) Economic Freedom of the World (EFW) histogram for 908 data points, i.e., when available for all (140) countries and for all (7) years; (**b**) cumulative probability density for the EFW and normal distribution fit with mean $\mu = 6.49$ and variance $\sigma^2 = 0.96$.

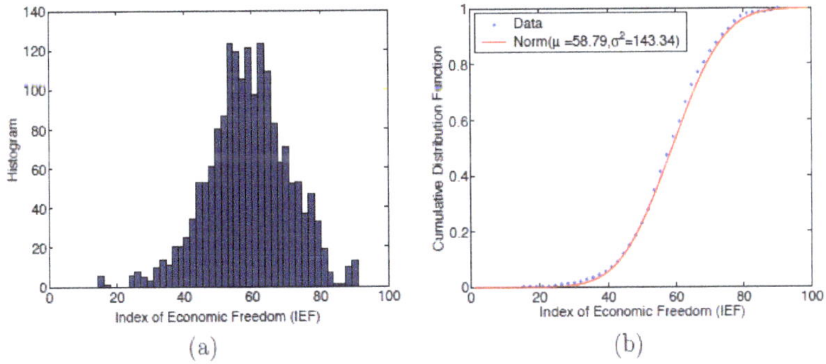

Figure 2. (**a**) Index of Economic Freedom (IEF) histogram for 1784 data points; (**b**) cumulative probability density for the IEF and normal distribution fit with mean μ = 58.79 and variance σ^2 = 143.34.

Figures 1 and 2 suggest that both indices follow a normal law slightly displaced to the right, i.e., to values greater than the median values, whence the negative skewness. This impression is reinforced by the average values of the indices: 6.49 for the EFW index and 58.79 for IEF, see Table 1. These two averages are greater than the corresponding median values: 5 in the case of EFW and 50 in the IEF. The skewness is negative for both indices: −0.3567 for the EFW index and −0.2373 for the IEF, see Table 1, confirming that the probability densities are no longer important for values above the median. These features show that the economies of the studied countries are generally more free than constrained.

Table 1. Summary of (rounded) main statistical characteristics of the economic freedom indicators distributions, i.e., the Economic Freedom of the World (EFW) index and Index of Economic Freedom (IEF), according to the examined time interval ΔT for the number N of data points.

	ΔT (yrs)	N	Mean (μ)	St.Dev. (σ)	Var. (σ^2)	CoV. (σ/μ)	Skewn.	Kurt.
EFW	7	908	6.49	0.98	0.96	0.151	−0.3567	3.3670
IEF	12	1784	58.79	11.97	143.34	0.2036	−0.2373	3.5416

In order to confirm that the distributions follow a normal law, a Kolmogorov–Smirnov (KS) test is performed. The results of the tests are shown in Table 2. The KS distances, DKS = 0.0399 for EFW and 0.0310 for IEF, are lower than the "critical values" of the normal distribution, 0.0449 for EFW and 0.0321 for the IEF. In addition, $p-$values, 0.1088 for the EFW index and 0.0633 for the IEF, are above the 5% significance level; thus the KS tests are considered to lead to statistically significant features. It is therefore possible to conclude that the EFW index and IEF values follow a normal law with μ = 6.49 and 58.79 and variance σ^2 = 0.96 and 143.34 respectively, i.e., the standard distribution (SD) is equal to 0.98 and 11.98, respectively.

3.2. EFW Index in Year 2006

For example, consider a specific year, 2006. Table 3 shows the EFW index values for the 20 freest countries for the year 2006. Hong Kong, Singapore, and New Zealand occupy the first 3 places. The rest of the top 20 is made up of the great Anglo-Saxon countries (USA, Canada, Australia) and European countries (Switzerland, United Kingdom, Ireland, Estonia, Iceland, Denmark, Finland, Austria, Netherlands, Germany, Slovakia). It should be noted that there is one South American country, Chile (in 6th position) and one country from the Arabian Peninsula, Kuwait (in 19-th position).

Table 2. Kolmogorov–Smirnov (KS) test for the adjustment of data from EFW and IEF to a normal distribution. The distances of KS (DKS) and the *p*-values indicate that KS tests are statistically significant. It is therefore allowed to conclude that the EFW and IEF values follow a normal law, with μ= 6.49 and 58.79 and variance σ^2 = 0.96 and 143.34, respectively.

Kolmogorov–Smirnov (KS) Test	EFW	IEF
p-value	0.1088	0.0633
Gaussian Distribution Critical Value	0.0449	0.0321
Significance Level	0.05	0.05
Number of data points	908	1784
DKS	0.0399	0.0310

In constrast, Table 4 shows the EFW index for the 21 least free countries in 2006. It is remarkable that the least free countries are mainly grouped in Africa: 16 out of the 21 last countries.

Table 3. 2006 Economic Freedom of the World (EFW) Index values for the 20 freest countries.

		2006 EFW Ranking			
Rank	Country	2006	Rank	Country	2006
1	Hong-Kong	8.94	11	Estonia	7.89
2	Singapore	8.57	12	Iceland	7.8
3	New Zealand	8.28	13	Denmark	7.78
4	Switzerland	8.20	14	Finland	7.69
5	United Kingdom	8.07	15	Austria	7.66
6	Chile	8.06	16	Netherlands	7.65
7	Canada	8.05	17	Germany	7.64
8	Australia	8.04	18	Taiwan	7.63
8	United States	8.04	19	Kuwait	7.62
10	Ireland	7.92	20	Slovak Rep.	7.61

Table 4. 2006 Economic Freedom of the World (EFW) Index values for the 21 least free countries. Unlike the 20 freest countries on the planet, the 21 least free countries are almost all in Africa (16 of the 21).

		2006 EFW Ranking			
Rank	Country	2006	Rank	Country	2006
121	Ethiopia	5.64	131	Burundi	5.23
121	Ukraine	5.64	131	Rwanda	5.23
123	Burkina Faso	5.63	133	Chad	5.12
124	Algeria	5.57	134	Central Africa Rep.	5.01
125	Syria	5.54	134	Guinea-Bissau	5.01
126	Malawi	5.42	136	Venezuela	4.76
127	Gabon	5.37	137	Niger	4.67
128	Nepal	5.35	138	Congo, Rep. of	4.64
129	Togo	5.33	139	Myanmar	4.19
130	Congo, Dem. Rep.	5.25	140	Angola	4.10
			141	Zimbabwe	2.67

3.3. *IEF in Year 2006*

Similarly, Tables 5 and 6 list the IEF values for the 20 freest countries and the 20 least free countries, respectively. The former British colonies still dominate the ranking. Hong Kong and Singapore occupy the top 2 places in the ranking. The big Anglo-Saxon (United States, United Kingdom, Australia, and Canada) countries are also in the top 20. Among all the regions of the world, Europe has the largest number of countries in the top 20 (9 of the 20 countries are European).

As in the case of the EFW index, a large majority of the "less free" countries are in Africa (10 out of 20 countries). The (last) Communist Countries (North Korea and Cuba) are appearing in the 2 last places of the ranking.

Table 5. 2006 Index of Economic Freedom (IEF) values for the 20 freest countries.

		2006 IEF Ranking			
Rank	**Country**	**2006**	**Rank**	**Country**	**2006**
1	Hong-Kong	88.6	11	Iceland	75.8
2	Singapore	88.0	12	Denmark	75.4
3	Ireland	82.2	12	Netherlands, The	75.4
4	New Zealand	82.0	14	Luxembourg	75.3
5	United States	81.2	15	Estonia	74.9
6	United Kingdom	80.4	16	Japan	73.3
7	Australia	79.9	17	Finland	72.9
8	Switzerland	78.9	18	Bahamas, The	72.3
9	Chile	78.0	19	Barbados	71.9
10	Canada	77.4	20	Cyprus	71.8

Table 6. 2006 Index of Economic Freedom (IEF) values for the 20 least free countries.

		2006 IEF Ranking			
Rank	**Country**	**2006**	**Rank**	**Country**	**2006**
138	Chad	50.0	148	Iran	45.0
139	Haiti	49.2	149	Venezuela	44.6
140	Nigeria	48.7	150	Turkmenistan	43.8
140	Burundi	48.7	150	Congo. Rep. of	43.8
140	Uzbekistan	48.7	152	Angola	43.5
143	Laos	47.5	153	Burma	40.0
143	Belarus	47.5	154	Zimbabwe	33.5
145	Togo	47.3	155	Libya	33.2
146	Guinea-Bissau	46.5	156	Cuba	29.3
147	Sierra Leone	45.2	157	Korea. North	4.0

3.4. Regional Evolution of Economic Freedom

In order to study the geographical distribution of economic freedom, it is possible to calculate an "average freedom value" for the six major continents (Africa, Asia, Europe, North America, Oceania, and South America). The distribution of countries by continent is carried out by following the geographical scheme of the United Nations Statistics Division [41]. This partition has been chosen because it has been developed with the aim of conducting statistical studies relevant to the various regions. However, the calculation of such an average selected is not a simple arithmetic mean. It does not make sense to give a similar weight to the United States and e.g., to Ecuador, to China, or to Vietnam. Instead, we consider that the weight should depend on the country's contribution to the world economy, for example through the GDP. Thereafter, the weight is given by

$$w_i = \frac{GDP_i}{\sum_{j=1}^{N} GDP_j} \tag{1}$$

where w_i represents the weight of the country i and GDP_j, the internal product country j.

The evolutions of the EF for the 6 continents, obtained by this method are reproduced in Figure 3 for the EFW index, and in Figure 4 for the IEF.

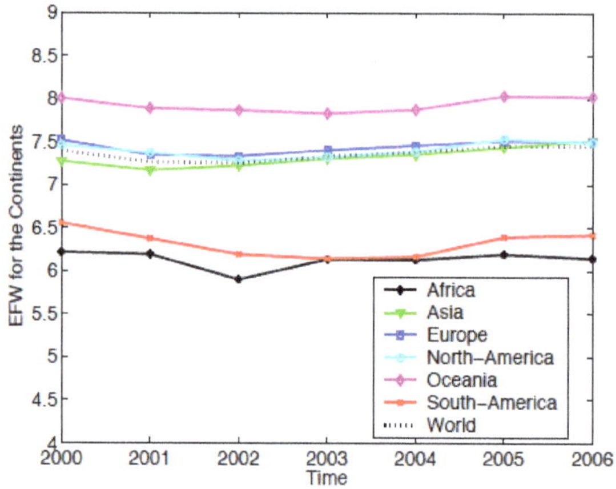

Figure 3. Yearly evolution of the Economic Freedom of the World (EFW) Index for the six continents (Africa, Asia, Europe, North America, Oceania, and South America). The index calculation for a region results from a weighted averaging of the indices of the countries belonging to the specific region. The weight of a country is the ratio of the GDP of the country to the GDP of the world economy.

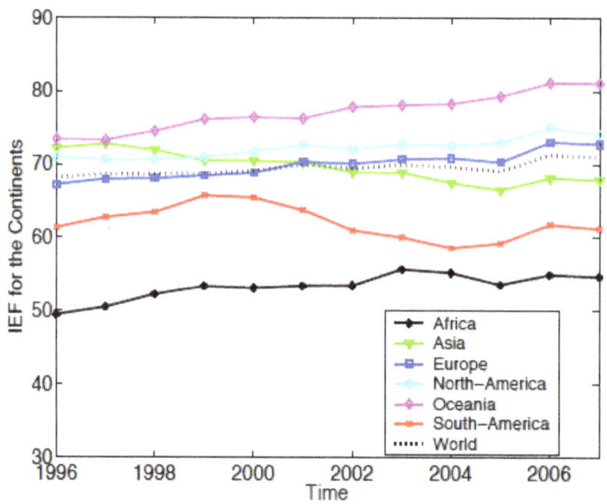

Figure 4. Yearly evolution of the Index of Economic Freedom (IEF) for the six continents (Africa, Asia, Europe, North America, Oceania, and South America). The index calculation for a region results from a weighted averaging of the indices of the countries belonging to the specific region. The weight of a country is in the ratio of the GDP of the country to the GDP of the world economy.

For the EFW index, Figure 3 shows that Oceania is the the freest of the six regions, with an index value $\simeq 8$, relatively stable of the 7 years. Europe, North America, and Asia are *ex aequo* with a value $\simeq 7.5$, which represents the world average value. Africa is the less free region and South America does not fare much better.

For the IEF, Figure 4 also shows that Oceania is the freest region with an ever increasing value. It goes from 73.36 in 1996 to 81 in 2007. Europe and North America follow the same evolution and have almost identical values. Asia regresses in terms of "economic freedom", even though there is a slight improvement in the last two years. It goes from 72.2 to 67.9 with a minimum value equal to 66.4 in 2005. Africa is again the least free region of the world, but progresses over the 12 years period. Overall, the world average freedom is rising from 68 in 1996 to 71 in 2007.

The "rate changes" appear to be different from one index to the other; this is due to the periods of study. Indeed, if the study period is restricted to 2000–2006 for the IEF, the results so obtained for both indices are almost identical. The slight differences are explained by the fact that the IEF is "more conservative" than the EFW; the IEF leads to values lower than EFW for a country. This topic is discussed further in Section 3.6.

3.5. Exponential Versus Power Law Behaviour

In this section, countries are ranked according to the value of the indices in a conventional order: a low ranking indicates that the country belongs to the group of the freest countries in the world. Conversely, a "high" rank means that the country has an index value, whence a low EF as compared to others.

The goal here is to determine, the so called "rank-size" law, once the countries are ranked, in particular whether the indices follow an exponential or a power law (These are the two most simple analytical functions carried over from statistical physics to econophysics; whence their mathematical origin is well known and not further discussed.), i.e.,

$$INDEX \sim e^{\lambda r} \tag{2}$$

or

$$INDEX \sim r^{\nu} \tag{3}$$

where r is the rank of the country; λ and ν are characteristic exponents. The latter equation corresponds to the (so called Zipf) rank-size law [42], if $\nu = -1$.

Figure 5a,c,e shows that the EFW has an exponential behaviour for countries with a rank higher than 20. The value of the exponent decreases a little bit more each year and ends up to stabilise at $\simeq -0.0049$ in 2005 and 2006 (see Table 7). The low error bars (less than 0.0001) and the high value regression coefficient (the regression coefficient is greater than 93%) confirm that the data perfectly follow the exponential law.

Figure 5b,d,f shows the power-law behaviour of the EFW. Table 8 reports the values of the exponent of the power law for the 6 studied years. It does not vary much between 2001 and 2004; it falls to -0.0743 in 2005 and -0.007 in 2006. Here again, the effectiveness of the regressions is high, between $\sim$89% and 93%. This indicates that the data follows a power law.

For the IEF, the semi-log graphs, see Figure 6a,c,e, indicates an exponential behavior according to the rank of countries. The exponent decreases every year, going down from -0.006 in 1996 to -0.0036 in 2007 (see Table 9). The regression coefficient shows that the exponential law has been "perfectly" followed since 2003, a year for which the efficiency of the regression exceeds 90%.

Unlike the EFW index, for which the data follow a power law for all ranks, Figure 6b,d,f shows a transition point between 2 different power law for the IEF, near rank 10. The exponent of the law for countries with a rank below 10 "increases" over the years, from -0.0931 in 1996 to -0.0518 in 2007. The exponent for countries with rank higher than 10 remains relatively stable $\simeq -0.016$ over the 12 years here studied (see Table 10).

It should be noted that countries with low EF (those which have a very high rank) follow neither a power law nor an exponential law; this feature holds for both indices. The difficulty of performing economic measures for these countries can explain that the index values are fraught with errors that are not possible to compensate. These countries are often those with a minimally developed economy, weakly connected to their outside world, apparently subject to the will of a dictator.

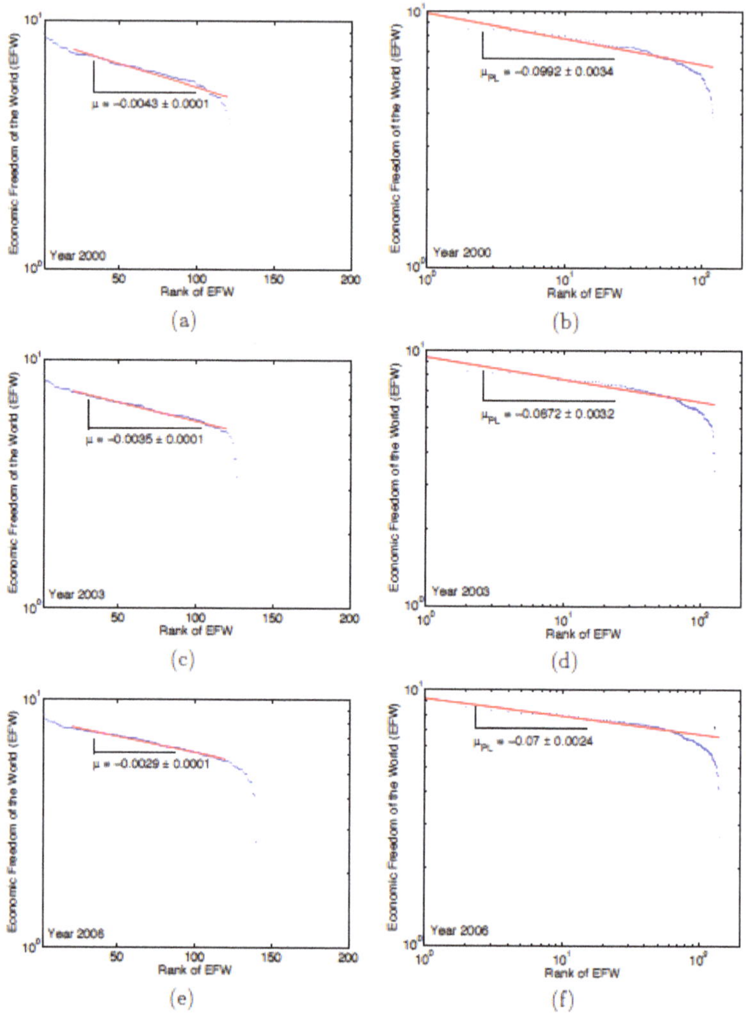

Figure 5. Examples of semi-log [(**a**,**c**,**e**)] and log–log [(**b**,**d**,**f**)] plots of the rank-size relation between the Economic Freedom of the World (EFW) index and the country rank for the years 2000, 2003, and 2006, respectively: the semi-log plots show that the relationship is exponential for countries of high rank (≥20); the log–log plots point to a behaviour close to a power law.

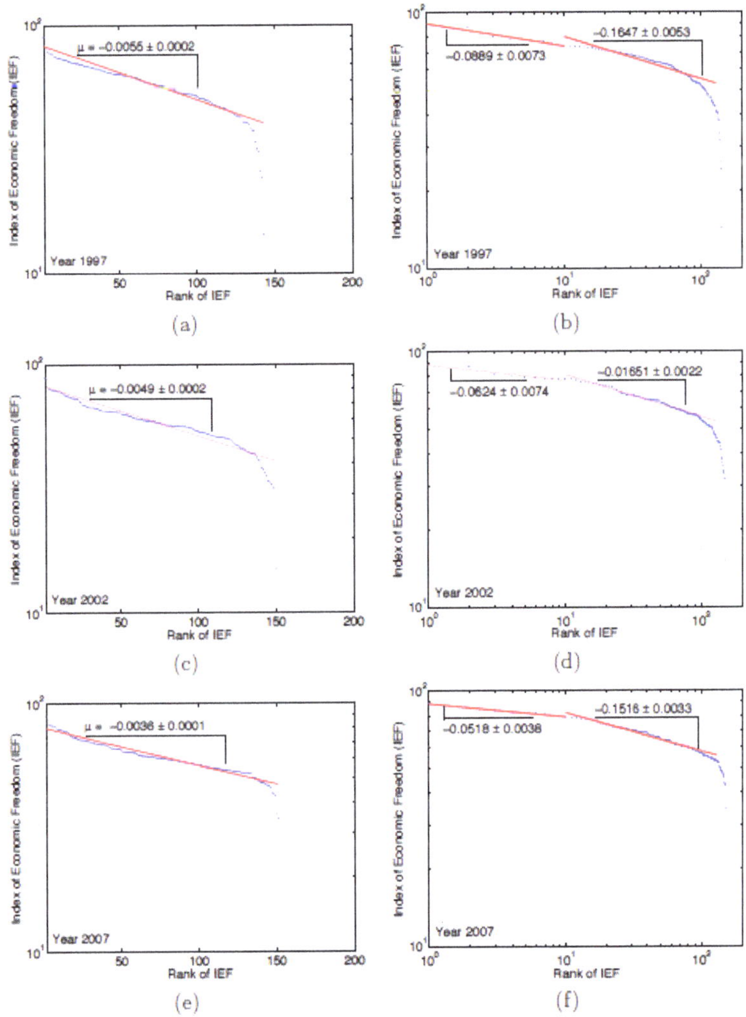

Figure 6. Examples of semi-log [(**a**,**c**,**e**)] and log–log [(**b**,**d**,**f**)] plots of the rank-size relation between the Index of Economic Freedom (IEF) and the country rank for the years 1997, 2002 and, 2007, respectively; the semi-log plots show that the IEF ranking has an exponential behaviour; the log–log plots point to the existence of a transitional point between two different power laws, i.e., near rank 10.

Table 7. Yearly evolution of the λ exponent in the assumed empirical exponential law between the EFW index and the rank (r), the standard error ($\Delta\lambda$), its relative value ($\Delta\lambda/\lambda$), and the efficiency (R^2) of the regression. The low error bar values (less than 0.0001) and the effectiveness of the regressions ($\geq$93%) confirm that the data are perfectly following the exponential law.

		EFW $\sim e^{\lambda r}$		
Year	λ	$\Delta\lambda$	$\Delta\lambda/\lambda$	R^2
2000	−0.0043	0.0001	0.0272	0.9316
2001	−0.0039	0.0001	0.0257	0.9388
2002	−0.0037	0.0001	0.0208	0.9591
2003	−0.0035	0.0001	0.0129	0.9839
2004	−0.0035	0.0001	0.0068	0.9954
2005	−0.0029	0.0001	0.0107	0.9889
2006	−0.0029	0.0001	0.0113	0.9876

3.6. Comparison of Both Indices

The purpose of this section is to compare the indices, whence it is necessary to restrict the observation "period" at the largest but common year interval. We should also take into account the countries common to both sets. That leaves 138 countries to be examined over a period extending from 2000 to 2006, i.e., 2 sets of 862 data points.

To have a meaningful comparison, it is best to "normalise" the index values in an observation interval; here we choose the interval to be [0, 1]. To do so, it is sufficient to divide the values of the EFW index by 10 and those of the IEF by 100.

The distributions of the 862 data points are reproduced in Figure 7 for both indices. The average of the EFW values is $\simeq$0.6542, while the average for the IEF is slightly lower at $\simeq$0.6118. This shows that the EFW gives, on average, an index value slightly greater than that given by the IEF for the same country (see below in Table 11).

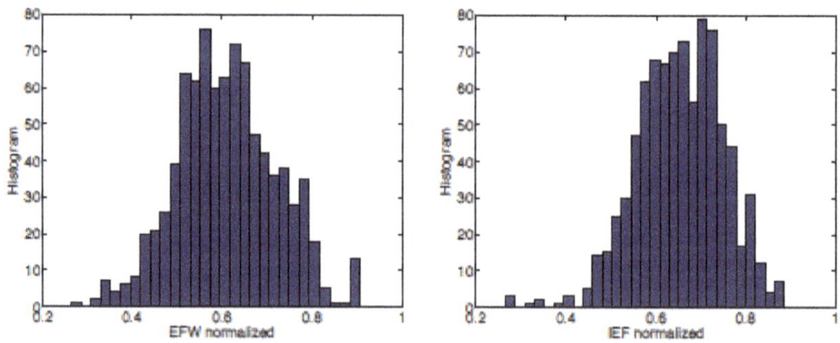

(a) (b)

Figure 7. Histogram of (**a**) Economic Freedom of the World (EFW) and (**b**) Index of Economic Freedom (IEF) values for the 862 data points, common to both indices, normalised over [0, 1].

In order to confirm that the IEF is more conservative than the EFW index, it is interesting to represent the EFW values according to the IEF values. This is done in Figure 8. By calculating the linear regression coefficient, the slope is found to be 0.7294. This value is markedly less than 1, whence confirming that the EFW gives index values greater than the IEF for a given country.

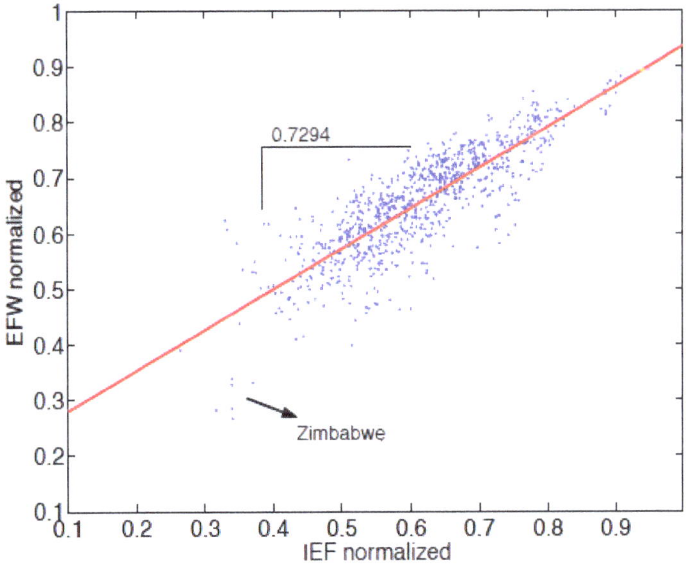

Figure 8. Scatter plot of the relationship between the Economic Freedom of the World (EFW) index and the Index of Economic Freedom (IEF) normalised values. The regression slope points to a linear relationship of ∼0.7294. This value, statistically significant, lower than 1, confirms that the IEF is "more conservative" than the EFW index. The worst EFW country (Zimbabwe) position is emphasised for framing of the data.

Table 8. Yearly evolution of the ν exponent in the empirical power law between the EFW and the rank (r), the standard error ($\Delta\nu$), its relative value ($\Delta\nu/\nu$), and the efficiency (R^2) of the regression. The low error bar values ($\Delta\nu/\nu \simeq 3\%$) and the effectiveness of the regressions confirm that the data is well following a power law.

		EFW $\sim r^{\nu}$		
Year	ν	$\Delta\nu$	$\Delta\nu/\nu$	R^2
2000	−0.0992	0.0034	0.0343	0.9161
2001	−0.0907	0.0029	0.0314	0.9285
2002	−0.0890	0.0029	0.0328	0.9226
2003	−0.0872	0.0032	0.0369	0.9038
2004	−0.0857	0.0034	0.0393	0.8924
2005	−0.0743	0.0023	0.0306	0.9319
2006	−0.0700	0.0024	0.0344	0.9154

4. Relationship between Economic Freedom and Wealth of Countries

As recalled here above, many studies show a strong relationship between economic freedom and the wealth of a country, i.e., between EF and the country gross domestic product (GDP). In this section, the goal is to evidence this relationship.

A graphic representation of EF according to the GDP, on Figures 9 and 10, shows that the relationship translates into a power law, i.e., thereby defining the exponent γ,

$$INDEX \simeq GDP^{\gamma} \,. \tag{4}$$

A positive exponent ($\gamma > 0$) indicates a "positive relationship" between EF and the GDP. This would mean that the freest countries are the richest ones. A negative exponent indicates a negative correlation: the freest countries would be the less rich ones.

The existence of this law is very important from an economic point of view. Indeed, it allows us to know the wealth which a country should have as a function of its level of economic freedom. By estimating the influence that a government decision will have on the economic freedom index of that country, it is possible to directly measure the impact of a government policy on the economy of the country. Moreover, the existence of this (simple) law will enable countries to be classified according to their position on the power law. Countries that are located above the law are countries that have a lower gross domestic product than they should for their level of economic freedom. These countries can be said to be 'underperforming'.

On the other hand, the countries that are located below the law are countries that have a gross domestic product greater than that which it should have. These countries are 'over-performing'.

On Table 9, we report the exponential law parameter (λ) between the IEF and the rank (r) of the IEF, the Standard Error ($\Delta\lambda$) and its Relative Error ($\Delta\lambda/\lambda$), together with the efficiency of the regression (R^2). The λ value decreases each year (in absolute value); it increases from -0.006 in 1996 to -0.0036 in 2007. The efficiency of the regression shows that the data follow an exponential law, rather perfectly since 2003, when the efficiency of the regression exceeds 90%.

Table 9. Yearly evolution of the λ exponent in the empirical exponential law between the IEF and the rank (r), the standard error ($\Delta\lambda$), its relative error ($\Delta\lambda/\lambda$), and the efficiency (R^2) of the regression. The low error bar values ($\Delta\lambda/\lambda \simeq 2$ to 4%) and the effectiveness of the regressions confirm that the data are closely following a power law.

		IEF $\sim e^{\lambda r}$		
Year	λ	$\Delta\lambda$	$\Delta\lambda/\lambda$	R^2
1996	−0.0060	0.0003	0.0422	0.8087
1997	−0.0055	0.0002	0.0405	0.8124
1998	−0.0057	0.0002	0.0385	0.8211
1999	−0.0054	0.0002	0.0416	0.7919
2000	−0.0051	0.0002	0.0382	0.8185
2001	−0.0050	0.0002	0.0345	0.8508
2002	−0.0049	0.0002	0.0381	0.8235
2003	−0.0044	0.0001	0.0212	0.9374
2004	−0.0043	0.0001	0.0241	0.9215
2005	−0.0041	0.0001	0.0246	0.9180
2006	−0.0037	0.0001	0.0238	0.9223
2007	−0.0036	0.0001	0.0227	0.9285

On Table 10, we report the (Zipf) rank-size law exponent (ν) between the IEF and the rank (r) of IEF, the Standard Error ($\Delta\nu$), the Relative Standard Error ($\Delta\nu/\nu$), and the yearly regression coefficients (R^2), for the observed different regimes. While the exponent for countries of rank below 10 decreases over the years the exponent for countries of rank higher than 10 remains relatively stable, near the value -0.016 over the 12 years of the study.

Table 10. Yearly evolution of the Zipf law exponent (ν) between the IEF and the rank (r) of IEF, the Standard Error ($\Delta\nu$), the Relative Standard Error ($\Delta\nu/\nu$), and the Regression Coefficient (R^2). While the exponent for countries of rank below 10 decreases over the years, the exponent for countries of rank higher than 10 remains relatively stable, near the value −0.016 over the 12 years of the study.

IEF $\sim r^{\nu}$								
	$r \leq 10$				$r \in [10-100]$			
Year	ν	$\Delta\nu$	$\Delta\nu/\nu$ (%)	R^2 (%)	ν	$\Delta\nu$	$\Delta\nu/\nu$ (%)	R^2 (%)
1996	−0.0931	0.0071	7.60	95.58	−0.1820	0.0056	3.09	92.16
1997	−0.0889	0.0073	8.26	94.82	−0.1647	0.0053	3.25	91.43
1998	−0.0808	0.0099	12.28	89.24	−0.1505	0.0044	2.92	92.96
1999	−0.0797	0.0079	9.90	92.73	−0.1477	0.0029	1.95	96.73
2000	−0.0807	0.0089	10.97	91.21	−0.1504	0.0030	1.98	96.63
2001	−0.0723	0.0058	8.02	95.11	−0.1634	0.0042	2.54	94.57
2002	−0.0624	0.0074	11.80	89.97	−0.1651	0.0022	1.36	98.38
2003	−0.0686	0.0070	10.17	92.35	−0.1704	0.0031	1.81	97.18
2004	−0.0690	0.0092	13.35	87.52	−0.1690	0.0024	1.45	98.16
2005	−0.0717	0.0095	13.26	87.67	−0.1678	0.0024	1.40	98.28
2006	−0.0564	0.0047	8.29	94.78	−0.1522	0.0022	1.45	98.17
2007	−0.0518	0.0038	7.33	95.88	−0.1516	0.0022	1.47	98.11

On Table 11, we report the main characteristics (average and standard deviation) of the normalised EFW and IEF data for the 138 countries out of the 7 years (i.e., 862 data points). The EFW mean is slightly higher than that for the EFW data. The coefficient of variation (σ/μ) shows a weak dispersion in both cases.

Table 11. Summary of (rounded) main statistical characteristics for the so called "normalized" EFW and IEF distributions of the economic freedom indicators, according to the number of countries N_c, the examined time interval ΔT, whence the number N of data points.

Variable	N_c	ΔT (Years)	N	Mean (μ)	StDev (σ)	CoV (σ/μ)
EFW	138	7	862	0.6542	0.0948	0.1449
IEF	138	7	862	0.6118	0.1094	0.1788

On Table 12, we list countries (The ISO 3166-1 code is used to facilitate the presentation of data) for which the EFW Index does not comply with the power law, i.e., the data points are located outside the area limited by twice the standard deviation from the power law.

Table 12. List of countries for which the EFW Index does not comply with the power law, i.e., are located outside the area limited by twice the standard deviation from the power law.

EFW	
Year	**Countries**
2000	DZA-COD-MMR-ZWE
2001	DZA-ZWE
2002	DZA-COD-MMR-VEN-ZWE
2003	DZA-MMR-VEN-ZWE
2004	DZA-COD-VEN-ZWE
2005	DZA-COD-VEN-ZWE
2006	AGO-COD-MMR-VEN-ZWE

Figures 9 and 10 clearly show that all countries, with a few exceptions obey the power law. The variation coefficient (σ/μ) shows a weak dispersion on both cases, because the countries are almost all in an interval corresponding to twice the standard deviation. For the EFW, the countries that pose a problem are Algeria, the Republic of Congo, Burma, and Zimbabwe, but also Venezuela since 2002. As regards the IEF, the problematic countries are more numerous: among these are Angola, Bosnia, Iran, Laos, Libya, and Zimbabwe. Venezuela is only an IEF problem since 2004. The lists of such countries are included in Tables 12 and 13 for each year of interest. In Table 12, we report the list of countries i for which the EFW Index values do not comply with the power law. In Table 13, we report the list of countries i for which the IEF Index does not comply with the power law.

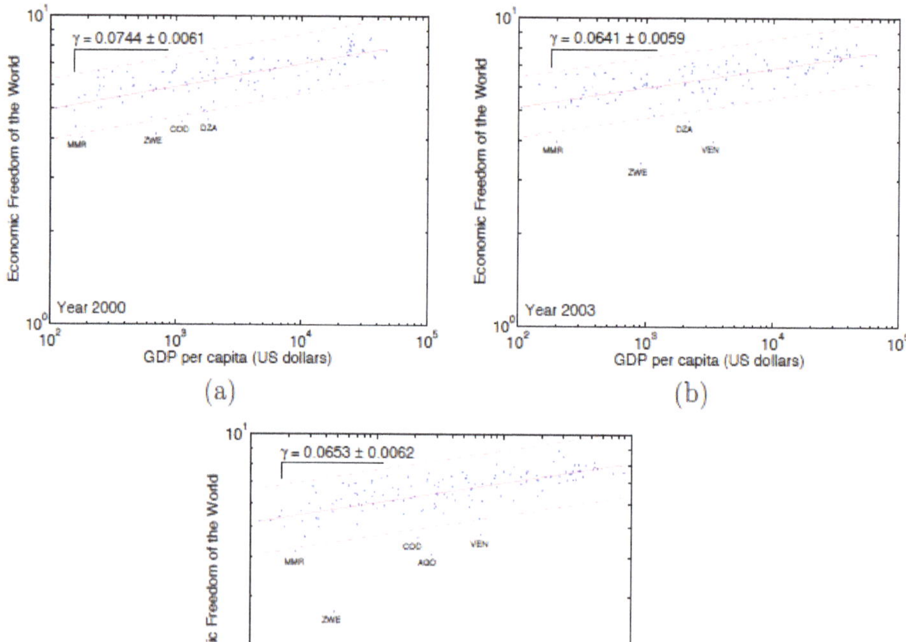

Figure 9. Examples of log–log plot of the Economic Freedom of the World (EFW) Index with respect to country's gross domestic product (GDP) for the years (**a**) 2000, (**b**) 2003, and (**c**) 2006. This relationship is characterised by a power law, with an exponent $\gamma \simeq 0.674$. The dotted lines encompass the region for which the data is within twice the standard deviation away from the trend.

The exponent γ values for the period 2000 to 2006 relationship between EFW and GDP are reported in Table 14, while the γ values for the IEF for the 1996 to 2007 period are shown in Table 15. In the case of the EFW (see Table 14), the exponent of the law in question remains stable on the 7 years with an average value $\simeq 0.0674$. Notice that the regressions coefficients for the EFW–GDP relation are not as high as in the case of the exponential and power (rank-size) laws. For the IEF, there are 3 periods on the 12 years during which the exponent holds different behaviours. For the 1996 to 2000 years, the exponent has an average value equal to 0.0948, which remains stable around this value over these 5 years. The second phase, which extends over the years 2001 to 2005, is a transition period during which the value of the exponent falls down. It ends up to some stabilisation around 0.0666 during the third period (2006–2007). The efficiency of the regressions is not very good, except for the third period during which R^2 is approaching 50%. Therefore, it may be conjectured that the IEF corrections, added in 2006, are bearing fruit.

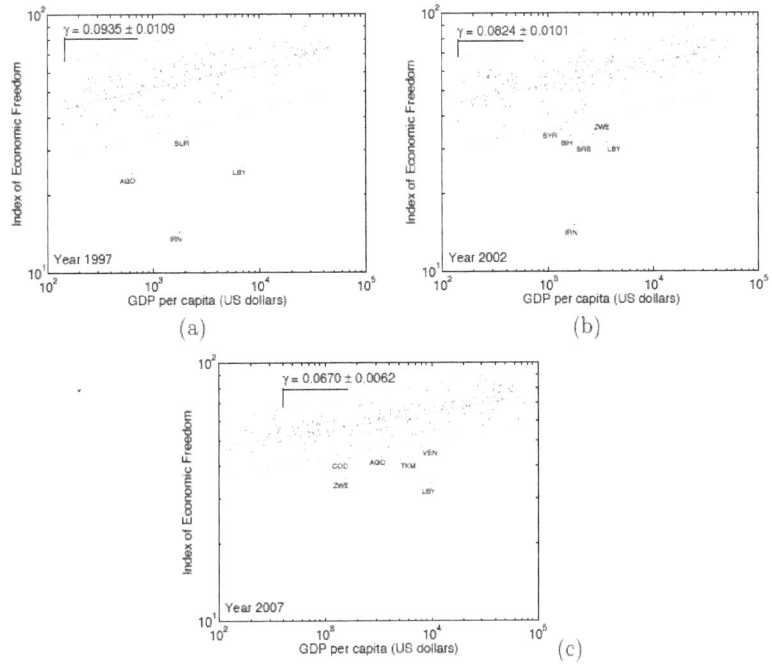

Figure 10. Examples of log–log plots of the Index of Economic Freedom (IEF) relationship to the country's gross domestic product (GDP) for the years (**a**) 1997, (**b**) 2002, and (**c**) 2007. This relationship is characterised by an evolutive power law. The dotted lines limit the region for which the data are located within a maximum distance equal to twice the standard deviation; the few outliers have been removed for calculating the power law exponent γ.

Table 13. List of countries fo which the IEF does not comply with the power law, i.e., are located outside the area limited by twice the standard deviation from the power law.

IEF	
Year	**Countries**
1996	AGO-AZE-IRN-LBY
1997	AGO-IRN-LBY-SUR
1998	AGO-BIH-IRN-LOA-LBY-UZB
1999	AGO-BIH-COG-IRN-LAO-LBY-UZB
2000	AGO-COG-IRN-LOA-LBY
2001	BLR-BIH-LOA-LBY
2002	BIH-IRN-LBY-SRB-SYR-ZWE
2003	BLR-BIH-LBY–SYR-ZWE
2004	BLR-LBY-SYR-VEN-ZWE
2005	LBY-VEN-ZWE
2006	AGO-COD-LBY-TKM-VEN-ZWE
2007	AGO-COD-LBY-TKM-VEN-ZWE

Table 14. Yearly evolution of the power law exponent (γ) between the EFW and GDP, the standard error ($\Delta\gamma$), the relative error bar ($\Delta\gamma/\gamma$) and the efficiency (R^2) of the regression. The power law exponent remains rather stable over the 7 years with an average value $\simeq$ 0.0674 ($\pm$0.004).

		EFW $\sim$ GDP$^\gamma$		
Year	γ	$\Delta\gamma$	$\Delta\gamma/\gamma$	R^2
2000	0.0744	0.0061	0.0824	0.5490
2001	0.0669	0.0061	0.0917	0.4959
2002	0.0636	0.0062	0.0978	0.4636
2003	0.0641	0.0059	0.0922	0.4847
2004	0.0705	0.0057	0.0814	0.5410
2005	0.0667	0.0062	0.0934	0.4540
2006	0.0653	0.0062	0.0952	0.4443

Table 15. Yearly evolution of the power law exponent (γ) between the IEF and GDP, the standard error ($\Delta\gamma$), the relative error ($\Delta\gamma/\gamma$), and the efficiency (R^2) of the regression. There are 3 periods to be noticed in which the exponent adopts different behaviours. For the years 1996 to 2000, the exponent has an average value $\simeq$0.0948 and remains stable ($\simeq$0.09) for about 5 years. The second phase spreads over the years 2001 to 2005, is a transitional period during which the value of the exponent falls down. It ends up stabilising around 0.0666 on the third and latest period (2006–2007). Notice that the regression coefficient (R^2) is not very high.

		IEF $\sim$ GDP$^\gamma$		
Year	γ	$\Delta\gamma$	$\Delta\gamma/\gamma$	R^2
1996	0.0940	0.0117	0.1248	0.3255
1997	0.0935	0.0109	0.1163	0.3439
1998	0.0994	0.0113	0.1140	0.3435
1999	0.0956	0.0112	0.1166	0.3261
2000	0.0915	0.0099	0.1086	0.3583
2001	0.0870	0.0098	0.1131	0.3472
2002	0.0824	0.0101	0.1224	0.3107
2003	0.0802	0.0075	0.0940	0.4332
2004	0.0773	0.0073	0.0947	0.4313
2005	0.0728	0.0070	0.0956	0.4267
2006	0.0662	0.0064	0.0961	0.4208
2007	0.0670	0.0062	0.0922	0.4414

In Table 15, we report the power law exponent (γ) between the IEF and GDP, the standard error ($\Delta\gamma$), the relative error ($\Delta\gamma/\gamma$), and the (R^2) regression coefficient. There are 3 periods to be noticed in which the exponent adopts different behaviours. For the years 1996 to 2000, the exponent has an average value $\simeq$0.0948 and remains stable for about 5 years. The second phase, which is spread over the years 2001 to 2005, is a transitional period during which the value of the exponent falls down. It ends up stabilising around 0.0666 on the third and latest period (2006–2007). Notice that the regression coefficient is not very high: $R^2 \sim 0.376$.

5. Conclusions

Let us recall the research questions: can one find an empirical law for describing the economic freedom (EF) of nations through the main measure indices, i.e., the Economic Freedom of the World (EFW) index [38] and the Index of Economic Freedom (IEF) [39]? What simple empirical laws can be found through a simple analysis of rank-size laws? Are such laws of interest for discussing the main determinant, according to the literature, i.e., each country's GDP?

We have taken some data pertaining to the 1997–2007 period, that is before 2008, thus before a recent "financial crisis", in order not to involve "multiple exaggerated developments" [43], but nevertheless in order to include a drastic turning point, 11 September 2001, following another geo-economico-political event, the fall of the Berlin wall. We have pointed out that the study of EF should develop over two distinct periods, at this time, mainly because the index's 2008 definition of economic freedom has been modified. In so doing we have selected data, leading to 138 countries examined over a period extending from 2000 to 2006, thus 2 sets of 862 data points.

We have found that the rank distributions obey either an exponential or a power law or a mixed behaviour. The EFW rank relationship is exponential for countries of high rank (≥20), but log-log plots point to a behaviour close to a power law when considering the whole sample. In contrast, the IEF overall ranking has an exponential behaviour. Interestingly, IEF rank-size rule log-log plots point to the existence of a transitional point between two different power laws, i.e., near rank 10.

Besides, the IEF appears to be "more conservative" than the EFW index.

Moreover, when searching for (analytical law) correlations between the country GDP and either EF indices (we have not looked for regressions between these macroeconomic variables and the various "pillars" of the indices, the literature is already abundant), we have distinguished regional aspects, i.e., defining six continents. We find that the EFW index relationship to country GDP is characterised by a power law, with a rather stable exponent ($\gamma \simeq 0.674$) as a function of time. In contrast, the IEF relationship with the country's gross domestic product points to a downward evolutive power law parameter as a function of time. Markedly the two studied indices provide different aspects of EF.

In so doing, we add numerical considerations to the literature, as should be somewhat expected by econophysics research, for this special issue, but presenting to others a different perspective. The rank-size law approach seems original for the present topics. It brings some information on the "statistical universality" of EF during the considered time interval. Thus we expect to open gates for rigorous approaches, i.e., stressing objectiveness in the modelling, rather than ideological bases.

Thereafter, suggestions for further research can be listed: among others, one could consider other time intervals; for example including the 2008 financial crisis, and nowadays considering the COVID-19 pandemic. This is left for our expected paper II. On the other hand, It would be nice to have more "economic considerations" and "historical considerations", looking at each pillar separately in more detail. For example, one could consider some renormalisation of the indices, taking into account, size (and type) of governments, size of country populations, inflation rates, foreign direct investments, health burden, etc., on one hand, and on the other hand, migration factors, religious factors, education levels, trade union strengths, pandemic constraints, local climate, etc., all of which presents quite a numerical challenge to econophysicists.

Author Contributions: All authors have contributed equally. All authors have read and agreed to submitting this version of the manuscript.

Funding: M.A. has been partially supported by a grant of the Romanian National Authority for Scientific Research and Innovation, CNDS-UEFISCDI, project number PN-III-P4-IDPCCF-2016-0084.

Institutional Review Board Statement: Not applicable.

Informed Consent Statement: Not applicable.

Data Availability Statement: Data sources are mentioned in the text and references; they are freely accessible.

Conflicts of Interest: The authors declare no conflict of interest.

References

1. Jolly, A. *OECD Economies and the World Today: Trends, Prospects and OECD Statistics*; Kogan Page Business Books: London, UK, 2003.
2. Smith, A. *The Wealth of Nations: An Inquiry into the Nature and Causes of the Wealth of Nations*; Harriman House Limited: Petersfield, UK, 2010.
3. Bastiat, F. *Economic Harmonies. Irvington-on-Hudson*; Foundation for Economic Education: New York, NY, USA, 1964.

4. Bastiat, F. *Harmonies of Political Economy*; Jazzybee Verlag Jürgen Beck: Altenmünster, Deutschland, 1944.
5. Gwartney, J.; Lawson, R.; Norton, S. *Economic Freedom of the World: 2008 Annual Report*; The Fraser Institute: Vancouver, BC, Canada, 2008.
6. Gollwitzer, S.; Quintyn, M. *The Effectiveness of Macroeconomic Commitment in Weak (er) Institutional Environments*; International Monetary Fund: Washington, DC, USA, 2010.
7. Friedman, M.; Friedman, R.D. *Two Lucky People: Memoirs*; University of Chicago Press: Chicago, IL, USA, 1998.
8. Krugman, P. Who Was Milton Friedman? *N. Y. Rev. Books* **2007**, *54*, 27.
9. Rothbard, M.N. *For a New Liberty: The Libertarian Manifesto*; Ludwig von Mises Institute: Auburn, AL, USA, 1978.
10. Rothbard, M.N. *Ethics of Liberty*; New York University Press: New York, NY, USA, 2015.
11. Weimer, D.L. *The Political Economy of Property Rights: Institutional Change and Credibility in the Reform of Centrally Planned Economies*; Cambridge University Press: Cambridge, UK, 1997; pp. 1–19.
12. De Soto, H. The Mystery of Capital. *Financ. Dev.* **2001**, *38*, 66–67.
13. Sen, A. Markets and Freedoms: Achievements and Limitations of the Market Mechanism in Promoting Individual Freedoms. *Oxf. Econ. Pap.* **1993**, *45*, 519–541. [CrossRef]
14. Goodin, R.E.; Rice, J.M.; Parpo, A.; Eriksson, L. *Discretionary Time: A New Measure of Freedom*; Cambridge University Press: Cambridge, UK, 2008.
15. Available online: http://www.fraserinstitute.rog/ (accessed on 30 October 2006).
16. Available online: http://www.freetheworld.com/index.html/ (accessed on 30 October 2006).
17. Lawson, R.A.; Murphy, R.; Powell, B. The determinants of economic freedom: A survey. *Contemp. Econ. Policy* **2020**, *38*, 622–642. [CrossRef]
18. Available online: http://www.heritage.org/ (accessed on 30 October 2006).
19. Available online: http://online.wsj.com/public/us (accessed on 30 October 2006).
20. Executive Summary, Index of Economic Freedom. 2008. Available online: http://www.heritage.org/index/PDF/2008/Index2008ExecSum.pdf (accessed on 30 October 2009).
21. Beach, W.W.; Kane, T. Methodology: Measuring the 10 Economic Freedoms, Index of Economic Freedom. 2008. Available online: http://thf$_{-}$media.s3.amazonaws.com/index/pdf/2008/Index2008$_{-}$Chapter4.pdf (accessed on 30 October 2009).
22. Dialga, I.; Vallée, T. The index of economic freedom: Methodological matters. *Stud. Econ. Financ.* **2021**, *38*, 529–561. [CrossRef]
23. Pei, M. *Political Institutions, Democracy, and Development, Democracy, Market Economics, and Development*; World Bank Publications: Herndon, VA, USA, 2001.
24. Easton, S.T.; Walker, M.A. Income, growth, and economic freedom. *Am. Econ. Rev.* **1997**, *87*, 328–332.
25. Ayal, E.B.; Karras, G. Components of economic freedom and growth: An empirical study. *J. Dev. Areas* **1998**, *32*, 327–338.
26. Scully, G. Economic Freedom, Government Policy, and the Trade-Off Between Equity and Economic Growth. *Public Choice* **2002**, *113*, 77–96. [CrossRef]
27. Berggren, N. Economic Freedom and Equality: Friends or Foes? *Public Choice* **1999**, *100*, 203–223. [CrossRef]
28. Holcombe, R.G. Entrepreneurship and Economic Growth. *Q. J. Austrian Econ.* **1998**, *1*, 45–62. [CrossRef]
29. De Haan, J.; Sturm, J.E. On the relationship between economic freedom and economic growth. *Eur. J. Political Econ.* **2000**, *16*, 215–241. [CrossRef]
30. Doucouliagos, C.; Ulubasoglu, M.A. Economic freedom and economic growth: Does specification make a difference? *Eur. J. Political Econ.* **2006**, *22*, 60–68 [CrossRef]
31. Hristova, K.D. Does Economic Freedom Determine Economic Growth? A Discussion of the Heritage Foundation's Index of Economic Freedom. Ph.D. Thesis, Mount Holyoke College, South Hadley, MA, USA, 2012.
32. Islam, S. Economic freedom, per capita income and economic growth. *Appl. Econ. Lett.* **1996**, *3*, 595–597. [CrossRef]
33. Rode, M.; Coll, S. Economic freedom and growth. Which policies matter the most? *Const. Political. Econ.* **2012**, *23*, 95–133. [CrossRef]
34. Cole, J.H. Contribution of Economic Freedom to World Economic Growth, 1980–1999. *Cato J.* **2003**, *23*, 189–198.
35. Dawson, J.W. Causality in the freedom-growth relationship. *Eur. J. Political. Econ.* **2003**, *19*, 479–495. [CrossRef]
36. Available online: http://www.dollarsandsense.org/archives/2005/0305miller.html (accessed on 30 October 2006).
37. Heckelman, J.C.; Stroup, M.D. Which Economic Freedoms contribute to growth? *Kyklos* **2000**, *53*, 527–544. [CrossRef]
38. Available online: http://www.freetheworld.com/datasets_-efw.html (accessed on 30 October 2006).
39. Available online: http://www.heritage.org/index/download.aspx (accessed on 30 October 2006).
40. Available online: http://www.imf.org/external/pubs/ft/weo/2007/01/data/index.aspx (accessed on 30 October 2009).
41. Available online: https://millenniumindicators.un.org/unsd/mdg/Data.aspx (accessed on 30 October 2009).
42. Zipf, G.K. *Human Behavior and the Principle of Least Effort: An Introduction to Human Ecology*; Addison Wesley: Cambridge, MA, USA, 1949.
43. Wigmore, B.A. *The Financial Crisis of 2008: A History of US Financial Markets 2000–2012*; Cambridge University Press: Cambridge, UK, 2021.

entropy

MDPI

Article

Plotting the Words of Econophysics

Gianfranco Tusset

Department of Economics and Management, University of Padua, via del Santo 33, 35123 Padua, Italy;
gianfranco.tusset@unipd.it

Abstract: Text mining is applied to 510 articles on econophysics to reconstruct the lexical evolution of the discipline from 1999 to 2020. The analysis of the relative frequency of the words used in the articles and their "visualization" allow us to draw some conclusions about the evolution of the discipline. The traditional areas of research, financial markets and distribution of wealth, remain central, but they are flanked by other strands of research—production, currencies, networks—which broaden the discipline by pushing towards a dialectical application of traditional concepts and tools drawn from statistical physics.

Keywords: lexical evolution of econophysics; text as data; correspondence analysis

Citation: Tusset, G. Plotting the Words of Econophysics. *Entropy* **2021**, 23, 944. https://doi.org/10.3390/e23080944

Academic Editor: Ryszard Kutner

Received: 29 May 2021
Accepted: 21 July 2021
Published: 23 July 2021

1. Introduction

The introduction in physics of a new kind of statistical law, or, better, simply a probabilistic law, which is hidden under the customary statistical laws, forces us to reconsider the basis of the analogy with the [. . .] statistical social laws. It is indisputable that the statistical character of social laws derives, at least in part from the manner in which the conditions for phenomena are defined. It is a generic manner, i.e., strictly statistical, allowing countless complexes of different concrete possibilities. On the other hand, [. . .] we are induced to ask ourselves whether there also exists here a real analogy with social facts, which are described with a somewhat similar language (p. 258) [1].

These words were written by a great theoretical physicist, Ettore Majorana, as a preamble to an article, *The Value of Statistical Laws in Physics and the Social Sciences*, on the convergence of natural and social sciences that Majorana wrote around 1930 before disappearing in 1938.

Majorana was hoping that physics and social sciences (including economics) would move in the direction of a shared language. If the social sciences, economics in particular, had always looked to classical physics as a model of scientific rigor, Majorana wanted the new physics and social sciences to converge on a common statistical field.

Majorana's message introduces the short journey we are about to make in the discipline of econophysics, that more than others have taken up the invitation to develop a research area in which natural and social sciences converge. Although there have been episodes that have anticipated some of its contents—from the far Bachelier random walk (1900) [2] and Pareto Law (1896–1897) [3] to the more recent Farjoun, and Machover *Laws of Chaos* (1983) [4], to name just a few—econophysics was born in the early nineties of the last century, with the celebrated article by Nunzio Mantegna on the *Lévy walks* (1991) [5]. Therefore, it has thirty years, maybe few to understand if it has been able to collect and develop Majorana's message, but enough for the definition of its own disciplinary identity.

Econophysics is a broad and magmatic field in terms of content and methods, as is well highlighted by at least a dozen highly scientific texts that deal in detail with the statistical, mathematical and theoretical facets of this new field. To understand if econophysics is moving in the direction desired by Majorana, if indeed that common language is on the horizon, we will consider the scientific articles on econophysics published during these years, analyzing them from a linguistic point of view, aware that words mean contents, methods, objectives.

The encounter between natural sciences and social sciences raises a theme that cannot be ignored and that goes beyond the very search for a common language: it is the theme of laws. In the world of relationships between individuals, of human behavior, are there social and economic laws that can be compared to the invariant laws that characterize the natural world? Econophysics does not ignore the problem, indeed it has made it a topic of discussion.

The linguistic reconstruction of econophysics will therefore be an opportunity to understand how positions are evolving on this point, to understand if the search for laws that characterizes physics represents a dominant feature also in the activity of econophysicists.

Section 2 presents the literature and our research methods. Section 3 illustrates the frequency of the main lexical cluster words identified in the texts considered as shining light on the evolution of the econophysics lexical corpus. Section 4 focuses on the possible correlation between the identified lexical clusters. Section 5 is devoted to the visual representation and analysis of econophysics words. Section 6 contains some concluding remarks.

2. Literature, Methods, and Results

Econophysics has known various moments in which it has discussed itself. The debate on empirical regularities that took place between the two components, economists and physicists, in 2006 [6,7], should be mentioned, as well as prolonged research on individual theoretical and methodological aspects of the discipline [8–13]. Also articles that periodically take stock of the state of econophysics should also not be ignored [14,15].

To try to understand the directions that econophysics is taking, we reconstructed its lexical development over the period from 1999 to 2020. The technique is that of "text as data" to define the frequency matrix of words used in econophysics articles. This matrix is then used for the realization of a scatterplot as a picture of the evolution of the lexicon of econophysics.

The approach to "text as data" proposed here can be defined as "bags of words" [16], i.e., the texts are broken down into words and short combinations of words (single or in segments of two or three), whose frequency is counted in relation to the year or quarter of publication, the latter considered 'active variables'. Therefore, the words and segments constitute the 'tokens', placed in a row of a large matrix that in each column presents the active variables chosen, in our case quarters and whole years. Thus, words and the text segment become the starting point of our analysis. In short, the words and segments that populate text are statistically analyzed to identify meta-trends and meta-behaviors that would not otherwise be immediately apparent.

For construction of the matrix and, thus, the scatterplot, 510 econophysics articles published between the years 1999 and 2020 were used, mainly in *Physica A* (287 or 56%), partly in *The European Physical Journal B* (165, 32%), and a minority (58, 12%) from other journals (*Physical Review E*, *Contemporary Physics* and few others). In the article only the words were used: therefore, the mathematical or statistical content is not taken into account.

Of course, this is only a fraction of the econophysics articles that have appeared since 1991. Nor would it be materially possible to analyze all the articles (each article must be cleaned to be included in the overall corpus). In selecting the articles, we used the following criteria.

First, we relied on the search engines of the sites of the two area journals considered *Phys. A* and *Eur. Phys. J. B*, which have been attentive to econophysics since the early days of its appearance. Rather than moving on the basis of a definition of econophysics, we relied on existing internal classifications.

For each of these journals, the choice of articles is proportional to the distribution of econophysics articles that have appeared in the various years (e.g., the 287 articles in *Phys. A* should be representative of the 1616 articles that the *Phys. A* website identifies as econophysics articles for the same period). The distribution of articles by subareas (the clusters) reflects the distribution of topics among the articles in the year examined.

The priority given to these two journals (*Phys. A* and *Eur. Phys. J. B*) stems from their emphasis on econophysics. Few articles published in other journals were included because they were particularly significant, often for the insights into the significance of the discipline they contained.

The impossibility, for now, of constructing a large corpus on the basis of most of the articles on econophysics published in journals of different subject areas limits the interpretative scope of an analysis constructed on the frequency of words. We believe, however, that, although not complete, the sample used here is sufficient to have a first significant result of the relative distribution of words and, therefore, of the sub-areas of research, during the period considered.

Words are certainly among the protagonists of our story, an idea that can be schematized in three steps. First, we treated the words and segments contained in the articles like our data, while the period of publication represents the variable under which words and segments are grouped. Thus, the early step was to construct a large matrix containing the frequencies of the overall words and articulated by quarter/year. The matrix or contingency tables contains 6915 rows (words/segments) and 88 columns (quarters from 1999 to 2020). All grammatical terms of 2 and 3 syllables and all words with fewer than 6 occurrences were removed from the corpus.

The analysis of the words contained in the matrix allows us to extract some lexical clusters that facilitate the modeling of the evolution of the econophysics lexical corpus. FIN* includes all words/segments concerning financial topics; DIST* the same for the broad area of distribution of wealth, income and other variables, often an object of sociophysics analysis. "Power law" is not included because considered more a tool than an object, as could be income or wealth; PROD* includes words/segments referring to the industrial and production world; CURR* refers to words concerning any kind of currency circulation, including cryptocurrency; and NETW* including all words concerning networks and complex networks.

To investigate the attitude towards the search for invariant laws, we introduced two other lexical clusters, STAT* and NONST*, which include words/segments related, the first one, to contents proper to statistical physics ("power law", "multifractality", "stationarity" are included here), more properly macro that do not imply the analysis of individual choices of agents; the second, to an analysis of 'rumors', of non-stationarity, of specificities often evident on the micro level ("minority game", "agent-based", "reflexivity", "non-stationarity" and so on) and emphasizing potential "noise", "instability" and similar phenomena.

The second step is to regress the time series organized into quarters from 1999 to 2020 regarding the lexical clusters above to determine the extent to which they are treated in isolation or together and how the STAT*-NONST* relationship of the debate is integrated into the treatment of other content. One can rightly question the use of VAR regression for time series regarding words, but it is more than an exact measure of potential causality, here we are interested in identifying trends to guide us in our treatment of the large topics above. The quantitative analysis is functional to the qualitative analysis developed in the next step.

Finally, the third step is dedicated to visualization and analysis of the words over the period considered here. We decided to adopt correspondence analysis as the most appropriate analytical method to visualize the words/segments contained in the papers in relation to the above active variables. This is an exploratory data processing technique belonging to multivariate statistics and designed to analyze the above matrixes containing frequencies, that is, measures of correspondence between rows and columns. Correspondence analysis was well suited to our purpose because our study lacked an *a priori* hypothesis to verify; it enabled us to identify systematic relationships between variables, without any prior expectations regarding the nature of these relationships [17,18].

Scatterplots showing the outcomes of this linguistic analysis are grounded on relative frequencies. Axes of the scatterplot were selected according to the level of inertia, i.e., the variance exhibited by the *active variables*. In other words, the active variables (in this case, whole years) were arranged according to the variance characterizing their own lexicon. The two pairs of active variables with the greatest distance in their lexicon identified the horizontal and vertical axes. Then we could also work with illustrative or *case variables*, i.e., words belonging to rows that can be pinpointed on the scatterplot showing the distribution of the dataset, and then associated with the active variables. This step helps to clarify the characteristics of the lexicon used by econophysicists.

The multiplicity of the active variables generates the multidimensionality of the data matrix. Exploratory factorial analysis enables this multidimensionality to be reduced by transforming data into noncorrelated variables and building factorial or semantic axes that constitute "points of view" on the phenomenon observed (p. 62) [19]. These points of view are contextual in that they display relationships across a broad corpus of texts by reducing the amount of information. Specific software is needed to analyze such a large dataset, so we used Automatic Lexical and Textual Processing for the Analysis of Content (*TALTAC*) and *R* to manage the corpus (both led to similar matrices), and *SPAD* to extract the figures relating to our study.

Briefly, the results of this lexical analysis. Econophysics tends to gradually widen its field of application, extending it to an increasing number of economic and social phenomena. This is a process which undoubtedly broadens the sphere of influence as well as the competence of the discipline. This process, however, pushes towards a dialectical, not dogmatic, application of the principles inherited from statistical physics: suffice it to mention the universality of the laws or the invariance of scale. This dialectical process, more common to the social sciences than to the natural sciences, does not weaken econophysics, on the contrary, it makes it more dynamic and alive. However, its application implies a challenge for econophysics, which remains, or aspires to remain, a natural science.

3. Measuring Lexical Clusters

To obtain a preliminary viewpoint, we reconstructed trends in the relative frequency of the five clusters. Figure 1 shows the results of this calculation in quarters since 1999. Since data are relative frequencies, the number of articles per quarter was used to calculate the total number of words on which to calculate the relative frequency of those words of interest to us. Clearly, we referred to the average number of words per article. Thus, each frequency is calculated on the total number of words appearing in the articles considered in that quarter.

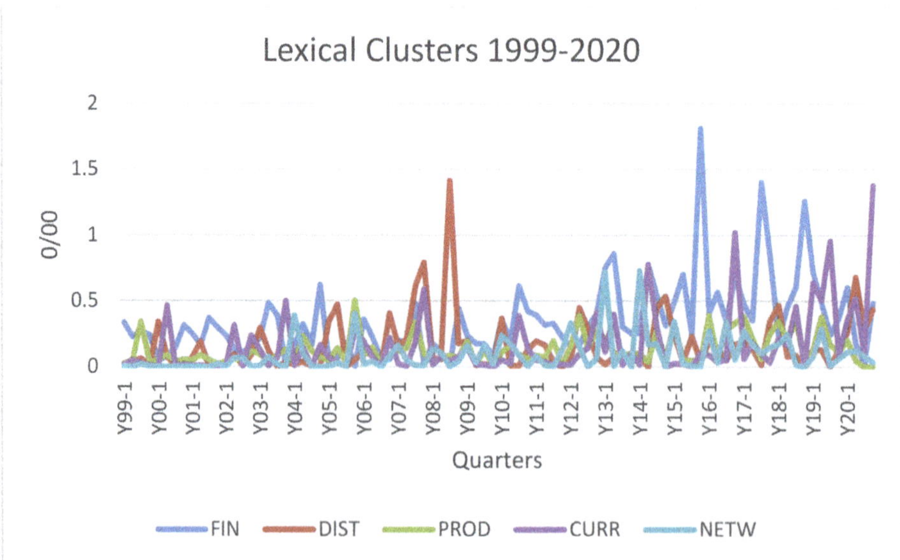

Figure 1. Econophysics main lexical clusters from 1999 to 2020.

Taking a quick look at Figure 1, what stands out is the prominence in terms of the relative frequency of the lexical cluster FIN* and partially of DIST*.

In particular, FIN*, which includes words referring to options, stocks, and all financial products, represents a constant in the interest of econophysicists but, contrary to what one might imagine when thinking about the financial origins of econophysics, it becomes dominant, from the perspective of lexical frequencies, from 2012 onwards, reaching various peaks, those of highest intensity in 2015, 2017 and 2019. DIST*, distribution of wealth and income, represents, since the early years of the period considered here, an important topic in the research of econophysicists, characterized by some peaks in different periods (the highest in the third quarter of 2008, and smaller ones in 2005, 2007, 2014 and 2020 respectively).

The trend of the PROD* cluster, including the reference to the real economy, is interesting. The attraction of econophysicists to industrial and production issues, without presenting relevant peaks, appears to present greater strength since mid-2014. As we will see later, these are the years in which interest in networks, financial and otherwise, grows.

The CURR* topic only exploded after 2014 and later, when cryptocurrency became the subject of analysis by econophysicists. Often treated as a financial asset, cryptocurrencies are also of interest as a means of circulation, an aspect that has prompted us to keep them separate from financial securities. Also included in this topic are all words that refer to monetary circulation, a recurring theme in the treatises on econophysics.

Although it indicates an approach rather than an area of economic/financial activity, we have also included here network, NETW*, whose prominence has grown, especially after the first years of the last decade, to the point of becoming an autonomous research area with respect to econophysics, an aspect that also explains its decreased frequency among the words of the discipline after 2015.

Stationarity or nonstationarity as well? To try to reconstruct the prevailing orientation among econophysicists, we have reported in Figure 2 the trend of two lexical clusters expressing the two possibilities. In the STAT* cluster we find those words/segments that indicate a preference for an econophysics faithful to physical statistics that analyzes the behavior of aggregates independently of that of individuals, searches for power laws and scale-invariance. The NONST* cluster, on the other hand, takes into account the development in the direction of nonstationarity, scale-dependence, reflexivity, behavior of individual agents not just aggregates, including agent-based computation.

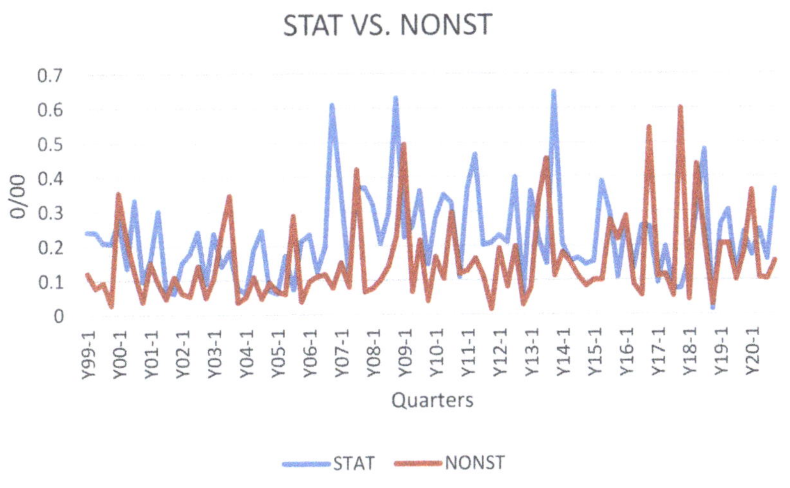

Figure 2. Lexical clusters on STAT* and NONST* approaches from 1999 to 2020.

The figure shows that econophysics is not solely the discipline of statistical physics devoted to aggregates. Interest in the two directions coexists showing various peaks of NONST* as well as STAT* peaks.

4. Correlation between Lexical Clusters

Is econophysics a discipline that deals primarily with financial markets? Or does it touch on a wide range of aspects of economic and financial life? Can relative frequencies tell us anything about the relationship between the topics (lexical clusters) that are the subject of econophysics work? Once the main lexical clusters were defined, we tried to study their evolution from 1999 to 2020. As a first step, we tried to understand whether the various research strands have, over time, constituted a single disciplinary corpus or have remained substantially separate, also in light of the debate on the macro or micro-orientation of the discipline. The idea is to test the existence of causality and correlations between lexical clusters represented by time series related to word frequencies.

To seek such hypothetical causalities or correlations, we start by testing Granger causality between the available time series: FIN*, DISTR*, PROD*, CURR* and NETW*. Adopting a level of confidence of 5 percent, we have identified the following outcomes (see Table A1 in Appendix A):

FIN*, DIST*, PROD* and NETW* are lexical clusters that have no causality or correlation between them. These sub-areas grow, expand, but within hypothetical sub-disciplinary boundaries. Thus, the lack of correlation between them should be read.

- Both DIST* and NETW* affect CURR*, which is equivalent to saying that the debate on currency circulation is influenced by the debates on distribution and production, mainly that on distribution, if we consider the two p-values (Table A1). As the CURR lexical cluster is the most recent in terms of development, the causality of which it is the subject is, perhaps, symptomatic of a lesser stiffening or closure of these sub-areas.
- Finally, the debate over NETW* is affected by the debate over STAT*.

The interesting fact is that FIN*, DIST*, PROD*, and NETW* represent a world unto themselves, not talking to each other or being influenced by other debates.

The two lexical clusters STAT* and NONST* concerning the more the approach than the content yielded the following outcomes.

- The STAT* orientation is conditioned by the FIN*, DIST*, NETW* and by the same NONST* cluster.
- The NONST* orientation does not affect any cluster, but is influenced by FIN*, DIST*, and NETW*.

The latter two causalities feed into the dialectical process above. Causality concerns not only STAT, which descends from statistical physics, but also NONST*, which instead challenges it. Hence the intertwining of the two lexical clusters, observable in Figure 2.

Finally, considering STAT* and NONST* in isolation, one cannot ignore the Granger causality from the latter to the former. Indeed, together with the observation of the absence of autocorrelation in the time series of the two variables, such causality induces an interpretation of this type: the centrality of statistical physics, power law and scale invariance need to be frequently reaffirmed in the face of the doubts evoked by NONST* words/segments.

These (few) causal relationships are only statistical hypotheses, however, and need to be validated. Consistently with the approach of our work, we opt for a textual validation: rather than seeing whether the above statistical hypotheses can be refuted or not, we use these hypotheses as a key to interpret how econophysics' lexicon changes over time.

5. Visualizing the Words of Econophysics

Recalling the English adage that 'a picture is worth a thousand words', our analysis of the texts produced during the crisis can be enriched by taking a further step and moving from number to image (image of words, in this case). The scatterplots presented here "can be regarded as maps" of the use of words and segments concerning topics in the econophysics corpus. The scatterplot simply "communicates [...] information" (p. 5) [17]. Correspondence analysis provides "ways for describing data, interpreting data, and generating hypotheses" without a theoretical model or preconceived hypothesis.

How can we interpret the scatterplot obtained by correspondence analysis? If a given word/segment is close to an active variable (a given year, for instance), this means that it characterizes speeches or discussion papers published at the time. On the other hand, words/segments that are common to most or all active variables (years) considered are to be found in the center (centroid) of the figure.

The scatterplot is constructed using a second matrix that differs from the one used for above figures solely because of the active variables (columns), the first quarters (88) and now years (22). In contrast, the words/segments (rows) remain the same (6915).

Our word/segment cloud lies in a $c-1$ dimensional space, where c is the number of active variables, the 22 years in our case. The choice of coordinates to be represented is such as to ensure the widest representation of words/segments consistent with their distance (in row and column) from the mean profiles located in the center of the plane. In short, the widest linguistic variability is guaranteed.

If $i = 1, \ldots, r$ are the words/segments considered here, $j = 1, \ldots, c$ the active variables i.e., the years analyzed, n the total of words/segments occurrences, $n_{i.}$ the total of the matrix i-row, $n_{.j}$ the total of the matrix j-column, we can express the distance, d, between two words/segments i and i' as Pearson chi-square distance (χ^2) in the form:

$$d^2(i,i') = \sum_{j=1}^{c} \frac{n}{n_{.j}} \left(\frac{n_{ij}}{n_{i.}} - \frac{n_{i'j}}{n_{i'.}} \right)^2 \quad (1)$$

The Euclidean distance weighting in Equation (1) results in a reassessment of the low-frequency components and a scaling of the high-frequency components. The very low frequencies (less than 6 occurrences) were removed to prevent them from weighing too heavily in the distance calculations due to the weighting (p. 107) [19].

Briefly, the scatterplot in Figure 3 shows the evolution of the vocabulary of econophysics articles. On the axes we find the inertia, which can be considered as an index of lexical change: the higher its value, the higher the variability of the words contained in the analyzed texts. In our case, it is quite low on both axes: 10.02 and 6.81 percent. This means that this representation explains only 16.83 percent of the total variability. By changing the combination of axes, we get lower values of total inertia. This result can be interpreted by stating that, in these twenty-two years, the vocabulary of econophysicists has changed little and very gradually. The movement can be read clockwise. Arranging the years in a sufficiently orderly sequence shows that the change has been gradual, but continuous. We will focus on the gradually introduced changes in the lexicon of econophysics, but the low variance makes it clear that previously used words and segments continue to be used. In other words, as new concepts are entered, the previous ones were retained. The most common words or segments, such as "Brownian motion", "statistical physics," "power law," found around the origins of the axes, are not shown because they were shared by most of the articles.

The lexicon used in econophysics in the period under consideration follows a sort of clockwise trajectory that goes from the left side of the axis to the fourth quadrant on the lower right, passing through the second and first quadrants. To make its interpretation easier, the lexical path has been divided into five phases, each of which is lexically characterized by marking a stage in the construction of the vocabulary of econophysics. The titles attributed to each phase look more to marginal novelty than to the main body of scholarship from that period, reiterating the interest in change at the margin.

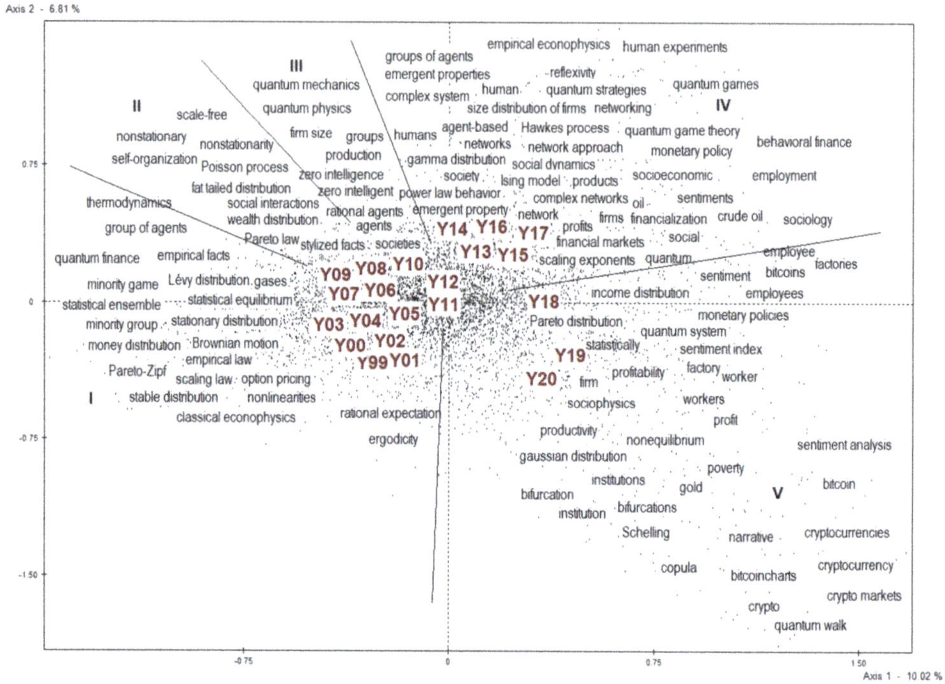

Figure 3. The vocabulary of econophysics between 1999 and 2020.

Briefly, reviewing the five phases will help us understand if and how the topics at the heart of econophysics have changed and how the orientation towards STAT* and NONST* has changed over time. Remember that words/segments are positioned in relation to years is done based on their respective relative frequencies, calculated on the set of words used in each year in the articles considered between 1999 and 2020.

About the distribution of words/segments in general, we can observe how it is rather spherical in the first four phases, signifying a rather weak inertia, while it shows a dilation in the fifth phase (2018–2020), proof that in the last years here considered the lexicon tends to show evident and not only gradual signs of change. The years 2019 and 2020 contribute 18.3 and 12.8 percent, respectively, to the formation of the horizontal axis and 12.2 and 31.3 percent, respectively, to the formation of the vertical axis. There are not many words/segments that characterize the fifth phase, but they show considerable weight in structuring the entire word/segment distribution.

5.1. Phase I—Statistical Aggregates

The first phase, subarea I of Figure 3, corresponds to the first six years, from 1999 to 2004. During this period, the new research area was presumably reinforcing its methodological and conceptual pillars drawn from statistical physics: "gases", "Brownian motion", "option pricing" and "Lévy distribution" testify that we are in the world of statistical physics applied mainly to financial markets. The idea that the theoretical properties of gases could be extended to a market composed of many agents, each operating as a particle, aroused great interest. The goods exchanged could be of any kind, including the income distributed throughout the economy as a whole. During this early phase, the discipline's focus is primarily on the outcome of the many unpredictable exchanges that occur in a market, not on what causes them or on the decision-making process of the agents. Statistical econophysics shows more interest in "predictions" of future prices or rather future price changes than in understanding how the market works. The direct challenge to economic theory anchored in individualism and the role of the representative agent is plain.

The aggregates of statistical physics produce distributions that cannot ignore what Mandelbrot [20,21] has shown, namely, that the distribution function of asset prices deviates significantly from Gaussian. Part of the subsequent development of econophysics is the result of this debate, including the need to normalize "stationary distributions."

About distribution, "Power law" appears in this first phase, but within econophysics, "power law" is something of a focal point that has allowed and contributed to the discipline's ability to stay within the confines of its inheritance from physics. Perhaps it is inappropriate to talk about power law science [22], but its popularity stems from the common belief that "small occurrences are extremely common, while large occurrences are extremely rare."

Taking a brief look at the stances taken on power law, can well represent the opinion of the early physicists engaging in socio-economic research: "Physicists are often fascinated by power laws. The reason for this is that complex, collective phenomena do give rise to power laws which are *universal*, that is, to a large degree independent of the microscopic details of the phenomenon. The power laws emerge from collective action and transcend individual specificities" (p. 105) [23]. However, power law models "contain multiplicative noise" and "lead to nonuniversal exponents that depend on the value of the parameters". It thus becomes necessary to model observations at the microscopic level to explain the decay of volatility correlations on this level (p. 112) [23]. Agent-based microscopic models were still advocated by Ausloos et al., for the same purpose, i.e., to determine "scaling exponents and universal laws" (p. 2) [24]. However, "although [power law] is probably not the universal law that some have claimed it to be, it is certainly a powerful and intriguing concept that potentially has applications to a variety of natural and man-made systems." (p. 346) [25].

Single agents, however, do not disappear in the aggregates of statistical physics, even in this first phase focused on statistical sets. The word "agents" itself weighs in at a significant 0.2 and 0.6 percent in determining the horizontal and vertical, respectively (the 100 percent is obtained by summing the individual contributions of the 6915 words/segments to the formation of the horizontal and vertical axes). Various types of noise can make their appearance in the study of aggregate phenomena or distributions. Typically, these noises are related to microscopic analyses of how markets work. This explains the presence in the first subarea of segments such as "minority game", a variant of Brian Arthur's El Farol bar problem [26], and "minority group", typical of computational models based on agents making decisions based on their memory of what happened in the past. Agent-based analysis, which has developed independently, is thus gradually being drawn into the galaxy of econophysics, given the need to explain microphenomena. Agent-based analysis has also generated an abundant literature on models based on assumptions very different from those that characterize statistical aggregates.

A sort of dialectic between macro and micro, between scale invariance and multiscale, between stationarity and non-stationarity makes its appearance since the first phase originating that causation between NONST* and STAT* mentioned above.

5.2. Phase II—Stationary or Nonstationary Processes?

Econophysics was a discipline that reached maturity in a few years. As Figure 3 shows, the period from 2005 to 2009 appears, in fact, characterized by those words that define its identity: "stylized facts", "Pareto law", "wealth distribution" and so on. Consistent with DIST's peak of the years 2006–2008 visible in Figure 1, "distribution" becomes a key word in econophysics, identifying an area of research, the distribution of wealth, that has begun to represent a specific field in the discipline. Since those years, it has been possible to state that wealth/income distribution analysis and financial market research have represented, not without overlap, the two main areas of research in the discipline.

When talking about income distribution and price changes, the notion (crucial for econophysics) of "stylized facts" is quite common. Simply put, these are phenomena that are primarily visible at the meso and macro levels, and usually lack a micro theoretical foundation. A stylized fact allows generalization without reference to time or spatial contextualization. Although the notion of "stylized fact" is widely accepted by statistical physicists aiming to explain aggregate or macro phenomena, it remains shrouded in a kind of vagueness, perhaps a legacy of its economic origin. Some recognized and universal stylized facts—such as distribution laws, option pricing and risk control—sit alongside less accepted stylized facts, such as trends in GDP or inflation. However, stylized facts also remain central to their use in the study of financial markets [27].

However, even at a stage when econophysics recognizes its roots in statistical physics, it does not fail to discuss them, thus making this discipline a living field of research.

It is a fact that deviations of price time series from random walk behavior and "price distribution" have been studied, moving also in the direction of stylized self-organizing facts. "Self-organizing" and "self-organization" together with "group of agents" highlight the novelty of this phase. A system characterized by self-organizing criticality is able to move towards a stable critical regime that is characterized by long-range correlations and free-scale power laws. From an economic perspective, we can look at the ability of markets to organize themselves by means of intermediate actors, such as groups of firms or sectors, or even uncoordinated agents [28]. If markets are able to converge toward stability, there is no need to analyze their internal or micro dynamics.

The transition from the microstate to the macrostate level or "phase transition" is part of the analysis of markets and socio-economic systems. "Self-organization" has a role in any phase transition. In 2007, Newman wrote: "There has been much excitement about self-organized criticality as a possible generic mechanism for explaining where power-law distributions come from [. . .] Self-organized critical models have been put forward not only for forest fires, but for earthquakes, solar flares, biological evolution, avalanches and many other phenomena" (p. 347) [25].

But self-organization does not necessarily mean homogeneity of agents. The models postulated the distinction between inactive agents ("chartists") and active agents ("fundamentalists"), and the feedback between price fluctuations and the number of active agents, implicitly admitting that agents can decide whether or not to enter the financial market based on their "predictions" regarding price changes. The choice involves a price dynamic that does not guarantee that the probability distribution will remain stationary over time. On the contrary, there may be a "nonstationary distribution" (p. 386) [29].

Not only that. The evolution of the income distribution debate has involved the assumption that agents have "saving propensities" [30] or saving parameters [31], which affect the volume of exchange between agents, viewed as particles colliding to exchange energy. When saving is allowed, the intensity of this exchange decreases, and the distribution consequently takes on a new shape (p. 166 ff) [32].

The ability of markets to organize themselves in a stable manner has been discussed. "Thermodynamics," which appears in the previous step, is connected to "stationarity" and "non-stationarity." The latter reminds us of what McCauley wrote in that very year: "There is no reliable analogue of energy in economics, and there are very good reasons why no meaningful thermodynamic analogy can be constructed" [7]. Thermodynamic equilibrium would require a stationary equilibrium, whereas markets and production are not stationary, nor are increases in the time series, with the consequence that growth processes can be understood by considering not only their variation over time, but also their initial conditions.

Time matters. With respect to financial markets, "non-stationarity" in time series could be caused by secular trends or other long-term factors that do not permanently characterize the observed phenomenon. In other words, the parameters of a process or distribution may change. This aspect distinguishes economics from physics. Clearly, nonstationary processes force us to set aside the ergodic condition and to reconsider "non-ergodicity" as a norm in economic processes (p. 3180) [33]. Are there concepts from physics that cannot be applied to economics? However, the parallel between natural and social sciences, rich in both similarities and differences, continues to be at the heart of econophysics.

5.3. Phase III—Zero-Intelligent Agents

We know that aggregates gave rise to empirical events because of so many causes that it was impossible to explain them by adopting a deductive approach. Phenomena were the product of too many causes to be investigated. Decision-making processes were ignored. However, are the interacting agents/particles that animate these phenomena incapable of making decisions? Are they zero intelligence [10,34]? Zero intelligence, the lexical protagonist of the third phase (2010–2012), must be conceived referring not to the decision-making capacity of agents, but to the inability to link the global outcome under observation to the behavior of the underlying microstructures. Agents are random factors, therefore assumptions about their behavior are not necessary to obtain stylized facts. The direction seems diametrically opposed to that of perfect rationality.

While minority game models have been proposed primarily to explain some stability and stationarity weaknesses at the aggregate level, zero-intelligence units are introduced into agent-based computation to assert "implicit microfoundations": individuals represent "black boxes" that are sources of unpredictable noise subject to objective constraints. Usually, microfoundations are explicit because the choice (optimization) mechanism is fully specified and functions as an essential explanatory factor. Here, agents are efficient even if their rationality is not explicit. What matters is the macro phenomenon, regardless of any individual rationality.

The point is not to assert that agents are purposeless and act randomly: zero-intelligence means that starting from individual behavior or rationality, macro phenomena cannot be predicted. In short, since rationality has no observable impact on market data, the rationality hypothesis may be superfluous.

This development of the macro-micro relationship, the true crux of econophysics, is not the only new element of this third phase, which is also distinguished by the prominence given to other fields. In fact, econophysics begins to be widely interested in the real economy in production and enterprises. A broad econophysics approach to production (the so-called "classical econophysics") has been proposed by Cockshott, Cottrell, Michaelson, Wright and Yakovenko, in a volume published in 2009, *Classical Econophysics* [33]. The title is explained in the following terms by the authors of the first part of the book: classical physics, from Galileo to Bohr plus classical economics, from Smith to Marx. We could say: econophysics devoted to work and energy on the one hand, and classical political economy focused on economic development on the other. The goal is actually even more ambitious than building an econophysics from classical physics and economics: the authors identify several categories that could unify the two disciplines, physics and economics.

Classical econophysics is close to the field of political economy, as highlighted by the treatment of "value"—a concept forgotten by neoclassical economics, and reinterpreted here based on "simulation data, empirical data, and statistical mechanics arguments" (p. 3) [35]. There is much interplay between physics and economics: from energy/value and energy/utility parallelism to fluid/monetary flow, to the common ground of technological innovation [36]. However, it is the relationship between thermodynamics and economics (hardly a new topic), with its burden of "entropy" and information, that remains at the heart of any econophysics view of production. In a nutshell, the point is: thermodynamics implies the conservation of energy, a principle that so far has not been confirmed in economic processes.

5.4. Phase IV—Emergent Properties

No concept is abandoned, but in the fourth phase (2013–2017) the frontiers of econophysics seem to be expanding, as highlighted by the repeated use of "complex systems." A "complex system" is "a system with a large number of mutually interacting parts, often open to their environment, that self-organize their internal structure and dynamics with new and sometimes surprising 'emergent' macroscopic properties" (p. 3196) [37]. The macroprospective is anchored in the idea that particles have "emergent properties," i.e., that [emergent properties] produce effects that are only visible at the macro level. Emergent properties originate from self-organization due to nonlinear interactions between humans or heterogeneous agents. It should be recognized that statistical econophysics does not provide a clear formulation for the occurrence of emergent properties. Econophysics looks at emergent properties because at the macro dimension. It is also interesting that physicists confess that they cannot predict the exact shape of these phenomena [38]: analysis of emergent properties requires tools other than those drawn from statistical physics.

The point is that the concept of "emergent property" was primarily devised by Keynes in economics, not physics, but has never been adequately developed in modern economics. Perhaps this is because, unlike econophysicists, economists base their reasoning on a movement from micro-level structures to complex global-level structures. Emergent properties involve phenomena that can only be observed at macro-level structures, where objects are irreducible to their components. They cannot be microfounded. Statistical physics states that it is not necessary to define the properties of particles or components. What matters are their effects at the macro level where the emergent properties are visible.

Emerging properties of systems are produced at the meso/macro level, the study of which requires new concepts: network is one of them. The occurrence of the words "network" peaked in 2014, after increasing considerably in 2012 and 2013 (weighing for the 0.2 per cent of horizontal axis and 0.7 percent of vertical one). The network shaped a real trend in econophysics studies during that period. The study of aggregates of indistinct particles/agents, followed by attention to the self-organizing capabilities of these particles/agents, paved the way for connections between agents and/or sets of agents, and their ability to build networks in financial and economic contexts. Graph theories provide the mathematical basis for the scientific description of networks. In 2014, Slanina wrote: "Numerous interdependences we find in society can be expressed in terms of a collection of networks, each of them mapping a certain aspect of pairwise interactions among humans or human collectives, or even products of human activities" (p. 222) [39].

Bargigli and Tedeschi wrote: "Network theory deals with the structure of interaction within a multiagent system. Consequently, it is naturally interested in the statistical equilibrium of these systems [...] Following this path, we come close to the idea [...] of reconstructing macroeconomics under the theoretical framework of statistical physics and combinatorial stochastic processes" (p. 2) [40]. The need to understand interactions at the meso and macro level fostered the growth of network analysis, which gradually became one of the foundations of "macroeconophysics". However, it is precisely the increased attention that has fostered its consolidation as an independent research area with respect to econophysics, as highlighted by the distribution of articles in *Physica A* in which "econophysics" and "network analysis" identify two distinct subareas.

One aspect of the explosion of attention to "network analysis" is a further broadening of financial market studies, as shown by the peaks in Figure 1. De Area Leão Pereira et al. (p. 258) [41] gave the first reason for this when they wrote: "The use of complex networks in financial markets has enabled a new view, mainly to measure the financial interaction between stock exchanges, assets, banks or companies. In this case, the nodes are usually assets, banks or countries." Complex networks add the interdependence of markets as a necessary condition for studying the fragility of financial systems. As with emergent properties and other topics, "network analysis" is brought back into the realm of statistical physics.

The segment "complex networks", which occurred more often than "complex system" in 2017, does not only refer to the financial world. It also includes production and business networks, reinforcing econophysics in the direction of the real economy as well as the financial economy. Econophysics was born with financial markets, and finance remains at the heart of this discipline. The question, however, is which econophysics is best suited to investigate production.

After rediscussing the temporal dimension, the other dimension to consider is space: terms such as "international network" and "macroeconomics" testify to a particular and gradual shift to great spaces. In 2016, Paul Ormerod wrote, "There is a great opportunity for econophysicists in the area of macroeconomics. Mainstream [DSGE] models are felt to be unsatisfactory, both by policy-makers and by mainstream economists" (p. 3288) [42]. Reference to communities of production networks [43] shifts econophysics to a spatial dimension that inevitably draws attention to the multiple connections that link productive or financial vertices at the international level. At these vertices we can find institutions, firms, industries, central banks, as well as agents. Econophysics is thus enriched by macro-econophysics, an important new field that opens up possibilities and raises challenges.

Consistently with these macro developments, the fourth is also the subarea comprising topics such as monetary and banking relationships, an operational field that, until then, has played a marginal role in econophysics [44].

5.5. *Phase IV—Cryptophysics*

Looking at Figure 1, one may wonder if there is a discontinuity between the fifth phase, which covers the years 2018 to 2020, and the previous phase. Some of the topics reported—bitcoin, cryptocurrency and sentiment analysis—seem far removed from the tradition of econophysics. "Sentiment" weighs 1 percent of the horizontal axis. A few trends can be detected.

First, it seems that econophysics is looking increasingly at macroeconomics and the real economy. The area is populated with words like "worker", "factory," "productivity" and "profit". It is decidedly interesting that a word as "profit" contributes in determining the axes (0.3 and 0.1 respectively). Are we facing a definitive shift to the real phenomena of the economy? Only in part. However, an "economic" strand of research seems to be consolidating, covering the firm [45], the price of crude oil (analyzed both financially and as a commodity) [46], capital income [47] and economic policy [48]. "Oil" weighs 1.1 percent in the horizontal axis and 0.8 percent in the vertical axis. "Crude oil" for 0.6 and 0.2 respectively. Innovations in methods and analytical tools are anchored in content with increasing areas of overlap with economics.

One may wonder why the attention to productive and industrial or economic-social issues does not explode, even if a growth of interest in real economy is undoubted. The doubt that arises is that econophysics remains tied to concepts and tools that, in a sense, prevent a decisive enlargement of the research area.

Second, the CURR* lexical cluster emerges strongly here. The word "bitcoin" alone contributes 3 percent of the horizontal axis and 2.3 percent of the vertical axis. Not only bitcoin, but "cryptocurrency" contributes 0.4 and 0.6 in structuring the axes and "cryptocurrencies" 0.3 and 0.5 respectively. Bitcoins are analyzed as financial assets and means of exchange [49–52]. According to this qualitative analysis, FIN* and CURR* converge. The same "gold" matters for 0.3 and 0.6 of horizontal and vertical axis. From the centrality of "option pricing" in the early years, to the relevance of "cryptocurrency pricing" in recent years [53,54].

Third, the consolidation of the "quantum walk" as a development of the now historical "random walk" opens new fields of application that, at least from a lexical point of view, seem to change econophysics. The "quantum communication" leads to "quantum cryptographic protocols" [55] (semi-quantum key distribution, among others), which seem to open to further enrichment of econophysics. In terms of content, the four mentioned above and CURR* in particular seem to be sufficient to contain also these developments that pertain mainly to the instrumental aspect.

All that being said, FIN* and DIST* remain the two central topics in econophysics [56–59], gradually joined by PROD*, CURR* and NETW*. The core of econophysics does not change, although the focus on specific phenomena induces continuous enlargement of the toolbox.

To conclude, Figure 3 shows that in the years 2018–2020 our word cloud undergoes a dilation and the distance between the words/segments that characterize those years and the core lexicon of econophysics tends to increase, as evidenced by the widespread presence at this stage of words/segments that weigh in the structuring of the scatterplot. This means that the use of the new words/segments has less need of the lexical apparatus typical of statistical mechanics, which is located around the origin of the axes, than was previously the case when new terms were introduced.

6. Concluding Remarks

A first conclusion of this lexical investigation is that econophysics can certainly be included among the attempts of synthesis between natural scientific language and economic and social language. Figure 3 speaks to both worlds, natural and social.

The words/segments legacy of statistical physics lie in the center (centroid) of the Figure 3, which, however, tells us that there is a dialectical relationship between this core of words/segments and words that over the years take over the scene, conditioning in some way the scientific debate within econophysics. It happened with the word "agents" in the first phase; with the word "network" but also "crude oil" in the fourth phase; with "bitcoin," "cryptocurrency," "sentiment," "gold" in the fifth and most recent phase. This is how the lexicon of econophysics evolves.

Does this dialectical process affect the propensity of econophysicists to seek "power laws" and invariant laws? Jovanovic and Schinckus stated: "The implicit disciplinary assumptions that econophysicists have regarding the identification of statistical laws come from the hypothesis of the universality of power laws. To put it in other words, econophysics inductively expects to identify a power law" (p. 37) [12]. However, the finding that linguistic variability increases in the last stages considered here (the fourth and fifth), together with the increased frequency of the NONST* lexical cluster from 2017 onwards (see Figure 2), leads us to conclude that the search for invariant laws is a fact that is far from being definitively established in econophysics.

The discipline seems to evolve on the basis of a different and less obvious point of attraction than "power law": the dialectical process that arises from the application and questioning of concepts and methods often drawn from statistical physics. The application of a complex, non-reductionist approach to observed phenomena seems to lead to the continued use of dialectical, if not contrasting concepts, as suggested by the oscillating values of the STAT* and NONST* time series.

In Figure 2 there is no bifurcation. The development of econophysics seems to depend on the intertwining and contamination between these conflicting concepts, rather than on the assertion of one orientation or the other, STAT* or NONST*.

The process of consolidation and enlargement of lexical clusters on the one hand reinvigorates the debate on stationarity and non-stationarity, in short, on the application of statistical physics to economic and social relations, on the other, it is the product of that debate.

Figure 3 and the lexical analysis of these twenty-two years show that the evolution of econophysics does not depend so much on the consolidation of certain principles, approaches or visions as on their continuous questioning and enrichment with other contents and areas.

To conclude, the effectiveness of "power law" does not seem to be a consequence of its universality, but rather of its non-dogmatic use which requires continuous verification. A conclusion that also seems relevant to the other pillars of econophysics—"scale invariance," "multifractality," and so on—and to the overall application of statistical physics to the social sciences.

Funding: This research received no external funding.

Institutional Review Board Statement: Not applicable.

Informed Consent Statement: Not applicable.

Data Availability Statement: Data sharing not applicable.

Acknowledgments: The author would like to thank Thomas Bassetti and anonymous referees for their helpful comments.

Conflicts of Interest: The authors declare no conflict of interest.

Appendix A

Table A1 reports a series of χ^2 tests through which we verify if the explanatory variable Granger causes the dependent one. In particular, for the sake of synthesis, we only consider those tests with a p-value below the usual 5% confidence level. More in general, the p-value displayed in the last column represents the confidence we have to reject the null hypothesis that the explanatory variable does not Granger cause the dependent one.

Table A1. Granger causality test on lexical clusters.

Dependent Variable	Explanatory Variable	χ^2	DF	p-Value
CURR	DIST	20.766	2	0
CURR	NETW	5.897	2	0.05
NETW	STAT	8.067	2	0.018
STAT	FIN	8.726	2	0.013
STAT	DIST	7.067	2	0.029
STAT	NETW	7.154	2	0.028
STAT	NONST	9.992	2	0.007
NONST	FIN	11.469	2	0.003
NONST	DIST	6.649	2	0.036
NONST	NETW	6.260	2	0.044

References

1. Mantegna, R.N. Presentation of the English translation of Ettore Majorana's paper: The value of statistical laws in physics and social sciences. *Quant. Financ.* **2005**, *5*, 133–140. [CrossRef]
2. Bachelier, L. Théorie de la Spéculation: Annales Scientifique de l'E.N.S. tome 17, 1900, p. 21–86. In *Louis Bachelier's Theory of Speculation. The Origins of Modern Finance*; Davis, M., Etheridge, E., Eds.; Princeton University Press: Princeton, NJ, USA; Oxford, UK, 2006.
3. Pareto, V. *Cours d'Economie Politique*; Bousquet, G.H., Busino, G., Eds.; New edition; Droz: Genève, Switzerland, 1964.
4. Farjoun, E.; Machover, M. *Laws of Chaos. A Probabilistic Approach to Political Economy*; Verso: London, UK, 1983.
5. Mantegna, R.N. Lévy walks and enhanced diffusion in Milan stock exchange. *Phys. A Stat. Mech. Appl.* **1991**, *179*, 232–242. [CrossRef]
6. Gallegati, M.; Keen, S.; Lux, T.; Ormerod, P. Worrying trends in econophysics. *Phys. A Stat. Mech. Appl.* **2006**, *370*, 1–6. [CrossRef]
7. McCauley, J.L. Response to "Worrying Trends in Econophysics". *Phys. A Stat. Mech. Appl.* **2006**, *371*, 601–609. [CrossRef]
8. Roehner, B.M. *Patterns of Speculation. A Study in Observational Econophysics*; Cambridge University Press: Cambridge, UK, 2002.
9. Schinckus, C. Is econophysics a new discipline? The neopositivist argument. *Phys. A Stat. Mech. Appl.* **2010**, *389*, 3814–3821. [CrossRef]
10. Schinckus, C. Introduction to econophysics: Towards a new step in the evolution of physical sciences. *Contemp. Phys.* **2013**, *54*, 17–32. [CrossRef]
11. Gingras, Y.; Schinckus, C. The institutionalization of econophysics in the shadow of physics. *J. Hist. Econ. Thought.* **2012**, *34*, 109–130. [CrossRef]
12. Jovanovich, F.; Schinckus, C. *Econophysics and Financial Economics. An Emerging Dialogue*; Oxford University Press: Oxford, UK, 2017.
13. Dash, K.C. *The Story of Econophysics*; Cambridge Scholars Publishing: Newcastle Upon Tyne, UK, 2019.
14. McCauley, J.; Roehner, B.; Stanley, E.; Schinckus, C. Editorial: The 20th anniversary of econophysics: Where we are and where we are going. *Int. Rev. Financ. Anal.* **2016**, *47*, 267–269. [CrossRef]
15. Kutner, R.; Ausloos, M.; Grech, D.; Di Matteo, T.; Schinckus, C.; Stanley, H.E. Econophysics and sociophysics: Their milestones & challenges. *Phys. A Stat. Mech. Appl.* **2019**, *516*, 240–253. [CrossRef]
16. Zhang, Y.; Jin, R.; Zhou, Z.-H. Understanding bag-of-words model: A statistical framework. *Int. J. Mach. Learn. Cybern.* **2010**, *1*, 43–52. [CrossRef]
17. Greenacre, M. *Correspondence Analysis in Practice*; Chapman & Hall: Boca Raton, NJ, USA, 2007.
18. Beh, E.J.; Lombardo, R. *Correspondence Analysis. Theory, Practice and New Strategies*; Wiley: Chichester, UK, 2014.
19. Bolasco, S. *Analisi Multidimensionale dei Dati*; Carocci: Rome, Italy, 2013.
20. Mandelbrot, B. The Pareto-Lévy Law and the Distribution of Income. *Int. Econ. Rev.* **1960**, *1*, 79–106. [CrossRef]
21. Mandelbrot, B. New Methods in Statistical Economics, 1963. In *Vilfredo Pareto. Critical Assessments of Leading Economists*; McLure, M., Wood, J.C., Eds.; Routledge: London, UK; New York, NY, USA, 1999; Volume IV, pp. 241–263.
22. Allen, P.; Maguire, S.; McKelvey, B. The Sage Handbook of Complexity and Management. In *The Sage Handbook of Complexity and Management*; Allen, P., Maguire, S., McKelvey, B., Eds.; SAGE: Los Angeles, CA, USA, 2011.
23. Bouchaud, J.-P. Power laws in economics and finance: Some ideas from physics. *Quant. Financ.* **2001**, *1*, 105–112. [CrossRef]

24. Ausloos, M.; Clippe, P.; Miśkiewicz, J.; Pekalski, A. A (reactive) lattice-gas approach to economic cycles. *Phys. A Stat. Mech. Appl.* **2004**, *344*, 1–7. [CrossRef]
25. Newman, M.E.J. Power laws, Pareto distributions and Zipf's law. *Contemp. Phys.* **2007**, *46*, 323–351. [CrossRef]
26. Arthur, B.W. Complexity and the economy. *Science* **1999**, *284*, 107–109. [CrossRef]
27. Zheng, B.; Qiu, T.; Ren, F. Two-phase phenomena, minority games, and herding models. *Phys. Rev. E* **2004**, *69*, 1–6. [CrossRef]
28. Samanidou, E.; Zschischang, E.; Stauffer, D.; Lux, T. Agent-based models of financial markets. *Rep. Prog. Phys.* **2007**, *70*, 409–450. [CrossRef]
29. Alfi, V.; Cristelli, M.; Pietronero, L.; Zaccaria, A. Minimal agent based model for financial markets I. *Eur. Phys. J. B* **2009**, *67*, 385–397. [CrossRef]
30. Chatterjee, A.; Sen, P. Agent dynamics in kinetic models of wealth exchange. *Phys. Rev. E* **2010**, *82*, 1–6. [CrossRef]
31. Patriarca, M.; Heinsalu, E.; Chakraborti, A. Basic kinetic wealth-exchange models: Common features and open problems. *Eur. Phys. J. B* **2010**, *73*, 145–153. [CrossRef]
32. Chakrabarti, B.K.; Chakraborti, A.A.; Chatterjee, A. (Eds.) *Econophysics and Sociophysics*; Wiley: Weinheim, Germany, 2006.
33. Gallegati, M. Beyond econophysics (not to mention mainstream economics). *Eur. Phys. J. Speec. Top.* **2016**, *225*, 3179–3185. [CrossRef]
34. Ponta, L.; Raberto, M.; Cincotti, S. A multi-assets artificial stock market with zero-intelligence traders. *EPL* **2011**, *93*, 28002. [CrossRef]
35. Cottrell, A.F.; Cockshott, P.; Michaelson, G.J.; Wright, I.P.; Yakovenko, V.M. *Classical Econophysics*; Routledge: London, UK; New York, NY, USA, 2009.
36. Chen, S.-H.; Li, S.-P. Econophysics: Bridges over a turbulent current. *Int. Rev. Financ. Anal.* **2012**, *23*, 1–10. [CrossRef]
37. Huber, T.A.; Sornette, D. Can there be a physics of financial markets? Methodological reflections on econophysics. *Eur. Phys. J. Speec. Top.* **2016**, *225*, 3187–3210. [CrossRef]
38. Schinckus, C. Ising model, econophysics and analogies. *Phys. A Stat. Mech. Appl.* **2018**, *508*, 95–103. [CrossRef]
39. Slanina, F. *Essentials of Econophysics Modelling*; Oxford University Press: Oxford, UK, 2013.
40. Bargigli, L.; Tedeschi, G. Interaction in agent-based economics: A survey on the network approach. *Phys. A Stat. Mech. Appl.* **2014**, *399*, 1–15. [CrossRef]
41. Pereira, E.J.D.A.L.; da Silva, M.F.; Pereira, H. Econophysics: Past and present. *Phys. A Stat. Mech. Appl.* **2017**, *473*, 251–261. [CrossRef]
42. Ormerod, P. Ten years after "Worrying trends in econophysics": Developments and current challenges. *Eur. Phys. J. Spec. Top.* **2016**, *225*, 3281–3291. [CrossRef]
43. Ikeda, Y.; Ayoama, H.; Iyetomi, H.; Mizuno, T.; Ohnishi, T.; Sakamoto, Y.; Watanabe, T. *Econophysics Point of View of Trade Liberalization: Community Dynamics, Synchronization, and Controllability as Example of Collective Motions*; RIETI Discussion Papers Series; Research Institute of Economy, Trade and Industry: Tokyo, Japan, 2016.
44. Hazan, A. Volume of the steady-state space of financial flows in a monetary stock-flow-consistent model. *Phys. A Stat. Mech. Appl.* **2017**, *473*, 589–602. [CrossRef]
45. Baaquie, B.E. A statistical model of the firm. *Phys. A Stat. Mech. Appl.* **2019**, *524*, 392–411. [CrossRef]
46. Bonaccolto, G.; Caporin, M.; Gupta, R. The dynamic impact of uncertainty in causing and forecasting the distribution of oil returns and risk. *Phys. A Stat. Mech. Appl.* **2018**, *507*, 446–469. [CrossRef]
47. Tempere, J. An equilibrium-conserving taxation scheme for income from capital. *Eur. Phys. J. B* **2018**, *91*, 1–6. [CrossRef]
48. Dai, P.-F.; Xiong, X.; Zhou, W.-X. Visibility graph analysis of economy policy uncertainty indices. *Phys. A Stat. Mech. Appl.* **2019**, *531*, 1–8. [CrossRef]
49. Begušić, S.; Kostanjčar, Z.; Stanley, H.E.; Podobnik, B. Scaling properties of extreme price fluctuations in Bitcoin markets. *Phys. A Stat. Mech. Appl.* **2018**, *510*, 400–406. [CrossRef]
50. Da Cunha, C.; Da Silva, R. Relevant stylized facts about bitcoin: Fluctuations, first return probability, and natural phenomena. *Phys. A Stat. Mech. Appl.* **2020**, *550*, 124155. [CrossRef]
51. Fang, W.; Tian, S.; Wang, J. Multiscale fluctuations and complexity synchronization of Bitcoin in China and US markets. *Phys. A Stat. Mech. Appl.* **2018**, *512*, 109–120. [CrossRef]
52. Alvarez-Ramirez, J.; Rodriguez, E.; Ibarra-Valdez, C. Long-range correlations and asymmetry in the Bitcoin market. *Phys. A Stat. Mech. Appl.* **2018**, *492*, 948–955. [CrossRef]
53. Azqueta-Gavaldón, A. Causal inference between cryptocurrency narratives and prices: Evidence from a complex dynamic ecosystem. *Phys. A Stat. Mech. Appl.* **2020**, *537*, 122574. [CrossRef]
54. Ghazani, M.M.; Khosravi, R. Multifractal detrended cross-correlation analysis on benchmark cryptocurrencies and crude oil prices. *Phys. A Stat. Mech. Appl.* **2020**, *560*, 125172. [CrossRef]
55. Ishak, N.I.; Muniandy, S.V.; Chong, W.Y. Scaling exponent analysis and fidelity of the tunable discrete quantum walk in the noisy channel. *Phys. A Stat. Mech. Appl.* **2020**, *559*, 125124. [CrossRef]
56. Di Vita, A. On avalanche-like perturbations of relaxed power-law distributions: Richardson's law of warfare as a consequence of the relaxation to a Pareto-like distribution of wealth. *Eur. Phys. J. B* **2020**, *93*, 27. [CrossRef]
57. Néda, Z.; Gere, I.; Biró, T.S.; Tóth, G.; Derzsy, N. Scaling in income inequalities and its dynamical origin. *Phys. A Stat. Mech. Appl.* **2020**, *549*, 124491. [CrossRef]

58. Safari, M.A.M.; Masseran, N.; Ibrahim, K.; Al-Dhurafi, N.A. The power-law distribution for the income of poor households. *Phys. A Stat. Mech. Appl.* **2020**, *557*, 124893. [CrossRef]
59. Tian, S.; Liu, Z. Emergence of income inequality: Origin, distribution and possible policies. *Phys. A Stat. Mech. Appl.* **2020**, *537*, 122767. [CrossRef]

entropy

MDPI

Article

Development of Econophysics: A Biased Account and Perspective from Kolkata

Bikas K. Chakrabarti [1,2,3,*] **and Antika Sinha** [1,4]

1 Saha Institute of Nuclear Physics, Kolkata 700064, India; antikasinha@gmail.com
2 S. N. Bose National Center for Basic Sciences, Kolkata 700106, India
3 Economic Research Unit, Indian Statistical Institute, Kolkata 700108, India
4 Department of Computer Science, Asutosh College, Kolkata 700026, India
* Correspondence: bikask.chakrabarti@saha.ac.in

Abstract: We present here a somewhat personalized account of the emergence of econophysics as an attractive research topic in physical, as well as social, sciences. After a rather detailed storytelling about our endeavors from Kolkata, we give a brief description of the main research achievements in a simple and non-technical language. We also briefly present, in technical language, a piece of our recent research result. We conclude our paper with a brief perspective.

Keywords: traveling salesman problem; simulated annealing technique; kinetic exchange model; Gini index; Kolkata index; minority game; Kolkata Paise Restaurant problem

Citation: Chakrabarti, B.K.; Sinha, A. Development of Econophysics: A Biased Account and Perspective from Kolkata. *Entropy* **2021**, *23*, 254. https://doi.org/10.3390/e23020254

Academic Editor: Ryszard Kutner

Received: 28 January 2021
Accepted: 19 February 2021
Published: 23 February 2021

1. Introduction

Countless attempts and research studies, mostly in physics, to model and comprehend the economic systems are about a century old. For the last three or four decades, major endeavor have been made and some successes have been achieved and published, notably under the general title 'Econophysics'. The term was coined at a Kolkata conference held in 1995 by Eugene Stanley, who later in an interview said " ... So, he (Bikas) started to have meetings on econophysics and I think the first one was probably in 1995 (he decided to start it in 1993–1994). Probably the first meeting in my life on this field that I went to was this meeting. In that sense Kolkata is — you can say — the nest from which the chicken was born ..." [1]. The entry on Econophysics by Berkeley Rosser in the New Palgrave Dictionary of Economics (2nd Edition [2]) starts with the sentence "According to Bikas Chakrabarti (...), the term 'econophysics' was neologized in 1995 at the second Statphys-Kolkata conference in Kolkata (formerly Calcutta), India, by the physicist H. Eugene Stanley ..." See also Figure 1 (and Reference [3]). It may be mentioned here that in a more generalized sense, the term 'Sociophysics' was introduced more than a decade earlier by Serge Galam and coworkers [4] (also see Reference [5]).

As we will discuss in the next section, economics, like all the natural sciences (physics, chemistry, biology, geology, etc.), are, epistemologically speaking, knowledge or truth acquired through induction from observations (natural or controlled in the laboratories) using inductive logic and analyzed or comprehended using deductive logic (like mathematics). The divisions of natural sciences between the streams, like physics, chemistry, biology, and geology, are for convenience and are man-made. 'Truth' established in one branch or stream of natural science does not become 'false' or wrong in another; only the importance often vary. This helps in the growth of an younger branch of science through interdisciplinary fusion of established knowledge from another older established branch; astrophysics, geophysics, biophysics, and biochemistry had been earlier examples. Econophysics has been the latest one, and this special issue of Entropy attempts to capture the history, success and future prospect of econophysics research studies.

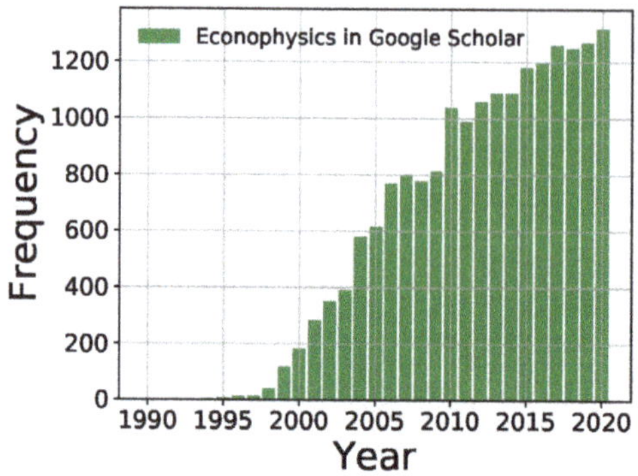

Figure 1. Histogram plot of yearwise numbers of entries containing the term econophysics against the corresponding year. The data are taken from Google Scholar (dated 31 December 2020). It may also be noted from Google Scholar that, while this 25-year old econophysics has today typical yearly citation frequency of order 1.3×10^3, more than 100-year old subjects, like astrophysics (Meghnad Saha published his thermal ionization equation for solar chromosphere in 1920), biophysics (Karl Pearson coined the term in his 1892 book 'Grammar of Science'), and geophysics (Issac Newton explained planetary motion, origin of tides, etc., in 'Principia Mathematica', 1687), today (31 December 2020) have typical yearly citation frequencies of order 32.5×10^3, 26.8×10^3, and 38.6×10^3, respectively.

In the next five sections (essentially following the structure and section-titles, suggested by the Editors of this special issue), we discuss in non-technical language, allowing them to be accessible to the uninitiated younger students and researchers (except in Section 4, where we present some new result of our research). True to the spirit suggested by the Editors, the second section has been presented in the form of a 'Dialogue' using the format of questions and answers between us, the coauthors.

2. What Attracted You to Econophysics?

As mentioned earlier, this section is formatted in the form of a dialogue (question and answer) between the coauthors.

AS: What attracted you to Econophysics? Can you briefly recount?

BKC: Meghnad Saha, founder of Saha Institute of Nuclear Physics (named so after his death in 1956), had been a pioneering Astrophysicist (known for the Saha Ionization formula in astrophysics), had also thought deeply about the scientific foundation of many social issues (see, e.g., Reference [6]). In early seventies, our undergraduate-level text book on heat and thermodynamics had been 'Treatise on Heat' [7], written by Saha together with Biswambhar Nath Srivastava (first published in 1931). This turns out to be the earliest textbook where the students were encouraged, in the section on Maxwell-Boltzmann distribution in kinetic theory of ideal gas, to compare it with the anticipated 'gamma' distribution of income in the society (see the page from Reference [7] reproduced in Reference [6]). If taken seriously, it asks the students to model the income distribution in a society, which maximizes the entropy (assuming stochastic market transactions)!

AS: Before you go further, let me ask why should one think of applying statistical physics to society in the first place?

BKC: One Robinson Crusoe in an island cannot develop a running market or a functional society. A typical thermodynamic system, like a gas, contains Avogadro number (about 10^{23}) of atoms (or molecules). Compared to this, the number (N) of 'social atoms' or agents in any market or society is of course very small (say, about 10^2 for a village market to about 10^9 in a global market). Still such many-body dynamical systems are statistical in nature and statistical physical principles should be applicable. Remember, each constituents particle in a gas follows some well-defined equations of motion (say, Newton's equation for classical gases or Schrödinger's equation for quantum gases), yet for the collective behavior of gases (or liquids or solids) we need to average over the 'appropriate' statistics for their stochastic behavior in such 'many-body' systems and calculate the emerging collective or thermodynamic properties of the entire system. Motivation to go for the 'appropriate' statistics to estimate the collective behavior or response of the society comes, therefore, very naturally. In the case of human agents in a society, the corresponding equations governing individual behavior are much more difficult and still unknown and unpredictable, yet many collective social behavior are quite predictable; ask any traffic engineer or engineers designing stadium evacuation in panic situation.

AS: Which problem of economics did Saha and Srivastava try to analyze using Maxwell-Boltzmann distribution or statistical mechanics of ideal gas?

BKC: As can be seen from the example they had put to the readers, they indicated to the students the problem of income and wealth inequalities (they assumed Gamma-function-like income distribution in Reference [7]; reproduced in Reference [6]). They suggested to them that the 'entropy maximization' principle, along with conservation of money (or wealth), across the market (with millions of transactions between the agents, buyers, and sellers) must be at work in such 'many-body' social or market systems. This will result in the consequent and inevitable inequality (equal distributions being entropically unstable against stochastic fluctuations, leading to steady state unequal distributions).

AS: That is quite interesting. Can you elaborate a bit more and explain a bit of statistical physics specifically for the classical ideal gas?

BKC: Let me try. One can present the derivation of the energy distribution among the constituent (Newtonian) particles of a (classical) ideal gas in equilibrium at a temperature T as follows: If $n(\epsilon)$ represents the number density of particles (atoms or molecules of a gas) having energy ϵ, then one can write $n(\epsilon)d\epsilon = g(\epsilon)f(\epsilon, T)d\epsilon$. Here, $g(\epsilon)$ denotes the 'density of states' giving $g(\epsilon)d\epsilon$ as the number of dynamical states possible for any of the free particles of the gas, having kinetic energy between ϵ and $\epsilon + d\epsilon$ (as counted by the different momentum vectors $\vec{p}$ corresponding to the same kinetic energy: $\epsilon = |p|^2/2m$, where m denotes the mass of the particles).

Since the momentum $\vec{p}$ is a three-dimensional vector, $g(\epsilon)d\epsilon \sim |p|^2 d|p| \sim (\sqrt{\epsilon})d\epsilon$. This is obtained purely from mechanics. For completely stochastic (ergodic) many-body dynamics or energy exchanges, maintaining the the energy conservation, the energy distribution function $f(\epsilon, T)$ should satisfy $f(\epsilon_1)f(\epsilon_2) = f(\epsilon_1 + \epsilon_2)$ for any arbitrary choices of ϵ_1 and ϵ_2. This suggests $f(\epsilon) \sim \exp(-\epsilon/\Delta)$, where the factor Δ can be identified from the equation of state for the gas (positive sign in the exponential is neglected because of the observation that the number decreases with increasing energy). This gives $n(\epsilon) = g(\epsilon)f(\epsilon) \sim (\sqrt{\epsilon})\exp(-\epsilon/KT)$. Knowing this $n(\epsilon)$, one can estimate the average pressure P the gas exerts on the walls of the container having volume V at equilibrium temperature T and compare with the ideal gas equation of state $PV = NKT$ (K denoting the Boltzmann constant). The gas pressure can be estimated from the average rate of momentum transfer by the atoms on the container wall, and one can compare with that obtained from the aforementioned equation of state and identify different quantities; in particular, one identifies $\Delta = KT$.

AS: How does one then extend this to the markets?

BKC: Yes, let us consider the trading markets, where there is no production (growth) or decay. In addition, the total amount of money (considered equivalent to energy) and number of traders (or agents, considered as particles or 'social atoms') remain fixed or constant throughout the trades. Since in the market money remains conserved as no one can print money or destroy money (will end-up in jail in both cases) and the exchange of money in the market is completely random, one would again expect, for any buyer-seller transaction in the market, $f(m_1)f(m_2) = f(m_1 + m_2)$, where $f(m)$ denotes the equilibrium or steady state distribution of money m among the traders in the market. This then, in a similar way, suggests $f(m) \sim \exp(-m/\Delta')$, where Δ' is a constant. Since there cannot be any equivalent of the particle momentum vector for the agents in the market, the density of states $g(m)$ here is a constant (any real-number value of m corresponds to one market state). Hence, the number $n(m)$ of traders or agents having money m will be given by $n(m) = c\exp(-m/\Delta')$, where c is a constant. One must also have $\int_0^M n(m)dm = N$, the total number of traders in the market, and $\int_0^M mn(m)dm = M$, the total amount money in circulation in the market (or country). This gives, the effective 'temperature' of the economy $\Delta' = M/N$, the average available money per trader or agent in this closed-economy (as no growth, migration of laborers, etc., are considered). This gives exponentially decaying (or Gibbs-like) distribution of money in the market (unlike the Maxwell-Boltzmann or Gamma distribution of energy in the ideal gas), where most of the people become pauper ($n(m)$ is maximum at $m = 0$).

AS: Is this exponentially decaying income or wealth distribution realistic for any economy?

BKC: That discussion will take us to the recent studies by econophysicists and data comparisons. We will defer those to the next section (Section 3). Indeed, some success of the model (sketched above) in capturing the real data has been explored extensively by Victor Yakovenko and his group from Maryland University. We, in Kolkata, explored what could make the distribution more like a Gamma distribution, as Saha and Srivastava indicated in their book [6] to be an observed phenomenon. We also tried to capture the Pareto tail of such a distribution. Avoiding detailed discussion here, we only refer here to three popular papers [8–10] in this context. The model sketched above essentially follows [6,8]. In this model, the exchanged money or wealth in each trade (equivalent to any of the particle-particle collision in Ideal gas) is completely random, subject to conservation of money (or wealth). A trader, acquiring a lot in earlier trades may lose the entire amount of money or wealth in the next trade as the total money (wealth) will be conserved if the partner trader gets that. If one introduces a saving propensity of each trader, so that each trader saves a fraction of their individual money (wealth) before the trade and exchanges randomly the respective rest amount in the trade (keeping total money or wealth again conserved) the resulting steady state distributions capture the above mentioned desirable features. One can easily see that, unlike in the Kinetic-exchange model described above, the possibility for any trader (with non-vanishing saving propensity) to become an absolute popper vanishes, as that will require that trader to lose in every trade. Consequently, the exponential distribution becomes unstable with effect to any non-vanishing saving propensity and the stable distribution will become Gamma-like for uniform saving propensity of the traders [9] and initially Gamma-like but crossing over to Pareto-like power-law decay when traders have non-uniform saving propensities [10]. These results are non-perturbative results; any non-vanishing saving propensity will induce these features; the saving propensity magnitudes only determine the most-probable income (wealth) or the income (wealth) crossover point for Pareto tail of the distribution.

AS: Can we come back to your journey towards econophysics? Apart from Saha-Srivastava's book, any influence from other books, especially from economics?

BKC: After Graduation and Post-Graduation from Calcutta University, I joined, in early 1975, the Saha Institute of Nuclear Physics as a Research Fellow in Condensed Matter Statistical Physics for my Ph.D. degree. By that time I had a huge personal collection of (mostly cheap editions, reprinted in India), general books, text books, other books and monographs in subjects outside physics; primarily in philosophy and economics. I had attempted closer studies of some them including: *The Problems of Philosophy*, Bertrand Russell (Cambridge Univ.), Oxford Univ. Press, Oxford (1959); *Mathematical Logic & the Foundations of Mathematics: An Introductory Survey*, Geoffrey Thomas Kneebone (Univ. London), D. van Nostrand Co. Ltd., London, UK, (1963); *The Problems of Philosophy*, Satischandra Chatterjee (Univ. Calcutta), Calcutta Univ. Press, Kolkata (1964); *The Philosophy of Wittgenstein*, George Pitcher (Princeton Univ.), Prentice-Hall Inc., New Delhi, India, (1964); *An Introduction to Philosophical Analysis*, John Hospers (Univ. Southern California), Prentice-Hall Inc., New Delhi (1971); Economics, Paul A. Samuelson (MIT), Tata-McGraw Hill, New Delhi (1971); and *Economic Theory & Operations Analysis*, William J. Baumol (Princeton Univ.), Prentice-Hall Inc., New Delhi (1978).

I tried to go through some of the isolated chapters or sections of these books, which I could understand, enjoyed, or liked most. Occasionally, I got excited and tried my own analysis, following them, on some interesting problems or discussions coming in my way. One such piece was a paper on 'Indeterminism and Freedom' by Bernard Berofsky of Columbia University, published in 1975, perhaps in Philosophical Quarterly. Among others, it also alluded to quantum physics in defending his thesis on freedom. I wrote a note detailing my criticisms and posted that to the author. The author, from the Department of Philosophy, Philosophy Hall, Columbia University in the City of New York, wrote to me the following letter on 17 June 1975 (see Figure 2):

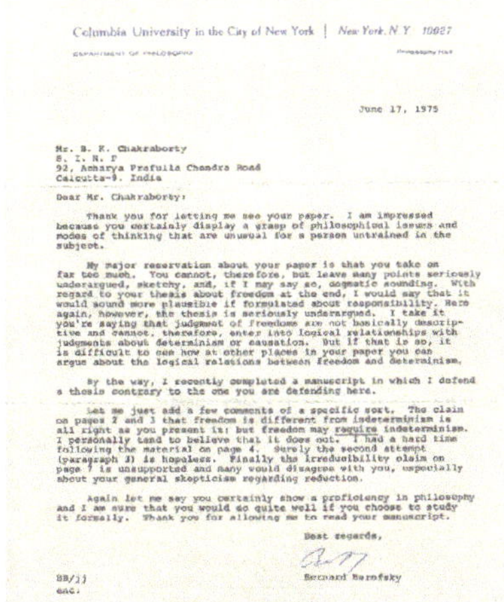

Columbia University in the City of New York | New York, N.Y. 10027

June 17, 1975

Mr. B. K. Chakraborty
S. I. N. P
92, Acharya Prafulla Chandra Road
Calcutta-9. India

Dear Mr. Chakraborty:

Thank you for letting me see your paper. I am impressed because you certainly display a grasp of philosophical issues and modes of thinking that are unusual for a person untrained in the subject.

My major reservation about your paper is that you take on far too much. You cannot, therefore, but leave many points seriously underargued, sketchy, and, if I may say so, dogmatic sounding. With regard to your thesis about freedom at the end, I would say that it would sound more plausible if formulated about responsibility. Here again, however, the thesis is seriously underargued. I take it you're saying that judgment of freedoms are not basically descriptive and cannot, therefore, enter into logical relationships with judgments about determinism or causation. But if that is so, it is difficult to see how at other places in your paper you can argue about the logical relations between freedom and determinism.

By the way, I recently completed a manuscript in which I defend a thesis contrary to the one you are defending here.

Let me just add a few comments of a specific sort. The claim on pages 2 and 3 that freedom is different from indeterminism is all right as you present it: but freedom may require indeterminism. I personally tend to believe that it does not. I had a hard time following the material on page 4. Surely the second attempt (paragraph 3) is hopeless. Finally the irreducibility claim on page 7 is unsupported and many would disagree with you, especially about your general skepticism regarding reduction.

Again let me say you certainly show a proficiency in philosophy and I am sure that you would do quite well if you choose to study it formally. Thank you for allowing me to read your manuscript.

Best regards,

Bernard Berofsky

BB/jj
enc.

Dear Mr. Chakraborty:

Thank you for letting me see your paper. I am impressed because you certainly display a grasp of philosophical issues and modes of thinking that are unusual for a person untrained in the subject.

My major reservation about your paper is that you take on far teo much. You cannot, therefore, but leave many points seriously under-argued, sketchy, and, if I may say so, dogmatic sounding. With regards to your thesis about freedom at the end, I would say that it would sound more plausible if formulated about responsibility. There again, however, the thesis is seriously underargued. I take it you're saying that judgment of freedoms are not basically descriptive and cannot, therefore, enter into logical relationships with judgements about determinism or causation. But if that is so. it is difficult to see how at other places in your paper you can argue about the logical relations between freedom and determinism.

By the way, I recently have completed a manuscript in which I defend a thesis contrary to the one you are defending here.

Let me just add a few comments of specific sort. The claim on pages 2 and 3 that freedom is different from indeterminism is all right as you present it. But freedom may REQUIRE indeterminism. I personally tend to believel that it does not. I had a hard time following the material on page 4. Surely the second attempt (paragra-ph 3) is hopeless. Finally the irreducibility claim on page 7 is unsupported and many would disagree with you, especially about your general skepticism regarding reduction.

Again let me say you certainly show a proficiency in philosophy and I am sure that you would do quite well if you choose to study it formally. Thank you for allowing me to read your manuscript.

Best regards,
Bernard Berofsky

Figure 2. Reply (dated 17 June 1975) from Bernard Berofsky of the Philosophy Department of Columbia University to BKC on his criticisms of Bernard's paper on 'Indeterminism and Freedom'.

AS: Obviously, you did not follow his suggestion, in fact, cordial invitation, to switch over to Philosophy. Why did you not?

BKC: Though I was seriously thinking of switching over to philosophy in a formal way, following Bernard's suggestion, some quick apparent success in my physics research publications with the newly developed Renormalization Group theory in those days kind of blinded me and left me with two minds. Somehow, I failed to take a decision and continued with my physics research until I practically forgot about the other choice! In late 1978, I submitted my Ph.D. thesis in Condensed Matter Physics to the University of Calcutta and got the degree in 1979, and, by the end of that year, I left for post-doctoral research studies in the Theoretical Physics Department of the University of Oxford and the Institute of Theoretical Physics, University of Cologne.

I came back and joined the Saha Institute of Nuclear Physics as Lecturer in 1983, and I started my research in statistical physics with four Ph.D. students joining me simultaneously (including Subhrangshu Sekhar Manna, who later developed the 'Manna Model', belonging to the 'Manna Universality Class'). Soon the statistical physics research in our group became so engaging and happening (with sixteen Ph.D. students, so far, getting their Ph.D. degrees and several of them becoming quite well known later for their pioneering research studies and still collaborating with me), I did not get much time until early nineties when I decided to try some research on 'economics-inspired physics'. I went back to the problem Saha and Srivastava addressed in their textbook mentioned above and I co-organized a conference in January 1995, together with some established Indian statistical physicists and (reluctant!) economists as participants. In the Proceedings of the Conference, I published (together with an economist Sugata Marjit) my first paper [11] dealing with statistical physics of Income distribution and related problems.

By the end of the year, as a part of the StatPhys-Kolkata II (series of International Conferences organized by us in Kolkata every 3–4 years, latest event StatPhys-Kolkata X, held end of 2019), we had organized a special session on 'Economics-Inspired Physics' and Eugene Stanley in his talk coined the term 'Econophysics' and had put that in the title of his paper [12] published in the Proceedings of the conference in Physica A, vol 224 (1996).

Though econophysics was quite risky as a topic of Ph.D. research in the late nineties (even today; still no faculty position in econophysics in our country, or for that matter, hardly exists elsewhere in the world), two brave students (Anirban Chakraborti and Arnab Chatterjee) expressed forcefully their desire to join the research on eventual topic of 'econophysics'. I was also fortunate, my colleague Sitabhra Sinha also joined us in such investigations. In the last 25 years, since that conference, significant developments have taken place in the subject, and many of them will be covered this special issue of Entropy.

AS: We will come back to those developments later. I understand, most of the papers on econophysics research studies are published in physics journals and not in economics journals. What is the cultural level of appreciation by the intellectuals today?

BKC: This is indeed very difficult to answer. To tell very frankly, the response so far is not very supportive or encouraging! Although, it must be mentioned, the term econophysics has now entered in dictionaries of economics (see, e.g., Reference [2]) and Encyclopedias of social science and philosophy (see, e.g., Reference [13]). That brings me to an interesting, rather recent, correspondence with my old philosopher 'guide' Bernard Berofsky in January 2013, after thirty-seven years! This was quite accidental, when I came across in my internet search a new book published by him. I contacted him (giving a link to my homepage) saying, "sorry, I could not follow your advice so far and had been very shy to contact you. Now that I have become sixty, I acquired sufficient courage and ...". Bernard immediately responded (see Figure 3) praising the development of econophysics due to the philosophical impulses of physicists.

-- Original Message --
Subject: Re: Hope you still remeber me!
From: "Bernard Berofsky" <...>
Date: Mon, February 11, 2013 3:59 am
To: bikask.chakrabarti@saha.ac.in

Dear Bikas,

It looks as if you were very wise not to pick up on my suggestion to study philosophy. You have established an incredible career for yourself as a physicist with the synthesizing impulses of a philosopher. So you have used your philosophical impulses in a most creative and fecund manner.

I am amazed and flattered that you have retained that letter. You must have quite a storehouse of memories.

We have in common that we have both published books by Cambridge and Oxford. My 2012 book by Oxford (Nature's Challenge to FreeWill) represents the culmination of my thinking about free will. (That is probably inaccurate; I cannot help but continue to think about it. But I am sure I will never undertake a major venture of that sort.)

It is very nice to hear from you and congratulations on all your wonderful achievements.

Best wishes,
Bernie

Figure 3. Mail from Bernard Berofsky of the Philosophy Department of Columbia University, in response to BKC's surprise contact mail in 2013 (after almost thirty-seven years!), appreciating and identifying the development of econophysics as one due to the "physicist(s) with synthesizing impulses of a philosopher ... (using) philosophical impulses in a most creative and fecund manner".

AS: Do you see really a philosophy behind econophysics?

BKC: Yes, indeed. I wrote about it earlier also (see, e.g., References [13,14]). I am not aware of all the documents on the mutual connection between philosophy and econophysics. I mentioned earlier about the entry on Econophysics in the Encyclopedia of Philosophy and Social Sciences [13]. I came to know of a rather recent entry on Social Ontology in The Stanford Encyclopedia of Philosophy [15] which, in the context of 'social atomism', writes "The idea is to model societies as large aggregates of people, much as liquids and gases are aggregates of molecules, ...". Then, after introducing the readers to two historical examples of Quetelet in 1848 [Adolphe Quetelet, 1848, Du système social et des lois qui le régissent, Paris: Guillaumin] and of Spencer in 1895 [Herbert Spencer, 1895, The Principles of Sociology, New York: Appleton] it says "Contemporary representatives include models in sociophysics and econophysics (see Chakrabarti et al., 2007) ... [which] take a society or market to be an aggregate of these interacting individuals [Bikas K. Chakrabarti, Anirban Chakraborti and Arnab Chatterjee, 2007, Econophysics and Sociophysics: Trends and Perspectives, Hoboken, NJ: Wiley]".

Let me now go back to the main to the main discussion and reiterate my basic argument in favor of considering economics as a natural science. Our knowledge about truth can, epistemologically speaking, be either deductive or inductive. Mathematics is an usual example of the deductive knowledge (though not all of it can be deduced from axiomatic logic); mathematical truths do not require any laboratory test or 'observational' support from the 'nature' to prove or validate them. Linguistically speaking, it is like the tautology "A bachelor does not have wife". One does not need to check each and every bachelor to confirm truth of the statement—the first part of the sentence confirms the second part. The same is true about the statement "two plus two equals four". Mathematical truths are analytical truths; left-hand side equals (in every intention and content) the right-hand side.

Mathematics, therefore, is not directly a natural science [16–18], though it has been at the root of the logical structure of many natural sciences, particularly physics. Natural sciences, however, are basically inductive in nature. They are based on natural or (controlled) laboratory observations. The statement "The sun rises every twenty four hours on the east in the morning" is not a tautology nor an analytical truth. Though east may be defined as the direction, and morning may be defined by the time of sunrise, that it rises every twenty four hours is an inductive (or empirically observed) truth and, therefore, tentative (and not like mathematical truths, which are analytical and certain). Natural sciences start with observations and end in observations, using both inductive reasoning or logic; in-between, they often employ the tools of deductive logic, mathematics (as most condensed form of deductive logic).

AS: So, the tools of mathematics and logic are employed to find and establish relationships among these 'natural' observations to develop natural sciences. Where does then economics belong to?

BKC: That is the crucial question. Intermediate analysis using mathematics is just applied mathematics, and can not be considered as (pure) mathematics. Any branch of natural science does that. Economics has been and will be a part of natural science, where natural observations, not much of controlled or laboratory observations, need to be analyzed employing deductive logic and mathematics. Economics, therefore, should naturally belong to natural sciences!

AS: Agreed. But why econophysics?

BKC: You see, in natural sciences today, there are several branches or disciplines, like physics, chemistry, biology, geology, etc. The differences are not natural and certainly nature did not create them: they are human creations. The demarcations among these disciplines are not always clear. As we mentioned earlier, there are clear differences (in the nature of logic employed) between mathematics and natural sciences. But that does not extend to the branches of natural sciences. In a white light spectrum, our color perception continuously change from violet to red (without any sharp boundary) as the wavelength changes in this collection of electromagnetic waves. Similar are the cases of the different branches of natural science. They are not strictly differentiable; are historical in origin and continued by us for our own convenience (during upbringing; like perhaps religion; both are man-made). Of course, it is hard today to be an expert in the whole of even one branch of natural science. We, therefore, try to learn and acquire expertise in one sub-branch or a sub-sub-branch of natural science. An unique feature of the sub- or sub-sub-branches of natural science is that an established 'truth' or a 'fundamental law' in one branch does not become 'false' or 'wrong' in another; only importance varies from discipline to discipline; quantum physics or gravity laws do not become invalid or wrong in chemistry or biology or mineralogy. Only gravity laws may be less important in chemistry or biology or mineralogy, and vice-versa. Models of geomagnetism in earth science cannot be built upon a law contradicting Maxwell's laws of electromagnetism. Developments in younger branches in science, therefore, profitably utilized earlier established laws or ideas in older branches of natural science. Many of the early successful scientists (even some mathematicians) happen to have been identified as physicists, and, consequently, physics has become like an 'elder brother' among natural sciences, and it is now equipped with a huge armory of ideas, laws and models to comprehend the nature. Economics as a relatively newer entrant to natural science can, therefore, expect gainful advantages from such econophysical attempts!

AS: I remember you once told me that the concept of modeling dynamics of physical systems and of economics systems are fundamentally different. Can you elaborate the point in this context?

BKC: I do not remember which point we had been discussing. However, there is a typical one which may be discussed. Modeling dynamics of a physical system, like a particle, using, say, the Newton's equation of motion, gives its dynamical state at a later time t by solving the equation of motion and utilizing the information regarding its dynamical state at an earlier time (say, at $t = 0$; called initial conditions). Exact solutions may not be possible as in the thermodynamic or many-body systems, but based on the statistical characterization of the state of the system at an earlier time, the dynamical formulation helps solving the statistics of the system for any future time. The economic agents or organizations, even under nominally identical economic situation, may have (continually upgradeable) anticipation about the future and the model dynamics need to accommodate, along with their initial economic state, such anticipatory factors (which are continually adjusted or learned through the ongoing dynamics itself!) to solve for the future. Such self-consistent 'learning' dynamics of physical systems are not typical, though some recent many-body game theoretic models, with iterative learning for optimal use of scarce resources as in the binary-choice Minority Game (see, e.g., Reference [19]), or many-choice Kolkata Paise Restaurant Problem (see, e.g., Reference [20]) naturally incorporate such evolving learning features in the self-correcting dynamics themselves. We will discuss some details of the later problem here. In any case, these studies are new and still very limited in scope.

AS: To summarize, though many of you had started your econophysics research studies more than twenty five years back, since Gene Stanley coined the term econophysics in 1996 (in his publication [12] in the Proceedings of the second StatPhys-Kolkata Conference), and many more physicists joined after that, the subject is not established yet.

BKC: You are partly right. In fact, physicists have long been trying to formulate and comprehend various problems of economics. As mentioned before, since 1931, the statistical physics modeling of income and wealth distributions are being tried. However, these older physics attempts had been sporadic and isolated ones; physicists, successful in such attempts, like Jan Tinbergen (Economics Nobel Prize winner in 1969; had Ph.D. in statistical Physics under Paul Ehrenfest of Leiden University), had to migrate to economics department. Since 1996 (more correctly perhaps since 1991, when Rosario Mantegna published his paper [21] on Milan stock exchange data modeling), however, the situation has changed considerably. Physicists are now investigating economics problems along with their students and colleagues from the same department and are publishing their econophysical research papers in physics journals (in around 2000, Econophysics had been assigned the Physics and Astronomy Classification Scheme (PACS) number 89.65Gh by the American Institute of Physics).

I personally think, however, that an intensive and successful branch of econophysics research started with Scott Kirkpatrick and coworkers in 1983 when they proposed [22] the idea and technique of 'Simulated Annealing' (or 'Classical Annealing') to get practical solutions of the computationally hard multi-variable optimization problems, like the (managerial) economics problem of the Traveling Salesman Problem (TSP), using tuning (annealing) of Boltzmann-type fluctuations (simulating thermal ones) to escape from the local minima to reach eventually one of the (degenerate) global minima of the cost function (travel distance). This is a very successful story of the application of (statistical) physics to solve a problem which in nature and basic intent a (financial) economics problem involving multi-variable optimization. It may be noted in this connection that the technique has since been applied to all branches of science, as well as technology, and the original paper [22] has received major attention of scientists and engineers (so far having received more than 48,000 citations, according to Google Scholar). This idea still continues leading to a very intriguing and active domain of research in computationally hard problems of optimizations, using statistical physics and physics of spin glasses. This eventually led to the concept and technique of 'Quantum Annealing' (or of 'Stochastic Quantum Computing'), where simulated quantum fluctuations (instead of simulated thermal fluctuations) are profitably used to tunnel through high but narrow local barriers [23], separating the global minima or solutions (see, e.g., Reference [24] for a brief review on solving TSP using quantum annealing). As I discussed earlier in my Econophysics-Kolkata Story [25], we started in 1986 (see Section 3.1) investigations on the statistical physics of the TSP. Soon my student Parangama Sen joined the effort [26]. (She eventually concentrated more on

Sociophysics and developed, among others, the Biswas-Chatterjee-Sen model, see, e.g., Reference [27], for collective opinion formation together with our students Soumyajyoti Biswas and Arnab Chatterjee. In this connection, let me take the opportunity to acknowledge the contributions of my other students, Srutarshi Pradhan, Asim Ghosh and Sudip Mukherjee, Suchismita Banerjee, and, of course, you, Antika, and of my colleagues in the Kolkata-econophysics group, namely Anindya Sundar Chakrabarti, Manipushpak Mitra, and Satya Ranjan Chakravarty, allowing us to make some significant contributions to econophysics, which we are going to summarize in the next section.)

AS: So, you think that successful research studies in econophysics already started with the Simulated Annealing paper by Kirkpatrick et al. in 1983, although econophysics research studies on more popular economics problems started in 1990s and, more specifically, after Stanley coined the term in 1996?

BKC: Yes, you are right. We will discuss in little more details (in the next section; Section 3.1) the impact of statistical physics in developing the Simulated (Classical or Quantum) Annealing techniques for the financial computation problems involving multi-variable optimization of the Traveling Salesman type. The inspiring success of the classical annealing technique, initiated by the Simulated Annealing method, has led to several intriguing developments in statistical physics and to many applications in computer science. Further potential extension in the context of solving NP-hard problems using quantum annealing has become one of the core research topic today in quantum many-body (statistical) physics and in quantum computation. Indeed I consider this outstanding development of simulated (classical or quantum) annealing techniques (starting with Kirkpatrick et al. [22]; also see Reference [23]) for the Traveling Salesman type multi-variable optimization problems to be a landmark achievement in the true spirit of econophysics. Of course, the present phase of econophysics research activities stemmed from several influential papers, analyzing financial market fluctuations, by Rosario Mantegna and Eugene Stanley and in particular following the publication of Kolkata Conference Proceedings paper [12] by Stanley et al. in 1996.

3. Major Achievements and Publications of the 'Kolkata School'

Physicist Victor Yakovenko and economist J. Barkley Rosser in their pioneering interdisciplinary collaborative review [28] in the Reviews of Modern Physics (2009) on econophysics of income and wealth distributions, discussed about some of the 'influential' and 'elegant' papers from the 'Kolkata School'. We will briefly summarize in this section some of our major research studies in econophysics (including those on wealth distributions).

3.1. *Traveling Salesman Problem and Simulated (Classical & Quantum) Annealing*

As already discussed, the Traveling Salesman Problem or TSP is, in its intent and structure, a computationally involved financial management problem (see, e.g., References [29,30]). The problem can be easily defined as a geometric one. Suppose in an unit square area there are N random dots, representing the cities. The salesman has to make a visit to all the cities and come back with minimum travel cost. The travel cost to visit all these cities will depend on the total travel distance of the tour. Each component of the travel distance between any two cities can be taken as the Euclidean distance (or as appropriate for the spatial metric, say Cartesian) between them. One can easily check that there are $N!/2$ (growing faster than exponential in N) distinct tours or trips to visit all the N cities. Obviously, all of these trips do not have identical value ('cost') for the total travel length (D), and the problem is to find the trips(s) which will correspond to the minimum travel distance D. Searching over all the possible trips soon becomes impossible as N becomes large, and there is no perturbative way to improve on any randomly chosen travel path to reach the global solution. At any point or city on a tour, there are N order choices for the next move or visit and the optimization problem of the total travel distance is truly a multi-variable one. It may be noted that the problem becomes trivial in one dimension (homes or offices placed randomly on a straight road), where the salesman can start from one end of the road and finish at the other end). Generally, for two dimension onwards, search time for such a minimum 'cost' (from among $\exp(N)$ number of trips or configurations), cannot be bounded by any (deterministic) polynomial in N (NP-hard problem).

From now onwards, let us concentrate our discussion on TSP in two dimension. The scale of the total travel distance, however, can be easily guessed using a 'mean field' picture. If N randomly placed points (cities) fill an unit (normalized) area, then the 'average' or 'mean' area per city is $1/N$, giving nearest neighbor distance to be of order $1/\sqrt{N}$ and total travel distance $D = \Omega\sqrt{N}$. Numerical estimates suggests $\Omega \simeq 0.71$ [31].

The problem is truly global in nature. Choice of the next city to visit depends on the position of even the farthest city in the country. However, one can approximately solve the problem (see References [32,33]) by reducing it to an effective one-dimensional problem where the country (unit square) is divided into hypothetical parallel strips of width w and the salesman visits the cities within each strip in a 'directed' way and the total travel distance D is optimized with respects to single variable w (optimal value then grows as $\sqrt{N}$) and gives (see, e.g., References [32,33]) $\Omega \simeq 0.92$. Another way is to put the cities randomly with concentration ρ on the lattice sites of, say, an unit square lattice. The lattice constraint can help then the calculation of the optimal travel distance. The optimal (normalized) travel path length then scales as $D = \Omega\sqrt{\rho}$. At $\rho = 1$, the lattice constraints would immediately imply that the global search problem reduce to a local one and all the space filling Hamiltonian walks would correspond to optimized tour with $\Omega = 1$. In the approximate single variable solution (minimization of D with respect to w) indicated above, the strip width w grows as $1/\sqrt{\rho}$ as ρ decreases. For $\rho \to 0$, however, the lattice constraints disappear, and the problem reduces to TSP on continuum as defined earlier (NP-hard, $w \to \infty$, with $\Omega \simeq 0.71$ [31]). Where does the problem become NP-hard? This study was initiated by us (see References [26,34–37]) and they indicated (also Reference [32]) that the TSP on dilute lattices becomes NP-hard only at $\rho \to 0$ (though this is not settled yet and some arguments support that it crosses over to NP-hardness at $\rho = 1_-$ or as soon as ρ becomes less than unity).

As already mentioned earlier (in Section 2), a major computational breakthrough of TSP and other such multi-variable optimization problems came from the 1983 seminal paper on Simulated Annealing' [22] by Kirkpatrick et al., who proposed a novel stochastic technique, inspired by the metallurgical annealing process and statistical physics of frustrated systems.

Imagine a bowl on the table, and you need to 'locate' its bottom point. Of course, one can calculate the local depths (from a reference height) everywhere along the inner volume of the bowl and search for the point where the local depth is maximum. However, as every one would easily guess, a much simpler and practical method would be to allow rolling of a marble ball along the inner surface of the bowl and wait for locating its resting position. Here, the physics of the forces of gravity and friction allows us to 'calculate' the location of the bottom point in an analog way! Algorithm-wise, it is simple. For any possible move, if the changed 'cost' function has lower value, one should accept the move and reject it otherwise. Success for the search of the minimum is guaranteed. In principle, a similar trick would work for cases where the bowl becomes larger and its internal surface gets modulated, as long as the surface contour or 'landscape' has valleys all tilted towards the same bottom point location. Computationally hard problems arise when these valleys are separated by 'barriers', which are (macroscopically) high. The simulated annealing suggests a way out to overcome (at least for finite height barriers) by allowing moves costing higher to have (Gibbs-like) lower probability of acceptance.

To search for the optimized cost (travel distance in TSP or energy of the ground state(s) in spin glasses) at eventually vanishing level of noise (or 'simulated temperature'), one starts from a high noise (temperature) 'melt' phase, and tune slowly the noise level. In this 'simulated' process, the (classical) noise at any intermediate level of annealing allows for the acceptance of the changed 'costs' ΔD in distance or energy D: 100% acceptance of the move if $\Delta D < 0$ and acceptance of the move with a Gibbs-like probability $\sim \exp(-\Delta D/T)$

for moves with increased in cost ($\Delta D > 0$)). As the noise level (T) is slowly reduced during the annealing process, the gradually decreasing probability of accepting higher cost values, allows the system to come out of the local minima valleys and settle eventually in (one of) the 'ground state' (with minimum D) of the system. For slow enough decrease of noise $T(t)$ with time t, one can estimate the quasi-equilibrium (thermal) average of the cost function $< D >$ at ant time t and derive the effective 'specific heat' value $\delta < D > / \delta T$ as a function of t. One needs to be very slow ($|dT/dt|$ very small) when the effective specific heat increases with decreasing T, indicating the 'glass' transition point and anneal at faster rates on both sides of the transition point.

As has been indicated in the earlier section, it has been a remarkably successful trick for 'practical' computational solutions of a large class of multi-variable optimization problems, as in most multi-city travel cost optimizations and similar multi-variable optimizations (see, e.g., References [29,30,38,39]).

Though some 'reasonable' optimization can be achieved very quickly using appropriate annealing schedules, the search time for reaching the lowest cost state or configuration for NP-hard problems, however, grows still as $\exp(N)$. The bottleneck could be identified soon. Extensive study of the dynamics of frustrated random systems, like the N spin (two state Ising) glasses, particularly of the Sherrington-Kirkpatrick variety (see, e.g., Reference [23] also for a TSP version of the quantum annealing), showed that its (free) energy landscape (in the 'glass' phase), is extremely rugged, and the barriers, separating the local valleys, often become N order implying the search for the degenerate ground states from 2^N (or $N!/2$) states is NP-hard (for the N-city TSP). In the macroscopic size limit (N approaching infinity), therefore, such systems effectively become non-ergodic or localized, and the classical (thermal) fluctuations, like that in the simulated annealing, fail to help the system to come out of such high barriers (at random locations or configurations, not dictated by any symmetry) as the escape probability is of order $\exp(-N/T)$ only. Naturally, the annealing time (inversely proportional to the escape probability), to get the ground state of the N-spin Sherrington-Kirkpatrick model, cannot be bounded by any polynomial in N.

The idea proposed by Ray et al. [40] was that quantum fluctuations in the Sherrington-Kirkpatrick model can perhaps lead to some escape routes to ergodicity or quantum fluctuation induced delocalization (at least in the low temperature region of the spin glass phase) by allowing tunneling through such macroscopically tall but thin (free energy or cost functions) barriers which are difficult to scale using classical fluctuations. This is based on the observation that escape probability due to quantum tunneling, from a valley with single barrier of height N and width $\tilde{w}$, scales as $\exp(-\sqrt{N}\tilde{w}/\Gamma)$, where Γ represents the quantum (or tunneling) fluctuation strength (see Figure 4). This extra handle through the barrier width $\tilde{w}$ (absent in the classical escape probability of order $\exp(-N/T)$) can help in a major way in its vanishing limit. Indeed, for a single narrow ($\tilde{w} \to 0$) barrier of height N, when Γ is slowly tuned to zero, the annealing time to reach the ground state or optimized cost, will become N independent (even in the $N \to \infty$ limit; δ-function barriers are transparent to quantum fluctuations, while classical or thermal annealing to escape from such a barrier is impossible)! It has led to some important clues. Of course, complications (localization) may still arise for many such barriers at random 'locations'. In any case, with this observation and some more developments, the quantum annealing technique was finally launched through the subsequent publications of a series of landmark papers (both theoretical and experimental; see Reference [23]) and through a remarkable practical realization of the quantum annealers by the D-wave Group [41].

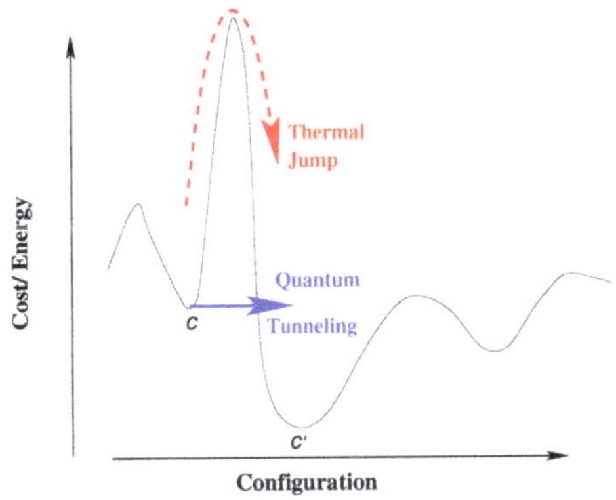

Figure 4. While optimizing the cost function of a computationally hard problem (like the minimum travel distance for the Traveling Salesman Problem (TSP)), one has to get out of a shallower local minimum, like the configuration C (travel route), to reach a deeper minimum C′. This requires jumps or tunneling, like fluctuations, in the dynamics. Classically, one has to jump over the energy or the cost barriers separating them, while quantum mechanically one can tunnel through the same. If the barrier is high enough, thermal jump becomes very difficult. However, if the barrier is narrow enough, quantum tunneling often becomes quite easy. Indeed, assuming the tall barrier to be of height N and width $\tilde{w}$, one can estimate (see, e.g., Reference [42]) the tunneling probability through the barrier to be of order $\exp[-(\tilde{w}\sqrt{N})/\Gamma]$, where Γ denotes the strength of quantum fluctuations (instead of the the classical escape probability of order $\exp[-N/T]$, T denoting the thermal or classical fluctuation strength).

Let us now conclude this subsection. Simulated Annealing technique, invented by Kirkpatrick et al. in 1983 [22], has since been applied extensively also to solve problems of collective decision making in economics and social sciences (see, e.g., Reference [43] for a recent review). As mentioned earlier [25], our group started investigations on statistical physics of TSP in 1986. The intriguing physics of Simulated Annealing inspired us to explore the possible further advantages of quantum tunneling (to allow escape through macroscopically tall but thin barriers in some NP-hard cases), where classical annealing (using thermal fluctuations) fails. This led finally to the quantum extension or to the invention of the quantum annealing technique, where our initial contributions (References [23,40]) are considered to be important and pioneering. See, e.g., Reference [24] for a brief review and Reference [44] for some recent discussions on the advantages of applying quantum annealing method to solve TSP. Quantum annealing is a very active research field today in quantum statistical physics and computation (see, e.g., References [45,46] for recent reviews).

3.2. *Social Inequality Measure and Kolkata Index*

Social inequality, particularly income or wealth inequality in, are ubiquitous. There are several indices or coefficients, used to measure them, the oldest and most popular one being the Gini index [47].

It is based on the Lorenz curve or function [48] $L(x)$, giving the cumulative fraction of (total accumulated) income or wealth possessed by the fraction (x) of the population, when counted from the poorest to the richest (see Figure 5). If the income (wealth) of every one would be identical, then $L(x)$ would be a straight line (diagonal) passing through the origin. This diagonal is called the equality line. The Gini coefficient (g) is given by the area between the Lorenz curve and the equality line (normalized by the area under the equality line: $g = 0$ corresponds to equality and $g = 1$ corresponds to extreme inequality.

We proposed [49] the Kolkata index or k-index given by the ordinate value of the intersecting point of the Lorenz curve and the diagonal perpendicular to the equality line (also see References [50–54]). By construction, $1 - L(k) = k$, saying that k fraction of wealth is being possessed by $(1-k)$ fraction of the richest population. As such, it gives a quantitative generalization of the approximately established (phenomenological) 80–20 law of Pareto [55], saying that, in any economy, typically about 80% wealth is possessed by only 20% of the richest population. Defining the complementary Lorenz function $L^{(c)}(x) \equiv [1 - L(x)]$, one gets k as its (nontrivial) fixed point (while Lorenz function $L(x)$ itself has trivial fixed points at $x = 0$ and 1). k-index can also be viewed as the normalized h-index [56] for social inequality; h-index is given by the fixed point value of the nonlinear citation function against the number of publications of individual researchers. We have studied the mathematical structure of k-index in Reference [53] (see Reference [54] for a recent review) and its suitability, compared with the Gini and other inequality indices or measures, in the context of different social statistics, in References [49–52]. In addition, see Reference [57] for redefining a generalized Gini index and Reference [58] for a recent application in characterizing the statistics of the spreading dynamics of COVID-19 pandemic in congested towns and slums of the developing world.

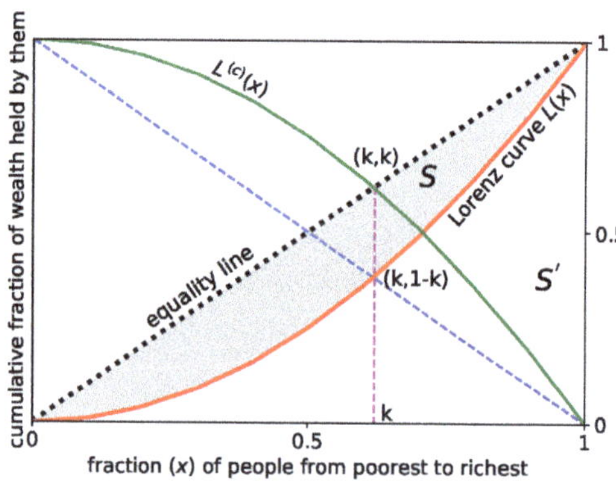

Figure 5. Lorenz curve (in red) or function $L(x)$ here represents the fraction of accumulated wealth against the fraction x of people possessing that, when arranged from the poorest to the richest. The diagonal from the origin represents the equality line. The Gini index (g) can be measured by the area (S) between the Lorenz curve and the equality line (shaded region), normalized by the total area $(S + S' = 1/2)$ under the equality line: $g = 2S$. The complementary Lorenz function $L^{(c)}(x) \equiv 1 - L(x)$ is shown by the green line. The Kolkata index (k) can be measured by the ordinate value of the intersecting point of the Lorenz curve and the diagonal perpendicular to the equality line. By construction, $L^{(c)}(k) = 1 - L(k) = k$, saying that k fraction of wealth is being possessed by $(1-k)$ fraction of richest population.

In summary, inspired by the observations of richer structure (self-similarity) of the (nonlinear) Renormalization Group equations near the fixed point (see, e.g., Reference [59]), or of the nonlinear chaos-driving maps near the fixed point (see e.g., Reference [60]) and noting that inequality functions, such as the Lorenz function $L(x)$ or the Complementary Lorenz function $L^{(c)}(x)$, to be generally nonlinear, we studied their nontrivial fixed points. As mentioned earlier, Lorenz function $L(x)$ has trivial fixed points (at $x = 0$ and 1), while the Complementary Lorenz function $L^{(c)}(x) \equiv 1 - L(x)$ has a nontrivial fixed point at $x = k$, the Kolkata index [49]. It also offers a tangible interpretation: k-index gives the fraction k of the total wealth possessed by the rich $(1-k)$ fraction of the population and gives a quantitative generalization of the Pareto's 80–20 law [55]. As discussed earlier, it can also be viewed as a normalized h-index of social inequality. Some unique features of Kolkata Index (k) may be noted: (a) Gini and other indices are mostly some average quantities based on the Lorenz function $L(x)$, which has trivial fixed points. k is a fixed point of the Complementary Lorenz function $L^{(c)}(x)$ and, if one considers the simplest form of Lorenz function $L(x) = x^2$, then $k = (\sqrt{5} - 1)/2$, inverse of the Golden Ratio [51]. (b) k

gives the fraction of wealth possessed by the rich $1 - k$ fraction of the population. As such, it provides a quantitative generalization of the Pareto's 80-20 law (see, e.g., Reference [55]). The observed values of k index for most of the cases of social inequalities [50–54] seem to fall in the range 0.80-0.86 (though have much smaller values today for world economies, presumably because of various welfare measures). (c) k-index is equivalent to a normalized version of the Hirsch-Index (h). While h corresponds to the fixed point of the publication success rate (measured by the integer numbers of citations) falling off nonlinearly with number of papers by individual academicians, k corresponds to the fixed point (fraction) of $1 - L(x)$, where $L(x)$ gives the nonlinearly varying fraction of cumulative wealth possessed by the increasing (from poor to rich) fraction of population in any society.

3.3. *Kinetic Exchange Model of Income and Wealth Distributions*

We have discussed already in Section 2 some details of the Kinetic Exchange model of income and wealth distributions. In an ideal gas, in thermal equilibrium, the number density $n(\epsilon)$ of (Newtonian) particles (atoms or molecules) having kinetic energy ϵ is given by

$$n(\epsilon) = g(\epsilon)f(\epsilon, T) \sim \sqrt{(\epsilon)}\exp(-\epsilon/\Delta), \quad (1)$$

where Δ is a constant, the density of states $g(\epsilon) \sim \sqrt{\epsilon}$ (coming from the counting of three-dimensional momenta vectors which correspond to the same kinetic energy) and $f(\epsilon) \sim \exp(-\epsilon/\Delta)$ (coming from stochastic, or entropy maximizing, scatterings between the particles, conserving their total kinetic energy). As discussed already in Section 2, to get the ideal gas equation of sate $PV = NKT$, where P and V denote, respectively, the pressure and volume of the gas at absolute temperature T, by calculating the pressure from the average transfer of momentum of the particles per unit area of the container (using Equation (1), Reference [7]), one identifies, $\Delta = KT$.

Following similar arguments [7,8] (also see Reference [28]), one gets (as discussed in Section 2) for the steady state distribution of the number $n(m)$ of agents in the market with income or wealth m.

$$n(m) = g(m)f(m) = c\exp(-m/\Delta'), \quad (2)$$

where $f(m) \sim \exp(-m/\Delta')$, with $g(m)$ a constant c (unlike in expression (1)), and Δ' as constants for the trading market. This is because, in a trading market, there is no production (growth) or decay, and the total amount of money (equivalent to energy in Kinetic theory of ideal gas), as well as the number of traders (buyers and sellers), remain fixed. Stochastic money exchanges in the trades involving indistinguishable buyers and sellers (who change their roles in different trades), keeping the buyer-seller total money in any trade to remain constant, lead to a distribution given by expression (2). This is also because there cannot be any equivalent of the particle momenta vectors for the agents in the market and hence the density of states $g(m)$ here is a constant. One must also have $\int_0^M n(m)dm = N$, the total number of traders in the market, and $\int_0^M mn(m)dm = M$, the total amount of money. These give the effective temperature of the market $\Delta' = M/N$, the average money in circulation in the market (economy).

As documented in several books and reviews (see, e.g., References [7,61,62]), the income or wealth distributions in any society have a Gamma function-like dip near zero income or wealth (unlike in the exponential distribution case discussed above, where the number density of pauper is the maximum). In addition, as is well known [62,63], the tail end of the distribution is known to be much more fat, described by the Pareto power law, and not by the thin exponentially decaying distribution.

As mentioned in the earlier section, following Saha and Srivastava's indication in their book [7], we explored how the kinetic theory of trading markets indicated above could be extended to accommodate a Gamma-like distribution at the least and explore further to capture the Pareto tail of such a distribution, as well.

We noted that many of the economics text books, in their chapters on Trades, discuss the saving propensity of the traders (habit of saving a fraction of the income or wealth possessed by the trader and do trade with the rest). We immediately realized [9,10], if one introduces the saving propensity of each trader, so that each trader saves a fraction of their individual money (wealth) before the trade and allows (random) exchanges of the rest amount in the trade (keeping the total money or wealth, including the saved portions, conserved), the traders will never become paupers. Unlike in the random exchange case (as in kinetic theory of gases, where one trader may lose its entire amount of money or wealth to the other in any trade), here, to lose the entire amount of money acquired at any point of time, the trader has to lose every time after that as the trader continues the successive trades (and, consequently, the saved portion becomes infinitesimal). The number density of paupers (having zero wealth) will become zero for any non-vanishing saving fraction of the traders and the exponential distribution will become unstable and the resulting steady state distributions will capture the above mentioned desirable features. This is a non-perturbative result; any amount of saving by the traders will induce this feature.

With uniform saving, the exponential distribution collapses and the stable distribution becomes Gamma-like [9] (also see Reference [64] for a micro-economic derivation of the kinetic exchange equations from the Cobb-Dauglas utility maximization with money saving propensity of the traders, and Reference [65] for extended microeconomic formulation of Kinetic exchange models, having economic growths, by incorporating additional saving of the production in the utility maximization equation). The steady state distribution becomes initially Gamma-like but crossing over to Pareto-like power-law decay when traders have non-uniform saving propensities [10]. The saving propensity magnitudes determine the most-probable income (wealth) and the income (wealth) crossover point for Pareto tail of the distribution (see References [63,66,67] for details).

It may be mentioned in this connection that one kind of saving by the traders, considered early by our group (including the students Anirban Chakrabarti and Srutarshi Pradhan) can, in fact, lead to wealth condensation or extreme inequality. When two randomly selected traders agree to trade (in the so-called 'Yard Sale' trade mode), such that the richer one among them will retain or save the extra money or wealth compared to that of her trade partner, the dynamics will eventually lead to aggregation of the entire amount of money or wealth in the hand of one trader, and the dynamics will stop. This happens because once any trader becomes pauper (loses entire amount of money or wealth), no other trader (with money) will engage in trade with her. Although this Yard Sale model has this uninteresting wealth condensation feature, it showed some interesting slow dynamics, and Anirban published that result [68]. Later, it was shown that inclusion of tax in the model, in the sense that a fraction of money is collected by the Government (non-playing member of the system) in every trade and, after some period of collections, redistributes the money among all (by investing on general social facilities, like road, hospital, etc., constructions, used equally by all in the society). Because of this general upliftment, the paupers come back to the trades and interesting steady state money distribution can emerge and such models of wealth distribution have become an active area of research (see, e.g., Reference [69] for a popular review on this development).

Kinetic model of gases and the kinetic theory is the first and extremely successful many-body theory in physics. Economic systems, markets in particular, are intrinsically many-body dynamical systems. Kinetic exchange models of markets may, therefore, be expected to provide the most successful models of market systems. In the kinetic exchange model, when one of the trader of a randomly chosen pair of traders is deliberately the poorest one at that instant of time (trade), the dynamics induces a self-organization in the market such that a 'poverty line' is spontaneously developed so that none of the trader remains below the emerged (self-organized) poverty threshold (see References [70,71] and references therein).

3.4. *Statistics of the Kolkata Paise Restaurant Problems*

Kolkata had, long back, very cheap fixed price 'Paise Restaurants' (also called 'Paise Hotels'; Paise is, rather was, the smallest Indian coin). These 'Paise Restaurant' were very popular among the daily laborers in the city. During lunch hours, these laborers used to walk down (to save the transport costs) from their place of work to one of these restaurants. These restaurants would prepare every day a (small) number of such dishes, sold at a fixed price (Paise). If several groups of laborers would arrive any day to the same restaurant, only one group would get their lunch, and others would miss their lunch that day. There were no cheap communication means in those days (like mobile phones) for mutual communications, for deciding the respective restaurants. Walking down to the next restaurant would mean failing to report back to work on time! To complicate this collective learning and decision-making problem, there were indeed some well-known rankings of these restaurants, as some of them would offer tastier items compared to the others (at the same cost, Paise, of course), and people would prefer to choose the higher rank of the restaurant, if not crowded! This 'mismatch' of the choice and the consequent decision not only creates inconvenience for the prospective customers (going without lunch), would also mean 'social wastage' (excess unconsumed food, services, or supplies somewhere).

A similar problem arises when the public administration plans and provides hospitals (beds) in different localities, but the local patients prefer 'better' perceived hospitals elsewhere. These 'outsider' patients would then have to choose other suitable hospitals elsewhere. Unavailability of the hospital beds in the over-crowded hospitals may be considered as insufficient service provided by the administration, and, consequently the unattended potential services will be considered as social wastage.

This kind of games [72] (see References [20,73] for recent reviews), anticipating the possible strategies of the other players and acting accordingly, is very common in society. Here, the number of choices need not be very limited (as in standard binary-choice formulations of most of the games, for example, in Minority Games [19,20,73]), and the number of players can be truly large! In addition, these are not necessarily one shot games, rather the players can learn from past mistakes and improve on their selection strategies for choosing the next move. These features make the games extremely intriguing and also versatile, with major collective or emerging social structures, not comparable to the standard finite choice, non-iterative games among finite number of players. Such repetitive collective social learning for a community sharing past knowledge for the individual intention to be in minority choice side in successive attempts are modeled by the 'Kolkata Paise Restaurant' (KPR) problem or, in short, by the 'Kolkata Restaurant' problem.

KPR is a repeated game, played among a large number of players or agents having no simultaneous communication or interaction among themselves. In KPR, the prospective players (customers/agents) choose from restaurants each day (time) in parallel decision mode, based on the past (crowd) information and their own (evolved or learned) strategies. There is no budget constraint to restrict the choice (and hence the solutions). Each restaurant has the same price for a meal but having a different rank, agreed upon by all the customers or players.

For simplicity, we may assume that each restaurant can serve only one customer (generalization to any fixed number of daily services for each would not change the complexion of the problem or game). If more than one customer arrives at any restaurant on any day, one of them is randomly chosen and is served, and the rest do not get meal that day. Information regarding the prospective customer or crowd distributions for the earlier days (up to a finite memory size) is made available to everyone. Each day, based on own learning and the developed (often mixed) strategies, each customer chooses a restaurant independent of the others. Each customer wants to go to the restaurant with the highest possible rank while avoiding a crowd so as to be able to get the meal there. Both from individual success and social efficiency perspective, the goal is to 'learn collectively' to utilize effectively the available resources.

The KPR problem seems to have a trivial solution: suppose that somebody, say a dictator (who is not a player), assigns a restaurant to each person the first day and asks them to shift to the next restaurant cyclically, on successive days. The fairest and most efficient solution: each customer gets food on each day (if the number of plates or choices is the same as that of the customers or players) with the same share of the rankings as others, and that too from the first day (minimum evolution time). This, however, is NOT a true solution of the KPR problem, where each customer or agent decides on his or her own every day, based on complete information about past events. In KPR, the customers try to evolve learning strategies to eventually to arrive at the best possible solution (close to the dictated solution indicated above). The time for this evolution needs also to be optimized; for example, a very efficient strategy, having convergence time which grows with the number of players (even linearly), is unsuitable for most of the social games, as our life-span is finite, and (in a democracy) the number of players or competitors cannot be restricted or bounded.

There have been many limiting formulations and studies using tricks from statistical physics and quantum physics (see, e.g., References [20,73–81]) and generalizations in computer science (see, e.g., Reference [82]) and mobility (vehicle on hire) markets (see, e.g., References [83,84]). We will present briefly in the next section some specific results of a new study on the nature phase transition and resource utilization in KPR with number of customers less (still very large) than the number of restaurants.

4. Some New Results for Statistics of the KPR Problem

Here, we consider the case where λN agents decide to choose among N available resources (for $\lambda < 1$). Every day each restaurant prepares one dish for lunch and serve it to the visitor. If, on some day, any restaurant is visited by more than one agent, then one of them is randomly chosen and served the prepared dish; the rest leave and have to starve for that day. Thus, every agent is required to make her choice such that the chosen restaurant will be visited alone by her (at most one agent arriving each restaurant) to assure her lunch that day. As λ is less than unity here, a fraction $(1-\lambda)$ of restaurants will any way go vacant any day. Additionally, a fraction $(1-f(t))$ of restaurants will go vacant on day (t) because of overcrowding at some restaurants due to fluctuations in choices of the prospective customers. On any day t, the average social success factor f for the agents, can be measured as

$$f(t) = \sum_{i=1}^{N}[\delta(n_i(t))/\lambda N], \quad (3)$$

with $\delta(n) = 1$ for $n = 1$ and $\delta(n) = 0$ otherwise; $n_i(t)$ denotes the number of agents arriving at the ith (rank) restaurant on day t. $[1-f(t)]$ gives the fraction of wastage due to fluctuation of choices and $[(1-\lambda)+(1-f(t)]$ gives the fraction of restaurants not visited by any agent on day t. The goal is to achieve $f(t) = 1$ preferably in finite convergence time (τ), i.e., for $t \geq \tau$, or at least as $t \to \infty$.

As usual, a dictated solution is extremely simple and efficient: A dictator asks everyone to form a queue for visiting the restaurants in order of their respective positions in the queue and then asks them to shift their positions by one step (rank) in the next day (assuming periodic boundary condition). Everyone gets the food and the steady state (t-independent) social utilization fraction $f = 1$. This is true even when the restaurants have ranks (agreed by all the agents or customers).

However, in democratic set-up, this dictated solution is not acceptable and the agents or players are expected to evolve their strategy to make the best minority choice independently (without presence of any dictator), using the publicly available information about the past record of crowd sizes in different restaurants, such that each arrives alone there in the respective restaurant and gets the dish. The more successful such collective learning, the more is the aggregated utilization fraction f. Earlier studies (see e.g., References [20,72,74,85–88] strategies for KPR game. Recently authors in Reference [81] have proposed two such stochastic strategies (strategy I and strategy II) where the agents collectively learn to make their decisions utilizing the publicly available history of crowd size of the last day's chosen restaurant. Below, we briefly discuss them.

Strategy I:

On day t, an agent goes back to her last day's visited restaurant k with probability

$$p_k^{(I)}(t) = [n_k(t-1)]^{-\alpha},\ \alpha > 0. \tag{4}$$

If $n_k(t-1) > 1$, each of the $n_k(t-1)$ agents or players try to arrive at the same k-th restaurant next day t with the above probability and chooses a different one $(k' \neq k)$ among any of the neighboring restaurants n_r on day t, with probability

$$p_{k'}^{(I)}(t) = (1 - p_k^{(I)}(t))/n_r. \tag{5}$$

Strategy II:

On day t, an agent tries to go back to the same restaurant as chosen the earlier day $(t-1)$ with probability

$$p_k^{(II)}(t) = 1, \text{ if } n_k(t-1) = 1 \text{ and} \tag{6}$$

$$p_{k'}^{(II)}(t) = p < 1, \text{ if } n_k(t-1) > 1 \tag{7}$$

for choosing any of the n_r neighboring restaurants $(k' \neq k)$.

4.1. Numerical Results

We have numerically studied the steady state dynamics of the KPR game where every day λN agents decide which restaurant to choose and visit among N restaurants following both the strategies I and II. We consider here infinite dimensional arrangement for restaurants, where the number of nearest neighboring restaurants n_r to each is $(N-1)$, and the cost to visit any of them is the same for all the time. The maximum social utilization f obtained from Equation (3) (from the point of view of agents or players) will be denoted further by f^a. Each day (iteration), parallel choice decisions by each are processed (following either strategy I or II) and used to compute f^a. Steady state is identified as the state when f^a does not change (within a predefined error margin) for the next (say, hundred) iteration.

On day t, $n_i(t-1)$, agents decide to revisit last day's visited restaurant (i) with probability $p_k^{(I)}(t)$ (Equation (4)) or probability $p_k^{(II)}(t)$ (Equation (6)), or else choose any other $(k' \neq k)$ from among any of the $(N-1)$ neighboring restaurants for both the strategies (Equations (5) and (7)). After the system stabilizes, ($f^a(t)$ becomes practically independent of t, the average statistics of $f^a(t)$ are noted as $[f^{a(I)}]$ or $[f^{a(II)}]$, respectively, for strategies I and II. We find the power law fits for the steady state wastage fraction $(1 - f^{a(I)}) \sim (1 - f^{a(II)}) \sim (\lambda - \lambda_c(N))^{\beta}$ with $\beta = 1.0 \pm 0.05$ (see Figures 6 and 7) and $\tau^{(I)} \sim \tau^{(II)} \sim (\lambda_c(N) - \lambda)^{-\gamma}$ with $\gamma = 0.5\pm\ 0.07$ (see Figures 8 and 9) in both of the strategies I and II. Varying λ, the steady state results of f^a, τ for different system sizes ($N = 500, 1000, 2000$), with $\alpha = 0.05, 0.25, 0.5, 1.0$ in strategy I or $p = 0.2, 0.4, 0.6, 0.8$ in strategy II are considered here. All simulations are done taking maximum $N = 2000$ with numbers of iteration/run of order 10^6. For finite system sizes, the effective critical points $\lambda_c(N)$ (where f^a becomes unity or τ reaches its peak value) obtained numerically for different system sizes (N) and are analyzed using finite size scaling method in Figure 10.

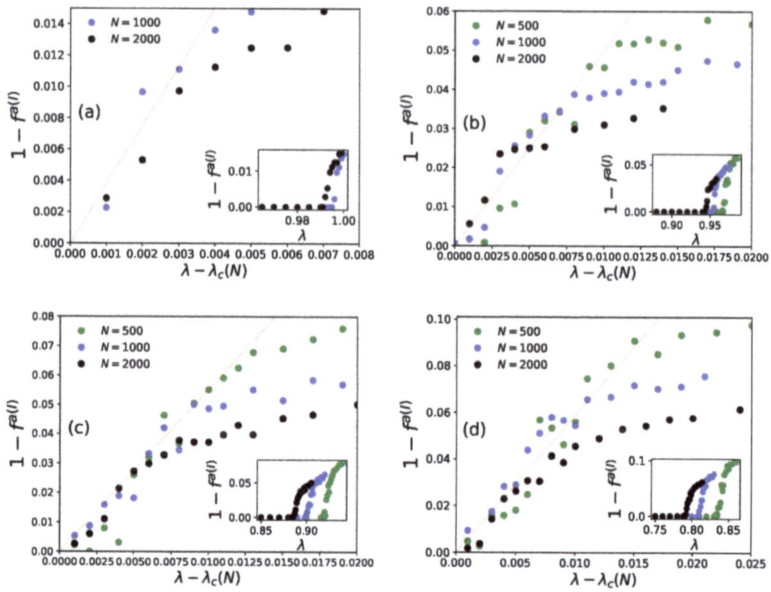

Figure 6. Plots of $(1 - f^{a(I)})$ against $\lambda - \lambda_c(N)$ following strategy I at (**a**) $\alpha = 0.05$, (**b**) $\alpha = 0.25$, (**c**) $\alpha = 0.5$, (**d**) $\alpha = 1.0$. A power law holds for $(1 - f^{a(I)}) \sim (\lambda - \lambda_c(N))^{\beta}$, where $\beta = 1.0 \pm 0.05$. The insets show direct relationship between $(1 - f^{a(I)})$ and λ (for strategy I).

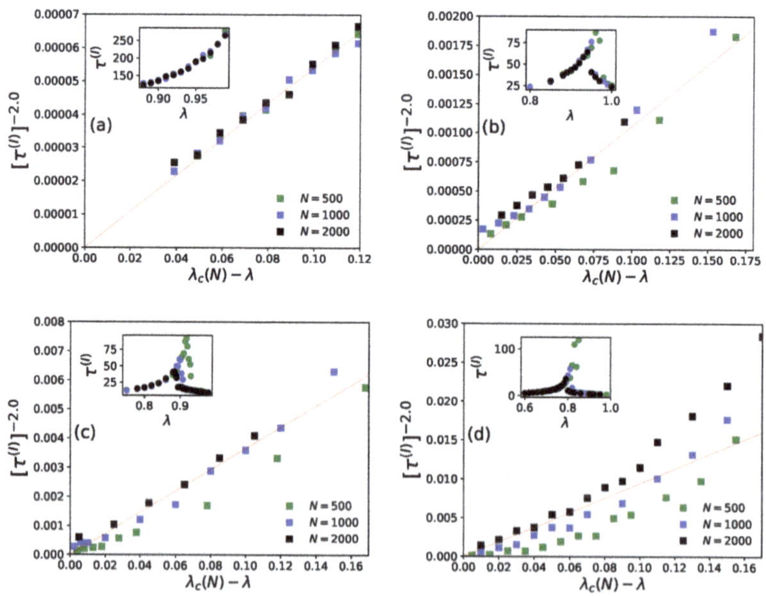

Figure 7. Plots of steady state convergence time $\tau^{(I)}$ from strategy I against $\lambda_c(N) - \lambda$ at (**a**) $\alpha = 0.05$, (**b**) $\alpha = 0.25$, (**c**) $\alpha = 0.5$, (**d**) $\alpha = 1.0$. A power law holds for $\tau^{(I)} \sim (\lambda_c(N) - \lambda)^{-\gamma}$, where $\gamma = 0.5 \pm 0.05$. The insets plot direct relationship between $\tau^{(I)}$ and λ for different system sizes (for strategy I), also showing the variation of λ as α increases.

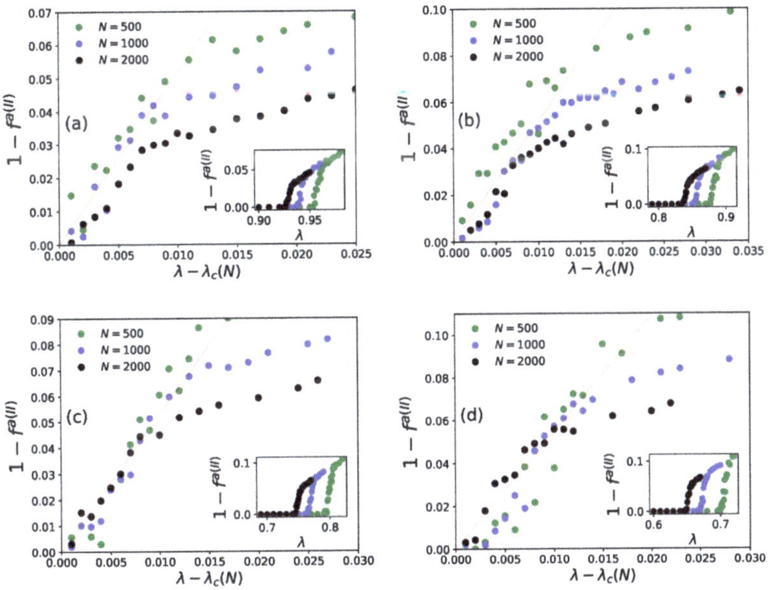

Figure 8. Plots of $(1 - f^{a(II)})$ versus $\lambda - \lambda_c(N)$ following strategy II at (**a**) $p = 0.8$, (**b**) $p = 0.6$, (**c**) $p = 0.4$, (**d**) $p = 0.2$. A power law holds for $(1 - f^{a(II)}) \sim (\lambda - \lambda_c(N))^{\beta}$ with $\beta = 1.0 \pm 0.05$. The insets show direct relationship between variations of $(1 - f^{a(II)})$ against λ (for strategy II).

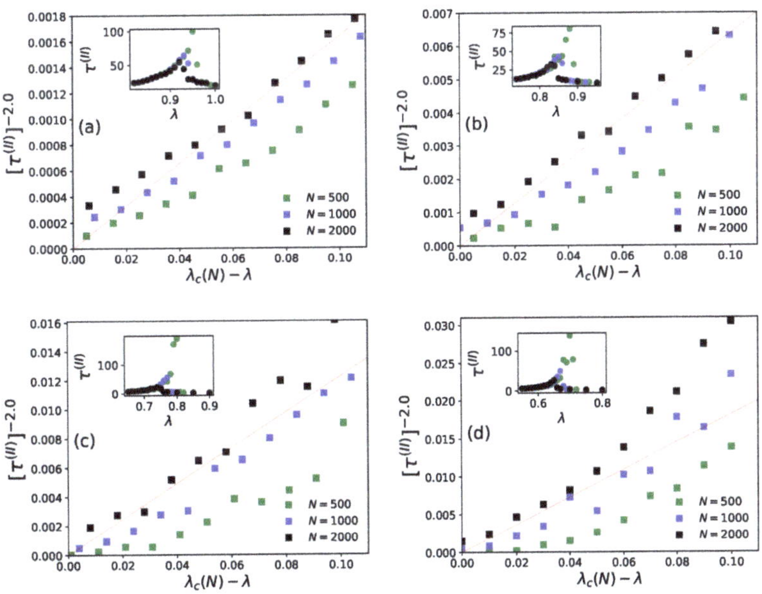

Figure 9. Plots of steady state convergence time $\tau^{(II)}$ against $\lambda_c(N) - \lambda$ following strategy II at (**a**) $p = 0.8$, (**b**) $p = 0.6$, (**c**) $p = 0.4$, (**d**) $p = 0.2$. A power law holds for $\tau^{(II)} \sim (\lambda_c(N) - \lambda)^{-\gamma}$ with $\gamma = 0.5 \pm 0.07$. The insets give direct relationship between $\tau^{(II)}$ and λ for different system sizes (for strategy II), also showing the variation of λ as p decreases.

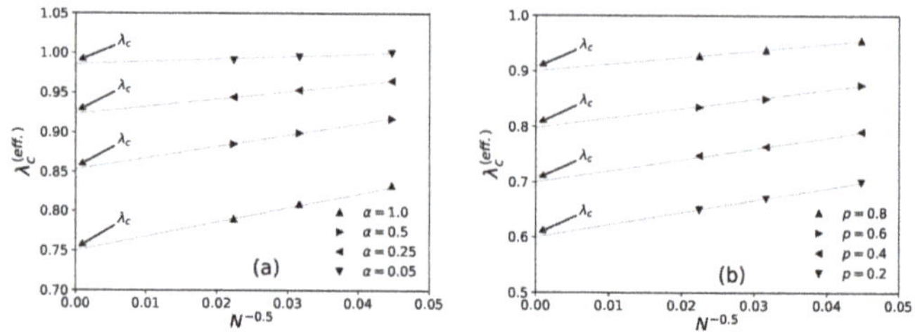

Figure 10. Extrapolation study of the effective finite size critical density of agents $\lambda_c(N)$. The system size dependence is numerically fitted to $\frac{1}{\sqrt{N}}$ and we estimate λ_c from $\lambda_c \equiv \lambda_c(N \to \infty)$. The extrapolated values of λ_c are $0.99, 0.92, 0.85, 0.75$ for $\alpha = 0.05, 0.25, 0.5, 1.0$ (strategy I) (**a**), and are $0.9, 0.8, 0.7, 0.6$ for $p = 0.8, 0.6, 0.4, 0.2$ (strategy II) (**b**).

It may be mentioned that, in general, for the estimation of errors in the exponents β and γ in Figures 6–9, we tried linear fits (without any intercept) for $\log y$ versus $\log x$, using best fits mostly for the intermediate range data points for all N values until they start deviating (due to extreme fluctuations near $\lambda = \lambda_c$ and towards their saturation values for λ approaching unity [74,81]) and anticipating their universal mean field values in this infinite dimensional system. From the slopes of these best fit lines for different α or p values, we extract the universal exponent values and their standard deviations. We give this higher error in the estimate of the unified (and universal) estimate of γ.

4.2. Summary

KPR is a multi-agent multi-choice repeated game where players try to learn from their past successes or failures, utilizing publicly available information on the crowd sizes at different restaurant in the past to decide which restaurant to visit that day such that she would be alone there for being served the only prepared dish. Here, asymmetric case such that λN ($\lambda < 1$) agents are considered against N restaurants, for sufficiently large N. End of each day (iteration), we have measured social utilization for agents $f^a(t) = \sum_{i=1}^{N}[\delta(n_i(t))/\lambda N]$ where $n_i(t)$ denotes number of customers visiting ith restaurant on day t.

As shown in Figure 6 (for strategy I) and Figure 8 (for strategy II), the social wastage fraction $(1 - f^a)$ vanishes at the effective critical point $\lambda_c(N)$ with the critical exponent β value near unity. In addition, from Figure 7 (for strategy I) and Figure 9 (for strategy II), we see that the the convergence or relaxation time t, required for f^a to stabilize, divergence near the same critical points $\lambda_c(N)$ for the respective strategies, with the exponent γ value about 1/2. Additionally, the finite size scaling analysis $\lambda_c(N) \sim \lambda_c + const.N^{-1/(d\nu)}$, where λ_c corresponds to $\lambda_c(N)$ for N going to infinity and d corresponds to the effective dimension, suggests the effective correlation length exponent $d\nu$ value to be around 2 for both the strategies, as expected for such mean field (infinite range systems).

In Reference [81], we have studied the dynamics of the KPR game following the same two strategies for the case $\lambda = 1$. For $\lambda = 1.0$, where the critical points λ_c (for both the strategies) vanish, the universality class (values of the critical exponents β and γ were observed to be distinctly different, and this point needs further investigations. We may, however, note that, since at $\lambda = 1$, the number of both agents and restaurants are same (N), full social utilization (where $f^a = 1 = f^r$, occurring at $\alpha = 0_+$ for strategy I and at $p = 1_-$ for strategy II) induces an additional frustrating constraint in the collective choice dynamics involved here.

The KPR game models have been extended already and used to study real life problems, like resource allocation in Internet of Things [82], vehicle for hire [83], matching in mobility markets [84], etc. We hope the KPR game models will be utilized much more effectively in the context of much wider practical areas of collective learning dynamics and choices.

5. Future of Econophysics: Some Perspective

One often says that the main purpose of economic activity is to optimize the limited funds of labor and capital, natural and technical resources and capital resources, to satisfy our (practically) unlimited needs. "Economic science is therefore the science of efficiency, and as such, it is a quantitative science." [89] (also see Reference [90]). We have already argued [14] in Section 2 that epistemologically economics belongs to natural science (and not mathematics). It begins with observation which are to be analyzed using logic or mathematics and eventually should end in observation, as in all natural sciences. Since 1990s, most Universities of the world offer Science Graduation degrees (Bachelor of Science or Master of Science degrees) in economics (in addition to Bachelor of Arts or Master of Arts from Fine Arts or Humanities Departments).

Robert Solow [91] pointed out that, in the 1940s, economics had been basically a descriptive and institutional subject for a 'gentleman scholar'. The textbooks of those days were 'civilized' and discursive. ... "Formal analysis were minimal and it made economics the domain of intuitive economists". He concluded his summary of the state of economics near the end of the 20th century "with a paraphrase of Oscar Wilde's description of a fox hunt - 'the unspeakable in pursuit of the inedible'-saying that perhaps economics was an example of 'the over-educated in pursuit of the unknowable'." [91]. Despite the ongoing controversies today in the field of economics, the "New Millennium economists are far more comfortable with what they do after the changes in the structure and content of economics over the last half century" [92]. The root cause of these changes have been identified by Colander [92] to be due to the rise of Complexity Science since early 1980s. In fact, concepts from physics had continually been absorbed into the main stream economic formulation of ideas and models. As Venkat Venkatasubramanian noted in his recent book [93], "Concepts such as equilibrium, forces of supply and demand, and elasticity reveal influence of classical mechanics on economics. The analytical model of utility-based preferences can be traced back to Daniel Bernoulli, the great Swiss mathematical physicist from nineteenth century. One of the founders of neoclassical economics, Irving Fisher, was trained under the legendary Yale physicist, Jisiah Willard Gibbs, a co-founder of the discipline of statistical mechanics. Similarly, Jan Tinbergen, who shared the first Nobel Prize in Economics in 1969, was the doctoral student of the great physicist Paul Ehrenfest at Leiden University".

Indeed, more specifically as discussed in Section 2, we would like to correlate these changes to occur following the successful development in econophysics of the Simulated Annealing technique [22] in 1983 for Traveling Salesman type multi-variable optimization problems, and other successive developments in econophysics of analyzing correlations in stock prices (see, e.g., Reference [3,21]) or the kinetic exchange modelings of income and wealth distributions (see, e.g., References [28,63]). The statistical physics of TSP, as an example of successful developments in econophysics, had already been introduced in our 2010 econophysics textbook [32], which has been the only 'suggested textbook' (since inception in 2012) of the formal course on econophysics, offered (by Diego Garlaschelli) at the Physics Department of the Leiden University (see the course prospectus for 2012–2013 through that of 2020–2021 [94]), where one of the first Nobel-laureates in economics Jan Tinbergen came from.

Econophysics has come as an exceptional development in interdisciplinary sciences (see, e.g., Reference [95] for a popular exposition on this development). Historically, economics, more specifically social sciences, belonged to the Humanities departments and not of Science. For earlier interdisciplinary developments of Astrophysics, Biophysics, or Geophysics, the scenario and ambiance had been quite different. The mother departments had been parts of the same science schools and even the corresponding resources, like books, journals, and also the faculty, had strong overlaps and could be shared. The marriage negotiations for Econophysics have been difficult, though extremely desirable and natural; as the saying goes: "marriage between the King of natural sciences with the Queen of social sciences!"

Regular interactions and collaborations between the communities of natural scientists and social scientists are, however, rare, even today! Though, as mentioned already, interdisciplinary research papers on econophysics and sociophysics are regularly being published at a steady and healthy rate, and a number of universities (including Universities of Bern, Leiden, London, Paris, and Tufts University) are offering the interdisciplinary courses on econophysics and sociophysics, not many clearly designated professor positions, or other faculty positions for that matter, are available yet (except for econophysics in Universities of Leiden and London). Neither are there designated institutions on these interdisciplinary fields, nor separate departments or centers of studies for instance. Of course, there have been several positive and inspiring attempts and approaches from both economics and finance side (see, e.g., References [96,97], along with a number of those [66,67,98–100] from physics, which have already been appreciated in the literature). Indeed, the thesis [101] in August 2018, Department of History and Philosophy of Science, University of Cambridge, by financial economist Christophe Schinckus (one of the co-editors of this special issue), says that "In order to reconstruct the subfield of econophysics, I started with the group of the most influential authors in econophysics and tracked their papers in the literature using the Web of Science database of Thomson-Reuters (The sample is composed of: Eugene Stanley, Rosario Mantegna, Joseph McCauley, Jean-Pierre Bouchaud, Mauro Gallegati, Benoît Mandelbrot, Didier Sornette, Thomas Lux, Bikas Chakrabarti and Doyne Farmer). These key authors are often presented as the fathers of econophysics simply because they contributed significantly to its early definition and development. Because of their influential and seminal works, these scholars are actually the most quoted authors in econophysics. Having the 10 highest quoted fathers of econophysics as a sample sounds an acceptable approach to define bibliometrically the core of econophysics". In addition, the entry on 'Social Ontology' in The Stanford Encyclopedia of Philosophy [15], as discussed in Section 2, confirms positive impact of such econophysics and sociophysics research studies on the overall modern philosophical outlook of social sciences.

We may note, however, a recently published highly acclaimed massive (580 page) book [96] on economics ('landmark volume', said E. Roy Weintraub, 'creative, elegant and brilliant work', said W. Brian Arthur and 'written by master economists', said D. Colander) by (Late) Martin Shubik (Ex-Seymour Knox Professor Emeritus of Mathematical Institutional Economics, Yale University and Santa Fe Institute) and Eric Smith (Santa Fe Institute) discussed extensively on econophysics approaches and in general on the potential of interdisciplinary research studies inspired by the developments in natural sciences. Getting somewhat excited, I wrote to Martin Shubik in late 2016 that their book can also serve as an outstanding 'white-paper' document in favor of a possible Proposal for an International Center for Interdisciplinary Studies on Complexity in Social Sciences. He immediately responded and gave his impression about the difficulties involved and indicated very briefly about the minimal financial and structural requirements (both my letter to him and his response is appended below (Figures 11 and 12).

-- Original Message --
Subject: RE: For your comments and suggestions
From: "Shubik, Martin" <martin.shubik@yale.edu>
Date: Sun November 27, 2016 11:26 pm
To: "bikask.chakrabarti@saha.ac.in" <bikask.chakrabarti@saha.ac.in>
Cc: "Smith Eric" <...>

Dear Professor Chakrabarti

Thank you for your kind comments. They are much appreciated. Interdisciplinary institutes are very difficult to organize and hard to fund. Unfortunately given the combination of my age (90) and an auto-immunesystem disease (Inclusion Body Myositis) I do not have the energy to look for the seed money that I estimate at least at $5,000,000. It is worth trying and to some extent the Santa Fe Institute succeeded. You need at least 4-5 founding members in different disciplines.

With regards
Martin Shubik

Figure 11. The first part of the email conversation between (late) Martin Shubik and BKC. Second part (email from BKC; appended to this part) is continued in Figure 12. The precise suggestions made in this immediate response indicate Shubik's prior plan for such 'interdisciplinary institutes' in economics.

-- Original Message --
From: Bikas K. Chakrabarti <bikask.chakrabarti@saha.ac.in>
Sent: Sunday, November 27, 2016 3:43 AM
To: Shubik, Martin <martin.shubik@yale.edu>
Subject: For your comments and suggestions

Professor Martin Shubik
Professor Emeritus of Mathematical Institutional Economics
Yale University.

Dear Professor Shubik,

Let me first introduce myself. I am a Professor of Physics at Saha Institute of Nuclear Physics, Kolkata, India (also a Visiting Prof. of Economics at the Indian Statistical Institute, Kolkata).

I found your recent book (with Eric Smith), The Guidance of an Enterprise Economy, MIT Press (2016),wonderful and extremely exciting. In particular, your specific addresses (and extensive discussions thought the book) towards multidisciplinary audience, including physicists and biologists, are truly inspiring and will be of far reaching consequences (I am particularly delighted to see five of our publications cited in your esteemed book!). This address, from an economist and scientist of your stature, will surely make major impact in the near future.

Indeed, it seems it is time to try for an international center for interdisciplinary studies on complexity in social and natural sciences. The model of the Abdus Salam International Centre for Theoretical Physics (ICTP), Trieste (funded in 1964 by UNESCO and IAEA), could surely be a helpful guide here. In view of the extreme interdisciplinary nature of, say econophysics and sociophysics, such efforts may be strengthened by instituting an international visiting center (may be under the aegis of UNESCO) where natural (physical, biological, etc) and social scientists and students from different universities and institutions of the world can meet for extended periods to participate in interdisciplinary workshops, discuss and interact on various interdisciplinary issues and collaborate for such researches and developments. Unlike in centers/ departments where the faculty members already initiated/ trained in such researches can work, such an international visiting center, where stalwarts in each of the basic disciplines can be invited to interact through collaborations/workshops/discussions with interested uninitiated younger researchers.

This is in brief the point on which I would like to receive your comments, whatever they may be, and suggestions, if any.

Thanking you, with kind regards.
Bikas K. Chakrabarti

Figure 12. Email conversation in the end of 2016 between (late) Martin Shubik and BKC regarding interdisciplinary developments in economics and the possibility of setting up an International Center for Interdisciplinary Studies on Complexity in Social Sciences. This email from BKC was appended to the response email (Figure 11) from Shubik. The (Yale) date and time mark in the mail-header (and that for BKC's in Figure 11, on arrival in Kolkata) indicate hardly any time gap between the two and the readiness with the precise suggestions indicate Shubik's prior thinking in similar line.

This ready and specific comments by Shubik clearly suggests that he actually had thought about the need of such an International Center for fostering interdisciplinary research which needs to be more inclusive than, for example, the Santa Fe Institute. The model of the Abdus Salam International Center for Theoretical Physics (ICTP), Trieste (funded by UNESCO and IAEA), was considered to provide helpful guidance for us here. It was contemplated, if an ICTP-type interdisciplinary research institute could be initiated for research studies on econophysics and sociophysics (see, e.g, Reference [102]). Though Shubik (who died in 2018 at the age of 92) agreed also to be one of its founding members, we could not make any progress yet. We may also note that Dirk Helbing and colleagues have been trying for an European Union funded 'Complex Techno-Socio- Economic Analysis Center' or 'Economic and Social Observatory' for the last decade or so (see Ref. [103] containing the White Papers arguing for their proposed project). We are also aware that Indian Statistical Institute had taken a decision to initiate a similar Center in India (see 'Concluding Remarks' in Reference [104]).

Hope, some such international visiting centers will come up soon and with them the spread of such interdisciplinary ideas will achieve more coherence and will lead to major success in such research studies.

Author Contributions: Data curation, A.S.; Writing—review & editing, B.K.C. and A.S. All authors have read and agreed to the published version of the manuscript.

Funding: This research received no external funding.

Institutional Review Board Statement: Not applicable.

Informed Consent Statement: Not applicable.

Data Availability Statement: Not applicable.

Acknowledgments: We acknowledge all our colleagues (mentioned by name in Section 2) for the collaborations. BKC is grateful to J.C. Bose National Fellowship (DST, Govt. of India) grant for support.

Conflicts of Interest: The authors declare no conflict of interest.

References

1. Gangopadhyay, K. Interview with Eugene H. Stanley. *IIM Kozhikode Soc. Manag. Rev.* **2013**, *2*, 73–78. [CrossRef]
2. Rosser, J.B., Jr. Econophysics. In *New Palgrave Dictionary of Economics*; Durlauf, S.N., Blume, L.E., Eds.; Palgrave Macmillan: London, UK, 2008; Volume 2, pp. 729–732.
3. Mantegna, R.N.; Stanley, H.E. *An Introduction to Econophysics*; Cambridge University Press: Cambridge, UK, 2000.
4. Galam, S.; Gefen, Y.; Shapir, Y. Sociophysics: A mean behavior model for the process of strike. *J. Mathe. Sociol. Scimago* **2000**, *9*, 1–13.
5. Galam, S. *Sociophysics: A Physicist's Modeling of Psycho-Political Phenomena*; Springer: New York, NY, USA, 2012.
6. Chakrabarti, B.K. Econophysics as conceived by Meghnad Saha. *Sci. Cult. Indian Sci. News Assoc.* **2018**, *84*, 365–369.
7. Saha, M.N.; Srivastava, B.N. *A Treatise on Heat*; Indian Press: Allahabad, India, 1931; p. 105.
8. Dragulescu, A.; Yakovenko, V.M. Statistical mechanics of money. *Eur. Phys. J. B-Condens. Matter Complex Syst.* **2000**, *17*, 723–729. [CrossRef]
9. Chakraborti, A.; Chakrabarti, B.K. Statistical mechanics of money: How saving propensity affects its distribution. *Eur. Phys. J. B-Condens. Matter Complex Syst.* **2000**, *17*, 167–170. [CrossRef]
10. Chatterjee, A.; Chakrabarti, B.K.; Manna, S.S. Pareto law in a kinetic model of market with random saving propensity. *Phys. A Stat. Mech. Appl.* **2004**, *335*, 155–163. [CrossRef]
11. Chakrabarti, B.K.; Marjit, S. Self-organisation and complexity in simple model systems: Game of life and economics. *Indian J. Phys. IACS* **1995**, *69B*, 681–698.
12. Stanley, H.E.; Afanasyev, V.; Amaral, L.A.N.; Buldyrev, S.V.; Goldberger, A.L.; Havlin, S.; Leschhorn, H.; Maass, P.; Mantegna, R.N.; Peng, C.-K. Anomalous fluctuations in the dynamics of complex systems: From DNA and physiology to econophysics. *Phys. A Stat. Mech. Appl.* **1996**, *224*, 302–321. [CrossRef]
13. Chakrabarti, B.K. Econophysics. In *Encyclopedia of Philosophy and the Social Sciences*; Kaldis, B., Ed.; Sage: Thousand Oaks, CA, USA, 2013; Volume 1, pp. 229–230.
14. Chakrabarti, B.K. Can economics afford not to become natural science? *Eur. Phys. J. Spec. Top.* **2016**, *225*, 3121–3125. [CrossRef]

15. Epstein, B. Social Ontology. In *The Stanford Encyclopedia of Philosophy*; Stanford University: Stanford, CA, USA, 2018. Available online: https://plato.stanford.edu/entries/social-ontology/ (accessed on 8 January 2021).
16. Whitehead, A.N.; Russell, B. *Principia Mathematica*; Cambridge University Press: Cambridge, UK, 1910; Volume I.
17. Whitehead, A.N.; Russell, B. *Principia Mathematica*; Cambridge University Press: Cambridge, UK, 1912; Volume II.
18. Whitehead, A.N.; Russell, B. *Principia Mathematica*; Cambridge University Press: Cambridge, UK, 1913; Volume III.
19. Challet, D.; Marsili, M.; Zhang, Y.-C. *Minority Games: Interacting Agents in Financial Markets*; Oxford University Press: Oxford, UK, 2005.
20. Chakrabarti, B.K.; Chatterjee, A.; Ghosh, A.; Mukherjee, S.; Tamir, B. *Econophysics of the Kolkata Restaurant Problem and Related Games: Classical and Quantum Strategies for Multi-Agent, Multi-Choice Repetitive Games*; Springer: Cham, The Netherland, 2017.
21. Mantegna, R.N. Lévy walks and enhanced diffusion in Milan stock exchange. *Phys. A Stat. Mech. Appl.* **1991**, *179*, 232–242. [CrossRef]
22. Kirkpatrick, S.; Gelatt, C.D.; Vecchi, M.P. Optimization by simulated annealing. *Science* **1983**, *220*, 671–680. [CrossRef]
23. Das, A.; Chakrabarti, B.K. Colloquium: Quantum annealing and analog quantum computation. *Rev. Modern Phys.* **2008**, *80*, 1061–1081. [CrossRef]
24. Santoro, G.E.; Tosatti, E. Optimization using quantum mechanics: Quantum annealing through adiabatic evolution. *J. Phys. A Math. Gener.* **2006**, *39*, R393–R431. [CrossRef]
25. Chakrabarti, B.K. Econophys-Kolkata: A short story. In *Econophysics of Wealth Distributions*; Chatterjee, A., Yarlagadda, S., Chakrabarti, B.K., Eds.; Springer: Milan, Italiy, 2005; pp. 225–228.
26. Sen, P.; Chakrabarti, B.K. Travelling salesman problem on dilute lattices: Visit to a fraction of cities. *J. Phys.* **1989**, *50*, 255–261. [CrossRef]
27. Sen, P.; Chakrabarti, B.K. *Sociophysics: An Introduction*; Oxford University Press: Oxford, UK, 2014.
28. Yakovenko, V.M.; Rosser, J.B., Jr. Colloquium: Statistical mechanics of money, wealth, and income. *Rev. Modern Phys.* **2009**, *81*, 1703. [CrossRef]
29. Orman, A.J. Williams, H.P. A survey of different integer programming formulations of the travelling salesman problem. *Optim. Econom. Financ. Anal.* **2006**, *9*, 93–108.
30. Rasmussen, R. TSP in Spreadsheets–a Guided Tour. *Int. Rev. Econom. Educ.* **2011**, *10*, 94–116. [CrossRef]
31. Percus, A.G.; Martin, O.C. Finite size and dimensional dependence in the Euclidean traveling salesman problem. *Phys. Rev. Lett.* **1996**, *76*, 1188–1191. [CrossRef] [PubMed]
32. Sinha, S.; Chatterjee, A.; Chakraborti, A.; Chakrabarti, B.K. *Econophysics: An Introduction*; John Wiley & Sons: New York, NY, USA, 2010.
33. Beardwood, J.; Halton, J.H.; Hammersley, J.M. The shortest path through many points. In *Mathematical Proceedings of the Cambridge Philosophical Society*; Cambridge Press: Cambridge, UK, 1959; Volume 55, pp. 299–327. [CrossRef]
34. Chakrabarti, B.K. Directed travelling salesman problem. *J. Phys. A Math. Gen.* **1986**, *19*, 1273–1275. [CrossRef]
35. Dhar, D.; Barma, M. Chakrabarti, B.K.; Taraphder, A. The travelling salesman problem on a randomly diluted lattice. *J. Phys. A Math. Gen.* **1987**, *20*, 5289–5298. [CrossRef]
36. Ghosh, M.; Manna, S.S.; Chakrabarti, B.K. The travelling salesman problem on a dilute lattice: A simulated annealing study. *J. Phys. A Math. Gen.* **1988**, *21*, 1483–1486. [CrossRef]
37. Chakraborti, A.; Chakrabarti, B.K. The travelling salesman problem on randomly diluted lattices: Results for small-size systems. *Eur. Phys. J. B-Condens. Matter Complex Syst.* **2000**, *16*, 677–680. [CrossRef]
38. Bonomi, E.; Lutton, J.-L. The N-city travelling salesman problem: Statistical mechanics and the Metropolis algorithm. *SIAM Rev.* **1984**, *26*, 551–568. [CrossRef]
39. Zhou, A.-H.; Zhu, L.-P.; Hu, B.; Deng, S.; Song, Y.; Qiu, H.; Pan, S. Traveling-salesman-problem algorithm based on simulated annealing and gene-expression programming. *Information* **2019**, *10*, 7. [CrossRef]
40. Ray, P.; Chakrabarti, B.K.; Chakrabarti, A. Sherrington-Kirkpatrick model in a transverse field: Absence of replica symmetry breaking due to quantum fluctuations. *Phys. Rev. B* **1989**, *39*, 11828–11832. [CrossRef] [PubMed]
41. Johnson, M.W.; Amin, M.H.S.; Gildert, S.; Lanting, T.; Hamze, F.; Dickson, N.; Harris, R.; Berkley, A.J.; Johansson, J.; Bunyk, P. Quantum annealing with manufactured spins. *Nature* **2011**, *473*, 194–198. [CrossRef] [PubMed]
42. Mukherjee, S.; Chakrabarti, B.K. Multivariable optimization: Quantum annealing and computation. *Eur. Phys. J. Spec. Top.* **2015**, *224*, 17–24. [CrossRef]
43. Lucas, A. Ising formulations of many NP problems. *Front. Phys.* **2014**, *2*, 1–15. [CrossRef]
44. Dong, Y.; Huang, Z. An Improved Noise Quantum Annealing Method for TSP. *Int. J. Theor. Phys.* **2020**, *59*, 3737–3755. [CrossRef]
45. Tanaka, S.; Tamura, R.; Chakrabarti, B.K. *Quantum Spin Glasses, Annealing and Computation*; Cambridge University Press: Cambridge, UK, 2017.
46. Albash, T.; Lidar, D.A. Adiabatic quantum computation. *Rev. Modern Phys.* **2018**, *90*, 015002. [CrossRef]
47. Gini, C. Measurement of inequality of incomes. *Econ. J.* **1921**, *31*, 124–126. [CrossRef]
48. Lorenz, M.O. Methods of measuring the concentration of wealth. *Publ. Am. Stat. Assoc.* **1905**, *9*, 209–219. [CrossRef]
49. Ghosh, A.; Chattopadhyay, N.; Chakrabarti, B.K. Inequality in societies, academic institutions and science journals: Gini and *k*-indices. *Phys. A Stat. Mech. Appl.* **2014**, *410*, 30–34. [CrossRef]
50. Ghosh, A.; Chatterjee, A.; Inoue, J.; Chakrabarti, B.K. Inequality measures in kinetic exchange models of wealth distributions. *Phys. A Stat. Mech. Appl.* **2016**, *451*, 465–474. [CrossRef]

51. Chatterjee, A.; Ghosh, A.; Chakrabarti, B.K. Socio-economic inequality: Relationship between Gini and Kolkata indices. *Phys. A Stat. Mech. Appl.* **2017**, *466*, 583–595. [CrossRef]
52. Sinha, A.; Chakrabarti, B.K. Inequality in death from social conflicts: A Gini & Kolkata indices-based study. *Phys. A Stat. Mech. Appl.* **2019**, *527*, 121185.
53. Banerjee, S.; Chakrabarti, B.K.; Mitra, M.; Mutuswami, S. On the Kolkata index as a measure of income inequality. *Phys. A Stat. Mech. Appl.* **2020**, *545*, 123178. [CrossRef]
54. Banerjee, S.; Chakrabarti, B.K.; Mitra, M.; Mutuswami, S. Social Inequality Measures: The Kolkata index in comparison with other measures. *Front. Phys.* **2020**, *8*, 562182. [CrossRef]
55. Available online: https://en.wikipedia.org/wiki/Pareto_principle (accessed on 28 January 2021).
56. Hirsch, J.E. An index to quantify an individual's scientific research output. *Proc. Natl. Acad. Sci. USA* **2005**, *102*, 16569–16572. [CrossRef] [PubMed]
57. Subramanian, S. More tricks with the Lorenz curve. *Econ. Bull.* **2015**, *35*, 580–589.
58. Sahasranaman, A.; Jensen, H.J. Spread of Covid-19 in urban neighbourhoods and slums of the developing world. *arXiv* **2020**, arXiv:2010.06958.
59. Fisher, M.E. Renormalization group theory: Its basis and formulation in statistical physics. *Rev. Modern Phys.* **1998**, *70*, 653–681. [CrossRef]
60. Feigenbaum, M.J. Universal behavior in nonlinear systems. *Phys. D Nonlinear Phenom.* **1983**, *7*, 16–39. [CrossRef]
61. Chatterjee, A.; Yarlagadda, S.; Chakrabarti, B.K. *Econophysics of Wealth Distributions*; Springer: Milano, Italiy, 2005.
62. Chatterjee, A.; Chakrabarti, B.K. Kinetic exchange models for income and wealth distributions. *Eur. Phys. J. B* **2007**, *60*, 135–149. [CrossRef]
63. Chakrabarti, B.K. Chakraborti, A.; Chakravarty, S.R.; Chatterjee, A. *Econophysics of Income and Wealth Distributions*; Cambridge University Press: Cambridge, UK, 2013.
64. Chakrabarti, A.S.; Chakrabarti, B.K. Microeconomics of the ideal gas like market models. *Phys. A Stat. Mech. Appl.* **2009**, *388*, 4151–4158. [CrossRef]
65. Quevedo, D.S.; Quimbay, C.J. Non-conservative kinetic model of wealth exchange with saving of production. *Eur. Phys. J. B* **2020**, *93*, 186. [CrossRef]
66. Pareschi, L.; Toscani, G. *Interacting Multiagent Systems: Kinetic Equations and Monte Carlo Methods*; Oxford University Press: Oxford, UK, 2013.
67. Ribeiro, M.B. *Income Distribution Dynamics of Economic Systems: An Econophysical Approach*; Cambridge University Press: Cambridge, UK, 2020.
68. Chakraborti, A. Distributions of money in model markets of economy. *Int. J. Modern Phys. C World Sci.* **2002**, *13*, 1315–1321. [CrossRef]
69. Boghosian, B.M. Is Inequality Inevitable? *Sci. Am.* **2019**, *321*, 70–77. [CrossRef]
70. Iglesias, J.R. How simple regulations can greatly reduce inequality. *arXiv* **2010**, arXiv:1007.0461.
71. Ghosh, A.; Basu, U.; Chakraborti, A.; Chakrabarti, B.K. Threshold-induced phase transition in kinetic exchange models. *Phys. Rev. E* **2011**, *83*, 061130. [CrossRef] [PubMed]
72. Chakrabarti, A.S.; Chakrabarti, B.K.; Chatterjee, A.; Mitra, M. The Kolkata Paise Restaurant problem and resource utilization. *Phys. A Stat. Mech. Appl.* **2009**, *388*, 2420–2426. [CrossRef]
73. Chakraborti, A.; Challet, D.; Chatterjee, A.; Marsili, M.; Zhang, Y.-C.; Chakrabarti, B.K. Statistical mechanics of competitive resource allocation using agent-based models. *Phys. Rep.* **2015**, *552*, 1–25. [CrossRef]
74. Ghosh, A.; Chatterjee, A.; Mitra, M.; Chakrabarti, B.K. Statistics of the kolkata paise restaurant problem. *New J. Phys.* **2010**, *12*, 075033. [CrossRef]
75. Sharif, P.; Heydari, H. Quantum solution to a three player Kolkata restaurant problem using entangled qutrits. *arXiv* **2011**, arXiv:1111.1962.
76. Sharif, P.; Heydari, H. An introduction to multi-player, multi-choice quantum games: Quantum minority games & kolkata restaurant problems. In *Econophysics of Systemic Risk and Network Dynamics*; Abergel, F., Ed.; Springer: Milano, Italy, **2013**; pp. 217–236.
77. Ghosh, D.; Chakrabarti, A.S. Emergence of distributed coordination in the Kolkata paise restaurant problem with finite information. *Phys. A Stat. Mech. Appl.* **2017**, *483*, 16–24. [CrossRef]
78. Banerjee, P.; Mitra, M.; Mukherjee, K. The economics of the Kolkata Paise Restaurant problem. *Sci. Cult. Indian Sci. News Assoc.* **2018**, *84*, 26–30.
79. Sharma, K.; Anamika; Chakrabarti, A.S. Chakraborti, A.; Chakravarty, S. The Saga of KPR: Theoretical and experimental developments. *Sci. Cult. Indian Sci. News Assoc.* **2018**, *84*, 31–36.
80. Tamir, B. Econophysics and the Kolkata Paise Restaurant Problem: More is different. *Sci. Cult. Indian Sci. News Assoc.* **2018**, *84*, 37–47.
81. Sinha, A.; Chakrabarti, B.K. Phase transition in the Kolkata Paise Restaurant problem. *Chaos Interdiscip. J. Nonlinear Sci.* **2020**, *30*, 083116. [CrossRef] [PubMed]

82. Park, T.; Saad, W. Kolkata paise restaurant game for resource allocation in the Internet of Things. In Proceedings of the 2017 51st Asilomar Conference on Signals, Systems, and Computers, IEEE Xplore, Pacific Grove, CA, USA, 29 October–1 November 2017; pp.1774–1778.
83. Martin, L. Extending Kolkata Paise Restaurant Problem to Dynamic Matching in Mobility Markets. *Jr. Manag. Sci.* **2019**, *4*, 1–34.
84. Martin, L.; Karaenke, P. The Vehicle for Hire Problem: A Generalized Kolkata Paise Restaurant Problem. In *Workshop on Information Technology and Systems*; Technical University of Munich: Seoul, Korea, 2017.
85. Ghosh, A. Chakrabarti, A.S.; Chakrabarti, B.K. Kolkata Paise Restaurant problem in some uniform learning strategy limits. In *Econophysics and Economics of Games, Social Choices and Quantitative Techniques*; Springer: Berlin, Germany, 2010; pp. 3–9.
86. Ghosh, A.; De Martino, D.; Chatterjee, A.; Marsili, M.; Chakrabarti, B.K. Phase transitions in crowd dynamics of resource allocation. *Phys. Rev. E* **2012**, *85*, 021116. [CrossRef] [PubMed]
87. Ghosh, A.; Chatterjee, A.; Chakrabarti, A.S.; Chakrabarti, B.K. Zipf's law in city size from a resource utilization model. *Phys. Rev. E* **2014**, *90*, 042815. [CrossRef]
88. Chakrabarti, B.K. Kolkata restaurant problem as a generalised el farol bar problem. In *Econophysics of Markets and Business Networks*; Springer: Milan, Italy, 2007; pp. 239–246.
89. Allais, M. Economics as a Science. *Cah. Vilfredo Pareto* **1968**, *6*, 5–24. Available online: https://www.jstor.org/stable/40368894?seq=1 (accessed on 28 January 2021).
90. Frey, B.S. *Economics As a Science of Human Behaviour: Towards a New Social Science Paradigm*, 2nd ed.; Springer: New York, NY, USA, 1999.
91. Solow, R.M. How did economics get that way and what way did it get? *Daedalus* **1997**, *126*, 39–58. [CrossRef]
92. Colander, D. New millennium economics: How did it get this way, and what way is it? *J. Econ. Perspect.* **2000**, *14*, 121–132. [CrossRef]
93. Venkatasubramanian, V. *How Much Inequality Is Fair? Mathematical Principles of a Moral, Optimal, and Stable Capitalist Society*; Columbia University Press: New York, NY, USA, 2017.
94. Leiden University. Econophysics e-Prospectuses for 2012–2013. 2020–2021. Available online: https://studiegids.universiteitleiden.nl/en/courses/34804/econophysics or https://studiegids.universiteitleiden.nl/courses/99643/econophysics (accessed on 28 January 2021).
95. Dash, K.C. *The Story of Econophysics*; Cambridge Scholars Publishing: Newcastle Upon Tyne, UK, 2019.
96. Shubik, M.; Smith, E. *The Guidance of an Enterprise Economy*; MIT Press: Cambridge, MA, USA, 2016.
97. Jovanovic, F.; Schinckus, C. *Econophysics and Financial Economics: An Emerging Dialogue*; Oxford University Press: Oxford, UK, 2017.
98. Richmond, P.; Mimkes, J.; Hutzler, S. *Econophysics and Physical Economics*; Oxford University Press: Oxford, UK, 2013.
99. Slanina, F. *Essentials of Econophysics Modelling*; Oxford University Press: Oxford, UK, 2013.
100. Aoyama, H.; Fujiwara, Y.; Ikeda, Y.; Iyetomi, H.; Souma, W. *Macro-Econophysics: New Studies on Economic Networks and Synchronization*; Cambridge University Press: Delhi, India; Cambridge, UK, 2017.
101. Schinckus, C. *When Physics Became Undisciplined An Essay on Econophysics*; University of Cambridge: Cambridge, UK, 2018. Available online: https://www.repository.cam.ac.uk/bitstream/handle/1810/279683/Chris_Thesis_FINAL.png?sequence=5&isAllowed=y (accessed on 28 January 2021).
102. Chakrabarti, B.K. International Center for Social Complexity, Econophysics and Sociophysics Studies: A Proposal. In *New Perspectives and Challenges in Econophysics and Sociophysics*; New Economic Windows Series; Abergel, F., Ed.; Springer: Cham, The Newtherland, 2019; pp. 259–267.
103. Helbing, D.; Balietti, S.; Bishop, S.; Lukowicz, P. Understanding, creating, and managing complex techno-socio-economic systems: Challenges and perspectives (Visioneer White Papers). *Eur. Phys. J. Spec. Top.* **2011**, *195*, 165–186. [CrossRef]
104. Ghosh, A. Econophysics Research in India in the last two Decades. *IIM Kozhikode Soc. Manag. Rev.* **2013**, *2*, 135–146. [CrossRef]

entropy

MDPI

Perspective

Radical Complexity

Jean-Philippe Bouchaud [1,2]

[1] Capital Fund Management, 75007 Paris, France; jean-philippe.bouchaud@cfm.fr
[2] Académie des Sciences, 75006 Paris, France

Abstract: This is an informal and sketchy review of five topical, somewhat unrelated subjects in quantitative finance and econophysics: (i) models of price changes; (ii) linear correlations and random matrix theory; (iii) non-linear dependence copulas; (iv) high-frequency trading and market stability; and finally—but perhaps most importantly—(v) "radical complexity" that prompts a scenario-based approach to macroeconomics heavily relying on Agent-Based Models. Some open questions and future research directions are outlined.

Keywords: financial markets; covariance matrices; copulas; high-frequency trading; market stability; agent-based models

Citation: Bouchaud, J.-P. Radical Complexity. *Entropy* **2021**, *23*, 1676. https://doi.org/10.3390/e23121676

Academic Editors: Ryszard Kutner, H. Eugene Stanley and Christophe Schinckus

Received: 9 November 2021
Accepted: 3 December 2021
Published: 14 December 2021

1. From Random Walks to Rough (Multifractal) Volatility

Since we will never really know *why* the prices of financial assets move, we should at least make a model of *how* they move. This was the motivation of Bachelier in 1900 [1] when he wrote in the introduction of their thesis that *contradictory opinions in regard to (price) fluctuations are so diverse that at the same instant buyers believe the market is rising and sellers that it is falling*. He went on to propose the first mathematical model of prices: the Brownian motion. He then built an option pricing theory that he compared to empirical data available to him, which already revealed, quite remarkably, what is now called the volatility smile! (Looking at his table on p. 30, one clearly see a smile that flattens with the maturity of the options, as routinely observed nowadays. As we now understand, this flattening comes from the slow convergence of returns towards Gaussian random variables as the time-lag increases, see, e.g., [2]).

After 120 years of improvements and refinements, we are closing in on a remarkably realistic model, which reproduces almost all known stylised facts of financial price series. However, are we there yet? As Benoît Mandelbrot once remarked: *In economics, there can never be a "theory of everything". However, I believe each attempt comes closer to a proper understanding of how markets behave.* In order to close the gap and justify the modern mathematical apparatus that has slowly matured, we will need to understand the interactions between the behaviour of zillions of traders—each with their or her own investment style, trading frequency, risk limits, etc. and the price process itself. Interestingly, recent research strongly suggests that markets self organise in a subtle way, as to be poised at the border between stability and instability. This could be the missing link—or the holy grail—that researchers have been looking for.

For many years, the only modification to Bachelier's proposal was to consider that log-prices, not prices themselves, are described by a Brownian motion. Apart from the fact that this modification prevents prices from becoming negative, none of the flaws of the Bachelier model were seriously tackled. Notwithstanding, the heyday of Brownian finance came when Black and Scholes published their famous 1973 paper, with the striking result that perfect delta-hedging is possible. However, this is because, in the Black–Scholes world, price jumps are absent and crashes impossible. This is, of course, a very problematic assumption, especially because the fat-tailed distribution of returns had been highlighted as soon as 1963 by Mandelbrot, who noted, in the same paper, that *large changes tend to be followed by large changes, of either sign, and small changes tend to be followed by small changes*, an effect now commonly referred to as "volatility clustering" and captured by the extended family of GARCH models.

It took the violent crash of October 1987, exacerbated by the massive impact of Black–Scholes' delta-hedging, for new models to emerge. The Heston model, published in 1993, is among the most famous post-Black–Scholes models, encapsulating volatility clustering within a continuous time, Brownian motion formalism. However, similar to GARCH, the Heston model predicts that volatility fluctuations decay over a single time scale; in other words, periods of high or low volatility have a rather well defined duration. This is not compatible with market data: volatility fluctuations have no clear characteristic time scale; volatility bursts can last anything between a few hours and a few years.

Mandelbrot had been mulling about this for a long while and actually, in 1974, proposed a model to describe a very similar phenomenon in turbulent flows called "multifractality". He adapted their theory in 1997 to describe currency exchange rates, before Bacry, Muzy and Delour formulated a more convincing version of the model in 2000, which they called the *Multifractal Random Walk* (MRW) [3]. With a single extra parameter (interpreted as a kind of volatility of volatility), the MRW satisfactorily captures many important empirical observations: fat-tailed distribution of returns and long-memory of volatility fluctuations. In 2014, Gatheral, Jaisson and Rosenbaum introduced their now famous "Rough Volatility" model [4], which can be seen as an extension of the MRW with an extra parameter allowing one to tune at will the roughness of volatility, while it is fixed in stone in the MRW model. Furthermore, indeed, empirical data suggest that volatility is slightly less rough than what the MRW posits. Technically, the Holder regularity of the volatility is $H = 0$ in the MRW and found to be $H \approx 0.1$ when calibrated within the Rough Volatility specification.

The next episode of the long saga came in 2009 when Zumbach noticed a subtle, yet crucial aspect of empirical financial time series: they are not statistically invariant upon time reversal [5]. The past and future are not equivalent, whereas almost all models to that date, including the MRW, did not distinguish the past from future. More precisely, past price trends, whether up or down, lead to higher future volatility but not the other way round. In 2019, following some work by P. Blanc, J. Donier and myself [6], A. Dandapani, P. Jusselin and M. Rosenbaum proposed to describe financial time series with what they called a "Quadratic Rough Heston Model" [7], which is a synthesis of all the ideas reviewed above. It is probably the most realistic model of financial price series to date. In particular, it provides a natural solution to a long standing puzzle, namely the joint calibration of the volatility smile of the S&P 500 and VIX options, which had eluded quants for many years [8]. The missing ingredient was indeed the Zumbach effect [9].

Is this the end of the saga? From a purely engineering point of view, the latest version of the Rough Volatility model is probably hard to beat. However, the remaining challenge is to justify how this particular model emerges from the underlying flow of buy and sell trades that interacts with market makers and high-frequency traders. Parts of the story are already clear; in particular, as argued by Jaisson, Jusselin and Rosenbaum in a remarkable series of papers, the Rough Volatility model is intimately related to the proximity of an instability [10] (see also [11]) that justifies the rough, multi-timescale nature of volatility. However, what is the self-organising mechanism through which all markets appear to settle close to such a critical point? Could this scenario allow one to understand why financial time series all look so much alike; stocks, futures, commodities, exchange rates, etc., share very similar statistical features, in particular in the tails. Beyond being the denouement of a 120-year odyssey, we would be allowed to believe that the final model is not only a figment of our mathematical imagination, but a robust, trustworthy framework for risk management and derivative pricing. The next step will be to generalise these models in a multivariate setting, capturing the various channels through which price fluctuations propagate between different stocks and asset classes. The description of linear correlations is already a headache (see next section), but they are in fact not enough to capture the complexity of *non-linear* dependence in financial markets, which we discuss in Section 3.

2. Random Matrix Theory to the Rescue

Harry Markowitz famously quipped that diversification is the only free lunch in finance. This is nevertheless only true if correlations are known and stable over time. Markowitz' optimal portfolio offers the best risk-reward tradeoff, for a given set of predictors, but requires the covariance matrix—of a potentially large pool of assets—to be known and representative of the future realised correlations. However, the empirical determination of large covariance matrices is fraught with difficulties and biases. Interestingly, the vibrant field of the "Random Matrix Theory" has provided original solutions to this big data problem and suggests droves of possible applications in econometrics, machine learning or other large dimensional models.

However, even for the simplest two-asset bond/equity allocation problem, the knowledge of the forward looking correlation has momentous consequences for most asset allocators in the planet. Will this correlation remain negative in the years to come, as it has been since late 1997, or will it jump back to positive territories? However, compared to volatility, our understanding of correlation dynamics is remarkably poor, and, surprisingly, the hedging instruments allowing one to mitigate the risk of bond/equity correlation swings are nowhere as liquid as the VIX itself.

Thus, there are two distinct problems in estimating correlation matrices. One is lack of data; the other one is time non-stationarity. Consider a pool of N assets, with N large. We have at our disposal T observations (say daily returns) for each of the N time series. The paradoxical situation is this: even though each individual off-diagonal covariance is accurately determined when T is large, the covariance matrix as a whole is strongly biased unless T is much larger than N itself. For large portfolios, with N of a few thousands, the number of days in the sample should be in the tens of thousands—say 50 years of data. This is simply absurd: Amazon and Tesla did not even exist 25 years ago. Maybe use 5-minute returns then, increasing the number of data points by a factor 100? Yes, except that 5-minute correlations are not necessarily representative of the risk of much lower frequency strategies, with other possible biases creeping in the resulting portfolios.

So in what sense are covariance matrices biased when T is not very large compared to N? The best way to describe such biases is in terms of eigenvalues. One finds that the smallest eigenvalues are way too small and the largest eigenvalues are too large. This results, in the Markowitz optimisation program, in a substantial over-allocation on a combination of assets that happened to have a small volatility in the past, with no guarantee that this will persist looking forward. The Markowitz construction can therefore lead to a considerable under-estimation of the realised risk in the next period.

Out-of-sample results are of course always worse than expected, but Random Matrix Theory (RMT) offers a guide to (partially) correct these biases when N is large. In fact, RMT gives an optimal, mathematically rigorous, recipe to tweak the value of the eigenvalues so that the resulting "cleaned" covariance matrix is as close as possible to the "true" (but unknown) one in the absence of any prior information on the direction of the eigenvectors. Such a result, first derived by Ledoit and Péché in 2011 [12], is already a classic and has been extended in many directions (see, e.g., [13,14]. Its operational implementation and the quality of out-of-sample predictions were extensively reviewed in [15–17]. The underlying mathematics, initially based on abstract "free probabilities", are now in a ready-to-use format, very similar to Fourier transforms or Ito calculus (see [17] for an introductory account). One of the exciting and relatively unexplored directions is to add some financially motivated prior, such as industrial sectors or groups, to improve upon the default "agnostic" recipe.

Now the data problem is solved as best as possible, but the stationarity problem pops up. Correlations, similar to volatility, are not fixed in stone but evolve with time. Even the sign of correlations can suddenly flip, as was the case for the S&P500/Treasuries

during the 1997 Asian crisis. After 30 years of correlations staunchly in positive territory (1965–1997), bonds and equities have been in a "flight-to-quality" mode (i.e., equities down and bonds up) ever since. More subtle, but significant, changes of correlations can also be observed between single stocks and/or between sectors in the stock market. For example, a downward move of the S&P500 leads to an increased average correlation between stocks. Here again, RMT provides powerful tools to describe the time evolution of the full covariance matrix [18,19].

As I discussed in the previous section, stochastic volatility models have made significant progress recently and, now, encode feedback loops that originate at the microstructural level, see also Section 4. Unfortunately, we are very far from having a similar theoretical handle to understand correlation fluctuations, although Matthieu Wyart and I had proposed a self-reflexive mechanism in 2007 to account for correlation jumps, such as the one that took place in 1997 [20]. Parallel to the development of descriptive and predictive models, the introduction of standardised instruments that hedge against such correlation jumps would clearly serve a purpose. This is especially true in the current environment [21] where inflation fears could trigger another inversion of the equity/bond correlation structure, which would be possibly devastating for many strategies that—implicitly or explicitly—rely on persistent negative correlations. Markowitz diversification free lunch can sometimes be poisonous!

3. My Kingdom for the Right Copula

As I just discussed, assessing linear correlations between financial assets is already hard enough. What about *non-linear* correlations then? If financial markets were kind enough to abide to Gaussian statistics, non-linear correlations would be entirely subsumed by linear ones. However, this is not the case: genuine non-linear correlations pervade the financial world and are quite relevant, both for the buy side and the sell side. For example, tail correlations in equity markets (i.e., stocks plummeting simultaneously) are notoriously higher than bulk correlations. Another apposite context is the Gamma-risk of large option portfolios, the management of which requires an adequate description of quadratic return correlations of the underlying assets.

In order to deal with non-linear correlations, mathematics has afforded us with a seemingly powerful tool—"copulas" [22]. Copulas are supposed to encapsulate all possible forms of multivariate dependence. However, in the zoo of all conceivable copulas, which one should one choose to faithfully represent financial data?

Following an unfortunate but typical pattern of mathematical finance, the introduction of copulas twenty years ago has been followed by a calibration spree, with academics and financial engineers alike frantically looking for copulas to best represent their pet multivariate problem. However, instead of first developing an intuitive understanding of the economic or financial mechanisms that suggest some particular dependence between assets and construct adequate copulas accordingly, the methodology has been to brute-force calibrate copulas straight out from statistics handbooks. The "best" copula is then decided from some quality-of-fit criterion, irrespective of whether the copula makes any intuitive sense at all.

This is reminiscent of local volatility models for option markets: although these models make no intuitive sense and cannot describe the actual dynamics of the underlying asset, it is versatile enough to allow the calibration of almost any option smile. Unfortunately, a blind calibration of some unwarranted model (even when the fit is perfect) is a recipe for disaster. If the underlying reality is not captured by the model, it will most likely derail in rough times—a particularly bad feature for risk management (recall the use of Gaussian copulas to price CDOs before the 2008 crisis). Another way to express this point is to use a Bayesian language: there are families of models for which the "prior" likelihood is clearly extremely small because no plausible scenarios for such models emerge from market mechanisms. Statistical tests are not enough—the art of modelling is precisely to recognise that not all models are equally likely.

The best way to foster intuition is to look at data before cobbling up a model and come up with a few robust "stylised facts" that you deem relevant and that your model should capture. In the case of copulas, one interesting stylised fact is the way the probability that two assets have returns simultaneously smaller than their respective medians depends on the linear correlation between the said two assets. Such a dependence exists clearly and persistently in stocks, and strikingly, it cannot be reproduced by most "out-of-a-book" copula families.

In particular, the popular class of so-called "elliptical" copulas is ruled out by such an observation. Elliptical copulas assume, in a nutshell, that there is a common volatility factor for all stocks: when the index becomes more or less volatile, all stocks follow suit. A moment of reflection reveals that this assumption is absurd since one expects that volatility patterns are at least industry-specific. However, this consideration also suggests a way to build copulas specially adapted to financial markets. In Ref. [23], R. Chicheportiche and I showed how to weld the standard factor model for returns with a factor model for volatilities. Perhaps surprisingly, the common volatility factor is not the market volatility, although it contributes to it. With a relatively parsimonious parameterisation, most multivariate "stylised facts" of stock returns can be reproduced, including the non-trivial joint-probability pattern alluded to above.

I have often ranted against the over-mathematisation of quant models, favouring theorems over intuition and convenient models over empirical data. The reliance on rigorous but misguided statistical tests is also plaguing the field. As an illustration related to the topic of copulas, let me consider the following question: is the univariate distribution of standardised stock returns *universal*, i.e., independent of the considered stock? In particular, is the famous "inverse-cubic law" [24,25] for the tail of the distribution indeed common to all stocks?

A standard procedure for rejecting such an hypothesis is the Kolmogorov–Smirnov (or Anderson–Darling) statistics. Furthermore, lo and behold, the hypothesis is strongly rejected. However, wait—the test is only valid if returns can be considered as independent, identically distributed random variables. Whereas returns are close to being uncorrelated, non-linear dependencies along the time axis are strong and long-ranged. Adapting the Kolmogorov–Smirnov test in the presence of long-ranged "self-copulas" is possible [26] and now leads to the conclusion that the universality hypothesis *cannot* be rejected by such a test. Intuitively, this is because the presence of long-range correlations in volatility drastically limit the *effective* size of the data set. We have much less independent data than we think.

Here again, thinking about the problem before blindly applying standard recipes is of paramount importance to get it right. Furthermore, of course, if the "inverse-cubic law" is indeed universal, as again recently advocated in [25], we should try to understand why. Despite many efforts in that direction, it is fair to say that there is no consensus on the underlying mechanism responsible for such a critical-like behaviour, see Sections 1 and 4.

The finer we want to hone in on the subtleties of financial markets, the more we need to rely on making sense of empirical data and to remember what the great Richard Feynman used to say: *It does not matter how beautiful your theory is, it does not matter how smart you are. If it does not agree with experiment, it is wrong.*

4. High-Frequency Trading and Market Stability

In the midst of the first COVID lockdown, the 10th anniversary of the infamous May 6th, 2010 "Flash Crash" went unnoticed. At the time, fingers were pointed at High-Frequency Trading (HFT), accused of both rigging the markets and destabilising them. Research has since then confirmed that HFT results in significantly lower bid-ask spread costs and, after correcting for technological glitches and bugs, does *not* increase the frequency of large price jumps. In fact, recent models explain why market liquidity is intrinsically unstable: managing the risk associated to market-making, whether by humans or by computers, unavoidably creates destabilising feedback loops. In order to make markets more resilient, research should focus on better market design and/or smart regulation that nip nascent instabilities in the bud.

Since orders to buy or to sell arrive at random times, financial markets are necessarily most of the time unbalanced. In such conditions, market-makers play a crucial role in allowing smooth trading and continuous prices. They act as liquidity buffers that absorb any temporary surplus of buy orders or sell orders. Their reward for providing such a service is the bid-ask spread—systematically buying a wee lower and selling a wee higher and pocketing the difference.

What is the fair value of the bid-ask spread? Well, it must at least compensate the cost of providing liquidity, which is *adverse selection*. Indeed, market-makers must post prices that can be picked up if deemed advantageous by traders with superior information. The classic Glosten–Milgrom model provides an elegant conceptual framework to rationalise the trade-off between adverse selection and bid-ask spread but fails to give a quantitative, operational answer (see, e.g., [27] for a recent discussion). In a 2008 study [28], we came up with a remarkably simple answer: the fair value of the bid-ask spread is equal to the ratio of the volatility to the square-root of the trade frequency. This simple rule of thumb has many interesting consequences.

First, it tells us that for a fixed level of volatility, increasing the trade frequency allows market-makers to reduce the spread and, hence, the trading costs for final investors. The logic is that trading smaller chunks more often reduces the risk of adverse selection. This explains in part the rise of HFT as modern market-making and the corresponding reduction in the spreads. Throughout the period 1900–1980, the spread on US stocks hovered around a whopping 60 basis points, whereas it is now only a few basis points [29]. In the meantime, volatility has always wandered around 40% per year—with of course troughs and occasional spikes, as we discuss below. In other words, investors were paying a much higher price for liquidity before HFT, in spite of wild claims that nowadays electronic markets are "rigged". In fact, after a few prosperous years before 2010, high-frequency market-making has become extremely competitive and average spreads are now compressed to minimum values.

From this point of view, the economic rents available to liquidity providers have greatly decreased since the advent of HFT. However, has this made markets more stable, or has the decrease in the profitability of market-making also made them more fragile? The second consequence of our simple relation between spread and volatility relates to this important question. The point is that this relation can be understood in a two-way fashion: clearly, when volatility increases, the impact of adverse selection can be dire for market-makers who mechanically increase their spreads. Periods of high volatility can however be quite profitable for HFT since competition for liquidity providing is then less fierce.

However, in fact, higher spreads by themselves lead to higher volatility since transactions generate a larger price jump—or even a crash when liquidity is low and the order book is sparse. Thus, we diagnose a fundamental destabilising feedback loop, intrinsic to any market-making activity:

$$\text{volatility} \longrightarrow \text{higher spreads and lower liquidity} \longrightarrow \text{more volatility.}$$

Such a feedback loop can actually be included in stochastic order book models (such as the now commonly used family of "Hawkes processes" [30]). As the strength of the feedback increases, one finds a phase transition between a stable market and a market prone to *spontaneous liquidity crises*, even in the absence of exogenous shocks or news [31].

This theoretical result suggests that when market-makers (humans or machines) react too strongly to unexpected events, liquidity can enter a death spiral. However, it is difficult to blame them since they are at risk of losing a full year of profit in a single adverse jump. As an old saying goes, *liquidity is a coward, it is never there when it is needed*. Liquidity can only be gossamer.

Such a paradigm allows one to understand why a large fraction of price jumps occur without any significant news—rather, they result from endogenous, unstable feedback loops [32,33]. Empirically, the frequency of 10-sigma daily moves of US stock prices has been fairly constant in the last 30 years, with no significant change between the pre-HFT epoch and more recent years [27]. Even the 6th May 2010 Flash Crash has a pre-HFT counterpart: on May 28th 1962, the stock market plunged 9% within a matter of minutes, for no particular cause, before recovering—much of the same weird price trajectory as in 2010. Our conjecture: markets are intrinsically unstable and have always been so. As noted in Section 1 above, this chronic instability may lie at the heart of the turbulent, multiscale nature of financial fluctuations and the universal power-law of the distribution of returns [24,25].

Can one engineer a smart solution that make markets less prone to such dislocations? From our arguments above, we know that the task would be to crush the volatility/liquidity feedback loop by promoting liquidity provision when it is on the verge of disappearing. One idea would be to introduce dynamical make/take fees, which would make cancellations more costly and limit order posting more profitable depending on the current state of the order book. These fees would then funnel into HFT's optimisation algorithms and (hopefully) drive the system away from the regime of recurrent endogenous liquidity crisis.

5. Radical Complexity and Scenario Based Macro-Economics

Good science is often associated with accurate, testable predictions. Classical economics has tried to conform to this standard and developed an arsenal of methods to come up with precise numbers for next year's GDP, inflation and exchange rates, among (many) other things. Few, however, will disagree with the fact that the economy is a complex system, with a large number of heterogeneous interacting units of different categories (firms, banks, households, public institutions) and very different sizes. In such complex systems, even qualitative predictions are hard. Thus, maybe we should abandon our pretense of exactitude and turn to another way to do science based on scenario identification. Aided by qualitative (agent based) simulations, swans that appear black to the myopic eye may in fact be perfectly white.

The main issue in economics is precisely about the emergent organisation, cooperation and coordination of a motley crowd of micro-units. Treating them as a unique representative firm or household risks throwing the baby out with the bathwater. Understanding and characterising such emergent properties is however difficult: genuine surprises can appear from micro- to macro-. One well-known example is the Schelling segregation model: even when all agents prefer to live is mixed neighbourhoods, myopic dynamics can lead to completely segregated ghettos [34]. In this case, Adam Smith's invisible hand badly fails.

More generally, slightly different micro-rules/micro-parameters can lead to very different macro-states: this is the idea of "phase transitions"; sudden discontinuities (aka crises) can appear when a parameter is only slightly changed. Because of feedback loops of different signs, heterogeneities and non-linearities, these surprises are hard, if not impossible, to imagine or anticipate, even aided with the best mathematical apparatus.

This is what I would like to call "Radical Complexity". Simple models can lead to unknowable behaviour, where "Black Swans" or "Unknown Unknowns" can be present, even if all the rules of the model are known in detail. In these models, even probabilities are hard to pin down, and rationality is de facto limited. For example, the probability of rare events can be exponentially sensitive to the model parameters and, hence, unknowable in practice [35]. In these circumstances, precise quantitative predictions are unreasonable. However, this does not imply the demise of the scientific method. For such situations, one should opt for a more qualitative, scenario-based approach, with emphasis on mechanisms, feedback loops, etc., rather than on precise but misleading numbers.

Establishing the list of possible (or plausible) scenarios is itself difficult. We need numerical simulations of Agent-Based Models (ABMs). While it is still cumbersome to experiment on large-scale human systems (although more and more possible using web-based protocols), experimenting with ABMs is easy and fun and indeed full of unexpected phenomena. These experiments in silico allow one to elicit scenarios that would be nearly impossible to imagine because of said feedback loops and non-linearities. Think, for example, of the spontaneous synchronisation of fireflies (or of neuron activity in our brains). It took nearly 70 years to come up with an explanation. Complex endogenous dynamics is pervasive but hard to guess without appropriate tools.

Experimenting with Agent-Based Models is interesting on many counts. One hugely important aspect is, in my opinion, that it allows to teach students in a playful, engaging way how complex social and economic systems work. Such simulations would foster their intuition and their imagination, much like lab experiments train the intuition of physicists about the real world beyond abstract mathematical formalism. Of course, similar to physics curriculums, experimenting with ABMs should be taught in parallel to, and not instead of, standard analytical models.

Creating one's own world and seeing how it unfolds clearly has tremendous pedagogical merits. It is also an intellectual exercise of genuine value: if we are not able to make sense of an emergent phenomenon within a world in which we set all the rules, how can we expect to be successful in the real world? We have to train our minds to grasp these collective phenomena and to understand how and why some scenarios can materialise and others not. The versatility of ABMs allows one to include ingredients that are almost impossible to accommodate in standard economic models and explore their impact on the dynamics of the systems [36,37], in particular the inability of some of these a priori well-behaved economic models to ever reach equilibrium [38]. For a recent review of macroeconomic ABMs, see, e.g., [39].

ABMs are often spurned because they are generally hard to calibrate, and therefore, the numbers they spit out cannot and should not be taken at face value. (For an interesting discussion of why ABMs are not yet part of mainstream economics, see [40,41]). They should rather be regarded as all-purpose *scenario generators*, allowing one to shape one's intuition about phenomena to uncover different possibilities and reduce the realm of Black Swans. The latter are often the result of our lack of imagination or of the simplicity of our models, rather than being inherently impossible to foresee.

Expanding the study of toy-models of economic complexity will create a useful corpus of scenario-based, qualitative macroeconomics [36,42,43], perhaps boosted by the recent Nobel prize of Giorgio Parisi. Instead of aiming for precise numerical predictions based on unrealistic assumptions, one should make sure that models rely on plausible causal mechanisms and encompass all plausible scenarios, even when these scenarios cannot be fully characterised mathematically. A qualitative approach to complexity economics should be high on the research agenda. As Keynes said, *it is better to be roughly right than exactly wrong*.

Funding: This research received no external funding.

Acknowledgments: I want to warmly thank my collaborators on these topics, especially: R. Allez, F. Benaych-Georges, R. Benichou, M. Benzaquen, P. Blanc, J. Bonart, F. Bucci, J. Bun, R. Chicheportiche, J. Donier, Z. Eisler, A. Fosset, M. Gould, S. Gualdi, S. Hardiman, A. Karami, Y. Lempérière, F. Lillo, R. Marcaccioli, I. Mastromatteo, F. Morelli, M. Potters, P. A. Reigneron, P. Seager, D. Sharma, M. Tarzia, B. Toth, M. Wyart, and F. Zamponi. I also want to pay tribute to various people with whom I had exciting and enlightening discussions, in particular: R. Bookstaber, D. Farmer, X. Gabaix, J. Gatheral, J. Guyon, A. Kirman, C. Lehalle, J. Moran, M. Rosenbaum, N. Taleb and G. Zumbach. Finally, I am deeply indebted to Mauro Cesa for encouraging me to put my thoughts together.

Conflicts of Interest: The author declares no conflict of interest.

References

1. Bachelier, L. Theory of Speculation, 1900. Available online: https://www.investmenttheory.org/uploads/3/4/8/2/34825752/emhbachelier.pdf (accessed on 13 December 2021).
2. Bouchaud, J.P.; Potters, M. Theory of financial risks. In *From Statistical Physics to Risk Management*; Cambridge University Press: Cambridge, UK, 2003.

3. Muzy, J.F.; Delour, J.; Bacry, E. Modelling fluctuations of financial time series: From cascade process to stochastic volatility model. *Eur. Phys. J. B-Condens. Matter Complex Syst.* **2000**, *17*, 537–548. [CrossRef]
4. Gatheral, J.; Jaisson, T.; Rosenbaum, M. Volatility is rough. *Quant. Financ.* **2018**, *18*, 933–949. [CrossRef]
5. Zumbach, G. Time reversal invariance in finance. *Quant. Financ.* **2009**, *9*, 505–515. [CrossRef]
6. Blanc, P.; Donier, J.; Bouchaud, J.P. Quadratic Hawkes processes for financial prices. *Quant. Financ.* **2017**, *17*, 171–188. [CrossRef]
7. Dandapani, A.; Jusselin, P.; Rosenbaum, M. From quadratic Hawkes processes to super-Heston rough volatility models with Zumbach effect. *arXiv* **2019**, arXiv:1907.06151.
8. Guyon, J. The joint S&P 500/VIX smile calibration puzzle solved. *Risk April* **2020**. [CrossRef]
9. Gatheral, J.; Jusselin, P.; Rosenbaum, M. The quadratic rough Heston model and the joint S&P 500/VIX smile calibration problem. *arXiv* **2020**, arXiv:2001.01789.
10. Jaisson, T.; Rosenbaum, M. Rough fractional diffusions as scaling limits of nearly unstable heavy tailed Hawkes processes. *Ann. Appl. Probab.* **2016**, *26*, 2860–2882. [CrossRef]
11. Hardiman, S.J.; Bercot, N.; Bouchaud, J.P. Critical reflexivity in financial markets: A Hawkes process analysis. *Eur. Phys. J. B* **2013**, *86*, 1–9. [CrossRef]
12. Ledoit, O.; Péché, S. Eigenvectors of some large sample covariance matrix ensembles. *Probab. Theory Relat. Fields* **2011**, *151*, 233–264. [CrossRef]
13. Bun, J.; Allez, R.; Bouchaud, J.P.; Potters, M. Rotational invariant estimator for general noisy matrices. *IEEE Trans. Inf. Theory* **2016**, *62*, 7475–7490. [CrossRef]
14. Benaych-Georges, F.; Bouchaud, J.P.; Potters, M. Optimal cleaning for singular values of cross-covariance matrices. *arXiv* **2019**, arXiv:1901.05543.
15. Bun, J.; Bouchaud, J.P.; Potters, M. Cleaning large correlation matrices: tools from random matrix theory. *Phys. Rep.* **2017**, *666*, 1–109. [CrossRef]
16. Ledoit, O.; Wolf, M. Nonlinear shrinkage of the covariance matrix for portfolio selection: Markowitz meets Goldilocks. *Rev. Financ. Stud.* **2017**, *30*, 4349–4388. [CrossRef]
17. Potters, M.; Bouchaud, J.P. *A First Course in Random Matrix Theory: For Physicists, Engineers and Data Scientists*; Cambridge University Press: Cambridge, UK, 2020. [CrossRef]
18. Reigneron, P.A.; Allez, R.; Bouchaud, J.P. Principal regression analysis and the index leverage effect. *Phys. A Stat. Mech. Its Appl.* **2011**, *390*, 3026–3035. [CrossRef]
19. Karami, A.; Benichou, R.; Benzaquen, M.; Bouchaud, J.P. Conditional Correlations and Principal Regression Analysis for Futures. *Wilmott* **2021**, *2021*, 63–73. [CrossRef]
20. Wyart, M.; Bouchaud, J.P. Self-referential behaviour, overreaction and conventions in financial markets. *J. Econ. Behav. Organ.* **2007**, *63*, 1–24. [CrossRef]
21. Breedt, A.; Seager, P. Available online: https://www.cfm.fr/insights/bond-equity-correlations-are-the-times-a-changin/ (accessed on 13 December 2021).
22. Mikosch, T. Copulas: Tales and facts. *Extremes* **2006**, *9*, 3–20. [CrossRef]
23. Chicheportiche, R.; Bouchaud, J.P. A nested factor model for non-linear dependencies in stock returns. *Quant. Financ.* **2015**, *15*, 1789–1804. [CrossRef]
24. Gabaix, X.; Gopikrishnan, P.; Plerou, V.; Stanley, H.E. Institutional investors and stock market volatility. *Q. J. Econ.* **2006**, *121*, 461–504. [CrossRef]
25. Watorek, M.; Kwapien, J.; Drozdz, S. Financial Return Distributions: Past, Present, and COVID-19. *Entropy* **2021**, *23*, 884. [CrossRef]
26. Chicheportiche, R.; Bouchaud, J.P. Goodness-of-fit tests with dependent observations. *J. Stat. Mech. Theory Exp.* **2011**, *2011*, P09003. [CrossRef]
27. Bouchaud, J.P.; Bonart, J.; Donier, J.; Gould, M. *Trades, Quotes and Prices: Financial Markets under the Microscope*; Cambridge University Press: Cambridge, UK, 2018.
28. Wyart, M.; Bouchaud, J.P.; Kockelkoren, J.; Potters, M.; Vettorazzo, M. Relation between bid–ask spread, impact and volatility in order-driven markets. *Quant. Financ.* **2008**, *8*, 41–57. [CrossRef]
29. Jones, C.M. A Century of Stock Market Liquidity and Trading Costs. *SSRN* **2002**. [CrossRef]
30. Bacry, E.; Mastromatteo, I.; Muzy, J.F. Hawkes processes in finance. *Mark. Microstruct. Liq.* **2015**, *1*, 1550005. [CrossRef]
31. Fosset, A.; Bouchaud, J.P.; Benzaquen, M. Endogenous liquidity crises. *J. Stat. Mech. Theory Exp.* **2020**, *2020*, 063401. [CrossRef]
32. Joulin, A.; Lefevre, A.; Grunberg, D.; Bouchaud, J.P. Stock price jumps: News and volume play a minor role. *arxiv* **2008**, arXiv:0803.1769 [CrossRef]
33. Marcaccioli, R.; Bouchaud, J.P.; Benzaquen, M. Exogenous and Endogenous Price Jumps Belong to Different Dynamical Classes. *SSRN* **2021**. doi:10.2139/ssrn.3866131
34. Grauwin, S.; Bertin, E.; Lemoy, R.; Jensen, P. Competition between collective and individual dynamics. *Proc. Natl. Acad. Sci. USA* **2009**, *106*, 20622–20626. [CrossRef]
35. Morelli, F.G.; Benzaquen, M.; Tarzia, M.; Bouchaud, J.P. Confidence collapse in a multihousehold, self-reflexive DSGE model. *Proc. Natl. Acad. Sci. USA* **2020**, *117*, 9244–9249. [CrossRef]

36. Gualdi, S.; Tarzia, M.; Zamponi, F.; Bouchaud, J.P. Tipping points in macroeconomic agent-based models. *J. Econ. Dyn. Control* **2015**, *50*, 29–61. [CrossRef] [PubMed]
37. Sharma, D.; Bouchaud, J.P.; Gualdi, S.; Tarzia, M.; Zamponi, F. V–, U–, L–or W–shaped economic recovery after Covid-19: Insights from an Agent Based Model. *PLoS ONE* **2021**, *16*, e0247823. [CrossRef]
38. Dessertaine, T.; Morán, J.; Benzaquen, M.; Bouchaud, J.P. Out-of-Equilibrium Dynamics and Excess Volatility in Firm Networks. *SSRN* **2020**. doi:10.2139/ssrn.3745898
39. Dosi, G.; Roventini, A. More is different ... and complex! the case for agent-based macroeconomics. *J. Evol. Econ.* **2019**, *29*, 1–37. [CrossRef]
40. Haldane, A.G.; Turrell, A.E. Drawing on different disciplines: Macroeconomic agent-based models. *J. Evol. Econ.* **2019**, *29*, 39–66;
41. Haldane, A.G.; Turrell, A.E. An interdisciplinary model for macroeconomics. *Oxf. Rev. Econ. Policy* **2018**, *34*, 219–251. [CrossRef]
42. Bookstaber, R. *The End of Theory*; Princeton University Press: Princeton, NJ, USA, 2017. [CrossRef]
43. Mounfield, C.C. *The Handbook of Agent Based Modelling*; Cambridge University Press: Cambridge, UK, 2021.

entropy

MDPI

Review

Three Decades in Econophysics—From Microscopic Modelling to Macroscopic Complexity and Back

Alex Smolyak * and Shlomo Havlin

Department of Physics, Bar-Ilan University, Ramat-Gan 52900, Israel; havlin@ophir.ph.biu.ac.il
* Correspondence: alex.smolyak@gmail.com

Abstract: We explore recent contributions to research in Econophysics, switching between Macroscopic complexity and microscopic modelling, showing how each leads to the other and detailing the everyday applicability of both approaches and the tools they help develop. Over the past decades, the world underwent several major crises, leading to significant increase in interdependence and, thus, complexity. We show here that from the perspective of network science, these processes become more understandable and, to some extent, also controllable.

Keywords: econophysics; dynamics of complex networks; cascading failure; network science

Citation: Smolyak, A.; Havlin, S. Three Decades in Econophysics—From Microscopic Modelling to Macroscopic Complexity and Back. *Entropy* **2022**, *24*, 271. https://doi.org/10.3390/e24020271

Academic Editors: Ryszard Kutner, Christophe Schinckus and H. Eugene Stanley

Received: 6 January 2022
Accepted: 11 February 2022
Published: 14 February 2022

1. Introduction

Historically, physical science deals with everything that surrounds us, from the smallest to the largest objects of our universe, with the small exception of life, which is mostly explored by biology, and human life specifically, which is handled by psychology, economics and many other sciences trying to find regularities, causalities and in general better understand our daily lives. The general implicit guiding principle of physics, reductionism, impeded physicists researching domains that are (at least to some extent) irreducible. The winds started to change about half a century ago, with the understanding that emergence is an important property in many realistic systems [1], and the mathematical apparatus developed in statistical physics is very useful in modelling and analyzing many everyday phenomena [2,3].

Written in the late 18th century, Adam Smith's Wealth of Nations [4] is considered the starting point of economic theory. Since then, theory broke into micro and macro, as well as a multitude of schools and approaches, from the simple to extremely complicated, from linear to partial differential equations. From the microscopic modelling perspective, the one looking at asset prices, the prevailing assumption was that prices [5] or price changes [6] follow a Gaussian random walk. This assumption means, on one hand, that the future could not be predicted from the present, and more importantly, risk from movement of assets was easily quantifiable and manageable. The macroscopic view deals with national income, gross domestic product (GDP), employment, production and typically does not concern itself with individual constituents. While it is clear that the macro is made up of the micro, the scales and layers between the individual economic agent or the single change in price of an asset and the contribution of a certain sector to the next year's GDP make it impossible to deduce one based on the other, or even assess their mutual dependence.

A major difference between physics and economics is the difficulty to set up controlled experiments. In physics, observation of nature will typically lead to hypotheses that could be translated into experiments to test them. In economics, with the possible exception of behavioural economics often tested on college students [7,8], observation is the only possible way to evaluate theory. While the macro-level view provides limited chances to assess accuracy, on the micro level it is to some extent easier. Asset prices, for example, are easily observable, with fine-grained information available and accessible. Those lend themselves to in-depth analysis, if not actual experimentation. Several real-life extreme events, such as Black Monday, a single day in October 1987 that saw the main index of the US stock market plunge over 20%, the collapse of a Nobel laureate backed fund, Long Term Capital Management L.P. in 1998 following several local crises and the economic meltdown following the housing market bubble together with predatory lending and a large web of financial derivatives tied to the housing market, gave experimental evidence to the immense interconnectedness of the micro and the macro, and the severe underestimation of risk by many of the prevalent economic theories.

These two aspects benefited greatly from research conducted by physicists over the last several decades under the common topic of a relatively new field of Econophysics [9–12]. In this brief review we shortly discuss a sample of several studies that focus on the micro- and macroscopic modelling and analysis to familiarize the interested reader in the usefulness and potential of both perspectives. We wish to highlight the close relationship between the micro and the macro, specifically showing how and where models from physics applied to economic and financial entities bring about the emergent properties that make the field of econophysics so challenging and interesting. From the early insights and models [13], their extensions [14,15] through what has become known as stylized facts [16] of the financial markets, including behaviour of prices and their volatility, and to the inherent connectivity driving global risk [17], tools and methods from physics and complexity sciences [18,19], such as phase transitions [2,3,20], fractal and multifractal analysis [21] and network science [22–25], all help to understand the intricacies of our economic and financial lives. The following sections will move back and forth between the micro and macro perspective highlighting some recent research driving econophysics forward.

2. Macro-Complexity, or the Interconnectedness of All Things

The physical infrastructure that surrounds us is a good starting point for the macro to micro journey and back. While it may not seem obvious, interdependence in infrastructure is critical to its continuous operations and resilience [26,27]. Figure 1 shows a schematic of various dependencies between different elements of such infrastructure. Those may be immediate (water, electric power) or longer-term (fuel, long-range transportation) but it is clear that efficient operation of every element depends either directly or indirectly on every other element. A prototypical example of interdependent network and connected infrastructures exhibit non-trivial and seemingly unpredictable transitions from operational to failed [28,29], unexpected critical junctions and positive, as well as negative, feedback loops.

Various theoretical models have been developed to analyze the resilience and onset of failure on such networks over the past decade or so [23,30–45], shedding light on the importance of degree distributions, intra- and inter-layer connectivity and mitigation strategies. We note here this list is far from exhaustive and focuses mainly on contributions from network science. Extensive literature exists from an economics perspective, and we refer to publications such as [46–48] and references therein as important examples of a complimentary view. As with the system under investigation, those models often take a probabilistic, generating function approach as their starting point. In its most fundamental form, a degree generating function is defined as $G_0(x) = \sum_{k=0}^{\infty} p_k x^k$ [49], where taking the kth derivative and equating x to zero gives the probability of having a degree of k. Taking this idea further, various connectivity constraints can be built into the generating functions and probabilities can be calculated to represent objects of interest, such as a surviving component size, given a defined connectivity and initial failure. Thus, if we have an initial model of how our networks are set up and connected, and we can specify how they fail, we can determine what they will look like when the process of failure plays out.

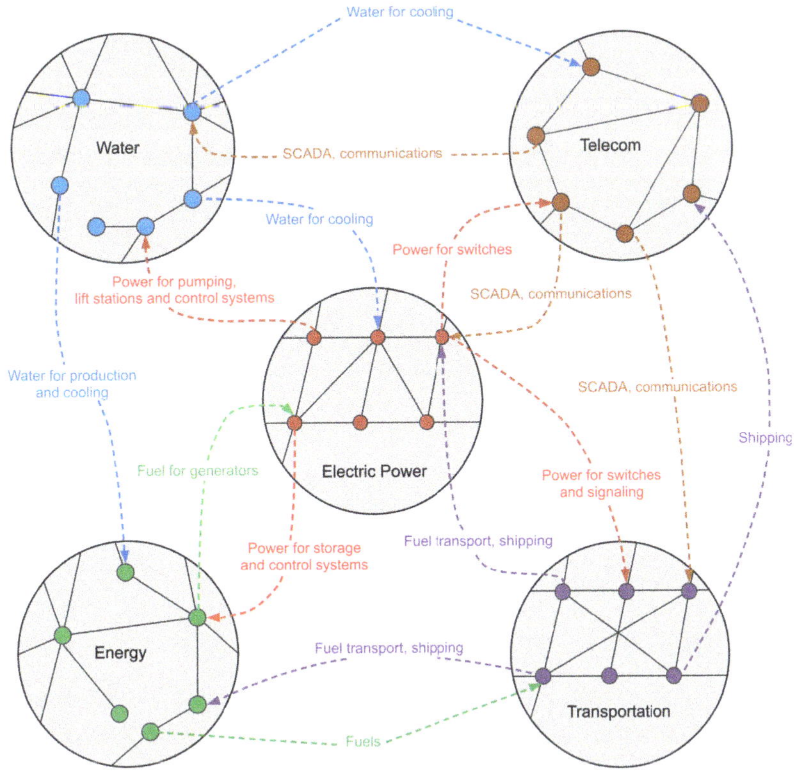

Figure 1. Interdependence in critical infrastructure. This schematic, after [50], shows the rich dependency coupling between different networks. Each circle is a complex network in its own right, with its degree distribution, connectivity and potential for random or intentional failure. These dynamics are greatly exacerbated due to the introduction of inter-network dependence. Each of the networks described requires some critical resource from one or more of the other networks to operate smoothly with failure in networks providing said critical resources may propagate failures to other networks.

These models are very powerful in the sense that they allow us to predict the infinite-time states of systems under cascading failures in the presence of complex interactions. The solution of the model showed that, due to dependencies, a microscopic failure of a single node can yield a macroscopic cascade and an abrupt collapse of the system [51,52]. They do suffer from some drawbacks, however. From the technical perspective, generating functions become very cumbersome for non-trivial or, worse yet, empirical distributions. We can solve them for simple cases such as regular, random (Erdos-Renyi, ER) networks, and for scale free (SF, power-law) distributed ones. However, we know these mathematically convenient constructs do not represent real networks. For example, they can not consider spatial embedding constraints. Nevertheless, it was shown by Bashan et al., both analytically and via simulations, that spatially embedded interdependent networks are far more vulnerable to microscopic failures [52]. Moreover, a recent analytical study that considered spatiality and cascading failures in modular interdependent networks was carried out by Vaknin et al. [53]. Lastly, because they deal with probabilities over the entire system, these macroscopic models have a hard time dealing with specificities such as individual nodes or small systems. When looking for answers at the level of the individual power plant or internet hub, macroscopic models are less helpful.

3. Microscopic Failure and Recovery Models—From the Lab to the Exchanges

To gain better insight into the behaviour of our system, more specifically, to enable analysis of its dynamical properties in addition to the long-term state, we turn to specifying individual node dynamics, which may include probabilities for failure [54] as well as recovery depending on its internal state and the state of its neighbours [55,56]. The methods described in the previous section are still useful and allow us to calculate steady state solutions for various parameters and detect non-trivial transitions between those states, but only when we switch our view to a microscopic one, and let a simulated system run its course can we discover the rich dynamics. Importantly, analytical solutions typically assume infinite time and size, while in life the actual time scales may be very important and systems are more often than not small in thermodynamic terms.

In particular, we may specify our model such that nodes in our network may fail spontaneously, fail under the influence of their neighbours or, if they are in a failed state, spontaneously recover with a certain probability density over a time period. As shown in Figure 2, a system under a simple failure-recovery model behaves highly non-trivially over time, not only spending long periods of time in its "active" or "failed" states, but also exploring transitions from one to the other without completing them due to the hysteresis region. As discussed in [55], this behaviour and the resulting bi-modal distribution of the system is very similar to the observed in financial markets when evaluating the fraction of companies listed in various indices with positive vs. negative returns.

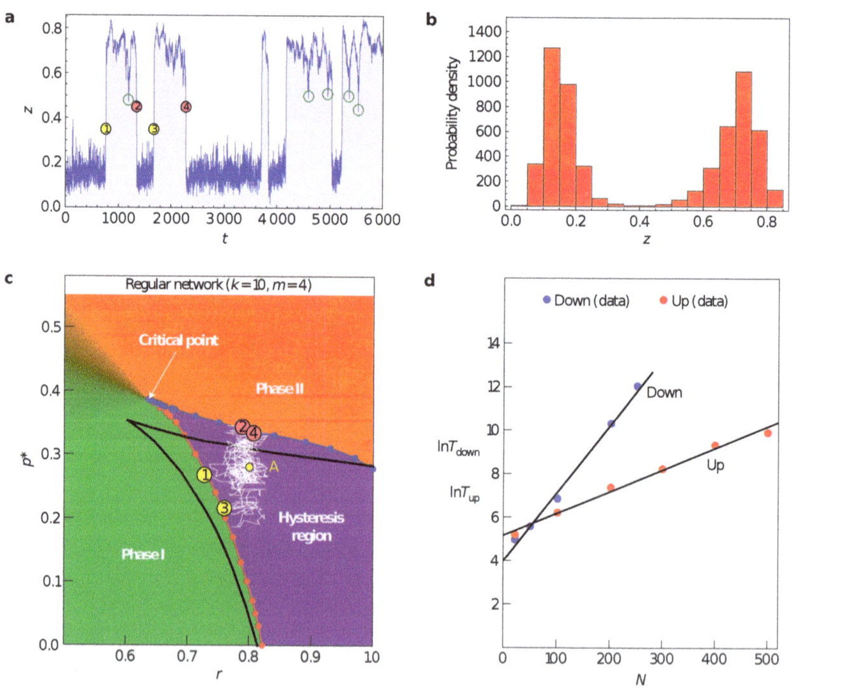

Figure 2. The importance of microscopic modelling. A simulation of a model with spontaneous recovery from failure on a regular network with k = 10 made of 100 nodes. (**a**) A realization of the time evolution starting from point A in panel (**c**) with switching between the failed and active states marked by yellow circles, and failure to complete a transition by green circles along the time series. (**b**) The probability density function of (**a**). (**c**) The phase space for the specified system, with the while line showing the path realized in (**a**), yellow and red circles matching the above. (**d**) The expected life time of the system in each state given a network size. After [55].

Temporal dynamics, small scales and fluctuations around the analytical solutions all highlight the importance and practical use of microscopic modelling. An instructive example was developed in Ref. [57]. The model specifies a risk-propagation mechanism on a bipartite Bank-Asset network, Figure 3. A bipartite network of banks and the assets on their books is set up, along with several parameters governing initial shock and levels of propagation at a node level. Then, an initial shock to an asset results in loss of value for a bank holding that asset. Enough such shocks may lead to a failure of a bank, leading to subsequent devaluation of the assets on its books. Importantly, the simple model allows one to build back up from the micro to the macro. In Ref. [57], the authors analyze the sensitivity of the whole system to various loan classes showing close relations with the events of real life. Below we discuss how such models can be taken a step further.

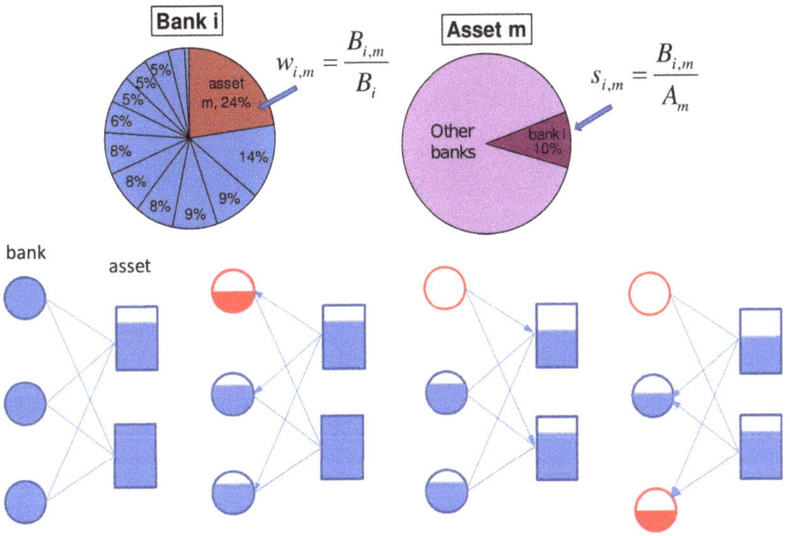

Figure 3. A microscopic model of a bank–asset risk propagation model. The relatively simple and intuitive model is very instructive and facilitates further modelling and analysis. The Bank's holdings are distributed between multiple assets and the assets are held by various banks. An initial reduction in the value of an asset leads to an impact to the values of all the banks exposed to it. For some banks the exposure is large enough to cause the bank to fail leading to further sell-off of its assets inducing a potential cascade. After [57].

4. Failure and Immunization in Real, Macroscopic Networks

The microscopic model presented in the previous section gives us powerful tools to stress-test our system against various potential failures and estimate their relative importance. There is, however, more to be done. As discussed in Ref. [58], given a specification of the failure process, the microscopic model can reveal nodes in the network that, more than others, propagate failure. These nodes do not display any high centrality values, and yet ensuring their protection from failure helps keep the network intact relatively cheaply in terms of number of nodes protected. Importantly, those nodes can be identified with only knowledge of their local neighborhood, without complete information of the entire network. Partial knowledge does not interfere with the performance of the suggested method. All that is needed is knowledge of the failure mechanism in order to devise an efficient mitigation strategy.

It is now possible to expand the local node-level insights up to a network-level view. Both in simulations and, more importantly, in real networks, we can use the microscopic model for macroscopic benefit. Figure 4 shows just that. The left panel visualizes a network of banks (right) and sovereign debt (left). The top right panel shows the fragility of the network under the default of various states given a certain threshold (colour bars). Most countries failing lead to cascading failure across the entire system for the lowest threshold, and many cause significant damage even with a high one. Protecting the nodes, the microscopic model highlights enable the entire network to remain mostly intact under most failing conditions.

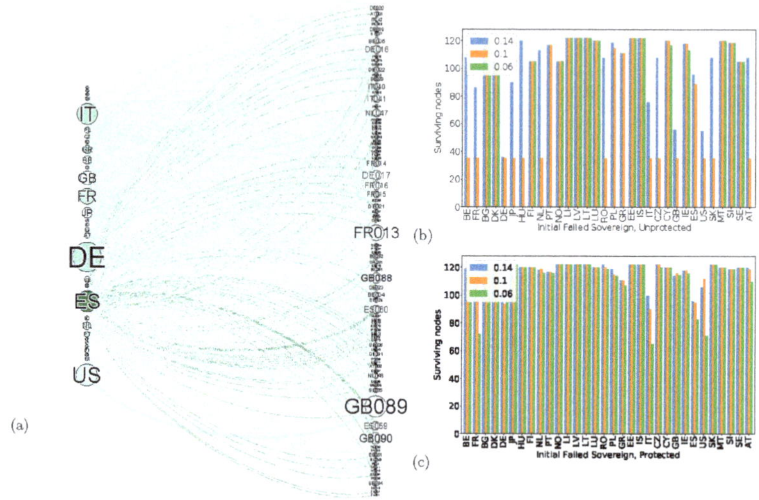

Figure 4. A macroscopic experiment showing the structure and cascading failure dynamics of a real Bank-Asset network. (**a**) The bipartite network, left part are the sovereign assets, right are the holding banks. Node size reflects the capital (either held or invested), edge colour shows size of holding, the darker the larger. (**b**) Survival of the unprotected network. The x-axis shows the initial failing sovereign debt asset, while the y-axis shows the number of remaining nodes. The colours show different levels of sensitivity. The higher the sensitivity, the bigger the impact needed to cause a failure. With the exception of very small countries, most failing sovereign debts cause a cascading failure of the entire network, some even for relatively resilient conditions. (**c**) The same network as (**b**) but with the defense mechanism in place, even the most sensitive networks do not undergo a complete cascading failure with the protection, leading to a much more stable overall economic environment. After [58].

Extending the model to other networks, failure mechanisms and topologies allow to protect many types of networks, bipartite or otherwise, from cascading failure stemming from unknown source. That feature is very important due to the inherent unknowability of every source of risk. Once the mitigation strategy is able to not care where the cascade starts, we know it will serve us well no matter the manifestation of risk. Mitigation and recovery models discussed here offer possible paths to alleviation of systemic financial risk. Those models highlight vulnerabilities and potential protection or recovery methods but are not yet applied (to our knowledge) in decision making by regulators. Thus, the practical ability of such approaches to positively impact economics is yet unknown.

The networks discussed here are created from exposure between various entities. Econophysics allows us to explore and analyze networks whose impact on the economy goes beyond the financial. Next, we explore another network constructed from the micro that gives tools to understand the economy as a whole. For that, we turn to mobility networks.

5. Micro-Mobility to Macroeconomics—From Cell Phones to GDP Estimates during a Pandemic

The year 2008 taught us important lessons about risk management, and made us understand that diversification, previously thought to be a strong risk-immunization technique, can fail catastrophically. We found out to which extent various financial and economic entities were interconnected, how that connectivity was often hidden from sight and how extreme its ramifications can be. Excess greed in the lending market for houses in the United States led to a credit crunch that almost toppled the entire global economic system. The abyss was avoided only by massive government bail-out plans, echoes of which are still with us today, more than a decade later. The year 2019 brought about a very different type of stress factor—this time a pandemic, a virus first recorded in Wuhan, China, and quickly spread throughout the globe. Response varied greatly between different countries, with many different non-pharmaceutical interventions (NPIs) taking place. Some pursued tight lockdown measures while others refrained from implementing strict limitations, striving for social distancing and herd immunity. Estimating the success of these measures is beyond the scope of this review, but the measures taken had indubitable consequences from an economic perspective. Global and local limitations on mobility led multi-national companies as well as individuals to find themselves disconnected, without air travel or public transportation, and with most businesses closed.

In recent years, proliferation of cellular devices with accurate GPS sensors, coupled with multiple companies that gather and process the data, exposed large amounts of fine-grained mobility data to researchers [59]. Several directions of research explored effects of lockdown and similar restrictions through the perspective of mobility networks [60–64]. Many of those deal with the effects of restrictions on epidemic spreading and vice versa, but some focus on the economic and social ramifications of the restrictions.

In Ref. [63], the authors start from the microscopic mobility patterns of individuals and construct a country-wide network of mobility in Italy, before, during and after the first lockdown. Various parameters of the emerging networks are analyzed, from the perspectives of scaling, dynamics and resilience. Interestingly, strong relations are found with two major economic indicators. Statically, i.e., using a historical point in time, regional mobility levels correlate strongly with official levels of regional GDP. Further, using a fast-moving estimate of the GDP based on several measures (and on its own shown to match the actual GDP when available), they show how well changes in levels of mobility correspond to changes of GDP, and from that they derive the local (regional) estimates of GDP (Figure 5), in near-real time, by far faster than official calculations. This estimate, easily calculated from available data, could allow decision makers to gauge the magnitude of economic impact to various regions following planned restrictions. Thus, again, a microscopic model gives macroscopic visibility into system behaviour and stability.

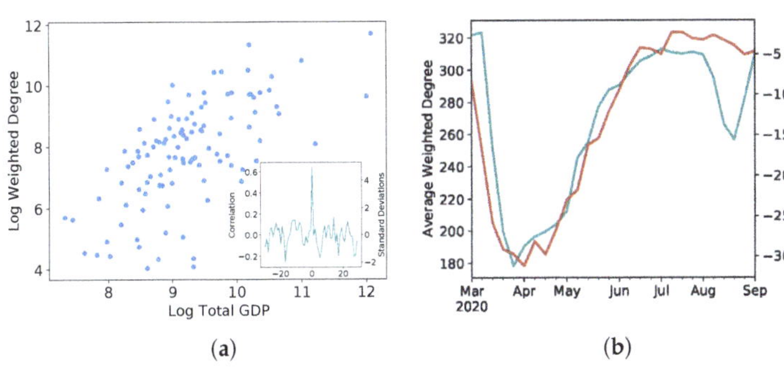

Figure 5. *Cont.*

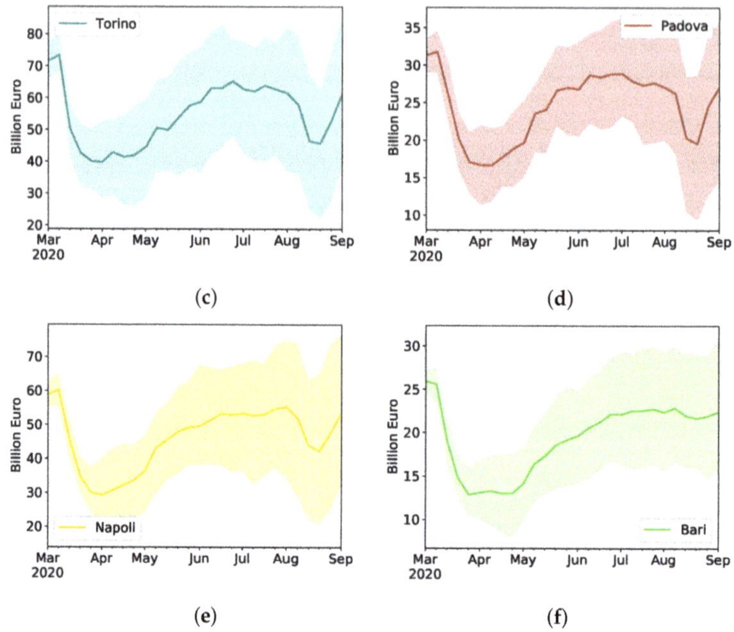

Figure 5. From micro to macro, inferring the economy from mobility, after [63]. (**a**) The correlation between the GDP levels of individual Italian provinces for 2017 and the weighted province degree. Log of values is shown to highlight the relation holds regardless the strongest province economies. Inset: shifted correlation to validate the significance of the relationship. (**b**) The Average weighted degree, teal, left axis; and the estimated real-time GDP, red, right axis. (**c**–**f**) The Province-based GDP forecast based on mobility data for Turin, Padua, Como and Pisa Provinces. The thick line is the average forecast, with the shaded area showing the 25–75 percentile range.

6. Trading, Failure and Centrality—From Local Thresholds to Global Importance

Finally, we bring together some of the methodologies described before to compare different countries and sectors and their importance to the global economy. The authors in Ref. [65] apply a model very similar to the one described in Section 3 after [57], only instead of banks and assets the network is comprised of various industries trading with each other. They then use sensitivity of the network to failure of a network's constituent to assess its relative importance. That is, similarly to the threshold shown in Figure 4, the higher the threshold above which the network undergoes cascading failure, the more important a sector or a country is. Figure 6a shows the progression of failure in the described network in a different setting than a bank–asset network. Figure 6b shows the results of the analysis. Surprisingly, through the lens of sensitivity to failure and examining the top one through eight sectors, the main drivers of economic activities that historically were in the United States have shifted to other geographies, with China taking the lead for the top component nearly two decades ago and surpassing in all eight top components around 2010.

It is noteworthy to highlight the contrast between the approach taken in [58] as mentioned in Section 4 and the one presented in [65]. While both approaches center around the onset of cascading failure given a specific failure mechanism, the next step is almost exactly opposite. The former approach designs a defense mechanism that, when functioning properly, is agnostic the specifics of the failure's origin. The latter, on the other hand, uses exactly that magnitude of impact to determine relative importance. Both models can be expanded and refined, each on its own, but the unifying theme, tying together different sections discussed above, highlights the strength of micro- and macroscopic modelling in the application of network science and econophysics for analysis of actual economic systems, their strengths and weaknesses.

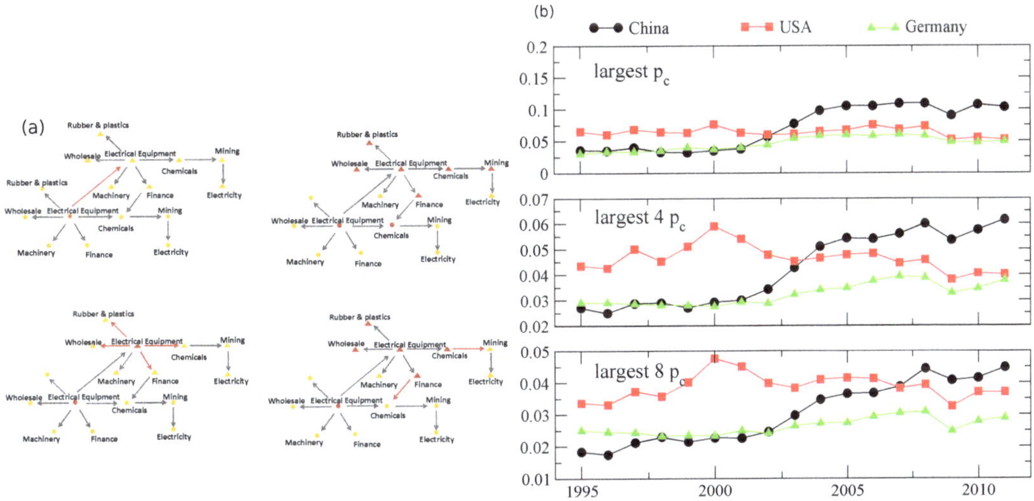

Figure 6. Tying things together, after [65]. (**a**) The propagation of cascading failure over a network of interacting industries and countries. Triangles and circles are different countries with interaction between and within countries facilitating the cascade. (**b**) A critical threshold exists beyond which the network fails. The higher the threshold, the more sensitive the network is to that failure. Looking at the top 1, 4 and 8 such critical thresholds a pattern emerges whereby Chinese sectors take a more central position in the networks.

7. Discussion

Throughout this paper we reviewed several research contributions over the last few years showcasing the immediate applicability of modelling and simulation methods expanded from statistical physics and network theory to various aspects of economics, finance and everyday life. From the big picture is system-level appreciation of the interdependence and nontrivial relations that are present in our most fundamental infrastructure, through fine-grained, local models of interaction that mirror high frequency trading asset behaviour to estimation of local and global economic behaviour in normal and highly anomalous times. These, and many other contributions, give us both new perspectives on known phenomena and mitigation tools in order to manage them such that they do not lead to massive damage. Even with this wide range of topics covered and approaches demonstrated, this is but a grain in the wide landscape of research conducted over the past decades, and evolving still, bridging the gap between methods and discoveries from physics to social and economic life.

Our economy, a textbook case of complex, adaptive system, is comprised of multiple time scales, types of interactions and agents driven by fear, greed, hope and desire, aiming to improve their state compared to themselves and others, poses a great challenge not typically encountered in physics, where particles' knowledge of governing rules does not change their behaviour. Yet, many patterns emerge over time that show that, nonetheless, there are regularities and laws governing some of the observable outcomes. There are arguably many more open questions, from big to small, yet to be answered pertaining to everyday life in its various aspects (economics, epidemiology, mobility, transportation and many more) compared to"classical" physics, and the growing body of research is a strong corroboration of that. The development of network science, econophysics and sociophysics has laid the directions, but the journey is far from over.

Author Contributions: Conceptualization, S.H. and A.S.; writing—original draft preparation, A.S. writing—review and editing, S.H. All authors have read and agreed to the published version of the manuscript.

Funding: This research received no external funding.

Institutional Review Board Statement: Not applicable.

Informed Consent Statement: Not applicable.

Conflicts of Interest: The authors declare no conflict of interest.

References

1. Anderson, P.W. More is different. *Science* **1972**, *177*, 393–396. [CrossRef]
2. Bak, P. *How Nature Works: The Science of Self-Organized Criticality*; Springer Science & Business Media: Berlin/Heidelberg, Germany, 2013.
3. Sornette, D. *Critical Phenomena in Natural Sciences: Chaos, Fractals, Selforganization and Disorder: Concepts and Tools*; Springer Science & Business Media: Berlin/Heidelberg, Germany, 2006.
4. Smith, A. *The Wealth of Nations*(1776); W. Strahan and T. Cadell: London, UK, 1937, Volume 11937.
5. Bachelier, L. Théorie de la spéculation. *Ann. Sci. L'ÉCole Norm. Supérieure* **1900**, *17*, 21–86. [CrossRef]
6. Black, F.; Scholes, M. The Pricing of Options and Corporate Liabilities. *J. Political Econ.* **1973**, *81*, 637–654. [CrossRef]
7. Kahneman, D.; Tversky, A. Prospect Theory: An Analysis of Decision under Risk. *Econometrica* **1979**, *47*, 263–292. [CrossRef]
8. Tversky, A.; Kahneman, D. Judgment under uncertainty: Heuristics and biases. *Science* **1974**, *185*, 1124–1131. [CrossRef]
9. Mantegna, R.N.; Stanley, H.E. *Introduction to Econophysics: Correlations and Complexity in Finance*; Cambridge University Press: Cambridge, UK, 1999.
10. Sinha, S.; Chatterjee, A.; Chakraborti, A.; Chakrabarti, B.K. *Econophysics: An Introduction*; John Wiley & Sons: Hoboken, NJ, USA, 2010.
11. Jackson, M.O. *Social and Economic Networks*; Princeton University Press: Princeton, NJ, USA, 2010.
12. Kutner, R.; Ausloos, M.; Grech, D.; Di Matteo, T.; Schinckus, C.; Stanley, H.E. Econophysics and Sociophysics: Their Milestones & Challenges. *Phys. A Stat. Mech. Appl.* **2019**, *516*, 240–253.
13. Mantegna, R.N. Lévy walks and enhanced diffusion in Milan stock exchange. *Phys. A Stat. Mech. Appl.* **1991**, *179*, 232–242. [CrossRef]
14. Werner, T.R.; Gubiec, T.; Kutner, R.; Sornette, D. Modeling of super-extreme events: An application to the hierarchical Weierstrass-Mandelbrot Continuous-time Random Walk. *Eur. Phys. J. Spec. Top.* **2012**, *205*, 27–52. [CrossRef]
15. Gubiec, T.; Klamut, J.; Kutner, R. Multi-phase Long-Term Autocorrelated Diffusion: Stationary Continuous-Time Weierstrass Walk Versus Flight. In *Simplicity of Complexity in Economic and Social Systems*; Springer Proceedings in Complexity; Grech, D., Miśkiewicz, J., Eds.; Springer: Cham, Switzerland, 2021; pp. 55–88.
16. Cont, R. Empirical properties of asset returns: Stylized facts and statistical issues. *Quant. Financ.* **2001**, *1*, 223. [CrossRef]
17. Battiston, S.; Puliga, M.; Kaushik, R.; Tasca, P.; Caldarelli, G. Debtrank: Too central to fail? financial networks, the fed and systemic risk. *Sci. Rep.* **2012**, *2*, 541. [CrossRef]
18. Kwapień, J.; Drożdż, S. Physical approach to complex systems. *Phys. Rep.* **2012**, *515*, 115–226. [CrossRef]
19. Klamut, J.; Kutner, R.; Struzik, Z.R. Towards a universal measure of complexity. *Entropy* **2020**, *22*, 866. [CrossRef]
20. Stanley, H. *Introduction to Phase Transitions and Critical Phenomena*; Oxford University Press: Oxford, UK, 1971.
21. Jiang, Z.-Q.; Xie, W.-J.; Zhou, W.-X.; Sornette, D. Multifractal analysis of financial markets: A review. *Rep. Prog. Phys.* **2019**, *12*, 125901. [CrossRef]
22. Barabási, A.L.; Albert, R. Emergence of scaling in random networks. *Science* **1999**, *286*, 509–512. [CrossRef]
23. Albert, R.; Barabási, A.L. Statistical mechanics of complex networks. *Rev. Mod. Phys.* **2002**, *74*, 47. [CrossRef]
24. Cohen, R.; Havlin, S. *Complex Networks: Structure, Robustness and Function*; Cambridge University Press: Cambridge, UK, 2010.
25. Newman, M. *Networks*; Oxford University Press: Oxford, UK, 2018.
26. Reed, D.A.; Kapur, K.C.; Christie, R.D. Methodology for assessing the resilience of networked infrastructure. *IEEE Syst. J.* **2009**, *3*, 174–180. [CrossRef]
27. Buldyrev, S.V.; Parshani, R.; Paul, G.; Stanley, H.E.; Havlin, S. Catastrophic cascade of failures in interdependent networks. *Nature* **2010**, *464*, 1025–1028. [CrossRef]
28. Di Muro, M.A.; Valdez, L.D.; Rêgo, H.A.; Buldyrev, S.; Stanley, H.; Braunstein, L.A. Cascading failures in interdependent networks with multiple supply-demand links and functionality thresholds. *Sci. Rep.* **2017**, *7*, 15059. [CrossRef]
29. Duan, D.; Lv, C.; Si, S.; Wang, Z.; Li, D.; Gao, J.; Havlin, S.; Stanley, H.E.; Boccaletti, S. Universal behavior of cascading failures in interdependent networks. *Proc. Natl. Acad. Sci. USA* **2019**, *116*, 22452–22457. [CrossRef]
30. Shao, J.; Buldyrev, S.V.; Havlin, S.; Stanley, H.E. Cascade of failures in coupled network systems with multiple support-dependence relations. *Phys. Rev. E* **2011**, *83*, 036116. [CrossRef]
31. Gao, J.; Buldyrev, S.V.; Stanley, H.E.; Xu, X.; Havlin, S. Percolation of a general network of networks. *Phys. Rev. E* **2013**, *88*, 062816. [CrossRef]

32. Cui, P.; Zhu, P.; Wang, K.; Xun, P.; Xia, Z. Enhancing robustness of interdependent network by adding connectivity and dependence links. *Phys. A Stat. Mech. Appl.* **2018**, *497*, 185–197. [CrossRef]
33. Zhang, H.; Zhou, J.; Zou, Y.; Tang, M.; Xiao, G.; Stanley, H.E. Asymmetric interdependent networks with multiple-dependence relation. *Phys. Rev. E* **2020**, *101*, 022314. [CrossRef]
34. Han, Y.; Guo, C.; Ma, S.; Song, D. Modeling cascading failures and mitigation strategies in PMU based cyber-physical power systems. *J. Mod. Power Syst. Clean Energy* **2018**, *6*, 944–957. [CrossRef]
35. Cohen, R.; Erez, K.; ben Avraham, D.; Havlin, S. Resilience of the Internet to Random Breakdowns. *Phys. Rev. Lett.* **2000**, *85*, 4626–4628. [CrossRef]
36. Callaway, D.S.; Newman, M.E.; Strogatz, S.H.; Watts, D.J. Network robustness and fragility: Percolation on random graphs. *Phys. Rev. Lett.* **2000**, *85*, 5468. [CrossRef]
37. Dorogovtsev, S.N.; Dorogovtsev, S.N.; Mendes, J.F. *Evolution of Networks: From Biological Nets to the Internet and WWW*; Oxford University Press: Oxford, UK, 2003.
38. Gallos, L.K.; Cohen, R.; Argyrakis, P.; Bunde, A.; Havlin, S. Stability and topology of scale-free networks under attack and defense strategies. *Phys. Rev. Lett.* **2005**, *94*, 188701. [CrossRef]
39. Boccaletti, S.; Latora, V.; Moreno, Y.; Chavez, M.; Hwang, D.U. Complex networks: Structure and dynamics. *Phys. Rep.* **2006**, *424*, 175–308. [CrossRef]
40. Parshani, R.; Buldyrev, S.V.; Havlin, S. Interdependent networks: Reducing the coupling strength leads to a change from a first to second order percolation transition. *Phys. Rev. Lett.* **2010**, *105*, 048701. [CrossRef]
41. Gao, J.; Buldyrev, S.V.; Havlin, S.; Stanley, H.E. Robustness of a network of networks. *Phys. Rev. Lett.* **2011**, *107*, 195701. [CrossRef]
42. Baxter, G.; Dorogovtsev, S.; Goltsev, A.; Mendes, J. Avalanche collapse of interdependent networks. *Phys. Rev. Lett.* **2012**, *109*, 248701. [CrossRef]
43. Gao, J.; Buldyrev, S.V.; Stanley, H.E.; Havlin, S. Networks formed from interdependent networks. *Nat. Phys.* **2012**, *8*, 40–48. [CrossRef]
44. Bianconi, G. *Multilayer Networks: Structure and Function*; Oxford University Press: Oxford, UK, 2018.
45. Amini, H.; Cont, R.; Minca, A. Resilience to contagion in financial networks. *Math. Financ.* **2016**, *26*, 329–365 [CrossRef]
46. Glasserman, P.; Young, H.P. Contagion in financial networks. *J. Econ. Lit.* **2016**, *54*, 779–831 [CrossRef]
47. Gai, P.; Kapadia, S. Contagion in financial networks. *Proc. R. Soc. A Math. Phys. Eng. Sci.* **2010**, *466*, 2401–2423.
48. Acemoglu, D.; Ozdaglar, A.; Tahbaz-Salehi, A. Systemic risk and stability in financial networks. *Am. Econ. Rev.* **2015**, *105*, 564–608. [CrossRef]
49. Newman, M.E. The structure and function of complex networks. *SIAM Rev.* **2003**, *45*, 167–256. [CrossRef]
50. Gao, J.; Li, D.; Havlin, S. From a single network to a network of networks. *Natl. Sci. Rev.* **2014**, *1*, 346–356. [CrossRef]
51. Zhou, D.; Bashan, A.; Cohen, R.; Berezin, Y.; Shnerb, N.; Havlin, S. Simultaneous first-and second-order percolation transitions in interdependent networks. *Phys. Rev. E* **2014**, *90*, 012803. [CrossRef]
52. Bashan, A.; Berezin, Y.; Buldyrev, S.V.; Havlin, S. The extreme vulnerability of interdependent spatially embedded networks. *Nat. Phys.* **2013**, *9*, 667–672. [CrossRef]
53. Vaknin, D.; Bashan, A.; Braunstein, L.A.; Buldyrev, S.V.; Havlin, S. Cascading failures in anisotropic interdependent networks of spatial modular structures. *New J. Phys.* **2021**, *23*, 113001. [CrossRef]
54. Watts, D.J. A simple model of global cascades on random networks. *Proc. Natl. Acad. Sci. USA* **2002**, *99*, 5766–5771. [CrossRef]
55. Majdandzic, A.; Podobnik, B.; Buldyrev, S.V.; Kenett, D.Y.; Havlin, S.; Stanley, H.E. Spontaneous recovery in dynamical networks. *Nat. Phys.* **2014**, *10*, 34–38. [CrossRef]
56. Majdandzic, A.; Braunstein, L.A.; Curme, C.; Vodenska, I.; Levy-Carciente, S.; Stanley, H.E.; Havlin, S. Multiple tipping points and optimal repairing in interacting networks. *Nat. Commun.* **2016**, *7*, 10850. [CrossRef]
57. Huang, X.; Vodenska, I.; Havlin, S.; Stanley, H.E. Cascading failures in bi-partite graphs: Model for systemic risk propagation. *Sci. Rep.* **2013**, *3*, 1219. [CrossRef]
58. Smolyak, A.; Levy, O.; Vodenska, I.; Buldyrev, S.; Havlin, S. Mitigation of cascading failures in complex networks. *Sci. Rep.* **2020**, *10*, 16124. [CrossRef]
59. Gonzalez, M.C.; Hidalgo, C.A.; Barabasi, A.L. Understanding individual human mobility patterns. *Nature* **2008**, *453*, 779–782. [CrossRef]
60. Bonaccorsi, G.; Pierri, F.; Cinelli, M.; Flori, A.; Galeazzi, A.; Porcelli, F.; Schmidt, A.L.; Valensise, C.M.; Scala, A.; Quattrociocchi, W.; et al. Economic and social consequences of human mobility restrictions under COVID-19. *Proc. Natl. Acad. Sci. USA* **2020**, *117*, 15530–15535. [CrossRef]
61. Arenas, A.; Cota, W.; Gómez-Gardeñes, J.; Gómez, S.; Granell, C.; Matamalas, J.T.; Soriano-Paños, D.; Steinegger, B. Modeling the spatiotemporal epidemic spreading of COVID-19 and the impact of mobility and social distancing interventions. *Phys. Rev. X* **2020**, *10*, 041055. [CrossRef]
62. Schlosser, F.; Maier, B.F.; Jack, O.; Hinrichs, D.; Zachariae, A.; Brockmann, D. COVID-19 lockdown induces disease-mitigating structural changes in mobility networks. *Proc. Natl. Acad. Sci. USA* **2020**, *117*, 32883–32890. [CrossRef]
63. Smolyak, A.; Bonaccorsi, G.; Flori, A.; Pammolli, F.; Havlin, S. Effects of mobility restrictions during COVID19 in Italy. *Sci. Rep.* **2021**, *11*, 21783. [CrossRef]

64. Gross, B.; Zheng, Z.; Liu, S.; Chen, X.; Sela, A.; Li, J.; Li, D.; Havlin, S. Spatio-temporal propagation of COVID-19 pandemics. *EPL (Europhys. Lett.)* **2020**, *131*, 58003. [CrossRef]
65. Li, W.; Kenett, D.Y.; Yamasaki, K.; Stanley, H.E.; Havlin, S. Ranking the economic importance of countries and industries. *J. Netw. Theory Financ.* **2017**, *3*, 1–17. [CrossRef]

entropy

MDPI

Review

Valuing the Future and Discounting in Random Environments: A Review

Jaume Masoliver [1,2,*], Miquel Montero [1,2,*], Josep Perelló [1,2,*] and J. Doyne Farmer [3,4,5,*] and John Geanakoplos [5,6]

1 Departament de Física de la Matèria Condensada, Universitat de Barcelona, 08028 Barcelona, Spain
2 Universitat de Barcelona Institute of Complex Systems (UBICS), 08028 Barcelona, Spain
3 Institute for New Economic Thinking at the Oxford Martin School, Oxford OX1 3UQ, UK
4 Mathematical Institute, University of Oxford, Oxford OX2 6GG, UK
5 Santa Fe Institute, Santa Fe, NM 87501, USA; john.geanakoplos@yale.edu
6 Department of Economics, Yale University, New Heaven, CT 06511, USA
* Correspondence: jaume.masoliver@ub.edu (J.M.); miquel.montero@ub.edu (M.M.); josep.perello@ub.edu (J.P.); doyne.farmer@inet.ox.ac.uk (J.D.F.)

Abstract: We address the process of discounting in random environments, which allows valuation of the future in economic terms. We review several approaches to the problem regarding different well-established stochastic market dynamics in the continuous-time context and include the Feynman–Kac approach. We also review the relation between bond-pricing theory and discounting and introduce both the market price of risk and the risk neutral measure from an intuitive point of view devoid of excessive formalism. We provide the discount for each economic model and discuss their key results. We finally present a summary of our previous empirical studies for several countries on the long-run discount problem.

Keywords: discounting; bond pricing; real interest rates; econophysics

Citation: Masoliver, J.; Montero, M.; Perelló, J.; Farmer, J.D.; Geanakoplos, J. Valuing the Future and Discounting in Random Environments: A Review. *Entropy* **2022**, *24*, 496. https://doi.org/10.3390/e24040496

Academic Editor: Geert Verdoolaege

Received: 3 February 2022
Accepted: 25 March 2022
Published: 1 April 2022

1. Introduction

The introduction around three decades ago of the view and methods of statistical physics into economics and finance signaled the appearance of a new interdisciplinary aspect of physics, which is sometimes called "econophysics" [1–3]. The fact that financial prices are random with sudden and uncontrollable ups and downs has been long-known; however, the first step towards a systematic mathematical analysis of price randomness was taken by Bachelier in 1900, who proposed a model for the market dynamics in which the prices follow ordinary Brownian motion [4].

However, Bachelier's model is not completely satisfactory because, in such a representation, prices can be either positive or negative, contradicting one of the most fundamental tenets of economics, the "principle of limited liability", which affirms that prices cannot attain negative values. This limitation in Bachelier's model was remedied more than six decades later by Osborne [5] by assuming the geometric Brownian motion where prices are described by the exponential of the ordinary Brownian motion and, hence, they can never attain negative values.

Let us denote by $S(t)$ a speculative price (or an economic index) at time t. In the continuous time framework, the geometric Brownian motion assumes that

$$S(t) = S_0 e^{x(t)}, \tag{1}$$

where $S_0 = S(t_0)$ is the price at some initial time t_0 and $x(t)$, the so-called return, is described by the ordinary Brownian motion, that is to say, by the stochastic differential equation

$$dx(t) = m dt + \sigma dW(t), \tag{2}$$

where $W(t)$ is the standard Wiener process with zero mean and unit variance. Note that the return is assumed to be a diffusion process with constant drift m and diffusion coefficient σ. In this model, the return is a Gaussian process with the mean and variance given, respectively, by m and σ^2. The price is, hence, a log-normal process, and the geometric Brownian motion is also called the log-normal model.

Despite the log-normal model being used in countless financial applications, it has certain limitations [6], which have given rise to several generalizations. One of them assumes that the return is a more complex diffusion process obeying a stochastic equation of the form

$$dx(t) = f(x,t)dt + g(x,t)dW(t), \tag{3}$$

which is interpreted in the sense of Itô. In this case, returns, and hence prices, are driven by an external "force" and by multiplicative noise, which, in the most general case, depend explicitly on the time and on the return level. Function $f(x,t)$ drives prices, and function $g(x,t)$ modulates the intensity of the fluctuations around the deterministic motion set by $f(x,t)$. In any case, and regardless of the values taken by $x(t)$, the prices given by Equation (1) are always nonnegative, thus, keeping the principle of limited liability.

Another significant shortcoming of the geometric Brownian motion model is the absence of both "fat tails" and skewness in the distribution of log-prices (i.e., returns). Indeed, empirical distributions of log-prices not only show fat tails—meaning that extreme losses and profits have a higher probability than those of the log-normal model—but also an asymmetric shape in the sense that losses are usually more probable than profits [6]. In order to address these and other problems, intense research has been prompted both in mathematics and physics, which, among others, may involve the use of the Lévy process as driving noise (instead of the Wiener process). One of the most popular alternative financial models—proposed by Mandelbrot [7] and Fama [8] in the early 1960s—is provided by substituting the Wiener process by the Lévy process, which can take into account the appearance of fat tails in the probability distribution of prices, a widely accepted empirical fact [6]. A major inconvenience of the non-Poissonian Lévy jump processes is, however, their lack of finite moments apart from (at most) the first one, which does not seem to be case in empirical data [6]. or models in which the variance σ^2 (or the noise intensity g) is a random process, such as in the so-called "stochastic volatility models" [9–11].

In economics and finance, one of the most consequential developments is that of "discounting", which essentially attempts to answer the crucial question of what the price will be in the future. In other words, discounting weighs the future relative to the present. Traditionally, the weighting procedure has been performed through a decreasing exponential. Thus, under a constant interest rate r, continuously compounded, a dollar invested today at time $t = 0$ yields e^{rt} dollars at time t [12]; hence, one dollar in any future time t is worth e^{-rt} today. This statement is true under constant and fixed rates; however, in real life, rates are random, and this uncertainty makes it completely unrealistic to represent rates by constant quantities or even by deterministic functions of time, and, as a consequence, random models for rates must be addressed.

The problem of discounting is widely known in finance where it has been thoroughly studied closely related to bond pricing particularly over short periods of time [13]. Discounting, particularly in the long run, is of importance not only in the context of finance but to many other aspects of the global economy. For instance, we may consider the long-term environmental planning, which is certainly sensitive in relation to climate action. Thus, in an oversimplified way, an environmental problem, which costs X to fix at a time t is worth an investment of $e^{-rt}X$ today.

This analysis assumes that the interest rate remains constant between today and the distant future t. The rate r becomes a key magnitude to decide whether it becomes more beneficial to take action today with a significant investment or whether the discount gives negligible value to today's investment. The choice of discount rate is perhaps the biggest factor influencing the debate on the urgency of the response to global warming as it relates today's investments with potential climate-related losses in future [14]. No wonder that, in recent years, obtaining a long-run discount rate valid for decades ahead has been the object of intense controversy.

Thus, Nicholas Stern, in an influential report commissioned by the UK government, advocated for a long-run discount of 1.4% [15], which on a 100-year horizon implies a present value of 25% (meaning that the future is worth 25% as much as the present). On the other hand, Willian Nordhaus proposed a discount rate of 4% [16] (implying a present value of 2%) and even the higher value of 6% [17], which implies a present value of 0.3%. Stern has been widely criticized for using such a low rate [18–20], and the question is far from being settled.

Economists present a variety of reasons for discounting, particularly for environmental problems in the long run. These reasons include, among others, ethical considerations [21,22], impatience, economic growth [23] and arguments based on the maximization of utility functions that are mostly chosen for mathematical convenience [24], all of them ingrained in a phenomenological expression called the Ramsey formula [25], which constitutes the standard approach to discounting in the economics literature (particularly in the long-run) [14].

From an empirical point of view, any practical economist involved in environmental debates might consider the average of historical interest rates, which occurred in the last 200 hundred years to estimate the forward discount rate (which is 2.7% in the less unstable countries [26]) or take the average of Wall Street forward looking models, with price bonds of maturity as long as 30 years. Unfortunately, due to historical fluctuations of real interest rates, the appropriate rate is considerably below such averages [26].

In econophysics, the problem of discounting, despite its relevance, is virtually unknown. The main purpose of this paper is to offer a survey of the problem devoid of excessive formalism and abstraction as well as to review some of our recent work on the problem [26–30]

2. The Process of Discounting—Fundamentals

Let us denote by $M = M(t)$ a given quantity of wealth at time t. In economics, the increment of $M(t)$ is assumed to be proportional to the quantity itself and the duration of the variation. For a continuous and instantaneous infinitesimal variation, this can be written as

$$dM(t) \propto M(t)dt. \tag{4}$$

This starting phenomenological law is built on the empirical observation that the larger $M(t)$, the greater its variation at a given time along with the simpler assumption that such a variation is linear in $M(t)$ and not, for instance, quadratic.

2.1. Definitions and General Setting

We define the interest rate as the relative time derivative

$$r(t) \equiv \frac{1}{M(t)}\frac{dM(t)}{dt} = \frac{d\ln M(t)}{dt}, \tag{5}$$

i.e., the rate is the time derivative of the logarithm of wealth. Let us incidentally note that the linearity shown in Equation (4) is equivalent to assume that $r(t)$ is independent of $M(t)$.

In the simplest case, the law (4) represents a direct proportionality, that is to say, $r(t)$ is constant and and from (5), we see that

$$dM(t) = rM(t)dt, \tag{6}$$

where r is the constant interest rate that has units of 1/(time) (wealth is assumed to be dimensionless). By direct integration, we have the usual exponential law [12]

$$M(t) = e^{r(t-t_0)}M(t_0), \tag{7}$$

which connects wealth at some time t_0, for instance, today (which, without loss of generality, we can take equal to zero) with wealth at some future time $t > t_0$.

The growth law (6) appears in many branches of natural and social sciences. Thus, in radioactivity, if $N(t)$ represents the number of active nuclei at time t, the usual hypothesis is that this number decreases as

$$dN(t) = -\lambda N(t)dt.$$

where $\lambda > 0$ is the decay constant. Similar considerations apply to many other situations, as in chemical reactions or population dynamics, to name a few.

As we mentioned above, discounting is the procedure of linking wealth at different times. This is done through the discount function defined as

$$\delta(t) \equiv \frac{M(0)}{M(t)}, \tag{8}$$

where $M(0)$ is today's wealth. In the case of constant rates, we see from Equation (7) that this function is given by the decreasing exponential

$$\delta(t) = e^{-rt}. \tag{9}$$

The assumption of constant rates is actually unrealistic. A first generalization would be to assume that rates are known functions of time $r = r(t)$. In such a case, the growth law (6) would be given by

$$dM(t) = r(t)M(t)dt, \tag{10}$$

which, after integrating, yields

$$\delta(t) = \exp\left(-\int_0^t r(t')dt'\right). \tag{11}$$

However, the assumption of rates being given by constants or by deterministic functions of time is unjustified, in particular over long periods of time. Financial interest rates are typically described as random, as the many models for stochastic interest rates appearing in the literature [13,31,32]. In other words, $r(t)$ is a random function of time, and in consequence the discount function $\delta(t)$ given by Equation (11) is a stochastic process. The effective discount function is then defined as the average of $\delta(t)$,

$$D(t) = \mathbb{E}\left[\exp\left(-\int_0^t r(t')dt'\right)\right], \tag{12}$$

taken over all possible realizations of $r(t)$. The function $r(t)$ can, in principle, be any random process. However, the most natural and simplifying assumption is that rates are Markovian processes with continuous paths—that is, they are diffusion processes [13]. This approach was proposed after the success of taking an identical approach to model stock prices with log-normal process in 1959 and contrasting with empirical data [5]. The first interest rate model was proposed by O. Vasicek in 1977 [33] and during the same decade when stochastic differential equations became crucial to obtain European option prices. Therefore, rates are solutions to stochastic differential equations of the form

$$dr = f(r)dt + g(r)dW(t), \tag{13}$$

where $W(t)$ is the Wiener process and the stochastic differential equation is interpreted in the sense of Itô. We assume that drift $f(r)$ and noise term $g(r)$ do not depend explicitly on time, that is to say, the time dependence is only implicit through $r = r(t)$, which means that the interest rate process is time homogeneous and may be stationary [34]. This is certainly an idealization because real markets do not appear to be time-homogeneous, at least over long periods of time [35].

On the other hand, explicit expressions for $f(r)$ and $g(r)$ should be proposed based on characteristics obtained by actual data, which is observed to have a reversion toward a mean value, and it is thus claimed to attain a stationary regime in contrast with, for instance, stock market price evolution where no stationary behavior is observed. We return to the comparison between models and empirical data in Section 4.

In order to obtain an operative expression for the effective discount function (12), we define the additional random process

$$x(t) = \int_0^t r(t')dt'. \tag{14}$$

The interpretation of $x(t)$ is apparent after substituting Equation (5) into Equation (14) and integrating. We find

$$x(t) = \ln \frac{M(t)}{M(0)} \qquad \Rightarrow \qquad M(t) = M(0)e^{x(t)},$$

which can be taken as an alternative definition of the accumulated return $x(t)$.

Substituting Equation (14) into Equation (12), we see that the effective discount function can be written as

$$D(t) = \mathbb{E}\left[e^{-x(t)}\right],$$

which implies that, in terms of the probability density function (PDF) $p(x, r, t|r_0)$ of the bidimensional diffusion process $(x(t), r(t))$, we can write

$$D(t|r_0) = \int_{-\infty}^{\infty} dr \int_{-\infty}^{\infty} e^{-x} p(x, r, t|r_0) dx, \tag{15}$$

where we have included the dependence on the initial rate, $r_0 = r(0)$, in the discount function $D(t|r_0)$.

From Equations (13) and (14), we see that the bidimensional process $(x(t), r(t))$ is defined by the following pair of stochastic differential equations

$$\begin{aligned} dx &= rdt, \\ dr &= f(r)dt + g(r)dW(t). \end{aligned} \tag{16}$$

Therefore, the joint density obeys the (forward) Fokker–Planck equation (FPE) [34]

$$\frac{\partial p}{\partial t} = -r\frac{\partial p}{\partial x} - \frac{\partial}{\partial r}[f(r)p] + \frac{1}{2}\frac{\partial^2}{\partial r^2}[g^2(r)p], \tag{17}$$

with the initial condition

$$p(x, r, 0|r_0) = \delta(x)\delta(r - r_0). \tag{18}$$

After solving the initial-value problem (17)—(18) and obtaining the joint PDF $p(x, r, t|r_0)$, the discount function follows from Equation (15). There are, however, two different approaches for achieving it. One of them, which is standard in the financial literature, is based on the backward Fokker–Planck equation, and this is called the *Feynman–Kac approach* [13]. A second procedure is based on Fourier analysis [27]. We will explain both approaches next.

2.2. The Feynman–Kac Approach

Using this method, one obtains a partial differential for the discount function $D(t|r_0)$, which is based on the backward Fokker–Planck equation for the joint density $p(x, r, t|r_0)$. In what follows, we assume that $t_0 \neq 0$ and denote $x_0 = x(t_0)$. By definition, $x_0 = 0$ (cf.

Equation (14)). However, we temporally keep $x_0 \neq 0$ and set $x_0 = 0$ at the end of the calculation when needed.

The backward FPE for the PDF $p(x, r, t|x_0, r_0, t_0)$ that corresponds to the bidimensional process (16) is [34]

$$\frac{\partial p}{\partial t_0} = -r_0 \frac{\partial p}{\partial x_0} - f(r_0) \frac{\partial p}{\partial r_0} - \frac{1}{2} g^2(r_0) \frac{\partial^2 p}{\partial r_0^2}, \tag{19}$$

with the *final condition* as $t_0 \to t$,

$$p(x, r, t|x_0, r_0, t) = \delta(x - x_0)\delta(r - r_0). \tag{20}$$

Let us observe that the problem (19)—(20) is invariant under translations of both time and x_0. We thus define the new variables

$$t' = t - t_0, \qquad x' = x - x_0, \tag{21}$$

so that

$$\frac{\partial p}{\partial t_0} = -\frac{\partial p}{\partial t'}, \qquad \frac{\partial p}{\partial x_0} = -\frac{\partial p}{\partial x'},$$

and Equation (19) reads

$$\frac{\partial p}{\partial t'} = -r_0 \frac{\partial p}{\partial x'} + f(r_0) \frac{\partial p}{\partial r_0} + \frac{1}{2} g^2(r_0) \frac{\partial^2 p}{\partial r_0^2}. \tag{22}$$

Under this change of variables, we also have

$$p = p(x, r, t|x_0, r_0, t_0) = p(x, r, t|x - x', r_0, t - t') = p(x', r, t'|r_0),$$

where the last equality comes from the invariance under time and x translations, that is,

$$p(x, r, t|x_0, r_0, t_0) = p(x - x_0, r, t - t_0|r_0).$$

Consequently, the final condition (20) becomes the initial condition

$$p(x', r, t' = 0|r_0) = \delta(x')\delta(r - r_0). \tag{23}$$

Having set the backward FPE in the form given by Equation (22), we next obtain the equation satisfied by the effective discount $D(t|r_0)$. To this end, we multiply Equation (22) by $e^{-x'}$ and integrate over x' and r, we have

$$\begin{aligned} \frac{\partial}{\partial t'} \int_{-\infty}^{\infty} dr \int_{-\infty}^{\infty} e^{-x'} p dx' &= -r_0 \int_{-\infty}^{\infty} dr \int_{-\infty}^{\infty} e^{-x'} \frac{\partial p}{\partial x'} dx' \\ &+ \left[f(r_0) \frac{\partial}{\partial r_0} + \frac{1}{2} g^2(r_0) \frac{\partial^2}{\partial r_0^2} \right] \int_{-\infty}^{\infty} dr \int_{-\infty}^{\infty} e^{-x'} p dx'. \end{aligned} \tag{24}$$

From Equation (15), we see that

$$\int_{-\infty}^{\infty} dr \int_{-\infty}^{\infty} e^{-x'} p(x', r, t'|r_0) dx' = D(t'|r_0). \tag{25}$$

On the other hand, integrating by parts, the first integral on the right hand side of Equation (24) and using (25), we have

$$\int_{-\infty}^{\infty} dr \int_{-\infty}^{\infty} e^{-x'} \frac{\partial p}{\partial x'} dx' = \int_{-\infty}^{\infty} dr \int_{-\infty}^{\infty} e^{-x'} p(x', r, t'|r_0) dx' = D(t'|r_0), \tag{26}$$

where we have considered the boundary condition (otherwise implicit in the definition of D given in Equation (15))

$$\lim_{x'\to+\infty}\left[e^{-x'}p(x',r,t'|r_0)\right]=0.$$

Substituting Equations (25) and (26) into Equation (24) and setting $t_0=0$, which implies $t'=t$ [cf. Equation (21)], we finally obtain

$$\frac{\partial D}{\partial t}=-r_0D+f(r_0)\frac{\partial D}{\partial r_0}+\frac{1}{2}g^2(r_0)\frac{\partial^2 D}{\partial r_0^2},\tag{27}$$

with the initial condition (cf. Equations (23) and (25))

$$D(0|r_0)=1.\tag{28}$$

The method for obtaining the discount function $D(t|r_0)$ by solving the initial-value problem (27)—(28) is called the *Feynman–Kac approach*, and Equation is (27) the *Feynman–Kac equation*. In some applications (see, for instance, Section 3), it is convenient to consider $t_0\neq 0$ so that $t'=t-t_0\neq t$. In these cases, it is appropriate to denote $D=D(t|r_0,t_0)$ and the Feynman–Kac Equation (27) reads

$$\frac{\partial D}{\partial t_0}=r_0D-f(r_0)\frac{\partial D}{\partial r_0}-\frac{1}{2}g^2(r_0)\frac{\partial^2 D}{\partial r_0^2},\tag{29}$$

with the final condition $D(t|r_0,t)=1$.

2.3. The Fourier Transform Approach

An alternative method for obtaining the discount function is based on the joint characteristic function—that is, on the Fourier transform of the joint density,

$$\tilde{p}(\omega_1,\omega_2,t|r_0)=\int_{-\infty}^{\infty}e^{-i\omega_2 r}dr\int_{-\infty}^{\infty}e^{-i\omega_1 x}p(x,r,t|r_0)dx.\tag{30}$$

One of the chief advantages of working with the characteristic function is that obtaining the effective discount is straightforward. Indeed, comparison of Equation (15),

$$D(t|r_0)=\int_{-\infty}^{\infty}dr\int_{-\infty}^{\infty}e^{-x}p(x,r,t|r_0)dx,$$

with Equation (30) shows that

$$D(t|r_0)=\tilde{p}(\omega_1=-i,\omega_2=0,t|r_0).\tag{31}$$

Therefore, in order to obtain the discount function, we only need to know the joint characteristic function of the bidimensional process (x,r). The procedure is quite advantageous in linear cases. In a forthcoming section, we will apply this approach to some standard models of interest rates.

2.4. Adding Risk Aversion

As we will see in the next section, the process of discounting just described is very closely related to an important problem in finance called bond pricing. In the context of bond pricing, there can be two kinds of investors. For one hand, if investors are *risk neutral*, then bond prices can be modeled based on the data generating measure p, which is the solution of the Fokker–Planck Equation (17) with the initial condition (18). This is sometimes called *the Local Expectation Hypothesis* (LEH) [36,37]. Nonetheless, a more general assumption is that investors are sensitive to risk.

In such a case, bonds are somewhat more accurately priced using an artificial density p^* usually called a *risk-neutral (or risk-correcting) probability measure*. Both magnitudes, the data generating measure p and the risk-neutral measure p^*, are related through a quantity that is denoted by $q(r,t)$ and called *market price of risk*, which, as described in the next section, is the extra return per unit of risk that investors demand to bear risk. This additional return is thus determined by a function $q = q(r,t)$ that, in its most general form, may depend on the rate r and current time t, although the most usual assumption is that $q = q(r)$ only depends on the rate [33]. Following a standard procedure for bond pricing [33,38], which we will present in Section 3, one takes risk into account by replacing the drift $f(r)$ by $f^*(r)$,

$$f(r) \to f^*(r),$$

where

$$f^*(r) = f(r) + g(r)q(r), \tag{32}$$

and $q(r) \geq 0$ is the market price of risk. The form of $q(r)$ is, in principle, unknown and has to be conjectured. The simplest and most common assumption is that $q(r) = q$ is constant, in such a case, the value of q may be more easily estimated from empirical data. Now, the risk-neutral measure $p^*(x,r,t|r_0)$ is given by the Fokker–Planck Equation (17) with $f(r)$ replaced by $f^*(r)$; that is,

$$\frac{\partial p^*}{\partial t} = -r\frac{\partial p^*}{\partial x} - \frac{\partial}{\partial r}\Big[\big[f(r) + g(r)q(r)\big]p^*\Big] + \frac{1}{2}\frac{\partial^2}{\partial r^2}\big[g^2(r)p^*\big], \tag{33}$$

with the initial condition given by

$$p^*(x,r,0|r_0) = \delta(x)\delta(r - r_0). \tag{34}$$

In an analogous way, the discount function adjusted for risk will now be given by the Feynman–Kac Equation (27) with $f(r)$ replaced by $f^*(r)$. Or, using the Fourier method, the discount function will be given in terms of the risk-neutral characteristic function, $\tilde{p}^*(\omega_1,\omega_2,t|r_0)$, by (cf. Equation (31))

$$D(t|r_0) = \tilde{p}^*(\omega_1 = -i, \omega_2 = 0, t|r_0). \tag{35}$$

3. Pricing Bonds—The Term Structure of Interest Rates

Pricing bonds is a traditional objective in finance and intimately related to the problem of discounting. It constitutes a vast subject with countless studies, many of them rather abstract, which have appeared in the mathematical finance literature over the last decades. We present a short and intentionally simple, yet rigorous, introduction to the subject devoid as much as possible of technicalities and mathematical subtleties and refer the interested reader to more specialized works for further information [13].

A bond is a financial instrument that one purchases now and that provides a payment in the future. From a more technical point of view, we say that a (discount) bond is a default-free claim on a specified sum of money to be delivered at a given future date called the maturity time. Such claims are bought and issued by investors. Let us denote by $B(t_0,t)$ the price at time t_0 of a discount bond maturing at time $t \geq t_0$, with unit maturity value,

$$B(t,t) = 1.$$

Let us incidentally note that, if the final maturity price is not 1 (say, $B(t,t) = \beta$) then the price of the bond at t_0 would be $\beta B(t_0,t)$.

Bonds are classified according to the *time interval to maturity* τ defined as

$$\tau = t - t_0.$$

Thus, if $\tau = 10$ years, we talk about a 10-year bond that is traded at t_0 (for instance, today) with price $B(t_0, t_0 + 10)$ and that, after 10 years, has unit value. Similarly for a 3-year bond, 3-month bond, etc.

The central question is to know the *backward evolution* of the bond price, from unit maturity to the initial purchasing price $B(t_0, t)$. Note that the problem is virtually identical to the problem of discounting discussed in Sect. II, with the sole difference that, in discounting, we look for the forward evolution from a known initial value to an unknown final value, while, in bond pricing, the situation is reversed, since we know the final value but not the initial one.

In order to proceed further, we define the *instantaneous rate of return* $b(t_0, t)$ (also called *forward rate*) as the relative time variation of the bond price (compare with Equation (5))

$$b(t_0, t) \equiv \frac{1}{B(t_0, t)} \frac{\partial B(t_0, t)}{\partial t_0} = \frac{\partial \ln B(t_0, t)}{\partial t_0}. \tag{36}$$

The knowledge of the forward rate $b(t_0, t)$ allows us to relate the initial price $B(t_0, t)$ and the maturing price $B(t, t) = 1$. Indeed, the integration of the above equation directly leads to

$$B(t_0, t) = \exp\left[-\int_{t_0}^{t} b(t_0', t) dt_0'\right]. \tag{37}$$

The close analogy between bond pricing and discounting is now apparent. Indeed, the comparison of Equation (37) with Equation (11) shows that $B(t_0, t)$ is the equivalent of the discount function $\delta(t)$ and that the forward rate $b(t_0, t)$ is the equivalent of the discount rate $r(t)$. However, in what follows, we will use the notation $r(t)$ not for the forward rate $b(t_0, t)$ but for the so-called spot rate (also called nominal rate), which we define in Equation (39).

Another quantity of interest is the *yield to maturity* $y(t_0, \tau)$ defined by

$$y(t_0, \tau) \equiv -\frac{1}{\tau} \ln B(t_0, t_0 + \tau) \qquad \Rightarrow \qquad B(t_0, t_0 + \tau) = e^{-\tau y(t_0, \tau)}. \tag{38}$$

From (37), we see that

$$y(t_0, \tau) = \frac{1}{\tau} \int_{t_0}^{t_0+\tau} b(t_0', t) dt_0',$$

that is to say, the yield is the time average of the forward rate over the maturity period τ.

A final quantity is needed, the *spot or nominal rate*, which is defined as the limit of the yield when the maturity tends to 0. In dealing with bonds, one sometimes uses, for the nominal rate, the notation $n(t_0)$ instead of $r(t_0)$—the later reserved for real interest rates, which can be negative due to inflation (see Section 3).

$$r(t_0) \equiv \lim_{\tau \to 0} y(t_0, \tau) = \lim_{\tau \to 0} \left[\frac{1}{\tau} \int_{t_0}^{t_0+\tau} b(t_0', t) dt_0'\right]. \tag{39}$$

Solving the indeterminacy by expanding the integral in powers of τ, we see that the spot rate is given in terms of the forward rate by

$$r(t_0) = b(t_0, t_0). \tag{40}$$

In other words, the spot rate is the instantaneous forward rate.

Let us finally note that a loan of amount M subscribed at time t_0 with an interest rate $r(t_0)$ (the spot rate) will, at time $t_0 + dt_0$, increase in value to $M + dM$, where

$$dM = r(t_0) M dt_0. \tag{41}$$

Indeed, at any time t_0, the value of the spot rate $r(t_0)$ is the instantaneous increase of the loan value, that is, $r(t_0) = d \ln M(t_0)/dt_0$ (compare with Equation (36)). All of this clearly heightens the close similarities with discounting mentioned above.

However, subsequent values of the spot rate are not necessarily certain. We will see next, the consequences of this fact on the time evolution of the bond price $B(t_0, t)$.

3.1. Dynamics of the Bond Price

Suppose the spot rate $r(t_0)$ is not deterministic but random. In such a case, and analogously to discounting, the usual assumption is that $r_0 = r(t_0)$ is a Markovian random process with continuous trajectories; that is, a diffusion process obeying a stochastic differential equation of the form

$$dr_0 = f(r_0)dt_0 + g(r_0)dW(t_0), \tag{42}$$

where $W(t_0)$ is the standard Wiener process. We assumed that the drift and the noise intensity are independent of time, thus, the time dependence of these coefficients is implicit through $r_0 = r(t_0)$. We know that this implies invariance under time translations, and we can set $t_0 = 0$ when needed without loss of generality.

We will now obtain the time evolution of the bond price $B(t_0, t)$ from the purchasing time t_0 to maturity t, and to this end, we follow Oldrich Vasicek [33]. Let us first observe that the most natural hypothesis consists in assuming that the bond price B is a function of the initial spot rate $r_0 = r(t_0)$ and write

$$B = B\big[t_0, t|r(t_0)\big]. \tag{43}$$

In this way, $B(t_0, t|r_0)$ represents the price of a bond issued at time t_0 and maturing at time t, given that the initial interest rate is $r_0 = r(t_0)$. The infinitesimal variation of the bond price is then defined by

$$dB = B\big[t_0 + dt_0, t|r(t_0 + dt_0)\big] - B\big[t_0, t|r(t_0)\big].$$

We expand in Taylor series up to second order

$$\begin{aligned} B\big[t_0 + dt_0, t|r(t_0 &+ dt_0)\big] = B\big[t_0, t|r(t_0)\big] + \frac{\partial B}{\partial t_0}dt_0 + \frac{\partial B}{\partial r_0}dr_0 \\ &+ \frac{1}{2}\left[\frac{\partial^2 B}{\partial t_0^2}dt_0^2 + \frac{\partial^2 B}{\partial r_0^2}dr_0^2 + 2\frac{\partial^2 B}{\partial t_0 \partial r_0}dt_0 dr_0\right] + \cdots . \end{aligned}$$

Substituting for Equation (42) and taking into account that $dW(t_0) = O(dt_0^{1/2})$ [34,39], we write

$$dB = \left[\frac{\partial B}{\partial t_0} + f(r_0)\frac{\partial B}{\partial r_0}\right]dt_0 + g(r_0)\frac{\partial B}{\partial r_0}dW(t_0) + \frac{1}{2}g^2(r_0)\frac{\partial^2 B}{\partial r_0^2}\left[dW(t_0)\right]^2 + O(dt_0^{3/2}).$$

However, $\left[dW(t_0)\right]^2 = dt_0$ (in mean square sense) [34,39], and, up to the first order in dt_0, we obtain

$$dB = \left[\frac{\partial B}{\partial t_0} + f(r_0)\frac{\partial B}{\partial r_0} + \frac{1}{2}g^2(r_0)\frac{\partial^2 B}{\partial r_0^2}\right]dt_0 + g(r_0)\frac{\partial B}{\partial r_0}dW(t_0). \tag{44}$$

Defining

$$\mu(t_0, t|r_0) \equiv \frac{1}{B}\left[\frac{\partial B}{\partial t_0} + f(r_0)\frac{\partial B}{\partial r_0} + \frac{1}{2}g^2(r_0)\frac{\partial^2 B}{\partial r_0^2}\right], \tag{45}$$

and

$$\sigma(t_0, t|r_0) \equiv -\frac{1}{B}g(r_0)\frac{\partial B}{\partial r_0}, \tag{46}$$

we see from (44) that the bond price satisfies the stochastic differential equation

$$\frac{dB}{B} = \mu(t_0, t|r_0)dt_0 - \sigma(t_0, t|r_0)dW(t_0), \tag{47}$$

showing that the bond price is also a diffusion process.

Averaging Equation (47) and recalling that $\mathbb{E}[dW(t_0)] = 0$, we see that

$$\mu(t_0, t|r_0) = \mathbb{E}\left[\frac{1}{B}\frac{dB}{dt_0}\right],$$

which proves that $\mu(t_0, t|r_0)$ is the average of the instantaneous rate of return [cf. Equation (36)] at time t_0 on a bond with maturing date t, given that the current spot rate is r_0. In an analogous way, one can easily show that $\sigma^2(t_0, t|r_0)$ is the variance [33].

We therefore see from the above development that the bond price is a random quantity but with a final fixed price (the maturity price). The question is: what is the (initial) price that an investor has to buy (or sell) a bond at time t_0 maturing at time $t = t_0 + \tau$ with the current spot rate r_0? One possible answer would be proceeding as in discounting to take the average over all possible realizations of the bond price. However, this procedure implies that the expected rate of return of a bond is invariant under risk variation—that is, under changes of the variance $\sigma^2(t_0, t|r_0)$—a fact that investors always have in mind.

We explain next a procedure resulting in a deterministic bond price, which takes into account the risk aversion of investors (in practice this is only true to some extend because the mathematical procedure assumes that the market is driven by Gaussian white noise—that is, the Wiener process, which is an idealized noise presenting, among other shortcomings, no fat tails, a key characteristic of real markets [6]).

3.2. The Market Price of Risk

Consider an investor who, at time t_0, sells an amount M_1 of a bond maturing at time t_1 and, at the same time, buys an amount M_2 of another bond with a different maturing date t_2. The total worth of the *portfolio*, thus, constructed is $M = M_2 - M_1$. Note that each amount M_i $(i = 1, 2)$ is a multiple of the bond price $B(t_0, t_i|r_0)$ $(i = 1, 2)$ and, hence, they also obey the stochastic differential Equation (47). That is,

$$\frac{dM_i}{M_i} = \mu(t_0, t_i|r_0)dt_0 - \sigma(t_0, t_i|r_0)dW(t_0).$$

As a consequence, the infinitesimal variation $dM = dM_2 - dM_1$ of the worth of the portfolio changes over time according to

$$\begin{aligned} dM &= \left[\mu(t_0, t_2|r_0)M_2 - \mu(t_0, t_1|r_0)M_1\right]dt_0 \\ &\quad - \left[\sigma(t_0, t_2|r_0)M_2 - \sigma(t_0, t_1|r_0)M_1\right]dW(t_0). \end{aligned} \tag{48}$$

Suppose we choose the amounts M_1 and M_2 such that

$$M_1 = \frac{M}{\sigma_1 - \sigma_2}\sigma_2, \qquad M_2 = \frac{M}{\sigma_1 - \sigma_2}\sigma_1, \tag{49}$$

where $M = M_2 - M_1$ and $\sigma_i = \sigma(t_0, t_i|r_0)$ $(i = 1, 2)$. Hence M_1 is proportional to σ_2, while M_2 is proportional to σ_1. With this choice, we have

$$\sigma_2 M_2 - \sigma_1 M_1 = \sigma_2\frac{\sigma_1 M}{\sigma_1 - \sigma_2} - \sigma_1\frac{\sigma_2 M}{\sigma_1 - \sigma_2} = 0,$$

and the random term in Equation (48) vanishes. This renders the portfolio composed of such amounts of the two bonds instantaneously riskless:

$$dM = \frac{M}{\sigma_1 - \sigma_2}(\mu_2\sigma_1 - \mu_1\sigma_2)dt_0, \tag{50}$$

where $\mu_i = \mu(t_0, t_i|r_0)$. The rate of return r_M of this portfolio is

$$r_M \equiv \frac{1}{M}\frac{dM}{dt_0} = \frac{\mu_2\sigma_1 - \mu_1\sigma_2}{\sigma_1 - \sigma_2}.$$

In order to avoid *arbitrage opportunities*—that is, making profits without taking any risk—the rate r_M must be equal to the spot rate r_0. If not, the portfolio can be purchased by taking funds borrowed at the spot rate, or otherwise sold and the profits lent out to accomplish a riskless arbitrage [33]. Therefore (compare also Equation (41) with Equation (50))

$$r_0 = \frac{\mu_2\sigma_1 - \mu_1\sigma_2}{\sigma_1 - \sigma_2}.$$

Rearranging terms, we find $(\mu_1 - r_0)/\sigma_1 = (\mu_2 - r_0)/\sigma_2$, so that

$$\frac{\mu(t_0, t_1|r_0) - r_0}{\sigma(t_0, t_1|r_0)} = \frac{\mu(t_0, t_2|r_0) - r_0}{\sigma(t_0, t_2|r_0)}.$$

This equation is valid for arbitrary maturities $t_1, t_2, \ldots,$ it then follows that the ratio $[\mu(t_0, t|r_0) - r_0]/\sigma(t_0, t|r_0)$ *must be independent of the maturity time t*.

Let us denote by $q(t_0|r_0)$ the common value of such a ratio for a bond of any maturity date, given that the current spot rate (at time t_0) is r_0,

$$q(t_0|r_0) \equiv \frac{\mu(t_0, t|r_0) - r_0}{\sigma(t_0, t|r_0)}, \qquad (t \geq t_0). \tag{51}$$

The quantity $q(t_0|r_0)$ is called the *market price of risk*, as it gives the variation of the expected rate of return on a bond (specified by the *risk premium* $\mu - r_0$) per an additional unit risk (specified by the standard deviation σ). The market price of risk $q(t_0|r_0)$ is the so-called Sharpe ratio [40] of the excess return $\mu - r_0$.

Note that, if $q = 0$, the spot rate $r_0 = r(t_0)$ and the average rate of return μ coincide.

$$\mu(t_0, t|r_0) = r(t_0)$$

$(t = t_0 + \tau)$ meaning that the expected instantaneous rates of return on bonds are the same for all maturities.

3.3. *The Term Structure Equation and the Risk-Neutral Measure*

The introduction of the market price of risk implies a non-random bond price $B = B(t_0, t|r_0)$, which, in turn, allows a deterministic equation for B. In effect, rewriting Equation (51) as

$$\mu(t_0, t|r_0) - r_0 = \sigma(t_0, t|r_0)q(t_0|r_0),$$

and substituting μ and σ for their definitions given in Equations (45) and (46), we have

$$\frac{1}{B}\left[\frac{\partial B}{\partial t_0} + f(r_0)\frac{\partial B}{\partial r_0} + \frac{1}{2}g^2(r_0)\frac{\partial^2 B}{\partial r_0^2}\right] - r_0 = -q(t_0|r_0)\frac{1}{B}g(r_0)\frac{\partial B}{\partial r_0},$$

which, after rearranging terms, yields

$$\frac{\partial B}{\partial t_0} = r_0 B - [f(r_0) + g(r_0)q(t_0|r_0)]\frac{\partial B}{\partial r_0} - \frac{1}{2}g^2(r_0)\frac{\partial B}{\partial r_0^2}. \tag{52}$$

This equation, called the term structure equation, is a partial differential equation for $B(t_0, t|r_0)$, that is obtained once we know the random character of the spot rate process $r(t)$ (through its drift f and noise intensity g) and once the market price of risk $q(t_0|r_0)$ is specified. Bond prices are thus obtained after solving the deterministic Equation (52) with the final condition:

$$B(t, t|r_0) = 1. \tag{53}$$

Let us observe that the term structure Equation (52) for the bond price B is identical to the Feynman–Kac Equation (27) for the discount function D as long as we make the following change of drift

$$f(r_0) \longrightarrow f(r_0) + g(r_0)q(t_0|r_0). \tag{54}$$

On the other hand, as we have seen in Section 2, the solution of the Feynman–Kac Equation (27) for the discount function $D(t|r_0)$ is written as the average (cf. Equation (15))

$$D(t|r_0, t_0) = \int_{-\infty}^{\infty} dr \int_{-\infty}^{\infty} e^{-x} p(x, r, t|r_0, t_0) dx,$$

where $p(x, r, t|r_0, t_0)$ is the probability density function of the bidimensional diffusion process $(x(t), r(t))$ defined by Equation (16),

$$dx = rdt, \qquad dr = f(r)dt + g(r)dW(t).$$

Now the analogy between the term structure Equation (52) and the Feynman–Kac Equation (29) suggests that we can write the bond price $B(t_0, t|r_0)$ as an average over the different realizations of the spot rate $r(t_0)$. However, this averaging procedure is taken using a modified PDF called the *risk-free measure*. Thus, it can be proven in a more rigorous way that [32,33]

$$B(t_0, t|r_0) = \int_{-\infty}^{\infty} dr \int_{-\infty}^{\infty} e^{-x} p^*(x, r, t|r_0, t_0) dx, \tag{55}$$

where $p^*(x, r, t|r_0, t_0)$ is the risk-free measure that is the PDF of the bidimensional process $(x(t_0), r(t_0))$ defined by the following pair of stochastic differential equations that include the market price of risk (see Equation (54)):

$$\begin{aligned} dx &= rdt, \\ dr &= [f(r) + g(r)q(t|r)]dt + g(r)dW(t). \end{aligned} \tag{56}$$

That is, p^* is the solution to the FPE

$$\frac{\partial p^*}{\partial t} = -r\frac{\partial p^*}{\partial x} - \frac{\partial}{\partial r}\Big[[f(r) + g(r)q(t|r)]p^*\Big] + \frac{1}{2}\frac{\partial^2}{\partial r^2}[g^2(r)p^*], \tag{57}$$

with the initial condition

$$p^*(x, r, t_0|r_0, t_0) = \delta(x)\delta(r - r_0). \tag{58}$$

Since, as we have shown in Section 2.2, the Feynman–Kac approach to discounting is equivalent to the Fourier method described in Section 2.3, we can apply the latter to directly obtain the bond price knowing only the risk neutral PDF, without having to solve the Feynman–Kac Equation (52) with condition (23). Indeed, the characteristic function of the risk neutral density p^* is the joint Fourier transform

$$\tilde{p}^*(\omega_1, \omega_2, t|r_0, t_0) = \int_{-\infty}^{\infty} e^{-i\omega_2 r} dr \int_{-\infty}^{\infty} e^{-i\omega_1 x} p^*(x, r, t|r_0, t_0) dx,$$

which, after comparing with Equation (55), yields

$$B(t_0, t|r_0) = \tilde{p}^*(\omega_1 = -i, \omega_2 = 0, t|r_0, t_0). \tag{59}$$

Finally, once we know the bond price, the yield to maturity $y(t_0, \tau|r_0)$ (also called *the term structure of interest rates*) is readily evaluated from Equation (38):

$$y(t_0, \tau|r_0) = -\frac{1}{\tau} \ln B(t_0, t_0 + \tau|r_0). \quad (60)$$

The graphic representations of $y(t_0, \tau|r_0)$ as a function of t_0 and for different values of the maturity interval τ are called *yield curves* and are of prime importance for practitioners.

4. Standard Models

Throughout the above development, it is clear that, in order to proceed further in the discounting process (as well as in pricing bonds), we need to identify the specific diffusion process chosen for modeling rates. Such a choice is mostly based on the analysis of empirical data [26,29]. Clearly, in any proposed market model, there are idealizations, otherwise a complete treatment of the problem would be very problematic, not to say impossible, not only analytically but computationally as well.

In addition to assuming a diffusive behavior for the market, the first of such idealizations is supposing that the market is stationary, i.e., the structural conditions of the market do not change over time. However, and in particular after the 1980s, market circumstances have largely changed due to a great increase of transaction volumes along with transparency and, to a lesser extend, changes in investor perspectives. In this review, we only address stationary models, although we are working on new models dealing with some non-stationary features of the market [29,35], and we refer the interested reader to these works for further information.

On the other hand, there is a property of the market that appears to be well founded on empirical grounds. This is the property of *mean reversion* meaning that prices tend to return to some fundamental value, called the *normal level*, which is typically identified as the long-time (i.e., stationary) mean value. The simplest method of introducing this feature in the diffusion market model is to assume a linear drift of the form $f(r) = -\alpha(r - m)$, where $\alpha > 0$ is the strength of the reversion to the normal level, identified by m.

In such a case, the drift acts like a linear restoring force driving $r(t)$ towards m as time increases. Despite that the introduction of mean reversion might create some arbitrage opportunities, the property of mean reversion is widely accepted in the literature [6], and we previously discussed this issue in the context of option pricing when considering the Ornstein–Uhlenbeck model [41].

4.1. Bonds and Real Rates

Before proceeding with the introduction of some standard models for the market evolution, we briefly explain the link between bonds and (real) interest rates.

Financial economists have developed a large number of models of interest rate processes to enable them to price bonds and other cash flows. In these models, interest rates are described by positive random processes since financial interest rates rarely take negative values. Although the models could be, in principle, extended to arbitrary horizons, they have only been studied carefully over time horizons of up to 30 years, since bonds are seldom issued for periods longer than this.

On the other hand, environmental economists are interested in the real behavior of the economic growth over longer horizons, in contrast to financial economists who are typically more interested in nominal rates over shorter periods of time. The behavior of real and nominal rates usually differ as, due to inflation, real rates can take on negative values. In this way, real rates $r(t)$ are generally defined by the so-called Fisher procedure:

$$r(t) = n(t) - i(t), \quad (61)$$

where $i(t)$ is the inflation rate that is usually generated from consumer price indexes as we will explain in the next section. The quantity $n(t)$ represents nominal rates, which are typically constructed out of government bonds and are usually positive (even though, in recent years, nominal rates have taken slightly negative values). In order to explain the close relationship between nominal rates and bonds, let us first recall that nominal rates were called spot rates in the previous section on bond pricing where we used the notation $r(t)$ instead of $n(t)$. We thus define the spot (i.e., nominal) rate as (see Equations (39) and (40))

$$n(t) \equiv \lim_{\tau \to 0} \left[\frac{1}{\tau} \int_t^{t+\tau} b(t', t) dt' \right] = b(t, t), \tag{62}$$

where $b(t', t)$ is the forward rate for bonds defined in Equation (36) (see also Equation (37)), that is

$$b(t', t) = \frac{\partial \ln B(t', t)}{\partial t'} \qquad \Rightarrow \qquad B(t', t) = \exp\left[- \int_{t'}^{t} b(t'', t) dt'' \right]$$

where $B(t', t)$ is the price at time t' of a (government) bond maturing at time $t \geq t'$. Let us recall the definition of the yield to maturity $y(t, \tau)$ given in Equation (38),

$$y(t, \tau) \equiv -\frac{1}{\tau} \ln B(t, t + \tau) \qquad \Rightarrow \qquad B(t, t + \tau) = e^{-\tau y(t, \tau)}$$

($\tau \geq 0$), so that

$$y(t, \tau) = \frac{1}{\tau} \int_t^{t+\tau} b(t', t) dt'$$

and comparing with Equation (62), we see that, in terms of the yield nominal rates, $n(t)$ can be defined as

$$n(t) = \lim_{\tau \to 0} y(t, \tau). \tag{63}$$

Thus, for empirical analysis, the yield can be used as an estimator of the nominal rates:

$$n(t) \sim y(t, \tau), \tag{64}$$

and the accuracy of such an estimator increases as $\tau \to 0$. We will return to this discussion in the next section.

In this way, taking nominal rates corrected by inflation as a proxy of economic growth, we recently demonstrated [26,29,30] through a detailed empirical study on many countries that real interest rates are negative around 25 % of the time (see next section). To understand how discounting depends on the random process used to characterize interest rates, we focused on three different models and obtained exact analytical expressions for the discount function [27]. The three models describe to, varying degrees, a number of relevant characteristics observed in the data, while being simple enough to allow for complete analytical treatment. The main results are summarized in Table 1.

The first model is based on the Ornstein–Uhlenbeck (OU) process—also called the Vasicek model in the financial literature [13]—which allows for negative rates and is, therefore, suitable for pricing environmental problems. The model has a stationary probability distribution and exhibits reversion to the mean, which means that the process tends to return to its average stationary value. We will review this model below.

The second and third models that we considered are given by the Feller and log-normal processes, respectively. For these processes, the rates cannot be negative. The Feller process—also known as the Cox–Ingersoll–Ross (CIR) model [42]—has reversion to the mean and stationary probability distribution.

This is one of the most popular models in finance [13], and we recently reviewed the main properties of the Feller process in previous works [27,43]. A third model, also implying positive rates, is the log-normal process (occasionally called the Dotham model in the financial literature [44]). The model does not have reversion to the mean nor a stationary distribution. Despite these shortcomings, the log-normal process has also been used in the financial literature mainly because it is positive and allows for analytical treatment [13]. We refer the interested reader to our previous work [27] for details on this model.

As remarked in the introduction, we are primarily interested in valuing the far future for environmental problems rather than the short time discount of finance, the latter implying positive interest rates, while the former involves positive as well as negative rates. For this reason, we next review in more detail the Vasicek model allowing for both positive and negative rates than the CIR and log-normal models of which we only present a sketched review.

Table 1. Key statistical features for three standard models: the Vasicek (Ornstein–Uhlenbeck), the Cox–Ingersoll–Ross (Feller) and the log-normal models. The average and variance are provided in terms of the model parameters to better compare the asymptotic behavior of $D(t)$. The asymptotic discount is provided by showing an exponential decay with a long-run rate of discount r_∞ for the Vasicek and the Cox–Ingersoll–Ross models and also in the log-normal case for a specific combination of parameters ($k^2/2 < \alpha$, mild fluctuations). The parameter δ is defined in Equation (111).

Model	$\mathbb{E}[r(t)]$	**Var**$[r(t)]$	$D(t \to \infty)$		r_∞
Vasicek	m	k^2/α	$\exp(-r_\infty t)$		$m - k^2/2\alpha^2$
Feller	m	$mk^2/(2\alpha)$	$\exp(-r_\infty t)$		$\frac{2m}{1+\sqrt{1+2k^2/\alpha}}$
Log-normal	$r_0 e^{\alpha t}$	$r_0^2 e^{2\alpha t}[e^{k^2 t} - 1]$	constant	$(k^2/2 > \alpha)$	$--$
			$\exp(-r_\infty t)$	$(k^2/2 < \alpha)$	$(\alpha - k^2/2)/\delta$
			$t^{-1/2}$	$(k^2/2 = \alpha)$	$--$

4.2. *The Vasicek (Ornstein–Uhlenbeck) Model*

In this model, the rates are described by the Ornstein–Uhlenbeck process [33], which is a diffusion model with linear drift and constant noise intensity:

$$dr(t) = -\alpha[r(t) - m] + kdW(t), \tag{65}$$

where $r(t)$ is the rate and $W(t)$ is the Wiener process. The parameter m (the normal level) is the mean value to which rates revert, $k > 0$ is the amplitude of fluctuations, and $\alpha > 0$ is the strength of the reversion to the mean. These parameters have to be estimated from empirical data.

In this case, the Fokker–Planck equation for the joint density $p(x, r, t|r_0)$ of the bidimensional process $(x(t), r(t))$, given by Equation (17), reads

$$\frac{\partial p}{\partial t} = -r\frac{\partial p}{\partial x} + \alpha\frac{\partial}{\partial r}[(r - m)p] + \frac{1}{2}k^2\frac{\partial^2 p}{\partial r^2}, \tag{66}$$

with the initial condition

$$p(x, r, 0|r_0) = \delta(x)\delta(r - r_0). \tag{67}$$

The joint Fourier transform of these equations results in a simpler initial-value problem for the joint characteristic function $\tilde{p}(\omega_1, \omega_2, t|r_0)$, which can be readily solved to yield the Gaussian density [27]

$$\tilde{p}(\omega_1, \omega_2, t|r_0) = \exp\Big\{-A(\omega_1, t)\omega_2^2 - B(\omega_1, t|r_0)\omega_2 - C(\omega_1, t|r_0)\Big\}, \tag{68}$$

where $A(\omega_1, t)$, $B(\omega_1, t)$ and $C(\omega_1, t)$ are given by [27]

$$A(\omega_1, t) = \frac{k^2}{4\alpha}\Big(1 - e^{-2\alpha t}\Big), \tag{69}$$

$$B(\omega_1, t|r_0) = ir_0 e^{-\alpha t} + \frac{k^2\omega_1}{2\alpha^2}\Big(1 - 2e^{-\alpha t} + e^{-2\alpha t}\Big) + im\Big(1 - e^{-\alpha t}\Big), \tag{70}$$

and

$$C(\omega_1,t|r_0) = i\omega_1 r_0 \frac{1}{\alpha}(1-e^{-\alpha t}) + \frac{k^2\omega_1^2}{2\alpha^3}\left[\alpha t - 2(1-e^{-\alpha t}) + \frac{1}{2}\left(1-e^{-2\alpha t}\right)\right] + im\omega_1\left[t - \frac{1}{\alpha}(1-e^{-\alpha t})\right]. \quad (71)$$

The characteristic function of the rate $r(t)$ is obtained by setting $\omega_2 = 0$ in Equation (68), which also results in the Gaussian density

$$\tilde{p}(\omega_2,t|r_0) = \exp\left\{-\frac{k^2}{4\alpha}\left(1-e^{-2\alpha t}\right)\omega_2^2 - i\left[r_0 e^{-\alpha t} + m(1-e^{-\alpha t})\right]\omega_2\right\}, \quad (72)$$

and in the stationary state ($t \to \infty$), we have

$$\tilde{p}_{st}(\omega_2) = e^{-(k^2/4\alpha)\omega_2^2 - im\omega_2} \qquad \Longrightarrow \qquad p_{st}(r) = \left(\frac{\alpha}{\pi k^2}\right)^{1/2} e^{-\alpha(r-m)^2/k^2}, \quad (73)$$

which proves that the normal level m is the stationary mean value,

$$m = \mathbb{E}[r(t)]. \quad (74)$$

It can also be shown that the correlation function of the process, defined as the average

$$C(\tau) = \mathbb{E}\big[r(t+\tau)r(t)\big] - \big[\mathbb{E}\big[r(t)\big]\big]^2,$$

($\tau \geq 0$) in the stationary state reads [27]

$$C(\tau) = (k^2/2\alpha)e^{-\alpha\tau}, \quad (75)$$

which means that α^{-1} is the correlation time, τ_c, of the rate. Indeed,

$$\tau_c \equiv \frac{1}{C(0)}\int_0^\infty C(\tau)d\tau = \alpha^{-1}.$$

Let us observe that the volatility, $\sigma^2 = C(0)$, is independent of the normal level and given by

$$\sigma^2 = k^2/2\alpha. \quad (76)$$

The discount function $D(t|r_0)$ is also obtained from Equations (68)–(71) although, in this case, after setting $\omega_1 = -i$ and $\omega_2 = 0$ (cf. Equation (31)). We have

$$\ln D(t) = -\frac{r_0}{\alpha}(1-e^{-\alpha t}) + \frac{k^2}{2\alpha^3}\left[\alpha t - 2(1-e^{-\alpha t}) + \frac{1}{2}\left(1-e^{-2\alpha t}\right)\right] - m\left[t - \frac{1}{\alpha}(1-e^{-\alpha t})\right],$$

which, after rearranging terms, can be written as

$$\ln D(t) = -\left(m - \frac{k^2}{2\alpha^2}\right)t + \frac{1}{\alpha}\left[m - r_0 - \frac{k^2}{4\alpha^2}(3-e^{-\alpha t})\right](1-e^{-\alpha t}), \quad (77)$$

where $r_0 = r(0)$ is the initial rate. Note that, as $t \to \infty$ (in fact when $t \gg \alpha^{-1}$, i.e., for times much greater than the correlation time α^{-1}) Equation (77) shows at once that the discount function of the Vasicek model has the typical exponential decay

$$D(t) \simeq e^{-r_\infty t}, \quad (78)$$

where

$$r_\infty = m - k^2/2\alpha^2, \tag{79}$$

is the long-run discount rate. Let us note that the long-run rate can be defined as the limit

$$r_\infty = -\lim_{t\to\infty} \frac{\ln D(t)}{t}, \tag{80}$$

as long as the limit exists. Let us also note the important fact that r_∞ is smaller than the mean value of the return given by the normal level m. This reduction is quantified by the "noise-to-signal" ratio k/α, which means that either a long persistence (recall that this is equivalent to long correlation time, i.e., α small) or an increase of the noise fluctuations (i.e., k large) reduce the long-run discount rate as compared with the average rate m.

Finally, we easily see from Equation (77) that, as $t \to 0$, the discount function approximates to $D(t) \simeq e^{-r_0 t}$, which would correspond to a fixed interest rate without random fluctuations or deterministic changes.

Risk Aversion

As mentioned above, risk aversion is taken into account by introducing the market price of risk $q(r)$ and changing the drift according to Equation (32). For the Vasicek model, in which $f(r) = -\alpha(r - m)$ and $g(r) = k$, we have

$$f^*(r) = -\alpha(r - m) + kq(r), \tag{81}$$

and assuming $q(r) = q$ to be a constant independent of r, we write

$$f^*(r) = -\alpha(r - m^*), \tag{82}$$

where

$$m^* = m + \frac{qk}{\alpha}. \tag{83}$$

Since the modified drift $f^*(r)$ has the same form that $f(r)$, we conclude that the adjusted-for-risk discount function will be given by Equation (77) after the replacement $m \to m^*$. In particular, the adjusted long-run discount now reads (cf. Equation (79))

$$r^*_\infty = m + \frac{qk}{\alpha} - \frac{k^2}{2\alpha^2}. \tag{84}$$

We thus see that the long-run discount depends on the historical rate m; however, this is shifted by two terms. The first term raises the long-run rate due to the market price of risk. The second shift lowers it by an amount given by the ratio of uncertainty (as measured by k) and persistence (as measured by α). We rewrite Equation (84) as

$$r^*_\infty = m + \frac{k}{\alpha}\left(q - \frac{k}{2\alpha}\right). \tag{85}$$

This shows that the overall shift in the long-run discount rate will be positive or negative depending on the size of the market price of risk and on the noise-to-signal ratio between the volatility parameter and the reversion rate.

It is not surprising that the market price of risk raises the long term rate; however, it is not so obvious that uncertainty and persistence can lower it. Indeed, for any given mean interest rate m, by varying k and α, the long-run discount rate r_∞ can take on any value less than m, including negative values, while, at the same time, the standard deviation σ can also be made to take on any arbitrary positive value.

A negative long-run rate is due to the amplification of negative real interest rates $r(t)$. Computation of the discount function involves an average over exponentials, rather than the exponential of an average. As a result, periods where interest rates are negative are amplified and can easily dominate periods where interest rates are large and positive, even if the negative rates are rarer and weaker. It does not take many such periods to substantially reduce the long run interest rate.

To summarize, in the Vasicek model, and even taking into account risk aversion, the long-run discounting rate can be much lower than the mean and, indeed, can correspond to low interest rates that are rarely observed.

4.3. The Cox–Ingersoll–Ross (Feller) Model

In the financial literature, one of the most accepted models for interest rates is the Cox–Ingersoll–Ross (CIR) model [42] where rates follow the Feller process described by drift and noise intensity given, respectively, by [45]

$$f(r) = -\alpha(r-m), \qquad g(r) = k\sqrt{r}. \tag{86}$$

The Feller model is thus a diffusion process described by the stochastic differential equation

$$dr(t) = -\alpha\big[r(t) - m\big]dt + k\sqrt{r(t)}dW(t), \tag{87}$$

where $W(t)$ is the standard Wiener process, and, as in the OU process, $m > 0$ represents the mean stationary rate (the *normal level*), and α^{-1} is the correlation time [27]. Let us note that, in one-dimensional diffusions, the diffusion coefficient is given by the square of the noise intensity, and we thus see that the Feller process has a linear diffusion vanishing at the origin. This turns the origin into a singular boundary, which results in significant properties for the process [43].

As in the Vasicek model, the linear drift results in a restoring force, which, in the absence of noise, makes the process decay toward the normal level m. On the other hand, the state-dependent noise intensity $k\sqrt{r}$ for large values of r magnifies the effect of noise, while when r goes to zero, this effect vanishes. Therefore, as the process approaches the origin, the drift drags r towards m. Hence, since $m > 0$, starting at some positive value $r_0 > 0$ the process cannot attain negative values with the overall result that *the Feller process always remains positive*.

Previous works [27,43] that we reviewed rather thoroughly presented the properties of the Feller process, and we refer the reader to these works for more detailed information. The process is not Gaussian, and the stationary PDF as $t \to \infty$ is the Gamma distribution [27]

$$p_{st}(r) = \frac{(2\alpha/k^2)^\theta}{\Gamma(\theta)} r^{\theta-1} e^{-(2\alpha/k^2)r}, \tag{88}$$

where

$$\theta = \frac{2\alpha m}{k^2} \tag{89}$$

is a positive and dimensionless constant that combines all the parameters of the model into a single expression. As mentioned above, a major characteristic of the Feller process is that $r(t)$ cannot attain negative values, which makes the model a convenient tool for pricing bonds, which are never negative [13].

In the Feller model, the joint density of the discounting process $(x(t), r(t))$ defined in Equation (16) obeys the Fokker–Planck equation (FPE) (cf. Equations (17) and (18))

$$\frac{\partial p}{\partial t} = -r\frac{\partial p}{\partial x} + \alpha\frac{\partial}{\partial r}[(r-m)p] + \frac{k^2}{2}\frac{\partial^2}{\partial r^2}(rp), \tag{90}$$

with the initial condition

$$p(x, r, 0|r_0) = \delta(x)\delta(r - r_0). \tag{91}$$

The joint Fourier transform, Equation (30), turns Equations (90)–(91) into a more manageable problem:

$$\frac{\partial \tilde{p}}{\partial t} = \left(\omega_1 - \alpha\omega_2 - i\frac{k^2}{2}\omega_2^2\right)\frac{\partial \tilde{p}}{\partial \tilde{\omega}_2} - i\alpha m\omega_2\tilde{p}, \tag{92}$$

$$\tilde{p}(\omega_1, \omega_2, 0|r_0) = e^{-i\omega_2 r_0}. \tag{93}$$

Equation (92) is a linear partial differential equation of first order whose solution can be obtained by the method of characteristics, and we refer the interested reader to our work [27] for a detailed information. Once we know the solution $\tilde{p}(\omega_1, \omega_2, t|r_0)$, the discount function is then obtained through Equation (31) with the result [27]

$$D(t) = \left[\frac{2\lambda e^{-(\lambda-\alpha)t/2}}{(\lambda+\alpha)+(\lambda-\alpha)e^{-\lambda t}}\right]^{\theta} \exp\left\{-\frac{2(1-e^{-\lambda t})r_0}{(\lambda+\alpha)+(\lambda-\alpha)e^{-\lambda t}}\right\}, \tag{94}$$

where θ is defined in Equation (89) and

$$\lambda = \sqrt{\alpha^2 + 2k^2}. \tag{95}$$

Notice that $\lambda > \alpha$ and the time scale represented by λ^{-1} is smaller than the correlation time α^{-1}.

In this case, the long-run discount rate, defined by the limit (cf. Equation (80))

$$r_\infty = -\lim_{t\to\infty}\frac{\ln D(t)}{t},$$

is directly obtained from Equation (94) with the result

$$r_\infty = \frac{1}{2}(\lambda - \alpha)\theta, \tag{96}$$

and, as in the Vasicek model, the effective discount reduces to the expected exponential decay

$$D(t) \simeq e^{-r_\infty t} \qquad (t \to \infty). \tag{97}$$

Substituting into Equation (96) the expressions for θ and λ given in Equations (89) and (95), we write

$$r_\infty = \frac{2m}{1+\sqrt{1+2k^2/\alpha^2}}, \tag{98}$$

which clearly shows that the long-run discount rate is always smaller than the stationary average rate:

$$r_\infty < m.$$

Figure 1 shows the discount function $D(t)$ along with the quantity $-\ln D(t)/t$ (cf. Equation (80)) and compare them with the Vasicek model with equivalent parameters.

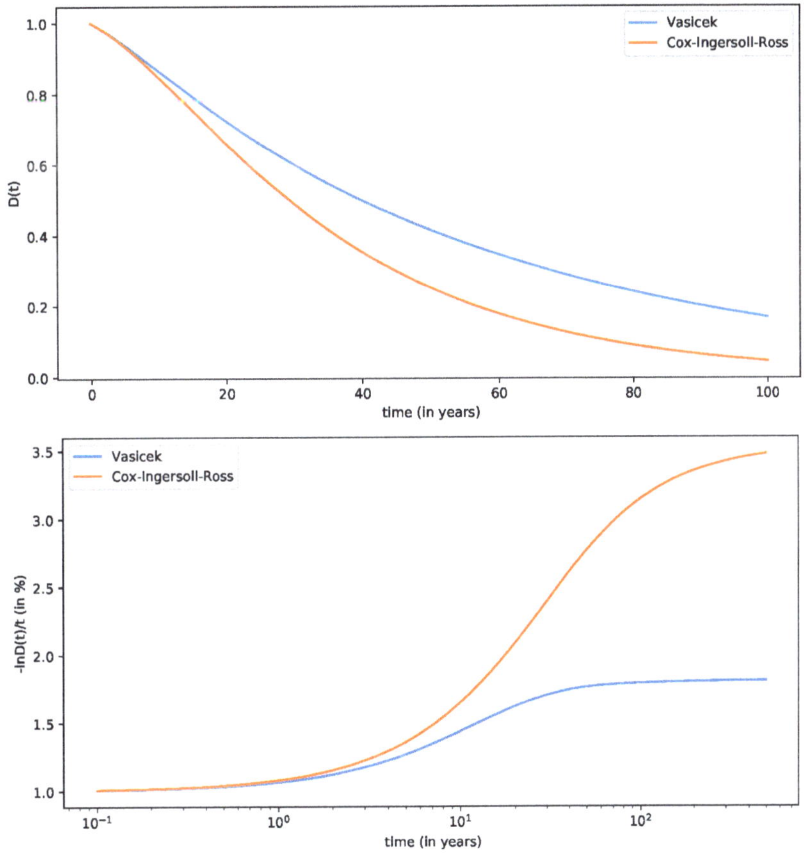

Figure 1. The Vasicek and Cox–Ingersoll–Ross discount functions. The parameters used are those corresponding to the United States and are provided by Table 5 of Ref. [29] (see Section 5). In the top figure, we plot the discount function $D(t)$, while in the bottom figure, we plot the log ratio $-\ln D(t)/t$. In the top figure, we observe the asymptotic exponential decay of the discount after more than a hundred years, while in the bottom figure, we clearly see the existence of a long-run discount rate for the Vasicek model (cf. Equation (80)). The initial rate r_0 is arbitrarily taken to be 1%. In both models, we assume no market price of risk $q(r) = 0$ (the Local Expectation Hypothesis).

Risk Aversion

For the Feller process, the adjusted drift for risk defined in Equation (54) reads

$$f^*(r) = -\alpha(r-m) + kq(r)\sqrt{r}, \tag{99}$$

where $q(r)$ is the market price of risk as discussed in the previous section. For any function $q(r)$ (including a constant market price of risk q), this adjusted drift leads to an unsolvable Fokker–Planck equation with no analytical expression for the adjusted discount and the long-run discount rate. It is, nonetheless, possible to obtain analytical expressions for these quantities if the market price of risk has the following functional form

$$q(r) = q\sqrt{r}, \tag{100}$$

where $q \geq 0$ is a positive quantity. In such a case, we may write

$$f^*(r) = -\alpha^*(r - m^*), \tag{101}$$

where

$$\alpha^* = \alpha - kq, \qquad m^* = \frac{\alpha m}{\alpha - kq}. \tag{102}$$

The adjusted drift has the same form as $f(r)$. Therefore, the adjusted discount function will be given Equation (94) with the replacements $\alpha \to \alpha^*$ and $m \to m^*$, and the long-run discount is (cf. Equation (98))

$$r^*_\infty = \frac{2m^*}{1 + \sqrt{1 + 2k^2/\alpha^{*2}}}. \tag{103}$$

From the definitions of α^* and m^*, we easily see that $\alpha^* \leq \alpha$ and $\alpha^* m^* = \alpha m$. Hence, writing r^* as

$$r^*_\infty = \frac{2\alpha^* m^*}{\alpha^* + \sqrt{\alpha^{*2} + 2k^2}} \geq \frac{2\alpha m}{\alpha + \sqrt{\alpha^2 + 2k^2}} = r_\infty,$$

so that $r^*_\infty \geq r_\infty$, and, if the market price of risk has the form given in Equation (100), then, in the CIR model, risk always increases the long-run discount rate regardless of the noise intensity and persistence.

4.4. *The Log-Normal Model*

In this model, rates are described by the the geometric Brownian motion (log-normal process), and the model is determined by the stochastic differential equation

$$\frac{dr}{r} = \alpha dt + kdW(t), \tag{104}$$

where r is the interest rate, α and k are constant parameters. α may be positive or negative, whereas k is always positive, and $W(t)$ is the standard Wiener process. Equation (104) can be integrated at once yielding

$$r(t) = r_0 \exp\left\{\left(\alpha - \frac{k^2}{2}\right)t + kW(t)\right\}, \tag{105}$$

showing that $r(t)$ is never negative ($r_0 > 0$). Therefore, the log-normal model is more suited for modeling nominal interest rates and bonds in finance than for the long-run real rates of environmental economics. Contrary to the OU and Feller processes, the log-normal process does not show reversion to the mean. Indeed, as t increases, we see from Equation (105) that the rate either diverges when $\alpha > 0$ or goes to zero if $\alpha < 0$. In an equivalent way, one can also show from Equation (105) that the mean and variance of the process are [27]

$$\langle r(t) \rangle = r_0 e^{\alpha t}, \qquad \mathrm{Var}[r(t)] = r_0^2 e^{2\alpha t}\left(e^{k^2 t} - 1\right).$$

The discount associated with the log-normal process model was studied in 1978 by L. U. Dothan [44], and, in finance, it is sometimes refereed to as the Dothan model. As it allows for analytical treatment, it is one of the models used in the literature [13]. For this model, the FPE for the joint density of the discounting processes $(x(t), r(t))$ is given by (cf. Equation (17))

$$\frac{\partial p}{\partial t} = -r\frac{\partial p}{\partial x} - \alpha\frac{\partial}{\partial r}(rp) + \frac{1}{2}k^2\frac{\partial^2}{\partial r^2}(r^2 p), \tag{106}$$

with the usual initial condition given by Equation (18). The Fourier transform of this expression leads to the following equation for the characteristic function $\tilde{p}(\omega_1, \omega_2, t|r_0)$

$$\frac{\partial \tilde{p}}{\partial t} = (\omega_1 + \alpha\omega_2)\frac{\partial \tilde{p}}{\partial \tilde{\omega}_2} + \frac{1}{2}k^2\omega_2^2\frac{\partial^2 \tilde{p}}{\partial \tilde{\omega}_2^2} \tag{107}$$

and the initial condition (91). Equation (106) is a partial differential equation of second order, which cannot be solved by the method of characteristics, and we refer the interested reader to our work [27] for more information on how to solve Equation (107) using the time–Laplace transform. Hence—and contrary to Vasicek and CIR models where it is possible to obtain exact expressions for the discount function $D(t)$—for the log-normal case, we can only achieve the exact expression of its Laplace transform,

$$\hat{D}(s) = \int_0^\infty e^{-st}D(t)dt.$$

The resulting formula—written as an integral of special functions, the Kummer function—is rather intricate, and we will not write it here (see [27] for more information). However, from the exact expression for $\hat{D}(s)$, we can obtain asymptotic expressions as $t \to \infty$ of the discount function $D(t)$ in real time. This is done using the so-called Tauberian theorems, which relate the small s behavior of $\hat{D}(s)$ with the long-time behavior of $D(t)$ [46,47]. The final result is the following asymptotic expression for the discount function $D(t)$ in the long run as $t \to \infty$ [27]

$$D(t) \sim \begin{cases} \text{constant} & k^2/2 > \alpha, \\ e^{-r_\infty t} & k^2/2 < \alpha, \\ t^{-1/2} & k^2/2 = \alpha. \end{cases} \tag{108}$$

The asymptotic form of the discount function thus depends on the values taken by the ratio α/k^2 between the strength of the constant deterministic drift α and the amplitude of fluctuations given by $k^2/2$ (which can be considered the "signal-to-noise ratio" of this model).

(i) The case $k^2/2 > \alpha$ corresponds to strong fluctuations, where the noise intensity $k^2/2$ is greater than the drift parameter α. In this case, the discount tends to a constant value (for the actual value of this constant, see [27]).

(ii) The case $k^2/2 < \alpha$ corresponds to mild fluctuations for which the deterministic drift is stronger than noise. In such a case, the discount function has the expected exponential decay

$$D(t) \sim e^{-r_\infty t}, \tag{109}$$

with a long-run rate of discount given by [27]

$$r_\infty = \frac{1}{\delta}\left(\alpha - \frac{k^2}{2}\right), \tag{110}$$

where $\delta > 1$ is a positive numerical factor that only depends on the ratio $2\alpha/k^2$ and reads

$$\delta = \psi\left(2\alpha/k^2\right) + \frac{1}{2\alpha/k^2 - 1}, \tag{111}$$

where $\psi(\cdot)$ is the digamma function.

Let us write Equation (109) in a more characteristic form. Indeed, from Equation (105), we see that

$$\mathbb{E}\left[\ln\frac{r(t)}{r_0}\right] = \left(\alpha - \frac{k^2}{2}\right)t,$$

and, with the help of Equation (109), we write Equation (109) as

$$D(t) \sim \exp\left\{-\frac{1}{\delta}\mathbb{E}\left[\ln\frac{r(t)}{r_0}\right]\right\}, \tag{112}$$

($t \to \infty$ and $k^2/2 < \alpha$). Note that the average $\mathbb{E}[\ln r(t)/r_0]$ is what a practitioner would take as an estimate of the discount rate up to time t within the log-normal model. Since $\delta > 1$, the analytical result (112) shows that the actual long-run rate of the model is a fraction of the average rate. We indicated elsewhere that the long-run discount rate is at most 73 % of the average rate [26].

In this way, when $2\alpha/k^2 > 1$, the log-normal model follows a similar pattern to that of the OU and Feller models: In all of them, the long-run rate is smaller than the average rate. This general statement is a direct consequence of Jensen's inequality, which states that the average of a convex function is greater than or equal to the function of the average; that is, $\mathbb{E}[f(X)] \geq f(\mathbb{E}[X])$. Assuming f to be the decreasing exponential and X as the cumulative process $x(t)$ defined in Equation (14), it follows immediately that the long-run rate r_∞ must be always less than or equal to the average rate. Nonetheless, our procedure quantifies the difference among averages [27].

(iii) The critical case $\alpha = k^2/2$, in which deterministic motion and fluctuations are balanced, leads to the hyperbolic discount function as obtained by Farmer and Geanakoplos [48,49]. The hyperbolic $D(t)$ is substantially greater than any exponential decaying function, showing that there is no long-run rate of interest in this case. In fact, the long-run rate of interest is 0; however, that does not convey as precise information as saying that $D(t)$ is approximately $k/\sqrt{t}$ for all large t. Since the sum (i.e., the integral) of all these $D(t)$ is infinite, such $D(t)$ assigns infinite value to any permanent positive flow of consumption: the infinite future is infinitely valuable.

Risk Aversion

Let us very briefly comment on the inclusion of risk aversion in the Dothan model. For the log-normal process $f(r) = \alpha r$ and $g(r) = kr$ and

$$f^*(r) = [\alpha + kq(r)]r.$$

Assuming a constant market price of risk, $q(r) = q \geq 0$, we have

$$f^*(r) = \alpha^* r, \qquad \alpha^* = \alpha + q.$$

Again, $f^*(r)$ has the same form than $f(r)$, and all previous results will apply after making the replacement $\alpha \to \alpha + q$.

5. Some Empirical Results

In order to choose an appropriate model for rates that would allow us to obtain realistic long-run discount functions, we performed a rather complete empirical study on interest rates combined with inflation. Our study follows the line partly initiated by Newell and Pizer [50] (see also [51]). To our knowledge, there are few empirical studies on real rates with some exceptions. We remark here the recent and excellent survey by Giglio et al. on the housing market in London and Singapore [52–54], which allowed for a rather realistic estimation of long-run discount rates.

Our first concern was knowing how the discount process depended on the underlying random process that characterizes interest rates. To this end, we collected data for the nominal interest rates and inflation of fourteen countries over time spans ranging from 87 to 318 years [26]. The countries in our sample are Argentina (ARG, 1864–1960), Australia (AUS, 1861–2012), Chile (CHL, 1925–2012), Germany (DEU, 1820–2012), Denmark (DNK, 1821–2012), Spain (ESP, 1821–2012), United Kingdom (GBR, 1694–2012), Italy (ITA, 1861–2012), Japan (JPN, 1921–2012), Netherlands (NLD, 1813–2012), Sweden (SWE, 1868–2012), the United States (USA, 1820–2012) and South Africa (ZAF, 1920–2012). The data are summarized in Table 2.

Table 2. List of the countries analyzed. CPI stand for Consumer Price Index. Data has different specificities, particularly in terms of empty records as has been reported elsewhere [26,29]. *, We have taken the discount (ID) rate since the government bond yield data was not available.

Country	CPI	Bond Yield	From	To	Records
Italy	CPITAM	IGITA10 annual from 12/31/1861 quarterly from 12/31/1919	12/31/1861 quarterly	09/30/2012	565
Chile	CPCHLM	IDCHLM quarterly	03/31/1925 quarterly	09/30/2012	312
Canada	CPCANM	IGCAN10 quarterly	12/31/1913 quarterly	09/30/2012	357
Germany	CPDEUM	IGDEU10 annual from 12/31/1820 quarterly from 12/31/1869	12/31/1820 quarterly	09/30/2012	729
Spain	CPESPM	IGESP10 annual from 12/31/1821 quarterly from 12/31/1920	12/31/1821 quarterly	09/30/2012	709
Argentina	CPARGM	IGARGM annual from 12/31/1864 quarterly from 12/31/1932	12/31/1864 quarterly	03/31/1960	342
Netherlands	CPNLDM	IGNLD10D annual	12/31/1813 annual	12/31/2012	189
Japan	CPJPNM	IGJPN10D quarterly	12/31/1921 quarterly	12/31/2012	325
Australia	CPAUSM	IGAUS10 annual from 12/31/1861 quarterly 12/31/1991	12/31/1861 quarterly	09/30/2012	564
Denmark	CPDNKM	IGDNK10 annual from 12/31/1821 quarterly from 12/31/1914	12/31/1821 quarterly	09/30/2012	725
South Africa	CPZAFM	IGZAF10 quarterly	12/31/1920 quarterly	09/30/2012	329
Sweden	CPSWEM	IGSWE10 annual	12/31/1868 annual	09/30/2012	135
United Kingdom	CPGBRM	IDGBRD * annual	12/31/1694 annual	12/31/2012	309
United States	CPUSAM	TRUSG10M annual	12/31/1820 annual	10/30/2012	183

Since all but two of our nominal interest rate processes are for 10-year government bonds, which pay out over a 10-year period, we smoothed out inflation rates with a 10-year moving average and subtracted the annualized inflation index from the annualized nominal rate to compute the real interest rate, as explained in the previous section by means of the Fisher's procedure (cf. Equation (61)),

$$r(t) = n(t) - i(t),$$

where $n(t)$ is the nominal rate and $i(t)$ is the inflation rate. The particular case of the United States is plotted in Figure 2.

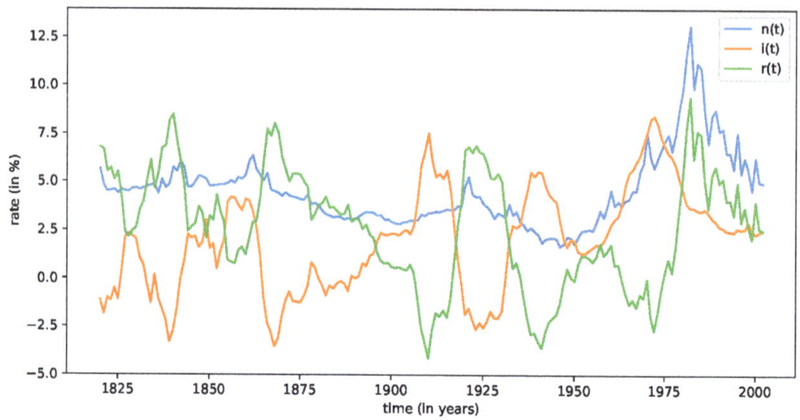

Figure 2. The construction of real interest rates $r(t)$ in terms of the nominal rates $n(t)$ and inflation $i(t)$ (Fisher's procedure). Large fluctuations and negative rates are shown here for the United States (USA).

In our empirical analysis, the nominal rates are determined by IG rates constructed from the 10-year Government Bond Yield $y(t,\tau)$ with τ= 10 years. Thus, looking at Equations (63) and (64), we estimate the nominal rates by

$$n(t) \sim y(t, \tau = 10 \text{ years}).$$

Let us recall that denoting, by $B(t, t+\tau)$, the government bond issued at time t and maturing at time $t+\tau$ with unit maturity, $B(t,t) = 1$, the yield $y(t,\tau)$ is defined as (cf. Equation (38))

$$y(t,\tau) \equiv -\frac{1}{\tau} \ln B(t, t+\tau) \qquad \Longrightarrow \qquad B(t, t+\tau) = e^{-\tau y(t|\tau)}.$$

One can argue that $\tau = 10$ years is not a short period of time in order to consider $y(t, \tau = 10 \text{ years})$ a very accurate estimator of $n(t)$ (cf. Equations (63) and (64)). Although this may be true, we must bear in mind that 10 year bonds are the shortest bonds available for most of the countries analyzed.

The inflation rate is estimated through the Consumer Price Index (CPI) as

$$i(t) \sim \frac{1}{\tau} \ln\big[I(t+\tau)/I(t)\big],$$

where $I(t)$ is the aggregated inflation up to time t, and τ =10 years. The relation between $I(t)$ and the Consumer Price Index (CPI) is

$$I(t+\tau) = I(t) \prod_{j=0}^{\tau-1} \big[1 + C(t+j)\big],$$

where $C(t)$ is the time series of the empirical CPI. The instantaneous rate of inflation $i(t)$ is, therefore, estimated by the quantity $i(t+\tau)$, which is written in terms of the CPI reads

$$i(t) \sim i(t+\tau) = \frac{1}{\tau} \sum_{j=0}^{\tau-1} \ln\big[1 + C(t+j)\big].$$

A remarkable characteristic observed for all countries is that real interest rates frequently become negative as the real interest rates are mostly dominated by inflation $i(t) > 0$

(see Figure 2). In some cases, as we can see in Table 3 (see also Figure 2), negative real rates show high frequency and long periods of time, and, on average, real interest rates are negative one quarter of the time.

Table 3. The OU (Vasicek) model parameter estimation in yearly units using stationary averages. "Neg RI" provides the time percentage and the number of years with negative real interest rates. The columns $\hat{m}$, $\hat{k}$ (in %) and $\hat{\alpha}$ are estimates from the country time series; $\hat{r}_\infty$ (in %) is evaluated from Equation (79). The Min and Max columns give reasons regarding the level of robustness of the estimation as they provide the minimum and the maximum values of the parameter estimation for four data blocks of equal length. The parameter α is estimated by fitting the empirical correlation function to an exponential (cf. Equation (75)) after using the whole data block. Countries in boldface are those considered historically more stable with positive long-run rates $\hat{r}_\infty > 0$.

Country	Neg RI	$\hat{m}$	Min	Max	$\hat{k}$	Min	Max	$\hat{\alpha}$	$\hat{r}_\infty$
Italy	28% (40y)	−0.3	−9.1	5.6	6.9	0.8	10.1	0.22	−5.4
Chile	56% (43y)	−6.8	−20.2	12.0	25.2	5.6	44.1	0.40	−26
Canada	22% (20y)	**2.9**	0.1	6	**2.3**	1.1	2.0	0.26	**2.5**
Germany	14% (25y)	−10.7	−51.0	4.0	33.9	0.9	61.4	0.20	−160
Spain	25% (45y)	5.7	−0.5	13.5	2.9	1.2	3.6	0.06	−6.4
Argentina	20% (17y)	**2.4**	−2.9	6.8	**6.2**	2.8	6.7	0.39	**1.1**
Netherlands	17% (33y)	**3.2**	0.8	5.4	**1.6**	0.8	2.2	0.14	**2.4**
Japan	33% (26y)	−2.2	−7.8	4.0	9.7	1.1	13.2	0.24	−10
Australia	23% (33y)	**2.6**	−0.7	4.9	**2.3**	0.7	2.8	0.19	**1.9**
Denmark	18% (33y)	**3.2**	1.5	4.3	**2.3**	1.1	2.9	0.23	**2.7**
South Africa	43% (36y)	**1.8**	−2.2	5.5	**2.5**	1.2	2.0	0.21	**1.1**
Sweden	28% (38y)	**2.3**	−0.3	3.9	**2.5**	0.6	3.4	0.25	**1.9**
United Kingdom	14% (45y)	**3.3**	1.4	4.3	**1.9**	1.0	2.4	0.19	**2.8**
United States	31% (36y)	**2.6**	1.0	4.0	**1.8**	1.2	2.1	0.18	**2.1**
Stable countries	23% (33y)	**2.7**	−0.14	5.0	**2.6**	1.04	2.94	0.23	**2.1**
Unstable counntries	31% (36y)	**−2.9**	17.7	1.8	**16**	1.9	26.5	0.22	**−42**

This makes the Feller and log-normal models—as well as any other model assuming positive interest rates [13]—less interesting or at least less appropriate to model real interest $r(t) = n(t) - i(t)$ instead of solely considering nominal rates $n(t)$. It is, however, necessary to remember the fact that nominal rates can indeed become negative as has recently been observed in Western economies. We, therefore, confined the empirical work to the OU (Vasicek) model and then assumed the Local Expectation Hypothesis [36–38], according to which, we live in a risk neutral world, and the market price of risk is zero. Let us recall, as explained in Section 3, that the market price of risk $q = q(r,t)$ may be any function of the rate and time. There is, hence, no unique expression for it. Thus, in Section 4, we presented several expressions of the long-run rate, which include risk in different forms for all market models analyzed. The usual assumption in the literature [33,38] is that the price of risk is a constant that is independent of time and the value of the rate but without any empirical justification. This is a sensitive issue since data is quite scarce, particularly in environmental applications, for obtaining a credible estimation of q. Moreover, to our knowledge, in environmental problems, the estimations of the long-run rates do not take into account, nor even mention, the market price of risk [14–16,50,54]. In any case, we do not lessen the importance of taking into account some kind of risk in estimating log-run rates; however, unfortunately, with the data available to us, we cannot make any reliable estimation of q. For this reason, we have not taken into account the market price of risk, assuming risk-neutral investors and following the Local Expectation Hypothesis. In any case, the question is under consideration).

We can estimate the parameters m, α and k of the Vasicek model to each of the data series. There are several possible procedures. One of the possible methods is to deal with stationary averages. The parameter m can be estimated through the stationary mean value of the rate (cf. Equation (74))

$$m = \mathbb{E}[r(t)].$$

Parameters α and k can be estimated via the correlation function of the Ornstein–Uhlenbeck process. Thus, from Equation (75), we have

$$C(t-t') = \frac{k^2}{2\alpha}e^{-\alpha|t-t'|}.$$

The empirical correlation can then be fitted by an exponential, which in turn allows us to estimate α (measured in 1/year units) for each country. The parameter k is obtained from the empirical standard deviation $\sigma^2 = \mathbb{E}[|r(t)-m|^2]$, and for the Vasicek model, it is given by

$$k = \sigma\sqrt{2\alpha}.$$

The resulting parameters are shown in Table 3. The minimum and maximum values for each country allows us to show that parameters may indeed fluctuate over different periods of time.

Finally, the long-run discount rate can be evaluated from Equation (79),

$$r_\infty = m - k^2/2\alpha^2.$$

For this calculation, we neglected the market price of risk as mentioned above.

The countries studied can be divided into two groups. Nine countries have long-run positive rates (boldface in Table 3). The average historical rate for these nine countries is $\overline{m} = 2.7$ % while the average long-run rate is $\overline{r}_\infty = 2.1$ %, which, on average, is 29 % lower than $\overline{m}$. Five countries with less stable behavior have long-run negative rates and an exponentially increasing discount.

Four cases of this group have a negative average rate m due to at least one period of runaway inflation; the exception is Spain, which has a (highly positive) mean real interest rate but still has a long-run negative rate. Convergence in this case to the long-run rate happens within 30 years and typically within less than a decade. This contrasts with other treatments, which assume that short term rates are always (or nearly always) positive and predict that the decrease in the discounting rate happens over a much longer timescale, which can be measured in hundreds of years [50,51,55–58].

Alternatively, we can estimate parameters using the well-established maximum likelihood procedure. For the Vasicek model, the maximum likelihood estimation is extensively documented in the financial mathematics literature (see for instance [13]). The approach differs from the previous one as it focuses attention on two consecutive steps of our time series (generally consecutive years) and takes the conditional probability to perform the estimation. Table 4 shows that the most inaccurate estimator is $\hat{\alpha}$, an unsurprising fact since the estimation of α is known to be quite difficult for the Vasicek model [59]. The last two columns in Table 4 include the long-run interest rate estimator $\hat{r}_\infty$ and its error calculated through error propagation.

Only four countries (the Netherlands, Sweden, the United Kingdom and the United States) show a positive long-run rate, $r_\infty > 0$. This estimation procedure leads to more negative r_∞. This feature can be attributed to the fact that, in most of the countries, estimating α via the maximum likelihood brings smaller values, which in turn leads to more drastic corrections to the long-run rate as r_∞ is inversely proportional to α (remember that $r_\infty = m - k^2/2\alpha^2$). This effect is particularly relevant in most turbulent countries during last century (e.g., Germany) thus signaling a more intense lack of stationarity in empirical data. The averaged r_∞ over all countries estimated via maximum likelihood is also sensitively smaller.

However, if we focus the attention on stable countries (with $r_\infty > 0$) both estimation procedures bring quite similar results (see, for instance, the United States case in

Tables 3 and 4, 2.1% versus 1.8%). As in the previous estimation procedure, we also neglected the effects of risk aversion and the market price of risk.

The Vasicek model is therefore the only one among the three classic models allowing for negative rates, and for this reason, both the Feller and the log-normal models have been excluded from our analysis. However, for the Cox–Ingersoll–Ross (Feller) model, it is possible to redefine the model by shifting the process $y = r - r_{min}$ where $r_{min} < 0$.

The estimation through the maximum likelihood procedure and its error analysis is then possible [59], and Figure 1 includes the shifted Cox–Ingersoll–Ross discount and compares it with the equivalent result assuming the Vasicek model. We demonstrated in Ref. [29] how to redefine the Feller process and how maximum likelihood estimation could be possible.

Table 4. Maximum likelihood estimation of the long-run interest rate for the Vasicek model. $\hat{m}$ estimates of the mean real interest rate in 1/years (in %). $\hat{\alpha}$ estimates the characteristic reversion time in 1/years. The squared root of $\hat{k}^2$ is given in terms of $1/(\text{year})^3$ (multiplied by 10^4 to be comparable with the results in Table 3). These estimators are accompanied by the square root of the variance of each estimator. $\hat{r}_\infty$ estimates the long-run real interest rate with 1/year (in %). Negative values of $\hat{r}_\infty$ imply that the discount function is asymptotically increasing. The standard error is obtained through error propagation. The last two rows show the average over all countries with the more stable countries ($r_\infty > 0$) and the less stable countries ($r_\infty < 0$). The error provided corresponds to the standard deviation of the $\hat{r}_\infty$ for the different countries.

Country	$\hat{m}$	$\sigma_{\hat{m}}$	$\hat{\alpha}$	$\sigma_{\hat{\alpha}}$	$\hat{k^2}$	$\sigma_{\hat{k^2}}$	$\hat{r}_\infty$	$\sigma_{\hat{r}_\infty}$
Italy	1.97	15.95	0.0056	0.0089	0.1146	0.068	−177.8	19.2
Chile	−5.79	31.46	0.0201	0.0227	31.07	2.49	−391.7	44.2
Canada	2.66	3.91	0.0142	0.0178	0.275	0.021	−4.15	3.94
Germany	−9.45	66.95	0.0071	0.0089	41.72	2.19	−4094	228
Spain	6.71	6.92	0.0167	0.0137	2.371	0.126	−35.78	7.28
Argentina	3.15	7.09	0.0228	0.0231	2.240	0.171	−18.31	7.27
Netherlands	**5.99**	0.78	0.1648	0.0550	**1.797**	0.243	**5.66**	0.78
Japan	5.02	24.68	0.0053	0.0114	1.396	0.109	−243.1	31.4
Australia	3.97	4.50	0.0089	0.0112	0.223	0.013	−10.29	4.58
South Africa	2.69	4.72	0.0154	0.0193	0.435	0.034	−6.49	4.77
Sweden	**2.79**	1.66	0.0676	0.0317	1.692	0.206	**0.95**	1.67
Denmark	4.10	2.59	0.0161	0.0133	0.315	0.017	−1.97	2.61
United Kingdom	**3.42**	0.62	0.1635	0.0326	**3.137**	0.253	**2.83**	0.62
United States	**3.19**	1.23	0.0603	0.0257	**1.003**	0.105	**1.81**	1.24
Stable countries	**3.85**	1.07	0.1140	0.0362	**1.907**	0.202	**2.81**	1.08
Unstable countries	1.50	16.86	0.0132	0.0150	8.120	0.523	−498.4	35.3

6. Discussion

We reviewed one of the most important aspects of economics and finance, i.e., the problem of discount, which weights the future relative to the present. The problem is clearly very relevant in finance over relatively short time spans; however, it is even more crucial for long-run planning in addressing environmental problems on how to act now with measures to mitigate the effects of climate change.

To our knowledge, this is a rather unknown issue to the econophysics community, and this review is particularly intended for this group. We thus addressed the problem with a simple approach and yet with a high level of rigor and generality. In this way, we also developed the traditional method used in mathematical finance to address the problem, i.e., the Feynman–Kac approach. In addition, we reviewed the bond pricing theory and its close similarity with discounting and presented a short introduction to the term structure of interest rates along with the market price of risk.

We obtained quantitative results on the problem by studying, in some detail, three standard models for the dynamical evolution of rates. These models are based on the

Ornstein–Uhlenbeck process (the Vasicek model), thus, allowing for both positive and negative rates and also on the Feller and log-normal processes for positive rates. We presented the exact results for the discount function and asymptotic expressions as $t \to \infty$ leading to the long-run discount rate, and we discussed the modifications of these expressions when the market price of risk is taken into account.

An important conclusion is that, for all models, the long-run discount rate is always less than the long-time average rate. This is a conclusion that necessarily has to have consequences in any long-run economic planning. Finally, we reviewed our recent empirical study on 14 different countries, which obtained numerical values for the parameters that appear in the Vasicek model. We demonstrated two different estimation procedures and briefly discussed their differences and similarities.

Author Contributions: Conceptualization, J.M., M.M., J.P., J.D.F. and J.G.; methodology, J.M., M.M., J.P., J.D.F. and J.G.; formal analysis, J.M., M.M., J.P., J.D.F. and J.G.; investigation, J.M., M.M., J.P., J.D.F. and J.G.; resources, J.M., M.M., J.P., J.D.F. and J.G.; writing—original draft preparation, J.M.; writing—review and editing, J.M., M.M., J.P., J.D.F. and J.G.; funding acquisition, M.M. and J.P. All authors have read and agreed to the published version of the manuscript.

Funding: This research was funded by MINEICO (Spain), Agencia Estatal de Investigación (AEI) grant number PID2019-106811GB-C33 (AEI/10.13039/501100011033) (JM, MM and JP); by Generalitat de Catalunya grant number 2017 SGR 608 (JM, MM and JP); by National Science Foundation grant 0624351 (JG); and by the Institute for New Economic Thinking (JDF).

Data Availability Statement: The study reports data described and analyzed in Refs. [26–30].

Conflicts of Interest: The authors declare no conflict of interest. The funders had no role in the design of the study; in the collection, analyses or interpretation of data; in the writing of the manuscript or in the decision to publish the results.

References

1. Mantegna, R.N.; Stanley, H.E. *Introduction to Econophysics: Correlations and Complexity in Finance*; Cambridge University Press: Cambridge, UK, 1999.
2. Bouchaud, J.-P.; Potters, M. *Theory of Financial Risk and Derivative Pricing: From Statistical Mechanics to Risk Management*; Cambridge University Press: Cambridge, UK, 2011.
3. Bouchaud, J.P. Econophysics: Still fringe after 30 years? *Europhys. News* **2019**, *50*, 24–27. [CrossRef]
4. Bachelier, L. Théorie de la spéculation. *Ann. Sci. École Norm. Sup.* **1900**, *17*, 21–86. Reprinted in Cootner, P.H. (Ed.) *The Random Character of Stock Market Prices*; MIT Press: Cambridge, MA, USA, 1964. [CrossRef]
5. Osborne, M.F.M. Brownian motion in stock markets. *Oper. Res.* **1959**, *7*, 145–173. Reprinted in Cootner, P.H. (Ed.) *The Random Character of Stock Market Prices*; MIT Press: Cambridge, MA, USA, 1964. [CrossRef]
6. Cont, R. Empirical properties of asset returns: Stylized facts and statistical issues. *Quant. Financ.* **2001**, *1*, 223–236. [CrossRef]
7. Mandelbrot, B. The variation of certain speculative prices. *J. Bus.* **1963**, *35*, 394–419. [CrossRef]
8. Fama, E. Mandelbrot and the stable Paretian hypothesis. *J. Bus.* **1963**, *35*, 420–429. [CrossRef]
9. Stein, E.; Stein, J. Stock price distributions with stochastic volatility: An analytic approach. *Rev. Financ. Stud.* **1991**, *4*, 727–752. [CrossRef]
10. Fouque, J.-P.; Papanicolau, G.; Sircar, K.R. *Derivatives in Financial Markets with Stochastic Volatility*; Cambdrige University Press: Cambridge, UK, 2000.
11. Masoliver, J.; Perelló, J. Multiple time scales and the exponential Ornstein–Uhlenbeck stochastic volatility model. *Quant. Financ.* **2006**, *6*, 423–433. [CrossRef]
12. Samuelson, P. A note on measurement of utility. *Rev. Econ. Stud.* **1937**, *4*, 155–161. [CrossRef]
13. Brigo, D. ;Mercurio, F. *Interest Rate Models—Theory and Practice*; Springer: Berlin, Germany, 2006.
14. Arrow, K.J.; Cropper, M.L.; Gollier, C.; Groom, B.; Heal, G.M.; Newell, R.G.; Nordhaus, W.D.; Pindyck, R.S.; Pizer, W.A.; Portney, P. R.; et al. Determining benefits and costs for future generations. *Science* **2013**, *341*, 349–350. [CrossRef]
15. Stern, N. *The Economics of Climate Change: The Stern Review*; Cambridge University Press: Cambridge, UK, 2006.
16. Nordhaus, W.D. The Stern Review on the economics of climate change. *J. Econ. Lit.* **2007**, *45*, 687–702. [CrossRef]
17. Nordhaus, W.D. Critical assumptions in the Stern Review on Climate Change. *Science* **2007**, *317*, 201–202. [CrossRef] [PubMed]
18. Dasgupta, P. *Comments on the Stern Review's Economics of Climate Change*; Cambridge University Press: Cambridge, UK, 2006.
19. Weitzman, M.L. A review on the Stern review on the economics of climate change. *J. Econ. Lit.* **2007**, *45*, 703–724. [CrossRef]
20. Nordhaus, W.D. *A Question of Balance*; Yale University Press: New Haven, CT, USA, 2008.
21. Stern, N. Ethics, equity and the economics of climate change. Paper 1. *Sci. Philos. Econ. Philos.* **2014**, *30*, 397. [CrossRef]

22. Stern, N. Ethics, equity and the economics of climate change. Paper 2. *Sci. Philos. Econ. Philos.* **2014**, *30*, 445. [CrossRef]
23. Drupp, M.A.; Freeman, M.C.; Groom, B.; Nesje, F. Discounting disentangled. *Am. Econ. J. Econ. Policy* **2018**, *10*, 109–134. [CrossRef]
24. Heal, G.M.; Millner, A. Agreeing to disagree on climate policy. *Proc. Natl. Acad. Sci. USA* **2014**, *111*, 3695–3698. [CrossRef]
25. Ramsey, F.P. A mathematical theory of saving. *Econ. J.* **1928**, *38*, 543–559. [CrossRef]
26. Farmer, J.D.; Geanakoplos, J.; Masoliver, J.; Montero, M.; Perelló, J. Discounting the Distant Future. University of Yale, Cowles Foundation Discussion Paper No. 1951. 2014. Available online: http://ssrn.com/Abstract=1448811 (accessed on 24 March 2022).
27. Farmer, J.D.; Geanakplos, J.; Masoliver, J.; Montero, M.; Perelló, J. Value of the future: Discounting in random environments. *Phys. Rev. E* **2015**, *91*, 052816. [CrossRef] [PubMed]
28. Masoliver, J. The value of the distant future: Discounting in random environments. In Proceedings of the 3er Congrés d'Economia i Empresa de Catalunya, Col·legi d'Economistes de Catalunya, Barcelona, Spain, 17 May 2018.
29. Perelló, J.; Montero, M.; Masoliver, J.; Farmer, J.D.; Geanakoplos, J. Statistical analysis and stochastic interest rate modeling for valuing the future with implications in climate change mitigation. *J. Stat. Mech.* **2020**, 0432110. [CrossRef]
30. Farmer, J.D.; Geanakoplos, J.; Masoliver, J.; Montero, M.; Perelló, J.; Richiardi, M.G. Discounting the distant future: What do historical bond prices imply about the long term discount rate? *J. Math. Econ.* 2021, *to appear*.
31. Andersen, L.B.G.; Piterbarg, V.V. *Interest Rate Modeling*; Atlantic Financial Press: London, UK, 2010; Volume I–III.
32. Duffie, D. Credit risk modeling with affine processes. *Bank. Financ.* **2005**, *29*, 2751–2802. [CrossRef]
33. Vasicek, O. An equilibrium characterization of the terms structure. *J. Financ. Econ.* **1977**, *5*, 177–188. [CrossRef]
34. Masoliver, J.*Random Processes, First-Passage and Escape*; World Scientific: Singapore, 2018.
35. Masoliver, J.; Montero, M.; Perelló, J. Valuing the future under random structural conditions: Non-stationary models for discounting. 2021, *in preparation*.
36. Cox, C.; Ingersoll, J.E.; Ross, S.A. A re-examination of the traditional hypothesis about the term structure of interest rates. *J. Financ.* **1981**, *35*, 769–799. [CrossRef]
37. Gilles, C.; Leroy, S.F. A note on the local expectation hypothesis. *J. Financ.* **1986**, *41*, 975–979. [CrossRef]
38. Piazzesi, M. Affine term structure models. In *The Handbook of Financial Econometrics*; Sahala, Y.A., Hansen, M.P., Eds.; Elsevier: Amsterdam, The Netherlands, 2009; pp. 691–766.
39. Gardiner, C.W. *Handbook of Stochastic Methods*; Springer: Berlin, Germany, 1986.
40. Sharpe, W.F. Mutual fund performance. *J. Business* **1966**, *39*, 119–138. [CrossRef]
41. Perelló, J.; Masoliver, J. Option pricing and perfect hedging on correlated stocks. *Physica A* **2003**, *330*, 622–652. [CrossRef]
42. Cox, J.C.; Ingersoll, J.E.; Ross, S.A. A theory of the term structure of interest rate. *Econometrica* **1985**, *53*, 385. [CrossRef]
43. Masoliver, J.; Perelló, J. First-passage and escape problems in the Feller process. *Phys. Rev. E* **2012**, *86*, 041116. [CrossRef] [PubMed]
44. Dothan, L.U. On the term structure of interest rates. *J. Financ. Econ.* **1978**, *6*, 59–69. [CrossRef]
45. Feller, W. Two singular diffusion processes. *Ann. Math.* **1951**, *54*, 173. [CrossRef]
46. Pitt, H.R. *Tauberian Theorems*; Oxford University Press: London, UK, 1958.
47. Handelsman, R.A.; Lew, J.S. Asymptotic expansion of Laplace convolutions for large argument and tail densities for certain sums of random variables. *SIAM J. Math. Anal.* **1974**, *5*, 425. [CrossRef]
48. Farmer, J.D.; Geanakoplos, J. Hyperbolic Discounting Is Rational: Valuing the Far Future with Uncertain Discount Rates. Cowles Foundation Discussion Paper No. 1719. 2009. Available online: http://ssrn.com/abstract=1448811 (accessed on 24 March 2022).
49. Geanakoplos, J.; Sudderth, W.; Zeitouini, O. Asymptotic behavior of stochastic discount rates. *Ind. J. Stat.* **2014**, *76 A*, 150. [CrossRef]
50. Newell, R.; Pizer, N. Discounting the distant future: How much do uncertain rates increase valuations? *J. Environ. Econ. Manag.* **2003**, *46*, 52–71. [CrossRef]
51. Gollier, C.; Koundouri, P.; Pantelidis, T. Declining discount rates: Economic justifications and implications for long-run policy. *Econ. Policy* **2008**, *23*, 757–795. [CrossRef]
52. Giglio, S.; Maggiori, M.; Stroebel, J. Very long-run discount rates. *Q. J. Econ.* **2015**, *130*, 1–53. [CrossRef]
53. Giglio, S.; Maggiori, M.; Stroebel, J. No-bubble conditions: Model-free test in housing markets. *Econometrica* **2016**, *84*, 1047–1091. [CrossRef]
54. Giglio, S.; Maggiori, M.; Rao, K.; Stroebel, J.; Weber, A. Climate change and long-run discount rates: Evidence from real estate. *Rev. Financ. Stud.* **2021**, *34*, 3527–3571. [CrossRef]
55. Weitzman, M.L. Why the far-distant future should be discounted at its lowest possible rate. *J. Environ. Econ. Manag.* **1998**, *36*, 201–208. [CrossRef]
56. Groom, B.; Koundouri, P.; Panopoulou, E.; Pantelidis, T. Discounting distant future: How much selection affect the certainty equivalent rate. *J. Appl. Econom.* **2007**, *22*, 641–656. [CrossRef]
57. Hepburn, C.; Koundouri, P.; Panopoulou, E.; Pantelidis, T. Social discounting under uncertainty: A cross-country comparison. *J. Environ. Econ. Manag.* **2007**, *57*, 140–150. [CrossRef]
58. Freeman, M.C.; Groom, B.; Panopoulou, E.; Pantelidis, T. Declining discount rates and the Fisher Effect: Inflated past, discounted future? *J. Environ. Econ. Manag.* **2015**, *73*, 32–49. [CrossRef]
59. Tang, C.Y.; Chen, S.X. Parameter estimation and bias correction for diffusion processes. *J. Econom.* **2009**, *149*, 65–81. [CrossRef]

entropy

MDPI

Article

Relationship between Continuum of Hurst Exponents of Noise-like Time Series and the Cantor Set

Maria C. Mariani [1], William Kubin [2], Peter K. Asante [2], Joe A. Guthrie [1] and Osei K. Tweneboah [3,*]

[1] Department of Mathematical Sciences, University of Texas at El Paso, El Paso, TX 79902, USA; mcmariani@utep.edu (M.C.M.); jguthrie@utep.edu (J.A.G.)
[2] Computational Science Program, University of Texas at El Paso, El Paso, TX 79902, USA; wkubin@miners.utep.edu (W.K.); pkasante@miners.utep.edu (P.K.A.)
[3] Department of Data Science, Ramapo College of New Jersey, Mahwah, NJ 07430, USA
* Correspondence: otwenebo@ramapo.edu

Abstract: In this paper, we have modified the Detrended Fluctuation Analysis (DFA) using the ternary Cantor set. We propose a modification of the DFA algorithm, Cantor DFA (CDFA), which uses the Cantor set theory of base 3 as a scale for segment sizes in the DFA algorithm. An investigation of the phenomena generated from the proof using real-world time series based on the theory of the Cantor set is also conducted. This new approach helps reduce the overestimation problem of the Hurst exponent of DFA by comparing it with its inverse relationship with α of the Truncated Lévy Flight (TLF). CDFA is also able to correctly predict the memory behavior of time series.

Keywords: Cantor set; fractals; homeomorphism; detrended fluctuation analysis; Hurst exponent

Citation: Mariani, C.M.; Kubin, W.; Asante, P.K.; Guthrie, J.A.; Tweneboah, O.K. Relationship between Continuum of Hurst Exponents of Noise-like Time Series and the Cantor Set. *Entropy* **2021**, *23*, 1505. https://doi.org/10.3390/e23111505

Academic Editors: Ryszard Kutner, Christophe Schinckus and H. Eugene Stanley

Received: 20 October 2021
Accepted: 10 November 2021
Published: 13 November 2021

1. Introduction

A ternary Cantor set is a set built by removing the middle part of a series when divided into three parts and repeating this process with the remaining shorter segments. It is the prototype of a fractal [1]. A fractal is a geometric object that has similar statistical properties to itself on all scales. If a fractal object is successively magnified, it looks similar or exactly like the original shape of the fractal. A similar pattern exhibited at increasingly smaller scales is often known in fractal mathematics as self-similarity [2,3]. In time series, self-similar phenomena describe the event in which the dependence in the time series decays more slowly than an exponential decay. Typically, it follows a power-like decay [4]. Scaling methods exist for quantifying the power-law exponent of the decay function such as Rescaled Range Analysis (R/S), Detrended Fluctuation Analysis (DFA) and the Truncated Lévy Flight (TLF).

The Rescaled Range Analysis (R/S) method by Hurst subdivides integrated time series into adjacent segment sizes and examines the range (R) of the integrated fluctuations. Then, a measure of dispersion, usually standard deviation (S), is determined as a function of segment size. A power law governs the approximate relationship between the Rescaled Range Analysis' statistic (R/S) and the segment size [5].

The Detrended Fluctuation Analysis (DFA) by Peng et al. (1994) is a technique that quantifies the same power-law exponent of the R/S method. Addressing difficulties in determining correct power-law exponents of the R/S method in non-stationary time series resulted in the introduction of the DFA. Unlike the R/S method, the DFA uses a local detrending approach (usually linear regression) in the segments of the integrated series. For time series with higher-order trends, polynomial fit replaces the linear regression approach of the DFA [6]. This provides its power-law exponents' protection against effects of nonstationarity and pollution of time series by external signals while eliminating spurious detection of long memory [7]. Empirical evidence has shown that the DFA performs well compared to other variance scaling methods including the R/S methods when estimating power-law exponents.

Usually, characterizing stochastic processes empirically requires the study of determining asymptotic probability density distributions (pdf) and temporal correlations. Brownian motion models the evolution of a particle's position over time with the assumption that the movement of the particle follows a diffusive process with Gaussian distribution. This model did not describe accurately real-world time series because kurtosis of associated pdf is greater than that of the Gaussian distribution [8,9]. The Truncated Lévy Flight (TLF) model originated as a means to address the difficulties of the Brownian motion for working in long-range correlation scales. The scaling exponent $(0 < \alpha \leq 2)$ of the TLF measures the memory behavior in time series that follows a diffusion process with Gaussian and non-Gaussian distributions [10].

In [11], a clear comparison was made between DNA and economics by the authors, showing the underlining similarities that allow researchers to model seemingly different phenomena using the same or slightly modified models. In the same manner, these variance scaling models have the added advantage of being used to model long memory effects in different fields where stochastic processes occur [2,7]. Thus, be it DNA sequencing, financial markets, geophysical time series etc., scaling methods have been used to detect long/short memory behaviors.

Scaling approaches serve as means of characterizing the dependence of observations separated in time series dominated by stochastic properties. Applications with DFA have been done in DNA sequences [6,12,13], neural oscillations [14], detection of speech pathology [15], heartbeat fluctuation in different sleep stages [16], describing cloud breaking [17], gearbox fault diagnosis [18], analysis of fetal cardiac data [19], streamflow in the Yellow River Basin in China [20], evaluation near infrared spectra of green and roasted coffee samples [21], just to mention a few.

Empirical evidence has shown that the DFA has the tendency of overestimating the scaling exponent [2,22]. We have not come accross any literature at the moment that describes a definite approach in the segment division step of the DFA algorithm. However, we observe that estimates of power-law exponents are influenced by the scale of choice [23,24]. Our goal in this work is to propose a definite non-overlapping segment division approach in the DFA algorithm (CDFA) that utilizes the theory of the ternary Cantor set. We show that using this approach we are able to rightly determine the correct scaling exponent to detect the memory behavior of the time series as well as reducing the over-fitting nature of the DFA. This approach has the advantage of generalizing the segment division step in the DFA algorithm. The Hurst exponents obtained from the CDFA method are then compared with the exponents of the DFA and the TLF on real time series.

In Section 2, we present proof of the relationship between the continuum of Hurst exponents of the DFA and Cantor set. We also present the scaling methods TLF, DFA and CDFA in this section. In Section 3, we present results and discussions from our investigation noting that for noise-like time series, anti-persistence, white noise and persistence behavior in time series imply $0 \leq H < 0.5$, $H = 0.5$ and $0.5 < H \leq 1$ respectively. The over-estimation of DFA's Hurst exponent decreasing with the Cantor scales is also discussed in this section. Section 4 concludes the paper.

2. Methods

2.1. The Truncated Lévy Flight (TLF)

We provide a brief overview of the Truncated Lévy Flight (TLF) model in this subsection. The most general representation of the Lévy stable distribution is denoted by the characteristic function:

$$\mathcal{K}(q,\alpha) = \exp\{i\mu q \;-\; \sigma^{\alpha} \mid q \mid^{\alpha} [1 \;+\; i\beta.sign(q).\phi(q,\alpha)]\} \tag{1}$$

where,

$$\phi(q,\alpha) = \begin{cases} (2/\pi)\ln(q), & \alpha = 1 \\ -\tan(\pi\alpha/2), & \alpha \neq 1. \end{cases}$$

The stability exponent $\alpha \in (0,2]$ defines the asymptotic decay of the pdf. $\sigma \in (0,\infty)$ measures dispersion. Skewness parameter $\beta \in [-1,1]$ measures asymmetry of the distribution. $\mu \in (-\infty,\infty)$ is a scalar which determines the "location" or shift of the distribution. The sign x is the signum function of $x \in \mathbb{R}$ defined as $sign(x) = x/|x|$. The problem is that the variance of the distribution in (1) is finite but is not stable. This is because, large cut-off l results in slow convergence and a smaller cut-off l may result in abrupt tail [8]. In [25], the author generated a TLF to address the convergence problem by using a decreasing exponential cut-off function. Thus, the process in Equation (1) is truncated to obtain the TLF given by:

$$\mathcal{T}(q,\alpha) = \begin{cases} c\,\mathcal{K}(q,\alpha), & |\,q\,| \leq l \\ 0, & |\,q\,| > l \end{cases} \tag{2}$$

for some normalizing constant c, stability exponent $\alpha \in (0,2]$ and cut-off length l. The characteristic function of the TLF in Equation (2) is given by

$$\ln[\mathcal{T}(q,\alpha)] = \frac{2\pi A l^{-\alpha} t \left[1 - ((ql/\sigma)^2 + 1)^{\alpha/2} \cos(\alpha \arctan(ql/\sigma))\right]}{\alpha\Gamma(\alpha)\sin(\pi\alpha)}. \tag{3}$$

To determine the best scaling exponent (α) from characteristic equation in (3), we adjust the values of A, the cut-off parameter l and the scaling exponent α simultaneously to fit the characteristic function to the data.

2.2. Detrended Fluctuation Analysis (DFA)

Given the noise-like time series ψ, we find the integrated series

$$Y = \sum_k (\psi_k - <\psi>). \tag{4}$$

to determine the Root Mean Squared Fluctuations (RMSF) from Equation (5) below

$$F(s) = \left\{ \frac{1}{N} \sum_j \left[Y_j - Y_j^s \right]^2 \right\}^{1/2} \tag{5}$$

A log–log plot of the RMSF against the series length s produces a directly proportional relation given by

$$F(s) \propto s^H$$
$$logF(s) - Hlog(s) = K, \tag{6}$$

where $H :=$ Hurst exponent of the DFA and $H_{min} \leq H \leq H_{max}$ [4].

2.3. Cantor Detrended Fluctuation Analysis (CDFA)

In this subsection, we prove that the subspace $[H_{min}, H_{max}]$ of Hurst exponents is homeomorphic to [0, 1] of the Cantor set. We also present an illustration of the Cantor set and the algorithm for the CDFA.

Theorem 1. *A map $f : [H_{min}, H_{max}] \to [0,1]$ between the topological spaces of Hurst exponents of noise-like time series and the Cantor set is a homeomorphism if it has the following properties:*

- f is a bijection;
- f is continuous;
- the inverse function f^{-1} is continuous.

If two topological spaces admit a homeomorphism between them, we say they are homeomorphic: they are essentially the same topological space.

Proof. Let $H_{min} \leq H \leq H_{max}$ and $0 \leq y = f(H) \leq 1$, then the map $f : [H_{min}, H_{max}] \rightarrow [0,1]$ gives

$$H_{min} - H_{min} \leq H - H_{min} \leq H_{max} - H_{min} \tag{7}$$

$$0 \leq \frac{H - H_{min}}{H_{max} - H_{min}} \leq 1. \tag{8}$$

Thus,

$$y = f(H) = \frac{H - H_{min}}{H_{max} - H_{min}}. \tag{9}$$

Now, we need to prove that the map f is homeomorphic to the Cantor set.

The map $f(H)$ is said to be bijective if and only if $f(a) = f(b)$ for all a, b implies that $a = b$. From

$$f(a) = \frac{a - H_{min}}{H_{max} - H_{min}} \quad and \quad f(b) = \frac{b - H_{min}}{H_{max} - H_{min}},$$

$$f(a) = f(b)$$

$$\implies a - H_{min} = b - H_{min}$$

$$\implies a = b.$$

Thus, the map $f(H)$ is a bijection.

The map $f(H)$ is continuous at some value c in its domain if $f(c)$ is defined, the limit of f as H approaches c exists and the function value of f at c equals the limit of f as H approaches c. The function $f(c)$ is defined as

$$f(c) = \frac{c - H_{min}}{H_{max} - H_{min}}. \tag{10}$$

The limit of f as H approaches c equals

$$\lim_{H \to c^+} f(H) = \lim_{H \to c^-} f(H) = \frac{c - H_{min}}{H_{max} - H_{min}}. \tag{11}$$

The left- and right-sided limits are equal from (11). Therefore,

$$\lim_{H \to c} f(H) = \frac{c - H_{min}}{H_{max} - H_{min}}. \tag{12}$$

Hence we observe that the right hand side of Equation (10) is equal to right hand side of Equation (12). Thus, it follows that

$$\lim_{H \to c} f(H) = f(c) = \frac{c - H_{min}}{H_{max} - H_{min}}.$$

Thus, the map f is continuous at some value $H = c$ for a differentiable fractal.

The inverse function of f (i.e., $f^{-1}(H)$) exists.

$$y = f(H) = \frac{H - H_{min}}{H_{max} - H_{min}} \tag{13}$$

$$(H_{max} - H_{min})y = H - H_{min} \tag{14}$$

$$H = H_{min} + (H_{max} - H_{min})y \tag{15}$$

Interchanging H and y gives

$$y = f^{-1}(H) = H_{min} + (H_{max} - H_{min})H, \tag{16}$$

the inverse function of $f(H)$.

The inverse map f^{-1} is continuous at some value s in its domain if $f^{-1}(s)$ is defined, the limit of f^{-1} as H approaches s exists and the function value of f^{-1} at s equals the limit of f^{-1} as H approaches s. $f^{-1}(s)$ is defined as

$$f^{-1}(s) = (1-s)H_{min} + sH_{max}. \tag{17}$$

The limit of f^{-1} as H approaches s equals

$$\lim_{H\to s^+} f^{-1}(H) = \lim_{H\to s^-} f^{-1}(H) = H_{min} + (H_{max} - H_{min})s. \tag{18}$$

$$\Rightarrow \quad \lim_{H\to s} f^{-1}(H) = H_{min} + (H_{max} - H_{min})s. \tag{19}$$

Since the right hand side of Equation (17) equals the right hand side of Equation (19) it implies that,

$$\lim_{H\to s} f^{-1}(H) = f^{-1}(s) = (1-s)H_{min} + sH_{max}.$$

Thus, the inverse map f^{-1} exists and is continuous at some value $H = s$.

Therefore, the map $f(H)$ is a homeomorphism and $H \in [H_{min}, H_{max}]$ is homeomorphic to $[0,1]$ of the Cantor set for noise-like time series. □

2.3.1. Illustration of the Cantor Set

In this subsection, we take real-world noise-like time series and remove middle thirds up to four (4) levels so that it is similar to the Cantor set. This phenomenon is depicted in Figure 1 [26]. It shows that the segments appear the same at different scales in successive magnifications of the Cantor set from levels C_0 to C_6. C_0 depicts the original time series with no missing parts and C_6 represents the remaining time series after removing middle thirds for the sixth time. For the sake of experimentation, we limit our scope to levels from C_0 to C_3.

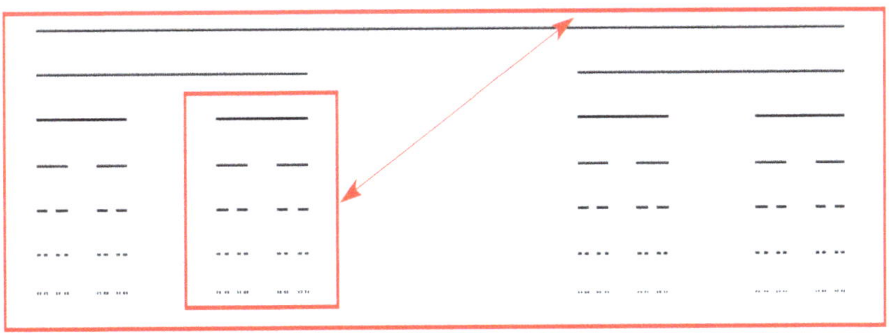

Figure 1. Fractal behavior of a ternary Cantor set.

2.3.2. Definition

The subset of intervals of the Cantor set is defined recursively as:

1. $C_0 = [0,1]$;
2. $C_1 = \left(\frac{1}{3}, \frac{2}{3}\right)$;
3. $C_n = \frac{C_{n-1}}{3} \cup \left(\frac{2}{3} + \frac{C_{n-1}}{3}\right)$ for $n \geq 2$.

The ternary Cantor set is defined as $C = [0,1] \setminus (\cup_{n=1}^{\infty} C_n)$. The level C_0 indicates the interval we begin with. For C_1, $[0,1]$ is divided into 3 sub-intervals and the middle sub-interval $\left(\frac{1}{3}, \frac{2}{3}\right)$ is removed. For C_2, each of the remaining intervals from C_1 are divided into 3 sub-intervals and their middle sub-intervals $\left(\frac{1}{9}, \frac{2}{9}\right)$ and $\left(\frac{7}{9}, \frac{8}{9}\right)$ are removed. This procedure can continue indefinitely by removing open middle third sub-interval of each interval obtained in the previous level. Due to issues with the dimension of the Cantor sets (i.e., dimension of $0.631 < 1$), we rescale the integrated series ψ_t by dividing each observation by the maximum data point:

$$\psi_t = \frac{\Psi_t}{\max(\Psi_t)}.$$

s.t. $\psi_t \in [0,1]$.

2.3.3. Algorithm of the CDFA

Here, we present a modification of the DFA algorithm called CDFA to generalize the segment division step of the DFA. The CDFA algorithm consists of four (4) main steps:

1. given the time series ψ_t of length N, find the integrated series shifted by the mean $< \psi >$,

$$Y_j = \sum_{i=1}^{j} (\psi_i - < \psi >).$$

2. the cumulatively summed series Y_j is then segmented into equal non-overlapping segments of various sizes Δs. Δs is based on the Cantor set theory scale ($\Delta s = 3^n$, $n \geq 0$). The number of non-overlapping segments is calculated as:

$$N_{\Delta s} \equiv int\left(\frac{N}{\Delta s}\right) = int\left(\frac{N}{3^n}\right).$$

The Cantor set scaling function is computed for multiple segments to highlight both slow- and fast-evolving fluctuations that control the structure of the time series.

3. Root Mean Squared Fluctuation (RMSF) is computed for multiple scales of the integrated series:

$$F(\Delta s) \equiv \left\{ \frac{1}{2N_{\Delta s}} \sum_{j=1}^{2N_{\Delta s}} \left[Y_j - Y_j^{\Delta s} \right]^2 \right\}^{1/2}$$

where j denotes the sample size of segments $N_{\Delta s}$. We compute RMSF from $j = 1$ to $2N_{\Delta s}$ not $N_{\Delta s}$. We sum from beginning to end and from end to beginning, then an average of the values is calculated so that every data point is considered. Conversely, the large segments interweave many local periods with both small and large fluctuations and therefore average out their differences in magnitude.

4. the least squares regression fit of $F(\Delta s)$ versus the Cantor scales Δs on a log–log scale produces the power-law notation computed for multiple scales:

$$F(\Delta s) \propto (\Delta s)^{H^c}$$

$$log(F(\Delta s)) = H^c log(\Delta s) + log(C),$$

where $H^c :=$ Hurst exponent of the CDFA which measures memory behavior in the noise-like time series.

2.3.4. Real Time Series

In Figure 2, the time series multifractal (upper panel), monofractal (middle panel) and white noise (lower panel) used in the experiment are noise-like biomedical time series with 8000 rescaled sample data points each [27]. The red trajectory depicts the random walk of the respective series. Observe that the fractal depicted by the multifractal time series at the peak looks very similar to the entire monofractal time series. Thus, comparing the series in the upper panel to the middle panel, the multifractal series has many fractals compared to the one for the monofractal series. We determine DFA's Hurst exponents for the remaining series after removing the middle thirds of each series at each level. It should be noted that the white noise time series has a structure independent of time with Hurst exponent close to $H = 0.5$ whereas noise-like monofractal and multifractal time series exhibit persistence behavior s.t. $0.5 < H \leq 1$.

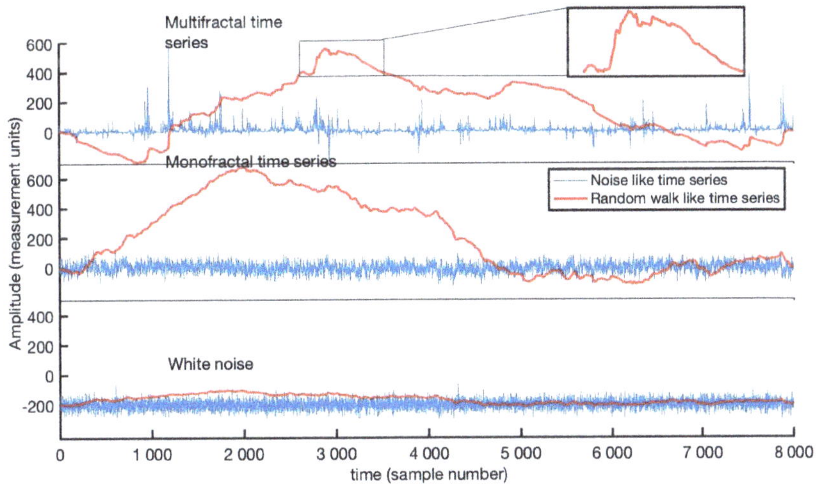

Figure 2. Biomedical time series plots.

3. Results

Below are tables of results of power-law exponents from the implementation of the DFA algorithm on the respective time series. Note from Figure 1 that

1. C_0 denotes the entire time series (with 8000 data points) and produces one Hurst exponent H_1 shown in row 1 of Tables 1–3;
2. C_1: after removing middle third of the time series for the first time, we have two(2) set of data with respective data points 2666 and 2667 and corresponding Hurst exponents H_1 and H_2 as shown in row 2 of Tables 1–3;
3. C_2: we have four (4) data sets remaining after deleting middle third for the second time. The data sets have 888, 889, 888 and 889 observations, respectively. Thus, we obtain four(4) Hurst exponents, namely, H_1, H_2, H_3 and H_4 as shown in row 3 of Tables 1–3;
4. C_3: deleting middle third for the third time produces eight (8) partitions of data with 296 data points each. Each of the data sets produces a Hurst exponent resulting in eight (8) exponents H_1, H_2, H_3, H_4, H_5, H_6, H_7, and H_8 in total. These exponents are shown in the last rows of Tables 1–3.

It should be noted that the same data points from partitioning correspond to the white noise, monofractal, and multifractal time series. The Hurst exponents also follow respectively.

Table 1. DFA's Hurst Exponents of White noise time series.

Levels	Hurst Exponents
C_0	$H_1 = 0.50$
C_1	$H_1 = 0.50,\ H_2 = 0.45$
C_2	$H_1 = 0.54,\ H_2 = 0.45,\ H_3 = 0.52,\ H_4 = 0.42$
C_3	$H_1 = 0.50,\ H_2 = 0.54,\ H_3 = 0.4,\ H_4 = 0.49,\ H_5 = 0.59,\ H_6 = 0.43,\ H_7 = 0.42,$ $H_8 = 0.57$

From Table 1, we observe closeness of the Hurst exponents of the white noise series to $H = 0.5$ for all levels from C_0 to C_3. This confirms the phenomena that are exhibited in the fractal nature of the Cantor set in white noise time series. No matter how many sections of a white noise series are removed, the left-over series still exhibits similar characteristics as the whole white noise series.

Table 2. DFA's Hurst exponents of monofractal time series.

Levels	Hurst Exponents
C_0	$H_1 = 0.79$
C_1	$H_1 = 0.80,\ H_2 = 0.68$
C_2	$H_1 = 0.81,\ H_2 = 0.69,\ H_3 = 0.74,\ H_4 = 0.68$
C_3	$H_1 = 0.65,\ H_2 = 0.80,\ H_3 = 0.63,\ H_4 = 0.72,\ H_5 = 0.78,\ H_6 = 0.68,$ $H_7 = 0.67,\ H_8 = 0.79$

Table 2 present Hurst exponents between 0.5 and 1 $(0.5 < H \leq 1)$ for long memory monofractal time series for levels C_0, C_1, C_2 and C_3. The phenomena exhibited in the monofractal time series from the table above are similar to the fractal nature of the Cantor set. The series left behind after removing the middle thirds of the monofractal time series exhibits similar statistical properties as the whole.

Table 3. DFA's Hurst exponents of multifractal time series.

Levels	Hurst Exponents
C_0	$H_1 = 0.86$
C_1	$H_1 = 0.86,\ H_2 = 0.75$
C_2	$H_1 = 0.75,\ H_2 = 0.88,\ H_3 = 0.70,\ H_4 = 0.78$
C_3	$H_1 = 0.82,\ H_2 = 0.69,\ H_3 = 0.77,\ H_4 = 0.97,\ H_5 = 0.69,\ H_6 = 0.70,$ $H_7 = 0.91,\ H_8 = 0.90$

Hurst exponents of the multifractal time series lie within the range $0.5 < H \leq 1$ for all levels C_0, C_1, C_2 and C_3 from Table 3. This illustrates the fractal phenomena depicted by the Cantor set where successive magnification of the Cantor produces a copy of itself. This can be seen in Figure 1. Thus, self-similar behavior persists after removing the middle thirds of the whole series up to the level C_3. Results from Tables 1–3 confirm that successive magnification of noise-like time series shows a similar pattern at increasingly smaller scales. Thus, the statistical characteristics of part of noise-like series are similar to that of the whole. This phenomenon is commonly known in fractals as self-similarity.

Figures 3–8 shows the log–log fits of RMSF and scales of the white noise, monofracal and multifractal bio-medical series using the DFA and the CDFA. The first two (2) plots (i.e., Figures 3 and 4) present fits of the white noise using the DFA and CDFA. The next two (2) plots (i.e., Figures 5 and 6) illustrate the fit of monofractal series using the DFA and CDFA. The last two(2) plots (i.e., Figures 7 and 8) show fits of the multifractal series using the DFA and the CDFA.

Table 4 above has six (6) columns of results in total. The first column (H) represents the Hurst exponents of the DFA, the second column (H^c) denotes the Hurst exponents of the CDFA and the difference between the exponents in the first two columns are found in the third column. The column for α denotes the scaling exponents of the TLF. The last two columns represent the multiplication of the Hurst exponents of the DFA (H) and the scaling exponent of the TLF (α), as well as the multiplication of the Hurst exponents of the CDFA (H^c) and the scaling exponents of the TLF. Upon investigating Hurst exponents of white noise, monofractal and multifractal time series using the DFA and CDFA, we observe differences in their exponents, as shown in Table 4. Hurst exponent of white noise time series changes slightly but that of the monofractal and multifractal time series changes about 1%. The slight changes in the exponents are a result of subdividing the time series as multiples of 3 (ternary base) at each level using the CDFA. This helps to curb the problem of overestimation associated with DFA. Notwithstanding the differences between the exponents, they still depict the same processes modeled herein (i.e., noise-like time series).

The exponent of the white noise is close to 0.5 whereas that of the noise-like monofractal and multifractal series lie within the range $0.5 < H \leq 1$, depicting long-memory behavior.

Table 1. Comparison of scaling exponents of DFA(H) & CDFA(H^c) & TLF (α) on noise-like time series.

Time Series	H	H^c	Difference	α	$H\alpha$	$H^c\alpha$
White noise	0.5	0.4997	0.0003	1.97	0.985	0.9844
Monofractal	0.79	0.781	0.009	1.28	1.0112	0.9997
Multifractal	0.86	0.851	0.009	1.17	1.0062	0.9976

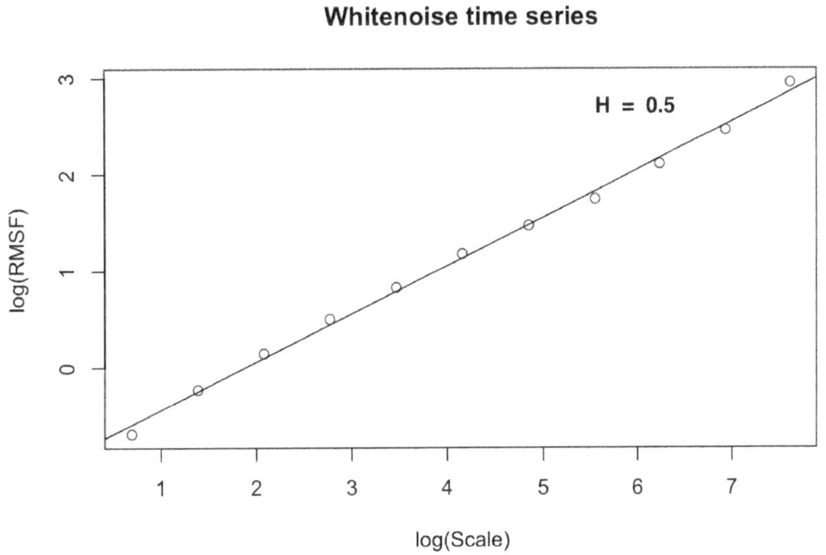

Figure 3. Log–log fit of white noise time series using DFA.

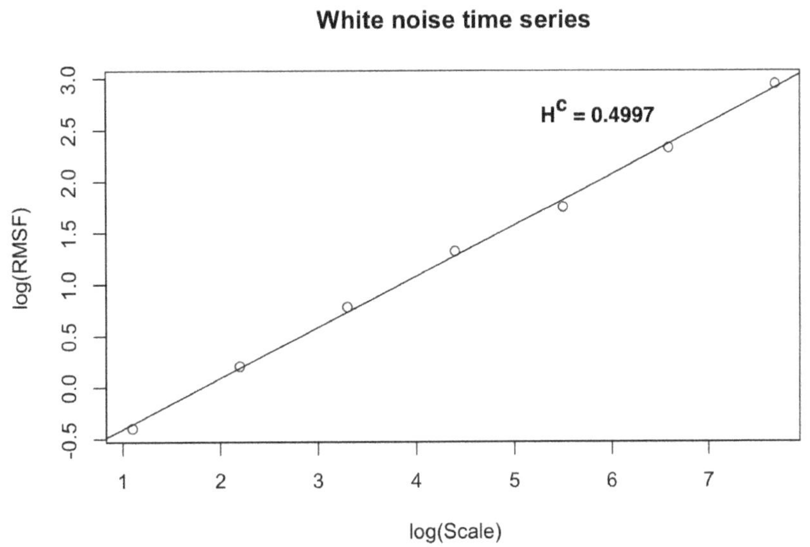

Figure 4. Log–log fit of white noise time series using CDFA.

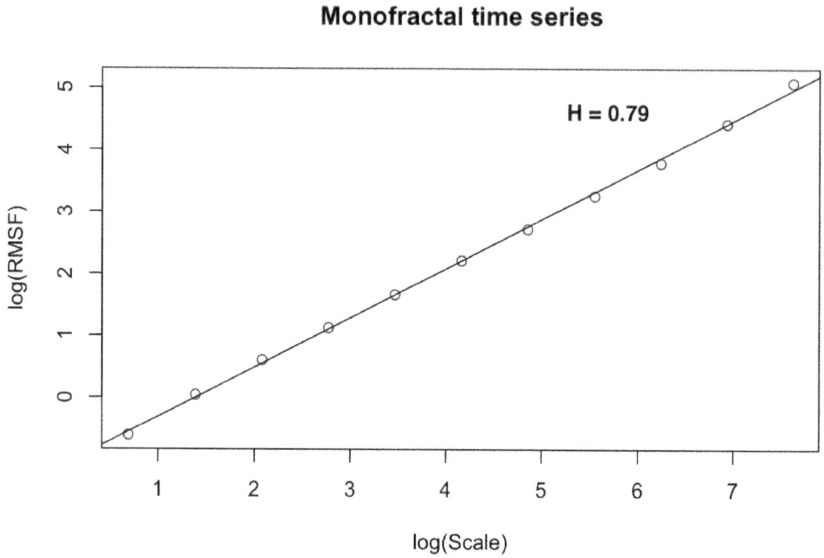

Figure 5. Log–log fit of monofractal time series using DFA.

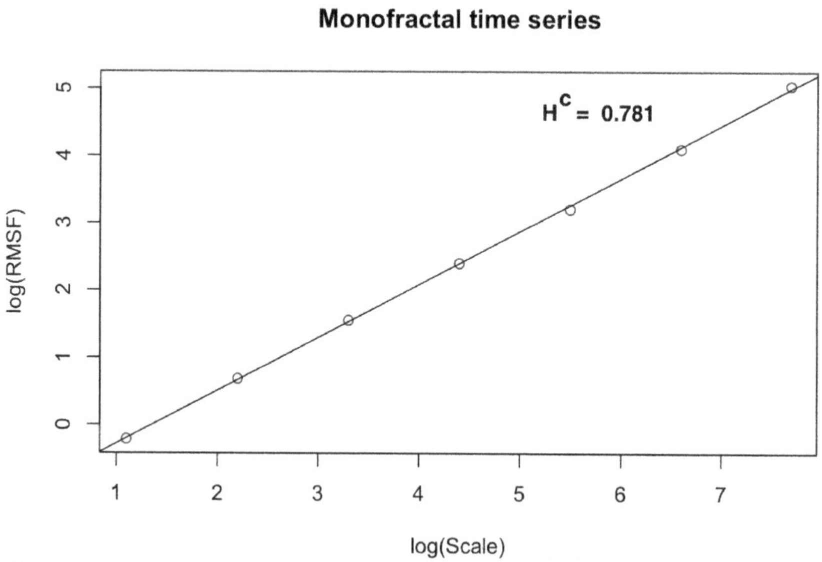

Figure 6. Log–log fit of monofractal time series using CDFA.

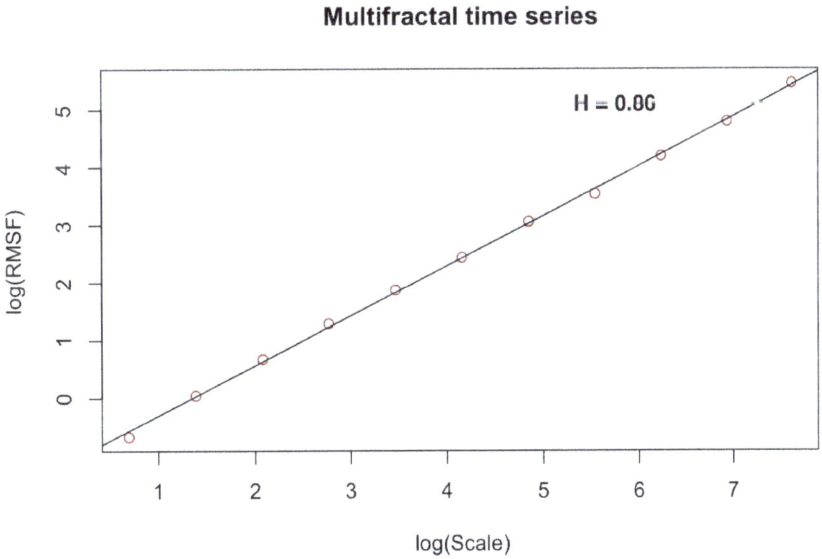

Figure 7. Log–log fit of multifractal time series using DFA.

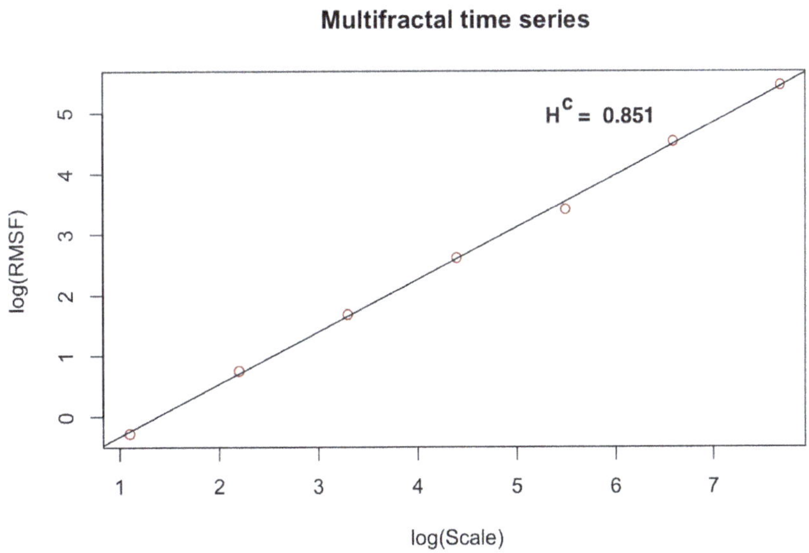

Figure 8. Log–log fit of multifractal time series using CDFA.

4. Discussion

The results obtained from the Tables 1–4 suggest that segment size may not always be hard-coded in the DFA algorithm based on the length of the time series in question. Especially for time series with odd lengths, the process can be automated using the fractal phenomena of the Cantor set to obtain equal segment sizes and satisfactory Hurst exponents.

Furthermore, multiplying Hurst exponents of the DFA and CDFA with the scaling exponents (α) of the Truncated Lévy flight (TLF) suggests that H^c is a better estimate. For the monofractal and multifractal noise-like time series, we observe that $H^c\alpha$ is approximately equal to 1 while $H\alpha$ exceeds 1. This deviates from the inverse relationship between the Hurst exponents and the scaling exponents of the TLF for Gaussian noise as discussed in the paper [4]. This highlights the overestimation of the Hurst exponent of the DFA approach that happens in practice.

5. Conclusions

In this work, we have proposed a modification to the DFA algorithm by utilizing the theory of the tenary Cantor set in the segment division step. The Cantor DFA (CDFA) has been compared to the α exponent of the truncated Lévy model and the Hurst exponent of the DFA. We have in addition proved that the interval of the Hurst exponent of the DFA is homeomorphic to the Cantor set. We confirm the results from the proof by illustrating the fractal phenomena exhibited by the Cantor set using real-world time series in Tables 1–3.

Our results from numerical simulations show that the CDFA generates better estimates of Hurst exponents. The CDFA proposed in this work automates the segment sizes in the DFA algorithm using the number base 3 theory of the Cantor set, where the time series is divided into multiples of 3 at each level. This modification helps to curb the overestimation problem of the Hurst exponent (H) of the DFA by determining segment sizes based on the fractal phenomena depicted by the Cantor set while correctly predicting the memory behavior of the series in question.

The results are shown in Table 4 where the Hurst exponent of the CDFA is compared with that of the DFA and the scaling exponents (α) of the TLF. In [4], a relationship was established between the Hurst exponent of the DFA and the α exponent of the TLF. The CDFA is also shown to satisfy this relation, thus making it possible to extract the α exponent of the TLF from the Hurst exponent of the CDFA.

The CDFA approach can be applied to time series with odd lengths, time series whose lengths are not easily divisible by even numbers, time series whose lengths do not permit equal segmentation, etc. These kinds of series exist in several industries, including financial, geophysics, health and the like. Another application of the CDFA would be to act as a control model for the ordinary DFA to reduce the chances of overestimation of the Hurst exponent.

Since this is a modification of the DFA, there is the need to simulate CDFA with different data sets having varying characteristics for which the DFA has been shown to correctly detect their scaling behavior. An example will be DNA sequences, financial markets, etc., for further comparison of the model performance against the DFA.

For future work, we seek to investigate the robustness of the CDFA as stated earlier by simulating the model with data sets from different fields, including, but not limited to, DNA sequences, financial markets and geophysical data. In the case of "big data", we seek to extend the CDFA by "parallelizing" the sequential code of the CDFA (PCDFA) to improve its efficiency in simulation.

Author Contributions: Conceptualization, M.C.M., W.K., P.K.A. and O.K.T.; methodology, M.C.M., W.K., P.K.A., J.A.G. and O.K.T.; software, W.K.; validation, M.C.M. and O.K.T.; formal analysis, W.K.; data curation, W.K.; writing—original draft preparation, W.K.; writing—review and editing, W.K., P.K.A. and O.K.T.; visualization, W.K. and P.K.A.; supervision, M.C.M. and O.K.T. All authors have read and agreed to the published version of the manuscript.

Funding: The study was funded partially by the National Institute on Minority Health and Health Disparities (NIMHD) grant (U54MD007592).

Institutional Review Board Statement: Not applicable.

Informed Consent Statement: Not applicable.

Data Availability Statement: The data used in this work can be accessed at https://www.ntnu.edu/inb/geri/software.

Conflicts of Interest: The authors declare no conflict of interest.

Abbreviations

Abbreviations

The following abbreviations are used in this manuscript:

DFA	Detrended Fluctuation Analysis
CDFA	Cantor Detrended Fluctuation Analysis
TLF	Truncated Lévy Flight
R/S	Rescaled Range Analysis

References

1. Guthrie, J.A.; Nymann, J.E. The topological structure of the set of subsums of an infinite series. *Colloquium Math.* **1988**, *55*, 323–327. Available online: http://yadda.icm.edu.pl/yadda/element/bwmeta1.element.desklight-a3fbdce0-a55c-413d-a53f-e1228d61fd02/c/cm55_2_15.pdf (accessed on 10 August 2021). [CrossRef]
2. Mariani, M.C.; Asante, P.K.; Bhuiyan, M.A.M.; Beccar-Varela, M.P.; Jaroszewicz, S.; Tweneboah, O.K. Long-Range Correlations and Characterization of Financial and Volcanic Time Series. *Mathematics* **2020**, *8*, 441. [CrossRef]
3. Mariani, M.C.; Kubin, W.; Asante, P.K.; Tweneboah, O.K.; Beccar-Varela, M.P. Multifractal Analysis of Daily US COVID-19 Cases. In Proceedings of the 10th Annual AHSE, STEM/STEAM and Education Conference, Honolulu, HI, USA, 9–11 June 2021; pp. 2333–4908. Available online: https://huichawaii.org/wp-content/uploads/2021/07/Mariani-Maria-C.-2021-HUIC.pdf (accessed on 12 August 2021).
4. Mariani, M.C.; Kubin, W.; Asante, P.K.; Tweneboah, O.K.; Beccar-Varela, M.P.; Jaroszewicz, S.; Gonzalez-Huizar, H. Self-Similar Models: Relationship between the Diffusion Entropy Analysis, Detrended Fluctuation Analysis and Lévy Models. *Mathematics* **2020**, *8*, 1046. [CrossRef]
5. Hurst, H.E. Long term storage capacity of reservoirs. *Trans. Am. Soc. Eng.* **1951**, *116*, 770–799. [CrossRef]
6. Kantelhardt, J.W.; Koscielny-Bunde, E.; Rego, H.H.A.; Havlin, S.; Bunde, A. Detecting Long-range Correlations with Detrended Fluctuation Analysis. *Phys. A* **2001**, 295, 441–454. [CrossRef]
7. Peng, C.K.; Buldyrev, S.V.; Havlin, S.; Simons, M.; Stanley, H.E.; Goldberger, A.L. Mosaic organization of DNA nucleotides. *Phys. Rev. E* **1994**, *49*, 1685–1689. [CrossRef] [PubMed]
8. Mantegna, R.N.; Stanley, H.E. An introduction to Econophysics: Correlations and Complexity in Finance. *Phys. Today* **2000**, *53*, 148. [CrossRef]
9. Gabaix, X.; Gopikrishnan, P.; Plerou, V.; Stanley, H.E. A theory of power-law distributions in financial market fluctuations. *Nature* **2003**, *423*, 267–270. [CrossRef]
10. Beccar-Varela, M.; Gonzalez-Huizar, H.; Mariani, M.C.; Serpa, L.F.; Tweneboah, O.K. Chile2015: Lévy Flight and Long-Range Correlation Analysis of Earthquake Magnitudes in Chile. *Pure Appl. Geophys.* **2016**, *173*, 2257–2266. [CrossRef]
11. Schinckus, C. From DNA to Economics: Analogy in Econobiology. *Rev. Contemp. Philos.* **2018**, *17*, 31–42. [CrossRef]
12. Buldyrev, S.V.; Goldberger, A.L.; Havlin, S.; Mantegna, R.N.; Matsa, M.E.; Peng, C.K.; Simons, M.; Stanley, H.E. Long-range correlation properties of coding and noncoding DNA sequences: GenBank analysis. *Phys. Rev. E* **1995**, *51*, 5084–5091. [CrossRef]
13. Bunde, A.; Havlin, S. *Fractals and Disordered Systems*; Springer: Berlin/Heidelberg, Germany; New York, NY, USA, 1996.
14. Hardstone, R.; Poil, S.S.; Schiavone, G.; Jansen, R.; Nikulin, V.V.; Mansvelder, H.D.; Linkenkaer-Hansen, K. Detrended Fluctuation Analysis: A Scale-Free View on Neuronal Oscillations. *Front. Physiol.* **2012**, *3*, 450. [CrossRef]
15. Little, M.; McSharry, P.; Moroz, I.; Roberts, S. Nonlinear, Biophysically-Informed Speech Pathology Detection. *IEEE Int. Conf. Acoust. Speed Signal Process.* **2006**, *2*, 1080–1083. [CrossRef]
16. Bunde, A.; Havlin, S.; Kantelhardt, J.W.; Penzel, T.; Peter, J.H.; Voigt, K. Correlated and Uncorrelated Regions in Heart-Rate Fluctuations during Sleep. *Phys. Rev. Lett.* **2000**, *85*, 3736–3739. [CrossRef] [PubMed]
17. Ivanova, K.; Ausloos, M. Application of the detrended fluctuation analysis (DFA) method for describing cloud breaking. *Phys. A Stat. Mech. Its Appl.* **1999**, *274*, 349–354. [CrossRef]
18. de Moura, E.P.; Vieira, A.P.; Irmao, M.A.S.; Silva, A.A. Applications of detrended-fluctuation analysis to gearbox fault diagnosis. *Mech. Syst. Signal Process.* **2009**, *23*, 682–689. [CrossRef]
19. Govindan, R.B.; Wilson, J.D.; PreiBl, H.; Eswaran, H.; Campbell, J.Q.; Lowery, C.L. Detrended fluctuation analysis of short datasets: An application to fetal cardiac data. *Phys. D Nonlinear Phenom.* **2007**, *226*, 23–31. [CrossRef]
20. Li, E.; Mu, X.; Zhao, G.; Gao, P. Multifractal Detrended Fluctuation Analysis of Streamflow in the Yellow River Basin, China. *Water* **2015**, *7*, 1670–1686. [CrossRef]
21. Benes, E.; Fodor, M.; Kovacs, S.; Gere, A. Application of Detrended Fluctuation Analysis and Yield Stability Index to Evaluate Near Infrared Spectra of Green and Roasted Coffee Samples. *Processes* **2020**, *8*, 913. [CrossRef]
22. Scafetta, N. *An Entropic Approach to the Analysis of Time Series*; University of North Texas, ProQuest Dissertations Publishing: Denton, TX, USA, 2001.
23. Weron, R. Measuring long-range dependence in electricity prices. In *Empirical Science of Financial Fluctuations*; Takayasu, H., Ed.; Springer: Tokyo, Japan, 2002. [CrossRef]

24. Kristoufek, L. *Long-Range Dependence in Returns and Volatility of Central European Stock Indices*; IES Working Paper 3/2010; IES FSV, Charles University: Prague, Czech Republic, 2010.
25. Koponen, I. Analytic approach to the problem of convergence of truncated Lévy flights towards the Gaussian stochastic process. *Phys. Rev. E* **1995**, *52*, 1197–1199. [CrossRef]
26. Available online: https://wiki.math.ntnu.no_media/tma4225/2015h/cantor_set_function.pdf (accessed on 13 September 2021).
27. Ihlen, E. Introduction to Multifractal Detrended Fluctuation Analysis in Matlab. *Front. Physiol.* **2012**, *3*, 141. [CrossRef] [PubMed]

entropy

MDPI

Article

Financial Return Distributions: Past, Present, and COVID-19

Marcin Wątorek [1], Jarosław Kwapień [2,*] and Stanisław Drożdż [1,2]

[1] Faculty of Computer Science and Telecommunications, Cracow University of Technology, ul. Warszawska 24, 31-155 Kraków, Poland; marcin.watorek@pk.edu.pl (M.W.); Stanislaw.Drozdz@ifj.edu.pl (S.D.)
[2] Complex Systems Theory Department, Institute of Nuclear Physics, Polish Academy of Sciences, ul. Radzikowskiego 152, 31-342 Kraków, Poland
* Correspondence: jaroslaw.kwapien@ifj.edu.pl

Abstract: We analyze the price return distributions of currency exchange rates, cryptocurrencies, and contracts for differences (CFDs) representing stock indices, stock shares, and commodities. Based on recent data from the years 2017–2020, we model tails of the return distributions at different time scales by using power-law, stretched exponential, and q-Gaussian functions. We focus on the fitted function parameters and how they change over the years by comparing our results with those from earlier studies and find that, on the time horizons of up to a few minutes, the so-called "inverse-cubic power-law" still constitutes an appropriate global reference. However, we no longer observe the hypothesized universal constant acceleration of the market time flow that was manifested before in an ever faster convergence of empirical return distributions towards the normal distribution. Our results do not exclude such a scenario but, rather, suggest that some other short-term processes related to a current market situation alter market dynamics and may mask this scenario. Real market dynamics is associated with a continuous alternation of different regimes with different statistical properties. An example is the COVID-19 pandemic outburst, which had an enormous yet short-time impact on financial markets. We also point out that two factors—speed of the market time flow and the asset cross-correlation magnitude—while related (the larger the speed, the larger the cross-correlations on a given time scale), act in opposite directions with regard to the return distribution tails, which can affect the expected distribution convergence to the normal distribution.

Keywords: return distributions; power-law tails; stretched exponentials; q-Gaussians; financial markets; COVID-19

Citation: Wątorek, M.; Kwapień, J.; Drożdż, S. Financial Return Distributions: Past, Present, and COVID-19. *Entropy* **2021**, *23*, 884. https://doi.org/10.3390/e23070884

Academic Editor: Ryszard Kutner

Received: 15 June 2021
Accepted: 9 July 2021
Published: 12 July 2021

1. Introduction

A proper risk assessment is one of the key prerequisites of any prospective financial investment. Even for an asset of moderate volatility, underestimating the probability of occurrence of an event of a given magnitude can lead to severe outcomes. Among the methods of dealing with risk assessment is the determination of a correct probability distribution function for asset price fluctuations in order to construct an adequate model of that asset's price dynamics. This issue has been of central interest since the early years of econometrics. It was Bachelier who proposed a model of the stock option price dynamics based on an uncorrelated random walk with a Gaussian distribution of fluctuations [1]. Later, it was found that the Gaussian noise hypothesis was only a poor approximation of the empirical data, which shows non-vanishing higher moments of the fluctuation distributions, i.e., skewness and positive excess kurtosis. Based on an observation of the cotton price dynamics, Mandelbrot proposed to model the logarithmic price increments (returns) with a process of Lévy flights, which is described by a heavy-tailed probability distribution function that is stable [2,3]. These distributions are defined by their characteristic function as they do not have a closed analytic form. However, their tails decrease as a power law in the limit of large x:

$$L_\alpha(x) \sim \frac{1}{|x|^{1+\alpha}}, \quad x \to \pm\infty, \tag{1}$$

where $0 < \alpha < 2$.

According to Mandelbrot [4], such a process can account for the absence of a convergence of the aggregated return distribution to the normal distribution as expected by the central limit theorem (CLT). The heavy tails are thus viewed as a natural limit of the aggregated independent or weakly dependent factors provided they are described by the stable distributions. However, this hypothesis has a weak point because the empirical data cannot exhibit the infinite variance required to maintain the distribution stability under aggregation. After the pioneering work of Mandelbrot, many researchers investigated financial time series in order to verify his outcomes. For example, Fama reported that the daily returns of stocks are better approximated by the infinite-variance distribution than the normal distribution or a mixture of the normal distributions [5]. The Lévy stability of the return pdfs in their central parts was also confirmed, among others, by Blume ($\alpha \approx 1.7 - 1.8$) [6], Teichmoeller ($\alpha \approx 1.6$) [7], and Blattberg and Gonedes ($\alpha \approx 1.6$) [8]. Some reports pointed out that, although central parts of the return distribution can be approximated by the stable distributions, the same cannot be said about the distant parts of their tails, which decay faster than expected. Officer found that the tails of the daily and monthly return distributions are no doubt thicker than Gaussian but at the same time thinner than Lévy-stable [9]. Barnea et al. observed that the daily return distributions for some stocks are well approximated by stable distributions, while for other stocks, they are not [10]. Much later, Young and Graff reported that the real-estate annual return distribution can be fitted by a stable function using $\alpha \approx 1.5$ [11].

Along with the research on empirical data, much effort was devoted to developing models that could mimic the market dynamics. Among such models, the subordinated stochastic processes do not require an assumption of the Lévy-stable character of the underlying dynamics and assume that the price movement is a Brownian motion that takes place in time, which itself is a stochastic process with positive increments and finite variance (e.g., a lognormal process) [12]. In practice, the subordinating process is assumed to be volume or transaction number. As an alternative, Engle proposed that the distribution tails are heavy because of the heteroskedasticity of the return-generating process, in which large returns are caused by a locally large variance of the process [13]. Mantegna and Stanley found a dual structure of the stock index return distribution (S&P500 index during the years 1984–1989), with its central part being in agreement with a Lévy-stable distribution and with exponentially decaying distant tails [14]:

$$L^{\text{tr}}_{\alpha,\gamma}(x) \sim \frac{1}{|x|^{1+\alpha}} e^{-\gamma|x|}, \quad \gamma > 0. \tag{2}$$

While considering the aggregated returns at different time horizons, they did not find any trace of a convergence to the normal distribution. Based on these findings, they proposed a new model for the price return dynamics: a truncated Lévy flight process. They also showed that the heteroskedastic model (GARCH) does not fit the data well [14]. This type of distribution ($\alpha \approx 1.6$–1.7) was also reported from an analysis of the same S&P500 index recorded over a longer interval (1986–2000). In contrast, the aggregated returns showed a crossover to a CLT regime around a time scale of 20 days [15].

Plerou et al. and Gopikrishnan et al. presented two parallel, comprehensive studies of the stock market high-frequency data representing stock price returns for 1000 American companies and S&P500 index returns [16,17], in which they observed the cumulative distribution function tails obtained from aggregated returns over a substantial spectrum of time scales from 5 min (stocks) and 1 min (index) to 4 years. They found that the return distributions have power law tails, with the exponent $2.5 < \alpha < 4$ depending on a stock. However, despite the fact that they did not fit the Lévy-stable domain ($\alpha < 2$), these distributions were invariant under a change in the time scale up to $\Delta t = 16$ days. Only for the sampling intervals longer than 16 days, a slow transition to a normal distribution was observed [16]. An analogous invariance of the return distribution shape with the power exponent $\alpha \approx 3$ under the time-scale change was observed for the S&P500 index, but in that case, the crossover occurred earlier at $\Delta t = 4$ days. Only for the time scales longer than 4 days, a slow convergence to a Gaussian distribution was seen. A similar behavior was found in the indices from other stock markets (Nikkei & Hang-Seng) [17]. This surprising behavior of the stock markets led the authors to formulate the so-called "inverse cubic law"—a conjecture that the power-law tails of the return distributions with the scaling exponent $\alpha \approx 3$ are a universal property of all stock markets at short and medium time scales [18]. Indeed, similar statistical characteristics were found by other researchers in data collected from other stock markets [19–35], Forex [36], commodity markets [36,37], and the cryptocurrency market [36,38–40].

The only possible explanation of this result is that the analyzed data violated the assumptions of the central limit theorem, i.e., the returns were significantly correlated. Indeed, the cross-correlations among the stock returns representing different companies are an obvious characteristics of all stock markets [34,41–45]. It was shown that the inter-stock cross-correlation strength has a strong impact on the index return distributions and can even modify their tail behavior, leading to a kind of alternation between different power-law regimes: stable and unstable [43]. On the other hand, the cross-correlations between different stock markets can also induce a significant regime change [21,22]. The existence of autocorrelation in returns is a more delicate issue: while the returns reveal some short-term memory lasting for a few minutes, the existence of long-term memory is doubtful [16,17,24,46,47], even though there were reports stating that the returns can show some autocorrelation or persistence over long terms [48–52]. On the other hand, there is consensus over a fact that the long-term autocorrelation is present in absolute returns (volatility) and in some more fundamental observables such as fluctuations in stock market orders, transaction size, and market liquidity [53–55]. The existence of a return autocorrelation can be considered an important factor that can destroy market efficiency [56,57]. These ubiquitous manifestations of the inverse cubic scaling in the financial data encouraged Gabaix et al. to propose a model that was able to account for this phenomenon [58]. According to this model, the inverse-cubic return fluctuations were a result of two processes: the volume fluctuation that forms a probability distribution function with the tail index 1.5 and a specific square-root form of the price impact function, which together produce a tail index equal to 3 [58]. However, Farmer and Lillo pointed out that the price impact function is specific to individual markets and even to individual stocks; thus, it cannot produce any universal behavior. Also, the dependence on transactions is slower than the square root and the volumes are not power-law distributed, so they cannot lead to a power-law behavior of the returns with $\alpha \approx 3$ [59]. The price changes are driven by more factors than simply volume and transaction number fluctuations—it can be the order book structure, for example [60,61]. Moreover, there is plenty of published evidence that various financial assets either do not have the power-law distribution tails [29,62–67] or their scaling exponent α differs from 3 even for the short time scales [36,68–73]. Given these results together, the inverse cubic scaling cannot be considered a universal property of financial returns and, thus, cannot be called "a law". However, it manifests itself sufficiently often to allow us to view it as one of the possible reference models describing the empirical return distribution tails (there is a plethora of volatility models, which takes into account various factors; a review of such models can be found in Poon and Granger [74]).

The power law tails of the return distributions, which are among the financial stylized facts, can be reproduced with a broad range of the scaling exponent by means of various models based on stochastic processes [63,75–83], including multiplicative processes [84–86], the minority game and other agent-related dynamics [87–91], as well as spin dynamics [23,92].

Apart from power-law functions, the tail behavior of the return distributions can also be approximated by exponential functions and stretched exponential functions [93]. The latter are defined by the following expression:

$$f(x) \sim \exp x^{-\beta}, \qquad 0 < \beta < 1. \tag{3}$$

Such a functional form allows for the stretched exponents to locally resemble the power laws. There were many published studies in which the return distributions were approximated successfully by the exponents, and some researchers advocate using these functions instead of the power laws [23,29,31,63–67,69,80,94,95]. Another type of exponential function that is sometimes considered in the context of financial data is the Laplace distribution function $p(x) \sim \exp(-|x|)$. This function can also demonstrate heavy tails. It was observed that some empirical return distributions can be approximated by this function [62,96].

The functions that have been discussed so far do not exhaust the possible models that can be used to approximate the empirical return distributions. In a financial context, a particularly important class is the q-Gaussian functions. They were derived as a part of the formalism of nonextensive statistical mechanics based on the Tsallis nonadditive entropy [97]:

$$S_q = k_{\mathrm{B}} \frac{1 - \int [p(x)]^q dx}{q - 1}, \tag{4}$$

where $p(x)$ is some probability distribution and k_{B} is a positive constant. Under certain conditions, this entropy is maximized by a family of q-Gaussian distributions given by

$$G_q(x) \sim \exp_q[-\mathcal{B}_q(x - \mu_q)^2], \tag{5}$$

where

$$\exp_q x = [1 + (1 - q_x]^{\frac{1}{1-q}}, \quad \mathcal{B}_q = [(3 - q)\sigma_q^2]^{-1}, \tag{6}$$

provided that $0 < q < 3$ and that μ_q and σ_q^2 are q-mean and q-variance, respectively. The q-Gaussians generalize both the normal distribution ($q = 1$) and the Lévy distributions ($5/3 < q < 3$). Their attractiveness comes from the fact that, for the correlated random variables, the q-Gaussians become stable distributions. Moreover, their tail behavior can also resemble the power laws [98]. As the price returns are correlated, one can expect that these functions can describe the statistical properties of returns. Indeed, there is a growing evidence that the q-Gaussian distributions can approximate the empirical return distributions [30,32,66,99–101].

The q-Gaussians are among the functions borrowed from the nonextensive statistical mechanics that were exploited in this context. Another example is the q-exponent given by Equation (6), which was also reported to fit the empirical returns from a stock market [102]. Finally, some researchers consider the normal-inverse Gaussian function to be a prospective model that can successfully be fitted to the data [71].

This short review of the return distribution modeling approaches shows that there is a cornucopia of the reported results that were even contradictory sometimes. The only firm observation that is shared by all the studies is that the return distributions reveal heavy tails, at least at short time scales. On medium and long time scales, the situation depends strongly on a data set, a market, and a financial instrument. Drożdż et al. attempted to resolve this problem by noticing that the most well-known results regarding the return distributions, i.e., Mandelbrot's Lévy stability ($\alpha < 2$) [4]; Mantegna and Stanley's truncated Lévy flights [14]; Plerou and Gopikrishnan's unstable power-law tails ($\alpha \approx 3$), which are persistent under aggregation of the returns until the time scales of days or even a month [16,17]; and their own results with the $\alpha \approx 3$ regime already breaking at the time scale of hours [24], were based on the data covering different epochs: 1816–1958 (Mandelbrot), 1984–1989 (Mantegna), 1926–1995 (Plerou and Gopikrishnan), and 1998–1999 (Drożdż). One can follow the whole historical process of the financial market development, introduction of new financial instruments, technological innovations, transition from the classic "floor-based" markets to the digital markets, computing power increase, telecommunication revolution, etc. from past to present. This inevitably leads to the constantly increasing number of investors, transactions, and pieces of information that arrive at the market. These are accompanied by the increasing amount of money and information processing speed, which, if taken together, result in an overall acceleration of the market time flow. Any unit of time nowadays corresponds to a much longer interval in the past. From this perspective, the market properties once observed, say, at a daily scale, now can be observed at scales of hours, minutes, or even seconds. This may be the very reason why Mandelbrot observed the Lévy-stable distributions that are hardly seen today and why Plerou and Gopikrishnan reported the crossover to the CLT-related convergence of distribution tails at the time scale of many days, while today, such a behavior is observed within hours or minutes. This hypothesis formulated by Drożdż et al. was later supported by other analyses as well [24,30,32,69,72].

However, based on data covering a given time interval, one can observe an analogous phenomenon by considering, e.g., the stocks representing companies with different capitalization. Since there is statistically a relation between the capitalization of a company and the number of transactions involving its stock shares, the highly capitalized stocks "feel" that time flows faster than their lower-capitalized counterparts. In consequence, the properties of the corresponding return distributions substantially differ between both groups, with the former displaying thicker tails than the latter [24,34,35,100,103]. Qualitatively similar observation can be made by comparing the distributions for the data from the markets of different developmental stage, e.g., the mature markets and the emerging markets. The former are characterized by higher liquidity and a higher transaction number than the latter; therefore, generally, the situation is parallel to the previous cases. Studies of the data from the emerging markets report thick tails with small scaling exponents more frequently than the mature markets [25,26,28,52,66,94,104–110].

Another issue related to return distributions is their asymmetry between positive and negative parts. It was investigated in various works as it is also an important factor in investment risk assessments (the gain–loss asymmetry). Typically, this property was tested by means of the third moment (skewness) of return distributions, in which a negative value means a higher probability of a significant gain with respect to a significant loss while a positive value means the opposite. The negative skewness is associated, thus, with a positive tail of the distribution being heavier than the negative tail. There are mixed outcomes of the empirical data reported in the literature, including indications of either positive, negative, or neutral skewness as well as the scaling exponent difference between the left and the right tails (in the case of power-law tails) dependent on the analyzed time intervals, markets, and securities (e.g., References [14,16,17,20,28,31,36,62,71,94,111–119]). However, even though a difference between the positive and negative tails exists in the data, it has a much weaker impact on the distribution shape and the related investment risk than the heavy tails. Therefore, in many studies reported in the literature, only absolute returns are considered, neglecting their actual signs (e.g., References [16,17,24,30,32,39]). As our study is focused on an investigation of the tail exponent stability with respect to the time scale Δt and, based on literature and our previous experience, we expect larger effects due to the time-scale change than due to the left–right tail asymmetry, we neglect the return sign and consider both tails together by analyzing the absolute return values. In fact, our major new finding is that, in recent years, the market's "internal" time stopped accelerating with respect to our ordinary "clock" time. Other factors also affect the convergence of return distributions to the Gaussian with increasing Δt, especially those that cause extreme volatility and strong cross-correlations between assets such as COVID-19. We discuss the interplay of these two factors in the following sections.

The remainder of our paper is organized as follows: in Section 2, we present the data sets that were analyzed; in Section 3, we discuss the results; and in Section 4, we collect the main conclusions of our study.

2. Data

We analyzed recent tick-by-tick recordings of the contracts for differences (CFDs) representing (1) six major stock market indices, CAC40 (Euronext), DAX30 (Deutsche Börse), FTSE100 (London SE), DJIA (New York SE), S&P500 (New York SE & NASDAQ), and NASDAQ100 (NASDAQ); (2) 240 U.S. stock shares and 30 stock shares with the highest capitalization from Germany, France, and the U.K. (see Appendix A for their list); (3) four commodities, U.S. crude oil (CL), high grade copper (HG), silver (XAG), and gold (XAU); (4) the currency exchange rates (not CFDs) involving five major currencies, USD, EUR, GBP, CHF, and JPY; and (5) two cryptocurrencies, bitcoin (BTC) [120] and ethereum (ETH) [121]. The commodity CFD prices are expressed in U.S. dollars. The data comes from Dukascopy (the index, stock share, commodity CFDs, and currency exchange rates) [122] and Kraken exchange (cryptocurrencies) [123] and covers 4 years from January 2017 to December 2020 (except for the stock share CFDs that cover a shorter interval starting from January 2018). Different instrument types have different trading hours, with the stock market index and commodity CFDs quoted from Monday to Friday (00:00–23:00 hours CET, daylight saving time-adjusted), the stock share CFDs quoted from Monday to Friday (U.S.: 15:30–22:00 CET, European: 09:00–17:30 CET), the currency exchange rates quoted around the clock from Monday to Friday, and the cryptocurrency exchange rates quoted continuously 24/7.

Price $P(t)$ of an asset is defined at the moment of transaction only and remains undefined otherwise. Therefore, in order to construct an evenly sampled time series of the price quotations, we assume that the price remains constant between the consecutive transactions, which is standard practice. The quotations of all the instruments were sampled with Δt= 1 s, 10 s, 1 min, 10 min, and 1 h frequency and transformed into the normalized logarithmic returns $r_{\Delta t}$ according to

$$r_{\Delta t} = (R_{\Delta t} - \mu_R)/\sigma_R, \quad R_{\Delta t}(t) = \log(P(t + \Delta t)) - \log(P(t)), \tag{7}$$

where μ_R and σ_R are the mean and standard deviation of $R_{\Delta t}(t)$, respectively, and Δt is a sampling interval. For each asset, we obtained five time series representing the returns for different time scales Δt. Figure 1 shows the evolution of $P(t)$ for various assets that are analyzed in our work. The COVID-19 outburst in the U.S. in March and April 2020 that had a strong impact on all financial markets has been distinguished by vertical lines. A few corresponding time series of the normalized returns $r_{\Delta t}(t)$ with Δt = 1 min are shown in Figure 2 together with a simulated Gaussian noise of the same length.

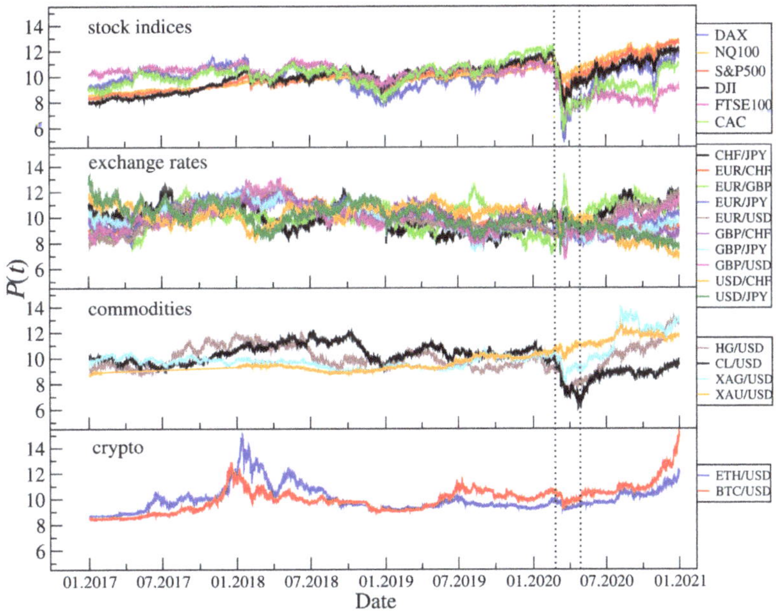

Figure 1. Evolution of the CFD price and exchange rate quotations of various assets over the 4 year interval 2017–2020 (data source: Dukascopy [122]) and the cryptocurrency prices (data source: Kraken [123]). The quotations have been standardized in order to facilitate comparison. The vertical dashed lines indicate the COVID-19 outburst in March–April 2020.

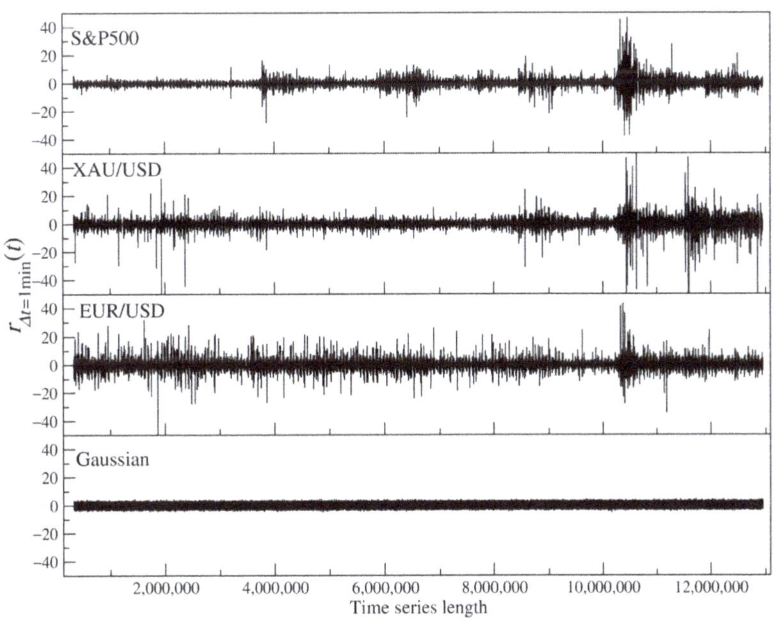

Figure 2. Time series of the standardized 1-min returns of sample financial instruments, S&P500 CFDs, gold CFDs (XAU/USD), and EUR/USD, together with a Gaussian noise of the same length. Note the leptokurtic character of the empirical data.

3. Results

For each individual time series of the absolute normalized returns, we created a cumulative distribution function and investigated how fast its tail decays. In order to quantify the tail behavior, we fit the empirical histograms with selected models that are of the highest significance in this context: the power-law function, the stretched exponential function, and the q-Gaussian function. Figure 3 shows sample return distributions with the three best-fitted models of interest. We refer the reader to the specific subsections for a discussion on the asset statistical properties; here, we consider only the fits. In each panel, it is evident that the power-law function (dashed line) is able to reproduce the empirical histograms in their far-tail region while it fails to describe the central part of the distributions completely. The stretched exponential and q-Gaussian functions perform much better in the central parts, while only the latter works well in the tails. However, as the q-Gaussian and power-law functions converge to the same behavior in the tail regions and as both the parameters α and q are related with each other via a relation,

$$q = \frac{3+\alpha}{1+\alpha}, \quad (8)$$

henceforth, we omit the q-Gaussian fit parameter q and explicitly give the fitted values of α and β only. For simplicity, we also omit the acronym "CFD" and use the asset names only, but one has to realize that the CFD contracts and the assets they refer to are not the same financial entities and that the statistical properties of the former may not necessarily reflect the properties of the latter.

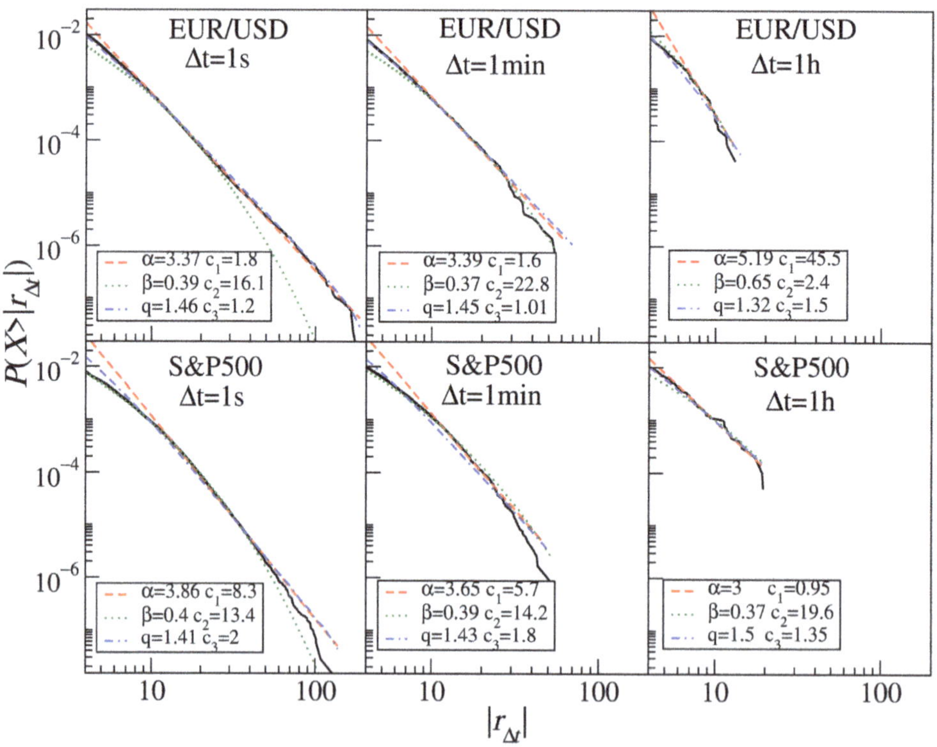

Figure 3. The least-square best fits of the power-law function (red dashed), the stretched exponential (green dotted), and the q-Gaussian function (blue dash-dotted). Sample cumulative distribution functions of the returns for the EUR/USD exchange rate (**top**) and the S&P500 index CFDs (**bottom**) are shown with different sampling intervals Δt from 1 s to 1 h.

3.1. Stock Market Indices

Let us start with the cumulative return distributions of the stock market index CFDs representing six principal indices, NASDAQ100, DJIA, S&P500, DAX30, FTSE100, and CAC40, and the five time scales, 1 s, 10 s, 1 min, 10 min, and 1 h. Figure 4 shows that the return distribution for the three U.S. indices (NASDAQ100, S&P500, and DJIA) does not show an inverse cubic decay. Dow Jones is the closest, but this may be due to the fact that this index has the least number of aggregated stocks (30 vs. 100 and 500). As the time scale Δt increases, we observe a gradual decrease in the thickness of the distribution tails, but this decrease is not so large that a convergence to the normal distribution could firmly be involved. The best power-law fits for $\Delta t = 1$ s are $\alpha \approx 3.9$ (S&P500), 3.8 (NASDAQ100), and 3.6 (DJIA). For a complete record of the fitted power-law and stretched exponential function parameters, see Table 1. For longer time scales, the tails appear to be significantly thinner only for NASDAQ100, and for $\Delta t = 1$ h, they reach $\alpha \approx 4.6$. For DJIA and S&P500, we do not observe any convergence to the normal distribution, and therefore, we assume that there is no such convergence for the scales up to 1 h. A strong discrepancy between the inverse cubic and the empirical distribution is also visible in the case of DAX30. For the returns with $\Delta t = 1$ s, we obtain the power law with $\alpha \approx 3.5$, and for the higher scales, we have a trace of $\alpha \to 4$; however, this is by no means a monotonous increase.

Table 1. Estimated tail exponent α and stretched exponent parameter β for the aggregated distributions of the CFD returns for select stock market indices.

Index	Param.	$\Delta t = 1$ s	$\Delta t = 10$ s	$\Delta t = 1$ min	$\Delta t = 10$ min	$\Delta t = 1$ h
DAX30	α	3.5	3.7	3.9	3.7	2.7
	β	0.37	0.64	0.63	0.45	0.38
CAC40	α	3.6	3.8	3.7	3.6	4.8
	β	0.38	0.62	0.40	0.42	0.63
FTSE100	α	2.8	3.4	3.7	3.5	4.6
	β	0.51	0.39	0.81	0.68	0.52
DJIA	α	3.6	3.7	3.3	3.3	3.0
	β	0.37	0.41	0.68	0.41	0.37
S&P500	α	3.9	3.9	3.6	3.5	3.0
	β	0.4	0.47	0.39	0.56	0.37
NASDAQ100	α	3.8	4.0	3.8	3.6	4.5
	β	0.41	0.44	0.36	0.42	0.47

The return distribution for the FTSE100 and CAC40 indices are different. Especially in the case of the latter, we observe an approximate inverse cubic decay $\alpha \approx 3$ for $\Delta t = 1$ s. It is also clearer than in the previous cases that the tails become much thinner with increasing scale, and for $\Delta t = 1$ h, we see $\alpha \to 5$. In the case of FTSE100, we do not deal with a homogeneous distribution but, rather, with two or more different distributions imposed. This is visible especially for the shortest time scale, where $\alpha < 3$. As the scale increases, we see a behavior similar to that of CAC40, although it is even more pronounced due to the thicker tail at 1 s.

These results can be compared to those obtained for the high-frequency data from 1998–1999, which included both DJIA and DAX30 [24]. The distributions for the shortest scale analyzed ($\Delta t = 5$ min) displayed tails close to those of the inverse cubic ones (even more in the case of DJIA than DAX30), but a crossover was visible for the scales $\Delta t > 2$ h for DJIA and $\Delta t > 30$ min for DAX30. Due to the limited maximum scale considered in the present study, we cannot conclude what the DJIA distributions for the 2 h scale look like, but it seems that, for the shorter scales, these distributions are slightly thinner than before. With regards to the results for the S&P500 and DAX30 data from the years 2004–2006 [32], the tail slope decrease was power-law starting from $\alpha \approx 4$ for $\Delta t = 1$ min to $\alpha \approx 6$ for $\Delta t = 1$ h for the American index and from $\alpha \approx 3.5$ to $\alpha \approx 5$ for the German index, respectively. These results differ from what we obtained here for the years 2017–2020. It seems that the tendency of the inverse cubic scaling regime to shift towards shorter time scales has at least stopped. The distribution tails also scale worse now than before. However, one has to notice that the years 2004–2006 were characterized by much lower volatility than the years 2017–2020, with a lack of comparably significant, dramatic events, which can have some impact on the results.

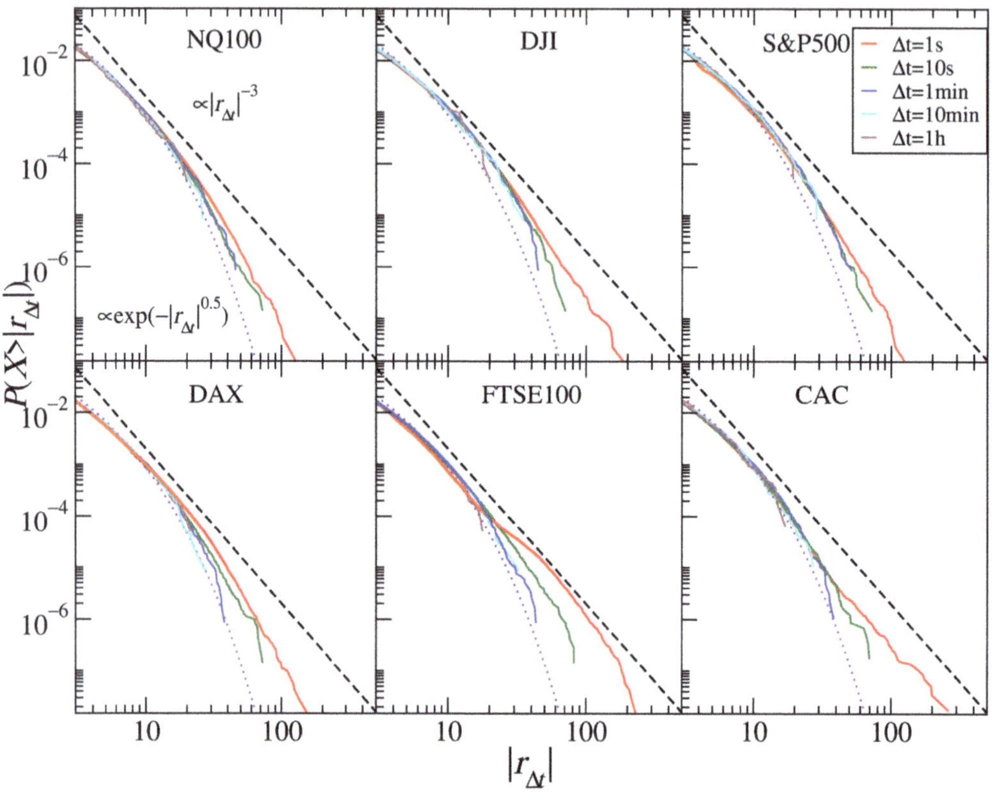

Figure 4. Cumulative distribution functions of the CFD returns for stock market indices NQ100 (NASDAQ), DJIA (New York SE), S&P500 (both New York SE and NASDAQ), DAX30 (Deutsche Börse), FTSE100 (London SE), and CAC40 (Euronext). Different sampling intervals (time scales) are shown from 1 s to 1 h. The inverse cubic scaling $\alpha = 3$ (dashed line) and the stretched exponential with $\beta = 0.5$ (dotted line) are shown in each panel to serve as a guide.

3.2. Individual Stocks

The return distributions for all individual stocks collected from four mature markets: the U.S., German, British, and French ones are shown in Figure 5. For the shortest time scale analyzed, three markets display approximate inverse cubic scaling of their tails: $\alpha \approx 3.2$ (U.K. and France) and $\alpha \approx 3.3$ (Germany), while the U.S. market shows a larger exponent: $\alpha \approx 3.6$ (see Table 2). With increasing Δt, the distribution tail becomes thinner, and already for $\Delta t = 10$ s, the exponent reaches $\approx$3.5 (the European markets) and $\approx$4.0 (the U.S. market). This seems to be the quickest departure from the $\alpha \approx 3$ behavior observed so far for individual stocks. The scaling index increases gradually up to $\Delta t = 10$ min, but for 1 h, this picture is altered and only the U.S. stocks show a further increase ($\alpha \approx 5.0$), while the exponent either stops—Germany—or even decreases—the U.K. and France (for these two longest scales, the stretched exponential function fits the empirical distribution better). This makes the situation less clear, but such a non-monotonous behavior was also observed for some scales in [16] despite a much larger set of stocks considered there (1000). In that study (the years 1994–1995), $\alpha \approx 5.0$ was observed for the returns sampled every 50–70 trading days. Later studies reported that the scaling regime with $\alpha \approx 3$ already broke at $\Delta t = 2$ h for 30 DJIA stocks and at $\Delta t = 5$ min for 30 DAX stocks [24] (1998–1999) and then that $\alpha \approx 3$ was valid up to $\Delta t = 1$ min and $\alpha \approx 5$ was reached for $\Delta t = 2$ h [32] (1000 U.S. stocks, 1998–1999). In fact, even though in Table 2 we do not observe a convincing convergence of the empirical distributions for the European stocks towards the normal distribution, our results show that contemporary stocks experienced an accelerated time flow compared with the past. Of course, since we analyzed the CFD contracts instead of the stock share spot quotations as in [16,24,32], we have to be careful in drawing decisive conclusions from the comparison between these two asset types.

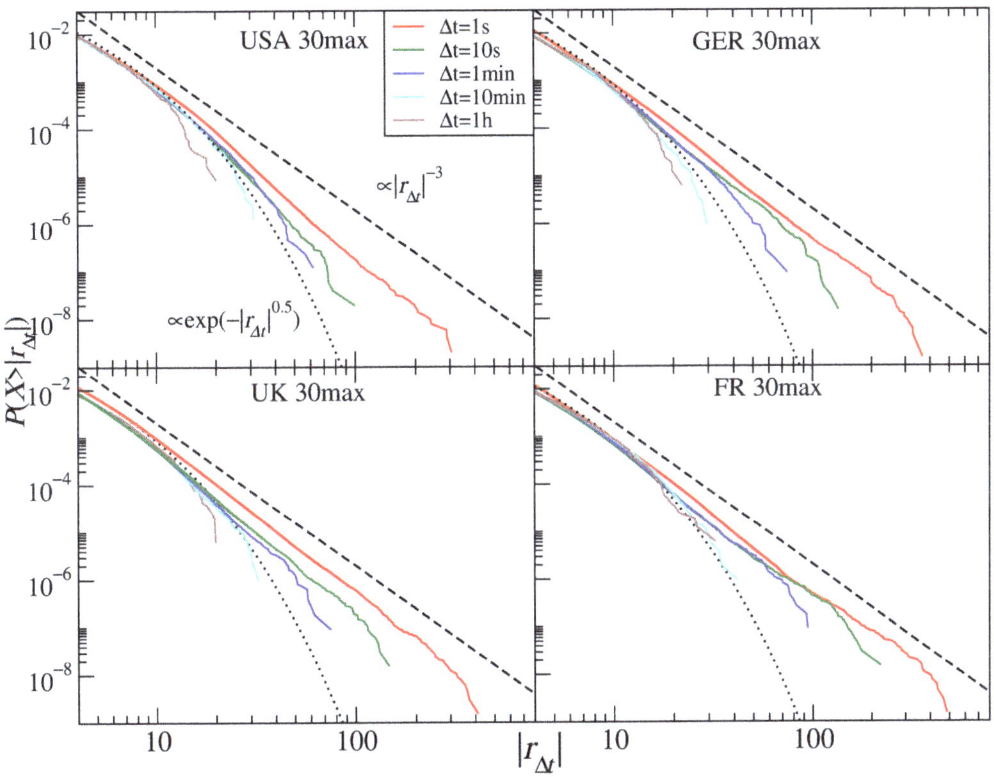

Figure 5. Cumulative distribution functions of the CFD returns for the stock shares representing different markets: the U.S. market (USA), the German market (GER), the U.K. market (UK), and the French market (FR). In each case, the aggregated distributions for 30 stocks with the largest capitalization are shown. Different time scales are shown from 1 s to 1 h. The inverse cubic scaling $\alpha = 3$ (dashed line) and the stretched exponential with $\beta = 0.5$ (dotted line) are shown in each panel to serve as a guide.

The faster convergence of aggregated returns nowadays, compared with a more or less distant past, is among others a consequence of a decreasing autocorrelation time [16,24,32]. On the other hand, somehow, an opposite process is the increase in the cross-correlation magnitude among different stocks, which leads to stronger violation of the CLT assumption about random variable independence and thickening of distribution tails for the stock market indices, which can cause a later crossover to the CLT regime. For the shortest scales available for analysis of the order of an inter-trade interval, the cross-correlations are relatively weak due to strong noise and a longer time needed for information to spread over the market. However, by increasing Δt, we also increase the cross-correlation magnitude, which can eventually reach a saturation level with a magnitude dependent on the stocks considered (the same industrial sector vs. different sectors, whether the stocks are included in the same index, etc.) [34,41,44,124]. If we review the available results on this problem, we can see that, in 1971, the saturation of the cross-correlation coefficient for the stocks of the largest capitalization was reached at $\Delta t \approx 1$ day [41], while it was 1/2 h for 1998–1999 for the largest companies and a few hours for the medium-sized companies [44]. In order to learn how much time is needed for the cross-correlation magnitude to saturate nowadays, we calculated the Pearson cross-correlation coefficients $C_{ij}(\Delta t)$ $(i,j = 1,\ldots,30)$ for all pairs of stocks within each of the markets studied here. Figure 6 shows the results for the mean coefficient $\langle C_{ij} \rangle$ together with a mean length of the zero-return sequences in the analyzed time series and the largest eigenvalue of the 30×30 correlation matrix $\mathbf{C}(\Delta t)$ in which the elements are C_{ij} for a given Δt and a given market. We also added two sets of U.S. stocks that represent medium-sized and small capitalization stocks. For all sets of stocks, a trace of saturation is observed already at the time scales of a few minutes, which is much less than the numbers presented above that from earlier works. This validates our statement that the market time "felt" by the assets accelerates. There is also a clear dependence of the mean cross-correlation coefficient on stock capitalization: the larger the capitalization, the stronger the correlation Figure 6.

Table 2. Estimated tail exponent α for the aggregated distributions of the CFD returns for 30 U.S. stocks with the largest, medium, and small capitalizations and 30 stocks representing selected European markets.

Market	Param.	$\Delta t = 1$ s	$\Delta t = 10$ s	$\Delta t = 1$ min	$\Delta t = 10$ min	$\Delta t = 10$ h
U.S. large	α	3.7	4.0	3.9	4.0	5.0
	β				0.47	0.54
U.S. medium	α	3.7	4.1	3.8	4.0	4.0
	β				0.41	0.48
U.S. small	α	3.5	3.7	3.8	3.9	4.5
	β				0.46	0.56
Germany	α	3.3	3.6	3.8	4.2	4.2
	β				0.49	0.56
U.K.	α	3.2	3.5	3.9	4.2	3.6
	β				0.51	0.48
France	α	3.2	3.4	3.5	3.7	3.5
	β				0.50	0.44

It is well-known that the cross-correlations are not stationary and that they strongly fluctuate across time [34,125,126]. Figure 7 displays the evolution of $\langle C_{ij}(t) \rangle$ calculated in 30-day windows over the years 2018–2020. Two time scales are considered: 1 s and 1 h. $\langle C_{ij}(t) \rangle$ fluctuates with a larger amplitude for $\Delta t = 1$ h than for $\Delta t = 1$ s. One of the periods associated with the largest values of $\langle C_{ij}(t) \rangle$ is 9–27 March 2020 (the COVID-19 pandemic outburst in the U.S.), when the markets underwent strong turbulence [39]. As the cross-correlations were particularly strong during that period, we suspect that it could contribute substantially to the tail shape of the stock return distributions.

To verify this hypothesis, we removed this period from the time series and constructed artificial stock indices by aggregating the returns for all stocks belonging to the same set. Figure 8 shows both the complete distributions and the resultant no-COVID ones. After removing the COVID-19 outburst period, the distribution tails became substantially thinner, which is particularly evident for $\Delta t = 1$ h (see Table 3). This supports our hypothesis that strong cross-correlations among the stocks can prevent stock indices from showing CLT convergence for short time scales. In this case, the stretched exponential function fits the empirical distribution better than the power-law function (Table 3). The numbers in this table illustrate how the stock-stock correlation strength can influence the stock index returns. While the stretching parameter β is comparable for each group of the U.S. stocks at $\Delta t = 1$ s, it becomes significantly different at $\Delta t = 1$ h, where the medium and small companies have thinner tails than the large companies. This is because the former are less cross-correlated than the latter and the distributions can more easily converge towards a Gaussian in this case, even though the medium and small companies should experience a slower time flow than the large ones, which acts towards tail thickening. From this example, we can see that both effects compete against each other and that the actual tail behavior depends on the interplay of both factors.

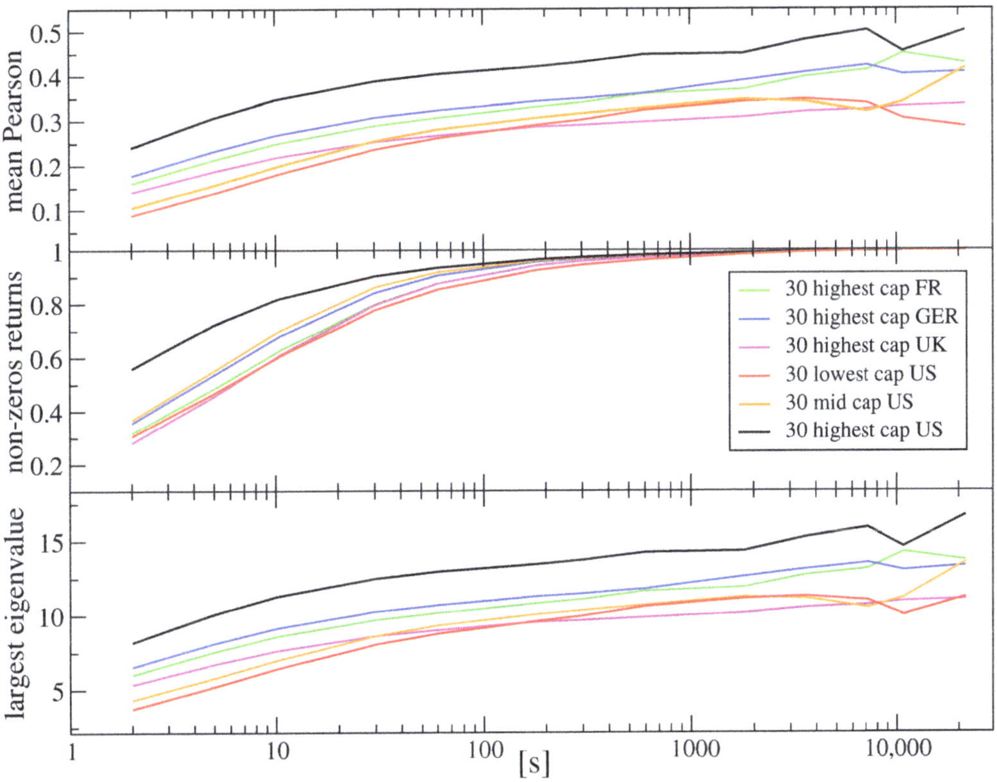

Figure 6. (**Top**) The mean Pearson cross-correlation coefficient $\langle C_{ij}\rangle(\Delta t)$ for the CFD returns as a function of time scale Δt for 30 companies, with the largest capitalization representing four stock markets, French (FR), German (GER), British (UK), and American (US), and for 30 companies with medium and small capitalization from the American market. Averaging was carried out over all pairs i, j with $i > j$ and $i, j = 1, \ldots, 30$. (**Middle**) The same was performed as above, but here, the zero returns were filtered out before calculating the correlation coefficients. (**Bottom**) The largest eigenvalue of the correlation matrix $\mathbf{C}(s)$ constructed from the Pearson cross-correlation coefficients $C_{ij}(s)$ for the same sets of stock share CFDs.

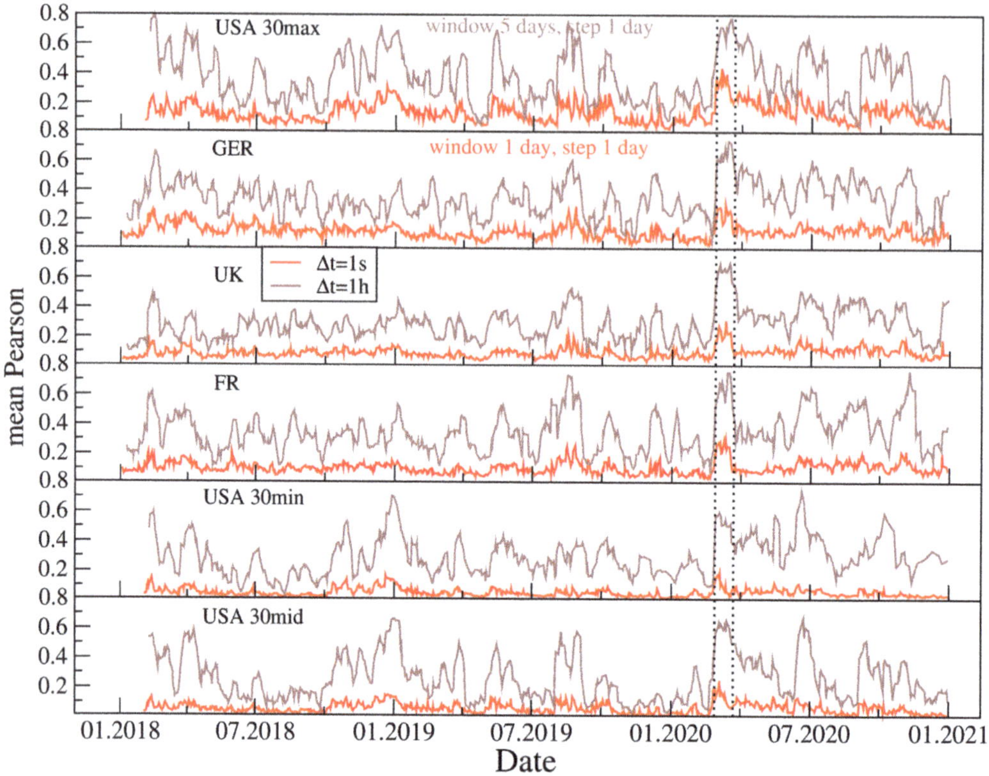

Figure 7. Evolution of the mean Pearson cross-correlation coefficient $\langle C_{ij}\rangle(\Delta t)(t)$ for the CFD returns of 30 companies, with the largest capitalization representing four stock markets, French (FR), German (GER), British (UK), and American (US), and for 30 companies with medium and small capitalization from the American market. The coefficient was calculated in a moving window of length of 30 days, and averaging was carried out over all pairs i, j with $i > j$ and $i, j = 1, \ldots, 30$. Two time scales with $\Delta t = 1$ s and $\Delta t = 1$ h are shown in each case.

Table 3. Estimated tail exponent α and stretched exponent parameter β for the aggregated distributions of the CFDs returns for 30 artificial stock indices representing different markets and different stock capitalization groups from the U.S. market. The results for the complete data and the data without the COVID-19 outburst in March 2020 (denoted nC) are shown for comparison.

Market	# Stocks	Param.	$\Delta t = 1$ s	$\Delta t = 1$ s (nC)	$\Delta t = 1$ h	$\Delta t = 1$ h (nC)
U.S. large	30	α	5.5	5.1	3.0	5.7
		β	0.51	0.49	0.49	0.56
U.S. mid	30	α	4.7	4.4	2.7	6.6
		β	0.43	0.41	0.48	0.72
U.S. small	30	α	4.4	4.4	5.2	5.5
		β	0.43	0.41	0.74	0.76
Germany	30	α	3.7	3.8	2.3	3.9
		β	0.45	0.43	0.32	0.50
U.K.	30	α	4.0	4.6	2.7	4.7
		β	0.41	0.42	0.46	0.66
France	30	α	3.9	4.0	2.7	6.2
		β	0.41	0.42	0.41	0.74

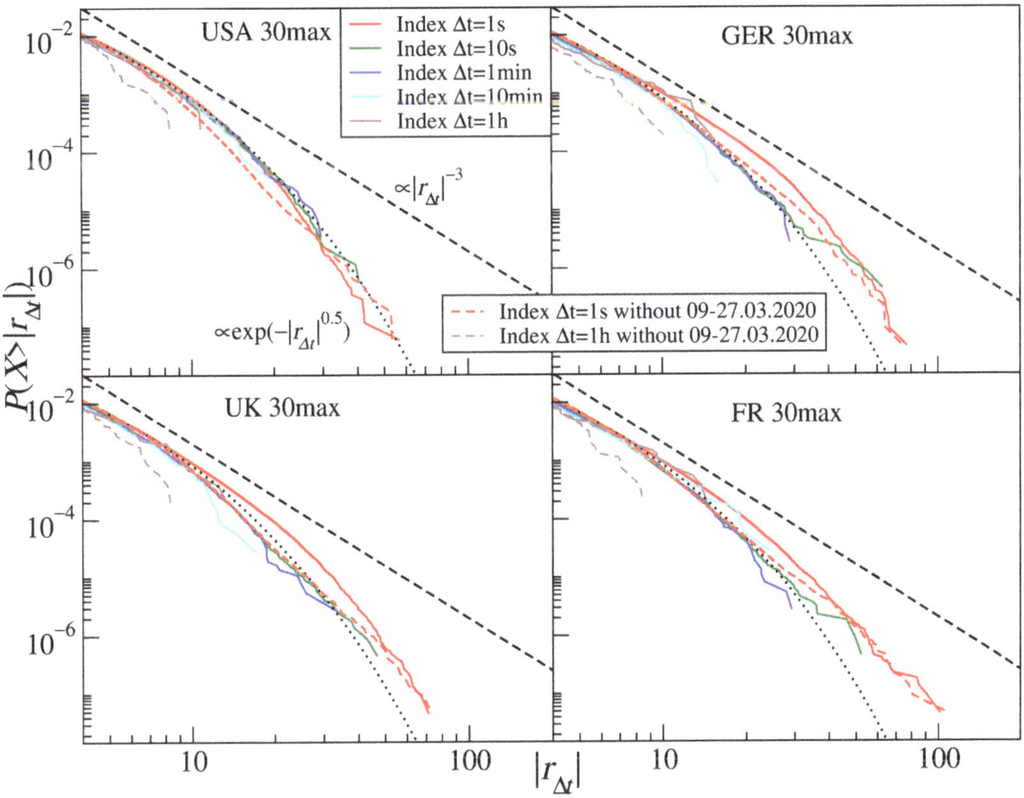

Figure 8. Cumulative distribution functions of the returns of an artificial index constructed as a sum of the stock share CFD quotes $I(t) = \sum_i P_i(t)$ for the 30 largest companies representing four stock markets, the U.S. market (USA), the German market (GER), the U.K. market (UK), and the French market (FR). Two time scales with $\Delta t = 1$ s and $\Delta t = 1$ h are shown and denoted by solid lines. In addition, the analogous distributions constructed from the CFD return time series after removing the COVID-19 outburst period corresponding to the strongest cross-correlations among the stock shares (9–27 March 2020) are denoted by dashed lines. The inverse cubic scaling $\alpha = 3$ (dashed line) and the stretched exponential with $\beta = 0.5$ (dotted line) are shown in each panel to serve as a guide to the eye.

3.3. Currencies

Unlike the stock indices, the return distributions for the exchange rates show the presence of increasingly thinner tails if the time scales increase, and hence, a faster convergence towards the normal distribution (Figure 9). If fitted by a power function, the differences between the individual exchange rates are smaller than those of the indices and generally exhibit an inverse cubic decay for smaller Δts: for $\Delta t = 1$ s, they vary from $\alpha \approx 3.0$ (USD/JPY, GBP/USD) to $\alpha \approx 3.4$ (EUR/USD) and, for $\Delta t = 1$ h, from $\alpha \approx 4.2$ (EUR/JPY) to $\alpha \approx 5.5$ (GBP/CHF) with the exception of GBP/JPY ($\alpha \approx 2.8$). The numbers are collected in Table 4. The scaling exponent $\alpha \approx 3$ was observed in many studies of the Forex data, including References [39,116,124,127,128]. If the stretched exponential function is used, the best-fitted parameter β reads for $\Delta t = 1$ s from $\beta = 0.37$ (EUR/JPY) to $\beta = 0.48$ (GBP/JPY, USD/CHF) and, for $\Delta t = 1$ h, from $\beta = 0.49$ (EUR/USD, USD/JPY) to $\beta = 0.87$ (GBP/USD). The mean values of the scaling exponent for the analyzed time scales are $\bar{\alpha} = 3.2$ (1 s), $\bar{\alpha} = 3.1$ (10 s), $\bar{\alpha} = 3.2$ (1 min), $\bar{\alpha} = 3.7$ (10 min), and $\bar{\alpha} = 5.0$ (1 h). The inverse cubic scaling can therefore now be identified for scales shorter than 10 min. These scale are longer than those in the years 2004–2008 for the 1-min scale $\bar{\alpha} = 3.9$ [124]. (It has to be noted, however, that those earlier results were obtained by fitting the q-Gaussian functions instead of the power-law functions, which may make it difficult to compare the results properly even though the relation given by Equation (8) holds). We thus observe a slower convergence to the normal distribution now than before for $\Delta t = 1$ h: $\bar{\alpha} = 5.9$ (2004–2008) vs. $\bar{\alpha} = 4.8$ (2017–2020). However, if the obtained results are compared with those from the study [116] for the years 1987–1993, the acceleration becomes visible: $\bar{\alpha} = 3.9$ (1987–1993) vs. $\bar{\alpha} = 4.8$ (2017–2020). In that case, the inverse cubic scaling was still visible for the 30-min scale, which is much longer than both in 2004–2008 and 2017–2020. This can be interpreted as the acceleration of the market time resulting in a faster convergence to the normal distribution in the 2000s, but later, this phenomenon of the effective time scale shortening disappeared.

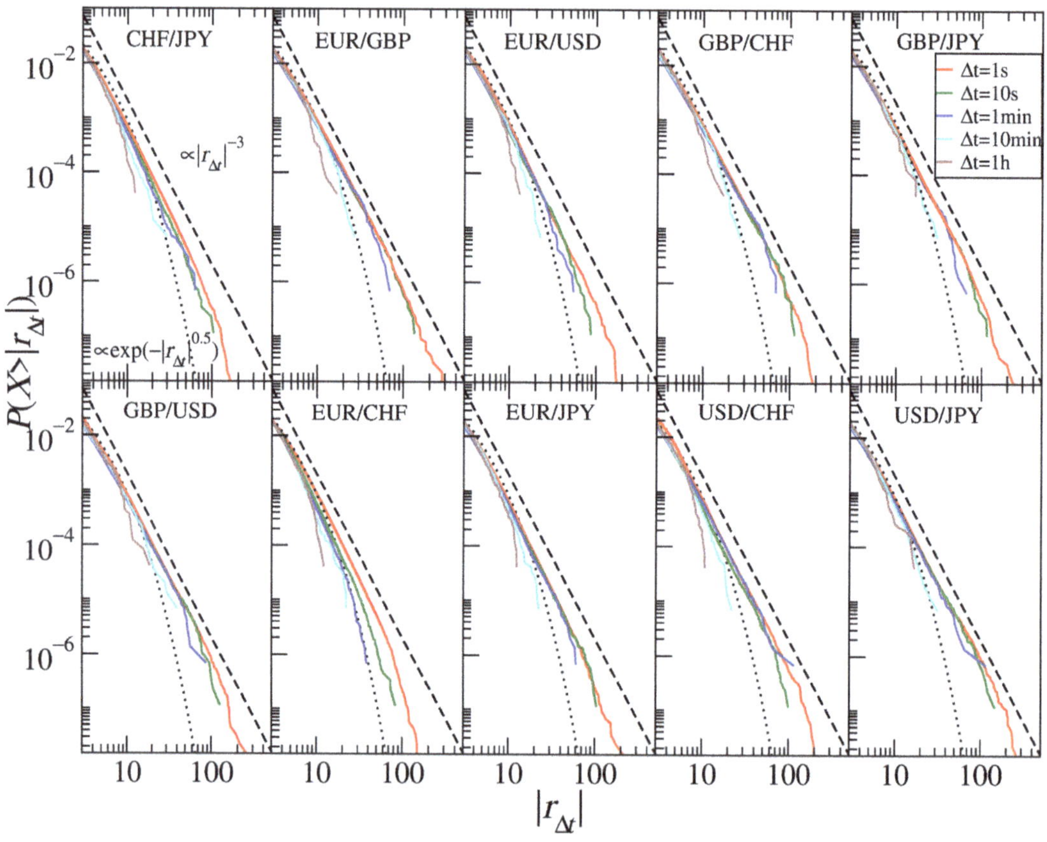

Figure 9. Cumulative distribution functions of the major currency exchange rate returns: Swiss franc (CHF), euro (EUR), British pound (GBP), Japanese yen (JPY), and the U.S. dollar (USD). Different time scales are shown from 1 s to 1 h. The inverse cubic scaling $\alpha = 3$ (dashed line) and the stretched exponential with $\beta = 0.5$ (dotted line) are shown in each panel to serve as a guide.

Table 4. Estimated tail exponent α and stretched exponent parameter β for the aggregated return distributions for the selected currency exchange rates and the cryptocurrency prices: BTC/USD and ETH/USD.

Exchange Rate	Param.	$\Delta t = 1$ s	$\Delta t = 10$ s	$\Delta t = 1$ min	$\Delta t = 10$ min	$\Delta t = 1$ h
CHF/JPY	α	3.2	3.3	3.4	3.6	5.2
	β	0.48	0.37	0.51	0.41	0.61
EUR/CHF	α	3.3	3.6	4.0	3.9	5.2
	β	0.40	0.39	0.41	0.40	0.78
EUR/GBP	α	3.1	2.9	2.9	3.6	5.1
	β	0.42	0.34	0.33	0.48	0.75
EUR/JPY	α	3.3	3.1	3.1	4.0	4.2
	β	0.37	0.29	0.39	0.64	0.58
EUR/USD	α	3.4	3.2	3.4	4.4	5.2
	β	0.39	0.35	0.37	0.55	0.65
GBP/CHF	α	3.1	2.9	2.9	3.5	5.5
	β	0.40	0.36	0.40	0.42	0.50
GBP/JPY	α	3.1	2.9	3.0	3.4	2.8
	β	0.48	0.34	0.36	0.54	0.37
GBP/USD	α	3.0	2.9	2.9	3.3	5.1
	β	0.40	0.33	0.34	0.39	0.87
USD/CHF	α	3.2	3.1	3.3	4.1	5.2
	β	0.45	0.52	0.38	0.48	0.62
USD/JPY	α	3.0	2.9	3.0	3.5	5.2
	β	0.40	0.34	0.36	0.56	0.54
BTC/USD	α	2.9	3.1	3.2	3.2	3.7
	β					
ETH/USD	α	2.8	3.1	3.2	3.3	4.3
	β					0.50

3.4. Cryptocurrencies

The cryptocurrency market is strongly related to Forex and its significance has risen steadily since its beginning [38,39]. The most important assets traded on this market in terms of their capitalization and volume are bitcoin (BTC) and ethereum (ETH). Their return distributions are shown in Figure 10. For Δt up to 10 min, the power-law function approximates the data well, with the tail exponent displaying the same inverse cubic scaling for BTC and ETH: $\alpha \approx 2.8$ (1 s), $\alpha \approx 3.1$ (10 s), $\alpha \approx 3.2$ (1 min), $\alpha \approx 3.3$ (10 min). In contrast, for $\Delta t = 1$ h, the crossover is observed and the tail exponent rises to $\alpha \approx 3.7$ for BTC and to 4.2 for ETH. (The stretched exponential function does not fit the data in the tail region, except on the 1 h scale for ETH) A good agreement between the empirical distributions of the cryptocurrency price returns expressed in major regular currencies and the inverse cubic scaling paradigm has already been reported [38–40,128], and it was interpreted as a sign of maturation (it used to be even more heavy-tailed with $\alpha \approx 2.2$ before 2014 [38,73]). A crossover for the same scale $\Delta t = 1$ h has also been reported [128].

A more time-resolution-oriented analysis [129] showed that the BTC dynamics can actually be compounded with alternating phases of fluctuations with different statistical properties. For example, during the COVID-19 outburst in March 2020, the BTC returns were characterized by $\alpha \approx 1.8$, which corresponds to the Lévy stability, but typically, the scaling index resided between 2.0 and 3.5 during the years 2019–2020 [129]. In contrast, the Lévy-stable distribution of daily returns with $\alpha \approx 1.3$ was reported to fit the BTC empirical data (the years 2011–2017) in [130], but there, the model was fitted to the whole distributions, not only the tails, and the analyzed period also covered the early years of the cryptocurrency market when it was immature. These may be a source of the discrepancy mentioned above.

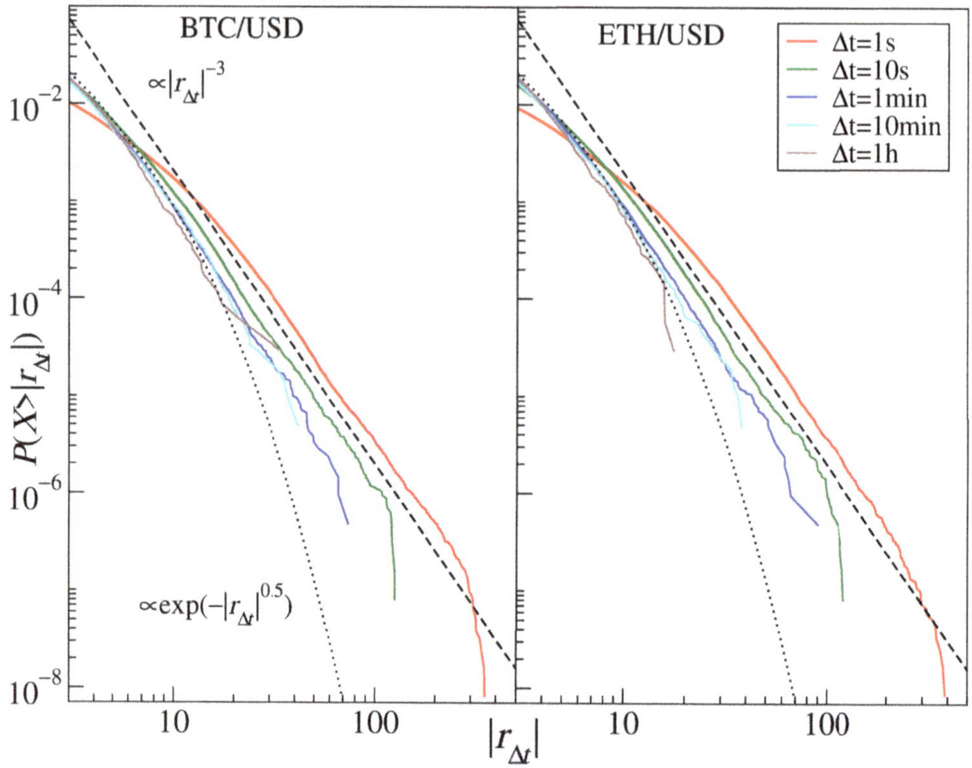

Figure 10. Cumulative distribution functions of the bitcoin-dollar exchange rate returns (BTC/USD) and the ethereum-dollar exchange rate returns (ETH/USD). Different time scales are shown from 1 s to 1 h. The inverse cubic scaling $\alpha = 3$ (dashed line) and the stretched exponential with $\beta = 0.5$ (dotted line) are shown in each panel to serve as a guide.

3.5. Commodities

Figure 11 shows the return distributions for the commodity CFDs (see also Table 5). Out of the commodities considered in our study, gold (XAU) has the strongest decay with increasing Δt: $\alpha \approx 2.6$ (1 s), $\alpha \approx 3.1$ (10 s), $\alpha \approx 3.6$ (1 min), $\alpha \approx 3.6$ (10 min), and $\alpha \approx 4.2$ (1 h). This has to be compared to $\alpha \approx 2.5$ for daily returns covering the years 1969-1999 [37] and 5-min returns covering the years 2012–2018 [36]. It is instructive to address why the tail for $\Delta t = 1$ s is so thick that α falls significantly below 3. We therefore identified a period of the largest gold price fluctuations, which occurred at the COVID-19 outburst in the U.S. during March and April, and removed it from the time series. The resulting distributions are shown in Figure 12 (left panel) for $\Delta t = 1$ s, 1 min, and 1 h (dashed lines) together with the complete ones (solid lines). It is evident that, for $\Delta t = 1$ s, the thickest part of the tail becomes thinner $\alpha \approx 3.2$, while no significant alternation is observed for $\Delta t = 1$ min ($\alpha \approx 3.6$) and 1 h ($\alpha \approx 4.2$). The distribution tails roughly agree in this case with the inverse cubic scaling on time scales that become increasingly short with time.

Compared to gold, silver (XAG) shows stronger invariance under the time scale change: α increases from 2.8 (1 s) to 3.0 (1 h), while it was 2.5 (positive tail) and 2.8 (negative tail) for the daily returns over the years 1969–1999 [37]. High-grade copper return distributions (HG) reveal the most wandering behavior: its scaling exponent goes from $\alpha \approx 3.7$ (1 s) through $\alpha \approx 3.6$ (10 s), $\alpha \approx 3.1$ (1 min), and $\alpha \approx 3.9$ (10 min) to $\alpha \approx 2.7$ (1 h) compared to $\alpha \approx 2.6$ (negative tail; daily data) and $\alpha \approx 2.8$ (positive tail) for the years 1971–1999 [37].

Table 5. Estimated tail exponent α and stretched exponent parameter β for the aggregated distributions of the CFDs returns for the selected commodities: high-grade copper (HG), crude oil (CL), silver (XAG), and gold (XAU).

Commodity	Param.	$\Delta t = 1$ s	$\Delta t = 10$ s	$\Delta t = 1$ min	$\Delta t = 10$ min	$\Delta t = 1$ h
HG	α	3.8	3.6	3.1	3.9	2.7
	β	0.42	0.49	0.36	0.45	0.25
CL	α	2.6	2.3	2.3	2.0	2.5
	β	0.48	0.29	0.28	0.41	0.43
XAG	α	2.8	2.9	2.9	3.0	3.0
	β	0.47	0.36	0.37	0.38	0.43
XAU	α	2.6	3.1	3.6	3.6	4.2
	β	0.49	0.62	0.42	0.51	0.85

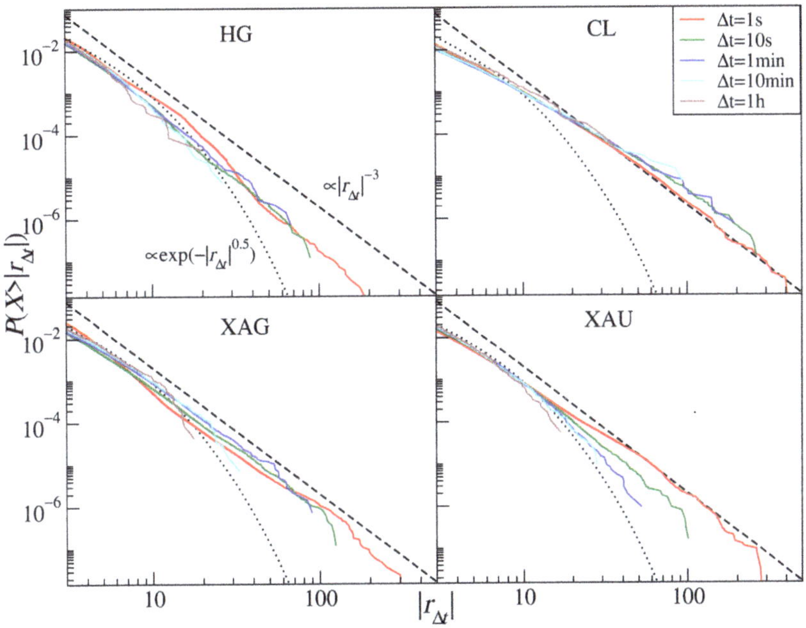

Figure 11. Cumulative distribution functions of the CFD returns for commodities: high-grade copper (HG), the U.S. crude oil (CL), silver (XAG), and gold (XAU). Different time scales are shown from 1 s to 1 h. The inverse cubic scaling $\alpha = 3$ (dashed line) and the stretched exponential with $\beta = 0.5$ (dotted line) are shown in each panel to serve as a guide.

Crude oil return distributions (CL) have the thickest tails with the scaling exponent $2.0 \leq \alpha \leq 2.6$ and with no signature of the CLT convergence. In parallel to what was observed for gold, we removed the COVID-19 outburst period from the time series and calculated the distributions again—see Figure 12 (right panel). Such incomplete signals are characterized by $\alpha = 3.0$ for $\Delta t = 1$ s and $\alpha \approx 2.7$ for $\Delta t = 1$ h. These numbers have to be compared with $\alpha \approx 2.9$ (negative tail) and $\alpha \approx 3.1$ (positive tail) reported for the WTI oil daily returns covering the years 1988–1998, with $\alpha \approx 2.0$ (negative tail) and $\alpha \approx 2.8$ (positive tail) reported for the crude oil daily returns covering the years 1983–1999 [37], and with $\alpha \approx 3.0$ (negative tail) and $\alpha \approx 3.1$ (positive tail) for the WTI oil 5-min returns covering the years 2012–2018 [36]. As those values do not differ much from each other, there is no evidence that the crude oil returns change their global dynamics over time. However, during the market turbulence that happened during the COVID-19 pandemic, the dynamics did change considerably, which was manifested by thickening the distribution tail for all considered time scales. It is noteworthy that the crude oil was the asset that was affected the strongest by the pandemic: in April 2020, the price of the May series of WTI oil futures even dropped below 0.

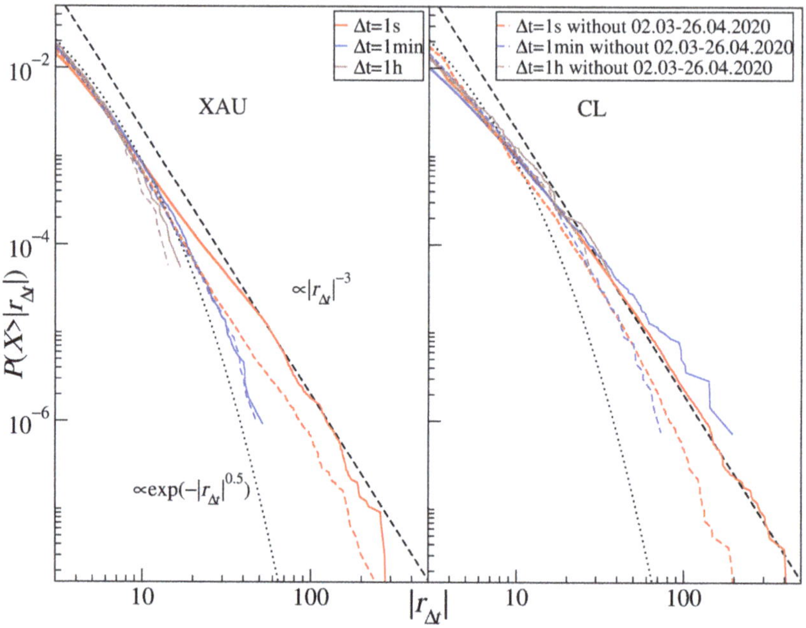

Figure 12. Cumulative distribution functions of the CFD returns for gold (XAU) and the U.S. crude oil (CL) after removing the COVID-19 outburst period (March–April 2020). Different time scales are shown from 1 s to 1 h. The inverse cubic scaling $\alpha = 3$ (dashed line) and the stretched exponential with $\beta = 0.5$ (dotted line) are shown in each panel to serve as a guide.

4. Summary

In this study, we analyzed high-frequency quotations of the CFD contracts associated with the stock market indices, the stocks themselves, and the selected commodities as well as with the most frequently traded currency exchange rates and the cryptocurrency prices. All of the data sets covered the years 2017–2020 except for the stock share CFDs, which covered the years 2018–2020. We analyzed the returns at a few different time scales from 1 s to 1 h and constructed the return distributions in order to investigate their tails. Our principal objective was to compare the tail behavior of the distributions derived from contemporary data with the behavior of the distribution tails in the past for the same assets. We applied the power-law function and the stretched exponential one to model the empirical distributions. A hypothesis that we planned to verify was the one formulated in [24,32,34], which states that, together with the acceleration of the information flow and processing across the financial markets, we can observe a significant change in the statistical properties of the returns at a particular time scale related to an effective acceleration of the market time with all of the possible consequences of this fact.

The results are mixed. On the one hand, the stock market indices (DJIA, DAX30, and S&P500, for which the present results can be compared directly with earlier works) do not show any further signatures of the time acceleration compared with the data from 1998–1999 and 2004–2006. It seems that the acceleration that was reported in [24,32] stopped or was only a temporary effect. Such effects were already reported before for Asian markets [35,69] as well as in this work regarding the stocks, so they may be a source of the observed behavior. On the other hand, the results for the individual stock groups show that the market time acceleration can still be ongoing, but it is masked at the level of indices owing to the cross-correlations among the stocks that are now stronger and developing faster than even during the years 2004–2006 [32]. That particular time interval (2004–2006) was characterized by a volatility much smaller than in recent years, which witnessed large market events such as the flash crash on 5 February 2018, the coronavirus-related unsteadiness in early 2020 and the subsequent rally ending with new record highs of S&P500 in August, the oil price drop in April 2020, etc. Large events, especially large falls, elevate the market correlation level, which can influence the statistical properties of data, including the distribution of returns. The auto- and cross-correlations are involved in an interesting interplay between two opposite-acting factors. The first factor is the market time flow speed, which works for market efficiency by shortening the period when the market autocorrelations are admissible. This factor shifts gradually the low-α behavior and the central limit theorem's realm to ever shorter time scales. The second factor is the asset cross-correlation strength, which causes thickening of the tails and decreases in α and β. It also violates the assumption of random variable independence and prevents the CLT from affecting the aggregated returns. This interplay and its consequences are interesting enough to be worthy of some more attention in future analyses. In particular, they can be responsible for the reported behavior of the return distributions in different time periods and suppressing the effects of the market time acceleration.

Currency exchange rates also no longer feel the market time acceleration such as that during 2004–2006 [32], but now, not only is there no further time scale shortening but also a moderate step backwards is observed: the inverse cubic scaling is seen at longer time scales than in 2004–2006 but is still significantly shorter than that during the years 1987–1993 [116]. The cryptocurrencies (BTC and ETH) show the same crossover scale as before—equal to 1 h [131]. Since this market is relatively young, it underwent a phase of strong market time acceleration after 2013, and now, it seems to be stabilized. It is still the market that shows the most exemplary inverse cubic scaling behavior across different scales out of all the markets analyzed in this work. Gold price CFDs show a clear difference between the present results and the distribution tails over the years 1969–1999 [37] and 2012–2018 [36] with increased tail slope during the recent years. In contrast, there is no clear change in the tail slope regarding silver, high-grade copper, and crude oil.

It should be noted, however, that the CFD contract price quotations analyzed here are not precisely the same as the related asset spot price quotations, which the authors of other works dealt with. This difference may partially account for the difference in the outcomes. Finally, the COVID-19 pandemic outburst that took place in March–April 2020 in the U.S. constituted a strong perturbation to all the markets, caused large-amplitude price fluctuations, and led to a strong increase in the cross-correlations among many assets. For example, it resulted in decreasing distribution tail slopes for the CFD returns for crude oil and gold. Even more significant were the bitcoin fluctuations, which become Lévy stable for the pandemic-outburst period.

In general, our results indicate that the monotonous shift in the time scales at which different types of dynamics can be observed in the financial data as well as the related continuously accelerating market time from past to present are oversimplified. In fact, there can be an underlying long-term trend of this type, but it is "decorated" with short-term phases of abrupt acceleration and, then, deceleration and stagnation. Our results indicate that the real market dynamics consists of continuous alternation of different regimes with different statistical properties that can form the overall impression of the market evolution direction. Together with the aforementioned problem of how the asset cross-correlations and the shortening autocorrelations compete against each other in shaping the statistical properties of data, it opens an intriguing direction for future work.

Author Contributions: Conceptualization, S.D., J.K. and M.W.; methodology, S.D., J.K. and M.W.; software, M.W.; validation, S.D., J.K. and M.W.; formal analysis, S.D., J.K. and M.W.; investigation, S.D., J.K. and M.W.; resources, M.W.; data curation, M.W.; writing—original draft preparation, J.K.; writing—review and editing, J.K. and M.W.; visualization, M.W.; supervision, S.D. All authors have read and agreed to the published version of the manuscript.

Funding: This research received no external funding.

Institutional Review Board Statement: Not applicable.

Informed Consent Statement: Not applicable.

Data Availability Statement: Publicly available datasets were analyzed in this study. This data can be found at [122,123].

Conflicts of Interest: The authors declare no conflict of interest.

Appendix A

Table A1. List of stocks that are included in the stock sets considered in Section 3.2. Capitalization is given in billions (10^9) of currency units.

U.S. Large		U.S. Medium		U.S. Small		U.K. Large		German Large		French Large	
Ticker	USD	Ticker	USD	Ticker	USD	Ticker	GBP	Ticker	EUR	Ticker	EUR
AAPL	2000	BIDU	68.6	CAG	18.7	RDSB	107.4	VOW3	140.6	VK	330.9
MSFT	1784	NSC	68.1	MGM	18.4	ULVR	105.7	SAP	126	MC	288
AMZN	1537	CL	67.7	CAH	18.3	AZN	94.2	SIE	112.5	OR	181.3
GOOGL	1378	SHW	67	CTL	18.1	RIO	88.7	ALV	89.5	SAN	105.2
FB	805	SO	65.9	ULTA	17.6	HSBA	86.4	DTE	81.8	FP	102.8
BABA	625	APD	63.1	OMC	16.3	DGE	70.4	DAI	80.6	AIR	78.9
BRKB	593	ICE	62.6	TIF	16	BATS	62.3	BAS	65.3	KER	74.8
TSM	592.8	D	60.9	DVN	14.8	BP	59	MRK	62.3	SU	73.8
TSLA	582	ADSK	59.5	AAL	14.7	RB	46.3	DPW	57.6	AI	65.9
JPM	473	ADI	56.6	WYNN	14.5	BLT	44.2	BMW	57.3	BNP	65.1
V	458	ILMN	56.2	WHR	14.1	PRU	40.6	VNA	56.7	CS	54.9
JNJ	438.5	PGR	55.9	L	13.8	AAL	39.5	ADS	52.8	SAF	51
WMT	383.5	VRTX	55.8	SJM	13.7	GLEN	38.1	BAYN	52.3	DG	50.8
MA	360	BSX	54.8	TEVA	12.7	VOD	37.7	IFX	47.6	RI	42.1
UNH	358.2	EMR	54.7	IPG	11.5	REL	35.5	HEN3	38.5	BN	37.8
DIS	337	NOC	53.7	DVA	11.5	LSE	35.3	MUV2	37	ACA	36.3
PG	336	HUM	53.3	NWL	11.4	BARC	31.7	PAH3	28.7	EDF	35.4
BAC	331	MET	53.1	GPS	11.3	NG	30.7	DB1	26	VIV	30.4
HD	324	PBR	53	TAP	11.3	LLOY	30.3	EOAN	26	ENGI	29.3
NVDA	318	REGN	51	CF	9.7	CPG	26.7	RWE	23.2	ORA	27.8
PYPL	277	TWTR	50	NRG	9.2	CRH	26.3	CON	22.8	SGO	26.8
INTC	261	KHC	49.9	KSS	9.1	EXPN	23.4	DBK	21.2	CAP	24.9
CMCSA	252.6	NEM	49.8	MRO	8.7	RBS	22.4	FRE	20.9	LR	21.3
XOM	243	F	49	X	6.9	AHT	20.1	BEI	20.5	UG	19.4
VZ	243	DG	48.6	MAT	6.9	WOS	20.1	FME	18.4	GLE	19.2
KO	231.6	ITUB	48.4	APA	6.8	ABF	19.4	HEI	15.3	ALO	16.2
NFLX	228	MAR	48	AA	6.1	CCL	18.1	TKA	7.1	EN	13
ADBE	224	FCX	47.7	JWN	6	TSCO	17.6	CBK	6.6	PUB	12.7
CSCO	222	KMB	47	M	5.1	LGEN	16.9	LHA	6.6	VIE	12.6
T	217.9	LVS	46.8	EQT	5.1	ANTO	16.7	LXS	5.5	CA	12.3

References

1. Bachelier, L. Théorie de spéculation. *Ann. Sci. l'Ecole Norm. Supér.* **1900**, 3, 21–86. [CrossRef]
2. Lévy, P. *Calcul des Probabilités*; Gauthier-Villars: Paris, France, 1925.
3. Paul, W.; Baschnagel, J. *Stochastic Processes: From Physics to Finance*; Springer: Berlin/Heidelberg, Germany, 1999.

4. Mandelbrot, B.B. The variation of certain speculative prices. *J. Bus.* **1963**, *36*, 394–419. [CrossRef]
5. Fama, E.F. The behavior of stock-market prices. *J. Bus.* **1965**, *38*, 404–419. [CrossRef]
6. Blume, M.E. Portfolio theory: A step towards its practical application. *J. Bus.* **1970**, *43*, 152–173. [CrossRef]
7. Teichmoeller, J. A note on the distribution of stock price changes. *J. Am. Stat. Assoc.* **1970**, *66*, 282–284. [CrossRef]
8. Blattberg, R.C.; Gonedes, N.J. A comparison of the stable and Student distributions as statistical models for stock prices. *J. Bus.* **1974**, *47*, 245–280. [CrossRef]
9. Officer, R.R. The distribution of stock returns. *J. Am. Stat. Assoc.* **1972**, *67*, 807–812. [CrossRef]
10. Barnea, A.; Downes, D.H. A reexamination of the empirical distribution of stock price changes. *J. Am. Stat. Assoc.* **1973**, *68*, 348–350. [CrossRef]
11. Young, M.S; Graff, R.A. Real estate is not normal: A fresh look at real estate return distributions. *J. Real Estate Financ. Econ.* 1995, *10*, 225–259. [CrossRef]
12. Clark P.K. A subordinated stochastic process model with finite variance for speculative prices. *Econometrica* **1973**, *41*, 135–155. [CrossRef]
13. Engle, R.F. Autoregressive conditional heteroscedasticity with estimates of the variance of United Kingdom inflation. *Econometrica* **1982**, *50*, 987–1007. [CrossRef]
14. Mantegna, R.N.; Stanley, H.E. Scaling behaviour in the dynamics of an economic index. *Nature* **1995**, *376*, 46–49. [CrossRef]
15. Couto Miranda, L.; Riera, R. Truncated Lévy walks and an emerging market economic index. *Phys. A Stat. Mech. Its Appl.* **2003**, *297*, 509–520. [CrossRef]
16. Plerou, V.; Gopikrishnan, P.; Amaral, L.A.N.; Meyer, M.; Stanley, H.E. Scaling of the distribution of price fluctuations of individual companies. *Phys. Rev. E* **1999**, *60*, 6519–6529. [CrossRef] [PubMed]
17. Gopikrishnan, P.; Plerou, V.; Amaral, L.A.N.; Meyer, M.; Stanley, H.E. Scaling of the distribution of fluctuations of financial market indices. *Phys. Rev. E* **1999**, *60*, 5305–5316. [CrossRef] [PubMed]
18. Gopikrishnan, P.; Meyer, M.; Amaral, L.A.N.; Stanley, H.E. Inverse cubic law for the distribution of stock price variations. *Eur. Phys. J. B* **1998**, *3*, 139–140. [CrossRef]
19. Lux, T. The stable Paretian hypothesis and the frequency of large returns: An examination of major German stocks. *Appl. Financ. Econ.* **1996**, *6*, 463–475. [CrossRef]
20. Makowiec, D.; Gnaciński, P. Fluctuations of WIG-the index of Warsaw Stock Exchange. Preliminary studies. *Acta Phys. Pol. B* **2001**, *32*, 1487–1500.
21. Drożdż, S.; Kwapień, J.; Grümmer, F.; Ruf, F.; Speth, J. Quantifying the dynamics of financial correlations. *Phys. A Stat. Mech. Its Appl.* **2001**, *299*, 144–153. [CrossRef]
22. Lillo, F.; Bonanno, G.; Mantegna, R.N. Variety of stock returns in normal and extreme market days: The August 1998 crisis. In *Empirical Science of Financial Fluctuations*; Takayasu, H., Ed.; Springer: Berlin/Heidelberg, Germany, 2002; pp. 77–89.
23. Kaizoji, T.; Bornholdt, S.; Fujiwara, Y. Dynamics of price and trading volume in a spin model of stock markets with heterogeneous agents. *Phys. A Stat. Mech. Its Appl.* **2002**, *316*, 441–452. [CrossRef]
24. Drożdż, S.; Kwapień, J.; Grümmer, F.; Ruf, F.; Speth, J. Are the contemporary financial fluctuations sooner converging to normal? *Acta Phys. Pol. B* **2003**, *34*, 4293–4306.
25. Kim, K.; Yoon, S.-M. Dynamical behavior of continuous tick data in futures exchange market. *Fractals* **2003**, *11*, 131–136. [CrossRef]
26. Kim, K.; Yoon, S.-M.; Kim, Y. Herd behaviors in the stock and foreign exchange markets. *Phys. A Stat. Mech. Its Appl.* **2004**, *341*, 526–532. [CrossRef]
27. Coronel-Brizio, H.F.; Hernándex-Montoya, A.R. On fitting the Pareto–Lévy distribution to stock market index data: Selecting a suitable cutoff value. *Phys. A Stat. Mech. Its Appl.* **2005**, *354*, 437–449. [CrossRef]
28. Sinha, S.; Pan, R.K. The power (law) of Indian markets: Analysing NSE and BSE trading statistics. In *Econophysics of Stock and Other Markets. Proc. Econophys-Kolkata II*; Chatterjee, A., Chakrabarti, B.K., Eds.; Springer: Berlin/Heidelberg, Germany, 2006.
29. Oh, G.; Kim, S.; Um, C.-J. Statistical properties of the returns of stock prices of international markets. *arXiv* **2006**, arXiv:physics/0601126.
30. Rak, R.; Drożdż, S.; Kwapień, J. Nonextensive statistical features of the Polish stock market fluctuations. *Phys. A Stat. Mech. Its Appl.* **2007**, *374*, 315–324. [CrossRef]
31. Gu, G.-F.; Zhou, W.-X. Statistical properties of daily ensemble variables in the Chinese stock markets. *Phys. A Stat. Mech. Its Appl.* **2007**, *383*, 497–506. [CrossRef]
32. Drożdż, S.; Forczek, M.; Kwapień, J.; Oświęcimka, P.; Rak, R. Stock market return distributions: From past to present. *Phys. A Stat. Mech. Its Appl.* **2007**, *383*, 59–64. [CrossRef]
33. Wang, F.; Shieh, S.-J.; Havlin, S.; Stanley, H.E. Statistical analysis of the overnight and daytime return. *Phys. Rev. E* **2009**, *79*, 056109. [CrossRef] [PubMed]
34. Kwapień, J.; Drożdż, S. Physical approach to complex systems. *Phys. Rep.* **2012**, *515*, 115–226. [CrossRef]
35. Eom, C.; Kaizoji, T.; Scalas, E. Fat tails in financial return distributions revisited: Evidence from the Korean stock market. *Phys. A Stat. Mech. Its Appl.* **2019**, *526*, 121055. [CrossRef]
36. Wątorek, M.; Drożdż, S.; Oświęcimka, P.; Stanuszek, M. Multifractal cross-correlations between the world oil and other financial markets. *Energy Econ.* **2019**, *81*, 874–885. [CrossRef]
37. Matia, K.; Amaral, L.A.N.; Goodwin, S.P.; Stanley, H.E. Different scaling behaviors of commodity spot and future prices. *Phys. Rev. E* **2002**, *66*, 045103(R). [CrossRef] [PubMed]

38. Drożdż, S.; Gębarowski, R.; Minati, L.; Oświęcimka, P.; Wątorek, M. Bitcoin market route to maturity? Evidence from return fluctuations, temporal correlations and multiscaling effects. *Chaos* **2018**, *28*, 071101. [CrossRef]
39. Wątorek, M.; Drożdż, S.; Kwapień, J.; Minati, L.; Oświęcimka, P.; Stanuszek, M. Multiscale characteristics of the emerging global cryptocurrency market. *Phys. Rep.* **2021**, *901*, 1–82. [CrossRef]
40. Takaishi, T. Recent scaling properties of Bitcoin price returns. *J. Phys. Conf. Ser.* **2021**, *1730*, 012124. [CrossRef]
41. Epps, T.W. Comovements in stock prices in the very short run. *J. Am. Stat. Assoc.* **1979**, *74*, 291–298.
42. Plerou, V.; Gopikrishnan, P.; Rosenow, B.; Amaral, L.A.N.; Guhr, T.; Stanley, H.E. Random matrix approach to cross correlations in financial data. *Phys. Rev. E* **2002**, *65*, 066126. [CrossRef] [PubMed]
43. Kwapień, J.; Drożdż, S.; Speth, J. Alternation of different fluctuation regimes in the stock market dynamics. *Phys. A Stat. Mech. Its Appl.* **2003**, *330*, 605–621. [CrossRef]
44. Kwapień, J; Drożdż, S.; Speth, J. Time scales involved in emergent market coherence. *Phys. A Stat. Mech. Its Appl.* **2004**, *337*, 231–242. [CrossRef]
45. Kwapień, J.; Drożdż, S.; Górski, A.Z.; Oświęcimka, P. Asymmetric matrices in an analysis of financial correlations. *Acta Phys. Pol. B* **2006**, *37*, 3039–3048.
46. Lo, A.W. Long term memory in stock market prices. *Econometrica* **1991**, *59*, 1279–1313. [CrossRef]
47. Mills, T.C. Is there long-term memory in UK stock returns? *Appl. Financ. Econ.* **1993**, *3*, 303–306. [CrossRef]
48. Fama, E.F.; French, K.R. Permanent and temporary components in stock prices. *J. Polit. Econ.* **1988**, *96*, 246–273. [CrossRef]
49. Wright, J.H. Long memory in emerging market stock returns. *Emerg. Mark. Quart.* **2001**, *5*, 50–55.
50. Henry, Ó.T. Long memory in stock returns: Some international evidence. *J. Appl. Financ. Econ.* **2002**, *12*, 725–729. [CrossRef]
51. Podobnik, B.; Fu, D.; Jagric, T.; Grosse, I.; Stanley, H.E. Fractionally integrated processes for transition economies. *Phys. A Stat. Mech. Its Appl.* **2006**, *362*, 465–470. [CrossRef]
52. Hull, M.; McGroarty, F. Do emerging markets become more efficient as they develop? Long memory persistence in equity indices. *Emerg. Mark. Rev.* **2014**, *18*, 45–61. [CrossRef]
53. Ding, Z.; Granger, C.W.J.; Engle, R.F. A long memory property of stock market returns and a new model. *J. Empir. Financ.* **1993**, *1*, 83–106. [CrossRef]
54. Baillie, R. Long memory processes and fractional integration in econometrics. *J. Econ.* **1996**, *73*, 5–59. [CrossRef]
55. Lillo, F.; Farmer, J.D. The long memory of the efficient market. *Stud. Nonlinear Dyn. Econom.* **2004**, *8*, 1–29. [CrossRef]
56. Fama, F.F. Efficient capital markets: A review of theory and empirical work. *J. Financ.* **1970**, *25*, 383–417. [CrossRef]
57. Alvarez-Ramirez, J.; Rodriguez, E.; Espinosa-Paredes, G. Is the US stock market becoming weakly efficient over time? Evidence from 80-year-long data. *Phys. A Stat. Mech. Its Appl.* **2012**, *391*, 5643–5647. [CrossRef]
58. Gabaix, X.; Gopikrishnan, P.; Plerou, V.; Stanley, H.E. A theory of power-law distributions in financial market fluctuations. *Nature* **2003**, *423*, 267–270. [CrossRef] [PubMed]
59. Farmer, J.D.; Lillo, F. On the origin of power law tails in price fluctuations. *Quant. Financ.* **2004**, *4*, 7–11. [CrossRef]
60. Gillemot, L.; Farmer, J.D.; Lillo, F. There's more to volatility than volume. *Quant. Financ.* **2006**, *6*, 371–384. [CrossRef]
61. Taranto, D.E.; Bormetti, G.; Bouchaud, J.-P.; Lillo, F.; Tóth, B. Linear models for the impact of order flow on prices II: The mixture transition distribution model. *Quant. Financ.* **2018**, *18*, 917–931. [CrossRef]
62. Kaizoji, T.; Kaizoji, M. Exponential laws of stock price index and a stochastic model. *Adv. Compl. Syst.* **2003**, *6*, 303–312. [CrossRef]
63. Silva, A.C.; Prange, R.E.; Yakovenko, V.M. Exponential distribution of financial returns at mesoscopic time lags: A new stylized fact. *Phys. A Stat. Mech. Its Appl.* **2004**, *344*, 227–235. [CrossRef]
64. Malevergne, Y.; Pisarenko, V.; Sornette, D. Empirical distributions of stock returns: Between the stretched exponential and the power law? *Quant. Financ.* **2005**, *5*, 379–401. [CrossRef]
65. Malevergne, Y.; Pisarenko, V.; Sornette, D. On the power of generalized extreme value (GEV) and generalized Pareto distribution (GPD) estimators for empirical distributions of stock returns. *Appl. Financ. Econ.* **2006**, *16*, 271–289. [CrossRef]
66. Cortines, A.A.G.; Riera, R. Non-extensive behavior of a stock market index at microscopic time scales. *Phys. A Stat. Mech. Its Appl.* **2007**, *377*, 181–192. [CrossRef]
67. Ren, F.; Gu, G.-F.; Zhou, W.-X. Scaling and memory in the return intervals of realized volatility. *Phys. A Stat. Mech. Its Appl.* **2009**, *388*, 4787–4796. [CrossRef]
68. Mart, T.; Surya, Y. Statistical properties of the Indonesian stock exchange index. *Phys. A Stat. Mech. Its Appl.* **2004**, *344*, 198–202. [CrossRef]
69. Yang, J.-S.; Chae, S.; Jung, W.-S.; Moon, H.-T. Dynamics of the return distribution in the Korean financial market. *Phys. A Stat. Mech. Its Appl.* **2006**, *363*, 377–382. [CrossRef]
70. Scalas, E.; Kim, K. The art of fitting financial time series with Lévy stable distributions. *arXiv* **2007**, arXiv:physics/0608224.
71. Suárez-García, P.; Gómez-Illate, D. Scaling, stability and distribution of the high-frequency returns of the IBEX35 index. *Phys. A Stat. Mech. Its Appl.* **2013**, *392*, 1409–1417. [CrossRef]
72. Rak, R.; Drożdż, S.; Kwapień, J.; Oświęcimka, P. Stock returns versus trading volume: Is the correspondence more general? *Acta Phys. Pol. B* **2013**, *44*, 2035–2050. [CrossRef]
73. Begušić, S.; Konstanjčar, Z.; Stanley, H.E.; Podobnik, B. Scaling properties of extreme price fluctuations in Bitcoin markets. *Phys. A Stat. Mech. Its Appl.* **2018**, *510*, 400–406. [CrossRef]
74. Poon, S.-H.; Granger C. Forecasting Volatility in Financial Markets: A Review. *J. Econ. Lit.* **2003**, *41*, 478–539. [CrossRef]

75. Solomon, S. Stochastic Lotka-Volterra systems of competing auto-catalytic agents lead generically to truncated Pareto power wealth distribution, truncated Levy distribution of market returns, clustered volatility, booms and craches. In *Decision Technologies for Computational Finance*; Springer: Boston, MA, USA, 1998; pp. 73–86. [CrossRef]
76. Solomon, S.; Richmond, P. Power laws of wealth, market order volumes and market returns. *Phys. A Stat. Mech. Its Appl.* **2001**, *299*, 188–197.
77. Lux, T.; Sornette, D. On rational bubbles and fat tails. *J. Money Credit Bank.* **2002**, *34*, 589–610. [CrossRef]
78. Sornette, D.; Malevergne, Y. From rational bubbles to crashes. *Phys. A Stat. Mech. Its Appl.* **2001**, *299*, 40–59. [CrossRef]
79. Sornette, D. "Slimming" of power law tails by increasing market returns. *Phys. A Stat. Mech. Its Appl.* **2002**, *309*, 403–418. [CrossRef]
80. Drăgulescu, A.A.; Yakovenko, V.M. Probability distribution of returns in the Heston model with stochastic volatility. *Quant. Financ.* **2002**, *2*, 443–453. [CrossRef]
81. Alejandro-Quiñones, Á.L.; Bassler, K.E.; Field, M.; McCauley, J.L.; Nicol, M.; Timofeyev, I.; Török, A.; Gunaratne, G.H. A theory of fluctuations in stock prices. *Phys. A Stat. Mech. Its Appl.* **2006**, *363*, 383–392. [CrossRef]
82. Bormetti, G.; Cazzola, V.; Montagna, G.; Nicrosini, O. Probability distribution of returns in the exponential Ornstein-Uhlenbeck model. *J. Stat. Mech.* **2008**, *2008*, P11013. [CrossRef]
83. Gerig, A.; Vicente, J.; Fuentes, M.A. Model for non-Gaussian intraday stock returns. *Phys. Rev. E* **2009**, *80*, 065102(R). [CrossRef]
84. Ghashghaie, S.; Breymann, W.; Peinke, J.; Talkner, P.; Dodge, Y. Turbulent cascades in foreign exchange markets. *Nature* **1996**, *381*, 767–770. [CrossRef]
85. Mandelbrot, B.B.; Fisher, A.; Calvet, L. Multifractal Model of Asset Returns. Cowles Foundation Discussion Paper no. 1164. 1997. Available online: https://papers.ssrn.com/sol3/papers.cfm?abstract_id=78588 (accessed on 26 May 2021). [CrossRef]
86. Breymann, W.; Shashghaie, S.; Talkner, P. A stochastic cascade model for FX dynamics. *Int. J. Theor. Appl. Financ.* **1996**, *3*, 357–360.
87. Bak, P.; Paczuski, M.; Shubik, M. Price variations in a stock market with many agents. *Phys. A Stat. Mech. Its Appl.* **1997**, *246*, 430–453. [CrossRef]
88. Caldarelli, G.; Marsili, M.; Zhang, Y.C. A prototype model of stock exchange. *EPL* **1997**, *40*, 479–484. [CrossRef]
89. Cont, R.; Bouchaud, J.-P. Herd behavior and aggregate fluctuations in financial markets. *Macroecon. Dyn.* **2000**, *4*, 170–196. [CrossRef]
90. Zhang, Z.-F. Self-organized model for information spread in financial markets. *Eur. Phys. J. B* **2000**, *16*, 379–385. [CrossRef]
91. Challet, D.; Marsili, M.; Zhang, Y.-C. Stylized facts of financial markets and market crashes in minority game. *Phys. A Stat. Mech. Its Appl.* **2001**, *294*, 514–524.
92. Bornholdt, S. Expectation bubbles in a spin model of markets: Intermittency from frustration across scales. *Int. J. Mod. Phys. C* **2001**, *12*, 667–674. [CrossRef]
93. Laherrère, J.; Sornette, D. Stretched exponential distributions in nature and economy: "Fat tails" with characteristic scales. *Eur. Phys. J. B* **1998**, *2*, 525–539. [CrossRef]
94. Matia, K.; Pal, M.; Salunkay, H.; Stanley, H.E. Scale-dependent price fluctuations for the Indian stock market. *Europhys. Lett.* **2004**, *66*, 909–914. [CrossRef]
95. Pisarenko, V.F.; Sornette, D. New statistic for financial return distributions: Power-law or exponential? *Phys. A Stat. Mech. Its Appl.* **2006**, *366*, 387–400. [CrossRef]
96. Linden, M. A model for stock return distribution. *Int. J. Financ. Econ.* **2001**, *6*, 159–169. [CrossRef]
97. Tsallis, C. Possible generalization of the Boltzmann-Gibbs statistics. *J. Stat. Phys.* **1988**, *52*, 479–487. [CrossRef]
98. Tsallis, C. *Introduction to Nonextensive Statistical Mechanics: Approaching a Complex World*; Springer: Berlin/Heidelberg, Germany, 2009. [CrossRef]
99. Michael, F.; Johnson, M.D. Financial market dynamics. *Phys. A Stat. Mech. Its Appl.* **2003**, *320*, 525–534.
100. Mu, G.-H.; Zhou, W.-X. Nonuniversal distributions of stock returns in an emerging market. *Phys. Rev. E* **2010**, *82*, 066103. [CrossRef]
101. Drożdż, S.; Kwapień, J.; Oświęcimka, P.; Rak, R. Quantitative features of multifractal subtleties in time series. *EPL* **2009**, *88*, 60003. [CrossRef]
102. Queirós, S.M.D. On anomalous distributions in intra-day financial time series and non-extensive statistical mechanics. *Phys. A Stat. Mech. Its Appl.* **2004**, *344*, 279–283. [CrossRef]
103. Lo, A.W.; MacKinlay, C. Stock market prices do not follow random walks: Evidence from a simple specification test. *Rev. Financ. Stud.* **1988**, *1*, 41–66. [CrossRef]
104. Bekaert, B.; Erb, C.B.; Harvey, C.R.; Viskanta, T.E. Distributional characteristics of emerging market returns and asset allocation. *J. Portf. Manag.* **1998**, *24*, 102–116. [CrossRef]
105. Lee, K.E.; Lee, J.W. Scaling properites of price changes for Korean stock indices. *arXiv* **2004**, arXiv:cond-mat/0407418. [CrossRef]
106. Sarkar, A.; Barat, P. Scaling analysis on Indian foreign exchange market. *Phys. A Stat. Mech. Its Appl.* **2006**, *364*, 362–368.
107. Vicente, R.; de Toledo, C.M.; Leite, V.B.P.; Caticha, N. Underlying dynamics of typical fluctuations of an emerging market price index: The Heston model from minutes to months. *Phys. A Stat. Mech. Its Appl.* **2006**, *361*, 272–288. [CrossRef]
108. Alfonso, L.; Mansilla, R.; Terrero-Escalante, C.A. On the scaling of the distribution of daily price fluctuations in Mexican financial market index. *Phys. A Stat. Mech. Its Appl.* **2012**, *391*, 2990–2996. [CrossRef]

109. Gang, G.-J.; Xie, C. Cross-correlations Between WTI Crude Oil Market and U.S. Stock Market: A Perspective from Econophysicss. *Acta Phys. Pol. B* **2012**, *45*, 2021–2037. [CrossRef]
110. Gang, G.-J.; Xie, C. Cross-correlations between the CSI 300 spot and futures markets. *Nonlinear Dyn.* **2013**, *73*, 1687–1696. [CrossRef]
111. Samuelson, P. The fundamental approximation theorem of portfolio analysis in terms of means, variances and higher moments. *Rev. Econ. Stud.* **1970**, *25*, 65–86. [CrossRef]
112. Kane, A. Skewness preference and portfolio choice. *J. Financ. Quant. Anal.* **1977**, *17*, 15–25. [CrossRef]
113. Friend, W.E.; Westerfield, R. Co-skewness and capital asset pricing. *J. Financ.* **1980**, *35*, 897–914.
114. Kon, S.J. Models of stock returns. A comparison. *J. Financ.* **1984**, *39*, 147–165. [CrossRef]
115. Müller, U.A.; Dacorogna, M.; Olsen, R.B.; Pictet, O.V.; Schwarz, M.; Morgenegg, C. Statistical study of foreign exchange rates, empirical evidence of a price change scaling law and intraday analysis. *J. Bank. Financ.* **1990**, *14*, 1189–1208. [CrossRef]
116. Guillaume, D.M.; Dacorogna, M.M.; Davé, R.R.; Müller, U.A.; Olsen, R.B.; Pictet, O.V. From the bird's eye to the microscope: A survey of new stylized facts of the intra-day foreign exchange markets. *Financ. Stoch.* **1997**, *1*, 95–130. [CrossRef]
117. Coronel-Brizio, H.F.; Hernández-Montoya, A.R.; Huerta-Quintanilla, R.; Rodríguez-Achach, M. Assessing symmetry of financial returns series. *Phys. A Stat. Mech. Its Appl.* **2007**, *383*, 5–9.
118. Derksen, M.; Kleijn, B.; de Vilder, R. Heavy tailed distributions in closing auctions. *arXiv* **2020**, arXiv:2012.10145. [CrossRef]
119. Miśkiewicz, J. Network analysis of cross-correlations on Forex market during crises. Globalisation on Forex market. *Entropy* **2021**, *23*, 352. [CrossRef]
120. Nakamoto, S. Bitcoin: A Peer-to-Peer Electronic Cash System. 2008. Available online: https://git.dhimmel.com/bitcoin-whitepaper/ (accessed on 26 May 2021). [CrossRef]
121. Ethereum. Available online: http://www.ethereum.org (accessed on 26 May 2021).
122. Dukascopy. Available online: https://www.dukascopy.com (accessed on 26 May 2021). [CrossRef]
123. Kraken. Available online: http://www.kraken.com (accessed on 26 May 2021).
124. Drożdż, S.; Kwapień, J.; Oświęcimka, P.; Rak, R. The foreign exchange market: Return distributions, multifractality, anomalous multifractality and the Epps effect. *New J. Phys.* **2010**, *12*, 105003.
125. Drożdż, S.; Grümmer, F.; Ruf, F. Speth, J. Towards identifying the world stock market cross-correlations: DAX versus Dow Jones, *Phys. A Stat. Mech. Its Appl.* **2001**, *294*, 226–234.
126. Drożdż, S.; Grümmer, F.; Ruf, F. Speth, J. Dynamics of correlations in the stock market. In *Empirical Science of Financial Fluctuations*; Takayasu, H., Ed.; Springer: Tokyo, Japan, 2002; p. 43.
127. Gębarowski, R.; Oświęcimka, P.; Wątorek, M.; Drożdż, S. Detecting correlations and triangular arbitrage opportunities in the Forex by means of multifractal detrended cross-correlations analysis. *Nonlinear Dyn.* **2019**, *98*, 2349–2364. [CrossRef]
128. Drożdż, S.; Minati, L.; Oświęcimka, P.; Stanuszek, M.; Wątorek, M. Signatures of crypto-currency market decoupling from the Forex. *Future Internet* **2019**, *11*, 154. [CrossRef]
129. Drożdż, S.; Kwapień, J.; Oświęcimka, P.; Stanisz, T.; Wątorek, M. Complexity in economic and social systems: Cryptocurrency market at around COVID-19. *Entropy* **2020**, *22*, 1043.
130. Muvunza, T. An α-stable approach to modelling highly speculative assets and cryptocurrencies. *arXiv* **2020**, arXiv:2002.09881. [CrossRef]
131. Drożdż, S.; Minati, L.; Oświęcimka, P.; Stanuszek, M.; Wątorek, M. Competition of noise and collectivity in global cryptocurrency trading: Route to a self-contained market. *Chaos* **2020**, *30*, 023122. [CrossRef]

entropy

MDPI

Article

Continuous Time Random Walk with Correlated Waiting Times. The Crucial Role of Inter-Trade Times in Volatility Clustering

Jarosław Klamut * and Tomasz Gubiec

Institute of Experimental Physics, Faculty of Physics, University of Warsaw, Pasteura 5, 02-093 Warsaw, Poland; Tomasz.Gubiec@fuw.edu.pl
* Correspondence: jaroslaw.klamut@fuw.edu.pl

Abstract: In many physical, social, and economic phenomena, we observe changes in a studied quantity only in discrete, irregularly distributed points in time. The stochastic process usually applied to describe this kind of variable is the continuous-time random walk (CTRW). Despite the popularity of these types of stochastic processes and strong empirical motivation, models with a long-term memory within the sequence of time intervals between observations are rare in the physics literature. Here, we fill this gap by introducing a new family of CTRWs. The memory is introduced to the model by assuming that many consecutive time intervals can be the same. Surprisingly, in this process we can observe a slowly decaying nonlinear autocorrelation function without a fat-tailed distribution of time intervals. Our model, applied to high-frequency stock market data, can successfully describe the slope of decay of the nonlinear autocorrelation function of stock market returns. We achieve this result without imposing any dependence between consecutive price changes. This proves the crucial role of inter-event times in the volatility clustering phenomenon observed in all stock markets.

Keywords: continuous time random walk; intertrade times; volatility clustering

Citation: Klamut, J.; Gubiec, T. Continuous Time Random Walk with Correlated Waiting Times. The Crucial Role of Inter-Trade Times in Volatility Clustering. *Entropy* **2021**, *23*, 1576. https://doi.org/10.3390/e23121576

Academic Editors: Geert Verdoolaege and Rosario Nunzio Mantegna

Received: 17 October 2021
Accepted: 20 November 2021
Published: 26 November 2021

1. Introduction

In many physical, biological, and economic systems we can identify elementary events occurring irregularly in time . Additionally, the times between those events can be interdependent in a non-trivial manner, which can lead to complex behavior. Therefore, it is no surprise that point processes are of high interest to researchers and their applications are widely studied [1,2]. Two of the most popular models are autoregressive conditional duration (ACD) [3] and the Hawkes model [4,5]. The canonical versions of both models include short-range dependencies (for ACD see [3,6–11]; for Hawkes see [12–20]). Both of them, however, have been extended to describe long-range memory (for ACD see [21–31]; for Hawkes see [32–42]).

Real-world stochastic processes have numerous features which can be associated with elementary events. For instance, in the transaction data from a stock market we observe the events—the transactions occurring in specific moments—and their features: the price and volume of each transaction. Inter-trade times from stock market transaction data are a perfect example of a point process. However, in order to describe the price of transactions, which we do below, one must go beyond the framework of point processes, which does not incorporate features of the elementary events. A natural generalization is the continuous-time random walk (CTRW).

The CTRW was the first proposed formalism to describe the dynamics of a variable changing its value in unevenly spaced points in time. Point processes extended to fit this phenomenon are called marked point processes [16]. Moreover, the distribution of time intervals between those points can be arbitrary. This formalism was introduced in 1965 by Montroll and Weiss [43] and since then it has been applied in a broad range of fields, ranging from astrophysics to economics and the social sciences. For a detailed review, see [44]. In the canonical CTRW, both increments of the observed process and waiting times (inter-event times) are i.i.d. random variables. An exemplary trajectory of such a process is shown in Figure 1.

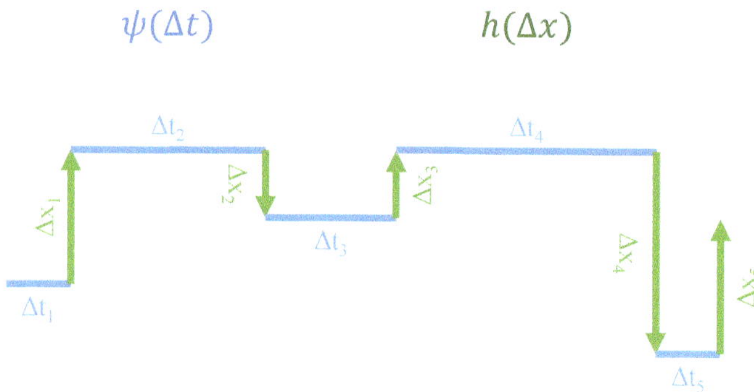

Figure 1. The example trajectory of the continuous-time random walk (CTRW), consisting of jumps of process values Δx_n preceded by waiting times Δt_n. In the canonical CTRW, Δt_n and Δx_n are i.i.d. random variables drawn from the distributions $\psi(\Delta t_n)$ and $h(\Delta x_n)$, respectively. In this paper, we consider the CTRW model with long-term dependence in the series of waiting times $\Delta t_1, \Delta t_2, \ldots, \Delta t_n$.

All kinds of random walks, starting with normal diffusion, through anomalous diffusion (both subdiffusion and superdiffusion) to Levy flights, can be described within the CTRW formalism. This can be achieved by using specific distributions of waiting times or increments (especially with heavy tails) and by considering memory in waiting times, increments, or coupling between them. The CTRW models with correlated increments were initially proposed to study lattice gases [45–47]. More recently, they have been used to model high-frequency financial data [48–60]. On the other hand, CTRW models with correlated waiting times are not well-studied. With the exception of a few recent attempts [52,61,62], these models have not been analyzed nor used to model empirical data. This fact is surprising in light of the recent popularity of point processes such as ACD and the Hawkes process. The aim of this work is to fill this gap. We propose a new CTRW model which incorporates dependencies of inter-event times. Our intention is to model long-range memories in the sequence of waiting times, an aim inspired by numerous empirical examples [63–69]. Our model is simple yet general enough to explain the properties of empirical data. That makes it a perfect candidate for future applications and a relevant reference point for future work.

The paper is organized as follows. In Section 2, we present the motivation behind the model, with correlated waiting times based on financial data. Next, in Section 3 we propose a way to include dependencies between the waiting times, in particular the long-range memory. In Section 4, we solve the CTRW model with correlated waiting times by calculating its propagator, moments, and the autocorrelation function (ACF) of increments. We then fit our model to tick-by-tick transaction data from the Warsaw Stock Exchange in Section 5. Finally, we provide a summary of our work in Section 6. Two appendices at the end provide a clarification of the mathematical methods that we have used.

2. Motivation

Models with interdependent waiting times are used to describe electron transfer [63], the firing of a single neuron [64], interhuman communication [65], and the modeling of earthquakes [66–69]. An excellent example of a process with correlated inter-event times that we will describe in this manuscript is tick-by-tick transaction price data from the stock market [70]. These data are very convenient to use, as they are of high quality and easily accessible in large amounts.

Firstly, let us recall two basic stylized facts observed in the majority of stock markets [71].

- In the ACF of time-dependent log-returns, we observe short-term negative autocorrelation.
- However, we observe slowly decaying positive autocorrelation for the ACF of absolute values of time-dependent log-returns.

The latter is considered to be reminiscent of the volatility clustering phenomenon.

Of course, these are not the only or the most significant stylized facts, but these two do not directly depend on the log-return distribution. The list should also contain the broad distribution of log-returns [72]; multi-fractality [73,74]; universal scaling of the distribution of times between large jumps [75,76]; and the slow, power-law decay of the correlation between these times. We will further discuss the latter in this manuscript. Usually, the CTRW models used to describe high-frequency stock market data consider waiting times Δt_n as inter-transaction times, and process increments Δx_n as logarithmic returns between consecutive transactions. Taking into account the so-called bid-ask bounce phenomenon allows CTRW processes to reproduce the first stylized fact of short-term negative autocorrelation [58,77,78]. In this type of models, waiting times Δt_n are i.i.d. variables and only the dependence between Δx_n and Δx_{n-1} is considered. Unfortunately, models considering only this type of dependencies turned out to be unable to describe the time ACF of absolute values of price changes [60]. Technically, it is possible to obtain a CTRW model reproducing both stylized facts, but it requires a power-law waiting-time distribution $\psi(\Delta t)$. However, this solution is not satisfying as we can obtain waiting-time distribution directly from the empirical data of inter-transaction times. It turns out that this distribution is far from a power-law one [58]. These results suggest that the source of the second stylized fact is not in the distributions of increments $h(\Delta x)$ and waiting times $\psi(\Delta t)$, but in the dependence between consecutive Δx and Δt values.

Let us start with an empirical analysis of the step ACF of series Δt_n and $|\Delta x_n|$. We observe approximately power-law memories in waiting times and absolute values of price changes; see Figure 2a. For a lag (in the number of steps) $\lesssim 3$, the autocorrelation of $|\Delta x_n|$ is higher than the autocorrelation of Δt_n, but for a lag > 3 it is otherwise. This result suggests that in the limit of long times, the dependence between waiting times may be more critical than dependence between price changes. To verify this hypothesis we perform a shuffling test. We compare the time ACF of price changes' absolute values for four samples of time series. The first one is the original time series of tick-by-tick transaction data. The second time series keeps the price changes Δx_n in the original order but shuffles the order of waiting times Δt_n. This way, we obtained a time series keeping all dependencies between price changes Δx_n, but without any dependencies between waiting times Δt_n. In the third time series, we kept the original waiting times Δt_n but shuffled the price changes Δx_n. In the last, fourth time series, both Δt_n and Δx_n were shuffled. Let us emphasise that all four time series have the same, unchanged distributions $\psi(\Delta t_n)$ and $h(\Delta x_n)$. The results are shown in Figure 2b. As expected, we observe the slow, almost power-law decay of the time ACF for the first empirical time series. Surprisingly, removing dependencies between waiting times does not change the time ACF in the limit of $t \to 0$, but significantly increases its slope of decay in the long-term. On the other hand, removing dependencies between price changes decreases the time ACF, dividing it by an almost constant factor but does not change the slope of the decay. The removal of all dependencies still leads to a positive time ACF, resulting from the non-exponential empirical distribution of waiting times.

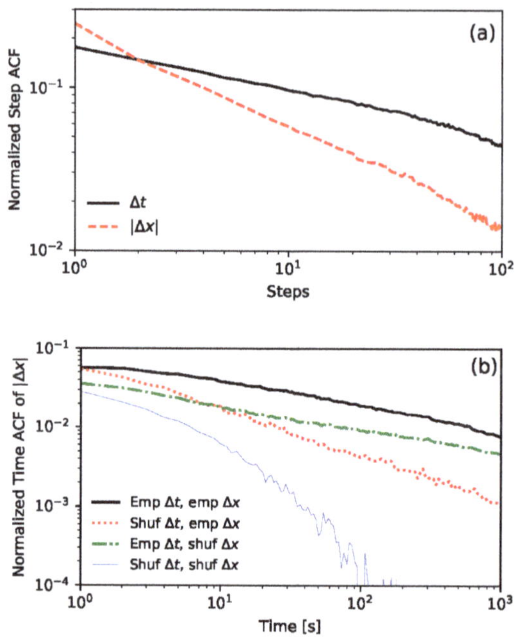

Figure 2. Figures 2 and 3 were prepared using transaction data for KGHM (one of the most liquid Polish stocks) from period of January 2013 to July 2017. Both figures are on a log-log scale. (**a**) The plot of normalized empirical step ACF of Δt and $|\Delta x|$. Both functions decay like a power-law. For lag = 1, the autocorrelation of $|\Delta x|$ is higher. However, it decays faster, and for long times the memory in waiting times is stronger. (**b**) The plot of the normalized time ACF of $|\Delta x|$ for four time series. The presented lines are for empirical data (thick black), empirical price changes, and intra-daily shuffled waiting times (dotted red); intra-daily shuffled price changes and empirical waiting times (dash-dotted green); and intra-daily independently shuffled price changes and waiting times (thin blue). Considering only empirical dependencies of waiting times reproduces the ACF, which decays with almost the same slope as the empirical one.

The empirical observations presented above convinced us that it is necessary to consider long-range dependencies between waiting times within CTRW to reproduce the slowly decaying ACF of price changes' absolute values observed in the financial data.

Please note that in Figure 2, we analyzed the step ACF for lags up to 100 and the time ACF for times up to 1000 s. The procedure used to estimate the time ACF was presented in [58] and is a modification of the classical slotting technique introduced in [79]. Such limits were chosen due to the length of trading sessions (around 8 h or 1000 trades). Unfortunately, these limits are not long enough to detect power-law dependencies. The only way to increase these limits is by joining all sessions into one sequence. In this procedure, we merge the end of one session with the beginning of the following one (we omit overnight price changes). These two periods of the sessions are different, as we observe intraday activity in financial data [80]. The session begins with short inter-transaction times and a high standard deviation of price changes. Usually, up to the middle of the session, average inter-trade times increase, and the standard deviation of price changes decreases. The situation reverts again close to the end of the session. This phenomenon is called the *lunch effect* [81]. We use the canonical method to remove intraday non-stationarity by dividing each waiting time by the corresponding average waiting time, depending on the time that has elapsed since the beginning of the session for each day of the week separately [82,83]. The comparison of the step ACFs of waiting times for non-stationarized and stationarized data is presented in Figure 3a. As a result of this procedure, we obtain the power-law decay over four orders of magnitude of lag. In Figure 3b, we present the time ACF of price changes' absolute values for stationarized data, which also exhibit power-law decay over four orders of magnitude of time lag. It is now reasonable to ask what the relationship is between the decay exponents of these autocorrelations. Fortunately, the model studied in this paper gives a strict answer to this question.

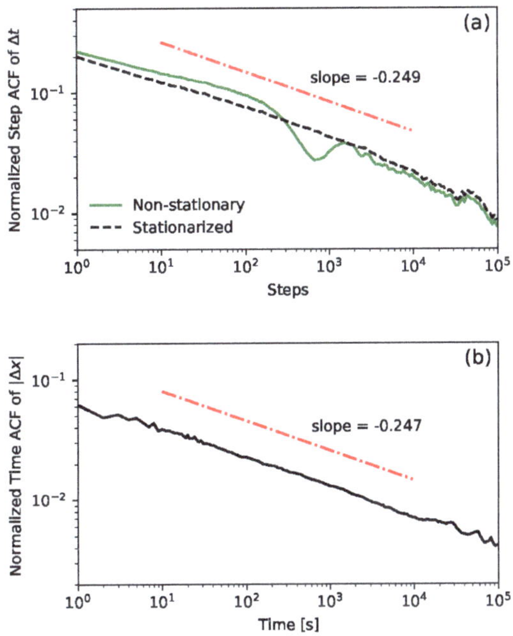

Figure 3. All intraday data (waiting times and corresponding price changes) are joined into one data set. (**a**) The plot shows the normalized step ACF of Δt for non-stationary and stationarized cases. The stationarizing procedure is described in the main text. (**b**) The plot of the normalized time ACF of $|\Delta x|$ with stationarized waiting times. Both stationarized autocorrealations decay like a power-law with similar slope.

3. Process of Waiting Times

Let us now focus on the sequence of inter-transaction times $\Delta t_1, \Delta t_2, \ldots, \Delta t_n, \ldots$. We are now looking for the point process to describe this series, which will be suitable for use in CTRW. For this reason, we need analytically solvable models. Moreover, we would like to use the empirical distribution of inter-event times $\psi(\Delta t_n)$ and observe the power-law step ACF, as shown in Figure 3a. Even these two simple conditions exclude ACD models and Hawkes processes from our considerations. We are not interested in ACD models, as the power-law ACF can be obtained only within the fractional extension. In the Hawkes process, both the waiting time distribution and autocorrelation depend on the memory kernel [15,84]. Therefore, they cannot be set independently. As the Hawkes process is defined solely by its kernel, both waiting time distribution and autocorrelation depend on it. Thus, it would be difficult (if it is possible at all) to reproduce both empirical WTD and ACF at the same time. This feature of the Hawkes process hampers its use in the description of empirical data.

As the solution to our search, we propose a simple point process in which waiting times Δt are repeated. In a very general sense, our proposition can be interpreted as a discretized version of CTRW, adapted to the role of the point process. Let us briefly describe this analogy. Within the canonical CTRW, values of the process are represented by a spatial variable, and the time is continuous. The spatial variable remains constant for a given period of continuous waiting time. Now, we define the point process by the series of waiting times. Here, the number of repetitions ν_i of the same value of waiting time is the analog of waiting time in the canonical CTRW. The exemplary realization of such an adapted process of waiting times is shown in Figure 4.

We require the waiting times Δt_n (values of the process in the discrete subordinated time n) to come from the distribution $\psi(\Delta t_n)$ ($\Delta t_n > 0$), with a finite mean $\langle \Delta t \rangle$. We define ν_i as the number of repetitions of the same waiting times (drawn independently for each series of repetitions). Let ν_i be the i.i.d. random variables with the distribution $\omega(\nu_i)$. In general, it can be any distribution, but to recreate the power-law step ACF of waiting times we will focus on a fat-tailed distribution with a finite first moment $\langle \nu \rangle$. In particular, we use the zeta distribution with parameter ρ

$$\omega_\rho(k) = k^{-\rho}/\zeta(\rho); \quad \zeta(\rho) = \sum_{i=1}^{\infty} i^{-\rho}, \ \rho > 1, \tag{1}$$

where $\zeta(\rho)$ is Riemann's zeta function. Its expected value is equal to $\langle \omega \rangle = \frac{\zeta(\rho-1)}{\zeta(\rho)}$ for $\rho > 2$ and the variance is finite for $\rho > 3$. The cumulative distribution function is given by $\frac{H_{k,\rho}}{\zeta(\rho)}$, where $H_{k,\rho} = \sum_{i=1}^{k} i^{-\rho}$ is the generalized harmonic number. Let us introduce $\Omega(k) = \sum_{i=k}^{\infty} \omega(i)$ as a sojourn probability. We have $\Omega(k) = 1 - \frac{H_{k-1,\rho}}{\zeta(\rho)}$ for the zeta distribution.

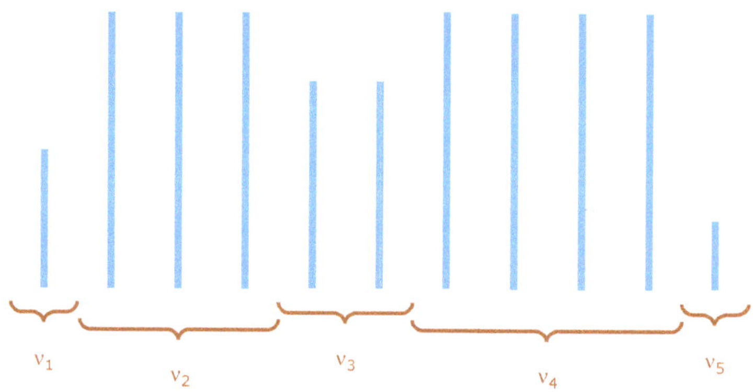

Figure 4. The example realization of the process of waiting times, the values of which correspond to the waiting times Δt_n of the point process used in the primary CTRW process. Process values $\Delta t_1, \Delta t_2, \ldots, \Delta t_n$ come from the values $\Delta t^1, \Delta t^2, \ldots, \Delta t^k$ repeated $\nu_1, \nu_2, \ldots, \nu_k$ times, respectively. Number of repetitions ν_i are drawn from the distribution $\omega(\nu_i)$. In the example above: $\nu_1 = 1, \nu_2 = 3, \nu_3 = 2, \ldots$ and $\Delta t_1 = \Delta t^1, \Delta t_2 = \Delta t_3 = \Delta t_4 = \Delta t^2, \Delta t_5 = \Delta t_6 = \Delta t^3, \ldots$.

We define a soft propagator of the process of times $\mathcal{P}(\Delta t; n | \Delta t_0, 0)$, which is the conditional probability density that the waiting time, which was initially (at $n = 0$) in the origin value ($\Delta t = \Delta t_0$), is equal to Δt after n steps. The soft propagator can be expressed by

$$\mathcal{P}(\Delta t; n | \Delta t_0, 0) = \delta(\Delta t - \Delta t_0)\Omega^{\text{first}}(n) + [1 - \Omega^{\text{first}}(n)]\psi(\Delta t), \tag{2}$$

where $\Omega^{\text{first}}(n)$ is the sojourn probability obtained from $\omega^{\text{first}}(n)$, which is the stationarized distribution of the repetition of the first waiting time:

$$\begin{aligned} \omega^{\text{first}}(n) &= \frac{\sum_{n'=1} \omega(n+n')}{\sum_{n''=0} \sum_{n'=1} \omega(n''+n')} = \frac{\sum_{n'=1} \omega(n+n')}{\sum_{n=1} n\omega(n)} = \frac{\sum_{n'=n+1} \omega(n')}{\langle \omega \rangle}, \\ \Omega^{\text{first}}(n) &= \frac{\sum_{i=n} \sum_{n'=i+1} \omega(n')}{\langle \omega \rangle} = \frac{\sum_{i=1} i\omega(i+n)}{\langle \omega \rangle} = \frac{\langle \omega \rangle - n\Omega(n+1) - \sum_{i=1}^{n} i\omega(i)}{\langle \omega \rangle}. \end{aligned} \tag{3}$$

The first term of the right-hand side of Equation (2) is the probability that the process value will stay constant (equal Δt_0) after n jumps. The second term indicates that there will be a process value jump with probability $1 - \Omega^{\text{first}}(n)$, so new process values will be completely independent, drawn from the distribution $\psi(\Delta t)$.

Restricting ourselves to $\omega(n)$ in the form of the zeta distribution, we can obtain

$$\Omega^{\text{first}}(n) = 1 - \frac{n}{\langle \omega \rangle} + \frac{n H_{n,\rho}}{\zeta(\rho-1)} - \frac{H_{n,\rho-1}}{\zeta(\rho-1)}, \tag{4}$$

and hence the propagator given by Equation (2). The step autocovariance of waiting times Δt_n can be expressed as

$$\text{cov}(n) = \langle \Delta t_i \Delta t_{i+n} \rangle - \langle \Delta t_i \rangle \langle \Delta t_{i+n} \rangle = \langle \Delta t_i \Delta t_{i+n} \rangle - \langle \Delta t \rangle^2, \tag{5}$$

where symbol $\langle \ldots \rangle$ means taking the average. Note that $\Delta t_{i+n} = \Delta t_i$ with probability $p = \Omega^{\text{first}}(n)$. With probability $1-p$, the Δt_i is independent. This leads to

$$\text{cov}(n) = p \langle \Delta t^2 \rangle + (1-p) \langle \Delta t \rangle^2 - \langle \Delta t \rangle^2 = \sigma_{\Delta t}^2 p = \sigma_{\Delta t}^2 \Omega^{\text{first}}(n). \tag{6}$$

We are interested in the asymptotic form of autocorrelation for $n \gg 1$. We can use following approximation (Theorem 12.21 from [85])

$$\zeta(\rho) - H_{n,\rho} \approx \frac{n^{1-\rho}}{\rho - 1}. \tag{7}$$

Finally, we obtain the normalized step ACF

$$\text{corr}(n) = \frac{\text{cov}(n)}{\text{cov}(0)} \approx \frac{n^{-(\rho-2)}}{\zeta(\rho-1)(\rho-2)(\rho-1)}. \tag{8}$$

The step ACF of waiting times decays like a power-law and the decay exponent is $\rho - 2$. It is worth emphasizing that even considering only $\rho > 2$, required for the existence of a finite average number of repetitions, we can obtain any value of the decay exponent.

4. The Primary Process

Now we are ready to define the primary CTRW process with repeating waiting times. This process is characterized by two key properties:

- changes of the process value Δx_n are i.i.d. random variables from the distribution $h(\Delta x)$, with finite variance σ_x^2 (and thus finite first two moments μ_1 and μ_2),
- waiting times Δt_n come from the process described in Section 3.

Note that we do not assume any dependence within the series of consecutive changes of the process value $\Delta x_1, \Delta x_2, \ldots, \Delta x_n$. We do not make any further assumptions about the shape of distributions $h(\Delta x)$. The memory in this process is present only in the sequence of waiting times.

Let us start the analysis of the properties of this process with the following observation. As the changes Δx_n are independent, the changes above any given threshold occur independently. Knowing the result in (8), we can calculate the autocorrelation of the series of inter-occurrence times between changes above or below any threshold. The details of the derivation are presented in Appendix B. It turns out that we also obtain power-law decay with the exponent $-(\rho-2)$, the same as in (8).

Moreover, we managed to obtain the soft propagator of the primary CTRW process and the characteristics derived from it. The details of calculations can be found in Appendix A. Here we present selected results, namely, the first two moments and the time autocorrelation of changes, in the limit of long times ($t \to \infty$). We consider analytical terms ($t, t^2, t^3, \ldots$) and the most significant power-law term when ρ is non-integer.

Using results from Appendix A, the first moment of the process for $t \to \infty$ can be approximated as

$$m_1(t) = \mathcal{L}^{-1}\left[-i\frac{\partial\tilde{P}(k;s)}{\partial k}\bigg|_{k=0}\right](t) \approx \frac{\mu_1}{\langle\Delta t\rangle}t + \mu_1\frac{\alpha\{\psi\}}{\Gamma(4-\rho)}t^{3-\rho}, \quad \rho \in (2;4), \tag{9}$$

where $\mathcal{L}^{-1}[\cdot](t)$ is the inverse Laplace transform, $\tilde{P}(k;s)$ is the propagator of the process in the Fourier–Laplace domain, $\Gamma(\cdot)$ is Euler's gamma function, and $\alpha\{\psi\}$ is a complex functional of ψ, which has to be calculated separately for each ψ. The most important term is typical, linear behavior, but we observe an additional power-law term. The second moment can be written in the form

$$\begin{aligned} m_2(t) &= \mathcal{L}^{-1}\left[-\frac{\partial^2\tilde{P}(k;s)}{\partial k^2}\bigg|_{k=0}\right](t) \\ &\approx \mu_1^2\left(\frac{t}{\langle\Delta t\rangle}\right)^2 + \sigma_x^2\frac{t}{\langle\Delta t\rangle} + \mu_1^2\beta\{\psi\}\frac{t}{\langle\Delta t\rangle} + \mu_1^2\frac{\gamma\{\psi\}}{\Gamma(5-\rho)}t^{4-\rho}, \quad \rho \in (2;5), \end{aligned} \tag{10}$$

where $\beta\{\psi\}$, $\gamma\{\psi\}$ are complex functionals of ψ, which have to be calculated separately for each ψ. From the first two moments of the process, we calculate the process variance (still considering only analytical and the most important power-law term)

$$\sigma^2(t) = m_2(t) - m_1^2(t) \approx \left(\sigma_x^2 + \mu_1^2\beta\{\psi\}\right)\frac{t}{\langle\Delta t\rangle} + \mu_1^2\frac{\gamma\{\psi\}}{\Gamma(5-\rho)}t^{4-\rho}, \quad \rho \in (2;5). \tag{11}$$

It is worth mentioning that for variance the power-law term from the second moment is more important than the power-law term from the first moment. We can observe normal diffusion for $\rho > 3$. However, there is superdiffusion in the case of $\rho \in (2;3)$. We obtain ballistic diffusion in the limit $\rho \to 2$.

Having the first two moments, one can calculate velocity ACF, which is equivalent to normalized ACF of changes for fixed sampling for the stationary process

$$C(t) = \frac{1}{2}\frac{\partial^2 m_2(t)}{\partial t^2} - \left(\frac{\partial m_1(t)}{\partial t}\right)^2 \Rightarrow C(t) \approx \mu_1^2\frac{1}{\Gamma(3-\rho)}\kappa\{\psi\}t^{2-\rho}, \tag{12}$$

where $\kappa\{\psi\} = \left(\frac{\gamma\{\psi\}}{2} - \frac{2\alpha\{\psi\}}{\langle\Delta t\rangle}\right)$, for $\rho \in (2;4)$. In the limit of $t \to \infty$ and $\mu_1 \neq 0$ we observe a power-law decay of ACF with the exponent $\rho - 2$. In the case of $\mu_1 = 0$, it can be proven that this exponent is $\rho - 1$, so the decay is faster (A5).

It is crucial to emphasize that in Equations (9)–(12) for ρ exceeding the mentioned range, there is still a power-law term with the same dependence on μ_1 and the same time exponent. However, the dependence of the amplitude on ρ takes a different, more complex form.

5. Empirical Results

We use the constructed process to investigate the role of correlated inter-trade times in the volatility clustering effect. We consider this process as a toy model, describing high-frequency financial data. The value of the process represents the logarithm of the stock price. We can treat transactions as events that change the price. Therefore, the inter-transaction times correspond to waiting times in our model. The jumps represent the difference in the logarithmic prices of consecutive transactions, which are logarithmic returns [52].

The CTRW formalism allows us to obtain the autocorrelation of price returns. Moreover, the same formalism can be used to obtain the nonlinear ACF of absolute increments. This can be achieved by using different jump distributions $h(\Delta x)$. To model the process of price changes in time, we should use the symmetric distribution $h(\Delta x)$, as the empirical distribution of returns is symmetrical. As a result, we obtain the vanishing mean $\mu_1 = 0$ and the quickly decaying ACF of returns. To derive the nonlinear ACF of absolute returns, we define the new CTRW process, and by calculating its linear ACF, we obtain the nonlinear ACF of price increments. Following [60], if as $h(\Delta x)$ we use only the positive half of the previous distribution multiplied by 2, we deal with the case of non-zero drift and obtain an artificial, monotonically increasing process. As $\mu_1 \neq 0$, we obtain the slow power-law decay of the autocorrelation of absolute returns, as in the empirical results presented as a solid black line in Figure 2b.

Since we assumed only one type of memory in our model, introduced by the distribution $\omega(\nu)$, we cannot expect that the model will be able to reproduce exact values of the empirical nonlinear ACF of the absolute returns. The model, however, should be able to reproduce its slope (as in Figure 2b, in which the green dash-dotted line reproduces the slope of the solid black line). The theoretical slope is obtained analytically and is equal $2-\rho$. It is worth emphasizing that the slope does not depend on the distribution of price changes $h(\Delta x)$ or waiting times $\psi(\Delta t)$ and is fully determined by the single parameter ρ, characterizing the distribution $\omega(\nu)$. This fact significantly simplifies the comparison with the empirical data, as we are required to estimate only one parameter ρ. On the other hand, the assumption of repeated waiting time is a technical method introducing memory. We cannot expect to observe such a phenomenon in the empirical time series. The parameter ρ is a measure of the memory present in the sequence of consecutive waiting times. Therefore, we estimate this parameter using the slope of the step ACF of waiting times, which is equal to $2-\rho$ in the model. It is a surprising and potentially essential fact that the exponent of the decay of the nonlinear time ACF is the same as in the step ACF of waiting times. This result motivates us to compare these two values for empirical financial data. Of course, in the empirical data we also observe a long-term positive step ACF of $|\Delta x|$, which was not included in our model. Therefore, we can expect that the slope of time ACF of $|\Delta x|$ should be slightly higher than the slope of the step ACF of Δt. Since a long-term nonlinear autocorrelation is usually interpreted as a reminiscence of the volatility clustering phenomenon, it is interesting to check what part of the observed volatility clustering effect can be explained only by memory between inter-trade times. We present the results for the five most traded stocks from the Warsaw Stock Exchange in Table 1 (ordered by the number of transactions), with the average inter-trade time not being greater than 30 s.

Table 1. Table with fitted slopes of the empirical stationarized step ACF of waiting times and the time ACF of price changes' absolute values for the five most liquid stocks from the WSE. The time ACF slopes are close to the corresponding step ACF slopes. The analysis was performed on the tick-by-tick market data from the public domain database [70]. The data covers the period from 3 January 2013 to 14 July 2017. For instance, the data set for KGHM contains 3,096,625 transactions.

Company	Step ACF Δt Slope	Time ACF $\lvert\Delta x\rvert$ Slope
KGHM	-0.25 ± 0.04	-0.25 ± 0.02
PKOBP	-0.33 ± 0.08	-0.30 ± 0.02
PZU	-0.26 ± 0.03	-0.28 ± 0.04
PGE	-0.33 ± 0.07	-0.36 ± 0.03
PEKAO	-0.33 ± 0.04	-0.37 ± 0.04

We see that our model can estimate the slope of time ACF with an accuracy of around 10%. Moreover, our model can successfully reproduce the power-law decay of the autocorrelation of inter-occurrence times between changes below or above any given threshold reported in [75,76]. Please note that the decay exponent predicted by our model $-(\rho - 2)$, with empirical values presented in the Table 1, is close to 0.31, as reported in [76].

6. Conclusions

We introduced a new continuous-time random walk (CTRW) model with long-term memory within a sequence of waiting times. We use a simple model of repeating waiting times instead of commonly-used point processes such as the ACD and the Hawkes process. Despite its simplicity, our model of repeating waiting times has a few valuable properties. It is stationary, can be treated analytically, and the distribution of waiting times and memory in its series can be set independently.

As we observe many phenomena with dependencies between waiting times, possible applications of this family of CTRW models go beyond the exemplary application presented here.

However, in this manuscript, we applied the proposed model to describe high-frequency financial time series. We asked ourselves which commonly known properties of the financial time series can be reproduced by the long-term memory introduced in our model, only by means of the repeating waiting times. We have to emphasize that part of these properties, known as stylized facts, depend on the waiting time distribution $\psi(\Delta t)$ and price change distribution $h(\Delta x)$. As we are not trying to study the general ability of continuous-time random walk to describe the high-frequency financial time series, we have not studied the broad distribution of log-returns [72], multi-fractality [73,74], or universal scaling of the distribution of times between large jumps [75,76]. We have analyzed the decay of the nonlinear time autocorrelation function of log-returns and the decay of the step autocorrelation function of times between large jumps. Although we considered only memory in a sequence of waiting times, we managed to show that long-term dependencies in waiting times are crucial in explaining the volatility clustering effect and results in the power-law decay of both measures mentioned above.

Our results indicate that the dependence between consecutive price changes is not the primary carrier of long-range memory in the volatility clustering phenomenon. To verify these results, we conducted another simulation. We prepared autocorrelated series of waiting times according to the Fourier filtering method (for example, described in [86]). Similarly, as in our model, both the slopes of the step ACF of WTs and the time ACF of absolute returns were the same. This verification confirms our conclusion and indicates that it is general, independently of the origin of the autocorrelation between inter-trade intervals.

Author Contributions: Data curation, J.K.; formal analysis, J.K. and T.G.; investigation, J.K.; methodology, T.G.; software, J.K.; supervision, T.G; validation, T.G.; visualization, J.K.; writing—original draft, J.K. and T.G.; writing—review and editing, J.K. and T.G. All authors have read and agreed to the published version of the manuscript.

Funding: This research received no external funding.

Institutional Review Board Statement: Not applicable.

Informed Consent Statement: Not applicable.

Data Availability Statement: No new data were created or analyzed in this study. Data sharing is not applicable to this article.

Acknowledgments: We want to thank Ryszard Kutner and Tomasz Raducha for their helpful remarks and comments on the manuscript.

Conflicts of Interest: The authors declare no conflict of interest.

Appendix A

In this Appendix, we sketch the solution for calculating the moments of the process in the limit of long times. All increments of the process Δx are independent, so first we will focus only on the number of jumps. We calculate the probability $P_n(t)$ for $n \geq 0$, which is the probability that it will be exactly n jumps up to time t, $P_0(t)$ can be obtained directly from the definition, as the probability of no jumps in the time t is

$$P_0(t) = \Psi(t) \Rightarrow \tilde{P}_0(s) = \tilde{\Psi}(s), \tag{A1}$$

where $\Psi(\cdot)$ is the sojourn probability for the waiting time distribution. For $n \geq 1$ the process will be described by the number of series of waiting times k, the waiting times in each series Δt^i , and the number of repetitions of waiting times in each series ν_i.

Of course, in each series the number of repetitions has to be at least one, so the number of series k cannot be larger than the number of jumps n. Particularly, the cumulative number of jumps in all k series must equal the total number of jumps: $\nu_1 + \nu_2 + \cdots + \nu_k = n$. All considered jumps have to happen before looking time t, so the total time elapsed has to be smaller than t: $\nu_1 \Delta t^1 + \nu_2 \Delta t^2 + \cdots + \nu_k \Delta t^k \leq t$. Let us call the period between t and the last jump $\delta t = t - \sum \Delta t^i \nu_i > 0$. For the simplicity of notation, let us redefine $\Omega(\nu) = \sum_{n=\nu+1}^{\infty} \omega(n)$.

Looking at the process at time t, we can be in one of two situations. There has been k series of WTs so far. The next $k+1$-th series has just begun, or we could still be in the k-th series. Therefore, the soft propagator $P_n(t)$ for $n \geq 1$ can be written as a sum of two parts:

1. The k-th series of waiting times Δt^k repeated ν_k times, ended before time t, and the process is still in the same position (the next waiting time will be from the new series). The probability of an individual event can be written as the product of these probabilities:
   (a) probabilities $\psi(\Delta t^i)$ that there was waiting time Δt^i in the i-th series for each series $1, \ldots, k$;
   (b) probabilities $\omega(\nu_i)$ that the repetition number was ν_i in the i-th series for each series $1, \ldots, k$;
   (c) probability $\Psi(\delta t)$ that the new WT Δt^{k+1} is larger than δt, because there was no jump in this period;
   (d) probability $\Omega(0) = 1$ that the new repetition number ν_{k+1} is at least one. By definition, this probability is one;

   To calculate the total probability, we need to consider the aggregated probability of all individual events. This means considering all possible sets of variables k, ν_i, Δt^i. To do this, we need to sum over all possible k and ν_i and integrate over all possible Δt^i.
2. The process is, during the series of WTs Δt^k, repeated ν_k times so far. The probability of an individual event can be written as the product of these probabilities:
   (a) probabilities $\psi(\Delta t^i)$ that there was waiting time Δt^i in the i-th series for each series $1, \ldots, k$;
   (b) probabilities $\omega(\nu_i)$ that the repetition number was ν_i in the i-th series for each series $1, \ldots, k-1$;
   (c) probability $\Omega(\nu_k)$ that there will be at least one more repetition of the k-th WT. This means that the total number or repetitions in this series is larger than ν_k observed so far;

   As before, we need to consider the aggregated probability of all individual events to calculate the total probability. In this case, we have another constraint for Δt^k , which has to be larger than δt, because in other cases, there should be another jump before time t.

Summing the above requirements, we can write the formula for the soft propagator:

$$P_n(t) = \sum_{k=1}^{n} \sum_{\substack{\nu_1,\dots,\nu_k \\ \nu_1+\dots+\nu_k=n}} \int_{\substack{\Delta t^1,\dots,\Delta t^k \\ 0<\delta t}} \psi(\Delta t^1)\dots\psi(\Delta t^k)\Psi(\delta t)\omega(\nu_1)\dots\omega(\nu_k)d\Delta t^1\dots d\Delta t^k$$
$$+\sum_{k=1}^{n} \sum_{\substack{\nu_1,\dots,\nu_k \\ \nu_1+\dots+\nu_k=n}} \int_{\substack{\Delta t^1,\dots,\Delta t^k \\ 0<\delta t<\Delta t^k}} \psi(\Delta t^1)\dots\psi(\Delta t^k)\omega(\nu_1)\dots\omega(\nu_{k-1})\Omega(\nu_k)d\Delta t^1\dots d\Delta t^k. \tag{A2}$$

Next, we calculate the Laplace transform ($t \to s$) and Z transform ($n \to z$) to obtain

$$\tilde{P}(z;s) = \tilde{P}_0(s) + \tilde{P}_z(s) = \frac{1}{s}\frac{1}{1-\tilde{f}(z;s)}\left[1+\tilde{F}(z;s) - z(\tilde{F}(z;s)+\tilde{f}(z;s))\right], \tag{A3}$$

where $\tilde{f}(z;s) = \sum_{\nu=1}^{\infty} z^{-\nu}\tilde{\psi}(s\nu)\omega(\nu)$ and, analogically, $\tilde{F}(z;s) = \sum_{\nu=1}^{\infty} z^{-\nu}\tilde{\psi}(s\nu)\Omega(\nu)$. Notice that the full soft propagator with included jumps can be easily expressed as the Z transform of $\tilde{P}_n$ at the point $z = \tilde{h}(k)^{-1}$

$$\begin{aligned}\tilde{P}(k;s) &= \sum_{n=0}^{\infty} \tilde{P}_n \tilde{h}^n(k) = \tilde{P}(z;s)\Big|_{z=\tilde{h}(k)^{-1}} \\ &= \frac{1}{s}\frac{1+\tilde{F}(\tilde{h}(k)^{-1};s) - \tilde{h}(k)^{-1}(\tilde{F}(\tilde{h}(k)^{-1};s)+\tilde{f}(\tilde{h}(k)^{-1};s))}{1-\tilde{f}(\tilde{h}(k)^{-1};s)}.\end{aligned} \tag{A4}$$

The first two moments of the process can be calculated as derivatives of the propagator at the point $k = 0$:

$$\begin{aligned}\tilde{m}_1(s) &= -i\frac{\partial \tilde{P}(k;s)}{\partial k}\Big|_{k=0} = \frac{\mu_1}{s}\frac{J_0+j_0}{1-j_0}, \\ \tilde{m}_2(s) &= -\frac{\partial^2 \tilde{P}(k;s)}{\partial k^2}\Big|_{k=0} \\ &= \frac{2\mu_1^2}{s}\frac{j_1(J_0+j_0)+(1-j_0)(J_1+j_1-J_0-j_0)}{(1-j_0)^2} \\ &+ \frac{\mu_2}{s}\frac{J_0+j_0}{1-j_0},\end{aligned} \tag{A5}$$

where we introduced

$$j_n = j(n;s) = \sum_{\nu=1}^{\infty} \nu^n\, \tilde{\psi}(s\nu)\, \omega(\nu), \quad J_n = J(n;s) = \sum_{\nu=1}^{\infty} \nu^n\, \tilde{\psi}(s\nu)\, \Omega(\nu). \tag{A6}$$

Next, we focus on the specific power-law memory. We set the distribution of the number of repeats to be zeta distribution with the parameter ρ: $\omega(\nu) = \frac{\nu^{-\rho}}{\zeta(\rho)}$, $\rho > 2$. The parameter ρ has to be bigger than two because the distribution of the number of repeats must have a finite mean not to break ergodicity. Furthermore, we expand the moments into series, assuming very small s (so for long times). To do that, we need expansions of $j(n;s)$ and $J(n;s)$ for $n = \{0,1\} < (\rho - 1)$. One can express $j(n;s)$ as the power-law sum

$$j(n;s) = \frac{1}{\zeta(\rho)}\sum_{\nu=1}^{\infty} \tilde{\psi}(s\nu)\, \nu^{-(\rho-n)} = \frac{s^{\rho-n-1}}{\zeta(\rho)} \underbrace{\sum_{\nu=1}^{\infty} \tilde{\psi}(s\nu)\, (s\nu)^{-(\rho-n)}\, s}_{I}. \tag{A7}$$

The behaviour of sum I can be estimated by integrals

$$\int_{2s}^{\infty} \tilde{\psi}(x)x^{-(\rho-n)}dx < I < \int_{s}^{\infty} \tilde{\psi}(x)x^{-(\rho-n)}dx. \tag{A8}$$

Therefore, we can approximate sum I into the series and finally obtain

$$j(n;s) = C_n s^{\rho-n-1} + C_n^0 + C_n^1 s + C_n^2 s^2 + C_n^3 s^3 + \cdots . \tag{A9}$$

One can calculate

$$C_n^0 = j(n;0) = \frac{1}{\zeta(\rho)} \sum_{\nu=1}^{\infty} \nu^{-(\rho-n)} = \frac{\zeta(\rho-n)}{\zeta(\rho)} \geq 1. \tag{A10}$$

Moreover, we can notice that

$$\frac{C_1^0}{C_0^1} = -\frac{1}{\langle \Delta t \rangle}. \tag{A11}$$

Similarly, we approximated

$$J(n;s) = D_n s^{\rho-n-2} + D_n^0 + D_n^1 s + D_n^2 s^2 + D_n^3 s^3 + \cdots . \tag{A12}$$

Constant terms are:

$$D_0^0 = \frac{\zeta(\rho-1)}{\zeta(\rho)} - 1 = C_1^0 - C_0^0, \quad D_1^0 = \frac{\zeta(\rho-2) - \zeta(\rho-1)}{2\zeta(\rho)} = \frac{C_2^0 - C_1^0}{2}. \tag{A13}$$

This gives the form of the first moment

$$\tilde{m}_1(s) \approx \frac{\mu_1}{s}\left(C_1^0 + D_0 s^{\rho-2}\right) \frac{\frac{C_0}{C_0^1} s^{\rho-2} - 1}{s C_0^1} = -\frac{\mu_1}{s^2}\frac{C_1^0}{C_0^1} - \frac{\mu_1}{s^{4-\rho}} \frac{D_0 + \frac{C_0 C_1^0}{C_0^1}}{C_0^1}, \tag{A14}$$

concerning only terms increasing with time ($s^{-\alpha}, \alpha > 1$)—the analytical and the most important power-law one. Switching to time variables, we obtain:

$$m_1(t) = \mathcal{L}^{-1}[\tilde{m}_1(s)] \approx \frac{\mu_1}{\langle \Delta t \rangle} t - \mu_1 \frac{D_0 + \frac{C_0}{\langle \Delta t \rangle}}{C_0^1 \Gamma(4-\rho)} t^{3-\rho}. \tag{A15}$$

The second moment can be expressed as

$$\begin{aligned} \tilde{m}_2(s) \approx\ & \frac{2\mu_1^2}{\langle \Delta t \rangle^2} s^{-3} - \mu_1^2 \frac{4C_0^2 + 3C_0^1 \langle \Delta t \rangle + 2C_1^1 \langle \Delta t \rangle + 2D_0^1 \langle \Delta t \rangle + C_2^0 \langle \Delta t \rangle^2}{2C_0^1 \langle \Delta t \rangle^2} s^{-2} \\ & - \mu_1^2 \frac{D_0 + C_1 - D_1 + 2\frac{C_0}{\langle \Delta t \rangle}}{C_0^1 \langle \Delta t \rangle} s^{\rho-5} + \frac{\mu_2}{\langle \Delta t \rangle} s^{-2}. \end{aligned} \tag{A16}$$

This gives us the variance in the time domain presented in the main text.

Appendix B

In our model, we can consider waiting times between extreme events. We define an extreme event as an event occurring on average every $\langle N \rangle$ steps. The autocovariance of waiting times between n extreme events $COV(n)$ can be written as

$$COV(n) = \sum_{W=n-1}^{\infty} \sum_{K_1=1}^{\infty} \sum_{K_2=1}^{\infty} \sum_{i=1}^{K_1} \sum_{j=1}^{K_2} NB(W;n-1) NB(K_1;1) NB(K_2;1) cov(j+i+W-1), \tag{A17}$$

where $cov(\cdot)$ is defined by Equation (5) and $NB(k,n)$ is a negative binomial distribution with k trials, given n successes and the probability of success $\frac{1}{\langle N \rangle}$:

$$NB(k,n) = \binom{k-1}{n-1} \left(\frac{1}{\langle N \rangle}\right)^n \left(1 - \frac{1}{\langle N \rangle}\right)^{k-n}. \tag{A18}$$

The simpler form of this autocovariance can be derived thus:

$$COV(n) = \left(\frac{1}{\langle N\rangle}\right)^{n-1}\sum_{x=0}^{\infty}\left(1-\frac{1}{\langle N\rangle}\right)^{x} cov(x+n)\binom{x+n}{n}. \tag{A19}$$

Next, we calculate its Z-transform

$$\widehat{COV}(z) = \langle N\rangle\, \widehat{cov}\left(\frac{1}{\frac{\langle N\rangle-1}{\langle N\rangle}+\frac{1/\langle N\rangle}{z}}\right), \tag{A20}$$

where

$$\widehat{cov}(z) = \frac{1}{\zeta(\rho-1)}\frac{z}{(z-1)^2}\left[(z-1)\zeta(\rho-1)-\zeta(\rho)+Li_{\rho,z^{-1}}\right] \tag{A21}$$

and $Li_{\rho,z^{-1}}$ is a polylogarithm function. Setting $z=\exp(s)$, we can show that the most important power-law term is $s^{\rho-3}$, which corresponds to a power-law decay $COV(n)\sim n^{-(\rho-2)}$, analogically to Equation (8).

References

1. Embrechts, P.; Klüppelberg, C.; Mikosch, T. *Modelling Extremal Events: For Insurance and Finance*; Stochastic Modelling and Applied Probability; Springer: Berlin/Heidelberg, Germany, 2013.
2. Daley, D.J.; Vere-Jones, D. *An Introduction to the Theory of Point Processes*, 2nd ed.; Springer: Berlin/Heidelberg, Germany, 2003.
3. Engle, R.F.; Russell, J.R. Autoregressive Conditional Duration: A New Model for Irregularly Spaced Transaction Data. *Econometrica* **1998**, *66*, 1127–1162. [CrossRef]
4. Hawkes, A. Spectra of some self-exciting and mutually-exciting point processes. *Biometrika* **1971**, *58*, 83. [CrossRef]
5. Hawkes, A. Point spectra of some mutually-exciting point processes. *J. R. Stat. Soc. B* **1971**, *33*, 438. [CrossRef]
6. Dufour, A.; Engle, R.F. Time and the Price Impact of a Trade. *J. Financ.* **2000**, *55*, 2467–2498. [CrossRef]
7. Engle, R.F.; Lange, J. Predicting VNET: A model of the dynamics of market depth. *J. Financ. Mark.* **2001**, *4*, 113–142. [CrossRef]
8. Engle, R.F.; Russell, J.R. Forecasting the frequency of changes in quoted foreign exchange prices with the autoregressive conditional duration model. *J. Empir. Financ.* **1997**, *4*, 187–212. [CrossRef]
9. Bauwens, L.; Giot, P. The Logarithmic ACD Model: An Application to the Bid-Ask Quote Process of Three NYSE Stocks. *Ann. D'éConomie Stat.* **2000**, *60*, 117–149. [CrossRef]
10. Grammig, J.; Maurer, K.O. Non-monotonic hazard functions and the autoregressive conditional duration model. *Econom. J.* **2000**, *3*, 16–38. [CrossRef]
11. Pacurar, M. Autoregressive conditional duration models in finance: A survey of the theoretical and empirical literature. *J. Econ. Surv.* **2008**, *22*, 711–751. [CrossRef]
12. Hawkes, A.G. Cluster models for earthquakes—Regional comparisons. *Bull. Int. Stat. Inst.* **1973**, *45*, 454.
13. Ogata, Y. The asymptotic behaviour of maximum likelihood estimators for stationary point processes. *Ann. Inst. Stat. Math.* **1978**, *30*, 243. [CrossRef]
14. Brillinger, D. The identification of point process systems. *Ann. Probab.* **1975**, *3*, 909. [CrossRef]
15. Oakes, D. The Markovian self-exciting process. *J. Appl. Probab.* **1975**, *12*, 69. [CrossRef]
16. Daley, D.J.; Vere-Jones, D. *An Introduction to the Theory of Point Processes. Volume I*, 2nd ed.; Probability and its Applications; Springer: Berlin/Heidelberg, Germany, 2003; pp. 22–469.
17. Bowsher, C. Modelling security market events in continuous time: Intensity based, multivariate point process models. *J. Econom.* **2007**, *141*, 876. [CrossRef]
18. Chavez-Demoulin, V.; Davison, A.C.; McNeil, A.J. Estimating value-at-risk: A point process approach. *Quant. Financ.* **2005**, *5*, 227–234. [CrossRef]
19. Hewlett, P. Clustering of order arrivals, price impact and trade path optimisation. In Proceedings of the Workshop on Financial Modeling with Jump Processes, Palaiseau, France, 6–8 September 2006.
20. Large, J. Measuring the resiliency of an electronic limit order book. *J. Financ. Mark.* **2007**, *10*, 1. [CrossRef]
21. Beran, J.; Feng, Y.; Ghosh, S. Modelling long-range dependence and trends in duration series: An approach based on EFARIMA and ESEMIFAR models. *Stat. Pap.* **2015**, *56*, 431–451. [CrossRef]
22. Jasiak, J. Persistence in Intertrade Durations. *Finance* **1999**, *19*, 166–195. [CrossRef]
23. Karanasos, M. The Statistical Properties of Exponential ACD Models. *Quant. Qual. Anal. Soc. Sci.* **2008**, *2*, 29–49.
24. Beran, J.; Feng, Y. Iterative Plug-in Algorithms for SEMIFAR Models: Definition, Convergence, and Asymptotic Properties. *J. Comput. Graph. Stat.* **2002**, *11*, 690–713. [CrossRef]
25. Beran, J.; Feng, Y. SEMIFAR models—A semiparametric approach to modelling trends, long-range dependence and nonstationarity. *Comput. Stat. Data Anal.* **2002**, *40*, 393–419. [CrossRef]

26. Deo, R.; Hsieh, M.; Hurvich, C.M. Long memory in intertrade durations, counts and realized volatility of NYSE stocks. *J. Stat. Plan. Inference* **2010**, *140*, 3715–3733. [CrossRef]
27. Deo, R.; Hurvich, C.M.; Soulier, P.; Wang, Y. Conditions for the propagation of memory parameter from durations to counts and realized volatility. *Econom. Theory* **2009**, *25*, 764–792. [CrossRef]
28. Sun, W.; Rachev, S.; Fabozzi, F.J.; Kalev, P.S. Fractals in trade duration: Capturing long-range dependence and heavy tailedness in modeling trade duration. *Ann. Finance* **2008**, *4*, 217–241. [CrossRef]
29. Ghysels, E.; Jasiak, J. GARCH for Irregularly Spaced Financial Data: The ACD-GARCH Model. *Stud. Nonlinear Dyn. Econom.* **1998**, *2*, 4. [CrossRef]
30. Hautsch, N. *Econometrics of Financial High-Frequency Data*; Springer: Berlin/Heidelberg, Germany, 2012; pp. 13–371.
31. Bhogal, S.K.; Thekke Variyam, R. Conditional duration models for high-frequency data: A review on recent developments. *J. Econ. Surv.* **2019**, *33*, 252–273. [CrossRef]
32. Hawkes, A.G. Hawkes processes and their applications to finance: A review. *Quant. Financ.* **2018**, *18*, 193–198. [CrossRef]
33. Bacry, E.; Mastromatteo, I.; Muzy, J.F. Hawkes Processes in Finance. *Mark. Microstruct. Liq.* **2015**, *1*, 1550005. [CrossRef]
34. Daley, D.J.; Vere-Jones, D. *An Introduction to the Theory of Point Processes. Volume II*, 2nd ed.; Probability and its Applications; Springer: Berlin/Heidelberg, Germany, 2008; pp. 18–573.
35. Chavez-Demoulin, V.; McGill, J. High-frequency financial data modeling using Hawkes processes. *J. Bank. Financ.* **2012**, *36*, 3415. [CrossRef]
36. Bacry, E.; Dayri, K.; Muzy, J.F. Non-parametric kernel estimation for symmetric Hawkes processes. Application to high frequency financial data. *Eur. Phys. J. B* **2012**, *85*, 157. [CrossRef]
37. Bacry, E.; Muzy, J.F. Hawkes model for price and trades high-frequency dynamics. *Quant. Financ.* **2014**, *14*, 1147–1166. [CrossRef]
38. Filimonov, V.; Sornette, D. Quantifying reflexivity in financial markets: Toward a prediction of flash crashes. *Phys. Rev. E* **2012**, *85*, 056108. [CrossRef] [PubMed]
39. Filimonov, V.; Sornette, D. Apparent criticality and calibration issues in the Hawkes self-excited point process model: Application to high-frequency financial data. *Quant. Financ.* **2015**, *15*, 1293–1314. [CrossRef]
40. Hardiman, S.J.; Bercot, N.; Bouchaud, J.P. Critical reflexivity in financial markets: A Hawkes process analysis. *Eur. Phys. J. B* **2013**, *86*, 442. [CrossRef]
41. Hardiman, S.J.; Bouchaud, J.P. Branching-ratio approximation for the self-exciting Hawkes process. *Phys. Rev. E* **2014**, *90*, 062807. [CrossRef]
42. Jaisson, T.; Rosenbaum, M. Limit theorems for nearly unstable Hawkes processes. *Ann. Appl. Probab.* **2015**, *25*, 600–631. [CrossRef]
43. Montroll, E.W.; Weiss, G.H. Random Walks on Lattices. II. *J. Math. Phys.* **1965**, *6*, 167–181. [CrossRef]
44. Kutner, R.; Masoliver, J. The continuous time random walk, still trendy: Fifty-year history, state of art and outlook. *Eur. Phys. J. B* **2017**, *90*, 50. [CrossRef]
45. Kutner, R. Correlated hopping in honeycomb lattice: Tracer diffusion coefficient at arbitrary lattice gas concentration. *J. Phys. Solid State Phys.* **1985**, *18*, 6323. [CrossRef]
46. Kehr, K.; Kutner, R.; Binder, K. Diffusion in concentrated lattice gases. Self-diffusion of noninteracting particles in three-dimensional lattices. *Phys. Rev. B* **1981**, *23*, 4931–4945. [CrossRef]
47. Haus, J.W.; Kehr, K.W. Random walk model with correlated jumps: Self-correlation function and frequency-dependent diffusion coefficient. *J. Phys. Chem. Solids* **1979**, *40*, 1019–1025. [CrossRef]
48. Scalas, E.; Gorenflo, R.; Mainardi, F. Fractional calculus and continuous-time finance. *Phys. Stat. Mech. Its Appl.* **2000**, *284*, 376–384. [CrossRef]
49. Mainardi, F.; Raberto, M.; Gorenflo, R.; Scalas, E. Fractional calculus and continuous-time finance II: The waiting-time distribution. *Phys. Stat. Mech. Its Appl.* **2000**, *287*, 468–481. [CrossRef]
50. Raberto, M.; Scalas, E.; Mainardi, F. Waiting-times and returns in high-frequency financial data: An empirical study. *Phys. Stat. Mech. Its Appl.* **2002**, *314*, 749–755. [CrossRef]
51. Scalas, E.; Gorenflo, R.; Mainardi, F. Uncoupled continuous-time random walks: Solution and limiting behavior of the master equation. *Phys. Rev. E* **2004**, *69*, 011107. [CrossRef]
52. Scalas, E. The application of continuous-time random walks in finance and economics. *Phys. Stat. Mech. Its Appl.* **2006**, *362*, 225–239. [CrossRef]
53. Kutner, R.; Świtała, F. Stochastic simulations of time series within Weierstrass - Mandelbrot walks. *Quant. Financ.* **2003**, *3*, 201–211. [CrossRef]
54. Masoliver, J.; Montero, M.; Weiss, G.H. Continuous-time random-walk model for financial distributions. *Phys. Rev. E* **2003**, *67*, 021112. [CrossRef]
55. Repetowicz, P.; Richmond, P. Modeling share price evolution as a continuous time random walk (CTRW) with non-independent price changes and waiting times. *Phys. Stat. Mech. Its Appl.* **2004**, *344*, 108–111. [CrossRef]
56. Masoliver, J.; Montero, M.; Perelló, J.; Weiss, G.H. The continuous time random walk formalism in financial markets. *J. Econ. Behav. Organ.* **2006**, *61*, 577–598. [CrossRef]
57. Masoliver, J.; Montero, M.; Perello, J.; Weiss, G.H. The CTRW in finance: Direct and inverse problems with some generalizations and extensions. *Phys. Stat. Mech. Its Appl.* **2007**, *379*, 151–167. [CrossRef]

58. Gubiec, T.; Kutner, R. Backward jump continuous-time random walk: An application to market trading. *Phys. Rev. E* **2010**, *82*, 046119. [CrossRef] [PubMed]
59. Gubiec, T.; Kutner, R. Continuous-Time Random Walk with multi-step memory: An application to market dynamics. *Eur. Phys. J. B* **2017**, *90*, 228. [CrossRef]
60. Klamut, J.; Gubiec, T. Directed continuous-time random walk with memory. *Eur. Phys. J. B* **2019**, *92*, 69. [CrossRef]
61. Tejedor, V.; Metzler, R. Anomalous diffusion in correlated continuous time random walks. *J. Phys. Math. Theor.* **2010**, *43*, 082002. [CrossRef]
62. Chechkin, A.V.; Hofmann, M.; Sokolov, I.M. Continuous-time random walk with correlated waiting times. *Phys. Rev. E* **2009**, *80*, 031112. [CrossRef]
63. Dasenbrook, D.; Hofer, P.P.; Flindt, C. Electron waiting times in coherent conductors are correlated. *Phys. Rev. B* **2015**, *91*, 195420. [CrossRef]
64. Karsai, M.; Kaski, K.; Barabási, A.L.; Kertész, J. Universal features of correlated bursty behaviour. *Sci. Rep.* **2012**, *2*, 397. [CrossRef]
65. Wu, Y.; Zhou, C.; Xiao, J.; Kurths, J.; Schellnhuber, H.J. Evidence for a bimodal distribution in human communication. *Proc. Natl. Acad. Sci. USA* **2010**, *107*, 18803–18808. [CrossRef]
66. Livina, V.N.; Havlin, S.; Bunde, A. Memory in the Occurrence of Earthquakes. *Phys. Rev. Lett.* **2005**, *95*, 208501. [CrossRef]
67. Lennartz, S.; Livina, V.N.; Bunde, A.; Havlin, S. Long-term memory in earthquakes and the distribution of interoccurrence times. *EPL (Europhys. Lett.)* **2008**, *81*, 69001. [CrossRef]
68. Zhang, Y.; Fan, J.; Marzocchi, W.; Shapira, A.; Hofstetter, R.; Havlin, S.; Ashkenazy, Y. Scaling laws in earthquake memory for interevent times and distances. *Phys. Rev. Res.* **2020**, *2*, 013264. [CrossRef]
69. Zhang, Y.; Zhou, D.; Fan, J.; Marzocchi, W.; Ashkenazy, Y.; Havlin, S. Improved earthquake aftershocks forecasting model based on long-term memory. *New J. Phys.* **2021**, *23*, 042001. [CrossRef]
70. Dom Maklerski Banku Ochrony Środowiska. Available online: https://bossa.pl/ (accessed on 30 April 2020).
71. Cont, R. Empirical properties of asset returns: Stylized facts and statistical issues. *Quant. Financ.* **2001**, *1*, 223–236. [CrossRef]
72. Mantegna, R.N.; Stanley, H.E. *Introduction to Econophysics: Correlations and Complexity in Finance*; Cambridge University Press: Cambridge, UK, 1999. [CrossRef]
73. Kantelhardt, J.W.; Zschiegner, S.A.; Koscielny-Bunde, E.; Havlin, S.; Bunde, A.; Stanley, H. Multifractal detrended fluctuation analysis of nonstationary time series. *Phys. Stat. Mech. Its Appl.* **2002**, *316*, 87–114. [CrossRef]
74. Jiang, Z.Q.; Xie, W.J.; Zhou, W.X.; Sornette, D. Multifractal analysis of financial markets: A review. *Rep. Prog. Phys.* **2019**, *82*, 125901. [CrossRef]
75. Ludescher, J.; Tsallis, C.; Bunde, A. Universal behaviour of interoccurrence times between losses in financial markets: An analytical description. *EPL (Europhys. Lett.)* **2011**, *95*, 68002. [CrossRef]
76. Ludescher, J.; Bunde, A. Universal behavior of the interoccurrence times between losses in financial markets: Independence of the time resolution. *Phys. Rev. E* **2014**, *90*, 062809. [CrossRef]
77. Montero, M.; Masoliver, J. Nonindependent continuous-time random walks. *Phys. Rev. E* **2007**, *76*, 061115. [CrossRef]
78. Gubiec, T.; Kutner, R. Share Price Evolution as Stationary, Dependent Continuous-Time Random Walk. *Acta Phys. Pol. A* **2010**, *117*, 669–672. [CrossRef]
79. Mayo, W.T. Spectrum Measurements with Laser Velocimeters. In *Proceedings of the Dynamic Flow Conference 1978 on Dynamic Measurements in Unsteady Flows*; Hansen, B.W., Ed.; Springer: Dordrecht, The Netherlands, 1978; pp. 851–868.
80. Gubiec, T.; Wiliński, M. Intra-day variability of the stock market activity versus stationarity of the financial time series. *Phys. Stat. Mech. Its Appl.* **2015**, *432*, 216–221. [CrossRef]
81. Gençay, R.; Dacorogna, M.; Muller, U.; Pictet, O.; Olsen, R. *An Introduction to High-Frequency Finance*; Elsevier Science: Amsterdam, The Netherlands, 2001.
82. Dacorogna, M.M.; Müller, U.A.; Nagler, R.J.; Olsen, R.B.; Pictet, O.V. A geographical model for the daily and weekly seasonal volatility in the foreign exchange market. *J. Int. Money Financ.* **1993**, *12*, 413–438. [CrossRef]
83. Tsay, R. *Analysis of Financial Time Series*; Wiley Series in Probability and Statistics; Wiley-Interscience: Hoboken, NJ, USA, 2005.
84. Saichev, A.; Sornette, D. "Universal" Distribution of Interearthquake Times Explained. *Phys. Rev. Lett.* **2006**, *97*, 078501. [CrossRef]
85. Apostol, T.M. *Introduction to Analytic Number Theory*; Springer: Berlin/Heidelberg, Germany, 1976; pp. 12–338.
86. Makse, H.A.; Havlin, S.; Schwartz, M.; Stanley, H.E. Method for generating long-range correlations for large systems. *Phys. Rev. E* **1996**, *53*, 5445–5449. [CrossRef]

entropy

Article

Understanding Changes in the Topology and Geometry of Financial Market Correlations during a Market Crash

Peter Tsung-Wen Yen [1,†], Kelin Xia [2] and Siew Ann Cheong [3,*,†]

1 Center for Crystal Researches, National Sun Yet-Sen University, No. 70, Lien-hai Rd., Kaohsiung 80424, Taiwan; peter.yen@mail.nsysu.edu.tw
2 Division of Mathematical Sciences, School of Physical and Mathematical Sciences, Nanyang Technological University, 21 Nanyang Link, Singapore 637371, Singapore; xiakelin@ntu.edu.sg
3 Division of Physics and Applied Physics, School of Physical and Mathematical Sciences, Nanyang Technological University, 21 Nanyang Link, Singapore 637371, Singapore
* Correspondence: cheongsa@ntu.edu.sg; Tel.: +65-6513-8084
† These authors contributed equally to this work.

Abstract: In econophysics, the achievements of information filtering methods over the past 20 years, such as the minimal spanning tree (MST) by Mantegna and the planar maximally filtered graph (PMFG) by Tumminello et al., should be celebrated. Here, we show how one can systematically improve upon this paradigm along two separate directions. First, we used topological data analysis (TDA) to extend the notions of nodes and links in networks to faces, tetrahedrons, or k-simplices in simplicial complexes. Second, we used the Ollivier-Ricci curvature (ORC) to acquire geometric information that cannot be provided by simple information filtering. In this sense, MSTs and PMFGs are but first steps to revealing the topological backbones of financial networks. This is something that TDA can elucidate more fully, following which the ORC can help us flesh out the geometry of financial networks. We applied these two approaches to a recent stock market crash in Taiwan and found that, beyond fusions and fissions, other non-fusion/fission processes such as cavitation, annihilation, rupture, healing, and puncture might also be important. We also successfully identified neck regions that emerged during the crash, based on their negative ORCs, and performed a case study on one such neck region.

Keywords: econophysics; financial markets; correlation filtering; minimal spanning tree; planar maximally filtered graph; topological data analysis; SGX; TAIEX

Citation: Yen, P.T.-W.; Xia, K.; Cheong, S.A. Understanding Changes in the Topology and Geometry of Financial Market Correlations during a Market Crash. *Entropy* **2021**, *23*, 1211. https://doi.org/10.3390/e23091211

Academic Editor: Ryszard Kutner

Received: 19 July 2021
Accepted: 6 September 2021
Published: 14 September 2021

1. Introduction

At the turn of the 20th century, Bachelier suggested in his PhD thesis that stock prices follow geometric Brownian motions and worked out some of the consequences [1]. This was a major breakthrough at that time, when few expected any theoretical understanding of the stock market. In his thesis, Bachelier assumed that the prices of term contracts follow a normal distribution. Osborn then proposed that it is the rate of return that follows a normal distribution [2]. Later, Mandelbrot and Fama independently found early evidence to suggest that this is not true, and the return distribution has fat tails better fitted by a Levy stable distribution with $b = 1.7$ [3,4]. Mandelbrot then proposed modeling financial returns using fractional Brownian motion [5] and, later, multifractals [6]. Parallel efforts to understand the complexity of financial markets using agent-based models and evolutionary computing were also undertaken at the Santa Fe Institute by Palmer et al. [7]. Up until this point in time, physicists studied economics problems sporadically, and this body of knowledge was not yet known as econophysics.

Widely recognized to be the start of econophysics are the 1991 paper by Mantegna [8] and the 1992 paper by Takayasu and his co-workers [9]. Then, in 1995, Stanley coined the name *econophysics* during the Statphys-Kolkata conference at Kolkata, India [10]. This marked a watershed moment in the field. After 1995, more physicists worked on economic and financial problems, publishing their results and findings in physics journals. These events ushered in the field of econophysics, where physicists (and mathematicians, as well as computer scientists) brought insights from their own fields to the study of economics and finance. Over the next two decades, econophysicists witnessed several breakthroughs. The earliest success of econophysics is the application of random matrix theory (RMT, which is a statistical theory developed to explain the energy spectra of heavy nuclei) to the stock market [11–14]. In RMT, one treats noise as a kind of symmetry, and thus information represents some form of symmetry breaking. This allows physicists to discriminate between noise and signal in financial markets. The next significant milestone in econophysics was a more compelling demonstration of fat tails in return distributions by Mantegna and Stanley [15,16], and also by Mittnik et al. [17]. These two groups estimated $b = 1.4$ for the Levy stable distribution.

Many other breakthroughs then followed, including the fitting of price time series to a log-periodic power-law (LPPL), which allowed precise predictions of market crashes [18,19], as well as the discovery of dragon kings [20] by Sornette, understanding and modeling of the Gibbs–Pareto distribution of wealth and income by Chakrabarti et al. [21] and Yakovenko [22], characterization of the actual Brownian motion in the price fluctuations [23,24], and the development of the DebtRank metric for measuring systemic risk in financial networks [25]. Other network approaches have also started appearing in econophysics recently. These include recurrence networks (RNs), visibility graphs (VGs), and transition networks (TNs). Recurrence networks were proposed by Marwan, Donner and their co-workers in 2009 [26,27] and are used to study the statistical properties of daily exchange rates [27]. Since the seminal work by Lacasa in 2008 [28], many groups have started using VGs to analyze financial time series, including exchange rates [29], stock indices across different countries [30], the macroeconomics series of China [31], and market indices in the US [32]. A recent article by Antoniades et al. [33] used the TN to investigate the Vosvrda macroeconomic model, but thus far no one has tested the approach on real financial time series data.

Other recent breakthroughs include the application of inverse statistics (IS) in finance. IS, which is deeply rooted in fluid dynamics, and related in particular to the phenomenon of turbulence, is an old yet challenging problem. For the last two decades, many concepts have been borrowed from past studies on turbulence and applied to financial problems. One of them was the use of forward statistics, which aims to answer the question "given a fixed time horizon, what are the typical returns that an investor will realize in that period?". In addition, Jensen [34] proposed the inverse statistics, by turning the question around, to ask "for a given return on an investment, what is the typical time required to realize it?". This latter question is no less pertinent and is more relevant to practical financial management. If IS such as the above can be computed, investors could earn market-beating profits.

Using the IS as a probe, Jensen, Simonsen, and Johansen, published a series of papers starting in the mid-2000s [35–38] to study many economic phenomena. They focused particularly on the Gain-Loss Asymmetry (GLA) in financial markets. GLA refers to the observation that, in a financial market, positive prices have different dynamics from the negative ones. After testing stock indices in the US such as the DJIA [35], Nasdaq, and the S&P 500 [37,39], those in other countries such as Austria [40], Korea [41], and 40 other world indices [39], and other instruments such as FOREX [38], mutual funds [42], it was found empirically that negative returns took shorter average times to realize compared to positive returns of the same magnitude. To explain how GLA occurs in real markets, models with a fear factor have been developed [43–46]. However, factors other than fear of loss might also explain the GLA [47]. A comprehensive survey on IS can be found in the review article by Ahlgren et al. [48].

In this Special Issue, we celebrate the breakthrough that is one of Mantegna's crowning achievements, which is the application of the *minimal spanning tree* (MST) to unravel hierarchical structures in financial markets [49]. We will start by reviewing the essence of Mantegna's insight, and the body of works that followed him (including the systematic embedding of cross correlations onto a hierarchy of surfaces with different genera [50]). We then describe attempts to overcome the limitations of the MST by going to hypergraph approaches [51–54]. A hypergraph is a natural extension of a graph, where instead of having each edge join only two nodes, an edge can join any number of nodes. Unfortunately, the hypergraph approach is difficult to implement starting from pairwise correlations, so we argue that the more promising approach to extract deeper insights into the hierarchical structure in financial markets is through *topological data analysis* (TDA) [55–58]. In TDA, the idea is to go beyond the concepts of nodes (0-simplex), links (1-simplex), and the network that they form to a *simplicial complex*, which can contain $(k > 1)$-simplices as constituents.

In a recent paper [59], we demonstrated how TDA can be used to understand the topological changes that accompany market crashes. For such extreme events in financial markets, one of the key questions not well answered through the use of MSTs or planar maximally filtered graphs (PMFGs) is how the hierarchy of cross correlations between stocks re-organizes itself. In particular, an important class of topological changes is the merging between disjoint clusters (or their time reversal—the splitting of a cluster into disjoint clusters). We found, by tracking how the Betti numbers β_0, β_1, and β_2 change over market crashes, that β_0 (the number of connected components) is small at the beginning of a market crash and increases as the market crash progresses. This tells us that we have a giant connected component in the market just before the crash, and as the market crashed, this broke up into many smaller components. The nature of this breaking up can be understood in greater detail through β_1 (the number of "holes" in the connected components), and β_2 (the number of "voids" in the connected components) (see Figure 1). Based on β_1 and β_2, we realized that a particular crash occurred in two stages. In the first stage, the topology of the giant connected component became more complex, as some "voids" grew outwards to become "holes". In the second stage, the number of "holes" decreased precipitously, presumably the result of handle-breaking events. These handle-breaking events are not simple, because the number of "voids" increases in this stage. Finally, the giant connected component broke up completely into many connected components that have simple topologies (few "holes" and "voids").

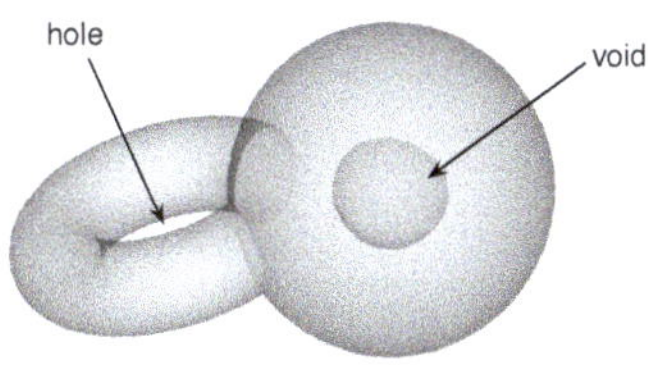

Figure 1. A manifold with a "hole" as well as a "void".

In addition to the TDA, we found another promising approach for extending the information filtering paradigm of MSTs and PMFGs. This is through calculating discrete versions of the Ricci curvature, either the Ollivier-Ricci curvature (ORC) for networks, or the Forman-Ricci curvature (FRC) for simplicial complexes. To identify which stocks in a network or simplicial complex make up the neck or bridge region between two densely connected clusters, the naive approach would be to identify them visually. Naturally, this is laborious and inefficient. It turns out the ORC is ideal for this task, because links in the neck regions have negative ORC. More importantly, the breaking up of a manifold into two involved the stretching and narrowing of the neck region through a process called *Ricci flow*. Physical fission processes closely resemble Ricci flow, even when the objects undergoing fragmentation are networks or simplicial complexes. In such discrete Ricci flows, the ORC or FRC become more negative over time to produce finite-time singularities. Our motives in computing the ORC are threefold: First, we would like to identify the neck regions by looking for where in the network the ORCs are negative. Second, by looking at how the negative ORCs are changing, we would like to predict when we run into finite-time singularities. These are when the fissions occur. Finally, from the natures of the singularities, we would like to understand the drivers for the different fissions.

To make the case for TDA and Ricci curvature analysis, we organized our paper as follows. In Section 2, we will review applications of the MST in econophysics. In Section 3, we will explain how the PMFG can provide more details on correlations between stocks, by keeping more links than in the MST. In fact, there is a hierarchy of maximally filtered networks on closed surfaces with increasing genera (the PMFG being the simplest, on a sphere with genus $g = 0$) that we can explore to understand the structure of correlations between stocks. Unfortunately, the algorithms for obtaining higher-order filtered networks become increasingly difficult to implement, which explains why the PMFG is not as popular as the MST. In fact, we found only one previous work that demonstrated how to filter the weighted links of an artificial complex network onto a torus (with genus $g = 1$) [60]. In Section 4, we describe the ideas behind TDA, and suggest that this is the natural extension going beyond MST and PMFG. To make our case, we explore four toy models for fusions and fissions, and thereafter use their TDA signatures to explain non-trivial topological changes observed in the cross correlations between stocks during a market crash in the Taiwan Stock Exchange (TWSE). In Section 5, we define what Ricci curvature is for smooth surfaces, and describe how this can be generalized to discrete networks and simplicial complexes, in the form of Ollivier-Ricci curvature and Forman-Ricci curvature, respectively. We then explain why we need Ricci curvature analysis to distinguish between different stages of fission processes that are topologically equivalent, before demonstrating this power for one of the toy models. Finally, we use the Ollivier-Ricci curvature to analyze a sequence of PMFGs obtained from the cross correlations of TWSE stocks in overlapping time windows leading up to the market crash of interest, before ending with a comparative case study of two neck regions. In Section 6, we present the conclusions.

2. The Minimal Spanning Tree

In Figure 2, we show the matrix of Pearson cross correlations

$$C_{ij} = \frac{\frac{1}{T}\sum_{t=1}^{T}(x_{i,t} - \bar{x}_i)(x_{j,t} - \bar{x}_j)}{\sqrt{\frac{1}{T-1}\sum_{t'=1}^{T}(x_{i,t'} - \bar{x}_i)^2}\sqrt{\frac{1}{T-1}\sum_{t''=1}^{T}\left(x_{j,t''} - \bar{x}_j\right)^2}} \tag{1}$$

between 561 stocks in the Singapore Exchange (SGX) within the period January 2008 to December 2009. In Equation (1), the time series $\mathbf{x}_i = (x_{i,1}, \ldots, x_{i,t}, \ldots, x_{i,T})$ and $\mathbf{x}_j = (x_{j,1}, \ldots, x_{j,t}, \ldots, x_{j,T})$ with average $\bar{x}_j = \frac{1}{T}\sum_{t=1}^{T} x_{j,t}$ can be the daily prices, daily price differences (also known as the daily returns), or daily log-returns (which are practically identical to the daily fractional returns) of stocks i and j. Their time averages are $\bar{x}_i = \frac{1}{T}\sum_{t=1}^{T} x_{i,t}$ and $\bar{x}_j = \frac{1}{T}\sum_{t=1}^{T} x_{j,t}$. In Section 4.3, we used the daily returns for our topological data analysis. This is acceptable for short time periods, e.g., six months, because the price levels do not change by much. For longer time periods, for example, two years, as in the example associated with Figure 2, we used the daily fractional returns, so that we do not have the problem of increasing weights when the price levels become significantly higher at the end of the time period.

Before the rows and columns are reordered, it is impossible to discern any correlational structures in the SGX stocks. After reordering the rows and columns, we find the strong correlations organized into diagonal blocks, with weaker correlations between them. We also see that within the largest diagonal block in Figure 2b, the correlations are not uniform, but are further organized into diagonal sub-blocks. In hindsight, doing the reordering of rows and columns to reveal these correlational structures in the SGX was a straightforward task, since they have been shown to exist in other markets [61–65]. Mantegna was the first to suspect such hierarchical organizations exist in stock markets and proposed methods to elucidate such structures. Like us, Mantegna employed hierarchical clustering methods to carry out the reordering of rows and columns. However, clustering methods are based on pairwise distances, so the first problem that he had to solve was mapping the conventional Pearson cross correlations, which do not satisfy the three axioms of a distance metric, to pairwise distances. After discussions with Sornette (see Ref. 14 in [49]), Mantegna adopted the mapping

$$D_{ij} = \sqrt{2(1 - C_{ij})} \tag{2}$$

going from a cross correlation C_{ij} between stock i and stock j to a pairwise distance D_{ij}, which satisfies the *strong triangle inequality* $D_{ij} \leq \max\{D_{ik}, D_{kj}\}$. Mantegna then investigated the correlational structures in the component stocks of the Dow Jones Industrial Average (DJIA) and Standards and Poors 500 (S&P 500) indices, using single-linkage hierarchical clustering. Based on these results, Mantegna argued that US stocks do not react equally strongly to the various economic factors, but do so in groups synonymous with those discovered by random matrix theory [66]. This corroboration between Mantegna's 1999 MST paper and Plerou et al.'s 1999 RMT paper was an important discovery at that time.

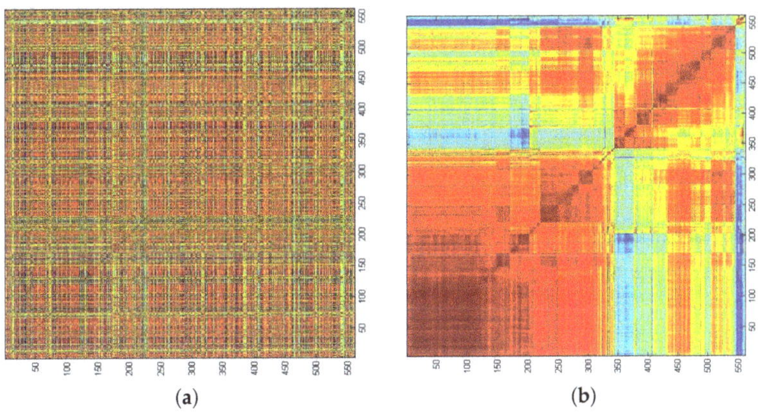

Figure 2. (**a**) The cross-correlation matrix for 561 stocks in the SGX from January 2008 to December 2009. In this figure, red correlations are strongly positive, blue correlations are strongly negative, while green correlations are close to zero. No structures can be discerned in this figure, because the stocks are arranged in alphabetical order. (**b**) After reordering the rows and columns of the cross-correlation matrix, we found strong correlations organized into diagonal blocks, with weaker correlations between them. Material from: Teh et al., Cluster fusion-fission dynamics in the Singapore stock exchange, Euro. Phys. J. B, published 2015 [67], Springer Nature Switzerland AG.

However, the greatest impact of this 1999 paper was the use of the minimal spanning tree (MST) as a caricature of the correlational structures between stocks. A *tree* is a graph with no cycles, and the MST was introduced as early as the 1950s as a special subgraph of a weighted graph containing cycles. In Figure 3a, we show the algorithm attributed to Kruskal [68] for constructing an MST, as well as an example in Appendix A. Following Mantegna's lead, many others (including ourselves) started publishing papers on the MSTs of different markets in, for example, the US [69–75], UK [76], Korea [77,78], Japan [79], China [80], India [81], Indonesia [82], and Africa [83]. We also find the MST applied to different classes of financial instruments: market indices [81,84–86], bonds and interest rates [87–89], currencies [90–95], commodities [96–101], overnight loans in an interbank network [102], housing market indices of different countries [103], to name just a few. Beyond Mantegna's test of the temporal stability of the MST representation (where he changed the time period slightly, recomputed the cross correlations, and drew the MST again) [69],

Onnela et al. also used the MST to visualize the progression of a market crash [70,104]. Other applications include Sun et al. [105,106] and Jiang et al. [107] using the MST to detect insider trading in stock markets, as well as Onnela et al. [70,73], Tola et al. [108], and Coelho et al. [109] using the MST for portfolio selection. The popularity of the MST in econophysics should be clear from this quick survey, and interested readers can refer to the reviews [110,111] for even more references.

(a)

```
Algorithm Kruskal's algorithm
procedure Build MST ({d_ij}; (1 ≤ i, j ≤ N))
    ▷ Start with a fully disconnected graph G = (V, E)
    E ← ∅
    V ← {i}_{1≤i≤N}
    ▷ Add edges with increasing distances
    for (i, j) ∈ V^2 ordered by increasing d_ij do
        ▷ Verify that i and j are not in a predetermined path
        if there is no path between (i, j) then
            ▷ connect i, j
            E ← E ∪ {(i, j)}
    G is the resulting MST
    return G(V, E)
```

(b)

```
Algorithm Planar maximally filtered graph algorithm
procedure Build PMFG ({d_ij}; (1 ≤ i, j ≤ N))
    ▷ Start with a maximum correlation C_ij, and draw a link
between i, and j, i.e. (i, j)
    ▷ Go to the next largest C_ij,
    if a new link preserve the planarity of the graph, do
        ▷ E ← E ∪ {(i, j)}
    else
        ▷ reject (i, j)
    if all nodes are incorporated into a simple graph, do
        ▷ Stop
    else
        ▷ Go to step 3
    return G(V, E)
```

Figure 3. Pseudo codes for (**a**) minimal spanning tree and (**b**) a planar maximally filtered graph.

3. The Planar Maximally Filtered Graph (PMFG)

The successes of the MST in econophysics inspired many other network studies. For example, to understand the same finance and economics problems, many groups experimented with other types of networks [112–119]. Others, such as Chen et al. [120], experimented with artificial markets on small world networks, scale-free networks, and multilayer networks, to find noticeable differences in market sentiments on these different networks. We even found work focusing on developing complex network metrics that can be used to track the evolution of financial markets across different states (for further information, see the review by Kennett and Havlin [121]). Working more or less separately from network scientists, economists approach the network structure of financial markets from the broader perspective of *market microstructure*. The National Bureau of Economic Research has a market microstructure research group that, it says, "··· is devoted to theoretical, empirical, and experimental research on the economics of securities markets, including the role of information in the price discovery process, the definition, measurement, control, and determinants of liquidity and transactions costs, and their implications for the efficiency, welfare, and regulation of alternative trading mechanisms and market structures" [122]. According to a quant school [123], market microstructure deals with issues of market structure and design, price formation and price discovery, transaction and timing cost, volatility, information and disclosure, and market maker and investor behavior. In short, market microstructure is a sub-field of economics that assumes a network structure as a given in financial markets, but introduces additional economic metrics that would help policy makers regulate market dynamics. To this end, we see more network metrics used in economics, and they have become more widely accepted by traditional economists. For example, in a recent paper, Tellez et al. distinguished between secured and unsecured interbank loans, and concluded that the Katz centrality and DebtRank are appropriate measures of systemic risk for the unsecured interbank network, while PageRank is more correlated with the interest rate spread in a secured interbank network [124].

Against this backdrop, one of the most important developments following the popularization of the MST was by Tumminello et al., who took the correlation filtering approach one step further. In econophysics, the MST is typically constructed starting from the cross-correlation matrix, which has $N(N-1)/2$ independent components. However, the MST only keeps $N-1 \ll N(N-1)/2$ of these. These $(N-1)$ MST links are clearly important, but we may also wonder whether some of the discarded links might be just as important. Tumminello et al. realized that we can obtain a hierarchy of filtered graphs by projecting the strongest cross correlations onto surfaces with different genera g [50]. The simplest such projection onto a sphere ($g = 0$) is the *planar maximally filtered graph* (PMFG). This keeps $3N-6$ links, which is more than in the MST but still small. In fact, all the MST links are contained in the PMFG. One advantage of using the MST (which is also true for the PMFG) is that we keep exactly the same number of links for the same number of nodes. This can be less biased than using a correlation threshold value because a small change in the threshold value may lead to a large change in the number of links kept. After the PMFG was introduced, we found the following econophysics papers applying it [86,125–129]. Unfortunately, the PMFG algorithm (see Figure 3b for said algorithm, and an example in Appendix A) is difficult to parallelize. Therefore, for larger data sets, Massara et al. developed a related algorithm called the *triangulated maximally filtered graph* (TMFG) [130].

4. Topological Data Analysis

In this section, we explain how to go beyond MSTs and PMFGs in our understanding of complex dynamics in financial markets by making use of methods developed for topological data analysis (TDA). In Section 4.1, we explain what the shortcomings of MSTs and PMFGs are, what we can understand and what we cannot, and why it is natural to turn to TDA. Following this, in Section 4.2 we briefly explain the ideas behind different TDA methods. We also describe three contrasting toy models for two manifolds to merge together, and a fourth toy model that is like a combination of the first three in Appendix B, before using the TDA signatures for each toy model to understand a real-world market crash in Section 4.3.

4.1. Why Topological Data Analysis?

In most of the MST and PMFG papers, econophysicists merely correlated the topologies of the networks obtained with events in the market, with little or no further explanation. When the MSTs or PMFGs of two successive time periods were compared, analysis is in terms of links created or deleted, but the market may have different numbers of connected components in the two time periods. A more thorough analysis would be to superimpose these connected components and the filtered graphs, to better understand the underlying reasons for link-level changes to the networks. However, when we project market cross correlations onto a MST or a PMFG, we always worry that we may be throwing out important information. Furthermore, by focusing on link-level changes, we are also implicitly assuming that changes to cross correlations can be understood in terms of pairwise interactions between stocks. Already, there are suggestions on the existence of important complex system dynamics that have to be described in terms of many-body interactions. For example, in gene expression networks, there are signs that important functions involve interactions between three or more genes [131–134]. The same is possibly also true for financial markets, but to identify such interactions, we must go beyond network descriptions of such systems.

The explanation for complex topological changes to the cross correlations between stocks lies ultimately with overlapping portfolios [135–140]. Simply put, each entity on the market owns multiple stocks, and because there are more entities than there are stocks, their portfolios necessarily overlap. Even for this bipartite system of entities and stocks, a network description would be an over-simplification. Based on the signals it receives and is capable of processing, an entity periodically optimizes its portfolio by buying and selling stocks. These trading activities generate signals for other entities in the market, who then react to optimize their own portfolios. These interactions at the portfolio level are not open information, but we can observe changes to the prices of stocks, and hence the cross correlations that these interactions produce. Over time, portfolios may accumulate so many changes that the cross correlations between stocks at different times become topologically distinct. Signatures of these topological changes can be seen in the MST [73,104,141] and PMFG [75,86,142] representations.

In their seminal work, Tumminello et al. explained how to project more and more cross correlations onto the surfaces of manifolds with increasing genera [50]. By keeping more links, we keep more of the information in the cross correlations. At the same time, we admit more complex groups (such as simplices) of cross correlations. Then, instead of asking about degree distributions and hubs, we can examine the distribution of k-simplices in the network, and how different simplices are connected to each other. The network obtained from the projection of cross correlations to a manifold with a large genus g should then be treated as a *simplicial complex*, i.e., a connected graph of simplices. In fact, in one of the PMFG papers [130], the authors pointed out that MSTs and PMFGs should already be recognized as simplicial complexes. There is thus potential for an improved understanding of the topological structure of cross correlations in terms of simplices, but somehow Massara et al. did not pursue it further to the natural TDA conclusion.

Recently, we published a TDA paper in the *Frontier in Physics* Special Issue "From Physics to Econophysics back to Physics: Methods and Insights" [59]. In this paper, we worked out the TDA signatures for (1) coalescing spheres, (2) torus to horn torus to spindle torus to sphere, and (3) sphere to ellipsoids, and used these toy models to develop a hypothesis on market crashes corresponding to the fragmentation of a multiply connected manifold with a non-zero genus. In this hypothesis, we have the creation of holes as well as handle-breaking events that accompany fragmentations associated with market crashes. We then presented preliminary evidence confirming the existence of hole creation and handle-breaking events. In this paper, we would like to go deeper to understand how a handle breaks, or its time-reversed event, which is how two disjoint manifolds fuse with each other.

4.2. What Is Topological Data Analysis?

TDA is a suite of mathematical tools developed by Edelsbrunner, Zomorodian, Carlsson, and Singh to analyze the topological properties of complex data sets [55–57]. Built on the foundations of topology [143–147], group theory [148,149], linear algebra [150,151], and graph theory [152–154], TDA has since became a popular field in applied mathematics, and has also found many applications in data analytics [58]. For more information on the history and developments of TDA, readers can consult these review articles [155–158].

In its simplest terms, TDA is a novel way to unravel the topological features of raw data, which can be in the form of point clouds, distance matrices, networks, or digital images. To perform a TDA, we first imagine a control parameter called the *proximity parameter* or *filtration parameter* ϵ. This is the radius of an imaginary ball centered at each of the data points, which we call 0-simplices. When we increase ϵ, the balls will grow outwards and eventually overlap with other balls. When the balls of data points i and j overlap, we draw a link between i and j, and say that the two data points now form a 1-simplex $\{i, j\}$. As ϵ increases further, there will be more overlaps, and if the $k+1$ data points $\{i_1, i_2, \ldots, i_{k+1}\}$ are such that the balls of i_α and i_β overlap, for all pairs of (i_α, i_β) in the set, then we say that $\{i_1, i_2, \ldots, i_{k+1}\}$ forms a k-simplex. The topological information contained in the data set can then be expressed in terms of the distribution of k-simplices, $k = 0, 1, \ldots$, and how they are connected to each other into a *simplicial complex*. For different ϵ, we have different connected subsets of the simplicial complex. The *homology group H* of a simplicial complex summarizes, in a group-theoretic way, the connectivities between k-simplices of different dimensions. As we analyze H_n, the n-dimensional subgroup of H, for $n = 0, 1, \ldots$ over the filtration process, we will discover simplices that remain the same over a large range of ϵ, as well as those that exist fleetingly over very small ranges of ϵ. We call the former the persistent homology of the data set, and based on these we construct useful TDA metrics such as barcodes, persistent diagrams, persistent landscapes, and also persistent Betti numbers. In addition, we combine these to design other tools, such as persistence-weighted kernels, or persistent entropy, and other persistent functions. To allow readers to more easily to grasp the general idea, we show cartoons in Figure 4 to demonstrate how TDA can be applied to a data cloud.

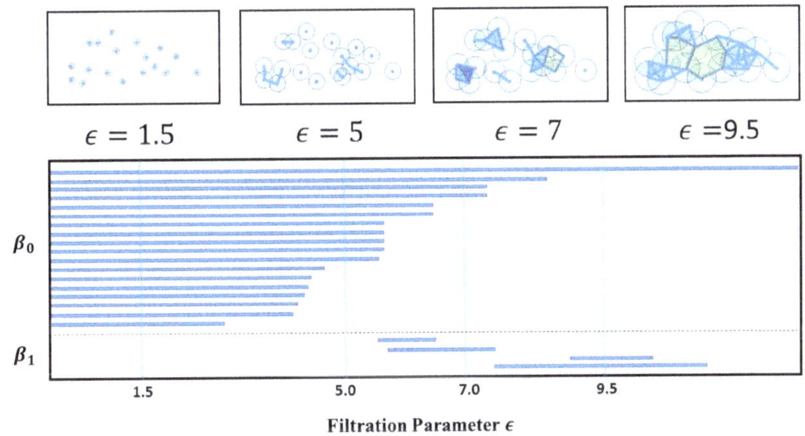

Figure 4. In the top row, we show a schematic diagram showing a data cloud and how the filtration process results in various overlapping outcomes for balls of different proximity parameters ϵ. In the bottom row, we show the barcodes obtained by scanning through the full range of ϵ. In this figure, we partition the barcodes into those for 0-dim simplices, which we indicate using the Betti number β_0, and those for 1-dim simplices, which we indicate using the Betti number β_1. In a barcode diagram, the barcode for a 0-dim simplex (a node) always starts at $\epsilon_1 = 0$, since all points are present at the start of the filtration process. The barcode of a 0-dim simplex ends at $\epsilon_2 > \epsilon_1$, when the point is incorporated into a higher-dimensional simplex. In contrast, the barcode of a 1-dim simplex (a link) starts at $\epsilon_1 > 0$, when two balls of radius ϵ_1 touch. The barcode of this 1-dim simplex then ends at $\epsilon_2 > \epsilon_1$, when a third ball with radius ϵ_2 (which may be part of another simplex) touches the first two. At the values of ϵ shown in the top row, we can also see β_0 going from $18 \to 11 \to 4 \to 1$, and β_1 going from $0 \to 0 \to 1 \to 2$, respectively.

Recently, TDA has found applications in many areas. These include computer network structures [159–161], computational biology [162–168], image analysis [169–172], vision [170], data analysis [58,173–176], shape recognition [177], and amorphous material structures [178,179]. More recently, we found the use of TDA in the reconstruction of brain functional networks [180,181], the analysis of financial markets [182,183], and haze detections [184,185]. In fact, TDA has become so much of a cottage industry that many softwares have become available for non-experts. These include Javaplex [186], Dipha [187], jHoles [188], Simpers [189], R-TDA [190], GUDHI [191], PHAT [192], Persus [193], Dinoysus [194], Ripser [195], as well as those reviewed by Otter et al. [155], and Pun et al. [158].

To the best of our knowledge, so far only very few works [182,183,196,197] had applied persistent homology and TDA to the study of trading networks, banking systems, and market crashes. The work closest to ours is that by Gidea and Katz in 2018, who treated the daily log-returns of S&P 500, DJIA, NASDAQ, and Russell 2000 as a four-dimensional data point [183]. They then slided a w-day time window one day at a time to create a sequence of point-cloud data sets that covered the Dotcom Crash of 2000, as well as the Global Financial Crisis of 2007–2009. The topological features they identified from the filtration process are high-order temporal correlations at various time scales. They then devised an L_p norm that can differentiate between persistent landscapes in two time windows, revealing early warning signals preceding crashes. Building on top of this work by Gidea and Katz, as well as the econophysics literature on MSTs and PMFGs, we will report in this paper an understanding of market crashes at levels of detail never before accomplished.

4.3. Using TDA to Understand Market Crashes

In this subsection, we examine the cross-correlation matrices of 671 stocks in the Taiwan Stock Exchange (TWSE) in successive six-month time windows that are seven days apart, and attempt to use the toy-model results in Appendix B to understand the fusion and fission processes associated with the March 2020 crash in greater details. In particular, we would like to ask "how many of each kind of processes do we find?" and "are there combinations of more than one kind of processes?" To answer these questions, we first organize in Table 1 the Betti numbers read off at the largest filtration parameters, for time windows between 1 August 2019 and 31 March 2020. Here, we see that, over the four time windows of August 2019, we have $\beta_0 = 1$, $\beta_1 = 6.5$, and $\beta_2 = 33.75$ on average. Then, in the first two time windows of September 2019, while $\beta_0 = 1$ and $\beta_2 = 40$ remained similar to those in August 2019, β_1 changed from an average of $\beta_1 = 6.5$ to $\beta_1 = 1.5$. For the next three time windows, the topological changes appear to have accelerated. Using the same $\epsilon_{\max} = 1.1$ over the five time windows, we found that the number of simplices increased dramatically from 10 million in the first time window of September 2019 to 85 million in the last time window of September 2019. In this last time window of September 2019, the Javaplex program failed to return any Betti numbers. It was only when we decreased the maximum filtration parameter from $\epsilon_{\max} = 1.1$ to $\epsilon_{\max} = 1.0$, that the number of simplices was reduced to 17 million, giving us $\beta_0 = 6$, $\beta_1 = 19$, and $\beta_2 = 17$.

Table 1. The calculated Betti numbers up to $k = 2$, total links, $\epsilon_{\max}$, and total number of simplices for TWSE during 1 August 2019 to 31 March 2021, which covers the COVID-19 crash with a sliding window of seven days.

Date	β_0	β_1	β_2	Links	$\epsilon_{\max}$	Simplices
(1 August 2019–31 January 2020)	1	6	23	27,675	1.1	3,444,963
(8 August 2019–8 February 2020)	1	9	31	37,696	1.1	15,194,973
(15 August 2019–15 February 2020)	1	5	33	43,708	1.1	31,321,288
(22 August 2019–22 February 2020)	1	6	48	46,507	1.1	41,178,428
(1 September 2019–01 March 2020)	1	2	40	46,944	1.1	42,079,525
(7 September 2019–8 March 2020)	1	1	40	47,482	1.1	44,068,045
(15 September 2019–15 March 2020)	2	8	19	58,201	1.1	39,877,266
(22 September 2019–22 March 2020)	65	17	1	36,504	0.75	40,871,885
(1 October 2019–31 March 2020)	91	8	1	37,640	0.65	57,119,884

To put these β_0 changes in the proper context, let us recall that we analyzed 671 stocks in the TWSE. When $\epsilon = 0$, none of these would be within the ϵ-ball of each other, and thus we found $\beta_0 = 671$. As ϵ increases, links start to form between stocks, and β_0 would decrease. After some point, the change in β_0 would be dominated by the gaps between clusters of stocks. If there are three such clusters, we would find $\beta_0 = 3$ over a wide range of ϵ, before it drops to 2, and then eventually to 1. This is the picture we should have in mind when we say that the persistent Betti number is $\beta_0 = 3$. However, for the last few time windows, we cannot be sure that the β_0 found by Javaplex are its persistent values. Physically, it is meaningful to compare persistent Betti numbers. It is also meaningful (but less so) to compare Betti numbers for a given value of ϵ. However, it is meaningless to compare Betti numbers obtained with different filtration parameters if they are not all persistent. Based on our past experience, there seems to be an analogy between the filtration parameter and the temperature of a thermodynamic system. Normally, a 10% change in the filtration parameter ϵ results in a corresponding 10% change in the number of simplices (akin to the number of arrangements whose logarithm gives us the entropy [181,198]), and hardly any changes to the Betti numbers, if we have already arrived at their persistent values. However, when the system is close to a critical point, a small change in temperature can produce a large change in the number of accessible states (analogous to simplices). To put it simply, our analysis of the Betti numbers suggests that the cross correlations in the first six time windows were more or less similar topologically, whereas for (15 September 2019, 15 March 2020) and subsequent time windows, the Betti numbers became extremely sensitive to ϵ over a broad range of ϵ, suggesting a non-trivial topological transition over the last three time windows. Another signature of this topological transition is the *persistence weakening* phenomenon that we observed in our earlier paper [59], where we found first a slow increase in the number of simplices, and then a rapid increase in the number of simplices after some threshold.

With the above in mind, let us consider the topological changes going from the second time window to the third time window, where we are confident that the Betti numbers obtained are persistent. Between these two time windows, we found that $\Delta\beta_0 = 0$, $\Delta\beta_1 = -4$, and $\Delta\beta_2 = +2$. Comparing these against the results of Appendix B, we realized that there were no fusions ($\Delta\beta_0 < 0$) or fissions ($\Delta\beta_0 > 0$), and therefore, none of the toy models we considered in Appendix B would be able to explain the changes to β_1 and β_2. In fact, for these first few time windows, the changes to β_1 appeared to be independent of changes to β_2, i.e., the creation/annihilation of holes seems to be independent of the creation/annihilation of voids. Some possible mechanisms for doing so are shown in Figure 5a–f. Although these time windows were still far from the crash, the picture of the market dynamics they suggest is more complex than we expected. We might need to go to higher-order Betti numbers to fully elucidate this dynamics.

In Appendix C, we showed that it is possible to have persistent Betti numbers (and thus equally meaningful pictures) at different scales. However, this makes the identification of the persistent β_0 more difficult, because we need to identify the filtration parameter values at which the lifetimes change most rapidly. Frequently, these are close to the largest scale, and cannot be easily seen from a full barcode (see Supplementary Material). To perform this multiscale analysis, we need to restrict ourselves to the longest-living barcodes, as shown in Figure 6 for the seventh of our nine time periods, i.e., (15 September 2019, 15 March 2020). In this figure, we find four persistent β_0 values at different scales. For the lowest of these four scales, from $0.91 \leq \epsilon \leq 0.93$, we have $\beta_0 = 19$. Thereafter, from $0.955 \leq \epsilon \leq 097$, we have $\beta_0 = 11$, and then $\beta_0 = 6$ for $0.985 \leq \epsilon \leq 1.015$, and $\beta_0 = 4$ for $1.02 \leq \epsilon \leq 1.04$). In Figure 6, the filtration parameter ends at $\epsilon = 1.05$. If we continue to increase ϵ, it is likely that we would find another persistent $\beta_0 = 2$ at a higher scale. Unlike for the seventh time period, which illustrated our ideas in Figure A6 very well, similar analyses for the eighth (22 September 2019, 22 March 2020) and ninth (1 October 2019, 31 March 2020) time periods would not yield equally convincing results, because the number of simplices at $\epsilon \approx 1$ is far too large for Javaplex to handle. We also do not expect to find strongly persistent β_0 to emerge at the scales of $\epsilon_{\max} = 0.76$ for the eighth time window, and $\epsilon_{\max} = 0.65$ for the ninth time window.

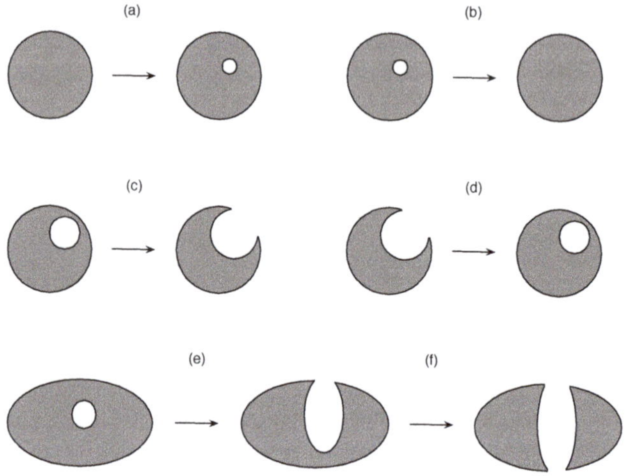

Figure 5. Topological changes that does not involve fusion or fission: (**a**) *cavitation*, in which a void forms within a manifold, (**b**) *annihilation*, in which a void within a manifold disappears, (**c**) *rupture*, in which a void breaks through the surface of the manifold, (**d**) *healing*, in which the surface of the manifold closes over a cavity to form a void, (**e**) another example of rupture, with the growing cavity proceeding to (**f**) *puncture* the manifold, forming a hole. The Betti number signatures of these changes are: (**a**) $\Delta\beta_2 = +1$, (**b**) $\Delta\beta_2 = -1$, (**c**) $\Delta\beta_2 = -1$, (**d**) $\Delta\beta_2 = +1$, (**e**) $\Delta\beta_2 = -1$, and (**f**) $\Delta\beta_1 = +1$.

To wrap up this section, we now understand that it is only meaningful to compare persistent Betti numbers or the Betti numbers at a fixed ϵ. However, we also realized from our analysis in Figure A6 that persistent Betti numbers can emerge at multiple scales, and the way to find them is to check where the lifetimes change most rapidly in the barcodes. Although we could not elucidate the persistent Betti number changes for the eighth and ninth time periods (for the March 2020 crash in the TWSE), our analyses of the first few time periods, as well as the seventh time period, are already a testament to the power of TDA. Without TDA, we would not have even guessed the roles of non-fission processes. Certainly, analyses based on the MST and PMFG would not be able to detect nucleation, rupture, and puncture events. Naturally, we need to ask whether such events are important, since these topological changes are not as drastic as fusion or fission. In any case, we must first be able to detect these events before we can evaluate how important they are relative to fusions and fissions. One hint that they might not be of negligible importance is the observed sequences of changes to β_2 and then to β_1 before β_0 changes. Therefore, any method that can detect $\Delta\beta_1$ and $\Delta\beta_2$ has the potential to provide early warning for fusion or fission events with $|\Delta\beta_0| > 0$.

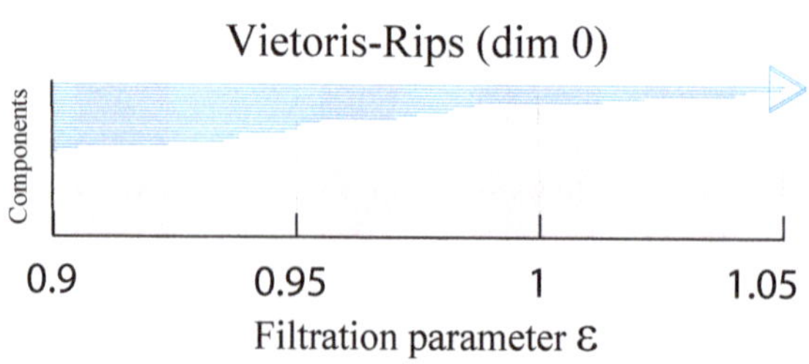

Figure 6. Visualization of the H_0 barcodes of TWSE in the seventh period (15 September 2019, 15 March 2020). We restrict our attention to $0.9 \le \epsilon \le 1.05$, so that we can inspect the finer details. In this figure, the persistent β_0 are highlighted as the pink shaded regions.

5. Going Beyond TDA: Ricci Curvature

Up to this point, we have a very detailed picture of global topological changes to the TWSE over the March 2020 crash. However, metrics such as Betti numbers cannot tell us which stocks participate in which stage of the changes. We can of course manually inspect the output of Javaplex to identify persistent simplices and then track their changes over the time windows. As can be imagined, this is extremely laborious. We would certainly like to have a metric that would automatically pick out not all persistent simplices, but those in the midst of rapid changes. It turns out that such a metric exists, after applied mathematicians recently adapted the idea of Ricci curvature to networks.

5.1. Ricci Curvature and Ricci Flow

To understand Ricci curvature and Ricci flow, we need to start with the *Riemannian metric* $g_{\mu\nu}$, which allows us to specify the distance $d(x,y)$ between any two points x, y on a surface. The Riemannian metric $g_{\mu\nu}$ is also important for the calculation of area. To explain this, let us introduce a disk $B(x,r)$ of radius r centered at x. This is the set of all points y whose distance $d(x,y)$ to x is less than r. On a Euclidean plane, the area $|B(x,r)|$ of $B(x,r)$ would be πr^2. On a Riemannian surface, however, this area can deviate from πr^2. To understand this deviation, let us imagine a disk on the surface of a sphere. Through elementary calculus, we can show that the area of such a disk is a little less than πr^2, and we can understand this deficit as due to the *scalar curvature*

$$R(x) := \lim_{r\to 0} \frac{\pi r^2 - |B(x,r)|}{\pi r^4/24} \tag{3}$$

on the surface of the sphere. One of the main disadvantages of using the scalar curvature is that we do not know whether the curvatures along different directions are the same, or different. Therefore, we extend the notion of the scalar curvature to directional curvatures by defining the *Ricci curvature* as

$$Ric(x)(\nu,\nu) := \lim_{r\to 0}\lim_{\theta\to 0} \frac{\frac{1}{2}\theta r^2 - |A(x,r,\theta,\nu)|}{\theta r^4/24}, \tag{4}$$

for an angular sector $A(x,r,\theta,\nu)$ inside a small disk $B(x,r)$, which has a small angular aperture θ (measured in radians) centered around some direction ν (a unit vector) emanating from x. Here, $|A(x,r,\theta,\nu)|$ is the area of the small angular sector, and $Ric(x)(\nu,\nu)$ is the inner product of the Ricci curvature tensor along the ν direction. If $Ric(x)$ has the same value for all ν, we say that the curvature of the surface is isotropic at x. Otherwise, the curvature at x is anisotropic.

The definition in Equation (4) allows the Ricci curvature to be computed intrinsically, i.e., without embedding the surface in a higher-dimensional space. This property is important when we generalize the Ricci curvature to networks. Going back to surface of a sphere, we will find that $Ric(x)$ has the same positive value at every x, and for every direction ν. Therefore, the Ricci curvature on the surface of a sphere is not only isotropic, it is also positive. For a planar surface, the Ricci curvature is also isotropic at all points, but its value is zero. For an arbitrary two-dimensional surface, the Ricci curvature at a given point will vary from some maximum value to some minimum value. These two values are called the *principal curvatures* of the surface at the given point. In general, for an n-dimensional surface, the Ricci curvature will vary between n principal curvatures, each of which can be positive, negative, or zero. A highly readable explanation can be found in Terence Tao's blog [199].

The Ricci curvature plays an important role in Einstein's theory of general relativity [200]. Even though Einstein worked through the Riemann curvature tensor $R^{\rho}_{\mu\sigma\nu}$, to get to the Ricci curvature $R_{\mu\nu} = \sum_{\rho,\sigma} R^{\rho}_{\mu\sigma\nu}$ and the scalar curvature $R = \sum_{\mu,\nu} R_{\mu\nu}$, we note that the Riemann curvature tensor is coordinate-dependent, while the Ricci curvature and scalar curvature are both coordinate-independent. It therefore makes perfect sense that only coordinate-independent quantities can enter Einstein's field equations. Another application of the Ricci curvature is its use to measure the growth of volumes of distance balls, transportation distances between balls, divergence of geodesics, and meeting probabilities in coupled random walks [201].

Due to its intrinsic character, the Ricci curvature is also the central concept behind the theory of *Ricci flow*,

$$\frac{d}{dt}g = -2Ric, \tag{5}$$

which is the mathematical theory that describes how manifolds deform. Informally, Ricci flow is the process of stretching the Riemannian metric g (increasing distance between points) in directions of negative Ricci curvature, and contracting g (decreasing distance between points) in directions of positive Ricci curvature. The stronger the curvature, the faster the stretching or contracting of the metric. In principle, one can use this equation to perform Ricci flow on a manifold for as long a period of time as one wished. In practice, however, it is possible for a manifold to develop singularities (where the curvature becomes infinite) during the Ricci flow. In three dimensions, many complicated singularities are possible. For instance, one can have a neck pinch, in which a cylinder-like "neck" of the manifold shrinks under Ricci flow, until at one or more places along the neck, the cylinder has tapered down to a point.

In pure mathematics, the theory of Ricci flow was instrumental in the proof of the Poincare conjecture (see Appendix D). So how does Ricci flow connect to what we care about in complex systems, or econophysics in particular? Conventionally, before one looks into the dynamics of a complex system, the first parsimonious step will always be to examine only the backbone (the "topology") of the dynamics. From this perspective, MSTs, PMFGs, graphs, networks, manifolds, or simplicial complexes are different constructs to inform us what this backbone is like. After constructing the backbone, and making sure that it is roughly correct, we then add the "geometry" of the dynamics in as a natural second step, and a natural and coordinate-independent way to quantify this would be to use the Ricci curvature. Therefore, it is important not to go directly into the geometry, before getting a perspective on the topological panorama, because the same curvature value can often mean different things when they are put onto different topologies. For example, the n-sphere and the n-torus are topologically different manifolds, but they could still have similiar average curvatures. Therefore, the use of curvature alone cannot distinguish them. This is also why the correct procedure should always work on the topologies first, before putting the curvatures back, to acquire the correct geometrical information.

From the viewpoint of theorists, the use of differentiable manifolds to describe complex system dynamics is rigorous. In real-world problems, however, manifold constructs are difficult to implement, due to computational limitations. Hence, our plan B often involves the coarse-graining of smooth manifolds. The way this works is to first collect real-world time series cross-section data and calculate their correlation matrices, before visualizing them in terms of networks or simplicial complexes to extract their topological characteristics. From this perspective, we are interested in the breaking of bridges, or the fusion of clusters. For differentiable manifolds, the different types of singularities that can be encountered in two-dimensional Ricci flow have been completely worked out [202], and partially so for Ricci flow in three dimensions [203,204]. For higher dimensions, these are still poorly understood [205]. For networks and simplicial complexes, we need to start with discrete versions of the Ricci curvature. These are the Ollivier-Ricci curvature (ORC) [206,207] and the Forman-Ricci curvature (FRC) [208,209]. The former is applied to networks, whereas the latter is devised for simplicial complexes. It has been found that the ORC is "related to" various graph invariants, ranging from local measures, such as the node degree and clustering coefficient, to global measures, such as betweenness centrality and network connectivity [210]. Thus far, ORC has been used to broadly investigate properties of the internet [210], gene expression networks related to cancer [211], and the structural connectivity of an animal brain [212], as well as to assist in specific tasks such as community detection [213,214], the measurement of market fragility, and the estimation of systemic risk [215]. In this work, we focus on using the ORC, and defer the application of the FRC to future works.

To define the ORC in mathematical terms, we start with an unweighted graph $G = (V, E)$ with vertex set $V = \{x_i\}_{i=1,\ldots,n}$ and edge set $E = \{e_k(x_{i_k}, x_{j_k})\}_{k=1,\ldots,m; i_k, j_k \in V}$, where n is the total number of vertices and m is the total number of edges. Let $\mathcal{N}_x$ be the neighborhood of a vertex $x \in V$. To introduce a curvature measure on a graph, Ollivier associated curvature with transport processes, much like the original concept of curvature being related to the parallel transport of one tangent vector along another. On a graph, the natural transport process to consider is a random walk, and the natural analog of parallel transport is how the hopping probabilities $\mu_x(x')$ from a vertex $x \in V$ to its neighbors $x' \in \mathcal{N}_x$ change to the hopping probabilities $\mu_y(y')$ from a vertex $y \in V$ to its neighbors $y' \in \mathcal{N}_y$ as we move the geodesic distance $d(x, y)$ from x to y. This change can be quantified by the *first Wasserstein distance*, also known as the *earth mover distance*

$$W_1(\mu_x, \mu_y) = \inf \sum_{x' \in \mathcal{N}_x} \sum_{y' \in \mathcal{N}_y} d(x', y') \xi^{xy}(x', y') \tag{6}$$

where inf is the infimum, and $\xi^{xy}(x', y')$ represent the amount of "mass" moved from x' to y', so that, after all movements, the hopping probabilities change from μ_x to μ_y. In the original paper by Ollivier, and others after him, the hopping probabilities μ_x are defined as

$$\mu_x(x') = \begin{cases} \frac{1}{|\mathcal{N}_x|}, & \text{if } x' \in \mathcal{N}_x; \\ 0, & \text{otherwise,} \end{cases} \tag{7}$$

where $|\mathcal{N}_x|$ is the total number of neighbors in $\mathcal{N}_x$. In the eighth example of his 2009 paper [207], Ollivier considered a *lazy random walk*, and used a modified set of hopping probabilities

$$\mu_x(x') = \begin{cases} \frac{1}{2}, & \text{if } x' = x; \\ \frac{1}{2|\mathcal{N}_x|}, & \text{if } x' \in \mathcal{N}_x; \\ 0, & \text{otherwise.} \end{cases} \tag{8}$$

This modification is useful, because in general, random walk on a graph G does not always lead to a stationary probability distribution, whereas a lazy random walk always do. Finally, in terms of $W_1(\mu_x, \mu_y)$, the ORC can be defined as

$$\mathrm{ORC}(x, y) := 1 - \frac{W_1(\mu_x, \mu_y)}{d(x, y)}. \tag{9}$$

This can be obtained using a linear programming procedure to optimize $W_1(\mu_x, \mu_y)$, as shown in Appendix E. In this Appendix, we computed $W_1(\mu_x, \mu_y)$ for two arbitrary nodes x and y on the network G, but in Equation (9), as part of the definition for $\mathrm{ORC}(x, y)$, we compute $W_1(\mu_{x_{i_k}}, \mu_{x_{j_k}})$ only for edges $e_k(x_{i_k}, x_{j_k}) \in E$. For the more common graphs, i.e., tree graphs, grid graphs, complete graphs, or bipartite graphs, $\mathrm{ORC}(x, y)$ can be evaluated in simple mathematical forms.

5.2. Ollivier-Ricci Curvature Analysis of TWSE

After confirming using the toy model in Appendix F the utility of negative ORCs to identify neck regions, we turn our attention to the TWSE March 2020 crash. The neck regions in the simplicial complexes across this market crash should also be thin and weakly connected parts that are most likely to be associated with rapid changes. Since the ORC computation requires a graph as input, we have to produce one starting from the Pearson cross correlations. Therefore, in the first part of this subsection, we limited ourselves to one time period (1 October 2019, 31 March 2020), and explore different ways to create the input graph.

Naively, we can create a complete network in which all links are present, but with different weights. However, such a network will always look like a fur ball when visualized, making it difficult for us to discern the various neck regions. Therefore, the first thing we tried is *threshold filtering*, i.e., we draw a link between stocks i and j if $C_{ij} > C_0$. The Python function that computes the ORC can accept as input disconnected graphs, but we adjusted C_0 until we obtain a fully connected graph. Unfortunately, for this (1 October 2019, 31 March 2020) time period, the fully connected graph obtained for the TWSE has 166,831 links. After visualization it still looks like a fur ball, impeding our investigations of topological changes in the network.

The next thing we tried is the *minimal spanning tree*, which can be constructed using the Kruskal algorithm shown in Figure 3a. Compared to a fur ball, the MST is more informative, especially when we used the *force atlas layout* [216]. In this layout, shown in Figure 7, nodes that are connected by short links have strong Pearson cross correlations, whereas those that are connected by long links have weak Pearson cross correlations. This geometrical feature of the layout allows us to discern clusters of strongly correlated nodes, separated from each other by weak correlations with bridging nodes. However, in the MST only, $N-1$ links are retained for N nodes. These are very few, so we checked how many important links were rejected by the MST in the nine periods, using two measures of importance: (1) correlations larger than the minimum correlation incorporated into the MST, and (2) correlations larger than the minimum correlation associated with the hub of the MST. These are shown in Table 2. Indeed, a large number of cross correlations larger than (1) were rejected in all nine time periods, and especially in the last two time periods. However, the importance measure (1) may be too strict, since we know that for all nodes to be connected in the MST, we frequently have to incorporate weak cross correlations. Based on importance measure (2), which is the correlation level set by the hub, the number of rejected cross correlations is significantly fewer, except during the fifth, seventh, and ninth time periods.

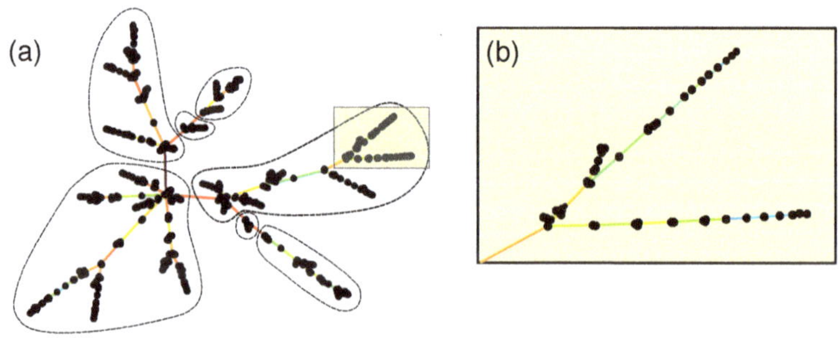

Figure 7. (**a**) The minimal spanning tree of 671 stocks on the TWSE, computed from their Pearson cross correlations between 1 October 2019 and 31 March 2020. In this figure, the black nodes represent stocks, while the colored links represent the most important cross correlations between stocks discovered using the Kruskal algorithm. If a link is red, it has negative ORC, whereas if a link is blue, it has positive ORC. We also sketched the seven clusters in the minimal spanning tree, using links with the most negative ORCs as a guide. (**b**) Enlarging the highlighted region in the minimal spanning tree shown in (**a**), we find that the links between closely spaced nodes have positive ORCs (and thus are shown in blue).

Table 2. In this table on the nine time periods of the TWSE, we show how many stronger cross correlations were rejected in favour of weaker ones, because they would lead to the inclusion of cycles in the MSTs.

Period	1	2	3	4	5	6	7	8	9
$C_{\min}$	0.442	0.429	0.420	0.416	0.444	0.436	0.429	0.428	0.368
Links Rejected	40,519	61,601	76,327	83,437	72,813	76,895	98,935	251,601	333,363
$C_{\min}^{(hub)}$	0.707	0.781	0.689	0.823	0.545	0.729	0.697	0.912	0.730
Links Rejected	2357	977	7971	661	38,811	5607	11,389	977	121,433

Therefore, the last filtering we tried is the *planar maximally filtered graph* (PMFG), adapting the Python example in https://gmarti.gitlab.io/networks/2018/06/03/pmfg-algorithm.html, accessed on 5 May 2021. As we have described in Section 3, this information filtering method was first proposed by Tumminello et al. [50]. We should add that in recent implementations, the Boyer–Myrvold planarity test [217] has replaced the Kuratowski theorem [153] for checking that the graph remains planar at different stages. The resulting PMFG is shown in Figure 8i. By allowing cycles, clusters are more compact in the PMFG. Additionally, $3(N-2)$ links were kept. This is an intermediate number that is still easy to visualize, and contains more of the important cross correlations. In particular, we observed that the cluster at the bottom of the visualization is connected to the rest of the network through two necks (instead of one). However, if we use the same two measures of link importance as for the MST, we see in Table 3 even more important cross correlations rejected in the PMFGs.

After deciding to use the PMFG visualization across all time periods, we tried to identify neck regions that persisted over several time periods to better understand how the market crash proceeded. Therefore, we used the final layout of the first time period as the initial layout of the second time period, the final layout of the second time period as the initial layout of the third time period, and so on and so forth. We had hoped that the PMFGs for successive periods would be sufficiently similar that we could identify features across them. Unfortunately, as we can see from Figure 8, this is not the case, even when we reduced the number of iterations to 100 for the force atlas layout algorithm.

Table 3. In this table on the nine time periods of the TWSE, we show that many stronger cross correlations were rejected in favour of weaker ones, because they would lead to the loss of planarity in the PMFGs.

Period	1	2	3	4	5	6	7	8	9
$C_{\min}$	0.018	0.103	0.097	0.012	−0.209	−0.044	−0.044	−0.079	−0.124
Links Rejected	245,846	224,312	237,712	286,892	380,440	313,430	340,352	415,946	429,102
$C_{\min}^{(hub)}$	0.407	0.401	0.459	0.448	0.545	0.597	0.454	0.600	0.705
Links Rejected	49,414	70,804	59,990	69,216	37,290	24,406	86,074	149,254	103,006

Due to this problem, we abandoned our original ambitious plan to automatically identify all neck regions and their changes. Instead, we manually analyzed the neck region that changed the most dramatically over the market crash. To begin, we first plotted in Figure 9 the number of links with strongly negative ORCs (<-0.5) over the time periods. As we can see, the number of such links increased as we approached the March 2020 market crash, but the number also increased in the third and fourth periods before falling back to levels close to the first and second periods. By checking the number of links with ORC <-0.45 and the number of links with ORC <-0.55, we see that these features are robust and associated with strongly negative ORCs. It appears therefore that a rising number of links with strongly negative ORCs is also an early warning indicator of a market crash. This was first observed by Sandhu et al. [215].

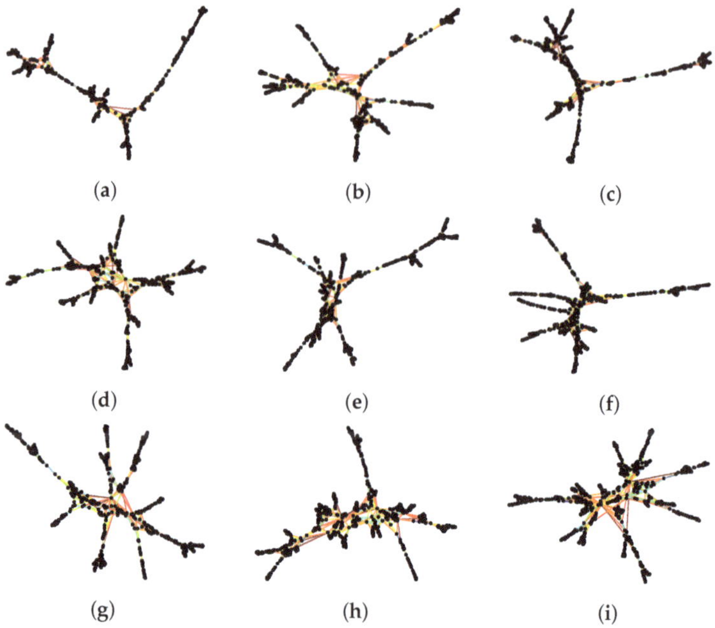

Figure 8. Sequence of PMFGs of 671 stocks on the TWSE, computed from their Pearson cross correlations for the time periods (**a**) 1 August 2019–31 January 2020, (**b**) 8 August 2019–8 February 2020, (**c**) 15 August 2019–15 February 2020, (**d**) 22 August 2019–22 February 2020, (**e**) 1 September 2019–1 March 2020, (**f**) 8 September 2019–8 March 2020, (**g**) 15 September 2019–15 March 2020, (**h**) 22 September 2019–22 March 2020, (**i**) 1 October 2019–31 March 2020. In this figure, the black nodes represent stocks, while the colored links represent the most important cross correlations between stocks discovered using the Kruskal algorithm. The links are colored according to their ORCs, with red being negative, green being approximately zero, and blue being positive.

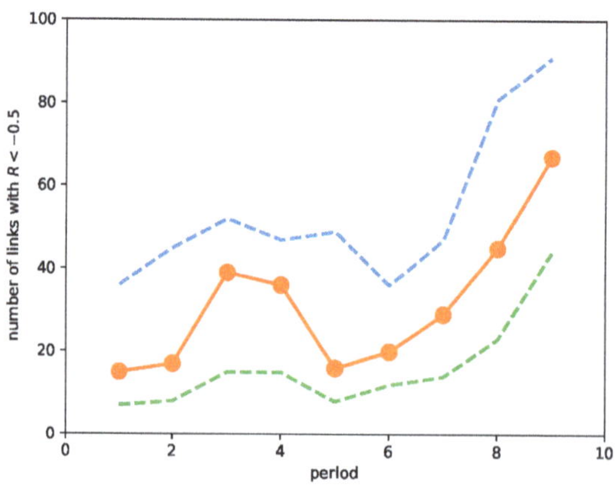

Figure 9. Number of links with ORC < -0.5 over the different time periods (shown in orange): (1) 1 August 2019–31 January 2020, (2) 8 August 2019–8 February 2020, (3) 15 August 2019–15 February 2020, (4) 22 August 2019–22 February 2020, (5) 1 September 2019–1 March 2020, (6) 8 September 2019–8 March 2020, (7) 15 September 2019–15 March 2020, (8) 22 September 2019–22 March 2020, (9) 1 October 2019–31 March 2020. Additionally, the number of links with ORC < -0.45 (blue dashed lines) and the number of links with ORC < -0.55 (green dashed lines) are also shown.

After inspecting the lists of links with ORC < -0.5, we focused on two links, (176, 193) and (176, 393), which (1) appeared frequently in the PMFGs across the nine time periods, and (2) had consistently negative curvatures. These three stocks are: (176) Tung Thih Electronic Co., Ltd., Taoyuan City, Taiwan (3552.TWO); (193) C-Tech United Corp., New Taipei City, Taiwan (3625.TWO); and (393) Taiwan Semiconductor Co., Ltd., New Taipei City, Taiwan (5425.TWO). (176) Tung Thih Electronic Co., Ltd. is a large company in the Auto Parts industry, with market capitalization 15.11 billion TWD, whereas (193) C-Tech United Corp. is a medium-size company in the Electrical Equipments & Parts industry, with market capitalization of 1.45 billion TWD. The last company, (393) Taiwan Semiconductor Co., Ltd., is another large company in the Semiconductors industry, with a market capitalization of 10.5 billion TWD. To put the sizes of these companies into the proper perspective, we compare them against TSMC (2330.TWO), the largest chip maker in the world and one of the largest companies in Taiwan, with a market capitalization of 14.73 trillion TWD. The ORCs of these links over the nine periods are shown in Table 4. Over the period of study, there are no PMFG links between 193 and 393.

From Table 4 we see that ORC$(176, 393)$ is less strongly negative, and change more slowly than ORC$(176, 193)$. Since the link $(176, 193)$ did not appear in the last time period, we suspect that the cluster associated with 193 has completely broken off from the cluster associated with 176. More importantly, comparing Table 4 and Figure 9, we see that the appearance of $(176, 193)$ in the PMFG coincided with the periods when the number of links with strongly negative curvature was increasing. This suggests that the link $(176, 193)$ might have formed in the third time period, broke off in the fifth time period and thereafter reformed in the sixth time period, before finally breaking up in the last time period. Such a sequence of events would surely be interesting to elucidate, but a detailed story might be better suited for a future study that we hope to do using the Generalized Forman-Ricci curvature [218] to more closely track how these fusions and fissions unfold. To wrap this paper up, let us visualize the clusters that these three nodes participate in over the last three time periods.

Table 4. Ollivier-Ricci curvatures of the links $(176, 193)$ and $(176, 393)$ in the PMFGs over the nine time periods. If the curvature value is left blank, the two nodes are not connected in the PMFG.

Period	ORC (176, 193)	ORC (176, 393)
1 August 2019–31 January 2020		
8 August 2019–8 February 2020		
15 August 2019–15 February 2020	−0.61	−0.53
22 August 2019–22 February 2020	−0.64	−0.41
1 September 2019–1 March 2020		−0.39
8 September 2019–8 March 2020	−0.59	−0.39
15 September 2019–15 March 2020	−0.55	
22 September 2019–22 March 2020	−0.37	−0.28
1 October 2019–31 March 2020		−0.07

In network science, in addition to global layout algorithms for visualizing entire networks, we also find ego-centric visualizations centered on a node that are of interest. In Figure 10, we chose 176, 193, and 393 to be the three centers we would like to visualize around. Then, we included all nodes in the immediate neighborhoods of 176, 193, and 393, and colored the links they have with 176, 193, and 393 red if they have ORC < -0.2 (strongly negative), green if $-0.2 \leq$ ORC ≤ 0.2 (roughly zero), and blue if ORC > 0.2 (strongly positive). Next, we drew only green and blue links between the neighbors of 176, 193, and 393, omitting red links between them. Finally, we colored simplices bound by green or blue links yellow. In this way, we keep the number of nodes and number of links to be visualized in Figure 10 to manageable numbers.

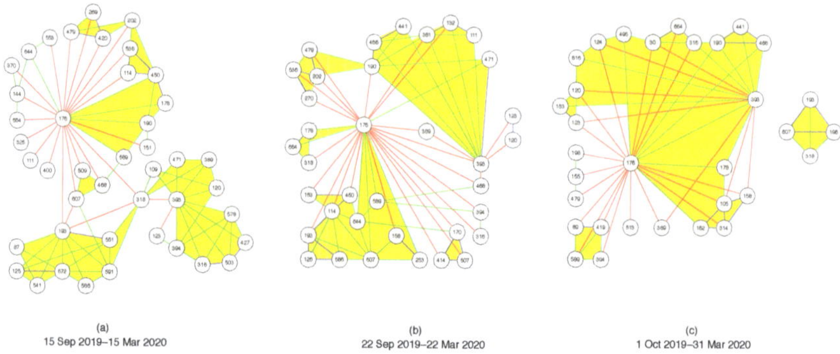

Figure 10. Rough sketch of the fission sequence in the TWSE from time window (**a**) 15 September 2019–15 March 2020 to time window (**b**) 22 September 2019–22 March 2020 to time window (**c**) 1 October 2019–31 March 2020. In this figure, we include the nodes 176, 193, and 393 for all three time windows. In each time window, we include all the nearest neighbors of 176, 193, and 393. We show all the links between these nodes with 176, 193, and 393, and color them red if their ORC < -0.2, green if their $-0.2 \leq$ ORC $\leq +0.2$, and blue if their ORC $> +0.2$. Finally, we show all green and blue links between these nearest-neighbor nodes, and color all simplices bound by green or blue links, to help visualize the clusters in the neighborhoods of 176, 193, and 393. Note that the members of these clusters are dynamic, suggesting strong mixing of cross correlations in the TWSE.

From this figure, we see in the seventh time period (15 September 2019–15 March 2020) that 176, 193, and 393 lied at the peripheries of the clusters they belong to respectively. We know that these nodes were at the peripheries of their respective clusters, because in their ego-centric visualizations, they would be surrounded by mostly green or blue links if they were part of the cores of their clusters. In this time period, 176 and 193 were connected directly through a red link, but the two of them were connected to 393 through 318 (Asia Electronic Material Co., Ltd., Zhubei, Taiwan (4939.TWO), Electronic Components). In the eighth time period (22 September 2019–22 March 2020), we see that the clusters containing 176 and 193 had merged, even though the two nodes were still at the fringe of this merged cluster, and still connected by a red link. 193 remained unlinked to 393, but 176 had "robbed" 318 from 393, but was now directly linked to 393 through a red link, as well as via 389 (AVY Precision Technology Inc., Taipei City, Taiwan (5392.TWO), Electronic Components) and 468 (Netronix Inc., Hsinchu City, Taiwan (6143.TWO), Communication Equipment). 176 also had other red links with the cluster 393 belonged to. Finally, in the last time period (1 October 2019–31 March 2020), we see that the cluster 193 belonged to had completely broken off from 176 (within the PMFG visualization for the entire network). Interestingly, 193 retained its link to 607 (Firich Enterprises Co., Ltd., New Taipei City, Taiwan (8076.TWO), Computer Hardware) from the eighth time period, and regained its connection to 318, at the same time made a new connection to 196 (Newmax Technology Co., Ltd., Taichung, Taiwan (3630.TWO), Electronic Components). Going from the eighth time period to the ninth time period, the biggest change (related to 176, 193, and 393) would be the clusters associated with 176 and 393 merging into a giant cluster. In this giant cluster, 176 and 393 were still peripheral nodes, but there was now a green link between them. In addition, 176 and 393 were also connected by green links through 178 (eGalax_eMPIA Technology Inc., Taipei City, Taiwan (3556.TWO), Semiconductors) and 190 (AimCore Technology Co., Ltd., Hsinchu City, Taiwan (3615.TWO), Electronic Components). In the eighth time period, 178 was connected to 176 by a green link, but not connected to 393, whereas 190 was connected to 393 by a green link, and to 176 by a red link in this time period.

To summarize, changes to the neck regions between 176, 193, and 393 appeared to be very sudden, even when we slid the time window by only seven days. This suggests the need to slide the time window through a smaller time step, to properly track changes to the network of stocks. However, to use such small time steps meaningfully, we will have to use intra-day time series data, instead of the daily data that we used in this paper. Furthermore, none of the fusions and fissions in Figure 10 resemble the simple toy models A or C (single neck, with fixed or varying dimensionality) described in the Appendix, even though it appears that these do start at the periheries of clusters. However, some aspects (multiple distant necks) of these events are similar to what happens in toy models B or D. In this sense, we are starting to understand why re-organizations of cross correlations in the financial market lead frequently to topological features such as voids.

6. Conclusions

Over the past 20 years, state-of-the-art information filtering methods such as the MST and the PMFG have revolutionized the field of econophysics, and also made contributions to other closely related disciplines. In this paper, we suggested two related directions to extend this information-filtering paradigm. The first is through topological data analysis (TDA), and the second is through the calculation of Ollivier-Ricci curvature. The former improves our understanding of the topological backbones of financial networks, whereas the latter puts the geometrical information back onto the topological backbones.

In the TDA, we explored four toy models of fusions, namely (1) the merging of two ellipsoidal surfaces, (2) the merging of two biconvex surfaces, (3) the merging of two anisotropic ellipsoidal surfaces through a sequence of higher-dimensional connections, and finally (4) the merging of two random irregular surfaces. By applying the insights extracted from this exploration to a recent crash in the TWSE, we found the number of simplices increasing slowly with increasing filtration parameter ϵ half a year before the market crash, and rapidly with increasing ϵ close to the crash. This suggests a non-trivial topological transition accompanied the market crash. However, we found that the four fusion/fission models proposed were not able to fully explain the topological changes, and additional processes (cavitation, annihilation, rupture, healing, and puncture) that do not involve fusion or fission, were needed to explain the changes in Betti numbers.

Moving beyond TDA, we used the Ollivier-Ricci curvature to quantify the distribution of curvatures in PMFGs constructed from the correlation matrices of the TWSE. We explained that positive ORCs correspond to stock components deep within a cluster, whereas negative ORCs pinpointed the neck (bridge) regions that connect distinct clusters. When we examined the PMFGs for nine periods between August 2019 and March 2020, we found dramatic topological changes between successive periods. This prevented us from systematically identifying all topological changes that were specifically associated with neck regions in the PMFGs. Instead, we look only at two neck regions—associated with the links (176, 193) and (176, 393)—that featured prominently during this period. These three nodes are: (176) Tung Thih Electronic Co., Ltd, (193) C-Tech United Corp., and (393) Taiwan Semiconductor Co., Ltd. During the last time period, (176, 193) was no longer found in the PMFG, while the curvature of (173, 393) became nearly zero. Using ego-network visualizations of these three nodes and selective visualization of links between them, we saw that all three nodes lie on the peripheries of the clusters they belonged to. In the seventh time period, all three clusters were distinct. In the eighth time period, the cluster containing 176 merged with the cluster containing 193. Finally, in the ninth time period, this cluster broke up into a small cluster containing 193, while the larger cluster containing 176 proceeded to merge with the cluster containing 393.

Supplementary Materials: MATLAB and Python scripts for TDA are available at https://doi.org/10.21979/N9/8XMZGF, accessed on 21 February 2021, whereas MATLAB and Python scripts and data files for Ricci curvature analysis are available at https://doi.org/10.21979/N9/EO5QON, accessed on 19 July 2021. The barcodes for the four toy models in Appendix B and the barcodes for the nine time periods of the TWSE in Section 4.3 are also available online at https://www.mdpi.com/1099-4300/23/9/1211/s1.

Author Contributions: Conceptualization, S.A.C. and K.X.; methodology, S.A.C. and P.T.-W.Y.; software, K.X.; writing—original draft preparation, S.A.C. and P.T.-W.Y.; writing—review and editing, S.A.C., K.X. and P.T.-W.Y.; visualization, S.A.C. and P.T.-W.Y. All authors have read and agreed to the published version of the manuscript.

Funding: This research received no external funding.

Institutional Review Board Statement: Not applicable.

Informed Consent Statement: Not applicable.

Data Availability Statement: All Python and Matlab scripts are provided, along with instructions on how to use them. This will download the raw data from Yahoo! Finance and perform the necessary computations to give the final results. See the links listed in Supplementary Material.

Acknowledgments: P.T.-W.Y. thanks the National Center for High-Performance Computing, Hsinchu, Taiwan for their technical support on providing high performance computing resources on Taiwania 1 server.

Conflicts of Interest: The authors declare no conflict of interest.

Appendix A. Construction of MST and PMFG

For readers interested in how to put the algorithms in Figure 3 into practice, let us use the Pearson cross correlations between 10 Asian stock market indices as an example. These 10 stock market indices are shown in Table A1, and their ordered cross correlations are shown in Table A2.

To construct the MST for these 10 Asian indices, we start from the largest cross correlation $C_{2,7} = 0.705$ in Table A2, to draw a link between node 2 (Hong Kong) and node 7 (South Korea). We then proceed to go through the cross correlations within Table A2 in decreasing order, to add a link between node 2 and node 10 (Singapore), node 7 and node 8 (Taiwan), node 1 (Japan) and node 7, ..., until we reach $C_{2,3} = 0.472$ between node 2 and node 3. To arrive at the tree graph shown in Figure A1a, we have rejected eight links, because they would result in cycles if we accepted them.

Table A1. List of ten Asian stock market indices.

i	Country/Region	Index
1	Japan	Nikkei 225 Index
2	Hong Kong	Hang Seng
3	China	Shanghai Stock Market Composite Index
4	Thailand	SET Index
5	India	BSE Sensex Index
6	Indonesia	Jakarta Stock Index
7	South Korea	KOSPI Index
8	Taiwan	TSE Index
9	Malaysia	Kuala Lumpur Composite Index
10	Singapore	Straits Times Index

According to Table A2, $C_{2,4} = 0.466$ is the next cross correlation to be considered. Indeed, if we add a link between node 2 and the new node 4 (Thailand), no cycles are created. Therefore, we accept this new link, to end up with the tree graph shown in Figure A1b.

After adding the link between node 2 and node 4, the next link we should consider is $C_{4,5} = 0.447$ according to Table A2. However, as we can see, adding this link that is colored red in Figure A1c, accepting this link between node 4 and node 5 (India) leads to a cycle in the network. Therefore, we reject the link between node 4 and node 5. Thereafter, we reject six more links, before adding a link between node 6 (Indonesia) and node 8 to complete the MST in Figure A1d.

Table A2. Ordered list of Pearson cross correlations between the ten Asian indices shown in Table A1.

i	j	C_{ij}	i	j	C_{ij}	i	j	C_{ij}
2	7	0.705	2	4	0.466	5	9	0.331
2	10	0.669	4	5	0.447	5	6	0.316
7	8	0.642	8	9	0.438	1	4	0.304
1	7	0.616	5	8	0.426	4	8	0.295
2	5	0.608	7	9	0.421	3	7	0.277
2	8	0.598	4	7	0.413	4	6	0.275
1	2	0.574	2	9	0.412	3	8	0.267
8	10	0.554	1	5	0.410	3	10	0.258
7	10	0.551	6	8	0.405	4	9	0.253
5	7	0.509	1	9	0.395	3	9	0.228
1	10	0.506	4	10	0.385	3	5	0.227
9	10	0.500	2	6	0.383	1	6	0.222
5	10	0.495	6	10	0.382	3	4	0.199
1	8	0.474	6	9	0.378	1	3	0.197
2	3	0.472	6	7	0.364	3	6	0.129

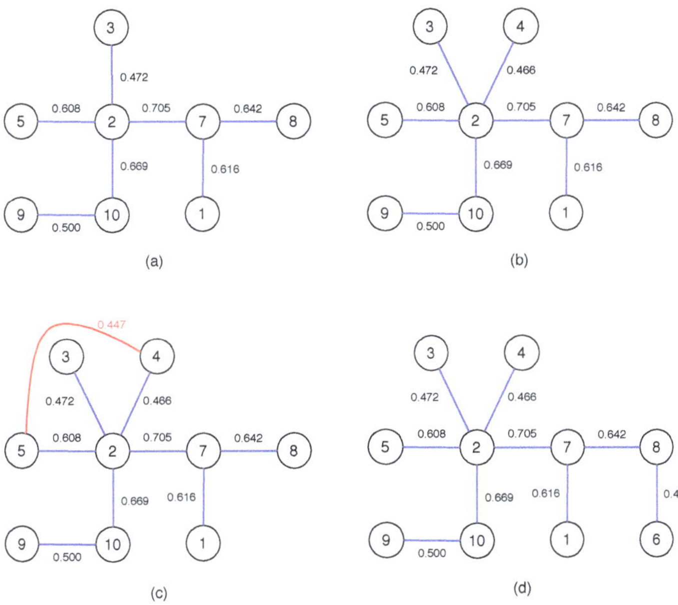

Figure A1. Intermediate and final stages in constructing the MST of 10 Asian indices: (**a**) after adding a link between node 2 and node 3; (**b**) after adding a link between node 2 and node 4; (**c**) after rejecting the red link between node 4 and node 5, due to the appearance of the cycle (5, 2, 4) in the network. Finally, (**d**) all 10 nodes are linked in the MST, after adding the link between node 6 and node 8.

Next, we construct the PMFG for the 10 Asian indices. Again, we start from the largest cross correlation $C_{2,7} = 0.705$ in Table A2, to draw a link between node 2 and node 7. Then, we go through the cross correlations within Table A2 in decreasing order, and arrive at the network shown in Figure A2a after adding a link between node 8 and node 10, without rejecting any stronger links.

According to Table A2, the next link that we should add is between node 7 and node 10. When we first draw this in Figure A2b, the link between node 7 and node 10 overlaps the link between node 1 and node 2. However, this does not necessarily imply that the link must be rejected. If we move node 1 to the other side of 5–2–7–8, but keeping it within the loop formed by the 2–8 link, as shown in Figure A2c, we find no overlaps between links.

Similarly, when we add the next link between node 5 and node 7 in Figure A2d, the new link overlaps with the link between node 2 and node 10. Just like when we add the link between node 7 and node 10, we do not immediately reject this new link, but check if the network obtained thus far can be redrawn to accommodate the new link. Indeed, by moving node 5 into the triangle formed by nodes 2, 7, and 10, we show that the network with the new link continues to be planar, as shown in Figure A2e.

Finally, when we attempt to add the next link between node 5 and node 10, as shown in Figure A2f, we see that there is no rearrangement of the existing nodes and links that we can do, for the new link to not overlap with existing links (and remain planar). Therefore, this link has to be rejected.

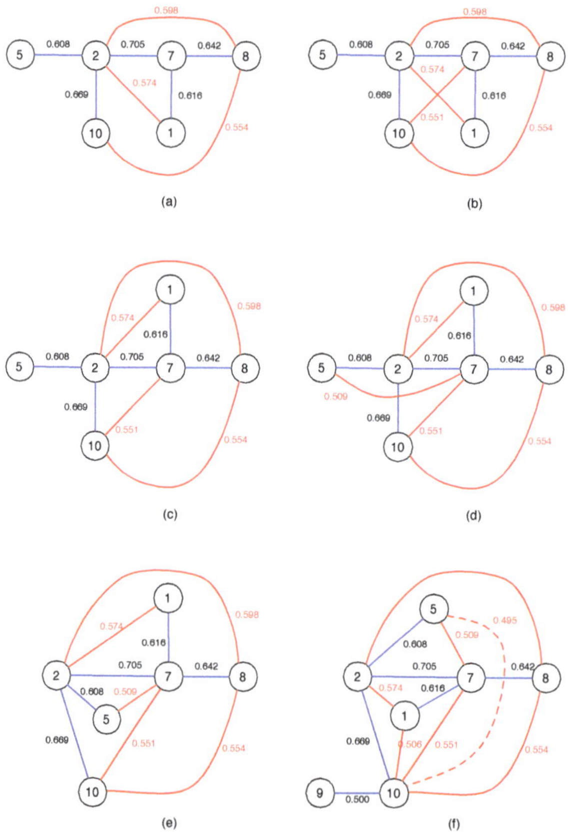

Figure A2. Intermediate stages in constructing the PMFG of 10 Asian indices: (**a**) after adding the link between node 8 and node 10; (**b**) after adding the link between node 7 and node 10, before checking for planarity; (**c**) after adding the link between node 7 and node 10, and after checking for planarity; (**d**) after adding the link between node 5 and node 7, before checking for planarity; (**e**) after adding the link between node 5 and node 7, and after checking for planarity; (**f**) after adding the link between node 5 and node 10, before checking for planarity. After checking for planarity, this link between node 5 and node 10 is rejected. In this figure, MST links are colored blue, PMFG links are colored red, while links that are rejected are shown as red dashed curves.

Appendix B. Topological Data Analysis of Toy Models of Fusions

Before we attempt to understand the complex events underlying a market crash, we should first understand the simplest topological changes associated with the fusion between two manifolds, or the breaking up of a manifold into multiple pieces. In this paper, we worked out the TDA signatures of four toy models (see Figure A3): (1) fusion through a single point of contact, (2) fusion through a 1-dimensional set, (3) fusion through a set with increasing dimensionality, and (4) random. For these toy models, we first normalize the distance matrices by dividing them by the largest distance, so that when we do filtrations, the threshold will always terminate at $\epsilon = 1$. According to some researchers in this field, TDA may detect artifacts or noise due to specific data sampling methods. To eliminate these artifacts due to data sampling, we randomized and reshuffled the data points. This helps us focus on the more meaningful topological features in the data set. For each toy model, we used the Javaplex software to compute the persistent Betti numbers at each stage of the fusion. The barcodes for all stages of all toy models are also shown in the Supplementary Material, and the persistent Betti numbers are read off from the barcodes at the largest filtration parameter used.

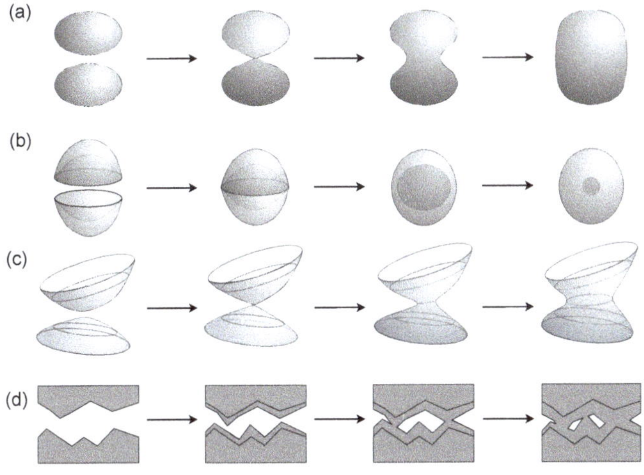

Figure A3. The four fusion models investigated: (**a**) two elliptical manifolds merging by first touching at a point, before developing a hyperbolic neck, and eventually having positive curvature everywhere; (**b**) two biconvex manifolds merging through their rims touching to form a void that shrinks in size; (**c**) a variation of (**a**), where the two anisotropic elliptical manifolds merging with their principal axes not aligned. After touching at a point, the neck between the two merging manifolds first becomes quasi-one-dimensional, before becoming fully two-dimensional; (**d**) the merging between two rough surfaces with random irregularities, which can be thought of as combinations of all of the above, plus emergent features like a hole that eventually becomes a void.

We show the results in Figure A4. For model (A), which is the reverse of the one studied by Santos et al. [181], we found in Figure A4(A1) that while the two ellipsoids are well separated, $\beta_0 = 2$ (two distinct objects), $\beta_1 = 0$ (no irreducible loops on either ellipsoids), and $\beta_2 = 2$ (one void enclosed by each ellipsoid). Then, in Figure A4(A2), the two deformed ellipsoids are just touching, and the point of contact has the form of a Dirac cone, which we expect to have non-trivial topological signatures of its own. As expected, the fusing of the two ellipsoids into a single manifold is reflected in $\beta_0 = 1$. We also found $\beta_1 = 0$, which tells us that there are no irreducible loops. Finally, the Dirac point kept the two voids separated, and hence $\beta_2 = 2$. As shown in Figure A4(A3), the fusion is complete, but the neck region joining the original two ellipsoids has negative curvature. In this situation, we found that $\beta_0 = 1$ (one distinct object), $\beta_1 = 0$, and $\beta_2 = 1$

(the two voids have merged). This is indeed what we expected, and it is indistinguishable from Figure A4(A4), which is the last stage of the whole coalescence process. The neck region has completely disappeared, and the curvature on the fused manifold is everywhere positive. Indeed, we found that $\beta_0 = 1$, $\beta_1 = 0$, and $\beta_2 = 1$. These Betti numbers are identical to those of a sphere.

For model (B), we found in Figure A4(B1) that when the two biconvex shells are separated, we have $\beta_0 = 2$ (two distinct objects), $\beta_1 = 0$ (no irreducible loops), and $\beta_2 = 0$. We were surprised that $\beta_2 = 0$ since each shell should still enclose a void. We increased the number of data points for this case, but β_2 remains zero, presumably because the typical gap in the voids is small, and the filtration procedure connects points across the voids. To test this hypothesis, we sliced the point cloud of (B3) into the top and bottom halves, moved them apart, and covered the gap in the equatorial plane between the inner and outer shells (see Figure A5a). For the top half, we found that $\beta_0 = 1$, $\beta_1 = 0$, and $\beta_2 = 1$. In terms of symmetry, this tells us that for the two parts together, we will have $\beta_0 = 2$, $\beta_1 = 0$, and $\beta_2 = 2$. This means that a void can indeed be identified if its narrowest part is distinctly larger than the typical distance between data points. In the next stage of the fusion, the two biconvex shells just touched in Figure A4(B2) to form an inner spherical shell and an outer ellipsoidal shell. As with (A2), the two two-dimensional shells also touched on a set of measure zero (even though the set is a one-dimensional ring, instead of a zero-dimensional point). This probably explains why $\beta_0 = 2$ (i.e., the shells remain topologically distinct). The other Betti numbers also remained the same as in (B1). Thereafter, with regard to (B3) and (B4), we expected them to have identical topological features, and found indeed that $\beta_0 = 1$ (one distinct object), $\beta_1 = 0$ (no irreducible loops), and $\beta_2 = 2$ (two voids enclosed) for the two cases. An observation we feel compelled to share is that for (B4), the density of red points (outer ellipsoid) and the density of blue points (inner sphere) are initially not the same, because we kept roughly the same number of blue and red points over the whole series of point clouds from (B1) to (B4). We then found the TDA calculations for (B4) were not complete even after four days (in contrast to (B1) to (B3), which took on average of two days). We suspected that this was because in the filtration process, the dense blue points at larger ϵ led to the emergence of very high-dimensional k-simplices. With every increase in dimension, the total number of simplices grew exponentially, and Javaplex slowed down. Indeed, when we performed an alternate TDA calculation in which we reduced the number of blue points, so that their density is comparable to the density of red points, the calculation became much faster: from longer than four days expected for the former, to roughly four hours in the latter.

Next, for model (C), we emulated the change in dimensionality during the fusion of two initially separated parabolic shells (Figure A4(C1)) approaching each other, by first connecting them with a one-dimensional line (Figure A4(C2)). As the fusion progressed, we connected the two parabolic shells with a two-dimensional rectangular sheet (Figure A4(C3)), and finally connected them with a two-dimensional cylindrical shell (Figure A4(C4)). In (C4), we removed points from the original parabolic shells that are within the neck region formed by the cylindrical shell to form the single surface with a channel shown in Figure A5b. We created this toy model as an alternative to model (A), in which the neck formation is locally isotropic. In physics, we know of materials and processes which are anisotropic, i.e., there are easy as well as hard axes. In these materials or processes, the neck formation is expected to go through a sequence of dimensionality changes. For (C1) in this model, we found that $\beta_0 = 2$, $\beta_1 = 0$, and $\beta_2 = 0$. It is understandable that $\beta_0 = 2$, because there are two distinct parabolic shells. We also understand why $\beta_1 = 0$ (because there are no irreducible closed loops) and $\beta_2 = 0$ (because the shells do not enclose any void). When we reached (C2), the two shells became connected. Therefore, we were not surprised to find that $\beta_0 = 1$ (one distinct object). We were also not surprised to find $\beta_1 = 0$ and $\beta_2 = 0$ remaining the same as for (C1). As we moved to (C3) and (C4), we found $\beta_0 = 1$, $\beta_1 = 0$, and $\beta_2 = 0$, the same as in (C2). We had expected differences between (C2), (C3), and (C4), but it seems that these differences do not manifest themselves in the persistent Betti numbers, but require us to look more closely at the Betti curves.

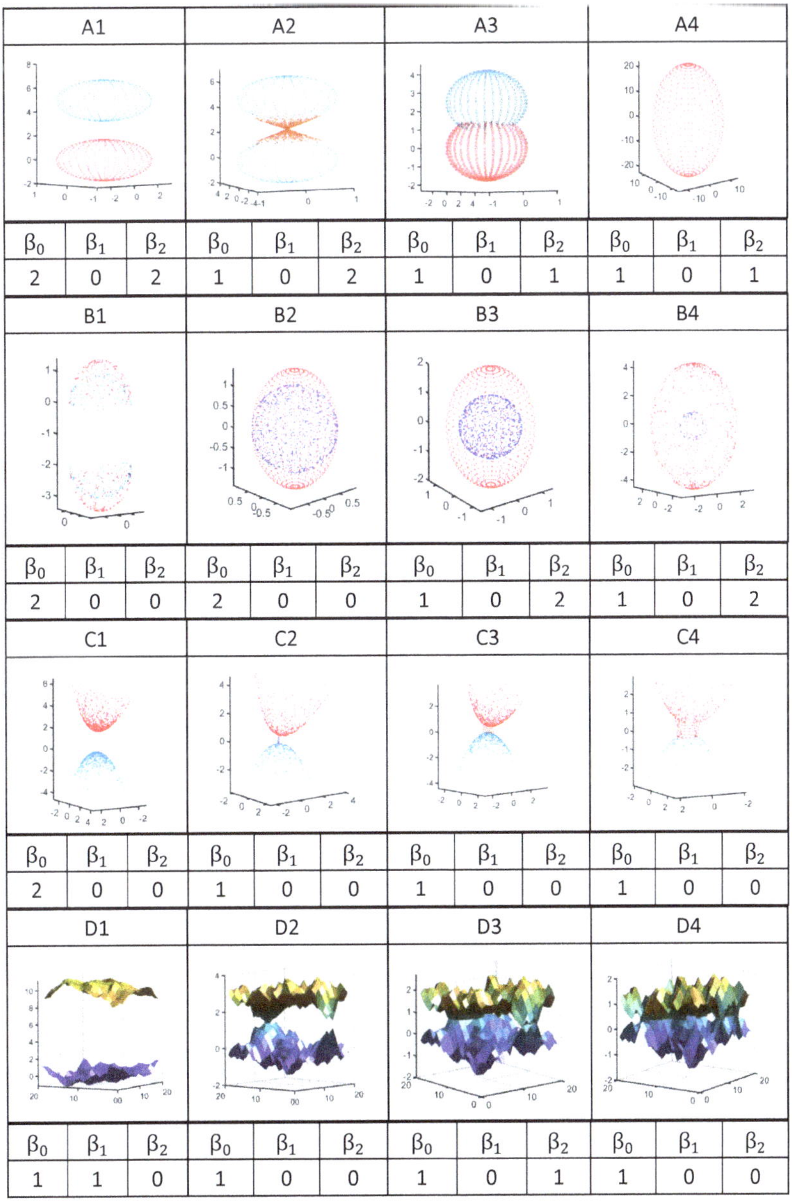

	β_0	β_1	β_2
A1	2	0	2
A2	1	0	2
A3	1	0	1
A4	1	0	1
B1	2	0	0
B2	2	0	0
B3	1	0	2
B4	1	0	2
C1	2	0	0
C2	1	0	0
C3	1	0	0
C4	1	0	0
D1	1	1	0
D2	1	0	0
D3	1	0	1
D4	1	0	0

Figure A4. Point clouds generated using MATLAB for different stages (1) to (4) of the four fusion models (**A–D**), and their associated Betti numbers. For different cases, the number of data points are different. For example, (**A2**) has more data points than the other stages of model (**A**), because we implemented it as a combination of two spheroids, plus a pair of Dirac cones.

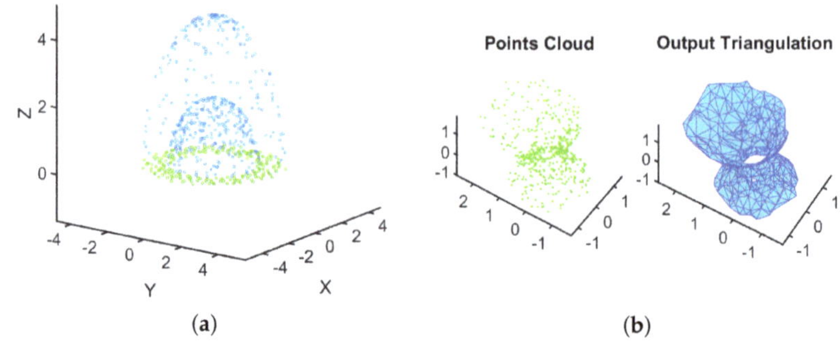

Figure A5. (**a**) To study how the typical gaps between data points affect the value of β_2 TDA produces, we split the point cloud in (B3) into top and bottom halves (blue points), moved them apart, and cover the gap in the equatorial plane between the inner and outer shells with green points. Such a structure would have a well-defined void whose average length scale is much larger than the distance between data points. (**b**) To show the channel formed in the neck region of (C4), we zoomed in to the green data points around the neck, and also rotated the view so that the channel formed by the two connected shells can be seen clearly, especially after the data points are meshed.

Finally, for model (D), we first generated two rugged surfaces on a 20×20 square grid. In Figure A4(D1), the two rugged surfaces were initially far apart, and then gradually moved closer, until they overlap at some parts, as shown in Figure A4(D2). We did not remove data points from the top rugged surface that penetrated the bottom rugged surface, or data points from the bottom rugged surface that penetrated the top rugged surface. In (D1), we have $\beta_0 = 1$ (as opposed to $\beta_0 = 2$ that we expected), $\beta_1 = 1$ (as opposed to $\beta_1 = 0$ that we expected), and $\beta_2 = 0$ (which is what we expected). In contrast, for (D2), we have $\beta_0 = 1$, $\beta_1 = 0$, and $\beta_2 = 0$, which are all as we expected. As the two rugged surfaces were brought closer and started to have more overlaps as in Figure A4(D3), their Betti numbers became $\beta_0 = 1$, and $\beta_1 = 0$, which were as expected, but we now had $\beta_2 = 1$ (a void has been formed between the two surfaces.). As we made the two surfaces merge further, we witnessed the change of β_2 from 1 to 0 (see Figure A4(D4)). This implies that the wrapped void formed by the two rugged surfaces in (D3), had disappeared in (D4). We believe this observation is the result of scale-dependence of β_2, discussed at the end of Appendix C.

Appendix C. Multiscale Analysis of Persistent Betti Numbers

We cannot directly apply the results of Appendix B, and Figure 5 to analyze the TDA results of the last three time windows, because we cannot be sure the Betti numbers are persistent. To better understand under what conditions we can have persistent Betti numbers, we visualize two contrasting mechanisms for fusion and fission in Figure A6. In Figure A6a–e, fusion occurs when thresholds decrease. Since $D_{ij} = \sqrt{2(1 - C_{ij})}$, these events correspond to increases in cross correlations. To put it more simply, we started out having two clusters with strong intra-cluster correlations, and weak inter-cluster correlations. When the inter-cluster correlations also become strong, the two clusters merge into one larger cluster. In contrast, for Figure A6f–i, fusion occurs when thresholds increase. These events are the result of correlations within the two original clusters weakening, to become comparable to the intra-cluster correlations. Without strong intra-cluster correlations to distinguish between their members, the two clusters effectively merged into a large but weakly correlated cluster. In the first sequence (Figure A6a–e), the persistent Betti number β_0 is the number of 0-simplexes where the lifetime changes most rapidly in the barcode. As we can see from Figure A6a–e, it is not difficult to identify the persistent β_0 for this sequence.

In the second sequence, the situation is more interesting. Instead of just one set of persistent β_0, we find two sets of persistent β_0 emerging at different scales. This means that we have one picture at the lower scale, but an equally meaningful picture at the higher scale. This also means that we need to be cautious when interpreting changes in β_0 as the result of fusions and fissions.

Similar scale-dependence can also occur for β_2. Consider the situation shown in Figure A7, where we can visually discern three voids in a (three-dimensional) point cloud. The three voids have different sizes, and are such that the smallest void occurs in the densest part of the point cloud, while the biggest void occurs in the sparsest part of the point cloud. When we perform TDA on this point cloud, the smallest void would be the first to emerge, when the filtration parameter ϵ_1 is just large enough to create a 2-simplex that encapsulates this smallest void. At this value of ϵ_1, the points around the medium void and the biggest void are too far apart for the simplices formed around them to fully encapsulate. Therefore, over a fairly large range of filtration parameters, this is the only persistent void we will find, and thus $\beta_2 = 1$.

When we continue to increase the filtration parameter, we will eventually reach the value ϵ_2, which is just large enough to create a 2-simplex encapsulating the medium void. At ϵ_2, the smallest void remains intact, while the sparse data points around the biggest void are still too far apart for the void to be encapsulated. If we continue to increase the filtration parameter, then at some point $\epsilon_{1'}$, the filtration parameter would become comparable to the size of the smallest void, and this void disappears. Therefore, over the range $\epsilon_2 < \epsilon < \epsilon_{1'}$, there are two persistent voids, and hence $\beta_2 = 2$.

Finally, at some other point $\epsilon_{1'} < \epsilon_3 < \epsilon_{2'}$, the biggest void become fully encapsulated and emerge as a topological feature. Later on, when the filtration parameter becomes large enough, we will find the medium void disappearing at $\epsilon_{2'}$, and the biggest void disappearing at $\epsilon_{3'}$. Therefore, depending on what scale ϵ we are at, the persistent β_2 may be 1 or 2, before dropping down to 1, and then eventually becoming 0.

This phenomenon of β_2 being scale-dependent explains the small values of β_2 in (D3) and (D4), where we expect from the rugged surfaces approaching each together to produce lots of small voids. In other words, by the time these small voids are produced, the filtration parameter would become so large that points on opposite sides of the void will be linked, and thus, the small voids will disappear.

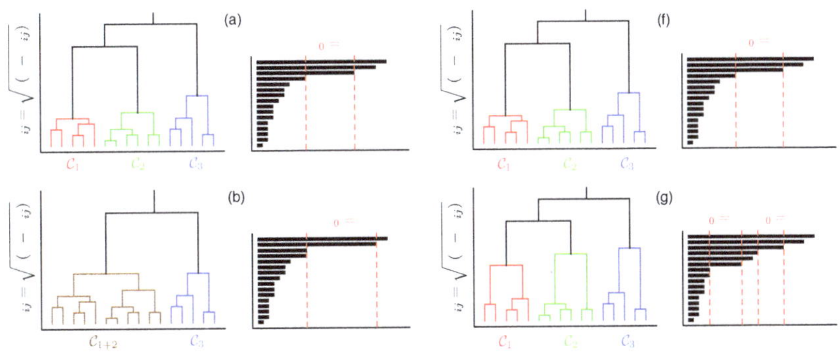

Figure A6. *Cont.*

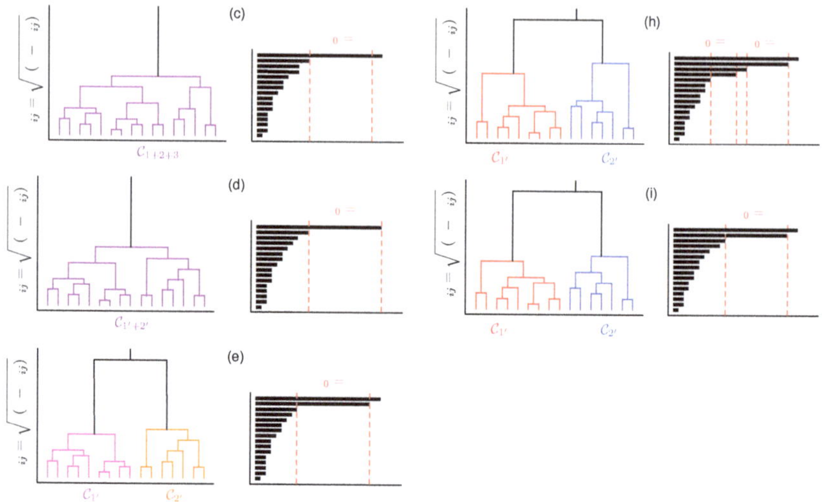

Figure A6. Using hierarchical clustering dendrograms and their corresponding H_0 barcodes to understand two fusion-and-fission sequences. In the first sequence, we first have the merging threshold between clusters $\mathcal{C}_1$ and $\mathcal{C}_2$ decreased going from (**a–b**), and thereafter the merging threshold between clusters $\mathcal{C}_{1+2}$ and $\mathcal{C}_3$ decreased going from (**b–c**). There is then a rearrangement $\mathcal{C}_{1+2+3} \to \mathcal{C}_{1'+2'}$ in (**d**), before the merging threshold between clusters $\mathcal{C}_{1'}$ and $\mathcal{C}_{2'}$ increased going from (**d–e**). The fusion $(1,2,3) \to (1+2,3) \to (1+2+3)$ thus occurs with decreasing threshold, while the fission $(1'+2') \to (1',2')$ occurs with increasing threshold. In the second sequence, the thresholds of $\mathcal{C}_1$, $\mathcal{C}_2$, and $\mathcal{C}_3$ first increased going from (**f–g**), before a rearrangement $\mathcal{C}_{1+2+3} \to \mathcal{C}_{1'+2'}$ occurs in (**h**) with comparable thresholds. Finally, the thresholds of $\mathcal{C}_{1'}$ and $\mathcal{C}_{2'}$ decreased in (**i**) to give two distinct clusters. In this figure, a large distance D_{ij} is associated with a small cross correlation C_{ij}, and vice versa. For each dendrogram, we also show the persistent Betti number β_0 in the corresponding barcode.

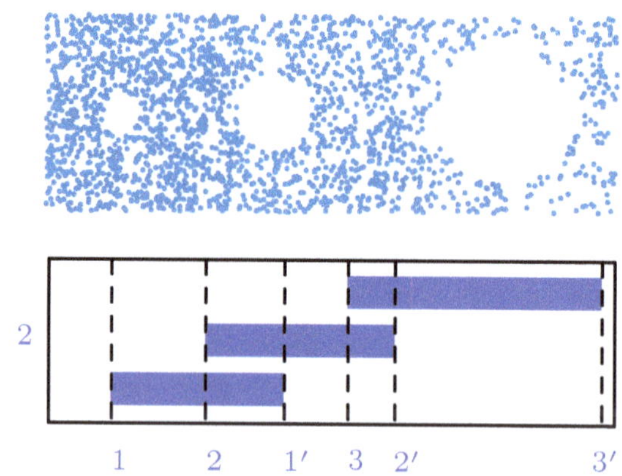

Figure A7. In this stylized depiction of a three-dimensional point cloud, we can visually discern three (persistent) voids. The smallest of these occurs in the densest part of the point cloud, whereas the biggest of these occurs in the sparsest part of the point cloud. (bottom) The barcodes associated with the emergences and disappearances of the three voids.

Appendix D. Ricci Flow and the Poincare Conjecture

In addition to its role in helping us understand dynamical reorganizations within complex systems, Ricci flow was also at the center of a major recent breakthrough in pure mathematics. In 1904, the French mathematician Poincare conjectured that "any closed simply connected 3-manifold is diffeomorphic to the standard 3-dimensional sphere S^3" [219], but later believed that the same is true in any dimension. Between 1960 and 1980, the Poincare conjectures for all dimensions were proven [220–223], except for three dimension. This problem was recognized as mathematically hard, but also of fundamental importance, that it was included as one of the seven Millenium Prize problems identified by the Clay Mathematics Institute in 2000. Later, the Poincare conjecture for three dimensions was stated in a more general form by Thurston, who conjectured that "any closed orientable 3-manifold can be canonically cut along embedded 2-spheres and 2-tori so as to decompose into eight different geometrical pieces" [224]. In response to this challenge, Hamilton crafted the Ricci flow model [225], and initiated a program to apply Ricci flow to solve Thurston's *geometrization conjecture* [224]. After the Poincare conjecture in three dimensions was stated in more general terms, the original problem and Thurston's generalized geometrization conjecture went unsolved for 20 more years, until Perelman cracked the final puzzle. With his original and ingenious "Ricci flow with surgery" approach, he successfully expunged the singular region as a work around for solving the geometrization conjecture, and ultimately resolved the Poincare conjecture in three dimensions [226,227]. This achievement not only won him the 2007 Fields Medal, but also made him the recipient of the first Millennium Prize. Since 2007, many mathematicians started dabbling into Ricci flow and related fields, creating something of a "gold rush" in this field over the past 15 years.

Appendix E. Computing the First Wasserstein Distance Using Linear Programming

In Equation (6), we defined the first Wasserstein distance (also known as the earth mover distance) conceptually, so that it can be used in the definition of the Ollivier-Ricci curvature. In this appendix, let us describe how the first Wasserstein distance on a network can be computed using linear programming.

First, consider a network G with five nodes. To compute the first Wasserstein distance between the lazy random walk probability distribution μ_2 (Figure A8a) and the lazy random walk probability distribution μ_5 (Figure A8b), let us consider all possible redistributions ξ_{ij} from $\mu_2(i)$ to $\mu_5(j)$, such that

$$\mu_2(i) = \sum_{j=1}^{8} \xi_{ij}, \quad \mu_5(j) = \sum_{i=1}^{8} \xi_{ij}. \tag{A1}$$

These are our constraints.

Next, let d_{ij} be the geodesic distance between node i and node j. On an unweighted network such as the one shown in Figure A8, the geodesic distance d_{ij} is the number of hops needed to go from node i to node j, and can therefore only take on integer values from 0 to 3. If there is more than one way to get from node i to node j, d_{ij} is the smallest number of hops. With this, we can write down the matrix of geodesic distances as

$$D = \begin{bmatrix} 0 & 1 & 2 & 2 & 3 \\ 1 & 0 & 1 & 1 & 2 \\ 2 & 1 & 0 & 2 & 1 \\ 2 & 1 & 2 & 0 & 1 \\ 3 & 2 & 1 & 1 & 0 \end{bmatrix}. \tag{A2}$$

Following this, let us define the cost of moving ξ_{ij} from node i to node j to be $d_{ij}\xi_{ij}$. The total cost to make μ_2 into μ_5 is thus

$$C(\mu_2, \mu_5) = \sum_{i=1}^{5}\sum_{j=1}^{5} d_{ij}\xi_{ij}. \tag{A3}$$

Our goal is to minimize $C(\mu_2, \mu_5)$, subject to the constraints in Equation (A1).

This constrained minimization problem can be recasted into a linear programming problem, if we change the indices from i and j to a fused index $k = (i, j)$. In terms of this fused index, the cost function can be rewritten as

$$C(\mu_2, \mu_5) = \sum_{k=1}^{25} d_k\xi_k, \tag{A4}$$

while the constraints can be written as

$$\sum_{k=1}^{25} a_{ik}\xi_k = \mu_2(i), \quad \sum_{k=1}^{25} b_{jk}\xi_k = \mu_5(j). \tag{A5}$$

In this formulation, $a_{ik} = 1$ if the first index in k is i, and $a_{ik} = 0$ otherwise, whereas $b_{jk} = 1$ if the second index in k is j, and $b_{jk} = 0$ otherwise.

The standard method for solving a high-dimensional linear programming problem is the *simplex method*, which can be found in numerous textbooks [228–230]. There are many variants for the simplex method, depending on whether the optimization is a maximization or minimization, and whether we are dealing with equality or inequality constraints. In general, the simplex method and its variants introduce auxiliary variables called *slack variables*, *excess variables*, or *artificial variables*. To illustrate how we can compute the first Wasserstein distance described above using the simplex method, we will also have to introduce artificial variables. Therefore, we should first eliminate redundant variables from with the list of 25 $\xi_k = \xi_{ij}$.

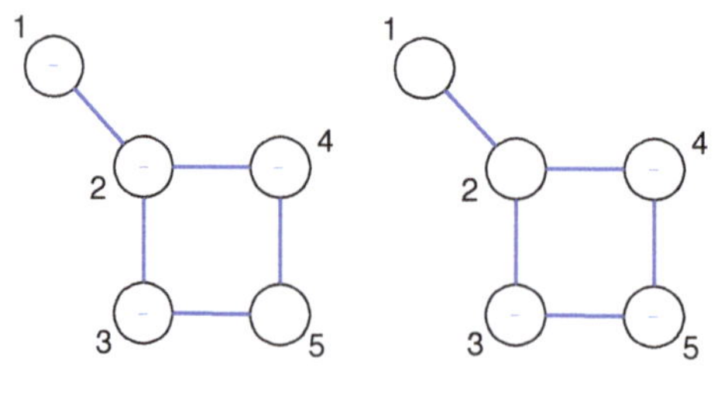

Figure A8. (**a**) The probability distribution μ_2 for lazy random walk starting from node 2, and (**b**) the probability distribution μ_5 for lazy random walk starting from node 5, on a network with five nodes. In (**a**), the probability of hopping from $2 \to 2$ is $\mu_2(2) = \alpha = \frac{1}{2}$, while the probability of hopping from 2 to its neighbors 1, 3, 4 are $\mu_2(1) = \mu_2(3) = \mu_2(4) = \frac{1}{6}$. These are shown in blue within the destination nodes. To complete the probability distribution μ_2, we also need to specify $\mu_2(5)$. Since it is not possible for the lazy random walk from node 2 to reach node 5 in a single hop, we set $\mu_2(5) = 0$, and show this probability in red within the destination node. In (**b**), we continue to have $\mu_5(5) = \frac{1}{2}$, but since node 5 has only two neighbors, we find $\mu_5(3) = \mu_5(4) = \frac{1}{4}$, and show these in blue within the destination nodes. We also show $\mu_5(1) = \mu_5(2) = 0$ in red within the destination nodes.

First of all, $\mu_2(5) = 0$. This cannot be broken down any further for redistribution to the various nodes in μ_5, and so we can just drop the variables ξ_{5j}, for $1 \leq j \leq 5$. Additionally, $\mu_5(1) = \mu_5(2) = 0$, so none of the nodes in μ_2 can make contributions to $j = 1, 2$. Hence, we can drop the variables ξ_{i1} and ξ_{i2}, for $1 \leq i \leq 5$. After eliminating these variables, we are then left with the 12 variables

$$\xi_{13}, \xi_{14}, \xi_{15}, \xi_{23}, \xi_{24}, \xi_{25}, \xi_{33}, \xi_{34}, \xi_{35}, \xi_{43}, \xi_{44}, \xi_{45}. \tag{A6}$$

Next, in terms of these remaining variables, we write the cost function as

$$C = 2\xi_{13} + 2\xi_{14} + 3\xi_{15} + \xi_{23} + \xi_{24} + 2\xi_{25} + 3\xi_{26} + 2\xi_{34} + \xi_{35} + 2\xi_{43} + \xi_{45}. \tag{A7}$$

We also write the seven equality constraints out explicitly as

$$\xi_{13} + \xi_{14} + \xi_{15} = \frac{1}{6}, \tag{A8}$$

$$\xi_{23} + \xi_{24} + \xi_{25} = \frac{1}{2}, \tag{A9}$$

$$\xi_{33} + \xi_{34} + \xi_{35} = \frac{1}{6}, \tag{A10}$$

$$\xi_{43} + \xi_{44} + \xi_{45} = \frac{1}{6}, \tag{A11}$$

$$\xi_{13} + \xi_{23} + \xi_{33} + \xi_{43} = \frac{1}{4}, \tag{A12}$$

$$\xi_{14} + \xi_{24} + \xi_{34} + \xi_{44} = \frac{1}{4}, \tag{A13}$$

$$\xi_{15} + \xi_{25} + \xi_{35} + \xi_{45} = \frac{1}{2}. \tag{A14}$$

To solve the minimization problem iteratively, we need to start from an initial feasible solution. This is not easy to guess, if we limit ourselves to the 12 variables. Therefore, we introduce one artificial variable for each of the seven equality constraints, such that

$$\xi_{13} + \xi_{14} + \xi_{15} + a_1 = \frac{1}{6}, \tag{A15}$$

$$\xi_{23} + \xi_{24} + \xi_{25} + a_2 = \frac{1}{2}, \tag{A16}$$

$$\xi_{33} + \xi_{34} + \xi_{35} + a_3 = \frac{1}{6}, \tag{A17}$$

$$\xi_{43} + \xi_{44} + \xi_{45} + a_4 = \frac{1}{6}, \tag{A18}$$

$$\xi_{13} + \xi_{23} + \xi_{33} + \xi_{43} + a_5 = \frac{1}{4}, \tag{A19}$$

$$\xi_{14} + \xi_{24} + \xi_{34} + \xi_{44} + a_6 = \frac{1}{4}, \tag{A20}$$

$$\xi_{15} + \xi_{25} + \xi_{35} + \xi_{45} + a_7 = \frac{1}{2}. \tag{A21}$$

This allows us to set $a_1 = \frac{1}{6}$, $a_2 = \frac{1}{2}$, $a_3 = \frac{1}{6}$, $a_4 = \frac{1}{6}$, $a_5 = \frac{1}{4}$, $a_6 = \frac{1}{4}$, $a_7 = \frac{1}{2}$, and all the ξ variables to zero as our initial solution. However, as their names imply, a_1 to a_9 are artificial variables, so we must be able to set their values to zero at the end of our minimization. To ensure that this will happen, let us also add $M(a_1 + a_2 + \cdots + a_7)$ to the cost function

$$\begin{aligned} C = {} & 2\xi_{13} + 2\xi_{14} + 3\xi_{15} + \xi_{23} + \xi_{24} + 2\xi_{25} + 2\xi_{34} + \xi_{35} + 2\xi_{43} + \xi_{45} + \\ & M(a_1 + a_2 + a_3 + a_4 + a_5 + a_6 + a_7), \end{aligned} \tag{A22}$$

where M is a large number. For this reason, this variant of the simplex method is called the *big M method*.

After years of teaching the simplex method to undergraduates, mathematicians have learned how to present the procedure in simple matrix/tabular form. The starting point is shown Table A3, where the coefficients of unknowns in the equality constraints have been organized into the first seventh rows of a matrix, and the coefficients of unknowns in the cost function organized into the eighth row. Initially, the columns V and r are not populated. To populate column V, we inspect the coefficient matrix to identify the *basic variables*. A basic variable is one that appears in one and only one row. At this start of the optimization, it should be clear that the basic variables are the artificial variables a_1 to a_7.

Starting from $a_1 = \frac{1}{6}$, $a_2 = \frac{1}{2}$, $a_3 = \frac{1}{6}$, $a_4 = \frac{1}{6}$, $a_5 = \frac{1}{4}$, $a_6 = \frac{1}{4}$, $a_7 = \frac{1}{2}$, and all the ξ variables set to zero, we want to ultimately be able to obtain a value of C that does not depend on the artificial variables a_1 to a_7. This can be achieved by subtracting M times the rows where the artificial variables appear from the last row where C appears, to produce the table shown in Table A4. In the last row of Table A4, all the coefficients associated with the ξ variables are negative, so this is not a feasible solution. To improve the solution, we need one of the ξ values to replace one of a as a basic variable. Looking for the most negative entry in the last row of Table A4, we find that this is $-2M$, in the columns associated with ξ_{33} and ξ_{44}. We can pick either one as the *entering variable* to replace one of the a's, which will be the *leaving variable*. Since we will surely pick ξ_{44} in the next iteration, let us for concreteness pick ξ_{33} as the entering variable. The final solution will not depend on the order we pick our entering variables. For ξ_{33} to become an entering variable, we determine the leaving variable by dividing the constants b by the coefficients in the ξ_{33} column, if this is possible, and populate the last column r. Since only two coefficients in the ξ_{33} column are non-zero, we compute r only along these two rows, to obtain the ratios $\frac{1}{6}$ and $\frac{1}{4}$. Since $\frac{1}{6}$ is the smaller of the two, we choose a_3 to be the leaving variable.

However, for ξ_{33} to replace a_3 as a basic variable, it must appear only in one row. In Table A4 it also appears in the row associated with a_5, as well as the last row. Therefore, we must perform elementary row operations to eliminate ξ_{33} from the a_5 row, as well as to eliminate ξ_{33} from the last row. After this is completed, we end up with the table shown in Table A5. Here we see also that the cost function has improved from $-2M$ to $-\frac{5M}{3}$. Since there are still negative coefficients in the last row, we know that we can continue to improve the cost function. In Table A5, the most negative coefficient is $-2M$, which appears in the column associated with ξ_{44}. We skipped ξ_{44} in favor of ξ_{33} the last iteration, so this is the right time to choose ξ_{44} as the entering variable. Thereafter, we divide the constants b's by the coefficients in the row associated with ξ_{44}, if this is possible, and look for the smallest ratio. In this case, we find only two ratios, $\frac{1}{6}$ and $\frac{1}{4}$, and the smallest ratio is associated with a_4, which will therefore be our new leaving variable.

Again, for ξ_{44} to become a basic variable, we must zero its coefficient in all other rows using elementary row operations. The matrix of coefficients then become the one shown in Table A6. We are making progress, because the cost function has improved to $-\frac{4M}{3}$, and many of the coefficients associated with ξ's have become positive in the last row. To find the next entering variable, we continue to look in the last row for the most negative coefficient. This would be $1 - 2M$, which appears in the column associated with ξ_{23}, as well as the column associated with ξ_{24}. As explained earlier, choosing ξ_{23} or ξ_{24} as the next entering variable will not change the solution, so let us make ξ_{23} the next entering variable. Then, from the ratios of the constants divided by the coefficients in the row associated with $\xi 23$, we see that a_5 will become the leaving variable.

Table A3. Matrix of coefficients at the start of the big M method to solve the linear programming problem for the 12 redistribution variables ξ_{ij}. In this table, the column V consists of basic variables, the column b consists of constraint values, while the column r consists of constraint values divided by elements in the column of the entering variable.

V	ξ_{13}	ξ_{14}	ξ_{15}	ξ_{23}	ξ_{24}	ξ_{25}	ξ_{33}	ξ_{34}	ξ_{35}	ξ_{43}	ξ_{44}	ξ_{45}	a_1	a_2	a_3	a_4	a_5	a_6	a_7	C	b	r
a_1	1	1	1	0	0	0	0	0	0	0	0	0	1	0	0	0	0	0	0	0	$\frac{1}{6}$	
a_2	0	0	0	1	1	1	0	0	0	0	0	0	0	1	0	0	0	0	0	0	$\frac{1}{2}$	
a_3	0	0	0	0	0	0	1	1	1	0	0	0	0	0	1	0	0	0	0	0	$\frac{1}{6}$	
a_4	0	0	0	0	0	0	0	0	0	1	1	1	0	0	0	1	0	0	0	0	$\frac{1}{6}$	
a_5	1	0	0	1	0	0	1	0	0	1	0	0	0	0	0	0	1	0	0	0	$\frac{1}{4}$	
a_6	0	1	0	0	1	0	0	1	0	0	1	0	0	0	0	0	0	1	0	0	$\frac{1}{4}$	
a_7	0	0	1	0	0	1	0	0	1	0	0	1	0	0	0	0	0	0	1	0	$\frac{1}{2}$	
	2	2	3	1	1	2	0	1	2	2	0	1	M	M	M	M	M	M	M	1	0	

Table A4. Matrix of coefficients after subtracting M times the constraint equations from the last row. In this table, the column V refers to basic variables, the column b refers to constraint values, while the column r refers to constraint values divided by elements in the column of the entering variable. In this table, the most recent changes are highlighted in bold.

V	ξ_{13}	ξ_{14}	ξ_{15}	ξ_{23}	ξ_{24}	ξ_{25}	ξ_{33}	ξ_{34}	ξ_{35}	ξ_{43}	ξ_{44}	ξ_{45}	a_1	a_2	a_3	a_4	a_5	a_6	a_7	C	b	r
a_1	1	1	1	0	0	0	0	0	0	0	0	0	1	0	0	0	0	0	0	0	$\frac{1}{6}$	
a_2	0	0	0	1	1	1	0	0	0	0	0	0	0	1	0	0	0	0	0	0	$\frac{1}{2}$	
a_3	0	0	0	0	0	0	1	1	1	0	0	0	0	0	1	0	0	0	0	0	$\frac{1}{6}$	$\boldsymbol{\frac{1}{6}}$
a_4	0	0	0	0	0	0	0	0	0	1	1	1	0	0	0	1	0	0	0	0	$\frac{1}{6}$	
a_5	1	0	0	1	0	0	1	0	0	1	0	0	0	0	0	0	1	0	0	0	$\frac{1}{4}$	$\boldsymbol{\frac{1}{4}}$
a_6	0	1	0	0	1	0	0	1	0	0	1	0	0	0	0	0	0	1	0	0	$\frac{1}{4}$	
a_7	0	0	1	0	0	1	0	0	1	0	0	1	0	0	0	0	0	0	1	0	$\frac{1}{2}$	
	$2-2M$	$2-2M$	$3-2M$	$1-2M$	$1-2M$	$2-2M$	$-2M$	$1-2M$	$2-2M$	$2-2M$	$-2M$	$1-2M$	**0**	**0**	**0**	**0**	**0**	**0**	**0**	1	$-2M$	

Table A5. Matrix of coefficients after elementary row operations to subtract appropriate multiples of the row associated with ξ_{33} from the row associated with a_5 and the last row. In this table, the column V refers to basic variables, the column b refers to constraint values, while the column r refers to constraint values divided by elements in the column of the entering variable. In this table, the most recent changes are highlighted in bold.

V	ξ_{13}	ξ_{14}	ξ_{15}	ξ_{23}	ξ_{24}	ξ_{25}	ξ_{33}	ξ_{34}	ξ_{35}	ξ_{43}	ξ_{44}	ξ_{45}	a_1	a_2	a_3	a_4	a_5	a_6	a_7	C	b	r
a_1	1	1	1	0	0	0	0	0	0	0	0	0	1	0	0	0	0	0	0	0	$\frac{1}{6}$	
a_2	0	0	0	1	1	1	0	0	0	0	0	0	0	1	0	0	0	0	0	0	$\frac{1}{2}$	
ξ_{33}	0	0	0	0	0	0	1	1	1	0	0	0	0	0	1	0	0	0	0	0	$\frac{1}{6}$	
a_4	0	0	0	0	0	0	0	0	0	1	1	1	0	0	0	1	0	0	0	0	$\frac{1}{6}$	$\mathbf{\frac{1}{6}}$
a_5	1	0	0	1	0	0	$\mathbf{0}$	$\mathbf{-1}$	$\mathbf{-1}$	1	0	0	0	0	$\mathbf{-1}$	0	1	0	0	0	$\mathbf{\frac{1}{12}}$	
a_6	0	1	0	0	1	0	0	1	0	0	1	0	0	0	0	0	0	1	0	0	$\frac{1}{4}$	$\mathbf{\frac{1}{4}}$
a_7	0	0	1	0	0	1	0	0	1	0	0	1	0	0	0	0	0	0	1	0	$\frac{1}{2}$	
	$2-2M$	$2-2M$	$3-2M$	$1-2M$	$1-2M$	$2-2M$	$\mathbf{0}$	$\mathbf{1}$	$\mathbf{2}$	$2-2M$	$-2M$	$1-2M$	0	0	$\mathbf{2M}$	0	0	0	0	1	$\mathbf{-\frac{5M}{3}}$	

Table A6. Matrix of coefficients after elementary row operations to subtract appropriate multiples of the row associated with ξ_{44} from the row associated with a_6 and the last row. In this table, the column V refers to basic variables, the column b refers to constraint values, while the column r refers to constraint values divided by elements in the column of the entering variable.

V	ξ_{13}	ξ_{14}	ξ_{15}	ξ_{23}	ξ_{24}	ξ_{25}	ξ_{33}	ξ_{34}	ξ_{35}	ξ_{43}	ξ_{44}	ξ_{45}	a_1	a_2	a_3	a_4	a_5	a_6	a_7	C	b	r
a_1	1	1	1	0	0	0	0	0	0	0	0	0	1	0	0	0	0	0	0	0	$\frac{1}{6}$	
a_2	0	0	0	1	1	1	0	0	0	0	0	0	0	1	0	0	0	0	0	0	$\frac{1}{2}$	$\mathbf{\frac{1}{2}}$
ξ_{33}	0	0	0	0	0	0	1	1	1	0	0	0	0	0	1	0	0	0	0	0	$\frac{1}{6}$	
$\boldsymbol{\xi_{44}}$	0	0	0	0	0	0	0	0	0	1	1	1	0	0	0	1	0	0	0	0	$\frac{1}{6}$	
a_5	1	0	0	1	0	0	0	-1	-1	1	0	0	0	0	-1	0	1	0	0	0	$\frac{1}{12}$	$\mathbf{\frac{1}{12}}$
a_6	0	1	0	0	1	0	0	1	0	$\mathbf{-1}$	$\mathbf{0}$	$\mathbf{-1}$	0	0	0	$\mathbf{-1}$	0	1	0	0	$\mathbf{\frac{1}{12}}$	
a_7	0	0	1	0	0	1	0	0	1	0	0	1	0	0	0	0	0	0	1	0	$\frac{1}{2}$	
	$2-2M$	$2-2M$	$3-2M$	$1-2M$	$1-2M$	$2-2M$	0	1	2	$\mathbf{2}$	$\mathbf{0}$	$\mathbf{1}$	0	0	$2M$	$\mathbf{2M}$	0	0	0	1	$\mathbf{-\frac{4M}{3}}$	

Performing elementary row operations to make ξ_{44} a proper basic variable, we obtained the table shown in Table A7. Again, the cost was improved, and more coefficients in the last row became positive. At this stage in the iterations, the most negative coefficient is $1-2M$, appearing in the column associated with ξ_{24}, which we postponed handling in the previous iteration. Therefore, we choose the next entering variable to be ξ_{24}. Thereafter, dividing the constants by coefficients in the column associated with ξ_{24}, we find the next leaving variable to be a_6.

After another round of elementary row operations, we obtained the table shown in Table A8. In this table, the most negative coefficient in the last row is $2-2M$, appearing in the columns associated with ξ_{25}, and ξ_{45}. Let us choose ξ_{25} to be the entering variable. For this variable, we find the ratio $\frac{1}{3}$ for the row associated with a_2, and $\frac{1}{2}$ in the row associated with a_7. Since $\frac{1}{3}<\frac{1}{2}$, we identify a_2 as the leaving variable.

After elementary row operations to make ξ_{25} a basic variable, we obtained the table shown in Table A9. The steady improvement to the cost function is obvious, and there are also fewer negative coefficients in the last row. To choose the next entering variable, let us inspect those columns with $3-2M$ in the last row. These are associated with ξ_{13}, ξ_{14}, and ξ_{15}. Since node 5 is the most important destination node, let us choose ξ_{15} as the next entering variable. The ratios are $\frac{1}{6}$ associated with a_1, and $\frac{1}{6}$ associated with a_7. Since they are the same, let us choose a_1 to be our next leaving variable.

After one last round of elementary row operations, making ξ_{15} a basic variable, we obtained the table shown in Table A10. Now, none of the coefficients in the last row are negative. This means that we have found our solution, and the iterations can end. From Table A10, we can read off the basic variables as

$$\xi_{15}=\frac{1}{6},\quad \xi_{23}=\frac{1}{12},\quad \xi_{24}=\frac{1}{12},\quad \xi_{25}=\frac{1}{3},\quad \xi_{33}=\frac{1}{6},\quad \xi_{44}=\frac{1}{6}. \tag{23}$$

We can check that $\xi_{23}+\xi_{24}+\xi_{25}=\frac{1}{12}+\frac{1}{12}+\frac{1}{3}=\frac{1}{2}$, satisfying the second constraint. Additionally, since $\xi_{33}=\frac{1}{6}$, this means that $\xi_{34}=\xi_{35}=0$. Furthermore, $\xi_{33}+\xi_{34}+\xi_{35}=\frac{1}{6}+0+0=\frac{1}{6}$, satisfying the third constraint. Similarly, since $\xi_{44}=\frac{1}{6}$, we also have $\xi_{43}=\xi_{45}=0$, and $\xi_{43}+\xi_{44}+\xi_{45}=0+\frac{1}{6}+0=\frac{1}{6}$ satisfies the fourth constraint. As in the cases of the third and fourth constraints, $\xi_{15}=\frac{1}{6}$ satisfies the first constraint by itself, meaning that $\xi_{13}=\xi_{14}=0$. More importantly, $\xi_{15}+\xi_{25}=\frac{1}{6}+\frac{1}{3}=\frac{1}{2}$, which already satisfy the seventh constraint, so we must have $\xi_{13}=\xi_{14}=0$. We also have $\xi_{13}+\xi_{23}+\xi_{33}+\xi_{43}=0+\frac{1}{12}+\frac{1}{6}+0=\frac{1}{4}$, satisfying the fifth constraint, and $\xi_{14}+\xi_{24}+\xi_{34}+\xi_{44}=0+\frac{1}{12}+0+\frac{1}{6}=\frac{1}{4}$ satisfying the sixth constraint.

Let us observe that the above solution implies that $a_1=a_2=a_3=a_4=a_5=a_6=a_7=0$. Let us also observe that $\xi_{13}, \xi_{14}, \xi_{34}, \xi_{35}, \xi_{43}, \xi_{45}$ did not make it onto the list of basic variables, and are thus non-basic variables. In the simplex method, non-basic variables are automatically set to zero at the end of the optimization. We also observe that we started out with seven artificial variables as basic variables, and iteratively replaced them with the redistribution variables. This means that at most seven redistribution variables can become basic, i.e., take on non-zero values at the end of the optimization. In this example, the optimization stopped after six redistribution variables became basic. The remaining six redistribution variables remained non-basic, and were thus set to zero, as dictated by the simplex method.

Finally, let us observe that after we prepared the coefficient matrix for the constraints and the cost function, including however many auxiliary variables as we need, the steps involved in each iteration are mechanical and easy to automate. In fact, most variants of the simplex method have been implemented in MATLAB, Python SciPy, and R as functions that users can call with minimal preparations.

Table A7. Matrix of coefficients after elementary row operations to subtract appropriate multiples of the row associated with ξ_{23} from the row associated with a_2 and the last row. In this table, the column V refers to basic variables, the column b refers to constraint values, while the column r refers to constraint values divided by elements in the column of the entering variable.

V	ξ_{13}	ξ_{14}	ξ_{15}	ξ_{23}	ξ_{24}	ξ_{25}	ξ_{33}	ξ_{34}	ξ_{35}	ξ_{43}	ξ_{44}	ξ_{45}	a_1	a_2	a_3	a_4	a_5	a_6	a_7	C	b	r
a_1	1	1	1	0	0	0	0	0	0	0	0	0	1	0	0	0	0	0	0	0	$\frac{1}{6}$	
a_2	$\mathbf{-1}$	0	0	**0**	1	1	0	**1**	**1**	$\mathbf{-1}$	0	0	0	1	**1**	0	$\mathbf{-1}$	0	0	0	$\frac{5}{12}$	$\frac{5}{12}$
ξ_{33}	0	0	0	0	0	0	1	1	1	0	0	0	0	0	1	0	0	0	0	0	$\frac{1}{6}$	
ξ_{44}	0	0	0	0	0	0	0	0	0	1	1	1	0	0	0	1	0	0	0	0	$\frac{1}{6}$	
$\boldsymbol{\xi_{23}}$	1	0	0	1	0	0	0	−1	−1	1	0	0	0	0	−1	0	1	0	0	0	$\frac{1}{12}$	
a_6	0	1	0	0	1	0	0	1	0	−1	0	−1	0	0	0	−1	0	1	0	0	$\frac{1}{12}$	$\frac{1}{12}$
a_7	0	0	1	0	0	1	0	0	1	0	0	1	0	0	0	0	0	0	1	0	$\frac{1}{2}$	
	1	$2-2M$	$3-2M$	**0**	$1-2M$	$2-2M$	0	$\mathbf{2-2M}$	$\mathbf{3-2M}$	$\mathbf{2M+1}$	0	1	0	0	**1**	$2M$	$\mathbf{2M-1}$	0	0	1	$-\frac{7M}{6}-\frac{1}{12}$	

Table A8. Matrix of coefficients after elementary row operations to subtract appropriate multiples of the row associated with ξ_{24} from the row associated with a_2 and the last row. In this table, the column V refers to basic variables, the column b refers to constraint values, while the column r refers to constraint values divided by elements in the column of the entering variable.

V	ξ_{13}	ξ_{14}	ξ_{15}	ξ_{23}	ξ_{24}	ξ_{25}	ξ_{33}	ξ_{34}	ξ_{35}	ξ_{43}	ξ_{44}	ξ_{45}	a_1	a_2	a_3	a_4	a_5	a_6	a_7	C	b	r
a_1	1	1	1	0	0	0	0	0	0	0	0	0	1	0	0	0	0	0	0	0	$\frac{1}{6}$	
a_2	−1	$\mathbf{-1}$	0	0	**0**	1	0	**0**	1	**0**	0	**1**	0	1	1	**1**	−1	$\mathbf{-1}$	0	0	$\frac{1}{3}$	$\frac{1}{3}$
ξ_{33}	0	0	0	0	0	0	1	1	1	0	0	0	0	0	1	0	0	0	0	0	$\frac{1}{6}$	
ξ_{44}	0	0	0	0	0	0	0	0	0	1	1	1	0	0	0	1	0	0	0	0	$\frac{1}{6}$	
ξ_{23}	1	0	0	1	0	0	0	−1	−1	1	0	0	0	0	−1	0	1	0	0	0	$\frac{1}{12}$	
$\boldsymbol{\xi_{24}}$	0	1	0	0	1	0	0	1	0	−1	0	−1	0	0	0	−1	0	1	0	0	$\frac{1}{12}$	
a_7	0	0	1	0	0	1	0	0	1	0	0	1	0	0	0	0	0	0	1	0	$\frac{1}{2}$	$\frac{1}{2}$
	1	**1**	$3-2M$	0	**0**	$2-2M$	0	**1**	$3-2M$	**2**	0	$\mathbf{2-2M}$	0	0	1	**1**	$2M-1$	$\mathbf{2M-1}$	0	1	$\mathbf{-M-\frac{1}{6}}$	

Table A9. Matrix of coefficients after elementary row operations to subtract appropriate multiples of the row associated with ξ_{25} from the row associated with a_7 and the last row. In this table, the column V refers to basic variables, the column b refers to constraint values, while the column r refers to constraint values divided by elements in the column of the entering variable.

V	ξ_{13}	ξ_{14}	ξ_{15}	ξ_{23}	ξ_{24}	ξ_{25}	ξ_{33}	ξ_{34}	ξ_{35}	ξ_{43}	ξ_{44}	ξ_{45}	a_1	a_2	a_3	a_4	a_5	a_6	a_7	C	b	r
a_1	1	1	1	0	0	0	0	0	0	0	0	0	1	0	0	0	0	0	0	0	$\frac{1}{6}$	$\frac{1}{6}$
ξ_{25}	−1	−1	0	0	0	1	0	0	1	0	0	1	0	1	1	1	−1	−1	0	0	$\frac{1}{3}$	
ξ_{33}	0	0	0	0	0	0	1	1	1	0	0	0	0	0	1	0	0	0	0	0	$\frac{1}{6}$	
ξ_{44}	0	0	0	0	0	0	0	0	0	1	1	1	0	0	0	1	0	0	0	0	$\frac{1}{6}$	
ξ_{23}	1	0	0	1	0	0	0	−1	−1	1	0	0	0	0	−1	0	1	0	0	0	$\frac{1}{12}$	
ξ_{24}	0	1	0	0	1	0	0	1	0	−1	0	−1	0	0	0	−1	0	1	0	0	$\frac{1}{12}$	
a_7	**1**	**1**	1	0	0	**0**	0	0	**0**	0	0	**0**	0	−1	−1	−1	1	1	1	0	$\frac{1}{6}$	$\frac{1}{6}$
	3 − 2M	**3 − 2M**	3 − 2M	0	0	**0**	0	1	1	2	0	**0**	0	**2M − 2**	**2M − 1**	**2M − 1**	1	1	0	1	$-\frac{M}{3}-\frac{5}{6}$	

Table A10. Matrix of coefficients after elementary row operations to subtract appropriate multiples of the row associated with ξ_{15} from the row associated with a_7 and the last row. In this table, the column V refers to basic variables, the column b refers to constraint values, while the column r refers to constraint values divided by elements in the column of the entering variable.

V	ξ_{13}	ξ_{14}	ξ_{15}	ξ_{23}	ξ_{24}	ξ_{25}	ξ_{33}	ξ_{34}	ξ_{35}	ξ_{43}	ξ_{44}	ξ_{45}	a_1	a_2	a_3	a_4	a_5	a_6	a_7	C	b	r
ξ_{15}	1	1	1	0	0	0	0	0	0	0	0	0	1	0	0	0	0	0	0	0	$\frac{1}{6}$	
ξ_{25}	−1	−1	0	0	0	1	0	0	1	0	0	1	0	1	1	1	−1	−1	0	0	$\frac{1}{3}$	
ξ_{33}	0	0	0	0	0	0	1	1	1	0	0	0	0	0	1	0	0	0	0	0	$\frac{1}{6}$	
ξ_{44}	0	0	0	0	0	0	0	0	0	1	1	1	0	0	0	1	0	0	0	0	$\frac{1}{6}$	
ξ_{23}	1	0	0	1	0	0	0	−1	−1	1	0	0	0	0	−1	0	1	0	0	0	$\frac{1}{12}$	
ξ_{24}	0	1	0	0	1	0	0	1	0	−1	0	−1	0	0	0	−1	0	1	0	0	$\frac{1}{12}$	
a_7	**0**	**0**	**0**	0	0	0	0	0	0	0	0	0	**−1**	−1	−1	−1	1	1	1	0	**0**	
	0	**0**	**0**	0	0	0	0	1	1	2	0	0	**2M − 3**	2M − 2	2M − 1	2M − 1	1	1	0	1	$-\frac{4}{3}$	

F. Ollivier-Ricci Curvature Analysis of a Toy Model Sequence of Fusion

In Section 4.3, we saw that the Betti numbers—being topological quantities—are not able to distinguish between parts of some models. There is, therefore, the need to go beyond TDA. Most notably, (A3) and (A4) have the same Betti numbers, but whereas the curvature is positive everywhere in (A4), the neck region in (A3) has negative curvature. In this subsection, we will use model (A) to illustrate the power of the ORC.

To compute the ORC for the different phases of model (A), we created wire meshes for the surfaces involved from (A1) to (A4) (see Figure 9), as instances of the `Graph` class from `networkx`. We then installed `GraphRicciCurvature` and all the Python packages that this depends on, before creating instances of the `OllivierRicci` class using the `networkx` graphs as inputs, using $\alpha = 0.5$ for the self-transition parameter. Finally, we called on the `compute_ricci_curvature()` method in the `OllivierRicci` class to compute the ORCs of all edges in the graphs.

As expected, the curvature ranged from slightly negative to slightly positive to significantly positive in (A1). Negative curvatures started appearing in (A2), at the Dirac point, and intensified in (A3). Finally, we have the curvature again ranging from slightly negative to slightly positive to significantly positive in (A4), thereby allowing us to distinguish between (A2), (A3), and (A4). The ORC changed most rapidly between (A2) and (A3).

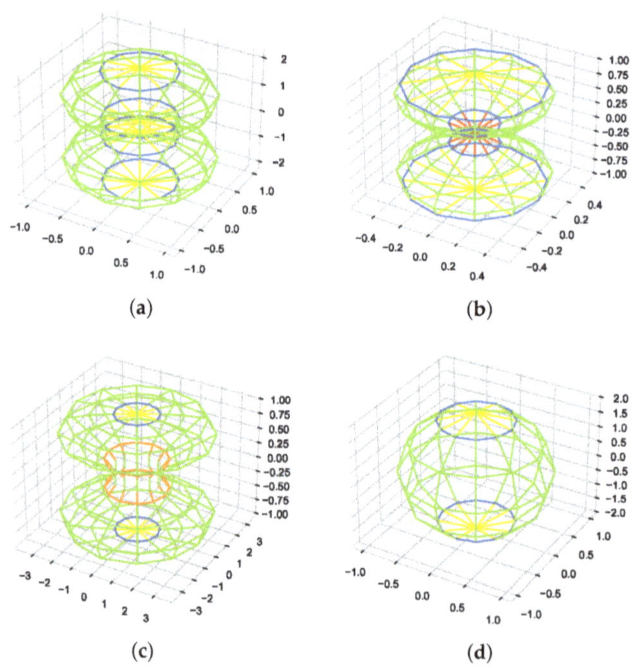

Figure 9. The sequence of topological changes going from (**a**) two disjoint spheres, to (**b**) them touching at a point to form a Dirac cone, to (**c**) them fusing to form a smoothed neck, to (**d**) eventually rounding off to an ellipsoid with no neck. In these figures, the links are colored according to their Ollivier-Ricci curvatures, standardized across the four scenarios from ORC $= -0.5$ (deep red) to ORC ≈ 0 (green (slightly negative) and yellow (slightly positive)) to ORC $= +0.5$ (deep blue).

References

1. Bachelier, L. Théorie de la spéculation. In *Annales Scientifiques de l'École Normale Supérieure*; Société Mathématique de France: Marseille, France, 1900; Volume 17, pp. 21–86.

2. Osborne, M.F. Brownian motion in the stock market. *Oper. Res.* **1959**, *7*, 145–173. [CrossRef]
3. Mandelbrot, B.B. The variation of certain speculative prices. In *Fractals and Scaling in Finance*; Springer: Berlin/Heidelberg, Germany, 1997; pp. 371–418.
4. Fama, E.F. The behavior of stock-market prices. *J. Bus.* **1965**, *38*, 34–105. [CrossRef]
5. Mandelbrot, B.B.; Van Ness, J.W. Fractional Brownian motions, fractional noises and applications. *SIAM Rev.* **1968**, *10*, 422–437. [CrossRef]
6. Mandelbrot, B.B.; Fisher, A.J.; Calvet, L.E. *A Multifractal Model of Asset Returns*; Cowles Foundation: New Haven, CT, USA, 1997.
7. Palmer, R.G.; Arthur, W.B.; Holland, J.H.; LeBaron, B.; Tayler, P. Artificial economic life: A simple model of a stockmarket. *Phys. D Nonlinear Phenom.* **1994**, *75*, 264–274. [CrossRef]
8. Mantegna, R.N. Lévy walks and enhanced diffusion in Milan stock exchange. *Phys. A Stat. Mech. Its Appl.* **1991**, *179*, 232–242. [CrossRef]
9. Takayasu, H.; Miura, H.; Hirabayashi, T.; Hamada, K. Statistical properties of deterministic threshold elements-the case of market price. *Phys. A Stat. Mech. Its Appl.* **1992**, *184*, 127–134. [CrossRef]
10. Stanley, H.E.; Afanasyev, V.; Amaral, L.A.N.; Buldyrev, S.V.; Goldberger, A.L.; Havlin, S.; Leschhorn, H.; Maass, P.; Mantegna, R.N.; Peng, C.K.; et al. Anomalous fluctuations in the dynamics of complex systems: From DNA and physiology to econophysics. *Phys. A Stat. Mech. Its Appl.* **1996**, *224*, 302–321. [CrossRef]
11. Laloux, L.; Cizeau, P.; Bouchaud, J.P.; Potters, M. Noise dressing of financial correlation matrices. *Phys. Rev. Lett.* **1999**, *83*, 1467. [CrossRef]
12. Plerou, V.; Gopikrishnan, P.; Rosenow, B.; Amaral, L.A.N.; Stanley, H.E. Universal and nonuniversal properties of cross correlations in financial time series. *Phys. Rev. Lett.* **1999**, *83*, 1471. [CrossRef]
13. Plerou, V.; Gopikrishnan, P.; Rosenow, B.; Amaral, L.N.; Stanley, H.E. A random matrix theory approach to financial cross-correlations. *Phys. A Stat. Mech. Its Appl.* **2000**, *287*, 374–382. [CrossRef]
14. Junior, L.S.; Franca, I.D.P. Correlation of financial markets in times of crisis. *Phys. A Stat. Mech. Its Appl.* **2012**, *391*, 187–208.
15. Mantegna, R.N.; Stanley, H.E. Stochastic process with ultraslow convergence to a Gaussian: The truncated Lévy flight. *Phys. Rev. Lett.* **1994**, *73*, 2946. [CrossRef] [PubMed]
16. Mantegna, R.N.; Stanley, H.E. Scaling behaviour in the dynamics of an economic index. *Nature* **1995**, *376*, 46–49. [CrossRef]
17. Mittnik, S.; Rachev, S.T.; Paolella, M.S. Stable Paretian modeling in finance: Some empirical and theoretical aspects. In *A Practical Guide to Heavy Tails*; Birkhäuser: Basel, Switzerland, 1998; pp. 79–110.
18. Sornette, D.; Sammis, C.G. Complex critical exponents from renormalization group theory of earthquakes: Implications for earthquake predictions. *J. Phys. I* **1995**, *5*, 607–619. [CrossRef]
19. Sornette, D.; Johansen, A.; Bouchaud, J.P. Stock market crashes, precursors and replicas. *J. Phys. I* **1996**, *6*, 167–175. [CrossRef]
20. Sornette, D. Dragon-kings, black swans and the prediction of crises. *arXiv* **2009**, arXiv:0907.4290.
21. Chatterjee, A.; Chakrabarti, B.K.; Manna, S. Money in gas-like markets: Gibbs and Pareto laws. *Phys. Scr.* **2003**, *2003*, 36. [CrossRef]
22. Dragulescu, A.; Yakovenko, V.M. Statistical mechanics of money. *Eur. Phys. J.-Condens. Matter Complex Syst.* **2000**, *17*, 723–729. [CrossRef]
23. Yura, Y.; Takayasu, H.; Sornette, D.; Takayasu, M. Financial brownian particle in the layered order-book fluid and fluctuation-dissipation relations. *Phys. Rev. Lett.* **2014**, *112*, 098703. [CrossRef]
24. Yura, Y.; Takayasu, H.; Sornette, D.; Takayasu, M. Financial Knudsen number: Breakdown of continuous price dynamics and asymmetric buy-and-sell structures confirmed by high-precision order-book information. *Phys. Rev. E* **2015**, *92*, 042811. [CrossRef]
25. Battiston, S.; Puliga, M.; Kaushik, R.; Tasca, P.; Caldarelli, G. Debtrank: Too central to fail? financial networks, the fed and systemic risk. *Sci. Rep.* **2012**, *2*, 541. [CrossRef]
26. Marwan, N.; Donges, J.F.; Zou, Y.; Donner, R.V.; Kurths, J. Complex network approach for recurrence analysis of time series. *Phys. Lett. A* **2009**, *373*, 4246–4254. [CrossRef]
27. Donner, R.V.; Donges, J.F.; Zou, Y.; Marwan, N.; Kurths, J. Recurrence-based evolving networks for time series analysis of complex systems. In Proceedings of the International Symposium on Nonlinear Theory and its Applications (NOLTA), Krakow, Poland, 5–8 September 2010; IEICE: Tokyo, Japan, 2010.
28. Lacasa, L.; Luque, B.; Ballesteros, F.; Luque, J.; Nuno, J.C. From time series to complex networks: The visibility graph. *Proc. Natl. Acad. Sci. USA* **2008**, *105*, 4972–4975. [CrossRef]
29. Yang, Y.; Wang, J.; Yang, H.; Mang, J. Visibility graph approach to exchange rate series. *Phys. A Stat. Mech. Its Appl.* **2009**, *388*, 4431–4437. [CrossRef]
30. Qian, M.C.; Jiang, Z.Q.; Zhou, W.X. Universal and nonuniversal allometric scaling behaviors in the visibility graphs of world stock market indices. *J. Phys. A Math. Theor.* **2010**, *43*, 335002. [CrossRef]
31. Wang, N.; Li, D.; Wang, Q. Visibility graph analysis on quarterly macroeconomic series of China based on complex network theory. *Phys. A Stat. Mech. Its Appl.* **2012**, *391*, 6543–6555. [CrossRef]
32. Stephen, M.; Gu, C.; Yang, H. Visibility graph based time series analysis. *PLoS ONE* **2015**, *10*, e0143015. [CrossRef]
33. Antoniades, I.; Stavrinides, S.; Hanias, M.; Magafas, L. Complex network time series analysis of a macroeconomic model. In *Dynamics on and of Complex Networks III*; Springer Science and Business Media LLC: Cham, Switzerland, 2020; pp. 135–147.

34. Jensen, M.H. Multiscaling and structure functions in turbulence: An alternative approach. *Phys. Rev. Lett.* **1999**, *83*, 76. [CrossRef]
35. Simonsen, I.; Jensen, M.H.; Johansen, A. Optimal investment horizons. *Eur. Phys. J.-Condens. Matter Complex Syst.* **2002**, *27*, 583–586. [CrossRef]
36. Jensen, M.H.; Johansen, A.; Simonsen, I. Inverse statistics in economics: The gain–loss asymmetry. *Phys. A Stat. Mech. Its Appl.* **2003**, *324*, 338–343. [CrossRef]
37. Johansen, A.; Simonsen, I.; Jensen, M.H. Optimal investment horizons for stocks and markets. *Phys. A Stat. Mech. Its Appl.* **2006**, *370*, 64–67. [CrossRef]
38. Jensen, M.H.; Johansen, A.; Petroni, F.; Simonsen, I. Inverse statistics in the foreign exchange market. *Phys. A Stat. Mech. Its Appl.* **2004**, *340*, 678–684. [CrossRef]
39. Zhou, W.X.; Yuan, W.K. Inverse statistics in stock markets: Universality and idiosyncracy. *Phys. A Stat. Mech. Its Appl.* **2005**, *353*, 433–444. [CrossRef]
40. Karpio, K.; Załuska-Kotur, M.A.; Orłowski, A. Gain–loss asymmetry for emerging stock markets. *Phys. A Stat. Mech. Its Appl.* **2007**, *375*, 599–604. [CrossRef]
41. Lee, C.Y.; Kim, J.; Hwang, I. Inverse statistics of the Korea composite stock price index. *J. Korean Phys. Soc.* **2008**, *52*, 517–523. [CrossRef]
42. Grudziecki, M.; Gnatowska, E.; Karpio, K.; Orlowski, A.; Zaluska-Kotur, M. New results on gain-loss asymmetry for stock markets time series. *Acta Phys.-Pol.-Ser. Gen. Phys.* **2008**, *114*, 569. [CrossRef]
43. Donangelo, R.; Jensen, M.H.; Simonsen, I.; Sneppen, K. Synchronization model for stock market asymmetry. *J. Stat. Mech. Theory Exp.* **2006**, *2006*, L11001. [CrossRef]
44. Simonsen, I.; Ahlgren, P.T.H.; Jensen, M.H.; Donangelo, R.; Sneppen, K. Fear and its implications for stock markets. *Eur. Phys. J. B* **2007**, *57*, 153–158. [CrossRef]
45. Ahlgren, P.T.H.; Jensen, M.H.; Simonsen, I.; Donangelo, R.; Sneppen, K. Frustration driven stock market dynamics: Leverage effect and asymmetry. *Phys. A Stat. Mech. Its Appl.* **2007**, *383*, 1–4. [CrossRef]
46. Siven, J.; Lins, J.; Hansen, J.L. A multiscale view on inverse statistics and gain/loss asymmetry in financial time series. *J. Stat. Mech. Theory Exp.* **2009**, *2009*, P02004. [CrossRef]
47. Sornette, D. Stock market speculation: Spontaneous symmetry breaking of economic valuation. *Phys. A Stat. Mech. Its Appl.* **2000**, *284*, 355–375. [CrossRef]
48. Ahlgren, P.T.H.; Dahl, H.; Jensen, M.H.; Simonsen, I. What Can Be Learned from Inverse Statistics? In *Econophysics Approaches to Large-Scale Business Data and Financial Crisis*; Springer: Berlin/Heidelberg, Germany, 2010; pp. 247–270.
49. Mantegna, R.N. Hierarchical structure in financial markets. *Eur. Phys. J.-Condens. Matter Complex Syst.* **1999**, *11*, 193–197. [CrossRef]
50. Tumminello, M.; Aste, T.; Di Matteo, T.; Mantegna, R.N. A tool for filtering information in complex systems. *Proc. Natl. Acad. Sci. USA* **2005**, *102*, 10421–10426. [CrossRef]
51. Berge, C.; Minieka, E. (translator) *Graphs and Hypergraphs*; North-Holland Publishing Company: Amsterdam, The Netherlands, 1976.
52. Gallo, G.; Longo, G.; Pallottino, S.; Nguyen, S. Directed hypergraphs and applications. *Discret. Appl. Math.* **1993**, *42*, 177–201. [CrossRef]
53. Zhou, D.; Huang, J.; Schölkopf, B. Learning with hypergraphs: Clustering, classification, and embedding. *Adv. Neural Inf. Process. Syst.* **2006**, *19*, 1601–1608.
54. Klamt, S.; Haus, U.U.; Theis, F. Hypergraphs and cellular networks. *PLoS Comput. Biol.* **2009**, *5*, e1000385. [CrossRef]
55. Zomorodian, A.; Carlsson, G. Computing persistent homology. *Discret. Comput. Geom.* **2005**, *33*, 249–274. [CrossRef]
56. Singh, G.; Mémoli, F.; Carlsson, G.E. Topological methods for the analysis of high dimensional data sets and 3d object recognition. *SPBG* **2007**, *91*, 100.
57. Edelsbrunner, H.; Letscher, D.; Zomorodian, A. Topological persistence and simplification. In Proceedings of the 41st Annual Symposium on Foundations of Computer Science, Redondo Beach, CA, USA, 12–14 November 2000; IEEE: Piscataway, NJ, USA, 2000; pp. 454–463.
58. Carlsson, G. Topology and data. *Bull. Am. Math. Soc.* **2009**, *46*, 255–308. [CrossRef]
59. Yen, P.T.W.; Cheong, S.A. Using topological data analysis (TDA) and persistent homology to analyze the stock markets in Singapore and Taiwan. *Front. Phys.* **2021**, *9*, 20. [CrossRef]
60. Aste, T.; Di Matteo, T.; Hyde, S. Complex networks on hyperbolic surfaces. *Phys. A Stat. Mech. Its Appl.* **2005**, *346*, 20–26. [CrossRef]
61. Chakraborty, A.; Kichikawa, Y.; Iino, T.; Iyetomi, H.; Inoue, H.; Fujiwara, Y.; Aoyama, H. Hierarchical communities in the walnut structure of the Japanese production network. *PLoS ONE* **2018**, *13*, e0202739. [CrossRef] [PubMed]
62. Fujiwara, Y.; Aoyama, H. Large-scale structure of a nation-wide production network. *Eur. Phys. J. B* **2010**, *77*, 565–580. [CrossRef]
63. Iino, T.; Iyetomi, H. Community structure of a large-scale production network in japan. In *The Economics of Interfirm Networks*; Springer: Berlin/Heidelberg, Germany, 2015; pp. 39–65.

64. Chakraborty, A.; Krichene, H.; Inoue, H.; Fujiwara, Y. Characterization of the community structure in a large-scale production network in Japan. *Phys. A Stat. Mech. Its Appl.* **2019**, *513*, 210–221. [CrossRef]
65. Krichene, H.; Chakraborty, A.; Inoue, H.; Fujiwara, Y. Business cycles's correlation and systemic risk of the Japanese supplier-customer network. *PLoS ONE* **2017**, *12*, e0186467. [CrossRef]
66. Plerou, V.; Gopikrishnan, P.; Rosenow, B.; Amaral, L.A.N.; Guhr, T.; Stanley, H.E. Random matrix approach to cross correlations in financial data. *Phys. Rev. E* **2002**, *65*, 066126. [CrossRef]
67. Teh, B.K.; Cheong, S.A. Cluster fusion-fission dynamics in the Singapore stock exchange. *Eur. Phys. J. B* **2015**, *88*, 1–14. [CrossRef]
68. Kruskal, J.B. On the shortest spanning subtree of a graph and the traveling salesman problem. *Proc. Am. Math. Soc. USA* **1956**, *7*, 48–50. [CrossRef]
69. Bonanno, G.; Lillo, F.; Mantegna, R.N. High-frequency cross-correlation in a set of stocks. *Quant. Finance* **2001**, *1*, 96–104. [CrossRef]
70. Onnela, J.P.; Chakraborti, A.; Kaski, K.; Kertiész, J. Dynamic asset trees and portfolio analysis. *Eur. Phys. J.-Condens. Matter Complex Syst.* **2002**, *30*, 285–288. [CrossRef]
71. Miccichè, S.; Bonanno, G.; Lillo, F.; Mantegna, R.N. Degree stability of a minimum spanning tree of price return and volatility. *Phys. A Stat. Mech. Its Appl.* **2003**, *324*, 66–73. [CrossRef]
72. Bonanno, G.; Caldarelli, G.; Lillo, F.; Mantegna, R.N. Topology of correlation-based minimal spanning trees in real and model markets. *Phys. Rev. E* **2003**, *68*, 046130. [CrossRef]
73. Onnela, J.P.; Chakraborti, A.; Kaski, K.; Kertesz, J.; Kanto, A. Dynamics of market correlations: Taxonomy and portfolio analysis. *Phys. Rev. E* **2003**, *68*, 056110. [CrossRef] [PubMed]
74. Brida, J.G.; Risso, W.A. Multidimensional minimal spanning tree: The Dow Jones case. *Phys. A Stat. Mech. Its Appl.* **2008**, *387*, 5205–5210. [CrossRef]
75. Zhang, Y.; Lee, G.H.T.; Wong, J.C.; Kok, J.L.; Prusty, M.; Cheong, S.A. Will the US economy recover in 2010? A minimal spanning tree study. *Phys. A Stat. Mech. Its Appl.* **2011**, *390*, 2020–2050. [CrossRef]
76. Coronnello, C.; Tumminello, M.; Lillo, F.; Micciche, S.; Mantegna, R.N. Sector identification in a set of stock return time series traded at the London Stock Exchange. *arXiv* **2005**, arXiv:cond-mat/0508122
77. Jung, W.S.; Chae, S.; Yang, J.S.; Moon, H.T. Characteristics of the Korean stock market correlations. *Phys. A Stat. Mech. Its Appl.* **2006**, *361*, 263–271. [CrossRef]
78. Eom, C.; Oh, G.; Kim, S. Topological properties of the minimal spanning tree in Korean and American stock markets. *arXiv* **2006**, arXiv:physics/0612068.
79. Cheong, S.A.; Fornia, R.P.; Lee, G.H.T.; Kok, J.L.; Yim, W.S.; Xu, D.Y.; Zhang, Y. The Japanese economy in crises: A time series segmentation study. *Econ. Open-Access Open-Assess. E-J.* **2012**, *6*, 1–81. [CrossRef]
80. Zhuang, R.; Hu, B.; Ye, Z. Minimal spanning tree for Shanghai-Shenzhen 300 stock index. In Proceedings of the 2008 IEEE Congress on Evolutionary Computation (IEEE World Congress on Computational Intelligence), Hong Kong, China, 1–6 June 2008; IEEE: Piscataway, NJ, USA, 2008; pp. 1417–1424.
81. Bonanno, G.; Vandewalle, N.; Mantegna, R.N. Taxonomy of stock market indices. *Phys. Rev. E* **2000**, *62*, R7615. [CrossRef]
82. Lee, G.S.; Djauhari, M.A. Network topology of Indonesian stock market. In Proceedings of the 2012 International Conference on Cloud Computing and Social Networking (ICCCSN), Bandung, Indonesia, 26–27 April 2012; IEEE: Piscataway, NJ, USA, 2012; pp. 1–4.
83. Majapa, M.; Gossel, S.J. Topology of the South African stock market network across the 2008 financial crisis. *Phys. A Stat. Mech. Its Appl.* **2016**, *445*, 35–47. [CrossRef]
84. Coelho, R.; Gilmore, C.G.; Lucey, B.; Richmond, P.; Hutzler, S. The evolution of interdependence in world equity markets-Evidence from minimum spanning trees. *Phys. A Stat. Mech. Its Appl.* **2007**, *376*, 455–466. [CrossRef]
85. Gilmore, C.G.; Lucey, B.M.; Boscia, M. An ever-closer union? Examining the evolution of linkages of European equity markets via minimum spanning trees. *Phys. A Stat. Mech. Its Appl.* **2008**, *387*, 6319–6329. [CrossRef]
86. Song, D.M.; Tumminello, M.; Zhou, W.X.; Mantegna, R.N. Evolution of worldwide stock markets, correlation structure, and correlation-based graphs. *Phys. Rev. E* **2011**, *84*, 026108. [CrossRef]
87. Di Matteo, T.; Aste, T. How does the eurodollar interest rate behave? *Int. J. Theor. Appl. Financ.* **2002**, *5*, 107–122. [CrossRef]
88. Dias, J. Sovereign debt crisis in the European Union: A minimum spanning tree approach. *Phys. A Stat. Mech. Its Appl.* **2012**, *391*, 2046–2055. [CrossRef]
89. Dias, J. Spanning trees and the Eurozone crisis. *Phys. A Stat. Mech. Its Appl.* **2013**, *392*, 5974–5984. [CrossRef]
90. McDonald, M.; Suleman, O.; Williams, S.; Howison, S.; Johnson, N.F. Detecting a currency's dominance or dependence using foreign exchange network trees. *Phys. Rev. E* **2005**, *72*, 046106. [CrossRef]
91. Mizuno, T.; Takayasu, H.; Takayasu, M. Correlation networks among currencies. *Phys. A Stat. Mech. Its Appl.* **2006**, *364*, 336–342. [CrossRef]
92. Górski, A.; Drozdz, S.; Kwapien, J. Minimal spanning tree graphs and power like scaling in FOREX networks. *arXiv* **2008**, arXiv:0809.0437.
93. Jang, W.; Lee, J.; Chang, W. Currency crises and the evolution of foreign exchange market: Evidence from minimum spanning tree. *Phys. A Stat. Mech. Its Appl.* **2011**, *390*, 707–718. [CrossRef]

94. Wang, G.J.; Xie, C.; Han, F.; Sun, B. Similarity measure and topology evolution of foreign exchange markets using dynamic time warping method: Evidence from minimal spanning tree. *Phys. A Stat. Mech. Its Appl.* **2012**, *391*, 4136–4146. [CrossRef]
95. Wang, G.J.; Xie, C.; Chen, Y.J.; Chen, S. Statistical properties of the foreign exchange network at different time scales: Evidence from detrended cross-correlation coefficient and minimum spanning tree. *Entropy* **2013**, *15*, 1643–1662. [CrossRef]
96. Sieczka, P.; Hołyst, J.A. Correlations in commodity markets. *Phys. A Stat. Mech. Its Appl.* **2009**, *388*, 1621–1630. [CrossRef]
97. Barigozzi, M.; Fagiolo, G.; Garlaschelli, D. Multinetwork of international trade: A commodity-specific analysis. *Phys. Rev. E* **2010**, *81*, 046104. [CrossRef]
98. Tabak, B.; Serra, T.; Cajueiro, D. Topological properties of commodities networks. *Eur. Phys. J. B* **2010**, *74*, 243–249. [CrossRef]
99. Kristoufek, L.; Janda, K.; Zilberman, D. Correlations between biofuels and related commodities before and during the food crisis: A taxonomy perspective. *Energy Econ.* **2012**, *34*, 1380–1391. [CrossRef]
100. Zheng, Z.; Yamasaki, K.; Tenenbaum, J.N.; Stanley, H.E. Carbon-dioxide emissions trading and hierarchical structure in worldwide finance and commodities markets. *Phys. Rev. E* **2013**, *87*, 012814. [CrossRef]
101. Kazemilari, M.; Mardani, A.; Streimikiene, D.; Zavadskas, E.K. An overview of renewable energy companies in stock exchange: Evidence from minimal spanning tree approach. *Renew. Energy* **2017**, *102*, 107–117. [CrossRef]
102. Iori, G.; De Masi, G.; Precup, O.V.; Gabbi, G.; Caldarelli, G. A network analysis of the Italian overnight money market. *J. Econ. Dyn. Control* **2008**, *32*, 259–278. [CrossRef]
103. Wang, G.J.; Xie, C. Correlation structure and dynamics of international real estate securities markets: A network perspective. *Phys. A Stat. Mech. Its Appl.* **2015**, *424*, 176–193. [CrossRef]
104. Onnela, J.P.; Chakraborti, A.; Kaski, K.; Kertesz, J. Dynamic asset trees and Black Monday. *Phys. A Stat. Mech. Its Appl.* **2003**, *324*, 247–252. [CrossRef]
105. Sun, X.Q.; Cheng, X.Q.; Shen, H.W.; Wang, Z.Y. Distinguishing manipulated stocks via trading network analysis. *Phys. A Stat. Mech. Its Appl.* **2011**, *390*, 3427–3434. [CrossRef]
106. Sun, X.Q.; Shen, H.W.; Cheng, X.Q.; Zhang, Y. Detecting anomalous traders using multi-slice network analysis. *Phys. A Stat. Mech. Its Appl.* **2017**, *473*, 1–9. [CrossRef]
107. Jiang, Z.Q.; Xie, W.J.; Xiong, X.; Zhang, W.; Zhang, Y.J.; Zhou, W.X. Trading networks, abnormal motifs and stock manipulation. *Quant. Financ. Lett.* **2013**, *1*, 1–8. [CrossRef]
108. Tola, V.; Lillo, F.; Gallegati, M.; Mantegna, R.N. Cluster analysis for portfolio optimization. *J. Econ. Dyn. Control* **2008**, *32*, 235–258. [CrossRef]
109. Coelho, R.; Hutzler, S.; Repetowicz, P.; Richmond, P. Sector analysis for a FTSE portfolio of stocks. *Phys. A Stat. Mech. Its Appl.* **2007**, *373*, 615–626. [CrossRef]
110. Iori, G.; Mantegna, R.N. Empirical analyses of networks in finance. In *Handbook of Computational Economics*; Elsevier: Amsterdam, The Netherlands, 2018; Volume 4, pp. 637–685.
111. Marti, G.; Nielsen, F.; Bińkowski, M.; Donnat, P. A review of two decades of correlations, hierarchies, networks and clustering in financial markets. *arXiv* **2017**, arXiv:1703.00485.
112. Onnela, J.P.; Kaski, K.; Kertész, J. Clustering and information in correlation based financial networks. *Eur. Phys. J. B* **2004**, *38*, 353–362. [CrossRef]
113. Gao, Y.C.; Zeng, Y.; Cai, S.M. Influence network in the Chinese stock market. *J. Stat. Mech. Theory Exp.* **2015**, *2015*, P03017. [CrossRef]
114. Kenett, D.Y.; Tumminello, M.; Madi, A.; Gur-Gershgoren, G.; Mantegna, R.N.; Ben-Jacob, E. Dominating clasp of the financial sector revealed by partial correlation analysis of the stock market. *PLoS ONE* **2010**, *5*, e15032. [CrossRef] [PubMed]
115. Kenett, D.Y.; Preis, T.; Gur-Gershgoren, G.; Ben-Jacob, E. Dependency network and node influence: Application to the study of financial markets. *Int. J. Bifurc. Chaos* **2012**, *22*, 1250181. [CrossRef]
116. Billio, M.; Getmansky, M.; Lo, A.W.; Pelizzon, L. Econometric measures of connectedness and systemic risk in the finance and insurance sectors. *J. Financ. Econ.* **2012**, *104*, 535–559. [CrossRef]
117. Výrost, T.; Lyócsa, Š.; Baumöhl, E. Granger causality stock market networks: Temporal proximity and preferential attachment. *Phys. A Stat. Mech. Its Appl.* **2015**, *427*, 262–276. [CrossRef]
118. Tu, C. Cointegration-based financial networks study in Chinese stock market. *Phys. A Stat. Mech. Its Appl.* **2014**, *402*, 245–254. [CrossRef]
119. Tumminello, M.; Micciche, S.; Lillo, F.; Piilo, J.; Mantegna, R.N. Statistically validated networks in bipartite complex systems. *PLoS ONE* **2011**, *6*, e17994. [CrossRef]
120. Chen, S.H.; Chang, C.L.; Tseng, Y.H. Social networks, social interaction and macroeconomic dynamics: How much could Ernst Ising help DSGE? *Res. Int. Bus. Financ.* **2014**, *30*, 312–335. [CrossRef]
121. Kenett, D.Y.; Havlin, S. Network science: A useful tool in economics and finance. *Mind Soc.* **2015**, *14*, 155–167. [CrossRef]
122. National Bureau of Economic Research. Market Microstructure. Available online: https://web.archive.org/web/20080722025938/http://www.nber.org/workinggroups/groups_desc.html (accessed on 29 June 2021).
123. Schools, Q. Market Microstructure. Available online: https://www.quantschools.co.uk/module/market-microstructure/ (accessed on 29 June 2021).
124. Téllez-León, I.E.; Martínez-Jaramillo, S.; Escobar-Farfán, L.O.; Hochreiter, R. How are network centrality metrics related to interest rates in the Mexican secured and unsecured interbank markets? *J. Financ. Stab.* **2021**, *55*, 100893. [CrossRef]

125. Aste, T.; Shaw, W.; Di Matteo, T. Correlation structure and dynamics in volatile markets. *New J. Phys.* **2010**, *12*, 085009. [CrossRef]
126. Pozzi, F.; Di Matteo, T.; Aste, T. Spread of risk across financial markets: Better to invest in the peripheries. *Sci. Rep.* **2013**, *3*, 1665. [CrossRef] [PubMed]
127. Wang, G.J.; Xie, C.; Chen, S. Multiscale correlation networks analysis of the US stock market: A wavelet analysis. *J. Econ. Interact. Coord.* **2017**, *12*, 561–594. [CrossRef]
128. Musmeci, N.; Aste, T.; Di Matteo, T. Relation between financial market structure and the real economy: Comparison between clustering methods. *PLoS ONE* **2015**, *10*, e0116201. [CrossRef]
129. Wen, F.; Yang, X.; Zhou, W.X. Tail dependence networks of global stock markets. *Int. J. Financ. Econ.* **2019**, *24*, 558–567. [CrossRef]
130. Massara, G.P.; Di Matteo, T.; Aste, T. Network filtering for big data: Triangulated maximally filtered graph. *J. Complex Netw.* **2016**, *5*, 161–178. [CrossRef]
131. Li, Z.; Pinson, S.; Marchetti, M.; Stansel, J.; Park, W. Characterization of quantitative trait loci (QTLs) in cultivated rice contributing to field resistance to sheath blight (Rhizoctonia solani). *Theor. Appl. Genet.* **1995**, *91*, 382–388. [CrossRef]
132. Tong, A.H.Y.; Lesage, G.; Bader, G.D.; Ding, H.; Xu, H.; Xin, X.; Young, J.; Berriz, G.F.; Brost, R.L.; Chang, M.; et al. Global mapping of the yeast genetic interaction network. *Science* **2004**, *303*, 808–813. [CrossRef] [PubMed]
133. Taylor, M.B.; Ehrenreich, I.M. Higher-order genetic interactions and their contribution to complex traits. *Trends Genet.* **2015**, *31*, 34–40. [CrossRef]
134. Kuzmin, E.; VanderSluis, B.; Wang, W.; Tan, G.; Deshpande, R.; Chen, Y.; Usaj, M.; Balint, A.; Usaj, M.M.; Van Leeuwen, J.; et al. Systematic analysis of complex genetic interactions. *Science* **2018**, *360*, 128800. [CrossRef]
135. Cifuentes, R.; Ferrucci, G.; Shin, H.S. Liquidity risk and contagion. *J. Eur. Econ. Assoc.* **2005**, *3*, 556–566. [CrossRef]
136. Huang, X.; Vodenska, I.; Havlin, S.; Stanley, H.E. Cascading failures in bi-partite graphs: Model for systemic risk propagation. *Sci. Rep.* **2013**, *3*, 1–9.
137. Caccioli, F.; Shrestha, M.; Moore, C.; Farmer, J.D. Stability analysis of financial contagion due to overlapping portfolios. *J. Bank. Financ.* **2014**, *46*, 233–245. [CrossRef]
138. Caccioli, F.; Farmer, J.D.; Foti, N.; Rockmore, D. Overlapping portfolios, contagion, and financial stability. *J. Econ. Dyn. Control* **2015**, *51*, 50–63. [CrossRef]
139. Corsi, F.; Marmi, S.; Lillo, F. When micro prudence increases macro risk: The destabilizing effects of financial innovation, leverage, and diversification. *Oper. Res.* **2016**, *64*, 1073–1088. [CrossRef]
140. Guo, W.; Minca, A.; Wang, L. The topology of overlapping portfolio networks. *Stat. Risk Model.* **2016**, *33*, 139–155. [CrossRef]
141. Yan, X.G.; Xie, C.; Wang, G.J. Stock market network's topological stability: Evidence from planar maximally filtered graph and minimal spanning tree. *Int. J. Mod. Phys. B* **2015**, *29*, 1550161. [CrossRef]
142. Eryiğit, M.; Eryiğit, R. Network structure of cross-correlations among the world market indices. *Phys. A Stat. Mech. Its Appl.* **2009**, *388*, 3551–3562. [CrossRef]
143. Hatcher, A. *Algebraic Topology*; Cambridge University Press: Cambridge, UK, 2002.
144. Edelsbrunner, H.; Harer, J. *Computational Topology: An Introduction*; American Mathematical Society: Providence, RI, USA, 2010.
145. Ghrist, R.W. *Elementary Applied Topology*; Createspace: Seattle, WA, USA, 2014.
146. Eilenberg, S.; Steenrod, N. *Foundations of Algebraic Topology*; Princeton University Press: Princeton, NJ, USA, 2015.
147. Munkres, J.R. *Elements of Algebraic Topology*; CRC Press: Boca Raton, FL, USA, 2018.
148. Cotton, F.A. *Chemical Applications of Group Theory*; John Wiley & Sons: Hoboken, NJ, USA, 2003.
149. Dresselhaus, M.S.; Dresselhaus, G.; Jorio, A. *Group Theory: Application to the Physics of Condensed Matter*; Springer Science & Business Media: Berlin/Heidelberg, Germany, 2007.
150. Strang, G. *Linear Algebra and Its Applications*, 4th ed.; Cengage Learning: Boston, MA, USA, 2006.
151. Lay, D.; Lay, S.; McDonald, J. *Linear Algebra and Its Applications*, 5th ed.; Pearson: London, UK, 2014.
152. Barabási, A.; PÃ3sfai, M. *Network Science*; Cambridge University Press: Cambridge, UK, 2016.
153. West, D.B. *Introduction to Graph Theory*; Prentice Hall: Upper Saddle River, NJ, USA, 2001; Volume 2.
154. Newman, M. *Networks: An Introduction*; OUP Oxford: Oxford, UK, 2010.
155. Otter, N.; Porter, M.A.; Tillmann, U.; Grindrod, P.; Harrington, H.A. A roadmap for the computation of persistent homology. *EPJ Data Sci.* **2017**, *6*, 1–38. [CrossRef] [PubMed]
156. Gholizadeh, S.; Zadrozny, W. A short survey of topological data analysis in time series and systems analysis. *arXiv* **2018**, arXiv:1809.10745.
157. Salnikov, V.; Cassese, D.; Lambiotte, R. Simplicial complexes and complex systems. *Eur. J. Phys.* **2018**, *40*, 014001. [CrossRef]
158. Pun, C.S.; Xia, K.; Lee, S.X. Persistent-Homology-based Machine Learning and its Applications—A Survey. *arXiv* **2018**, arXiv:1811.00252.
159. De Silva, V.; Ghrist, R.; Muhammad, A. Blind Swarms for Coverage in 2-D. In *Robotics: Science and Systems*; MIT Press: Cambridge, MA, USA, 2005; pp. 335–342.
160. Horak, D.; Maletić, S.; Rajković, M. Persistent homology of complex networks. *J. Stat. Mech. Theory Exp.* **2009**, *2009*, P03034. [CrossRef]

161. Lee, H.; Kang, H.; Chung, M.K.; Kim, B.N.; Lee, D.S. Persistent brain network homology from the perspective of dendrogram. *IEEE Trans. Med. Imaging* **2012**, *31*, 2267–2277.
162. Kasson, P.M.; Zomorodian, A.; Park, S.; Singhal, N.; Guibas, L.J.; Pande, V.S. Persistent voids: A new structural metric for membrane fusion. *Bioinformatics* **2007**, *23*, 1753–1759. [CrossRef]
163. Yao, Y.; Sun, J.; Huang, X.; Bowman, G.R.; Singh, G.; Lesnick, M.; Guibas, L.J.; Pande, V.S.; Carlsson, G. Topological methods for exploring low-density states in biomolecular folding pathways. *J. Chem. Phys.* **2009**, *130*, 04B614. [CrossRef] [PubMed]
164. Xia, K.; Wei, G.W. Persistent homology analysis of protein structure, flexibility, and folding. *Int. J. Numer. Methods Biomed. Eng.* **2014**, *30*, 814–844. [CrossRef] [PubMed]
165. Gameiro, M.; Hiraoka, Y.; Izumi, S.; Kramar, M.; Mischaikow, K.; Nanda, V. A topological measurement of protein compressibility. *Jpn. J. Ind. Appl. Math* **2015**, *32*, 1–17. [CrossRef]
166. Xia, K.; Feng, X.; Tong, Y.; Wei, G.W. Persistent homology for the quantitative prediction of fullerene stability. *J. Comput. Chem.* **2015**, *36*, 408–422. [CrossRef] [PubMed]
167. Xia, K.; Wei, G.W. Multidimensional persistence in biomolecular data. *J. Comput. Chem.* **2015**, *36*, 1502–1520. [CrossRef]
168. Wang, B.; Wei, G.W. Object-oriented persistent homology. *J. Comput. Phys.* **2016**, *305*, 276–299. [CrossRef]
169. Carlsson, G.; Ishkhanov, T.; De Silva, V.; Zomorodian, A. On the local behavior of spaces of natural images. *Int. J. Comput. Vis.* **2008**, *76*, 1–12. [CrossRef]
170. Singh, G.; Memoli, F.; Ishkhanov, T.; Sapiro, G.; Carlsson, G.; Ringach, D.L. Topological analysis of population activity in visual cortex. *J. Vis.* **2008**, *8*, 11. [CrossRef]
171. Bendich, P.; Edelsbrunner, H.; Kerber, M. Computing robustness and persistence for images. *IEEE Trans. Vis. Comput. Graph.* **2010**, *16*, 1251–1260. [CrossRef] [PubMed]
172. Pachauri, D.; Hinrichs, C.; Chung, M.K.; Johnson, S.C.; Singh, V. Topology-based kernels with application to inference problems in Alzheimer's disease. *IEEE Trans. Med. Imaging* **2011**, *30*, 1760–1770. [CrossRef] [PubMed]
173. Niyogi, P.; Smale, S.; Weinberger, S. A topological view of unsupervised learning from noisy data. *SIAM J. Comput.* **2011**, *40*, 646–663. [CrossRef]
174. Wang, B.; Summa, B.; Pascucci, V.; Vejdemo-Johansson, M. Branching and circular features in high dimensional data. *IEEE Trans. Vis. Comput. Graph.* **2011**, *17*, 1902–1911. [CrossRef]
175. Liu, X.; Xie, Z.; Yi, D. A fast algorithm for constructing topological structure in large data. *Homol. Homotopy Appl.* **2012**, *14*, 221–238. [CrossRef]
176. Rieck, B.; Mara, H.; Leitte, H. Multivariate data analysis using persistence-based filtering and topological signatures. *IEEE Trans. Vis. Comput. Graph.* **2012**, *18*, 2382–2391. [CrossRef]
177. Di Fabio, B.; Landi, C. A Mayer–Vietoris formula for persistent homology with an application to shape recognition in the presence of occlusions. *Found. Comput. Math.* **2011**, *11*, 499–527. [CrossRef]
178. Hiraoka, Y.; Nakamura, T.; Hirata, A.; Escolar, E.G.; Matsue, K.; Nishiura, Y. Hierarchical structures of amorphous solids characterized by persistent homology. *Proc. Natl. Acad. Sci. USA* **2016**, *113*, 7035–7040. [CrossRef]
179. Saadatfar, M.; Takeuchi, H.; Robins, V.; Francois, N.; Hiraoka, Y. Pore configuration landscape of granular crystallization. *Nat. Commun.* **2017**, *8*, 15082. [CrossRef] [PubMed]
180. Reimann, M.W.; Nolte, M.; Scolamiero, M.; Turner, K.; Perin, R.; Chindemi, G.; Dłotko, P.; Levi, R.; Hess, K.; Markram, H. Cliques of neurons bound into cavities provide a missing link between structure and function. *Front. Comput. Neurosci.* **2017**, *11*, 48. [CrossRef] [PubMed]
181. Santos, F.A.; Raposo, E.P.; Coutinho-Filho, M.D.; Copelli, M.; Stam, C.J.; Douw, L. Topological phase transitions in functional brain networks. *Phys. Rev. E* **2019**, *100*, 032414. [CrossRef]
182. Gidea, M. Topological data analysis of critical transitions in financial networks. In *International Conference and School on Network Science*; Springer: Berlin/Heidelberg, Germany, 2017; pp. 47–59.
183. Gidea, M.; Katz, Y. Topological data analysis of financial time series: Landscapes of crashes. *Phys. A Stat. Mech. Its Appl.* **2018**, *491*, 820–834. [CrossRef]
184. Zulkepli, N.; Noorani, M.; Razak, F.; Ismail, M.; Alias, M. Haze detection using persistent homology. In *AIP Conference Proceedings*; AIP Publishing LLC: Melville, NY, USA, 2019; Volume 2111, p. 020012.
185. Zulkepli, N.F.S.; Noorani, M.S.M.; Razak, F.A.; Ismail, M.; Alias, M.A. Topological characterization of haze episodes using persistent homology. *Aerosol Air Qual. Res.* **2019**, *19*, 1614–1624. [CrossRef]
186. Adams, H.; Tausz, A.; Vejdemo-Johansson, M. JavaPlex: A research software package for persistent (co) homology. In *International Congress on Mathematical Software*; Springer: Berlin/Heidelberg, Germany, 2014; pp. 129–136.
187. Bauer, U.; Kerber, M.; Reininghaus, J. Distributed computation of persistent homology. In *2014 Proceedings of the Sixteenth Workshop on Algorithm Engineering and Experiments (ALENEX)*; SIAM: Philadelphia, PA, USA, 2014; pp. 31–38.
188. Binchi, J.; Merelli, E.; Rucco, M.; Petri, G.; Vaccarino, F. jholes: A tool for understanding biological complex networks via clique weight rank persistent homology. *Electron. Notes Theor. Comput. Sci.* **2014**, *306*, 5–18. [CrossRef]
189. Dey, T.K.; Fan, F.; Wang, Y. Computing topological persistence for simplicial maps. In Proceedings of the Thirtieth Annual Symposium on Computational Geometry, Kyoto, Japan, 8–11 June 2014; ACM: New York, NY, USA, 2014; pp. 345–354.
190. Fasy, B.T.; Kim, J.; Lecci, F.; Maria, C. Introduction to the R package TDA. *arXiv* **2014**, arXiv:1411.1830.

191. Maria, C. Filtered Complexes. Available online: https://gudhi.inria.fr/doc/3.4.1/group__simplex__tree.html (accessed on 20 April 2020).
192. Bauer, U.; Kerber, M.; Reininghaus, J.; Wagner, H. Phat–persistent homology algorithms toolbox. *J. Symb. Comput.* **2017**, *78*, 76–90. [CrossRef]
193. Nanda, V. Perseus, the Persistent Homology Software. Available online: http://www.sas.upenn.edu/~vnanda/perseus (accessed on 20 April 2020).
194. Dionysus. Dionysus: The Persistent Homology Software. Available online: https://mrzv.org/software/dionysus2/ (accessed on 20 April 2020).
195. Bauer, U. Ripser: A Lean C++ Code for the Computation of Vietoris-Rips Persistence Barcodes. 2017. Available online: https://github.com/Ripser/ripser (accessed on 20 April 2020).
196. Schauf, A.; Cho, J.B.; Haraguchi, M.; Scott, J.J. *Discrimination of Economic Input-Output Networks Using Persistent Homology*; The Santa Fe Institute CSSS Working Paper; The Santa Fe Institute: Santa Fe, NM, USA, 2016.
197. de la Concha, A.; Martinez-Jaramillo, S.; Carmona, C. Multiplex financial networks: Revealing the level of interconnectedness in the banking system. In Proceedings of the International Conference on Complex Networks and their Applications, Lyon, France, 29 November–1 December 2017; Springer: Berlin/Heidelberg, Germany, 2017; pp. 1135–1148.
198. Santos, F.; Da Silva, L.; Coutinho-Filho, M. Topological approach to microcanonical thermodynamics and phase transition of interacting classical spins. *J. Stat. Mech. Theory Exp.* **2017**, *2017*, 013202. [CrossRef]
199. Tao, T. Ricci Flow. Available online: https://terrytao.files.wordpress.com/2008/03/ricci.pdf (accessed on 29 June 2021).
200. Albert, E.; Perrett, W.; Jeffery, G. The foundation of the general theory of relativity. *Ann. Der Phys.* **1916**, *49*, 769–822.
201. Samal, A.; Sreejith, R.; Gu, J.; Liu, S.; Saucan, E.; Jost, J. Comparative analysis of two discretizations of Ricci curvature for complex networks. *Sci. Rep.* **2018**, *8*, 18650. [CrossRef]
202. Isenberg, J.; Mazzeo, R.; Sesum, N. Ricci flow in two dimensions. *arXiv* **2011**, arXiv:1103.4669.
203. Topping, P. *Lectures on the Ricci Flow*; Cambridge University Press: Cambridge, UK, 2006; Volume 325.
204. Brendle, S. *Ricci Flow and the Sphere Theorem*; American Mathematical Society: Providence, RI, USA, 2010; Volume 111.
205. Máximo, D. On the blow-up of four-dimensional Ricci flow singularities. *J. Für Die Reine Angew. Math. (Crelles J.)* **2014**, *2014*, 153–171. [CrossRef]
206. Ollivier, Y. Ricci curvature of metric spaces. *Comptes Rendus Math.* **2007**, *345*, 643–646. [CrossRef]
207. Ollivier, Y. Ricci curvature of Markov chains on metric spaces. *J. Funct. Anal.* **2009**, *256*, 810–864. [CrossRef]
208. Forman, R. Bochner's method for cell complexes and combinatorial Ricci curvature. *Discret. Comput. Geom.* **2003**, *29*, 323–374. [CrossRef]
209. Sreejith, R.; Mohanraj, K.; Jost, J.; Saucan, E.; Samal, A. Forman curvature for complex networks. *J. Stat. Mech. Theory Exp.* **2016**, *2016*, 063206. [CrossRef]
210. Ni, C.C.; Lin, Y.Y.; Gao, J.; Gu, X.D.; Saucan, E. Ricci curvature of the internet topology. In Proceedings of the 2015 IEEE Conference on Computer Communications (INFOCOM), Hong Kong, China, 26 April–1 May 2015; IEEE: Piscataway, NJ, USA, 2015; pp. 2758–2766.
211. Sandhu, R.; Georgiou, T.; Reznik, E.; Zhu, L.; Kolesov, I.; Senbabaoglu, Y.; Tannenbaum, A. Graph curvature for differentiating cancer networks. *Sci. Rep.* **2015**, *5*, 12323. [CrossRef] [PubMed]
212. Farooq, H.; Chen, Y.; Georgiou, T.T.; Tannenbaum, A.; Lenglet, C. Network curvature as a hallmark of brain structural connectivity. *Nat. Commun.* **2019**, *10*, 4937. [CrossRef]
213. Ni, C.C.; Lin, Y.Y.; Luo, F.; Gao, J. Community detection on networks with ricci flow. *Sci. Rep.* **2019**, *9*, 9984. [CrossRef]
214. Sia, J.; Jonckheere, E.; Bogdan, P. Ollivier-ricci curvature-based method to community detection in complex networks. *Sci. Rep.* **2019**, *9*, 9800. [CrossRef] [PubMed]
215. Sandhu, R.S.; Georgiou, T.T.; Tannenbaum, A.R. Ricci curvature: An economic indicator for market fragility and systemic risk. *Sci. Adv.* **2016**, *2*, e1501495. [CrossRef]
216. Bastian, M.; Heymann, S.; Jacomy, M. Gephi: An open source software for exploring and manipulating networks. In Proceedings of the International AAAI Conference on Weblogs and Social Media, San Jose, CA, USA, 17–20 May 2009; pp. 361–362.
217. Boyer, J.M.; Myrvold, W.J. Simplified O (n) Planarity by Edge Addition. *Graph Algorithms Appl.* **2006**, *5*, 241.
218. Wee, J.; Xia, K. Ollivier Persistent Ricci Curvature-Based Machine Learning for the Protein–Ligand Binding Affinity Prediction. *J. Chem. Inf. Model.* **2021**, *61*, 1617–1626. [CrossRef] [PubMed]
219. Poincaré, M. Cinquième complément à l'analysis situs. *Rendiconti del Circolo Matematico di Palermo (1884–1940)* **1904**, *18*, 45–110. [CrossRef]
220. Smale, S. Generalized Poincare?'s conjecture in dimensions greater than four. *Matematika* **1962**, *6*, 139–155.
221. Freedman, M. The topology of four-differentiable manifolds. *J. Diff. Geom* **1982**, *17*, 357–453.
222. Zeeman, E. The generalised Poincaré conjecture. *Bull. Am. Math. Soc.* **1961**, *67*, 270. [CrossRef]
223. Stallings, J. The piecewise-linear structure of Euclidean space. *Math. Proc. Camb. Philos. Soc.* **1962**, *58*, 471–488. [CrossRef]
224. Thurston, W.P. Three dimensional manifolds, Kleinian groups and hyperbolic geometry. *Bull. Am. Math. Soc.* **1982**, *6*, 357–381. [CrossRef]
225. Hamilton, R.S. Three-manifolds with positive Ricci curvature. *J. Differ. Geom* **1982**, *17*, 255–306. [CrossRef]
226. Perelman, G. The entropy formula for the Ricci flow and its geometric applications. *arXiv* **2002**, arXiv:math/0211159 .

227. Perelman, G. Ricci flow with surgery on three-manifolds. *arXiv* **2003**, arXiv:math/0303109.
228. Bertsimas, D.; Tsitsiklis, J.N. *Introduction to Linear Optimization*; Athena Scientific: Belmont, MA, USA, 1997; Volume 6.
229. Hurlbert, G. *Linear Optimization: The Simplex Workbook*; Undergraduate Texts in Mathematics; Springer: New York, NY, USA, 2010.
230. Metei, A.; Jain, V. *Optimization Using Linear Programming*; Mercury Learning & Information: Dulles, VA, USA, 2019.

entropy

MDPI

Article

Understanding the Nature of the Long-Range Memory Phenomenon in Socioeconomic Systems

Rytis Kazakevičius *,†, **Aleksejus Kononovicius** †, **Bronislovas Kaulakys** † and **Vygintas Gontis** *,†

Institute of Theoretical Physics and Astronomy, Vilnius University, Sauletekio al. 3, 10257 Vilnius, Lithuania; aleksejus.kononovicius@tfai.vu.lt (A.K.); Bronislovas.Kaulakys@tfai.vu.lt (B.K.)

* Correspondence: rytis.kazakevicius@tfai.vu.lt (R.K.); vygintas@gontis.eu (V.G.); Tel.: +370-698-12384 (V.G.)

† These authors contributed equally to this work.

Abstract: In the face of the upcoming 30th anniversary of econophysics, we review our contributions and other related works on the modeling of the long-range memory phenomenon in physical, economic, and other social complex systems. Our group has shown that the long-range memory phenomenon can be reproduced using various Markov processes, such as point processes, stochastic differential equations, and agent-based models—reproduced well enough to match other statistical properties of the financial markets, such as return and trading activity distributions and first-passage time distributions. Research has lead us to question whether the observed long-range memory is a result of the actual long-range memory process or just a consequence of the non-linearity of Markov processes. As our most recent result, we discuss the long-range memory of the order flow data in the financial markets and other social systems from the perspective of the fractional Lèvy stable motion. We test widely used long-range memory estimators on discrete fractional Lèvy stable motion represented by the auto-regressive fractionally integrated moving average (ARFIMA) sample series. Our newly obtained results seem to indicate that new estimators of self-similarity and long-range memory for analyzing systems with non-Gaussian distributions have to be developed.

Keywords: long-range memory; $1/f$ noise; absolute value estimator; anomalous diffusion; ARFIMA; first-passage times; fractional Lèvy stable motion; Higuchi's method; mean squared displacement; multiplicative point process

Citation: Kazakevičius, R.; Kononovicius, A.; Kaulakys, B.; Gontis, V. Understanding the Nature of the Long-Range Memory Phenomenon in Socioeconomic Systems. *Entropy* **2021**, *23*, 1125. https://doi.org/10.3390/e23091125

Academic Editor: Ryszard Kutner

Received: 4 August 2021
Accepted: 25 August 2021
Published: 29 August 2021

1. Introduction

Many empirical data sets and theoretical models have been investigated using the tool of spectral analysis. Many researchers across different fields find the power spectral density (abbr. PSD) of the $1/f^\beta$ form (with $0.5 \lesssim \beta \lesssim 1.5$) to be of a particular interest [1–10], both because of its apparent omnipresence and the implication of slowly decaying autocorrelation, which indicates the presence of the long-range memory phenomenon. Long-range memory is also one of the established stylized facts of the financial markets [11–19]. Consequently, as our group was investigating $1/f$ noise [20–23], we have become naturally interested in the rapidly growing field of econophysics. The term "econophysics" was coined by H. E. Stanley in the Statphys conference in Kolkata in 1995 [24]. Over the last three decades, econophysics has matured both from the theoretical and the applied perspectives. Here, we review mostly our own and directly adjacent approaches, and we would like to recommend a couple of broader reviews, which can be found in [25,26].

Our first publications were devoted to the modeling of the financial markets [27,28]. In those works, we have considered trades occurring in the financial markets as point events driven by a point process proposed in [21–23]. Thanks to the organizers of the international conference Applications of Physics in Financial Analysis 4, held in Warsaw in 2003, we were able to present our findings to econophysicists. Our first results, inspired by interaction with the participants of the APFA 4 conference, have been published in [29,30]. We presented our ideas in a more general context of complex systems in [31,32].

Later, we took part in the COST Action P10 "Physics of Risk" and the follow-up COST Action MP0801 "Physics of Competition and Conflicts". Bronislovas Kaulakys and Vygintas Gontis were executive committee members of both COST Actions, while the other group members gave talks and poster presentations during the annual meetings and helped organize an annual action meeting in Vilnius in 2006. This COST action meeting has helped us embrace econophysics and be recognized as econophysicists.

While it may be natural to see trades in the financial markets as point events [27–30], modeling volatility and return as a point process was not as straightforward. We have developed our approach further by abstracting the point process away and considering a continuous framework of Langevin stochastic differential equations (abbr. SDEs). First, we have shown that the continuous interpretation of the point process model works well for trading activity [33]; thus, we have refined the SDE approach with model for volatility and return [34–38]. Interestingly, similar SDEs can be derived from a simple agent-based model (abbr. ABM) [39,40], too. With time, we have developed more complicated ABMs to account for the separation of time scales and order flow [41,42]. We have even branched out into sociophysics [43–46] as we have understood that the herding ABM we used to model the financial market is essentially equivalent to the well-known voter model [47–49].

For 10 months (in 2015 and 2016), Vygintas Gontis, with the support of the Baltic American Freedom Foundation, has stayed as a visiting researcher at the Center of Polymer Studies of Boston University. Discussions with the founding fathers of econophysics, H. E. Stanley, professors Sh. Havlin, B. Podobnik, and S. Buldyrev, resulted in a paper [50]. Together, we have considered volatility return intervals (term inspired by the studies [51–54]) of the financial time series at various time scales. In the paper, we have shown that the time intervals between large financial fluctuations is distributed according to a power–law probability density function (abbr. PDF) $p(\tau) \sim \tau^{-3/2}$ [50]. The same distribution arise in our models and from many other one-dimensional Markov processes [55], while the long-range memory process would exhibit a different distribution, such as $p(\tau) \sim \tau^{2-H}$, which is a well-known result for the fractional Brownian motion (abbr. FBM) [56].

Here, we provide an overview of our approach to understanding and modeling the long-range memory phenomenon in financial markets and other complex systems and share our most recent result. In Section 2, we introduce the original point process and discuss how to derive a non-linear SDE, which can reproduce the long-range memory phenomenon. We also discuss numerous extensions of both the point process model and non-linear SDE. Next, in Section 3, we show how we can obtain a similar SDE from a simple herding ABM. Following the overview, we also present a novel result, which concerns understanding the nature of the self-similarity and long-range memory phenomenon from the perspective of fractional Lèvy stable motion (abbr. FLSM) and auto-regressive fractionally integrated moving average (abbr. ARFIMA) time series. In Section 4, we tested various long-range memory estimators such as mean squared displacement, method of absolute value estimator, Higuchi's method, and burst and interburst duration analysis on fractional Lèvy stable motion (ARFIMA(0,d,0) time series). Finally, in Section 5, we share our future considerations.

2. The Multiplicative Point Process, the Class of Stochastic Differential Equations, and Their Applications

In this section, we overview how the physically motivated point process proposed in [21–23] was applied to model trading activity and absolute returns in the financial markets. We also discuss numerous extensions of the model into some related research topics, such as superstatistics and anomalous and non-homogeneous diffusion.

2.1. The Multiplicative Point Process Model

Let us consider signal $I(t)$ composed of pulses with profiles given by $A_k(x)$:

$$I(t) = \sum_k A_k(t - t_k), \quad (1)$$

where t_k is the event (pulse) time. There are many physical and social systems, which generate signals of such nature: electric current [57], music [58], human heartbeat [59], internet traffic [32], or trading activity [29] to name a few.

As most profiles of the pulses are brief, it is trivial that they would influence only high frequencies corresponding to the typical inverse pulse length. If we are interested in longer-term dynamics, it is sufficient to assume that the Kronecker delta function well approximates the profile, $A_k(x) = a_k\delta(x)$. Many such systems are driven by the flow of identical or similar objects, such as electrons, packets, or trades. This lets us simplify (1) and investigate it as a temporal point process with unit events. Such a process can be either described by the event times $\{t_k\}$ or by the inter-event times $\{\tau_k = t_{k+1} - t_k\}$.

The inter-event times are a far more convenient choice to model as they at least can give a semblance of the stationarity, while event times are obviously non-stationary as $\{t_k\}$ is monotonically increasing series. In [21–23], it was analytically shown that a relatively slow autoregressive AR(1) Brownian motion of τ_k yield $1/f$ fluctuations of the signal $I(t)$. The author of [29] has built upon this observation and introduced multiplicative point process for the inter-event time

$$\tau_{k+1} = \tau_k + \sigma^2\gamma\tau_k^{2\mu-1} + \sigma\tau_k^{\mu}\varepsilon_k. \tag{2}$$

In the above, it is assumed that inter-event time fluctuates due to exogenous perturbations. Perturbations are assumed to be standard uncorrelated Gaussian random variables, ε_k. The general rate of change is governed by σ, while γ is the damping constant. Multiplicativity, specified by μ, ensures that $I(t)$ is multifractal and has a power–law PDF. This point process model has found its use for the analysis of $1/f$ noise and long-range memory in many diverse phenomena such as musical rhythm spectra [58], human cognition [60], human interaction dynamics [61], turbulence [62], and few others [63–66]. Inspired by this model, [67] has shown under which conditions $1/f^{\beta}$ spectrum can arise from reversible Markov chains.

After closer examination, it should be evident that Equation (2) can be seen as an iterative solution of a certain SDE if Euler–Maruyama method was used [68]. Hence the corresponding Langevin SDE can be trivially recovered from the iterative relation (2):

$$\mathrm{d}\,\tau = \sigma^2\gamma\tau^{2\mu-1}\,\mathrm{d}\,k + \sigma\tau^{\mu}\,\mathrm{d}\,W_k. \tag{3}$$

Here W is uncorrelated standard Wiener process. Note that this SDE is in the event space (or k–space) and not in the real time. Further, this SDE must be solved by restricting the diffusion of the inter-event time τ to some arbitrary interval $[\tau_{\min}, \tau_{\max}]$ on the positive half-plane as otherwise this SDE may not have a stationary distribution. If stationary distribution exists, then the stationary PDF of τ is a power–law:

$$p_k(\tau) = \frac{\alpha+1}{\tau_{\max}^{\alpha+1} - \tau_{\min}^{\alpha+1}}\tau^{\alpha}, \quad \alpha = 2(\gamma - \mu). \tag{4}$$

Yet the main result of [29] is the power–law statistical properties of $I(t)$. In the limit $\tau_{\min} \to 0$ and $\tau_{\max} \to \infty$ PSD of $I(t)$ in arbitrarily long range of frequencies has a power–law slope:

$$S(f) \sim 1/f^{\beta}, \quad \beta = 1 + \frac{2(\gamma-\mu)}{3-2\mu}. \tag{5}$$

The number of events in a selected time window, for example number of trades per minute, also has a power–law distribution [29]:

$$p(N) \sim N^{-2(\gamma-\mu)-3}. \tag{6}$$

Formally, one could define the number of events in a window of length w as $N[t] = \int_t^{t+w} I(u)\,\mathrm{d}\,u$ (here the square brackets indicate that N is in discrete time). These analytical results can be confirmed by numerical simulation (see Figure 1).

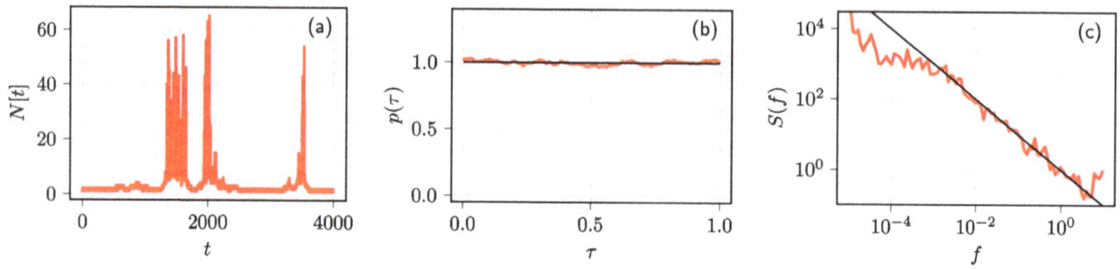

Figure 1. Statistical properties of the point process by numerically solving Equation (2): (**a**) sample fragment of corresponding $N[t]$ time series, (**b**) PDF of the inter-event times, and (**c**) PSD of the process. Red curves correspond to numerical results, while black curves are theoretical power–law fits with (**b**) $\alpha = 0$ and (**c**) $\beta = 1$. Model parameter values: $\gamma = 0$, $\mu = 0$, $\sigma = 0.1$, $w = 1$.

2.2. *The Class of Non-Linear Stochastic Differential Equations*

In [33,69–71], we have made a transition from k-space to real time and this enabled us to model trading activity and absolute returns in the financial markets not only qualitatively, but quantitatively, too. The transition from SDE in k-space, Equation (3), to real time is achieved by substitution $\mathrm{d}\,t = \tau\,\mathrm{d}\,k$, which yields:

$$\mathrm{d}\,\tau = \sigma^2\gamma\tau^{2\mu-2}\,\mathrm{d}\,t + \sigma\tau^{\mu-1/2}\,\mathrm{d}\,W. \tag{7}$$

Modeling inter-event time in real time makes less sense than in the k-space, so let us change the variable to the number of events per unit time $x = \frac{1}{\tau}$. Applying Itô transformation yields:

$$\mathrm{d}\,x = \sigma^2\left(\eta - \frac{\lambda}{2}\right)x^{2\eta-1}\,\mathrm{d}\,t + \sigma x^{\eta}\,\mathrm{d}\,W. \tag{8}$$

In the above, we have introduced a more convenient set of parameters:

$$\eta = \frac{5}{2} - \mu, \quad \lambda = 2(\gamma - \mu) + 3. \tag{9}$$

As far as SDE (8) corresponds to the point process defined by Equation (2), the results for stationary PDF and PSD should apply:

$$p(x) \sim x^{-\lambda}, \qquad S(f) \sim 1/f^{\beta}, \quad \beta = 1 + \frac{\lambda - 3}{2\eta - 2}. \tag{10}$$

The validity of these theoretical predictions was extensively checked numerically (see Figure 2 for a quick example) and also, in [72], proven analytically. The analytical proof provided in [72] allows interpreting the process modeled by SDE (8) in a more general context. In fact we can model any process possessing these power–law statistical properties, even processes, which make less sense from the perspective of the original point process.

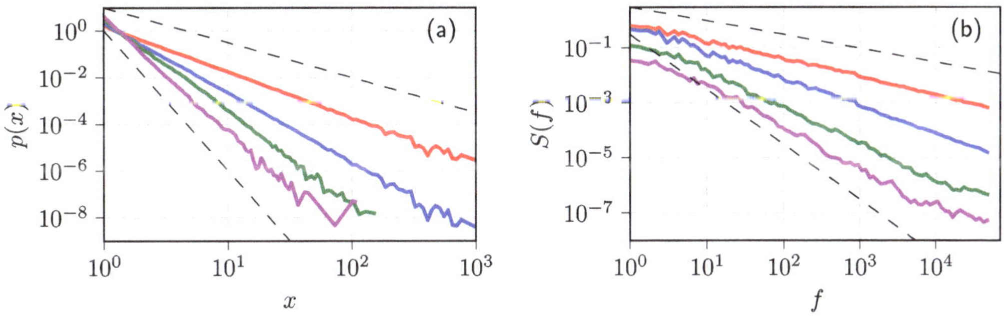

Figure 2. Various slopes of PDF (**a**) and PSD (**b**) reproduced by the numerical solutions of SDE (8). Model parameter values: $\sigma = 1$, $\eta = 2.5$ (all cases) and $\lambda = 2$ (red curves in both (**a**,**b**)), 3 (blue curves), 4 (green curves), and 5 (magenta curves). Black dashed lines correspond to (**a**) $p(x) \sim x^{-\lambda}$ with $\lambda = 1.5$ and $\lambda = 6$ (upper and lower curves), (**b**) $S(f) \sim 1/f^{\beta}$ with $\beta = 0.5$ and $\beta = 2$ (upper and lower curves).

Equation (8) and similar random walk models have been used to model the EUR/CHF exchange rate [73]. It has also lead to numerous modifications by our group, which we discuss in detail in the following subsections.

2.3. Reproducing the Long-Range Memory Using GARCH(1,1) Process

Autoregressive conditional heteroscedasticity (abbr. ARCH) family models [74–79] are quite popular forecasting tools among professional traders as well as researchers interested in the long-range memory phenomenon. Unlike SDEs, ARCH family models have explicitly built-in memory, which is built-in either via explicit dependence on the numerous previous states, infinitely many in the case of the ARCH(∞) model [80–82], or via fractional integration procedure, which introduces memory similar to the one present in the fractional Brownian motion, as in the fractionally integrated GARCH (abbr. FIGARCH) model [83–85]. In [86], we have shown that it is possible to modify the GARCH(1,1) model, which is Markovian in nature, to reproduce $1/f$ spectrum.

Generalized autoregressive conditional heteroskedasticity (abbr. GARCH) processes can be approximated by the diffusion processes. There are two competing approaches, which yield continuous approximations of GARCH processes using sets of SDEs. One of the approaches was proposed by Nelson [87] and the other by Kluppelberg et al. [88,89]. In the GARCH(1,1), Nelson's approach is easier to apply, but has a drawback that the resulting COGARCH(1,1) would be driven by two sources of noise, instead of the one in the GARCH(1,1). Yet, we can circumvent the problem by ignoring the observed heteroskedastic economic variable z_t and focusing on the approximation of the volatility process, σ_t^2, of GARCH(1,1):

$$z_t = \sigma_t \omega_t, \tag{11}$$

$$\sigma_t^2 = a + b z_{t-1}^2 + c\sigma_{t-1}^2 = a + b\sigma_{t-1}^2 \omega_{t-1}^2 + c\sigma_{t-1}^2. \tag{12}$$

In the above, ω_t is the noise, while a, b, and c are the GARCH(1,1) model parameters. For Nelson's approach to work, we need to compute first and second moments of change in volatility. With the usual GARCH(1,1) we obtain SDE for geometric Brownian motion [86].

Now lets introduce non-linearity into Equation (12). In [86], we have explored two such options:

$$\sigma_t^2 = a + b\sigma_{t-1}^{\mu} \omega_{t-1}^{\mu} + c\sigma_{t-1}^2, \tag{13}$$

$$\sigma_t^2 = a + b\sigma_{t-1}^{\mu} |\omega_{t-1}|^{\mu} + \sigma_{t-1}^2 - c\sigma_{t-1}^{\mu}. \tag{14}$$

Both of these options can be approximated by SDEs belonging to the class of SDEs (8) with $\lambda = \mu$ and $\eta = \mu/2$. Consequently both of these options reproduce $1/f$ spectrum with $\mu = 3$. Other parameters, a, b, and c, influence only the additional terms, which restrict the diffusion of σ_t^2. Setting these values too high shrinks the interval and the power–law distribution becomes extremely hard to observe.

2.4. Anomalous Diffusion in the Long-Range Memory Process

SDE (8) can be also seen to describe a heterogeneous diffusion in a non-linear potential. Such diffusion leads to anomalous growth in variance [90]

$$\left\langle [x(t) - \langle x(t) \rangle]^2 \right\rangle \sim t^{\theta}, \quad \theta = \frac{1}{1-\eta}. \tag{15}$$

This phenomenon is also known as anomalous diffusion [91–93]. If $\theta = 1$ then the process exhibits normal diffusion. Otherwise if $0 < \theta < 1$, the diffusion is slower than normal and is referred to as sub-diffusion. The diffusion may also be faster, if $1 < \theta < 2$, in that case it is called super-diffusion.

The anomalous diffusion can be obtained from SDE (8) only for specific parameter values such as $\lambda < 1$ and $\eta < 1/2$ [90]. Because power–law slope of the PSD, β, varies between 0 and 2, from Equation (10), it follows that anomalous diffusion and power–law noise can be observed at the same time only for negative parameter η values, specifically for $\eta < (\lambda - 1)/2$ and $\lambda < 1$; however, for these parameters values numerical simulation would become very slow and inefficient [72]; therefore, we have considered generalizing SDE (8) by considering non-Gaussian white noise.

In [94], we have considered Lévy α-stable noise. SDE equivalent to SDE (8), but with Lévy α-stable noise takes the following form:

$$\frac{\mathrm{d}\,x}{\mathrm{d}\,t} = \gamma(\eta, \lambda, \alpha) x^{\alpha(\eta-1)+1} + x^{\eta} \xi_{\alpha}(t). \tag{16}$$

Here, $\xi_{\alpha}(t)$ is a white noise, the intensity of which is distributed according to the symmetric Lévy α-stable distribution. The characteristic function of the noise intensity is given by:

$$\langle \exp(ik\xi_{\alpha}) \rangle = \exp\left(-\sigma^{\alpha} |k|^{\alpha}\right). \tag{17}$$

Here, α is the index of stability and σ is the scale parameter. We interpret SDE (16) in an Itô sense and it can also be written in the form

$$\mathrm{d}\,x = \gamma(\eta, \lambda, \alpha) x^{\alpha(\eta-1)+1} \,\mathrm{d}\,t + x^{\eta} \,\mathrm{d}\,L_t^{\alpha}. \tag{18}$$

Here, $\mathrm{d}\,L_t^{\alpha}$ stands for the increments of Lévy α-stable motion L_t^{α}. If SDE (16) is solved with reflective boundary conditions and

$$\gamma(\eta, \lambda, \alpha) = \frac{\sin\left[\pi\left(\frac{\alpha}{2} - \alpha\eta + \lambda\right)\right]}{\sin[\pi(\alpha(\eta - 1) - \lambda)]} \frac{\Gamma(\alpha\eta - \lambda + 1)}{\Gamma(\alpha(\eta - 1) - \lambda + 2)}, \tag{19}$$

then generalized SDE (16) generate time series with power–law steady-state PDF and power–law PSD:

$$p(x) \sim x^{-\lambda}, \qquad S(f) \sim \frac{1}{f^{\beta}}, \quad \beta = 1 + \frac{\lambda - 3}{\alpha(\eta - 1)}. \tag{20}$$

Extensive numerical simulations have shown that due to the presence of the multiplicative Lévy α-stable noise in Equation (16) both sub-diffusion and super-diffusion can be observed together with power–law noise even for positive η values [95]; however, no analytical expression for anomalous diffusion exponent dependence on SDE parameters has been derived yet.

In Figure 3, we show a sample series of the solutions of SDE (16) and the statistical properties of the series when the noise is Lévy α-stable noise with $\alpha = 1$. The other SDE (16) parameters were picked so $1/f$ spectrum would be reproduced. As can be seen in the subfigure (**a**), ongoing diffusion is disrupted by huge jumps, which are characteristic to Lévy flights.

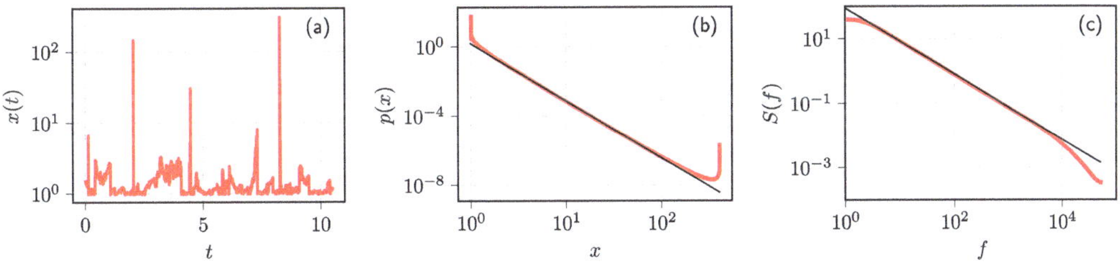

Figure 3. Statistical properties of the time series obtained by solving SDE with Lévy α-stable noise, Equation (16): (**a**) sample fragment of the time series, (**b**) PDF, and (**c**) PSD of time series. Red curves correspond to numerical results, while black curves are power–law best fits with exponents (**b**) $\lambda \approx 3.3$, (**c**) $\beta \approx 1$.

If we consider modeling only sub-diffusive processes, then we can study another generalization of SDE (8), originally proposed in [96]. If we start with a Markovian process described by the Itô SDE

$$\mathrm{d}\, x(\tau) = f[x(\tau)]\, \mathrm{d}\, \tau + g[x(\tau)]\, \mathrm{d}\, W(\tau). \tag{21}$$

The drift and diffusion functions of the above SDE are given by

$$f(x) = \sigma^2 \left(\eta - \frac{\lambda}{2} \right) x^{2\eta - 1}, \qquad g(x) = \sigma x^{\eta}. \tag{22}$$

We interpret the time τ as an internal (operational) time. For the trapping processes that have a distribution of the trapping times with power–law tails, the physical time $t = T(\tau)$ is given by the strictly increasing α_+-stable Lévy motion defined by the Laplace transform

$$\left\langle e^{-kT(\tau)} \right\rangle = e^{-\tau k^{\alpha_+}}. \tag{23}$$

Here, the parameter α_+ takes the values from the interval $0 < \alpha_+ < 1$. Thus, the physical time t obeys the SDE

$$\mathrm{d}\, t(\tau) = \mathrm{d}\, L^{\alpha_+}(\tau), \tag{24}$$

where $dL^{\alpha_+}(\tau)$ stands for the increments of the strictly increasing α_+-stable Lévy motion $L^{\alpha_+}(\tau)$. For such physical time t the operational time τ is related to the physical time t via the inverse α_+-stable subordinator

$$S(t) = \inf\{\tau : T(\tau) > t\}. \tag{25}$$

Such subordination leads to power spectral density

$$S(f) \sim \begin{cases} \frac{1}{\omega^{\beta}}, & 1 - \alpha_+ < \beta < 1 + \alpha_+, \\ \frac{1}{\omega^{1+\alpha_+}}, & \beta > 1 + \alpha_+. \end{cases}, \qquad \beta = 1 + \frac{\alpha_+(\lambda - 3)}{(\eta - 1)} \tag{26}$$

Proposed SDEs (8), (16), and (21) have served as a basis to study heterogeneous diffusion in a non-homogeneous medium [90,96,97] and time subordinated processes [98,99] as well as the effects of non-linear variable transformations [100,101].

In paper [98], we investigated the distinction between the internal time of the system and the physical time as a source of $1/f$ noise. We have introduced the internal (operational) time into the earlier point process [21–23] together with additional equations relating the internal time to the physical time. In this scenario, we can still recover power–law statistical features similar to the ones obtained by solving Equation (8). In the financial markets, the internal time could reflect the fluctuating human activity, e.g., trading activity, yielding the long-range correlations in the volatility. The effective approach for the solution of highly non-linear SDEs was proposed [98] by a suitable choice of the internal time and variable steps of integration.

The effects of non-linear variable transformations [100,101] suggest that long-range memory in certain cases can be just a measurement effect. As far as the non-linear transformation of the observable x to y

$$x = \frac{1}{y^{\delta}}, \tag{27}$$

with δ being the transformation exponent, yields SDE for the variable y of the same form such as Equation (8) for x.

2.5. Inverse Cubic Law for Long-Range Correlated Processes

The inverse cubic law is an established stylized fact stating that the cumulative distributions of various financial market time series such as the number of trades, the trade volume, or the return [12,14,15,19]. Thus, this law is as important for the modeling as the consideration of long-range memory and fractal scaling, which are also stylized facts [6,12,14,15,19]. We have in proposed [102] that the non-linear SDE yields both the power–law behavior of the PSD and the inverse cubic law of the cumulative distribution. This was achieved using the idea that when the market evolves from calm to violent behavior there is a decrease of the delay time of multiplicative feedback of the system in comparison to the driving noise correlation time. This results in a transition from the Itô to the Stratonovich sense of the SDE and yields a long-range memory process.

We start from a simple quadratic SDE

$$\mathrm{d}\,x = x^2 \circ_{\alpha} \mathrm{d}\,W \tag{28}$$

where α is the interpretation parameter, defining the α-dependent stochastic integral of the SDE (28),

$$\int_0^T f(x(t)) \circ_{\alpha} \mathrm{d}W_t \equiv \lim_{N\to\infty} \sum_{n=0}^{N-1} f(x(t_n))\Delta W_{t_n}. \tag{29}$$

Here, $t_n = \frac{n+\alpha}{N}T$ with $0 \le \alpha \le 1$. Natural choices of the parameter α are: (i) $\alpha = 0$, pre-point (Itô convention), (ii) $\alpha = 1/2$, mid-point (Stratonovich convention), and (iii) $\alpha = 1$, post-point (Hänggi–Klimontovich, kinetic, or isothermal convention) [103].

The quadratic SDE (28) is the simplest multiplicative SDE without the drift term symmetric for the positive and negative deviations of some observable x. More generally, the same process can be described by the delayed SDE [103]

$$\mathrm{d}\,x(t) = f(x(t))\,\mathrm{d}\,t + g(x(t-\delta))\zeta_t^{\tau}\,\mathrm{d}\,t. \tag{30}$$

Here, $f(x)$ represents arbitrary deterministic drift of the observable x, while $g(x)$ effectively controls the diffusion as ζ_t^{τ} is the noise term, which is assumed to have correlation time τ. Note that the diffusion function depends on the delayed value of the observable x (by time interval δ).

It may be shown [103] that in the limit $\delta \to 0$ and $\tau \to 0$ (under the condition $\delta/\tau = const$) SDE (30) can be transformed into

$$\mathrm{d}\,x = f(x(t))\,\mathrm{d}\,t + g(x(t)) \circ_{\alpha} \mathrm{d}\,W \tag{31}$$

with the interpretation parameter being determined by

$$\alpha\left(\frac{\delta}{\tau}\right) \simeq \frac{1}{2(1+\delta/\tau)}. \tag{32}$$

Under the perturbation by the white noise, in a case of $\tau \ll \delta$, even for a short delay in feedback δ, we achieve the Itô outcome, because there is no correlation between the sign of the noise ζ_t and the time-derivative of the feedback $g(x)$. On the contrary, under the perturbation by the correlated noise, $\tau \gg \delta$, a correlation emerges between the sign of ζ_t and the time-derivative of $g(x)$. In this case the correlation yields the Stratonovich outcome [103].

In general, the value of α may depend on the coordinate x and/or other system' parameters. SDE (28) with $\alpha \neq 0$ may be transformed into SDE in Itô sense

$$\mathrm{d}\,x = 2\alpha x^3\,\mathrm{d}\,t + x^2\,\mathrm{d}\,W. \tag{33}$$

This SDE is a particular case of the general Itô Equation (8) yielding the power–law steady-state PDF and the power–law PSD (10). These SDEs become identical for $\eta = 2$ and $\lambda = 4(1-\alpha)$.

Let us note that $1/f^{\beta}$ noise emerges due to the large fluctuations in the time series, while the finite time studies reveal the commonly observed magnitudes of the observable. The common fluctuations can be modeled by the familiar in the financial application's Itô SDEs. On the other hand, the large rapid fluctuations of the violent market arise due to the strong correlated influences; the processes of such a market are fast, all durations become short in comparison to the herding correlation time, and, consequently, the market should be modeled by the Stratonovich version of SDE.

For the modeling of such dynamics, we generalize Equations (28) and (33) with x-dependent parameter $\alpha(x)$. Let

$$\mathrm{d}\,x = 2\alpha(x)x^3\,\mathrm{d}\,t + x^2\,\mathrm{d}\,W, \tag{34}$$

with, e.g.,

$$\alpha(x) = \frac{1}{2}\left[1 - \exp\left\{-\left(\frac{x}{x_c}\right)^2\right\}\right], \tag{35}$$

where x_c is the Itô to Stratonovich interpretations crossover parameter. Equations (34) and (35) represent transition from Itô to Stratonovich convention with an increase in the variable x and decrease of the delay time of multiplicative feedback for larger x, according to the Wong–Zakai theorem [103]. Detailed numerical analysis of the model represented by Equations (34) and (35) is presented in paper [102].

2.6. $1/f^{\beta}$ *Noise with Distributions Other Than Power–Law*

Solutions of the SDE (8) will always have power–law statistical properties of the (10) form; however, often noise with $1/f^{\beta}$ PSD is distributed according to PDF, which is not power–law, but Gaussian or some other distribution. Here, we review two different approaches, which allow for other distributions to be observed in time series with $1/f^{\beta}$ spectrum: superstatistical and coupled SDE approaches.

In [104], it was suggested that the Poissonian-like process with the slowly changing average inter-event time may be represented as the superstatistical process exhibiting $1/f$ noise. It was assumed that the inter-event time τ_k, obtained by solving Equation (2), represents not the actual (observed) inter-event time, but its average (reciprocal of the event rate). In this setup, the actual inter-event time $\hat{\tau}_k$ would be given by the conditional probability

$$\varphi(\hat{\tau}_k|\tau_k) = \frac{1}{\tau_k}e^{-\hat{\tau}_k/\tau_k}, \tag{36}$$

similar to the non-homogeneous Poisson process. This additional randomization has no influence on the lower frequencies of the PSD and the intensity of the signal.

The PDF of the observed inter-event time $\hat{\tau}_k$ may be derived from the superstatistical model,

$$p(\hat{\tau}_k) = \int_0^\infty \varphi(\hat{\tau}_k|\tau_k) p_k(\tau_k) \, \mathrm{d}\, \tau_k. \tag{37}$$

Equations (36) and (37) generate the q-exponential distribution used in the non-extensive statistical mechanics and many real systems [105]. Detailed analytical derivations and the numerical verification were presented in [104].

In the paper [38], a similar superstatistical approach was taken with respect to the intensity of the signal x, obtained by solving SDE (8). The observed series $\hat{x}$ is assumed to be generated from x series by applying exogenous noise, which is described by an arbitrary conditional distribution $\varphi(\hat{x}|x)$. In this approach, the steady-state distribution of $\hat{x}$ is given by

$$p(\hat{x}) = \int_0^\infty \varphi(\hat{x}|x) p(x) \, \mathrm{d}\, x. \tag{38}$$

Analytical and numerical analysis of inter-trade duration, the trading activity, and the return using the superstatistical method with the exponential and normal distributions of the local signal, driven by the stochastic process, were discussed in detail in [38].

In later sections of this paper, we show that the superstatistical approach is not the only approach that allows us to change the observed signal PDF. The coupled SDE approach, proposed in [99], allows for more flexibility and easier interpretation of how the statistical properties become independent of each other. The general form of the set of coupled SDEs was derived from the scaling properties needed for the realization of $1/f^\beta$ noise [99]

$$\mathrm{d}\, x = f(x) y^{2\eta} \, \mathrm{d}\, t + g(x) y^\eta \, \mathrm{d}\, W_1, \tag{39}$$

$$\mathrm{d}\, y = \sigma^2 \left(\eta + 1 - \frac{\lambda}{2} \right) y^{2\eta+1} \, \mathrm{d}\, t + \sigma y_t^{\eta+1} \, \mathrm{d}\, W_2. \tag{40}$$

Here, $f(x)$ and $g(x)$ are arbitrary drift and diffusion functions, which determine the stationary PDF of x; W_1 and W_2 are uncorrelated standard Wiener processes. The first equation describes the changes in the intensity of the signal, while the second equation represents fluctuations in the rate of change. These coupled SDEs allow for $1/f^\beta$ spectrum to be reproduced together with arbitrary steady-state PDF of the observed value x. It was shown that the power–law slope of the PSD, β, of the time series of x generated by solving SDEs (39) and (40) depends on the parameters η and λ as follows

$$\beta = 1 + \frac{\lambda - 1}{2\eta}. \tag{41}$$

In Figure 4, we show that one can obtain a Gaussian distribution of x (subfigure (**b**)) together with $1/f$ spectrum (subfigure (**c**)). In subfigure (**a**), one can visually see the impact of the variations in the rate of change.

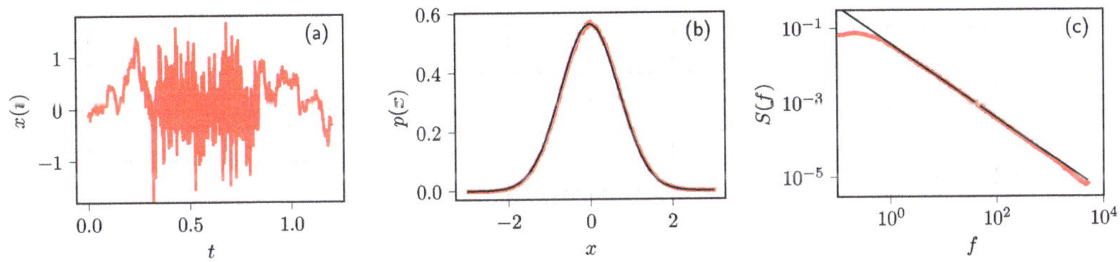

Figure 4. Statistical properties of the time series obtained by solving coupled SDEs (39) and (40): (**a**) sample fragment of $x(t)$ time series, (**b**) PDF of the externally observed values x, and (**c**) PSD of $x(t)$. Red curves correspond to numerical results, while black curves are theoretical fits: (**b**) standard Gaussian PDF, (**c**) $S(f) \sim 1/f^{\beta}$.

2.7. Reproducing Statistical Properties of the Financial Markets

While qualitatively, the trading activity and the absolute returns have power–law distributions and exhibit long-range memory property [14,19], corresponding empirical statistical properties have a finer structure. In order to reproduce the empirical statistical properties in detail, some modifications to the SDE are needed.

The author of [13] has determined that Hurst exponents of the trading activity time series of 1000 US stocks are remarkably close: $H \approx 0.85$. This implies that the PSD of the trading activity should have a power–law slope $\beta = 2H - 1 \approx 0.7$. The author of [13] has also discovered the that slope of the PDFs of the trading activity also has a power–law tail with exponent $\lambda \approx 4.4$. It would be impossible to reproduce such values by using SDE (8), because Equation (10) implies that if $\lambda > 3$, then $\beta > 1$. In our analysis of 26 US stocks [106], we have confirmed the slope of the PDF, but we have observed a more complicated PSD, with two slopes instead of one ($\beta < 1$ for both slopes).

Both of these issues are resolved by a modified SDE for trade intensity, n [33]:

$$\mathrm{d}\, n = \sigma^2 \left[\eta - \frac{\lambda}{2} + \left(\frac{n_0}{n} \right)^2 \right] \frac{n^{2\eta - 1}}{(n\epsilon + 1)^2} \mathrm{d}\, t + \sigma \frac{n^{\eta}}{n\epsilon + 1} \mathrm{d}\, W. \tag{42}$$

The problem of the two PSD slopes is resolved, because this SDE has two different effective η values. For $n \gg \epsilon^{-1}$ the effective η is equal to the specified parameter value (in the numerical simulations we have used $\eta = 5/2$, thus $\hat{\eta}_1 = 5/2$). For $n \ll \epsilon^{-1}$ the effective η is one smaller than the specified parameter value $\hat{\eta}_2 = \eta - 1 = 3/2$). The slope of the PDF increases from the value predicted in Equation (10) due to integration, as trading activity is defined as number of trades per time window w, or in the current parametrization, an integral of trade intensity: $N[t] = \int_t^{t+w} n(u)\, \mathrm{d}\, u$.

Parameter n_0 and the related term in the drift function ensure that n would not become very small as the term causes the potential to rapidly grow for $n < n_0$. This helps us avoid negative trade intensities, which are impossible by definition, as well as ensure some level of minimal trading activity, which in our experience may differ for different stocks and different markets [37,106].

In Figure 5, we have shown that the stochastic model can match statistical properties of MMM stock traded on NYSE. While the matches are not perfect, some of the noticeable differences can be explained by the fact that the stochastic model does not take into account intraday seasonalities.

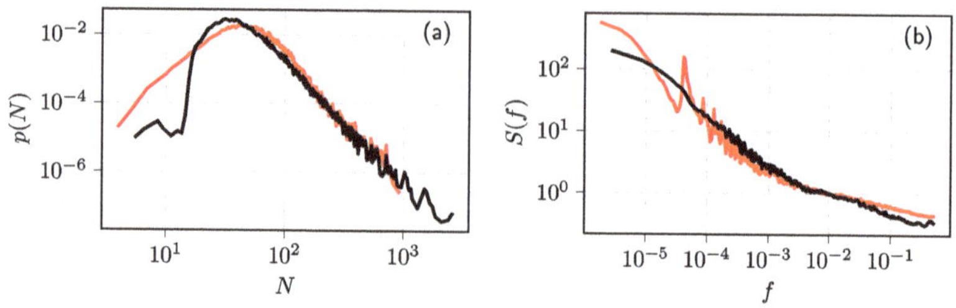

Figure 5. Trading activity (**a**) PDF and (**b**) PSD for MMM stock traded on NYSE (red curve) and the numerical solutions of SDE (42). Model parameters values: $\eta = 2.5$, $\lambda = 4.3$, $\sigma^2 = 0.045$, $\epsilon = 0.36$, $n_0 = 0.14$. Empirical and numerical PDF was obtained by considering trades in the 300 s time window.

Reproducing statistics of absolute return requires another modification of the SDE [36]. Our empirical analysis, confirmed by the other authors [105], indicated that the q-Gaussian distribution [38,107] seems to be a good fit for the empirical absolute return, defined as the log–price difference, distribution. This is achieved by:

$$\mathrm{d}\,x = \sigma^2\left[\eta - \frac{\lambda}{2} - \left(\frac{x}{x_{\max}}\right)^2\right]\frac{(1+x^2)^{\eta-1}}{\left(1+\epsilon\sqrt{1+x^2}\right)^2}x\,\mathrm{d}\,t + \sigma\frac{(1+x^2)^{\frac{\eta}{2}}}{1+\epsilon\sqrt{1+x^2}}\,\mathrm{d}\,W. \tag{43}$$

To reproduce the full complexity of the empirical data, another ingredient is needed, namely external noise, which can be understood as an effect of news flow or the distortions caused by the discrete order flow:

$$r_t = \xi\left\{r_0 = 1 + \frac{2}{w}\left|\int_{t-w}^{t} x(u)\,\mathrm{d}\,u\right|, q = 1 + 2/\lambda_2\right\}. \tag{44}$$

This relation was inspired by the superstatistical approach (discussed in Section 2.6) and determined by trying to fit the empirical data as best we can. We have empirically determined that the best fit is obtained when ξ is a process that generates uncorrelated random variates from a q-Gaussian distribution with $q \approx 1.4$ ($\lambda_2 \approx 5$) and r_0 being one minute ($w \approx 60$ s) moving average filter of the solutions of SDE (43). Using this model, we were able to reproduce empirical statistical properties of stock from New York (abbr. NYSE) and Vilnius stock exchanges (abbr. VSE) [36,37].

In Figure 6, we have demonstrated that the stochastic model reproduces empirical data reasonably well from NYSE and VSE. Some of the noticeable differences can be observed because we do not take into account the intraday seasonality, and we do not directly take into account that VSE had relatively low liquidity (many one minute time intervals have zero returns). Differing liquidity is a likely explanation for the differences seen between NYSE and VSE, too.

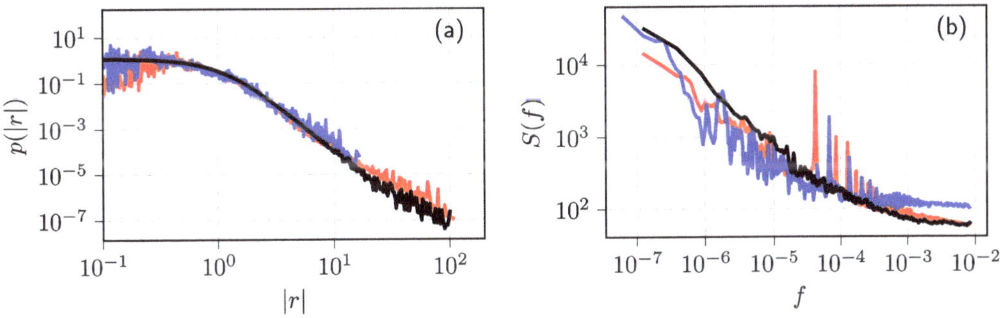

Figure 6. Comparison of empirical (**a**) PDFs and (**b**) PSDs of absolute one minute return as observed in NYSE (red curves) and VSE (blue curves) stocks. Empirical results are compared against the model, generated by the SDE (43) and exogenous noise Equation (44), (black curves). Model parameter values: $\eta = 2.5$, $\lambda = 3.6$, $\epsilon = 0.017$, $x_{\max} = 10^3$, $\lambda_2 = 5$.

2.8. Variable Step Method for Solving Non-Linear Stochastic Differential Equations

Note that SDEs (8), (42), and (43) are not Lipshitz continuous [68]; thus, they have to be solved by imposing boundary conditions, which would prevent the explosion of the solutions. An alternative way to achieve Lipshitz continuity is to include additional terms for restricting diffusion, which would have no detrimental effects on the PSD and PDF of the time series. Such is the role of the n_0 term in SDE (42) and $x_{\max}$ term in SDE (43).

Lacking Lipshitiz continuity causes another complication in solving the SDEs: the standard Euler–Maruyama or Milsten methods [68] do not yield good results with reasonable step sizes. This complication is resolved by using a variable step size. The core idea is to use a larger step size whenever the anticipated changes would be small and use the smaller step size whenever significant changes are coming. The mathematical form of the variable step size is often unique to the SDE being solved, but a good rule of thumb would be to linearize the drift and the diffusion functions. See [69,70] for more details.

For example, SDE (8) in our works is solved by the following set of difference equations:

$$x_{i+1} = x_i + \kappa^2 \left(\eta - \frac{\lambda}{2} \right) x_i + \kappa x_i \varepsilon_i, \tag{45}$$

$$t_{i+1} = t_i + \kappa^2 x^{2-2\eta}. \tag{46}$$

In the above κ is a small number that acts as an error tolerance parameter. The smaller it becomes, the better x_i reproduces desired statistical properties given by Equation (10), but at the expense of numerical computation time.

Similarly, this variable step method can be also applied to SDEs with α-stable Lévy noise. For example, we can solve SDE (16) numerically by using the following set of difference equations

$$x_{k+1} = x_k + \kappa^\alpha \gamma x_k + \frac{\kappa}{\sigma} x_k \zeta_k^\alpha, \tag{47}$$

$$t_{k+1} = t_k + \frac{\kappa^\alpha}{\sigma^\alpha} x_k^{-\alpha(\eta-1)}, \tag{48}$$

where ζ_k^α is a random variable having α-stable Lévy distribution. This set of difference equations should be solved only with the reflective boundaries at $x = x_{\min}$ and $x = x_{\max}$ using the projection method [108]. In nutshell, if the variable x_{k+1} acquires the value outside of the interval $[x_{\min}, x_{\max}]$ then the value of the nearest reflective boundary is assigned to x_{k+1}. Iterative equations for SDEs (42) and (43) are a bit more complicated [36,106], but they still remain qualitatively the same.

Note that the introduction of the variable time step into the numerical solution of an SDE is equivalent to introducing the subordination scheme directly into the SDE, when internal time and physical time are related by a non-linear transformation [98].

3. Agent-Based Model of the Long-Range Memory in the Financial Markets

In the previous section, we have discussed how our group has started from the physically motivated point process model and arrived at the general class of SDEs reproducing long-range memory phenomenon; however, this generality has its drawback: microscopic mechanisms of the modeled systems are ignored. We then tried to investigate some existing financial ABMs for the possibility to derive SDE of a similar form to SDE (8). We have failed to do so with some prominent yet complicated ABMs, such as the ones proposed in [109,110] (for more prominent ABMs of the time, which include some other candidates we have tried, see [111]); however, we have found success with Kirman's herding model, initially proposed in [112] and later analyzed in financial market context by [113,114].

3.1. Kirman's Herding Model

Kirman's herding model can be defined via two one-step transition probabilities in a system with two possible states:

$$p(X \to X+1) = (N-X)[\sigma_1 + hX]\Delta t, \tag{49}$$

$$p(X \to X-1) = X[\sigma_2 + h(N-X)]\Delta t. \tag{50}$$

In the above, X is the number of agents in state 1 and N is the total number of agents within the system. Total number of agents is conserved, so the number of agents in state 2 is trivially given by $N - X$. Here, Δt is a short time window during which only one transition should be likely. Transitions may occur either due to independent behavior (governed by parameters σ_i), or due to recruitment (governed by parameter h). Using birth–death process formalism [115] it is easy to find SDE corresponding to Kirman's herding model with $x = X/N$:

$$\mathrm{d}\,x = [(1-x)\sigma_1 - x\sigma_2]\,\mathrm{d}\,t + \sqrt{2hx(1-x)}\,\mathrm{d}\,W. \tag{51}$$

3.2. Kirman's Herding Model for the Financial Markets

Evidently, SDE (51) is not of the same form as SDE (8), but we have not yet discussed the meaning of states 1 and 2. In many financial ABMs of the time, it was a common choice to assume that agents represent chartist and fundamentalist traders [111]. Assuming that chartist traders trade based on the wide variety of technical trading tools, which often produce conflicting predictions, their excess demand (difference between the supply and demand generated by the group as a whole) is given by:

$$D_c = r_0 X_c(t)\xi(t), \tag{52}$$

where $X_c(t)$ is the number of chartist traders and $\xi(t)$ is their average mood (describing average sentiment to buy or sell). The relative impact of the chartists' traders in comparison to fundamentalist traders is given by r_0. Fundamentalist traders on the other hand are often assumed to trade based on the quantity known as a fundamental price, P_f, with the expectation that the price, $P(t)$, in the long run, will converge towards the fundamental price. Under this assumption, their excess demand is given by:

$$D_f = X_f(t)\ln\frac{P_f}{P(t)}. \tag{53}$$

Using the excess demand functions of the both groups, we can use Walras law [116] to obtain the expression for the price [40,113]:

$$P(t) = P_f \exp\left[r_0 \frac{X_c(t)}{X_f(t)} \xi(t)\right]. \quad (54)$$

The log–return of the price is evidently given by:

$$r_w(t) = \ln P(t) - \ln P(t-w) = r_0 \frac{x_c(t)}{x_f(t)} \zeta_w(t). \quad (55)$$

In the above, $\zeta_w(t)$ is the mood change function over time window w. As the mood changes on a very short time scale and we are interested in the long-term dynamics, we can simply assume that $\zeta_w(t)$ is some kind of uncorrelated noise and consider only a more slowly varying ratio between fractions of chartists and fundamentalists. As the total number of agents is fixed, we can define long-term component of return, modulating return, as:

$$y(t) = \frac{x(t)}{1 - x(t)}. \quad (56)$$

SDE for the modulating return is given by:

$$\mathrm{d}\,y = [\sigma_1 + (2 - \sigma_2)y](1 + y)\,\mathrm{d}\,t + \sqrt{2hy}(1 + y)\,\mathrm{d}\,W, \quad (57)$$

which is roughly similar to the SDE (8) with $\eta = 3/2$ and $\lambda = \frac{\sigma_2}{h} + 1$.

This SDE can be generalized by introducing variable event rate $\tau(y) = y^{-\alpha}$. This addition can be explained by the fact that it is well known that returns and trading volume correlate and the best correlation is achieved between squared returns and volume [16–18,117], hence suggesting that $\alpha = 2$ is a likely candidate. With this extension and when considering only the highest powers of y (as the large y tend to influence the PSD), we obtain [40]:

$$\mathrm{d}\,y = h(2 - \sigma_2)y^{2+\alpha}\,\mathrm{d}\,t + \sqrt{2hy^{3+\alpha}}\,\mathrm{d}\,W. \quad (58)$$

Now this SDE is completely equivalent to the SDE (8) with $\eta = \frac{3+\alpha}{2}$ and $\lambda = \frac{\sigma_2}{h} + \alpha + 1$. Consequently PSD of y will have a frequency range in which:

$$S_y(f) \sim 1/f^{\beta}, \quad \beta = 1 + \frac{\frac{\sigma_2}{h} + \alpha - 2}{1 + \alpha}. \quad (59)$$

In the later papers, we modified this herding ABM until it was able to reproduce the absolute return PDF and PSD close to the empirical absolute return PDFs and PSDs. In [118], we have shown that considering mood dynamics can help in reproducing fractured PSD. In [41], we have reliably introduced the exogenous noise, much similar to what was achieved with the SDE driven model in [36], into this ABM, thus producing a consentaneous model. In [119,120], we have explored the opportunities to control the fluctuations in the artificial financial markets driven by the herding ABM, showing that the random trading, control strategy suggested in [121], may also destabilize the market. In [42], we have removed the assumption about the exogenous noise and replaced it with order book dynamics, thus presenting another possible explanation for fracture in the PSD: it also arises due to market price lagging behind the changes in the equilibrium price, Equation (54). Notably, the order book version of the model was able to reproduce both trading activity and absolute return statistical properties at the same time.

In Figure 7, we have reproduced one of the figures from [41] to show how well the ABM can reproduce the empirical data from New York, Vilnius, and Warsaw stock exchanges (abbr. WSE). Here, we have shown that the model was able to reproduce 10 min absolute return PDFs and PSDs from the different stock exchanges, but in the original article, more intraday time scales are covered, and seasonality was also taken into account.

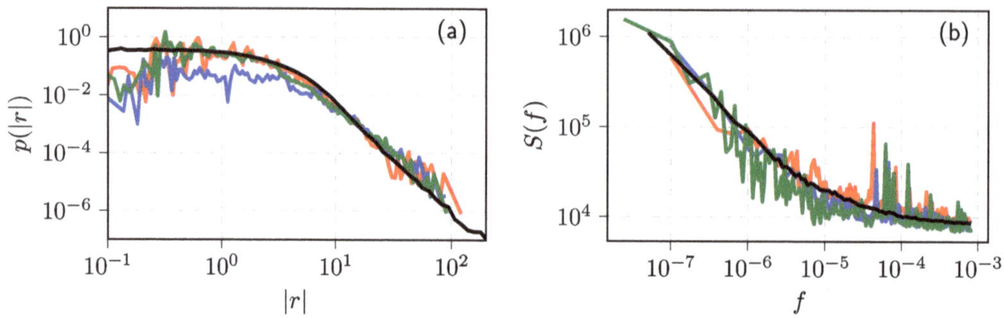

Figure 7. Comparison of empirical (**a**) PDFs and (**b**) PSDs of absolute ten minute return as observed in NYSE (red curves), VSE (blue curves), and WSE (green curves) stocks. Empirical results are compared against the consentaneous model, defined in [41]. Model parameter values are the same as in Figure 2 of [41].

3.3. Kirman's Herding Model, Voter Model, and the Opinion Dynamics Context

Attentive reader with a background in opinion dynamics will likely notice that Kirman's model is remarkably similar to the well-known voter model [47–49]. They are identical, which has prompted us to question whether the voter model is truly a model for voters, which Fernandez–Garcia et al. in [122] also raised. This has lead us to explore and model statistical properties of spatially heterogeneous electoral data [43]. As we have noticed segregation effects in the electoral data, we have continued our investigation by considering the migratory nature of census and electoral data [44]. Similar approaches were taken by others as well. Sano and Mori [123] have looked into spatiotemporal Japanese election data in their model, assuming a noticeable fraction of stubborn voters who do not allow for the party's popularity to drop below a certain threshold. Braha et al. [124] have considered spatiotemporal US election data and have also emphasized the role of opinion leaders and spatial variability of external influences. Fenner et al. [125,126] have started from a generative model inspired by survival analysis, but in later works transition to the SDE framework [127,128]. Michaud and Szilva [129] have fixed issues with the model originally proposed by Fernandez–Garcia et al. [122], mainly, they have redefined how the noise term is handled so that the model would be more mathematically well-posed. Marmani et al. [130] have provided a similar empirical analysis of Italian electoral data and provided an additional perspective from the point of view of Shannon entropy.

As is common in opinion dynamics [47–49], we have also explored the influence of network topologies on the statistical properties of Kirman's herding model. Namely, we have demonstrated [131] a continuous transition from extensive case, characterized by localized interactions, Gaussian distributions, and Boltzmann entropy, to a non-extensive case, characterized by global interactions, q-Gaussian distribution, and Tsallis entropy. Similar results were demonstrated earlier by Alfarano and Milakovic [132], who have explored how Kirman's herding model works on random, Barabasi–Albert, and small-world network topologies. Similar observations were also made in [133], but Carro et al. have used the so-called annealed approximation, which takes into account network structures better than the usual mean-field approximation.

Recently, we have also used the noisy voter model to model parliamentary presence [45]. A paper by Vieira et al. [134] has inspired us to look into the Lithuanian parliamentary presence data. Unlike Vieira et al., we have observed not a ballistic diffusion regime but superdiffusive behavior; however, both of these regimes can be obtained from the noisy voter model with imperfectly acting agents. Namely, agents can internally intend to attend the parliamentary session or skip, but the action itself may be random despite being conditioned on the intended action. As Vieira et al. have used fractional diffusion equation as a model, this result implies that it may be possible to fake long-range memory encoded in the fractional diffusion equation by using Markov models employing non-linear transformations of the voter model [101].

The classical voter model incorporates only a recruitment mechanism, despite other responses to social interaction being possible. For example, diamond model [135] posits that independence and anti-conformity mechanism may be important to understanding human social behaviors. Similarly, Latane social impact theory [136] predicts the importance of supportive interactions—namely, individuals strengthening the conviction of their like-minded peers. While this theory was recently studied in the opinion dynamics context [137,138], it has not been combined with the voter model. One could also consider majority-vote models [139–141] and q-voter models [142,143] as implementing some kind of support by the like-minded agents. In majority-vote models, recruitment is only possible if a majority of agents have opposing opinions (therefore, the majority becomes harder to convince, but the minority remains as susceptible to change). In most q-voter models, a group of q agents must share an opinion to convince a single agent. We have implemented supportive interactions by decreasing the transitions rates of the agents by an amount proportional to the number of like-minded agents. In some cases, these modifications cause the transition rates go to zero, which freezes the system state. Similar qualitative behavior is observed in works, which consider non-Markovian mechanisms, such as implicit opinion freezing or aging [144–147]. This serves as another example that highly non-linear Markovian models can lead to similar dynamics as the dynamics generated by the non-Markovian models.

4. Searching for the True Long-Range Memory Test

We have reviewed our experience of modeling long-range memory phenomena using Markovian models in the earlier sections. We have shown numerous examples of non-linearity causing behaviors and dynamics reminiscent of the models with true long-range memory (such as delayed feedback, aging, freezing, and fractional dynamics). In this section, we present our latest endeavor to find a statistical test, which would distinguish whether the real-life systems possess true or spurious long-range memory. We proposed a test earlier, based on the specific first-passage times, which we refer to as the burst and interburst duration analysis (abbr. BDA) [148–151].

Investigating empirical PDF of burst and interburst duration compared with the model properties, we have interpreted the observed long-range memory in the financial markets by ordinary non-linear SDEs representing multifractal stochastic processes with non-stationary increments [152,153]. One has to take into account the interplay of endogenous and exogenous fluctuations in the financial markets to build a comprehensive model of this complex system [154]. Non-linear SDEs might be applicable in the modeling of other social systems, where models of opinion or population dynamics lead to the macroscopic description by these equations [148–151]. The description by SDEs is an alternative to the modeling incorporating fractional dynamics, if power–law statistical properties are observed in the empirical data.

The BDA employs the dependence of first-passage time PDF on Hurst exponent H for the fractional Brownian motion [56,152,153,155].

FBM, FLSM, and ARFIMA [156–158] form the theoretical background of long-range memory and self-similar processes. These processes, first of all, served for the modeling of systems with anomalous diffusion and expected fractional dynamics [159]. We can consider fractional models possessing true long-range memory as they have correlated increments. Self-similar processes with non-Gaussian stable increments are essential for the modeling of social systems as well. In the financial markets, power–law distributions of noise often interplay with autocorrelations [160–162]. In [163], we implemented BDA for the order disbalance time series seeking to confirm or reject the long-range memory in the order flow. Further, we analyzed the same LOBSTER data of order flow in the financial markets [164] from the perspective of FLSM and ARFIMA models seeking to identify the impact of increment distributions and correlations on estimated parameters of self-similarity [165]. The revealed peculiarities of non-Gaussian fractional dynamics in this financial system raise new questions about whether used sample estimators are reliable. In this section, we test various long-range memory estimators such as mean squared displacement, absolute value estimator, Higuchi's method, and BDA on discrete fractional Lèvy stable motion represented by the ARFIMA sample series.

4.1. Fractional Processes with Non-Gaussian Noise

FBM serves as a model of the correlated time series with stationary Gaussian increments and generalizes the classical Brownian motion [1]. One can define FBM, $B_H(t)$, of the index H (Hurst parameter) in the interval $0 < H < 1$ as the Itô integration over classical Brownian motion B

$$B_H(t) = \int_{-\infty}^{\infty} \left((t-u)_+^d - (-u)_+^d \right) \mathrm{d}\, B(u), \tag{60}$$

where $d = H - 1/2$, $(x)_+ = \max(x, 0)$. The parameter H in FBM quantifies fractal behavior, long-range memory, and anomalous diffusion. This is not the case for the other more general stochastic processes. Thus, in this contribution the Hurst parameter H is responsible only for the fractal properties of the trajectories. We will consider fractional Lèvy stable motion as more general process with non-Gaussian distribution $L_H^\alpha(t)$ representing an integrated process of independent and stable stationary increments $\mathrm{d}\, L^\alpha(u)$ [156]

$$L_H^\alpha(t) = \int_{-\infty}^{\infty} \left((t-u)_+^d - (-u)_+^d \right) \mathrm{d}\, L^\alpha(u), \tag{61}$$

where parameter d depends on H and parameter of stable distribution α, $d = H - 1/\alpha$. The parameter α characterizes special class of stable, invariant under summation, distributions [166], useful in the modeling both super and sub-diffusion [159]. Here, we are interested in the symmetric zero mean, stable distribution defined by the stability index in the region $0 < \alpha < 2$. This new parameter is responsible for the power–law tails of the new PDF $P(x) \sim |x|^{-1-\alpha}$.

FBM and FLSM exhibit identical self-similar scaling behavior in statistical sense,

$$B_H(ct) \sim c^H B_H(t), \qquad L_H^\alpha(ct) \sim c^H L_H^\alpha(t), \tag{62}$$

where $x \sim y$ means that x and y have identical distributions. One can establish the relation with the fractal dimension of trajectories $D = 2 - H$ [167]. In analogy to the notions used in fractal geometry, these types of processes can be considered self-similar.

Mean squared displacement (abbr. MSD) is another important statistical property of various complex systems. Mathematically it was introduced as an ensemble average of the possible microscopic trajectories $x(t)$ [159]

$$\langle (x(t) - x(0))^2 \rangle \sim t^\lambda, \quad \lambda = 2d + 1. \tag{63}$$

Note that Equation (63) is valid for the FBM, while the ensemble average of FLSM diverges [156]. For the FBM $d = H - 1/2$, while for the FLSM λ is not defined. When $d < 0$, one observes dynamics as sub-diffusion and for $d > 0$ as super-diffusion.

In experimental or empirical data analysis, one usually deals with discrete-time sample data series $\{X_i\}$. It is challenging to decide which model to apply in the description of empirical data when diffusion is anomalous $d \neq 0$, as observed dynamics in the sample data can originate from the long-range memory or power–law of the noise. We will use the sample MSD defined as

$$M_N(k) = \frac{1}{N-k+1} \sum_{i=0}^{N-k} (X_{i+k} - X_k)^2. \tag{64}$$

Let us also introduce increment process $\{Y_i = X_i - X_{i-1}\}$, which is extracted from the sample data series. In the case of the FBM increment process, it is called fractional Gaussian noise (abbr. FGN), and in the case of FLSM, it is called fractional Lèvy stable noise (abbr. FLSN). The authors in [156] provide evidence of FLSM non-ergodicity and that $M_N(k) \sim k^{\lambda}$, where $\lambda = 2d + 1$, for large N, k, and N/k. Thus, the MSD sample analysis of time series with FLSM assumption becomes very important providing estimation of the memory parameter d. The long-range memory usually is defined through the divergence of autocovariance $\rho(k)$, $\sum_{k=1}^{\infty} \rho(k) = \infty$, [11]

$$\rho(k) = \frac{1}{N-k+1} \sum_{i=1}^{N-k+1} Y_i Y_{i+k} = 2^{-1}\{(k+1)^{2H} - 2k^{2H} + |k-1|^{2H}\} \tag{65}$$
$$\sim H(2H-1)k^{-\gamma}, \quad k \to \infty.$$

For the FGN, the exponent of autocorrelation is defined by the Hurst parameter $\gamma = 2 - 2H$. We see that FBM is an essential long-range memory process with various statistical properties defined by the Hurst parameter. Thus, researchers use an extensive choice of statistical estimators to determine H and evaluate memory effects even when investigated time series deviate from the Gaussian distribution.

Accepting a more general FLSM approach, one has to reevaluate previously used estimators [163], as we now have more independent parameters. The stability index $0 < \alpha < 2$ and the memory parameter d both contribute to the observed sample properties. Since in the Lèvy stable case, the second moment is infinite the measure of noise autocorrelation, e.g., the co-difference [166,168], is used instead of covariance

$$\tau(k) =\sim k^{-(\alpha - \alpha H)}. \tag{66}$$

Note that the parameter $\gamma = \alpha - \alpha H = \alpha - \alpha d - 1$, has a strong dependency on α, when for the Gaussian processes, it was considered just as the indicator of long-range memory. Consequently, the previously used sample power spectral density analysis, the rescaled range analysis [169–171], or multifractal detrended fluctuation analysis [172,173] has to be reevaluated from the perspective of FLSM [163,165].

Earlier, we have introduced the burst and interburst duration analysis (BDA) as one more method to quantify the long-range memory through the evaluation of H [149,152,153,163]. For the one dimensional bounded sample time series, any threshold divides these series into a sequence of burst T_j^b and interburst T_j^i duration, $j = 1, ...N_b$. The notion of burst and interburst duration follows from the threshold first-passage problem initiated at the nearest vicinity of the threshold. The burst duration is the first-passage time from above and interburst from below the threshold, see [149,152,153,163] for more details. The empirical (sample) PDF (histogram) of T_j gives us the information about H, as the power–law part of this PDF should be T^{2-H} [56]. We have to revise the method of BDA from the more general perspective of FLSM [165], as the question of which properties can be recovered using this method is open and has to be investigated.

The method of absolute value estimator (abbr. AVE) works correctly even for the time series with infinite variance [11,167,168,174]. The method is based on mean value δ_n calculated from sample series Y_i and evaluating its scaling with length of sub-series n. Divide the increment series Y_i into blocks of size n, so that $m \cdot n = N$, and average within each block to obtain the aggregated series $Y_j^{(n)} = \frac{1}{n} \sum_{i=(j-1)n+1}^{jn} Y_i$. Calculate δ_n

$$\delta_n = \frac{1}{m} \sum_{j=1}^{m} |Y_j^{(n)} - \langle Y \rangle|, \tag{67}$$

where $\langle X \rangle$ is the overall series mean. Then the absolute value scaling parameter H_{AV} can be evaluated from the scaling relation

$$\delta_n \sim n^{H_{AV}-1}. \tag{68}$$

One more almost equivalent estimator of scaling properties regarding the FLSM is Higuchi's method [11,175]. It relies on finding fractional dimension D of the length of the path. The normalized path length L_n in this method is defined as follows

$$L_n = \frac{N-1}{n^3} \sum_{i=1}^{n} \frac{1}{m-1} \sum_{j=1}^{m-1} |X_{i+jn} - X_{i+(j-1)n}|, \tag{69}$$

and $L_n \sim n^{-D}$, where $D = 2 - H$.

We investigate four methods: AVE, Higuchi's, MSD, and BDA for the analysis of ARFIMA time series as a test sample of FLSM.

4.2. Numerical Exploration of the Accumulated ARFIMA(0,d,0) Time Series

Let us consider the discrete process $\{X_i\}$ defined as a cumulative sum,

$$X_{i+1} = X_i + Y_i, \tag{70}$$

of correlated increments $\{Y_i\}$. Let the increments be generated by the ARFIMA(0,d,0) process [158,176]:

$$Y_i = \sum_{j=0}^{\infty} \frac{\Gamma(j+d)}{\Gamma(d)\Gamma(j+1)} Z_{i-j}, \tag{71}$$

with random Z_{i-j} from the domain of attraction of an α-stable law with $0 < \alpha \leq 2$. One can calculate the sum in Equation (71) using the fast Fourier transform algorithm. The approximate relation between FLSM and ARFIMA can be derived using Riemann-sum approximation, see [176] for details.

Seeking to generate comparable time series with that analyzed in [165], the order disbalance time series of the financial markets we choose is $N = 7 \times 10^6$, nine values of $d = \{-0.4, -0.3, -0.2, -0.1, 0.0, 0.1, 0.2, 0.3, 0.4\}$ and four values of $\alpha = \{2, 1.5, 1.25, 1.0\}$. The sample time series for any set of parameters have been evaluated using four estimators described above: MSD, AVE, Higuchi's estimator, and BDA. We evaluate H as described in the previous subsection. First of all, we partition time series Y_i in subsets with 5×10^5 time steps and accumulate them to obtain 14 subseries X_i. Then, the exponent λ or the Hurst parameter are evaluated for each subseries using MSD, AVE, and Higuchi's sample estimators. Finally, we calculate the mean and standard deviation of defined 14 λ and H sets. Estimated d we calculate using $d = H - 1/\alpha$ or $d = (\lambda - 1)/2$ in MSD case. The graphs in Figure 8 of estimated d versus used ARFIMA model d serve as a good test of used estimators.

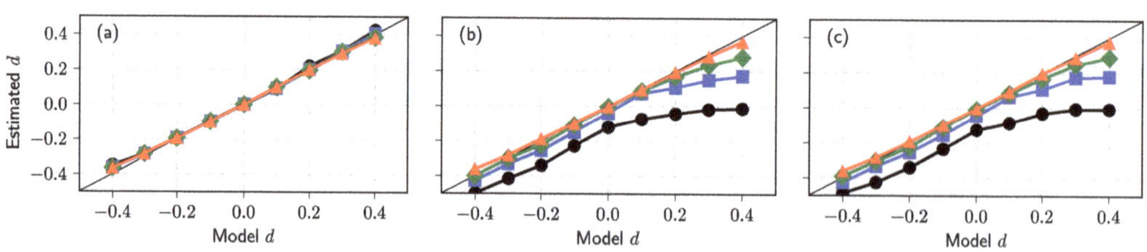

Figure 8. Comparison of the MSD (**a**), AVE (**b**), and Higuchi (**c**) estimator performance when estimating d from the accumulated ARFIMA(0,d,0) series in the unbounded case, $\{X_i\}$ generated by Equation (70). Different curves correspond to the different values of the noise distribution stability parameter: $\alpha = 2$ (red triangles), 1.5 (green diamonds), 1.25 (blue squares), and 1 (black circles).

Our numerical result given in subfigure (**a**) confirms the theoretical prediction for the sample MSD $M_N(k) \sim k^{2d+1}$ [156] as estimated d using this relation almost coincides with model d for all values of α. It is accepted that two estimators, absolute value and Higuchi's, are almost equivalent and should be applicable for the analysis of fractional processes with stable distribution [11,167,168,174]. Indeed, the results of our numerical investigation, see (**b**) and (**c**) subfigures in Figure 8b,c, confirm the equivalence of these estimators. Nevertheless, the estimated values of memory parameter d deviate considerably from its model value, when $\alpha \to 1$, and these deviations are much more prominent for the super-diffusion case $d > 0$. These deviations do not arise as a computational effect, as the estimated relative standard deviation decreases from 0.15 to 0.02 for the evaluated H in the investigated interval of d. Fortunately, this result does not contradict the study [165], where we used these estimators to evaluate d in empirical order disbalance time series exhibiting sub-diffusion.

It is important to note that the estimators, MSD, AVE, and Higuchi's should work well only for the unbounded time series when the most physical systems and processes are of finite size and duration. In all such cases, boundary effects might become important, and one must choose or propose more reliable estimators [167]. The BDA considered in our previous work [149,152,153,163], probably, can serve as an alternative approach. This method works better for the bounded time series, where more intersections of series with the threshold can be expected. Thus, in this contribution for the BDA, we restrict the diffusion of X_i to the interval $[-X_{max}, X_{max}]$ (in our analysis we use $X_{max} = (10^5)^{2d+1}$). This restriction is implemented as a soft boundary condition:

$$X_{i+1} = \max(\min(X_i + Y_i, X_{max}), -X_{max}). \tag{72}$$

This iterative relation replaces Equation (70) in the $\{X_i\}$ series generation algorithm. We define the PDF of the burst and interburst duration T_j for the whole set of time steps $N = 7 \times 10^6$ and the series threshold equal to zero mean. Note that only in this symmetric case PDF's of burst and interbust duration coincide. Seeking to understand how the diffusion restriction mechanism impacts the results of other estimators, we use the same restriction mechanism for the 14 subseries obtained after the partition procedure. We present the results of this analysis in Figure 9.

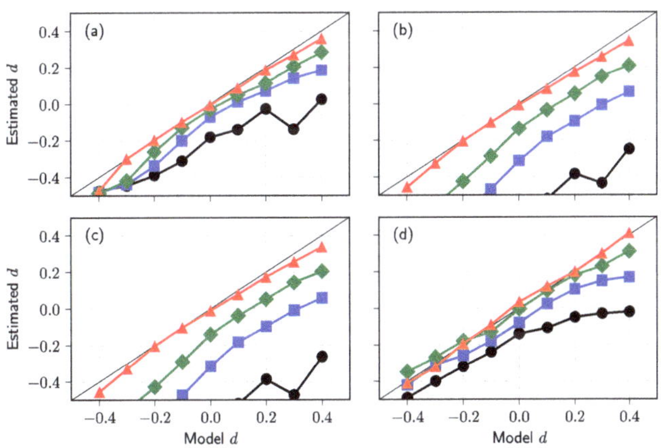

Figure 9. Comparison of the MSD (**a**), AVE (**b**), Higuchi (**c**), and BDA (**d**) estimator performance when estimating d from the accumulated ARFIMA(0,d,0) series in the bounded case, $\{X_i\}$ generated by Equation (72). Different curves correspond to the different values of the noise distribution stability parameter: $\alpha = 2$ (red triangles), 1.5 (green diamonds), 1.25 (blue squares), and 1 (black circles).

Though the used diffusion restriction is relatively soft and changes the direction of movement in the limited number of trajectories points, the results of MSD, AVE, and Higuchi's estimators changed very considerably—compare subfigures (**a**–**c**) with the corresponding results in Figure 9. Contrary, the results obtained using H defined by BDA, see subfigure (**d**), resembles AVE (**b**) and Higuchi's estimator (**c**) subfigures from unbounded series Figure 9. Further investigation is needed to define the best methods and sample estimators for evaluating parameters of fractional time series impacted by various diffusion restrictions. The vast amount of data available from the financial markets can serve as empirical time series considered from the perspective of FLSM.

5. Future Considerations

Here, we have reviewed our approaches to modeling the long-range memory phenomenon and power–law statistics in a variety of complex systems. Our approach differs from the usual approach taken by mathematicians in that we have used Markovian models instead of the non-Markovian alternatives. We were able to reproduce similar behaviors due to our models being driven by various non-linear dependencies. In the case of SDEs, non-linearity may cause the increments of the stochastic process to be non-stationary and, by consequence, cause spurious long-range memory [177,178]. The many models we have built over the years are not models of true long-range memory; however, the critical question is whether our models capture the memory as observed in the financial markets and possibly other socioeconomic complex systems. Section 4, which describes our most recent endeavor, hints at three components that are needed to provide an answer.

The first component is a statistical test, which should distinguish between spurious and true long-range memory. Currently, we are considering the BDA method [148–151], which performs reasonably well in comparison to the alternatives. The core idea of the method is that for any one-dimensional Markovian random walk first-passage time PDF should be a power–law with exponent $-3/2$ at least for some of the duration. Deviations from this law could indicate the presence of true long-range memory. Though the method may fail when the stochastic process is not one-dimensional, the study of what happens in the multidimensional case, e.g., as in [99], is pending. Other challenges may also arise, as discussed in Section 4.

The second component would be a selection of models exhibiting both spurious and true long-range memory. Our prior research has introduced a variety of models of spurious long-range memory; hence, the next steps would be formulating comparable alternative models and studying properties of the existing long-range memory models. Here, we have focused on estimating long-range memory in the fractional Lévy stable motion (modeled using ARFIMA(0,d,0) discrete process), which is a generalization of the fractional Brownian motion; however, in general, other models could also be considered, for example, the multiplicative point process (see Section 2) could be generalized by replacing uncorrelated Gaussian noise with fractional Gaussian noise. Other correlation structures or variable pulse duration could also be considered as an extension [179]. Other notable alternatives and extensions include continuous-time random walk [180] and complex contagion frameworks [181,182].

The third component would be a variety of data from socioeconomic complex systems. Many of our earlier approaches relied on high-frequency absolute return and trading activity time series, but in our most recent works, we have shifted our attention to the order book data obtained from LOBSTER [164]. Order book data seem to invite a more general approach by understanding the data within FLSM or ARFIMA mindset for a broad class of anomalous diffusion processes [157,167,168]. The vast data in social and financial systems have to be investigated to identify and validate the fractional dynamics and long-range memory. Our first results in this direction [163,165] question the interpretation of long-range memory in the order flow data of financial markets. First of all, a prudent choice of estimators based on FLSM and ARFIMA assumptions are needed. After extensive analysis from this perspective, it would be possible to decide whether the investigated social system exhibits true long-range memory or observed power–law statistical properties are just the outcome of strong non-linear effects.

Research effort combining all these three components could yield a better understanding of the long-range memory phenomenon as it is observed in the variety of complex systems. The comprehensive interpretation of long-range memory observed in the financial and other social systems should considerably contribute to developing advanced analytical tools for applications in financial markets. Thus, we have focused on the description and explanation of the long-range memory phenomenon. Notably, a few more recent works refer to or use some of our results and are more application-minded. In [73] a non-linear SDE was derived, providing both physical and economic arguments, to study the performance of EUR/CHF exchange rate. The derived SDE belongs to the class described by (8). The author of [183] has considered the relationship between aging and long-range memory phenomena in a couple of physics experiments: blinking-quantum-dots, single-file diffusion, and Brownian motion in a logarithmic potential. The author of [184] has shown that SDE (8) applies to the modeling of the dynamics on microblogging networks. The author of [185] has considered the effects of perturbations on the stability of power–law distributions in general with an application to wealth distributions. The author of [186] tested the applicability of simple stochastic models to the modeling of non-stationary behavior of intraday tick-by-tick returns. The author of [187] has tested forecast robustness of non-linear GARCH model when time series exhibit high positive autocorrelation. Mean reversion phenomenon was studied in Karachi Stock Exchange data from the perspective of GARCH models in [188]. The authro of [189] has compared the performance of non-linear SDE models against Black and Scholes model, which is one of the models used by the practitioners. Various modifications of Heston model, another model favored by the practitioners, are also reminiscent of SDE (8) [190]. We hope to inspire and maybe take up more application–minded endeavors.

Author Contributions: Conceptualization, R.K., A.K., B.K. and V.G.; methodology, R.K., A.K., B.K. and V.G.; software, A.K. and V.G.; validation, R.K., A.K., B.K. and V.G.; formal analysis, B.K.; investigation, R.K., A.K., B.K. and V.G.; resources, R.K., A.K., B.K. and V.G.; data curation, R.K., A.K., B.K. and V.G.; writing—original draft preparation, R.K., A.K., B.K. and V.G.; writing—review and editing, R.K.; visualization, R.K., A.K., B.K. and V.G.; supervision, V.G.; project administration, R.K.; funding acquisition, R.K. and V.G. All authors have read and agreed to the published version of the manuscript.

Funding: This project was funded by the European Union (project No 09.3.3-LMT-K-712-19-0017) under the agreement with the Research Council of Lithuania (LMTLT).

Institutional Review Board Statement: Not applicable.

Informed Consent Statement: Not applicable.

Conflicts of Interest: The authors declare no conflict of interest.

Abbreviations

The following abbreviations are used in this manuscript:

ABM	agent-based model
APFA	Applications of Physics in Financial Analysis
ARCH	autoregressive conditional heteroscedasticity
ARFIMA	autoregressive fractionally integrated moving average
AVE	absolute value estimator
BDA	burst and interburst duration analysis
COST	European Cooperation in Science and Technology
FBM	fractional Brownian motion
FGN	fractional Gaussian noise
FIGARCH	fractionally integrated GARCH
FLSM	fractional Lèvy stable motion
FLSN	fractional Lèvy stable noise
GARCH	generalized ARCH
MSD	mean squared displacement
NYSE	New York stock exchange

PDF	probability density function
PSD	power spectral density
SDE	stochastic differential equation
VSE	Vilnius stock exchange
WSE	Warsaw stock exchange

References

1. Mandelbrot, B.; Van Ness, J.W. Fractional Brownian motions, fractional noises and applications. *SIAM Rev.* **1968**, *10*, 422–437. [CrossRef]
2. Press, W.H. Flicker noises in astronomy and elsewhere. *Comments Astrophys.* **1978**, *7*, 103–119.
3. Dutta, P.; Horn, P.M. Low-frequency fluctuations in solids: $1/f$ noise. *Rev. Mod. Phys.* **1981**, *53*, 497–516. [CrossRef]
4. Bak, P.; Tang, C.; Wiesenfeld, K. Self-organized criticality: An explanation of the 1/fnoise. *Phys. Rev. Lett.* **1987**, *59*, 381–384. [CrossRef]
5. West, B.J.; Shlesinger, M.F. On the ubiquity of $1/f$ nose. *Int. J. Mod. Phys. B* **1989**, *3*, 795–819. [CrossRef]
6. Mandelbrot, B.B. *Multifractals and 1/f Noise: Wild Self-Affinity in Physics*; Springer: New York, NY, USA, 1999.
7. Milotti, E. 1/f noise: A pedagogical review. *arXiv* **2002** , arXiv:physics/0204033.
8. Ward, L.; Greenwood, P. 1/f noise. *Scholarpedia* **2007**, *2*, 1537. [CrossRef]
9. Rodriguez, M.A. Complete spectral scaling of time series: Towards a classification of $1/f$ noise. *Phys. Rev. E* **2014**, *90*, 042122. [CrossRef]
10. Yadav, A.C.; Kumar, N. Scaling theory for the $1/f$ noise. *arXiv* **2021**, arXiv:2103.11608.
11. Taqqu, M.S.; Teverovsky, V.; Willinger, W. Estimators for long-range dependence: An empirical study. *Fractals* **1995**, *3*, 785–788. [CrossRef]
12. Gopikrishnan, P.; Meyer, M.; Amaral, L.; Stanley, H. Inverse cubic law for the distribution of stock price variations. *Eur. Phys. J. B* **1998**, *3*, 139–140. [CrossRef]
13. Plerou, V.; Gopikrishnan, P.; Nunes Amaral, L.A.; Gabaix, X.; Stanley, H.E. Economic fluctuations and anomalous diffusion. *Phys. Rev. E* **2000**, *62*, R3023–R3026. [CrossRef]
14. Cont, R. Empirical properties of asset returns: Stylized facts and statistical issues. *Quant. Financ.* **2001**, *1*, 1–14. [CrossRef]
15. Mantegna, R.N.; Stanley, H.E. *An Introduction to Econophysics: Correlations and Complexity in Finance*; Cambridge University Press: Cambridge, UK, 2000.
16. Gabaix, X.; Gopikrishnan, P.; Plerou, V.; Stanley, H.E. A theory of power law distributions in financial market fluctuations. *Nature* **2003**, *423*, 267–270. [CrossRef]
17. Farmer, J.D.; Gillemot, L.; Lillo, F.; Mike, S.; Sen, A. What really causes large price changes. *Quant. Financ.* **2004**, *4*, 383–397. [CrossRef]
18. Gabaix, X.; Gopikrishnan, P.; Plerou, V.; Stanley, H.E. Institutional investors and stock market volatility. *Q. J. Econ.* **2006**, 461–504. [CrossRef]
19. Alfi, V.; Cristelli, M.; Pietronero, L.; Zaccaria, A. Minimal agent based model for financial markets I: Origin and self-organization of stylized facts. *Eur. Phys. J. B* **2009**, *67*, 385–397. [CrossRef]
20. Kaulakys, B.; Vektaris, G. Transition to nonchaotic behaviour in randomly driven systems: intermittency and 1/f-noise. In *Noise in Physical Systems and 1/f Fluctuations, Proceedings of the 13th International Conference, Palanga, Lithuania, 29 May–3 June 1995*; World Scientific: Singapore, 1995; pp. 677–680.
21. Kaulakys, B.; Meskauskas, T. Modeling $1/f$ noise. *Phys. Rev. E* **1998**, *58*, 7013–7019. [CrossRef]
22. Kaulakys, B. Autoregressive model of 1/f noise. *Phys. Lett. A* **1999**, *257*, 37–42. [CrossRef]
23. Kaulakys, B.; Gontis, V.; Alaburda, M. Point process model of 1/f noise vs a sum of Lorentzians. *Phys. Rev. E* **2005**, *71*, 1–11. [CrossRef]
24. Ghosh, A. Econophysics research in India in the last two decades. *IIM Kozhikode Soc. Manag. Rev.* **2013**, *2*, 135–146. [CrossRef]
25. de Area Leao Pereira, E.J.; da Silva, M.F.; Pereira, H. Econophysics: Past and present. *Phys. A* **2017**, *473*, 251–261. [CrossRef]
26. Jovanovic, F.; Schinckus, C. *Econophysics and Financial Economics: An Emerging Dialogue*; Oxford University Press: Oxford, UK, 2017. [CrossRef]
27. Gontis, V. Modelling share volume traded in financial markets. *Lith. J. Phys.* **2001**, *41*, 551–555.
28. Gontis, V. Multiplicative stochastic model of the time interval between trades in financial markets. *Nonlinear Anal. Model. Control* **2002**, *7*, 43–54. [CrossRef]
29. Gontis, V.; Kaulakys, B. Multiplicative point process as a model of trading activity. *Phys. A* **2004**, *343*, 505–514. [CrossRef]
30. Gontis, V.; Kaulakys, B. Modeling financial markets by the multiplicative sequence of trades. *Phys. A* **2004**, *344*, 128–133. [CrossRef]
31. Gontis, V.; Kaulakys, B.; Alaburda, M.; Ruseckas, J. Evolution of complex systems and 1/f noise: From physics to financial markets. *Solid State Phenom.* **2004**, *97–98*, 65–70. [CrossRef]
32. Gontis, V.; Kaulakys, B.; Ruseckas, J. Point process models of 1/f noise and internet traffic. *AIP Conf. Proc.* **2005**, *776*, 144–149. [CrossRef]

33. Gontis, V.; Kaulakys, B. Modeling long-range memory trading activity by stochastic differential equations. *Phys. A* **2007**, *382*, 114–120. [CrossRef]
34. Gontis, V.; Kaulakys, B. Long-range memory model of trading activity and volatility. *J. Stat. Mech.* **2006**, *2006*, P10016. [CrossRef]
35. Gontis, V.; Ruseckas, J.; Kononovicius, A. A Non-linear stochastic model of return in financial markets. In *Stochastic Control;* Myers, C., Ed.; InTech : London, UK, 2010; pp. 559–580. [CrossRef]
36. Gontis, V.; Ruseckas, J.; Kononovicius, A. A long-range memory stochastic model of the return in financial markets. *Phys. A* **2010**, *389*, 100–106. [CrossRef]
37. Gontis, V.; Kononovicius, A. Nonlinear stochastic model of return matching to the data of New York and Vilnius stock exchanges. *Dyn.-Socio-Econ. Syst.* **2011**, *2*, 101–109.
38. Ruseckas, J.; Gontis, V.; Kaulakys, B. Nonextensive statistical mechanics distributions and dynamics of financial observables from the nonlinear stochastic differential equations. *Adv. Complex Syst.* **2012**, *15*, 1250073. [CrossRef]
39. Ruseckas, J.; Kaulakys, B.; Gontis, V. Herding model and 1/f noise. *EPL* **2011**, *96*, 60007. [CrossRef]
40. Kononovicius, A.; Gontis, V. Agent based reasoning for the non-linear stochastic models of long-range memory. *Phys. A* **2012**, *391*, 1309–1314. [CrossRef]
41. Gontis, V.; Kononovicius, A. Consentaneous agent-based and stochastic model of the financial markets. *PLoS ONE* **2014**, *9*, e102201. [CrossRef]
42. Kononovicius, A.; Ruseckas, J. Order book model with herding behavior exhibiting long-range memory. *Phys. A* **2019**, *525*, 171–191. [CrossRef]
43. Kononovicius, A. Empirical analysis and agent-based modeling of Lithuanian parliamentary elections. *Complexity* **2017**, *2017*, 7354642. [CrossRef]
44. Kononovicius, A. Compartmental voter model. *J. Stat. Mech.* **2019**, *2019*, 103402. [CrossRef]
45. Kononovicius, A. Noisy voter model for the anomalous diffusion of parliamentary presence. *J. Stat. Mech.* **2020**, *2020*, 063405. [CrossRef]
46. Kononovicius, A. Supportive interactions in the noisy voter model. *Chaos Solitons Fractals* **2021**, *143*, 110627. [CrossRef]
47. Castellano, C.; Fortunato, S.; Loreto, V. Statistical physics of social dynamics. *Rev. Mod. Phys.* **2009**, *81*, 591–646. [CrossRef]
48. Dong, Y.; Zhan, M.; Kou, G.; Ding, Z.; Liang, H. A survey on the fusion process in opinion dynamics. *Inf. Fusion* **2018**, *43*, 57–65. [CrossRef]
49. Noorazar, H. Recent advances in opinion propagation dynamics. *Eur. Phys. J. Plus* **2020**, *135*, 521. [CrossRef]
50. Gontis, V.; Havlin, S.; Kononovicius, A.; Podobnik, B.; Stanley, H.E. Stochastic model of financial markets reproducing scaling and memory in volatility return intervals. *Phys. A* **2016**, *462*, 1091–1102. [CrossRef]
51. Yamasaki, K.; Muchnik, L.; Havlin, S.; Bunde, A.; Stanley, H. Scaling and memory in volatility return intervals in financial markets. *Proc. Natl. Acad. Sci. USA* **2005**, *102*, 9424–9428. [CrossRef] [PubMed]
52. Wang, F.; Yamasaki, K.; Havlin, S.; Stanley, H. Scaling and memory of intraday volatility return intervals in stock market. *Phys. Rev. E* **2006**, *73*, 026117. [CrossRef]
53. Wang, F.; Yamasaki, K.; Havlin, S.; Stanley, H. Indication of multiscaling in the volatility return intervals of stock markets. *Phys. Rev. E* **2008**, *77*, 016109. [CrossRef]
54. Denys, M.; Gubiec, T.; Kutner, R.; Jagielski, M.; Stanley, H.E. Universality of market superstatistics. *Phys. Rev. E* **2016**, *94*, 042305. [CrossRef]
55. Redner, S. *A Guide to First-Passage Processes;* Cambridge University Press: Cambridge, UK, 2001.
56. Ding, M.; Yang, W. Distribution of the first return time in fractional Brownian motion and its application to the study of on-off intermittency. *Phys. Rev. E* **1995**, *52*, 207. [CrossRef]
57. Johnson, J.B. The Schottky effect in low frequency circuits. *Phys. Rev.* **1925**, *26*, 71–85. [CrossRef]
58. Levitin, D.J.; Chordia, P.; Menon, V. Musical rhythm spectra from Bach to Joplin obey a 1/f power law. *Proc. Natl. Acad. Sci. USA* **2012**, *109*, 3716–3720. [CrossRef]
59. Kobayashi, M.; Musha, T. 1/f fluctuation of heartbeat period. *IEEE Trans. Biomed. Eng.* **1982**, *29*, 456–457. [CrossRef] [PubMed]
60. Wagenmakers, E.J.; Farrell, S.; Ratcliff, R. Estimation and interpretation of $1/f^{\alpha}$ noise in human cognition. *Psychon. Bull. Rev.* **2004**, *11*, 579–615. [CrossRef] [PubMed]
61. Mathiesen, J.; Angheluta, L.; Ahlgren, P.T.H.; Jensen, M.H. Excitable human dynamics driven by extrinsic events in massive communities. *Proc. Natl. Acad. Sci. USA* **2013**, *110*, 17259–17262. [CrossRef] [PubMed]
62. Leonardis, E.; Chapman, S.C.; Foullon, C. Turbulent characteristics in the intensity fluctuations of a solar quiescent prominence observed by the Hinode Solar Optical Telescop. *Astrophys. J.* **2012**, *745*, 185. [CrossRef]
63. Meskauskas, T.; Kaulakys, B. 1/f noise in fractal quaternionic structures. *AIP Conf. Proc.* **2005**, *780*, 91–94. [CrossRef]
64. Ribeiro, L.C.; de Deus, L.G.; Loureiro, P.M.; Albuquerque, E.D.M. Profits and fractal properties: Notes on Marx, countertendencies and simulation models. *Rev. Political Econ.* **2017**, *29*, 282–306. [CrossRef]
65. Ribeiro, L.C.; Rapini, M.S.; Silva, L.A.; Albuquerque, E.M. Growth patterns of the network of international collaboration in science. *Scientometrics* **2018**, *114*, 159–179. [CrossRef]
66. Nakamura, T.; Small, M.; Tanizawa, T. Long-range correlation properties of stationary linear models with mixed periodicities. *Phys. Rev. E* **2019**, *99*, 022128. [CrossRef] [PubMed]
67. Erland, S.; Greenwood, P.E. Constructing $1/\omega^{\alpha}$ noise from reversible Markov chains. *Phys. Rev. E* **2007**, *76*, 031114. [CrossRef]

68. Kloeden, P.E.; Platen, E. *Numerical Solution of Stochastic Differential Equations*; Springer: Berlin, Germany, 1999.
69. Kaulakys, B.; Ruseckas, J. Stochastic nonlinear differential equation generating 1/f noise. *Phys. Rev. E* **2004**, *70*, 020101. [CrossRef] [PubMed]
70. Kaulakys, B.; Ruseckas, J.; Gontis, V.; Alaburda, M. Nonlinear stochastic models of 1/f noise and power-law distributions. *Phys. A* **2006**, *365*, 217–221. [CrossRef]
71. Kaulakys, B.; Alaburda, M. Modeling scaled processes and $1/f^{\beta}$ noise using non-linear stochastic differential equations. *J. Stat. Mech.* **2009**, *2009*, P02051. [CrossRef]
72. Ruseckas, J.; Kaulakys, B. 1/f noise from nonlinear stochastic differential equations. *Phys. Rev. E* **2010**, *81*, 031105. [CrossRef]
73. Lera, S.C.; Sornette, D. Currency target-zone modeling: An interplay between physics and economics. *Phys. Rev. E* **2015**, *92*, 062828. [CrossRef]
74. Engle, R. Autoregresive conditional heteroscedasticity with estimates of the variance of United Kingdom inflation. *Econometrica* **1982**, *50*, 987–1008. [CrossRef]
75. Bollerslev, T. Generalized autoregressive conditional heteroskedasticity. *J. Econom.* **1986**, *31*, 307–327. [CrossRef]
76. Engle, R.; Bollerslev, T. Modeling the persistence of conditional variances. *Econom. Rev.* **1986**, *5*, 1–50. [CrossRef]
77. Potters, M.; Cont, R.; Bouchaud, J.P. Financial markets as adaptive systems. *EPL* **1998**, *41*, 239–244. [CrossRef]
78. Giraitis, L.; Robinson, P.M.; Surgailis, D. A model for long memory conditional heteroscedasticity. *Ann. Appl. Probab.* **2000**, *10*, 1002–1024. [CrossRef]
79. Bollerslev, T. Glossary to ARCH (GARCH). *CREATES Res. Pap.* **2008**, *49*, 1–4 . [CrossRef]
80. Giraitis, L.; Leipus, R.; Surgailis, D. Recent advances in ARCH modelling. In *Long Memory in Economics*; Teyssiere, G., Kirman, A., Eds.; Springer: Berlin/Heidelberg, Germany, 2007; pp. 3–38.
81. Giraitis, L.; Leipus, R.; Surgailis, D. ARCH(∞) models and long memory. In *Handbook of Financial Time Series*; Anderson, T.G.; Davis, R.A.; Kreis, J.; Mikosh, T., Eds.; Springer Verlag: Berlin, Germany, 2009; pp. 71–84. [CrossRef]
82. Giraitis, L.; Surgailis, D.; Škarnulis, A. Stationary integrated ARCH(∞) and AR(∞) processes with finite variance. *Econom. Theory* **2018**, *34*, 1159–1179. [CrossRef]
83. Granger, C.W.J.; Joyeux, R. An introduction to long-memory time series models and fractional differencing. *J. Time Ser. Anal.* **1980**, *1*, 15–29. [CrossRef]
84. Baillie, R.T.; Bollerslev, T.; Mikkelsen, H.O. Fractionally integrated generalized autoregressive conditional heteroskedasticity. *J. Econom.* **1996**, *74*, 3–30. [CrossRef]
85. Tayefi, M.; Ramanathan, T.V. An overview of FIGARCH and related time series models. *Austrian J. Stat.* **2012**, *41*, 175–196. [CrossRef]
86. Kononovicius, A.; Ruseckas, J. Nonlinear GARCH model and 1/f noise. *Phys. A* **2015**, *427*, 74–81. [CrossRef]
87. Nelson, D.B. ARCH models as diffusion approximations. *J. Econom.* **1990**, *45*, 7–38. [CrossRef]
88. Kluppelberg, C.; Lindner, A.; Maller, R. A continuous-time GARCH process driven by a Levy process: Stationarity and second-order behaviour. *J. Appl. Probab.* **2004**, *41*, 601–622. [CrossRef]
89. Kluppelberg, C.; Maller, R.; Szimayer, A. The COGARCH: A Review, with News on Option Pricing and Statistical Inference. In *Surveys in Stochastic Processes*; EMS press: Berlin, Germany, 2010. [CrossRef]
90. Kazakevicius, R.; Ruseckas, J. Influence of external potentials on heterogeneous diffusion processes. *Phys. Rev. E* **2016**, *94*, 032109. [CrossRef]
91. Havlin, S.; Ben-Avraham, D. Diffusion in disordered media. *Adv. Phys.* **2002**, *51*, 187–292. [CrossRef]
92. ben Avraham, D.; Havlin, S. *Diffusion and Reactions in Fractals and Disordered Systems*; Cambridge University Press: Cambridge, UK, 2005.
93. Metzler, R.; Jeon, J.H.; Cherstvy, A.G.; Barkai, E. Anomalous diffusion models and their properties: Non-stationarity, non-ergodicity, and ageing at the centenary of single particle tracking. *Phys. Chem. Chem. Phys.* **2014**, *16*, 24128–24164. [CrossRef] [PubMed]
94. Kazakevičius, R.; Ruseckas, J. Lévy flights in inhomogeneous environments and 1/f noise. *Phys. A* **2014**, *411*, 95. [CrossRef]
95. Kazakevicius, R.; Ruseckas, J. Power-law statistics from nonlinear stochastic differential equations driven by Lévy stable noise. *Chaos Solitons Fractals* **2015**, *81*, 432–442. [CrossRef]
96. Kazakevicius, R.; Ruseckas, J. Anomalous diffusion in nonhomogeneous media: Power spectral density of signals generated by time-subordinated nonlinear Langevin equations. *Phys. A* **2015**, *438*, 210–222. [CrossRef]
97. Kazakevicius, R.; Ruseckas, J. Power law statistics in the velocity fluctuations of Brownian particle in inhomogeneous media and driven by colored noise. *J. Stat. Mech.* **2015**, *2015*, P02021. [CrossRef]
98. Ruseckas, J.; Kazakevicius, R.; Kaulakys, B. 1/f noise from point process and time-subordinated Langevin equations. *J. Stat. Mech.* **2016**, *2016*, 054022. [CrossRef]
99. Ruseckas, J.; Kazakevicius, R.; Kaulakys, B. Coupled nonlinear stochastic differential equations generating arbitrary distributed observable with 1/f noise. *J. Stat. Mech.* **2016**, *2016*, 043209. [CrossRef]
100. Kaulakys, B.; Alaburda, M.; Ruseckas, J. 1/f noise from the nonlinear transformations of the variables. *Mod. Phys. Lett. B* **2015**, *29*, 1550223. [CrossRef]
101. Kazakevicius, R.; Kononovicius, A. Anomalous diffusion in nonlinear transformations of the noisy voter model. *Phys. Rev. E* **2021**, *103*, 032154. [CrossRef] [PubMed]

102. Kaulakys, B.; Alaburda, M.; Ruseckas, J. Modeling of long-range memory processes with inverse cubic distributions by the nonlinear stochastic differential equations. *J. Stat. Mech.* **2016**, *2016*, 054035. [CrossRef]
103. Pesce, G.; McDaniel, A.; Hottovy, S.; Wehr, J.; Volpe, G. Stratonovich-to-Itô transition in noisy systems with multiplicative feedback. *Nat. Commun.* **2013**, *4*, 2733. [CrossRef]
104. Kaulakys, B.; Alaburda, M.; Gontis, V.; Ruseckas, J. Modeling long-memory processes by stochastic difference equations and superstatistical approach. *Braz. J. Phys.* **2009**, *39*, 453–456. [CrossRef]
105. Tsallis, C. Economics and finance: q-Statistical stylized features galore. *Entropy* **2017**, *19*, 457. [CrossRef]
106. Gontis, V.; Kaulakys, B.; Ruseckas, J. Trading activity as driven Poisson process: Comparison with empirical data. *Phys. A* **2008**, *387*, 3891–3896. [CrossRef]
107. Ruseckas, J.; Kaulakys, B. Tsallis distributions and 1/f noise from nonlinear stochastic differential equations. *Phys. Rev. E* **2011**, *84*, 051125. [CrossRef] [PubMed]
108. Pettersson, R. Approximations for stochastic differential equations with reflecting convex boundaries. *Stoch. Process. Appl.* **1995**, *59*, 295–308. [CrossRef]
109. Lux, T.; Marchesi, M. Scaling and criticality in a stochastic multi-agent model of a financial market. *Nature* **1999**, *397*, 498–500. [CrossRef]
110. Challet, D.; Marsili, M.; Zecchina, R. Statistical mechanics of systems with heterogeneous agents: Minority games. *Phys. Rev. Lett.* **2000**, *84*, 1824–1827. [CrossRef]
111. Cristelli, M.; Pietronero, L.; Zaccaria, A. Critical overview of agent-based models for economics. *arxiv* 2012, **arXiv**:1101.1847.
112. Kirman, A.P. Ants, rationality and recruitment. *Q. J. Econ.* **1993**, *108*, 137–156. [CrossRef]
113. Alfarano, S.; Lux, T.; Wagner, F. Estimation of agent-based models: The case of an asymmetric herding model. *Comput. Econ.* **2005**, *26*, 19–49. [CrossRef]
114. Alfarano, S.; Lux, T.; Wagner, F. Time variation of higher moments in a financial market with heterogeneous agents: An analytical approach. *J. Econ. Dyn. Control* **2008**, *32*, 101–136. [CrossRef]
115. van Kampen, N.G. *Stochastic Process in Physics and Chemistry*; North Holland: Amsterdam, The Netherlands, 2007.
116. Walras, L. *Elements of Pure Economics*; Routledge: London, UK, 2013.
117. Rak, R.; Drozdz, S.; Kwapien, J.; Oswiecimka, P. Stock returns versus trading volume: Is the correspondence more general? *Acta Phys. Pol. B* **2013**, *44*, 2035–2050. [CrossRef]
118. Kononovicius, A.; Gontis, V. Three state herding model of the financial markets. *EPL* **2013**, *101*, 28001. [CrossRef]
119. Kononovicius, A.; Gontis, V. Control of the socio-economic systems using herding interactions. *Phys. A* **2014**, *405*, 80–84. [CrossRef]
120. Kononovicius, A.; Gontis, V. Herding interactions as an opportunity to prevent extreme events in financial markets. *Eur. Phys. J. B* **2015**, *88*, 189. [CrossRef]
121. Biondo, A.E.; Pluchino, A.; Rapisarda, A.; Helbing, D. Stopping financial avalanches by random trading. *Phys. Rev. E* **2013**, *88*, 062814. [CrossRef]
122. Fernandez-Gracia, J.; Suchecki, K.; Ramasco, J.J.; San Miguel, M.; Eguiluz, V.M. Is the voter model a model for voters? *Phys. Rev. Lett.* **2014**, *112*, 158701. [CrossRef]
123. Sano, F.; Hisakado, M.; Mori, S. Mean field voter model of election to the house of representatives in Japan. In Proceedings of the JPS Conference Proceedings, The Physical Society of Japan, Kanazawa, Japan, 13–18 November 2017; Volume 16, p. 011016. [CrossRef]
124. Braha, D.; de Aguiar, M.A.M. Voting contagion: Modeling and analysis of a century of U.S. presidential elections. *PLoS ONE* **2017**, *12*, e0177970. [CrossRef]
125. Fenner, T.; Kaufmann, E.; Levene, M.; Loizou, G. A multiplicative process for generating a beta-like survival function with application to the UK 2016 EU referendum results. *Int. J. Mod. Phys. C* **2017**, *28*, 1750132. [CrossRef]
126. Fenner, T.; Levene, M.; Loizou, G. A multiplicative process for generating the rank-order distribution of UK election results. *Qual. Quant.* **2017**, *52*, 1069–1079. [CrossRef]
127. Fenner, T.; Levene, M.; Loizou, G. A stochastic differential equation approach to the analysis of the UK 2016 EU referendum polls. *J. Phys. Commun.* **2018**, *2*, 055022. [CrossRef]
128. Levene, M.; Fenner, T. A stochastic differential equation approach to the analysis of the 2017 and 2019 UK general election polls. *Int. J. Forecast.* **2021**, *37*, 1227–1234. [CrossRef]
129. Michaud, J.; Szilva, A. Social influence with recurrent mobility and multiple options. *Phys. Rev. E* **2018**, *97*, 062313. [CrossRef]
130. Marmani, S.; Ficcadenti, V.; Kaur, P.; Dhesi, G. Entropic analysis of votes expressed in Italian elections between 1948 and 2018. *Entropy* **2020**, *22*, 523. [CrossRef]
131. Kononovicius, A.; Ruseckas, J. Continuous transition from the extensive to the non-extensive statistics in an agent-based herding model. *Eur. Phys. J. B* **2014**, *87*, 169. [CrossRef]
132. Alfarano, S.; Milakovic, M. Network structure and N-dependence in agent-based herding models. *J. Econ. Dyn. Control* **2009**, *33*, 78–92. [CrossRef]
133. Carro, A.; Toral, R.; San Miguel, M. The noisy voter model on complex networks. *Sci. Rep.* **2016**, *6*, 24775. [CrossRef]
134. Vieira, D.S.; Riveros, J.M.E.; Jauregui, M.; Mendes, R.S. Anomalous diffusion behavior in parliamentary presence. *Phys. Rev. E* **2019**, *99*, 042141. [CrossRef] [PubMed]

135. Willis, H.R. Conformity, independence and anticonformity. *Hum. Relat.* **1965**, *18*, 373–388. [CrossRef]
136. Latane, B. The psychology of social impact. *Am. Psychol.* **1981**, *36*, 343–356. [CrossRef]
137. Bancerowski, P.; Malarz, K. Multi-choice opinion dynamics model based on Latane theory. *Eur. Phys. J.* **2019**, *92*, 219. [CrossRef]
138. Kowalska-Styczeń, A.; Malarz, K. Noise induced unanimity and disorder in opinion formation. *PLoS ONE* **2020**, *15*, e0235313. [CrossRef]
139. de Oliveira, M.J. Isotropic majority-vote model on a square lattice. *J. Stat. Phys.* **1992**, *66*, 273–281. [CrossRef]
140. Vilela, A.L.M.; Stanley, H.E. Effect of strong opinions on the dynamics of the majority-vote model. *Sci. Rep.* **2018**, *8*, 8709. [CrossRef]
141. Galesic, M.; Stein, D.L. Statistical physics models of belief dynamics: Theory and empirical tests. *Phys. A* **2019**, *519*, 275–294. [CrossRef]
142. Castellano, C.; Munoz, M.A.; Pastor-Satorras, R. The non-linear q-voter model. *Phys. Rev. E* **2009**, *80*, 041129. [CrossRef]
143. Jedrzejewski, A.; Sznajd-Weron, K. Statistical physics of opinion formation: Is it a SPOOF? *Comptes Rendus Phys.* **2019**, *20*, 244–261. [CrossRef]
144. Stark, H.U.; Tessone, C.J.; Schweitzer, F. Decelerating microdynamics can accelerate macrodynamics in the voter model. *Phys. Rev. Lett.* **2008**, *101*, 018701. [CrossRef]
145. Stark, H.U.; Tessone, C.J.; Schweitzer, F. Slower is faster: Fostering consensus formation by heterogeneous inertia. *Adv. Complex Syst.* **2008**, *11*, 551–563. [CrossRef]
146. Wang, Z.; Liu, Y.; Wang, L.; Zhang, Y.; Wang, Z. Freezing period strongly impacts the emergence of a global consensus in the voter model. *Sci. Rep.* **2014**, *4*, 3597. [CrossRef] [PubMed]
147. Artime, O.; Peralta, A.F.; Toral, R.; Ramasco, J.; San Miguel, M. Aging-induced continuous phase transition. *Phys. Rev. E* **2018**, *98*, 032104. [CrossRef]
148. Gontis, V.; Kononovicius, A.; Reimann, S. The class of nonlinear stochastic models as a background for the bursty behavior in financial markets. *Adv. Complex Syst.* **2012**, *15*, 1250071. [CrossRef]
149. Gontis, V.; Kononovicius, A. Spurious memory in non-equilibrium stochastic models of imitative behavior. *Entropy* **2017**, *19*, 387. [CrossRef]
150. Kononovicius, A.; Gontis, V. Approximation of the first passage time distribution for the birth-death processes. *J. Stat. Mech.* **2019**, *2019*, 073402. [CrossRef]
151. Gontis, V.; Kononovicius, A. Bessel-like birth-death process. *Phys. A* **2020**, *540*, 123119. [CrossRef]
152. Gontis, V.; Kononovicius, A. Burst and inter-burst duration statistics as empirical test of long-range memory in the financial markets. *Phys. A* **2017**, *483*, 266–272. [CrossRef]
153. Gontis, V.; Kononovicius, A. The consentaneous model of the financial markets exhibiting spurious nature of long-range memory. *Phys. A* **2018**, *505*, 1075–1083. [CrossRef]
154. Gontis, V. Interplay between endogenous and exogenous fluctuations in financial markets. *Acta Phys. Pol. A* **2016**, *129*, 1023–1031. [CrossRef]
155. Metzler, R.; Oshanin, G.; Redner, S. *First-Passage Phenomena and Their Applications*; World Scientific: Singapore, 2014.
156. Burnecki, K.; Weron, A. Fractional Levy stable motion can model subdiffusive dynamics. *Phys. Rev. E* **2010**, *82*, 021130. [CrossRef]
157. Burnecki, K.; Weron, A. Algorithms for testing of fractional dynamics: A practical guide to ARFIMA modelling. *J. Stat. Mech.* **2014**, *2014*, P10036. [CrossRef]
158. Burnecki, K.; Sikora, G. Identification and validation of stable ARFIMA processes with application to UMTS data. *Chaos Solitons Fractals* **2017**, *102*, 456–466. [CrossRef]
159. Klafter, J.; Lim, S.C.; Metzler, R. (Eds.) *Fractional Dynamics: Recent Advances*; World Scientific: New York, NY, USA, 2012.
160. Lillo, F.; Farmer, J.D. The long memory of the efficient market. *Stud. Nonlinear Dyn. Econom.* **2001**, *8*, 1–35. [CrossRef]
161. Bouchaud, J.P.; Gefen, Y.; Potters, M.; Wyart, M. Fluctuations and response in financial markets: The subtle nature of 'random' price changes. *Quant. Financ.* **2004**, *4*, 176–190. [CrossRef]
162. Toth, B.; Palit, I.; Lillo, F.; Farmer, J.D. Why is equity order flow so persistent? *J. Econ. Dyn. Control* **2015**, *51*, 218–239. [CrossRef]
163. Gontis, V. Long-range memory test by the burst and inter-burst duration distribution. *J. Stat. Mech.* **2020**, *2020*, 093406. [CrossRef]
164. Huang, R.; Polak, T. *LOBSTER: The Limit Order Book Reconstructor*; Technical Report; Discussion Paper School of Business and Economics; Humboldt Universitat zu Berlin: Berlin, Germany, 2011.
165. Gontis, V. Order Flow in the Financial Markets from the Perspective of the Fractional Lévy Stable Motion. *arXiv* **2021**, arXiv:2105.02057.
166. Smarodinsky, G.; Taqqu, M. *Stable Non-Gaussian Random Processes: Stochastic Models with Infinite Variance*; Chapman and Hall: London, UK, 1994. [CrossRef]
167. Magdziarz, M.; Slezak, J.K.; Wojcik, J. Estimation and testing of the Hurst parameter using p-variation. *J. Phys. Math. Theor.* **2013**, *46*, 325003. [CrossRef]
168. Weron, A.; Burnecki, K. Complete description of all self-similar models driven by Levy stable noise. *Phys. Rev. E* **2005**, *71*, 016113. [CrossRef]
169. Hurst, H.E. Long-term storage capacity of reservoirs. *Trans. Am. Soc. Civ. Eng.* **1951**, *116*, 770–799. [CrossRef]
170. Beran, J. *Statistics for Long-Memory Processes*; CRC press: Boca Raton, FL, USA, 1994.

171. Montanari, A.; Taqqu, M.S.; Teverovsky, V. Estimating long-range dependence in the presence of periodicity: An empirical study. *Math. Comput. Model.* **1999**, *29*, 217–228. [CrossRef]
172. Peng, C.K.; Buldyrev, S.V.; Havlin, S.; Simons, M.; Stanley, H.E.; Goldberger, A.L. Mosaic organization of DNA nucleotides. *Phys. Rev. E* **1994**, *49*, 1685–1689. [CrossRef] [PubMed]
173. Kantelhardt, J.W.; Zschiegner, S.A.; Koscielny-Bunde, E.; Havlin, S.; Bunde, A.; Stanley, H.E. Multifractal detrended fluctuation analysis of nonstationary time series. *Phys. A* **2002**, *316*, 87–114. [CrossRef]
174. Mercik, S.; Weron, K.; Burnecki, K.; Weron, A. Enigma of self-similarity of fractional Levy stable motions. *Acta Phys. Pol. B* **2003**, *34*, 3773–3791.
175. Higuchi, T. Approach to an irregular time series on the basis of the fractal theory. *Phys. D* **1988**, *31*, 277–283. [CrossRef]
176. Stoev, S.; Taqqu, M.S. Simulation methods for linear fractional stable motion and FARIMA using the Fast Fourier Transform. *Fractals* **2004**, *12*, 95. [CrossRef]
177. Bassler, K.E.; Gunaratne, G.H.; McCauley, J.L. Markov processes, Hurst exponents, and nonlinear diffusion equations: With application to finance. *Phys. A* **2006**, *369*, 343–353. [CrossRef]
178. McCauley, J.L.; Gunaratne, G.H.; Bassler, K.E. Hurst exponents, Markov processes, and fractional Brownian motion. *Phys. A* **2007**, *379*, 1–9. [CrossRef]
179. Ruseckas, J.; Kaulakys, B.; Alaburda, M. Modelling of $1/f$ noise by sequences of stochastic pulses of different duration. *Lith. J. Phys.* **2003**, *43*, 223–228.
180. Kutner, R.; Masoliver, J. The continuous time random walk, still trendy: Fifty-year history, state of art and outlook. *Eur. Phys. J. B* **2017**, *90*, 50. [CrossRef]
181. Baronchelli, A. The emergence of consensus: A primer. *R. Soc. Open Sci.* **2018**, *5*, 172189. [CrossRef] [PubMed]
182. Landry, N.; Restrepo, J.G. The effect of heterogeneity on hypergraph contagion models. *Chaos* **2020**, *30*, 103117. [CrossRef] [PubMed]
183. Leibovich, N.; Dechant, A.; Lutz, E.; Barkai, E. Aging Wiener-Khinchin theorem and critical exponents of $1/f\beta$ noise. *Phys. Rev. E* **2016**, *94*, 052130. [CrossRef]
184. Dmitriev, A.; Tsukanova, O.; Maltseva, S. Modeling of microblogging social networks: Dynamical system vs. Random dynamical system. *Procedia Comput. Sci.* **2017**, *122*, 812–819. [CrossRef]
185. Vita, A.D. On the response of power law distributions to fluctuations. *Eur. Phys. J. B* **2019**, *92*, 255. [CrossRef]
186. Ponta, L.; Trinh, M.; Raberto, M.; Scalas, E.; Cincotti, S. Modeling non-stationarities in high-frequency financial time series. *Phys. A* **2019**, *521*, 173–196. [CrossRef]
187. Emenogu, N.G.; Adenomon, M.O. Robustness of GARCH family models to high positive autocorrelation. *J. Niger. Stat. Assoc.* **2020**, *32*, 13–28.
188. Vveinhardt, J.; Streimikiene, D.; Rizwan, A.R.; Nawaz, A.; Rehman, A. Mean reversion: An investigation from Karachi stock exchange sectors. *Technol. Econ. Dev. Econ.* **2016**, *22*, 493–511. [CrossRef]
189. Lima, L.S.; Melgaço, J.H.C. Dynamics of stocks prices based in the Black & Scholes equation and nonlinear stochastic differentials equations. *Phys. A* **2021**, *581*, 126220. [CrossRef]
190. Benhamou, E.; Gobet, E.; Miri, M. Time dependent Heston model. *SIAM J. Financ. Math.* **2010**, *1*, 289–325. [CrossRef]

entropy

MDPI

Article

Are Mobility and COVID-19 Related? A Dynamic Analysis for Portuguese Districts

António Casa Nova [1], Paulo Ferreira [1,2,3,*], Dora Almeida [3], Andreia Dionísio [3] and Derick Quintino [4]

1 Instituto Politécnico de Portalegre, 7300-110 Portalegre, Portugal; casanova@ipportalegre.pt
2 VALORIZA—Research Center for Endogenous Resource Valorization, 7300-555 Portalegre, Portugal
3 CEFAGE-UE, IIFA, Universidade de Évora, Largo dos Colegiais 2, 7004-516 Évora, Portugal; dmfa1982@gmail.com (D.A.); andreia@uevora.pt (A.D.)
4 Department of Economics, Administration and Sociology, University of São Paulo, Piracicaba 13418-900, SP, Brazil; derickdq@usp.br
* Correspondence: pferreira@ipportalegre.pt

Abstract: In this research work, we propose to assess the dynamic correlation between different mobility indices, measured on a daily basis, and the new cases of COVID-19 in the different Portuguese districts. The analysis is based on global correlation measures, which capture linear and non-linear relationships in time series, in a robust and dynamic way, in a period without significant changes of non-pharmacological measures. The results show that mobility in retail and recreation, grocery and pharmacy, and public transport shows a higher correlation with new COVID-19 cases than mobility in parks, workplaces or residences. It should also be noted that this relationship is lower in districts with lower population density, which leads to the need for differentiated confinement policies in order to minimize the impacts of a terrible economic and social crisis.

Keywords: correlation coefficient; detrended cross-correlation analysis; COVID-19; mobility indices

Citation: Casa Nova, A.; Ferreira, P.; Almeida, D.; Dionísio, A.; Quintino, D. Are Mobility and COVID-19 Related? A Dynamic Analysis for Portuguese Districts. *Entropy* **2021**, *23*, 786. https://doi.org/10.3390/e23060786

Academic Editors: Ryszard Kutner, Christophe Schinckus and H. Eugene Stanley

Received: 20 May 2021
Accepted: 19 June 2021
Published: 21 June 2021

1. Introduction

The numbers of COVID-19 cases, both infections and casualties, are increasing daily all over the world, and concerns about their effects show no decrease. Even with the start of vaccination programs, it has not been possible to break the advance of the numbers, primarily because the speed of vaccination is asymmetric in different countries, but also because, contrarily to some respiratory diseases in the past, the spread between countries was higher [1–3]. With various negative economic and financial effects (see References [4–10]), COVID-19 also has several other consequences in people's lives, such as fear and depression [11,12], suicide trends [13] or in mental health [14,15].

The substantial effects of COVID-19 are related to the lockdowns that countries had to impose to control the spread of the disease. According to Reference [16], human behavior, among other factors, could contribute to respiratory viral infections, even more in a context where the superspreading conditions are not fully known [17]. However, it is crucial to reduce the number of social contacts, as complete vaccination programs are absent or not yet fully developed, and social-distancing measures could be the key in helping to solve the problem [18].

The spread of COVID-19 could be related to several factors. For example, Reference [19] identified several of these factors in assessing community risk factors in Catalonia, Spain, such as air pollution, population density, demographic and socioeconomic conditions, or even land use. In addition to these factors, which could affect the incidence of the disease in a general way, the authors also identify other factors related to the possible individual prevalence of the disease, such as the existence of comorbidities.

The existence of social contacts could be proxied by mobility data [20], with frameworks such as Google's Community Mobility Reports (CMR) being able to measure that mobility, as it measures citizens' mobility according to different types (for more details about CMR, see References [21,22]).

The use of CMR and its effects in COVID-19 has already been made using different approaches; see, for example, the studies of References [20,23–28], which, at a country level, found that the reduction of the mobility has a direct impact on the decrease of the infections. Reference [29] also confirms these trends and adds that reducing cases due to mobility restrictions has a very significant effect on a 2-week basis.

At a regional level, we can find the studies of References [30,31], both for the US. Although both find relevance in the effect of mobility on controlling the disease, Reference [30] finds differences between urban and rural locations, while Reference [31] identifies that population density has different implications in the reduction of mobility (higher density has more impact on the reduction of mobility, for example, in stores). In Poland, Reference [32] concluded that the restrictions helped control COVID-19, although with the difference between regions, related to the strictness of state restrictions.

During January 2021, Portugal was constantly in the news, as it was considered the worst country in the world regarding the infections and death rate (see https://www.politico.eu/article/portugal-coronavirus-rate-surge/, accessed on 19 May 2021). The lifting of some restrictions during the Christmas season may have compounded this tragic scenario. In this context, our purpose is to analyze, in a dynamic way, and based on daily data, the relationship between citizens' mobility and new COVID-19 infections, using regional-level data, in this case, for Portuguese districts. Our main objective is to assess the relevant relationship between the number of new infections of COVID-19 and citizen mobility. Moreover, we also want to distinguish between the different types of mobility. Differentiating the analysis between regions could give important insights for possible future decisions about new lockdowns or lifting of restrictions.

The implementation of non-pharmacological measures has a relevant impact on the control of the dissemination of COVID-19. In Portugal, the introduction of mandatory personal protective equipment (PPE) such as masks, or the instructions for frequent use of alcohol gel and washing hands, among others, started with the beginning of the pandemic in March/April 2020. Since then, the use of PPE has remained mandatory, and the non-pharmacological measures have not changed significantly.

In this paper, the mobility is measured considering Google CMR reports, and the relationship between mobility and new cases is assessed through the detrended cross-correlation analysis correlation coefficient. This non-linear framework has the ability to capture the relationship between variables for different timescales, which could give important information about the number of days needed to reduce infections. Moreover, we also propose the use of a sliding windows approach, which allows analysis of the evolution of the relationship over time.

Our main results corroborate that mobility is correlated with the number of new COVID-19 cases. However, the mobility correlation is not equal for the different typologies: for example, mobility in retail, recreation and groceries seems to have a higher correlation, while in general the mobility in workplaces shows little relationship. Despite the temporal evolution of the relationship, confirming that the lift of restrictions at Christmas had a highly significant impact on the increase of new COVID-19 cases, we also find that the impacts of the mobility are different across districts.

The remainder of the paper is organized as follows: in Section 2, both data and methodology are presented, with the results being present in Section 3, while Section 4 provides discussion and conclusions for the study.

2. Data and Methodology

Since the outbreak of COVID-19, and until 13 April 2021, almost 138 million cases were reported worldwide, with almost 3 million deaths. Portugal has about 828,000 cases and around 17,000 deaths. For cases of disease, information is available from the Portuguese Health Ministry, through Sistema Nacional de Vigilância Epidemiológica

(SINAVE), with the complete set of registered cases until 28 February 2021 (due to data availability). Until this day, Portugal has had a total of 805,140 cases. Intending to analyze the relationship between mobility and COVID-19 in the different Portuguese districts, we considered only the information which is registered in Portuguese mainland districts due to the availability of data about mobility. In total, the number of cases of the districts is 775,954. All the data were transformed in daily incidence for each district to perform the correlational analysis with the information from Google CMR. In these reports, it is possible to retrieve information about six distinct mobility indices: (i) retail and recreation (I1); (ii) groceries and pharmacies (I2), (iii) parks (I3), (iv) transit stations (I4), (v) workplaces (I5) and (vi) residential areas (I6). For more information about the indices and the places where mobility is referred to, see https://www.google.com/covid19/mobility/index.html?hl=en (accessed on 19 May 2021).

Daily data for these indices were retrieved for Portuguese districts from 15 February 2020 to 28 February 2021, in a total of 380 observations. Some districts do not have information for the mobility indices in some days of August and September 2020, implying that the sample is smaller for those districts (355 observations). The information about the number of cases and the number of observations for each district are identified in Table 1. Moreover, as some districts present missing information for some indices, the correlations were calculated for the remainder, where data are available.

Table 1. Total number of COVID-19 cases for each district and the number of observations considered in the analysis.

District	Total Cases	Observations	District	Total Cases	Observations
Aveiro	54,974	380	Leiria	24,647	380
Beja	7778	355	Lisbon	195,131	380
Braga	83,524	380	Portalegre	6936	355
Bragança	9787	355	Porto	160,398	380
Castelo Branco	10,914	355	Santarém	26,762	380
Coimbra	28,953	380	Setúbal	66,228	380
Évora	10,018	355	Viana do Castelo	16,920	355
Faro	19,594	380	Vila Real	13,972	355
Guarda	12,264	355	Viseu	27,154	355

To perform our correlational analysis, we use the detrended cross-correlation analysis coefficient ($\rho DCCA$), proposed by Reference [33] and derived from the work of Reference [34]. The DCCA measures the long-range cross-correlation between two series Y_i and X_i consisting on the sequence of $k = 1, 2, \ldots, N$ observations. The first step of the DCCA consists of the calculation of the profiles:

$$Y_k = \sum_{i=1}^{k} (y_i - \langle y \rangle) \text{ and } X_k = \sum_{i=1}^{k} (x_i - \langle x \rangle) \tag{1}$$

with $\langle . \rangle$ as the mean operator. Those profiles are then divided into $(N - n)$ overlapping boxes, from $n = 4$ to $n = N/4$ and for each box, based on the ordinary least squares, local trends $\widetilde{Y}_{k,i}$ and $\widetilde{X}_{k,i}$ are calculated, for future detrend of the profiles Y_k and X_k. With the local trends, the covariance of the residuals of each box is calculated as follows:

$$f_{xy}^2(n,i) = \frac{1}{(n+1)} \sum_{k=1}^{i+n} \left(X_k - \widetilde{X}_{k,i} \right) \left(Y_k - \widetilde{Y}_{k,i} \right). \tag{2}$$

Considering the information of all the set of $N - n$ boxes, the DCCA covariance is calculated as follows:

$$F_{xy}^2(n) = \frac{1}{(N-n)} \sum_{i=1}^{N-n} f_{xy}^2(n,i), \tag{3}$$

which was used by Reference [33] to obtain the correlation coefficient given by the following:

$$\rho_{DCCA} = \frac{F_{xy}^2(n)}{F_x^2(n)F_y^2(n)}. \quad (4)$$

The denominator of ρ_{DCCA} consists of the fluctuation functions of the detrended fluctuation analysis of Reference [35], which analyzes the long-range behavior of each time series individually.

The ρ_{DCCA} is a non-linear correlation coefficient, robust to the presence of non-stationarity, and confirms the property of $-1 \leq \rho_{DCCA} \leq 1$ according to [36–39] and is testable according to [40]. Moreover, this is a multiscale correlation coefficient, allowing for the analysis of the behavior between variables in different time periods. Despite the statistical properties previously referred to, the robustness of the correlation coefficient is confirmed by its use in different research areas (see, for example, [41–46], among others).

In this analysis, the ρ_{DCCA} will be calculated using a sliding windows approach to analyze the evolution of the correlation over time, using windows of 250 observations. In Table 2 we present the critical values to test the null hypothesis of absence of correlation, considering 250 observations, as it is the dimension of the samples used in the analysis.

Table 2. Critical values to test the ρ_{DCCA} considering time series of 250 observations and different timescales, considering a confidence level of 95% (source: Reference [40]).

Timescale	Critical Value
$n = 4$	0.137
$n = 8$	0.152
$n = 16$	0.193
$n = 32$	0.271
$n = 64$	0.383

3. Results

As previously stated, this study uses the DCCA correlation coefficient to assess the relationship between mobility indices and COVID-19 in Portuguese districts, also applying a sliding windows approach in order to evaluate the evolution of the correlation over time.

Figure 1 shows the behavior of the DCCA correlation coefficient between new COVID-19 cases and the six mobility indices, identifying the evolution over time for Portugal as a whole. Considering the multiscale feature of the measure and the temporal dynamics, a tri-dimensional analysis could be made. The information could be represented in different dimensions, as we can see in Figure A1, Appendix A. There, the results for Portugal as a whole are available, considering the correlation between the retail and recreation index and new COVID-19 cases, in three panels. Panel (a) reinforces the difference between time scales; in panel (b), the view is more about the evolution of the correlation over time; panel (c) adopts a panoramic view and is the one chosen for presentation the general results throughout the paper.

The results may be analyzed through different dimensions and perspectives, allowing an in-depth interpretation of the results.

Firstly, in general, the behavior of the correlation of retail and recreation, groceries and pharmacies and transit stations indices is qualitatively similar. In the very short run (lower timescales) the correlation coefficients are relatively high, meaning that mobility has a positive correlation with the number of new cases. However, there is a time-varying behavior, with a significant increase at the beginning of 2021, more marked in the case of groceries and pharmacies. Despite the continuous increase in the correlation, a peak can clearly be noticed after Christmas, probably related to the lifting of mobility restrictions in the country (in a season when environmental conditions could be more conducive to the development of respiratory problems). In the middle of January 2021, the Portuguese government took severe restrictive measures. Immediately afterwards, the correlation levels remained high, meaning that the mandatory restrictions to the mobility probably had a significant correlation with the reduction in the number of new COVID-19 cases. Over time, those measures could have had result on a progressive decline of the correlation levels, in agreement with References [20,23–28]. Another important feature is that the peak of the correlation is about the 7th/8th day, although in the groceries and pharmacies index it seems to be a little bit more, but it is remarkable that the duration of the correlations (red ones) is higher during the peak of the beginning of 2021. This means that lifting mobility restrictions or imposing new mobility restrictions could have an expected impact in about a week, which is consistent, for example, with the incubation period of the virus [47–49].

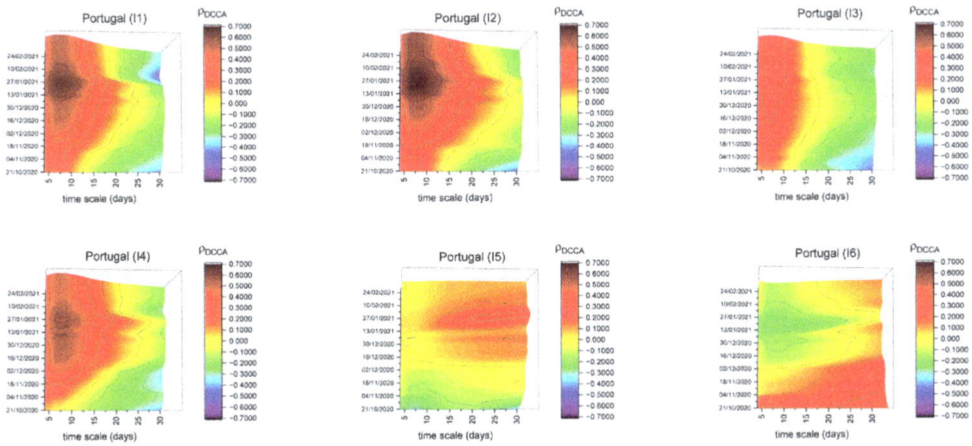

Figure 1. DCCA correlation coefficients between the different mobility indices and new COVID-19 cases in Portugal.

The results of the correlation of parks mobility index show different behavior and are more constant over time. Even though it seems relevant to explain the increase of new COVID-19 cases, the impact of this mobility type is not as high. As it measures mobility in open spaces, it should be related to a lower capacity of contagion in those spaces.

Finally, the correlation of the workplaces and residential areas indices presents different behavior, also considering the differences of the places to which they refer. Compared with the previously analyzed indices, the reduced correlations in workplaces mean that they seem to be relatively secure locals, probably due to the different measures taken by the employers. Despite the reduced levels of the correlations compared with the previously analyzed indices, the workplaces index seems to increase its correlation with COVID-19 cases over time during part of the sample, moving from negative to positive correlations in mid-December and continuing to increase during January and February. Moreover, it is important to highlight that at the beginning of 2021, the correlation is higher for higher timescales.

The results of the correlation of the workplaces index could firstly be justified with a period of a greater confluence of employees to their workplaces, especially before Christmas and New Year and, after this, the sharper increase may reflect the lifting of measures to restrict mobility during the Christmas and New Year period. Workplaces concern with the active population, that is, mostly between 30 and 50 years old. It is in this age group that asymptomatic cases are most significant. So, it could be a "domino effect": people left for Christmas, the "family bubbles" were broken, and when they returned to work, they infected others, which could justify the increased correlation and the impact even in longer timescales.

Regarding mobility in residential areas, as expected, it has higher moments of negative correlations, meaning that keeping people in their homes would decrease new infections. This finding is similar to References [30,50], both for the case of the US. However, it is noteworthy that some positive correlations are noted at the beginning of the analysis, although weaker than in the other mobility indices. This could happen because during the first months of the pandemic, most disease cases could have appeared in family circles.

As a final note referring to the statistical significance of the correlations, due to the multidimensional analysis, it is not feasible to introduce the information of the critical values in the figures. For this, it is necessary to identify the critical values from Table 2. Roughly, it is possible to say that, until n = 16, orange plans mean statistical significance, while, for higher timescales, darker oranges or blue plans are necessary.

In addition to the global analysis, we also aimed to analyze the relationship between mobility and COVID-19 in the different Portuguese districts. To do that, we made a similar analysis for each district, comparing it with the results presented for Portugal as a whole. Due to space limitation, we highlight non-similar patterns on the analysis of those indices (all the figures, organized by indices, are presented in Appendix B, in Figures A2–A7. The existence of significant differences across districts could lead to thinking that adopting different lockdown measures between districts should be a hypothesis to be considered.

If we consider the retail and recreation index (I1) (see Figure A2), in general, all districts in the country show a similar correlation pattern with the national results. This pattern is characterized by a lower correlation at the beginning of the sample period, increasing gradually until its peak at the beginning of 2021. Despite this similar pattern, it is important to mention districts such as Beja, Bragança, Évora, Faro, Guarda and Portalegre, in which the correlation intensity is lower than that found for Portugal, as seen in Figure 2. Excepting Faro, these districts are located in inland (and more rural) regions which have lower population density levels, in line with Reference [30]. Another district that we consider relevant to include is the Lisbon district. Between mid-November and early February, high correlation levels are observed for the different timescales. This evidence contrasts with that observed at the national level, which shows higher correlation levels for the same period, mainly for short timescales. This behavior may reflect the greater confluence of people in this type of space, not only in the period leading up to Christmas and New Year (for the traditional festive season shopping) but also in the period that followed (taking advantage, for example, of the traditional sales season). The fact that high levels of correlation are observed for longer timescales may indicate the need for restrictive measures to be adopted earlier.

These features lead us to think about the possibility of dichotomies between inland and coast, which could allow the conclusion that mobility restrictions could have been differentiated according to these dichotomies.

Figure 3 shows the correlation patterns between the groceries and pharmacies (I2) mobility index and new COVID-19 cases for Beja, Évora, Lisbon, Portalegre and Setúbal. For Beja, Évora and Portalegre, until nearly the end of 2020, this index presented a low correlation, close to zero, lower than that found for Portugal, indicating a lesser correlation between this type of mobility on the number of new cases for these districts. This empirical evidence may be justified by the smaller number of spaces available in these districts, which are still sufficiently available to serve the needs of their populations. From the beginning of 2021 and for short timescales, these correlations have increased, which may show that the frequency of these spaces could have a positive correlation with the emergence of new cases. This evidence may reflect the return home at the end of the holiday season and the onset of symptoms. These are all inland districts, with lower population density levels, as already stated, which reinforces the possibility of the adoption of differentiated confinement measures.

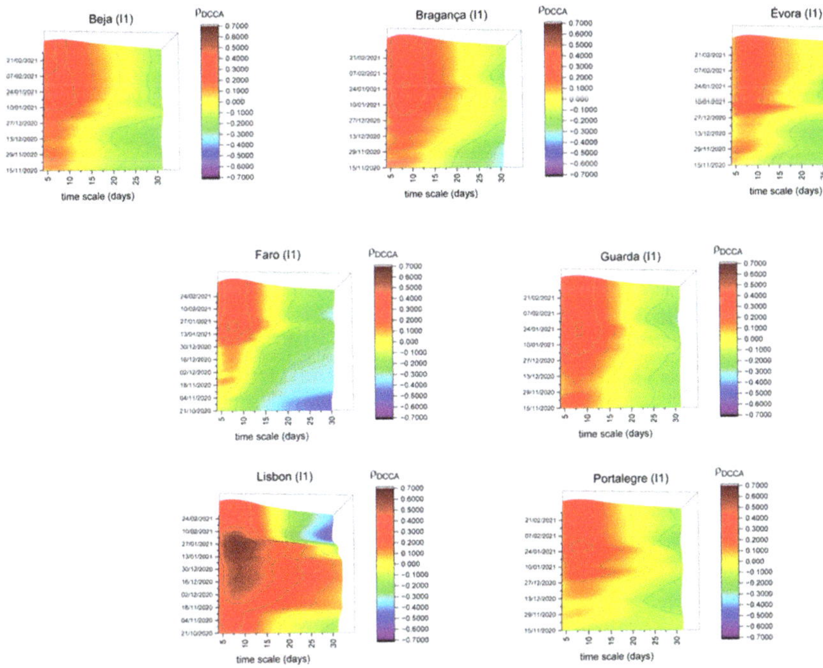

Figure 2. DCCA correlation coefficients between retail and recreation (I1) and new COVID-19 cases in Beja, Bragança, Évora, Faro, Guarda, Lisbon and Portalegre.

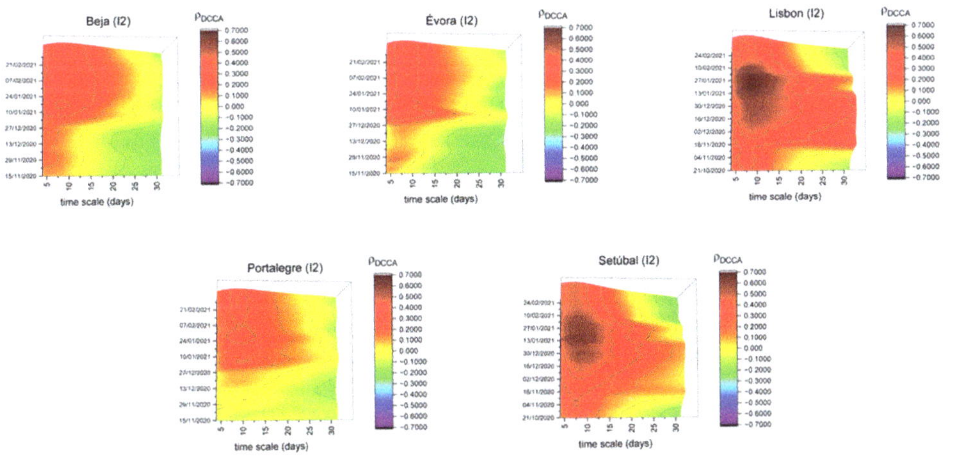

Figure 3. DCCA correlation coefficients between groceries and pharmacies (I2) and new COVID-19 cases in Beja, Évora, Lisbon, Portalegre and Setúbal.

Regarding Lisbon and Setúbal, between mid-November and early February, high correlation levels are observed for the different timescales. Contrary to that stated for the Beja, Évora and Portalegre districts, Lisbon and Setúbal are districts with high population density, which may justify the observed behavior. Furthermore, it could also be justified not only by the high number of this kind of space but also by the increase in the number of people who go to those places. This could lead us to think that the adoption of different measures (more restrictive in this case) should be considered.

In Figure 4, we have selected Beja, Coimbra, Évora, Faro and Portalegre because they present a different pattern compared to the parks index presented in Figure 1. This index has a lower correlation with the number of new cases, when compared to those found for Portugal. Parks refers to open spaces, where it is known that the propagation of the virus could be less significant. The low population density could also explain the differences of Beja, Évora and Portalegre, as was found by Reference [51] for US counties, while Faro's location, on the south coast of Portugal, and the extension of its beaches, could lead to different results (i.e., enjoying those type of open spaces cautiously could imply lower correlation levels). Regarding Coimbra, it is also a district that is close to beaches but also with some municipalities with reduced population density levels. It is also necessary to highlight that, in Faro, the sliding windows correlation coefficients until November show high negative values, meaning that the possibility of enjoying time in those open spaces was negatively correlated with the increase of new COVID-19 cases.

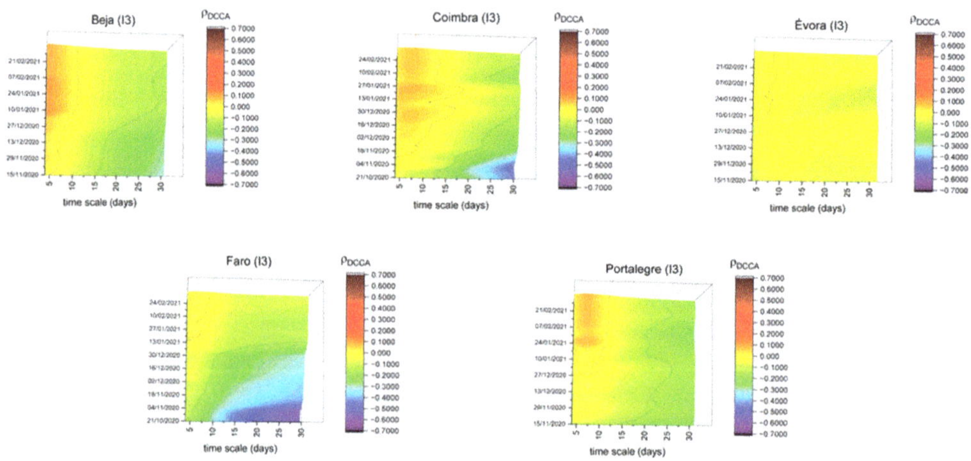

Figure 4. DCCA correlation coefficients between parks (I3) and new COVID-19 cases in Beja, Coimbra, Évora, Faro and Portalegre.

On the other hand, in the period following the adoption of the new confinement measures, an increase could be noted in the correlation between this index and the number of new cases, which may reflect the possibility of using the so-called "hygienic walks". Thus, the adoption of restrictive measures concerning the frequency of use of these spaces may seem counterproductive. In other words, the fact that some of these spaces closed completely (e.g., walled public gardens), may have led to the displacement of people to those where only circulation was allowed (and not staying there), having an impact on the increase in correlation, especially on short timescales.

Before we start our analysis about the transit stations (I4) index for some districts, we would like to state that this is the only index for which some districts do not have available information, which may be related to lesser presence of public transportation.

Figure 5 shows the correlation between the indices referring to the mobility in transit stations and new COVID-19 cases for five different districts, all located in the north region. In mid-January, new confinement measures were adopted by the government. They had a national impact, which could have led to the reduction of the correlation between this index and the number of new COVID-19 cases; however, there was no significant correlation reduction in Aveiro, Braga and Porto. This mobility index continued to show high correlations for short timescales with the number of new COVID-19 cases. Regarding

Coimbra, its correlation is lower over the entire sample period and for all timescales. On the one hand, it may indicate the security of the transport network or a lower rate of its usage in this district. Finally, in Vila Real, we can see higher correlations in the short-term, without significant change over time. This fact may reflect that transport habits in this district have remained unchanged.

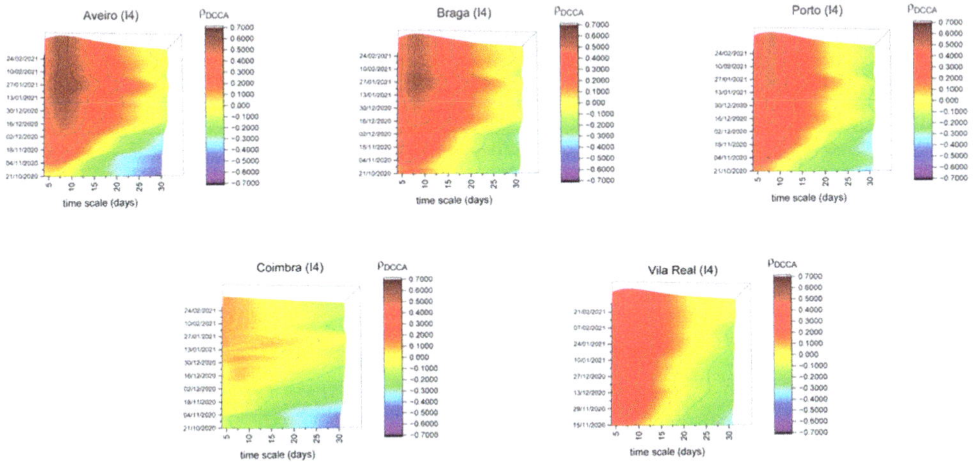

Figure 5. DCCA correlation coefficients between the mobility in transit stations (I4) and new COVID-19 cases in Aveiro, Braga, Porto, Coimbra and Vila Real.

Comparing the results in workplace mobility (I5), it is possible to distinguish a different pattern of correlations mainly in Évora and Castelo Branco, as represented in Figure 6. These are the only districts with significant differences throughout the period under analysis for the different timescales, showing a positive correlation between this index and new COVID-19 cases. This could be related to less efficient security measures in workplaces or the fact that they were adopted later.

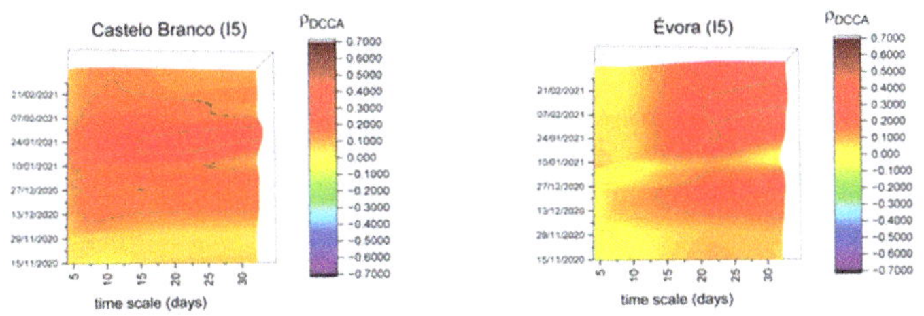

Figure 6. DCCA correlation coefficients between workplace mobility (I5) and new COVID-19 cases in Castelo Branco and Évora.

Finally, considering the residential areas (I6) index, Figure 7 shows the patterns registered in Aveiro, Braga, Castelo Branco and Lisbon, although with different patterns. Aveiro and Braga show the highest negative correlations after the confinement of the beginning of 2021, probably meaning that the success of the lockdown was greater in those districts. Regarding Castelo Branco and Lisbon, these districts are the only ones showing a positive correlation over the entire sample period, mainly in short timescales. This may indicate that, in these districts, the family nuclei could have caused an emergence of new COVID-19 cases, although with different possible explanations. Lisbon is the most populous district of the country and in some cases the quality or undersized dimensions of the habitations could promote the increase of contagion. On the other hand, in the case of

Castelo Branco, the situation could be related to an existent gap between the beginning of the cases in this district and the rest of the country. For example, when the first confinement occurred, Castelo Branco had practically no COVID-19 cases, meaning that people had no necessity to go to their houses, i.e., confinement could be considered unnecessary in the district.

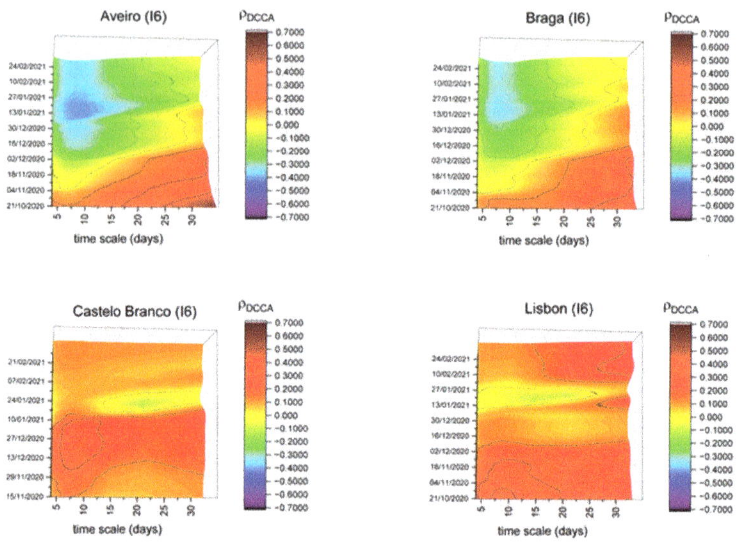

Figure 7. DCCA correlation coefficients between the residential areas (I6) index and new COVID-19 cases in Aveiro, Braga, Castelo Branco and Lisbon.

Taking the different patterns found for the correlations between some of the six indices and the spread of new COVID-19 cases, we would like to highlight that confinement measures do not have the same effect on all districts, which could indicate that the adoption of different measures in different districts could be desirable. We also highlight that there are locations that seem to present more risk of contagion (the ones related to retail and recreation, groceries and pharmacies and transit stations), while residential areas seem to present a lower risk factor of contagion, as expected. Applying the DCCA coefficient, an unexplored method to address this issue, allows us to analyze the behavior between each mobility index and the spread of new COVID-19 cases in different timescales and leads us to understand, for example, when peak correlations occurred and that not all the indices have the same peak correlation.

4. Discussion and Conclusions

In this research work, the intention was to assess the correlation between the number of contagions and the mobility indices of people. For this purpose, an approach based on the DCCA was used, which has the capacity to assess the global correlation between serial variables. Simultaneously, it presents robustness in the face of issues related to stationarity, non-linearity and non-normality of the data and also allows for the analysis of the evolution of the relationship over time. The whole sample under consideration was Portugal and its respective districts, with daily data on the variables under analysis. It should be noted that, despite Portugal being an interesting case study, as it was considered exemplary in the first phase of the pandemic crisis and was catalogued as the "worst country in the world" in January 2021, the truth is that the approach is robust and valid and can be successfully applied to any country or region.

The global results essentially indicate that a dynamic association exists between the different mobility indices and the new COVID-19 cases, with three main risk factors being identified in terms of mobility: retail and recreation (I1); groceries and pharmacies (I2) and public transport (I4). In addition to the considerations already taken, regarding the effectiveness of confinement to contain contagions, we can infer that some of these mobility factors may imply the non-use of a mask in certain situations, which may justify the values found for the retail and recreation and groceries and pharmacies indices. Take as an example recreation (cafes and restaurants) in which the consumption of food and drink goods prevents the use of a mask. In the case of public transport (I4), it could also be related to the fact that people may touch the same surfaces sequentially, with the respective risk of contagion.

When we perform the district analysis, for the majority of districts, we found similar behavior to that of the country as a whole. However, there are some distinct behaviors during the period under analysis and for different mobility risk factors. These differences may be related to the low population densities of some districts, especially those inland. Note that, for all the indices except residential areas, in general, the least densely populated districts were the ones showing lower correlations than those of the country as a whole, in line with the results found in Reference [31]. Regarding residential areas, Lisbon has a higher level of correlation than the average for Portugal, which may indicate that, in large cities, with a high population density and possibly weaker habitational conditions, residential mobility may be a significant contagion factor. Once again, it is the districts with the lowest population density that stand out (due to the lowest correlation) in this factor, also related to the difference between urban and rural areas, as identified by Reference [30].

Overall, and always bearing in mind that other factors could be related to the increasing number of new COVID-19 cases, as stated by Reference [19], we can conclude that some mobility indices are more likely than others to have correlation patterns with the contagion levels of COVID-19, which may be linked to the crowding of people, wearing masks and hand hygiene. In addition to this, we also concluded that population density might affect the correlation level of mobility indices with the new confirmed cases of COVID-19. It appears that districts with lower population density have lower correlations, which indicates that a different definition of confinement policies may be more appropriate for controlling the pandemic and simultaneously minimizing its effects in economic and social terms. Blindly imposing confinements leads to population revolt and the growth of states of anxiety and general impoverishment. It is increasingly important to understand which risk factors related to mobility most potentiate contagion and which regions and moments tougher measures are justified in terms of containment. These results are in line, for example, with the conclusions of Reference [28], where it is stated that re-arranging local restrictions can be much more effective in controlling the number of COVID-19 cases without causing unnecessary economic costs than local or country-wide mobility restrictions.

The results obtained in this study and the respective conclusions may be an important contribution to political decision-making about measures to be taken to contain the amount of contagion and, possibly taking measures which are differentiated by district and/or region, combining them with the available different non-pharmacological measures, which have been relatively stable during the period under analysis.

It is important to state again that the focus is on the method and respective abilities, which has proven to be robust and adequate, providing accurate and detailed information about the variables that have the greatest correlation with the number of COVID-19 infected persons. It is also relevant to highlight that the increased mobility in Portugal was made considering the break in social distancing, especially between family and close social meetings. Given this, we believe that there is a high probability that the increased mobility had a strong impact on the increase in numbers of people infected with COVID-19, given the tendency for breaking social distancing, especially in the Christmas period.

Author Contributions: Conceptualization, A.C.N., P.F., D.A., A.D. and D.Q.; formal analysis, A.C.N., P.F., D.A., A.D. and D.Q.; data curation, A.C.N., P.F., D.A., A.D. and D.Q.; writing—original draft preparation, A.C.N., P.F., D.A., A.D. and D.Q.; writing—review and editing, A.C.N., P.F., D.A., A.D. and D.Q. All authors have read and agreed to the published version of the manuscript.

Funding: Paulo Ferreira acknowledges the financial support of Fundação para a Ciência e a Tecnologia (grants UIDB/05064/2020 and UIDB/04007/2020). Andreia Dionísio acknowledges the financial support of Fundação para a Ciência e a Tecnologia (grant UIDB/04007/2020). Derick Quintino acknowledges the financial support of the Coordenação de Aperfeiçoamento de Pessoal de Nível Superior—Brasil (CAPES)—Finance Code 001.

Institutional Review Board Statement: The study was conducted according to the guidelines of the Declaration of Helsinki, and approved by Ethics Committee of Instituto Politécnico de Portalegre (3/2021, approved on 25 February 2021).

Informed Consent Statement: Not applicable.

Data Availability Statement: Restrictions apply to the availability of these data. Data were obtained from Sistema Nacional de Vigilância Epidemiológica (SINAVE) and are available with the permission of Sistema Nacional de Vigilância Epidemiológica (SINAVE).

Acknowledgments: The authors would like to acknowledge the Sistema Nacional de Vigilância Epidemiológica (SINAVE), of the Portuguese Health Ministry, for making available the data used in this paper.

Conflicts of Interest: The authors declare no conflict of interest.

Appendix A

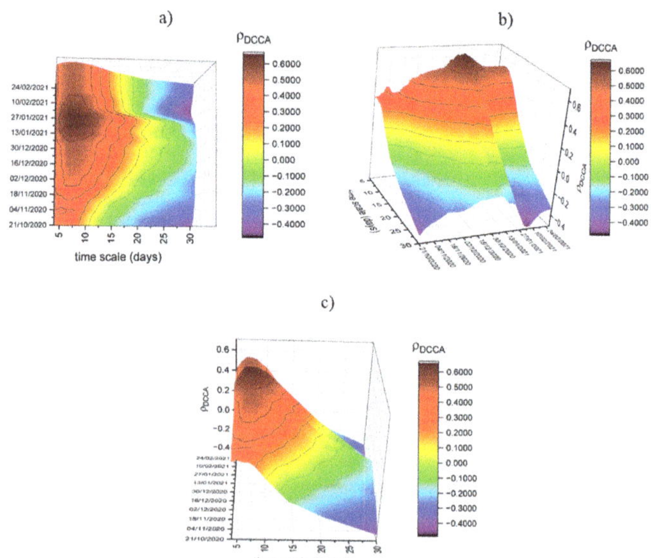

Figure A1. Representation of the DCCA correlation coefficient between the retail and recreation mobility index and new COVID-19 cases in Portugal. Panel (**a**) reinforces the analysis through the time-scale view; panel (**b**) reinforces the analysis through the view of the temporal evolution; panel (**c**) shows a panoramic view of the results.

Appendix B

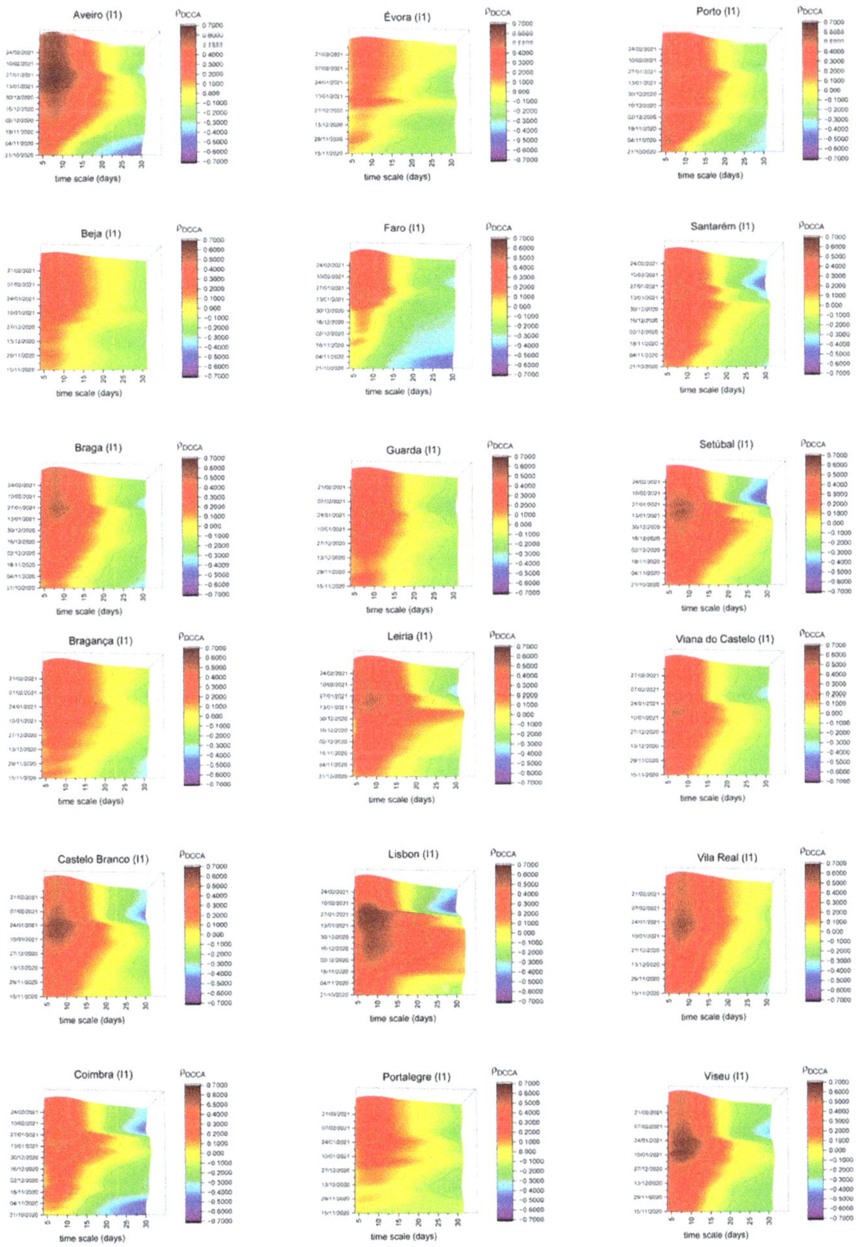

Figure A2. DCCA correlation coefficients between the retail and recreation index and new COVID-19 cases in the complete set of Portuguese districts.

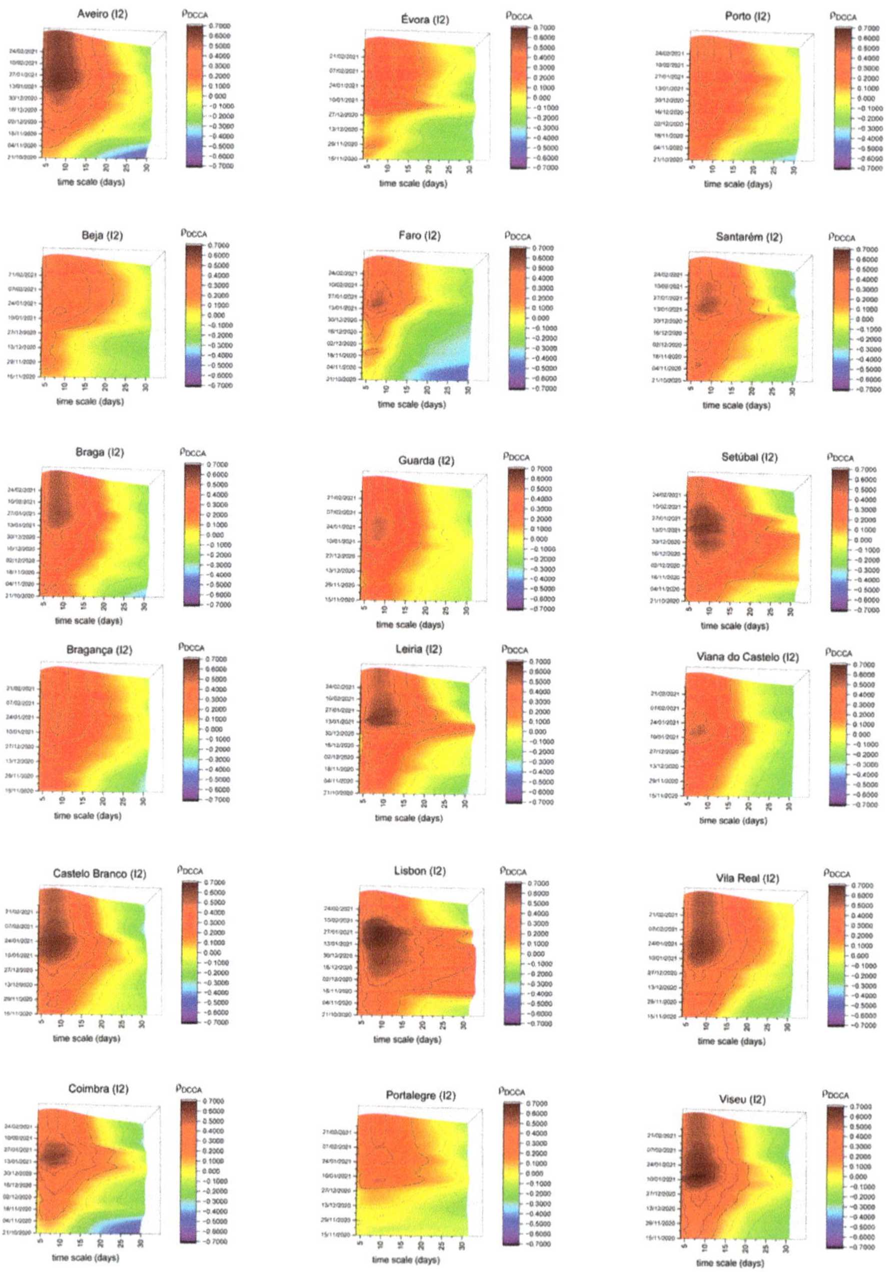

Figure A3. DCCA correlation coefficients between the groceries and pharmacies index and new COVID-19 cases in the complete set of Portuguese districts.

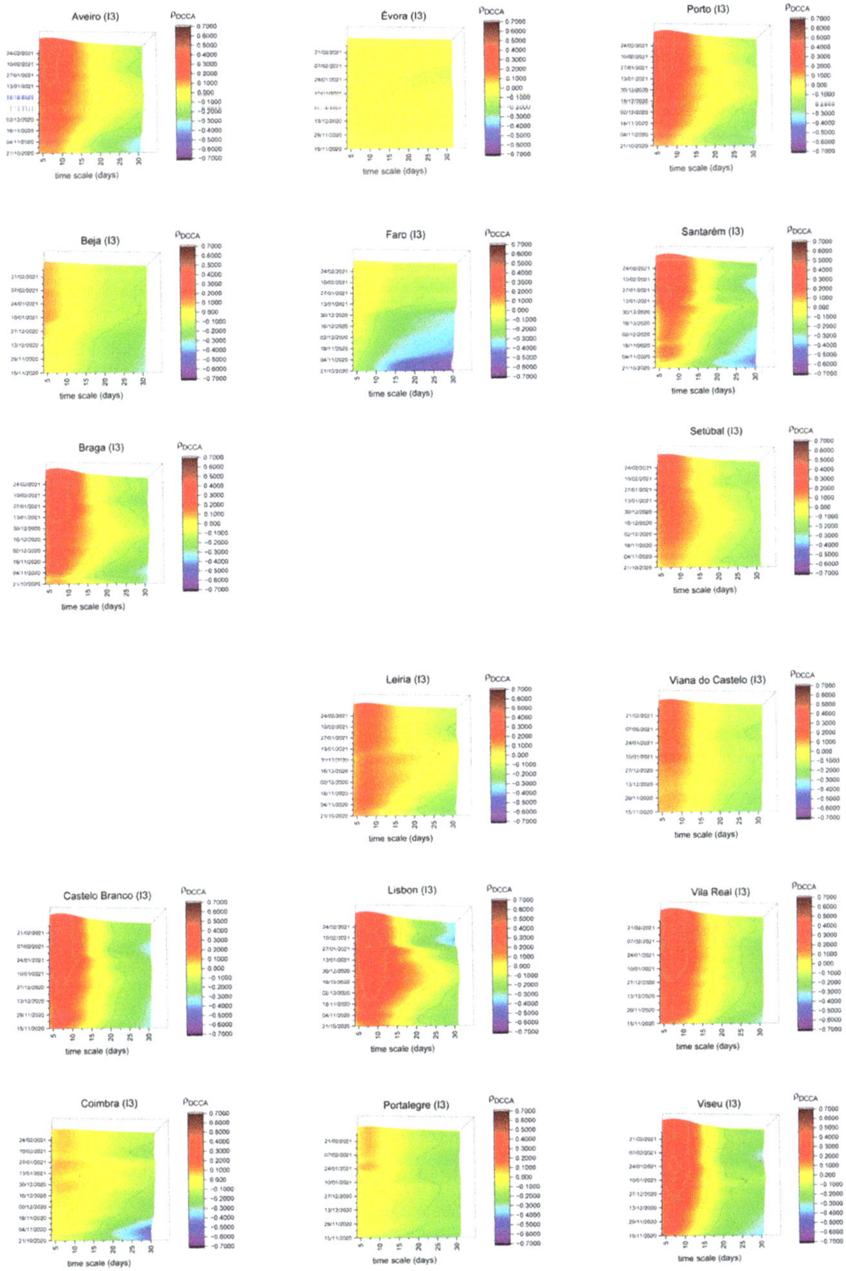

Figure A4. DCCA correlation coefficients between the parks index and new COVID-19 cases in the complete set of Portuguese districts.

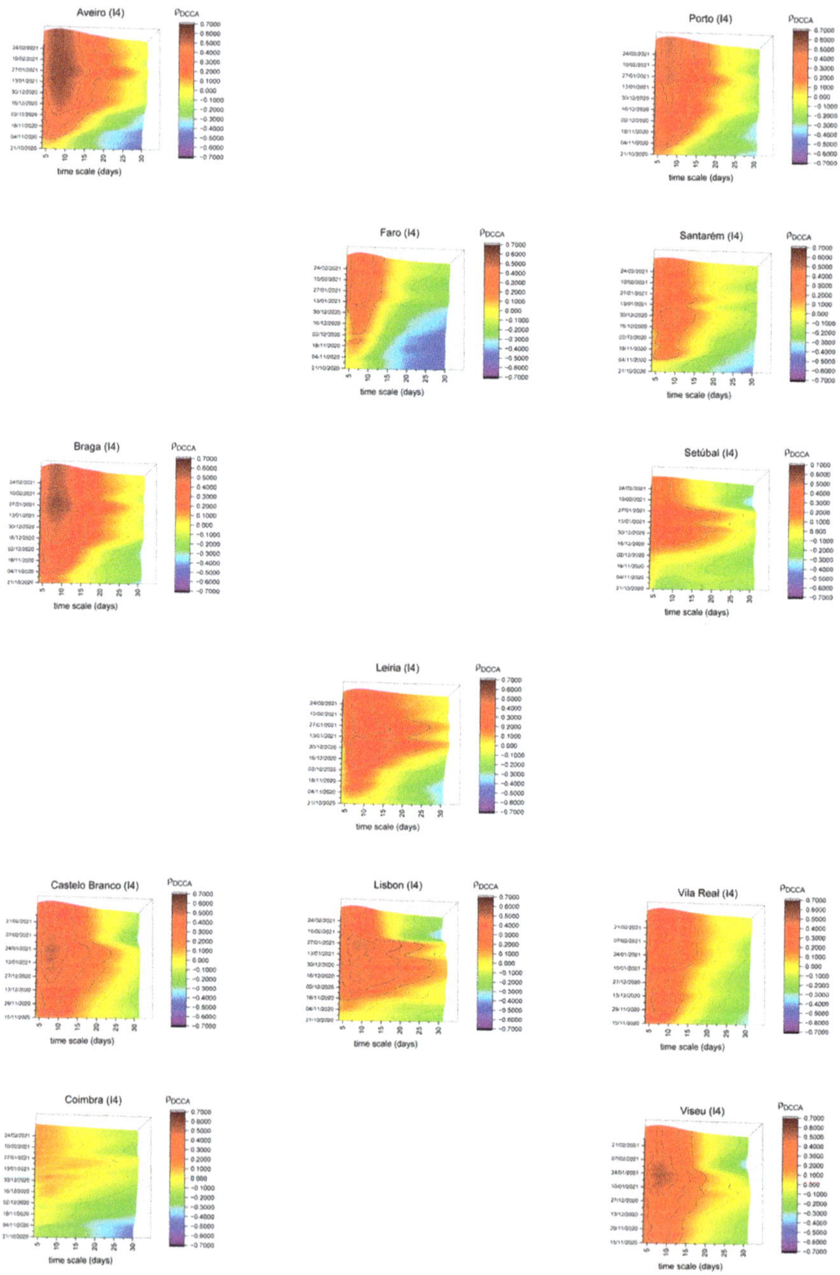

Figure A5. DCCA correlation coefficients between the transit stations index and new COVID-19 cases in the complete set of Portuguese districts.

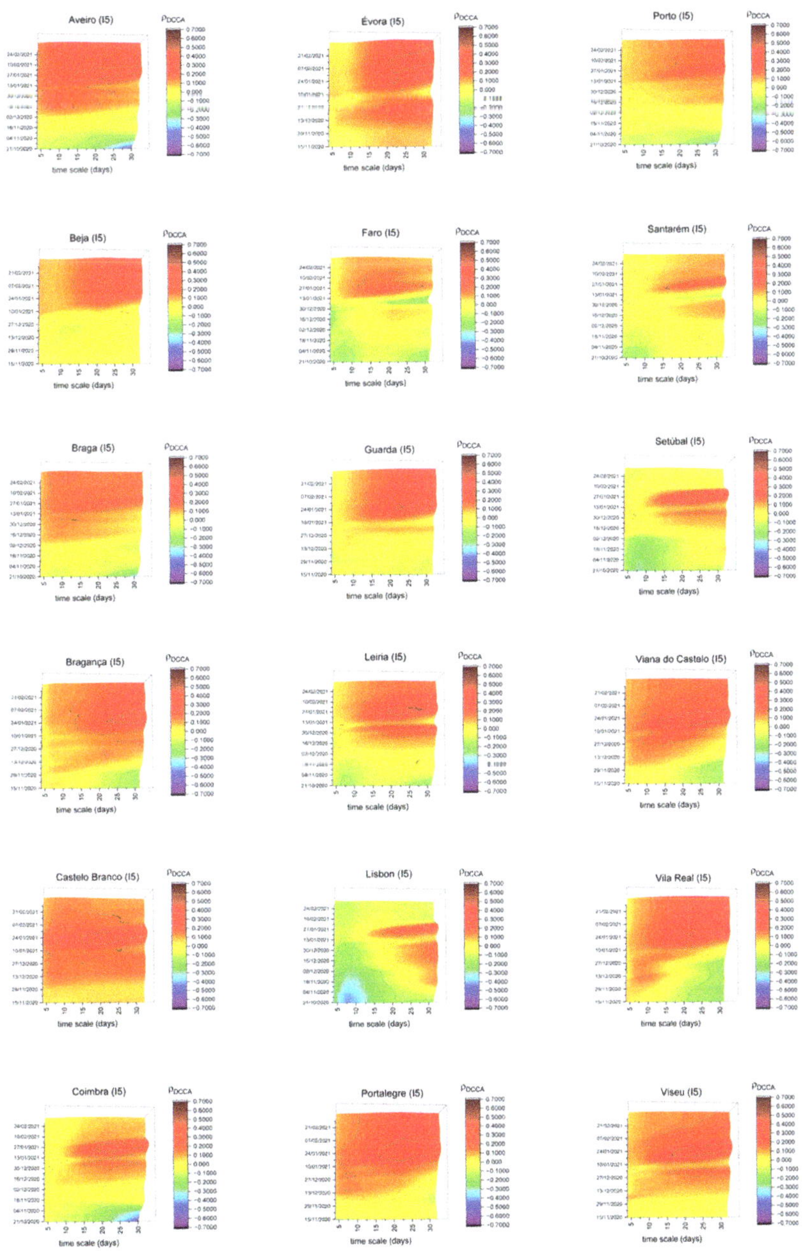

Figure A6. DCCA correlation coefficients between the workplaces index and new COVID-19 cases in the complete set of Portuguese districts.

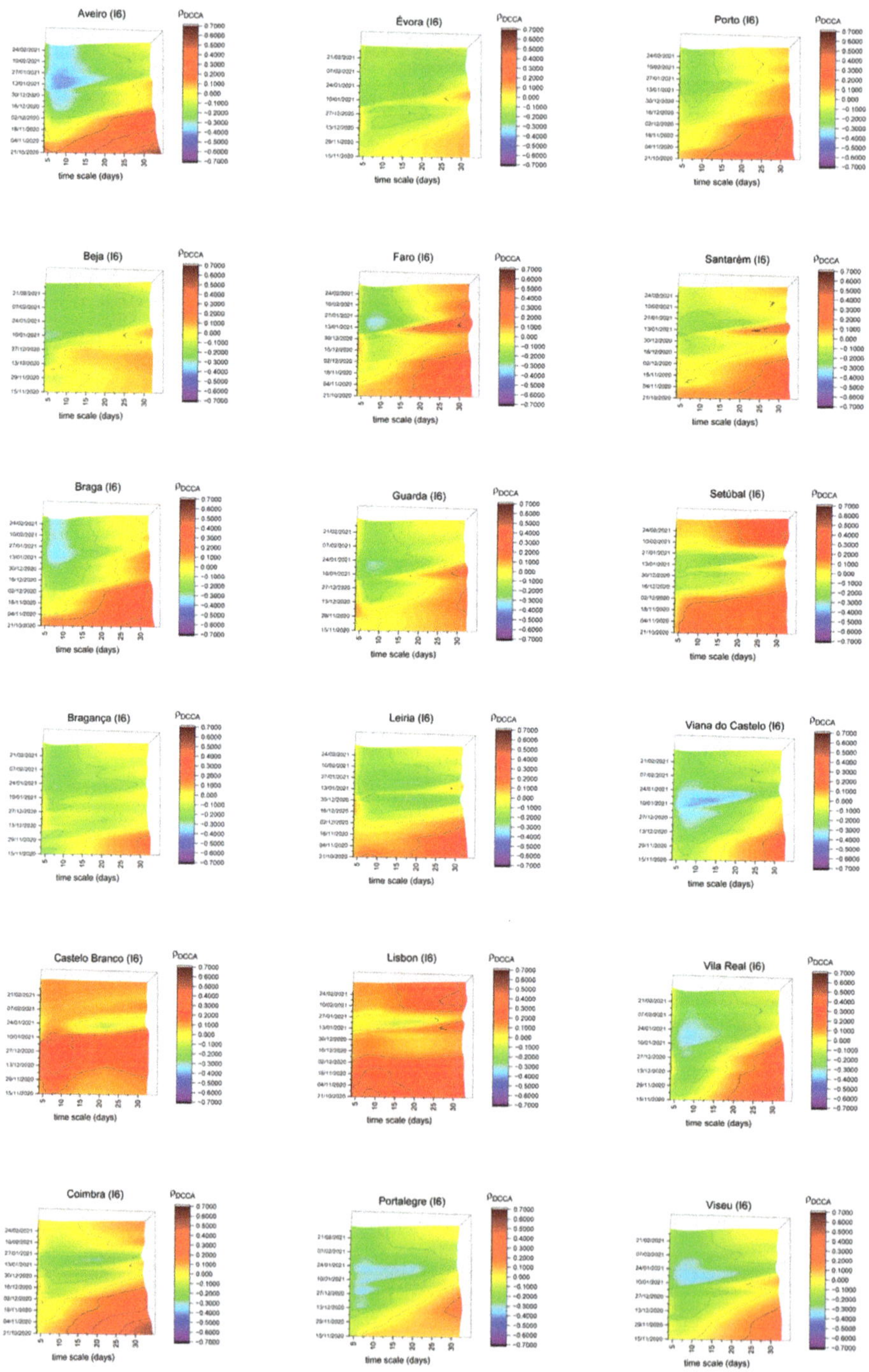

Figure A7. DCCA correlation coefficients between the residential areas index and new COVID-19 cases in the complete set of Portuguese districts.

References

1. Chen, M.H.; Jang, S.S.; Kim, W.G. The Impact of the SARS Outbreak on Taiwanese Hotel Stock Performance: An Event-Study Approach. *Int. J. Hosp. Manag.* **2007**, *26*, 200–212. [CrossRef]
2. Ezhilan, M.; Suresh, I.; Nesakumar, N. SARS-CoV, MERS-CoV and SARS-CoV-2: A Diagnostic Challenge. *Measurement* **2021**, *168*, 108335. [CrossRef]
3. Wu, J.T.; Leung, K.; Leung, G.M. Nowcasting and Forecasting the Potential Domestic and International Spread of the 2019-nCoV Outbreak Originating in Wuhan, China: A Modelling Study. *Lancet* **2020**, *395*, 689–697. [CrossRef]
4. Milani, F. COVID-19 outbreak, social response, and early economic effects: A global VAR analysis of cross-country interdependencies. *J. Popul. Econ.* **2021**, *34*, 223–252. [CrossRef]
5. Asahi, K.; Undurraga, E.A.; Valdés, R.; Wagner, R. The effect of COVID-19 on the economy: Evidence from an early adopter of localized lockdowns. *J. Glob. Health* **2021**, *11*, 05002. [CrossRef] [PubMed]
6. Albulescu, C. COVID-19 and the United States financial markets' volatility. *Finance Res. Lett.* **2021**, *38*, 101699. [CrossRef]
7. Baig, A.S.; Butt, H.A.; Haroon, O.; Rizvi, S.A.R. Deaths, panic, lockdowns and US equity markets: The case of COVID-19 pandemic. *Finance Res. Lett.* **2021**, *38*, 101701. [CrossRef]
8. Fernandes, N. *Economic Effects of Coronavirus Outbreak (COVID-19) on the World Economy*; Paper No. WP-1240-E; IESE Business School Working: Navarra, Spain, 2020; p. 33.
9. Baker, S.R.; Farrokhnia, R.A.; Meyer, S.; Pagel, M.; Yannelis, C. How does household spending respond to an epidemic? Consumption during the 2020 COVID-19 pandemic. *Rev. Asset Pricing Stud.* **2020**, *10*, 834–862. [CrossRef]
10. Blustein, D.; Duffy, R.; Ferreira, J.; Cohen-Scali, V.; Cinamon, R.; Allan, B. Unemployment in the time of COVID-19: A research agenda. *J. Vocat. Behav.* **2020**, *119*, 103436. [CrossRef]
11. Tsang, S.; Avery, A.R.; Duncan, G.E. Fear and depression linked to COVID-19 exposure: A study of adult twins during the COVID-19 pandemic. *Psychiatry Res.* **2021**, 113699. [CrossRef] [PubMed]
12. Schimmenti, A.; Billieus, J.; Starcevic, V. The four horsemen of fear: An integrated model of understanding fear experiences during the COVID-19 pandemic. *Clin. Neuropsychiatry* **2020**, *17*, 41–45.
13. Kawohl, W.; Nordt, C. COVID-19, unemployment, and suicide. *Lancet Psychiatr.* **2020**, *7*, 389–390. [CrossRef]
14. Ravens-Sieberer, U.; Kaman, A.; Erhart, M.; Devine, J.; Schlack, R.; Otto, C. Impact of the COVID-19 pandemic on quality of life and mental health in children and adolescents in Germany. *Eur. Child Adolesc. Psychiatry* **2021**. [CrossRef] [PubMed]
15. Fore, H. A wake-up call: COVID-19 and its impact on children's health and wellbeing. *Lancet Glob. Health* **2020**, *8*, e861–e862. [CrossRef]
16. Salzberger, H.; Buder, F.; Lampl, B.; Ehrenstein, B.; Hitzenbichler, F.; Holzmann, T.; Schmidt, B.; Hanses, F. Epidemiology of SARS-CoV-2. *Infection* **2020**, *8*, 1–7. [CrossRef] [PubMed]
17. Moryiama, M.; Hugentobler, W.; Iwasaki, A. Seasonality of Respiratory Viral Infections. *Annu. Rev. Virol.* **2020**, *7*, 83–101. [CrossRef] [PubMed]
18. Zhang, M.; Wang, S.; Hu, T.; Fu, X.; Wang, X.; Hu, Y.; Halloran, B.; Cui, Y.; Liu, H.; Liu, Z.; et al. Human mobility and COVID-19 transmission: A systematic review and future directions. *medRxiv* **2021**. [CrossRef]
19. Zaldo-Aubanell, Q.; López, F.; Bach, A.; Serra, I.; Olivet-Vila, J.; Saez, M.; Pino, D.; Maneja, R. Community Risk Factors in the COVID-19 Incidence and Mortality in Catalonia (Spain). A Population-Based Study. *Int. J. Environ. Res. Public Health* **2021**, *18*, 3768. [CrossRef]
20. Nouvellet, P.; Bhatia, S.; Cori, A.; Ainslie, K.E.; Baguelin, M.; Bhatt, S.; Donnelly, C.A. Reduction in mobility and COVID-19 transmission. *Nat. Commun.* **2021**, *12*, 1–9. [CrossRef]
21. Aktay, A.; Bavadekar, S.; Cossoul, G.; Davis, J.; Desfontaines, D.; Fabrikant, A.; Gabrilovich, E.; Gadepalli, K.; Gipson, B.; Wilson, R.; et al. Google COVID-19 Community Mobility Reports: Anonymization Process Description (Version 1.1). 2020. Available online: https://arxiv.org/abs/2004.04145 (accessed on 19 May 2021).
22. Drake, T.M.; Docherty, A.B.; Weiser, T.G.; Yule, S.; Sheikh, A.; Harrison, E.M. The effects of physical distancing on population mobility during the COVID-19 pandemic in the UK. *Lancet Digit. Health* **2020**, *2*, e385–e387. [CrossRef]
23. Sulyok, M.; Walker, M. Community movement and COVID-19: A global study using Google's Community Mobility Reports. *Epidemiol. Infect.* **2020**, *148*, 1–9. [CrossRef]
24. Yilmazkuday, H. Stay-at-home works to fight against COVID-19: International evidence from Google mobility data. *J. Hum. Behav. Soc. Environ.* **2021**, *31*, 210–220. [CrossRef]
25. Zhu, D.; Mishra, S.R.; Han, X.; Santo, K. Social distancing in Latin America during the COVID-19 pandemic: An analysis using the Stringency Index and Google Community Mobility Reports. *J. Travel Med.* **2020**, *27*, taaa125. [CrossRef] [PubMed]
26. Murphy, M.M.; Jeyaseelan, S.M.; Howitt, C.; Greaves, N.; Harewood, H.; Quimby, K.; Sobers, N.; Landis, R.; Rocke, K.; Hambleton, I.R. COVID-19 containment in the Caribbean: The experience of small island developing states. *Res. Glob.* **2020**, *2*, 100019. [CrossRef]
27. Wang, S.; Liu, Y.; Hu, T. Examining the change of human mobility adherent to social restriction policies and its effect on COVID-19 cases in Australia. *Int. J. Environ. Res. Pub. Health* **2020**, *17*, 7930. [CrossRef] [PubMed]
28. Kartal, M.T.; Depren, O.; Depren, S.K. The Relationship between Mobility and COVID-19 Pandemic: Daily Evidence from an Emerging Country by Causality Analysis. *Transp. Res. Interdiscip. Perspect.* **2021**, *10*, 100366.

29. McGrail, D.J.; Dai, J.; McAndrews, K.M.; Kalluri, R. Enacting national social distancing policies corresponds with dramatic reduction in COVID19 infection rates. *PLoS ONE* **2020**, *15*, e0236619. [CrossRef] [PubMed]
30. Li, X.; Rudolph, A.E.; Mennis, J. Association Between Population Mobility Reductions and New COVID-19 Diagnoses in the United States Along the Urban–Rural Gradient, February–April. *Prev. Chronic Dis.* **2020**, *17*, 200241. [CrossRef]
31. Hamidi, S.; Zandiatashbar, A. Compact development and adherence to stay-at-home order during the COVID-19 pandemic: A longitudinal investigation in the United States. *Landsc. Urban Plan.* **2020**, *205*, 103952. [CrossRef]
32. Wielechowski, M.; Czech, K.; Grzęda, Ł. Decline in Mobility: Public Transport in Poland in the time of the COVID-19 Pandemic. *Economies* **2020**, *8*, 78. [CrossRef]
33. Zebende, G.F. DCCA cross-correlation coefficient: Quantifying level of cross-correlation. *Physica A* **2011**, *390*, 614–618. [CrossRef]
34. Podobnik, B.; Stanley, H.E. Detrended cross-correlation analysis: A new method for analyzing two nonstationary time series. *Phys. Rev. Lett.* **2008**, *100*, 084102. [CrossRef]
35. Peng, C.K.; Buldyrev, S.V.; Havlin, S.; Simons, M.; Stanley, H.E.; Goldberger, A.L. Mosaic organization of DNA nucleotides. *Phys. Rev. E.* **1994**, *49*, 1685.
36. Kristoufek, L. Measuring correlations between non-stationary series with DCCA coefficient. *Physica A* **2014**, *402*, 291–298. [CrossRef]
37. Kristoufek, L. Detrending moving-average cross-correlation coefficient: Measuring cross-correlations between non-stationary series. *Physica A* **2014**, *406*, 169–175. [CrossRef]
38. Wang, G.J.; Xie, C.; Chen, Y.J.; Chen, S. Statistical properties of the foreign exchange network at different time scales: Evidence from detrended cross-correlation coefficient and minimum spanning tree. *Entropy* **2013**, *15*, 1643–1662. [CrossRef]
39. Zhao, X.; Shang, P.; Huang, J. Several fundamental properties of DCCA cross-correlation coefficient. *Fractals* **2017**, *25*, 1750017. [CrossRef]
40. Podobnik, B.; Jiang, Z.Q.; Zhou, W.X.; Stanley, H.E. Statistical tests for power-law cross-correlated processes. *Phys. Rev. E* **2011**, *84*, 066118. [CrossRef] [PubMed]
41. Brito, A.A.; Santos, F.R.; de Castro, A.P.N.; da Cunha Lima, A.T.; Zebende, G.F.; da Cunha Lima, I.C. Cross-correlation in a turbulent flow: Analysis of the velocity field using the ρ DCCA coefficient. *Europhys. Lett.* **2018**, *123*, 20011. [CrossRef]
42. Machado Filho, A.; da Silva, M.F.; Zebende, G.F. Autocorrelation and cross-correlation in time series of homicide and attempted homicide. *Physica A* **2014**, *400*, 12–19. [CrossRef]
43. Paiva, A.S.S.; Rivera-Castro, M.A.; Andrade, R.F.S. DCCA analysis of renewable and conventional energy prices. *Physica A* **2018**, *490*, 1408–1414. [CrossRef]
44. Zebende, G.F.; Brito, A.A.; Silva Filho, A.M.; Castro, A.P. ρDCCA applied between air temperature and relative humidity: An hour/hour view. *Physica A* **2018**, *494*, 17–26. [CrossRef]
45. Chen, Y.; Cai, L.; Wang, R.; Song, Z.; Deng, B.; Wang, J.; Yu, H. DCCA cross-correlation coefficients reveals the change of both synchronization and oscillation in EEG of Alzheimer disease patients. *Physica A* **2018**, *490*, 171–184. [CrossRef]
46. Marinho, E.; Sousa, A.; Andrade, R. Using Detrended Cross-Correlation Analysis in geophysical data. *Physica A* **2013**, *392*, 2195–2201. [CrossRef]
47. Qin, J.; You, C.; Lin, Q.; Hu, T.; Yu, S.; Zhou, X. Estimation of incubation period distribution of COVID-19 using disease onset forward time: A novel cross-sectional and forward follow-up study. *Sci. Adv.* **2020**, *6*, eabc1202. [CrossRef] [PubMed]
48. Rokni, M.; Ghasemi, V.; Tavakoli, Z. Immune responses and pathogenesis of SARS-CoV-2 during an outbreak in Iran: Comparison with SARS and MERS. *Rev. Med. Virol.* **2020**, *30*, e2107. [CrossRef] [PubMed]
49. Wassie, G.; Azene, A.; Bantie, G.; Dessie, G.; Arabaw, A. Incubation Period of Severe Acute Respiratory Syndrome Novel Coronavirus 2 that Causes Coronavirus Disease 2019: A Systematic Review and Meta-Analysis. *Curr. Ther. Res. Clin. Exp.* **2020**, *93*, 100607. [CrossRef] [PubMed]
50. Li, Y.; Li, M.; Rice, M.; Zhang, H.; Sha, D.; Li, M.; Su, Y.; Yang, C. The Impact of Policy Measures on Human Mobility, COVID-19 Cases, and Mortality in the US: A Spatiotemporal Perspective. *Int. J. Environ. Res. Pub. Health* **2021**, *18*, 996. [CrossRef] [PubMed]
51. Sy, K.T.L.; White, L.F.; Nichols, B.E. Population density and basic reproductive number of COVID-19 across United States countries. *PLoS ONE* **2021**, *16*, e0249271. [CrossRef]

entropy MDPI

Article

Evolving Network Analysis of S&P500 Components: COVID-19 Influence of Cross-Correlation Network Structure

Janusz Miśkiewicz [1,2,*] and Dorota Bonarska-Kujawa [2]

1 Institute of Theoretical Physics, University of Wrocław, 50-137 Wroclaw, Poland
2 Physics and Biophysics Department, Wrocław University of Environmental and Life Sciences, 50-375 Wroclaw, Poland; dorota.bonarska-kujawa@upwr.edu.pl
* Correspondence:janusz.miskiewicz@upwr.edu.pl

Abstract: The economy is a system of complex interactions. The COVID-19 pandemic strongly influenced economies, particularly through introduced restrictions, which formed a completely new economic environment. The present work focuses on the changes induced by the COVID-19 epidemic on the correlation network structure. The analysis is performed on a representative set of USA companies—the S&P500 components. Four different network structures are constructed (strong, weak, typically, and significantly connected networks), and the rank entropy, cycle entropy, averaged clustering coefficient, and transitivity evolution are established and discussed. Based on the mentioned structural parameters, four different stages have been distinguished during the COVID-19-induced crisis. The proposed network properties and their applicability to a crisis-distinguishing problem are discussed. Moreover, the optimal time window problem is analysed.

Keywords: network analysis; structural entropy; time series analysis; COVID-19

Citation: Miśkiewicz, J.; Bonarska-Kujawa, D. Evolving network analysis of S&P500 components: COVID-19 Influence of Cross-Correlation Network Structure. *Entropy* **2022**, *24*, 21. https://doi.org/10.3390/e24010021

Academic Editors: Ryszard Kutner, Christophe Schinckus, H. Eugene Stanley and Philip Broadbridge

Received: 31 October 2021
Accepted: 19 December 2021
Published: 23 December 2021

1. Introduction

1.1. Literature Review

The crucial and most obvious features of economic systems are cooperation and interaction, e.g., supply chains, ownership dependencies, cooperation networks, and financial networks. The companies and their activities form complex networks of relationships and dependencies [1–3]. The idea that the economy can be considered as a complex system can be found at the end of the XX century in economic literature, e.g., [4]. The features of the network depend on various parameters, such as technology, law, regulations, culture, climate, weather, resources, and many other parameters. Among various aspects of economic activity, the financial market plays a special role. Besides fundraising for the business, they allow monitoring the condition of enterprises and even whole economic sectors. The stock market indices are regularly published and are considered to be not only a measure of the change of companies' group values, but also a test of the economy's status [5]. Therefore, within this study, the cross-correlations among stock time series of S&P 500 index components are analysed. This index was chosen because it is based on quotations of 500 shares of the largest stocks that trade on the New York Stock Exchange and NASDAQ. Considering the importance of those companies, as well as their variety and number, this set can be considered to be representative of the USA's economy and, therefore, can be the basis of the analysis of cross-correlations among them. Particularly, these stocks can be analysed from the point of view of cross-correlation network structures formed by those companies. The cross-correlation analysis consists of two steps: the distance (usually based on linear correlation [6]) matrix calculation and its analysis by constructing a network of random matrix approaches [7–9]. The alternative cross-correlation analysis should be here mentioned. Recently, detrended cross-correlation analysis based on detrended fluctuation analysis (DFA) [10] and their modifications have been very popular; see the recent review [11], or the power law classification scheme [12]. However, the most popular strategy is based on the network construction and its structure analysis [13,14]. The most popular choice is the minimum spanning tree (MST) analysis [3,9,15–21]. MST application results primarily from the portfolio optimisation problem [22], but is also due to the proper recovering of the industrial sectors [3,6]. Besides MST analysis, the second most commonly used group of methods are those which construct networks based on assumed threshold [23]. This approach is used also in this paper—the distance matrix is filtered assuming that nodes are connected when the distance fulfills a given condition. Within this analysis, the properties of the most typical, weak, strong or significant correlations are investigated. The network generation procedures are described in detail in Section 3. The main aim of the paper is the description of the structural changes observed during the COVID-19 pandemic and their comparison to the changes in cross-correlation network properties during other, recently observed crises.

Another important aspect is the influence of the external parameters on the network structure. It is a truism to say that the state of the market depends on the macroeconomic situation. Particularly, the reaction of the market to crashes is also widely discussed from the point of view of network structure, e.g., [9,24,25], or in the analysis of globalization processes [26–28]. However, the present pandemic situation should not be considered as the typical shock but rather a change of the "economic environment". The most important fact is that the pandemic was expected and induced serious changes in the economy. The government, due to the pandemic situation, introduced special rules lasting a relatively long time. The restrictions form special conditions which are expected to change the structure of cross-correlations among companies. The network structure of shares' cross-correlations is the subject of interest of various studies [9,21,29]. A natural extension of time series analysis through the network methods is evolving network analysis, since the economy time series are non-stationary, and thus, should be described by an evolving network rather than a static network. Evolving network theory was initially applied to systems naturally described by network structures, such as social networks [30], scientific collaboration networks [31] and economy time series [8,25,32,33], to mention a few examples.

1.2. COVID-19 History

Although the COVID-19 epidemic is a contemporary event, for the convenience of future readers, a short description of the epidemic key points in the USA is presented below.

December 2019 The first known cases have been identified in Wuhan, China.
January 2020 The epidemic spreads to other provinces of China.
February 2020 Italy is affected with a rapidly growing number of infected and fatal cases.
March 2020 The USA overtakes China and Italy with the highest number of confirmed cases in the world.

The present situation is the subject of various studies. More detailed history and discussion on the influence of pandemic on stock markets from the standard time series analysis point of view can be found at [34–38].

1.3. Paper Structure

The present paper is organised as follows. Section 2 describes the data analysed. The Section 3 defines the methods used: the statistical distance, the network construction algorithms and the network parameters (node entropy, cycle entropy, averaged clustering coefficient and transitivity) calculated and analysed in the study. Section 4 presents the obtained results, including the evolution of the node and cycle entropy and the averaged clustering coefficient and transitivity. It is worth stressing that the parameters introduced here (the node and the cycle entropy) are sensitive to economic crises. Moreover, the performed analysis shows that the structure of the defined networks changes significantly in the crisis. The main outcome of the paper is the analysis of a representative group of USA companies and the observation of their reaction to changing economic situations. A very promising outcome of the study is that four periods during the COVID-19 pandemic are distinguished, which shows that the reactions to various factors are different and that this analysis is capable of observing this.

2. Data

The study is based on the S&P 500 index components' daily time series in the interval from 2016.01.04 to 2021.03.26. There are a total of 1315 data points in each of the time series. The S&P500 index consists of the largest stocks of the New York Stock Exchange and NASDAQ (National Association of Securities Dealers Automated Quotations). Considering the importance of stock indices for the assessment of the state of the economy, we can conclude that the entities on which the S&P 500 index is based constitute a representative group that allows the observation of important processes taking place in the economy. Furthermore, this index is based on a broad range of companies of different sectors; therefore, it can be considered representative for the USA economy. The time interval is chosen such that it contains a period before the COVID-19 pandemic. It is worth noticing that the pandemic period is before the broad availability of vaccination, so the observed changes are the effects of the institutional response to the pandemic situation. The inclusion of different periods is particularly important because the algorithm of the study is new and has never been tested.

The time series were downloaded from the web page Available online: https://stooq.pl (accessed on 28 March 2021). Although the index is based on 500 quotes, after inspection of the data, 432 time series were chosen for the analysis due to missing data. The list of quotations used for the analysis is presented in Appendix A.

The time series were converted into logarithmic daily return time series (so-called log-returns) according to Equation (1),

$$LogR(A)_i = \log \frac{a_i}{a_{i-1}}, \tag{1}$$

where A represents the time series, and a_i represents the i-th element of the time series A.

The evolution of the mean value of stocks included in the study is presented in Figure 1. By analysing the evolution of the mean quote in Figure 1, it should be noticed that it has the form of a visible growing trend with periods of significant fluctuations. The fluctuations correspond to periods of rapid growth followed by a significant drop in value—price bubbles or crises resulting from external factors. The major fluctuations which should be pointed out are the beginning of 2016 (which is the result of two shocks, the Chinese stock market and USA stock market selloffs), the third quarter of 2017, the first quarter of 2018, the second half of 2018 (the cryptocurrency crash), minor fluctuations in the middle of 2019 and a dramatic drop at the beginning of 2020 related to the COVID-19 pandemic. In general, the fluctuations seen in the mean value of quotes are also present in the averaged log-returns evolution graph seen in Figure 1, particularly in the left and right plots, respectively. It is worth noting that the range of fluctuations at the beginning of 2020 was ≈ 0.2, while the others observed in the period 2016–2019 did not exceed 0.07; therefore, the fluctuations resulting from the pandemic shock were approximately 3 times bigger than the other shocks. Within the analysed period, COVID-19 is the dominating factor in financial markets.

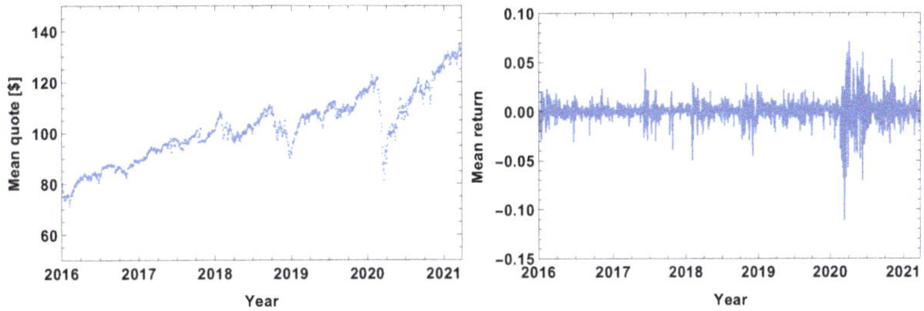

Figure 1. (**Left**) Evolution of the mean value of 432 quotes included in the study. (**Right**) Evolution of the averaged log-returns of 432 quotes included in the study.

3. Methods

Considering the fact that the evolution of the network structure is investigated in this study, the sliding window technique was applied. The essence of this method is that a fragment of a fixed length (the time window size) is selected from the time series. An analysis is performed for this fragment, and subsequently, the beginning and the end are shifted by one point and all calculations are repeated. The procedure is repeated until the time window reaches the end of the series.

The analysis carried out in this work can be divided into the following main steps:

1. Distance matrix calculations;
2. Network construction;
3. Network feature analysis.

3.1. Distance Matrix

The distance between the log-returns time series is calculated based on the ultrametric distance [26,27,39,40] as in Equation (2),

$$DM(A,B)_{t,T} = \sqrt{\frac{1}{2}(1-\rho(A,B)_{t,T})}, \tag{2}$$

where the correlation $\rho(A,B)_{t,T}$ is calculated using Pearson correlation coefficient, as in Equation (3):

$$\rho(A,B)_{t,T} = \frac{\langle AB\rangle_{t,T} - \langle A\rangle_{t,T}\langle B\rangle_{t,T}}{\sigma(A)_{t,T}\sigma(B)_{t,T}}. \tag{3}$$

where the indices $()_{t,T}$ denote the interval $(t, t+T)$. T stands for the time window size. The distance DM, when equal to zero, indicates a perfect linear correlation between time series, while a distance DM equal to one is obtained in the case of a lack of linear correlation (which does not mean that the time series are not correlated by other functions [39]).

In the literature, there is an alternative formulation of Equation (2), the ultrametric distance, which utilizes different normalization techniques [20]. Of course, the normalization does not affect the conclusions. The ultrametric distance DM is calculated for all possible pairs of time series, and the results are presented in the form of the distance matrix. The distance matrix DM is symmetrical due to the definition of the ultrametric distance Equation (2).

3.2. Network Construction

Considering the fact that each distance matrix contains $\frac{(n(n-1)}{2}$ different elements, here it gives 93096 different numbers. The analysis of the distance matrix requires the construction of higher-order structure—networks. Although in the literature the minimum spanning tree (MST) is one of the most popular structures [6,16,18,19,41,42], it imposes a very strong bias on the generated network. For example, due to the imposed tree structure, it is impossible to observe cliques, which are quite important elements of economic relationship analysis. In the case of MST analysis, with some additional effort, it is also possible to distinguish clusters [16], but such analysis is not straightforward due to the tree structure. MST analysis often distinguishes one prominent node, eg. [16,42], but in different network structures, the node could be a member of a clique and such a conclusion of its special role would be not possible.

Therefore, in this paper, the threshold method is used. The distances are categorised into defined groups, and, in each case, the network is constructed based on the appropriately filtered distance matrix.

Distance categorisation:

- Strongly connected time series—the companies are connected when the distance is shorter than the first quartile of the distances in the analysed distance matrix, so the network is built on a set of the 25% shortest links;

- Weakly connected time series—the companies are connected when the distance is longer than the third quartile of the distance in the analysed distance matrix, so the network is built on a set of the 25% longest links;
- The most typical connections—the companies are connected when the distance between them is longer than the first quartile and shorter than the third quartile of the distances in the analysed distance matrix, so the network is built on this set of 50% of the links;
- Significantly connected time series—the companies are connected in the network when the distance between them is shorter than the median of the distances in the analysed distance matrix, so the network is built on a set of 50% of the links.

The examples of the network generated in the study are presented in the Appendix B. Due to the huge number of graphs generated in the analysis (the time series length diminishes by the time window size) and the size of the networks, only a few examples are presented focusing on the state before the COVID-19 pandemic (July/August 2019) and two examples during the pandemic (March 2020 and August/September 2020).

On the other hand, the MST analysis allows the dominating node to be distinguished, usually with the highest number of links, eg. GE in [16]. However, this result partially depends on the imposed tree structure. In the threshold method, such situations are less probable, and a very high number of companies have a high number of links, so such prominent nodes are not observed.

3.3. *Network Analysis*

The last step of the analysis is the network parameter calculations. Considering the fact that, in the study, more than a thousand networks are constructed (due to the sliding window technique) and each network consists of 432 nodes, the direct analysis is tremendous. On the other hand, the general state of the system can be characterised by calculating appropriately chosen parameters.

The study aims to observe changes in the structure of the network of correlations. In the case of economic systems, some structures are of special interest. Usually one of the very first issues analysed is the leadership, or the presence of dominating companies, which are network hubs. The second most important structures are clusters that correspond to strongly cooperating companies or sets with strong mutual relationships, e.g., belonging to the same highly specialised sector, with the same ownership or sharing another common factor. The question of the presence of dominating companies is answered by the rank node analysis, which ranks nodes with respect to the number of links. It was shown in [22,32,42] that during crises, the dominating structure is a star-like network with a well-defined centre. On the other hand, in independently developing companies, one can expect that the statistical distances among time series would be similar (with some fluctuations). Moreover, the most interesting aspects, from the point of view of questions raised, are the changes in the network structure. Thus, the measure which properly exposes such structures and their changes is Shannon entropy. Therefore, the rank node distribution is characterised by information entropy; here it will be called **rank node entropy** and defined by Equation (4),

$$SN = -\sum_{i \in L} p_i \ln p_i \tag{4}$$

where L represents the list of all observed ranks, and p_i represents the probability of the i-th rank node.

The second feature investigated is the formation of particular structures, specifically triangles and cycles. The triangles expose the companies forming closely interacting groups; analogously, cycles are the groups with significant relationships (a chain of dependence). These two parameters are analysed by the calculation of transitivity and cycle entropy. The transitivity is defined as the fraction of all possible triangles in the graph.

$$T = 3\frac{\#triangles}{\#triads} \tag{5}$$

where *triad* indicates two edges with a shared vertex. **The cycle entropy** is defined as the information entropy of the cycle length distribution:

$$SC = -\sum_{i \in C} p_i \ln p_i \tag{6}$$

where C indicates the list of all observed cycle lengths, and p_i represents the probability of observing a cycle of the length i.

The last analysed network parameter is the clustering coefficient, which is the standard characteristic of the link density. Here, the averaged clustering coefficient is used, which is defined by Equation (7):

$$C = \frac{1}{n}\sum_{v \in G} c_v, \quad c_v = \frac{2T(v)}{deg(v)(deg(v)-1)} \tag{7}$$

where $T(v)$ is the number of triangles through node v, and $deg(v)$ represents the degree of node v.

The last element of the analysis procedure to define is the time window length. Considering the analysis of daily time series, three time window lengths have been chosen: 5 days, 20 days, and 60 days, which correspond to the week, month, and quarter periods, respectively.

A summary of the analysis algorithm is as follows:

1. Choose the representative set of companies (shares);
2. Verify the integrity of the time series and their length (should be identical);
3. Normalise the time series by converting them to the daily log-return time series;
4. Choose the time window size;
5. For each of the time series, starting at the beginning, take the interval of the time window length and calculate the time series correlation (distance) matrix;
6. Based on the correlation matrix, generate the network. Here, four possible strategies are considered: (i) strongly, (ii) weakly, (iii) most typical, (iv) significantly connected networks, so the following steps should be repeated for each network type;
7. Calculate the network's characteristics: rank entropy, cycle entropy, averaged clustering coefficient and transitivity;
8. Move the starting point by one point and repeat steps 5-8. Continue until the end of the time series length is reached.

Finally, the time evolution of the network characteristics is received and discussed.

4. Results

4.1. Week Size Time Window, $T = 5d$

The analysis begins with the shortest time window $T = 5d$. The evolutions of the strongly, weakly, most typically, and significantly connected network properties are presented in Figures 2–5. As was mentioned in Section 3.2, each of the structures is focused on different features of the system. The first network presented, which is of strongly connected companies, is built under the assumption that the companies are connected when the distance between them is shorter than the first quartile of the distances in the given distance matrix. The evolution of rank entropy, cycle entropy, averaged clustering coefficient, and the transitivity in the considered period are presented in Figure 2.

In the rank entropy evolution chart, one can distinguish the maximum state, which corresponds to the periods of "normal" trading, i.e., beyond crisis periods. Furthermore, similar observations can be made for the other rank entropy graphs independent of the time window size and the network structure considered; in all of them, stable maximum entropy is observed, suggesting that there exists a stable level of the rank distribution entropy. Besides the presence of the maximum rank entropy state, there are periods when the rank entropy is clearly smaller. At the beginning of 2016, which corresponds to the first fluctuation period in the considered interval, the rank entropy decreases from the value of 5 to 3. An analogous change is observed in the middle of 2017, the first quarter of 2018, and the crisis moment of 2019. The lowest values of rank node entropy are observed in the second quarter of 2020, which correlates with the development of the COVID-19 pandemic. Moreover, the evolution of the rank entropy reflects different stages of the reaction to the pandemic. At the end of 2019, it was obvious that the pandemic would spread all over the world, so in the beginning of January 2020 the first decrease in the rank entropy is observed as the result of news. Afterwords, the network structure began returning to the typical state. However, when the first cases were observed in USA, and consequently, the number of hospitalised persons began rapidly growing, the rank entropy decreased, reaching the lowest observed value and indicating significant changes in the network structure of the strongly connected companies.

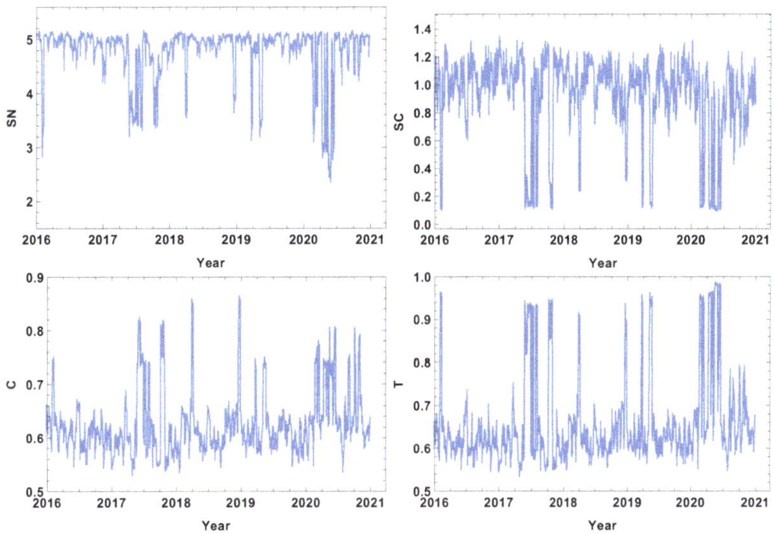

Figure 2. Network feature evolution in the case of the strongly connected companies, i.e., the distance between them is shorter than the first quartile of distances in the analysed distance matrix. The time window $T = 5d$. The top left figure represents rank entropy, the top right represents cycle entropy, the bottom left represents the averaged clustering coefficient, and the bottom right represents transitivity.

The cycle entropy is focused on the cycle distribution length. In contrast to the rank entropy, the week time window analysis of the cycle entropy (Figure 3) does not show a clear stable maximum state. The cycle entropy in the period between crushes takes a value in the interval between 0.7 and 1.3, but during crises, the cycle entropy decreases to the value 0.1 (high fluctuation periods). It seems that during crises, most of the cycles are broken and the cycle entropy takes a very low value. In contrast to the rank entropy, the cycle entropy does not allow the severity of crises to be measured, since the same level is obtained for the crises at the beginning of 2016, the middle of 2017, the first and second quarter of 2019 and the COVID-19-induced crisis in 2020. Therefore, the cycle entropy achieves the lowest observed values relatively faster.

Besides the new measures introduced here (rank entropy and cycle entropy), the standard network parameters—averaged clustering coefficient and transitivity—are also sensitive to the COVID-19-induced crisis. Particularly, the transitivity obtains a very high value at the beginning of 2020. However, similarly to the cycle entropy, the transitivity does not allow the severity of the crises to be measured. The COVID-19 crash is characterised by similar values as the other crashes. The averaged clustering coefficient seems to be a less useful parameter in measuring crises strength because the highest values are observed in the first quarters of 2018 and 2019. When analysing the results of the network of the strongly correlated companies, it should be taken into account that this network is based on 25% of the most correlated time series, so this assumption may induce a dichotomous state of the network structure: crises and not crises.

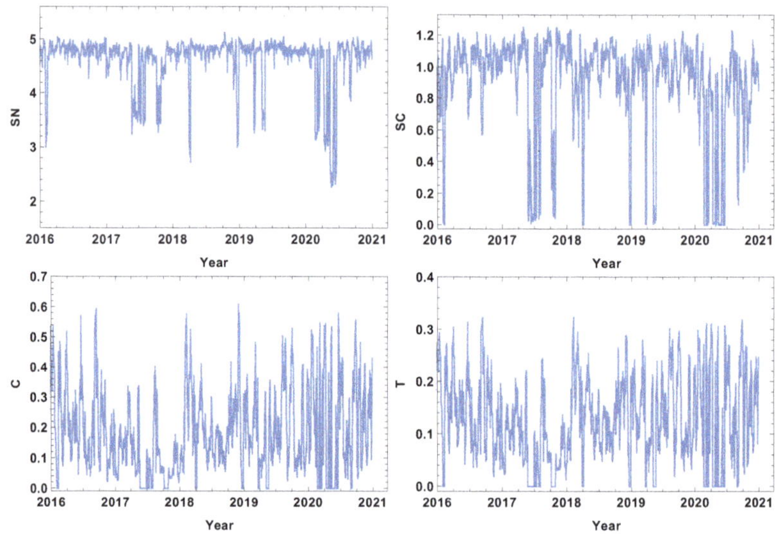

Figure 3. Network feature evolution in the case of the weakly connected companies, i.e., the distance between them is longer than the third quartile of the distances in the analysed distance matrix. The time window $T = 5d$. The top left figure represents rank entropy, the top right represents cycle entropy, the bottom left represents the averaged clustering coefficient, and the bottom right represents transitivity.

The evolution of the weakly connected companies' network parameters seems to be complementary to the strongly connected time series. In this case, the analysis is focused on the structure of the networks connected by a long statistical distance. The results for the week time window are presented in Figure 3. Analogously to the case of strongly connected companies, the rank entropy graph has a clear maximum value (≈5). The crash periods are correlated with a significant decrease in the rank entropy, and, similarly to the already discussed case, the lowest rank entropy is observed in the first and second quarter of 2020, which corresponds with pandemic development in the USA. The decrease in the rank entropy indicates the increase in differences in rank distributions, which is a quite natural process—during crises, a star-like network is dominating [22,32,41,42]. The two other coefficients analysed, i.e., the averaged clustering coefficient and transitivity, present a very noisy graph. In this very short time window, the fluctuations are dominating and do not allow any particular network features to be distinguished.

The results of the most typically connected companies are presented in Figure 4. In this case, the analysis is focused on the companies among which the distance is within the first and third quartile. Therefore, the network excludes extreme cases but shows the structure of typical connections among companies. The rank entropy graph, similar to the already discussed cases, is sensitive to significant price fluctuations. During crashes, the rank entropy value visibly decreases. On the other hand, the network during normal trading is characterized by rank entropy $SN \approx 5$. In contrast to the strongly and weakly correlated networks, after the large decrease in rank entropy related to the COVID-19 pandemic, the system does not return to the standard state at the level $SN \approx 5$, but is in the interval $SN \in (4, 4.5)$. Of course, the initial shock was the strongest one, particularly as it was followed by strong restrictions. However, in the second half of 2020, the situation did not return to the normal situation as the economy was still affected by the pandemic, and the rank entropy of the typically connected company network seems to be sensitive to this fact. The cycle entropy for the typically correlated company network, similar to the strongly and weakly correlated company networks, attains its minimal value at the time of large fluctuation periods $SC \approx 0.1$, indicating that crashes very strongly affect the cycle length structure. Averaged clustering coefficient evolution, in contrast to the weakly collected time series network, is sensitive to crises, having local maxima at the stock market crises. The essential feature of this result is that the highest local maximum is correlated with the COVID-19 period, when the largest fluctuation appeared. The latter observation is important because it shows that this network structure is sensitive not only to the presence of fluctuations but also to its magnitude. In the case of the last parameter, transitivity, the graph evolution seen in Figure 4 shows that, during crises, the structure of the network changes significantly (clearly distinguished local maxima), achieving transitivity twice as big compared to the standard fluctuation level.

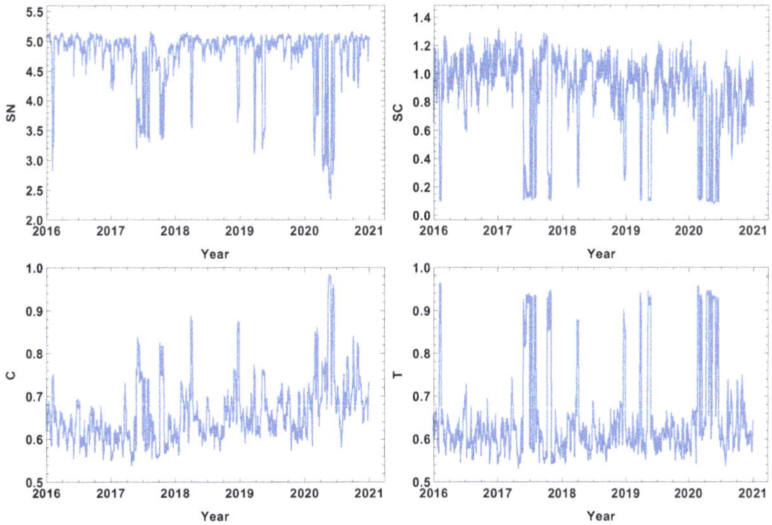

Figure 4. Network feature evolution in the case of the most typically connected companies, i.e., the distance between them is within the interval between the first and the third quartile of the distances in the analysed distance matrix. The time window $T = 5d$. The top left figure represents the rank entropy, the top right represents cycle entropy, the bottom left represents the averaged clustering coefficient, and the bottom right represents transitivity.

The evolution of the network parameters of significantly connected companies is presented in Figure 5. In this case, the analysis concentrates on highly correlated companies, including those which are the most correlated. This is also the network based on half of the correlations. The features of the significantly connected network are slightly surprising, since the local minima related to the COVID-19 pandemic period are not the deepest minima. Beginning the analysis of this type of network with the rank entropy graph, it is seen that the smallest values of SN are observed in the 2017 crisis. The difference between the local minima in 2017 and 2020 is ≈ 0.6, which is not very high when comparing it to the maximum level $SN \approx 5$. A similar observation is made on the cycle entropy graph, in which SC fluctuates significantly even beyond the periods of crises. This indicates that the time window length seems to be too short to smooth the system fluctuations. The averaged clustering coefficient graph of significantly connected companies at the periods of crises achieves a value close to zero, which means that during a crisis, the cliques are almost whipped out of the network. The transitivity of the significantly connected time series supports the observations made on the averaged clustering coefficient graph, while the minima correspond to crisis periods. However, the changes in the network structure are so significant that the transitivity nearly reaches zero, indicating that triangles of correlated companies are very rare during crises.

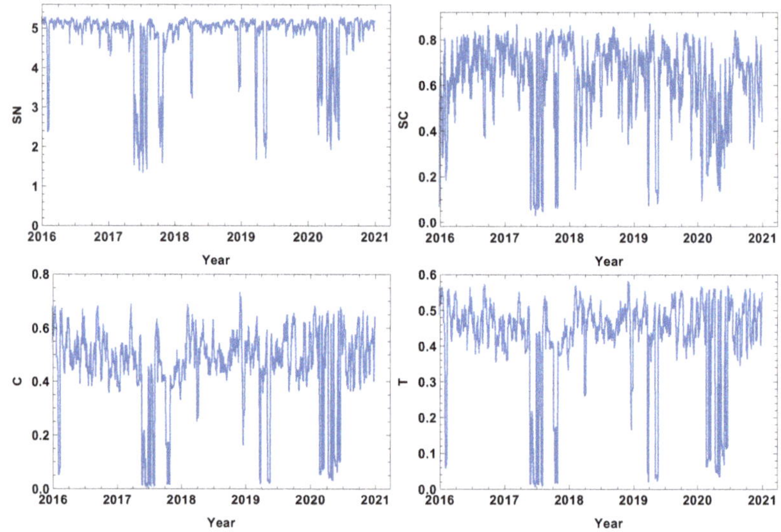

Figure 5. Network feature evolution in the case of the significantly connected companies, i.e., the distance between them is shorter than the median of the distances in the analysed distance matrix. The time window $T = 5d$. The top left figure represents rank entropy, the top right represents cycle entropy, the bottom left represents the averaged clustering coefficient, and the bottom right represents transitivity.

To summarise the week time window analysis, the COVID-19 effect is observed at the beginning of 2020, as seen in the decrease in the rank entropy value. It is worth noting that the reaction of the rank entropy evolution to the price fluctuation does not depend on the type of network considered. Of course, the obtained results differ in details, but for all considered cases, the rank entropy graph has a "maximum level" describing normal stock exchange market activity and significantly decreases with large fluctuations, indicating changes in the rank node distribution.

4.2. Month Size Time Window, $T = 20d$

The next considered time window size was the month size time window ($T = 20d$). The results are presented in Figures 6–9. The first and most visible observation of the month window size analysis is the reduction of local fluctuations compared to the results obtained for the week time window size, as seen in Figures 3–5.

The network features of the strongly correlated companies are presented in Figure 6. The most obvious observation is the decrease in the noise level compared to the week time window size analysis. The rank node entropy, as in the previous case, has a clear maximum level ($SN \approx 5$), which corresponds to the period of trading without significant price fluctuations. However, the intervals of decrease are not in the form of rapid and large oscillations, but have a shape of intervals, indicating that the change of structure was observed in the whole crisis period. The significant decrease in the rank entropy by 2.5 or more indicates a rapid and serious change of network structure. At the crash, the network reconstructs immediately to a new state characterised by much lower rank entropy. It can be seen that, for the month resolution analysis of the group of strongly correlated companies, the COVID-19-induced crises had three stages in which the rank entropy decreased abruptly. The three other network parameters, i.e., cycle entropy, averaged clustering coefficient, and transitivity, also decrease at this crisis; however, the value does not depend on the crisis severity, but they achieve the lowest possible value equal to zero indicating that during a crisis the higher-order structures, such as loops, triangles or clusters, do not exist in the network of the most-correlated companies.

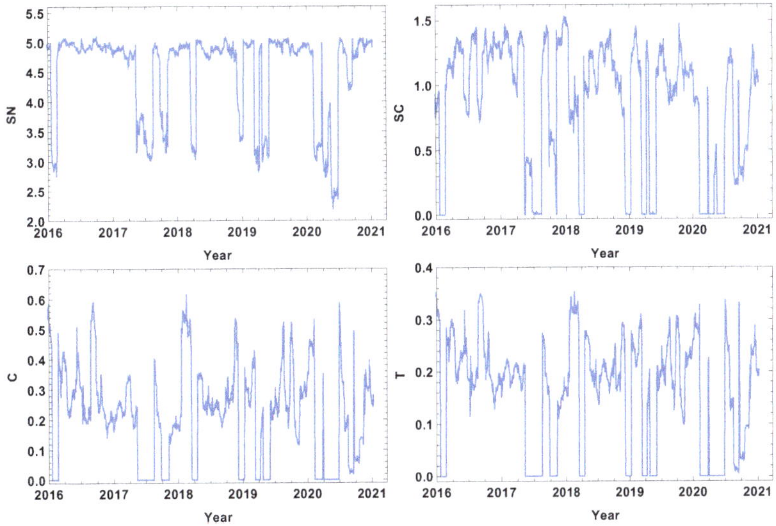

Figure 6. Network feature evolution in the case of the strongly connected companies, i.e., the distance between them is shorter than first quartile of distances in the analysed distance matrix. The time window $T = 20d$. The top left figure represents rank entropy, the top right represents cycle entropy, the bottom left represents the averaged clustering coefficient, and the bottom right represents transitivity.

The weakly connected time series network features are presented in Figure 7. Similarly to the strongly correlated network case, the rank entropy evolution for the network of the weakly correlated companies allows the crises intervals to be distinguished. The lowest rank entropy is observed during the COVID-19 crisis. The cycle entropy graph in the case of weakly connected companies also has a visible drop of cycle entropy value, but the lowest observed value is $SC \approx 0.1$, which indicates that only a few types of cycles are present in the network. In view of the already discussed cases, the averaged clustering coefficient evolution is very interesting. The weakly correlated companies' network structure follows a different pattern than the already discussed network structures; during crises, the averaged clustering coefficient takes a very high value, indicating the strong clustering of companies. The same observation can be made by the analysis of the transitivity evolution graph; the maximum value is obtained at the crises periods, so weakly correlated companies form a high number of triangles. This finding correlates with the results of the network of the strongly correlated companies, where, during crises, the complex structures disappear; they emerge in the weakly correlated network.

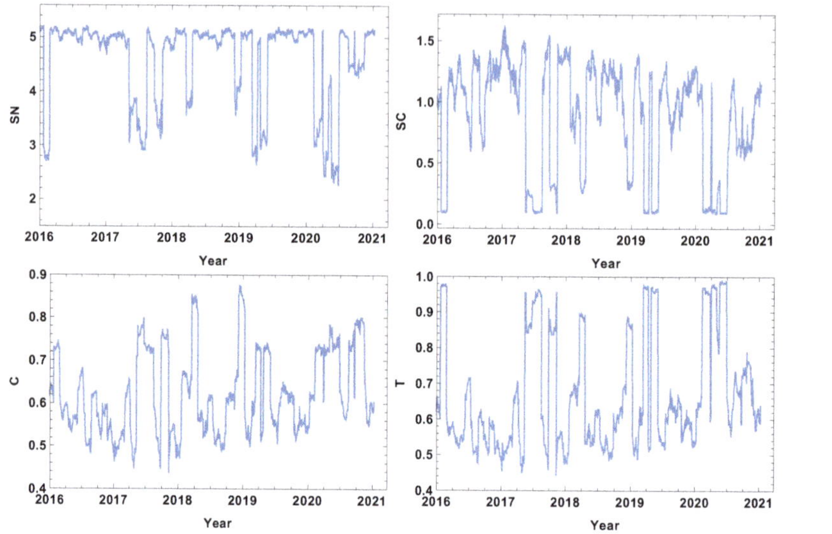

Figure 7. Network feature evolution in the case of the weakly connected companies, i.e., the distance between them is longer than the third quartile of the distances in the analysed distance matrix. The time window $T = 20d$. The top left figure represents rank entropy, the top right represents cycle entropy, the bottom left represents the averaged clustering coefficient, and the bottom right represents transitivity.

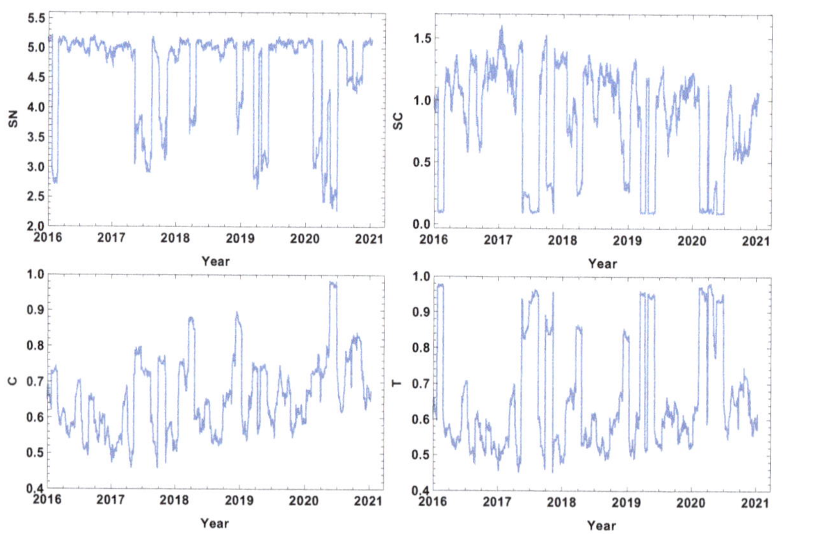

Figure 8. Network feature evolution in the case of the most typically connected companies, i.e., the distance between them is within the interval between the first and the third quartile of the distances in the analysed distance matrix. The time window $T = 20d$. The top left figure represents rank entropy, the top right represents cycle entropy, the bottom left represents the averaged clustering coefficient, and the bottom right represents transitivity.

The results of the analysis of the network of typically connected companies are presented in Figure 8. The rank entropy graph for the network of typically connected companies is similar to the already discussed cases, which properly indicates the crisis periods when the rank entropy decreases significantly. During the COVID-19-induced crisis, three stages of rank entropy value can be distinguished. However, they are not so separated as in the case of the strongly connected company network Figure 6. The cycle entropy graph supports the previous findings: during crises, cycles almost disappear from the network. On the other hand, the clustering coefficient is rather high in the crisis period. The highest averaged clustering coefficient is observed during the second quarter of 2020 (C is close to 1). An analogous observation can be made on the transitivity when the number of triangles is very high during crises. An interesting observation is made while comparing the averaged clustering graph with the transitivity plot. In the first half of 2020, the transitivity achieved a high value (the first and second quarter of 2020), while the averaged clustering coefficient takes a value close to one during the second quarter of 2020, at the time when the pandemic became very serious and stronger restrictions were imposed. This observation is meaningful since the transitivity is very sensitive to price fluctuations and immediately goes to a value close to one, while the clustering coefficient is more robust, allowing us to not only observe the fact that the network structure has changed but also relate the changes to the crisis severity.

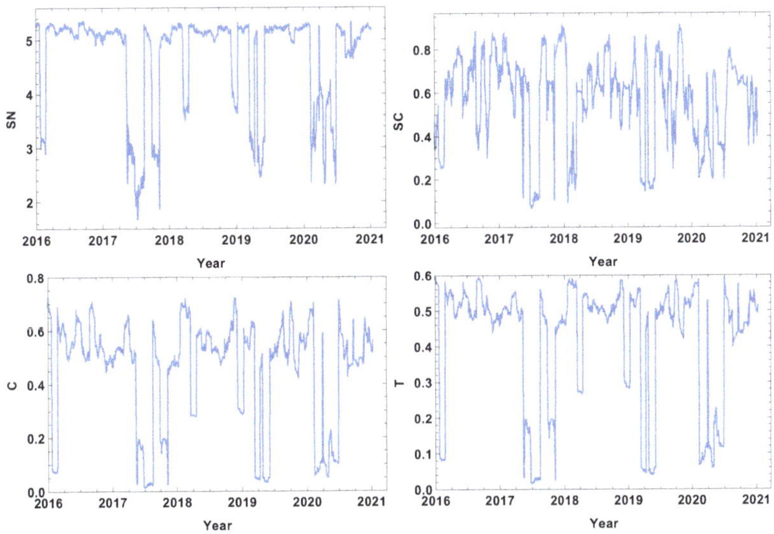

Figure 9. Network feature evolution in the case of the significantly connected companies, i.e., the companies are connected when the distance between them is shorter than the median of the distances in the analysed distance matrix. The time window $T = 20d$. The top left figure represents rank entropy, the top right represents cycle entropy, the bottom left represents the averaged clustering coefficient, and the bottom right represents transitivity.

The network results of the significantly connected companies are presented in Figure 9. Here, the companies are connected in the network when the distance between them is smaller than the median of the distance matrix, so the investigated group consists of relatively strongly correlated companies, and the analysis is based on half of the possible links. Comparing the results obtained in $T = 20d$ and in $T = 5d$, it can be observed that the fluctuation level is significantly reduced, but the main findings are also valid for this analysis. The first important observation is that, in the case of the network of significantly connected companies, the smallest value of rank entropy occurs in the middle of 2017. This means that, from the point of view of the strongly connected companies, the unexpected market fluctuations are much worse than even severe but predicted; the rank entropy decrease during the COVID-19 crisis is not so strong. The cycle entropy graph is still difficult to interpret because it is hard to indicate a clear relationship between cycle entropy value evolution and the crash history. It is probable that the time window is too short and the fluctuations of the system are still hiding the crisis influence. At the graph of the average clustering coefficient evolution, two states can be distinguished—the normal trading period when $C \in (0.45, 0, 75)$, and the states of significantly lower value $C \in (0, 0, 4)$, which corresponds to the crash interval. The market fluctuations observed at the beginning of 2016, the middle of 2017, the first two quarters of 2019, and COVID-19 had similar effects on the averaged clustering coefficient, which became nearly zero at those periods. The main features observed in the averaged clustering graph Figure 9 are also present in the transitivity graph in Figure 9, with the difference that the numerical values characterising normal trading periods and crashes are slightly different. The transitivity of the network of significantly connected companies for intervals without spectacular events are in the range $T \in (0.4, 0.6)$, while during crashes, these values decrease as far as zero. Similarly to the rank entropy and the averaged clustering coefficient 9, the transitivity in 2017 went even lower than during the COVID-19 crisis, indicating slightly weaker changes despite much more significant price fluctuations. Once again this raises the question about the importance of the shock expectations. The correlated companies could react similarly, so the network structure is not completely changed.

4.3. *Quarter Size Time Window, $T = 60d$*

The results of the analysis for the quarter time window size $T = 60d$ are presented in Figures 10–13. The quarter time window size is the longest time window considered in this study. As one can expect, the extension of the time window size filters the higher frequency fluctuations, allowing only the long-lasting correlations to remain, since each of the points in the graph is based on the cross-correlation distance calculated by the interval of 60 consecutive log-returns.

The network features of the strongly correlated companies are presented in Figure 10. The extension of the time window size resulted in clarifying the main features of the rank entropy graph, including the presence of a base level which describes the normal trading periods when the rank entropy is $SN \approx 5$. The crash periods demonstrate a significantly lower value of rank entropy of $SN \approx 3$. Considering the aim of the study, the most interesting evolution is the evolution of network parameters in 2020. The rank entropy plot shows that the network structure switched between three stages. The first stage was observed in the first quarter of 2020, when the pandemic was expected in the USA, and, due to the situation in China, the supply chain was affected. In the second quarter of 2020, the rank entropy increased to $SN \approx 4$, so the fact that the pandemic was expected resulted in some increase of the rank structure complexity when it came to the USA. However, as the situation developed and became severe in the first half of 2020, the rank entropy dropped up to $SN \approx$ 2–2.4. The interesting finding is that the system adapted to the present situation, and at the end of 2020, the rank entropy returned to the level of the normal trading time $SN \approx 5$. The cycle entropy graph, in contrast to the rank entropy plot, does not have a stable value for normal trading, but during high fluctuation periods, the cycle entropy drops as far as zero. Very similar observations can be made on the averaged clustering coefficient and transitivity, suggesting that in the group of the most correlated companies, the higher-order structures are not present during crises. This is a new observation not previously discussed in the literature. By further analysing the cycle entropy graph, Figure 10, during the COVID-19 pandemic, it can be observed that after the initial drop of the cycle entropy in the first quarter of 2020, in the second quarter, the cycle entropy increases up to a value of $SC \approx 0.5$, which corresponds to the temporary increase of the rank entropy in the same period. This supports the expressed idea that, despite objective difficulties, the system is trying to adapt to the new situations. After the short decrease in the middle of 2020, the cycle entropy increases, reaching a value at the beginning of 2021 of $SN \approx 1.3$. The averaged clustering plot and transitivity graph follow a similar sequence, differing only in minor details during the periods beyond crises, but during the crises, values of both parameters decrease up to zero indicating disappearing complex structures among the most correlated companies. It should be stressed that the latter finding is observed in a very long time window, which means that during crises, long correlations do not form a complex structure.

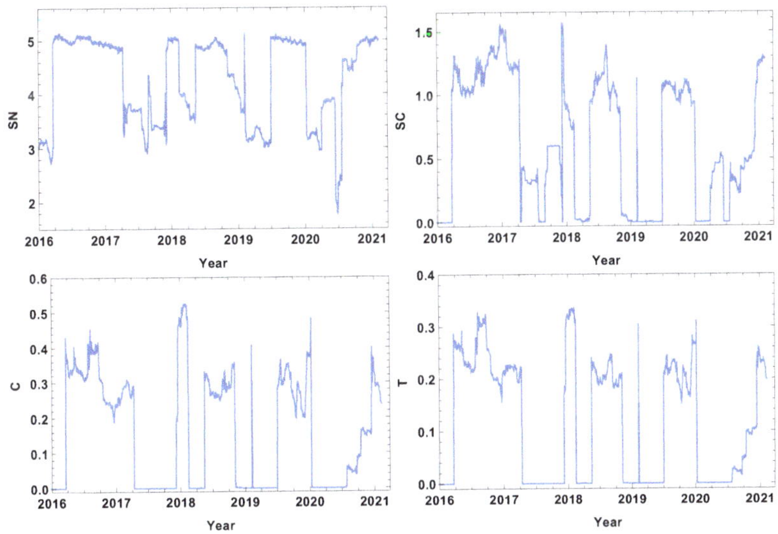

Figure 10. Network feature evolution in the case of the strongly connected companies, i.e., the distance between them is shorter than first quartile of the distances in the analysed distance matrix. The time window $T = 60d$. The top left figure represents rank entropy, the top right represents cycle entropy, the bottom left represents the averaged clustering coefficient, and the bottom right represents transitivity.

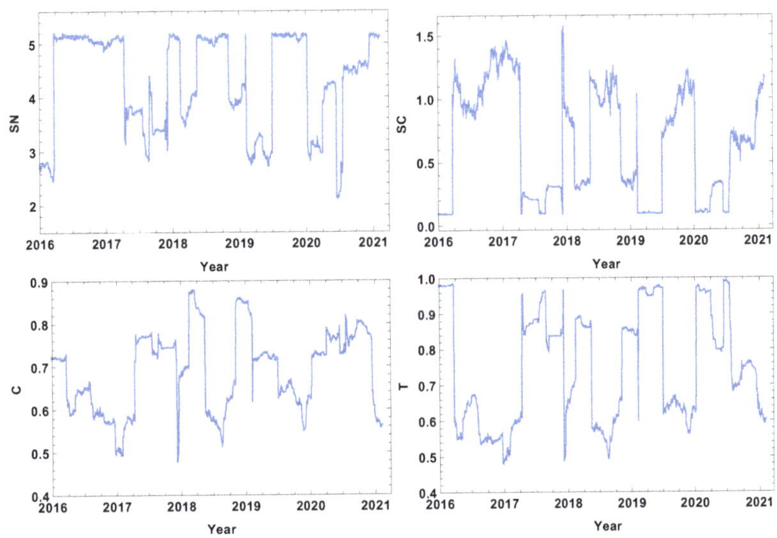

Figure 11. Network feature evolution in the case of the weakly connected companies, i.e., the distance between them is longer than the third quartile of the distances in the analysed distance matrix. The time window $T = 60d$. The top left figure represents rank entropy, the top right represents cycle entropy, the bottom left represents the averaged clustering coefficient, and the bottom right represents transitivity.

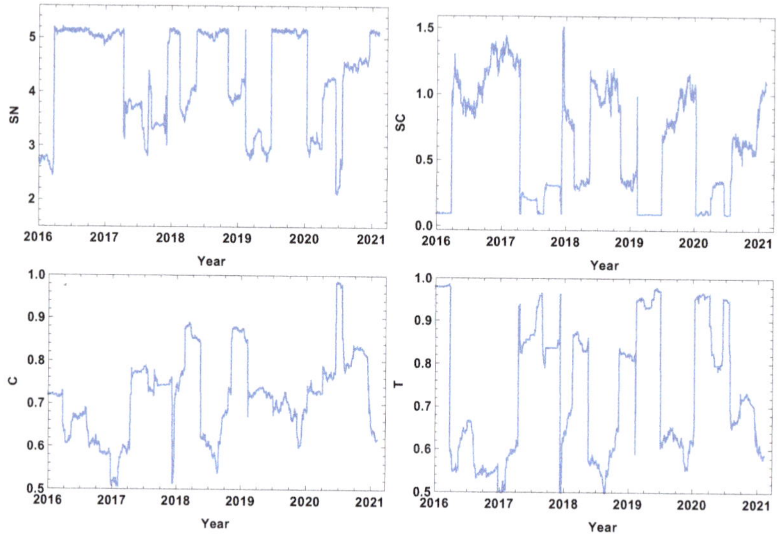

Figure 12. Network feature evolution in the case of the most typically connected companies, i.e., the distance between them is within the interval between the first and the third quartile of the distances in the analysed distance matrix. The time window $T = 60d$. The top left figure represents rank entropy, the top right represents cycle entropy, the bottom left represents the averaged clustering coefficient, and the bottom right represents transitivity.

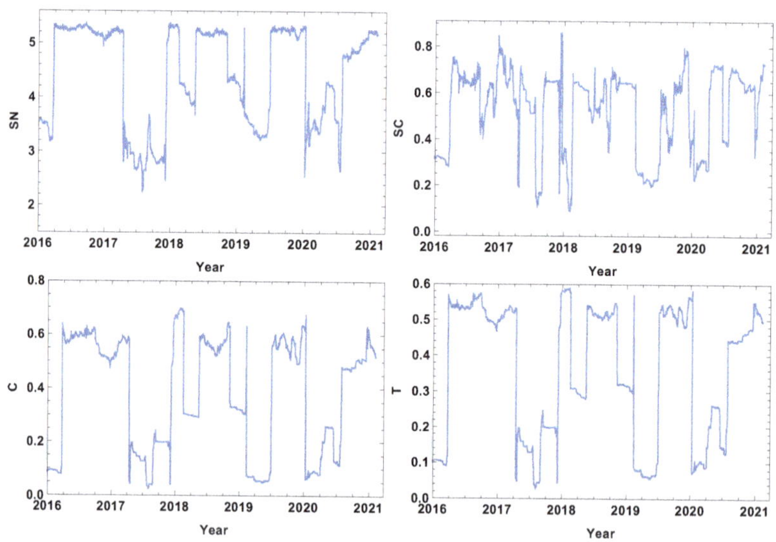

Figure 13. Network feature evolution in the case of the significantly connected companies, i.e., the distance is shorter than the median of the distances in the analysed distance matrix. The time window $T = 60d$. $SN \approx 0.9$.

The evolution of the network parameters of the weakly correlated companies is presented in Figure 11. The rank entropy evolution can be clearly divided into two categories: the crisis periods and the time beyond. During crises, the rank entropy decreases. The lowest value is observed at the beginning of 2016, when $SN \approx 2.5$, and in the middle of 2020, when $SN \approx 2$. Similar to the previous case discussed, the COVID-19 crash period can be divided into three stages: the change of SN from 5 to 3, then, in the second quarter, the rise to 4.1 and the significant drop in the middle of 2020 to 2. Afterwards, the rank entropy returns to its typical value. This observation supports the hypothesis that the network structure responded to the information of the incoming pandemic, then adopted to the situation and strongly reacted to the quick increase of affected cases and restrictions. The cycle entropy graph also supports the hypothesis of three COVID-19 stages. However, in SC, the first and the second stage is characterised by a very low value of cycle entropy $SC \approx 0.1$, with a short rise in SC during the second quarter of 2020 to the value $SC \approx 0.3$. The averaged clustering coefficient and transitivity show that, during crises, for the network of the weakly connected companies, the dominating structures are densely connected groups, since both parameters reach values close to the maximum, particularly for the transitivity where $T \approx 1$ is observed for all crashes in the analysed intervals. The long time window reduces the fluctuations present in shorter time windows ($T = 5d, T = 20d$) such that it is possible to analyse the evolution of the weakly correlated network structure.

The most typically connected time series network features are presented in Figure 12. The rank entropy graph shows that the most typically connected companies beyond the crises periods form networks, the rank entropy of which is $SN \approx 5$. During crises, the rank entropy decreases significantly, e.g., in the crisis of 2016 $SN \approx 2.4$, and during the COVID-19 crisis, $SN \approx 2$. In the evolution of SN in 2020, four stages can be distinguished. The first stage was in the first quarter of 2020, where $SN \approx 3$ as the stock market was scared by news from China. The second was in the second quarter of 2020, where $SN \approx 4.2$ when COVID-19 entered the USA. The third was observed in the middle of 2020, when $SN \approx 2$ when the situation worsened significantly and serious restrictions were imposed. The fourth stage lasted through the second half of 2020 when $SN \approx 4.6$. The cycle entropy graph in Figure 12 differs from the rank entropy in that, during crisis, cycle entropy is likely to reach a very low value of $SC \approx 0.1$. However, similar to the rank entropy, the four stages of the COVID-19 crisis can be distinguished by the difference that, during the first quarter in 2020 and in the third stage in half of 2020, the cycle entropy went to the same value $SC \approx 0.1$. Another important feature of the cycle entropy graph is that, in contrast to the rank entropy plot, the cycle entropy has a nontrivial evolution between crashes, showing the increasing complexity of the typically connected network. The two other parameters (averaged cluster coefficient and transitivity) show that, during crises, the most typically connected companies form a cluster or a structure close to it. Particularly high values of the averaged clustering coefficient of $SN \approx 0.9$ were observed in the middle of 2020 when the pandemic situation was significantly worsened. The transitivity graph in Figure 12 supports the observation of four stages in the COVID-19 crisis in 2020 made in the analysis of the rank entropy and cycle entropy of the most typically connected companies.

The analysis of the significantly connected companies over the quarter time window size of $T = 60d$ are presented in Figure 13. The rank entropy evolution for the network of significantly connected companies can be divided into two stages: the crisis period, when the rank entropy takes low values of $SN \in (2, 4.6)$, and the intervals beyond crises, when SN is relatively stable, such as $SN \in (5, 5.4)$. Although in 2020 the decrease resulting from the COVID-19 crisis is clearly visible, the stages observed in the cases of the networks of the strongly, weakly, and typically connected companies cannot be distinguished. The cycle entropy plot in Figure 13 shows that the cycle length distribution entropy is rather difficult to interpret in view of crisis presence and its severity. The opposite observation can be made for the averaged clustering coefficient and transitivity graphs. In these two graphs, Figure 13, the crises periods are characterised by a clear decrease of those parameters. During the COVID-19 crisis, four stages can be distinguished, similar to the networks of the strongly, weakly, and typically connected companies. The 2020 crisis began with a decrease in the averaged clustering coefficient from $C \approx 0.65$ to $C \approx 0.04$, and this coefficient remained at this value through the first quartile of 2020. Afterwards, it increased to 0.24 and kept this value until the middle of 2020, when it decreased to 0.1. Then, in the middle of the third quarter, it increased to the value $C \approx 0.45$. Analogous evolution is observed in the transitivity graph Figure 13 with slightly different values but with identical periods.

5. Conclusions

The presented study analysed the impact of the pandemic on the structure of cross-correlation networks among the most important companies on S&P500 components. The stock market crashes strongly influence cross-correlation network structure. Four different networks have been introduced and investigated: strongly, weakly, most typically, and significantly connected companies. The first 2 networks are based on 25% of links while the 2 latter networks are constructed on 50% of links. In general, all constructed networks are sensitive to large price fluctuations. Of particular interest was the crisis induced by the COVID-19 pandemic, where four stages of the market reaction were distinguished. It is worth stressing that the observed changes in the network structure can be related to particular features of the 2020 situation. The essential result is that the discussed changes can be quantified by rank entropy and cycle entropy measures, as well as the standard network parameters, such as the averaged clustering coefficient and the transitivity. The networks of strongly connected companies react in a different way to the crisis than the networks of weakly connected companies. Besides the type of the network and its features, the optimal size of the time window to calculate cross-correlation has been investigated. The optimal window size is a month, $T = 20d$. In the analysis based on the $T = 20d$ cross-correlation time window, the fluctuations are suppressed such that important trends can be seen and discussed. On the other hand, in the analysis performed for the shortest time window (T = 5 days), the averaged clustering coefficient and the transitivity for the network of the significantly connected companies decreases in the crises, while for the networks of strongly and typically connected companies, these parameters visibly increase. This situation might be the effect of the short window time wherein the Pearson correlation coefficient is calculated on the five data point sets. This observation supports the conclusion that the optimal time window for the analysis of the daily time series returns is a month period.

The presented results show that the proposed network structures are capable of describing and measuring the changes resulting from crises on the stock markets. Moreover, the introduced parameters, the rank network entropy and the cycle entropy, are useful parameters in the analysis of structure changes and crises recognition. Particularly, the rank entropy, which is capable of quantitatively characterising network structure changes and those parameters, might be useful in crash analysis. On the other hand, the introduced network structures, which are composed of strongly, significantly, typically and weakly correlated companies, do not introduce as strong of constraints as the frequently used MST structures. For example, the GE company, which is the centre of the MST in [16], is one of the highly connected companies here, but is not so prominent as in the MST structure. The number of links of GE is comparable to the median level of the number of links for a given network type.

Besides the main results of the paper, it has been observed that the rank entropy is likely to change its value in a step-like function, showing that, according to the market situation, the network will change to some well-established structures. This is a very intriguing hypothesis which deserves further study.

Author Contributions: Conceptualization, J.M.; methodology, J.M.; software, J.M.; investigation, J.M.; data verification, J.M.; results discussion, J.M., D.B.-K.; writing—original draft preparation, J.M.; writing—review and editing, D.B.-K.; visualization, J.M. All authors have read and agreed to the published version of the manuscript.

Funding: This research received no external funding.

Data Availability Statement: The data used in this study were downloaded from the web page www.stoog.pl (accessed on 28 March 2021).

Conflicts of Interest: The authors declare no conflict of interest.

Appendix A. Companies List

The analysis of this paper is based on the following companies quotes (using the standard abbreviations):

A, AAL, AAP, AAPL, ABBV, ABC, ABT, ACN, ADBE, ADI, ADM, ADP, ADS, ADSK, AEE, AEP, AES, AFL, AIG, AIV, AIZ, AJG, AKAM, ALB, ALGN, ALK, ALL, ALLE, ALXN, AMAT, AMD, AME, AMG, AMGN, AMP, AMT, AMZN, ANSS, ANTM, AON, AOS, APA, APD, APH, APTV, ARE, ARNC, ATVI, AVB, AVGO, AVY, AWK, AXP, AYI, AZO, BA, BAC, BAX, BBY, BDX, BEN, BFb, BHF, BIIB, BK, BKNG, BKR, BLK, BLL, BMY, BPYU, BRKb, BSX, BWA, BXP, C, CAG, CAH, CAT, CB, CBOE, CBRE, CCI, CCL, CDNS, CERN, CF, CFG, CHD, CHRW, CHTR, CI, CINF, CL, CLX, CMA, CMCSA, CME, CMG, CMI, CMS, CNC, CNP, COF, COG, COO, COP, COST, COTY, CPB, CPRI, CRM, CSCO, CSX, CTAS, CTSH, CTXS, CVS, CVX, D, DAL, DD, DE, DFS, DG, DGX, DHI, DHR, DIS, DISCA, DISCK, DISH, DLR, DLTR, DOV, DRE, DRI, DTE, DUK, DVA, DVN, DXC, EA, EBAY, ECL, ED, EFX, EIX, EL, EMN, EMR, EOG, EQIX, EQR, EQT, ES, ESS, ETN, ETR, EW, EXC, EXPD, EXPE, EXR, F, FAST, FB, FBHS, FCX, FDX, FE, FFIV, FIS, FISV, FITB, FL, FLIR, FLR, FLS, FMC, FRT, FTI, FTV, GD, GE, GILD, GIS, GL, GLW, GM, GOOG, GOOGL, GPC, GPN, GPS, GRMN, GS, GT, GWW, HAL, HAS, HBAN, HBI, HCA, HD, HES, HIG, HLT, HOG, HOLX, HON, HP, HPE, HPQ, HRB, HRL, HSIC, HST, HSY, HUM, IBM, ICE, IDXX, IFF, ILMN, INCY, INFO, INTC, INTU, IP, IPG, IPGP, IR, IRM, ISRG, IT, ITW, IVZ, J, JBHT, JCI, JEF, JNJ, JNPR, JPM, JWN, K, KDP, KEY, KHC, KIM, KKR, KLAC, KMB, KMI, KMX, KO, KR, KSS, KSU, L, LB, LEG, LEN, LH, LHX, LKQ, LLY, LMT, LNC, LNT, LOW, LRCX, LUMN, LUV, LYB, M, MA, MAA, MAC, MAR, MAS, MAT, MCD, MCK, MCO, MDLZ, MDT, MET, MGM, MHK, MKC, MMC, MMM, MNST, MO, MOS, MPC, MRK, MRO, MS, MSFT, MSI, MTB, MTD, MU, NAVI, NCLH, NDAQ, NEE, NEM, NFLX, NI, NKE, NLOK, NLSN, NOC, NOV, NRG, NSC, NTAP, NTRS, NVDA, NWL, NWS, NWSA, O, OKE, OMC, ORCL, ORLY, OXY, PAYX, PBCT, PCAR, PCG, PDCO, PEAK, PEG, PEP, PFE, PFG, PG, PGR, PH, PHM, PKG, PKI, PLD, PM, PNC, PNR, PNW, PPG, PPL, PRGO, PRU, PSA, PSX, PVH, PWR, PXD, PYPL, QCOM, QRVO, RCL, RE, REG, REGN, RF, RHI, RL, RMD, ROK, ROP, ROST, RRC, RSG, RTX, SBAC, SBUX, SCHW, SEE, SHW, SIG, SJM, SLB, SLG, SNA, SNPS, SO, SPG, SPGI, SRCL, SRE, STT, STX, STZ, SWK, SWKS, SYF, SYK, SYY, T, TAP, TDG, TEL, TFC, TGT, TJX, TMO, TNL, TRIP, TRV, TSCO, TSN, TXN, TXT, UA, UAA, UAL, UDR, UHS, ULTA, UNH, UNM, UNP, UPS, URI, USB, V, VAR, VFC, VIAC, VLO, VMC, VNO, VRSK, VRSN, VRTX, VTR, VTRS, VZ, WAT, WBA, WDC, WEC, WELL, WFC, WHR, WLTW, WM, WMB, WMT, WRK, WU, WY, WYNN, XEC, XEL, XLNX, XOM, XRAY, XRX, XYL, YUM, ZBH, ZION, ZTS.

Appendix B. Graph Examples

Here, a few of the network examples generated and analysed in the study are presented (Figures A1–A4). The figures were obtained using Mathematica 11 with the "SpringElectricalEmbeding" algorithm. This algorithm optimises the position of nodes with respect to its rank. However, due to the number of nodes and links, the graphs are a bit unclear, particularly in the case of the network of significantly connected companies in Figure A4. However, even a cursory observation shows that the proposed structures give significantly different results. Particularly, networks representing the state of the stock market during the pandemic vary significantly, even though loss of network connectivity is observed. It goes beyond the scope of this paper, but the detailed analysis of the network's evolution from the point of view of the role of a particular company or the reaction of a group of companies to the pandemic situation might be very interesting; however, this is left for another study.

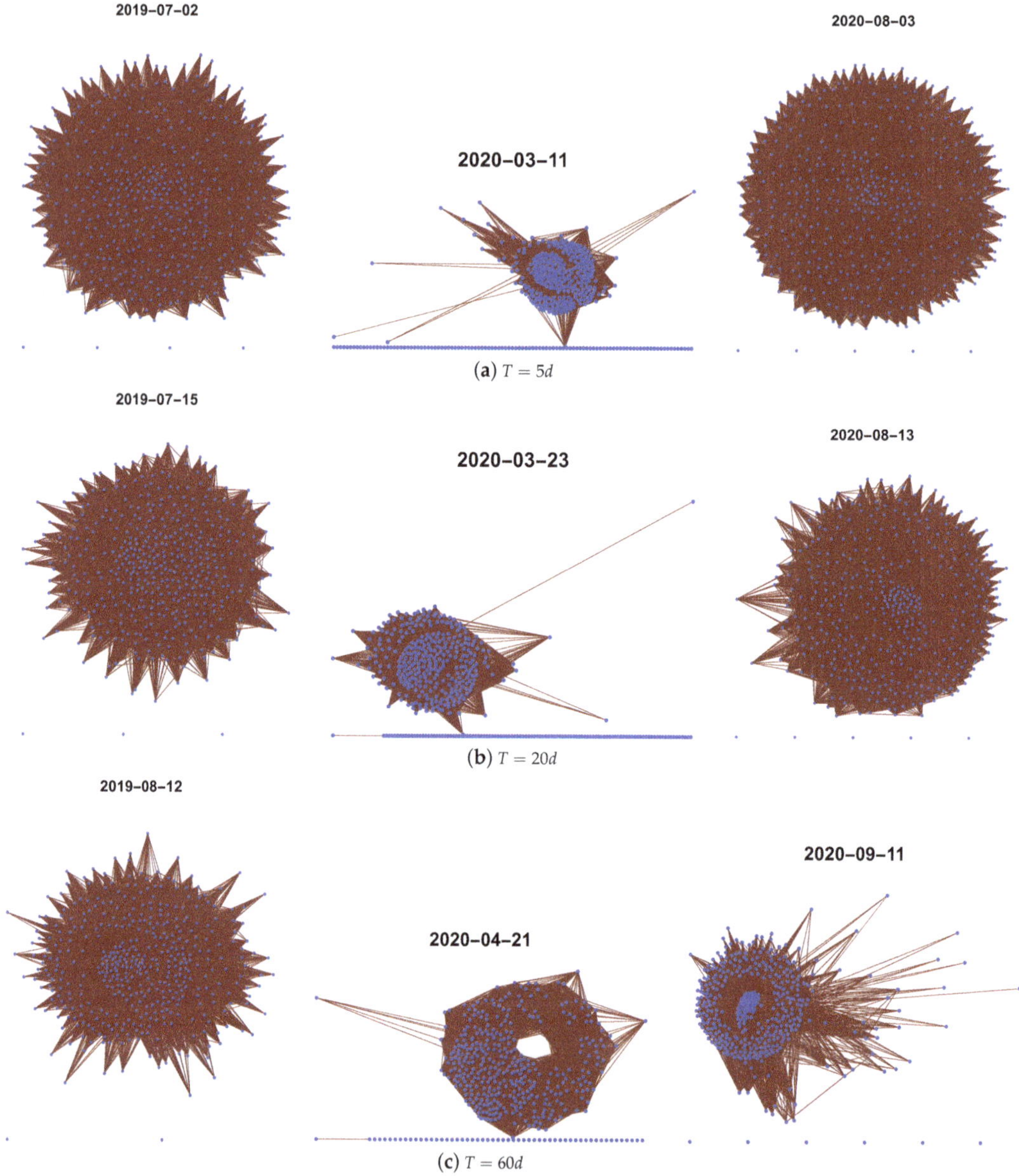

Figure A1. Examples of the graphs obtained for the network of strongly correlated companies. The presented graphs show the network state before and in the first and second stage of the COVID-19 pandemic. The top graphs correspond to the shortest time window $T = 5d$, the middle three graphs represent networks for the month time window and the bottom graphs present the examples for the quarter time window.

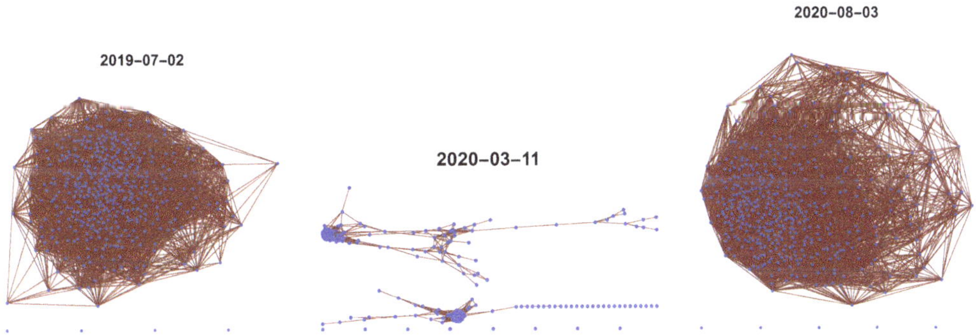

(**a**) $T = 5d$

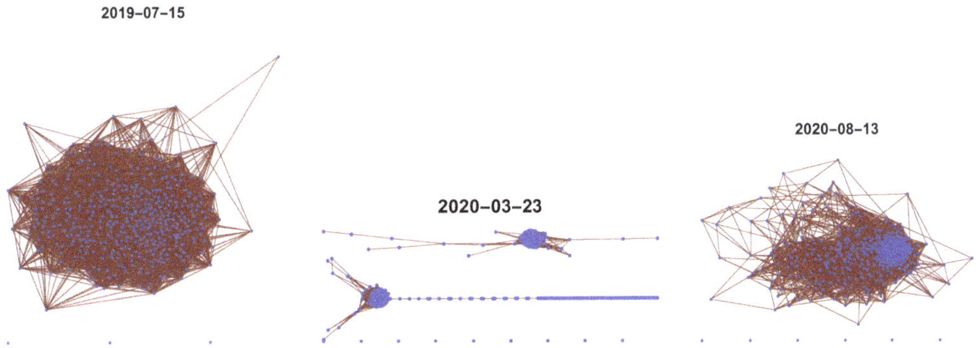

(**b**) $T = 20d$

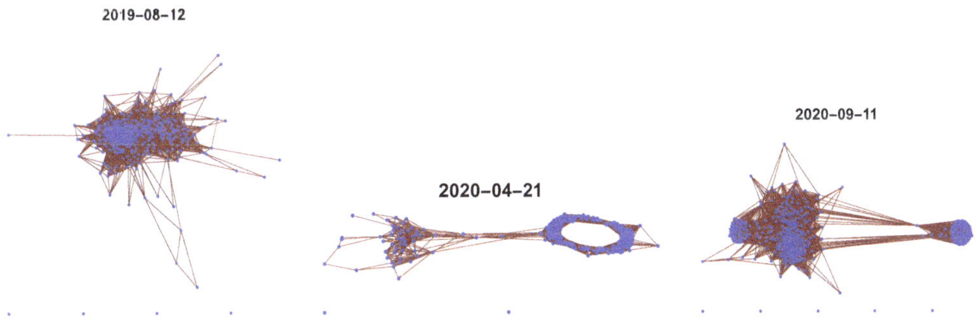

(**c**) $T = 60d$

Figure A2. Examples of the graphs obtained for the network of weakly connected companies. The presented graphs show the network before and in the first and second stage of the COVID-19 pandemic. The top graphs correspond to the shortest time window $T = 5d$, the middle three graphs represent networks for the month time window and the bottom graphs present the examples for the quarter time window.

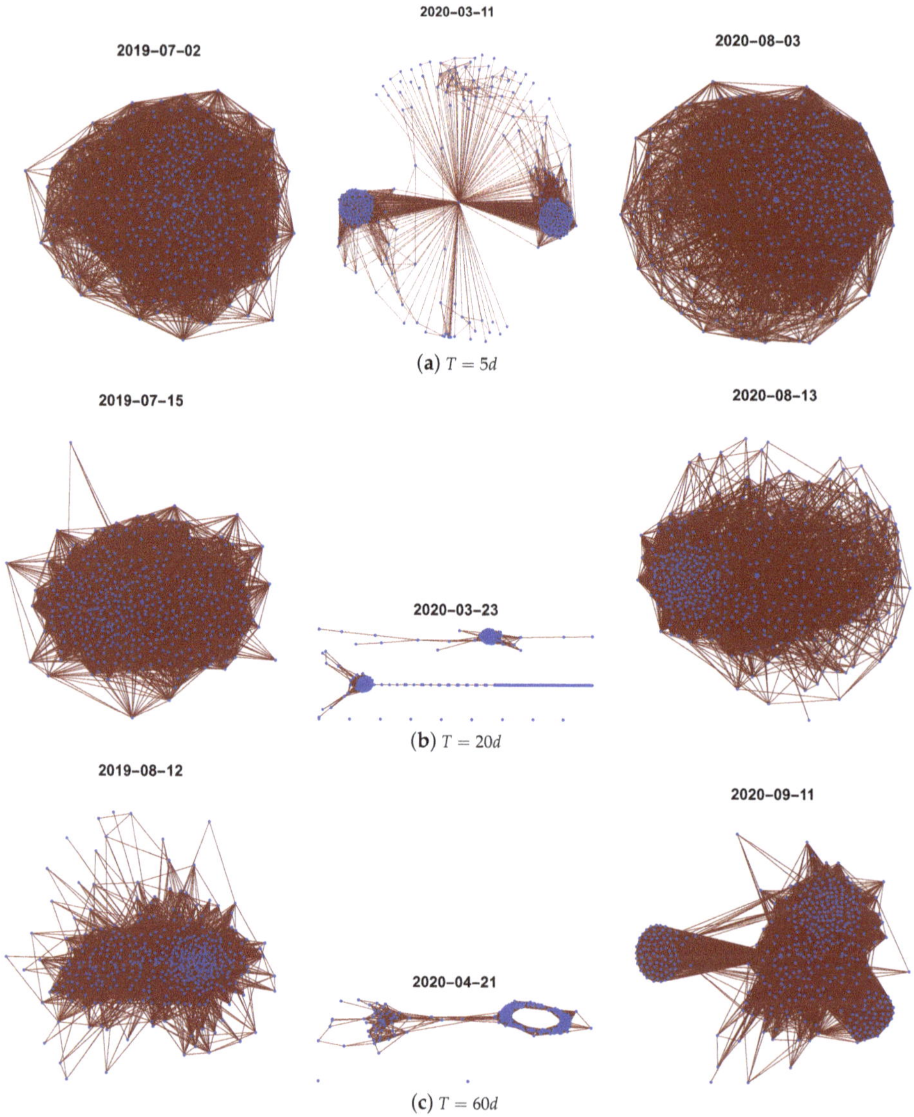

Figure A3. Examples of the graphs obtained for the network of the typically connected companies. The presented graphs show the network before and in the first and second stage of the COVID-19 pandemic. The top graphs correspond to the shortest time window $T = 5d$, the middle three graphs represent networks for the month time window and the bottom graphs present the examples for the quarter time window.

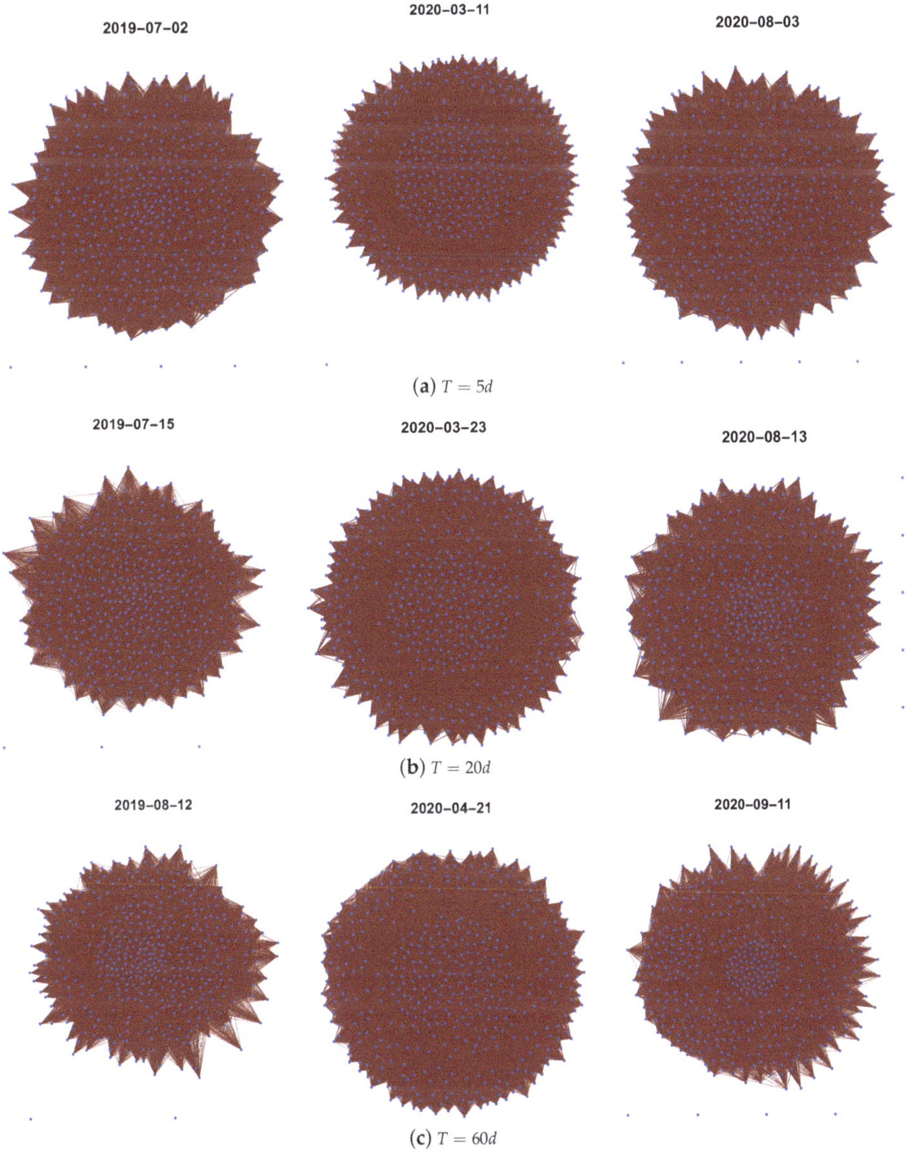

Figure A4. Examples of the graphs obtained for the network of significantly connected companies. The presented graphs show the network before and in the first and second stage of the COVID-19 pandemic. The top graphs correspond to the shortest time window $T = 5d$, the middle three graphs represent networks for the month time window and the bottom graphs present the examples for the quarter time window.

References

1. Jackson, M. *Social and Economic Networks*; Princeton University Press: Princeton, NJ, USA, 2008.
2. Souma, W.; Fujiwara, Y.; Aoyama, H. Complex networks and economics. *Phys. A Stat. Mech. Its Appl.* **2003**, *324*, 396–401. [CrossRef]
3. Bonanno, G.; Vandewalle, N.; Mantegna, R.N. Taxonomy of stock market indices. *Phys. Rev. E* **2000**, *62*, R7615–R7618. [CrossRef]

4. Kirman, A. The economy as an evolving network. *J. Evol. Econ.* **1997**, *7*, 339–353. [CrossRef]
5. Maysami, R.; Howe, L.; Rahmat, M. Relationship between Macroeconomic Variables and Stock Market Indices: Cointegration Evidence from Stock Exchange of Singapore's All-S Sector Indices. *J. Pengur. (UKM J. Manag.)* **2012**, *24*.
6. Mantegna, R.N. Hierarchical structure in financial markets. *Eur. Phys. J. B* **1999**, *11*, 193–197. [CrossRef]
7. Stosic, D.; Stosic, D.; Ludermir, T.B.; Stosic, T. Collective behavior of cryptocurrency price changes. *Phys. A Stat. Mech. Its Appl.* **2018**, *507*, 499–509. [CrossRef]
8. Ren, F.; Zhou, W.X. Dynamic Evolution of Cross-Correlations in the Chinese Stock Market. *PLoS ONE* **2014**, *9*, e97711. [CrossRef]
9. Wang, G.J.; Xie, C.; Chen, S.; Yang, J.J.; Yang, M.Y. Random matrix theory analysis of cross-correlations in the US stock market: Evidence from Pearson's correlation coefficient and detrended cross-correlation coefficient. *Phys. A Stat. Mech. Its Appl.* **2013**, *392*, 3715–3730. [CrossRef]
10. Podobnik, B.; Stanley, H.E. Detrended Cross-Correlation Analysis: A New Method for Analyzing Two Nonstationary Time Series. *Phys. Rev. Lett.* **2008**, *100*, 084102. [CrossRef]
11. Wątorek, M.; Drożdż, S.; Kwapień, J.; Minati, L.; Oświęcimka, P.; Stanuszek, M. Multiscale characteristics of the emerging global cryptocurrency market. *Phys. Rep.* **2021**, *901*, 1–82. [CrossRef]
12. Miśkiewicz, J. Power law classification scheme of time series correlations. On the example of G20 group. *Phys. A Stat. Mech. Its Appl.* **2013**, *392*, 2150–2162. [CrossRef]
13. Zou, Y.; Donner, R.V.; Marwan, N.; Donges, J.F.; Kurths, J. Complex network approaches to nonlinear time series analysis. *Phys. Rep.* **2019**, *787*, 1–97.
14. Silva, V.F.; Silva, M.E.; Ribeiro, P.; Silva, F. Time series analysis via network science: Concepts and algorithms. *Wiley Interdiscip. Rev. Data Min. Knowl. Discov.* **2021**, *11*, e1404. [CrossRef]
15. Mantegna, R. Information and hierarchical structure in financial markets. *Comput. Phys. Commun.* **1999**, *121–122*, 153–156. [CrossRef]
16. Kwapień, J.; Drożdż, S. Physical approach to complex systems. *Phys. Rep.* **2012**, *515*, 115–226.
17. Tumminello, M.; Lillo, F.; Mantegna, R.N. Correlation, hierarchies, and networks in financial markets. *J. Econ. Behav. Organ.* **2010**, *75*, 40–58. [CrossRef]
18. Brida, J.G.; Risso, W.A. Hierarchical structure of the German stock market. *Expert Syst. Appl.* **2010**, *37*, 3846–3852. [CrossRef]
19. Deviren, S.A.; Deviren, B. The relationship between carbon dioxide emission and economic growth: Hierarchical structure methods. *Phys. A Stat. Mech. Its Appl.* **2016**, *451*, 429–439. [CrossRef]
20. Bonanno, G.; Lillo, F.; Mantegna, R. High-frequency cross-correlation in a set of stocks. *Quant. Financ.* **2001**, *1*, 96–104. [CrossRef]
21. Xia, L.; You, D.; Jiang, X.; Guo, Q. Comparison between global financial crisis and local stock disaster on top of Chinese stock network. *Phys. A Stat. Mech. Its Appl.* **2018**, *490*, 222–230. [CrossRef]
22. Onnela, J.P.; Chakraborti, A.; Kaski, K.; Kertiész, J. Dynamic asset trees and portfolio analysis. *Eur. Phys. J. B-Condens. Matter Complex Syst.* **2002**, *30*, 285–288. [CrossRef]
23. Namaki, A.; Shirazi, A.; Raei, R.; Jafari, G. Network analysis of a financial market based on genuine correlation and threshold method. *Phys. A Stat. Mech. Its Appl.* **2011**, *390*, 3835–3841. [CrossRef]
24. Zheng, Z.; Podobnik, B.; Feng, L.; Li, B. Changes in Cross-Correlations as an Indicator for Systemic Risk. *Sci. Rep.* **2012**, *2*, 888. [CrossRef] [PubMed]
25. Sensoy, A.; Yuksel, S.; Erturk, M. Analysis of cross-correlations between financial markets after the 2008 crisis. *Phys. A Stat. Mech. Its Appl.* **2013**, *392*, 5027–5045. [CrossRef]
26. Miśkiewicz, J.; Ausloos, M. Has the world economy reached its globalization limit? *Phys. A Stat. Mech. Its Appl.* **2010**, *389*, 797–806. [CrossRef]
27. Miśkiewicz, J.; Ausloos, M. Correlation measure to detect time series distances, whence economy globalization. *Phys. A Stat. Mech. Its Appl.* **2008**, *387*, 6584–6594. [CrossRef]
28. Kali, R.; Reyes, J. The architecture of globalization: A network approach to international economic integration. *J. Int. Bus. Stud.* **2007**, *38*, 595–620. [CrossRef]
29. Tóth, B.; Kertész, J. Increasing market efficiency: Evolution of cross-correlations of stock returns. *Phys. A Stat. Mech. Its Appl.* **2006**, *360*, 505–515. [CrossRef]
30. Lin, J.; Ban, Y. The evolving network structure of US airline system during 1990–2010. *Phys. A Stat. Mech. Its Appl.* **2014**, *410*, 302–312. [CrossRef]
31. Barabasi, A.; Jeong, H.; Neda, Z.; Ravasz, E.; Schubert, A.; Vicsek, T. Evolution of the social network of scientific collaborations. *Phys. A: Stat. Mech. Its Appl.* **2002**, *311*, 590–614. [CrossRef]
32. Kullmann, L.; Kertész, J.; Kaski, K. Time-dependent cross-correlations between different stock returns: A directed network of influence. *Phys. Rev. E* **2002**, *66*, 026125. [CrossRef] [PubMed]
33. Eryigit, M.; Eryigit, R. Network structure of cross-correlations among the world market indices. *Phys. A Stat. Mech. Its Appl.* **2009**, *388*, 3551–3562. [CrossRef]
34. Baker, S.R.; Bloom, N.; Davis, S.J.; Kost, K.J.; Sammon, M.C.; Viratyosin, T. *The Unprecedented Stock Market Impact of COVID-19*; Technical Report; National Bureau of Economic Research, MA, USA, 2020.
35. Onali, E. COVID-19 and Stock Market Volatility. 2020. Available online: https://ssrn.com/abstract=3571453 (accessed on 20 October 2021).

36. Mazur, M.; Dang, M.; Vega, M. COVID-19 and the march 2020 stock market crash. Evidence from S&P1500. *Financ. Res. Lett.* **2021**, *38*, 101690. [CrossRef] [PubMed]
37. Harjoto, M.A.; Rossi, F.; Paglia, J.K. COVID-19: Stock market reactions to the shock and the stimulus. *Appl. Econ. Lett.* **2021**, *28*, 795–801. [CrossRef]
38. Ramelli, S.; Wagner, A. What the stock market tells us about the consequences of COVID-19. In *Mitigating COVID Economic Crisis: Act Fast and Do Whatever*; CEPR Press: London, UK, 2020; Volume 63.
39. Miśkiewicz, J. Analysis of time series correlation. The choice of distance metrics and network structure. In Proceedings of the 5th Symposium on Physics in Economics and Social Sciences, Warszawa, Poland, 25–27 November 2010.
40. Miśkiewicz, J. Distance matrix method for network structure analysis. In *Statistical Tools for Finance and Insurance*; Springer: Berlin/Heidelberg, Germany, 2011; pp. 251–289.
41. Vandewalle, N.; Brisbois, F.; Tordoir, X. Non-random topology of stock markets. *Quant. Financ.* **2001**, *1*, 372–374. [CrossRef]
42. Wiliński, M.; Sienkiewicz, A.; Gubiec, T.; Kutner, R.; Struzik, Z. Structural and topological phase transitions on the German Stock Exchange. *Phys. A Stat. Mech. Its Appl.* **2013**, *392*, 5963–5973. [CrossRef]

entropy

MDPI

Article

Effects of Vaccination Efficacy on Wealth Distribution in Kinetic Epidemic Models

Emanuele Bernardi [1], Lorenzo Pareschi [2,*], Giuseppe Toscani [1,3] and Mattia Zanella [1]

1 Department of Mathematics "F. Casorati", University of Pavia, 27100 Pavia, Italy; emanuele.bernardi01@universitadipavia.it (E.B.); giuseppe.toscani@unipv.it (G.T.); mattia.zanella@unipv.it (M.Z.)
2 Department of Mathematics and Computer Science, University of Ferrara, 44121 Ferrara, Italy
3 IMATI "E. Magenes", CNR, 27100 Pavia, Italy
* Correspondence: lorenzo.pareschi@unife.it

Abstract: The spread of the COVID-19 pandemic has highlighted the close link between economics and health in the context of emergency management. A widespread vaccination campaign is considered the main tool to contain the economic consequences. This paper will focus, at the level of wealth distribution modeling, on the economic improvements induced by the vaccination campaign in terms of its effectiveness rate. The economic trend during the pandemic is evaluated, resorting to a mathematical model joining a classical compartmental model including vaccinated individuals with a kinetic model of wealth distribution based on binary wealth exchanges. The interplay between wealth exchanges and the progress of the infectious disease is realized by assuming, on the one hand, that individuals in different compartments act differently in the economic process and, on the other hand, that the epidemic affects risk in economic transactions. Using the mathematical tools of kinetic theory, it is possible to identify the equilibrium states of the system and the formation of inequalities due to the pandemic in the wealth distribution of the population. Numerical experiments highlight the importance of the vaccination campaign and its positive effects in reducing economic inequalities in the multi-agent society.

Keywords: wealth distribution; kinetic models; wealth inequalities; compartmental epidemic modelling; vaccination campaign; COVID-19

Citation: Bernardi, E.; Pareschi, L.; Toscani, G.; Zanella, M. Effects of Vaccination Efficacy on Wealth Distribution in Kinetic Epidemic Models. *Entropy* **2022**, *24*, 216. https://doi.org/10.3390/e24020216

Academic Editors: Ryszard Kutner, H. Eugene Stanley and Christophe Schinckus

Received: 7 January 2022
Accepted: 25 January 2022
Published: 29 January 2022

1. Introduction

In the early 2020s, the spread of the COVID-19 pandemic highlighted the close link between economics and health in the context of emergency management. Because of this, assessing the impact of an epidemic phenomenon on a country's economy has emerged as one of the key aspects to consider in the context of containment strategies. From a mathematical point of view, a systematic approach to the study of the effects on the economies of countries facing a severe pandemic is a very complex problem and a mathematical model can only provide rough indications of the possible consequences, based on simplifying assumptions about the key parameters driving the pandemic evolution. The basic idea is to trace these phenomena back to the evolution of the so-called wealth distribution of a country, which measures how many people belong to increasing income levels.

A first attempt to understand changes in wealth distribution in the presence of epidemic spread was proposed in [1] by combining the classical SIR compartmental model of susceptible, infected and recovered individuals [2,3] with the kinetic model of wealth distribution introduced in [4], and assuming that, due to the presence of the pandemic, individuals in different compartments act differently in the economic process. Although the model was developed in a relatively simplified context, it has provided a general framework for socio-epidemiological modeling that can be easily extended to more complex dynamics, both in terms of economic transactions [5] and in terms of epidemic interactions [6,7]. We mention in this direction the recent survey reported in [8] and the seminal approaches proposed in [9–12] investigating the economic effects of infectious diseases, as well as the study presented in [13].

More precisely, according to [4], the financial transactions in [1] were based on the choice of two parameters. The first defines the so-called safeguard threshold, i.e., the maximum percentage of money that the individual is willing to employ in a transaction, and the second is the random risk inherent in the transaction, characterized by its variance through a spread proportional to the square of the individual's wealth. There, the time dependence of the variance was postulated by assuming that, in the presence of a significant epidemic spread, the variance of the risk tends to increase. This is in agreement with the financial market reactions that were often observed during the COVID-19 pandemic to announcements of rising numbers of infected people in several countries [14]. With the use of the model in [1], it was possible to qualitatively observe the effects of the pandemic in terms of a reduction in the middle class and the increase in social inequalities (see also [15,16]).

The possibility, starting in early 2021, of launching a widespread vaccination campaign has led to general optimism about the ability to improve economic performance while limiting the health consequences of the epidemic. However, it is clear that the reduction of economic consequences is closely linked to the effectiveness of the vaccine in containing infections.

In this paper we will focus, at the level of wealth distribution, on the economic improvements induced by the vaccination campaign in terms of its percentage of effectiveness. The interplay between the economic trend and the pandemic will be evaluated by resorting to a mathematical model joining a kinetic model of wealth distribution based on binary transactions with a compartmental epidemic model including vaccinated individuals (see also [17]). In particular, a fraction of vaccinated individuals, which is determined by the efficacy of the vaccine, may contract the disease. Without intending to review the extensive literature on this topic, we cite the recent papers [18–26] that highlight the possible partial immunity provided by vaccinations. Moreover, the emergence of viral variants means that the efficacy of the vaccine inherently non-constant and subject to collective compliance with non-pharmaceutical interventions.

The underlying theoretical framework we consider is that of kinetic models for collective social phenomena, which allows for the linking of microscopic agent-based behavior to emerging observable patterns [27]. In particular, mathematical modeling of wealth distribution has seen a marked development in recent decades [5,28–35], in which, at least partially, the essential economic mechanisms that are responsible for the formation of large-scale economic indicators such as the Pareto or Gini index have been understood [36,37].

The interplay between epidemic spread and the social economic background is described here as the result of interactions among a large number of individuals, each of which is characterized by the variable $w \in \mathbb{R}_+$, measuring the amount of wealth of a single agent. In this regard, as shown in [1,8,38,39], the fundamental tools of statistical physics allow the understanding of epidemiological dynamics by linking classical compartmental approaches with a statistical description of economic aspects. Indeed, the multiscale nature of kinetic theory allows for the determination of the macroscopic (or aggregate) and measurable features of disease evolution [27,40,41].

The rest of the paper is organized as follows. Section 2 introduces the SIR-type system of kinetic equations that includes vaccinated individuals and combines the dynamics of wealth evolution with the spread of infectious disease in a system of interacting agents. Next, in Section 3 we study the main mathematical properties of the system, and show that, through a suitable asymptotic procedure, the solution of the kinetic system tends to the solution of a system of Fokker–Planck-type equations, which exhibits explicit equilibria of the inverse Gamma type. Finally, in Section 4, we investigate numerically the solutions of the Boltzmann-type kinetic system, and its Fokker–Planck asymptotics, along with the evolution of the Gini index, characterizing the wealth inequalities. These simulations confirm the model's ability to describe phenomena that are characteristic of economic trends in situations compromised by the rapid spread of an epidemic, and their variations as a function of the effectiveness of the vaccination campaign.

2. Wealth Dynamics in Epidemic Phenomena

In this Section, we present an extension of the SIR-kinetic compartmental description of epidemic spreading introduced in [1], which additionally takes into account the population of vaccinated individuals. The model consists of a system of four kinetic equations describing the evolution of wealth in the presence of an infectious disease with partial efficacy of vaccination. The entire population is divided then into four compartments: susceptible individuals (S), who can contract the disease; identified infectious individuals (I), who are recognized to have contracted the disease and can transmit it; vaccinated individuals (V), who have received a vaccine, but can still be at least partially infected and contagious; and the recovered individuals (R), who are healed and immune. The model can be easily adapted to include disease-related mortality and other compartments of interest in terms of available data, such as records of hospitalized individuals. We refer to [3,6,7,42] and the references therein for possible developments in these directions. It should be noted that, since we are referring to an advanced epidemic situation in which we assume the existence of a vaccine, the dynamics of unidentified asymptomatic individuals, so significant in the early stages of the COVID-19 pandemic, has become less relevant thanks to mass screening programs. For this reason, we have chosen to employ only one compartment I related to the identified infected individuals. To measure the aggregate effects of vaccination over the whole population, we have considered the compartment V with a given vaccine efficacy.

The agents of each compartment are characterized uniquely by their wealth $w \geq 0$. Hence, we denote by $f_H(w,t)$, $H \in \{S, I, V, R\}$, the distributions of wealth at time $t \geq 0$ in each compartment, such that $f_H(w,t)dw$ denotes the fraction of agents belonging to the compartment J, which, at time $t \geq 0$, are characterized by wealth between w and $w + dw$. The total wealth distribution density is then defined by the sum of the distributions in all compartments

$$f(w,t) = f_S(w,t) + f_I(w,t) + f_V(w,t) + f_R(w,t), \qquad \int_{\mathbb{R}_+} f(w,t)dw = 1,$$

for all $t \geq 0$. Hence, the fractions of the population belonging to each compartment are given by

$$J(t) = \int_{\mathbb{R}_+} f_J(w,t)dw, \qquad J \in \{S, I, V, R\}.$$

We denote by $m_{J,\kappa}(t)$ the local momenta of order κ for the wealth distributions in each compartment

$$m_{\kappa,J}(t) = \frac{1}{J(t)} \int_{\mathbb{R}_+} w^\kappa f_J(w,t)dw, \tag{1}$$

and we denote with $m_\kappa(t)$ the moment of order $\kappa > 0$ of the wealth distribution $f(w,t)$

$$m_\kappa(t) = \int_{\mathbb{R}_+} w^\kappa f(w,t)dw = \sum_{J \in \{S,I,V,R\}} J(t)\, m_{\kappa,J}(t).$$

2.1. The Kinetic Model

Following [1], we assume that the evolution of the densities obeys an SIR-type compartmental model and that the wealth exchange process is influenced by the epidemic's dynamics. This gives a system of four kinetic equations for the unknown distributions $f_H(w,t)$, $H \in \{S, I, V, R\}$, expressed by

$$
\begin{aligned}
\partial_t f_S(w,t) &= -K(f_S,f_I)(w,t) - \alpha f_S(w,t) + \sum_{J\in\{S,I,V,R\}} Q_{SJ}(f_S,f_J)(w,t),\\
\partial_t f_I(w,t) &= K(f_S,f_I)(w,t) + (1-\zeta)K(f_V,f_I)(w,t) - \gamma_I f_I(w,t) + \sum_{J\in\{S,I,V,R\}} Q_{IJ}(f_I,f_J)(w,t),\\
\partial_t f_V(w,t) &= \alpha f_S(w,t) - (1-\zeta)K(f_V,f_I)(w,t) + \sum_{J\in\{S,I,V,R\}} Q_{VJ}(f_V,f_J)(w,t),\\
\partial_t f_R(w,t) &= \gamma_I f_I(w,t) + \sum_{J\in\{S,I,V,R\}} Q_{RJ}(f_R,f_J)(w,t),
\end{aligned}
\tag{2}
$$

where $\gamma \geq 0$ is the recovery rate for the infected compartment and $\alpha \in [0,1]$ is the vaccination rate of individuals, whereas the term $0 \leq 1-\zeta \leq 1$ quantifies the effectiveness of the vaccine, in such a way that high effectiveness corresponds to values close to one of the parameters, ζ. The operator $K(\cdot,\cdot)$ governs the transmission of the infection and is considered to be of the following form

$$
K(f_H,f_I)(w,t) = f_H(w,t)\int_{\mathbb{R}_+} \beta(w,w_*) f_I(w_*,t)\,dw_*,
\tag{3}
$$

for any $H \in \{S,I,V,R\}$. In (3) the function $\beta(w,w_*) \geq 0$ denotes the contact rate between people with wealth w and, respectively, w_*. A leading example for $\beta(w,w_*)$ is obtained by choosing analogously to [1]

$$
\beta(w,w_*) = \frac{\bar{\beta}}{(c+|w-w_*|)^{\nu}},
\tag{4}
$$

where $\bar{\beta} > 0$, $\nu > 0$ and $c \geq 0$. According to the above contact rate, agents with similar wealth are more likely to interact. The extrapolation of heterogeneous contact rates have been deeply studied in mathematical epidemiology; see [1,43–47] and the references therein.

Finally, the operators $Q_{HJ}(f_H,f_J)$, $H,J \in \{S,I,V,R\}$ characterize the evolution of the wealth in each compartment due to wealth exchange activities between agents of the same class, or between agents of different classes H and J. Their form follows the one originally proposed in the Cordier–Pareschi–Toscani model [4]. An interaction between two individuals in compartment H and J with wealth pair (w,w_*) leads to a wealth pair (w'_{JH}, w'_{HJ}) defined by relations

$$
\begin{aligned}
w'_{HJ} &= (1-\lambda_H)w + \lambda_J w_* + \eta_{HJ} w\\
w'_{JH} &= (1-\lambda_J)w_* + \lambda_H w + \eta_{JH} w_*,
\end{aligned}
\tag{5}
$$

with $H,J \in \{S,I,V,R\}$. In (5) the constants $\lambda_H, \lambda_J \in (0,1)$ are exchange parameters defining the saving propensities $1-\lambda_H$ and $1-\lambda_J$, i.e., the maximum percentage of money that individuals are willing to employ in a general monetary transaction. Note that the parameters are different in each compartment, underlining the differing behavior of agents in the presence of the pandemic. The choice $\lambda_V > \lambda_S$, for example, reflects the fact that susceptible non-vaccinated agents have reduced action in wealth exchanges due to various government restrictions with respect to vaccinated individuals.

Furthermore, $\eta_{JH} \geq -\lambda_H, \eta_{HJ} \geq -\lambda_J$ are independently centered random variables with the same distribution Θ such that $\mathrm{Var}(\eta_{HJ}) = \mathrm{Var}(\eta_{HJ}) = \sigma^2(t)$. The quantity $\sigma^2(t)$ represents the market risk, which is the same for the whole population and is influenced by the progress of the pandemic. This is in agreement with market reactions that have been observed during new epidemic waves; see, e.g., ref. [14]. It is convenient to express the operators $Q_{HJ}(f_H,f_J)$ in weak form, i.e., the way these operators act on observable quantities [27].

Let $\varphi(w)$ be a test function and let $\langle\cdot\rangle$ denote the expectation with respect to the pair of random variables η_{JH}, η_{HJ} in the interaction process (5). Then, for $H, J \in \{S, I, V, R\}$ we define the Boltzmann-type bilinear operators as follows

$$\int_{\mathbb{R}_+} \varphi(w) Q_{HJ}(f_H, f_J)(w,t) dw = \left\langle \int_{\mathbb{R}_+^2} (\varphi(w'_{HJ}) - \varphi(w)) f_H(w,t) f_J(w_*,t) dw\, dw_* \right\rangle \tag{6}$$

where $(w, w_*) \to (w'_{JH}, w'_{HJ})$ as in (5) and where $\langle\cdot\rangle$ denotes the expectation with respect to the independent random variables η_{HJ}, η_{HJ}.

Binary interactions between individuals (5) reflect the idea that wealth exchanges occur between pairs of agents who invest a fraction of their wealth in the presence of an equivalent good. In each case, such investments involve nondeterministic speculative risks that can provide additional wealth or a loss of wealth. The aggregate behavior of the population is then provided by the operators (6), from which we obtain the emerging macroscopic trends of the binary exchanges considered in each epidemiological compartment.

Remark 1. *In the kinetic epidemic model (2) the passage from susceptible to vaccinated is governed by a very simple dynamics that does not take into account possible vaccine limitations, as in the first phase of the vaccination campaign. In general, the vaccination rate α may depend on several factors such as the age and work status of individuals and time. It is worthwhile to observe that, in addition to the natural dependency of the recovery rate γ_I from age [8,22,48], we may also consider wealth-dependent recovery rates to take into account the fact that high wealth can provide access to better hospitals in some health systems, thus ensuring a higher chance of recovery [1]. We point the interested reader to [39] for a more detailed discussion based on the available data.*

2.2. Evolution of Macroscopic Quantities

In the following, we discuss the evolution of emerging macroscopic quantities from the kinetic model (2). Let $\varphi(w)$ be a test function. Choosing $\varphi(w) = 1$ in (6), we have

$$\sum_{J \in \{S,I,V,R\}} \int_{\mathbb{R}_+} \varphi(w) Q_{HJ}(f_H, f_J)(w,t) dw = 0,$$

which corresponds to mass conservation, i.e., the conservation of the number of agents. If $\varphi(w) = w$ in (6), we get the evolution of the average wealth in each compartment, corresponding to the first quantity not conserved in time:

$$\begin{aligned} \frac{d}{dt} m_{1,H}(t) &= \frac{1}{H(t)} \sum_{J \in \{S,I,V,R\}} \int_{\mathbb{R}_+^2} \langle w'_{HJ} - w \rangle f_H(w,t) f_J(w_*,t) dw dw_* \\ &= H(t) \sum_{J \in \{S,I,V,R\}} J(t)(\lambda_J m_{1,J}(t) - \lambda_H m_{1,H}). \end{aligned} \tag{7}$$

The total mean wealth is then conserved:

$$\frac{d}{dt} \sum_{H \in \{S,I,V,R\}} \int_{\mathbb{R}_+} w f_H(w,t) dw = \frac{d}{dt} m_1 = 0.$$

The evolution of mass fractions can be easily obtained from (2) via direct integration

$$
\begin{aligned}
\frac{d}{dt}S(t) &= -\int_{\mathbb{R}_+^2} \beta(w,w_*)f_S(w,t)f_I(w,t)dw\,dw_* - \alpha S(t),\\
\frac{d}{dt}I(t) &= \int_{\mathbb{R}_+^2} \beta(w,w_*)f_S(w,t)f_I(w,t)dw\,dw_* + (1-\zeta)\int_{\mathbb{R}_+^2} \beta(w,w_*)f_V(w,t)f_I(w,t)dw\,dw_* - \gamma_I I(t),\\
\frac{d}{dt}V(t) &= \alpha S(t) - (1-\zeta)\int_{\mathbb{R}_+^2} \beta(w,w_*)f_V(w,t)f_I(w,t)dw\,dw_*,\\
\frac{d}{dt}R(t) &= \gamma_I I(t).
\end{aligned}
\tag{8}
$$

To obtain a closed-form evolution of the macroscopic quantities, we consider a constant rate function, $\beta(w,w_*) = \bar{\beta} > 0$, obtained from (4) for $\nu = 0$, and a constant-in-time market risk $\sigma^2(t) = \sigma^2$. Under these assumptions, thanks tothe mass conservation of Boltzmann-type operators (6), we obtain a classical SIR model with vaccination

$$
\begin{aligned}
\frac{d}{dt}S(t) &= -\bar{\beta}S(t)I(t) - \alpha S(t),\\
\frac{d}{dt}I(t) &= \bar{\beta}S(t)I(t) + (1-\zeta)\bar{\beta}V(t)I(t) - \gamma_I I(t),\\
\frac{d}{dt}V(t) &= \alpha S(t) - (1-\zeta)\bar{\beta}V(t)I(t),\\
\frac{d}{dt}R(t) &= \gamma_I I(t).
\end{aligned}
\tag{9}
$$

As a consequence, for large times $t \to +\infty$, we have a disease-free equilibrium state, where $I(t) \to 0^+$, $S(t) \to 0^+$, $V(t) \to V^\infty$ and $R(t) \to R^\infty$ with $V^\infty + R^\infty = 1$ (see [3]).

The dynamics of mean wealth can be recovered from (7) as follows

$$
\begin{aligned}
S(t)\frac{d}{dt}m_{1,S}(t) &= S(t)(\bar{m}_1(t) - \lambda_S m_{1,S}(t)),\\
I(t)\frac{d}{dt}m_{1,I}(t) &= \bar{\beta}S(t)I(t)(m_{1,S} - m_{1,I}) + \bar{\beta}(1-\xi)V(t)I(t)(m_{1,V} - m_{1,I})\\
&\quad + I(t)(\bar{m}_1 - \lambda_I m_{1,I}),\\
V(t)\frac{d}{dt}m_{1,V}(t) &= \alpha S(t)(m_{1,S} - m_{1,V}) + V(t)(\bar{m}_1 - \lambda_V m_{1,V}),\\
R(t)\frac{d}{dt}m_{1,R}(t) &= \gamma_I I(t)(m_{1,R}(t) - m_{1,I}(t)) + R(t)(\bar{m}_1(t) - \lambda_R m_{1,R}(t)),
\end{aligned}
\tag{10}
$$

where we defined the weighted mean wealth as

$$
\bar{m}_1(t) = \sum_{J\in\{S,I,V,R\}} \lambda_J m_{1,J}(t)J(t). \tag{11}
$$

Therefore, based on (10), we can observe that the large time behavior of the mean wealth satisfies

$$
2\bar{m}_1^\infty - \lambda_V m_{1,V}^\infty - \lambda_R m_{1,R}^\infty = 0.
$$

Hence, we obtain

$$
\lambda_V m_{1,V}^\infty = \lambda_R m_{1,R}^\infty,
$$

together with the constraint $R^\infty m_{R,1}^\infty + V^\infty m_{V,1}^\infty = m$, based on the conservation of total mean wealth. Thanks to the latter equalities, we can observe that the asymptotic mean wealth in the compartments of vaccinated and recovered individuals is given by

$$
m_{1,V}^\infty = \frac{\lambda_R}{\lambda_R V^\infty + \lambda_V R^\infty}m, \qquad m_{1,R}^\infty = \frac{\lambda_V}{\lambda_R V^\infty + \lambda_V R^\infty}m. \tag{12}
$$

Likewise, we obtain the system for the the second moments

$$\begin{aligned}
S(t)\frac{d}{dt}m_{2,S}(t) &= (\lambda_S^2 - 2\lambda_S + \sigma^2)Sm_{2,S} + S(t)\bar{m}_2 + 2(1-\lambda_S)Sm_{1,S}\bar{m}_1, \\
I(t)\frac{d}{dt}m_{2,I}(t) &= \bar{\beta}SI(m_{2,S} - m_{2,I}) + (1-\zeta)\bar{\beta}VI(m_{2,V} - m_{2,I}) \\
&\quad + (\lambda_I^2 - 2\lambda_I + \sigma^2)Im_{2,I} + I\bar{m}_2 + 2(1-\lambda_I)Im_{1,I}\bar{m}_1, \\
V(t)\frac{d}{dt}m_{2,V}(t) &= \alpha S(m_{2,S} - m_{2,V}) + (\lambda_V^2 - 2\lambda_V + \sigma^2)Vm_{2,V} + V\bar{m}_2 \\
&\quad + 2(1-\lambda_V)Vm_{1,V}\bar{m}_1, \\
R(t)\frac{d}{dt}m_{2,R}(t) &= (\lambda_R^2 - 2\lambda_R + \sigma^2)Rm_{2,R} + R\bar{m}_2 + 2(1-\lambda_R)Rm_{1,R}\bar{m}_1,
\end{aligned} \tag{13}$$

where $\bar{m}_1$ has been defined in (11) and we have introduced the following notation

$$\bar{m}_2(t) = \sum_{J\in\{S,I,V,R\}} \lambda_J^2 m_{2,J}(t)J(t).$$

The evolution of the second moment for the whole system is governed by

$$\frac{d}{dt}m_2(t) = \bar{m}_2(t) + \sum_{J\in\{S,I,V,R\}} \Big(m_{J,2}(\lambda_J^2 - 2\lambda_J + \sigma) + 2(1-\lambda_J)m_{J,2}\bar{m}_1(t)\Big)J(t).$$

For large times, the second-order moment for susceptible and infected is such that $m_{2,S}, m_{2,I} \to 0^+$ for $t \to +\infty$. Therefore, $m_{2,V}^\infty$, $m_{2,R}^\infty$ are solutions to

$$\begin{aligned}
(\lambda_V^2 - 2\lambda_V + \sigma^2)m_{2,V}^\infty + \bar{m}_2^\infty + (1-\lambda_V)m_{1,V}^\infty\bar{m}_1^\infty &= 0, \\
(\lambda_R^2 - 2\lambda_R + \sigma^2)m_{2,R}^\infty + \bar{m}_2^\infty + (1-\lambda_R)m_{1,R}^\infty\bar{m}_1^\infty &= 0.
\end{aligned}$$

from which we get

$$\begin{aligned}
m_{2,R}^\infty &= \frac{\lambda_V^2(1-\lambda_V)V^\infty m_{1,V}^\infty\bar{m}_1^\infty - A_V(1-\lambda_R)m_{1,R}^\infty\bar{m}_1^\infty}{A_V(\lambda_R^2(1+R^\infty) - 2\lambda_R + \sigma^2) - \lambda_V^2\lambda_R^2V^\infty R^\infty} \\
m_{2,V}^\infty &= \frac{\lambda_R^2(1-\lambda_R)R^\infty m_{1,R}^\infty\bar{m}_1^\infty - A_R(1-\lambda_V)m_{1,V}^\infty\bar{m}_1^\infty}{A_R(\lambda_V^2(1+V^\infty) - 2\lambda_V + \sigma^2) - \lambda_R^2\lambda_V^2V^\infty R^\infty}
\end{aligned}$$

where

$$A_H = \lambda_V^2(1+H^\infty) - 2\lambda_V + \sigma^2, \qquad H \in \{V,R\},$$

and $\bar{m}_1^\infty = \lambda_V m_{1,V}^\infty V^\infty + \lambda_R m_{1,R}^\infty R^\infty$ and $m_{1,V}^\infty, m_{1,R}^\infty$ have been obtained in (12).

Remark 2. *In the general case where a non-constant incidence rate $\beta = \beta(w, w_*)$ is considered, the macroscopic system of equations is not closed. Depending on the specific choice of β and using the knowledge on the equilibrium states discussed in Section 3.1 it is possible, through the classical hydrodynamic closure of kinetic theory, to derive epidemic models where the dynamics, instead of being homogeneous as in classical compartmental modeling, is influenced by the heterogeneous wealth status of individuals. We refer to [8,38] for examples in this direction.*

3. Properties of the Kinetic Model

In this section we study the mathematical model (2) from an analytical point of view, by proving the well-posedness and convergence to equilibrium of the solution. To this end, we made suitable simplification assumptions on the contact rate by restricting to the case $\beta(w, w_*) = \bar{\beta}$. We resort to classical mathematical approaches for kinetic equations to characterize the trend to equilibrium [1,27]. In particular, taking into account methods for nonconservative systems—see, e.g., ref. [49]—we provide an existence and uniqueness result. Given a function $f(w) \in L^1(\mathbb{R}_+)$, we define its Fourier transform as follows

$$\hat{f}(z) = \int_{\mathbb{R}} e^{-iwz} f(w)dw.$$

According to the above assumption regarding the contact rate, we rewrite (2) in weak form:

$$\begin{aligned}
\partial_t \int_{\mathbb{R}_+} \varphi(w) f_S(w,t)dw &= -\bar{\beta} I(t) \int_{\mathbb{R}_+} \varphi(w) f_S(w,t)dw - \alpha \int_{\mathbb{R}_+} \varphi(w) f_S(w,t)dw \\
&\quad + \sum_{J\in\{S,I,V,R\}} \int_{\mathbb{R}_+} \varphi(w) Q_{SJ}(f_S, f_J)(w,t)dw, \\
\partial_t \int_{\mathbb{R}_+} \varphi(w) f_I(w,t)dw &= \bar{\beta} I(t) \int_{\mathbb{R}_+} \varphi(w) f_S(w,t)dw + (1-\zeta)\bar{\beta} I(t) \int_{\mathbb{R}_+} \varphi(w) f_V(w,t)dw \\
&\quad - \gamma_I \int_{\mathbb{R}_+} \varphi(w) f_I(w,t)dw + \sum_{J\in\{S,I,V,R\}} \int_{\mathbb{R}_+} \varphi(w) Q_{IJ}(f_I, f_J)(w,t)dw, \\
\partial_t \int_{\mathbb{R}_+} \varphi(w) f_V(w,t)dw &= \alpha \int_{\mathbb{R}_+} \varphi(w) f_S(w,t)dw - (1-\zeta)\bar{\beta} I(t) \int_{\mathbb{R}_+} \varphi(w) f_V(w,t)dw \\
&\quad + \sum_{J\in\{S,I,V,R\}} \int_{\mathbb{R}_+} \varphi(w) Q_{VJ}(f_V, f_J)(w,t)dw, \\
\partial_t \int_{\mathbb{R}_+} \varphi(w) f_R(w,t)dw &= \gamma_I \int_{\mathbb{R}_+} \varphi(w) f_I(w,t)dw + \sum_{J\in\{S,I,V,R\}} \int_{\mathbb{R}_+} \varphi(w) Q_{RJ}(f_R, f_J)(w,t)dw.
\end{aligned} \tag{14}$$

Hence, we consider $\varphi(w) = e^{-izw}$ in (14) to get

$$\begin{aligned}
\partial_t \hat{f}_S(z,t) &= -\bar{\beta} I(t) \hat{f}_S(z,t) - \alpha \hat{f}_S(z,t) + \sum_{J\in\{S,I,V,R\}} \hat{Q}_{SJ}(\hat{f}_S, \hat{f}_J)(z,t), \\
\partial_t \hat{f}_I(z,t) &= \bar{\beta} I(t) \hat{f}_S(z,t) + (1-\zeta)\bar{\beta} \hat{f}_I(z,t) \hat{f}_V(z,t) - \gamma_I \hat{f}_I(z,t) + \sum_{J\in\{S,I,V,R\}} \hat{Q}_{IJ}(\hat{f}_I, \hat{f}_J)(z,t), \\
\partial_t \hat{f}_V(z,t) &= \alpha \hat{f}_S(z,t) - (1-\zeta)\bar{\beta} \hat{f}_I(z,t) \hat{f}_V(z,t) + \sum_{J\in\{S,I,V,R\}} \hat{Q}_{VJ}(\hat{f}_V, \hat{f}_J)(z,t), \\
\partial_t \hat{f}_R(z,t) &= \gamma_I \hat{f}_I(z,t) + \sum_{J\in\{S,I,V,R\}} \hat{Q}_{RJ}(\hat{f}_R, \hat{f}_J)(z,t).
\end{aligned} \tag{15}$$

Similarly to [1] the operators $\hat{Q}_{HJ}(\hat{f}_H, \hat{f}_J)(z,t)$ may be rewritten as follows

$$\int_{\mathbb{R}_+} e^{-iwz} Q_{HJ}(f_H, f_J)dw = \langle \hat{f}_H(A_{HJ}z, t)\rangle \hat{f}_J(\lambda_J z, t) - J(t)\hat{f}_H(z,t),$$

where

$$A_{HJ} = 1 - \lambda_H + \eta_{HJ}.$$

We assume that the parameters of the trading activity satisfy the condition

$$\nu = \max_{H,J\in\{S,I,V,R\}} [\lambda_J^2 + \langle A_{HJ}^2\rangle] < 1. \tag{16}$$

Let $\mathcal{P}_s(\mathbb{R}_+)$ be the set of probability measures $f(w)$ with bounded $s-$moment, and, for any pair of densities f and g in $\mathcal{P}_s(\mathbb{R}_+)$, let us consider the class of metrics d_s defined by

$$d_s(f,g) = \sup_{z\in\mathbb{R}} \frac{|\hat{f}(z) - \hat{g}(z)|}{|z|^s}, \tag{17}$$

where $\hat{f}$ and $\hat{g}$ denote the Fourier transforms of f and g. Then, the distance (17) is well-defined and finite for any pair of probability measures with equal moments up to order

$[s]$ (where $[s]$ denotes the integer part of s), if s is a real number or up to $s-1$, if s is an integer [27].

Inequality (16), combined with a Fourier-based distance, allows one to obtain an exponential convergence to equilibrium for system (2). This condition is verified whenever

$$\sigma^2 < 2 \min_{J\in\{S,I,V,R\}} \lambda_J(1-\lambda_J),$$

namely, when the market risk is not too great in relation to the saving propensities. To study the large-time behavior of the solution to systems such as (15) we follow [1,27]. Then, we have the following result

Theorem 1. *Let $f_J(w,t)$ and $g_J(w,t)$, $J \in \{S,I,V,R\}$, be two solutions of the kynetic system (2), corresponding to the initial values $f_J(w,0)$ and $g_J(w,0)$ such that $d_2(f_J(w,0), g_J(w,0))$, $J \in \{S,I,V,R\}$, is finite. Then, if condition (16) holds, the Fourier-based distance $d_2(f_J(w,t), g_J(w,t))$ decays exponentially in time toward zero and the following holds:*

$$\sum_{J\in\{S,I,V,R\}} d_2(f_J(w,t), g_J(w,t)) < \sum_{J\in\{S,I,V,R\}} d_2(f_J(w,0), g_J(w,0)) \exp\{-(1-\nu)t\}. \quad (18)$$

The previous result and the Equation (18) give us the contractivity of the system in the d_2 metric, which will be the essential to prove the existence theorem. Theorem 1 allows us to further investigate the properties of the steady state $f_J^\infty(w)$, $J \in \{S,I,V,R\}$.
In order to obtain an existence result we need to introduce a subset of $\mathcal{P}_2(\mathbb{R})$

$$\mathcal{D}_{m_1,m_2} := \left\{ F \in \mathcal{P}_2(\mathbb{R}) : \int_{\mathbb{R}} v dF(v) = m_1, \int_{\mathbb{R}} v^2 dF(v) = m_2 \right\}. \quad (19)$$

Following [49], it is possible to prove that $\mathcal{D}_{m_1,m_2}$ is a metric Banach space with the $d_2(\cdot,\cdot)$ metric. Now, we define

$$\mathcal{D}^\infty := \mathcal{D}_{m^\infty_{V,1}, m^\infty_{V,2}} \times \mathcal{D}_{m^\infty_{R,1}, m^\infty_{R,2}}$$

as the product space of two sets such as (19), where the momenta are those of the steady states for the relative distributions $f_J(w)$, for $J \in \{V,R\}$ (we are only considering these two classes since for large time $I,S \to 0^+$). We also recall a variant of the metric used in Theorem 1

$$\bar{d}_2(f,g) := \sum_{J\in\{V,R\}} d_2(f_J(w,t), g_J(w,t)). \quad (20)$$

Now, we are able to prove the following theorem.

Theorem 2. *If the initial value $f_0(w) = f(w,0) \in \mathcal{D}^\infty$ and condition (16) holds, then the system*

$$\begin{aligned} \partial_t f_V(w,t) &= \sum_{J\in\{V,R\}} Q_{VJ}(f_V, f_J)(w,t), \\ \partial_t f_R(w,t) &= \sum_{J\in\{V,R\}} Q_{RJ}(f_R, f_J)(w,t), \end{aligned} \quad (21)$$

has a unique steady state $f^\infty(w)$, and it also belongs to $\mathcal{D}^\infty$.

Proof. Let us consider the flow map

$$T_t : \left(\mathcal{D}^\infty, \bar{d}_2\right) \to \left(\mathcal{D}^\infty, \bar{d}_2\right) \quad (22)$$

which , for any time $t > 0$, is given by $T_t(f_0(w)) = f(t) = (f_V(w,t), f_R(w,t))$, where $f(t)$ is the solution of (21) at time t with $f(w,0) = f_0(w) \in \mathcal{D}^\infty$. Thanks to (18) we have

$$\overline{d}_2(T_t(f_0(w)),\ T_t(g_0(w))) < \overline{d}_2(f_0(w),\ g_0(w)) \exp\{-(1-\nu)t\}$$

which is a strict contraction for (22) with constant $\exp\{-(1-\nu)t\} < 1$. Now, it is easy to see that $(\mathcal{D}^\infty, \overline{d}_2)$ is a Banach space and therefore the Banach fixed-point theorem ensures the existence and uniqueness for the steady state in $\mathcal{D}^\infty$. □

Remark 3. *Similar results may be obtained in the more realistic case $\beta(w, w_*) = \beta(w - w_*)$ since the transmission operator $K(\cdot,\cdot)$ defined in (3) possesses, in this case, a convolution structure, which naturally converts into a product in the Fourier space. We omit the details.*

3.1. Fokker–Planck Scaling and Steady States

In the general case, it is difficult to compute analytically the large-time behaviour of the compartmental kinetic system (2). A deeper insight into the steady states can be obtained through the so-called quasi-invariant limit procedure [1,4,27]. The goal is to derive a simplified Fokker–Planck model in which the study of the asymptotic properties is much easier. It is worth mentioning that this approach is inspired by the so-called grazing collision limit of the Boltzmann equation; see [50,51].

The driving idea is to scale interactions and trading frequency at the same time. As a consequence, the equilibrium of the wealth distribution is reached more quickly with respect to the time scale of the epidemic. Hence, given $\epsilon \ll 1$ we introduce the following scaling

$$\begin{aligned} &\lambda_S \to \epsilon\lambda_S, \quad \lambda_I \to \epsilon\lambda_I, \quad \lambda_V \to \epsilon\lambda_V, \quad \lambda_R \to \epsilon\lambda_R, \\ &\sigma^2 \to \epsilon\sigma^2, \quad \beta(w,w_*) \to \epsilon\beta(w,w_*), \quad \gamma_I \to \epsilon\gamma_I, \end{aligned} \tag{23}$$

together with the time scaling $t \to t/\epsilon$. We denote as $Q^\epsilon_{HJ}(\cdot,\cdot)$, $H, J \in \{S, I, V, R\}$, the scaled interaction terms. Using a Taylor expansion for small values of ϵ, we get [1]

$$\begin{aligned} &\frac{1}{\epsilon}\int_{\mathbb{R}_+} Q^\epsilon_{HJ}(f_H, f_J)(w,t)\varphi(w)dw \\ &\quad = \int_{\mathbb{R}_+} \left\{ -\varphi'(w)(w\lambda_H J - m_{1,J}\lambda_J) + \frac{\sigma^2}{2}\varphi''(w)w^2 J(t) \right\} f_H(w,t)dw + O(\epsilon). \end{aligned}$$

Integrating back by parts, in the limit $\epsilon \to 0$, we obtain the system of Fokker–Planck equations

$$\begin{aligned} \frac{\partial f_S(w,t)}{\partial t} &= -K(f_S, f_I)(w,t) - \alpha f_S(w,t) + \frac{\partial}{\partial w}\{[w\lambda_S - \bar{m}(t)]f_S(w,t)\} \\ &\quad + \frac{\sigma^2}{2}\frac{\partial^2}{\partial w^2}(w^2 f_S(w,t)), \\ \frac{\partial f_I(w,t)}{\partial t} &= K(f_S, f_I)(w,t) + (1-\zeta)K(f_V, f_I)(w,t) - \gamma_I f_I(w,t) \\ &\quad + \frac{\partial}{\partial w}\{[w\lambda_I - \bar{m}(t)]f_I(w,t)\} + \frac{\sigma^2}{2}\frac{\partial^2}{\partial w^2}(w^2 f_I(w,t)), \\ \frac{\partial f_V(w,t)}{\partial t} &= \alpha f_S(w,t) - (1-\zeta)K(f_V, f_I)(w,t) + \frac{\partial}{\partial w}\{[w\lambda_V - \bar{m}(t)]f_V(w,t)\} \\ &\quad + \frac{\sigma^2}{2}\frac{\partial^2}{\partial w^2}(w^2 f_V(w,t)), \\ \frac{\partial f_R(w,t)}{\partial t} &= \gamma_I(w,t) + \frac{\partial}{\partial w}\{[w\lambda_R - \bar{m}(t)]f_R(w,t)\} + \frac{\sigma^2}{2}\frac{\partial^2}{\partial w^2}(w^2 f_R(w,t)), \end{aligned} \tag{24}$$

where $\bar{m}$ has been defined in (11). The above Fokker–Planck system is complemented with the following boundary conditions

$$\frac{\partial}{\partial w}[w^2 g_J(w,t)]|_{w=0} = 0 \qquad [w\lambda_J - \bar{m}]g_J + \frac{\sigma}{2}\frac{\partial}{\partial w}(w^2 g_J)\bigg|_{w=0} = 0.$$

We can verify under suitable assumptions that the Fokker–Planck system (24) possesses an explicitly computable steady state [52]. Let us consider the case of a constant contact rate, i.e., $\beta(w, w_*) = \bar{\beta}$. Since for large times $S, I \to 0^+$ we find that the stationary states $f_V^\infty(w)$ and $f_R^\infty(w)$ solve the following equations:

$$\lambda_V \frac{\partial}{\partial w}\left[(w - m_V^\infty)f_V^\infty(w)\right] + \frac{\sigma^2}{2}\frac{\partial^2}{\partial w^2}[w^2 f_V^\infty(w)] = 0,$$
$$\lambda_R \frac{\partial}{\partial w}\left[(w - m_R^\infty)f_R^\infty(w)\right] + \frac{\sigma^2}{2}\frac{\partial^2}{\partial w^2}[w^2 f_R^\infty(w)] = 0.$$

Based on the above equalities, we find that the two steady states are inverse Gamma densities

$$f_V^\infty(w) = V^\infty \frac{\kappa^{\mu_V}}{\Gamma(\mu_V)} \frac{e^{-\frac{\kappa}{w}}}{w^{1+\mu_V}} \qquad f_R^\infty(w) = R^\infty \frac{\kappa^{\mu_R}}{\Gamma(\mu_R)} \frac{e^{-\frac{\kappa}{w}}}{w^{1+\mu_R}} \tag{25}$$

with Pareto indices defined as follows

$$\mu_V = 1 + 2\frac{\lambda_V}{\sigma^2}, \qquad \mu_R = 1 + 2\frac{\lambda_R}{\sigma^2},$$

$$\kappa = (\mu_V - 1)m_V^\infty = (\mu_R - 1)m_R^\infty = \frac{2\lambda_R\lambda_V}{\sigma^2(\lambda_R V^\infty + \lambda_V R^\infty)} m.$$

Consequently, the global steady state is a mixture of the inverse Gamma distribution

$$f^\infty(w) = f_V^\infty(w) + f_R^\infty(w), \tag{26}$$

which may present a bimodal shape with a different intensity.The formation of two peaks at the equilibrium is due to the fact that we have two different maxima corresponding to the points

$$\bar{w}_V = \frac{\kappa}{\mu_V + 1} = \frac{\lambda_R\lambda_V}{(\lambda_V + \sigma)(\lambda_R V^\infty + \lambda_V R^\infty)} m, \tag{27}$$

$$\bar{w}_R = \frac{\kappa}{\mu_R + 1} = \frac{\lambda_R\lambda_V}{(\lambda_R + \sigma)(\lambda_R V^\infty + \lambda_V R^\infty)} m, \tag{28}$$

for the vaccinated and for the recovered wealth distributions, respectively. In the next section we report on the resulting profiles for different choices of λ_V, λ_R, σ and V^∞, R^∞.

Remark 4. *The emergence of a multimodal equilibrium wealth distribution has been classically linked to the appearance of new inequalities in highly stressed societies; see, e.g., [15,35,53]. In these cases, the economic segregation of part of the society leads to the pauperization of substantial layers of the middle class. In the present case, the different economic impact played by agents in each compartment is capable of shaping the wealth distribution towards a bimodal distribution. Indeed, the trading propensities modeling personal responses to the economic scenario can be substantially modified by the progression of the epidemic and the vaccine efficacy.*

4. Numerical Results

In this section we study the impact of vaccination on the equilibrium of the kinetic system through several numerical simulations. This allows us to show the model's ability to describe different situations of wealth distribution in the presence of epidemic dynamics. In particular, we will adopt standard direct simulation Monte Carlo methods to simulate the system of kinetic Equation (2); see [27] and the references therein. In all the subsequent tests we will consider $N = 10^5$ agents and the densities are reconstructed through standard histograms.

In the first test, we verify numerically the convergence of the solution to the kinetic system (2) to the solution of the Fokker–Planck system (24) under the scaling (23). Then, we study the emergence of wealth inequalities, measured through the Gini index, in relation to the effectiveness of the vaccine. These results are obtained both in the case of a constant market risk variance σ^2 and in the case of a variance that depends on the current epidemic situation. Lastly, we introduce the possibility that the effectiveness of the vaccine is also affected by the number of positive cases. This situation mimics the realistic case of the diffusion of viral variants for which an up-to-date vaccine may be not immediately available.

4.1. Test 1: Long-Time Behavior and Convergence to Equilibrium

In this test, we want to observe the convergence of the numerical solution of the kinetic system (2) to the one of the Fokker–Planck system (24) in the quasi-invariant limit introduced in Section 3.1. We consider the simplified case where $\beta(w, w_*) = \bar{\beta} = 0.2$, $\gamma_I = 1/12$ and $\zeta = 0.9$, for which we obtained the steady distributions in (25). These values are representative of realistic dynamics during the beginning of the COVID-19 pandemic; see, e.g., [6–8,39,54].

At time $t = 0$ we consider an inverse Gamma distribution

$$f(w,0) = \frac{(\mu-1)^{\mu}}{\Gamma(\mu)} \frac{\exp\left(-\dfrac{\mu-1}{\mu}\right)}{w^{1+\mu}}, \tag{29}$$

where $\Gamma(\cdot)$ is the Gamma function and $\mu = 10$. The distributions of the epidemic compartments are

$$f_S(w,0) = \rho_S f(w), \quad f_I(w,0) = \rho_I f(w), \quad f_V(w,0) = \rho_V f(w), \quad f_R(w,0) = \rho_R f(w), \tag{30}$$

where the mass fractions are $\rho_I = 7.5 \times 10^{-3}$, $\rho_V = 0$, $\rho_R = 4 \times 10^{-2}$ and $\rho_S = 1 - (\rho_I + \rho_V + \rho_R)$. Furthermore, we consider the value $\sigma^2 = 0.02$ for the market risk. In Figure 1 we show the numerical solution at time $T = 300$ of (2) in the scaling regime (23) with $\epsilon = 1, 0.5, 10^{-3}$.

In particular, provided an epidemic dynamics such that $V^\infty = 0.51$ and $R^\infty = 0.49$, we give numerical evidence of the aforementioned convergence in two regimes expressing increasing safeguard thresholds $1 - \lambda_J$, $J \in \{S, I, V, R\}$, for non-vaccinated agents

(*i*) $\lambda_S = 0.15, \lambda_I = 0.10, \lambda_V = 0.30, \lambda_R = 0.20$
(*ii*) $\lambda_S = 0.10, \lambda_I = 0.05, \lambda_V = 0.30, \lambda_R = 0.15$

where the same values of V^∞ and R^∞ are unchanged. In particular, we assume that recovered individuals are characterized by a greater safeguard parameter. This is coherent with the possibility of reinfection, which will be investigated in the last numerical test.

We observe that, if $\epsilon \ll 1$, the Fokker–Planck asymptotic distribution is a consistent approximation of the equilibrium distribution of the Boltzmann-type model. In both cases, the global distribution is a mixture of inverse Gamma densities and in the righ-hand plot depicted in Figure 1, we can clearly observe a bimodal shape for the wealth distribution. To highlight this, we have drawn the maximum points of the distributions f_V^∞, f_R^∞, which are at $\overline{w}_V, \overline{w}_R$, defined in (27) and (28).

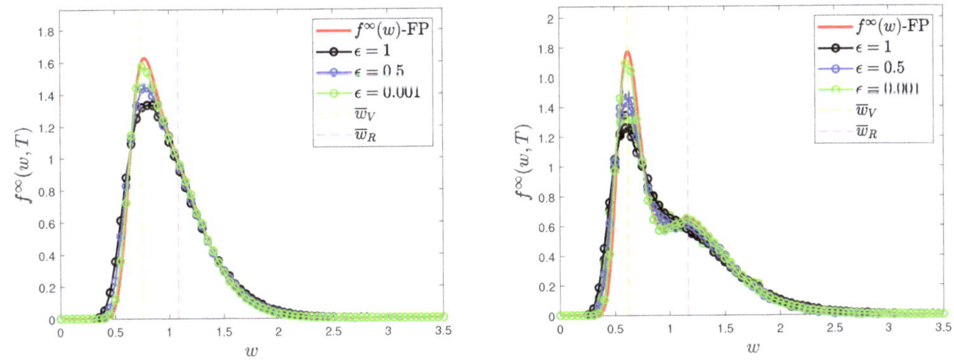

Figure 1. Test 1. Comparison of the wealth distributions at the end of the epidemic for the kinetic system (2) with the explicit Fokker–Planck asymptotics (26) with scaling parameters $\epsilon = 1, \frac{1}{2}, 10^{-3}$. (**Left**) $\lambda_S = 0.15$, $\lambda_I = 0.10$, $\lambda_V = 0.30$, $\lambda_R = 0.20$. (**Right**) $\lambda_S = 0.10$, $\lambda_I = 0.05$, $\lambda_V = 0.30$ $\lambda_R = 0.15$. In both cases we fixed $\bar{\beta} = 0.2$, $\gamma_I = 1/12$, $\alpha = 0.005$, $\zeta = 0.9$ and $\sigma^2 = 0.02$.

4.2. Test 2: Wealth Inequalities and Vaccination Campaign

In the second test case we analyze the emergence of wealth inequalities through the computation of the Gini index. In particular, we concentrated on the effects linked to the outbreak of the infection and on the impact of an effective vaccination campaign.

We fixed the epidemic parameters as follows: $\bar{\beta} = 0.15$, $\gamma_I = 1/12$ and a vaccination rate of $\alpha = 10^{-2}$. Furthermore, we considered two different vaccine efficacies $\zeta = 0.95$, corresponding to a high efficacy of the vaccine, and $\zeta = 0.55$ corresponding to a low efficacy of the vaccine. Since we are interested in the behavior of the system up to the conclusion of the epidemic phenomenon, the final time was fixed as $T = 810$, corresponding to a wide time-span. We kept the same values for the saving propensities and market risk as those defined for Section 4.1. Hence, we considered initial wealth distributions as in (29) and mass fractions as in (30), with $\rho_I = 7 \times 10^{-3}$, $\rho_V = 0$, $\rho_R = 4 \times 10^{-2}$ and $\rho_S = 1 - (\rho_I + \rho_V + \rho_R)$. The scaling coefficient was $\epsilon = 5 \times 10^{-2}$. The resulting epidemic dynamic is reported in Figure 2.

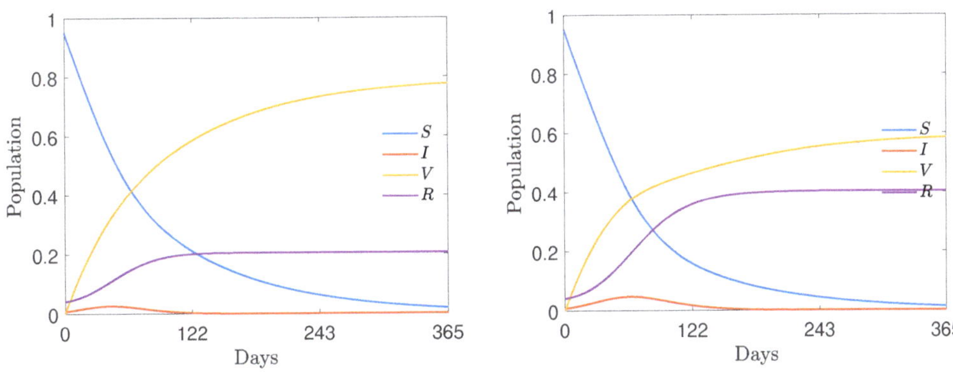

Figure 2. Test 2. Evolution of the epidemic dynamics from (9) for the choice of parameters $\bar{\beta} = 0.15$, $\gamma_I = 1/12$, $\alpha = 0.01$ and $\zeta = 0.95$ (**left**), $\zeta = 0.55$ (**right**).

We evaluated the Gini coefficient of the emerging equilibrium distributions. The Gini index is commonly computed from the Lorenz curve

$$L(F(w)) = \int_0^w f^\infty(w_*)w_* dw_*,$$

where $F(w) = \int_0^w f^\infty(w_*)dw_*$ and is defined as follows

$$G_1 = 1 - 2\int_0^1 L(x)dx.$$

This index should be understood as a measure of a country's wealth discrepancy and it varies in $[0,1]$, where in the case $G_1 = 0$ the country is in a situation of perfect equality, whereas $G_1 = 1$ indicates complete inequality. A reasonable value for this parameters is in the range $[0.2, 0.5]$ for most Western economies [36].

In Figure 3 we show the evolution of the Gini index with the parameters described above. We may observe that the epidemic peak leads to an increasing of inequalities that is then absorbed for later times in relation to the efficacy of the vaccine. Consequently, only when the vaccine is made available to the majority of the population does it actually contribute to reducing inequalities; otherwise, it may have the opposite effect. This reminds us of how, on a global level, the importance of making vaccines available to all countries should be seen not only in terms of epidemics, but also in terms of reducing economic inequalities. In all the considered cases, in the long term, the Gini index decreases thanks to the vaccine.

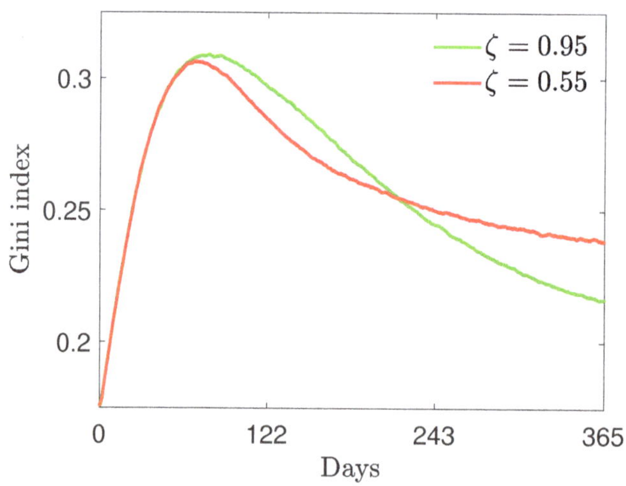

Figure 3. Test 2. Evolution of Gini index under the epidemic dynamics described in Figure 2 and for the choice of parameters $\lambda_S = 0.10$, $\lambda_I = 0.07$, $\lambda_V = 0.30$, $\lambda_R = 0.15$. Two vaccine efficacies were considered: 95% (green) and 55% (red). In both cases we considered $\sigma^2 = 0.02$.

Next, we consider the case where the market risk is related to the behavior of the epidemic's spread and where there is a linear relation between the market risk and the number of people infected. The introduction of a time-dependent market risk $\sigma^2(t)$ mimics an instantaneous influence of the pandemic on the volatility of a market economy, as is often observed. Therefore, we consider the following:

$$\sigma^2(t) = \sigma_0^2(1 + \mu I(t)) \quad (31)$$

where $\mu > 0$ expresses the effective influence of the epidemic dynamics on the market volatility and $\sigma_0^2 > 0$ is an ineradicable baseline risk.

In the following, we choose $\mu = 50$ and $\sigma_0^2 = 0.02$. In Figure 4 we represent the evolution of $\sigma^2(t)$ in the presence of an epidemic characterized by $\bar{\beta} = 0.15$, $\gamma_I = 1/10$. Furthermore, we compare the Gini index in the presence of two effectiveness rates of the vaccine, i.e., $\zeta = 0.95$ and $\zeta = 0.55$. We may easily observe how an increasing variability leads to a worsening of the Gini index and, therefore, of the inequalities. The long-term behavior of the Gini index depends, as before, on the vaccine efficacy ζ such that low efficacy leads to increasing inequalities in the long term. This is due to the fact that as $t \to +\infty$ we have $I \to 0^+$ and then $\sigma^2(t) \to \sigma_0^2$.

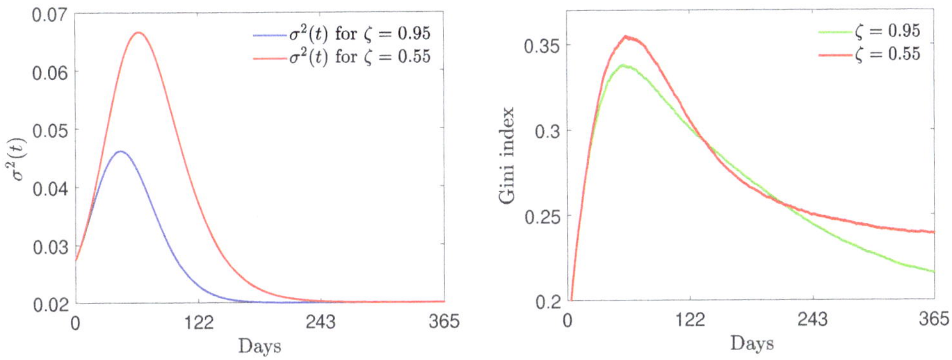

Figure 4. Test 2. (**Left**) evolution of the market risk $\sigma^2(t)$ as defined in (31) with $\mu = 50$ and $\sigma_0^2 = 0.02$ in case of two different vaccine efficacies. (**Right**) evolution of Gini index under the epidemic dynamics described in Figure 2 and epidemic-dependent market risk parameter (31).

Finally, in Figure 5 we present the evolution of the full kinetic density solution to (2) in the scaling $\epsilon = 5 \times 10^{-2}$ in the presence of fixed market risk σ^2 or with the epidemic-dependent $\sigma^2(t)$ discussed in (31).

4.3. Nonlinear Incidence Rate and Time-Varying Vaccine Efficacy

In this last test case, to model different frequencies of interactions between agents that belong to the same wealth class, we introduce a wealth-dependent contact rate $\beta(w, w_*)$ of the form

$$\beta(w, w_*) = \frac{\bar{\beta}}{(c + |w - w_*|)^\nu}, \tag{32}$$

where $\bar{\beta}, c, \nu > 0$. We have depicted the above contact rate in Figure 6.

We also introduce a time-dependent efficacy of the vaccine ζ of the form

$$\zeta(t) = \zeta_0 - \psi \int_0^t \int_{\mathbb{R}_+} f_I(w,t)\,dw\,ds = \zeta_0 - \psi \int_0^t I(s)\,ds, \tag{33}$$

with $\zeta_0 \in [0,1]$ indicating the initial efficacy of the vaccine and $0 < \psi \leq \zeta_0$. This time-dependence in vaccine coverage describes, in a simplified way, the fact that with more infected individuals it is more likely to encounter mutations of the original virus, for which the vaccine is less effective. In the following, we compare the evolution of the wealth inequalities in the presence of two different values ζ_0.

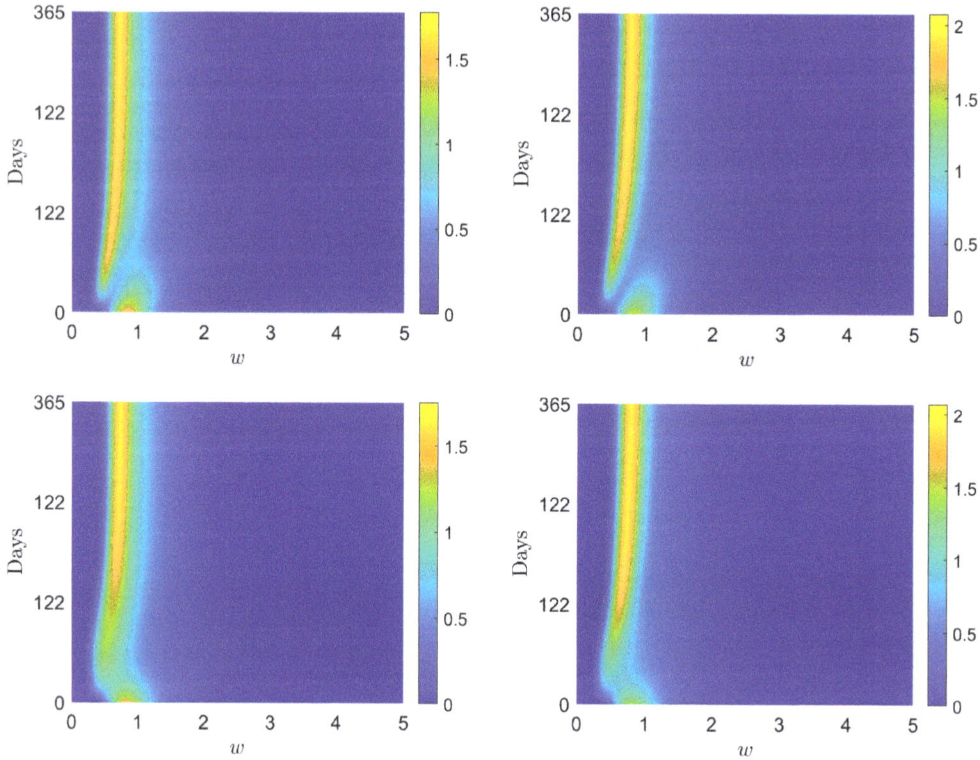

Figure 5. Test 2. Time evolution of the wealth distribution of the kinetic model (2) in the scaling $\epsilon = 5 \times 10^{-2}$ with vaccine efficacy $\zeta = 0.55$ (**left** column) or $\zeta = 0.95$ (**right** column) and with constant market risk $\sigma^2 = 0.02$ (top row) or $\sigma^2(t)$, defined in (31) with $\mu = 50$. In all the evolutions we considered $\lambda_S = 0.10$, $\lambda_I = 0.07$, $\lambda_V = 0.30$ and $\lambda_R = 0.15$. The initial distribution was defined in (29) and (30). In the left image, we can observe the evolution of the wealth distribution for the kinetic model (2) in the scaling parameter $\epsilon = 5 \times 10^{-2}$ with $\zeta = 0.95$, whereas, in the right image we have the comparison between the behaviors of the Gini index with vaccine effectiveness, equal to 95% (green line) and 65% (red line). In both images we considered a variable market risk (31) with $\sigma_0^2 = 0.02$ and $\mu = 50$ and $\lambda_S = 0.10$, $\lambda_I = 0.07$, $\lambda_V = 0.30$ and $\lambda_R = 0.15$.

Furthermore, to make the modeling more realistic, we assume the loss of immunity of the agents in the compartment R. To this end, we have to modify the first and last equations of the model (2) as follows

$$
\begin{aligned}
\partial_t f_S(w,t) &= -K(f_S, f_I)(w,t) - \alpha f_S(w,t) + \gamma_R f_R(w,t) + \sum_{J\in\{S,I,V,R\}} Q_{SJ}(f_S, f_J)(w,t) \\
\partial_t f_R(w,t) &= \gamma_I f_I(w,t) - \gamma_R f_R(w,t) + \sum_{J\in\{S,I,V,R\}} Q_{RJ}(f_R, f_J)(w,t),
\end{aligned}
\tag{34}
$$

where $\gamma_R \geq 0$ is the rate expressing the loss of immunity of recovered agents. Note that this latter assumption substantially changes the epidemic dynamics, since asymptotically, instead of a disease-free scenario, we have the emergence of endemic states [3].

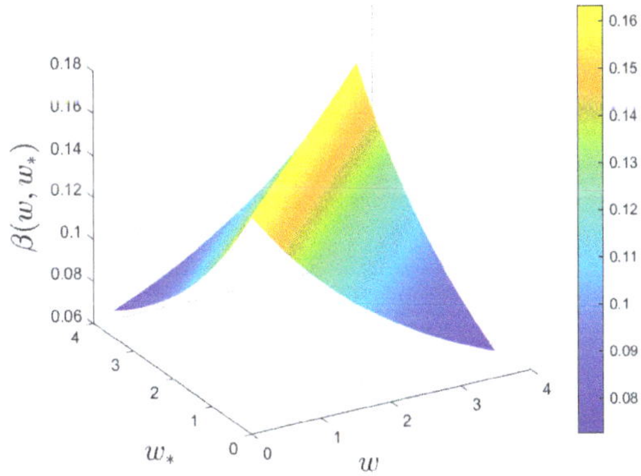

Figure 6. Test 3. Wealth-dependent contact rate $\beta(w, w_*)$ of the form (32) with $\bar{\beta} = 8$, $c = 7$, $\nu = 2$.

4.3.1. Test 3A: $\gamma_R = 0$

First, we consider model (2) without the modified relations (34) (or equivalently, in the absence of reinfection, i.e., $\gamma_R = 0$) and, as before, a fixed recovery rate $\gamma_I = 1/12$ and a vaccination rate $\alpha = 0.005$ with the same initial masses as those defined in (30). Furthermore, we fixed $\psi = 0.005$. In Figure 7, in the top row, we show the evolution for the fractions of the population in the case of $\zeta_0 = 0.95$ (left) and $\zeta_0 = 0.55$ (right). We may observe how a variable efficacy of the vaccine, affected by epidemic peaks, may strongly shape the immunity of the population, even in the presence of an initial high efficacy. Interestingly, in this latter case, a variable efficacy leads to the emergence of secondary peaks of infection. This is due to the presence of a smaller number of recovered persons who, unlike vaccinated people, maintain immunity.

In Figure 7, in the bottom-left row, we can observe the evolution of the resulting vaccine efficacy for $\zeta_0 = 0.95$, $\zeta_0 = 0.55$ and $\psi = 0.005$. The vaccine efficacy is degraded by the epidemic dynamics due to the increasing of the infected compartment, with a slower efficacy decay for high initial ζ_0.

For the same choice of coefficient, in the bottom-right plot of Figure 7, we show the evolution of the Gini coefficient in the case of variable efficacy as (33). With respect to a vaccine with constant efficacy, the efficacy decay forces the emergence of sharper inequalities, which is well evidenced by the evolution of the Gini coefficient.

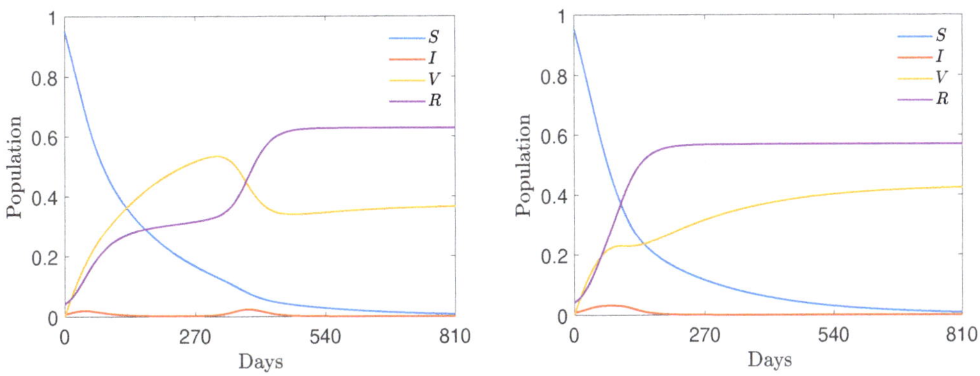

Figure 7. *Cont.*

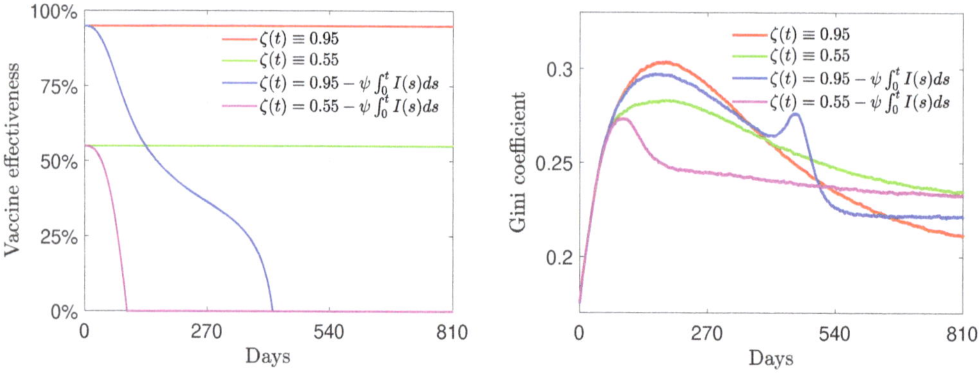

Figure 7. Test 3A. Top row: epidemic dynamics with wealth-dependent $\beta(w, w_*)$, defined in (32) with $\bar{\beta} = 8$, $c = 7$, $\nu = 2$, $\gamma_I = 1/12$, $\alpha = 0.005$ and variable ζ as in (33) with $\psi = 0.005$. We considered $\zeta_0 = 0.95$ (**left**) and $\zeta_0 = 0.55$ (**right**). The initial distribution is (29) with mass fractions (30). Bottom row: decline in vaccine efficacy due to the presence of a high number of infective people (**left**) and the evolution of the Gini index (**right**) for a variable infection rate $\beta(w, w_*)$ as in (32) and vaccine effectiveness $\zeta(t)$ as in (33). We considered $\lambda_S = 0.10$, $\lambda_I = 0.07$, $\lambda_V = 0.25$, $\lambda_R = 0.15$ and $\bar{\beta} = 8$, $c = 7$, $\nu = 2$ and $\psi = 0.005$.

4.3.2. Test 3B: $\gamma_R > 0$

Finally, we consider model (2) including the modified Equation (34), with a reinfection period of 180 days, i.e., $\gamma_R = 1/180$ and, as before, a fixed recovery rate $\gamma_I = 1/12$ and vaccination rate $\alpha = 0.005$ with the same initial masses as those defined in (30). In the first row of Figure 8, we present two epidemic dynamics with nonlinear contact rates (32) and the time-dependent efficacy $\zeta(t)$ defined in (33) with $\psi = 1.5 \times 10^{-4}$. In the left plot, we present the case of strong initial vaccine efficacy $\zeta_0 = 0.95$ and in the right plot the case of mild initial vaccine efficacy $\zeta_0 = 0.45$. The macroscopic dynamics present an endemic equilibrium due to the presence of the reinfection rate γ_R. Furthermore, in contrast to the previous case, in the case of reduced initial efficacy of the vaccine, a second infection wave is seen to emerge.

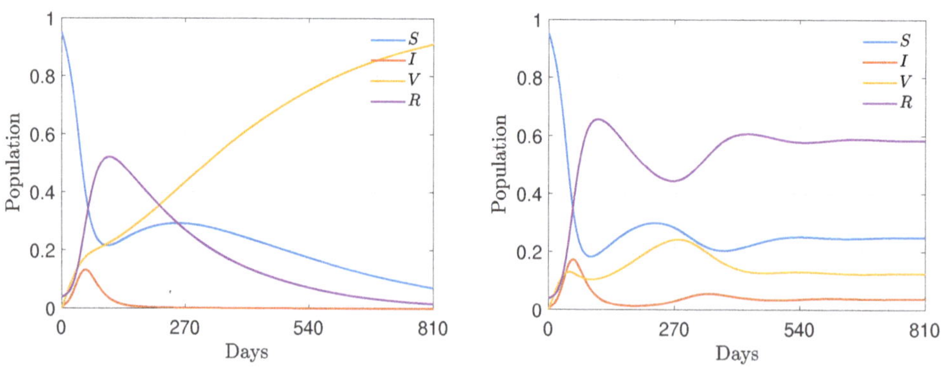

Figure 8. *Cont.*

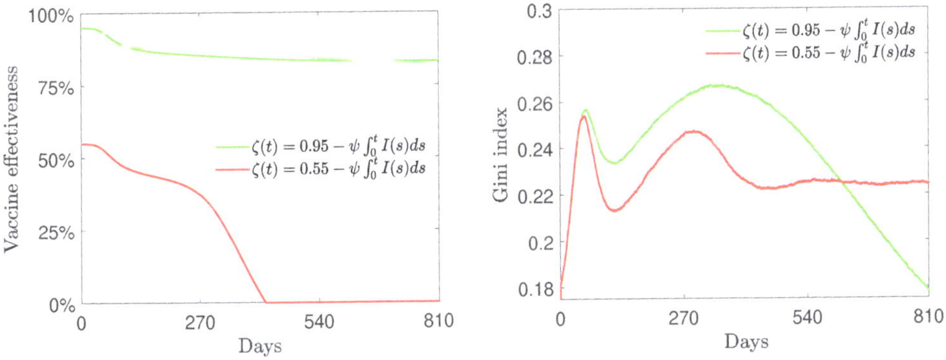

Figure 8. Test 3B. Top row: epidemic dynamics with wealth-dependent $\beta(w, w_*)$, defined in (32) with $\bar{\beta} = 8$, $c = 7$, $\nu = 2$, $\gamma_I = 1/12$, $\gamma_R = 1/180$, $\alpha = 0.005$ and variable ζ as in (33) with $\psi = 1.5 \times 10^{-4}$. We considered $\zeta_0 = 0.95$ (**left**) and $\zeta_0 = 0.55$ (**right**). The initial distribution is (29) with mass fractions (30). Bottom row: decline in vaccine efficacy due to the presence of a high number of infected people (**left**) and evolution of the Gini index (**right**). We considered $\lambda_S = 0.10$, $\lambda_I = 0.07$, $\lambda_V = 0.25$, $\lambda_R = 0.15$ and $\bar{\beta} = 8$, $c = 7$ and $\nu = 2$.

Looking at the bottom-left plot, we can observe that, in the present regime of parameters, a strong initial vaccine efficacy is robust with respect to the efficacy decay due to epidemic waves. On the other hand, mild initial efficacies can dissipate their positive influence on the evolution of the infection. At the level of the evolution of the Gini index, in the presence of reinfection, it appears even more evident that inequalities appear for large times in the presence of mild vaccinations. Nevertheless, in transient regimes, the higher possibility of investing wealth for vaccinated agents may create temporary inequalities.

5. Conclusions

The widespread vaccination campaign undertaken in Western countries to counteract the evolution of the COVID-19 epidemic and its economic effects depends in large part on the efficacy of vaccines. Mathematical models capable of predicting the evolution of the economy in relation to the effectiveness of the vaccination campaign can play a fundamental role in configuring possible scenarios and suggesting further measures to be taken by governments. In this paper we analyzed, at the level of wealth distribution, the economic improvements induced by the vaccination campaign in terms of its percentage of effectiveness. Following the ideas developed in [1,8], the interplay between the economic trend and the pandemic has been evaluated, resorting to a mathematical model combining a kinetic model for wealth exchanges based on binary interactions with a classical SIR compartmental epidemic model, including the compartment of vaccinated individuals. Extensions of the presented methodology are possible to include disease-related mortality and redistribution operators. Moreover, since a direct comparison of the results of similar compartmental kinetic models—in the case of social aspects related to the transience of the epidemic—outlined a good agreement with the actual data [8,38,39], we can assume that the present approach is able to follow the real evolution of the economic parameters of a country over a sufficiently long period of time. Indeed, even though the model introduced here necessarily represents a strong simplification of an extremely complex phenomenon, its qualitative behavior is capable of describing the essential features of the pandemic's impact on individuals' wealth. A key aspect of the model is, in fact, the possibility of obtaining explicit configurations of the stationary wealth distributions in the form of inverse Gamma densities, with the essential parameters depending on the percentage of vaccinated and recovered individuals, thus relating the effectiveness of the vaccination campaign to the formation of wealth inequalities. Several numerical experiments have also been conducted to quantify how a highly effective vaccination campaign has a direct effect on the decrease over time of the Gini coefficient, a classic measure of inequality in the distribution of wealth in Western societies.

Author Contributions: Conceptualization, L.P. and M.Z.; Data curation, E.B.; Investigation, E.B.; Methodology, L.P.; Project administration, G.T. and M.Z.; Supervision, L.P. and G.T.; Writing—original draft, E.B.; Writing—review & editing, M.Z. All authors have read and agreed to the published version of the manuscript.

Funding: This research was funded by Ministero dell'Università e della Ricerca, grant number 2020JLWP23 PRIN2020 and Università di Ferrara, grant number FIR2021.

Institutional Review Board Statement: Not applicable.

Informed Consent Statement: Not applicable.

Data Availability Statement: Not applicable.

Acknowledgments: This work was conducted within the activities of the GNFM and GNCS groups of INdAM (National Institute of High Mathematics). M.Z. acknowledges the partial support of MUR-PRIN2020 Project (No. 2020JLWP23) "Integrated mathematical approaches to socio-epidemiological dynamics". The research of M.Z. was partially supported by MIUR, Dipartimenti di Eccellenza Program (2018–2022), and Department of Mathematics "F. Casorati", University of Pavia. The research of L.P. was partially supported by FIR project "No hesitation. For effective communication of COVID-19 vaccination", University of Ferrara.

Conflicts of Interest: The authors declare no conflict of interest.

References

1. Dimarco, G.; Pareschi, L.; Toscani, G.; Zanella, M. Wealth distribution under the spread of infectious diseases. *Phys. Rev. E* **2020**, *102*, 022303. [CrossRef] [PubMed]
2. Brauer, F.; Castillo-Chavez, C.; Feng, Z. *Mathematical Models in Epidemiology*; Text in Applied Mathematics; Springer: Berlin/Heidelberg, Germany, 2019; Volume 69.
3. Hethcote, H.W. The mathematics of infectious diseases. *SIAM Rev.* **2000**, *42*, 599. [CrossRef]
4. Cordier, S.; Pareschi, L.; Toscani, G. On a kinetic model for a simple market economy. *J. Stat. Phys.* **2005**, *120*, 253. [CrossRef]
5. Chakraborti, A.S.; Chakrabarti, B.K. Microeconomics of the ideal gas like market models. *Phys. A* **2009**, *388*, 4151–4158. [CrossRef]
6. Gatto, M.; Bertuzzo, E.; Mari, L.; Miccoli, S.; Carraro, L.; Casagrandi, R.; Rinaldo, A. Spread and dynamics of the COVID-19 epidemic in Italy: Effects of emergency containment measures. *Proc. Natl. Acad. Sci. USA* **2020**, *117*, 10484–10491. [CrossRef] [PubMed]
7. Parolini, N.; Dedè, L.; Antonietti, P.F.; Ardenghi, G.; Manzoni, A.; Miglio, E.; Pugliese, A.; Verani, M.; Quarteroni, A. SUIHTER: A new mathematical model for COVID-19. Application to the analysis of the second epidemic outbreak in Italy. *Proc. R. Soc. A* **2021**, *477*, 20210027. [CrossRef]
8. Albi, G.; Bertaglia, G.; Boscheri, W.; Dimarco, G.; Pareschi, L.; Toscani, G.; Zanella, M. Kinetic modelling of epidemic dynamics: Social contacts, control with uncertain data, and multiscale spatial dynamics. In *Predicting Pandemics in a Globally Connected World*; Bellomo, N., Chaplain, M., Eds.; Springer: Berlin/Heidelberg, Germany, 2022; Volume 1.
9. Ashraf, B.N. Economic impact of government interventions during the COVID-19 pandemic: International evidence from financial markets. *J. Behav. Exp. Financ.* **2020**, *27*, 1003701. [CrossRef]
10. Bonaccorsi, G.; Pierri, F.; Cinelli, M.; Flori, A.; Galeazzi, A.; Porcelli, F.; Schmidt, A.L.; Valensise, C.M.; Scala, A.; Quattrociocchi, W.; et al. Economic and social consequences of human mobility restrictions under COVID-19. *Proc. Natl. Acad. Sci. USA* **2020**, *117*, 15530–15535. [CrossRef]
11. Gersovitz, M.; Hammer, J.S. The economical control of infectious diseases. *Econ. J.* **2004**, *114*, 1–27. [CrossRef]
12. Goenka, A.; Liu, L.; Nguyen, M.H. Infectious diseases and economic growth. *J. Math. Econ.* **2014**, *50*, 34. [CrossRef]
13. Gozzi, N.; Tizzoni, M.; Chinazzi, M.; Ferres, L.; Vespignani, A.; Perra, N. Estimating the effect of social inequalities on the mitigation of COVID-19 across communities in Santiago de Chile. *Nat. Commun.* **2021**, *12*, 2429. [CrossRef] [PubMed]
14. Zhang, D.; Hu, M.; Ji, Q. Financial markets under the global pandemic of COVID-19. *Financ. Res. Lett.* **2020**, *36*, 101528. [CrossRef] [PubMed]
15. Deaton, A. *COVID-19 and Global Income Inequality*; NBER Working Paper 28392; National Bureau of Economic Research: Cambridge, MA, USA, 2021.
16. von Braun, J.; Zamagni, S.; Sorondo, M.S. The moment to see the poor. *Science* **2020**, 368, 214. [CrossRef] [PubMed]
17. Ghostine, R.; Gharamti, M.; Hassrouny, S.; Hoteit, I. An extended SEIR model with vaccination for forecasting the COVID-19 pandemic in Saudi Arabia using an ensamble Kalman filter. *Mathematics* **2021**, *9*, 636. [CrossRef]

18. Bolzoni, L.; Bonacini, E.; Soresina, C.; Groppi, M. Time-optimal control strategies in SIR epidemic models. *Math. Biosci.* **2017**, *292*, 86–96. [CrossRef] [PubMed]
19. Buonomo, B.; Carbone, G.; D'Onofrio, A. Effect of seasonality on the dynamics of an imitation-based vaccination model with public health intervention. *Math. Biosci. Eng.* **2018**, *15*, 299–321. [PubMed]
20. Buonomo, B.; Lacitignola, D.; Vargas-De-León, C. Qualitative analysis and optimal control of an epidemic model with vaccination and treatment. *Math. Comput. Simul.* **2014**, *100*, 88–102. [CrossRef]
21. Buonomo, B.; Marca, R.D.; d'Onofrio, A.; Groppi, M. A behavioural modelling approach to assess the impact of COVID-19 vaccine hesitancy. *J. Theor. Biol.* **2022**, *534*, 110973. [CrossRef]
22. Colombo, R.M.; Garavello, M. Optimizing vaccination strategies in an age structured SIR model. *Math. Biosci. Eng.* **2020**, *17*, 1074–1089. [CrossRef]
23. Dolgin, E. COVID vaccine immunity is waning—How much does that matter? *Nature* **2021**, *597*, 606–607. [CrossRef]
24. Moore, S.; Hill, E.M.; Dyson, L.; Tildesley, M.J.; Keeling, M.J. Modelling optimal vaccination strategy for SARS-CoV-2 in the UK. *PLoS Comput. Biol.* **2021**, *17*, e1008849. [CrossRef] [PubMed]
25. Sun, D.; Li, Y.; Teng, Z.; Zhang, T.; Lu, J. Dynamical properties in an SVEIR epidemic model with age-dependent vaccination, latency, infection, and relapse. *Math. Meth. Appl. Sci.* **2021**, *44*, 12810–12834. [CrossRef]
26. Townsend, J.P.; Hassler, H.B.; Wang, Z.; Miura, S.; Singh, J.; Kumar, S.; Ruddle, N.H.; Galvani, A.P.; Dornburg, A. The durability of immunity against reinfection by SARS-CoV-2: A comparative evolutionary study. *Lancet* **2021**, *2*, E666–E675. [CrossRef]
27. Pareschi, L.; Toscani, G. *Interacting Multiagent Systems: Kinetic Equations & Monte Carlo Methods*; Oxford University Press: Oxford, UK, 2013.
28. Bouchaud, J.F.; Mézard, M. Wealth condensation in a simple model of economy. *Phys. A* **2000**, *282*, 536–545. [CrossRef]
29. Chakraborti, A.S.; Chakrabarti, B.K. Statistical mechanics of money: How saving propensity affects its distribution. *Eur. Phys. J. B* **2000**, *17*, 167–170. [CrossRef]
30. Chatterjee, A.; Chakrabarti, B.K.; Stinchcombe, R.B. Master equation for a kinetic model of trading market and its analytic solution. *Phys. Rev. E* **2005**, *72*, 026126. [CrossRef]
31. Drăgulescu, A.; Yakovenko, V.M. Statistical mechanics of money. *Eur. Phys. J. B* **2000**, *17*, 723–729. [CrossRef]
32. Garibaldi, U.; Scalas, E.; Donadio, S. Statistical equilibrium in simple exchange games I: Methods of solution and application to the Bennati-Dragulescu-Yakovenko (BDY) game. *Eur. Phys. J. B* **2006**, *53*, 267–272.
33. Ghosh, A.; Chatterjee, A.; Inoue, J.I.; Chakrabarti, B.K. Inequality measures in kinetic exchange models of wealth distributions. *Phys. A* **2016**, *451*, 465–474. [CrossRef]
34. Giordano, G.; Blanchini, F.; Bruno, R.; Colaneri, P.; Filippo, A.D.; Matteo, A.D.; Colaneri, M. Modelling the COVID-19 epidemic and implementation of population-wide interventions in Italy. *Nat. Med.* **2020**, *26*, 855–860. [CrossRef]
35. Gupta, A.K. Models of wealth distributions: A perspective. In *Econophysics and Sociophysics: Trends and Perspectives*; Chakrabarti, B.K., Chakraborti, A., Chatterjee, A., Eds.; Wiley VHC: Weinheim, Germany, 2006; pp. 161–190.
36. Düring, B.; Pareschi, L.; Toscani, G. Kinetic models for optimal control of wealth inequalities. *Eur. Phys. J. B* **2018**, *91*, 265. [CrossRef]
37. Mantegna, R.N.; Stanley, H.E. Scaling behaviour in the dynamics of an economic index. *Nature* **1995**, *376*, 46–49. [CrossRef]
38. Dimarco, G.; Perthame, B.; Toscani, G.; Zanella, M. Kinetic models for epidemic dynamics with social heterogeneity. *J. Math. Biol.* **2021**, *83*, 4. [CrossRef] [PubMed]
39. Zanella, M.; Bardelli, C.; Dimarco, G.; Deandrea, S.; Perotti, P.; Azzi, M.; Figini, S.; Toscani, G. A data-driven epidemic model with social structure for understanding the COVID-19 infection on a heavily affected Italian Province. *Math. Mod. Meth. Appl. Sci.* **2021**, *31*, 2533–2570. [CrossRef]
40. Bellomo, N.; Bingham, R.; Chaplain, M.A.J.; Dosi, G.; Forni, G.; Knopoff, D.A.; Lowengrub, J.; Twarock, R.; Virgillito, M.E. A multiscale model of virus pandemic: Heterogeneous interactive entities in a globally connected world. *Math. Mod. Meth. Appl. Scie.* **2020**, *30*, 1591–1651. [CrossRef]
41. Loy, N.; Tosin, A. A viral load-based model for epidemic spread on spatial networks. *Math. Biosci. Eng.* **2021**, *18*, 5635–5663. [CrossRef] [PubMed]
42. Liu, X.; Stechlinski, P. Infectious disease models with time-varying parameters and general nonlinear incidence rate. *Appl. Math. Model.* **2012**, *36*, 1974–1994. [CrossRef] [PubMed]
43. Britton, T.; Ball, F.; Trapman, P. A mathematical model reveals the influence of population heterogeneity on herd immunity to SARS-CoV-2. *Science* **2020**, *369*, 846–849. [CrossRef]
44. Fumanelli, L.; Ajelli, M.; Manfredi, P.; Vespignani, A.; Merler, S. Inferring the structure of social contacts from demographic data in the analysis of infectious diseases spread. *PLoS Comput. Biol.* **2012**, *8*, e1002673. [CrossRef]
45. Lunelli, A.; Pugliese, A.; Rizzo, C. Epidemic patch models applied to pandemic influenza: Contact matrix, stochasticity, robustness of predictions. *Math. Biosci.* **2009**, *220*, 24–33. [CrossRef] [PubMed]
46. McCarthy, Z.; Xiao, Y.; Scarabel, F.; Tang, B.; Bragazzi, N.L.; Nah, K.; Heffernan, J.K.; Asgary, A.; Murty, V.K.; Ogden, N.H.; et al. Quantifying the shift in social contact patterns in response to non-pharmaceutical interventions. *J. Math. Ind.* **2020**, *10*, 28. [CrossRef]

47. Mossong, J.; Hens, N.; Jit, M.; Beutels, P.; Auranen, K.; Mikolajczyk, R.; Massari, M.; Salmaso, S.; Tomba, G.S.; Wallinga, J.; et al. Social contacts and mixing patterns relevant to the spread of infectious diseases. *PLoS Med.* **2008**, *5*, 0381–0391. [CrossRef] [PubMed]
48. Colombo, R.M.; Garavello, M.; Marcellini, F. An age and space structured SIR model describing the COVID-19 pandemic. *J. Math. Ind.* **2020**,*10*, 22. [CrossRef] [PubMed]
49. Bisi, M.; Carrillo, J.A.; Toscani, G. Contractive metrics for a Boltzmann equation for granular gases: Diffusive equilibria. *J. Stat. Phys.* **2005**, *118*, 301–331. [CrossRef] [PubMed]
50. Cercignani, C. *The Boltzmann Equation and its Applications*; Springer: Berlin, Germany, 1988. [CrossRef]
51. Villani, C. On a new class of weak solutions to the spatially homogeneous Boltzmann and Landau equations. *Arch. Ration. Mech. Anal.* **1998**, *143*, 273–307.
52. Risken, H. *The Fokker-Planck Equation: Methods of Solution and Applications*; Springer Series in Synergetics; Springer: Berlin/Heidelberg, Germay, 1996; Volume 18. [CrossRef]
53. Ferrero, J.C. The monomodal, polymodal, equilibrium and nonequilibrium distribution of money. In *Econophysics of Wealth Distributions*; Chatterjee, A., Yarlagadda, S., Chakrabarti, B.K., Eds.; Springer: Cernusco, Italy, 2005; pp. 159–167.
54. Dolbeault, J.; Turinici, G. Social heterogeneity and the COVID-19 lockdown in a multi-group SEIR model. *Comput. Math. Biophys.* **2021**, *9*, 14–21.

entropy

Article

Asymmetric Relatedness from Partial Correlation

Carlos Saenz de Pipaon Perez [1], Andrea Zaccaria [2,3,*] and Tiziana Di Matteo [1,3,4]

1 Department of Mathematics, King's College London, The Strand, London WC2R 2LS, UK; cspipaon@gmail.com (C.S.d.P.P.); tiziana.di_matteo@kcl.ac.uk (T.D.M.)
2 Istituto dei Sistemi Complessi (ISC)—CNR, UoS Sapienza, P.le A. Moro, 2, 00185 Rome, Italy
3 Centro Ricerche Enrico Fermi, Piazza del Viminale, 1, 00184 Rome, Italy
4 Complexity Science Hub Vienna, Josefstädter Straße 39, A 1080 Vienna, Austria
* Correspondence: andrea.zaccaria@cnr.it

Abstract: Relatedness is a key concept in economic complexity, since the assessment of the similarity between industrial sectors enables policymakers to design optimal development strategies. However, among the different ways to quantify relatedness, a measure that takes explicitly into account the time correlation structure of exports is still lacking. In this paper, we introduce an asymmetric definition of relatedness by using statistically significant partial correlations between the exports of economic sectors and we apply it to a recently introduced database that integrates the export of physical goods with the export of services. Our asymmetric relatedness is obtained by generalising a recently introduced correlation-filtering algorithm, the partial correlation planar graph, in order to allow its application on multi-sample and multi-variate datasets, and in particular, bipartite temporal networks. The result is a network of economic activities whose links represent the respective influence in terms of temporal correlations; we also compute the statistical confidence of the edges in the network via an adapted bootstrapping procedure. We find that the underlying influence structure of the system leads to the formation of intuitively-related clusters of economic sectors in the network, and to a relatively strong assortative mixing of sectors according to their complexity. Moreover, hub nodes tend to form more robust connections than those in the periphery.

Keywords: complex systems; economic complexity; relatedness; products and services; planar graph; partial correlation

Citation: Saenz de Pipaon Perez, C.; Zaccaria, A.; Di Matteo, T. Asymmetric Relatedness from Partial Correlation. *Entropy* **2022**, *24*, 365. https://doi.org/10.3390/e24030365

Academic Editor: Stanisław Drożdż

Received: 10 January 2022
Accepted: 24 February 2022
Published: 3 March 2022

1. Introduction

In the past few years, the use of bipartite networks for the representation of real-world complex systems has become widespread in a variety of fields and applications. These networks are usually constructed using multi-sample, multi-variate structured data used to model complex systems such as biological networks (enzymes and reactions [1], genes and diseases [2], plants and pollinators [3]), movies and actors [4,5], authors and papers [5,6], board of directors members and companies [7,8], companies and technologies they patent [9], members of peer-to-peer networks and data provided [10], international NGO branches and cities hosting them [11], supreme court judges and their votes [12], and legislators and bills they sponsor [13].

A prominent example is the bipartite network formed by countries and the products they export. This type of data has been used extensively in the field of economic complexity (EC) [14,15] to assess various quantities of interest for the modelling of the economic development of countries. The first one is the competitiveness of countries and the sophistication of products [16–20], and the relatedness between products, countries, or between countries and products [21,22]. With respect to the datasets implemented in the literature up to now, the dataset we use in this paper adds the inclusion of services to the set of tangible products traditionally considered in the EC literature [23,24].

An agreed definition of relatedness still does not exist, despite the vast number of applications of this concept, that ranges from forecasting industrial upgrading [25] to its use as an explanatory variable in a number of different contexts (see [26] and references therein). In most cases, one computes a projection of the bipartite network (e.g., country-product) onto one of the two sets of nodes to obtain a monopartite network (e.g., product-product) [21,22,27]; the relatedness between the nodes of the target layer is given by the weights of the corresponding links. Since the information content of the projected network is always smaller than that in the bipartite network, the choice of the method employed to achieve this is highly non-trivial. The resulting network should be a meaningful representation of the bipartite network for the specific problem being tackled while minimising the information loss due to the projection. There are several methods available in the literature to carry out this task (see [26,28]); however, to the best of our knowledge, no one takes explicitly into account the temporal structure, with the possible exception of the time-delayed co-occurrences approach described in [23,29] which, however, does not take into account the correlation between the different time series involved. This is a key element, since a comprehensive unveiling of the complex interactions between industrial sectors clearly requires a dynamical perspective.

In this paper, we tackle this issue by quantifying the average influence between industrial sectors in terms of partial correlation. To do so we introduce a framework that generalises a network generation method based on correlation-filtering called the partial correlation planar graph (PCPG) algorithm [30] in order to allow for its use with multi-sample multi-variate datasets. Since this methodology is particularly suitable for bipartite networks such as the ones usually studied in EC, we have called our framework biPCPG. The PCPG is an adaptation of the Planar Maximally Filtered Graph (PMFG) [31] which is in turn a further step from the Minimum Spanning Tree (MST) [32]. Fruitfully applied to financial market dynamics [33], these methods are able to capture the heterogeneity of similarities usually found at different scales of correlation in complex systems thanks to them employing a hierarchical clustering approach rather than a thresholding approach. The advantage of the PMFG over the MST is that, due to its relaxed constraints, its output network contains loops and a larger amount of information than the MST by preserving all the hierarchical properties of the MST [31].

The PCPG [30] adapts the PMFG in order to capture asymmetric interactions among variables in the system, thus producing a directed network. The PCPG achieves this by employing an edge-weighting scheme based on partial correlations, which are a measure of how the correlation of two variables is affected by a third variable. More specifically, the so-called *influence* (the difference between correlation and partial correlation) is employed to measure the similarities in the system and is used as a metric to select the edges included in the network. In our case, this formulation of relatedness allows asymmetries to be detected in the system.

As a result, the PCPG network is a weighted, connected, directed network that includes the MST as a subgraph as well as allowing for other substructures such as loops and cliques of three and four elements which add to the information content of the graph [31]. The fact that the links present in the PCPG are mostly those which correspond to the largest correlations in the system ensures the statistical robustness of the network to a high extent [34].

The PCPG was originally developed for its use on multi-variate datasets of only one sample: the time series of different stocks. In our case we have the export time series, so not only many variables (the different products) but also many samples, one of each country. In this paper we propose an extension of the PCPG, that we call biPCPG, to allow its application on multi-sample and multi-variate datasets, e.g., the export time series, by product, of many countries.

Our proposed extension to the PCPG method involves the preparation of the multi-sample dataset in order to apply the PCPG algorithm. This is achieved by structuring the dataset into a set of correlation matrices among the time series of products exported by countries, averaging these, and applying the existing PCPG procedure. Following similar principles, we also adapt an existing bootstrapping procedure (see [34]) in order to determine the statistical reliability of the links present in the resulting network.

The contribution of this paper is many fold. Firstly, the biPCPG framework opens the possibility of the application of the PCPG algorithm to a wide variety of datasets with a multi-sample and multi-variate structure, including, but not limiting to, the ones usually analysed using the EC framework. Furthermore, the data-processing methodology introduced here could be utilised to apply other correlation-filtering algorithms for network generation (e.g., [31,33]).

Secondly, this paper introduces a network which describes the asymmetric relatedness among physical products (manufacturing) and services. This is an addition with respect to the networks usually present in the literature, such as the product space [21] and product taxonomy network [22], which are constituted only by products.

Thirdly, this paper introduces an adapted bootstrapping procedure to asses the reliability of the edges present in a network generated from multi-sample multi-variate datasets. Similarly to the network-generating framework, this bootstrapping procedure can be utilised to asses the reliability of edges in networks generated using alternative correlation-filtering methods with datasets with this structure.

Fourthly, in order to assess the information content of the biPCPG network we calculate two assortativity measures and run a community detection procedure, finding that meaningful clusters and connections emerge, as well as a relevant complexity-related assortativity. In summary, the biPCPG analysis unveils the average influence between industrial and service sectors, efficiently encapsulating the information about the correlation structure of the system.

Finally, we provide a Python package named "biPCPG" [35] with its documentation hosted in [36]. The 0.1.0 version of this package was used to perform all the calculations done in this paper, including the data-handling, biPCPG network generation, bootstrapping procedure and calculations done on the biPCPG network. It is worth noting that the package has a modular structure such that the data-handling and the generation of the biPCPG network are computed independently of each other. This allows the user to, for example, utilise the data-handling module to prepare a multi-sample multi-variate dataset for an alternative correlation-filtering method, or to implement the PCPG algorithm on a dataset of her choice, without the need for the dataset to have a multi-sample multi-variate structure. To the best of our knowledge, the PCPG module in the biPCPG package is the first publicly available Python implementation of the PCPG algorithm.

The rest of this paper is organised as follows. In Section 2, we describe the dataset used in this investigation and the cleaning procedure performed on it. In Section 3, we describe the set of methods to generate the biPCPG network and comment on the resulting network. In the result sections we describe the assortativity calculations and community detection procedure done on the biPCPG network and show the results obtained. Section 5 concludes.

2. Data Description and Preprocessing

The dataset used in this research project is an integration of the United Nations Commodity Trade Statistics Database (UN-COMTRADE—https://comtrade.un.org, accessed on 13 February 2019) and the International Monetary Fund's Balance of Payments data (BPM6) [37], relative to physical goods and service exports respectively. This integrated dataset was introduced in a World Bank working paper [23]. The UN-COMTRADE data consists of the amount of exports from each country per category of products (in USD). The categorisation of products is given by the World Customs Organization's (WCO) Harmonized System 2007 edition (HS2007) [38], which classifies products by using a hierarchical six-digit code depending on the category of the product. The IMF BPM6 dataset consists of the amount (in USD) of services provided abroad by each country and is collected according to the 6th edition of its manual, provided by the International Monetary

Fund (IMF). Henceforth, we will globally refer to the collection of products in COMTRADE and services in BPM6 as *sectors*.

The hierarchical structure of the HS classification allows for an aggregation from the most granular six-digit level, consisting in about 5000 different products, into a coarser two-digit level. A further aggregation of a few small (in terms of export quantities) two-digit sectors into a single two-digit sector was also performed in this dataset, leaving a total of 78, roughly homogeneous aggregated product sectors at the two-digit level. From the BPM6 part of the dataset, there are a further 22 service sectors at a comparable level of aggregation.

The aggregated dataset used in our study is therefore comprised of $78 + 22 = 100$ sectors of products and services, these are listed in Table A1 in Appendix D. The data span a total of 22 years, from 1995 to 2016. As there are missing data points in some years for several countries, we apply a sanitation procedure where only countries with complete data for all sectors throughout the 22 years are kept. This reduces the dataset to from 129 countries to 99 countries. The analysed dataset has a total $99 \times 100 = 9900$ time series of length 22, with no missing values, representing the amount of product exports or service provisions in USD for each country.

In order to perform specific calculations (see Section 4.2), the 100 sectors in the dataset must be aggregated one level further. The product sectors can be further aggregated using what the WCO refers to as *sections*. The WCO provides a total of 21 sections which are available at [38]. In this case, services sectors can be aggregated into a single "section". Thus, in our aggregated dataset we have a total of 22 sections of sectors—21 product sections arising from the HS2007 classification, and one additional section containing the service sectors from the BPM6 dataset.

Revealed Comparative Advantage Matrices

The raw data used to construct in this paper are the amount of exports $E^y_{c,p}$ (in USD) of a sector p (product or service) by a country c in year y. We compute the Revealed Comparative Advantage (RCA) [39] as

$$RCA^y_{c,p} = \frac{\text{ratio of } c\text{'s exports of } p \text{ to the total exports of } c \text{ in year } y}{\text{ratio of the world's exports of } p \text{ to the total world's exports of all sectors in year } y}$$
$$= \frac{E^y_{c,p} \Big/ \sum_{p' \in P} E^y_{c,p'}}{\sum_{c' \in C} E^y_{c',p} \Big/ \sum_{c' \in C, p' \in P} E^y_{c',p'}} \tag{1}$$

where P and C are the sets of unique sectors and unique countries in the dataset discussed above.

The use RCA is ubiquitous in the EC literature, because removes trivial dependencies from the sectors' and countries' size. When the $RCA^y_{c,p}$ is above 1, the country is said to have a revealed comparative advantage in exporting a given sector in that year. Conversely, when $RCA^y_{c,p}$ is below 1 the country can be thought of as not being very competitive in that particular sector. Finally, when $RCA^y_{c,p}$ is equal to 1 the country has the expected (average) share of the world's exports in the given sector and year.

Therefore, the dataset on which we perform the following calculations consists of time series $RCA_{c,p} = (RCA^y_{c,p} : y \in Y)$ for 99 countries and 100 sectors, where Y is the index set of years $[1995, 2016]$. The data is then shaped into a set of 22 matrices $\mathbf{RCA}^y$, one for each year, where each row represents a country, each column represents a sector and each entry is the corresponding $RCA^y_{c,p}$ value.

3. Methods: The biPCPG Framework

3.1. Methodology Description

Before discussing the detailed implementation of the biPCPG methodology, here we provide a summarised description of our procedure; a visual representation can be found in Figure 1.

Given the multi-sample nature of the dataset analysed, a series of data-preprocessing steps are needed before the application of PCPG. The PCPG algorithm takes a single correlation matrix as an input and outputs a network (see Section 3.5). In order to obtain our biPCPG network, along with reliability values for its edges from a multi-sample dataset, we need two main procedures, a "Network generating procedure" and a "Bootstrapping procedure".

The "Network generating procedure" is shown in the black box in Figure 1 and deals with the data handling necessary to obtain a PCPG network from a dataset with a multi-sample structure. In our case, we are interested in obtaining a biPCPG network where nodes are sectors, therefore the input matrix should describe the correlations between sectors.

To find this input correlation matrix, the initial step is to shape the dataset such that, for each country, we have a matrix where the columns are the relevant time series of each sector. We then compute a correlation matrix for each of these time series matrices. Finally, we average these correlation matrices over countries to obtain an average correlation matrix which serves as the input to the PCPG algorithm, i.e., the last step in the biPCPG framework. The output of the biPCPG algorithm is the network we refer to as G, as well as the weights of the edges in contains, i.e., the average influence between sectors.

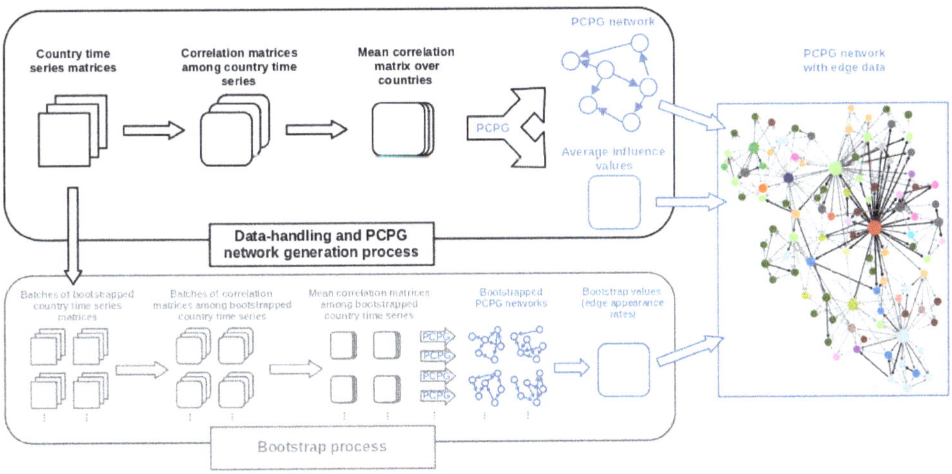

Figure 1. Flowchart of procedures and methods involved in obtaining the final biPCPG network.

The "Bootstrapping procedure" of our framework, shown in the grey box in Figure 1, deals with the bootstrapping procedure necessary to asses the reliability of the edges in the biPCPG network obtained. This starts from the country time series matrices, which are bootstrapped R times, obtaining a "batch" of replicates each time. Each of these batches contains C matrices, one for each country, where the rows have been drawn coherently from their corresponding original country matrices. This is done in order to randomise the time dimension while preserving the correlation structure across countries (see Section 3.6). We then replicate the "Network generating procedure" described above by treating each batch of replicates as a new dataset of country time series matrices and follow the steps to obtain a replicate biPCPG network. This means that, for each batch, we calculate a correlation matrix for every time series matrix, we then average across these correlation matrices and use the average correlation matrix as an input to the PCPG algorithm. Repeating this procedure for all R batches we obtain R replicate networks. We find the fraction of times each edge in G appears in the replicate networks, which is a measure of the reliability of the edge.

3.2. Partial Correlations and Average Influence: Definitions

As described in the original PCPG paper (see [30]), the starting point of our analysis is the *partial correlation*, which measures the effect that a random variable Z has on the correlation between two other random variables, X and Y. The partial correlation $\rho(X, Y : Z)$ is defined in terms of the Pearson correlations $\rho(\cdot, \cdot)$ between the three variables, formally

$$\rho(X, Y : Z) = \frac{\rho(X, Y) - \rho(X, Z)\rho(Y, Z)}{\sqrt{[1 - \rho^2(X, Z)][1 - \rho^2(Y, Z)]}}. \quad (2)$$

A small value of $\rho(X, Y : Z)$ may be ambiguous, as this could be due to the correlations among the three variables being small; or due to variable Z having a strong effect on the correlation between X and Y, which is generally the interesting case. In order to discriminate between these two cases the *correlation influence* or *influence* of variable Z on the pair of elements X and Y is used. This is defined as

$$d(X, Y : Z) \equiv \rho(X, Y) - \rho(X, Y : Z). \quad (3)$$

We define the *average influence* of variable Z on the correlations between X and all other variables in the system as follows:

$$d(X : Z) = \langle d(X, Y : Z) \rangle_{Y \neq X}. \quad (4)$$

We anticipate that the average influence will be the input of the network building algorithm also described in [30].

Note that, potentially, there could be certain values of *measured* correlations $\rho(X, Y)$, $\rho(X, Z)$ and $\rho(Y, Z)$ that lead to a *measured* partial correlation $\rho(X, Y : Z)$, to be out of its defined range $[-1, 1]$. In our analysis, this occurred in 0.02% of the partial correlations computed. In these cases, partial correlations were set to be undefined (*NaN* in programming terms) which in turn makes the influence values based on these partial correlations also undefined. Similarly to the undefined correlation values described above, these undefined influences are not included in calculation of average influence $d(X : Z)$.

Some of the values obtained for $\rho(X, Y)$, $\rho(X, Y : Z)$, $d(X, Y : Z)$ and $d(X : Z)$ in our dataset and their interpretation are discussed in Section 3.4. An important point is that, in general, $d(X : Z) \neq d(Z : X)$: the influence is asymmetric, and the largest among these two quantities indicates the main direction of influence between X and Z. For example, in our dataset when $X =$ Glass and $Z =$ Furniture, the average influence of Furniture on Glass $d(X : Z) = 0.03$ while the corresponding reverse average influence of Glass on Furniture $d(Z : X) = 0.29$, suggesting that the direction of influence is from Glass to Furniture and not vice-versa. This, however, is an example of a clear-cut case, where difference between the two average influence values is not small. In general, these differences tend to be much smaller. This can be an effect of the complex relationship and mutual interaction between the economic sectors, or a consequence of the noise present in the data. This makes a bootstrapping procedure necessary in order to asses the statistical confidence in the overall direction of influence, as well as the average influence values themselves. We will discuss the bootstrapping procedure in Section 3.6.

3.3. Average Correlation Matrix

The input to the PCPG algorithm is a correlation matrix [30]. In our procedure, to allow its use on our multi-sample dataset, this correlation matrix is replaced by an average correlation matrix over countries. In order to obtain this average correlation matrix, we reshape the 22 $\mathbf{RCA}^y$ matrices into a total of $C = 99$ matrices, one for each country, each consisting of $T = 22$ rows and $P = 100$ columns. We denote these $\mathbf{TS}_c$, $c \in 1, \ldots, C$.

In this way, the columns of each matrix $\mathbf{TS}_c$ are the $RCA_{c,p}$ time series of the given country c, where each column represents a sector p in the dataset.

In order to obtain the input matrix to the PCPG algorithm, we first find C correlation matrices denoted $\mathbf{K}_c$, $c \in 1, \ldots, C$ from the pair correlations between the columns of each matrix $\mathbf{TS}_c$. Thus the entries of the country correlation matrix $\mathbf{K}_c$ are given by

$$\left(K_c\right)_{p,p'} = \rho\left(\left(TS_c\right)_{*,p}, \left(TS_c\right)_{*,p'}\right) = \rho\left(RCA_{c,p}, RCA_{c,p'}\right) \tag{5}$$

where ρ is the Pearson correlation, the subscript $*, p$ denotes the column p of the matrix and $RCA_{c,p}$ is the RCA time series for country c and sector p.

For each correlation value we obtain p-value via a two-sided T-test procedure [40]. Given we are performing multiple tests, we apply a False Discovery Rate (FDR) correction to obtain *adjusted* p-values via the Benjamini–Hochberg (BH) procedure [41]. We choose the BH procedure since it ultimately allows the inclusion of more information in the biPCPG network than a more restrictive correction procedure such as the Bonferroni correction [42]. Note that the FDR correction has been extensively used in the literature for the statistical validation of networks and, in particular, it has been previously used to validate networks representing bipartite complex systems [43].

We reject non-statistically significant correlation samples when the adjusted p-value is above a critical value of 0.01. In these cases, the corresponding entries to the $\mathbf{K}_c$ matrix are marked as undefined. The same procedure for obtaining country correlation matrices was also performed without the FDR correction for the 0.01 and alternative critical values. This produced networks which have the same main features as the network presented below, including the main hub nodes, clusters of sectors and communities detected.

Once the country correlation matrices $\mathbf{K}_c$ are found, we then compute the element-wise mean of these matrices, obtaining the average correlation matrix $\bar{\mathbf{K}}$ with entries

$$\bar{K}_{p,p'} = \frac{1}{C}\sum_{c=1}^{C}\left(K_c\right)_{p,p'}, \tag{6}$$

where row and column indices p and p' denote economic sectors. Any undefined correlation is discarded during the averaging process.

Note that, using this notation, the correlations $\rho(\cdot,\cdot)$ mentioned in Section 3.2, are replaced by the average correlations $\bar{K}_{p,p'}$ described here. This leads to an equivalent expression for the partial correlation

$$\rho(p,p' : p'') = \frac{\bar{K}_{p,p'} - \bar{K}_{p,p''}\bar{K}_{p',p''}}{\sqrt{\left[1-(\bar{K}_{p,p''})^2\right]\left[1-(\bar{K}_{p',p''})^2\right]}}. \tag{7}$$

3.4. Partial Correlation and Average Influence: Empirical Analysis

In order to clarify the meaning of the intermediate quantities that are used to build the biPCPG network, we devote this subsection to the discussion of some empirical features.

Bearing in mind how the influence of a variable on the correlation of two other variables is defined (see Equation (3)), we explore four examples of the results obtained from these computations. Note that, in the description below, the variables X, Y and Z used in the definition of Equation (3), are replaced by sectors of our system. Thus, the partial correlation column in Table 1 describes the average correlation, $\bar{K}_{p,p'}$, between sectors p and p' accounting for the effect of a third sector p'', and similarly for the influence column. We therefore denote these quantities $\rho(p,p' : p'')$ and $d(p,p' : p'')$, respectively.

Example 1 shown in Table 1 is an example of the case described in Section 3.2, which shows a very small partial correlation due to all correlations among the three variables being small. By definition, this makes the resultant influence value is small, which reduces the average influence of the sector "Other textile" on the sector "Cereals", making the appearance of this edge in the network less probable.

Example 2 also shows a case where the partial correlation between p and p', accounting for the effect of p'', is small. However, contrary to the case in Example 1, this is due to p'' strongly affecting the correlation between p and p', i.e., $\rho(p, p') \sim \rho(p, p'')\rho(p', p'')$. Therefore, the resulting influence is relatively high, which increases the probability of an edge from "Cultural" to "Audiovisual" being present in the biPCPG network. In addition, note that the probability of an edge from "Cultural" to "Audiovisual" also increases under these results, due to the symmetry between the p and p' variables.

In Example 3, we have a case where the correlation between p and p' is relatively strong and variable p'' has a small effect on it. This is due to the similar values of the correlation $\rho(p, p')$ and the partial correlation $\rho(p, p' : p'')$. Therefore, the resulting influence of "Knitted clothing" on the correlation between the "Pigments" and "Aluminium" sectors is close to zero.

Finally, Example 4 shows a seemingly counter-intuitive case where the correlation between p and p' is small while their partial correlation given p'' is negative, yielding a high influence. A negative partial correlation occurs when the correlation between p and p' is small but both p and p' have a high correlation with p''. In this case, the influence of "Plastics" can be interpreted as preventing the correlation $\rho(p, p')$ between "Vehicles" and "Earths and stone" from being lower, or being negative.

Table 1. Examples of values used in the computations of influence $d(p, p' : p'')$.

	Variable & Sector		Corr. $\bar{K}_{p,p'}$	Corr. $\bar{K}_{p,p''}$	Corr. $\bar{K}_{p',p''}$	Partial Corr. $\rho(p, p' : p'')$	Influence $d(p, p' : p'')$
Ex. 1	p	Cereals	0.024388	−0.017268	0.028770	0.024899	−0.000511
	p′	Telecommunication					
	p″	Other textile					
Ex. 2	p	Audiovisual	0.283807	0.772049	0.368241	−0.000834	0.284641
	p′	Sea Transport					
	p″	Cultural					
Ex. 3	p	Pigments	0.602575	0.064069	0.040062	0.601727	0.000848
	p′	Aluminium					
	p″	Knitted clothing					
Ex. 4	p	Vehicles	0.025574	0.781281	0.542898	−0.760384	0.785958
	p′	Earths and stone					
	p″	Plastics					

It is important to note that the average influence values among sector pairs determine the structure of any PCPG network (see Section 3.5). Figure 2 displays a scatter plot that shows the correlation $\rho(\cdot, \cdot)$ and average influence $d(\cdot, \cdot)$ among all $N(N-1) = 9900$ pairs of sectors in our biPCPG network. Note that this includes data points for both $d(p : p'')$ and $d(p'' : p)$ influences at the same horizontal coordinate as the correlation between p and p'' is symmetric.

This plot shows that the average influence between a pair of sectors is highly correlated with the correlation between the same pair of sectors, showing a very narrow 95% confidence interval (barely visible as it is only slightly wider than the fit line). See Appendix B for details on the calculation of the confidence and prediction intervals shown in Figure 2.

This is not surprising given how the average influence is calculated; however, the relatively high coefficient of determination $R^2 = 0.58$ indicates that, generally, the partial correlation values obtained are relatively small. This may be due to there actually not being large influences between the sectors, or due to limitations of the dataset. For example, hidden influences between the sectors could potentially be detected in datasets with longer time series.

In Figure 2, we can observe that most of the correlations (around 80%) are positive. Around 10.7% of the pairs of sectors with positive correlations have an average influence below zero. This quantity is over an order of magnitude larger than its counterpart, the percentage of pairs of sectors with negative correlation but a positive average influence, which is around 0.47%.

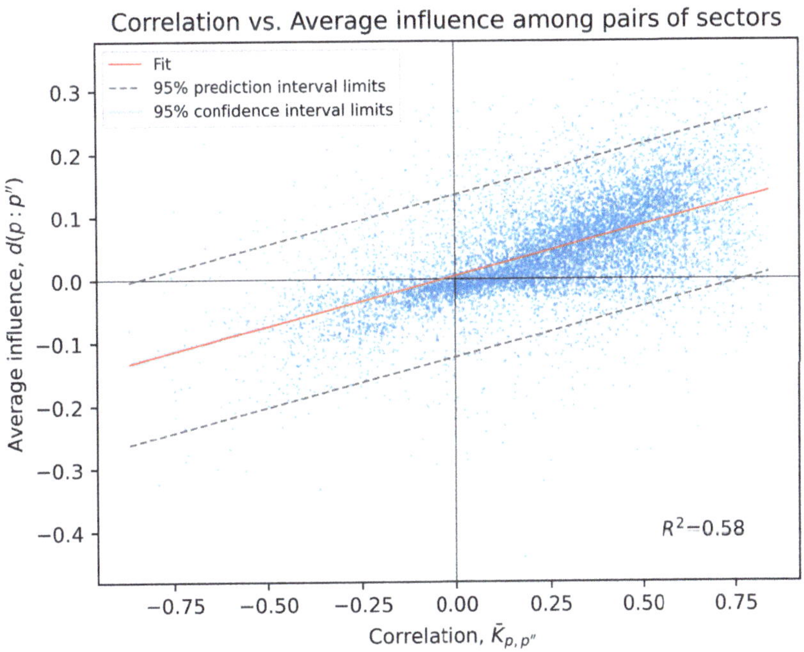

Figure 2. Plot showing correlation and average influence values among all 9900 pairs of sectors in the system. A line of best fit among the points is shown in red along with the coefficient of determination $R^2 = 0.58$, with the 95% confidence interval limits in light blue and the 95% prediction interval limits in dashed grey lines. Note the confidence interval is so narrow it is only visible at the edges of the red best fit line upon close inspection.

3.5. Network Construction

The construction algorithm of a PCPG network starts with a list of the $N(N-1)$ average influence values in decreasing order and an empty graph of N nodes and no edges, where N is the number of variables in the system. In our case, we have $N = 100$ economic sectors. We then cycle through the sorted list, starting with the largest average influence value found, e.g., $d(p : p'')$, where p and p'' are a given pair of products. The edge $p'' \to p$ is included in the network if and only if the resulting network is still planar and the edge $p \to p''$ has not been included already. We stop adding edges if adding the next edge in the list would break the planarity of the graph. This procedure ensures two things: (i) only the largest among $d(p : p'')$ and $d(p'' : p)$ will be included in the network, and (ii) the final network has $3(N-2)$ edges. It is important to note that for a given input correlation matrix of size $N \times N$ the PCPG network will always have $3(N-2)$ edges and that the identity of these edges solely depends on the correlation values in the input matrix.

The final result of this procedure is what we refer to as the biPCPG network, G. Naturally, we also obtain the average influence d associated to each edge in G, as well as the network's adjacency matrix $\mathbf{A}$ defined as

$$A_{p,p''} = \begin{cases} 1 & \text{if edge } p \to p'' \in G, \\ 0 & \text{otherwise.} \end{cases} \tag{8}$$

3.6. *biPCPG Bootstrapping*

To assess the reliability of the links in the biPCPG network, we adapt a bootstrapping procedure originally introduced in [34]. The aim is to obtain a bootstrap value for each link which is proportional to the reliability of the link.

We build R batches, where the matrices to be bootstrapped in each batch are the time series matrices of all countries $\mathbf{TS}_c\ \forall\ c \in 1,\ldots,C$. From each matrix $\mathbf{TS}_c$, a replicate time series matrix $\mathbf{TS}_c^r\ \forall\ r \in 1,\ldots,R$ is obtained, where $R = 1000$ is the total number of batches. An important feature of our procedure is how the null model, i.e., the replicate time series matrices, is generated. For each batch, the bootstrapping of the time series matrices is done coherently across countries. This means rows are drawn with repetition from each of the country matrices *jointly*—the same row indices are selected across the matrices. In addition, the new locations of the selected rows in their corresponding replicate matrices are exactly the same. This way, in the replicate time series matrices, $\mathbf{TS}_c^r$, the time structure of the time series is destroyed while preserving the country-level correlations.

Take, for example, the first batch, $r = 1$. In order to obtain the first batch of replicate matrices $\mathbf{TS}_c^1\ \forall\ c \in 1,\ldots,C$, we randomly select a sequence of $T = 22$ row indices, allowing repetitions. These row indices denote which rows from the original matrices $\mathbf{TS}_c$ are included in the corresponding replicates $\mathbf{TS}_c^1$ in this batch, as well as their order. This way, any row of a replicate matrix in this first batch will contain data points corresponding to the same year as rows of the same index in all the other replicate matrices in the batch.

After all the replicate matrices are obtained for all countries and batches, we calculate a replicate correlation matrix $\mathbf{K}_c^r$ for each of them, rejecting non-statistically significant samples as described in Section 3.3. We then find the element-wise mean of the replicate correlation matrices in each batch r, obtaining R replicate average correlation matrices $\bar{\mathbf{K}}^r$ where

$$\bar{K}^r_{p,p'} = \frac{1}{C}\sum_{c=1}^{C}\left(K_c^r\right)_{p,p'}. \tag{9}$$

Note that, similarly to the replicate time series matrices, in these replicate correlations matrices the time structure of the time series is destroyed while preserving the country-level correlations due to the way the bootstrapping has been performed.

We then apply the PCPG algorithm described in Section 3.5 to each matrix $\bar{\mathbf{K}}^r$, obtaining R replicate adjacency matrices, $\mathbf{A}^r\ \forall\ r \in 1,\ldots,R$.

To compute the bootstrap value, $b_{p,p''}$, for each link $p \to p''$, we evaluate the number of time the link appears in the replicate adjacency matrices $\mathbf{A}^r$, and normalise by the number of replicates R, formally

$$b_{p,p''} = \frac{\sum_{r=1}^{R} A^r_{p,p''}}{R} \tag{10}$$

Each bootstrap value is therefore some number in the interval [0-1] and is proportional to the reliability of the link.

4. Results

4.1. *Descriptive Analysis of the biPCPG Network*

The network G resulting from the application of the biPCPG method to our dataset is shown in Figure 3. This network displays some interesting results with a few distinct hub nodes. The most noticeable of these nodes are "Plastics", "Pigments" and "Vegetables" nodes. Hub nodes in the network also tend to have high average influence on other nodes in the network, this being displayed by the width of the edges stemming out of them. The colour of the edge represents its bootstrap value. We note that the hub nodes are also the source of most of the darker edges in the network, i.e., the most reliable edges, especially the "Plastics" node, whose edges bootstrap values are very high.

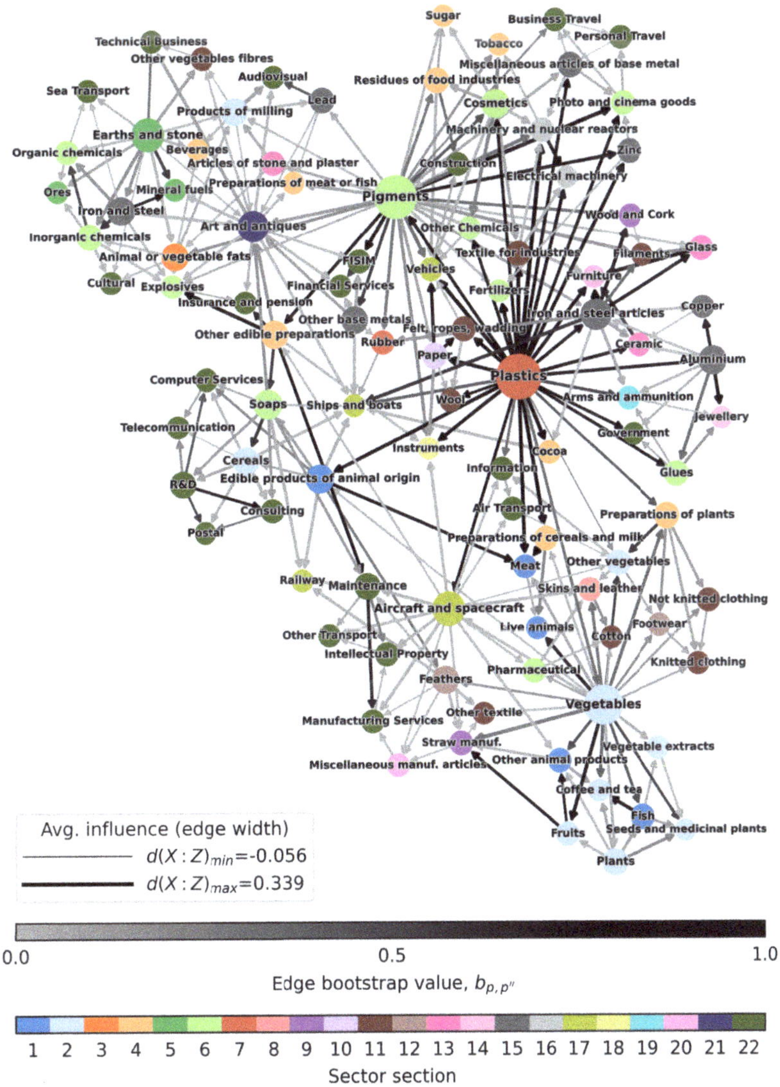

Figure 3. The biPCPG network. The widths of the edges are proportional to the average influence value, $d(p, p'')$ they represent. The colours of the edges are proportional to their bootstrap value, $b_{p,p''}$. The darker the edge, the more reliable it is. Node colours represent the sector section each product and service belong to. Node sizes are proportional to out-degree. The node layout was found using the ForceAtlas2 algorithm [44].

The resulting network also displays distinct clusters of intuitively related economic sectors. For example, the most recognisable "food and plant" cluster can be found at the bottom-right of the network, surrounding the "Vegetables" hub node. At the top-left of the network, we can observe another distinct cluster containing several sectors related to chemicals or raw materials. Finally, on the top-right of the network, surrounding the "Plastics" and "Pigments" nodes, one can find a "macro-cluster" formed mostly by industrial and manufacturing sectors.

It is worth noting that, while most edges connect intuitively related sectors, the are several cases of less-intuitive connections spread around the network. This causes the inclusion of some of these seemingly unrelated sectors in some of the clusters mentioned above. This is partially due the original construction of the PCPG algorithm, which ensures a fixed number of edges to be included in the network. Therefore, edges representing small influences among sectors could be forced to be included in the network. In our case, the biPCPG network obtained contains around 5% of edges representing Average influence values of 0.05 or smaller.

4.2. Assortativity Analysis

As described in Section 2, the 100 sectors in our dataset can be grouped into 22 groups of sectors called *sections*. Furthermore, a key metric within the field of economic complexity is the *complexity* of a product or service, which measures the capabilities needed by a country to produce it (see Appendix A). In order to better understand the structure of this network, and by extension the information contained in it, one can then investigate its *homophily* or *assortativity* according to these characteristics. Roughly speaking, this is the tendency for nodes belonging to the same group to be connected to each other. In this paper, we make use of two different assortativity metrics which we describe below. The motivation behind this analysis is to assess if our framework generates a meaningful network which is able to synthesise information about the system.

4.2.1. Assortativity by Unordered Characteristics

This quantity is used to measure the assortativity between, for example, nodes with an associated qualitative characteristic such as, in our case, sector sections, s (see Section 2). The *assortativity coefficient* is defined as [45]

$$s_s = \frac{\mathrm{Tr}\mathbf{F} - ||\mathbf{F}^2||}{1 - ||\mathbf{F}^2||} \tag{11}$$

where entries of the matrix $\mathbf{F}$ are the fractions of edges in the network that connect a vertex of section s to one of section s', and $||\mathbf{X}||$ is the sum of all elements of a matrix $\mathbf{X}$ [45]. Therefore the numerator is a quantity that measures the fraction of the edges in the network that connect vertices of the same type (i.e., within-section edges) minus the expected value of the same quantity in a network with the same community divisions but random connections between the vertices. The denominator is one minus the same expected value.

This formula gives $s_s = 0$ when there is no assortative mixing and $s_s = 1$ when there is perfect assortative mixing. For a perfectly disassortative network, the value is in the range $-1 \leq s_s < 0$ (see [45] for its interpretation). We evaluate this metric for the section of sectors described in Section 2, denoting this by the subscript s.

4.2.2. Assortativity by Scalar Characteristics

A measure of assortativity for numeric quantities associated with nodes can also be defined [45]. First, note that the entries of the matrix $\mathbf{F}$ are the fraction of all edges in a network that connect nodes with associated scalar values q and q'. Note that the values q and q' are discrete—in our case these are the *Complexity rank* [17] of sectors—computed by taking average complexity *value* of each product (across the available years in our dataset) and ranking these averages from highest to smallest. The complexity of a product or service is a well-known quantity in the economic complexity literature that describes the capabilities needed by a country to produce it, see Appendix A for its definition. The *numeric assortativity coefficient* is defined as

$$s_q = \frac{\sum_{q,q'} qq'\left(F_{q,q'} - a_q b_{q'}\right)}{\sigma_a \sigma_b} \tag{12}$$

where $a_q = \sum_{q'} F_{q,q'}$, $b_{q'} = \sum_q F_{q,q'}$ and σ_a and σ_b are the standard deviations of the distributions of a_q and $b_{q'}$, respectively. The value of s_q is in the range $-1 \leq s_q \leq 1$ with $s_q = 1$ indicating perfect assortativity and $s_q = -1$ indicating perfect disassortativity. Typically, assortativity values in the range 0.3–0.7 are considered to indicate a significant community structure in social networks (higher values are rare) [46,47].

4.2.3. Assortativity Results

The results for the two assortativity metrics defined above are as follows:

- assortativity by sector section = s_s = 0.08 (0.15 without FDR correction);
- assortativity by sector mean complexity rank = s_q = 0.19 (0.31 without FDR correction).

These results indicate that the structure of the resulting biPCPG network encodes information efficiently. Firstly, the *Assortativity by sector section*, $s_s = 0.15$, is positive, this means that sectors that belong to the same *section* (see Section 2) tend to be connected in the network, i.e., they influence each other. The section of each sector is reflected in Figure 3 by the colour of the node. The most evident clustering of sectors within the same section is found at the top of the plot where a highly connected cluster of service sectors is found.

Furthermore, the moderately high *Assortativity by sector mean complexity rank*, $s_q = 0.19$, indicates that sectors around the same level of complexity tend to influence each other. This makes sense intuitively since, according to the economic complexity literature, these tend to be connected in other networks that describe the relationship among products (e.g., product space network, product taxonomy network [21,22]).

4.3. Community Detection on the biPCPG Network

We apply a well-known community detection algorithm for directed networks based on spectral optimisation [48]. The modularity, or quality function, to be maximised is

$$Q^{\text{dir}} = \frac{1}{m} \sum_{p,p''} \left(A_{p,p''} - \frac{k_p^{\text{out}} k_{p''}^{\text{in}}}{m} \right) \delta\left(\nu_p, \nu_{p''}\right) \tag{13}$$

where $\mathbf{A}$ is the adjacency matrix, k_p^{in} and k_p^{out} are the weighted in-degree and out-degree of node p, m is the total edge weight in the network, ν_p is the community of node p and $\delta\left(\nu_p, \nu_{p''}\right) = 1$ if $\nu_p = \nu_{p''}$ and 0 otherwise. This method does not require any parameter choices relating to community size or number of communities; however, adaptations of this method that allow for these choices are available in the literature. It is worth pointing out that, for the analysis carried out in this paper, edge-weights are all set to 1. In Equation (13), this makes the weighted in-degree and out-degree simply the in- and out-degree as well as fixing $m = 294$, the total number of edges in the network.

Since there is no universal definition for communities in directed networks, we also apply the same community detection algorithm for the undirected version of the biPCPG network G^{und}. In this case, the modularity to be maximised is given by

$$Q^{\text{und}} = \frac{1}{2m} \sum_{p,p''} \left(A^{\text{und}}_{p,p''} - \frac{k_p k_{p''}}{2m} \right) \delta\left(\nu_p, \nu_{p''}\right) \tag{14}$$

where $\mathbf{A}^{\text{und}}$ is the undirected adjacency matrix which defines the undirected network G^{und}. This can be obtained from the adjacency matrix, $\mathbf{A}$, which defines the directed biPCPG network G as follows

$$A^{\text{und}}_{p,p''} = \begin{cases} 1 & \text{if } A_{p,p''} = 1 \text{ or } A_{p'',p} = 1, \\ 0 & \text{otherwise.} \end{cases} \quad (15)$$

This allows us to qualitatively assess if the structure of the biPCPG network is sufficient for reasonable communities to be detected, without the bias of the information contained in the average influence or bootstrap values associated to edges. We implement this algorithm via the *leidenalg* Python package (version 0.8.4) [49], an implementation of the *leiden* algorithm for modularity optimisation.

Note that optimising modularity is an NP-hard problem [50], and therefore heuristics have to be implemented for algorithms to be efficient. One of the steps in the *leiden* algorithm used here involves selecting a random community for a node to be added to. However, this randomness can be controlled via a *seed* to the random number generator. This makes the process deterministic such that the same communities are selected every time the algorithm is run on a given network using the same seed value. In our analysis, we tested several seed values finding that the detected communities varied only for a few nodes, with many seed values returning the exact same partitions. The results shown in Section 4.3 were found using 1 as the seed, as well as for many other seed values tested.

Furthermore, we compare the the communities obtained for the directed and undirected versions of the network for seed values 1, ..., 1000 via the *Adjusted Mutual Information* [51]. Take, for example, our set of P of N sectors and consider two partitions of P, namely $U = \{U_1, U_2, \ldots, U_J\}$ with J pairwise-disjoint clusters found by maximising Q^{und} for the undirected version of the network, and $V = \{V_1, V_2, \ldots, V_D\}$ with D pairwise-disjoint clusters found by maximising Q^{dir} for the directed version of the network. The AMI between the two partitions is then defined as

$$AMI(U,V) = \frac{MI(U,V) - E\{MI(U,V)\}}{\max\{H(U), H(V)\} - E\{MI(U,V)\}} \quad (16)$$

where $MI(U,V)$ is the mutual information between two partitions, $E\{MI(U,V)\}$ is the expected mutual information and $H(U)$ and $H(V)$ are the entropy values associated to partitions U and V respectively. The AMI equals 1 when two partitions are exactly the same and 0 when the MI between them equals its expected value and therefore serves as a similarity measure for the two partitions, for further details on its calculation see [51]. In Section 4.3, we give the result for the *average* AMI obtained for the 1000 seed values tested using the *scikit-learn 0.23* Python package.

Community Detection Results

The community detection procedure described above yielded 5 distinct communities when applied on the undirected biPCPG network, G^{und}, which we denote communities $\nu = 1, \ldots, 5$. These communities have 31, 22, 21, 13 and 13 sectors contained in each of them, respectively.

The detected communities in the network can be seen highlighted in Figure 4. When comparing with Figure 3, which shows the network highlighting the section of each sector, one can see that the detected communities partition the network into groups that contain intuitively related sectors. For example, communities 2, 3 and 5 contain mostly nodes related to industrial and chemical sectors, while community 1 captures the "food and plant" cluster described above as well as some service sectors. Finally, for community 4, it is slightly more difficult to find a common theme. However, it is worth noting that over half of the sectors it contains are service sectors.

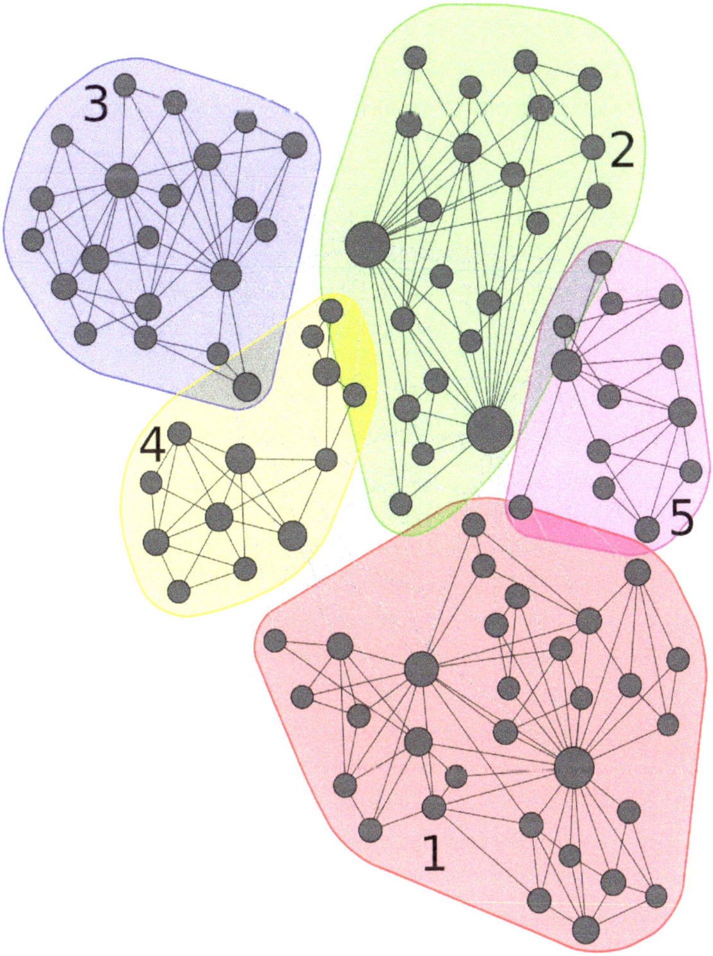

Figure 4. biPCPG network, G, resulting from the application of the PCPG algorithm on the mean correlation matrix $\bar{\mathbf{K}}$ between sectors' RCA time series. Nodes are grouped by their community, ν, found by maximising modularity in the network. The node layout was found using the ForceAtlas2 algorithm [44].

The information structure these communities contain can be seen when sorting rows and columns of the average correlation matrix $\bar{\mathbf{K}}$ and average influence matrix by community index as seen in Figures A2 and A3 in Appendix C. We can observe, for example, that brighter colours, meaning higher values, are generally found close the diagonal of the matrices (i.e., among sectors within the same community). This is especially noticeable for communities 1 and 2. We can also identify which rows and columns represent service sectors, as these tend to have a lower correlation and average influence values with non-service sectors (depicted in dark blue) and higher values among themselves.

The average *adjusted mutual information* obtained for the 1000 seed values tested is 0.90. This is a very high value which tells us that, on average, the partitions obtained for the directed and undirected versions of the network were very similar. This suggests that the community detection procedure is weakly dependent on the version of the network (directed vs. undirected) as well as the seed value used.

5. Discussion

In this paper we have introduced the biPCPG framework, a generalisation of the PCPG [30] algorithm to datasets with a multi-sample and multi-variable structure that allows a statistical significant and robust analysis, mainly by generating confidence bounds via an adapted bootstrapping procedure. We have then applied this new procedure to a recently introduced dataset that integrates the export of physical goods and services data. The proposed procedure allows the generation of a network of these economic sectors whose links represent the average influence in terms of temporal correlation. This can be seen as an an asymmetric formulation of relatedness [26,52]. The resulting network contains several hub nodes with high degree (namely Plastics, Pigments, Iron and steel articles, Preparations of cereals and milk and Aluminium) as well as distinct clusters of intuitively-related economic sectors (such as a food and plant cluster, a services cluster and manufacturing cluster). We find that, in this network, economic sectors display a relatively high assortativity according to their complexity rank and, to a lesser extent, their category.

6. Conclusions

In this work, we have introduced an asymmetric definition for relatedness by extending the PCPG methodology introduced in [30] for its use on bipartite datasets, which we call biPCPG. We apply this approach to a recently introduced dataset containing the exports of countries regarding both manufactured products and intangible services. We show that the biPCPG methodology is able to generate a statistically robust network of economic sectors which captures the underlying influence structure int erms of temporal correlations.

This work can be extended in a number of possible directions. First of all, the biPCPG framework can be applied to any temporal bipartite network, such as those of common use in economic complexity, such as the company-technology [9] or the country-scientific field network [29]. Moreover, the adapted bootstrapping procedure can be used to other network-generating techniques based on correlation-filtering to datasets with a multi-sample and multi-variable structure. These techniques include those based on threshold methods [53], the Minimum Spanning Tree [33] and the aforementioned PMFG [31], as well as more recent techniques based on a null-model approach [54]. This would be possible by replacing the last step in our procedure, the original PCPG algorithm, with the correlation-filtering technique of interest. Finally, it would also be particularly interesting to apply our procedure to datasets with the same structure but longer time series, such as financial datasets containing, for example, asset prices at the different exchanges where they are traded.

Author Contributions: Conceptualisation, A.Z. and T.D.M.; methodology, A.Z. and T.D.M.; software, C.S.d.P.P.; validation, C.S.d.P.P., A.Z. and T.D.M.; formal analysis, C.S.d.P.P.; investigation, C.S.d.P.P., A.Z. and T.D.M.; data curation, A.Z.; manuscript writing, review and editing, C.S.d.P.P., A.Z. and T.D.M. All authors have read and agreed to the published version of the manuscript.

Funding: This research received no external funding.

Institutional Review Board Statement: Not applicable.

Informed Consent Statement: Not applicable.

Data Availability Statement: Raw databases are available from (https://comtrade.un.org, accessed on 13 February 2019) and (https://data.imf.org, accessed on 13 February 2019).

Acknowledgments: The authors acknowledge Michele Tumminello for providing the PCPG Mathematica code.

Conflicts of Interest: The authors declare no conflict of interest.

Abbreviations

The following abbreviations are used in this manuscript:

AMI	Adjusted Mutual Information
BH	Benjamini–Hochberg
biPCPG	Bipartite Partial Correlation Planar Graph
BPM	International Monetary Fund's Balance of Payments data
EC	Economic Complexity
FDR	False Discovery Rate
HS	Harmonized System
IMF	International Monetary Fund
MST	Minimum Spanning Tree
PCPG	Partial Correlation Planar Graph
PMFG	Planar Maximally Filtered Graph
RCA	Revealed Comparative Advantage
UN-COMTRADE	United Nations Commodity Trade Statistics Database
USD	United States Dollar
WCO	World Customs Organization

Appendix A. Fitness and Complexity of Economic sectors

From the matrices containing $RCA_{c,p}$ time series, described in Section 3.3 we can derive the $\mathbf{M}^y$ matrix which has entries given by

$$M^y_{c,p} = \begin{cases} 1 & \text{if } RCA^y_{c,p} \geq 1, \\ 0 & \text{otherwise} \end{cases} \tag{A1}$$

where c represents a country, p represents a product (or service), and y represents a given year.

This matrix therefore summarises the countries having a comparative advantage at exporting the different products or services in a given year, or not. Two key quantities from the economic complexity literature are defined using this matrix, namely the *fitness* of countries and the *complexity* of products (or services) [17,55]. The intuition behind these quantities is that the higher the fitness of a country the higher its capability of exporting products of high complexity. It is therefore natural for the fitness to be proportional to the weighted sum of the products of which it is a competitive exporter. The definition of the complexity of a product is more subtle. In general terms, the complexity of a product should be inversely proportional to the number of countries exporting it. We should also note that more economically developed countries tend to have a highly diversified export basket, while less economically developed countries tend to have a much more limited diversification in their exports, and focused on low complexity products. Therefore, the upper bound of a product's complexity should be determined by the fitness of the countries' exporting it, with a strong bias towards lower fitness countries: if a product is exported by lower fitness countries, its complexity can not be high. The fitness F_c of a country and the complexity Q_p of a product (or service) are therefore defined using the following set of coupled iterative equations

$$\begin{cases} \tilde{F}_c^{(n)} = \sum_p M_{cp} Q_p^{(n-1)} \\ \tilde{Q}_p^{(n)} = \dfrac{1}{\sum_c M_{cp} \dfrac{1}{F_c^{(n-1)}}} \end{cases} \rightarrow \begin{cases} F_c^{(n)} = \dfrac{\tilde{F}_c^{(n)}}{\left\langle \tilde{F}_c^{(n)} \right\rangle_c} \\ Q_p^{(n)} = \dfrac{\tilde{Q}_p^{(n)}}{\left\langle \tilde{Q}_p^{(n)} \right\rangle_p} \end{cases} \tag{A2}$$

which are iterated until a fixed point is reached [56]. This fixed point has been shown to be stable and not dependent on the initial conditions, which are set to $\tilde{Q}_p^{(0)} = 1 \forall p$ and

$\tilde{F}_c^{(0)} = 1 \forall c$ [17]. We use the complexity of products and services in our dataset to calculate an assortativity metric on the network G as described in Section 4.2.

It is worth noting that the dataset analysed and similar datasets explored in the economic complexity literature exhibit a nested structure [56]. This nested structure is manifested as a triangular structure in the $\mathbf{M}^y$ matrices when countries (rows) and sectors (columns) are sorted by their fitness and complexity rank, respectively. This can be seen in Figure A1, which is the $\mathbf{M}^y$ matrix for the year $y = 2005$.

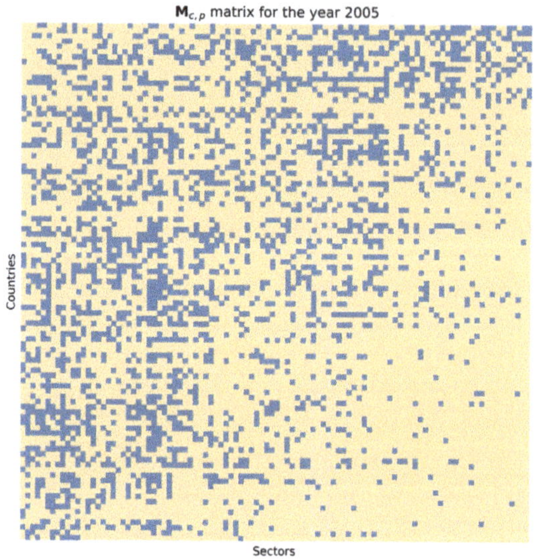

Figure A1. Binary matrix $\mathbf{M}^{2005}$ displaying high $RCA_{c,p}$ values for the year 2005. Blue indicates an entry of one and yellow an entry of zero. The triangular structure of the matrix implies a nestedness in the data.

Appendix B. Confidence and Prediction interval calculations

The 95% confidence interval around a linear fit $\hat{\mu}_{y|x_0}$ done on n data points (x_i, y_i) $n = 1, \ldots, n$ contains the mean response of new values $\mu_{y|x_0}$ at a given value x_0 with a 95% probability. This is given by

$$\left|\hat{\mu}_{y|x_0} - \mu_{y|x_0}\right| \leq T_{n-2}^{.975} \hat{\sigma} \sqrt{\frac{1}{n} + \frac{(x_0 - \bar{x})^2}{\sum_{i=1}^{n}(x_i - \bar{x})^2}} \tag{A3}$$

where $\hat{\mu}_{y|x_0} = a + bx_0$ is computed from the linear fit, $T_{n-2}^{.975}$ is the 97.5th percentile of the Student's t-distribution with $n - 2$ degrees of freedom and $\hat{\sigma}$ is the standard deviation of the residuals in the linear fit given by

$$\hat{\sigma} = \sqrt{\sum_{i=1}^{n} \frac{(y_i - \hat{y})^2}{n-2}}. \tag{A4}$$

The 95% prediction interval around a linear fit $\hat{y}_0$ is the interval within which a new observation, y_0, at a given value, x_0, is found, with 95% probability. This is given by

$$|\hat{y}_0 - y_0| \leq T_{n-2}^{.975} \hat{\sigma} \sqrt{1 + \frac{1}{n} + \frac{(x_0 - \bar{x})^2}{\sum_{i=1}^{n}(x_i - \bar{x})^2}} \tag{A5}$$

where $\hat{y}_0 = a + bx_0$ is computed from the linear fit. See [57] for a more detailed description.

Appendix C. Avg. Correlation and Avg. Influence Matrices Sorted by Community

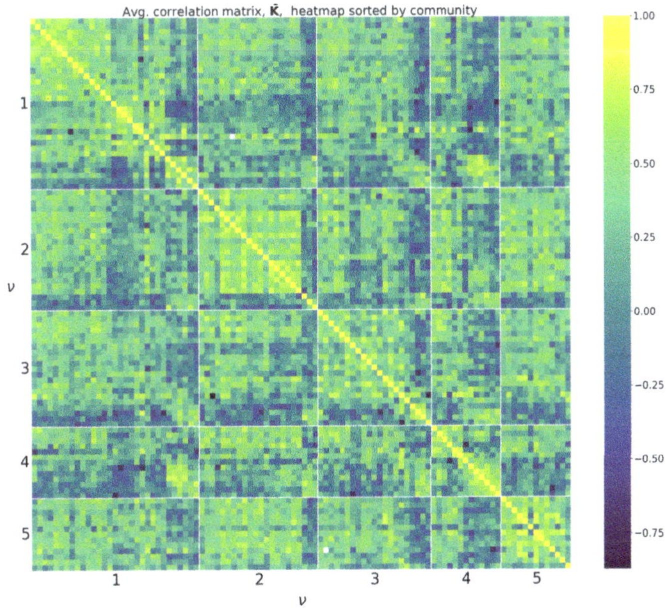

Figure A2. Average correlation matrix $\bar{\mathbf{K}}$ sorted by communities ν found by maximising modularity.

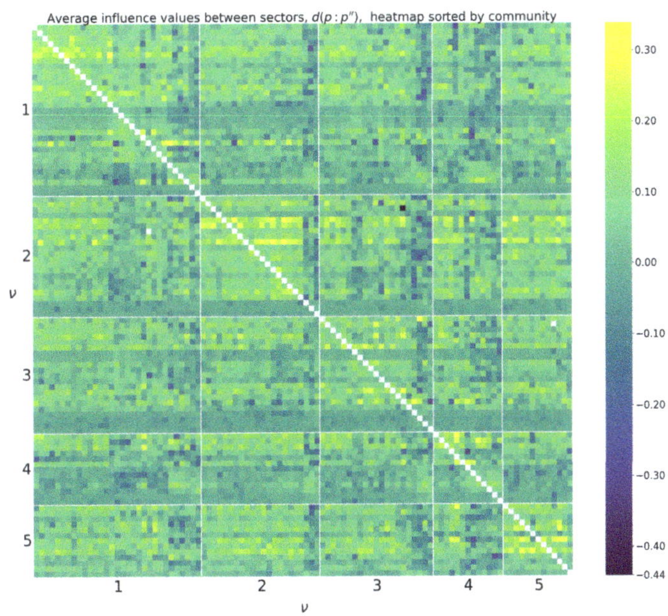

Figure A3. Matrix showing average influence values between products $d(p:p'')$ sorted by communities ν found by maximising modularity. Entries in white indicate that the average influence of a sector on itself is undefined.

Appendix D. Sector List

Table A1. List of product (HS2007) and service (IMF BP6) sector codes in the analysed dataset.

Sector Code	Sector Name	Sector Code	Sector Name
01	Live animals	61	Knitted clothing
02	Meat	62	Not knitted clothing
03	Fish	63	Other textile
04	Edible products of animal origin	64	Footwear
05	Other animal products	67	Feathers
06	Plants	68	Articles of stone and plaster
07	Vegetables	69	Ceramic
08	Fruits	70	Glass
09	Coffee and tea	71	Jewellery
10	Cereals	72	Iron and steel
11	Products of milling	73	Iron and steel articles
12	Seeds and medicinal plants	74	Copper
13	Vegetable extracts	76	Aluminium
14	Other vegetables	78	Lead
15	Animal or vegetable fats	79	Zinc
16	Preparations of meat or fish	81	Other base metals
17	Sugar	83	Miscellaneous articles of base metal
18	Cocoa	84	Machinery and nuclear reactors
19	Preparations of cereals and milk	85	Electrical machinery
20	Preparations of plants	86	Railway
21	Other edible preparations	87	Vehicles
22	Beverages	88	Aircraft and spacecraft
23	Residues of food industries	89	Ships and boats
24	Tobacco	90	Instruments
25	Earths and stone	93	Arms and ammunition
26	Ores	94	Furniture
27	Mineral fuels	96	Miscellaneous manuf. articles
28	Inorganic chemicals	97	Art and antiques
29	Organic chemicals	BXSM_BP6_USD	Manufacturing Services
30	Pharmaceutical	BXSOCN_BP6_USD	Construction
31	Fertilizers	BXSOFIEX_BP6_USD	Financial Services
32	Pigments	XSOFIFISM_BP6_USD	FISIM
33	Cosmetics	BXSOGGS_BP6_USD	Government
34	Soaps	BXSOIN_BP6_USD	Insurance and pension
35	Glues	BXSOOBPM_BP6_USD	Consulting
36	Explosives	BXSOOBRD_BP6_USD	R&D
37	Photo and cinema goods	BXSOOBTT_BP6_USD	Technical Business
38	Other Chemicals	BXSOPCRAU_BP6_USD	Audiovisual
39	Plastics	BXSOPCRO_BP6_USD	Cultural
40	Rubber	BXSORL_BP6_USD	Intellectual Property
41	Skins and leather	BXSOTCMC_BP6_USD	Computer Services
44	Wood and Cork	BXSOTCMM_BP6_USD	Information
46	Straw manuf.	BXSOTCMT_BP6_USD	Telecommunication
47	Paper	BXSR_BP6_USD	Maintenance
51	Wool	BXSTRA_BP6_USD	Air Transport
52	Cotton	BXSTROT_BP6_USD	Other Transport
53	Other vegetables fibres	BXSTRPC_BP6_USD	Postal
54	Filaments	BXSTRS_BP6_USD	Sea Transport
56	Felt, ropes, wadding	BXSTVB_BP6_USD	Business Travel
59	Textile for industries	BXSTVP_BP6_USD	Personal Travel

References

1. Burgos, E.; Ceva, H.; Hernández, L.; Perazzo, R.P.; Devoto, M.; Medan, D. Two classes of bipartite networks: Nested biological and social systems. *Phys. Rev. E-Nonlinear Soft Matter Phys.* **2008**, *78*, 046113. [CrossRef] [PubMed]
2. Kontou, P.I.; Pavlopoulou, A.; Dimou, N.L.; Pavlopoulos, G.A.; Bagos, P.G. Network analysis of genes and their association with diseases. *Gene* **2016**, *590*, 68–78. [CrossRef] [PubMed]
3. Domínguez-García, V.; Muñoz, M.A. Ranking species in mutualistic networks. *Sci. Rep.* **2015**, *5*, 8182. [CrossRef] [PubMed]
4. Watts, D.J.; Strogatz, S.H. Collective dynamics of 'small-world' networks. *Nature* **1998**, *393*, 440–442. [CrossRef]
5. Newman, M.E. Scientific collaboration networks. I. Network construction and fundamental results. *Phys. Rev. E-Phys. Plasmas Fluids Relat. Interdiscip. Top.* **2001**, *64*, 8. [CrossRef]
6. Newman, M.E.J. The structure of scientific collaboration networks. *Proc. Natl. Acad. Sci. USA* **2001**, *98*, 404–409. [CrossRef]
7. Conyon, M.J.; Muldoon, M.R. The small world of corporate boards. *J. Bus. Financ. Account.* **2006**, *33*, 1321–1343. [CrossRef]
8. Ramasco, J.J.; Dorogovtsev, S.N.; Pastor-Satorras, R. Self-organization of collaboration networks. *Phys. Rev. E-Phys. Plasmas Fluids Relat. Interdiscip. Top.* **2004**, *70*, 10. [CrossRef]
9. Pugliese, E.; Napolitano, L.; Zaccaria, A.; Pietronero, L. Coherent diversification in corporate technological portfolios. *PLoS ONE* **2019**, *14*, e0223403. [CrossRef]
10. Guillaume, J.L.; Latapy, M.; Le-Blond, S. Statistical Analysis of a P2P Query Graph Based on Degrees and Their Time-Evolution. In *Distributed Computing—IWDC 2004*, Lecture No ed.; Sen, A., Das, N., Das, S.K., Eds.; Springer: Berlin/Heidelberg, Germany, 2004; Volume 3326, pp. 126–137. [CrossRef]
11. Taylor, P.J. The new geography of global civil society: NGOs in the world city network. *Globalizations* **2004**, *1*, 265–277. [CrossRef]
12. Doreian, P.; Batagelj, V.; Ferligoj, A. Generalized blockmodeling of two-mode network data. *Soc. Netw.* **2004**, *26*, 29–53. [CrossRef]
13. Fowler, J.H. Legislative cosponsorship networks in the US House and Senate. *Soc. Netw.* **2006**, *28*, 454–465. [CrossRef]
14. Hidalgo, C.A. Economic complexity theory and applications. *Nat. Rev. Phys.* **2021**, *3*, 92–113. [CrossRef]
15. Pietronero, L.; Cristelli, M.; Gabrielli, A.; Mazzilli, D.; Pugliese, E.; Tacchella, A.; Zaccaria, A. Economic Complexity: "Buttarla in caciara" vs. a constructive approach. *arXiv* **2017**, arXiv:1709.05272.
16. Hidalgo, C.A.; Hausmann, R. The building blocks of economic complexity. *Proc. Natl. Acad. Sci. USA* **2009**, *106*, 10570–10575. [CrossRef] [PubMed]
17. Tacchella, A.; Cristelli, M.; Caldarelli, G.; Gabrielli, A.; Pietronero, L. A new metrics for countries' fitness and products' complexity. *Sci. Rep.* **2012**, *2*, 723. [CrossRef] [PubMed]
18. Liao, H.; Vidmer, A. A Comparative Analysis of the Predictive Abilities of Economic Complexity Metrics Using International Trade Network. *Complexity* **2018**, *2018*, 2825948. [CrossRef]
19. Pugliese, E.; Chiarotti, G.L.; Zaccaria, A.; Pietronero, L. Complex economies have a lateral escape from the poverty trap. *PLoS ONE* **2017**, *12*, e0168540. [CrossRef]
20. Angelini, O.; Di Matteo, T. Complexity of Products: The Effect of Data Regularisation. *Entropy* **2018**, *20*, 814. [CrossRef]
21. Hidalgo, C.A.; Klinger, B.; Barabasi, A.L.; Hausmann, R. The Product Space Conditions the Development of Nations. *Science* **2007**, *317*, 482–487. [CrossRef]
22. Zaccaria, A.; Cristelli, M.; Tacchella, A.; Pietronero, L. How the taxonomy of products drives the economic development of countries. *PLoS ONE* **2014**, *9*, e0113770. [CrossRef] [PubMed]
23. Zaccaria, A.; Mishra, S.; Cader, M.; Pietronero, L. Integrating Services in the Economic Fitness Approach. In *Policy Research Working Paper No. 8485.*; World Bank: Washington, DC, USA, 2018. Available online: https://openknowledge.worldbank.org/handle/10986/29938 (accessed on 13 February 2019).
24. Stojkoski, V.; Utkovski, Z.; Kocarev, L. The impact of services on economic complexity: Service sophistication as route for economic growth. *PLoS ONE* **2016**, *11*, e0161633. [CrossRef] [PubMed]
25. Albora, G.; Pietronero, L.; Tacchella, A.; Zaccaria, A. Product Progression: A machine learning approach to forecasting industrial upgrading. *arXiv* **2021**, arXiv:2105.15018.
26. Hidalgo, C.A.; Balland, P.A.; Boschma, R.; Delgado, M.; Feldman, M.; Frenken, K.; Glaeser, E.; He, C.; Kogler, D.F.; Morrison, A.; et al. The Principle of Relatedness. *Springer Proc. Complex.* **2018**, *1*, 451–457. [CrossRef]
27. Teece, D.J.; Rumelt, R.; Dosi, G.; Winter, S. Understanding corporate coherence: Theory and evidence. *J. Econ. Behav. Organ.* **1994**, *23*, 1–30. [CrossRef]
28. Zhou, T.; Ren, J.; Medo, M.; Zhang, Y.C. Bipartite network projection and personal recommendation. *Phys. Rev. E* **2007**, *76*, 046115. [CrossRef]
29. Pugliese, E.; Cimini, G.; Patelli, A.; Zaccaria, A.; Pietronero, L.; Gabrielli, A. Unfolding the innovation system for the development of countries: coevolution of Science, Technology and Production. *Sci. Rep.* **2019**, *9*, 16440. [CrossRef]
30. Kenett, D.Y.; Tumminello, M.; Madi, A.; Gur-Gershgoren, G.; Mantegna, R.N.; Ben-Jacob, E. Dominating clasp of the financial sector revealed by partial correlation analysis of the stock market. *PLoS ONE* **2010**, *5*, e0015032. [CrossRef]
31. Tumminello, M.; Aste, T.; Di Matteo, T.; Mantegna, R.N. A tool for filtering information in complex systems. *Proc. Natl. Acad. Sci. USA* **2005**, *102*, 10421–10426. [CrossRef]
32. Kruskal, J.B. On the shortest spanning subtree of a graph and the traveling salesman problem. *Proc. Am. Math. Soc.* **1956**, *7*, 48–50. [CrossRef]

33. Mantegna, R.N.; Stanley, H.E. *An Introduction to Econophysics: Correlations and Complexity in Finance*; Cambridge University Press: Cambridge, UK, 1999.
34. Tumminello, M.; Di Matteo, T.; Aste, T.; Mantegna, R.N. Correlation based networks of equity returns sampled at different time horizons. *Eur. Phys. J. B* **2007**, *55*, 209–217. [CrossRef]
35. Saenz de Pipaon Perez, C. biPCPG Python Package. Available online: http://www.github.com/cspipaon/biPCPG (accessed on 6 January 2022).
36. Saenz de Pipaon Perez, C. biPCPG Python Package Documentation. Available online: http://bipcpg.readthedocs.io (accessed on 6 January 2022).
37. International Monetary Fund Data. International Trade in Services and the Comparative Advantage of Nations. Available online: https://data.imf.org/ITS (accessed on 13 February 2019).
38. World Customs Organization. Harmonized System Nomenclature 2007 Edition. Available online: http://www.wcoomd.org/en/topics/nomenclature/instrument-and-tools/hs_nomenclature_previous_editions/hs_nomenclature_table_2007.aspx (accessed on 13 February 2019).
39. Balassa, B. Trade Liberalisation and "Revealed" Comparative Advantage. *Manch. Sch.* **1965**, *33*, 99–123. [CrossRef]
40. Student. Probable error of a correlation coefficient. *Biometrika* **1908**, *6*, 302–310. [CrossRef]
41. Benjamini, Y.; Hochberg, Y. Controlling the false discovery rate: A practical and powerful approach to multiple testing. *J. R. Stat. Soc. Ser. (Methodol.)* **1995**, *57*, 289–300. [CrossRef]
42. Miller, R.G. *Simultaneous Statistical Inference*, 2nd ed.; Springer: Berlin/Heidelberg, Germany, 1981; p. 166.
43. Tumminello, M.; Miccichè, S.; Lillo, F.; Piilo, J.; Mantegna, R.N. Statistically validated networks in bipartite complex systems. *PLoS ONE* **2011**, *6*, e0017994. [CrossRef]
44. Jacomy, M.; Venturini, T.; Heymann, S.; Bastian, M. ForceAtlas2, a continuous graph layout algorithm for handy network visualization designed for the Gephi software. *PLoS ONE* **2014**, *9*, e0098679. [CrossRef]
45. Newman, M.E. Mixing patterns in networks. *Phys. Rev. E-Phys. Plasmas Fluids Relat. Interdiscip. Top.* **2003**, *67*, 13. [CrossRef]
46. Newman, M.E.; Girvan, M. Finding and evaluating community structure in networks. *Phys. Rev. E-Nonlinear Soft Matter Phys.* **2004**, *69*, 026113. [CrossRef]
47. Clauset, A.; Newman, M.E.; Moore, C. Finding community structure in very large networks. *Phys. Rev. E-Phys. Plasmas Fluids Relat. Interdiscip. Top.* **2004**, *70*, 6. [CrossRef]
48. Leicht, E.A.; Newman, M.E. Community structure in directed networks. *Phys. Rev. Lett.* **2008**, *100*, 118703. [CrossRef]
49. Traag, V.A.; Waltman, L.; van Eck, N.J. From Louvain to Leiden: Guaranteeing well-connected communities. *Sci. Rep.* **2019**, *9*, 5233. [CrossRef]
50. Brandes, U.; Delling, D.; Gaertler, M.; Gorke, R.; Hoefer, M. On Modularity Clustering. *IEEE Trans. Knowl. Data Eng.* **2008**, *20*, 172–188. [CrossRef]
51. Vinh, N.X.; Epps, J.; Bailey, J. Information theoretic measures for clusterings comparison: Variants, properties, normalization and correction for chance. *J. Mach. Learn. Res.* **2010**, *11*, 2837–2854.
52. Tacchella, A.; Zaccaria, A.; Miccheli, M.; Pietronero, L. Relatedness in the Era of Machine Learning. *arXiv* **2021**, arXiv:2103.06017.
53. Onnela, J.P.; Kaski, K.; Kertész, J. Clustering and information in correlation based financial networks. *Eur. Phys. J. B* **2004**, *38*, 353–362. [CrossRef]
54. Kojaku, S.; Masuda, N. Constructing networks by filtering correlation matrices: A null model approach. *Proc. R. Soc. A Math. Phys. Eng. Sci.* **2019**, *475*, 12–14. [CrossRef]
55. Cristelli, M.; Gabrielli, A.; Tacchella, A.; Caldarelli, G.; Pietronero, L. Measuring the Intangibles: A Metrics for the Economic Complexity of Countries and Products. *PLoS ONE* **2013**, *8*, e0070726. [CrossRef]
56. Pugliese, E.; Zaccaria, A.; Pietronero, L. On the convergence of the Fitness-Complexity algorithm. *Eur. Phys. J. Spec. Top.* **2016**, *225*, 1893–1911. [CrossRef]
57. Montgomery, D.C.; Peck, E.A.; Vining, G.G. *Introduction to Linear Regression Analysis*, 5th ed.; Wiley: Hoboken, NJ, USA, 2012.

entropy

MDPI

Article

Network Analysis of Cross-Correlations on Forex Market during Crises. Globalisation on Forex Market

Janusz Miśkiewicz [1,2]

1 Institute of Theoretical Physics, University of Wrocław, 50-204 Wrocław, Poland; janusz.miskiewicz@uwr.edu.pl
2 Physics and Biophysics Department, Wrocław University of Environmental and Life Sciences, 50-375 Wrocław, Poland; janusz.miskiewicz@upwr.edu.pl

Abstract: Within the paper, the problem of globalisation during financial crises is analysed. The research is based on the Forex exchange rates. In the analysis, the power law classification scheme (PLCS) is used. The study shows that during crises cross-correlations increase resulting in significant growth of cliques, and also the ranks of nodes on the converging time series network are growing. This suggests that the crises expose the globalisation processes, which can be verified by the proposed analysis.

Keywords: time series analysis; cross-correlations; power law classification scheme; network analysis; globalisation; entropy

Citation: Miśkiewicz, J. Network Analysis of Cross-Correlations on Forex Market during Crises. Globalisation on Forex Market. *Entropy* **2021**, *23*, 352. https://doi.org/10.3390/e23030352

Academic Editor: H. Eugene Stanley

Received: 11 February 2021
Accepted: 11 March 2021
Published: 15 March 2021

1. Introduction

The economy is a human activity where interactions are particularly important. The mutual impacts are caused by an exchange of goods, services, and co-operation, but also competition, company overtaking, industrial espionage, etc. In result, one can observe a grouping among entities in the form of co-operation, branches, common interest, or competition on the market. These effects are the subject of many research fields, e.g., portfolio analysis [1–4], market structure analysis [5,6], globalisation researches [7–9], and many others.

The main tool for exploring the nature of interdependence among entities (companies, branches, shares, countries, etc.) is the cross-correlation analysis. In fact, this term is gathering a great variety of methods. Just mentioning the most often used: classical variance analysis and Pearson correlation coefficient [10–15], cointegration analysis [16–19], multifractal analysis [20–23], random matrix theory [24–27], power law classification scheme [28–30], or entropy-based methods [31,32].

The range of problems investigated by cross-correlation analysis is very broad, starting from sociology, economy, econophysics [20,23,33,34], transport [35,36], genome analysis, biology, food network, biochemistry network, science collaboration network [37], up to sport [38], and many others.

Within this study, the globalisation is analysed by the power law classification scheme (PLCS). In difference to other cross-correlation methods, such as detrended fluctuation analysis (DFA) [39–45] or the Pearson coefficient-based method [12,15,46,47], which are focused on noise correlation, PLCS is focused on trends. In the case of globalisation, trends seem to be more important, because they reflect similarities in evolution rather than mutual dependence and sensitivity to external impulses. Besides that, PLCS analysis allows for observing different features—medium-range correlations. On the other hand, the method is sensitive to long-term deterministic correlations that are related to "fundamental" effects [48]. The research analyses the currency exchange rate time series as an objective measure of mutual relationship and interactions among economies. The currency exchange rates are one of the most important parameters of the economy status. There are several platforms where the exchange of currencies occur. The best known and one of the most important from the global point of view is the Forex market, which is focused on the institutional market. Besides that, there are many other exchange platforms that are aimed at individuals, such as exchange office, banks, and Internet exchange systems. The present study focuses on the Forex exchange time series, since the main goal is the analysis of economy globalisation, particularly cluster formation during stock market crises.

2. Methods

The power law classification scheme (PLCS) is focused on correlations of trends [28]. The algorithm will be shortly described here for the clarity of presentation and convenience of the reader.

Let assume that there are two time series recorded simultaneously with the same length N. In the first step, the subseries from the initial point k are taken and the Manhattan distance between them calculated. The procedure is repeated for each $k \in \{1, \ldots, N\}$. At this point, the series of cumulative Manhattan distance is obtained. Each point of this series corresponds to a different "k". Finally, the power law function is fitted to the cumulative Manhattan distance series. The power of the fitted function diminished by one defines the correlation strength.

Example of Application

Let us assume that there are two time series that are generated by the linear functions:

$$f_1(t) = a_1 \cdot t, \quad f_2(t) = a_2 \cdot t.$$

The data are registered in equal intervals e.g., $t = 1, 2, \ldots, N$. The generated time series are denoted as f_1 and f_2. Subsequently, the cumulative series of the Manhattan distance between series f_1 and f_2 is equal to

$$MD(k) = \sum_{i=1}^{k} |a_1 - a_2| i = |a_1 - a_2| \frac{(1+k)k}{2},$$

so

$$MD(k) = \frac{|a_1 - a_2|}{2}(k + k^2).$$

The last step is the fitting of the power law function. The most popular method is fitting the linear function to the log-log transformed data e.g., $(ln(k), ln(MD(k)))$. Of course, the quality of the fit depends on the series length. In the case of the analysed functions f_1 and f_2, the fitted exponent for the first 100 data points is equal to 1.922, but, for 1000 data points, is equal to 1.982 and asymptotically approach 2. The observed uncertainity is the result of numerical limitations of the computer memory while calculating the logarithm. In order to obtain the correlation strength, one has to diminish the exponent of the fitted function by one and finally obtains 1. Of course, this result is in agreement with the linear relationship between the considered functions. Other examples and more detailed analysis can be found in [28–30].

The results of PLCS analysis can be classified into two categories:

$\alpha < 0$ when the correlation strength is smaller than zero—the distance between time series is decreasing, the time series are converging.

$\alpha > 0$ when the correlation strength is greater than zero—the distance between time series is increasing, and the time series are diverging.

The special case of $\alpha = 0$ is observed when the time series are overlapping [28].

In the present study, the *time evolution* of correlation strength is analysed; therefore, the additional correlation window parameter is introduced T_c. The correlation strength is calculated in a moving time window, so the appropriate subseries of the length T_c are taken and the correlation strength between them calculated; subsequently, the starting point is shifted by one day and the procedure is repeated.

The application of PLCS to a time series gives symmetrical correlation matrix with $\frac{N^2-N}{2}$ unique elements (N is the number of time series elements considered). Therefore, to conclude, it should be further analysed. The popular strategy is to construct a network, e.g., Minimum Spanning Tree or others. However, PLCS allows for distinguishing two types of cross-correlation: convergent and divergent time series. Therefore, in this paper, the following two networks are constructed:

- converging time series networks, i.e., only the nodes (representing the currency time series) with a correlation strength smaller than one are connected, and
- diverging time series network, i.e., only the nodes with a correlation strength greater than one are connected.

Clearly, the first type of network is focused on the time series approaching each other, while the second on the time series increasing differences.

In the presented study, the grouping of currencies was analysed, particularly the clique and community formation were investigated. Therefore, the following network features were calculated: the clique size evolution, the community number, the frequency of the connection on the graph, the evolution of the network node rank distribution, and the rank node entropy.

Clique size evolution is obtained by calculating the size of the biggest clique for each of the generated networks. The clique size evolution illustrates a process of unification of the market. Indeed, if the giant clique is observed, then one type of correlation is dominating on the market and, on the contrary, if the size of the biggest cluster is small, then the correlation matrix consists of a variety of correlation type.

Community number is obtained by measuring the number of community structure partitions that group nodes, such that there is a higher density of edges within the community than between them. This parameter is weaker than the clique number, but still allows observing grouping on Forex market.

The frequency of connection on the graph is the measure where the frequency of being connected on the graph is analysed. The most important feature of this measure is the ability to distinguish the most stable connections in the considered period.

Node rank distribution is the analysis where the most detailed information regarding the graph is obtained. The rank of nodes is an important feature allowing for observing the hierarchy of a network and is often used to determine network type [49–51]. This measure gives very detailed information regarding the graph. It may be considered as a quick overview of the network main features, e.g., if it is densely connected or whether each node is only connected with a small number of links.

Rank node entropy is the Shannon entropy that is defined in the standard way (Equation (1)), where the evolution of the entropy of node rank is calculated.

$$S = \sum_i -p_i \ln p_i, \tag{1}$$

where p_i is the probability of i-th rank. A summation is done over all ranks of nodes present in the network.

Those analyses are performed for both types of networks (diverging and converging).

3. Data

3.1. Data Source

The foreign exchange market (Forex) is a global network of brokers and computers that serves as a place of currency exchange. The market is active from Monday morning in Asia to Friday afternoon in New York and is active 24 h per day.

The most important feature of the Forex market (and very natural) is that the exchange is quoted in pairs in difference to stock markets, where each stokes has its value. It is important to mention that the arbitrage on Forex is possible in a short time scale [52–54]. This induces some bias on the analysis, because the choice of the base currency may influence the results, particularly on the very short time scale. On the other hand, one can distinguish a group of leading currencies, which are the most frequently traded: US dollar, euro, and Japanese yen, which are dominating in the market. The bias resulting from the arbitrage is reduced by PLCS feature—due to the averaging procedure. Moreover, in the present study, the euro, as the leading currency, has been chosen as a central currency and exchange rate time series investigated in this paper.

Within this study, the daily exchange rates registered on the Forex market were analysed. The data set consists of 34 time series with the euro as the base currency. The following exchange rates have been investigated: AR, CZK, AUD, DKK, BGN, EGP, BRL, HKD, CAD, HRK, CHF, HUF, IDR, CNY, ISK, JPY, KRW, MXN, MYR, NAD, NOK, NZD, PHP, PLN, RON, RUB, SEK, SGD, THB, TRY, TWD, UAH, USD, and ZAR. Standard abbreviations are used. The period is from 03.09.1996 until 05.02.2020, i.e., 1000 data points.

Within the considered period, one can distinguish several crises (on a regional and global scale). The crises are playing a special role in the presented analysis, because we can expect highlighting the globalisation processes. To mention the most serious crises within the considered interval: 1997—Asian financial crisis [55], 1998—Russian crisis [56], 1999—Argentine crisis [57], early 2000s recession [58], dot-com bubble [59], 2008 financial crisis [60], 2010 European sovereign debt crisis [61], national government debt-crises (Spanish, Greek, Russian, and Turkish), and others. Those crises are discussed in view of the performed analysis results.

3.2. Descriptive Statistics of the Series

The exchange rate time series were converted into return time series by Equation (2).

$$r_i(t) = \frac{a_i(t) - a_i(t-1)}{a_i(t-1)} \tag{2}$$

where a_i denotes the analysed time series.

Table 1 presents the statistical properties of the investigated time series. The mean value of the exchange rate returns of the considered time series was in the interval $(-0.629 \times 10^{-4}, 9.087 \times 10^{-4})$, so the average daily fluctuations are rather small, and they are close to zero. However, the range of observed returns is significant—the lowest noticed return was -0.282, while the greatest was 0.585. The next considered parameter—standard deviation—is particularly important, because it is broadly used as a measure of volatility. When comparing the values of standard deviation and the mean, one can notice that the dispersion is huge. The standard deviation is two orders of magnitude greater than the mean. Another important piece of information is given by skewness analysis. Many of the time series have skewness that is much different from zero, which means that the return distribution is asymmetric. The lowest skewness is observed for CHF exchange rate return, while the highest value is achieved for EGP. The last discussed statistical feature is the result of kurtosis, which is much bigger than one and are leptokurtic for all considered time series. The highest values are observed for EGP, CHF, AR, UAH, IDR, and RUB.

Table 1. Statistical properties of the exchange rate returns.

Currency	Mean $\cdot 10^{-4}$	Median $\cdot 10^{-4}$	Std $\cdot 10^{-2}$	Max	Min	Skewness	Kurtosis
AR	9.087	3.836	1.413	0.403	−0.126	11.147	261.7
CZK	−0.471	−0.740	0.486	0.093	−0.064	1.291	41.6
AUD	0.324	−2.245	0.760	0.079	−0.050	0.743	10.8
DKK	0.053	0	0.048	0.079	−0.009	−0.560	84.1
BGN	1.393	0.452	0.845	0.063	−0.060	0.324	6.4
EGP	3.513	0.665	1.261	0.586	−0.075	21.336	961.4
BRL	3.334	−0.718	1.182	0.129	−0.1108	0.513	15.8
HKD	−0.079	0	0.663	0.055	−0.070	−0.094	8.4
CAD	−0.140	−1.480	0.674	0.044	−0.043	0.201	5.7
HRK	0.351	0.135	0.492	0.049	−0.053	0.092	18.2
CHF	−0.629	0	0.468	0.088	−0.159	−6.186	304.3
HUF	1.335	0.289	0.595	0.070	−0.062	1.174	20.1
IDR	4.849	0	1.802	0.462	−0.207	5.287	134.0
CNY	−0.313	0.329	0.834	0.050	−0.062	−0.102	8.4
ISK	1.392	−0.991	0.876	0.145	−0.133	1.199	71.2
JPY	0.061	2.313	0.849	0.083	−0.116	−0.606	17.0
KRW	1.050	−1.545	1.084	0.158	−0.232	−0.678	78.1
MXN	1.944	0	0.904	0.068	−0.091	0.221	10.1
MYR	1.033	−0.253	0.773	0.068	−0.070	0.129	13.0
NAD	2.782	−0.375	1.100	0.184	−0.101	1.500	25.9
NOK	0.585	−0.937	0.530	0.050	−0.082	−0.350	23.5
NZD	0.162	−3.306	0.783	0.057	−0.051	0.341	6.3
PHP	1.367	1.125	0.799	0.111	−0.130	−0.039	29.5
PLN	0.615	−1.442	0.642	0.057	−0.048	0.609	9.6
RON	5.452	0.675	0.908	0.192	−0.096	3.521	76.2
RUB	6.074	1.393	1.651	0.347	−0.282	4.050	124.6
SEK	0.570	−0.450	0.475	0.036	−0.039	0.228	8.4
SGD	−0.159	0	0.595	0.043	−0.052	−0.154	7.0
THB	0.412	0.367	1.016	0.171	−0.067	1.045	25.0
TRY	9.005	4.949	1.177	0.267	−0.086	4.395	89.3
TWD	0.09	−0.232	0.654	0.068	−0.069	0.079	9.6
UAH	6.469	0	1.732	0.554	−0.215	8.250	258.0
USD	−0.082	0	0.672	0.077	−0.077	−0.046	11.6
ZAR	2.768	−1.856	1.113	0.121	−0.143	0.259	18.1

Additionally, the time evolution of the mean return exchange rate is presented in Figure 1. This graph allows for obtaining a general idea of Forex market evolution, particularly to distinguish the periods of instability of the market.

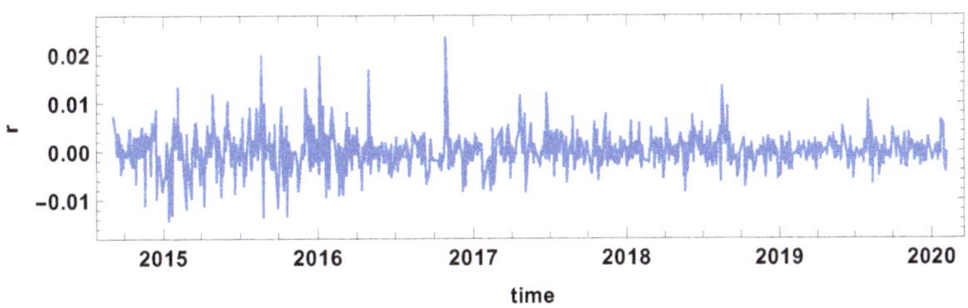

Figure 1. The mean value of the exchange rates return of the considered time series.

4. Results

The moving time window technique must be used to study the time evolution of cross-correlation. The results of the analysis depend on the correlation time window length. The long time window smooths the fluctuations and it can hide important system features. On the other hand, the short time window does not provide a good quality fit of the power law, and the fluctuations are more apparent in the analysis. Therefore, PLCS algorithm was applied for three time window lengths: $T_c \in (20, 60, 120)$, which correspond to a month, quarter, and half of the year period.

4.1. Month Time Window

The frequency of connection is the first parameter investigated here. This parameter informs how often the correlation strength was converging or diverging, so how stable was the correlation in the analysed period. In the case of the diverging correlation strength network, the result is presented in Figure 2.

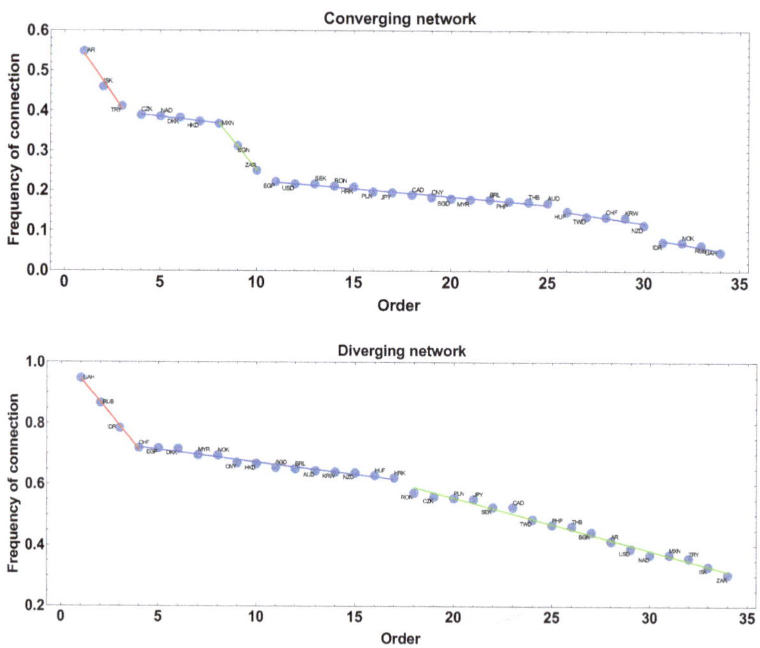

Figure 2. The frequency of connection presented in descending order. The time window $T_c = 20$ days. The blue line denotes a group of currencies of similar frequency of being connected on the network.

Applying the linear fit to the frequency rank allowed for distinguishing three groups of currencies. The first group is marked by the red line: UAH, RUB, and IDR. The second group is marked by the blue line: CHF, EGP, DKK, MYR, NOK, CNY, HKD, SGD, BRL, AUD, KRW, NZD, and HUF. The third group is marked by the green line: RON, CZK, PLN, JPY, SEK, CAD, TWD, PHP, THB, BGN, AR, USD, NAD, MXN, TRY, ISK, and ZAR.

In the case of the network construction based on the converging time series, i.e., the correlation strength $\alpha < 0$ the frequency of connection ranks are presented in Figure 2 and denoted as the converging network. In this case, six groups can be distinguished. Taking more detailed analysis into account, the following groups can be pointed out: the first, marked by the red line AR, ISK, and TRY, and the second, denoted by the blue line, consists of CZK, NAD, DKK, HKD, and MXN. The third group, marked by the green line consists of two members BGN, ZAR. The fourth group is the biggest EGP, USD, SEK, RON, HRK, PLN, JPY, CAD, CNY, SGD, MYR, BRL, PHP, THB, and AUD. The two other groups are formed by HUF, TWD, CHF, KRW, NZD and IDR, NOK, RUB, UAH. Although both graphs are, in some sense, complementary, divergent correlation graphs are constructed under the condition that on the graph there are currencies with $\alpha > 0$, while the divergent graph under condition $\alpha < 0$ the graphs in Figure 2 are not simple mirror images of each other. This is because, in the analysis, the whole correlation matrix is investigated and a given currency may be present on both types of graphs at the same time (it might be convergent with respect to one time series and divergent with concerning another). Particularly interesting are the groups denoted by the blue lines. These groups consist of currencies with similar frequency of being present in the network (divergent or convergent respectively), so the method introduces a natural categorization of time series.

Clique size evolution. In the context of correlation strength network, the cliques are special formations. The cliques are the fully connected group of currencies, with the same type of correlations. Figure 3 presents the clique size evolution graphs for both types of networks. The main advantage of the clique size evolution analysis is the possibility to observe the clique formation in time. The converging time series network that is presented in Figure 3 shows that the biggest clusters were formed in the fourth quarter in 2014, which can be interpreted as the moment when most of the time series were converging, so the differences were decreasing. The clique was formed by 24 currencies. At the other maxima, the formed clusters were not so large and they were in the interval 17–10 currencies. The local maxima were observed in mid-2015, the second and third quarter of 2016, the first quarter of 2017, the second and third quarter of 2018, and the second quarter of 2019. It is also worth noticing that the average level of clique size before 2017 was on the level of 12 currencies, whereas, afterwards, the average value becomes about five time series. Thus, the significant decrease of the clique size is noticeable.

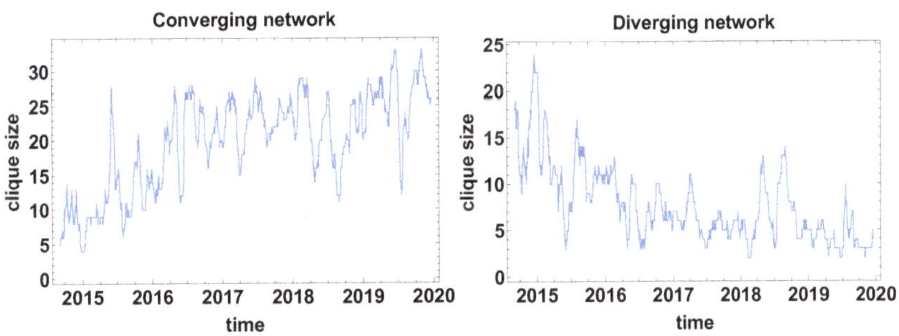

Figure 3. The biggest clique size evolution. Time window size $T_c = 20$ days.

The changes in the average size of the clique that are observed for the converging time series graph are supported by the analysis of the clique size evolution for the diverging time series graph Figure 3. In this case, the initial average size of the clusters was increased from the size of about 10 currencies to more than 23 currency time series. In the high frequency (short time window) analysis, the clique size in the diverging time series network is of high variance, which means that there is no stable tendency. The clusters are formed temporarily. However, the significant value of the cluster size suggests that the majority of the time series are divergent.

The structure analysis of the network was continued by calculating the number of communities that formed on the network. This structure community analysis is based on a weaker constraint than the clique search. Another difference to the biggest clique size is the number of communities is analysed instead of the biggest clique size. The number of communities algorithm looks for the subgraph group with nodes with a higher density of connections than the other part of the network (indifference to the clique that is a fully connected subgraph). Figure 4 presents the results of the number of community analysis.

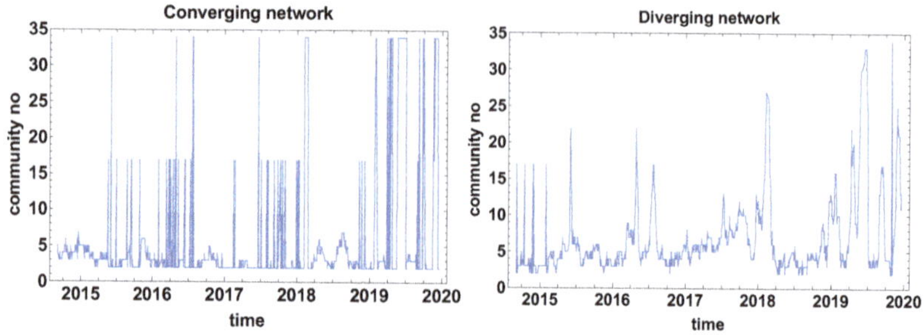

Figure 4. Evolution of the communities number. Time window size $T_c = 20$ days.

Figure 4 presents the evolution of the community number that is observed for both types of networks. Intriguing is the evolution of community number in the case of the converging network (the correlation strength $\alpha > 0$). Three levels of community number can be distinguished in Figure 4 for the converging network these are the ground level where a few communities are observed and two other states of 17 and 34 nodes. Such a big number of communities suggests that they are of very small size (one or at most a few nodes), additionally, the huge increment denotes that shift of the time window by one day has changed the situation significantly. This can be interpreted as either the period is extremely unstable or the correlation strength is approximately close to zero and small changes of the data set have affected the classification of the time series. This observation suggests that, in future applications of the method, it might be worth considering the introduction of an additional class of time series cross-correlation $\alpha \approx 0$. Besides the two-state period, the other local maxima are not spectacular, because they are not exceeding seven communities.

The graph presenting the evolution of the community size for the diverging network (Figure 4) differs significantly from the converging network. In this case, except for the initial part at the end of 2014, the two-level behaviour is not observed. Therefore, the diverging network seems to be more robust to the network switching effect. Similarly to the converging network, the "baseline" of the community number can be distinguished (2–5 communities). Several clear maximums can be distinguished in the case of the diverging network quantity of community evolution: June 2015, April and July 2016, February 2018, and several maxima in 2019. 2019 was the most unstable year out of those analysed when many times the network was split into a big number of small communities.

The evolution of the community number for the converging network might suggest that the time window size T_c is too short and fluctuations significantly influence the results of the analysis.

Figures 5 and 6 present the evolution of the node rank histogram for converging and diverging time series networks, respectively. When analysing the evolution of the node rank histogram for the converging time series network shown in Figure 5, it can be observed that in 2015 and 2016 the nodes with a significant number of links ($k > 20$) are dominating. Whereas, in 2017 and later, the nodes with the low number of links ($k < 15$) are dominating. A short exception is observed in 2018 (during the Chinese crisis) when nodes with a high number of links were clearly present in the network. In 2019 and later, the nodes with a small number of links are prevailing on the converging network.

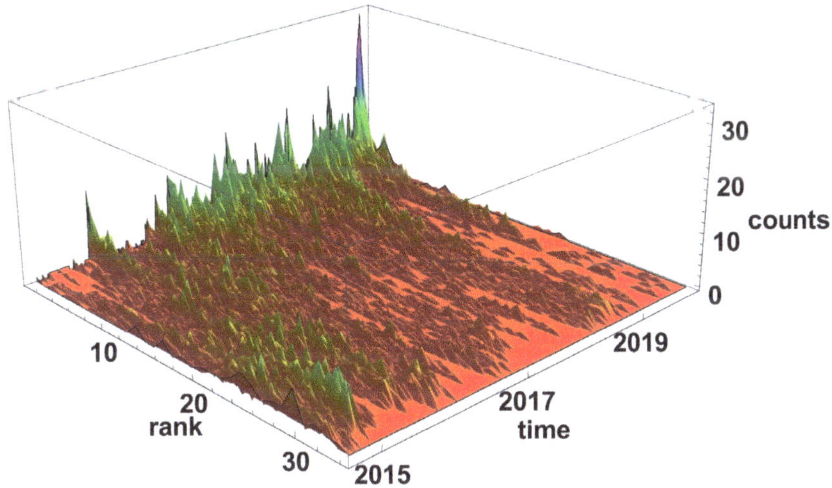

Figure 5. Evolution of the rank nodes histogram for converging network. The time window size $T_c = 20$ days. The counts denote how many times the node of given rank (number of links) was observed on the network.

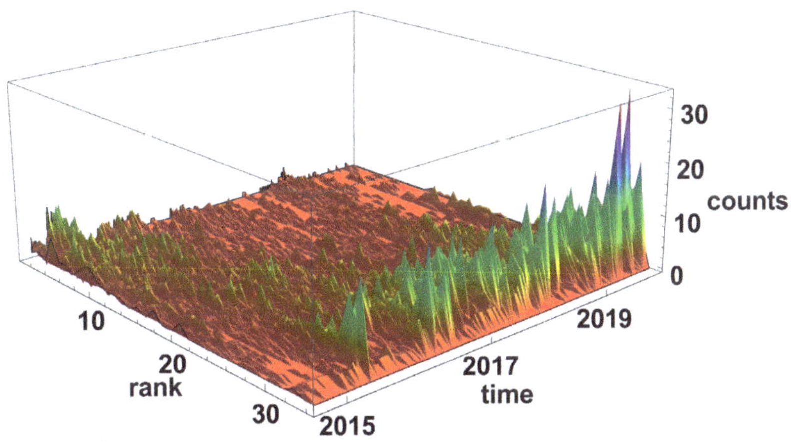

Figure 6. Evolution of the rank nodes histogram for diverging network. The time window size $T_c = 20$ days. Counts denotes how many times the node of given rank (number of links) was observed on the network.

The evolution of the diverging time series network histograms is presented in Figure 6. Initially, in 2015, the nodes with a small number of links are most evident, but, since 2016, the situation has changed and the nodes with a high number of connections are the most common on the network. It is particularly well seen at the end of 2019 and the beginning of 2020, when nodes with the degree $k > 30$ are dominating on the network.

Figure 7 presents the evolution of rank node entropy. There are no significant differences between the generated networks. Particularly interesting are the minima, which correspond to the situation where there is a significant group of nodes of the same rank. Although several minima can be distinguished, they do not form a clear evolution; this is due to the noise influence. This results indicate that the time window is rather too short to obtain a clear evolution of the system.

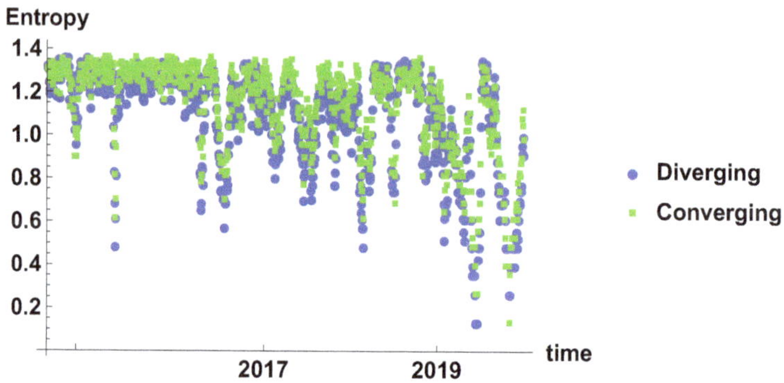

Figure 7. Evolution of the rank node entropy for diverging and converging networks. The time window size $T_c = 20$ days. The blue circles and green squares denote the entropy of diverging and convergent network, respectively.

4.2. Quarter Time Window

Extending the size of the time window T_c to 60 days results in filtering high frequency changes, which were observed in the one-month time window. Following the same scheme of network feature analysis shown in Section 4.1, the discussion starts from the frequency of being connected. The results are presented in Figure 8. In the case of networks constructed with the constraint of the converging time series, the most frequent connections are ISK and TRY, while, for the divergent time series network, the most frequent observed currencies are UAH and RUB, which are present in 94% and 93% of the constructed network. The blue line denotes the group of currencies with similar frequency of the network member. For the converging time series network, the biggest group has a frequency in the interval 27–3%, being rather low, while, in the second type of network considered here, the frequency is in the interval 82–51%, so the probability of connection is significantly higher.

Figure 9 shows the time evolution of the biggest clique size (so the clusters of a fully connected set of currencies). It can be noticed that the divergent and convergent time series networks results are on average complementary—the size of the cliques in convergent time series is growing in time, but in divergent time series are decreasing. Of course, the graphs differ in details. Moreover, the general similarity does not apply to the position and magnitude of extreme points. For the converging network, as in Figure 9, six local extremes can be distinguished. The local maxima are observed in April 2015, March 2016, May–June 2016, April–June 2017, January 2018 (which is the highest maxima of 30 nodes in one clique), and the local minimum in June 2018. The clique size evolution in the diverging time series network has approximately four local extremes. The first maximum is observed at the end of 2014, which is followed by a very deep minimum in April 2015. The decrease of the clique size is enormous, because, at the first maximum, there are 26 nodes in the clique, while at the minimum the biggest clique consists of 5 nodes, so the biggest clique size decreased by 21 nodes. Immediately after that minimum, the biggest clique is growing to achieve the size of 17 nodes in August 2015. Subsequently, the clique size is relatively slowly decreasing to the level of 4–7 nodes. The last maximum is observed in July 2018.

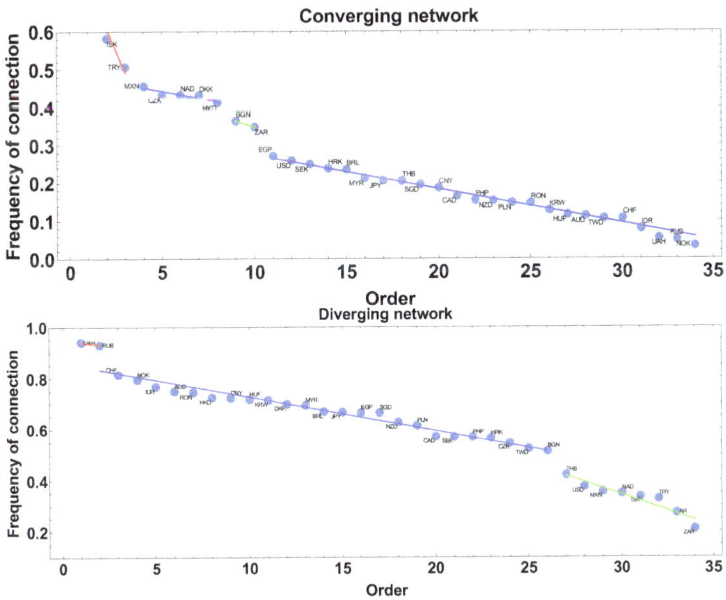

Figure 8. The frequency of connection presented in descending order. The time window $T_c = 60$ days. The blue line denotes group of currencies of similar frequency of being connected on the network.

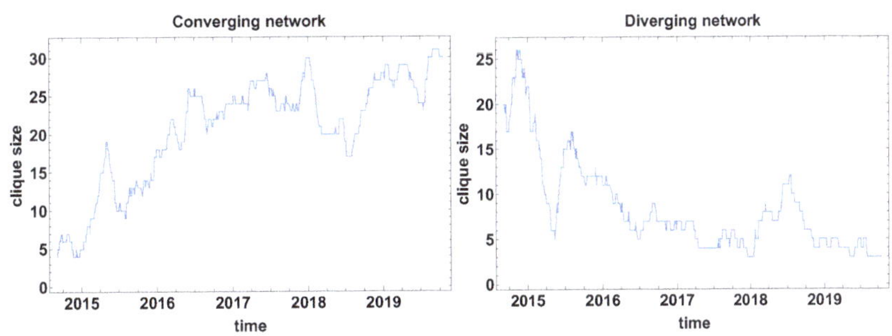

Figure 9. The biggest clique size evolution. Time window size $T_c = 20$ days.

Figure 10 presents the evolution of the community number. In the evolution of community number of converging networks, one can distinguish three levels: the ground state, where the community approximately 3–6 communities, the second level of 16–17 communities, and the third level of 34 communities. Because the border between converging and diverging time series is $\alpha = 0$, the bistable behaviour of the graph means that a significant group of currencies is at the border and a small shift of the time window position is changing their classification. A similar observation was made for the evolution of community number for $T_c = 20$ days. As it was already mentioned, the additional class of $\alpha \approx 0$ is not introduced here due to the clarity of the analysis, because the main aim of the study is to verify the properties of the algorithm. The additional class should be considered in such a case in, e.g., commercial analysis.

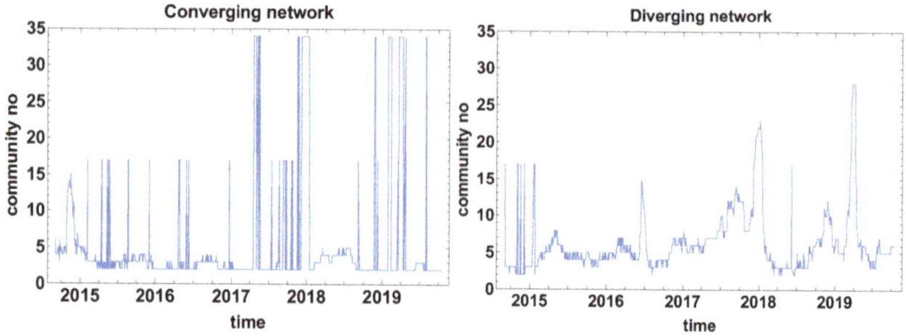

Figure 10. Evolution of the community number. The time window $T_c = 60$ days.

The bistable behaviour of the size of the community size is also observed in the diverging network shown in Figure 10 at the end of 2014. Afterwards, the bistable evolution is not observed and several clear maxima can be noticed: June 2016, at the end of 2017, and in April–May 2019. It can be observed that, due to the longer time window, the number of maxima has been reduced when compared to the previously discussed case, as in Figure 4.

The evolution of the node rank histograms for converging and diverging networks are presented in Figures 11 and 12, respectively. In both types of network, two periods can be distinguished: the most common is the high-rank nodes or the reverse situation—the low-rank nodes. The converging time series network, as in Figure 11, is, in general, complementary to the diverging network case, as in Figure 12. At the end of 2014, the low-rank nodes are dominating, while, in 2016, 2017, and 2019, the high-rank nodes are prevailing in the histograms. Combining the results of the rank node histograms evolution with the clique size analysis, where huge clusters are observed, as in Figure 9, it can be concluded that the generated networks are very close to a fully connected network.

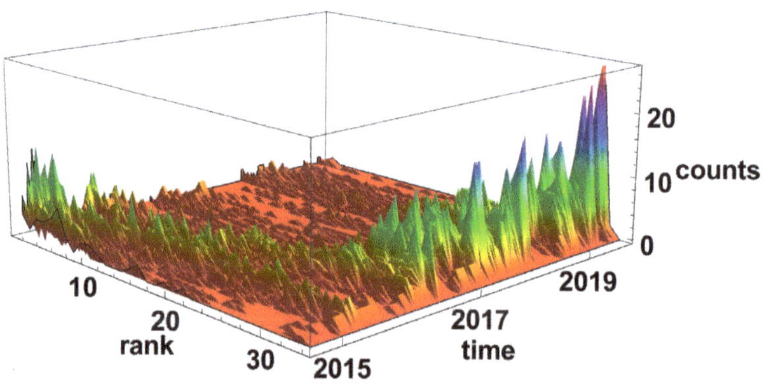

Figure 11. Evolution of the rank nodes histogram for converging network. The time window size $T_c = 60$ days. Counts denote how many times the node of given rank (number of links) was observed on the network.

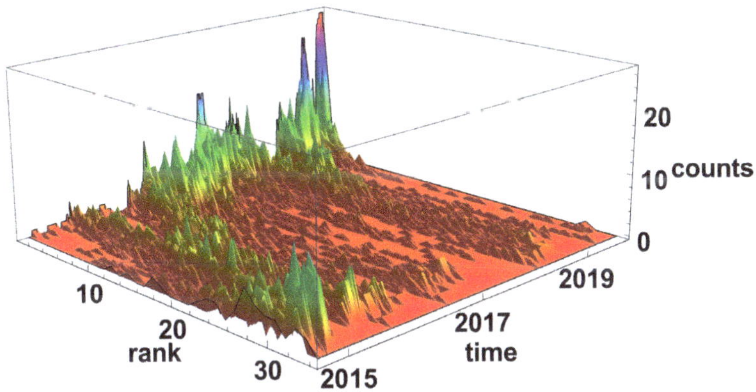

Figure 12. Evolution of the rank nodes histogram for diverging network. The time window size $T_c = 60$ days. Counts denote how many times the node of given rank (number of links) was observed on the network.

In the case of the diverging time series network, as in Figure 12, the nodes of high rank are observed at the end of 2014, at the end of 2015, and the beginning of 2016. A very special situation occurs at the beginning of 2015, when there is no dominating group of nodes, but all the ranks of nodes are present in the histogram. In 2017, at the end of 2018, and then the beginning of 2019, the networks are divided into small subgraphs. In the mid of 2018, the increase of high-order nodes is observed—this situation can be related to the Chinese crisis.

Figure 13 presents the entropy of the rank node distribution for the time window of $T_c = 20$ days. In this case, the influence of noise is significantly reduced. The different periods can be clearly distinguished. Initially, in 2015 the decrease of entropy is observed, which is the effect of domination of high rank nodes in the histograms. The period of stable high entropy follows, which lasts until the mid of 2016. Later, oscillation appears, which are combined with the decrease of the minimum value to achieve minimum in the beginning of 2018. In 2018, another period of maximum entropy is observed. It seems that level 1.4 is the maximum entropy observed in these networks and can be considered as a measure of the stability of the market. A significant lowering of the entropy may be considered as a signature of the crisis.

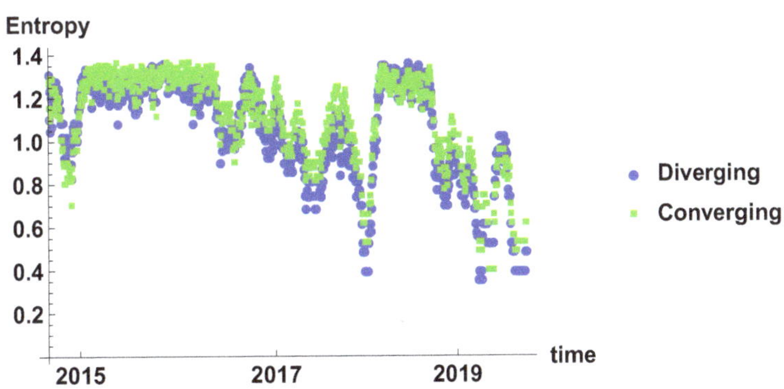

Figure 13. Evolution of the rank node entropy for diverging and converging networks. The time window size $T_c = 60$ days. The blue circles and green squares denote the entropy of diverging and convergent network respectively.

4.3. Half Year Time Window

This subsection contains the results obtained for the longest time window $T_c = 120$ days. Figure 14 presents the results of the frequency of connection of nodes to the network for both types of networks. In the case of the converging network, AR is the most frequent currency, which is present in 83% generated graphs. This node is separated and does not belong to any group. The first group, which can be distinguished in this analysis, consists of five currencies: ISK, TRY, MXN, HKD, and NAD. Currencies of this group are connected to others in 58–54% of networks. The second group consists of two currencies: ZAR and BGN. The last group is the biggest one—26 currencies. Within this group, the frequency of being connected is rather low: from 32% to 3%.

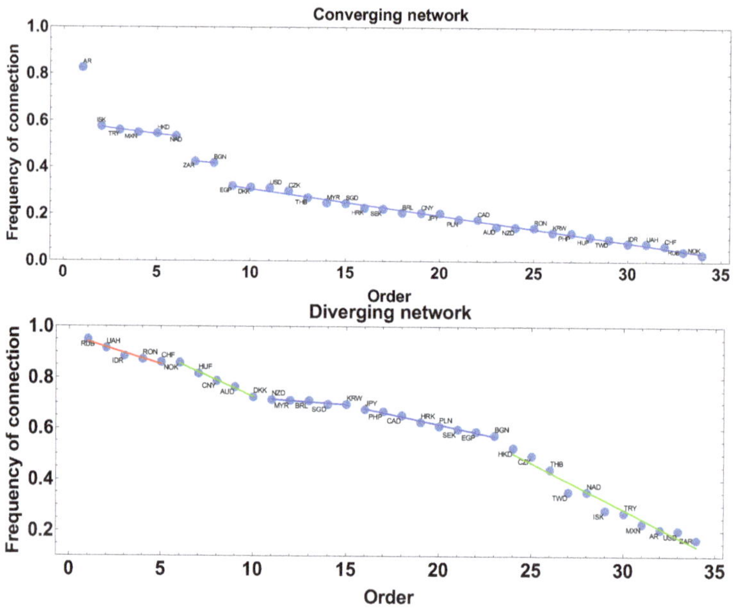

Figure 14. The frequency of connection presented in descending order. The time window $T_c = 120$ days. The blue line denotes group of currencies of similar frequency of being connected on the network.

The frequency of being connected on a divergent time series graph is slightly different because only two groups of similar frequency, i.e., without significant differences between consecutive elements, can be distinguished. The first group consists of five currencies: NZD, MYR, BRL, SGD, and KRW, and their frequency is varying from 72% to 70%. This group is followed by the second one: JPY, PHP, CAD, HRK, PLN, SEK, EGP, and BGN with the frequencies from 68% to 57%.

Figure 15 presents the biggest clique size evolution for the time window $T_c = 120$ days. When comparing to the previously discussed cases, i.e., $T_c = 20, 60$ days, the smoothing effect of the time window size is clearly visible. In this case, the biggest clique size for the converging time series network is asymptotically increasing with the exception in the middle of 2018, which can be related to the Chinese stock market crisis. An analogous maximum is observed in the graph presenting the biggest clique size evolution for the diverging network shown in Figure 15.

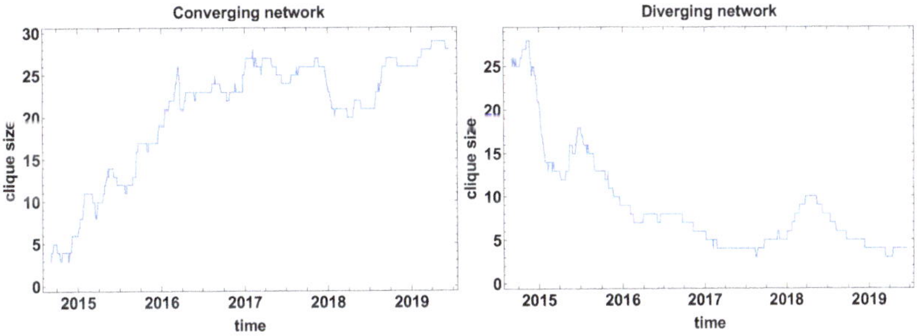

Figure 15. The biggest clique size evolution. Time window size $T_c = 120$ days.

Figure 16 presents the evolution of the community number on the graph for the time window $T_c = 120$ days. In the case of the converging network, the observed previously switching effect between two states for shorter time windows is also present in this case. However, in difference to the previous analyses, there is a period when the network brakes into separate nodes. This is the second and third quarter of 2017. At this time, in the community number of the diverging network graph, the maximum is reaching the value of 20 nodes. Simultaneously, the high number of communities is observed in diverging and converging networks this suggests that no clear tendency (or significant correlation) is present in the market. This finding agreed with the fact that, at this time, there was no serious global crisis.

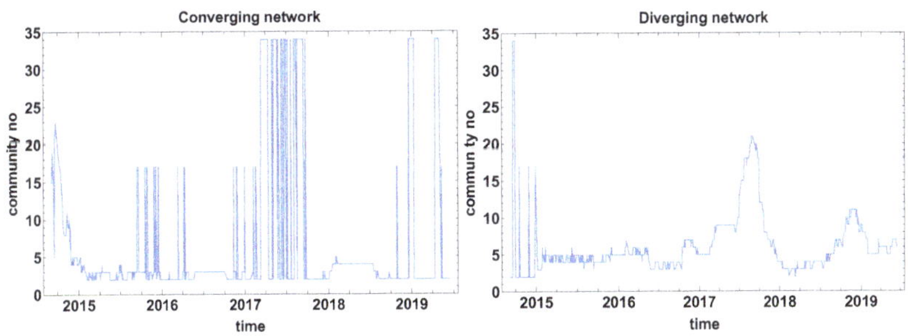

Figure 16. Evolution of the community number. The time window $T_c = 120$ days.

The node rank histogram evolution that were obtained for the time window $T_c = 120$ days are presented in Figures 17 and 18. In both graphs, the change node rank structure is clearly visible. In the case of converging time series network, as in Figure 17, at the beginning of the analysed period, i.e., at the end of 2014 and in the first quarter of 2015 the low-rank nodes are prevailing in the network, while, from 2016, the high-rank nodes are dominating. Differently from the already analysed rank histograms evolutions for shorter time windows ($T_c = 20$ and $T_c = 60$ days) in the case of $T_c = 120$ days, the process of network transition from domination of low-rank nodes to high-rank nodes, prevailing network is a kind of continuous process. The transformation process lasts approximately a year when the nodes are gaining connections. The significant shift of the maximum position of the low-rank nodes is observed in mid-2018, probably due to the Chinese stock market crisis.

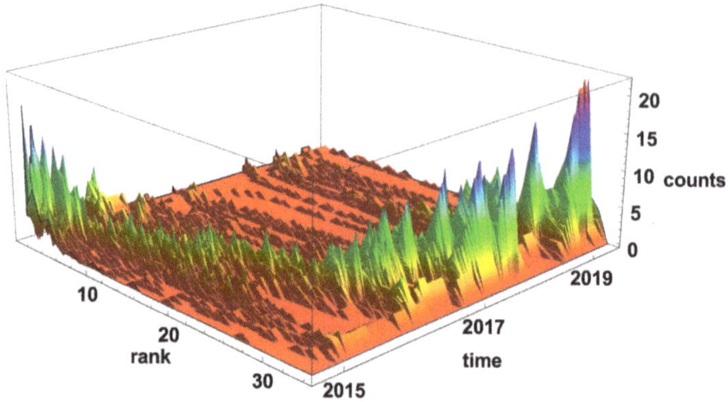

Figure 17. Evolution of the rank nodes histogram for converging network. The time window size $T_c = 120$ days. Counts denotes how many times the node of given rank (number of links) was observed on the network.

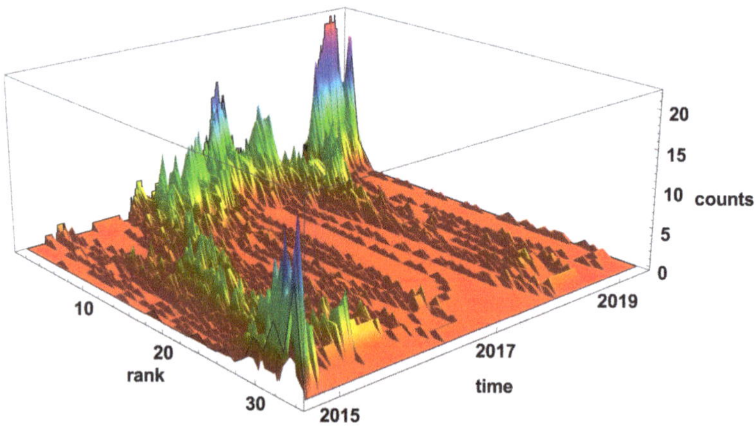

Figure 18. Evolution of the rank nodes histogram for diverging network. The time window size $T_c = 120$ days. Counts denote how many times the node of given rank (number of links) was observed on the network.

The diverging network rank node histogram evolution, as shown in Figure 18, is complementary to the converging series network. At the end of 2014, the high-rank nodes are prevailing in the histogram. During 2016, the node rank frequency of occurrence is evolving from high node rank domination to low-rank nodes prevailing in 2016. Finally, since 2016, the low-rank nodes have dominated the network except for mid-2018.

The rank node entropy evolution that is observed in the case of the time window $T_c = 120$ days is presented in Figure 19. The long time window results in significant filtering of the time series. In this case, the most stable effects can be observed. In the presented results, there are two such events—one in 2017 and the second in 2019. The outcomes of the analysis for the half-year time window confirm the previous observations that the crisis is characterised by a low value of the entropy of the rank node distribution.

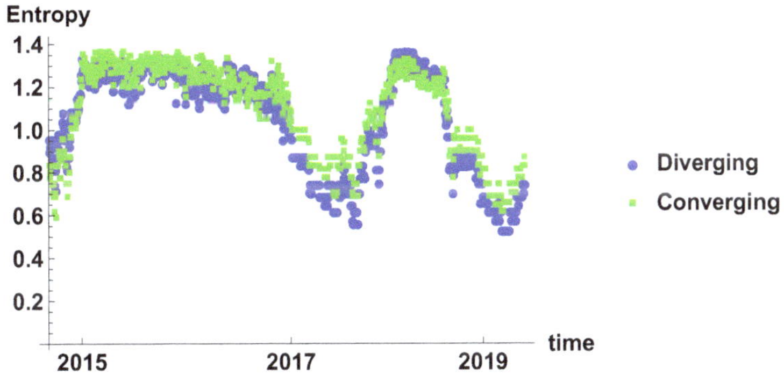

Figure 19. Evolution of the rank node entropy for diverging and converging networks. The time window size $T_c = 120$ days. The blue circles and green squares denote the entropy of diverging and convergent network respectively.

5. Conclusions

The presented study investigates the cross-correlations among currency exchange rates on Forex market by the PLCS algorithm, followed by network analysis. The PLCS method is focused on the trend correlations and unlike other methods, allows to observe cross-correlation of trends. The results of this paper show that crises influence trend correlations. The convergent and divergent networks are not simple mirrors of each other. Because the network is constructed with the cross-correlation matrix, the introduced constraint may reveal a different feature, e.g., the community number observed in the converging network presents a two-state evolution that is rarely observed in a diverging network. Particularly interesting is the biggest cluster size analysis, which is sensitive to crisis occurrence. Particularly, the change of the cluster size can expose the severity crisis. The third feature investigated here is the frequency of the connection, which verifies the stability of the connection. Currencies are forming groups concerning the frequency of connections to the network. It might give an opportunity to develop a new classification of currencies with respect to their relationship to the group. The last performed analysis—the rank node histogram evolution—provides the most detailed information about the structure and evolution of the cross-correlation among currencies. The analysis of the rank node entropy is particularly interesting. The obtained results suggest that entropy might be a synthetic measure of crisis. Of course, this conclusion needs further analysis, but the presented results are very promising.

A very special outcome of this analysis is that, in recent times, e.g., 2017, the structure of the observed networks has changed and depending on the type of the network (converging or diverging) the high or low-rank nodes are prevailing. It means that the cross-correlation in the Forex market has changed significantly. The observed changes in the biggest clique size and the number of communities are the results of globalisation, which are more transparent during crises. In this special condition, correlations and mutual dependence are exposed. Of course, the results depend on the choice of central currency and the analysis can be repeated for other central currencies. However, the main aim of this paper was establishing new analysis methods, so the detailed analysis of the role of the central currency choice is left for other studies. The additional results are the analysis of the role of the time window length. The presented results allow for estimating the window size with the requested quality of research. It is not recommended to use time windows shorter than 20 days. Of course, extending the size of the time window improves the quality of the results from the statistical point of view, and it filters the high frequency changes exposing the long-term proprieties. Although this aspect was not discussed here, longer time windows might be more appropriate for forecasting.

Funding: This research received no external funding.

Institutional Review Board Statement: Not applicable.

Informed Consent Statement: Not applicable.

Conflicts of Interest: The author declares no conflict of interest.

References

1. Chandra, P. *Investment Analysis and Portfolio Management*; McGraw-Hill Education: New York, NY, USA, 2017.
2. Briston, R.J. *The Stock Exchange and Investment Analysis*; Routledge: London, UK, 2017.
3. Yu, J.N. Research on Financial Portfolio Analysis in the New Era. *J. Phys. Conf. Ser.* **2020**, *1437*, 012055. [CrossRef]
4. Safitri, I.N.N.; Sudradjat, S.; Lesmana, E. Stock portfolio analysis using Markowitz model. *Int. J. Quant. Res. Model.* **2020**, *1*, 47–58. [CrossRef]
5. Auer, R.A.; Schoenle, R.S. Market structure and exchange rate pass-through. *J. Int. Econ.* **2016**, *98*, 60–77. [CrossRef]
6. Corbet, S.; Lucey, B.; Urquhart, A.; Yarovaya, L. Cryptocurrencies as a financial asset: A systematic analysis. *Int. Rev. Financ. Anal.* **2019**, *62*, 182–199. [CrossRef]
7. Levitt, T. The globalization of markets. In *Readings in International Business: A Decision Approach*; The MIT Press: Cambridge, MA, USA, 1993; Volume 249.
8. Beck, U. *What Is Globalization?* John Wiley & Sons: Hoboken, NJ, USA, 2018.
9. Scholte, J.A. *Globalization: A Critical Introduction*; Macmillan International Higher Education: Montgomery, AL, USA, 2005.
10. Wang, G.J.; Xie, C.; Stanley, H.E. Correlation structure and evolution of world stock markets: Evidence from Pearson and partial correlation-based networks. *Comput. Econ.* **2018**, *51*, 607–635. [CrossRef]
11. Piao, L.; Fu, Z. Quantifying distinct associations on different temporal scales: Comparison of DCCA and Pearson methods. *Sci. Rep.* **2016**, *6*, 36759. [CrossRef]
12. Mantegna, R.N. Hierarchical structure in financial markets. *Eur. Phys. J. B* **1999**, *11*, 193–197. [CrossRef]
13. Miśkiewicz, J.; Ausloos, M. Correlation measure to detect time series distances, whence economy globalization. *Phys. A Stat. Mech. Its Appl.* **2008**, *387*, 6584–6594. [CrossRef]
14. Miśkiewicz, J. Distance matrix method for network structure analysis. In *Statistical Tools for Finance and Insurance*; Springer: Berlin, Germany, 2011; pp. 251–289.
15. Mantegna, R.N.; Stanley, H.E. *Introduction to Econophysics: Correlations and Complexity in Finance*; Cambridge University Press: Cambridge, UK, 1999.
16. Granger, C.W. Causality, cointegration, and control. *J. Econ. Dyn. Control* **1988**, *12*, 551–559. [CrossRef]
17. Johansen, S. Statistical analysis of cointegration vectors. *J. Econ. Dyn. Control* **1988**, *12*, 231–254. [CrossRef]
18. Watson, M.W. Vector autoregressions and cointegration. *Handb. Econom.* **1994**, *4*, 2843–2915.
19. Adebola, S.S.; Gil-Alana, L.A.; Madigu, G. Gold prices and the cryptocurrencies: Evidence of convergence and cointegration. *Phys. A Stat. Mech. Its Appl.* **2019**, *523*, 1227–1236. [CrossRef]
20. Wang, J.; Shang, P.; Ge, W. Multifractal cross-correlation analysis based on statistical moments. *Fractals* **2012**, *20*, 271–279. [CrossRef]
21. El Alaoui, M.; Bouri, E.; Roubaud, D. Bitcoin price–volume: A multifractal cross-correlation approach. *Financ. Res. Lett.* **2019**, *31*. [CrossRef]
22. Oświęcimka, P.; Drożdż, S.; Forczek, M.; Jadach, S.; Kwapień, J. Detrended cross-correlation analysis consistently extended to multifractality. *Phys. Rev. E* **2014**, *89*, 023305. [CrossRef]
23. Pal, M.; Rao, P.M.; Manimaran, P. Multifractal detrended cross-correlation analysis on gold, crude oil and foreign exchange rate time series. *Phys. A Stat. Mech. Its Appl.* **2014**, *416*, 452–460. [CrossRef]
24. Ren, F.; Zhou, W.X. Dynamic Evolution of Cross-Correlations in the Chinese Stock Market. *PLoS ONE* **2014**, *9*, e97711. [CrossRef]
25. Utsugi, A.; Ino, K.; Oshikawa, M. Random matrix theory analysis of cross correlations in financial markets. *Phys. Rev. E* **2004**, *70*, 026110. [CrossRef]
26. Plerou, V.; Gopikrishnan, P.; Rosenow, B.; Amaral, L.A.N.; Guhr, T.; Stanley, H.E. Random matrix approach to cross correlations in financial data. *Phys. Rev. E* **2002**, *65*, 066126. [CrossRef] [PubMed]
27. Pharasi, H.K.; Sharma, K.; Chakraborti, A.; Seligman, T.H. Complex market dynamics in the light of random matrix theory. In *New Perspectives and Challenges in Econophysics and Sociophysics*; Springer: Berlin, Germany, 2019; pp. 13–34.
28. Miśkiewicz, J. Power law classification scheme of time series correlations. On the example of G20 group. *Phys. A Stat. Mech. Its Appl.* **2013**, *392*, 2150–2162. [CrossRef]
29. Miśkiewicz, J. Cross-correlations of the Forex market using power law classification scheme picture. *Acta Phys. Pol. A* **2016**, *129*, 917–921. [CrossRef]
30. Miśkiewicz, J.; Tadla, A.; Trela, Z. Does the monetary policy influenced cross-correlations on the main world stocks markets? Power Law Classification Scheme analysis. *Phys. A Stat. Mech. Its Appl.* **2019**, *519*, 72–81. [CrossRef]
31. Miśkiewicz, J. Entropy of Globalizing World Macroeconomy Time Series Analysis. *Acta Phys. Pol. A* **2020**, *138*, 25–30. [CrossRef]
32. Teng, Y.; Shang, P. Transfer entropy coefficient: Quantifying level of information flow between financial time series. *Phys. A Stat. Mech. Its Appl.* **2017**, *469*, 60–70. [CrossRef]

33. Ramchand, L.; Susmel, R. Volatility and cross correlation across major stock markets. *J. Empir. Financ.* **1998**, *5*, 397–416. [CrossRef]
34. Kristoufek, L. Detrending moving-average cross-correlation coefficient: Measuring cross-correlations between non-stationary series. *Phys. A Stat. Mech. Its Appl.* **2014**, *406*, 169–175. [CrossRef]
35. Jin, S.T.; Kong, H.; Sui, D.Z. Uber, public transit, and urban transportation equity: A case study in new york city. *Prof. Geogr.* **2019**, *71*, 315–330. [CrossRef]
36. Zaarane, A.; Slimani, I.; Hamdoun, A.; Atouf, I. Real-Time Vehicle Detection Using Cross-Correlation and 2D-DWT for Feature Extraction. *J. Electr. Comput. Eng.* **2019**, *2019*, 6375176. [CrossRef]
37. Hellsten, I.; Lambiotte, R.; Scharnhorst, A.; Ausloos, M. Self-citations, co-authorships and keywords: A new approach to scientists' field mobility? *Scientometrics* **2007**, *72*, 469–486. [CrossRef]
38. Ausloos, M. Rank–size law, financial inequality indices and gain concentrations by cyclist teams. The case of a multiple stage bicycle race, like Tour de France. *Phys. A Stat. Mech. Its Appl.* **2020**, *540*, 123161. [CrossRef]
39. Chen, Z.; Ivanov, P.C.; Hu, K.; Stanley, H.E. Effect of nonstationarities on detrended fluctuation analysis. *Phys. Rev. E* **2002**, *65*, 041107. [CrossRef] [PubMed]
40. Hu, K.; Ivanov, P.C.; Chen, Z.; Carpena, P.; Eugene Stanley, H. Effect of trends on detrended fluctuation analysis. *Phys. Rev. E* **2001**, *64*. [CrossRef] [PubMed]
41. Fan, Q.; Liu, S.; Wang, K. Multiscale multifractal detrended fluctuation analysis of multivariate time series. *Phys. A Stat. Mech. Its Appl.* **2019**, *532*, 121864. [CrossRef]
42. Kwapień, J.; Drożdż, S. Physical approach to complex systems. *Phys. Rep.* **2012**, *515*, 115–226. [CrossRef]
43. Bryce, R.M.; Sprague, K.B. Revisiting detrended fluctuation analysis. *Sci. Rep.* **2012**, *2*, 315. [CrossRef] [PubMed]
44. Höll, M.; Kiyono, K.; Kantz, H. Theoretical foundation of detrending methods for fluctuation analysis such as detrended fluctuation analysis and detrending moving average. *Phys. Rev. E* **2019**, *99*. [CrossRef]
45. Oświęcimka, P.; Kwapień, J.; Drożdż, S. Wavelet versus detrended fluctuation analysis of multifractal structures. *Phys. Rev. E* **2006**, *74*. [CrossRef]
46. Mantegna, R.N.; Palágyi, Z.; Stanley, H.E. Applications of statistical mechanics to finance. *Phys. A Stat. Mech. Its Appl.* **1999**, *274*, 216–221. [CrossRef]
47. Bonanno, G.; Lillo, F.; Mantegna, R. High-frequency cross-correlation in a set of stocks. *Quant. Financ.* **2001**, *1*, 96–104. [CrossRef]
48. Bouchaud, J.P.; Cont, R. A Langevin approach to stock market fluctuations and crashes. *Eur. Phys. J. B* **1998**, *6*, 543–550. [CrossRef]
49. Hassan, M.K.; Islam, L.; Haque, S.A. Degree distribution, rank-size distribution, and leadership persistence in mediation-driven attachment networks. *Phys. A Stat. Mech. Its Appl.* **2017**, *469*, 23–30. [CrossRef]
50. Bauer, B.; Jordán, F.; Podani, J. Node centrality indices in food webs: Rank orders versus distributions. *Ecol. Complex.* **2010**, *7*, 471–477. [CrossRef]
51. Hou, B.; Yao, Y.; Liao, D. Identifying all-around nodes for spreading dynamics in complex networks. *Phys. A Stat. Mech. Its Appl.* **2012**, *391*, 4012–4017. [CrossRef]
52. Gębarowski, R.; Oświęcimka, P.; Wątorek, M.; Drożdż, S. Detecting correlations and triangular arbitrage opportunities in the Forex by means of multifractal detrended cross-correlations analysis. *Nonlinear Dyn.* **2019**, *98*, 2349–2364. [CrossRef]
53. Mancini-Griffoli, T.; Ranaldo, A. *Limits to Arbitrage during the Crisis: Funding Liquidity Constraints and Covered Interest Parity*; Swiss National Bank: Bern, Switzerland, 2011.
54. Chen, K.S.; Chen, C.M.; Lee, C.C. Arbitrage, Covered Interest Parity and Cointegration Analysis on the NTD/USD Forex Market Revisited. *Int. J. Econ. Financ. Issues* **2017**, *7*, 420–428.
55. Wade, R. The Asian debt-and-development crisis of 1997-?: Causes and consequences. *World Dev.* **1998**, *26*, 1535–1553. [CrossRef]
56. Chiodo, A.J.; Owyang, M.T. A case study of a currency crisis: The Russian default of 1998. *Fed. Reserve Bank St. Louis Rev.* **2002**, *84*, 7. [CrossRef]
57. Bebczuk, R.; Galindo, A. Financial crisis and sectoral diversification of Argentine banks, 1999–2004. *Appl. Financ. Econ.* **2008**, *18*, 199–211. [CrossRef]
58. Imbs, J. The first global recession in decades. *IMF Econ. Rev.* **2010**, *58*, 327–354. [CrossRef]
59. Goodnight, G.T.; Green, S. Rhetoric, risk, and markets: The dot-com bubble. *Q. J. Speech* **2010**, *96*, 115–140. [CrossRef]
60. Luchtenberg, K.F.; Vu, Q.V. The 2008 financial crisis: Stock market contagion and its determinants. *Res. Int. Bus. Financ.* **2015**, *33*, 178–203. [CrossRef]
61. Lane, P.R. The European sovereign debt crisis. *J. Econ. Perspect.* **2012**, *26*, 49–68. [CrossRef]

entropy

Article

Neural Networks for Estimating Speculative Attacks Models

David Alaminos [1,*], Fernando Aguilar-Vijande [2] and José Ramón Sánchez-Serrano [3,4]

1 Department of Financial Management, Universidad Pontificia Comillas, 28015 Madrid, Spain
2 PhD in Economics and Business, Universidad de Málaga, 29071 Málaga, Spain; fernando.aguilar@uma.es
3 Department of Finance and Accounting, Universidad de Málaga, 29071 Málaga, Spain; joseramonsanchez@uma.es
4 Cátedra de Economía y Finanzas Sostenibles, Universidad de Málaga, 29071 Málaga, Spain
* Correspondence: dalaminos@icade.comillas.edu

Abstract: Currency crises have been analyzed and modeled over the last few decades. These currency crises develop mainly due to a balance of payments crisis, and in many cases, these crises lead to speculative attacks against the price of the currency. Despite the popularity of these models, they are currently shown as models with low estimation precision. In the present study, estimates are made with first- and second-generation speculative attack models using neural network methods. The results conclude that the Quantum-Inspired Neural Network and Deep Neural Decision Trees methodologies are shown to be the most accurate, with results around 90% accuracy. These results exceed the estimates made with Ordinary Least Squares, the usual estimation method for speculative attack models. In addition, the time required for the estimation is less for neural network methods than for Ordinary Least Squares. These results can be of great importance for public and financial institutions when anticipating speculative pressures on currencies that are in price crisis in the markets.

Keywords: speculative attacks; currency crisis; neural networks; deep learning; Quantum-Inspired Neural Network

Citation: Alaminos, D.; Aguilar-Vijande, F.; Sánchez-Serrano, J.R. Neural Networks for Estimating Speculative Attacks Models. *Entropy* **2021**, 23, 106. https://doi.org/10.3390/e23010106

Received: 7 December 2020
Accepted: 10 January 2021
Published: 13 January 2021

1. Introduction

A currency crisis is defined as the inability of the authorities of a country to defend a certain parity for the exchange rate. In turn, the exchange rate crisis will occur as a result of a speculative attack carried out by operators in the foreign exchange market, which causes a large and sudden increase in the ability to readjust the central parity [1]. The models of speculative attacks best known from the previous literature are the so-called first- and second-generation models. The first-generation models are based on the incompatibility between the economic policy of a government and its commitments to a fixed exchange rate, which ends up leading to a speculative attack on its currency and the collapse of the exchange regime. The first formulation of this type of model is due to Krugman [2]; second-generation ones incorporate private agents, their expectations, and interaction with economic policy, generating the possibility of multiple equilibria and self-generated crises. This second-generation model was built by the work of Obstfeld [3]. The experience of countries with exchange rate crises shows that they cause significant welfare losses for economic agents, insofar as they have generated falls in output and employment, and large losses in international reserves without neglecting significant fiscal problems. Hence the importance of having indicators that warn about events of excessive fragility is that they allow the authorities to act promptly to minimize the costs associated with the outcome of these episodes of speculative attacks in currency crises.

In the last decade, many countries have suffered a currency crisis that has led to high pressure against the price of their currency in financial markets [4]. This has been due to the significant deterioration of their balance of payments concerning international trade. However, the reasons why they have suffered these falls have been varied. Countries like

Russia and Iran suffered in recent years different important falls in the value of their currency due to the economic sanctions imposed by the United States and the European Union. This caused a drop in their commercial activity, and therefore, an abrupt deterioration in their international trade balances. Other African countries such as Namibia or South Africa have also recently suffered acute currency crises due to domestic political crises and continuing instability that has deteriorated their international image and their bilateral and trade relations with other countries. Lastly, Latin American countries such as Mexico or Argentina have suffered successive currency crises with consequences of speculative attacks due to their current account crises with failed economic policies.

Different authors have analyzed speculative attacks based on macroeconomic theory, being the object of continuous study and with strong consequences both in the economy and in the financial markets. However, in the last decade, we can find various works on speculative attacks with very specific objectives on the procedure in which they occur. Even so, these studies have not obtained a great repercussion, the first- and second-generation models created previously are currently of great importance [5–10]. Others that follow this line of speculative attack models stand out, such as those carried out by [11–17], where they have tried to explain the origins of speculative attacks and currency crises, managing to establish the theory that helps to explain these phenomena. This has also been studied in various works such as those of [15–21] discussing what type of exchange rate to establish or what type of economic policy to choose to reduce the chances of suffering a speculative attack. Despite this, recent previous literature has revealed difficulties in achieving a certain degree of predictive capacity [15–17,21]. The current complexity in economic decisions and especially in financial markets leads to the need to search for new methodologies that more accurately estimate the models of speculative attacks. These models on speculative attacks have always been estimated using the Ordinary Least Squares (OLS) method, as the most widely used statistical technique in estimating these models [7–15].

In order to cover this gap, and given the importance that currency trading problems continue to have for many countries, the present study develops different machine learning techniques for estimating the two main popular speculative attacks models that respond to the most current concerns of the financial situation of the currencies. To this end, the data have been used for the cases of Mexico and Thailand, two countries that in recent decades have shown difficulties with the price of their currencies, being targets of attacks by numerous agents in the foreign exchange market. Specifically, the neural networks of Perceptron Multilayer, Deep Recurrent Neural Networks, Deep Neural Decision Trees, and Quantum-Inspired Neural Networks have been used, to be compared with the usual OLS method. The quantum variant is the one that achieves the best results both outside the sample and also in the forecasts of final postestimations made. Besides, the computational methodologies used in this study improve the precision results obtained by the OLS method. These results are repeated for both the first-generation and second-generation models, as well as for the data used from Mexico and Thailand.

We make some contributions to the literature. We consider new estimation techniques for forecasting the speculative attacks through the first- and second-generation models, testing the precision and level of residuals obtained by each methodology. It has important implications for public institutions, governments, central banks, financial institutions, and other stakeholders concerned in the foreign exchange markets for the accurate estimation of speculative attacks.

The present study is organized as follows: Section 2 reviews the speculative models used in this study. In Section 3, the methods used are presented. In Section 4, the data and the variables used in the research are detailed and the results obtained are analyzed. Finally, the conclusions of the study and its implications are exposed.

2. Speculative Attacks Models

2.1. First Generation Model

The models of currency crisis or balance of payments crisis try to explain why and the logic of how a currency crisis is unleashed. Thus, the first-generation models were based, mainly, on the fact that exchange rate crises occur due to the existence of incompatibility in monetary and fiscal policies (both expansive) with the maintenance of a fixed exchange rate regime in the long term. In other words, these occur in a situation in which a government (central bank), which promised to keep the exchange rate fixed, is running constant fiscal deficits and these are monetized by its central bank. This situation creates an incompatibility that will mean that this exchange rate regime cannot be maintained for long. The reason why this regime will end up collapsing is that there is a surplus of the money supply over demand continuously and this surplus will be reduced by the central bank by selling reserves. Thus, the central bank will lose reserves in all periods to balance the money market. Faced with this situation of constant loss of reserves, investors, anticipating the natural disappearance of reserves, will carry out a speculative attack on the local currency that will lead to reserves decreasing to a "critical" value, a level that may be zero according to the Flood and Garber model [10] or that they reach a level below the critical value [1–3].

The first-generation basic model considers that private agents (investors or speculators) have perfect foresight on the future behavior of economic variables and work in continuous time. It is a model that assumes a small and open economy, where a single good is produced, and it is assumed that the Purchasing Power Parity (PPP) and the discovered interest parity are met. There are two types of assets, local and foreign money, and bonds, also local and foreign, the latter perfectly substitutes (this implies the existence of an interest rate). The model proposes a small country, where it produces a marketable good in the international market, whose price in the national territory (P) is defined by the exchange rate (TC) of the national currency expressed in terms of the foreign currency (s) multiplied by the price of the product in international markets (P *), as it appears in expression (1),

$$P = sP^*, \tag{1}$$

The hypothesis also assumed that the price of the good abroad P * is constant and equal to 1 ($P^* = 1$). So, the internal price of the product will be equal to the exchange rate ($P = s$).

The approach of Krugman is completed with flexible wages and prices, with production in full employment, and the trade balance, regardless of the role of the balance of payments in the current account model, will be the difference between production and expenditure:

$$B = Y - G - C(Y - T, W) \qquad C_1, C_2 > 0, \tag{2}$$

where B is the current account balance, Y is the level of production, G defines public spending, C represents private consumption, T is the tax variable, and W is total household wealth.

Regarding the asset market, the model establishes that investors can only choose between two assets: national currency (M), and foreign currency (F), with the nominal interest rate of both assets equal to zero. In this way, the real wealth of national residents (W) will be equal to the sum of holdings in the national currency (M) plus those of foreign currency (F) as defined in expression (3):

$$W = \frac{M}{P} + F. \tag{3}$$

Lastly, the model assumes that foreigners do not have a national currency, so (M) represents the national currency stock, and in equilibrium, it assumes that national residents must be willing to maintain said stock. The equilibrium condition of the portfolio establishes that asset holdings in national currency are equivalent to a proportion of residents' real wealth and that this, in turn, depends on the expected inflation rate (π). Furthermore, one of the assumptions of the model is that the domestic price level (P) corresponds to the exchange rate (s), and asset holdings in national currency depend on the expected depreciation rate of the currency, expressed in Equation (4):

$$\frac{M}{P} = L(\pi) \times W. \quad (4)$$

Krugman considers two different economic regimes: a system with a flexible exchange rate and a system with a fixed exchange rate. The behavior of the economy in the short term is different depending on the exchange rate system. An increase in the expected inflation rate under a flexible exchange rate regime produces an increase in the domestic price level, while when the exchange rate is fixed, an increase in the expected inflation rate implies an alteration in the composition of residents' wealth, increasing foreign currency assets (ΔF) and decreasing domestic currency assets. This situation causes a compensatory change in government reserves that decrease by the same amount as holdings of foreign currency in the hands of private residents increase:

$$\Delta R = -\Delta F = \Delta \frac{M}{P}. \quad (5)$$

Krugman also analyzes the dynamic behavior of the economy under both exchange rates. In the case of flexible TC, it is assumed that the creation of money depends solely on the financing needs of the government. Therefore, the growth of the money stock will be determined by the differences between the government's fiscal expenses and revenues, as expressed in Equation (6):

$$\frac{M}{P} = G - T. \quad (6)$$

Relating public spending and money supply, under the assumption of perfect forecasting of the inflation rate, Krugman shows that the demand for assets in national currency will depend exclusively on price growth and that national residents will only be willing to increase the proportion of national currency over foreign currency if there is a reduction in the price level.

In a fixed exchange rate regime, it is assumed that the government has a stock of reserves in foreign currency, which it uses to stabilize the exchange rate. This is equivalent to saying that the price level is constant, where $P = sP^*$ and $P^* = 1$, and therefore $P = s = 0$. The private sector can only acquire assets if it decreases its spending relative to its income and therefore, private sector savings are considered:

$$S = Y - T - C(Y - T, W). \quad (7)$$

In this case, and because the price level is constant, the growth of residents' wealth is equivalent to the savings of the private sector, that is:

$$\dot{W} = \frac{\dot{M}}{P} + \dot{F} = S. \quad (8)$$

In this way, the distribution of savings between assets denominated in national currency and assets in a foreign currency will be determined by the equilibrium condition of the trade balance. As long as investors trust the government to maintain the price level, the expected inflation will be zero, giving a stable relationship between wealth and deposits in national currency. If there is an increase in the wealth of residents, a proportion L will go to the national currency, given: $\frac{M}{P} = L(\pi) \times W$ and $(1 - L)$. It will be used for assets in foreign currency. The government will be able to cover its deficit by issuing new national currency or by using its foreign currency reserves (R). Therefore, the composition of the state budget can be expressed:

$$\frac{\dot{M}}{\dot{P}} + \dot{R} - G \quad T = g\left(\frac{M}{P}\right). \tag{9}$$

From this expression, it follows that if the government commits to maintaining the exchange rate, it has no control over how it finances its deficit. Over time, both private sector wealth and government reserves will vary. When the government runs a deficit, its reserves decrease, even though the private sector saving is zero. In a deficit situation, fixing the exchange rate is impossible regardless of the initial amount of reserves that the government had and the effect derived from said fixing will generate a balance of payments crisis, caused by a speculative attack at the moment in which the agents anticipate the depletion of reserves.

2.2. Second Generation Model

The second-generation models differ from the first generation because they are models of multiple equilibria, since they consider an interaction between the private sector and the behavior of the government, giving rise to multiple solutions. These second-generation models consider that in a country's economy, there is an interrelation between the behavior of the private sector and the decisions made by the public sector. Thus, a financial crisis under this relationship can take place when international financial operators have expectations about a possible devaluation of the currency, this situation is reflected in interest rates, which by rising try to attract national currency against the foreign currency. This scenario can lead the government to devalue due to the cost of debt service. On the contrary, if the private agents do not have expectations that the exchange rate will change, the interest rate remains low and the devaluation is less likely.

Second-generation models were developed by Flood and Marion [11] to understand crises in their self-fulfilling character. According to this mechanism, if the agents foresee a possible devaluation of the currency, this will be reflected in the salary negotiations, which will cause economic imbalances, including a rise in the country's price level. These imbalances can be corrected by the government through the exchange rate since it is set after wage negotiations. If the government decides not to devalue, it will correct economic imbalances avoiding an increase in inflation by reducing its control over the variables that define the level of production. If, on the contrary, the government decides to lean towards the flexible exchange rate, it will be feeding a process through which both the level of wages and prices in the country will increase. Both situations are reflected in Equation (10), which reflects the so-called cost of the exchange rate regime.

$$L_t = 0.5\theta(p_t - p_{t-1}) + 0.5(y_t - y^*)^2, \tag{10}$$

where p_t is the national price level, y_t is the country's output at time t, y^* is the output target set by economic policy, and θ is the weight associated with deviations in inflation from the political objective.

According to this approach, the government will decide to devalue its currency provided that the loss for leaving the fixed exchange rate system, together with the cost for the government of the loss of credibility of making this decision, is less than the loss obtained for not giving up under pressure and keep the exchange rate fixed. In this model, the existence of different levels of economic equilibrium stands out, where each level reflects the expectations that economic agents maintain about the economic policy that the government will carry out in the following period, since depending on the levels of devaluation expectations, the parameters of the equation will also be different, thus obtaining multiple results.

3. Neural Networks Methods

3.1. Multilayer Perceptron (MLP)

The multilayer perceptron (MLP) is a feed-forward, supervised artificial neural network model that is composed of a layer of input units, another layer of output, and several intermediate layers called hidden layers in so much so that they have no connections with the outside world. Each input sensor would relate to the units of the second layer, these in turn with those of the third layer, and so on. The network will aim to establish a correspondence between a set of input data and a set of desired outputs.

Moreover, [22] show that learning in MLP was a special case of a functional approach, where there is no assumption about the model underlying the data analyzed. This process involves finding a function that correctly represents the learning patterns, in addition to carrying out a generalization process that allows the efficient treatment of unanalyzed individuals during said learning. To do this, we proceed to adjust the W weights from the information from the sample set, considering that both the architecture and the network connections are known. The objective is to obtain those weights that minimize the learning error. Given, then, a set of pairs of learning patterns $\{(x_1, y_1), (x_2, y_2) \ldots (x_p, y_p)\}$, and an error function $\varepsilon(W, X, Y)$, the training process implies the search for the set of weights that minimizes the learning error $E\ (W)$, as expressed in (11).

$$\min_{w} E(W) = \min_{w} \sum_{i=1}^{p} \varepsilon(W, x_i, y_i). \tag{11}$$

Most of the analytical models used to minimize the error function use methods that require the evaluation of the local gradient of the $E(W)$ function and techniques based on second-order derivatives can also be considered [23,24].

3.2. Deep Recurrent Convolution Neural Network

Recurrent neural networks (RNN) have been successfully used in many fields for time-series prediction due to its huge prediction performance. For a simple neural network (NN), the inputs are assumed to be independent of each other. The common structure of RNN is organized by the output of which is depended on its previous computations [24,25]. Given an input sequence vector x, the hidden states of a recurrent layer s, and the output of a single hidden layer y, it can be calculated as appears in expressions (12) and (13):

$$s_t = \sigma(W_{xs}x_t + W_{ss}s_{t-1} + b_s) \tag{12}$$

$$y_t = o(W_{so}s_t + b_y) \tag{13}$$

where W_{xs}, W_{ss}, and W_{so} denote the weights from the input layer x to the hidden layer s, the hidden layer to itself, and the hidden layer to its output layer, respectively. b_s and b_y are the biases of hidden layer and output layer, respectively. σ and o are the activation functions. The Equation (14) represents the function of vibration signals.

$$STFT\{z(t)\}(\tau, \omega) \equiv T(\tau, \omega) = \int_{-\infty}^{+\infty} z(t)\omega(t-\tau)e^{-j\omega t}\, dt \tag{14}$$

where $z\ (t)$ is the vibration signals, $\omega\ (t)$ is the Gaussian window function focused around 0, and $T\ (\tau, \omega)$ is a complex function that describes the vibration signals over time and frequency.

When time-frequency features $\{T_i\}$ are used to estimate speculative attacks with RNN, the convolutional operation is conducted in the state transition. To calculate the hidden layers with a convolutional operation, the next Equations (15) and (16) are applied:

$$S_t = \sigma(W_{TS} \times T_t + W_{ss} \times S_{t-1} + B_s) \tag{15}$$

$$Y_t = o(W_{YS} \times S_t + B_y) \tag{16}$$

where the term W indicates the convolution kernels. The convolutional operation has been determined by local connections, weight sharing, and local grouping, which allow every unit to integrate time-frequency data in the current layer. The convolution is operated between weights and inputs and is performed in the transition of inputs to the hidden layers.

Recurrent Convolutional Neural Network (RCNN) can be heaped to establish a deep architecture, named "deep recurrent convolutional neural network" [25]. When DRCNN is used to estimate speculative attacks, the last part of the model is a supervised learning layer, which is determined as appears in Equation (17):

$$\hat{r} = \sigma(W_h \times h + b_h) \tag{17}$$

where W_h is the weight and b_h is the bias. The error between predicted observations and actual ones in the training data for speculative attacks estimation can be calculated and back propagated to train the model [25]. Considering that the actual data at time t is r, the loss function is determined as shown in the next Equation (18):

$$L(r, \hat{r}) = \frac{1}{2} \parallel r - \hat{r} \parallel_2^2 \tag{18}$$

Stochastic gradient descent is applied for optimization to learn the parameters. The gradient of loss function regarding parameters W_h and b_h are determined as follows in the Equations (19) and (20):

$$\frac{\partial L}{\partial W_h} = -(r - \hat{r})\sigma'(.)h \tag{19}$$

$$\frac{\partial L}{\partial b_h} = -(r - \hat{r})\sigma'(.) \tag{20}$$

3.3. Deep Neural Decision Trees (DNDT)

DNDT are DT models executed by deep-learning NNs, where a configuration of DNDT weightings corresponds to a specific decision tree and is thus interpretable [26]. The algorithm begins by implementing a soft binning function [27–29] to calculate the error rate for each node, making it possible to make decisions divided into DNDT. In general, the input of a binning function is a real scalar x, which generates an index of the containers to which x belongs. Assuming x is a continuous variable, group it into n + 1 intervals. This requires n cut-off points, which are trainable variables in this context. The cut-off points are denoted as [β1, β2, . . . , βn] and are strictly ascending such that β1 < β2 < . . . < βn.

The activation function of the DNDT algorithm is implemented based on the NN defined in Equation (21).

$$\pi = f_{w,b,\tau}(x) = \text{softmax}((wx + b)/\tau), \tag{21}$$

where w is a constant with value w = [1, 2, . . . , n + 1], $\tau > 0$ is a temperature factor, and b is defined in Equation (22).

$$b = [0, -\beta 1, -\beta 1 - \beta 2, \ldots, -\beta 1 - \beta 2 - \cdots - \beta n] \tag{22}$$

The NN defined in Equation (22) gives a coding of the binning function x. Additionally, if τ tends to 0 (often the most common case), the vector sampling is implemented using the Straight-Through (ST) Gumbel–Softmax method [30].

Given the binning function described above, the key idea is to build the DT using the Kronecker product, assuming we have an input instance $x \in R^D$ with D characteristics. Associating each characteristic x_d with its own NN $f_d(x_d)$, we can determine all the final nodes of the DT, in line with Equation (23).

$$z = f1(x1) \otimes f2(x2) \otimes \cdots \otimes f_d(x_d) \tag{23}$$

where z is now also a vector that indicates the index of the leaf node reached by instance x. Finally, we assume that a linear classifier on each leaf z classifies the instances that reach it.

However, the main drawback of the design is the use of the Kronecker product, which means it is not scalable in terms of the number of characteristics. In our current implementation, we avoid this problem using broad datasets and training a forest with random subspace [27–30]. This involves introducing multiple trees and training each with a subset with random characteristics. A better solution that does not require a forest of hard interpretability involves exploiting the dispersion of the binning function during the learning, since the number of nonempty leaves grows much slower than the total.

3.4. *Quantum-Inspired Neural Networks (QNN)*

The QNN is built from quantum computation techniques. These neural networks are inspired in quantum framework. The calculation unit of this model consists of quantum gates and their inputs and outputs are qubits. Any gate can calculate any local unit operation on the inputs. Quantum gates are interconnected by links. A quantum computational network is a computing machine that consists of quantum gates with synchronized steps. The calculation is done from left to right. The outputs of the gates are connected to the inputs of others. Some of the inputs are used as input to the network. Other inputs are connected to gates for 0 and 1 qubits. A few outputs are connected to sink gates, where arriving qubits are rejected [31,32]. An output qubit can be measured across the state $|0\rangle$ and $|1\rangle$, and is watched based on the probability amplitudes associated with the qubit [33–35]. Qubit is defined as the smallest unit of information in quantum computation, which is a probabilistic representation. A qubit may either be in the "1" or "0" or in any superposition of the two [36]. The state of the qubit can be defined as follows in the Equation (24):

$$|\psi\rangle = \alpha|0\rangle + \beta|1\rangle, \tag{24}$$

where α and β are the numbers that point out the amplitude of the corresponding states such that $|\alpha|_2 + |\beta|_2 = 1$. A qubit is defined as the smallest unit of information in quantum computation. It is determined as a pair of numbers $\begin{bmatrix} \alpha \\ \beta \end{bmatrix}$. An angle θ is a specification that represents geometrical aspects and is defined such that: $\cos(\theta) = |\alpha|$ *and* $\sin(\theta) = |\beta|$. Quantum gates may be applied for adjusting the probabilities because of weight upgrading [31,37]. An example of rotation gate can be: expressed as appears in the expression (25):

$$U(\Delta\theta) = \begin{bmatrix} \cos(\Delta\theta) & -\sin(\Delta\theta) \\ \sin(\Delta\theta) & \cos(\Delta\theta) \end{bmatrix} \tag{25}$$

A state of the qubit can be upgraded by applying the quantum gate explained previously. Application of rotation gate on a qubit is defined as follows in expression (26):

$$\begin{bmatrix} \alpha' \\ \beta' \end{bmatrix} = \begin{bmatrix} \cos(\Delta\theta) & -\sin(\Delta\theta) \\ \sin(\Delta\theta) & \cos(\Delta\theta) \end{bmatrix} \begin{bmatrix} \alpha \\ \beta \end{bmatrix} \tag{26}$$

The next hybrid quantum-inspired neural network is proposed for forecasting speculative attacks. The process is begun with a quantum hidden neuron from the state $|0\rangle$. The superposition expressed in the Equation (27) is prepared:

$$\sqrt{p}|0\rangle + \sqrt{1-p}\Big|1\Big\rangle \; with \; 0 \leq \Big|p\Big| \leq 1, \tag{27}$$

where p represents random probability of starting the system in the state $|0\rangle$. The classical neurons are initiated by random number generation. The output from the quantum neuron is determined as follows in the Equation (28):

$$v_j = f\left(\sum_{i=1}^{n} w_{ji} \times x_i\right) \tag{28}$$

where f is a problem-dependent sigmoid or Gaussian function. The output from the network is represented as appears in the Equation (29):

$$y_k = f\left(\sum_{j=1}^{l} w_{jk} \times v_j\right) \tag{29}$$

The desired output is the o_k. The squared error ($E^2{}_k$) is defined in the expression (30):

$$E^2{}_k = \frac{1}{2}|y_k - o_k|^2 \tag{30}$$

The learning follows the rules of the feed forward backpropagation algorithm. The upgrading of output layer weight is defined as follows in the Equation (31):

$$\Delta w_{jk} = \eta e_k f' \, v_j \tag{31}$$

Upgrading of quantum hidden layer weight in quantum backpropagation algorithm, the weights are upgraded by quantum gate conforming to Equation (26), so in this case, the equation would be as it appears in the Equation (32):

$$\begin{bmatrix} \alpha_{ij}' \\ \beta_{ij}' \end{bmatrix} = \begin{bmatrix} \cos(\Delta\theta) & -\sin(\Delta\theta) \\ \sin(\Delta\theta) & \cos(\Delta\theta) \end{bmatrix} \begin{bmatrix} \alpha_{ij} \\ \beta_{ij} \end{bmatrix} \tag{32}$$

where $\Delta\theta_{ij} = -\frac{\partial E}{\partial \theta_{ij}}$, the index i represents the number of outputs from quantum neuron and the index j defines the number of outputs from network, $\gamma_{ij}' = \gamma_{ij} + \eta\Delta\theta_{ij}$, and η is the learning rate [36,37]. This ratio usually takes the value of 0.1.

4. Data and Variables

The present study employs a sample of the quotations of the Mexican peso (MXN) and the Thai baht (THB). There have been two cases of currencies that have suffered speculative attacks in the past and analyzed by previous literature [1–3]. The period analyzed includes from 1995 to 2019, with the quotations of the currencies mentioned concerning the US dollar. In addition, the macroeconomic data of the current account balance, gross domestic product (GDP), consumption, total household wealth, inflation rate, assets in foreign currency, national savings, public spending, tax revenues, foreign currency reserves, quotation of the Mexican peso, the Thai baht against the US dollar, etc. have been used. These data have been obtained from Yahoo Finance, Federal Reserve Economic Data of St. Louis (FRED), and Open Data World Bank.

Besides, to check the reliability level of the models built, different test samples were created. This sample data set has been divided into mutually exclusive two groups, i.e., one for training (70% of the data) and another for testing (30% of the data). As is well known, the training data are used to fit the parameters of the models. For its part, the testing data are used to evaluate the built model and make predictions. The percentage of correctly classified cases (accuracy) and the root of the mean square error have been used for the evaluation. Furthermore, for the treatment of each of the three groups, the 10-fold cross-validation procedure has been applied with 500 iterations [33]. On the other hand, for our estimations, we used two four-core Intel Core i7-6500 processor as computing resources to make estimates. The code for the estimation of our methods has been performed by

Python (3.8 version), with the support of the libraries such as NumPy, PyTorch, and QisKit to create the mathematical routines, Deep Learning algorithms, and Quantum processing, respectively. The MLP and OLS models have been created with MATLAB code (MATLAB R2016b package).

5. Results

Tables 1 and 2, and Figures 1–3 show adjustment levels using accuracy, the mean square error (RMSE), and the mean absolute percentage error (MAPE). In all computational methods, the level of accuracy always exceeds 82.64% for testing data, while for OLS, it reaches 75.27% for Mexico and 77.41% for Thailand. For its part, the RMSE and MAPE levels are adequate. Therefore, computational methods improve OLS by a large margin, with QNN being the one that best adjusts the result in terms of residuals (with 91.62% accuracy), followed by DNDT (with 88.10%) for Mexico. In the case of Thailand, the results improve slightly, but the order of precision is the same since the best methodology is QNN with 92.84% in test data, followed by DNDT with 89.05%. Taken together, these results provide a level of accuracy far superior to that of previous studies. Thus, in the work of [7], an accuracy of around 78.2% is revealed. In the work of [9], it is close to 73.1%, and in the study of [12], it approaches 71%. Other studies such as [1–3,5,6] achieve a precision of even less than 70%. Therefore, the difference shown by the computational methodologies applied in this study far exceeds the precision shown by the previous literature.

Table 1. Results of accuracy evaluation: Mexico.

		First Generation Model		Second Generation Model	
		Training	Testing	Training	Testing
OLS	Accuracy (%)	78.45	75.27	80.02	77.41
	RMSE	1.12	1.20	1.01	1.10
	MAPE	0.57	0.61	0.41	0.47
MLP	Accuracy (%)	85.37	82.64	86.78	84.11
	RMSE	0.93	1.07	0.81	0.95
	MAPE	0.44	0.50	0.37	0.43
DRCNN	Accuracy (%)	90.04	84.30	91.95	86.18
	RMSE	0.67	0.84	0.59	0.80
	MAPE	0.27	0.33	0.24	0.31
DNDT	Accuracy (%)	92.15	88.10	93.62	89.05
	RMSE	0.46	0.67	0.42	0.65
	MAPE	0.18	0.27	0.16	0.23
QNN	Accuracy (%)	94.51	91.62	95.72	92.84
	RMSE	0.35	0.54	0.34	0.64
	MAPE	0.15	0.22	0.10	0.07

Table 2. Results of accuracy evaluation: Thailand.

		First Generation Model		Second Generation Model	
		Training	Testing	Training	Testing
OLS	Accuracy (%)	78.67	76.43	80.27	78.06
	RMSE	1.09	1.03	0.99	1.04
	MAPE	0.54	0.55	0.43	0.52

Table 2. *Cont.*

		First Generation Model		Second Generation Model	
		Training	Testing	Training	Testing
MLP	Accuracy (%)	87.81	85.01	89.27	86.52
	RMSE	0.87	1.00	0.76	0.89
	MAPE	0.41	0.47	0.34	0.40
DRCNN	Accuracy (%)	92.61	86.71	94.58	88.65
	RMSE	0.63	0.78	0.55	0.74
	MAPE	0.25	0.31	0.22	0.28
DNDT	Accuracy (%)	93.87	89.74	95.37	90.71
	RMSE	0.43	0.62	0.39	0.60
	MAPE	0.17	0.25	0.14	0.21
QNN	Accuracy (%)	96.27	93.32	97.50	94.57
	RMSE	0.32	0.50	0.32	0.60
	MAPE	0.13	0.21	0.10	0.06

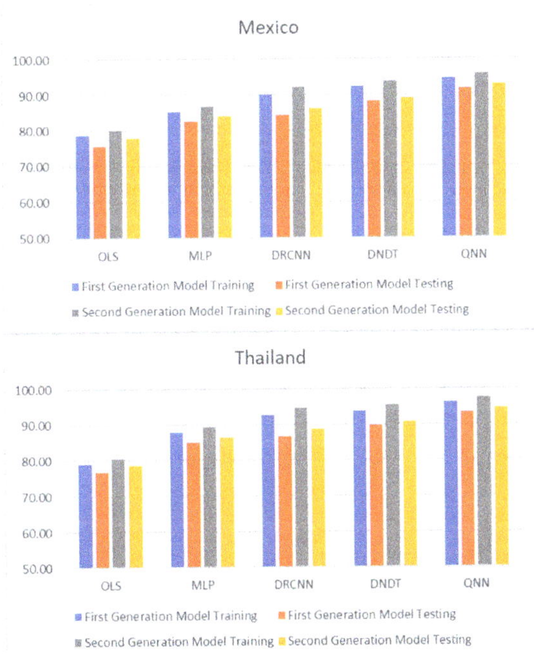

Figure 1. Results of accuracy evaluation: classification (%).

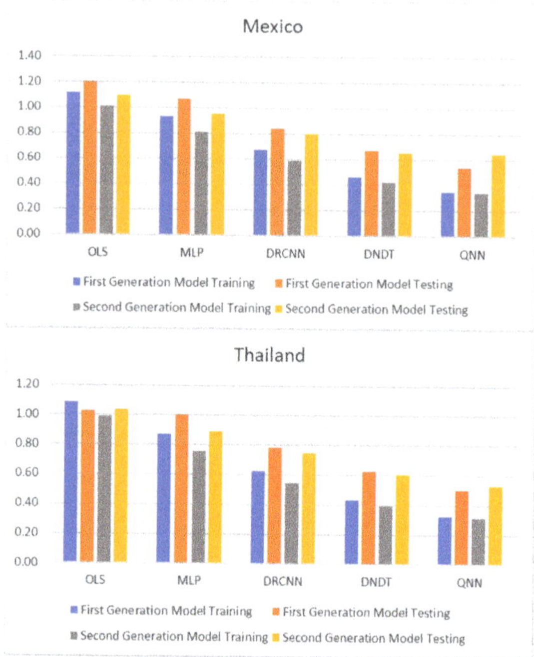

Figure 2. Results of accuracy evaluation: mean square error (RMSE).

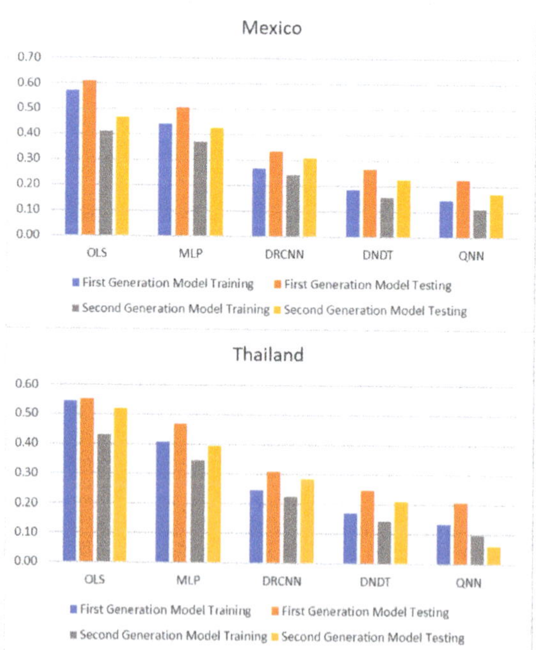

Figure 3. Results of accuracy evaluation: mean absolute percentage error (MAPE).

These results demonstrate the greater stability offered by the QNN model compared to the rest, especially in the light of the RMSE and MAPE results obtained for three other computational methods. The results of the QNN improve the results of the popular OLS, just as it improves the precision results shown in previous works such as [9–13]. This set of computational methods observed as highly accurate represents a group of novel methods that estimate the speculative attacks and therefore different from that shown in the previous literature.

To reinforce the superiority of neural network methodologies for estimating speculative attack models, the Diebold-Mariano (DM) and Harvey-Leybourne-Newbold (HLN) tests [38,39] have been applied to compare the methodologies used and the time elapsed to perform the estimation with each of the techniques. Table 3 reports the results of the DM test, showing that all the neural network methodologies used are better options than OLS. Like QNN, it is the best option compared to the rest, since the DM test ensures that the results that exceed 1.96/−1.96 do not reject the null hypothesis at 5% of significance, and therefore the differences observed between methodologies in the estimate are significant. On the same line, being the result with a negative sign means that the second option of the comparative is better than the second option. Likewise, the HLN test is adjusted version of DM test [39], which has better small-sample properties. Both DM and HLN tests show a significance difference between computational and statistical techniques, and the computational superiority over conventional methods. On the other hand, Figure 4 shows the average run time of the methodologies used for the estimation, where it is shown that neural network methodologies need a shorter estimation time, both for training and testing data, with QNN being the most common option efficient in terms of time use, needing 0.11 and 0.10 min to estimate with training and testing data, respectively, in the case of Mexico. For the case of Thailand, the estimate needs 0.13 and 0.11 min to estimate with training and testing data, respectively.

Table 3. Comparison of testing results using Diebold-Mariano (DM) and Harvey-Leybourne-Newbold (HLN) tests.

	First Generation Model		Second Generation Model	
	DM	**HLN**	**DM**	**HLN**
OLS vs. MLP	−2.42 **	−2.31 *	−2.57 **	−2.25 **
OLS vs. DRCNN	−2.86 **	−2.57 **	−2.93 **	−2.83 **
OLS vs. DNDT	−3.02 **	−2.84 **	−2.99 **	−2.67 **
OLS vs. QNN	−3.17 **	−2.99 **	−3.29 **	−3.06 **
MLP vs. DRCNN	−2.15 **	−2.03 *	−2.47 *	−2.41 *
MLP vs. DNDT	−2.34 *	−2.17 **	−2.63 **	−2.49 **
MLP vs. QNN	−2.76 **	−2.62 **	−3.20 **	−3.07 **
DRCNN vs. DNDT	−2.08 *	−1.93 *	−2.47 *	−2.36 *
DRCNN vs. QNN	−2.53 *	−2.14 *	−2.45 **	−2.28 *
DNDT vs. QNN	−2.11 *	−1.97 *	−2.46 *	−2.13 **

* Indicates significance at the 5% level. ** Indicates significance at the 10% level.

Postestimations

To perform multiple-step-ahead prediction to obtain greater robustness of results, we use the iterative strategy. For this, we have trained the models for prediction for one step and two forward steps, that is, for the moments $t + 1$ and $t + 2$ [38]. These forecasted data for $t + 1$ and $t + 2$ are included in the data sample as actual observations. Tables 4 and 5, and Figures 5–7 point out the accuracy and residual results (RMSE and MAPE) for one-year and two-year forecasting horizons. For $t + 1$, the range of precision for the four neural networks techniques is 83.07–90.94% overall, being in the model of QNN where the percentage of accuracy is higher (90.94%) for the Mexican case. With the OLS method, the accuracy decreases to 74.72–74.90%. On the same line, for the Thai case, the precision range has been 83.34–92.63%, with QNN being again the methodology with the highest precision (92.63%). With the OLS method, the accuracy decreases to 75.64–77.15%. For $t + 2$, this range of precision is 81.34–89.52%, being also the method of QNN in which the percentage of accuracy is higher (89.52%) for the Mexican estimations. For the OLS method, the accuracy decreases to the range of 72.78–73.81%. Moreover, in $t + 2$ for the Thai estimations, again confirms the predictive superiority of QNN (90.54%). These results show the high precision and great robustness of the NN techniques.

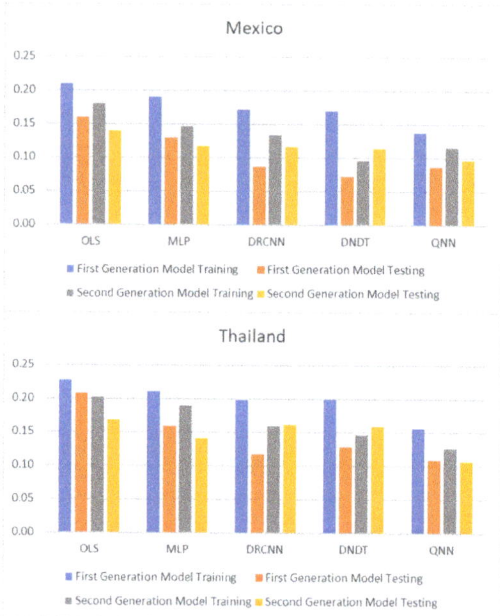

Figure 4. Results of time lapse for estimation.

Table 4. Multiple-step ahead forecasts in forecast horizon = $t + 1$ and $t + 2$ (Mexico).

		First Generation Model		**Second Generation Model**	
		$t + 1$	$t + 2$	$t + 1$	$t + 2$
OLS	Accuracy (%)	74.72	73.81	74.90	72.78
	RMSE	1.32	1.38	1.19	1.42
	MAPE	0.71	0.75	0.58	0.81
MLP	Accuracy (%)	83.07	81.34	84.51	80.89
	RMSE	1.00	1.15	0.87	1.02
	MAPE	0.47	0.54	0.40	0.46
DRCNN	Accuracy (%)	84.46	83.81	83.05	82.98
	RMSE	0.72	0.90	0.63	0.86
	MAPE	0.29	0.36	0.26	0.33
DNDT	Accuracy (%)	86.62	82.81	88.00	83.71
	RMSE	0.50	0.72	0.45	0.69
	MAPE	0.20	0.28	0.17	0.24
QNN	Accuracy (%)	89.78	87.04	90.94	89.52
	RMSE	0.37	0.58	0.37	0.53
	MAPE	0.16	0.24	0.11	0.15

Table 5. Multiple-step ahead forecasts in forecast horizon = $t + 1$ and $t + 2$ (Thailand).

		First Generation Model		**Second Generation Model**	
		$t+1$	$t+2$	$t+1$	$t+2$
OLS	Accuracy (%)	75.64	73.57	77.15	75.12
	RMSE	1.26	1.33	1.18	1.29
	MAPE	0.65	0.74	0.60	0.67
MLP	Accuracy (%)	83.34	81.58	87.16	83.94
	RMSE	0.93	1.07	0.81	0.95
	MAPE	0.44	0.50	0.37	0.42
DRCNN	Accuracy (%)	86.13	84.64	87.96	84.54
	RMSE	0.67	0.84	0.58	0.77
	MAPE	0.27	0.33	0.24	0.30
DNDT	Accuracy (%)	87.20	83.37	88.59	85.27
	RMSE	0.46	0.67	0.42	0.65
	MAPE	0.18	0.26	0.15	0.22
QNN	Accuracy (%)	91.45	88.66	92.63	90.54
	RMSE	0.35	0.54	0.32	0.48
	MAPE	0.14	0.22	0.11	0.14

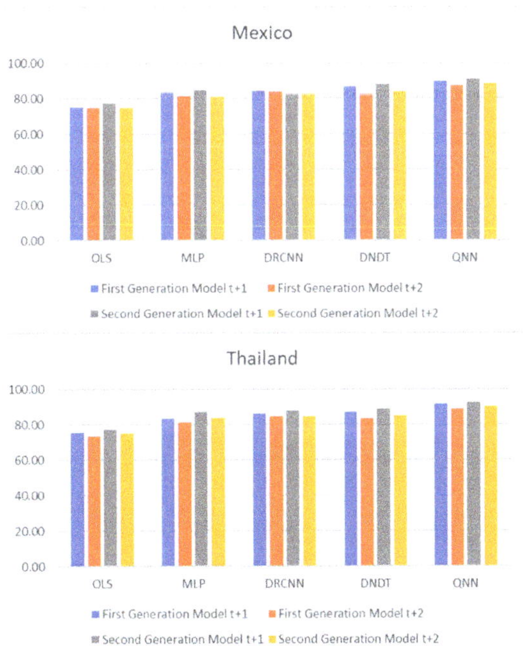

Figure 5. Multiple-step ahead forecasts in forecast horizon: accuracy.

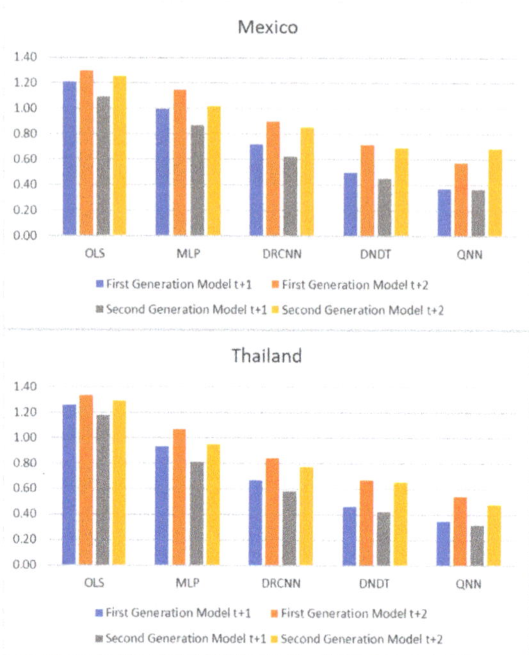

Figure 6. Multiple-step ahead forecasts in forecast horizon: RMSE.

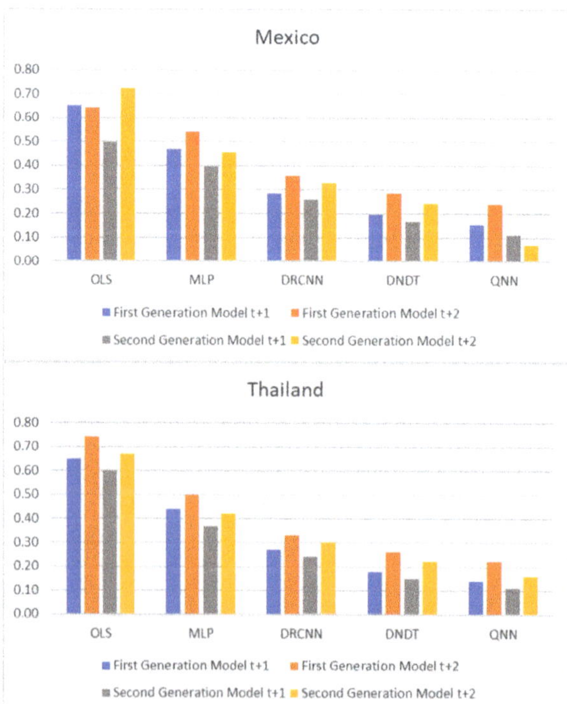

Figure 7. Multiple-step ahead forecasts in forecast horizon: MAPE.

6. Conclusions

This study has developed a new simulation of speculative attack models using machine learning techniques. Using data of period 1995–2019 for the cases of the currencies of Mexico and Thailand (Peso and Baht) and applying four different NN methods in the estimation of the first- and second-generation speculative attacks models to achieve a robust accuracy capacity, such as MLP, DRCNN, DNDT, and QNN. This last methodology is the one that has obtained the highest levels of precision. Most of the proposed NN methodologies have shown a low level of error and stability in the estimates made from speculative attack models, proving their interesting alternative to conventional statistical methods, such as OLS.

Besides, the target has been to improve the accuracy of previous studies using different methodologies. The results obtained in this research are higher than those obtained in the existing literature, with an accuracy range of 82.64–92.84% using the NN methods, while OLS method has only reached an accuracy range of 75.27–78.06%. It has also detected new significant variables to consider in speculative attacks models in weak currencies, allowing a high level of stability in the models developed over forecasting horizons of $t + 1$ and $t + 2$. In contrast to previous research, this study has been able to expand the estimation of speculative attacks in exchange rate attending to accuracy and error results. The results have identified a set of significant variables for each methodology applied and for each standard dependent variable. Furthermore, the time elapsed to make the estimates is less for the proposed NN techniques compared to the time needed for the OLS method. This makes an essential contribution to the field of computational macroeconomics and finance. The conclusions are relevant to public managers, financial analysts, central bankers, and other stakeholders in the foreign exchange markets, who are generally interested in knowing which indicators provide reliable, accurate, and potential forecasts of performance evolution. Our study suggests new explanatory significant variables to allow these agents to analyze the performance of speculative attack models. This research has also provided a new estimation analysis developed for speculative attacks using four NN methods, being the QNN the most accurate. Hence, this study attempts to contribute to existing knowledge in the field of machine learning. These new simulations of estimation can be used as a reference to improve decision-making in public and financial institutions.

In summary, this study provides a significant opportunity to contribute to the research line of currency crises and speculative attacks, since the results obtained have significant implications for the future decisions of public institutions, making it possible to avoid big negative changes of the trend of the exchange rate and the potential associated risks. It also helps these agents send warning signals to governments and central banks and avoid currency crisis losses derived from a huge decrease in the balance of payments. Further research could include speculative attack models with other new variables to take advantage of the benefits of machine learning techniques.

Author Contributions: This study has been designed and performed by all of the authors. D.A. collected the data. D.A., F.A.-V. and J.R.S.-S. analyzed the data. The introduction and literature review were written by D.A. and F.A.-V. All of the authors wrote the discussion and conclusions. All authors have read and agreed to the published version of the manuscript.

Funding: This research was funded by Universidad de Málaga, Spain, and Cátedra de Economía y Finanzas Sostenibles Universidad de Málaga, Spain.

Institutional Review Board Statement: Not applicable.

Informed Consent Statement: Not applicable.

Data Availability Statement: Data available on request due to restrictions.

Conflicts of Interest: The authors declare no conflict of interest.

References

1. Minsky, H. *The Financial Instability Hypothesis;* Columbia University Press: New York, NY, USA, 1975.

2. Krugman, P. A Model of Balance-of-Payments Crises. *J. Money Credit. Bank.* **1979**, *11*, 311–325. [CrossRef]
3. Obstfeld, M. Rational and Self-fulfilling Balance-of-Payments Crises. *Am. Econ. Rev.* **1986**, *76*, 72–81.
4. Laeven, L.; Valencia, F. Systemic Banking Crises Database II. *IMF Econ. Rev.* **2020**, *68*, 307–361. [CrossRef]
5. Obstfeld, M. Models of currency crises with self-fulfilling features. *Eur. Econ. Rev.* **1996**, *40*, 1037–1047. [CrossRef]
6. Eichengreen, B.; Rose, A.K. Contagious Currency Crises: Channels of Conveyance. NBER Chapters. In *Changes in Exchange Rates in Rapidly Developing Countries: Theory, Practice, and Policy Issues*; National Bureau of Economic Research: Washington, DC, USA, 1999; pp. 29–56.
7. Frankel, J.A.; Rose, A.K. *Currency Crashes in Emerging Markets: An Empirical Treatment*; International Finance Discussion Papers 534; Board of Governors of the Federal Reserve System (U.S.): Washington, DC, USA, 1996.
8. Kaminsky, G.; Lizondo, S.; Reinhart, C. Leading Indicators of Currency Crises. *IMF Staff Pap.* **1998**, *45*, 1–48. [CrossRef]
9. Berg, A.; Pattillo, C. Predicting Currency Crises: The Indicator Approach and an Alternative. *J. Int. Money Financ.* **1999**, *18*, 561–586. [CrossRef]
10. Flood, R.P.; Garber, P.M. Collapsing exchange-rate regimes: Some linear examples. *J. Int. Econ.* **1984**, *17*, 1–13. [CrossRef]
11. Flood, R.; Marion, N. The Size and Timing of Devaluation in Capital Controlled Economies. *J. Dev. Econ.* **1995**, *54*, 123–147. [CrossRef]
12. Jurek, M. Choosing the exchange rate regime–a case for intermediate regimes for emerging and developingeconomies. *Econ. Bus. Rev.* **2018**, *4*, 46–63. [CrossRef]
13. Macroeconomic regime switches and speculative attacks. *J. Econ. Dyn. Control* **2007**, *31*, 3321–3347. [CrossRef]
14. Broz, J.L.; Frieden, J.A. The political economy of international monetary relations. *Annu. Rev. Polit. Sci.* **2001**, *4*, 317–343. [CrossRef]
15. Benchimol, J.; Fourçans, A. Money and Monetary Policy in the Eurozone: An Empirical Analysis during Crises. *Macroecon. Dyn.* **2017**, *21*, 677–707. [CrossRef]
16. Cruz-Rodríguez, A. *Exchange Arrangements and Speculative Attacks: Is there a Link?* MPRA Paper 72359; University Library of Munich: Munich, Germany, 2016.
17. Afonso, J.R.; Eliane, C.A.; Fajardo, B.G. The role of fiscal and monetary policies in the Brazilian economy: Understanding recent institutional reforms and economic changes. *Q. Rev. Econ. Financ.* **2016**, *62*, 41–55. [CrossRef]
18. Smith, G.W. 2001. Speculative attacks with unpredictable or unknown foreign exchange reserves. *Can. J. Econ. Can. Econ. Assoc.* **2001**, *34*, 882–902.
19. Esaka, T. De facto exchange rate regimes and currency crises: Are pegged regimes with capital account liberalization really more prone to speculative attacks? *J. Bank. Financ.* **2010**, *34*, 1109–1128. [CrossRef]
20. Nkwatoh, L.S.; Cornelius, K. Is the CFA Franc prone to speculative attacks or a contagion effect: A stochastic-Markov transition analysis for Cameroon. *CBN J. Appl. Stat.* **2019**, *10*, 97–117. [CrossRef]
21. Himmels, C.; Kirsanova, T. Discretionary Policy in a Small Open Economy: Exchange Rate Regimes and Multiple Equilibria. *J. Macroecon.* **2018**, *56*, 53–64. [CrossRef]
22. He, H.; Zhao, J.; Sun, G. Prediction of MoRFs in Protein Sequences with MLPs Based on Sequence Properties and Evolution Information. *Entropy* **2019**, *21*, 635. [CrossRef]
23. Johnson Singh, K.; Thongam, K.; De, T. Entropy-Based Application Layer DDoS Attack Detection Using Artificial Neural Networks. *Entropy* **2016**, *18*, 350. [CrossRef]
24. Yeung, D.S.; Cloete, I.; Shi, D.; Ng, W.W.Y. *Sensitivity Analysis for Neural Networks*; Natural Computing Series; Springer: Berlin/Heidelberg, Germany, 2010.
25. Becerra-Vicario, R.; Alaminos, D.; Aranda, E.; Fernández-Gámez, M.A. Deep Recurrent Convolutional Neural Network for Bankruptcy Prediction: A Case of the Restaurant Industry. *Sustainability* **2020**, *12*, 5180. [CrossRef]
26. Yang, Y.; Garcia-Morillo, I.; Hospedales, T.M. Deep Neural Decision Trees. In Proceedings of the 2018 ICML Workshop on Human Interpretability in Machine Learning (WHI 2018), Stockholm, Sweden, 14 July 2018.
27. Alaminos, D.; Becerra-Vicario, R.; Fernández-Gámez, M.Á.; Cisneros Ruiz, A.J. Currency Crises Prediction Using Deep Neural Decision Trees. *Appl. Sci.* **2019**, *9*, 5227. [CrossRef]
28. Dougherty, J.; Kohavi, R.; Sahami, M. Supervised and unsupervised discretization of continuous features. *Mach. Learn. Proc.* **1995**, 194–202. [CrossRef]
29. Jang, E.; Gu, S.; Poole, B. Categorical reparameterization with Gumbel-Softmax. *arXiv* **2017**, arXiv:1611.01144.
30. Ho, T.K. The random subspace method for constructing decision forests. *IEEE Trans. Pattern Anal. Mach. Intell.* **1998**, *20*, 832–844. [CrossRef]
31. Gupta, S.; Zia, R.K.P. Quantum Neural Networks. *J. Comput. Syst. Sci.* **2020**, *63*, 355–383. [CrossRef]
32. Jia, Z.; Yi, B.; Zhai, R.; Wu, Y.; Guo, G.; Guo, G. Quantum Neural Network States: A Brief Review of Methods and Applications. *Adv. Quantum Technol.* **2019**, *2*, 1800077. [CrossRef]
33. Verdon, G.; Broughton, M.; McClean, J.R.; Sung, K.J.; Babbush, R.; Jiang, Z.; Neven, H.; Mohseni, M. Learning to learn with quantum neural networks via classical neural networks. *arXiv* **2019**, arXiv:1907.05415.
34. Jeswal, S.K.; Chakraverty, S. Recent Developments and Applications in Quantum Neural Network: A Review. *Arch. Comput. Methods Eng.* **2019**, *26*, 793–807. [CrossRef]

35. Alaminos, D.; Esteban, I.; Salas, M.B.; Callejón, A.M. Quantum Neural Networks for Forecasting Inflation Dynamics. *J. Sci. Ind. Res.* **2020**, *79*, 103–106.
36. Alaminos, D.; Esteban, I.; Fernández-Gámez, M.A. Financial Performance Analysis in European Football Clubs. *Entropy* **2020**, *22*, 1056. [CrossRef]
37. Lamothe-Fernández, P.; Alaminos, D.; Lamothe-López, P.; Fernández-Gámez, M.A. Deep Learning Methods for Modeling Bitcoin Price. *Mathematics* **2020**, *8*, 1245. [CrossRef]
38. Chen, H.; Wan, Q.; Wang, Y. Refined Diebold-Mariano Test Methods for the Evaluation of Wind Power Forecasting Models. *Energies* **2014**, *7*, 4185–4198. [CrossRef]
39. Harvey, D.; Leybourne, S.; Newbold, P. Testing the equality of prediction mean squared errors. *Int. J. Forecast.* **1997**, *13*, 281–291. [CrossRef]

entropy

MDPI

Article

What Drives Bitcoin? An Approach from Continuous Local Transfer Entropy and Deep Learning Classification Models

Andrés García-Medina [1,2,*] and Toan Luu Duc Huynh [3,4,5]

[1] Unidad Monterrey, Centro de Investigación en Matemáticas, A.C. Av. Alianza Centro 502, PIIT, Apodaca 66628, Mexico
[2] Consejo Nacional de Ciencia y Tecnología, Av. Insurgentes Sur 1582, Col. Crédito Constructor, Ciudad de México 03940, Mexico
[3] WHU—Otto Beisheim School of Management, 56179 Düsseldorf, Germany; toanhld@ueh.edu.vn
[4] UEH Institute of Innovation (UII), University of Economics Ho Chi Minh City, Ho Chi Minh City 70000, Vietnam
[5] IPAG Business School, 75006 Paris, France
* Correspondence: andres.garcia@cimat.mx

Abstract: Bitcoin has attracted attention from different market participants due to unpredictable price patterns. Sometimes, the price has exhibited big jumps. Bitcoin prices have also had extreme, unexpected crashes. We test the predictive power of a wide range of determinants on bitcoins' price direction under the continuous transfer entropy approach as a feature selection criterion. Accordingly, the statistically significant assets in the sense of permutation test on the nearest neighbour estimation of local transfer entropy are used as features or explanatory variables in a deep learning classification model to predict the price direction of bitcoin. The proposed variable selection do not find significative the explanatory power of NASDAQ and Tesla. Under different scenarios and metrics, the best results are obtained using the significant drivers during the pandemic as validation. In the test, the accuracy increased in the post-pandemic scenario of July 2020 to January 2021 without drivers. In other words, our results indicate that in times of high volatility, Bitcoin seems to self-regulate and does not need additional drivers to improve the accuracy of the price direction.

Keywords: local transfer entropy; long-short-term-memory; Bitcoin

Citation: García-Medina, A.; Luu Duc Huynh, T. What Drives Bitcoin? An Approach from Continuous Local Transfer Entropy and Deep Learning Classification Models. *Entropy* **2021**, *23*, 1582. https://doi.org/10.3390/e23121582

Academic Editors: Ryszard Kutner, Christophe Schinckus and H. Eugene Stanley

Received: 28 September 2021
Accepted: 23 November 2021
Published: 26 November 2021

1. Introduction

Currently, there is tremendous interest in determining the dynamics and direction of the price of Bitcoin due to its unique characteristics, such as its decentralization, transparency, anonymity, and speed in carrying out international transactions. Recently, these characteristics have attracted the attention of both institutional and retail investors. Thanks to technological developments, investor trading strategies are benefited by digital platforms; therefore, market participants are more likely to digest and create information for this market. Of special interest is its decentralized character, since its value is not determined by a central bank but, essentially, only by supply and demand, recovering the ideal of a free market economy. At the same time, it is accessible to all sectors of society, which breaks down geographic and particular barriers for investors. The fact that there are a finite number of coins and the cost of mining new coins grows exponentially has suggested to some specialists that it may be a good instrument for preserving value. That is, unlike fiat money, Bitcoin cannot be arbitrarily issued, so its value is not affected by the excessive issuance of currency that central banks currently follow, or by low interest rates as a strategy to control inflation. In other words, it has been recently suggested that bitcoin is a safe-haven asset or store of value, having a role similar to that once played by gold and other metals.

The study of cryptocurrencies and bitcoin has been approached from different perspectives and research areas. It has been addressed from the point of view of financial economics, econometrics, data science, and more recently by econophysics. In these approaches, various methodologies and mathematical techniques have been utilised to understand different aspects of these new financial instruments. These topics range from systemic risk, the spillover effect, autoscaling properties, collective patterns, price formation, and forecasting in general. Remarkable work in the line of multiscale analysis of cryptocurrency markets can be found in [1]. However, this paper is motivated by using the econphysics approach, incorporated with rigorous control variables to predict Bitcoin price patterns. We would like to offer a comprehensive review of the determinants of Bitcoin prices. The first pillar can be defined as sentiment and social media content. While Bitcoin is widely considered a digital financial asset, investors pay attention to this largest market capitalization by searching its name. Therefore, the strand of literature on Google search volume has become popular for capturing investor attention [2]. Concomitantly, not only peer-to-peer sentiment (individual Twitter accounts or fear from investors) [3,4] but also influential accounts (the U.S. President, media companies) [5–7] significantly contribute to Bitcoin price movement. Given the greatest debate on whether Bitcoin should act as a hedging, diversifying or safe-haven instrument, Bitcoin exhibits a mixture of investing features. More interestingly, uncertain shocks might cause changes in both supply and demand in Bitcoin circulation, implying a change in its prices [8]. Thus, the diverse stylized facts of Bitcoin, including heteroskedasticity and long memory, require uncertainty to be controlled in the model. While uncertainties represent the amount of risk (compensated by the Bitcoin returns) [9], our model also includes the price of risk, named the 'risk aversion index' [10]. These two concepts (amount of risk and the price of risk) demonstrate discount rate factors in the time variation of any financial market [11]. In summary, the appearance of these determinants could capture the dynamics of the cryptocurrency market. Since cryptocurrency is a newly emerging market, the level of dependence in the market structure is likely higher than that in other markets [12]. Furthermore, the contagion risk and the connectedness among these cryptocurrencies could be considered the risk premium for expected returns [13,14]. More importantly, this market can be driven by small market capitalization, implying vulnerability of the market [15]. Hence, our model should contain alternative coins (altcoins) to capture their movements in the context of Bitcoin price changes. Finally, investors might consider the list of these following assets as alternative investment, precious metals being the first named. They are not only substitute assets [16] but also predictive factors (for instance, gold and platinum) [17], which additionally include commodity markets (such as crude oil [18,19], exchange rate [20], equity market [21]), and Tesla's owner [22]). In summary, there are voluminous determinants of Bitcoin prices. In the scope of this study, we focus on the predictability of our model, especially the inclusion of social media content, representing the high popularity of information, on the Bitcoin market. However, the more control variables there are, the higher the accuracy of prediction. Our model thus may be a useful tool by combining the huge predictive factors for training and forecasting the response dynamics of Bitcoin to other relevant information.

This study approaches Bitcoin from the framework of behavioural and financial economics using an approach from econophysics and data science. In this sense, it seeks to understand the speculative character and the possibilities of arbitrage through a model that includes investor attention and the effect of the news, among other factors. For this, we will use a causality method originally proposed by Schreiber [23], and we will use the information as characteristics of a deep learning model. The current literature only focuses on specific sentiment indicators (such as Twitter users [3] or the number of tweets [24,25]), and our study crawled the original text from influential Twitter social media users (such as the President of United States, CEO of Tesla, and well-known organizations such as the United Nations and BBC Breaking News). Then, we processed language analyses to construct the predictive factor for Bitcoin prices. Therefore, our model incorporates a new perspective on Bitcoin's drivers.

In this direction, the work of [26] uses the effective transfer entropy as an additional feature to predict the direction of U.S. stock prices under different machine learning approaches. However, the approximation is discrete and based on averages. Furthermore, the employed metrics are not exhaustive to determine the predictive power of the models. In a similar vein, the authors of [27] perform a comparative analysis of machine learning methods for the problem of measuring asset risk premiums. Nevertheless, they do not take into account recurrent neural network models or additional nontraditional features. Furthermore, an alternative approach to study the main drivers of Bitcoin is discussed in [28], where the author explores wavelet coherence to examine the time and frequency domains between short- and long-term interactions. In the same vein, the recent studies employed the correlation networks and vector error correction models to explain the price prediction and exchange spillovers [29,30]. Of course, Bitcoin prediction is more likely to have sentimental and 'noise' factors differing from stock prediction.

On the other hand, there are methodologies to explain machine learning results known as eXplainable Artificial Intelligence (XAI). Among these, two of the most popular are Local Interpretable Model Agnostic [31] and Shapley Additive Explanation (SHAP) [32]. Both techniques are based on disturbing the model locally. The former assumes a linear model to obtain the score of the characteristics in terms of the importance of making predictions; the latter uses game theory concepts to find the best feature fitting in terms of predictive gain. In [33] these techniques are extended to include temporal dependencies and demonstrate the need to develop XAI techniques applicable to time series. In [34,35] is proposed an XAI method applicable to credit risk. In a similar vein, the authors of [36] mention the difficulty of estimating out-of-sample behavior in stress scenarios. An interesting work is [37], where it is considered a gradient boosting decision trees approximation to predict the drops of the S&P 500 markets using a large number of characteristics. The authors claim that retaining a small and carefully selected amount of features improves the learning model results.

However, as mentioned in the cornerstone work [31] it is not possible to explain a highly non-linear model through local perturbations. That is, there is a high instability derived from the characteristics of the inherent dynamical system. In addition, the examples of the articles mentioned above run in most cases in seconds or minutes. Therefore, the LIME and SHAP methods are appropriate mainly for machine learning models or simple deep learning scenarios [38]. In this spirit, it is not practical to follow the traditional XAI approach, given the computational demand derived from the number of hyperparameters and configurations to be implemented. However, our proposal to use transfer entropy in the variable selection process can be considered an alternative strategy to XAI. In particular, of interest for highly non-linear dependency conditions, such as bitcoin dynamics.

Our study embodied a wide range of Bitcoin's drivers from alternative investment, economic policy uncertainty, investor attention, and so on. However, social media is our main contribution to predictive factors. Specifically, we study the effect that a set of Twitter accounts belonging to politicians and millionaires has on the behaviour of Bitcoin's price direction. In this work, the statistically significant drivers of Bitcoin are detected in the sense of the continuous estimation of local transfer entropy (local TE) through nearest neighbours and permutation tests. The proposed methodology deals with non-Gaussian data and nonlinear dependencies in the problem of variable selection and forecasting. One main aim is to quantify the effects of investor attention and social media on Bitcoin in the context of behavioural finance. Another aim is to apply classification metrics to indicate the effects of including or not the statistically significant features in an LSTM's classification problem.

The next Section 2 introduce the local transfer entropy, the nearest neighbour estimation technique, the deep learning forecasting models, and the classification metrics. Section 3 describes the data and their main descriptive characteristics. Section 4 presents and highlights the main results. Finally, Section 5 highlights the implications of the results, and future work is proposed.

2. Materials and Methods

2.1. Transfer Entropy

Transfer Entropy (TE) [23] measures the flow of information from system Y over system X in a nonsymmetric way. Denote the sequences of states of systems X, Y in the following way: $x_i = x(i)$ and $y_i = y(i), i = 1, \dots, N$. The idea is to model the signals or time series as Markov systems and incorporate the temporal dependencies by considering the states x_i and y_i to predict the next state x_{i+1}. If there is no deviation from the generalized Markov property $p(x_{i+1}|x_i, y_i) = p(x_{i+1}|x_i)$, then Y has no influence on X. Hence, TE is derived using the last idea and defined as

$$T_{Y \to X}(k,l) = \sum p(x_{i+1}, x_i^{(k)}, y_i^{(l)}) \log \frac{p(x_{i+1}|x_i^{(k)}, y_i^{(l)})}{p(x_{i+1}|x_i^{(k)})}, \tag{1}$$

where $x_i^{(k)} = (x_i, \dots, x_{i-k+1})$ and $y_i^{(l)} = (y_i, \dots, y_{i-l+1})$.

TE can be thought of as a global average or expected value of a local transfer entropy at each observation [39]

$$T_{Y \to X}(k,l) = \left\langle \log \frac{p(x_{i+1}|x_i^{(k)}, y_i^{(l)})}{p(x_{i+1}x_i^{(k)})} \right\rangle \tag{2}$$

The main characteristic of the local version of TE is to be measured at each time n for each destination element X in the system and each causal information source Y of the destination. It can be either positive or negative for a specific event set $(x_{i+1}, x_i^{(k)}, y_i^{(l)})$, which gives the opportunity to have a measure of informativeness or noninformativeness at each point of a pair of time series.

On the other hand, there exist several approximations to estimate the probability transition distributions involved in TE expression. Nevertheless, there is not a perfect estimator. It is generally impossible to minimize both the variance and the bias at the same time. Then, it is important to choose the one that best suits the characteristics of the data under study. That is the reason finding good estimators is an open research area [40]. This study followed the Kraskov-Stögbauer-Grassberger) KSG estimator [41], which focused on small samples for continuous distributions. Their approach is based on nearest neighbours. Although obtaining insight into this estimator is not easy, we will try it in the following.

Let $\mathbf{X} = (x_1, x_2, \dots, x_d)$ now denote a d-dimensional continuous random variable whose probability density function is defined as $p : \mathbb{R}^d \to \mathbb{R}$. The continuous or differential Shannon entropy is defined as

$$H(\mathbf{X}) = -\int_{\mathbb{R}^d} p(\mathbf{X}) \log p(\mathbf{X}) d\mathbf{X} \tag{3}$$

The KSG estimator aims to use similar length scales for K-nearest-neighbour distance in different spaces, as in the joint space to reduce the bias [42].

To obtain the explicit expression of the differential entropy under the KSG estimator, consider N i.i.d. samples $\chi = \{\mathbf{X}(i)\}_{i=1}^N$, drawn from $p(\mathbf{X})$. Beneath the assumption that $\epsilon_{i,K}$ is twice the (maximum norm) distance to the k-th nearest neighbour of $\mathbf{X}(i)$, the differential entropy can be estimated as

$$\hat{H}_{KSG,K}(\mathbf{X}) \equiv \psi(N) - \psi(K) + \frac{d}{N} \sum_{i=1}^{N} \log \epsilon_{i,K}, \tag{4}$$

where ψ is known as the digamma function and can be defined as the derivative of the logarithm of the gamma function $\Gamma(x)$

$$\psi(K) = \frac{1}{\Gamma(K)} \frac{d\Gamma(K)}{dK} \tag{5}$$

The parameter K defines the size of the neighbourhood to use in the local density estimation. It is a free parameter, but there exists a trade-off between using a smaller or larger value of K. The former approach should be more accurate, but the latter reduces the variance of the estimate. For further intuition, Figure 1 graphically shows the mechanism for choosing the nearest neighbours at $K = 3$.

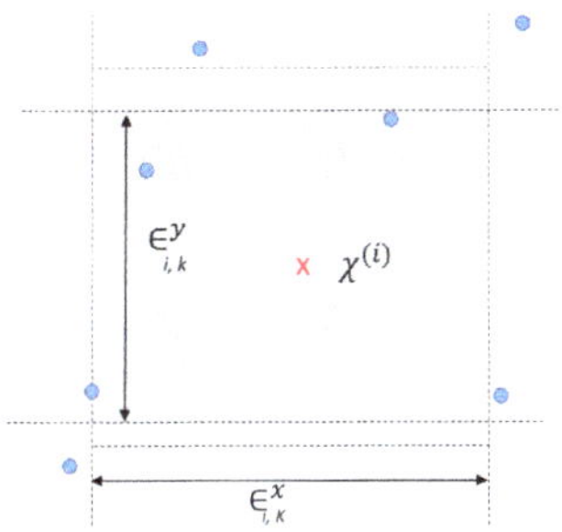

Figure 1. Graphical representation of nearest-neighbors selection. At a given sample point, $\mathbf{X}(i)$, the max-norm rectangle contains the $K = 3$ nearest-neighbors.

The KSG estimator of TE can be derived based on the previous estimation of the differential entropy. Yet, in most cases, as analysed in this work, no analytic distribution is known. Hence, the distribution of $T_{Y_s \to X}(k,l)$ must be computed empirically, where Y_s denotes the surrogate time series of Y. This is done by a resampling method, creating a large number of surrogate time-series pairs $\{Y_s, X\}$ by shuffling (for permutations or redrawing for bootstrapping) the samples of Y. In particular, the distribution of $T_{Y_s \to X}(k,l)$ is computed by permutation, under which surrogates must preserve $p(x_{n+1}|x_n)$ but not $p(x_{n+1}|x_n, y_n)$.

2.2. Deep Learning Models

We can think of artificial neural networks (ANNs) as a mathematical model whose operation is inspired by the activity and interactions between neuronal cells due to their electrochemical signals. The main advantages of ANNs are their non-parametric and nonlinear characteristics. The essential ingredients of an ANN are the neurons that receive an input vector x_i, and through the point product with a vector of weights w, generate an output via the activation function $g(\cdot)$:

$$f(x_i) = g(x_u \cdot w) + b, \tag{6}$$

where b is a trend to be estimated during the training process. The basic procedure is the following. The first layer of neurons or input layer receives each of the elements of the input vector x_i and transmits them to the second (hidden) layer. The next hidden layers calculate their output values or signals and transmit them as an input vector to the next layer until reaching the last layer or output layer, which generates an estimation for an output vector.

Further developments of ANNs have brought recurrent neural networks (RNNs), which have connections in the neurons or units of the hidden layers to themselves and are more appropriate to capture temporal dependencies and therefore are better models for time series forecasting problems. Instead of neurons, the composition of an RNN includes a unit, an input vector x_t, and an output signal or value h_t. The unit is designed with a recurring connection. This property induces a feedback loop, which sends a recurrent signal to the unit as the observations in the training data set are analysed. In the internal process, backpropagation is performed to obtain the optimal weights. Unfortunately, backpropagation is sensitive to long-range dependencies. The involved gradients face the problem of vanishing or exploding. Long-short-term memory (LSTM) models were introduced by Hochreiter and Schmidhuber [43] to avoid these problems. The fundamental difference is that LSTM units are provided with memory cells and gates to store and forget unnecessary information.

The final ANNs we need to discuss are convolutional neural networks (CNNs). They can be thought of as a kind of ANN that uses a high number of identical copies of the same neuron. This allows the network to express computationally large models while keeping the number of parameters small. Usually, in the construction of these types of ANNs, a max-pooling layer is included to capture the largest value over small blocks or patches in each feature map of previous layers. It is common that CNN and pooling layers are followed by a dense fully connected layer that interprets the extracted features. Then, the standard approach is to use a flattened layer between the CNN layers and the dense layer to reduce the feature maps to a single one-dimensional vector [44].

2.3. *Classification Metrics*

In classification problems, we have the predicted class and the actual class. The possible scenarios under a classification prediction are given by the confusion matrix. They are true positive (TP), true negative (TN), false positive (FP), and false negative (FN). Based on these quantities, it is possible to define the following classification metrics:

- Accuracy = $\frac{TP+TN}{TP+TN+FP+FN}$
- Sensitivity, recall or true positive rate (TPR) $= \frac{TP}{TP+FN}$
- Specificity, selectivity or true negative rate (TNR) $= \frac{TN}{TN+FP}$
- Precision or Positive Predictive Value (PPV) $= \frac{TP}{TP+FP}$
- False Omission Rate (FOR) $= \frac{FN}{FN+TN}$
- Balanced Accuracy (BA) $= \frac{TPR+TNR}{2}$
- F1 score $= 2\frac{PPV \times TPR}{PPV+TPR}$.

The most complex measure is the area under the curve (AUC) of the receiver operating characteristic (ROC), where it expresses the pair $(TPR_\tau, 1 - TNR_\tau)$ for different thresholds τ. Contrary to the other metrics, the AUC of the ROC is a quality measure that evaluates all the operational points of the model. A model with the aforementioned metric equal to 0.5 is considered a random model. Then, a value significantly higher than 0.5 is considered a model with predictive power, with a value of 1 the upper bound of this quantity.

3. Data

An important part of the work is the acquisition and preprocessing of data. We focus on the period of time from 1 January 2017 to 9 January 2021 at a daily frequency for a total of $n = 1470$ observations. As a priority, we consider the variables listed in Table 1 as potential drivers of the price direction of Bitcoin (BTC). Investor attention is considered Google Trends with the *query* = *"Bitcoin"*. Additionally, the number of mentions is properly scaled to make comparisons between days of different months because by default, Google Trends weighs the values by a monthly factor. Then, the log return of the resulting time series is calculated.

Table 1. Type of driver and variable name.

Type	Variables
Investor attention	Google Trends
Social media	BBC Breaking News
	Department of State
	United Nations
	Elon Musk
	Donald Trump
Twitter-EPU	Twitter-based Uncertainty Index
Risk Aversion	Financial Proxy to Risk Aversion and Economic Uncertainty
Cryptocurrencies	ETH
	LTC
	XRP
	DOGE
	TETHER
Financial indices	Gold
	Silver
	Palladium
	Platinum
	DCOILBRENTEU
	DCOILWTICO
	EUR/USD
	S&P 500
	NASDAQ
	VIX
	ACWI
	Tesla

The social media data are collected from the Twitter API (https://developer.twitter.com/en/docs/twitter-api, accessed on 15 January 2021). Nevertheless, the API of Twitter only enables downloading the latest 3200 tweets of a public profile, which generally was not enough to cover the period of study. Then, the dataset has been completed with the freely available repository of https://polititweet.org/ (accessed on 15 January 2021). In this way, the collected number of tweets was 21,336, 22,808, 24,702, 11,140, and 26,169 for each of the profiles listed on Table 1 in the social media type, respectively. The textual data of each tweet in the collected dataset are transformed to a sentiment polarity score through the VADER lexicon [45]. Then, the scores are aggregated daily for each profile. The resulting daily time series have missing values due to the inactivity of the users, and then a third-order spline is considered before calculating their differences. The last is to stationarize the polarity time series. It is important to remember that Donald Trump's account was blocked on 8 January 2021, so it was also necessary to impute the last value to have series of the same length.

The economic policy uncertainty index is a Twitter-based uncertainty index (Twitter-EPU). The creators of the index used the Twitter API to extract tweets containing keywords related to uncertainty ("uncertain", "uncertainly", "uncertainties", "uncertainty") and economy ("economic", "economical", "economically", "economics", "economies", "economist", "economists", "economy"). Then, we use the index consisting of the total number of daily tweets containing inflections of the words uncertainty and economy (Please consult https://www.policyuncertainty.com/twitter_uncert.html for further details of the index, accessed on 15 January 2021). The risk aversion category considers the financial proxy to risk aversion and economic uncertainty proposed as a utility-based aversion coefficient [10]. A remarkable feature of the index is that in early 2020, it reacted more strongly to the new COVID-19 infectious cases than did a standard uncertainty proxy.

As complementary drivers, it includes a set of highly capitalized cryptocurrencies and a heterogeneous portfolio of financial indices. Specifically, Ethereum (ETH), Litecoin (LTC), Ripple (XRP), Dogecoin (DOGE), and the stable coin TETHER are included from yahoo finance (https://finance.yahoo.com/, accessed on 15 January 2021). The components of the heterogeneous portfolio are listed in Table 1, which takes into account the Chicago Board Options Exchange's CBOE Volatility Index (VIX). This last information was extracted from Bloomberg (https://www.bloomberg.com/, accessed on 15 January 2021). It is important to point out that risk aversion and the financial indices do not have information that corresponds to weekends. The imputation method to obtain a complete database consisted of repeated Friday values as a proxy for Saturday and Sunday. Then, the log return of the resulting time series is calculated. This last transformation was also made for Twitter-EPU and cryptocurrencies. The complete dataset can be found in the Supplementary Material.

Usually, the econophysics and data science approaches share the perspective of observing data first and then modelling the phenomena of interest. In this spirit, and with the intention of gaining intuition on the problem, the standardized time series (target and potential drivers), as well as the cumulative return of the selected cryptocurrencies and financial assets are plotted in Figures 2 and 3. The former figure shows high volatility in almost all the studied time series around March 2020, which might be due to the declaration of the pandemic by the World Health Organization (WHO) and the consequent fall of the worlds main stock markets. The latter figure exhibits the overall best cumulative gains for BTC, ETH, LTC, XRP, DOGE, and Tesla. It is worth noting that the only asset with a comparable profit to that of the cryptocurrencies is Tesla, which reaches high cumulative returns starting at the end of 2019 and increases its uptrend immediately after the announcement of the worldwide health emergency.

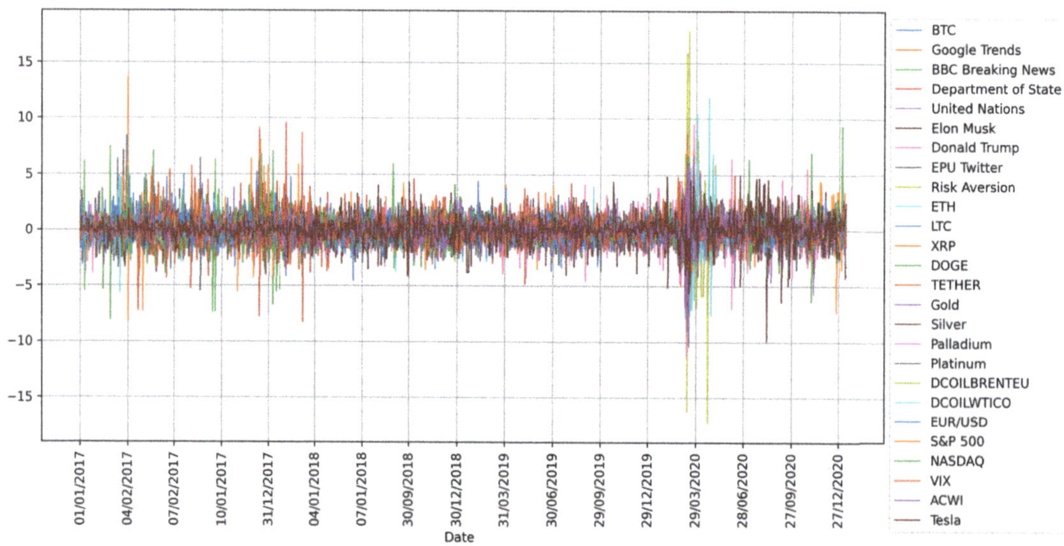

Figure 2. Standardized time series after preprocessing, as explained in the main text.

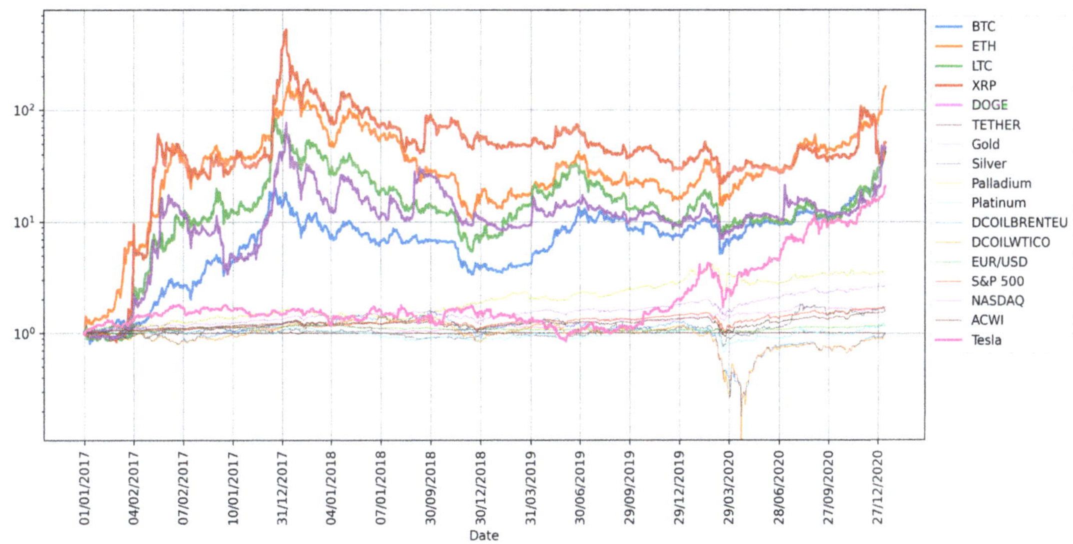

Figure 3. Cumulative returns of the selected cryptocurrencies and financial assets. The scale is logarithmic in the y-axis and starts in one to be financially interpreted as the gains.

Furthermore, Figure 4 shows the heatmap of the correlation matrix of the preprocessed dataset. We can observe the formation of certain clusters, such as cryptocurrencies, metals, energy, and financial indices, which tells us about the heterogeneity of the data. It should also be noted that the VIX volatility index is anti-correlated with most of the variables.

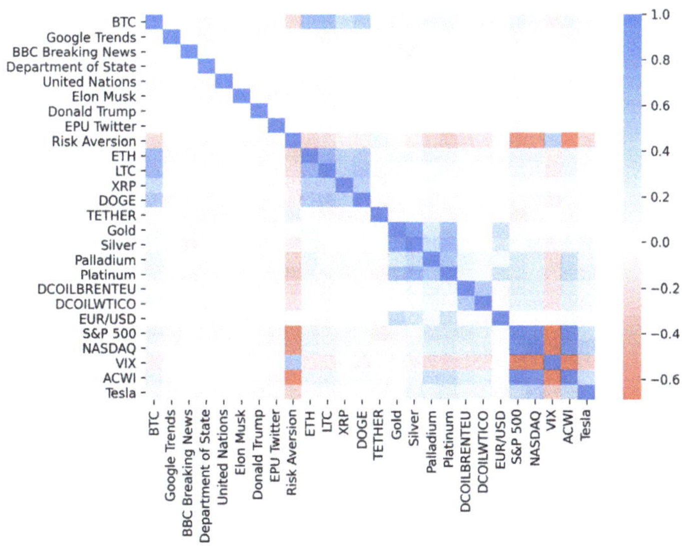

Figure 4. Correlation matrix of the preprocessed time series.

Additionally, the main statistical descriptors of the data are presented in Table 2. The first column is the variable's names or tickers. The subsequent columns represent the mean, standard deviation, skewness, kurtosis, Jarque Bera test (JB), and the associated p value of the test for each variable, i.e., target, and potential drivers. Basically, none of the time series passes the test of normality distribution, and most of them present a high kurtosis, which is indicative of heavy tail behaviour. Finally, stationarity was checked in the sense of Dickey-Fuller and the Phillips-Perron unit root tests, where all variables pass both tests.

Table 2. The symbols **, and *** denote the significance at the 5%, and 1% levels, respectively.

Variable	Mean	Std. Dev.	Skewness	Kurtosis	JB	p-Value
BTC	0.0025	0.0424	−0.8934	12.7470	10,073.5034	***
Google Trends	0.0018	0.1915	0.2611	6.8510	2868.6001	***
BBC Breaking News	0.0007	3.5295	−0.1789	15.5376	14,686.4748	***
Department of State	−0.0013	4.0610	0.0941	8.4698	4362.1014	***
United Nations	0.0007	3.2689	0.1748	0.2747	11.9228	**
Elon Musk	0.0035	1.8877	0.0672	3.8630	906.9805	***
Donald Trump	0.0001	4.8294	0.0461	5.2434	1670.4686	***
Twitter−EPU	0.0009	0.3001	0.3049	3.6071	812.4777	***
Risk Aversion	0.0001	0.0726	3.7594	165.2232	1,664,075.5884	***
ETH	0.0034	0.0566	−0.3991	9.7009	5759.1438	***
LTC	0.0025	0.0606	0.6919	10.5404	6870.4947	***
XRP	0.0027	0.0753	2.2903	36.2405	81,162.9262	***
DOGE	0.0026	0.0669	1.2312	15.0342	14,113.2759	***
TETHER	0.0000	0.0062	0.3255	20.0952	24,581.8501	***
Gold	0.0006	0.0082	−0.6595	5.5761	1995.0834	***
Silver	0.0005	0.0164	−1.1304	13.0841	10,720.4362	***
Palladium	0.0015	0.0197	−0.9198	20.5312	25,840.0775	***
Platinum	0.0004	0.0145	−0.9068	10.7274	7196.3125	***
DCOILBRENTEU	−0.0004	0.0374	−3.1455	81.2272	403,755.7220	***
DCOILWTICO	0.0006	0.0358	0.7362	38.4244	89,931.3161	***
EUR/USD	0.0002	0.0042	0.0336	0.8999	49.0930	***
S&P 500	0.0006	0.0125	−0.5714	20.5446	25,746.7436	***
NASDAQ	0.0008	0.0145	−0.3601	11.7771	8463.6169	***
VIX	−0.0061	0.0810	1.4165	8.4537	4833.8826	***
ACWI	0.0006	0.0115	−1.1415	20.4837	25,833.5682	***
Tesla	0.0017	0.0371	−0.3730	5.5089	1877.5034	***

4. Results

4.1. Variable Selection

The observed characteristics of the data in the previous section justify the use of a non-parametric approach to determine the explainable features to be employed in the predictive classification model. Therefore, the variable selection procedure consisted of applying the continuous transfer entropy from each driver to Bitcoin using the KSG estimation. Figure 5 shows the average transfer entropy when varying the Markov order k, l and neighbour parameter K from one to ten for a total of 1000 different estimations by each driver. The higher the intensity of the colour, the higher the average transfer entropy (measured in nats). The grey cases do not transfer information to BTC. In other words, these cases do not show a statistically significant flow of information, where the permutation test is applied to construct 100 surrogate measurements under the null hypothesis of no directed relationship between the given variables.

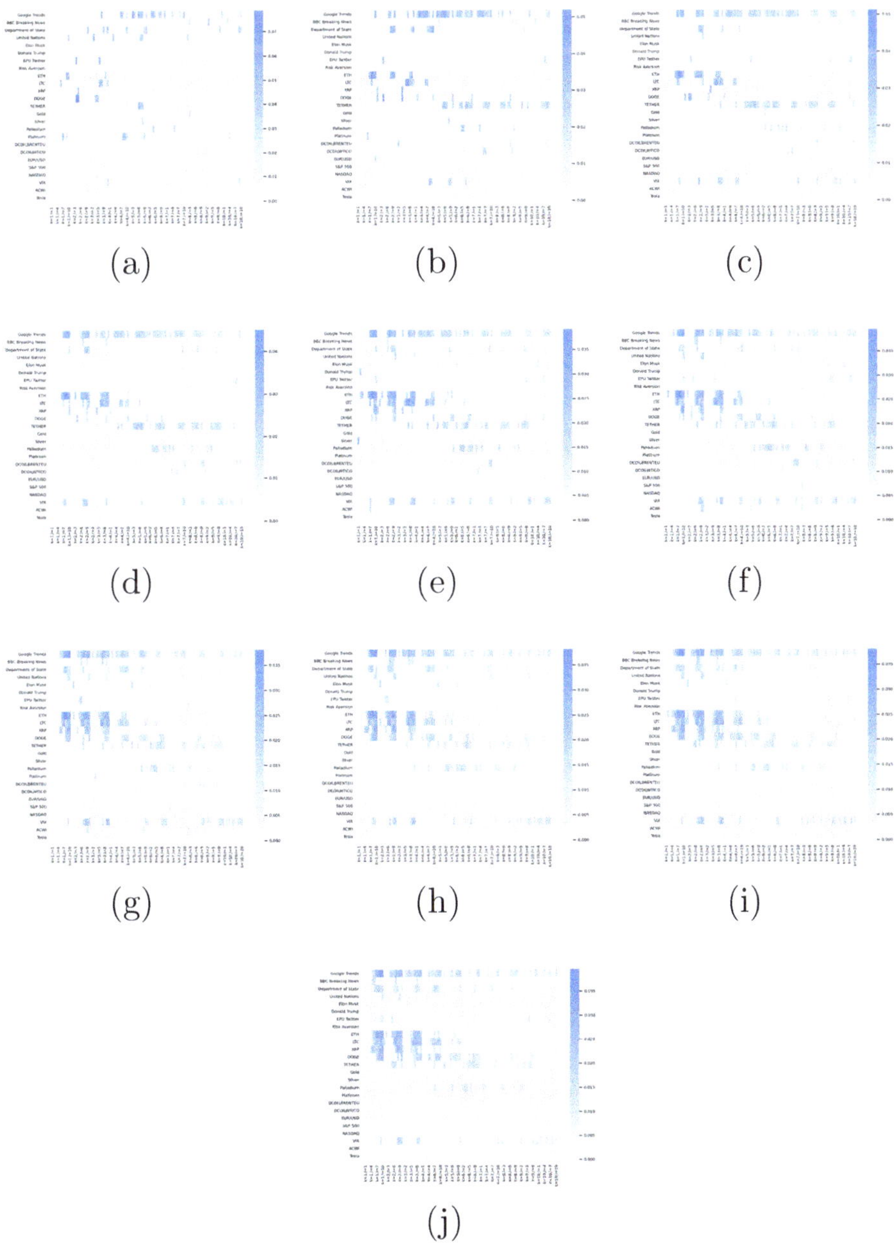

Figure 5. Average transfer entropy from each potential driver to BTC. The y-axis indicates the driver, and the x-axis indicates the Markov order pair k, l of the source and target. From (**a**) to (**j**), nearest neighbours K run from one to ten, respectively.

The tuple of parameters $\{k, l, K\}$ that give the highest average transfer entropy from each potential driver to BTC are considered optimal, and the associated local TE is kept as a feature in the classification model of Bitcoin's price direction. Figure 6 shows the local TE from each statistically significant driver to BTC at the optimal parameter tuple $\{k, l, K\}$. Note that the set of local TE time series is limited to 23 features. Consequently, the set of originally proposed potential drivers is reduced from 25 to 23. Surprisingly, NASDAQ and Tesla do not send significative information to BTC for any value of $\{k, l, K\}$ in the grid of the 1000 different configurations. The variations are smooth on K, but not on the Markov order k, l. It is also notorious the negligible amounts of the flow of information at $k = l = 1$.

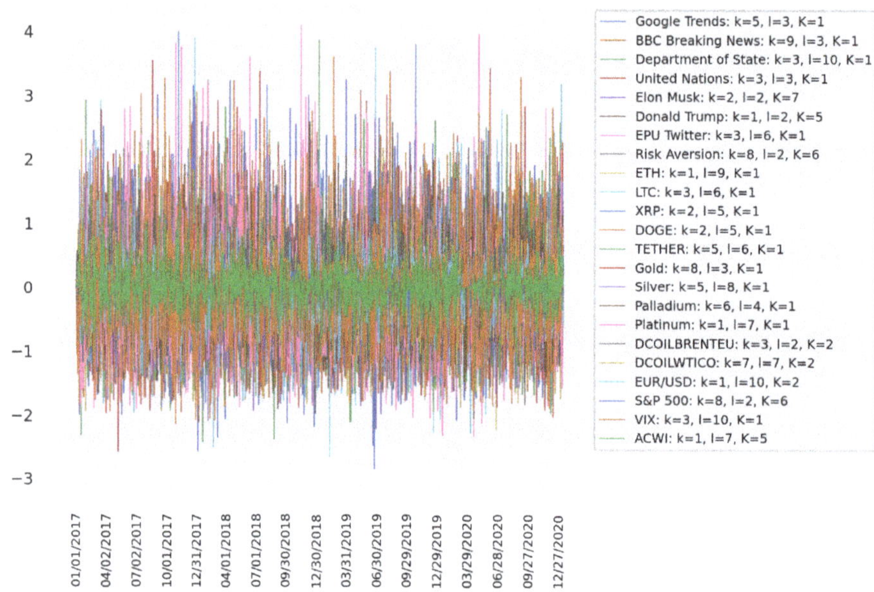

Figure 6. Local TE of the highest significant average values on the tuple $\{k, l, K\}$. NASDAQ and Tesla are omitted because they do not send significative information to BTC for any considered value on the grid of the tuple $\{k, l, K\}$.

4.2. *Bitcoin's Price Direction*

The task of detecting Bitcoin's price direction was done through a deep learning approach. The first step consisted of splitting the data into training, validation, and test datasets. The chosen training period runs from 1 January 2017 to 4 January 2020, or 75% of the original entire period of time, and is characterized as a prepandemic scenario. The validation dataset is restricted to the period from 5 January 2020 to 11 July 2020, or 13% of the original data, and is considered the pandemic scenario. The test dataset involves the postpandemic scenario from 12 July 2020 to 9 January 2021 and contains 12% of the complete dataset. Deep learning forecasting requires transforming the original data into a supervised data set. Here, samples of 74 historical days and a one-step prediction horizon are given to the model to obtain a supervised training dataset, with the first dimension being a power of two, which is important for the hyperparameter selection of the batch dimension. Specifically, the sample dimensions are 1024, 114, and 107 for training, validation, and testing, respectively. Because we are interested in predicting the direction of BTC, the time series are not demeaned and instead are only scaled by their variance when feeding the deep learning models. An important piece in a deep learning model is the selection of the activation function. In this work, the rectified linear unit (ReLU) was selected for the hidden layers. Then, for the output layer, the sigmoid function is chosen because we are dealing with a classification problem. In addition, an essential ingredient is the selection of the stochastic gradient descent method. Here, Adam optimization is selected based on adaptive estimation of the first- and second-order moments. In particular, we used version [46] to search for the long-term memory of past gradients to improve the convergence of the optimizer.

There exist several hyperparameters to take into account when modelling a classification problem under a deep learning approach. These hyperparameters must be calibrated on the training and validation datasets to obtain reliable results on the test dataset. The usual procedure to set them is via a grid search. Nevertheless, deeper networks with more computational power are necessary to obtain the optimal values in a reasonable amount of time. To avoid excessive time demands, we vary the most crucial parameters in a small grid and apply some heuristics when required. The number of epochs, is selected under the early stopping procedure. Another crucial hyperparameter is the batch, or the number of samples to work through before updating the internal weight of the model. For this parameter the selected grid was $\{32, 64, 128, 256\}$. Additionally, we consider the initial learning rates at which the optimizer starts the algorithm, which were $\{0.001, 0.0001\}$. As an additional method of regularization, the effect of dropping between consecutive layers is added. This value can take values from 0 to 1. Our grid for this hyperparameter is $\{0.3, 0.5, 0.7\}$. Finally, because of the stochastic nature of the deep learning models, it is necessary to run several realizations and work with averages. We repeat the hyperparameter selection with ten different random seeds for robustness. The covered scenarios are the following: *univariate* (S1), where bitcoin is self-driven; *all features* (S2), where all the potential drivers listed in Table 1 are included as features of the model; *significative features* (S3), only statistically significant drivers under the KSG transfer entropy approach are considered as features; *local TE*, only the local TE of the statistically significant drivers are included as a feature; and finally the *significative features + local TE* (S5) scenario, which combines scenarios (S3) and (S4). Finally, five different designs have been proposed for the architectures of the neural networks, which are denoted as *deep LSTM* (D1), *wide LSTM* (D2), *deep bidirectional LSTM* (D3), *wide bidirectional LSTM* (D4), and *CNN* (D5). The specific designs and diagrams of these architectures are displayed in Figure 7. In total, 6000 configurations or models were executed, which included the grid search for the optimal hyperparameters, the different scenarios and architectures, and the realizations on different seeds to avoid biases due to the stochastic nature of the considered machine learning models.

The computation was done in a workstation with the following characteristics: Alienware Aurora R7, Ubuntu 20.10, Processor i9-9900X 8 cores, 16 logic, 64 GB RAM, Dual NVIDIA RTX 2080 ti, 3TB HHD. On this equipment, the computational demand extends the execution to nearly 60 h of computation. Tables 3 and 4 present the main results for the validation and test datasets, respectively. Table 2 explicitly states the best value for the dropout, learning rate (LR), and batch hyperparameters. In both tables, the hashtag (#) column indicates the number of times the specific scenario gives the best score for the different metrics considered so far. Hence, the architecture design D3 for case S3 yields the highest number of metrics with the best scores in the validation dataset. In contrast, in the test dataset, the highest number of metrics with the best scores correspond to design D2 for case S1. Nevertheless, design D5 from case S5 is close in the sense of the # value, where it presents the best AUC and PPV scores. An important point to keep in mind is that only during the validation stage we find models with an AUC greater than 0.6, so this metric does not give evidence of predictive power in the testing stage.

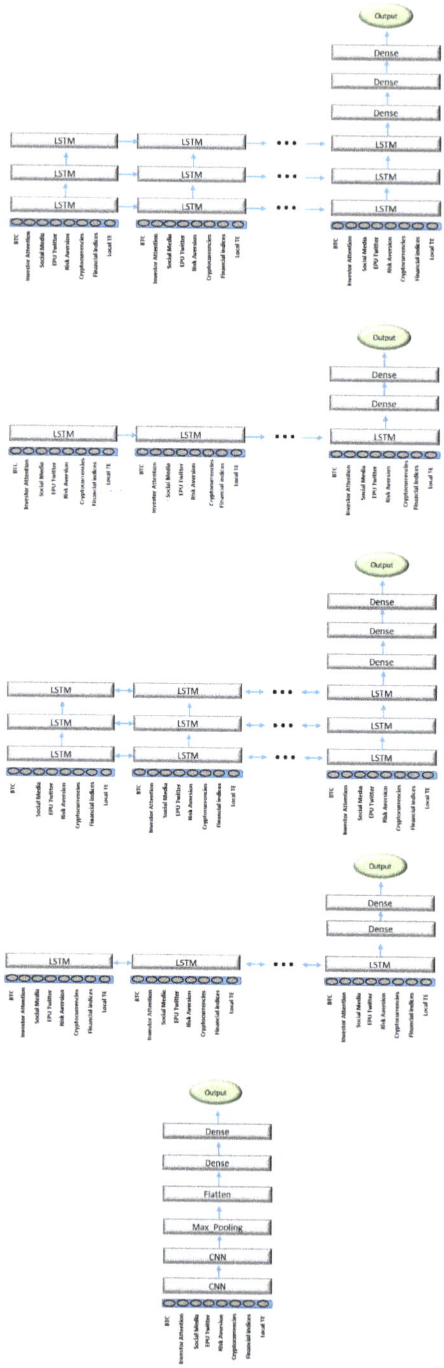

Figure 7. From (**top**) to (**bottom**): D1, D2, D3, D4, and D5.

Table 3. Classification metrics on the validation dataset.

Design	Case	Dropout	LR	Batch	Acc	AUC	TPR	TNR	PPV	FOR	BA	F1	#
D1	S1	0.3	0.001	32	57.11	0.5388	84.75	25.28	56.63	40.97	55.02	67.89	
	S2	0.3	0.001	128	57.28	0.5391	80.33	30.75	57.18	42.40	55.54	66.80	
	S3	0.7	0.001	128	58.07	0.5379	74.43	39.25	58.51	42.86	56.84	65.51	
	S4	0.3	0.001	256	57.98	0.5304	75.41	37.92	58.30	42.74	56.67	65.76	
	S5	0.3	0.001	256	57.19	0.5100	87.54	22.26	56.45	39.18	54.90	**68.64**	
D2	S1	0.3	0.001	64	59.82	0.5444	81.97	34.34	58.96	**37.67**	58.15	68.59	
	S2	0.3	0.001	32	61.14	0.5909	65.57	56.04	63.19	41.42	60.81	64.36	
	S3	0.3	0.001	128	**62.28**	**0.6062**	62.95	**61.51**	**65.31**	40.94	**62.23**	64.11	5
	S4	0.5	0.0001	32	55.44	0.4964	75.08	32.83	56.27	46.63	53.96	64.33	
	S5	0.7	0.001	32	58.07	0.5706	63.77	51.51	60.22	44.74	57.64	61.94	
D3	S1	0.3	0.001	128	56.23	0.4865	**88.52**	19.06	55.73	40.94	53.79	68.40	
	S2	0.3	0.001	64	59.65	0.5816	68.69	49.25	60.90	42.26	58.97	64.56	
	S3	0.3	0.001	128	60.09	0.5619	76.72	40.94	59.92	39.55	58.83	67.29	
	S4	0.3	0.001	32	58.16	0.5350	79.18	33.96	57.98	41.37	56.57	66.94	
	S5	0.3	0.001	256	59.47	0.5702	68.69	48.87	60.72	42.44	58.78	64.46	
D4	S1	0.5	0.001	32	57.28	0.5276	80.16	30.94	57.19	42.46	55.55	66.76	
	S2	0.7	0.001	128	58.68	0.5447	66.23	50.00	60.39	43.74	58.11	63.17	
	S3	0.7	0.001	64	58.25	0.5468	64.26	51.32	60.31	44.49	57.79	62.22	
	S4	0.5	0.001	256	57.11	0.5092	78.36	32.64	57.25	43.28	55.50	66.16	
	S5	0.7	0.0001	32	57.11	0.5328	70.33	41.89	58.21	44.91	56.11	63.70	
D5	S1	0.7	0.001	128	60.09	0.5834	72.13	46.23	60.69	40.96	59.18	65.92	
	S2	0.3	0.001	64	60.00	0.5683	67.70	51.13	61.46	42.09	59.42	64.43	
	S3	0.5	0.001	32	59.39	0.5648	68.03	49.43	60.76	42.67	58.73	64.19	
	S4	0.5	0.001	32	59.12	0.5572	75.57	40.19	59.25	41.16	57.88	66.43	
	S5	0.3	0.001	128	60.79	0.5825	70.33	49.81	61.73	40.67	60.07	65.75	

In a robustness discussion, we would like to compare our predictive feature with the existing approaches. While the current studies look at the conventional approach of econometrics [29,30], our study sheds light on the deep learning method. Accordingly, we had two samples (training sample and test group). Therefore, it allows us to validate our findings with different periods. The unique, comparable study that we have found in the area of learning models is due to [26]. However, they only show the results for two accuracy metrics when predicting the direction of the US markets. Even so, barely the metrics exceed the value of 0.6, and it is not clear if they are considering a test set.

Table 4. Classification metrics on the test dataset.

Design	Case	Acc	AUC	TPR	TNR	PPV	FOR	BA	F1	#
D1	S1	64.30	0.4981	90.99	11.67	67.01	60.38	51.33	77.18	
	S2	56.82	0.4946	74.08	22.78	65.42	69.17	48.43	69.48	
	S3	61.78	0.5188	79.58	26.67	68.15	**60.17**	**53.12**	73.42	2
	S4	51.96	0.4898	60.70	34.72	64.71	69.06	47.71	62.65	
	S5	60.47	0.4842	83.66	14.72	65.93	68.64	49.19	73.74	
D2	S1	60.75	0.4786	85.92	11.11	65.59	71.43	48.51	74.39	
	S2	52.06	0.4870	56.48	43.33	66.28	66.45	49.91	60.99	
	S3	53.46	0.4997	56.76	46.94	67.85	64.50	51.85	61.81	
	S4	55.70	0.4794	70.56	26.39	65.40	68.75	48.48	67.89	
	S5	50.93	0.4806	55.49	41.94	65.34	67.67	48.72	60.02	
D3	S1	**65.05**	0.5072	**95.21**	5.56	66.54	62.96	50.38	**78.33**	3
	S2	55.70	0.5248	63.38	40.56	67.77	64.04	51.97	65.50	
	S3	57.38	0.5176	67.32	37.78	68.09	63.04	52.55	67.71	
	S4	51.40	0.5051	52.96	48.33	66.90	65.75	50.65	59.12	
	S5	54.21	0.5094	60.56	41.67	67.19	65.12	51.12	63.70	
D4	S1	61.21	0.4831	86.48	11.39	65.81	70.07	48.93	74.74	
	S2	48.13	0.4718	48.17	48.06	64.65	68.02	48.11	55.21	
	S3	47.20	0.4771	43.24	**55.00**	65.46	67.05	49.12	52.08	
	S4	45.98	0.4359	50.99	36.11	61.15	72.80	43.55	55.61	
	S5	51.96	0.4743	55.49	45.00	66.55	66.11	50.25	60.52	
D5	S1	58.04	0.5017	78.59	17.50	65.26	70.70	48.05	71.31	
	S2	55.23	0.4942	62.39	41.11	67.63	64.34	51.75	64.91	
	S3	54.21	0.4994	63.10	36.67	66.27	66.50	49.88	64.65	
	S4	54.49	0.5269	62.39	38.89	66.82	65.60	50.64	64.53	
	S5	55.79	**0.5316**	61.83	43.89	**68.49**	63.17	52.86	64.99	2

5. Discussion

We start from descriptive statistics as a first approach to intuitively grasp the complex nature of Bitcoin, as well as its proposed heterogeneous drivers. As expected, the variables did not satisfy the normality assumption and presented high kurtosis, highlighting the need to use non-parametric and nonlinear analyses.

The KSG estimation of TE found a consistent flow of information from the potential drivers to Bitcoin through the considered range of K nearest neighbours. Even when, in principle, the variance of the estimate decreases with K, the results obtained with $K = 1$ do not change abruptly for larger values. In fact, the variation in the structure of the TE matrix for different Markov orders k, l is more notorious. Additionally, attention must be paid to the evidence about the order $k = l = 1$ through values near zero. Practitioners usually assume this scenario under Gaussian estimations. A precaution must then be made about the memory parameters of Markov, at least when working with the KSG estimation. The associated local TE does not show any particular pattern beyond high volatility, reaching values of four nats when the average is below 0.1. Thus, volatility might be a better proxy for price fluctuations in future studies.

In terms of intuitive explanations, we found that the drivers of Bitcoin might not truly capture its returns in distressed periods. Although we expected to witness that the predictive power of these determinants might play an important role across time horizons, it turns out that the prediction model of Bitcoin relies on a choice of a specific period. Thus, our findings also confirm the momentum effect that exists in this market [47]. Due to the momentum effect, the timing of market booms could not truly be supported much for further analysis by our models. In regard to our main social media hypothesis, the popularity of Bitcoin content still exists as the predictive component in the model. More noticeably, our study highlights that Bitcoin prices can be driven by momentum on social media [24]. However, the selection of training and testing periods should be cautious with the boom and burst of this cryptocurrency. Apparently, while the fundamental value of Bitcoin is still debatable [48], using behavioural determinants could have some merits in predicting Bitcoin. Thus, we believe that media content would support the predictability of Bitcoin prices alongside other financial indicators. Concomitantly, after clustering these factors, we found that the results seem better able to provide insights into Bitcoin's drivers.

On the other hand, the forecasting of Bitcoin's price direction improves in the validation set but not for all metrics in the test dataset when including significant drivers or local TE as a feature. Nonetheless, the last assertion relies on the number of metrics with the best scores. Although the test dataset having the best performance corresponds to the *deep bidirectional LSTM* (D3) for the scenario *univariate* (S3), this case only beat three of the eight metrics. The other five metrics are outperformed by scenarios including *significative features* (S3) and *significative features + local TE* (S5). Furthermore, the second-best performances are tied with two of the eight metrics with leading values. Interestingly, the last case shows the best predictive power on the CNN model using significant features as well as local TE indicators (D5–S5). In particular, it outperforms the AUC and PPV overall, yet AUC is in the border of a random model. To delve into the explainable aspect, a future work will seek to apply the Shapley-Lorentz decomposition proposed in [49,50]. There the authors develop a global methodology, which can be associated with a generalization of AUC-ROC.

Moreover, it is important to note that the selected test period is atypical in the sense of a bull period for Bitcoin as a result of the turbulence generated by the COVID-19 public health emergency; this might induce safe haven behaviour related to this asset and increase its price and capitalization. This atypical behaviour opens the door to propose future work to model Bitcoin by the self-exciting process of the Hawkes model during times of great turbulence.

We would like to end by emphasizing that we were not exhaustive in modelling classification forecasting. In contrast, our intention was to exemplify the effect of including the significant features and local TE indicators under different configurations of a deep learning model through a variety of classification metrics. Two methodological contributions to highlight are the use of nontraditional indicators such as market sentiment, as well as a continuous estimation of the local TE as a tool to determine additional drivers in the classification model. Finally, the models presented here are easily adaptable to high-frequency data because they are non-parametric and nonlinear in nature.

Supplementary Materials: The following are available online at https://www.mdpi.com/1099-4300/23/12/1582/, File S1: preprocessed data.

Author Contributions: Conceptualization, A.G.-M. & T.L.D.H.; Data curation, A.G.-M.; Formal analysis, A.G.-M.; Funding acquisition, A.G.-M.; Investigation, A.G.-M. & T.L.D.H.; Methodology, A.G.-M.; Writing original draf, A.G.-M. & T.L.D.H.; Writing—review & editing, A.G.-M. & T.L.D.H. All authors have read and agreed to the published version of the manuscript.

Funding: This research was funded Consejo Nacional de Ciencia y Tecnología (CONACYT, Mexico) through fund FOSEC SEP-INVESTIGACION BASICA (Grant No. A1-S-43514).

Institutional Review Board Statement: Not applicable.

Informed Consent Statement: Not applicable.

Data Availability Statement: The data is available as Supplementary Materials.

Acknowledgments: We thank Román A. Mendoza for support in the acquisition of the financial time series. To Rebeca Moreno for her kindness in drawing the diagrams of the Supplementary Materials. Also, it is necessary to thank the fruitful discussion with Victor Muñiz. T.L.D.H. acknowledges funding from the University of Economics Ho Chi Minh City (Vietnam) with registered project 2021-08-23-0530.

Conflicts of Interest: The authors declare no conflict of interest. The funders had no role in the design of the study; in the collection, analyses, or interpretation of data; in the writing of the manuscript, or in the decision to publish the results.

References

1. Wątorek, M.; Drożdż, S.; Kwapień, J.; Minati, L.; Oświęcimka, P.; Stanuszek, M. Multiscale characteristics of the emerging global cryptocurrency market. *Phys. Rep.* **2020**, *901*, 1–82. [CrossRef]
2. Urquhart, A. What causes the attention of Bitcoin? *Econ. Lett.* **2018**, *166*, 40–44. [CrossRef]
3. Shen, D.; Urquhart, A.; Wang, P. Does twitter predict Bitcoin? *Econ. Lett.* **2019**, *174*, 118–122. [CrossRef]
4. Burggraf, T.; Huynh, T.L.D.; Rudolf, M.; Wang, M. Do FEARS drive Bitcoin? *Rev. Behav. Financ.* **2020**, *13*, 229–258. [CrossRef]
5. Huynh, T.L.D. Does Bitcoin React to Trump's Tweets? *J. Behav. Exp. Financ.* **2021**, *31*, 100546. [CrossRef]
6. Corbet, S.; Larkin, C.; Lucey, B.M.; Meegan, A.; Yarovaya, L. The impact of macroeconomic news on Bitcoin returns. *Eur. J. Financ.* **2020**, *26*, 1396–1416. [CrossRef]
7. Cavalli, S.; Amoretti, M. CNN-based multivariate data analysis for bitcoin trend prediction. *Appl. Soft Comput.* **2021**, *101*, 107065. [CrossRef]
8. Gronwald, M. Is Bitcoin a Commodity? On price jumps, demand shocks, and certainty of supply. *J. Int. Money Financ.* **2019**, *97*, 86–92. [CrossRef]
9. Baker, S.R.; Bloom, N.; Davis, S.J. Measuring economic policy uncertainty. *Q. J. Econ.* **2016**, *131*, 1593–1636. [CrossRef]
10. Bekaert, G.; Engstrom, E.C.; Xu, N.R. The Time Variation in Risk Appetite and Uncertainty. Working Paper 25673, National Bureau of Economic Research, 2019. Available online http://www.nber.org/papers/w25673 (accessed on 15 January 2021).
11. Das, S.; Demirer, R.; Gupta, R.; Mangisa, S. The effect of global crises on stock market correlations: Evidence from scalar regressions via functional data analysis. *Struct. Chang. Econ. Dyn.* **2019**, *50*, 132–147. [CrossRef]
12. Lahiani, A.; Jlassi, N.B.; Jeribi, A. Nonlinear tail dependence in cryptocurrency-stock market returns: The role of Bitcoin futures. *Res. Int. Bus. Financ.* **2021**, *56*, 101351. [CrossRef]
13. Luu Duc Huynh, T. Spillover risks on cryptocurrency markets: A look from VAR-SVAR granger causality and student'st copulas. *J. Risk Financ. Manag.* **2019**, *12*, 52. [CrossRef]
14. Huynh, T.L.D.; Nguyen, S.P.; Duong, D. Contagion risk measured by return among cryptocurrencies. In Proceedings of the International Econometric Conference of Vietnam, Ho Chi Minh, Vietnam, 15–16 January 2018; pp. 987–998.
15. Huynh, T.L.D.; Nasir, M.A.; Vo, X.V.; Nguyen, T.T. "Small things matter most": The spillover effects in the cryptocurrency market and gold as a silver bullet. *N. Am. J. Econ. Financ.* **2020**, *54*, 101277. [CrossRef]
16. Thampanya, N.; Nasir, M.A.; Huynh, T.L.D. Asymmetric correlation and hedging effectiveness of gold & cryptocurrencies: From pre-industrial to the 4th industrial revolution. *Technol. Forecast. Soc. Chang.* **2020**, *159*, 120195.
17. Huynh, T.L.D.; Burggraf, T.; Wang, M. Gold, platinum, and expected Bitcoin returns. *J. Multinatl. Financ. Manag.* **2020**, *56*, 100628. [CrossRef]
18. Huynh, T.L.D.; Shahbaz, M.; Nasir, M.A.; Ullah, S. Financial modelling, risk management of energy instruments and the role of cryptocurrencies. *Ann. Oper. Res.* **2020**, 1–29. [CrossRef]
19. Huynh, T.L.D.; Ahmed, R.; Nasir, M.A.; Shahbaz, M.; Huynh, N.Q.A. The nexus between black and digital gold: Evidence from US markets. *Ann. Oper. Res.* **2021**, 1–26. [CrossRef]
20. Chu, J.; Nadarajah, S.; Chan, S. Statistical analysis of the exchange rate of bitcoin. *PLoS ONE* **2015**, *10*, e0133678.
21. Lahmiri, S.; Bekiros, S.; Bezzina, F. Multi-fluctuation nonlinear patterns of European financial markets based on adaptive filtering with application to family business, green, Islamic, common stocks, and comparison with Bitcoin, NASDAQ, and VIX. *Phys. A Stat. Mech. Its Appl.* **2020**, *538*, 122858. [CrossRef]
22. Ante, L. How Elon Musk's Twitter Activity Moves Cryptocurrency Markets. 2021. SSRN 3778844. Available online: https://papers.ssrn.com/sol3/papers.cfm?abstract_id=3778844 (accessed on 15 June 2021).
23. Schreiber, T. Measuring information transfer. *Phys. Rev. Lett.* **2000**, *85*, 461–464. [CrossRef] [PubMed]
24. Philippas, D.; Rjiba, H.; Guesmi, K.; Goutte, S. Media attention and Bitcoin prices. *Financ. Res. Lett.* **2019**, *30*, 37–43. [CrossRef]
25. Naeem, M.A.; Mbarki, I.; Suleman, M.T.; Vo, X.V.; Shahzad, S.J.H. Does Twitter Happiness Sentiment predict cryptocurrency? *Int. Rev. Financ.* **2020**. [CrossRef]
26. Kim, S.; Ku, S.; Chang, W.; Song, J.W. Predicting the Direction of US Stock Prices Using Effective Transfer Entropy and Machine Learning Techniques. *IEEE Access* **2020**, *8*, 111660–111682. [CrossRef]
27. Gu, S.; Kelly, B.; Xiu, D. Empirical asset pricing via machine learning. *Rev. Financ. Stud.* **2020**, *33*, 2223–2273. [CrossRef]

28. Kristoufek, L. What are the main drivers of the Bitcoin price? Evidence from wavelet coherence analysis. *PLoS ONE* **2015**, *10*, e0123923. [CrossRef] [PubMed]
29. Giudici, P.; Polinesi, G. Crypto price discovery through correlation networks. *Ann. Oper. Res.* **2021**, *299*, 443–457. [CrossRef]
30. Giudici, P.; Pagnottoni, P. Vector error correction models to measure connectedness of Bitcoin exchange markets. *Appl. Stoch. Models Bus. Ind.* **2020**, *36*, 95–109. [CrossRef]
31. Ribeiro, M.T.; Singh, S.; Guestrin, C. "Why should i trust you?" Explaining the predictions of any classifier. In Proceedings of the 22nd ACM SIGKDD International Conference on Knowledge Discovery and Data Mining, San Francisco, CA, USA, 13–17 August 2016; pp. 1135–1144.
32. Lundberg, S.M.; Lee, S.I. A unified approach to interpreting model predictions. In Proceedings of the 31st International Conference on Neural Information Processing Systems, Long Beach, CA, USA, 4–9 December 2017; pp. 4768–4777.
33. Schlegel, U.; Arnout, H.; El-Assady, M.; Oelke, D.; Keim, D.A. Towards a rigorous evaluation of xai methods on time series. In Proceedings of the 2019 IEEE/CVF International Conference on Computer Vision Workshop (ICCVW), Seoul, Korea, 27–28 October 2019; pp. 4197–4201.
34. Bussmann, N.; Giudici, P.; Marinelli, D.; Papenbrock, J. Explainable AI in fintech risk management. *Front. Artif. Intell.* **2020**, *3*, 26. [CrossRef] [PubMed]
35. Bussmann, N.; Giudici, P.; Marinelli, D.; Papenbrock, J. Explainable machine learning in credit risk management. *Comput. Econ.* **2021**, *57*, 203–216. [CrossRef]
36. Bracke, P.; Datta, A.; Jung, C.; Sen, S. Machine Learning Explainability In Finance: An Application To Default Risk Analysis. 2019. Available online: https://www.bankofengland.co.uk/working-paper/2019/machine-learning-explainability-in-finance-an-application-to-default-risk-analysis (accessed on 15 November 2021).
37. Ohana, J.J.; Ohana, S.; Benhamou, E.; Saltiel, D.; Guez, B. Explainable AI (XAI) models applied to the multi-agent environment of financial markets. In Proceedings of the International Workshop on Explainable, Transparent Autonomous Agents and Multi-Agent Systems, Virtual Event, 3–7 May 2021; Springer: Berlin/Heidelberg, Germany, 2021; pp. 189–207.
38. Hailemariam, Y.; Yazdinejad, A.; Parizi, R.M.; Srivastava, G.; Dehghantanha, A. An Empirical Evaluation of AI Deep Explainable Tools. In Proceedings of the 2020 IEEE Globecom Workshops GC Wkshps, Madrid, Spain, 7–11 December 2020; pp. 1–6.
39. Lizier, J.T.; Prokopenko, M.; Zomaya, A.Y. Local information transfer as a spatiotemporal filter for complex systems. *Phys. Rev. E* **2008**, *77*, 026110. [CrossRef]
40. Bossomaier, T.; Barnett, L.; Harré, M.; Lizier, J.T. *An introduction to Transfer Entropy*; Springer: Berlin/Heidelberg, Germany, 2016; pp. 78–82.
41. Kraskov, A.; Stögbauer, H.; Grassberger, P. Estimating mutual information. *Phys. Rev. E* **2004**, *69*, 066138. [CrossRef] [PubMed]
42. Gao, S.; Ver Steeg, G.; Galstyan, A. Efficient estimation of mutual information for strongly dependent variables. *arXiv* **2015**, arXiv:1411.2003.
43. Hochreiter, S.; Schmidhuber, J. Long Short-Term Memory. *Neural Comput.* **1997**, *9*, 1735–1780. [CrossRef] [PubMed]
44. Brownlee, J. *Deep Learning for Time Series Forecasting: Predict the Future with MLPs, CNNs and LSTMs in Python*; Machine Learning Mastery: 2018. Available online: https://machinelearningmastery.com/deep-learning-for-time-series-forecasting/ (accessed on 15 November 2021).
45. Hutto, C.; Gilbert, E. Vader: A parsimonious rule-based model for sentiment analysis of social media text. *Proc. Int. AAAI Conf. Web Soc. Media* **2014**, *8*, 216–225.
46. Reddi, S.J.; Kale, S.; Kumar, S. On the convergence of adam and beyond. *arXiv* **2018**, arXiv:1904.09237.
47. Grobys, K.; Sapkota, N. Cryptocurrencies and momentum. *Econ. Lett.* **2019**, *180*, 6–10. [CrossRef]
48. Chaim, P.; Laurini, M.P. Is Bitcoin a bubble? *Phys. A Stat. Mech. Its Appl.* **2019**, *517*, 222–232. [CrossRef]
49. Giudici, P.; Raffinetti, E. Shapley-Lorenz eXplainable artificial intelligence. *Expert Syst. Appl.* **2021**, *167*, 114104. [CrossRef]
50. Giudici, P.; Raffinetti, E. Lorenz model selection. *J. Classif.* **2020**, *37*, 754–768. [CrossRef]

entropy

MDPI

Article

Cryptocurrency Market Consolidation in 2020–2021

Jarosław Kwapień [1,*], Marcin Wątorek [2] and Stanisław Drożdż [1,2]

[1] Complex Systems Theory Department, Institute of Nuclear Physics, Polish Academy of Sciences, ul. Radzikowskiego 152, 31-342 Kraków, Poland; Stanislaw.Drozdz@ifj.edu.pl
[2] Faculty of Computer Science and Telecommunications, Cracow University of Technology, ul. Warszawska 24, 31-155 Kraków, Poland; marcin.watorek@pk.edu.pl
* Correspondence: jaroslaw.kwapien@ifj.edu.pl

Abstract: Time series of price returns for 80 of the most liquid cryptocurrencies listed on Binance are investigated for the presence of detrended cross-correlations. A spectral analysis of the detrended correlation matrix and a topological analysis of the minimal spanning trees calculated based on this matrix are applied for different positions of a moving window. The cryptocurrencies become more strongly cross-correlated among themselves than they used to be before. The average cross-correlations increase with time on a specific time scale in a way that resembles the Epps effect amplification when going from past to present. The minimal spanning trees also change their topology and, for the short time scales, they become more centralized with increasing maximum node degrees, while for the long time scales they become more distributed, but also more correlated at the same time. Apart from the inter-market dependencies, the detrended cross-correlations between the cryptocurrency market and some traditional markets, like the stock markets, commodity markets, and Forex, are also analyzed. The cryptocurrency market shows higher levels of cross-correlations with the other markets during the same turbulent periods, in which it is strongly cross-correlated itself.

Keywords: financial markets; cryptocurrencies; multiscale analysis; detrended cross-correlations; minimal spanning tree; COVID-19

Citation: Kwapień J.; Wątorek, M.; Drożdż, S. Cryptocurrency Market Consolidation in 2020–2021. *Entropy* **2021**, *23*, 1674. https://doi.org/10.3390/e23121674

Academic Editors: Ryszard Kutner, H. Eugene Stanley and Christophe Schinckus

Received: 18 November 2021
Accepted: 9 December 2021
Published: 13 December 2021

1. Introduction

Over the past few years, two processes have had a particularly strong impact on financial markets: the emergence of the cryptocurrency market [1–5] and the COVID-19 pandemic [6–12]. Each of these processes alone has already been a topic in numerous pieces of the scientific literature, but they also were studied together [5,13–21]. Of particular interest in this context is how the ongoing pandemic is changing the cryptocurrency market and how this market position among the other financial and commodity markets undergoes an accelerated evolution. The cryptocurrency market is an interesting object for analysis from the perspective of complex systems, as it is a unique financial market whose establishment and evolution was entirely spontaneous with no intervening government or other regulatory institution. Thus, a process of the market's self-organization can be traced from the very beginning until the present.

As the cryptocurrency market properties are constantly evolving and they are still far from being fully identified and understood, there is heavy ongoing related research that points in various directions (see, for example, [4] for comprehensive literature listing and pointing out several significant research voids). On the general level, the cryptocurrency markets are studied at an angle of trading security, the vulnerability to improper trading practices [22], and the formation of demand [23]. On the asset level, the fundamental aspects of the market processes that drive price discovery [24,25], price fluctuations [26–28], asset liquidity [29], and asset–asset correlations [30,31] are studied from the investor's perspective in order to facilitate the optimal portfolio construction both inside the cryptocurrency market and across different markets, including the cryptocurrency one. An associated important direction of research is the possibility of market forecasting, which includes the approach developed in econophysics that is based on a search for evidence of the exogeneous and endogeneous market shocks, speculative bubbles, crashes, and their precursors [32]. Among the voids, one can count the sparse analyses based on high-frequency data, the exaggerated focus on bitcoin (BTC) alone, and the insufficient attention paid to how different mining protocols can affect the related asset properties and how various legal regulations being (actually or potentially) imposed on the cryptocurrency markets can perturb both the mining and the trade [4].

From a perspective of their statistical and dynamical properties, the cryptocurrencies neither resemble regular currencies, like the US dollar (USD) or Chinese yuan (CNH), nor commodities, like gold or oil [33–35]. Among the major problems associated with cryptocurrencies is their significant volatility. In consequence, even the largest and the most capitalized cryptocurrency, BTC, is considered an asset that resides at the interface between a standard financial asset and a speculative one [36]. Most studies of the cryptocurrency market relations with the traditional markets reported in the literature point to relative independence of the cryptocurrencies (see, for example, [13,31,37]). However, there were also some reports concluding that there are temporary or stable cross-correlations or even causality between the major cryptocurrencies and some regular currencies, like TRY [34] and some Asian currencies, like BHT, CNH, and TWD [38], as well as between the cryptocurrencies and commodities [37].

As a new system, it took several years for the cryptocurrency market to reveal any signatures of maturity, like the market efficiency [39,40]. However, already prior to the crash of April 2018, its statistical properties became similar to the properties of the other markets, among which there were the financial stylized facts (the power-law tails of the return distributions, volatility clustering, etc.) [5,28,40] and some other complexity traits, like multifractality [26], and, in some aspects, it started to resemble Forex [26,41]. On the other hand, one of the interesting facts about the cryptocurrency market's inner structure is that, unlike other financial markets where, typically, the highly capitalized assets have spillover effects on the less capitalized ones, here the less capitalized assets are able to influence the evolution of the highly capitalized ones. This can lead to more a complex structure than a typical structure of the other markets, where causality is unidirectional [42–45].

These and other similarities and differences opened space for a concern, whether bitcoin and other cryptocurrencies may be considered as a safe haven during market turmoils or whether they may be used to hedge against the traditional assets. Although the literature on this issue is growing, the conclusions are mixed: BTC and the other major cryptocurrencies are sometimes indicated as good candidates for a safe haven [15,16,18,46,47] but the opposite can also be suggested [15,17,36,48–52], depending on the analyzed data. Sometimes the answer can even be conditional: "yes" to a safe haven, "no" to a hedge [53]. An important risk factor of BTC and other cryptocurrencies that acts against their use for hedging is their possible lack of fundamental value [54].

As regards the asset–asset correlations among the cryptocurrencies, it was shown that, besides a trend going towards the stronger market cross-correlations, the cryptocurrencies reveal a cyclic amplification of volatility connectedness during periods of economic instability or external shocks. However, BTC does not play a central role in driving market volatility [42]. A different study applying different methodologies (principal component analysis, cross-sectional dependence, and vector autoregression framework) confirmed this finding and extended it from volatility to returns [45]. Another work reported the increased cross-correlations among the cryptocurrencies after the bubble of 2017 as compared to the earlier period by using the detrended fluctuation analysis [44]. The highly capitalized cryptocurrencies show statistically significant time-lagged autocorrelations that may indicate substantial market inefficiency (although not necessarily usable for profit-making) [35]. All these works analyzed very small sets of assets, however, which significantly limited the market insight they were able to provide. A more comprehensive study, which considered over 50 cryptocurrencies, also brought more diversified results, and identified some assets that were statistically and dynamically different than the others (these were tether, holo, maker, NEM, and nexo) [20].

In our former publications we thoroughly analyzed the cryptocurrency market evolution from its early stages of development to the current, relatively mature phase. In the Ref. [55] we reported that the cryptocurrency dynamics over the years 2016–2019 displayed signatures of decoupling from dynamics of the regular currencies [55]. In the Refs. [5,13] we analyzed the cryptocurrency market properties during the pandemic onset (January 2019–October 2020). We showed that before the pandemic, over the years 2018–2019, the evolution of the cryptocurrency market was largely independent from the evolution of the traditional markets. We interpreted this independence as a consequence of a quiet period on the traditional markets and a disparity in the market capitalization: the cryptocurrency market was too small to perturb other markets, while they were too tranquil then to induce any turmoil among the cryptocurrencies. However, in the second half of January 2020, at the moment when the first COVID-19 case was reported in the United States, some cryptocurrencies responded and thus lost their independence. For example, BTC gained positive cross-correlation with JPY, CHF, and gold, which are considered as a financial safe haven, and negative cross-correlation with other major assets, while ETH preserved its independent dynamics longer. Later, during the outburst of the first wave of COVID-19 in April 2020, the cryptocurrencies underwent a crash together with all the major markets, except for a few regular currencies like JPY. This state of cross-market coupling continued in the months that followed, both at the moments of the subsequent pandemic waves and the market rallies. Our analyses ended in the middle of the third pandemic wave before the introduction of anti-COVID vaccines, thus we could not report on how the markets would respond to a decreased pandemic risk. From this angle, our present analysis can be viewed inter alia as a continuation of those previous works based on a new data set.

In the following, we will report on our study of a set of the most liquid cryptocurrencies whose high-frequency price quotes cover the last 21 months. We will apply the generalized detrended cross-correlation analysis [56–59] and study the spectral properties of a detrended correlation matrix, as well as the topological properties of its network representation. In the context of the current cryptocurrency research, our main objectives are (1) to look into the most recent data that have not been covered by other works yet, and compare results with the earlier ones, (2) to consider a set of assets that is wide as possible provided the available data quality, and (3) to apply a methodology that is rarely used in this context, that is, the q-dependent cross-correlation analysis that is able to filter data according to its magnitude. In Section 2 we will briefly recollect the related formalism, in Section 3 we present and discuss the main results, and in Section 4 we will present the summary and conclusions.

2. Methods

Data from the cryptocurrency market, which is characterized by volatility that exceeds volatility of the traditional markets, are not well-suited to being studied by means of the standard correlation formalism based on the Pearson correlation [60] that requires data stationarity. Thus, methods based on signal detrending are advised [56,61].

The q-dependent detrended correlation coefficient $\rho_q(s)$ was proposed in the Ref. [59] to quantify the detrended cross-correlations between two, typically non-stationary time series $\{x(i)\}_{i=1,\ldots,T}$ and $\{y(i)\}_{i=1,\ldots,T}$ of length T. Let these time series be divided into M_s boxes of length s starting from its opposite ends (thus, there are $2M_s$ boxes total). In each box, the data points are subject to integration and polynomial trend removal:

$$X_\nu(s,i) = \sum_{j=1}^{i} x(\nu s + j) - P^{(m)}_{X,s,\nu}(i), \qquad Y_\nu(s,i) = \sum_{j=1}^{i} x(\nu s + j) - P^{(m)}_{Y,s,\nu}(i), \tag{1}$$

where the polynomials $P^{(m)}$ of order m are applied. The next step is calculation of the local residual variances and covariance:

$$f_{XX}^2(s,\nu) = \sum_{i=1}^{s}(X_\nu(s,i) - \tilde{X}_\nu(s))^2, \qquad f_{YY}^2(s,\nu) = \sum_{i=1}^{s}(Y_\nu(s,i) - \tilde{Y}_\nu(s))^2, \tag{2}$$

$$f_{XY}^2(s,\nu) = \sum_{i=1}^{s}(X_\nu(s,i) - \tilde{X}_\nu(s))(Y_\nu(s,i) - \tilde{Y}(s)), \tag{3}$$

where $\tilde{X}$ and $\tilde{Y}$ denote the local mean of X and Y, respectively. These quantities are used to define a family of the fluctuation functions of order q:

$$F_{XX}^{(q)}(s) = \frac{1}{2M_s}\sum_{\nu=0}^{2M_s-1}\left[f_{XX}^2(s,\nu)\right]^{q/2}, \qquad F_{YY}^{(q)}(s) = \frac{1}{2M_s}\sum_{\nu=0}^{2M_s-1}\left[f_{YY}^2(s,\nu)\right]^{q/2}, \tag{4}$$

$$F_{XY}^{(q)}(s) = \frac{1}{2M_s}\sum_{\nu=0}^{2M_s-1}\mathrm{sign}\left[f_{XY}^2(s,\nu)\right]|f_{XY}^2(s,\nu)|^{q/2}. \tag{5}$$

The sign function in Equation (5) preserves the information that is otherwise lost after taking the modulus of $f_{XY}^2(s,\nu)$, while the modulus itself excludes a possibility of obtaining complex values of the covariance f_{XY}^2 raised to a real power $q/2$ [59,62]. The q-dependent detrended correlation coefficient is defined by the following formula:

$$\rho_q^{XY}(s) = \frac{F_{XY}^{(q)}(s)}{\sqrt{F_{XX}^{(q)}(s)F_{YY}^{(q)}(s)}}, \tag{6}$$

which generalizes for any q the standard ($q = 2$) detrended correlation coefficient ρ_{DCCA} [58]. The parameter q plays the role of a filter weighting the boxes ν in the sums in Equations (4) and (5) by their variance/covariance magnitudes. For $q > 2$, the boxes with large signal fluctuations are given higher weights with respect to the $q = 2$ case, while for $q < 2$ the boxes with small fluctuations contribute more than for $q = 2$. Therefore, by applying ρ_q, one can learn which fluctuations are the source of the observed detrended correlation of the time series.

For a set of N parallel time series indexed by i, the q-dependent correlation coefficient can be calculated for each time series pair (i,j) $(i,j = 1,...,N)$, and a q-dependent detrended correlation matrix $\mathbf{C}_q(s)$ with the entries $\rho_q^{(i,j)}(s)$ can be created, as well as a q-dependent metric distance matrix $\mathbf{D}_q(s)$ whose entries are

$$d_q^{(i,j)}(s) = \sqrt{2(1 - \rho_q^{(i,j)}(s))}. \tag{7}$$

The matrix $\mathbf{D}_q(s)$ can then be used to create a weighted graph, where nodes labelled by $i = 1,...,N$ represent the time series and $N(N-1)/2$ edges connecting the nodes i,j are attributed the weights equal to $d_q^{(i,j)}(s)$. A subset of the complete graph, consisting of all N nodes and only $N-1$ edges that minimize the weight sum, is a q-dependent detrended minimum spanning tree (qMST) [63]. This tree can be constructed by means of the Prim algorithm, for instance [64]. However, although the very algorithm is the same, such a tree differs from the standard approach that uses the Pearson correlation coefficient and a corresponding Pearson correlation matrix (see, for example, [55,65] for such a standard approach applied to the cryptocurrency market).

A data set of interest is the 1 min price quotations of the 80 cryptocurrencies that were among the most actively traded ones on the Binance platform [66] over the period from 1 January 2020 to 1 October 2021. The quotes are expressed in USD Tether (USDT) that is a stablecoin linked to the US dollar and its value is \$1.00 by design [67]. Each time series of the price quotations is 921,600 points long and covers 640 trading days (the Binance platform is active 24 hours a day and 7 days a week). All the assets used in this study are listed in Appendix A (Table A1).

3. Results and Discussion

The price quotation time series $p_i(t_m)$, where $m = 1, ..., T$ and i stands for a given cryptocurrency ticker, were first transformed to the time series of logarithmic returns $R_X(t_m) = \ln p_i(t_{m+1}) - \ln p_i(t_m)$ and then normalized to zero mean and unit variance, which is a standard procedure. Then, for each pair of cryptocurrencies (i, j), the q-dependent detrended cross-correlation coefficient $\rho_q^{(i,j)}(s)$ given by Equation (6) was determined for a number of time scales s from $s = 10$ min to $s = 360$ min and different values of the filtering parameter q. In what follows, we will present results obtained for $q = 1$, which corresponds to a situation where the small fluctuation period variances in Equations (4) and (5) are amplified relatively to the large ones, and for $q = 4$, which corresponds to the opposite situation. Thus, we can consider the asset cross-correlations for the quiet and turbulent periods in a separate manner.

Before we start a presentation of our results, in Figure 1 we show the historical data of the BTC price in USD in the years 2020–2021 together with the BTC share in the total cryptocurrency market capitalization over the same period. Among the most characteristic events for BTC was the crash on 13 March 2020 related to the COVID-19 pandemic onset in the United States, when BTC surged below 4107 USD, a long rally that started in October 2020 and ended on 14 April 2021 with then the all-time-high equal to 64,830 USD, a subsequent drop-down phase that ended on 20 July 2021 at 29,324 USD, and the next all-time-high on 20 October 2021 equal to 66,961 USD. As the BTC has been priced higher and higher, its share in the total market capitalization drops down steadily from about 70% in January 2020 to below 45% in October 2021, which seems to be inevitable if the number of the actively traded cryptocurrencies grows quickly.

Figure 1. Price evolution of bitcoin (BTC) expressed in US dollars (black) and the BTC share in the total cryptocurrency market capitalization (magenta) over the period from 1 January 2020 to 31 October 2021. Characteristic events are indicated by vertical dashed lines and Roman numerals: COVID-19 crash in March 2020 (event I), strong bull market on cryptocurrency valuation October 2020–April 2021 (event II), all-time high on 14 April 2021 (event III), the May crash on the cryptocurrency market (event IV), and recent rally with new all-time high on 20 October 2021 (event V).

Since for $N = 80$ cryptocurrencies there are $\mathcal{N} = N(N-1)/2 = 3160$ cryptocurrency pairs that have to be considered, it is convenient to analyze the whole set collectively by means of the spectral analysis of the $N \times N$ q-dependent detrended correlation matrix $\mathbf{C}_q(s)$, whose entries are the coefficients $\rho_q^{(i,j)}(s)$. We can diagonalize it and calculate its eigenvalues λ_i and eigenvectors $\mathbf{v}_i$ (with $i = 1, ..., N$):

$$\mathbf{C}_q(s)\mathbf{v}_i^{(q)}(s) = \lambda_i^{(q)}(s)\mathbf{v}_i^{(q)}(s). \tag{8}$$

The eigenvalues are ordered typically from the largest one ($i = 1$) to the smallest one ($i = N$). (For simplicity, from now on we will omit the parameters q and s when dealing with the eigenvalues and eigenvectors of $\mathbf{C}_q(s)$. Their value will be known from the context.)

For the financial markets, a typical eigenvalue spectrum of the Pearson-coefficient-based correlation matrix consists of a large λ_1 that is separated from the remaining eigenvalues by a considerable gap and corresponds to the average behaviour of the considered assets (the so-called market factor), a few elevated non-random eigenvalues that correspond to subsets of related assets (e.g., representing companies from the same industry or currencies from the same geographical region), and a bulk of mean eigenvalues that correspond to random fluctuations and, essentially, carry no genuine information. Here we use the detrended correlation coefficient ρ_q instead of the Pearson coefficient [60], but our experience shows that the corresponding matrix $\mathbf{C}_q$ reveals similar spectral properties [63]. The largest eigenvalue λ_1 is associated with a maximally delocalized eigenvector $\mathbf{v}_1$ with many significant components, while the eigenvectors representing smaller eigenvalues are more localized, that is, few components are significant. The eigenvector structure is usually expressed by the inverse participation ratio or the localization length [68], but here we apply the Shannon entropy defined by

$$H(\mathbf{v}_i) = -\sum_{j=1}^{N} p_i(j) \ln p_i(j), \tag{9}$$

with $p_i(j) = v_i^2(j)$ (the eigenvectors are normalized to unit length, so that $\sum_{j=1}^{N} v_i^2(j) = 1$). If the eigenvector is maximally delocalized and all its components are equal to each other, the Shannon entropy assumes its maximum value: $H(\mathbf{v}_i) = \ln N$, while if there is only a single non-zero component, the entropy vanishes: $H(\mathbf{v}_i) = 0$. Entropy can thus serve as a measure of vector localization.

In order to track the evolution of the asset–asset detrended cross-correlations, we apply a moving window of size of 7 days (10,080 data points), which was shifted by a daily step (1440 data points) along the time series. For each window position t, based on the 80 time series of price returns, we create a detrended correlation matrix $\mathbf{C}_q(s,t)$ for a few selected values of q ($q = 1$ and $q = 4$) and s ($s = 10$ min, $s = 60$ min, $s = 180$ min, and $s = 360$ min). Next we diagonalize $C_q(s,t)$ and derive a complete set of the eigenvalues $\lambda_i(t)$ and eigenvectors $\mathbf{v}_i(t)$. Figure 2 exhibits $\lambda_1(t)$, $H(\mathbf{v}_1(t))$, and the largest squared component $v_1^{(\max)}(t)$ of the eigenvector $\mathbf{v}_1(t)$ for different time scales s and different values of the filtering parameter q. By increasing s, we also obtain a systematically increasing $\lambda_1(t)$, which reflects the increasing strength of the mean asset–asset detrended cross-correlations for the longer time scales s. This is a well-known property of the financial and commodity markets and it is called the Epps effect [41,69–71]. This effect has already been observed on the cryptocurrency market and reported, for example, in the Ref. [5]. It is a consequence of the fact that what dominates the price evolution on short time scales is noise: it takes time to spread a piece of information among the assets, especially if the asset liquidity is small like in the case of the cryptocurrencies. Therefore, only on the sufficiently long time scales, the cross-correlations are able to be built up to a full extent.

Another observation is that the difference in correlation strength between $s = 10$ min and $s = 360$ min is much stronger for $q = 1$ than for $q = 4$; the correlation strength for large scales is also significant then. The behavior of λ_1 is also different: in the case of $q = 1$, periods with a large value of λ_1 are accompanied by periods of moderate value, but there are also few periods with relatively small values of the largest eigenvalue. In turn, for $q = 4$ the $\lambda_1(t)$ evolution consists of large, but short "bursts" separated by small background values. In the latter case, λ_1 is more sensitive to changes. Looking at the $\lambda_1(t)$ chart for $q = 1$ and the shorter s time scales, two characteristic epochs can be distinguished: (1) more or less until October 2020, we observe a horizontal trend, where the average value of λ_1 does not change much, and (2) from October 2020 to mid-2021, a strong upward trend is noticeable. This is confirmed by looking at the Shannon entropy panel, where the behavior of this quantity is very similar. This means that from the fall of 2020 to mid-2021, there was a gradual increase in the strength of the market correlation and more cryptocurrencies began to behave in a similar way. It can be said that the market has consolidated. In the third quarter of 2021, this trend was halted, λ_1 began to decrease slightly, and $H(\mathbf{v}_1)$ was saturated close to its maximum allowed value of approximately 4.38. Understandably, as the delocalization of the vector $\mathbf{v}_1$ increases, the value of its largest component $v_1^{(\max)}$ decreases (see Figure 2).

Figure 3 shows the changes over time of the second largest eigenvalue λ_2, the entropy of the components of the corresponding eigenvector $\mathbf{v}_2$ and the changes in the value of the largest component $v_2^{(\max)}$ of this vector. For both $q = 1$ and $q = 4$, the value of λ_2 is much lower than the value of λ_1, which results from a much smaller number of significant eigenvector components: entropy is lower than 4, and for $q = 1$, in the vast majority of windows, its value decreases as s increases, which is the opposite of the λ_1 case. For $q = 4$, we do not observe such an effect. With $q = 1$, the global maximum of λ_2 falls in July 2020, when its value more than doubled if compared to other time intervals. Simultaneously, λ_1 reached one of its lowest values, as did $H(\mathbf{v}_1)$. At the same time, the entropy for $\mathbf{v}_2$ did not change much from its typical value, but then and in the preceding period $H(\mathbf{v}_2$ was similar for different time scales. For large fluctuations ($q = 4$), the maximum λ_2 also occurred in the same period, but was not as unique as for the smaller fluctuations ($q = 1$), because λ_2 reached equally high magnitude in April and May 2021. However, λ_1 for $q = 4$ also had its maxima at the same moments. This means that briefly in July 2020, there was a strong correlation of a small group of cryptocurrencies, and this mainly concerned small and medium fluctuations in their price, while the market as a whole was in a decoupling stage. In turn, in April and May 2021 there was a stronger than usual correlation of the entire market, with large fluctuations being particularly strongly correlated. As for the largest component of the vector $\mathbf{v}_2$ and $q = 1$, we do not observe systematic changes in its value for the short time scales, while for the long ones, starting from autumn 2020, there is a growing trend that ends in mid-2021. This increase in $v_2^{(\max)}$ suggests that one of the cryptocurrencies increased its dominance over other cryptocurrencies at that time. This behavior differs from the behavior of the analogous measure described above in the case of the vector $\mathbf{v}_1$, where there was a clear decrease.

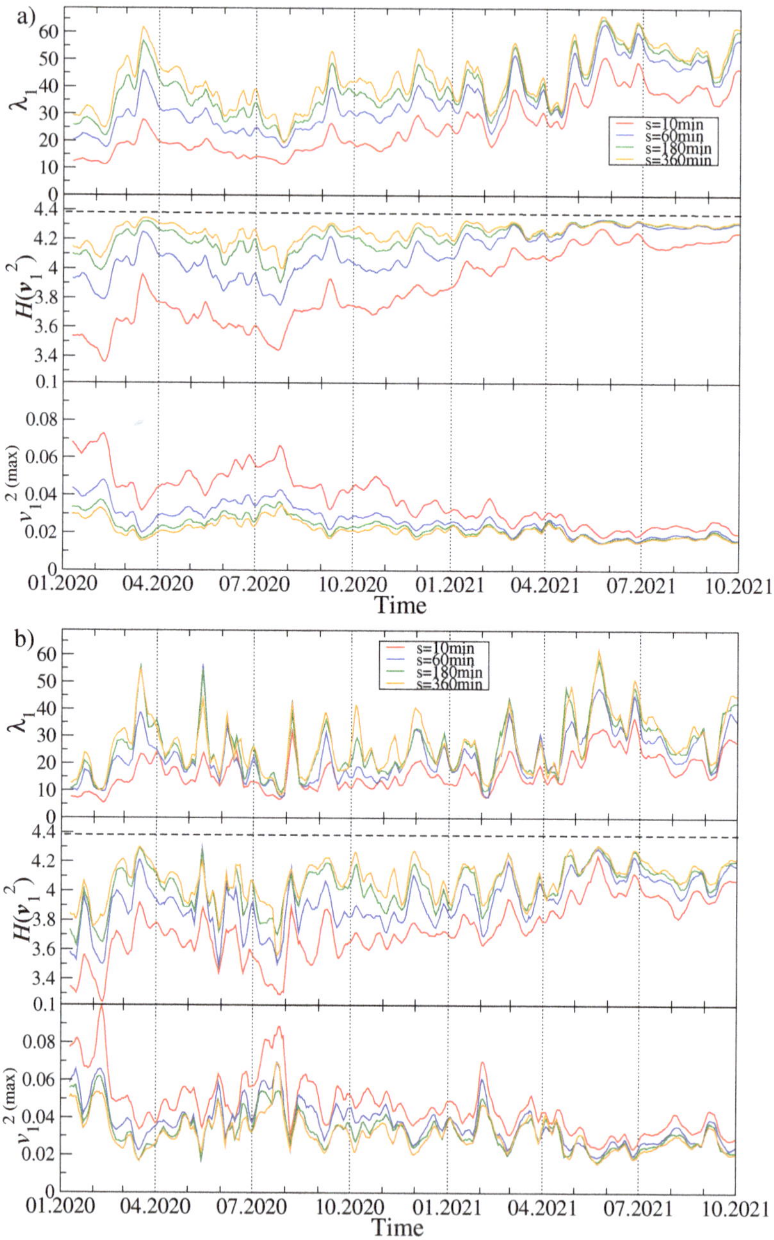

Figure 2. Time evolution of the selected spectral characteristics of the q-dependent detrended correlation matrix $\mathbf{C}_q(s)$ for $q = 1$ (**a**) and $q = 4$ (**b**). A moving window of a length of 7 days shifted by 1 day was applied for sample values of the scale: $s = 10$ min (red), $s = 60$ min (blue), $s = 180$ min (green), and $s = 360$ min (orange). The largest eigenvalue λ_1 (top panels in (**a**,**b**)), the Shannon entropy $H(\mathbf{v}_1)$ of the squared eigenvector components $v_1(j)$ with $j = 1, ..., N$ (middle panels), and the squared maximum component of the eigenvector $\mathbf{v}_1$ associated with λ_1 (bottom panels) are shown. The cryptocurrency prices are expressed in USDT.

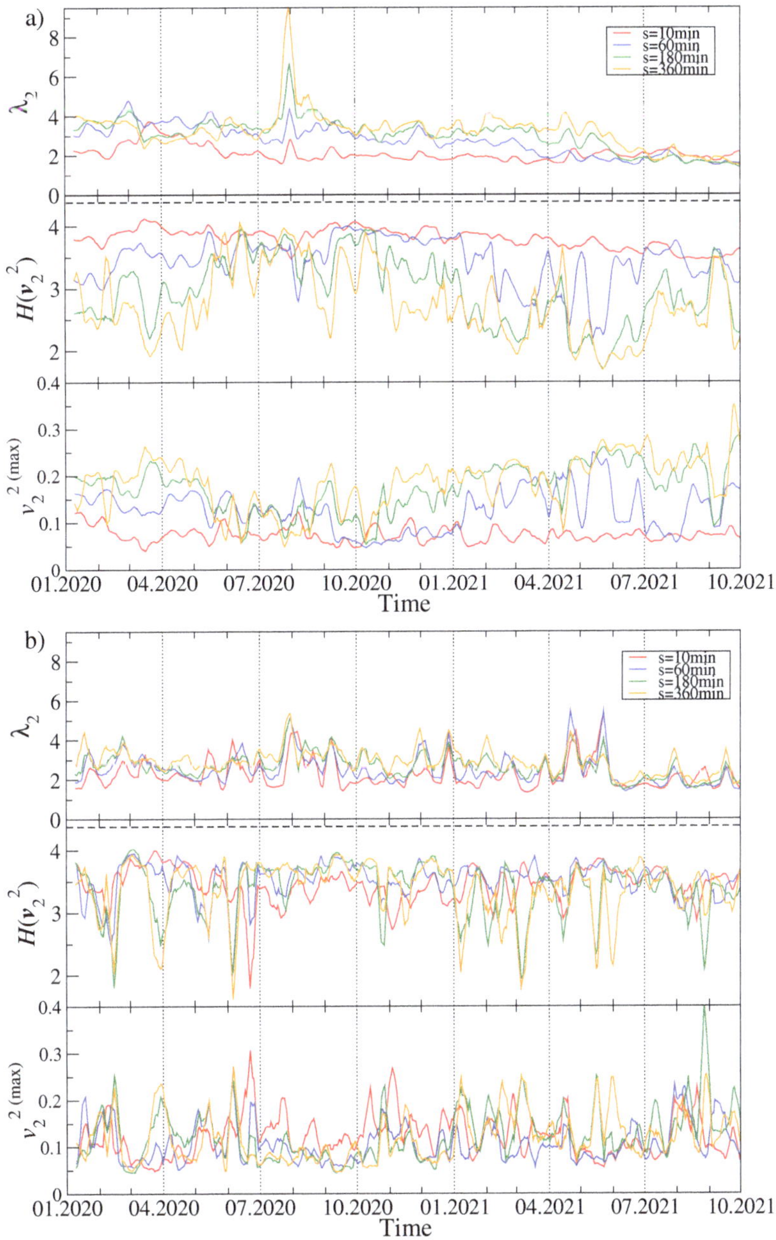

Figure 3. Time evolution of the selected spectral characteristics of $\mathbf{C}_q(s)$ (continuing). As in Figure 2, two cases are shown: $q = 1$ (**a**) and $q = 4$ (**b**). A moving window of length 7 days shifted by 1 day was applied for sample values of the scale: $s = 10$ min (red), $s = 60$ min (blue), $s = 180$ min (green), and $s = 360$ min (orange). The second largest eigenvalue λ_2 (top panels), the Shannon entropy $H(\mathbf{v}_2)$ of the squared eigenvector components $v_2(j)$ with $j = 1, ..., N$ (middle panels), and the squared maximum component of the eigenvector $\mathbf{v}_2$ associated with λ_2 (bottom panels) are shown.

Since a sum of all the eigenvalues must equal trace of $\mathbf{C}_q$ with $\mathrm{Tr}\mathbf{C}_q = N$, the high values of λ_1 take a significant part of each time series variance. This can suppress all the other eigenvalues with λ_2 in particular and can also have a strong impact on the eigenvector $\mathbf{v}_2$. We thus prefer to look at these quantities once more after removing the variance contribution of λ_1 from the original time series of returns. In order to accomplish this, we created an eigensignal representing λ_1 as a sum of the original time series weighted by the corresponding eigenvector components $z_1(t_m) = \sum_{j=1}^{N} v_1(j) r_j(t_m)$, where $r_j(t_m)$ are the normalized returns of a cryptocurrency j at time t_m, $m = 1, ..., T$. We then least-square fit the eigensignal $\{z_1(t_m)\}$ to each original time series $\{r_j(t_m)\}$ and subtract the fitted component from $\{r_j(t_m)\}$. What remains then is a residual signal $\{r_j^{(\mathrm{res})}\}$, which does not comprise any contribution from $\{z_1(t_m)\}$ and, thus, also from λ_1:

$$r_i^{(\mathrm{res})}(t_m) = r_i(t_m) - \alpha_i z_1(t_m) - \beta_i, \tag{10}$$

where α_i, β_i are the parameters of a linear fit. Finally, we calculate the coefficients $\rho_q^{(i,j)}(s)$ for all the cryptocurrency pairs (i, j) and form a residual q-dependent detrended correlation matrix $\mathbf{C}_q^{(\mathrm{res})}(s)$. After diagonalising it, we obtain its eigenvalues $\lambda_i^{(\mathrm{res})}$ and eigenvectors $\mathbf{v}_i^{(\mathrm{res})}$. We repeat this procedure a few times for different scales s and filtering parameters q. Figure 4 collects the results.

Now the largest eigenvalue $\lambda_1^{(\mathrm{res})}$, which inherits some information stored previously in λ_2 but without the former clear impact of λ_1, is not suppressed any more and, for $q = 1$, it shows richer behaviour with more fluctuations and more pronounced maxima (see Figure 4a). Interestingly, the large maximum of λ_2 observed in Figure 3a in July 2020 disappeared almost completely here and was replaced by a series of pronounced maxima in February, March, September, and December 2020, and a smaller one in May 2021. They are the more visible the longer time scale is considered. From a present perspective, the unique maximum of λ_2 in July 2020 might solely be a product of a relatively small value of λ_1 in that moment, which was unable to suppress λ_2 to its overall level of 4.

As regards the Shannon entropy, three phases can be distinguished: (1) from January to May 2020, (2) from May 2020 to April 2021, and (3) from May to October 2021. In the first and third phases there is no difference in $H(\mathbf{v}_1^{(\mathrm{res})})$ if we consider different scales s, while during the second phase, which largely overlapped with the bull market, the entropy fluctuates in time and increases with increasing s. However, its saturation level for $s = 360$ min in this phase is comparable with the analogous level in the other phases – this is because $H(\mathbf{v}_1^{(\mathrm{res})})$ for $s = 10$ min can be much smaller in phase (2) than in phases (1) and (3). Dissimilarity between the phases is observed also for $v_1^{(\mathrm{res})(max)}$: in phase (2) its value is substantially elevated as compared with the phases (1) and (2). These outcomes suggest that the eigenvector $\mathbf{v}_1^{(\mathrm{res})}$ became delocalised and some cryptocurrency used to contribute more to this eigenvector during phase (2) than the other cryptocurrencies did.

For $q = 4$ (Figure 4b), both $H(\mathbf{v}_1^{(\mathrm{res})})$ and $v_1^{(\mathrm{res})(\mathrm{max})}$ fluctuate over the whole analysed period more than it is observed for $q = 1$. The largest residual eigenvalue for $q = 4$ displays local maxima in the same moments as for $q = 1$, but their height varies. Apart from the maxima, typical fluctuations of $\lambda_1^{(\mathrm{res})}$ are smaller in 2021 than they used to be in 2020.

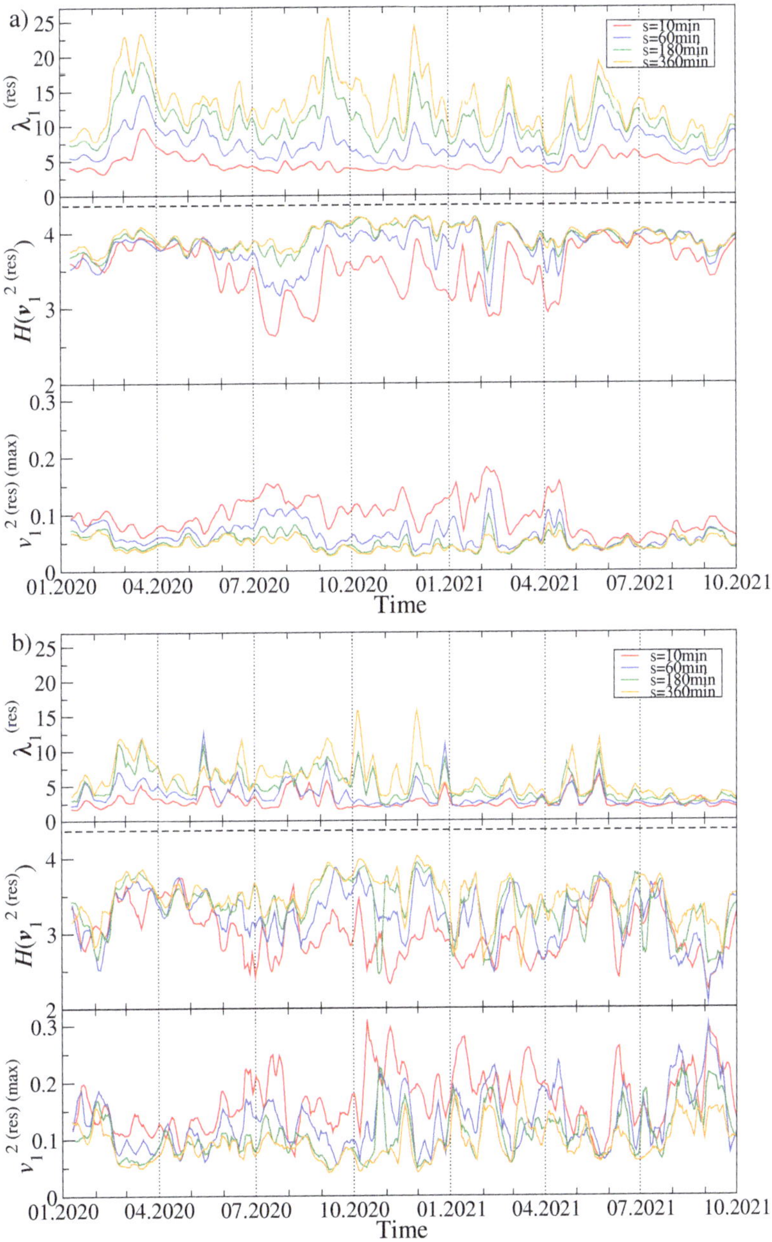

Figure 4. Time evolution of the selected spectral characteristics of the residual q-dependent detrended correlation matrix $\mathbf{C}_q^{(\mathrm{res})}(s)$ after filtering out the component corresponding to λ_1. As in Figure 2, two cases are shown: $q = 1$ (**a**) and $q = 4$ (**b**). A moving window of length 7 days shifted by 1 day was applied for sample values of the scale: $s = 10$ min (red), $s = 60$ min (blue), $s = 180$ min (green), and $s = 360$ min (orange). The largest residual eigenvalue $\lambda_1^{(\mathrm{res})}$ (top panels), the Shannon entropy $H(\mathbf{v}_1^{(\mathrm{res})})$ of the squared eigenvector components $v_1^{(\mathrm{res})}(j)$ with $j = 1, ..., N$ (middle panels), and the squared maximum component of the eigenvector $\mathbf{v}_1^{(\mathrm{res})}$ associated with $\lambda_1^{(\mathrm{res})}$ (bottom panels) are shown.

Some deeper insight into the cross-correlation structure of the cryptocurrency market can be gained by transforming the q-dependent detrened correlation matrix $\mathbf{C}_q(s)$ into a related distance matrix $\mathbf{D}_q(s)$, whose elements are defined by Equation (7). The latter is used as a basis for creating a minimum spanning tree, in which each node represents a particular cryptocurrency and each weighted edge represent the metric distance between a pair of assets or, equivalently, the detrended cross-correlation coefficient. To facilitate comprehension of the MST pictures, the edge weights between the nodes (i,j) are proportional to the coefficients $\rho_q^{(i,j)}(s)$ even though the metric distances $d_q^{(i,j)}(s)$ were used to determine the MST edges in this work.

We created an MST for each moving window position and for the same values of s and q as before. Owing to this, we are able to observe the evolution of the MST topology along the considered time span. The first topological characteristics we discuss here is the probability that a given node has a degree k. Its cumulative distributions $P(X \geq k)$ for a few sample window positions are shown in Figure 5 for $q = 1$ (top) and $q = 4$ (bottom) and for $s = 10$ min (red line) and $s = 360$ min (blue line). The MST topology expressed by these characteristics varies between different time intervals from a centralized graph with a single dominant node playing the role of a hub, that is, when there is a significant gap between the largest degree $k_{\max}$ and the second largest degree, to a distributed graph with a small $k_{\max}$ and a small difference in the degrees of the most connected nodes. The former situation is more typical for the short time scales ($s = 10$ min) and the periods with small return fluctuations ($q = 1$), while the latter situation occurs frequently for the long time scales ($s = 360$ min) and both the small and large fluctuation periods ($q = 1$ and $q = 4$); see Figure 5.

While increasing the scale from $s = 10$ min to $s = 360$ min, for $q = 1$ we observe a systematic change of the MST topology from centralized towards more distributed. For $q = 4$ there is no such a change and the topology is largely preserved. From the network perspective, this means that the detrended cross-correlations during the strong volatility periods are already well-developed at the 10-min time scale and, possibly, one has to consider even shorter scales to detect any topological transition (this would require a higher frequency of the price quotations than 1 min considered here, however). It is also worth noting that the cumulative probability distributions in some windows show a scale-free decay with k (the almost-straight lines in double logarithmic plots). This conclusion supports the results reported earlier for the data covering the years 2016–2019 [5] and 2017–2018 [72].

Topological changes of the MSTs while going from past to present can be expressed by the time evolution of the node degree $k_i(t)$ for the most connected nodes representing the cryptocurrencies i. The results for the MSTs created based on three distinct data sets are presented in Figure 6: (1) the original time series of the price returns, (2) the residual time series obtained after filtering out the contribution of λ_1 from the original data (both are based on the quotes given in USDT), and (3) the time series of the price returns based on the quotes given in BTC. The latter case allows us for effective filtering out the impact of BTC on the other assets' detrended cross-correlations.

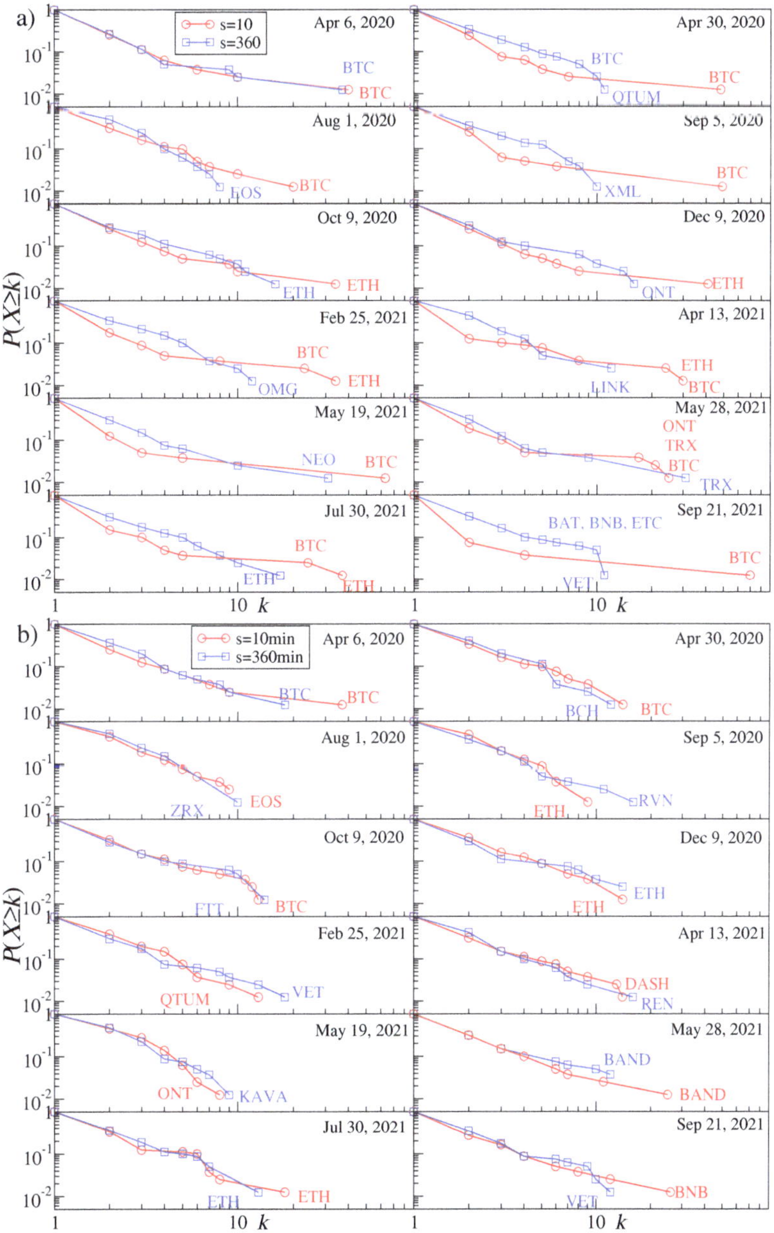

Figure 5. Node degree cumulative distribution $P(X \geq k)$ of the MSTs created for the cryptocurrency prices expressed in USDT. Results for sample moving windows are shown for $q = 1$ (**a**) and $q = 4$ (**b**). In each panel the distributions for two temporal scales are displayed: $s = 10$ min (red) and $s = 360$ min (blue). The nodes with the highest degree k are labelled by the corresponding cryptocurrency ticker.

There are the following observations:

(1) Exactly as expected from the above discussion related to Figure 5, for each data type, the degree of the most connected nodes tends to decrease with increasing s and the degree gap between k_{max} and the smaller values of k_i decreases as well. For longer time scales, the topology becomes less centralized and more "democratic" with a few hubs of a comparable connectivity.

(2) As the most capitalized cryptocurrency, BTC remains the most connected node over the longest time for $s = 10$ min and, to a lesser extent, for $s = 60$ min. However, for $s = 360$ min, it ceases to play such a role in August 2020, when the MST becomes decentralized permanently and the most connected node can be a cryptocurrency of moderate capitalization (see, for example, [73] for a similar observation).

(3) It happened for $s = 10$ min that the periods when ETH was the most connected node as frequently as BTC prevailed between September 2020 and February 2021. For $s = 60$ min also some other assets like ONT and TRX are represented by the most connected nodes from time to time, but it happens more because of a temporarily diminished degree of BTC and ETH than because of their own importance.

(4) In the residual data, BTC does not play so substantial role as in the original data, because its dominating role was largely wiped out by filtering out the λ_1 contribution. It remains, however, a hub with the second largest connectivity throughout the whole analyzed interval for $s = 10$ min. If s is increased to 60 min, BTC is degraded further on to be among a few secondary hubs with a few connections only. For both the scales, the most connected node is FTT, but its distinguished structural position vanishes almost completely after April 2021. For $s = 360$ min the MSTs always show a decentralized topology.

(5) If the prices are expressed in BTC, $k_{max}(t)$ is typically smaller ($k_{max} < 30$ out of 68) than when they are expressed in USDT ($k_{max} < 70$ out of 80). This is the expected property as BTC is the most connected hub in the case of the prices given in a stable coin. For any scale, a typical situation in this case is that there is frequent alternation of the most connected nodes: ETH, BNB, LINK, ONT, LTC, XRP, DASH, and so forth are among the assets that have the largest degree in certain time intervals, but none of them is able to substantially centralize the network. For long time scales, it even occurs that the largest degree nodes are switched almost random.

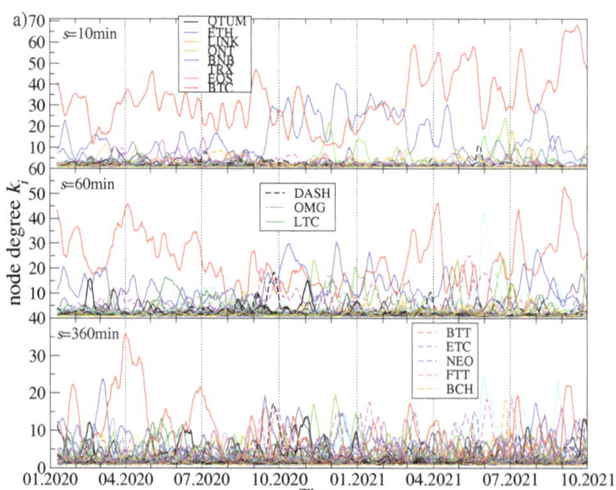

Figure 6. *Cont.*

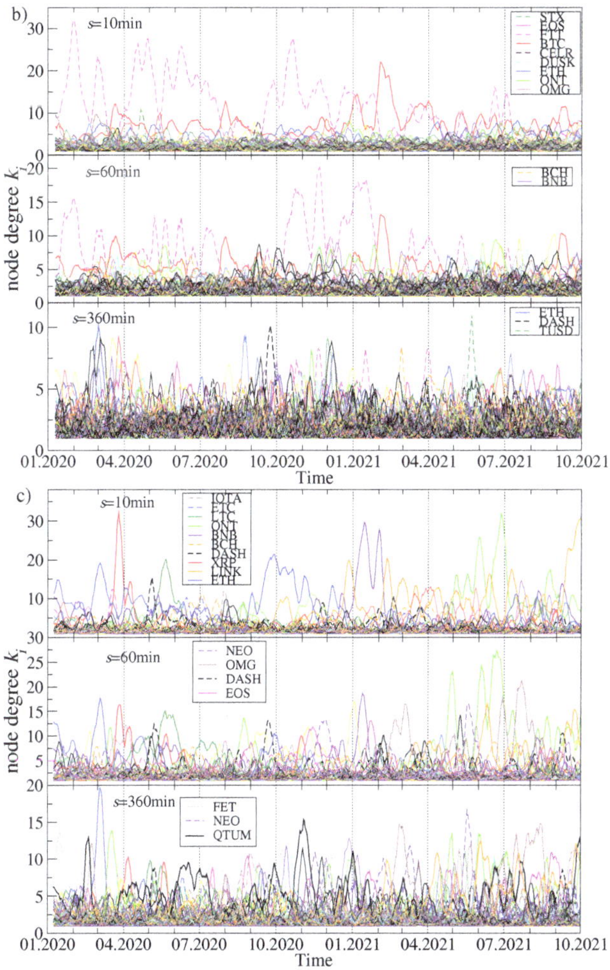

Figure 6. Evolution of the node degree k_i for the most connected nodes of the MST calculated in the seven-day-long moving window with a step of 1 day. For the prices expressed in USDT, two cases are shown: (**a**) the results for the complete data set without any filtering and (**b**) the results for the residual signals after filtering out a contribution from the component represented by the largest eigenvalue λ_1. The results for (**c**)—the prices expressed in BTC , which corresponds to filtering out any BTC-related contribution to other assets' evolution, are also shown. In each case, three exemplary scales are shown: $s = 10$ min (top graph in each panel), $s = 60$ min (middle graph), and $s = 360$ min (bottom graph). Different colors and line styles denote the node degree for different cryptocurrencies.

A variety of the MST topologies that can be observed in the cryptocurrency market in different periods is illustrated in Figures 7 and 8. The top left MST in Figure 7 has largely a star-like structure with BTC being its central node and ETH being a secondary hub. All other nodes are peripheral in respect to these two. The bottom left tree is also significantly centralized but now the most connected node is ETH, while BTC, FTT, and BAT are secondary hubs. A mixed type of topology is shown in the bottom right MST, where there are two primary hubs that are almost equivalent topologically (BTC and ETH) and a single secondary hub (BCH). However, despite this interesting dual centrality, the network has a part that is rather distributed. A largely distributed structure can be seen in the top right MST, in which only BTC possesses a significant number of the satellite nodes, while the overall network structure is distributed and almost random.

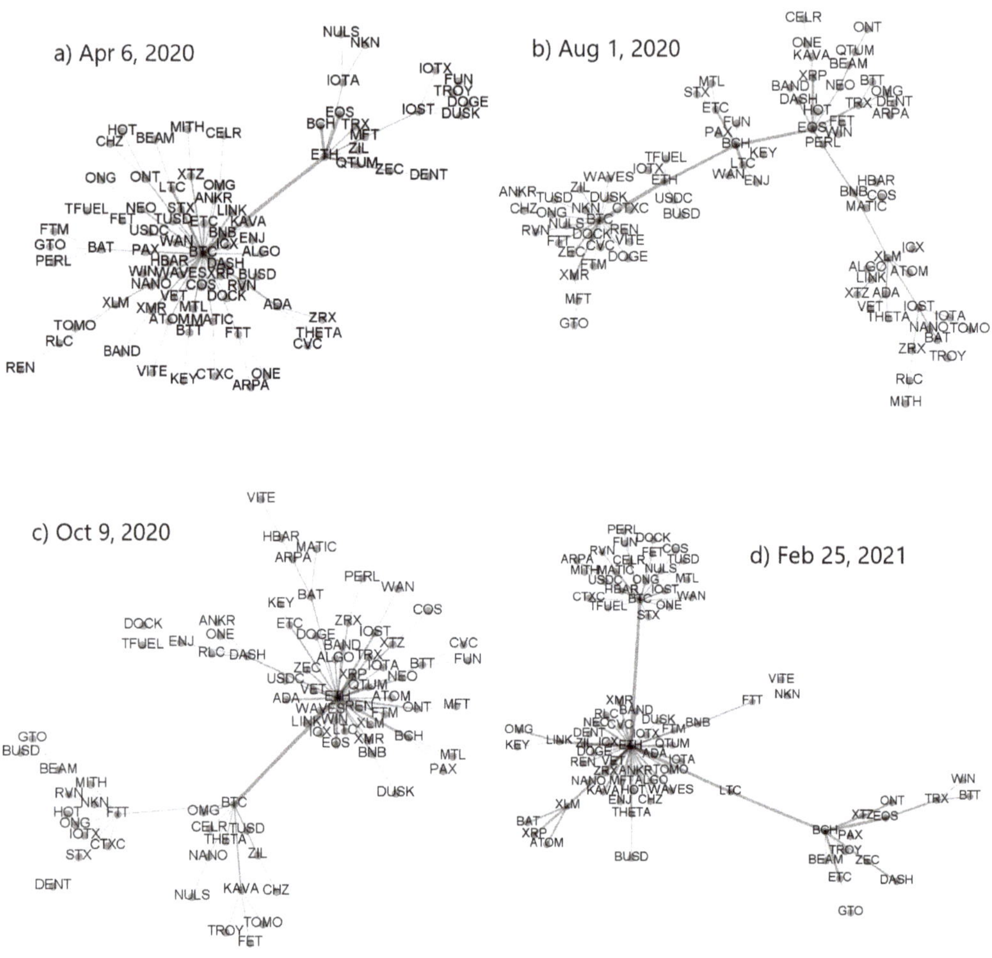

Figure 7. Minimal spanning trees calculated from a distance matrix $\mathbf{D}_q(s)$ based on $\rho_q(s)$ for $q = 1$ and $s = 10$ min. Each node represents a cryptocurrency and the edge widths are proportional to value of the corresponding coefficient $\rho_q(s)$. Each MST was created for moving window of length 7 days ended at specific dates: (**a**) 6 April 2020, (**b**) 1 August 2020, (**c**) 9 October 2020, and (**d**) 25 February 2021.

While the asset–asset correlation strength can be amplified by increasing scale s, Figure 8 shows that this operation weakens at the same time the centralized topology of the associated MST, which can show the signatures of a decentralized network. This can be seen by comparing the trees corresponding to the same windows in Figures 7 and 8.

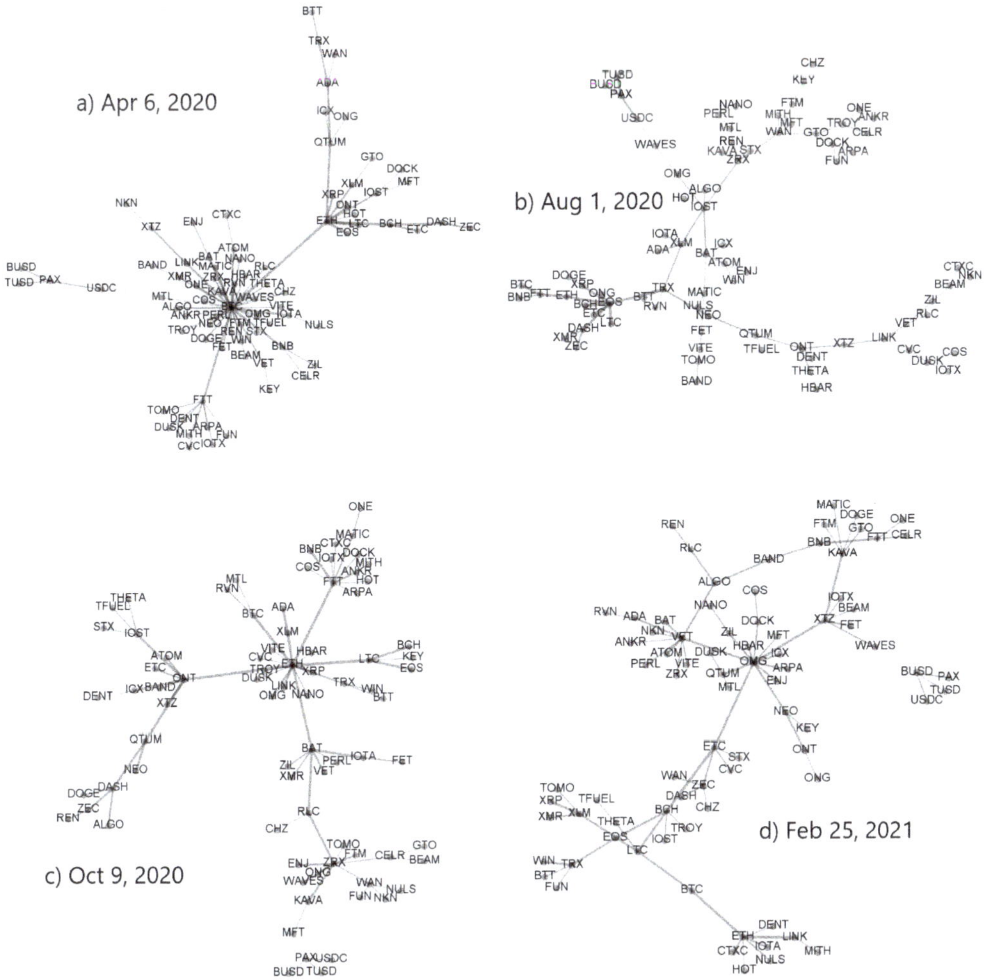

Figure 8. Minimal spanning trees calculated from a distance matrix $\mathbf{D}_q(s)$ based on $\rho_q(s)$ for $q = 1$ and $s = 360$ min. Each node represents a cryptocurrency and the edge widths are proportional to value of the corresponding coefficient $\rho_q(s)$. Each MST was created for moving window of a length of 7 days ended at specific dates: (**a**) 6 April 2020, (**b**) 1 August 2020, (**c**) 9 October 2020, and (**d**) 25 February 2021.

This conclusion receives additional support from the top panels of Figure 9a,b presenting the mean path length as a function of time. It is defined by the following formula:

$$\langle L(q,s,t) \rangle = \frac{1}{N(N-1)} \sum_{i=1}^{N} \sum_{j=i+1}^{N} L_{ij}(q,s,t), \tag{11}$$

where L_{ij} is the length of the path connecting nodes i and j. The larger $\langle L(q,s,t) \rangle$ is, the more distributed is the corresponding MST. Indeed, by considering a given window, this quantity systematically increases with increasing s. The smallest values of the mean path length ($2 < \langle L_{ij}(q,s,t) \rangle < 3$) can be seen in April-May 2020 (see also [74]), in August–September 2020, between March and May 2021, in May 2021, and in September–October

2021 for $s = 10$ min. These are the periods of the most centralised market, where a vast majority of the nodes is connected to a central hub. In each of these periods, the maximum node degree k_{max} assumes high values as well (see Figure 6a). In contrast, the elevated values of $\langle L_{ij}(q,s,t)\rangle$ ($L_{ij}(q,s,t) > 5$) are observed in February 2020, July 2020, and between February and May 2021.

The power-law exponent $\gamma(q,s,t)$ describing slope of the cumulative probability distribution of the node degree is shown in the middle panel of Figure 9a for $q = 1$ and it is accompanied by the standard error of its least-square fit (the lower panel). It is an unstable quantity that fluctuates between 0.5 and 2 (see also Figure 5) for the results in sample windows. The smaller $\gamma(q,s,t)$ is, the more distant k_{max} can be from the smaller values of k_i, but this relation does not always hold. The same quantities are shown in Figure 9b for the case of $q = 4$. Now we see smaller differences between the network characteristics for different time scales. This is the same rule as the one observed in Figure 5 for $q = 4$.

Topology of the MSTs representing the residual time series $\{r_i^{(\text{res})}(t_m)\}$ differs from the original time series $\{r_i(t_m)\}$ significantly. Because the removed component representing λ_1 is connected with the strength of the average detrended cross-correlation coefficient $\langle\rho_q(s)\rangle$, a lack of this component weakens the detrended cross-correlations and can thus destroy the star-like structures within the MST. This must obviously lengthen many inter-node paths and increase $\langle L_{ij}(q,s,t)\rangle$. In fact, Figure 10a,b shows that $\langle L_{ij}(q,s,t)\rangle > 5$ over almost the whole analyzed period for both $q = 1$ and $q = 4$. It happens sometimes that its value reaches 10, which indicates a distributed network topology. The slope exponent $\gamma(q,s,t)$ behaves even more erratically than for the original, complete data in Figure 9, and the standard error of the fitted values is much larger.

The same topological characteristics for the MSTs created from the time series of price quotations expressed in BTC are presented in Figure 11a,b. Their temporal evolution seems to be less random than in Figure 10 and resembles the picture for the USDT-based data shown in Figure 9. For $q = 1$, the mean path length fluctuates along a horizontal line at $\langle L_{ij}\rangle \approx 5$ until April 2021. Then the trend line starts to decrease towards a level of 4 or even below this value. This suggests that the MST topology has gradually become more centralized in the recent months. Such an effect is hardly visible for $q = 4$. A rather high values of $\gamma(q,s,t)$ above 1.5 for $q = 4$ confirm a more compact topology of the corresponding MSTs than in the case of the prices expressed in USDT.

Our study of the cryptocurrency network topology can be completed with an analysis of the network cluster structure. Obviously, in this case we have to consider the complete weighted networks defined by the matrix $\mathbf{C}_q(s)$ instead of the MSTs. In order to identify node clusters, we exploit the Louvain algorithm of community detection, whose performance is counted among the best methods [75].

For the most moving window positions, the algorithm detects a few cryptocurrency clusters, but their composition fluctuates among the windows. To show how the clusters vary in time, we select a few significant nodes and associate them with a set of nodes they share a given cluster with. Among the distinguished nodes that frequently play a role of the MST cluster centers are BTC, ETH, LINK, TRX, ONT, BNB, and others. In the case of BTC, we consider a network of all 80 cryptocurrencies expressed in USDT, while for the other nodes, we consider a limited set of 68 cryptocurrencies expressed in BTC and that are not pegged to US dollar. Some of the related clusters consist of a few nodes only throughout the whole period under study, but there are also clusters consisting of a variable number of nodes. Here we show the examples of the latter group of clusters: the clusters to which BTC, ETH, BNB, or ONT belong. It should be noted, however, that (1) a node representing a given cluster might not necessarily be its center in the MST representation, (2) some clusters are merged in some windows, while they remain separate in the other windows, and (3) the nodes can jump between clusters.

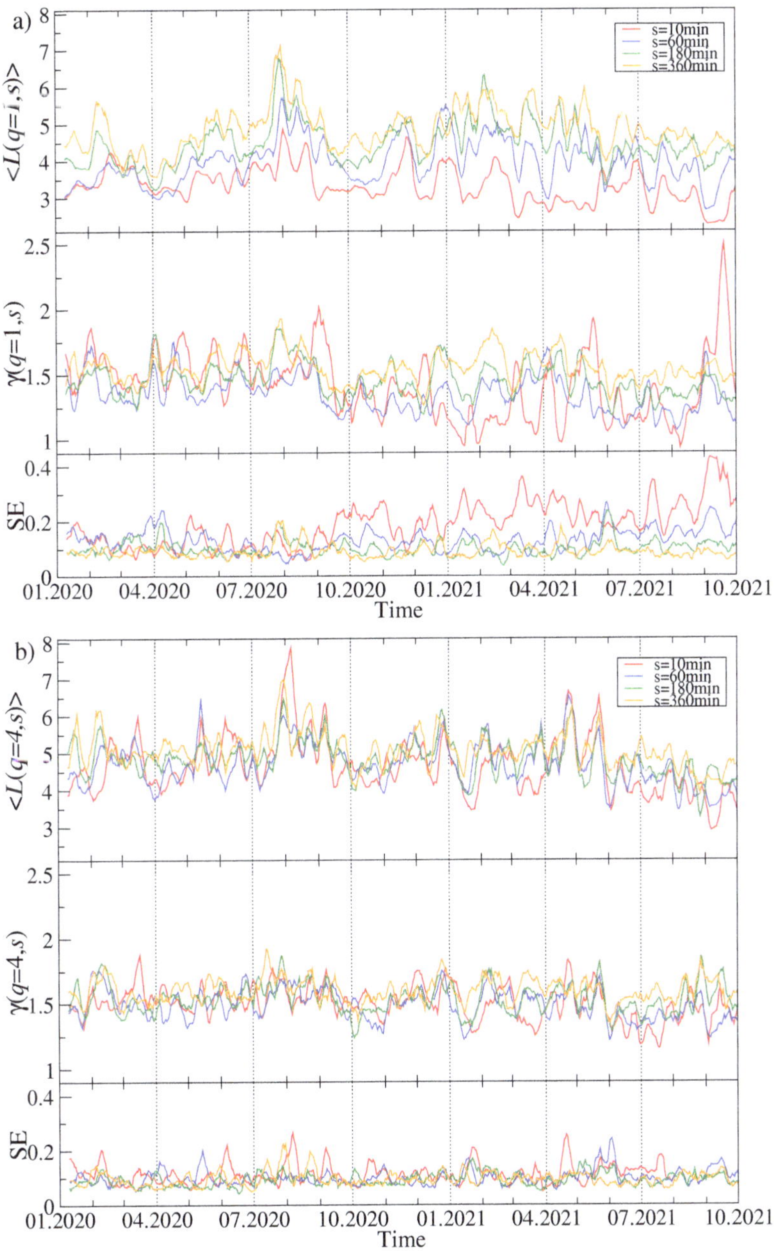

Figure 9. Time evolution of the selected network characteristics of the MST created from a distance matrix $\mathbf{D}_q(s)$. Two cases are shown: $q = 1$ (**a**) and $q = 4$ (**b**). In each case, a moving window of length 7 days shifted by 1 day was applied for the scales: $s = 10$ min (red), $s = 60$ min (blue), $s = 180$ min (green), and $s = 360$ min (orange). The mean path length $\langle L(q,s,t) \rangle$ (top panels), the node degree cumulative probability distribution $P(X \geq k)$ power-law slope exponent $\gamma(q,s,t)$ (middle panels) together with its standard error (SE, bottom panels). The cryptocurrency prices are expressed in USDT.

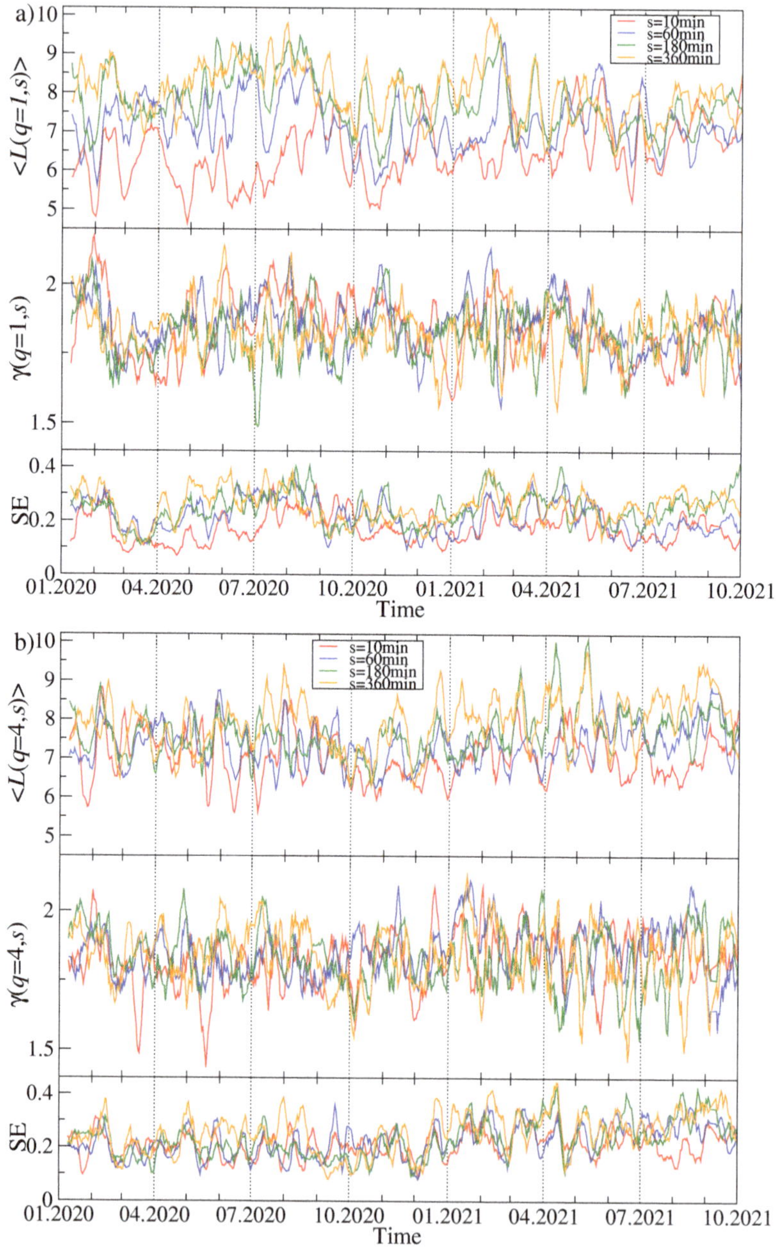

Figure 10. The same quantities as in Figure 9 but here obtained from the residual MSTs calculated for $\mathbf{D}_q^{(\mathrm{res})}(s)$ after filtering out the component corresponding to λ_1. Two cases are shown: $q = 1$ (**a**) and $q = 4$ (**b**). The cryptocurrency prices are expressed in USDT.

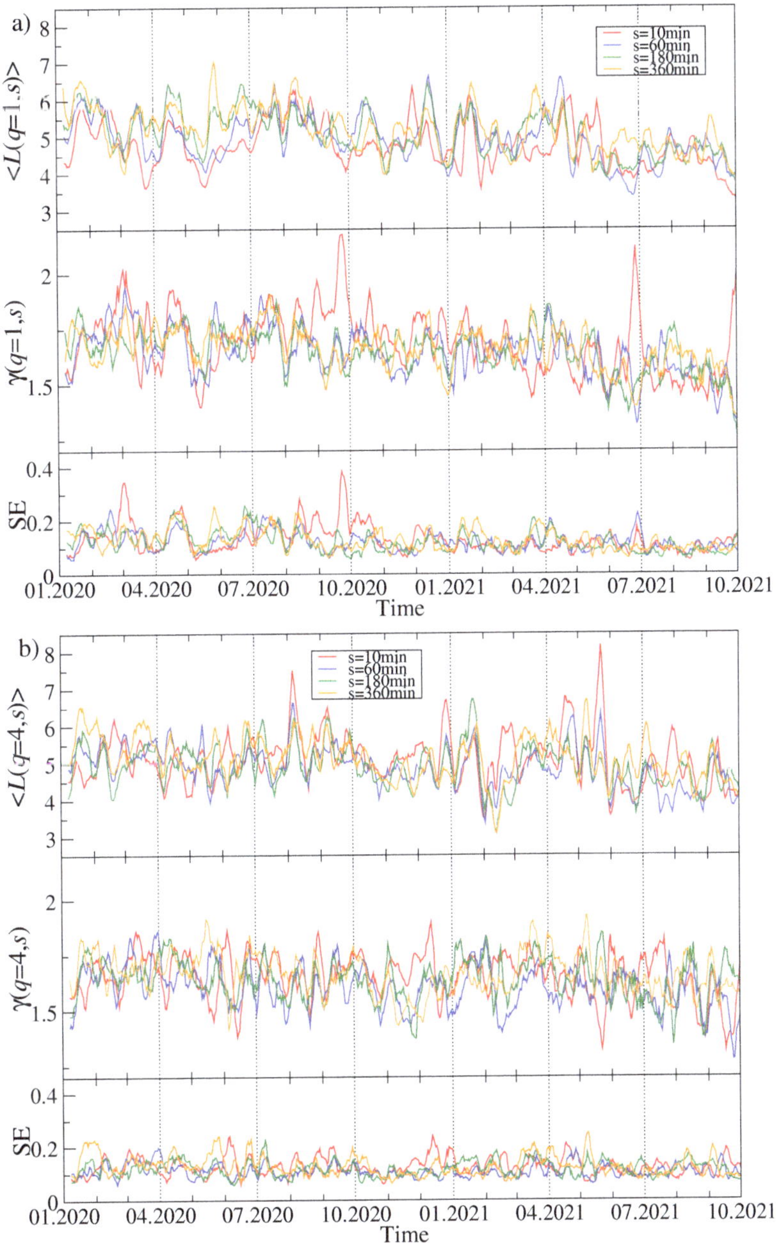

Figure 11. The same quantities as in Figures 9 and 10 but obtained from the cryptocurrency prices expressed in BTC. Two cases are shown: $q = 1$ (**a**) and $q = 4$ (**b**).

In Figures 12–15 we present the time evolution of the cluster composition for different time scales: $s = 10$ min, $s = 60$ min, and $s = 360$ min, and for $q = 1$. For example, a full point in the plot depicting the BTC cluster indicates that a respective cryptocurrency shares a cluster with BTC in a particular time window. The more dense points are seen along a horizontal line representing that cryptocurrency, the more stable is the coexistence of these two cryptocurrencies within the same cluster. On the other hand, the more numerous are the points along a vertical line, the larger is the cluster at that particular moment.

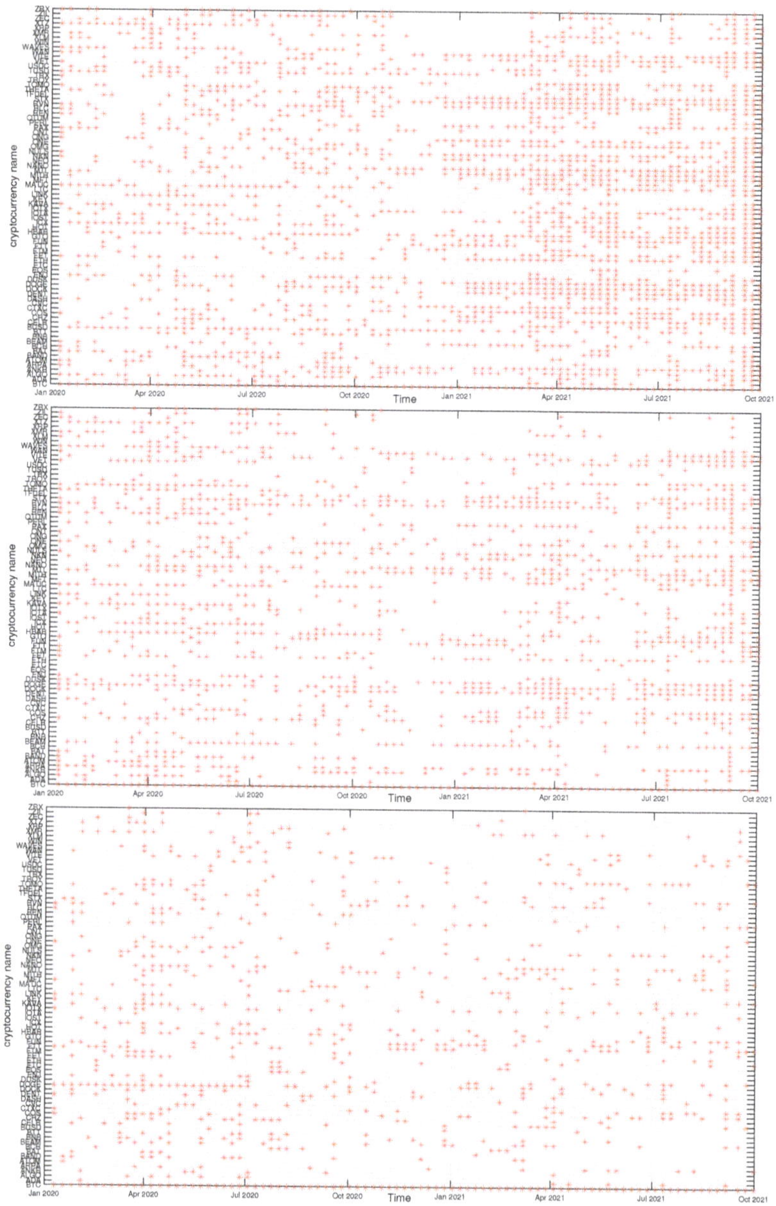

Figure 12. Composition of the BTC-related cryptocurrency cluster as a function of time for sample temporal scales: $s = 10$ min (**top**), $s = 60$ min (**middle**), and $s = 360$ min (**bottom**). Each point on the horizontal axis represents a non-overlapping seven-day-long moving window. Asset prices have been expressed in USDT.

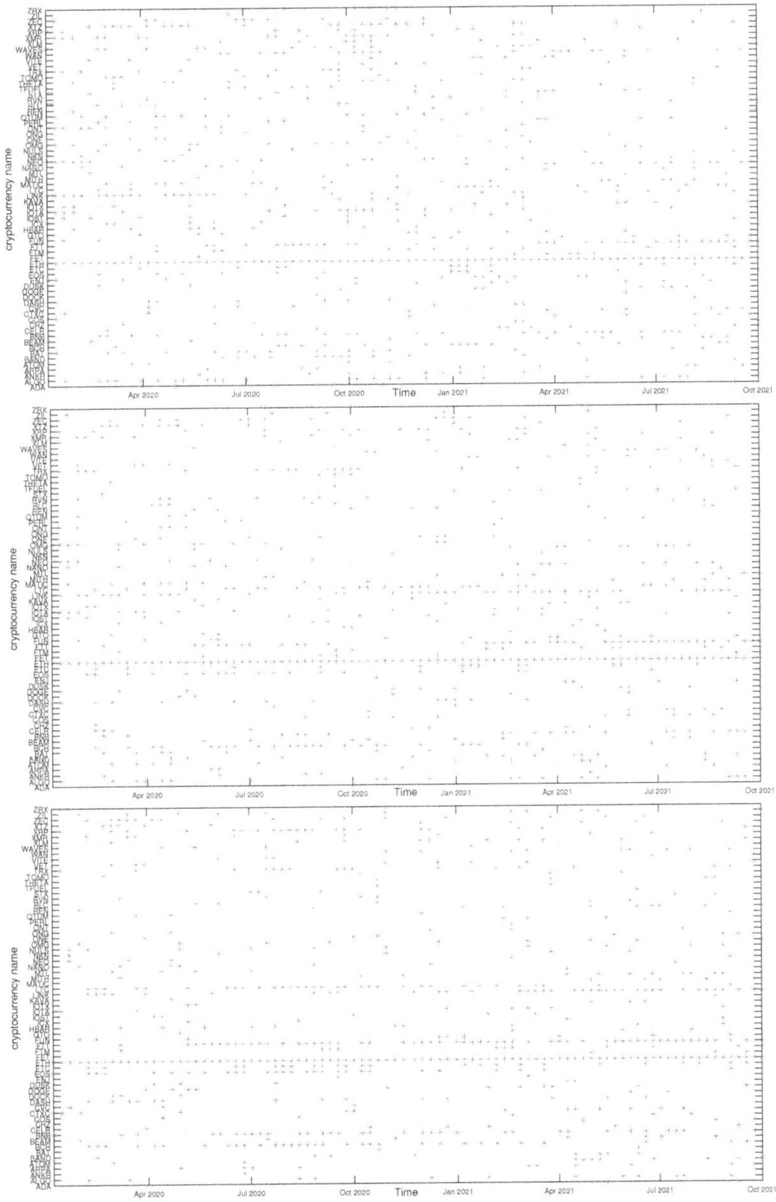

Figure 13. Composition of the ETH-related cryptocurrency cluster as a function of time for sample temporal scales: $s = 10$ min (**top**), $s = 60$ min (**middle**), and $s = 360$ min (**bottom**). Each point on the horizontal axis represents a non-overlapping seven-day-long moving window. Asset prices have been expressed in BTC, therefore any BTC-related contribution has been filtered out.

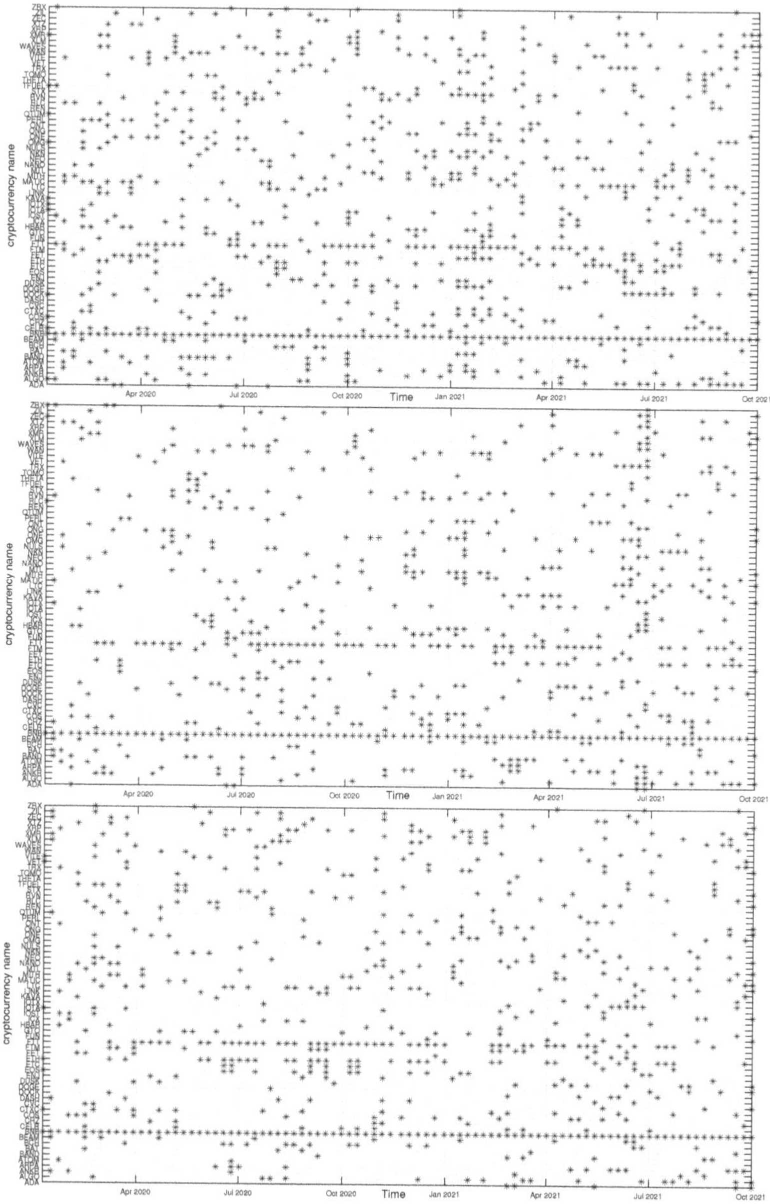

Figure 14. Composition of the BNB-related cryptocurrency cluster as a function of time for sample temporal scales: $s = 10$ min (**top**), $s = 60$ min (**middle**), and $s = 360$ min (**bottom**). Each point on the horizontal axis represents a non-overlapping seven-day-long moving window. Asset prices have been expressed in BTC, therefore any BTC-related contribution has been filtered out.

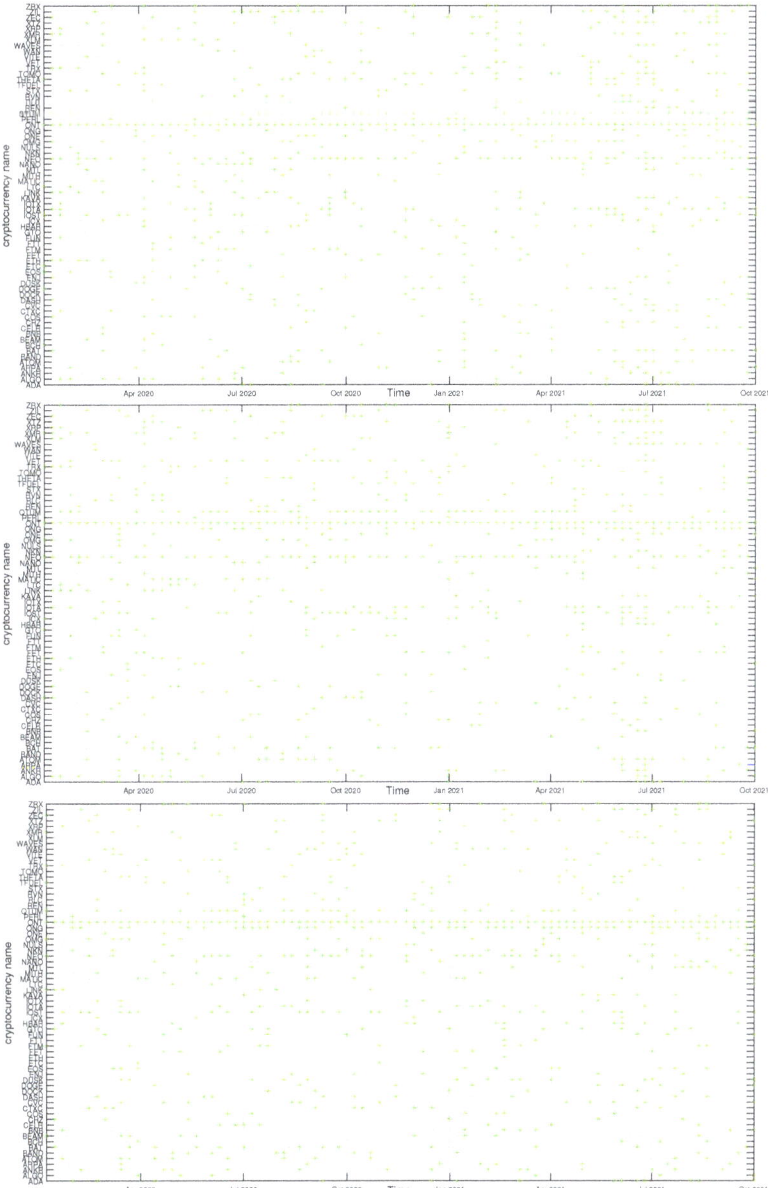

Figure 15. Composition of the ONT-related cryptocurrency cluster as a function of time for sample temporal scales: $s = 10$ min (**top**), $s = 60$ min (**middle**), and $s = 360$ min (**bottom**). Each point on the horizontal axis represents a non-overlapping seven-day-long moving window. Asset prices have been expressed in BTC, therefore any BTC-related contribution has been filtered out.

A cluster, to which BTC belongs, is typically the largest cluster in the network. By looking at Figure 12, we see that, on the shortest time scale of 10 min, the BTC cluster's size increases substantially in March 2021 and remains such till the end of the analyzed time interval. This is in agreement with the increase of the Shannon entropy $H(\mathbf{v}_1)$ observed in Figure 2 and it indicates that the market network has become more compact recently.

A situation looks different for $s = 60$ min, because apart from the BTC cluster growth observed in the $s = 10$ min case, only a slightly smaller cluster structure was seen before mid-2020. Thus, for $s = 60$ min the BTC cluster shrunk considerably over the period from July 2020 to February 2021 and it was larger outside that period. There are nodes that accompany BTC regularly, like STX, RVN, NANO, and BEAM, and there are nodes that fall into the BTC cluster only few times, like TRX and ETH, or even never do this, like ETH. For $s = 360$ min we do not detect any comparably large cluster and the BTC cluster is much smaller. It also tends to shrink even more after mid-2020.

The ETH cluster is much less numerous that the BTC one, which is partially due to a smaller number of the analyzed assets, but also to the properties of this cluster. Despite this, however, some long-term trend can be seen for $s = 10$ min that resulted in the temporary cluster growth in the latter half of 2020 followed by its shrinking that lasts till the end of the analyzed period. Such an effect cannot be noticed for the longer scales, where the density of points remains at the same level throughout the years 2020-2021. Among the nodes that frequently accompany ETH are BNB, LTC, BCH, and LINK.

The BNB cluster can be counted among the most numerous clusters on a par with the ETH cluster. For $s = 10$ min we also observe its interim growth between September 2020 and January 2021, which overlaps with the ETH growth phase. It also overlaps with the BTC cluster shrinking phase, which suggests that these events can be related with each other. No significant trends can be seen for $s = 60$ min and $s = 360$ min. The nodes that share the cluster with BNB most frequently are FTT and ETH.

Finally, the ONT cluster also shows its specific growth phase between May and July 2021 ($s = 10$ min and $s = 60$ min), outside of which no significant trend can be seen. NEO and IOTA are the nodes that appear the most frequently in the same cluster with ONT. In general, Figures 12–15 show highly unstable composition of the analyzed clusters. This outcome differs from the results of some earlier studies based on data from more a distant past that reported stability of the cryptocurrency clusters (e.g., [76]). Additionally, the identified community structure of the market differs from the result of another study, where a core-periphery structure was identified instead [72].

Our discussion hitherto is focused on the simultaneous time series without delays between them. However, there is an interesting question whether the most capitalized and liquid cryptocurrencies like BTC and ETH drive the remaining ones, which can generate the delayed cross-correlations that can be observable. In order to address this question, we calculated the coefficients $\rho_q^{(\mathrm{BTC,X})}(s,\tau)$ for all the cryptocurrency pairs (BTC,X) and (ETH,X), where X stands for any cryptocurrency other than BTC and ETH. A time lag τ that can assume two values: $\tau = -1$ min and $\tau = 1$ min, defines whether the BTC (ETH) time series is advanced or lagged relatively to the second time series. For these two cases, we calculate the average coefficients $\langle \rho_q(s,\tau) \rangle$ for BTC and ETH (the averaging is carried out over all other cryptocurrencies X).

Figure 16 shows the results for $q = 1$ and $q = 4$ and for the shortest scale $s = 10$ min (a potential effect of 1 minute delay can be too weak to be detectable on longer scales). If the time series of the BTC returns is considered, $\langle \rho_q(s,\tau) \rangle$ is significantly larger for $\tau = 0$ than for $\tau = \pm 1$. For $q = 1$ the advanced BTC time series produces larger $\langle \rho_q(s,\tau) \rangle$ than the lagged one. This difference is statistically significant. For $q = 4$ both shifted time series produce $\langle \rho_q(s,\tau) \rangle$ with comparable magnitude for a vast majority of windows with a few exceptions, where the advanced BTC time series produces slightly stronger cross-correlations than the lagged one does. The qualitatively similar results are obtained for the advanced and lagged ETH time series. We can therefore conclude that by shifting the time series representing BTC or ETH we still preserve some amount of the valid detrended cross-correlations. The relative dominance of the advanced ($\tau = -1$ min) time series over the lagged ($\tau = 1$ min) ones suggest that the remaining part of the market absorbs information that occurred first in the price fluctuations of BTC and ETH with a time needed for this absorption being as long as a minute. An opposite process of information transfer from the less liquid cryptocurrencies to BTC and ETH cannot be detected based on our data set. It must be noted, however, that both the BTC and ETH returns exhibit a detrended autocorrelation with the length of more than 1 min. Such an autocorrelation can artificially produce the delayed detrended cross-correlations which can manifest themselves in a way similar to that observed in Figure 16. We cannot therefore answer the formulated question decisively.

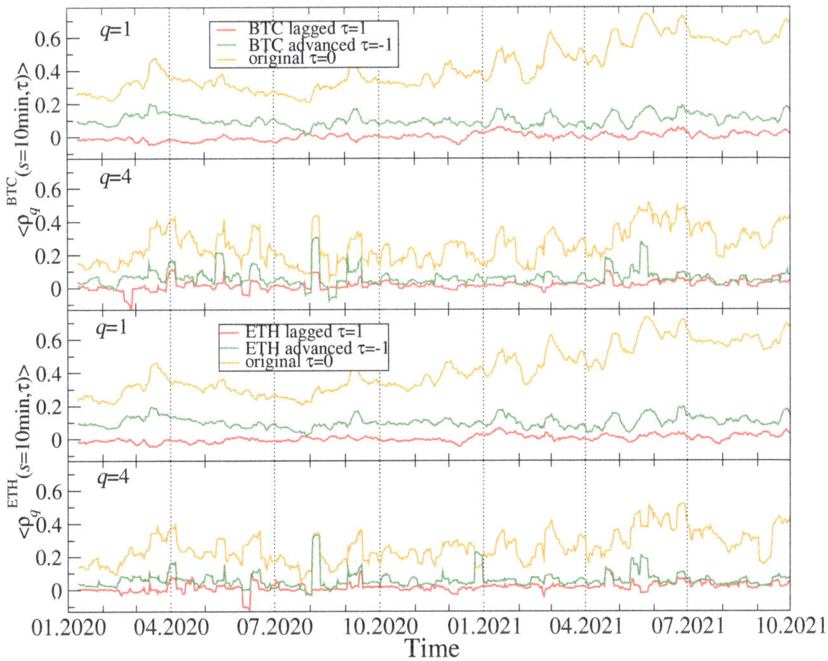

Figure 16. Mean lagged q-dependent detrended cross-correlation coefficient $\rho_q(s, \tau)$ as a function of time after averaging over all the considered cryptocurrencies other than BTC and ETH. Time series representing BTC and ETH returns have been advanced (green) or delayed (red) by $\tau = 1$ min and compared with the original non-shifted time series (orange). Two values of the filtering parameter q are shown: $q = 1$ (all fluctuations enter with the same weight, the first and third panels) and $q = 4$ (large fluctuations are amplified, the second and fourth panels).

Our former studies of the cryptocurrency market showed that, recently, it begun to be positively or negatively cross-correlated in some specific periods with the traditional financial markets like the stock market, the currency exchange market, and the commodity markets [5,13]. Among such periods of the statistically significant detrended cross-correlations there was the COVID-19 pandemic in the United States: the very first case on the US territory in the end of January 2020, the first COVID-19 wave outburst in April, and the second wave development in June–July, and the subsequent pandemic slowdown, which brought the across-market rally starting in September 2020. As we have already collected more contemporary data that end in October 2021, we are able to extend our analysis of the detrended cross-correlations between the cryptocurrencies and a few other financial assets. We consider the logarithmic price returns of a few basic cryptocurrencies (BTC, ETH, DASH, EOS, and XMR), the main regular currencies (AUD, CAD, CHF, CNH, CZK, EUR, GBP, JPY, MXN, NOK, NZD, PLN, and ZAR), sample commodities (crude oil, copper, silver, and gold), and the most important stock market indices (S&P500, NASDAQ100, Russel 2000, DJIA, FTSE, DAX, and NIKKEI). All the assets except the stock market indices are priced in US dollars (data from Dukascopy [77]).

Figure 17 shows the historical quotes of S&P500 and BTC together with the distinguished periods of the elevated detrended cross-correlations inside the cryptocurrency markets. One can see that these periods are associated with specific market events that are observed in the historical data: the all-market surge at the COVID-19 pandemic onset in March-April 2020, the second pandemic wave in June–July 2020, a market rally and the following drawdowns in September–October 2020, the cryptocurrency market rally in March–April 2021 and a surge and a subsequent rally in September–October 2021. Looking from a macroscopic perspective, in all these cases the coarse-grained behaviours of S&P500 and BTC were similar to each other at least for some period of time.

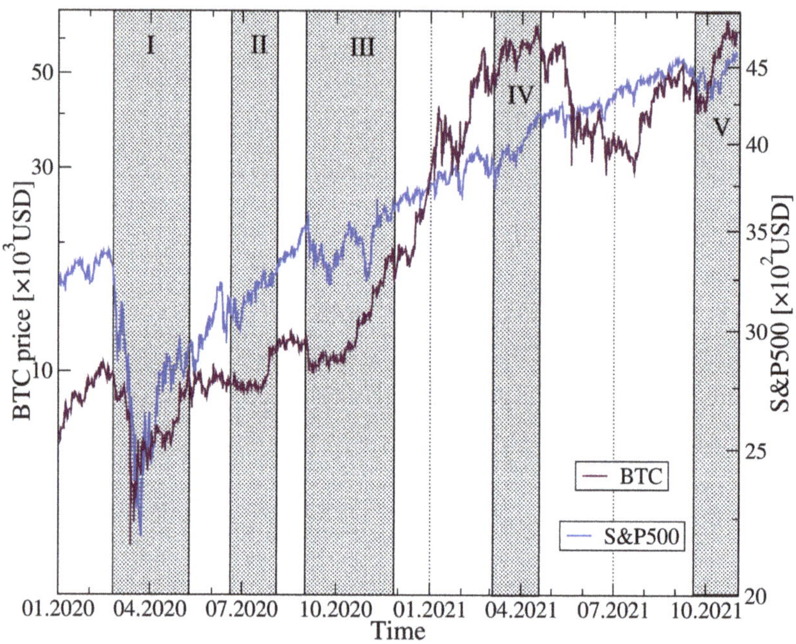

Figure 17. Temporal co-evolution of BTC price in USD (maroon) and the S&P500 index (blue) over the years 2020–2021. Periods, in which $\rho_q(s)$ calculated for these two assets exceed a threshold of 0.25 for $s = 360$ min and $q = 1$ (see Figure 18), are denoted by grey vertical strips. Specific market events are indicated by Roman numerals: I—the all-market surge at the COVID-19 pandemic onset in March–April 2020, II—the second pandemic wave in June–July 2020, III—a market rally and the following drawdowns in September–October 2020, IV—the cryptocurrency market rally in March–April 2021, and V—a surge and a subsequent rally in September–October 2021.

To inspect this issue in more detail, we calculated the q-dependent detrended cross-correlation coefficients for all the possible pairs of the considered assets. Before we did this, we had to concord all the time series by eliminating the gaps caused by different trading hours. The results for $q = 1$ and $q = 4$ and for $s = 10$ min and $s = 360$ min are shown in Figure 18. For both values of the filtering parameter q, the cross-correlations are stronger on the long time scale and weaker on the short one. Except for the maximum of $\rho_q(s)$ that occurred for $q = 4$ and $s = 10$ min in the end of June 2020, which is not present at all for $q = 4$ and $s = 360$ min and for $q = 1$, all the other periods of the amplified cross-correlations can be observed in each case. The maxima of $\rho_q(s)$ calculated for BTC and the traditional assets occur, roughly, over the same periods than the maxima of the inner cross-correlations on the cryptocurrency market.

Different traditional assets reveal different levels of the detrended cross-correlation with BTC: the strongest correlations can be detected for S&P500 and other stock indices, while the weaker but also significant ones for crude oil, copper, CAD and other regular currencies except for JPY and, to a much smaller extent gold. The Japanese currency is significantly anticorrelated with BTC in the periods, in which the other assets are positively cross-correlated. This means that JPY can be used for the hedging purposes while investing on the cryptocurrency market. After comparing the cross-correlation strength for $q = 1$ with that for $q = 4$, we may conclude that, during the large fluctuation periods, the traditional assets are less strongly cross-correlated with BTC than during the smaller fluctuation periods. They also need rather long time scales to be fully built up. What can be inferred from these results is that the detrended cross-correlations are weaker in 2021 than they used to be in 2020, but they are still stronger than the corresponding cross-correlations before the COVID-19 pandemic.

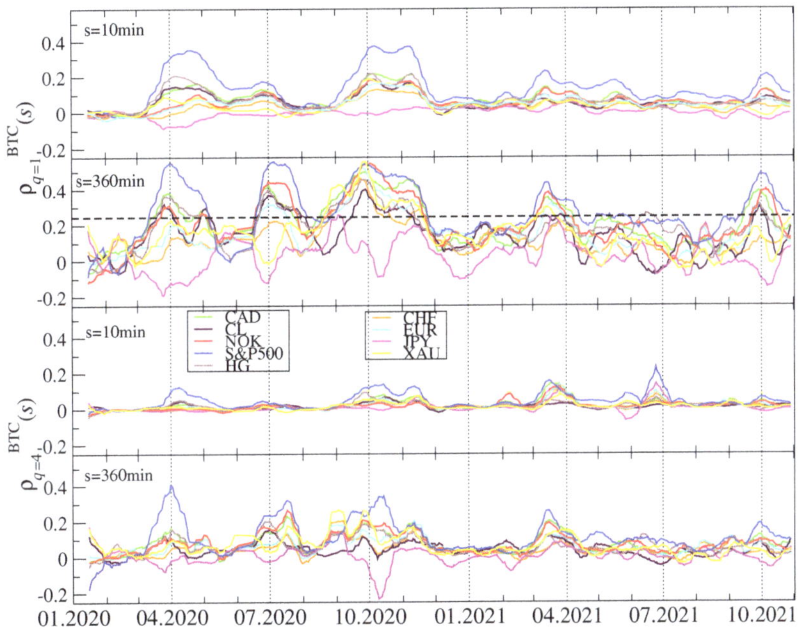

Figure 18. The q-dependent detrended cross-correlation coefficient $\rho_q(s)$ calculated in 10-day-long moving windows with a 1-day step for BTC and the traditional market assets: the S&P500 index (blue), crude oil price (CL, black), copper price (HG, brown), gold price (XAU, yellow), and a few regular currencies expressed in the US dollars: euro (EUR, cyan), Swiss franc (CHF, orange), Canadian dollar (CAD, light green), Japanese yen (JPY, magenta), and Norwegian krone (NOK, red). Two temporal scales s ($s = 10$ min in the first and third panels, and $s = 360$ min in the second and fourth panels) and two filtering parameter q values ($q = 1$ in the first and second panels, and $q = 4$ in the third and fourth panels) are shown. The horizontal dashed line at $\rho_q(s) = 0.25$ in the second panel denotes a discrimination threshold applied to determine the shaded regions in Figure 17.

Based on the coefficients $\rho_q^{(i,j)}(s)$, where i and j labels the cryptocurrencies and traditional assets, we created the related minimal spanning trees. A few sample trees for specific moving window positions are presented in Figure 19. It is easy to notice that the detrended cross-correlation strength between BTC and the traditional markets, the closest ones being the stock markets and not the currency markets is much smaller than the analogous strength among the traditional assets representing the same market type and even different market types. Topology of the MSTs is heterogeneous with both the significant hubs (S&P500, AUD, EUR, and some cryptocurrency) and the long branches.

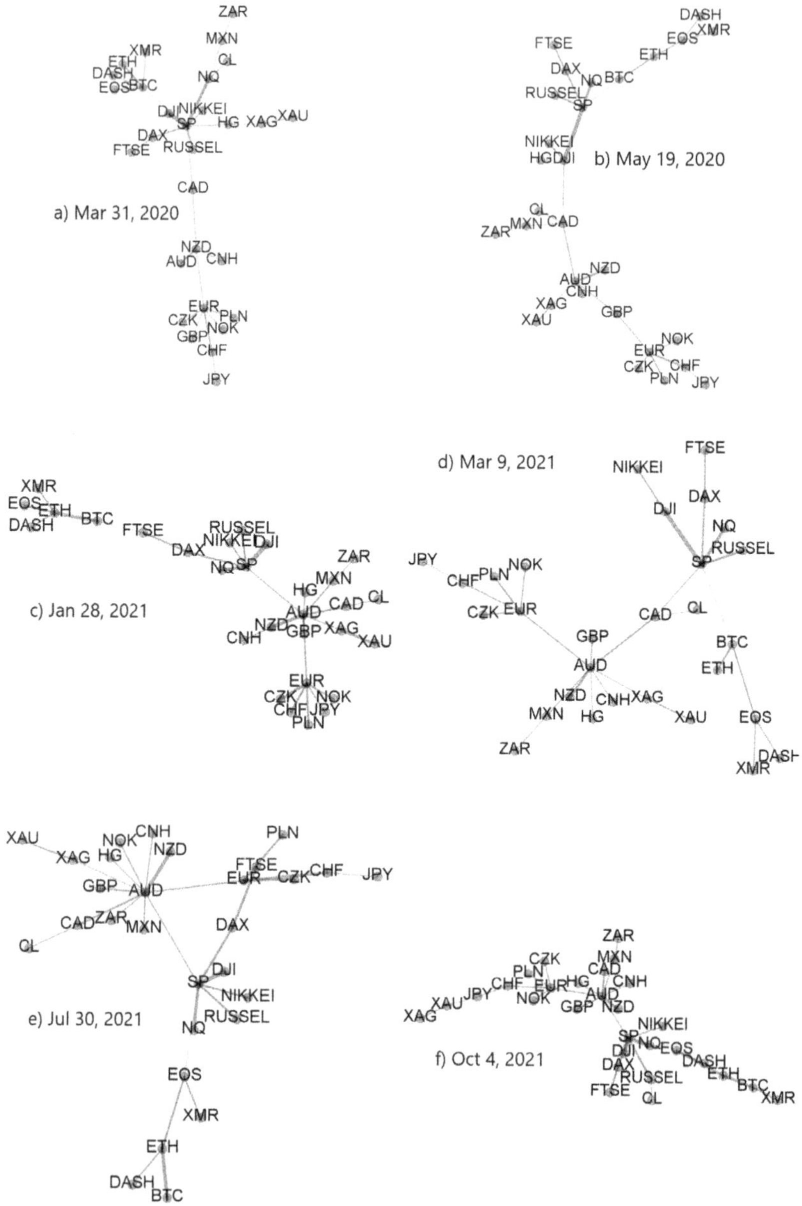

Figure 19. Minimal spanning trees calculated from a distance matrix $\mathbf{D}_q(s)$ based on $\rho_q(s)$ for $q = 1$ and $s = 10$ min. The data used to create MSTs consists of cryptocurrencies (BTC, ETH, DASH, EOS, and XMR), regular currencies (AUD, GBP, NZD, MXN, ZAR, CNH, EUR, CHF, JPY, CZK, NOK, CAD, and PLN), commodities (gold-XAU, silver-XAG, copper-HG, and crude oil-CL), as well as stock market indices (S&P500-SP, NASDAQ100-NQ, Russel 2000, FTSE, DAX, NIKKEI, and DJIA) in 10-day-long moving windows ended at specific dates: (**a**) 31 March 2020 (highly correlated markets during the pandemic onset in the United States), (**b**) 19 May 2020 (maximum cross-market correlations), (**c**) 28 January 2021 (the GameStop short squeeze related market turbulence accompanied by the cryptocurrency market decoupling), (**d**) 9 March 2021 (the elevated market cross-correlations), (**e**) 30 July 2021 (the cryptocurrencies starting a rally phase with minimum cross-market correlations), and (**f**) 4 October 2021 (the latest phase of the cross-market correlations).

4. Conclusions

In this paper, we studied the high-frequency time series of price returns representing 80 cryptocurrencies that were the most actively traded on the Binance platform. We focused on the detrended cross-correlation structure of the cryptocurrency market at different time intervals and calculated the q-dependent detrended cross-correlation coefficient $\rho_q(s)$ for all the cryptocurrency pairs and in different moving window positions. Based on these coefficients, we analyzed the spectral properties of the detrended correlation matrix and topology of the minimal spanning trees calculated from this matrix.

The main issue that has been pointed out is that our analysis comprises only a small fraction of all traded cryptocurrencies, whose number exceeds 7500 [78]. However, the less well-known and less capitalized a cryptocurrency is, the less liquid and less reliable are the related data. This is why restricting our analysis to the most capitalized ones was crucial. Another related issue was the MST construction, and it has already been mentioned in Section 3 that the exact connectivity of the MST links is prone to noise effects, which is the most significant source of possible errors. Fortunately, the more important these errors are, the weaker the correlations, while they are less effective if the correlations are strong (this is an issue that should be addressed independently in future work).

Our principal result is the observation that, over the last year, the cryptocurrency market has gradually become more compact from a topological perspective. This was achieved by the increasing market cohesion expressed by the rising average cross-correlation strength among the cryptocurrencies. Spectrally, it was manifested by the elevated magnitude of the largest correlation matrix eigenvalue λ_1 after mid-2020, as compared with the earlier periods. λ_1 is associated with an eigenvector that becomes more and more delocalised with time (as detected by the increasing entropy of its components). The largest component of this eigenvector is suppressed by the delocalisation, and its absolute value decreases significantly. These effects are observed if either the large or the small fluctuation intervals are filtered out by tuning the parameter q in $\rho_q(s)$. In addition, the detrended cross-correlations saturated faster than before (small difference between λ_1 for different time scales). This is a detrended counterpart of the classic Epps effect, which describes a process of the market consolidation due to the cross-correlations among the assets [69–71].

The topological properties of the MSTs are in agreement with the outcomes of the spectral analysis and show that the market becomes more centralized with time. On the short scales, the most connected node nowadays develops more connections to other nodes than it used to have before. The MST topology in this case is centralized and close to a star-like structure. Usually, the role of a stable central hub is played by BTC or ETH on the short time scales, but on the longer scales (e.g., an hour or longer), the hub is unstable and it frequently switches among the most liquid cryptocurrencies. The corresponding MST topology is distributed without any central hub. By increasing the scales, the mean path length also increases and it indicates that the structure for the longer time scales is more distributed and random than for the short scales. In this case, the market consolidates quickly on the short time scales (e.g., 10 min), but then the fine-grained community structure develops itself owing to the increasing cross-correlations and the average cross-correlation level rises across the network. The structure becomes less centralized, but at this point the market is already strongly coupled and compact.

We also calculated the detrended cross-correlation coefficients for BTC and some selected traditional assets like the stock market indices, commodity prices, and the regular currency exchange rates. We found that during the periods associated with the strongly correlated cryptocurrencies, the inter-market cross-correlations are also stronger than usual. Typically, the inter-market couplings rise in the periods of market instability like the COVID-19-pandemic-related events and fall in the more quiet times. However, even in such periods, the cryptocurrency market is more independent from the other markets than those markets are independent among themselves. As the pandemic becomes a normal component of our reality, the cross-correlations between the cryptocurrency market and the other markets tend to decrease, but this process is more prolonged now than the opposite process that occurred suddenly in early 2020. It is an open issue now whether the cryptocurrencies will at some point return to be an entirely independent market or the correlations that can occur from time to time will remain observable.

The main issue that has to be pointed out is that our analysis comprises only a small fraction of all traded cryptocurrencies, whose number exceeds 7500 [78]. However, the less well-known and less capitalized a cryptocurrency is, the less liquid and less reliable are the related data. This is why restricting our analysis to the most capitalized ones was crucial. Another issue is related to the MST construction, which has already been mentioned in Section 3: the exact connectivity of the MST links is prone to noise effects, which is the most significant source of possible errors. Fortunately, these errors are the more important, the weaker are the correlations, while they are less effective if the correlations are strong (this is an issue that should independently be addressed in future work).

Author Contributions: Conceptualization, S.D. and M.W.; methodology, S.D., J.K., and M.W.; software, M.W.; validation, S.D., J.K., and M.W.; formal analysis, M.W.; investigation, S.D., J.K., and M.W.; resources, M.W.; data curation, M.W.; writing—original draft preparation, J.K.; writing—review and editing, J.K. and M.W.; visualization, M.W.; supervision, S.D. and J.K. All authors have read and agreed to the published version of the manuscript.

Funding: This research received no external funding.

Institutional Review Board Statement: Not applicable.

Informed Consent Statement: Not applicable.

Data Availability Statement: Data available freely from Binance [66] and Dukascopy [77].

Conflicts of Interest: The authors declare no conflict of interest.

Appendix A

Table A1. List of tickers from Dukascopy and Binance.

Dukascopy		Binance			
Ticker	**Name**	**Ticker**	**Name**	**Ticker**	**Name**
BTC	bitcoin	BTC	bitcoin	LINK	chainlink
ETH	ethereum	ADA	cardano	LTC	litecoin
DASH	dash	ALGO	algorand	MATIC	polygon
EOS	eos	ANKR	ankr	MFT	hifi finance
XMR	monero	ARPA	arpa chain	MITH	mithril
AUD	Australian dollar	ATOM	cosmos	MTL	metal
EUR	euro	BAND	band protocol	NANO	nano
GBP	British pound	BAT	basic atention token	NEO	neo
NZD	New Zealand dollar	BCH	bitcoin cash	NKN	nkn
CAD	Canadian dollar	BEAM	beam	NULS	nuls
CHF	Swiss franc	BNB	binance coin	OMG	omg network
CNH	offshore renminbi	BTT	bittorrent	ONE	harmony
CZK	Czech krone	BUSD	binance USD	ONG	ontology gas
JPY	Japanese yen	CELR	celer network	ONT	ontology
MXN	Mexican peso	CHZ	chiliz	PAX	pax dollar
NOK	Norwegian krone	COS	contentos	PERL	perl
PLN	Polish zloty	CTXC	cortex	QTUM	qtum
ZAR	South African rand	CVC	civic	REN	ren
NIKKEI	Nikkei 225	DASH	dash	RLC	iexec
RUSSEL	Russell 2000	DENT	dent	RVN	ravencoin
DAX	DAX 30	DOCK	dock	STX	stacks
FTSE	FTSE 100	DOGE	dogecoin	TFUEL	theta fuel

Table A1. *Cont.*

Dukascopy		Binance			
Ticker	**Name**	**Ticker**	**Name**	**Ticker**	**Name**
DJI	Dow Jones Industrial Average	DUSK	dusk network	THETA	theta
SP	S&P 500	ENJ	enj coin	TOMO	tomochain
NQ	NASDAQ 100	EOS	eos	TROY	troy
XAG	silver	ETC	ethereum classic	TRX	tron
XAU	gold	ETH	ethereum	TUSD	trueusd
HG	high-grade copper	FET	fetch	USDC	USD coin
CL	crude oil	FTM	fantom	VET	vechain
		FTT	ftx token	VITE	vite
		FUN	funtoken	WAN	wanchain
		GTO	gifto	WAVES	waves
		HBAR	hedera	WIN	winklink
		HOT	holo	XLM	stellar
		ICX	icon	XMR	ripple
		IOST	iost	XRP	monero
		IOTA	miota	XTZ	tezos
		IOTX	iotex	ZEC	zcash
		KAVA	kava	ZIL	zilliqa
		KEY	key	ZRX	0x

References

1. Gerlach, J.-C.; Demos, G.; Sornette, D. Dissection of Bitcoin's multiscale bubble history from January 2012 to February 2018. *R. Soc. Open Sci.* **2019**, *6*, 180643. [CrossRef]
2. Corbet, S.; Lucey, B.; Urquhart, A.; Yarovaya, L. Cryptocurrencies as a financial asset: A systematic analysis. *Int. Rev. Financ. Anal.* **2019**, *62*, 182–199. [CrossRef]
3. Flori, A. Cryptocurrencies in finance: Review and applications. *Int. J. Theor. Appl. Finance* **2019**, *22*, 1950020. [CrossRef]
4. Bariviera, A. F.; Merediz-Solà, I. Where do we stand in cryptocurrencies economic research? A survey based on hybrid analysis. *J. Econ. Surv.* **2021**, *35*, 377–407. [CrossRef]
5. Wątorek, M.; Drożdż, S.; Kwapień, J.; Minati, L.; Oświęcimka, P.; Stanuszek, M. Multiscale characteristics of the emerging global cryptocurrency market. *Phys. Rep.* **2021**, *901*, 1–82. [CrossRef]
6. Zhang, D.; Hu, M.; Ji, Q. Financial markets under the global pandemic of COVID-19. *Fin. Res. Lett.* **2020**, *36*, 101528. [CrossRef]
7. Buszko, M.; Orzeszko, W.; Stawarz M. COVID-19 pandemic and stability of stock market - A sectoral approach. *PLoS ONE* **2021**, *16*, e0250938. [CrossRef]
8. James, N.; Menzies, M. Association between COVID-19 cases and international equity indices. *Physica D* **2021**, *417*, 132809. [CrossRef]
9. James, N.; Menzies, M. Efficiency of communities and financial markets during the 2020 pandemic. *Chaos* **2021**, *31*, 083116. [CrossRef]
10. Chahuán-Jiménez, K.; Rubilar, R.; de la Fuente-Mella, H.; Leiva, V. Breakpoint Analysis for the COVID-19 Pandemic and Its Effect on the Stock Markets. *Entropy* **2021**, *23*, 100. [CrossRef] [PubMed]
11. Maheu, J.M.; McCurdy, T.H. Song, Y. Bull and bear markets during the COVID-19 pandemic. *Fin. Res. Lett.* **2021** *42*, 102091. [CrossRef]
12. Song, R.; Shu, M.; Zhu, W. The 2020 global stock market crash: Endogenous or exogenous? *Physica A* **2022**, *585*, 126425. [CrossRef]
13. Drożdż, S.; Kwapień, J.; Oświęcimka, P.; Stanisz, T.; Wątorek, M. Complexity in economic and social systems: Cryptocurrency market at around COVID-19. *Entropy* **2020**, *22*, 1043. [CrossRef]
14. Mnif, E.; Jarboui, A.; Mouakhar, K. How the cryptocurrency market has performed during COVID-19? A multifractal analysis. *Financ. Res. Lett.* **2020**, *36* 101647. [CrossRef]
15. Conlon, T.; Corbet, S.; McGee, R.J. Are cryptocurrencies a safe haven for equity markets? An international perspective from the COVID-19 pandemic. *Res. Int. Bus. Financ.* **2020**, *54*, 101248. [CrossRef]
16. Demir, E.; Bilgin, M.H.; Karabulut, G.; Doker, A.C. The relationship between cryptocurrencies and COVID-19 pandemic. *Eurasian Econ. Rev.* **2020**, *10*, 349–360. [CrossRef]
17. Kristoufek, L. Grandpa, grandpa, tell me the one about Bitcoin being a safe haven: New evidence from the COVID-19 pandemic. *Front. Phys.* **2020**, *8*, 296. [CrossRef]
18. Goodell, J.W.; Goutte, S. Co-movement of COVID-19 and Bitcoin: Evidence from wavelet coherence analysis. *Fin. Res. Lett.* **2021**, *38*, 101625. [CrossRef]

19. James, N. Dynamics, behaviours, and anomaly persistence in cryptocurrencies and equities surrounding COVID-19. *Physica A* **2021**, *570*, 125831. [CrossRef]
20. James, N.; Menzies, M.; Chan, J. Changes to the extreme and erratic behaviour of cryptocurrencies during COVID-19. *Physica A* **2021**, *565*, 125581. [CrossRef]
21. Wątorek, M.; Kwapień, J.; Drożdż, S. Financial return distributions: Past, present, and COVID-19. *Entropy* **2021** *23*, 884. [CrossRef]
22. Gandal, N.; Hamrick, J.T.; Moore, T.; Oberman, T. Price manipulation in the bitcoin ecosystem. *J. Monetary Econ.* **2018**, *95* 86–96. [CrossRef]
23. de la Horra, L.P.; de la Fuente, G.; Perote, J. The drivers of bitcoin demand: A short and long-run analysis. *Int. Rev. Fin. Anal.* **2019**, *62*, 21–34. [CrossRef]
24. Urquhart, A. What causes the attention of bitcoin? *Econ. Lett.* **2018**, *166*, 40–44. [CrossRef]
25. Aalborg, H.A.; Molnr, P.; de Vries, J.E. What can explain the price, volatility and trading volume of bitcoin? *Fin. Res. Lett.* **2019**, *29*, 255–265. [CrossRef]
26. Drożdż, S.; Gębarowski, R.; Minati, L.; Oświęcimka, P.; Wątorek, M. Bitcoin market route to maturity? Evidence from return fluctuations, temporal correlations and multiscaling effects. *Chaos* **2018**, *28*, 071101. [CrossRef] [PubMed]
27. Gkillas, K.; Katsiampa, P. An application of extreme value theory to cryptocurrencies. *Econ. Lett.* **2018**, *164*, 109–111. [CrossRef]
28. Katsiampa, P. Volatility co-movement between Bitcoin and Ether. *Fin. Res. Lett.* **2019**, *30*, 221–227. [CrossRef]
29. Godfrey, K.R. Toward a model-free measure of market efficiency. *Pacific-Basin Fin. J.* **2017**, *44*, 97–112. [CrossRef]
30. Dyhrberg, A.H. Bitcoin, gold and the dollar—A GARCH volatility analysis. *Fin. Res. Lett.* **2016**, *16*, 85–92. [CrossRef]
31. Corbet, S.; Meegan, A.; Larkin, C.; Lucey, B.; Yarovaya, L. Exploring the dynamic relationships between cryptocurrencies and other financial assets. *Econ. Lett.* **2018**, *165*, 28–34. [CrossRef]
32. Fry, J.; Cheah, E.-T. Negative bubbles and shocks in cryptocurrency markets. *Int. Rev. Fin. Anal.* **2016**, *47*, 343–352. [CrossRef]
33. Baur, D.G.; Dimpfl, T.; Kuck, K. Bitcoin, gold and the US dollar—A replication and extension. *Fin. Res. Lett.* **2018**, *25*, 103–110. [CrossRef]
34. Manavi, S.A.; Jafari, G.; Rouhani, S.; Ausloos, M. Demythifying the belief in cryptocurrencies decentralized aspects. A study of cryptocurrencies time cross-correlations with common currencies, commodities and financial indices. *Physica A* **2020**, *556*, 124759. [CrossRef]
35. Ferreira, P.; Kristoufek, L.; Pereira, E.J.D.A.L. DCCA and DMCA correlations of cryptocurrency markets. *Physica A* **2020**, *545*, 123803. [CrossRef]
36. Kristoufek, L. What are the main drivers of the Bitcoin price? Evidence from wavelet coherence analysis. *PLoS ONE* **2015**, *10*, 1–15. [CrossRef] [PubMed]
37. Ji, Q.; Bouri, E.; Gupta, R.; Roubaud, D. Network causality structures among Bitcoin and other financial assets: A directed acyclic graph approach. *Quart. Rev. Econ. Financ.* **2018**, *70*, 203–213. [CrossRef]
38. Corelli, A. Cryptocurrencies and exchange rates: A relationship and causality Analysis. *Risks* **2018**, *6*, 111. [CrossRef]
39. Urquhart, A. The inefficiency of bitcoin. *Econ. Lett.* **2016**, *148*, 80–82. [CrossRef]
40. Bariviera, A.F. The inefficiency of Bitcoin revisited: A dynamic approach. *Econ. Lett.* **2017**, *161*, 1–4. [CrossRef]
41. Drożdż, S.; Minati, L.; Oświęcimka, P.; Stanuszek, M.; Wątorek, M. Signatures of crypto-currency market decoupling from the Forex. *Future Internet* **2019**, *11*, 154. [CrossRef]
42. Yi, S.; Xu, Z.; Wang, H.-J. Volatility connectedness in the cryptocurrency market: Is Bitcoin a dominant cryptocurrency? *Int. Rev. Fin. Anal.* **2018**, *60*, 98–114. [CrossRef]
43. Aste, T. Cryptocurrency market structure: Connecting emotions and economics. *Digital Fin.* **2019**, *1*, 5–21. [CrossRef]
44. Ferreira, P.; Pereira, É. Contagion effect in cryptocurrency market. *J. Risk Fin. Man.* **2019**, *12*, 115. [CrossRef]
45. Aslanidis, N.; Bariviera, A.F.; Perez-Laborda, A. Are cryptocurrencies becoming more interconnected? *Econ. Lett.* **2021**, *199*, 109725. [CrossRef]
46. Corbet, S.; Hou, Y.G.; Hu, Y.; Larkin, C.; Oxley, L. Any port in a storm: Cryptocurrency safe-havens during the COVID-19 pandemic. *Econ. Lett.* **2020**, *194*, 109377. [CrossRef]
47. Mariana, C.D.; Ekaputra, I.A.; Husodo, Z.A. Are Bitcoin and Ethereum safe-havens for stocks during the COVID-19 pandemic? *Fin. Res. Lett.* **2021**, *38*, 101798. [CrossRef]
48. Lahmiri, S.; Bekiros, S. The impact of COVID-19 pandemic upon stability and sequential irregularity of equity and cryptocurrency markets. *Chaos Solit. Fract.* **2020**, *138*, 109936. [CrossRef]
49. Grobys, K. When Bitcoin has the flu: On Bitcoin's performance to hedge equity risk in the early wake of the COVID-19 outbreak. *Appl. Econ. Lett.* **2021**, *28*, 860–865. [CrossRef]
50. Jiang, Y.; Lie, J.; Wang, J.; Mu, J. Revisiting the roles of cryptocurrencies in stock markets: A quantile coherency perspective. *Econ. Model.* **2021**, *95*, 21–34. [CrossRef]
51. Shahzad, S.J.H.; Bouri, E.; Roubaud, D.; Kristoufek, L. Safe haven, hedge and diversification for G7 stock markets: Gold versus bitcoin. *Econ. Model.* **2019**, *87*, 212–224. [CrossRef]
52. Conlon, T.; McGee, R. Safe haven or risky hazard? Bitcoin during the COVID-19 bear market. *Fin. Res. Lett.* **2020**, *35*, 101607. [CrossRef]
53. Wang, P.; Zhang, W.; Li, X.; Shen, D. Is cryptocurrency a hedge or a safe haven for international indices? A'comprehensive and dynamic perspective. *Fin. Res. Lett.* **2019**, *31*, 1–18. [CrossRef]

54. Cheah, E.T.; Fry, J. Speculative bubbles in bitcoin markets? An empirical investigation into the fundamental value of bitcoin. *Econ. Lett.* **2015**, *130*, 32–36. [CrossRef]
55. Drożdż, S.; Minati, L.; Oświęcimka, P.; Stanuszek, M.; Wątorek, M. Competition of noise and collectivity in global cryptocurrency trading: Route to a self-contained market. *Chaos* **2020**, *30*, 023122. [CrossRef]
56. Podobnik, B.; Stanley, H.E. Detrended cross-correlation analysis: A new method for analyzing two nonstationary time series. *Phys. Rev. Lett.* **2008**, *100*, 1–4. [CrossRef]
57. Zhou, W.-X. The components of empirical multifractality in financial returns. *EPL* **2009**, *88*, 28004. [CrossRef]
58. Zebende, G.F. DCCA cross-correlation coefficient: Quantifying level of cross-correlation. *Physica A* **2011**, *390*, 614–618. [CrossRef]
59. Kwapień, J.; Oświęcimka, P.; Drożdż, S. Detrended fluctuation analysis made flexible to detect range of cross-correlated fluctuations. *Phys. Rev. E* **2015**, *92*, 052815. [CrossRef]
60. Pearson, K. Note on regression and inheritance in the case of two parents. *Proc. R. Soc. Lond.* **1895**, *58*, 240–242.
61. Peng, C.-K.; Buldyrev, S.V.; Havlin, S.; Simons, M.; Stanley, H.E.; Goldberger, A.L. Mosaic organization of DNA nucleotides. *Phys. Rev. E* **1994**, *49*, 1685–1689. [CrossRef] [PubMed]
62. Oświęcimka, P.; Drożdż, S.; Forczek, M.; Jadach, S.; Kwapień, J. Detrended cross-correlation analysis consistently extended to multifractality. *Phys. Rev. E* **2014**, *89*, 023305. [CrossRef] [PubMed]
63. Kwapień, J.; Oświęcimka, P.; Forczek, M.; Drożdż, S. Minimum spanning tree filtering of correlations for varying time scales and size of fluctuations. *Phys. Rev. E* **2017**, *95*, 052313. [CrossRef] [PubMed]
64. Prim, R.C. Shortest connection networks and some generalizations. *Bell Syst. Tech. J.* **1957**, *36*, 1389–1401. [CrossRef]
65. Zięba, D.; Kokoszczyński, R.; Śledziewska, K. Shock transmission in the cryptocurrency market. Is Bitcoin the most influential? *Int. Rev. Financ. Anal.* **2019**, *64*, 102–125. [CrossRef]
66. Binance. Available online: https://www.binance.com/ (accessed on 10 November 2021).
67. Tether. Available online: https://tether.to (accessed on 10 November 2021).
68. Kwapień, J.; Drożdż, S. Physical approach to complex systems. *Phys. Rep.* **2012**, *515*, 115–226. [CrossRef]
69. Epps, T.W. Comovements in stock prices in the very short run. *J. Am. Stat. Assoc.* **1979**, *74*, 291–298.
70. Kwapień, J.; Drożdż, S.; Speth, J. Time scales involved in emergent market coherence. *Physica A* **2004**, *337*, 231–242. [CrossRef]
71. Drożdż, S.; Kwapień, J.; Oświęcimka, P.; Rak, R. The foreign exchange market: Return distributions, multifractality, anomalous multifractality and the Epps effect. *New J. Phys.* **2010**, *12*, 105003. [CrossRef]
72. Polovnikov, K.; Kazakov, V.; Syntulsky, S. Core-periphery organization of the cryptocurrency market inferred by the modularity operator. *Physica A* **2020**, *540*, 123075. [CrossRef]
73. Papadimitriou, T.; Gogas, P.; Gkatzoglou, F. The evolution of the cryptocurrencies market: A complex networks approach. *J. Comp. Appl. Math.* **2020**, *376*, 112831. [CrossRef]
74. García-Medina, A.; Hernández, J.B. Network analysis of multivariate transfer entropy of cryptocurrencies in times of turbulence. *Entropy* **2020**, *22*, 760. [CrossRef] [PubMed]
75. Blondel, V.D.; Guillaume, J.-L.; Lambiotte, R.; Lefebvre, E. Fast unfolding of communities in large networks. *J. Stat. Mech. Theory Exp.* **2008**, *10*, P10008. [CrossRef]
76. Stosić, D.; Stosić, D.; Ludermir, T.B.; Stosić, T. Collective behavior of cryptocurrency price changes. *Physica A* **2018**, *507*, 499–509. [CrossRef]
77. Dukascopy. Available online: https://www.dukascopy.com/ (accessed on 10 November 2021).
78. Statista. Available online: https://www.statista.com/statistics/863917/number-crypto-coins-tokens/ (accessed on 3 December 2021).

entropy

MDPI

Article

Analysis of Individual High-Frequency Traders' Buy–Sell Order Strategy Based on Multivariate Hawkes Process

Hiroki Watari [1], Hideki Takayasu [2,3] and Misako Takayasu [1,2,*]

[1] Department of Mathematical and Computing Science, School of Computing, Tokyo Institute of Technology, Yokohama 226-8502, Japan; watari.h.aa@m.titech.ac.jp
[2] Institute of Innovative Research, Tokyo Institute of Technology, Yokohama 226-8502, Japan; takayasu@csl.sony.co.jp
[3] Sony Computer Science Laboratories, Tokyo 141-0022, Japan
* Correspondence: takayasu.m.aa@m.titech.ac.jp

Abstract: Traders who instantly react to changes in the financial market and place orders in milliseconds are called high-frequency traders (HFTs). HFTs have recently become more prevalent and attracting attention in the study of market microstructures. In this study, we used data to track the order history of individual HFTs in the USD/JPY forex market to reveal how individual HFTs interact with the order book and what strategies they use to place their limit orders. Specifically, we introduced an 8-dimensional multivariate Hawkes process that included the excitations due to the occurrence of limit orders, cancel orders, and executions in the order book change, and performed maximum likelihood estimations of the limit order processes for 134 HFTs. As a result, we found that the limit order generation processes of 104 of the 134 HFTs were modeled by a multivariate Hawkes process. In this analysis of the EBS market, the HFTs whose strategies were modeled by the Hawkes process were categorized into three groups according to their excitation mechanisms: (1) those excited by executions; (2) those that were excited by the occurrences or cancellations of limit orders; and (3) those that were excited by their own orders.

Keywords: high-frequency trader; multivariate Hawkes process; econophysics; forex market

Citation: Watari, H.; Takayasu, H.; Takayasu, M. Analysis of Individual High-Frequency Traders' Buy–Sell Order Strategy Based on Multivariate Hawkes Process. *Entropy* **2022**, *24*, 214. https://doi.org/10.3390/e24020214

Academic Editors: Ryszard Kutner, Christophe Schinckus and H. Eugene Stanley

Received: 7 January 2022
Accepted: 28 January 2022
Published: 29 January 2022

1. Introduction

To gain a deeper understanding of the mechanisms of financial markets, it is necessary to clarify the order strategies of individual market participants. In financial markets, recent developments in information technology have made it possible to track the transactions of individual market participants in detail. These technological advances have led to the analysis of the trading strategies of individual market participants and how these strategies affect financial markets. For example, Odean [1], and Grinblatt and Keloharju [2], reported the relationship between historical returns and market participants' decisions to buy and sell stocks. The position management strategies of individual market participants were analyzed based on the data, which confirmed that these strategies actually affected market prices in the near future [3]. Individual strategies for placing buy and sell orders in response to market price changes were approximated using a simple mathematical model, and the basic statistical properties of financial Brownian motion were theoretically derived based on the kinetic theory in a manner parallel to traditional statistical physics [4–6].

In particular, high-frequency traders (HFTs) have recently attracted attention. HFTs are algorithmic traders who can react to market changes in milliseconds and place or cancel, buy and sell orders at high frequencies [7]. Because of the development of information technology, they have a large presence in financial markets around the world. In fact, HFTs accounted for 68.3% of the total trading volume in the stock market [8]. Furthermore, HFTs currently account for the majority of orders shown in the order book [9–11]. The availability of high-frequency trading data has triggered the academic study of HFTs [12,13]. Previous studies have generally agreed that HFTs make market spreads smaller and enhance market liquidity [8,14,15]. As an indicator for predicting the short-term volatility of market prices from the order book information, the volume-synchronized probability of informed trading (VPIN) has been proposed and actively studied [16–20]. In addition, informed trading using the advantage of information such as public news and confidential information has been studied using high-frequency data [21–25]. We believe that it is crucial to gain a deep understanding of their trading behavior in current financial markets, where HFTs provide most of the liquidity.

In this study, we used a multivariate Hawkes process to investigate the processes used by individual HFTs for generating sell and buy limit orders in the USD/JPY forex market, and clarified when each HFT placed buy–sell limit orders. The Hawkes process is a type of non-homogeneous Poisson process proposed by Hawkes [26]. As will be explained later, it is characterized by an intensity function which determines the probability of the occurrence of an event in a point process. It utilizes an excitation term that is affected by past events, and can describe a point process associated with past events. Similar ideas have been independently introduced for financial markets to explain the strong correlation to past events, such as the "autoregressive conditional duration model" [27] and "self-modulation processes" [28]. The Hawkes process is a useful model for interpreting financial phenomena, in which many factors interact to produce complex aspects. In this paper, we show that it is also useful for interpreting the behavior of HFTs. Specifically, we introduced a multivariate Hawkes process in which the process of generating HFTs' buy–sell limit orders is mutually excited by a total of eight events, such as the creation of limit orders, the cancellation of limit orders, and execution in the order book, showing that the order behaviors of many HFTs can be modeled by the Hawkes process.

Hawkes processes [29] have various applications in the financial field, such as those related to volatility clustering [30], market activity and risk [31–33], and market impact [34]. In particular, the Hawkes process has been actively employed as an approach to the dynamic description of order books, where a set of order types is specified and a multivariate Hawkes process is fitted to their timestamps [35–41]. However, there has been no study that used a multivariate Hawkes process to investigate the order generation processes of individual HFTs. In today's financial markets, where the majority of order books are made up of HFTs' orders, our empirical results provide new information from a more microscopic perspective. We believe that this study shed light on how HFTs provide liquidity to the market.

The remainder of this paper is organized as follows. Section 2 explains the datasets and describes the HFTs that were analyzed in this research. Section 3 introduces the multivariate Hawkes process and describes the method used for parameter estimation. In Section 4, a clustering analysis of 134 HFTs is introduced to categorize their strategies based on the estimated Hawkes' parameters. In Section 5, we discuss our results.

2. Data

First, we provide a basic description of the order data for the USD/JPY forex market (EBS market), along with individual trader IDs (see Section 2.1). We then define the HFTs in this market (see Section 2.2) and show some examples to explain how their buy–sell limit order generation is linked to changes in the order book (see Section 2.3).

2.1. EBS Market Data Description

In this study, we used high-frequency data for the USD/JPY forex market provided by the EBS. EBS is an interbank forex market and one of the largest financial platforms in the world. Because it is an interbank market, most market participants are professional traders from banks, hedge funds, and other financial institutions, and our forex dataset contains their trading data. Our dataset contains information from five days (from 21:00 GMT on 5 June 2016 to 21:00 GMT on 10 June 2016), with a total of approximately 2.8 million orders and a transaction volume of USD 68 billion corresponding to this period. Table 1 shows an example of the raw data we used. The data for each of the 2.8 million orders contained not only the order type, price, volume, and timestamp (in milliseconds), but also an anonymized trader ID that could identify who submitted the order. Using these trader IDs, we could track individual traders' full orders in milliseconds. In addition, the minimum price unit that a trader could submit was JPY 0.005, and the minimum transmission volume was USD 1 million.

Table 1. Examples of raw data. Each order datum is tied to an anonymized trader ID.

Date	Order Time	Trader ID	Order Type	USD/JPY	Volume	Deal Time
5 June 2016	21:00:12.946	578	Sell limit	106.515	1	–
5 June 2016	21:01:13.647	HT6	Buy cancel	105.390	2	–
5 June 2016	21:02:20.148	JR1	Buy limit	105.405	1	21:02:20.499
5 June 2016	21:02:20.499	HSH	Sell market	105.405	1	21:02:20.499
5 June 2016	21:03:00.950	7KP	Bid market	106.470	1	–
⋮	⋮	⋮	⋮	⋮	⋮	⋮
10 June 2016	20:59:20.148	HT6	Buy Limit	107.405	3	20:59:29.072

The EBS market is open 24 h a day from Monday morning to Friday at midnight, and trading is conducted via a double auction system in the order book. Figure 1 shows a schematic of the trading in the order book, where the horizontal axis is the price and the vertical axis is the volume. There are six order types for trading: buy/sell limit orders, buy/sell cancel orders, and buy/sell market orders. Limit orders are submitted at the trader's desired price and remain in the order book until traded or cancelled. Cancel orders are submitted by a trader to cancel a limit order that they previously submitted. Market orders are submitted at the current best limit price. Transactions that are executed by buy market orders are called hit sell transactions, and transactions executed by sell market orders are called hit buy transactions (see Figure 1). If the best price worsens (e.g., the best sell limit price becomes higher) before the market receives the market order at the best price, the market order is automatically invalidated. In fact, this study found that 79.5% of the market orders were invalidated without being executed.

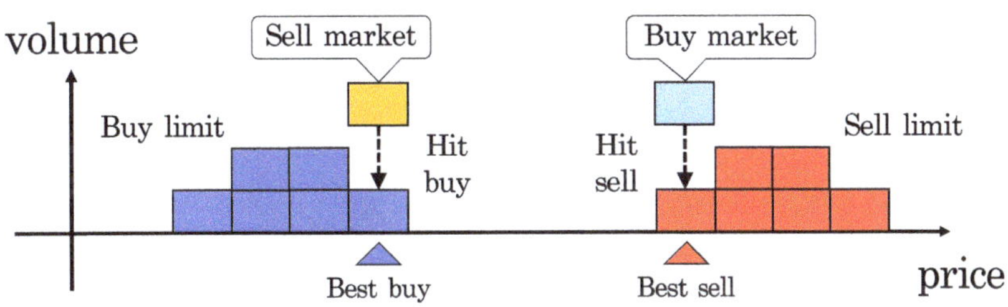

Figure 1. Schematic of trading in order book. In the EBS market, even a sell (buy) limit order becomes a hit buy (sell) if a buy (sell) limit order at the same price is already in the order book.

Figure 2a shows the average trading price per 10 min window over the 5 days we analyzed. During this period, there are no market crashes or spikes. Figure 2b shows the number of each type of order per day, which looks stable.

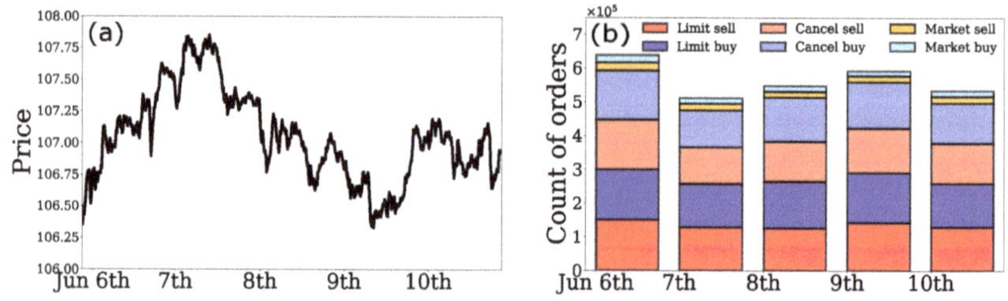

Figure 2. (**a**) Average trading price per 10 min window; (**b**) daily number of orders for the 6 types of orders in unit of 10^5.

2.2. Definition of HFTs

"HFTs" is a general term for traders who place and cancel orders at high speed and high frequency according to an algorithm, but there are various definitions. In this study, we define an HFT as a trader who places both buy and sell limit orders and presents an average of 500 or more limit orders per day following the previous report written by a researcher from EBS [9]. Based on this definition, the number of HFTs was 134 out of the 1031 traders included in this 5-day data set. These 134 HFTs accounted for 89.6% of the market's total number of limit orders.

Figure 3a shows the histogram of the minimum time interval between orders for each HFT. There is no description in the data to identify whether the ID is a human or a computer; however, Figure 3a shows that most of the intervals are within 0.1 s, which are difficult for a human to execute.

In Figure 3b, we plot the number of HFTs and non-HFTs participating in the market every hour, indicating that the number of HFTs is relatively stable compared to non-HFTs. Figure 3c shows the percentage of limit orders placed by HFTs every hour, demonstrating that the majority of the limit orders are provided by the HFTs.

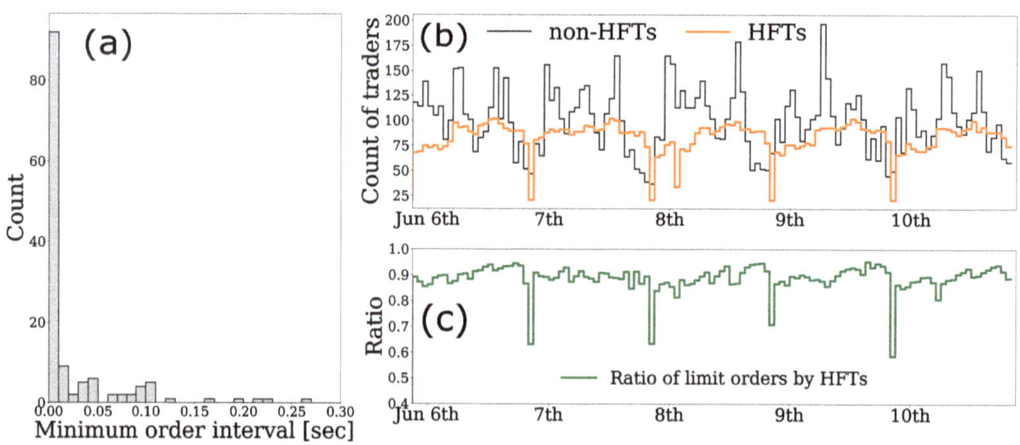

Figure 3. (**a**) Histogram of the minimum time intervals between orders for 134 HFTs individually; (**b**) hourly changes in the number of HFTs and non-HFTs participating in trading; and (**c**) hourly change of the percentage of limit orders provided by HFTs.

2.3. Basic Properties of HFTs

In this study, we focused on the limit order generation process of HFTs, which accounted for the majority of limit orders in the order book. Naturally, the order strategies (i.e., the processes used to submit limit orders) of HFTs differed from every algorithm. However, it is natural for them to see the quotes in the order book when submitting their limit orders. Figure 4a,b plot the numbers of buy–sell limit orders per 10 min window for three HFTs, respectively, and Figure 4c plots the numbers for six types of orders in the order book. From Figure 4, we can observe that the numbers of buy–sell limit orders from the three HFTs increased or decreased simultaneously and tended to be in sync. More interestingly, the numbers of these HFTs' buy–sell limit orders tend to be in sync with the numbers for each type of order in the order book where all market participants' orders are submitted. Since the above synchronization phenomenon was confirmed for many HFTs, we believe that many HFTs react instantaneously to some changes in the market when submitting limit orders.

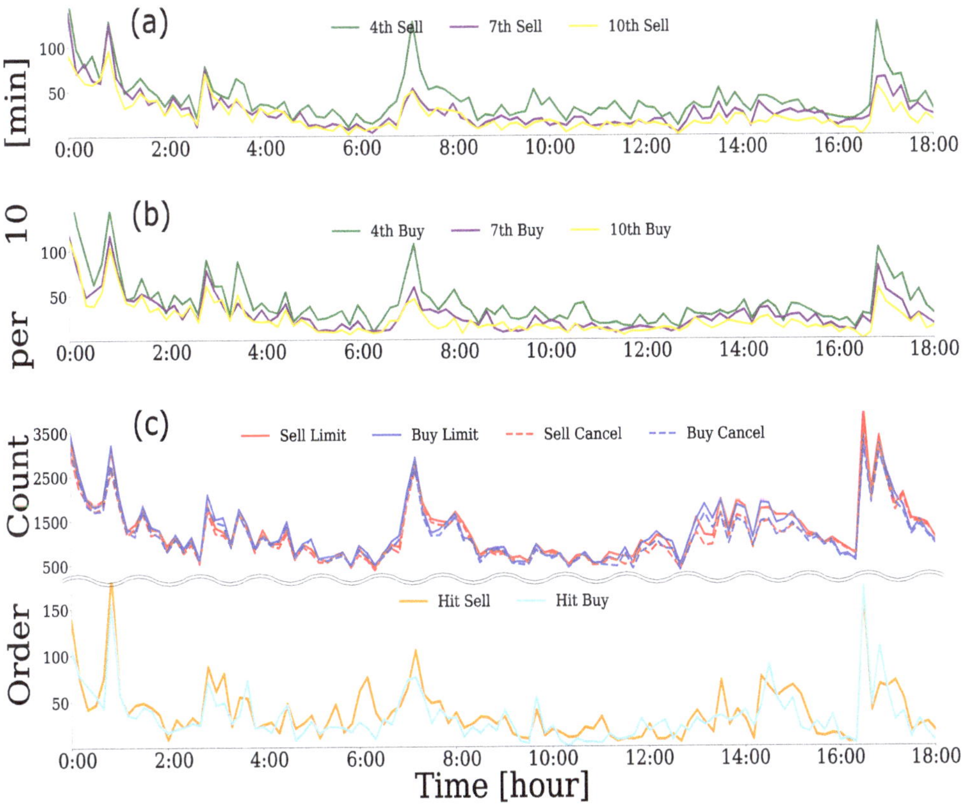

Figure 4. Numbers of (**a**) sell limit orders and (**b**) buy limit orders per 10 min window for three HFTs (green: HFT with 4th highest order frequency; purple: HFT with 7th highest order frequency; yellow: HFT with 10th highest order frequency). (**c**) Numbers for six types of orders per 10 min window in the order book (red: sell limit order; blue: buy limit order; red dotted line: sell cancel; blue dotted line: buy cancel; orange: hit sell; sky blue: hit buy). The vertical axis of each figure shows the number of each type of order per 10 min window, and the horizontal axis shows the time from 0:00 to 18:00 on 6 June 2016.

3. Method

The preceding section showed that the limit order generation processes of the HFTs tended to be in sync with traders' orders or orders in the order book. To clarify how 134 HFTs' buy–sell limit orders react to the other order events, their order processes are modeled by the multivariate Hawkes process. In this section, we introduce the multivariate Hawkes process that we used in this study (see Section 3.1) and explain the parameter estimation method based on maximum likelihood estimation (see Section 3.2). We then describe the validity of the estimation results (see Section 3.3).

3.1. Model

This section presents an overview of the Hawkes process and introduces our model.

3.1.1. Mathematical Notation

Let us consider point process $\{t_i\}$, which is a sequence of non-negative random variables such that $\forall i \in \mathbb{N}$, $t_i < t_{i+1}$. For point process $\{t_i\}$, the conditional intensity function is defined as follows [42]:

$$\lambda(t \mid \mathcal{H}_t) = \lim_{\Delta t \to 0} \frac{\mathbb{P}\{N(t+\Delta t) - N(t) = 1 \mid \mathcal{H}_t\}}{\Delta t} \tag{1}$$

where $\mathbb{P}\{A|B\}$ represents the probability of A under condition B, $N(t)$ is the cumulative number of event occurrences at time t (i.e., a counting process), and $\mathcal{H}_t$ is the history of the process up to time t containing a sequence of event times $\{t_i\}$ (i.e., a filtration). As can be seen from the definition, $\lambda(t \mid \mathcal{H}_t)\Delta t$ represents the probability of an event occurring in time interval $[t, t+\Delta t)$. Here, we use the shorthand notation $\lambda(t) \equiv \lambda(t \mid \mathcal{H}_t)$, assuming the history up to time t, and we call the Poisson parameter $\lambda(t)$ an intensity function.

3.1.2. Overview of Hawkes Process

The Hawkes process is a point process in which the intensity function is affected by the occurrence of past events. Let $\{t_i\}$ be a point process, and $N(t)$ be the associated counting process, and the intensity function of the generalized Hawkes process is defined as follows [43]:

$$\lambda(t) = c + \int_{-\infty}^{t} \phi(t-s)dN(s) = c + \sum_{t_i<t} \phi(t-t_i) \tag{2}$$

where c is a positive constant showing a base intensity, and $\phi(t)$ is a kernel function that expresses the effect of event t_i from the past on the current intensity [44]. There are various types of kernel functions, and their properties have been well studied. In this study, we applied an exponential kernel $\alpha e^{-\beta t}$, which is a popular kernel function that was originally proposed by Hawkes [26]. We call this Hawkes process a univariate Hawkes process because it is affected by its own events.

The Hawkes process can be extended to a multivariate model in which several types of point processes interact with each other. Let $\{\mathbf{t}_i\} \equiv \{\{t_{1,i}\}, \{t_{2,i}\}, \ldots, \{t_{M,i}\}\}$ be M-variable point processes, and $\mathbf{N}(t) = \{N_1(t), N_2(t), \ldots, N_M(t)\}$ be the associated counting process, the intensity function of a multivariate Hawkes process for point process $\{t_{n,i}\}$ is defined as follows [43]:

$$\lambda_n(t) = c_n + \sum_{m=1}^{M} \int_{-\infty}^{t} \phi_{n,m}(t-s)dN_m(s) = c_n + \sum_{m=1}^{M} \sum_{t_{m,i}<t} \phi_{n,m}(t-t_{m,i}) \tag{3}$$

As in the case of the univariate Hawkes process, c_n is a positive constant showing a base intensity, and $\phi_{n,m}(t)$ is a kernel function that expresses the effect of event $t_{m,i}$ from the past on the current intensity. In this case, $\phi_{n,m}(t) = \alpha_{n,m}e^{-\beta_{n,m}t}$, with positive constant parameters $\alpha_{n,m}$ and $\beta_{n,m}$, and the intensity function for point process $\{t_{n,i}\}$ is given as follows:

$$\begin{aligned}
\lambda_n(t) &= c_n + \sum_{m=1}^{M} \int_{-\infty}^{t} \alpha_{n,m}e^{-\beta_{n,m}(t-s)}dN_m(s) \\
&= c_n + \sum_{m=1}^{M} \sum_{t_{m,i}<t} \alpha_{n,m}\exp(-\beta_{n,m}(t-t_{m,i}))
\end{aligned} \tag{4}$$

Figure 5 is a schematic of this intensity function (Equation (4)). The intensity function, $\lambda_n(t)$, increases α_{nm} at event time $t_{n,i}$ and exponentially decays with a time constant of

$1/\beta_{nm}$. Thus, $\lambda_n(t)$ is excited not only by its own events, but also by other events, and the multivariate Hawkes process can represent such mutual interactions.

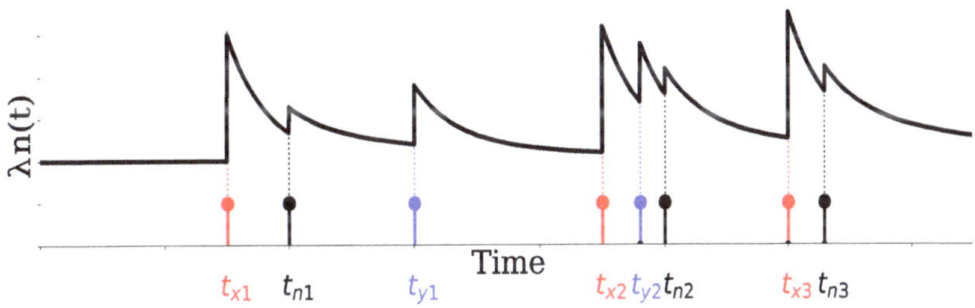

Figure 5. Schematic showing an example of the time evolution of the intensity function for a multivariate Hawkes process with the exponential kernel $\phi_{n,m}(t) = \alpha_{n,m}e^{-\beta_{n,m}t}$

The quantity, $\rho_{n,m}$, expressed in the following equation (Equation (5)) in the case of an exponential kernel is called the branching ratio [45,46]. This is the expectation of the number of occurrences of event n caused by the occurrence of event m. A larger value for this number represents a greater impact of event m on event n, and this value is an important quantity for interpreting the Hawkes process:

$$\rho_{n,m} \equiv \frac{\alpha_{n,m}}{\beta_{n,m}} = \int_{t_{m,i}}^{\infty} \alpha_{n,m} \exp(-\beta_{n,m}(t - t_{m,i}))dt \tag{5}$$

3.1.3. Trader Model

In this study, the buy and sell limit order processes of 134 HFTs are modeled by eight-variable Hawkes processes with exponential kernels that are excited by a total of eight-point processes. These eight types were the target HFTs' sell limit (TS) and buy limit (TB), and six types of orders in the order book: sell limit (SL); buy limit (BL); sell cancel (SC); buy cancel (BC); hit sell (HS); and hit buy (HB).

Let $\{\mathbf{t}_i\} \equiv \{\{t_{TS,i}\}, \{t_{TB,i}\}, \{t_{SL,i}\}, \{t_{BL,i}\}, \{t_{SC,i}\}, \{t_{BC,i}\}, \{t_{HS,i}\}, \{t_{HB,i}\}\}$ be an eight-variable point process, the intensity functions for $\{t_{TS,i}\}$ and $\{t_{TB,i}\}$ are given as follows:

$$\lambda_{TS}(t) = c_{TS} + \sum_{m \in \mathbb{M}} \sum_{t_{m,i}<t} \alpha_{TS,m} \exp(-\beta_{TS,m}(t - t_{m,i})) \tag{6}$$

$$\lambda_{TB}(t) = c_{TB} + \sum_{m \in \mathbb{M}} \sum_{t_{m,i}<t} \alpha_{TB,m} \exp(-\beta_{TB,m}(t - t_{m,i})) \tag{7}$$

where $\mathbb{M} \equiv \{TS, TB, SL, BL, SC, BC, HS, HB\}$. We assume that the above intensity functions (Equations (6) and (7)) represent the buy and sell limit order processes of each HFT and examine which events affect their order generation processes. Hereafter, the abbreviations listed in Table 2 are used for the order events.

Table 2. Abbreviations for eight types of order events. Note the six types of orders in the order book do not include the orders of target HFTs. Therefore, $\{\mathbf{t}_i\}$ differs for each HFT.

TS	:	Sell limit of the target HFT itself	**TB**	:	Buy limit of the target HFT itself
SL	:	Sell limit in order book	**BL**	:	Buy limit in order book
SC	:	Sell cancel in order book	**BC**	:	Buy cancel in order book
HS	:	Hit sell	**HB**	:	Hit buy

3.2. Parameter Estimation Using Maximum Likelihood Estimation

For Equations (6) and (7), we apply the maximum likelihood estimation method for the parameter estimation. Because the functional forms of the intensity functions are the same for $\{t_{TS,i}\}$ and $\{t_{TB,i}\}$, we solve the log-likelihood functions in the following manner.

For the point process $\{t_{TS,i}\}$, the log-likelihood function in time interval $[0,T]$ is given by the following [47]:

$$\begin{aligned}\log L(c_{TS},\boldsymbol{\alpha}_{TS},\boldsymbol{\beta}_{TS}) &= -\int_0^T \lambda_{TS}(t)dt + \sum_{i=1}^{n} \log \lambda_{TS}(t_i) \\ &= -c_{TS}T + \sum_{m\in\mathbb{M}} \frac{\alpha_{TS,m}}{\beta_{TS,m}}\left[\sum_{t_{m,i}<T} \{exp(-\beta_{TS,m}(T-t_{m,i})-1\}\right] \\ &\quad + \sum_{i=1}^{n} \log\left[c_{TS} + \sum_{m\in\mathbb{M}} \sum_{t_{m,j}<t_{TS,i}} \alpha_{TS,m} exp(-\beta_{TS,m}(t_{TS,i}-t_{m,j}))\right] \end{aligned} \quad (8)$$

The same formulation is also applied for $\log L(c_{TB},\boldsymbol{\alpha}_{TB},\boldsymbol{\beta}_{TB})$.

These log-likelihood functions are differentiable by each parameter, and we optimized them by Adam [48], which is a type of gradient descent method, to obtain the maximum likelihood estimators. Here, the initial values are $(c_{TS},\boldsymbol{\alpha}_{TS},\boldsymbol{\beta}_{TS}) = (c_{TB},\boldsymbol{\alpha}_{TB},\boldsymbol{\beta}_{TB}) = (0.1,\mathbf{0.1},\mathbf{10})$, and the various parameters required for Adam are set following the values in the original reference [48]. Here, $\boldsymbol{\alpha}_{TS}$, $\boldsymbol{\beta}_{TS}$, $\boldsymbol{\alpha}_{TB}$, and $\boldsymbol{\beta}_{TB}$ are vectors of eight variables (e.g., $\boldsymbol{\alpha}_{TS} = (\alpha_{TS,TS},\alpha_{TS,TB},\ldots,\alpha_{TS,HS})$).

It is also known that the computation of the gradients of the log-likelihood function of the Hawkes process usually requires the computation of $O(N^2)$. However, using a recursive formulation that can be used when the kernel function is an exponential function, we perform maximum likelihood estimation with $O(N)$ computational complexity (see Ogata [49] for the recursive formulation).

In addition, the HFTs do not always continuously place orders. Therefore, if an HFT did not place any orders for more than 15 min, we considered such period as not participating in the trade and performed the maximum likelihood estimation for each trader by ignoring such inactive periods.

3.3. Validity of Estimation Results

Based on a residual analysis [50], we assessed the goodness-of-fit of the point process model. Let the intensity function for point process $\{t_i\}$ be $\lambda(t)$, then the sequence, $\{\tau_i\}$, of random variables, where each element is transformed by $\tau_i \equiv \int_0^{t_i} \lambda(t)dt$, has the distribution of a stationary Poisson process with intensity 1, and the transformed residual $\Delta\tau_i \equiv \tau_{i+1} - \tau_i$ has an exponential distribution with the unit mean.

Therefore, if the estimated HFT intensities, $\hat{\lambda}_{TS}(t)$ and $\hat{\lambda}_{TB}(t)$, are good approximations of the true intensities, $\lambda_{TS}(t)$ and $\lambda_{TB}(t)$, respectively, then the transformed residuals, $\Delta\hat{\tau}_{TS,i}$ and $\Delta\hat{\tau}_{TB,i}$, are expected to follow the exponential distributions with the unit mean. Here, the transformed residuals are defined as $\Delta\hat{\tau}_{TS,i} \equiv \int_{t_{TS,i}}^{t_{TS,i+1}} \hat{\lambda}_{TS}(t)dt$ and $\Delta\hat{\tau}_{TB,i} \equiv \int_{t_{TB,i}}^{t_{TB,i+1}} \hat{\lambda}_{TB}(t)dt$, respectively, which can be derived in the case of $\Delta\hat{\tau}_{TS,i}$ as an example (Equation (9)) because the intensity function has the same form as described above:

$$\begin{aligned}\Delta\hat{\tau}_{TS,i} = \hat{c}_{TS}\Delta t_{TS,i} - \textstyle\sum_{m\in\mathbb{M}}\sum_{t_{m,i}<t_{TS,i+1}} \frac{\hat{\alpha}_{n,m}}{\hat{\beta}_{n,m}}(exp(-\hat{\beta}_{TS,m}(t_{TS,i+1}-t_{m,i})) \\ -exp(-\hat{\beta}_{TS,m}(t_{TS,i}-t_{m,i})))\end{aligned} \quad (9)$$

Figure 6a shows the cumulative distribution of the three HFTs' original residuals, $\Delta t_{TS,i}$, and Figure 6b shows the cumulative distribution of the same three HFTs' transformed residuals, $\Delta\hat{\tau}_{TS,i}$. We confirmed that the transformed residuals for the three HFTs approximately followed the exponential distribution with the unit mean, which implied that the intensities of $\{t_{TS,i}\}$ were properly approximated. Because not all of the HFTs' order generation processes could be modeled by the Hawkes process proposed here, the above operations were performed for the sell and buy order processes of the 134 HFTs.

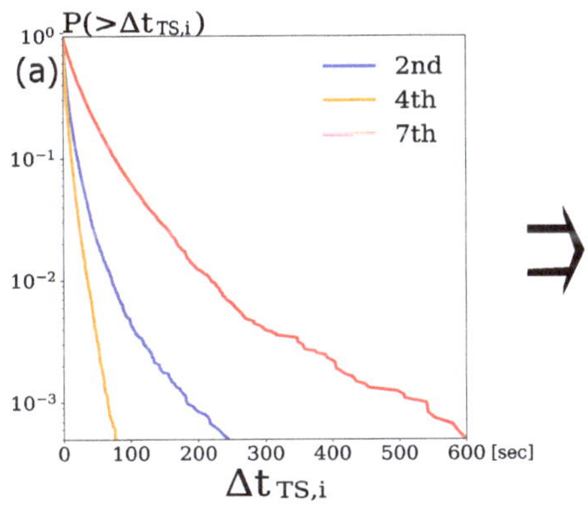

Figure 6. Cumulative distributions of $\Delta t_{TS,i}$ (**a**) and $\Delta \hat{\tau}_{TS,i}$ (**b**) for three typical HFTs (blue: HFT with 2nd highest order frequency; orange: HFT with 4th highest order frequency; and red: HFT with 7th highest order frequency).

Here, we apply the Kullback–Leibler divergence between the distribution of the transformed residuals and the exponential distribution with the unit mean as a criterion to determine whether the approximations of the intensities of each HFTs' $\{t_{TS,i}\}$ and $\{t_{TB,i}\}$ are appropriate. Since we estimate the intensity functions of the two-point processes for each trader, we calculate the sum of the respective Kullback–Leibler divergences (D^{KL}_{TS+TB}), as defined by Equation (10) below:

$$D^{KL}_{TS+TB} = \sum_j p_j(\Delta\hat{\tau}_{TS,i}) \log \frac{p_j(\Delta\hat{\tau}_{TS,i})}{q_j(\Delta\hat{\tau}_{TS,i})} + \sum_j p_j(\Delta\hat{\tau}_{TB,i}) \log \frac{p_j(\Delta\hat{\tau}_{TB,i})}{q_j(\Delta\hat{\tau}_{TB,i})} \tag{10}$$

Here, $q_j(\tau)$ is the discrete exponential distribution with the unit mean, and $p_j(\tau)$ is the sampling distribution. The bin size of the discrete distribution is assumed to be 1. The threshold of accepting D^{KL}_{TS+TB} error will be determined in the next section.

4. Results

In this section, we first show that the Hawkes process introduced here successfully approximated the intensity of the limit order generation process for 104 of the 134 HFTs by applying the method described in Section 3.3 (see Section 4.1). We then categorize the order generation processes of the 104 successfully estimated HFTs into three groups according to their excitation mechanisms, and explain how each group of HFTs places their orders (see Section 4.2).

4.1. D^{KL}_{TS+TB} Calculation Results for All HFTs

Figure 7 shows a histogram of the D^{KL}_{TS+TB} values for the 134 HFTs. From the histogram, there are many HFTs whose values of D^{KL}_{TS+TB} are very close to 0 and the plots are scattered for $D^{KL}_{TS+TB} > 0.05$, so we set the threshold as 0.05. The 104 HFTs who fell into the range of $D^{KL}_{TS+TB} < 0.05$ were considered to be traders whose order generation processes were well modeled, while the rest of the 30 HFTs who fell into the range of $D^{KL}_{TS+TB} \geq 0.05$ were considered to be traders whose order generation processes were poorly modeled by the Hawkes process introduced here.

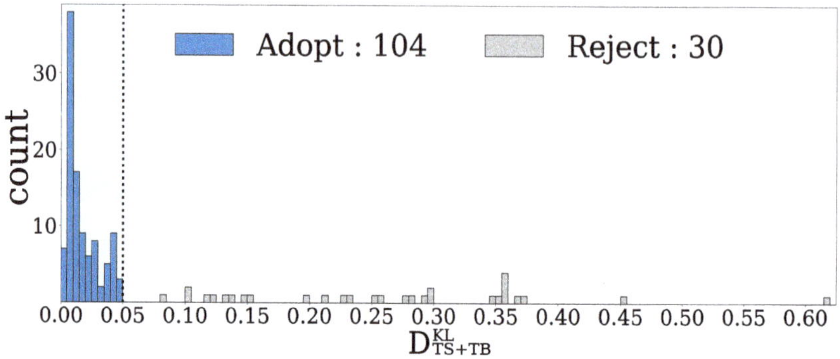

Figure 7. A histogram of D^{KL}_{TS+TB} values for all 134 HFTs. Out of 134 HFTs, 104 fell within the acceptable error threshold of 0.05, and the remaining 30 HFTs exceeded the threshold.

With the estimated parameters, the Hawkes process could be simulated using the thinning method [51]. For example, Figure 8 compares the time series of the number of orders per 10 min window for the real data of an HFT with $D^{KL}_{TS+TB} = 0.0238$ and a simulated time series. It can be confirmed that both sell limit orders (upper figure) and buy limit orders (lower figure) successfully reproduce the behavior of the real data.

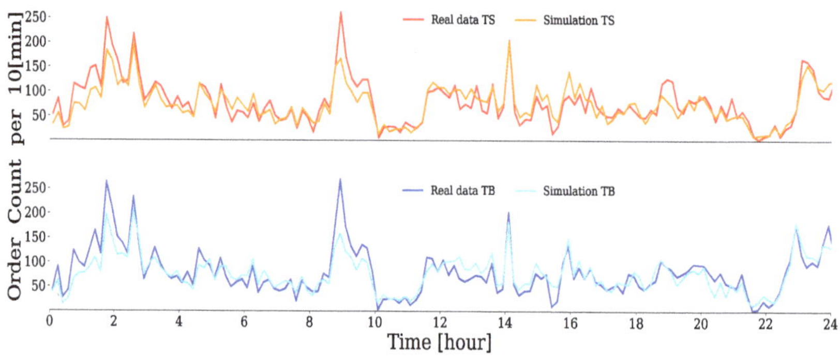

Figure 8. Comparison between simulations and real data of $\{t_{TS,i}\}$ and $\{t_{TB,i}\}$ for HFT with $D^{KL}_{TS+TB} = 0.0238$. The horizontal axis represents the time over a 24 h period, and the vertical axis represents the number of order occurrences per 10 min window.

On the other hand, Figure 9 shows a comparison of the simulated and real data for an HFT with $D^{KL}_{TS+TB} = 0.211$, which was judged not to be properly estimated by the Hawkes process. It can be seen that the deviations from the real data for both sell limit orders (upper figure) and buy limit orders (lower figure) are larger than in the case of Figure 8. The Hawkes process introduced here did not adequately explain the order generation process for this trader with a large error, D^{KL}_{TS+TB}. Because some traders could not be modeled by the Hawkes process, in the following, we report the results of our clustering analysis of the order generation processes of 104 HFTs after excluding 30 traders.

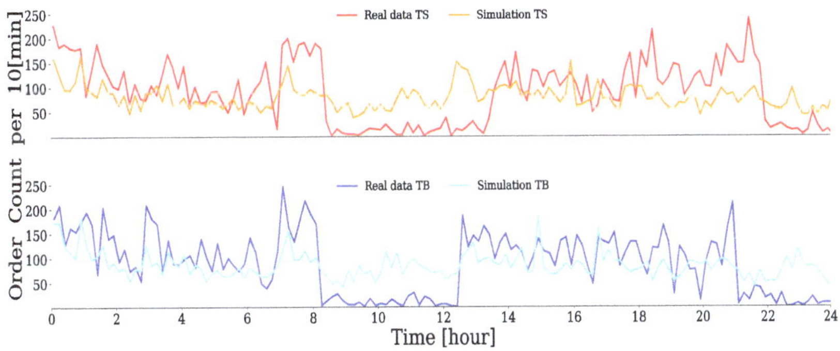

Figure 9. Comparison between simulations and real data of $\{t_{TS,i}\}$ and $\{t_{TB,i}\}$ for HFT with $D^{KL}_{TS+TB} = 0.211$. The horizontal axis represents the time during a 24 h period, and the vertical axis represents the number of order occurrences per 10 min window.

4.2. Results of Clustering Analysis

Here, we categorize the 104 HFTs whose limit order generation process was properly estimated according to the similarity of their excitation mechanisms (the branching ratio), and describe how each group of HFTs placed buy–sell limit orders and provided liquidity to the market. As defined in Equation (5) in Section 3.3, the branching ratio, $\rho_{n,m}$, is an absolute value that represents the expectation for the number of occurrences of event n caused by the occurrence of event m. To evaluate the excitations of $\{t_{TS,i}\}$ and $\{t_{TB,i}\}$ relative to each HFT, we introduced normalized branching ratios $\bar{\rho}_{TS,m}$ and $\bar{\rho}_{TB,m}$, which are defined by the following equations, so that the sum is equal to 1:

$$\bar{\rho}_{TS,m} \equiv \frac{\rho_{TS,m}}{\sum_{i \in \mathbb{M}} \rho_{TS,i}} \tag{11a}$$

$$\bar{\rho}_{TB,m} \equiv \frac{\rho_{TB,m}}{\sum_{i \in \mathbb{M}} \rho_{TB,i}} \tag{11b}$$

Because both $\{t_{TS,i}\}$ and $\{t_{TB,i}\}$ are 8-variable Hawkes processes, 16 normalized branching rates were defined for each HFT. Figure 10 shows a dendrogram of the hierarchical clustering of the 104 HFTs using these 16 variables. For this hierarchical clustering, we used the Ward method [52] to join clusters in the order of the decreasing sum of squares after joining. The vertical axis in Figure 10 represents the distance between clusters with an increase in the sum of squares when clusters A and B are joined, and is defined by the following equation:

$$\Delta(A,B) = \sum_{i \in A \cup B} ||\vec{x}_i - \vec{m}_{A \cup B}||^2 - \sum_{i \in A} ||\vec{x}_i - \vec{m}_A||^2 - \sum_{i \in B} ||\vec{x}_i - \vec{m}_B||^2 \tag{12}$$

where $||d||$ denotes the Euclidean distance, and $\vec{m}_j$ is the center of cluster j.

Based on the distance between the clusters, we found it reasonable to categorize the HFTs into three groups with the threshold distance around 3, as shown in Figure 10, and designated them as Group A, Group B, and Group C. There were 77 HFTs in Group A, 12 in Group B, and 15 in Group C. The number of clusters becomes larger for a lower threshold distance, however, we confirmed that properties of any smaller groups are quite similar to one of these three groups in the graphical representation of an interaction network to be discussed in the following.

The remainder of this section explains the order events that excited the HFTs in each group to place buy–sell limit orders based on the estimated Hawkes parameters.

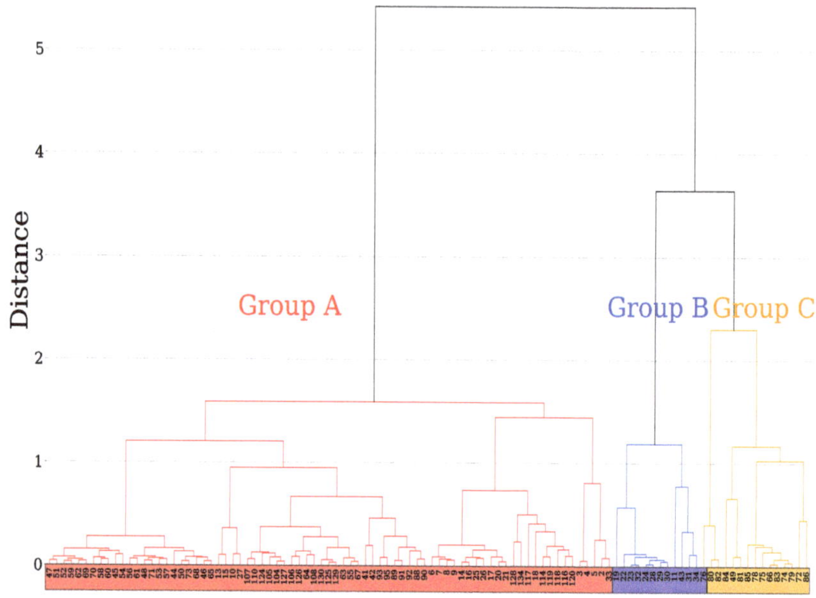

Figure 10. Dendrogram based on the Ward method of clustering for 104 HFTs that were successfully modeled by the Hawkes process. The vertical axis represents the distance between the clusters, as defined in Equation (12), and the horizontal axis shows the labels of the HFTs according to the order frequency (red: Group A with 77 HFTs; blue: Group B with 12 HFTs; yellow: Group C with 15 HFTs).

4.2.1. Group A

Group A is comprised of 77 HFTs. The total number of limit orders in the 5 days was approximately 850,000, which accounted for 62.5% of the total number of limit orders in the market. Figure 11 shows the quartiles and means of the 16 normalized branching ratios for these 77 HFTs, where (a) represents $\bar{\rho}_{TS,m}$ and (b) represents $\bar{\rho}_{TB,m}$. From Figure 11a, it can be seen that the generation of sell limit orders by the HFTs in Group A was most excited by hit sell, which greatly exceeded the excitation from other events. In contrast, Figure 11b shows that their buy limit order generation was most excited by hit buy, which also greatly exceeded the excitation from other events.

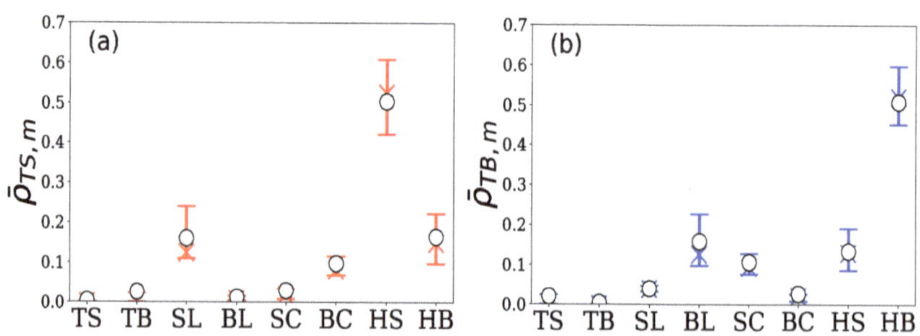

Figure 11. Percentile plot of normalized branching ratios (**a**) $\bar{\rho}_{TS,m}$ and (**b**) $\bar{\rho}_{TB,m}$ for 77 HFTs in group A. The vertical axis represents the normalized branching ratios by event m, and the horizontal axis represents element $m \in \mathbb{M}$ in both figures (top bar: 75th percentile; X symbol: median; bottom bar: 25th percentile; ◯ symbol: the mean).

Figure 12 illustrates the network graph of the buy and sell limit orders of the HFTs in Group A, along with all types of orders, using these normalized branching ratios.

The size of the directed edges of the network is proportional to the mean value of the normalized branching ratio, e.g., edges directed from HS to TS and from HB to TB represent strong excitations.

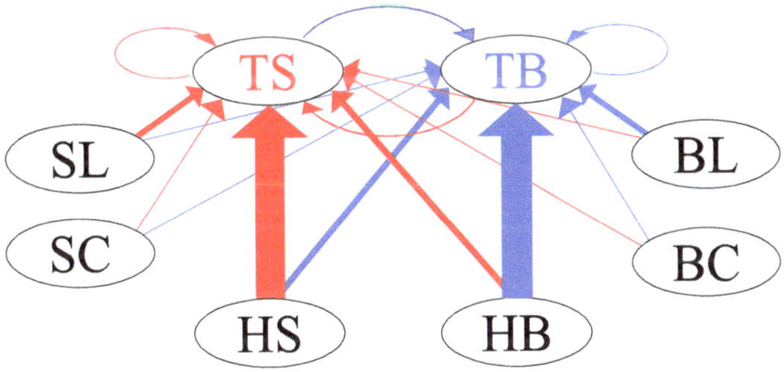

Figure 12. Network graph of interaction between buy–sell limit orders of HFTs in Group A and all types of orders in the order book.

In addition, because the kernel function of the Hawkes process is an exponential function, the time constant, which is a measure of the response speed of one excitation at a time, is given by $\beta_{n,m}^{-1}$. The mean values of the estimated time constant for the HFTs in Group A are summarized in Table 3. It is suggested that their reaction speed to an event is approximately 0.1 s, which is reasonable for HFTs who trade at very high speeds.

Table 3. Mean values of the estimated time constant $\hat{\beta}_{n,m}^{-1}$ (s) for the HFTs in Group A.

n \ m	TS	TB	SL	BL	SC	BC	HS	HB
TS	0.102	0.295	0.076	0.087	0.081	0.069	0.110	0.404
TB	0.118	0.099	0.114	0.071	0.066	0.091	0.148	0.111

4.2.2. Group B

Group B is comprised of 12 HFTs. The total number of limit orders in the 5 days was approximately 174,000, which accounted for 12.7% of the total number of limit orders in the market. Figure 13 shows the quartiles and means of the 16 normalized branching ratios for these 12 HFTs. From Figure 13a, we can see that their sell limit order generation was excited by sell limit and cancel buy, and from Figure 13b, we can see that their buy limit order generation was excited by buy limit and cancel sell. Unlike Group A, they did not react to execution events but were excited by the generation and cancellation of limit orders in the order book.

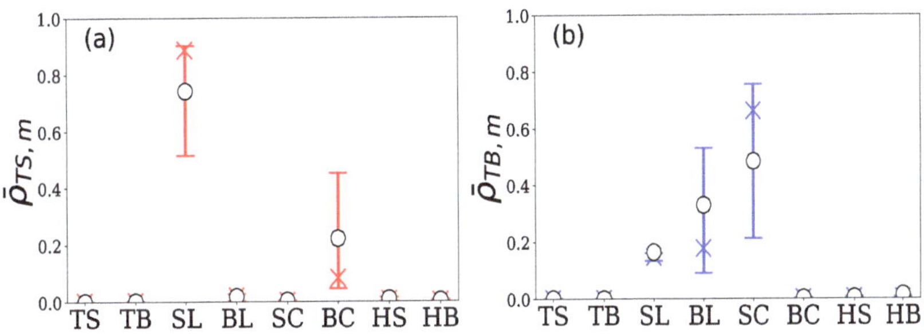

Figure 13. Percentile plot of normalized branching ratios (**a**) $\bar{\rho}_{TS,m}$ and (**b**) $\bar{\rho}_{TB,m}$ for 12 HFTs in Group B. The vertical and horizontal axes are the same as those in Figure 10 (top bar: 75th percentile; X symbol: median; bottom bar: 25th percentile; ◯ symbol: the mean).

Figure 14 shows the interaction network of the buy and sell limit orders of the HFTs in Group B, along with all types of orders, using the mean of the normalized branching ratios, as in Figure 12.

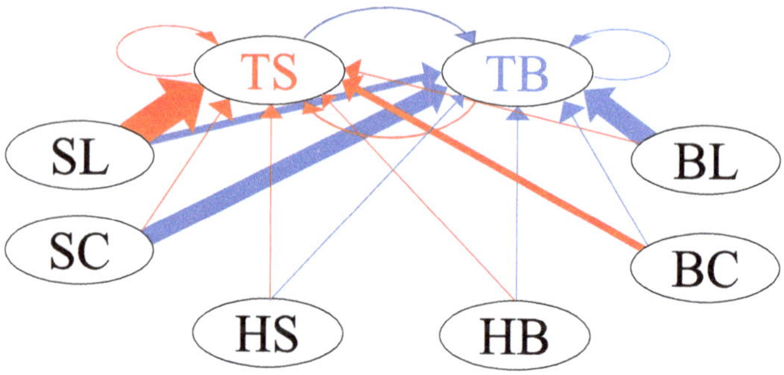

Figure 14. Network graph of interaction between the buy–sell limit orders of HFTs in Group B and all types of orders in the order book.

The mean values of the time constants for each event are summarized in Table 4. As in the case of HFTs in Group A, these values suggest that the reaction speed to events were to be measured in milliseconds.

Table 4. Sample means of 16 time constants, $\beta_{n,m}^{-1}$(S), for HFTs in Group B.

n \ m	TS	TB	SL	BL	SC	BC	HS	HB
TS	0.099	0.100	0.681	0.109	0.110	0.285	0.101	0.099
TB	0.099	0.100	0.259	0.373	0.558	0.105	0.099	0.101

4.2.3. Group C

Group C is comprised of 15 HFTs. Their total number of limit orders in the 5 days was approximately 95,000, which accounted for 6.9% of the total number of limit orders in the market. From Figure 15a, it can be seen that the sell limit order generation of the HFTs in Group C was most strongly excited by their own buy limit, and was also excited by the sell limit and cancel buy. On the other hand, Figure 15b shows that their buy limit order generation was most strongly excited by their own sell limit, but was also excited by buy limit and cancel buy.

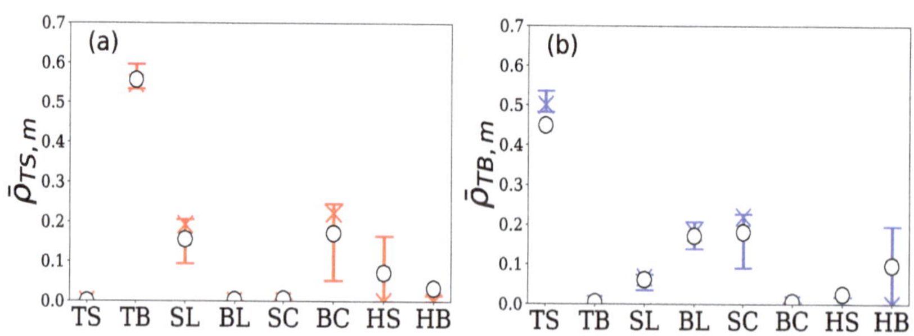

Figure 15. Percentile plot of normalized branching ratios (**a**) $\bar{\rho}_{TS,m}$ and (**b**) $\bar{\rho}_{TB,m}$ for 15 HFTs in Group C. The vertical and horizontal axes are the same as those in Figures 10 and 11 (top bar: 75th percentile; X symbol: median; bottom bar: 25th percentile; ○ symbol: the mean).

Figure 16 shows the interaction network of the buy–sell limit orders of the HFTs in Group C, along with all types of orders, as in Figures 12 and 14. The HFTs' sell/buy limit orders interacted with each other.

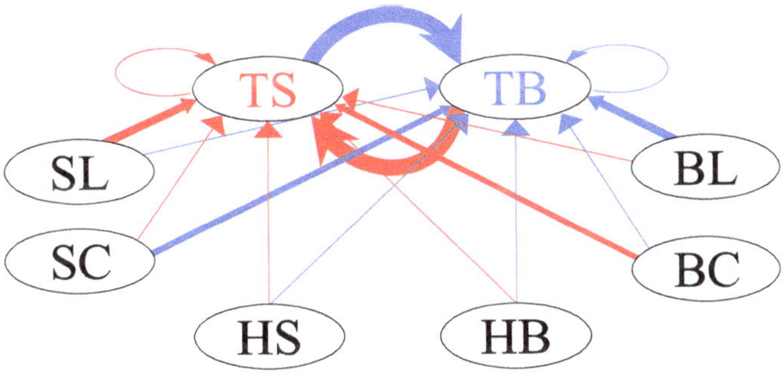

Figure 16. Network graph of interaction between buy–sell limit orders of HFTs in Group C and all types of orders in the order book.

The mean values of the time constants for each event are summarized in Table 5. The time constants of the excitations from TS to TB and from TB to TS were larger than 10 s, suggesting that the excitations were sustained for a very long time compared to those previously observed.

Table 5. Sample means of 16 time constants, $\beta_{n,m}^{-1}$(S), for HFTs in Group C.

n \ m	TS	TB	SL	BL	SC	BC	HS	HB
TS	0.134	17.781	0.266	0.099	0.096	0.359	0.109	0.154
TB	11.201	0.098	0.269	0.258	0.355	0.094	0.132	0.103

5. Conclusions

In summary, we introduced a multivariate Hawkes process to model the limit order generation processes of individual HFTs participating in the USD/JPY foreign exchange market for 5 days and analyzed their limit order generation mechanisms. First, we confirmed that an eight-variable Hawkes process, which consisted of each HFTs' own buy–sell limit orders and the six types of orders in the order book, could adequately model the limit order generation processes of 104 of the 134 HFTs. Then, we categorized the 104 properly modeled HFTs into three categories based on the similarity of the excitation mechanisms measured by the parameter values of the Hawkes process. As a result, we confirmed that the majority of the HFTs in our dataset reacted to the execution of trades, while 12 of the 134 HFTs only reacted to limit orders and 15 of the 134 HFTs reacted to their orders. By evaluating the time constants of the estimated excitations of individual HFTs, we found that many HFTs responded to the most recent change in the order book in a very short time, by placing or canceling new orders. Since HFTs currently account for the majority of limit orders shown in the order book, the results of this analysis provide more microscopic insight into the dynamics of the order book than previous studies.

The following issues will be studied in the future as a generalization of the present work. The first goal is to clarify the limit order generation processes of the remaining 30 HFTs who could not be adequately modeled by the present analysis. The Hawkes process adopted in this study only included the impact of the occurrence of a recent order event and ignored important financial market influences such as the volume of orders, market price fluctuations and trends, and the positions of the HFTs. We believe that the information ignored in this study could contain variables that would explain their order generation processes. Second, although this study only focused on the generation of limit orders by HFTs, it is also important to clarify the cancellation process for limit orders by HFTs and the generation of market orders. Third, we did not pay attention to profit and loss; however, practically, a key factor in an HFT strategy is the ability to make stable profits.

As the period of our data is very short, we did not observe any abnormal behavior in the market; however, we cannot deny the possibility that HFTs may overreact and result in serious synchronization during other periods or in other markets. Further studies of the relations among Hawkes parameters and the case of crashes are needed to prevent the excessive synchronization of biased orders of buy or sell. Our results are important since the model we derived in this paper provides a foundation for performing such studies through simulations. HFTs play a central role in providing liquidity to the market, and further detailed analyses of HFT strategies will contribute to the development of modern financial markets in general.

Author Contributions: Conceptualization, M.T.; formal analysis, H.W.; data curation, H.W.; investigation, H.W., H.T. and M.T.; methodology, H.W., H.T.; project administration, M.T.; resources, M.T.; software, H.W.; supervision, M.T.; writing—original draft, H.W.; writing—review and editing, H.T., M.T. All authors have read and agreed to submit this version of the manuscript.

Institutional Review Board Statement: Not applicable.

Informed Consent Statement: Not applicable.

Data Availability Statement: Restrictions apply to the availability of these data. Data were obtained from the EBS. We obtained permission for publication.

Acknowledgments: We appreciate EBS, NEX Group plc. for their provision of the EBS data.

Conflicts of Interest: The authors declare no conflict of interest.

References

1. Odean, T. Are investors reluctant to realize their losses? *J. Financ.* **1998**, *53*, 1775–1798. [CrossRef]
2. Grinblatt, M.; Keloharju, M. The investment behavior and performance of various investor types: A study of Finland's unique data set. *J. Financ. Econ.* **2000**, *55*, 43–67. [CrossRef]
3. Sueshige, T.; Sornette, D.; Takayasu, H.; Takayasu, M. Classification of position management strategies at the order-book level and their influences on future market-price formation. *PLoS ONE* **2019**, *14*, e0220645. [CrossRef]
4. Sueshige, T.; Kanazawa, K.; Takayasu, H.; Takayasu, M. Ecology of trading strategies in a forex market for limit and market orders. *PLoS ONE* **2018**, *13*, e0208332. [CrossRef] [PubMed]
5. Kanazawa, K.; Sueshige, T.; Takayasu, H.; Takayasu, M. Kinetic theory for financial Brownian motion from microscopic dynamics. *Phys. Rev. E* **2018**, *98*, 052317. [CrossRef]
6. Kanazawa, K.; Sueshige, T.; Takayasu, H.; Takayasu, M. Derivation of the Boltzmann equation for financial Brownian motion: Direct observation of the collective motion of high-frequency traders. *Phys. Rev. Lett.* **2018**, *120*, 138301. [CrossRef] [PubMed]
7. Biais, B.; Foucault, T.; Moinas, S. Equilibrium fast trading. *J. Financ. Econ.* **2015**, *116*, 292–313. [CrossRef]
8. Carrion, A. Very fast money: High-frequency trading on the NASDAQ. *J. Financ. Mark.* **2013**, *16*, 680–711. [CrossRef]
9. Schmidt, A.B. Ecology of the Modern Institutional Spot FX: The EBS Market in 2011. 2012. Available online: https://papers.ssrn.com/sol3/papers.cfm?abstract_id=1984070 (accessed on 7 January 2022).
10. Gerig, A. High-Frequency Trading Synchronizes Prices in Financial Markets. 2015. Available online: https://papers.ssrn.com/sol3/papers.cfm?abstract_id=2173247 (accessed on 7 January 2022).
11. Mukerji, P.; Chung, C.; Walsh, T. The impact of algorithmic trading in a simulated asset market. *J. Risk Financ. Manag.* **2019**, *12*, 68. [CrossRef]
12. O'Hara, M. High frequency market microstructure. *J. Financ. Econ.* **2015**, *116*, 257–270. [CrossRef]
13. Goldstein, M.A.; Kumar, P.; Graves, F.C. Computerized and high-frequency trading. *Financ. Rev.* **2014**, *49*, 177–202. [CrossRef]
14. Jones, C.M. What Do We Know about High-Frequency Trading? Columbia Business School Research Paper 13-11. 2013. Available online: https://papers.ssrn.com/sol3/papers.cfm?abstract_id=2236201 (accessed on 7 January 2022).
15. Menkveld, A.J. High frequency trading and the new market makers. *J. Financ. Mark.* **2013**, *16*, 712–740. [CrossRef]
16. Easley, D.; De Prado, M.M.L.; O'Hara, M. The microstructure of the "flash crash": Flow toxicity, liquidity crashes, and the probability of informed trading. *J. Portf. Manag.* **2011**, *37*, 118–128. [CrossRef]
17. Easley, D.; de Prado, M.M.L.; O'Hara, M. VPIN and the flash crash: A rejoinder. *J. Financ. Mark.* **2014**, *17*, 47–52. [CrossRef]
18. Andersen, T.G.; Bondarenko, O. VPIN and the flash crash. *J. Financ. Mark.* **2014**, *17*, 1–46. [CrossRef]
19. Andersen, T.G.; Bondarenko, O. Reflecting on the VPIN dispute. *J. Financ. Mark.* **2014**, *17*, 53–64. [CrossRef]
20. Andersen, T.G.; Bondarenko, O. Assessing measures of order flow toxicity and early warning signals for market turbulence. *Rev. Financ.* **2015**, *19*, 1–54. [CrossRef]

21. D'Souza, C. Where Does Price Discovery Occur in FX Markets? 2007. Available online: https://papers.ssrn.com/sol3/papers.cfm?abstract_id=966446 (accessed on 7 January 2022).
22. Gençay, R.; Gradojevic, N.; Olsen, R.; Selçuk, F. Informed traders' arrival in foreign exchange markets: Does geography matter? *Empir. Econ.* **2015**, *49*, 1431–1462. [CrossRef]
23. Gençay, R.; Gradojevic, N. Private information and its origins in an electronic foreign exchange market. *Econ. Model.* **2013**, *33*, 86–93. [CrossRef]
24. Gradojevic, N.; Erdemlioglu, D.; Gençay, R. Informativeness of trade size in foreign exchange markets. *Econ. Lett.* **2017**, *150*, 27–33. [CrossRef]
25. Elaut, G.; Frömmel, M.; Lampaert, K. Intraday momentum in FX markets: Disentangling informed trading from liquidity provision. *J. Financ. Mark.* **2018**, *37*, 35–51. [CrossRef]
26. Hawkes., A.G. Spectra of some self-exciting and mutually exciting point processes. *Biometrika* **1971**, *58*, 83–90. [CrossRef]
27. Engle, R.F.; Russell, J.R. Forecasting the frequency of changes in quoted foreign exchange prices with the autoregressive conditional duration model. *J. Empir. Financ.* **1997**, *4*, 187–212. [CrossRef]
28. Takayasu, M.; Takayasu, H. Self-modulation processes and resulting generic 1/f fluctuations. *Phys. A Stat. Mech. Its Appl.* **2003**, *324*, 101–107. [CrossRef]
29. Hawkes, A.G. Hawkes processes and their applications to finance: A review. *Quant. Financ.* **2018**, *18*, 193–198. [CrossRef]
30. Bowsher, C. Modelling security market events in continuous time: Intensity based, multivariate point process models. *J. Econom.* **2007**, *141*, 876–912. [CrossRef]
31. Filimonov, V.; Sornette, D. Quantifying reflexivity in financial markets: Toward a prediction of flash crashes. *Phys. Rev. E* **2012**, *85*, 056108. [CrossRef]
32. Hardiman, S.J.; Bercot, N.; Bouchaud, J.P. Critical reflexivity in financial markets: A Hawkes process analysis. *Eur. Phys. J. B* **2013**, *86*, 442. [CrossRef]
33. Hardiman, S.J.; Bouchaud, J.P. Branching-ratio approximation for the self-exciting Hawkes process. *Phys. Rev. E* **2014**, *90*, 062807. [CrossRef]
34. Bacry, E.; Muzy, J.F. First- and Second-Order Statistics Characterization of Hawkes Processes and Non-Parametric Estimation. *IEEE Trans. Inf. Theory* **2016**, *62*, 2184–2202. [CrossRef]
35. Bacry, E.; Jaisson, T.; Muzy, J.F. Estimation of slowly decreasing hawkes kernels: Application to high-frequency order book dynamics. *Quant. Financ.* **2016**, *16*, 1179–1201. [CrossRef]
36. Achab, M.; Bacry, E.; Muzy, J.F.; Rambaldi, M. Analysis of order book flows using a non-parametric estimation of the branching ratio matrix. *Quant. Financ.* **2018**, *18*, 199–212. [CrossRef]
37. Rambaldi, M.; Pennesi, P.; Lillo, F. Modeling foreign exchange market activity around macroeconomic news: Hawkes-process approach. *Phys. Rev. E* **2015**, *91*, 012819. [CrossRef] [PubMed]
38. Rambaldi, M.; Bacry, E.; Lillo, F. The role of volume in order book dynamics: A multivariate Hawkes process analysis. *Quant. Financ.* **2017**, *17*, 999–1020. [CrossRef]
39. Rambaldi, M.; Filimonov, V.; Lillo, F. Detection of intensity bursts using Hawkes processes: An application to high-frequency financial data. *Phys. Rev. E* **2018**, *97*, 032318. [CrossRef] [PubMed]
40. Lu, X.; Abergel, F. High-dimensional Hawkes processes for limit order books: Modelling, empirical analysis and numerical calibration. *Quant. Financ.* **2018**, *18*, 249–264. [CrossRef]
41. Fosset, A.; Bouchaud, J.P.; Benzaquen, M. Non-parametric estimation of quadratic Hawkes processes for order book events. *Eur. J. Financ.* **2021**. [CrossRef]
42. Rizoiu, M.A.; Lee, Y.; Mishra, S.; Xie, L. A tutorial on hawkes processes for events in social media. *arXiv* **2017**, arXiv:1708.06401.
43. Hawkes, A.G. Point Spectra of Some Mutually Exciting Point Processes. *J. R. Stat. Soc. Ser. B Methodol.* **1971**, *33*, 438–443. [CrossRef]
44. Bacry, E.; Mastromatteo, I.; Muzy, J.F. Hawkes processes in finance. *Mark. Microstruct. Liq.* **2015**, *1*, 1550005. [CrossRef]
45. Helmstetter, A.; Sornette, D. Importance of direct and indirect triggered seismicity in the ETAS model of seismicity. *Geophys. Res. Lett.* **2003**, *30*. [CrossRef]
46. Helmstetter, A.; Sornette, D. Subcritical and supercritical regimes in epidemic models of earthquake aftershocks. *J. Geophys. Res. Solid Earth* **2002**, *107*, ESE-10. [CrossRef]
47. Toke, I.M. An Introduction to Hawkes Processes with Applications to Finance. Lectures Notes from Ecole Centrale Paris, BNP Paribas Chair of Quantitative Finance. 2011; Volume 193. Available online: http://www.smallake.kr/wp-content/uploads/2015/01/HawkesCourseSlides.pdf (accessed on 7 January 2022).
48. Kingma, D.P.; Ba, J. Adam: A method for stochastic optimization. *arXiv* **2014**, arXiv:1412.6980.
49. Ogata, Y. On Lewis' simulation method for point processes. *IEEE Trans. Inf. Theory* **1981**, *27*, 23–31. [CrossRef]
50. Ogata, Y. Statistical models for earthquake occurrences and residual analysis for point processes. *J. Am. Stat. Assoc.* **1988**, *83*, 9–27. [CrossRef]
51. Lewis, P.W.; Shedler, G.S. Simulation of nonhomogeneous Poisson processes by thinning. *Nav. Res. Logist. Q.* **1979**, *26*, 403–413. [CrossRef]
52. Ward, J.H., Jr. Hierarchical grouping to optimize an objective function. *J. Am. Stat. Assoc.* **1963**, *58*, 236–244. [CrossRef]

entropy

MDPI

Article

The Stock Market Model with Delayed Information Impact from a Socioeconomic View

Zhiting Wang [1], Guiyuan Shi [2], Mingsheng Shang [3] and Yuxia Zhang [1,*]

[1] Physics and Photoelectricity School, South China University of Technology, Guangzhou 510640, China; 201820127594@mail.scut.edu.cn
[2] International Academic Center of Complex Systems, Beijing Normal University at Zhuhai, Zhuhai 519087, China; sgy@bnu.edu.cn
[3] Chongqing Institute of Green and Intelligent Technology, Chinese Academy of Sciences, Chongqing 400714, China; msshang@cigit.ac.cn
* Correspondence: zhangyux@scut.edu.cn

Abstract: Finding the critical factor and possible "Newton's laws" in financial markets has been an important issue. However, with the development of information and communication technologies, financial models are becoming more realistic but complex, contradicting the objective law "Greatest truths are the simplest." Therefore, this paper presents an evolutionary model independent of micro features and attempts to discover the most critical factor. In the model, information is the only critical factor, and stock price is the emergence of collective behavior. The statistical properties of the model are significantly similar to the real market. It also explains the correlations of stocks within an industry, which provides a new idea for studying critical factors and core structures in the financial markets.

Keywords: econophysics; financial complexity; collective intelligence; emergent property; stock correlation; detrended cross-correlation analysis

Citation: Wang, Z.; Shi, G.; Shang, M.; Zhang, Y. The Stock Market Model with Delayed Information Impact from a Socioeconomic View. *Entropy* **2021**, *23*, 893. https://doi.org/10.3390/e23070893

Academic Editors: Ryszard Kutner, Christophe Schinckus and H. Eugene Stanley

Received: 19 May 2021
Accepted: 10 July 2021
Published: 14 July 2021

1. Introduction

With the massive use of information and communication technologies, we can collect traceable data from almost anyone. The rise of network science [1] and computational social science [2] have provided opportunities for innovative research in econophysics and sociophysics. In particular, econophysics regards the financial market as a complex system and attempts to depict it more realistically, such as the interactions between investors by network dynamic evolution. Econophysics describes the economic system with many interacting heterogeneous entities (people, firms, institutions, etc.), and expects to find similar laws to the physical system. However, humans are not ideal gas molecules, it is unclear how many and which quantities would be needed for determining and anticipating a given macroscopic, in the sense of collective, observable [3]. Moreover, because human beings are adaptable, the study of economic systems is bound to be a difficult problem.

Researchers have proposed numerous different mechanisms to model the microstructure of financial markets. They pursued the most detailed descriptions, such as creating diverse agents and setting rules for interactions between agents and trading rules. Researchers collected data about investors' behavior through information technology to deal with the variables of different individuals. But individuals rely on different risk preferences and reference points. Even if we can reasonably describe the behavior of a single individual, we cannot directly generalize to a group. Investors' decisions in financial markets are not always rational; their buying and selling decisions are affected by emotion, personality, and bias [4,5]. People are different, and they are not rational to some extent. For individuals, faith may be stronger than reason, personal interest may be stronger than the good of the team, etc. Meanwhile, the COVID-19 disease is a new and dreaded event [6], and in the process of keeping the virus under control, people's cognitive functioning has been enhanced, and their behavior has been changed to some extent. For example, more people are willing to wear a mask after the epidemic outbreak and so on. In the stock market, there are so many unpredictable fluctuations. When new information is generated, what does it mean for the stock market? That would hardly be positive for the stock market because investors are different, and their cognitive processes and cognitive environment are fickle and changeable. In the face of changes in the information environment, different investors have different reaction capacities and speeds. After thinking about the information, even for a specific investor, they will understand the information from a new perspective and form their own judgment slowly. Thus, in a financial system, microstructure models are not enough to consider the variable adaptability of investors.

Although investors are different and unpredictable, research exhibits that pieces of statistical evidence remain stable, accordant to the stability of the statistical properties of particle motion in physics models [7,8]. Therefore, in the studies of financial markets, statistical results of different micro models exhibit universal characteristics. The classical percolation model [9–11] simulates herd behavior. For any pair of agents i and j, they are connected with a probability, and then agent i makes the buying or selling decision with another probability. The model explains the power-law distribution of stock price returns appropriately. The two-dimensional Ising model [12] considers investors' imitation of neighbors, the influence of public information, and personal traits. Here the influence of public information is a Gaussian distribution. The investor's decision function also has a probability form, and the returns of the model are "fat-tailed" [13,14]. The financial models with network topology [15] also produce the universal characteristics of real stock markets by setting the link probability of nodes and performing decision functions. These models share common features. First, they generate a stock trading environment in the form of probability. Second, investors make buy-sell decisions with probability or decision functions. More details are introduced to depict a more realistic financial market based on these basic models and their common features. Over the past century or so, stock trading information flow has changed from slow to intensive, investors' literacy from low to high, relationship from simple social relationships to complex social networks. Individual characteristics of investors and the market environment have dramatically changed. Stock trading rules also varied in different countries; for example, China has a 10% price limit [16]. Nevertheless, no matter what changed the environment or rules, it is observed that universal characteristics are robust on different timescales and in different stock markets. Therefore, in the study of the macro laws, statistical properties of the stock market, the critical factor should not be the relationship network of investors, the speed of information flow, or the level of literacy of investors, which researchers want to introduce. On the other hand, collective intelligence results from intelligence, which emerges out of collaboration and coordination of many individual agents [17]. Collective intelligence, which Wooley et al. [18] define as the ability of a group to perform a wide variety of tasks. They studied "collective intelligence" and demonstrated that the critical factor characterizing "collective intelligence" is not the group members' average or maximum individual intelligence. Here, we view the ability of investors to make buying and selling decisions. Investors gamble in the stock market, where supply and demand determine the stock price, i.e., the result of their behavior is reflected in the price of the stock. Investors' collective intelligence is the emergence of investors' collective behavior. In this paper, we abstract all the factors that impact the market to the only value of information. In given information, the behavior of investors emerging with probabilities results in the evolution of stock markets. Here, unlike the micro model that pursues a realistic and detailed structure, we discard individual features and interaction. We present a stock price evolution model with emergence properties in the given information in Section 2 and verify its rationality using real market data in Section 3. We aim to find the critical factor and capture stable macroscopic law in the ever-changing stock market.

The paper is organized as follows: Section 2: A detailed description of the stock market model with delayed information impact. Section 3: Statistical analysis and nonlinear behavior of the proposed model. Section 4: Correlation analysis between stocks in the industry.

2. Stock Price Model with Delayed Information Impact

The analysis of financial stock market prices has been found to exhibit some universal characteristics similar to those observed in physical systems with many interacting units, and several microscopic models have been developed to study them. Examples include percolation models, Ising models, network models, and their extensions to social interactions. Though these models are very different, they all can be used to simulate the stock market. Because the simulation results are consistent with the statistical properties of real market price fluctuations, these models may generate the "Newton's laws" in financial markets. Thus, we aim to find the possible "Newton's laws" in these models and try to prove it.

The classical percolation model is generated with the connection probability of neighbor nodes. The Ising model is a random field with a probability, and the evolution of the swing is closely related to the structure of space and initial state. The network model is also generated with a probability. We find the common feature that they generate is stock trading in the form of probability.

Mitchell and Mulherin [19] studied the relation between the number of news announcements reported daily by Dow Jones & Company and aggregate measures of securities market activity, including trading volume and market returns. They employed a distinctive proxy for the information, i.e., the number of announcements released daily by Dow Jones & Company. Meanwhile, the social sciences have obtained access to huge datasets based on the internet activity of millions of users all over the world. Among the most frequently utilized providers of data, social media such as Twitter and Facebook and search engines Google and Yahoo play the most important roles. For example, the frequency of searched terms has been shown to provide helpful information for forecasting various phenomena ranging from trading volumes [20] to consumer behavior [21] and finance [22]. In summary, information is too complicated to be considered fully in a theoretical model, let alone delayed information in stock markets. In previous studies, Albers et al. [23] studied "delayed information." In the paper, the time when relevant information is available and the time that a decision has an effect could be decoupled. Investors might not have access to the latest exchange rates or stock prices. They refer to this as the delayed information model. However, we define a new concept of delayed information here. In the stock market, information comes in various ways and at different influence levels. In general, there is a small amount of super good news and bad news. Most of the news is ordinary. In our model, the influence of information is an abstract concept. The influence of information will last for some time, and the disappearance time of influence will be delayed. This is what we refer to as delayed information.

We propose the stock price model of delayed information impact based on the common feature and abstract information. It includes two components, i.e., the generation and delay of market information and the emergence of collective decision-making in the given information.

2.1. Information Generation and Delay

- Suppose the initial stock price is P_0. The stock market environment is fickle daily and is influenced by a series of stochastic events, including supply and demand, macroeconomic, political factors, corporate finances and performance, market sentiment, etc. We coarse-grain all the stochastic events by information into just a single influence value. The impact of information is an abstract concept, which is a random variable that is normally distributed with mean 0 and standard deviation σ_1, here $\sigma_1 = \lambda P_0$. Any theoretical normal distribution has a maximum of infinity and a minimum of minus infinity. There is an infinite range. In our model, the impact of information is normally distributed, and it must be finite. Thus, there must be a truncation. The truncation interval should be large enough and reasonable. The information has an impact on the stock price, so the truncation interval has a relation with the stock price P_0. It cannot stand alone. Here, considering the extreme cases (terrible information, great information), we set the truncation interval to be $[-4\sigma_1, +4\sigma_1]$. New information sequence I_t can be obtained by random sampling from the truncated Gaussian distribution.

- In the stock market, the influence of information is in a state of change and eventually disappears. Thus, we introduce the delayed information. The progress of influence disappearance is a different matter from the memory deterioration. That the influence of information eventually disappears does not mean that the people forget the information; it is just that the information is a dead issue. Considering that significant events have a sustained impact on the investors, and the impact strength of the information will delay over time, we assume that the information influence I_t delays linearly with time simply, and the information influence after the i-th day I_i' is expressed as

$$I_i' = \begin{cases} I_t - ai, I_t > 0 \\ I_t + ai, I_t < 0 \end{cases} \tag{1}$$

where a is the delay coefficient.

2.2. Stock Price Evolution Process

The given information determines the theoretical stock price P_t'.

$$P_t' = P_{t-1} + I_t + \sum_{i=0}^{i=t-1} I_i' \tag{2}$$

Investors participate in the game and make decisions based on the given information. Their collective behaviors result in actual stock prices. As the investors vary from radicals or conservatives, daredevils, or followers, etc., statistical properties of the final actual stock price series are stable in the ever-changing stock market. The actual stock price P_t in day t has emergence properties of collective intelligence, which is a random sampling from a truncated Gaussian distribution $P_t \sim N(P_t', \sigma_2^2)$. As the price fluctuation is related to the information, here $\sigma_2 = \frac{1}{3} \times |P_t' - P_{t-1}|$. Considering the extremes, we set the truncation interval as $[-4\sigma_2, +4\sigma_2]$.

Figure 1 shows the simulated stock price series P_t and the corresponding return series r_t, $P_0 = 3000, \sigma_1 = 20, a = 5$. In Figure 1, volatility clustering is easily observable. High-volatility tends to follow high-volatility, and low-volatility tends to follow low-volatility.

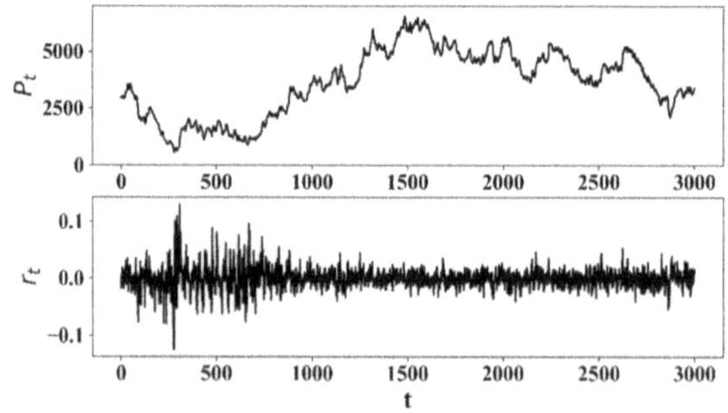

Figure 1. Stock price series of the proposed model and its corresponding return.

3. Descriptive Statistics and Nonlinear Behavior Analysis

This section discusses the stock price model's descriptive statistics and nonlinear behavior with delayed information impact and verifies the simulation results with the real stock market. We use real daily closing price data from 1 January 2010 to 3 December 2020 ($T \approx 2700$), including the SSE (Shanghai Composite Index), SZES (Shenzhen Stock Exchange Index), and S&P500 (S&P 500 Index) (https://finance.yahoo.com, 3 May 2021). The simulated data length $T = 3000$ matches with the real data ($T \approx 2700$).

3.1. Descriptive Statistics of Returns

The "Fat-tailed" characteristic of returns has been verified in extensive empirical studies [24–26]. It is an important criterion for the reasonableness of price dynamics in the stock model research. Here, the definition of price return is $r_t = lnP_t - lnP_{t-1}$ [27]. The probability density distributions of three simulated and real market returns are shown in Figure 2a. Simulated and real return distributions are almost identical. Compared to the Gaussian distribution, they both exhibit distinct "fat-tailed" characteristics. Table 1 shows the statistics: mean, standard deviation, maximum, minimum, skew, kurtosis, the results of Kolmogorov–Smirnov test (K-S test) and power-law fit, where the kurtosis of all returns is larger than three that is the kurtosis of the Gaussian distribution [28]. In the K-S test, all p-values are very small, and all the H-values are 1, so we reject the null hypothesis that the distribution follows the Gaussian distribution at a 5% significance level. Figure 2b shows that the cumulative probability distributions of simulated and real market returns follow power-law distribution $P(|r_t| > x) \sim x^{-\alpha}$, α is the power-law exponent. The corresponding power-law exponent values in Table 1 approximately equal to 3, it obeys the "Inverse cubic law" [29].

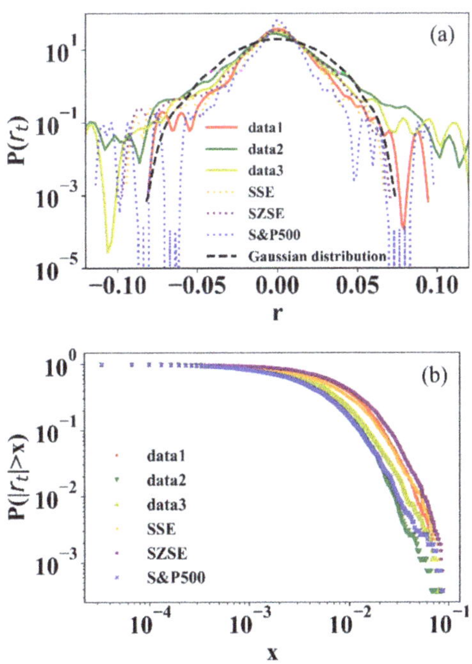

Figure 2. (**a**) The probability density distributions of simulated and empirical returns (semi-log); (**b**) The cumulative distributions of simulated and empirical returns (log-log).

Table 1. Descriptive statistics, power-law fit, and K-S test of returns.

Data	Mean	Std	Max	Min	Skew	Kurtosis	K-S Test		α
							p-Value	H	
x_1	0.00004	0.0172	0.1294	−0.1240	0.2987	6.6543	8.1208×10^{-9}	1	3.5784
x_2	−0.00002	0.0217	0.1601	−0.2125	−0.3848	9.3611	1.8554×10^{-10}	1	4.0968
x_3	0.00005	0.0182	0.1520	−0.1367	−0.0234	8.3799	4.2418×10^{-10}	1	3.8109
S&P500	0.00004	0.0111	0.0934	−0.1066	−0.9710	15.2922	4.0739×10^{-18}	1	3.4624
SSE	0.00002	0.0136	0.0060	−0.0887	−0.8969	6.1958	1.6704×10^{-10}	1	3.5277
SZSE	0.00001	0.0164	0.0625	−0.0895	−0.7368	3.7987	5.8053×10^{-7}	1	3.4777

3.2. *Nonlinear Statistical Analysis of Returns*

Some studies have investigated the nonlinear properties of financial markets [30–32]. Hsieh [30] discussed some of the methodological issues in detecting chaotic and nonlinear behavior. Alves et al. [31] focused on the Dow Jones Index to determine the chaotic dynamics. Zhu et al. [32] revealed the long-term memory of financial time series. Here, we compare the nonlinear behavior of the simulated return series with the real market series.

3.2.1. Correlation Dimension Analysis

The correlation dimension method measures the complexity of dynamical systems that distinguishes deterministic systems (including low-dimensional chaos) and stochastic systems [33]. According to the method of Grassberger et al. [34], the correlation dimension can be calculated when the appropriate embedding dimension m and time lag τ are selected for the phase space reconstruction. For an m-dimensional phase space, the correlation integral *C(r)* is calculated by

$$C(r) = \lim_{N \to \infty} \frac{2}{N(N-1)} \sum_{i,j=1, i \neq j}^{N} \Theta(r - |X_i - X_j|) \tag{3}$$

where Θ is the step function. The appropriate choice of r enables the correlation dimension of the system D to describe as

$$D = \lim_{r \to 0} \frac{log_2 C(r)}{log_2 r} \tag{4}$$

A common method is to fit the $log_2 C(r)$ and $log_2 r$ using least squares, and the slope is the correlation dimension D. For random sequences, D increases linearly with the embedding dimension m with no saturation. While for deterministic chaotic sequences, D increases with m to a certain position to reach saturation, and the saturation m is the correlation dimension D of the time series attractor. Figure 3 shows the correlation integral $log_2 C(r)$ and $log_2 r$ in different embedding dimensions m. Figure 4 shows the correlation dimension. It is observed that all correlation dimensions increase with m and reach saturation at a certain position. It can be seen that all the returns have deterministic noise, which means the systems are chaotic. The simulated data from the proposed model coincide with the real market data.

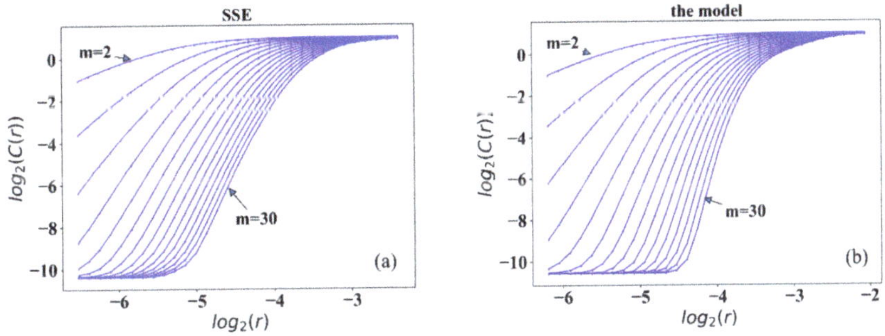

Figure 3. Correlation integral results of return series from SSE (**a**), the model (**b**).

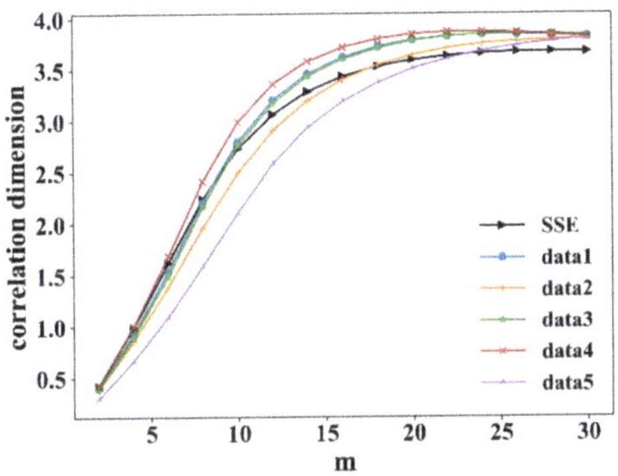

Figure 4. Correlation dimension of returns from SSE and five simulated data.

3.2.2. Lyapunov Exponent Analysis, Sample Entropy Analysis, and Hurst Exponent

We further compare the nonlinear behavior of simulated and empirical rates of return in this section. The maximal Lyapunov exponent (MLE) determines the predictability of a dynamical system. A positive MLE is usually taken as an indication that the system is chaotic. Consequently, any system with MLE > 0 is considered to be chaotic. We calculate the MLE of each stock price series using the algorithm of Rosenstein et al. [35]. In Table 2, the simulated and real returns have similar positive MLE, and indicate they are not totally stochastic. They have a similar chaotic property to some extent.

Table 2. The maximum Lyapunov exponent (m = 10), Sample Entropy (m = 2) and Hurst exponent of returns from the model and empirical market.

Data	MLE	Sample Entropy	Hurst Exponent
Data1	0.0778	1.7497	0.6281
Data2	0.0762	1.6832	0.6364
Data3	0.0773	1.7033	0.6478
Data4	0.0757	1.7401	0.6152
Data5	0.0575	1.4901	0.5840
SSE	0.0628	1.7889	0.5238
SZSE	0.0842	1.8750	0.5176
S&P500	0.0639	1.4902	0.5022

Hurst exponent is used as a measure of the "long memory" of a time series, which measures how the range of fluctuations in a time series varies over time. H ranges between 0 and 1 (excluding 0 and 1). Where H = 0.5, the time series indicates a completely uncorrelated series. When H > 0.5, the time series has long-term memory, and when H < 0.5, the time series has inverse persistence, it exhibits stronger fluctuations than totally random. We calculate the Hurst exponent by the rescaled range analysis [36]. In Table 2, the Hurst exponent is slightly larger than 0.5, which means that the simulated and real returns have similar long-term memory.

Sample entropy is a measure of the complexity of time series. The smaller the sample entropy, the higher the sequence self-similarity; the larger sample entropy, the more complex the sample sequence. We calculate the sample entropy method following Richman et al. [37]. In Table 2, the simulated and real returns have similar sample entropy values that indicate their similar complexity.

4. Correlation Analysis of Stocks

Portfolio theory is a framework for assembling a portfolio of assets such that the expected return is maximized and the level of risk is minimized. Investors can reduce risk by holding a portfolio of stocks that are not perfectly positively correlated. Diversification can help to construct optimal investment portfolios. Charu et al. [38] use mutual information for measuring stock correlations and construct the stock network. Sun et al. [39] applied DCCA coefficients to construct the correlation matrix of assets. Thus, the correlation between stocks is an important criterion to weigh the correlation of stock market risk level and portfolio rationality. Studies on the properties of stock correlation show that the stronger correlations between stocks are, the higher risk in the corresponding asset portfolio [40]. Usually, stocks belonging to the same industry are more correlated because they are influenced by the same external information, including natural climate, macro policies, raw materials, and other factors [41]. The stocks in an industry have strong positive correlations and risky portfolios, so sound investments usually cover different industries. In our model, stock rises or falls are affected by external information; thus, the model can be considered to study the correlation between stocks.

This section investigates the correlation of stock returns within per industry in China using the detrended cross-correlation analysis (DCCA) [42,43] and calculates their distributions. The DCCA coefficient measures the correlation level between non-stationary series such as financial series. ρ is the DCCA coefficient, $-1 \leq \rho \leq 1$. $\rho = 1$ indicates that two time series are perfectly correlated; $\rho = -1$ indicates that the two time series are perfectly anti-correlated; $\rho = 0$ indicates that the two time series are uncorrelated processes. There are 28 industries in the Shenwan Industry Classification Standard. We selected 16 industries from 1 January 2016 to 10 December 2020 ($T \approx 1200$), which contain a sufficient number of stocks (the number of stocks $N > 30$). We then simulated stock data in an industry: As the initial stock price is the same, to avoid the sensitivity to initial conditions, we selected the data from 6000 to 7500 steps in the simulation ($T = 1500$), then we obtained 100 stocks under the same historical information series.

Figure 5 shows the distribution of the correlation of stock returns within an industry. Figure 5a–c are three empirical data examples, and Figure 5d–f are three simulated ones that are generated in different historical information series. It can be seen in Figure 5 that ρ distributions within each of the 16 industries show a regular single-peaked distribution. The most probable correlation coefficients ρ_m are around 0.3, which indicates that the model is consistent with the real market, and most stocks have weak positive correlations within an industry. Figure 6 shows the most probable correlation coefficients ρ_m within the 16 industries and the three simulated data. The three simulated data peaks are 0.34, 0.33, and 0.32; all are lying within the peak range from 0.21 to 0.43 in the real market. Moreover, since each set of simulated data is generated in given the same historical information series, there is probable that the stock market evolution will recur when there is similar information series. In our model, the correlation of the simulated stock with the same historical information can be analogized statistically to the correlation of the stocks within China's industry. It is a supplement method of stock correlation research that helps investors obtain a better portfolio strategy.

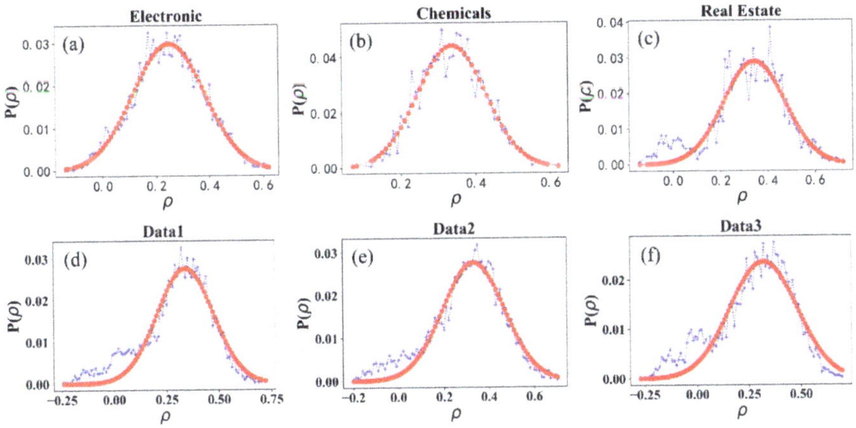

Figure 5. The distribution of ρ from Chemicals (**a**), Real Estate (**b**), Electronics (**c**), and three simulated data (**d–f**).

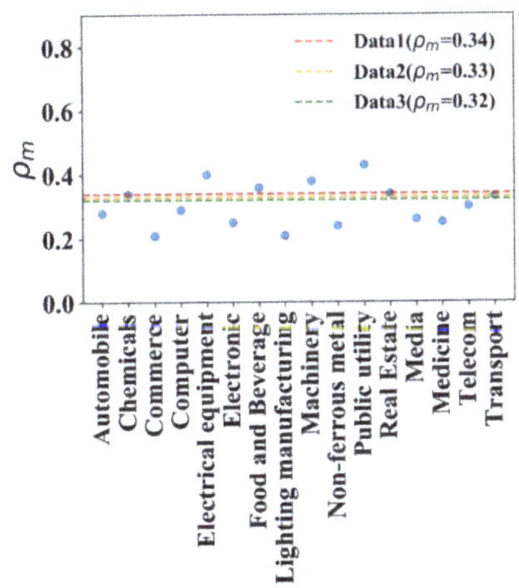

Figure 6. The ρ_m of stocks in 16 industries and three simulated data.

5. Conclusions

"Greatest truths are the simplest" is an objective law. The principles also apply to the stock market. With the development of the stock market, the spread of information is faster. It is easier to get information, the literacy level of the investors has improved. They are closer to each other; their relationships are more complicated than ever, society's wealth has increased, etc. The empirical studies show that no matter how the stock market environment changes, the universal characteristics (the crashes, the skewed distributions with specific kurtosis values, the fat tails, etc.) remain stable. It means that the "Greatest truths in stock market remain stable." In this paper, we aim to find the "Greatest truths in the stock market."

We analyze three typical models (the percolation model, Ising model, and network topology financial model) and their extensions that are used for stock market research. We find that these models can represent the universal characteristics successfully. It means that these models should contain the "Greatest truths in the stock market." We find that "they generate a stock trading in the form of probability."

The stock market environment is variable daily and is influenced by a series of stochastic events (supply and demand, macroeconomic, political factors, corporate finances and performance, market sentiment, etc.). We coarse-grain all the stochastic events by information just a single influence value. The information can influence investors' performance. The stock price is the result of all investors' performance. We model the progress in probability and find that it can represent the universal characteristics.

Our model is based on the idea of "Greatest truths in the stock market." Our results suggest that the investors' individual characteristic is not the critical factor; the stock market's micro-specialties are not the greatest truths. In the stock market, the critical factor is information, and the stock price is the emergence of collective performance of all investors. Besides, the model can generate different stock price series in the same historical information, analogous to the stocks in the same industry. Similar single-peaked distribution proving that the model can be effectively used in stock correlation research and history recur rules. It opens a new way of selecting rational portfolios, complementing current industry correlation research methods, and providing theoretical support. The paper provides a helpful framework for understanding stock price evolution through the emergence of collective performance. We find the possible critical factor and the essence of the financial market at a macro level.

Author Contributions: Writing—original draft preparation, Z.W.; writing—review and editing, Y.Z., G.S. and M.S. All authors have read and agreed to the published version of the manuscript.

Funding: This research was funded by the National Natural Science Foundation of China (Grant No. 61503140).

Data Availability Statement: Data sharing not applicable.

Acknowledgments: Z.W. is obliged to Jiawei Zhang for several enlightening discussions and suggestions for writing.

Conflicts of Interest: The authors declare no conflict of interest. The funders had no role in the design of the study; in the collection, analyses, or interpretation of data; in the writing of the manuscript, or in the decision to publish the results.

References

1. Barabási, A.-L. The Network Takeover. *Nat. Phys.* **2012**, *8*, 14–16. [CrossRef]
2. Lazer, D.; Pentland, A.; Adamic, L.; Aral, S.; Barabasi, A.-L.; Brewer, D.; Christakis, N.; Contractor, N.; Fowler, J.; Gutmann, M.; et al. Social Science. Computational Social Science. *Science* **2009**, *323*, 721–723. [CrossRef] [PubMed]
3. Caldarelli, G.; Wolf, S.; Moreno, Y. Physics of Humans, Physics for Society. *Nat. Phys.* **2018**, *14*, 870. [CrossRef]
4. Ackert, L.F.; Church, B.K.; Deaves, R. Emotion and Financial Markets. *Fed. Reserv. Bank Atlanta Econ. Rev.* **2003**, *88*, 33–41.
5. Sadi, R.; Asl, H.G.; Rostami, M.R.; Gholipour, A.; Gholipour, F. Behavioral Finance: The Explanation of Investors' Personality and Perceptual Biases Effects on Financial Decisions. *Int. J. Econ. Financ.* **2011**, *3*, 234–241. [CrossRef]
6. Kraemer, M.U.G.; Yang, C.-H.; Gutierrez, B.; Wu, C.-H.; Klein, B.; Pigott, D.M.; du Plessis, L.; Faria, N.R.; Li, R.; Hanage, W.P.; et al. The Effect of Human Mobility and Control Measures on the COVID-19 Epidemic in China. *Science* **2020**, *368*, 493–497. [CrossRef] [PubMed]
7. Castellano, C.; Fortunato, S.; Loreto, V. Statistical Physics of Social Dynamics. *Rev. Mod. Phys.* **2009**, *81*, 591. [CrossRef]
8. Perc, M. The Social Physics Collective. *Sci. Rep.* **2019**, *9*, 16549. [CrossRef]
9. Cont, R.; Bouchaud, J.-P. Herd Behavior And Aggregate Fluctuations In Financial Markets. *Macroecon. Dyn.* **2000**, *4*, 170–196. [CrossRef]
10. Eguiluz, V.M.; Zimmermann, M.G. Transmission of Information and Herd Behavior: An Application to Financial Markets. *Phys. Rev. Lett.* **2000**, *85*, 5659–5662. [CrossRef]
11. Ren, F.; Zheng, B. Generalized Persistence Probability in a Dynamic Economic Index. *Phys. Lett. A* **2003**, *313*, 312–315. [CrossRef]
12. Zhou, W.X.; Sornette, D. Self-Organizing Ising Model of Financial Markets. *Eur. Phys. J. B* **2007**, *55*, 175–181. [CrossRef]
13. Cont, R. Empirical Properties of Asset Returns: Stylized Facts and Statistical Issues. *Quant. Financ.* **2001**, *1*, 223–236. [CrossRef]
14. Maganini, N.D.; Filho, A.C.D.S.; Lima, F.G. Investigation of Multifractality in the Brazilian Stock Market. *Phys. A Stat. Mech. Appl.* **2018**, *497*, 258–271. [CrossRef]
15. Zhao, H.; Zhou, J.; Zhang, A.; Su, G.; Zhang, Y. Self-Organizing Ising Model of Artificial Financial Markets with Small-World Network Topology. *Europhys. Lett.* **2013**, *101*, 18001. [CrossRef]
16. Wan, Y.-L.; Wang, G.-J.; Jiang, Z.-Q.; Xie, W.-J.; Zhou, W.-X. The Cooling-off Effect of Price Limits in the Chinese Stock Markets. *Phys. A Stat. Mech. Appl.* **2018**, *505*, 153–163. [CrossRef]

17. Singh, V.K.; Gautam, D.; Singh, R.R.; Gupta, A.K. *Agent-Based Computational Modeling of Emergent Collective Intelligence BT—Computational Collective Intelligence. Semantic Web, Social Networks and Multiagent Systems*; Nguyen, N.T., Kowalczyk, R., Chen, S.-M., Eds.; Springer: Berlin/Heidelberg, Germany, 2009; pp. 240–251.
18. Woolley, A.W.; Chabris, C.F.; Pentland, A.; Hashmi, N.; Malone, T.W. Evidence for a Collective Intelligence Factor in the Performance of Human Groups. *Science* **2010**, *330*, 686–688. [CrossRef]
19. Mitchell, M.L.; Mulherin, J.H. The Impact of Public Information on the Stock Market. *J. Financ.* **1994**, *49*, 923–950. [CrossRef]
20. Preis, T.; Reith, D.; Stanley, H.E. Complex Dynamics of Our Economic Life on Different Scales: Insights from Search Engine Query Data. *Philos. Trans. R. Soc. A Math. Phys. Eng. Sci.* **2010**, *368*, 5707–5719. [CrossRef]
21. Carrière-Swallow, Y.; Labbé, F. Nowcasting with Google Trends in an Emerging Market. *J. Forecast.* **2013**, *32*, 289–298. [CrossRef]
22. Preis, T.; Moat, H.S.; Stanley, H.E. Quantifying Trading Behavior in Financial Markets Using Google Trends. *Sci. Rep.* **2013**, *3*, 1684. [CrossRef] [PubMed]
23. Albers, S.; Charikar, M.; Mitzenmacher, M. Delayed Information and Action in On-Line Algorithms. *Inf. Comput.* **2001**, *170*, 135–152. [CrossRef]
24. Gopikrishnan, P.; Plerou, V.; Nunes Amaral, L.A.; Meyer, M.; Stanley, H.E. Scaling of the Distribution of Fluctuations of Financial Market Indices. *Phys. Rev. E* **1999**, *60*, 5305–5316. [CrossRef] [PubMed]
25. Qiu, T.; Zheng, B.; Ren, F.; Trimper, S. Return-Volatility Correlation in Financial Dynamics. *Phys. Rev. E* **2006**, *73*, 65103. [CrossRef]
26. Zhang, J.W.; Zhang, Y.; Kleinert, H. Power Tails of Index Distributions in Chinese Stock Market. *Phys. A Stat. Mech. Appl.* **2007**, *377*, 166–172. [CrossRef]
27. Wang, Y.; Zheng, S.; Zhang, W.; Wang, J. Complex and Entropy of Fluctuations of Agent-Based Interacting Financial Dynamics with Random Jump. *Entropy* **2017**, *19*, 512. [CrossRef]
28. Balanda, K.P.; Macgillivray, H.L. Kurtosis: A Critical Review. *Am. Stat.* **1988**, *42*, 111–119.
29. Gopikrishnan, P.; Meyer, M.; Amaral, L.A.N.; Stanley, H.E. Inverse Cubic Law for the Distribution of Stock Price Variations. *Eur. Phys. J. B* **1998**, *3*, 139–140. [CrossRef]
30. Hsieh, D.A. Chaos and Nonlinear Dynamics: Application to Financial Markets. *J. Finance* **1991**, *46*, 1839–1877. [CrossRef]
31. Alves, P.R.L.; Duarte, L.G.S.; da Mota, L.A.C.P. Detecting Chaos and Predicting in Dow Jones Index. *Chaos Solitons Fractals* **2018**, *110*, 232–238. [CrossRef]
32. Zhu, H.; Zhang, W. Multifractal Property of Chinese Stock Market in the CSI 800 Index Based on MF-DFA Approach. *Phys. A Stat. Mech. Appl.* **2018**, *490*, 497–503. [CrossRef]
33. Grassberger, P.; Procaccia, I. Dimensions and Entropies of Strange Attractors from a Fluctuating Dynamics Approach. *Phys. D Nonlinear Phenom.* **1984**, *13*, 34–54. [CrossRef]
34. Grassberger, P.; Procaccia, I. Measuring the Strangeness of Strange Attractors. *Phys. D Nonlinear Phenom.* **1983**, *9*, 189–208. [CrossRef]
35. Rosenstein, M.T.; Collins, J.J.; De Luca, C.J. A Practical Method for Calculating Largest Lyapunov Exponents from Small Data Sets. *Phys. D Nonlinear Phenom.* **1993**, *65*, 117–134. [CrossRef]
36. Couillard, M.; Davison, M. A Comment on Measuring the Hurst Exponent of Financial Time Series. *Phys. A Stat. Mech. Appl.* **2005**, *348*, 404–418. [CrossRef]
37. Richman, J.S.; Moorman, J.R. Physiological Time-Series Analysis Using Approximate Entropy and Sample Entropy. *Am. J. Physiol. Circ. Physiol.* **2000**, *278*, H2039–H2049. [CrossRef]
38. Sharma, C.; Habib, A. Mutual Information Based Stock Networks and Portfolio Selection for Intraday Traders Using High Frequency Data: An Indian Market Case Study. *PLoS ONE* **2019**, *14*, e0221910. [CrossRef]
39. Sun, X.; Liu, Z. Optimal Portfolio Strategy with Cross-Correlation Matrix Composed by DCCA Coefficients: Evidence from the Chinese Stock Market. *Phys. A Stat. Mech. Appl.* **2016**, *444*, 667–679. [CrossRef]
40. Eom, C.; Park, J.W. Effects of Common Factors on Stock Correlation Networks and Portfolio Diversification. *Int. Rev. Financ. Anal.* **2017**, *49*, 1–11. [CrossRef]
41. Guo, X.; Zhang, H.; Tian, T. Development of Stock Correlation Networks Using Mutual Information and Financial Big Data. *PLoS ONE* **2018**, *13*, e0195941. [CrossRef]
42. Kristoufek, L. Measuring Correlations between Non-Stationary Series with DCCA Coefficient. *Phys. A Stat. Mech. Appl.* **2014**, *402*, 291–298. [CrossRef]
43. Ferreira, P.; Pereira, É.J.d.A.L.; Silva, M.F.d.; Pereira, H.B. Detrended Correlation Coefficients between Oil and Stock Markets: The Effect of the 2008 Crisis. *Phys. A Stat. Mech. Appl.* **2019**, *517*, 86–96. [CrossRef]

entropy

MDPI

Article

A Maximum Entropy Model of Bounded Rational Decision-Making with Prior Beliefs and Market Feedback

Benjamin Patrick Evans * and **Mikhail Prokopenko**

Centre for Complex Systems, The University of Sydney, Sydney, NSW 2006, Australia; mikhail.prokopenko@sydney.edu.au
* Correspondence: benjamin.evans@sydney.edu.au

Abstract: Bounded rationality is an important consideration stemming from the fact that agents often have limits on their processing abilities, making the assumption of perfect rationality inapplicable to many real tasks. We propose an information-theoretic approach to the inference of agent decisions under Smithian competition. The model explicitly captures the boundedness of agents (limited in their information-processing capacity) as the cost of information acquisition for expanding their prior beliefs. The expansion is measured as the Kullblack–Leibler divergence between posterior decisions and prior beliefs. When information acquisition is free, the homo economicus agent is recovered, while in cases when information acquisition becomes costly, agents instead revert to their prior beliefs. The maximum entropy principle is used to infer least biased decisions based upon the notion of Smithian competition formalised within the Quantal Response Statistical Equilibrium framework. The incorporation of prior beliefs into such a framework allowed us to systematically explore the effects of prior beliefs on decision-making in the presence of market feedback, as well as importantly adding a temporal interpretation to the framework. We verified the proposed model using Australian housing market data, showing how the incorporation of prior knowledge alters the resulting agent decisions. Specifically, it allowed for the separation of past beliefs and utility maximisation behaviour of the agent as well as the analysis into the evolution of agent beliefs.

Keywords: decision-making; bounded rationality; complexity economics; information-theory; maximum entropy principle; quantal response statistical equilibrium

JEL Classification: D91; G41; D83; C61; C60; C50

Citation: Evans, B.P.; Prokopenko, M. A Maximum Entropy Model of Bounded Rational Decision-Making with Prior Beliefs and Market Feedback. *Entropy* **2021**, *23*, 669. https://doi.org/10.3390/e23060669

Academic Editors: Ryszard Kutner, Christophe Schinckus and Eugene Stanley

Received: 21 April 2021
Accepted: 21 May 2021
Published: 26 May 2021

1. Introduction

Economic agents are often faced with partial information and make decisions under pressure, yet many canonical economic models assume perfect information and perfect rationality. To address these challenges, Simon [1] introduced bounded rationality as an alternate attribute of decision-making. Bounded rationality aims to represent partial access to information, with possible acquisition costs, and limited computational cognitive processing abilities of the decision-making agents.

Information theory offers several natural advantages in capturing bounded rationality, interpreting the economic information as the source data to be delivered to the agent (receiver) through a noisy communication channel (where the level of noise is related to the "boundedness" of the agent). This representation has spurred the creation of information-theoretic approaches to economics, such as Rational Inattention (R.I.) [2], and more recently, the application of R.I. to discrete choice [3]. Another approach represents decision-making as a thermodynamic process over state changes and employs the energy-minimisation principle to derive suitable decisions [4].

These approaches have shown how one can incorporate a priori knowledge into decision-making, but place no consideration to inferring these decisions based on observed macroeconomic outcomes (e.g., a distribution of profit rates within a financial market) and market feedback loops. Independently, another recent information-theoretic framework, Quantal Response Statistical Equilibrium (QRSE) [5], was developed aiming to infer least biased (i.e., "maximally noncommittal with regard to missing information" [6]) decisions through the maximum entropy principle, given only the macroeconomic outcomes (e.g., when the choice data is unobserved). However, the ways to incorporate prior knowledge into such a system remain mostly unexplored.

In this work, we provide a unification of these approaches, showing how to incorporate prior beliefs into QRSE in a generic way. In doing so, we provide a least biased inference of decision-making, given an agent's prior belief. Specifically, we show how the incorporation of prior beliefs affects the agent's resulting decisions when their individual choices are unobserved (as is common in many real-world economic settings). The proposed information-theoretic approach achieves this by considering a cost of information acquisition (measured as the Kullback-Leibler divergence), where this cost controls deviations from an agent's prior knowledge on a discrete choice set. When the cost of information acquisition is prohibitively high (i.e., when an agent is faced with limitations through time, cognition, cost, or other constraints), the agent falls back to their prior beliefs. When information acquisition is free, the agent becomes a perfect utility maximiser. The cost of information acquisition therefore measures the boundedness of the agent's decision-making.

The proposed approach is general, allowing the incorporation of any form of prior belief, while separating the agents' current expectations from their built-up beliefs. In particular, we show how incorporating prior beliefs into the QRSE framework allows for modelling decisions in a rolling way, when previous decisions "roll" into becoming the latest beliefs. Furthermore, we place the original QRSE in the context of related formalisms, and show that it is a special case of the general model proposed in our study, when the prior preferences (beliefs) are assumed to be uniform across the agent choices. Finally, we verify and demonstrate our approach using actual Australian housing market data, in terms of agent buying and selling decisions.

The remainder of the paper is organised as follows. Section 2 provides a background of information-theoretic approaches to economic decision-making, Section 3 describes QRSE and relevant decision-making literature. Section 4 outlines the proposed model, and Section 5 applies the developed model to the Australian housing market. Section 6 presents conclusions.

2. Background and Motivation

The use of statistical equilibrium (and more generally, information-theoretic) models remains a relatively new concept in economics [7]. For example, Yakovenko [8] outlines the use of statistical mechanics in economics. Scharfenaker and Semieniuk [9] detail the applicability of maximum entropy for economic inference, Scharfenaker and Yang [10] give an overview of maximum entropy and statistical mechanics in economics outlining the benefits of utilising the maximum entropy principle for rational inference, and Wolpert et al. [11] outline the use of maximum entropy for deriving equilibria with bounded rational players in game theory. Earlier, Dragulescu and Yakovenko [12] showed how in a closed economic system, the probability distribution of money should follow the Boltzmann-Gibbs law [13]. Foley [14] discusses Rational expectations and boundedly rational behaviour in economics. Harré [15] gives an overview of information-theoretic decision-theory and applications in economics, and Foley [16] analyses information-theory and results on economic behaviour.

Ömer [17] provides a comparison of "conventional" economic models and newly proposed ideas from complex systems such as maximum entropy methods and Agent-based models (ABM), which deviate from the assumption of homo economicus—a perfectly rational representative agent. Yang and Carro [18] discuss how a combination of agent-based modelling and maximum entropy models can be complementary, leveraging the analytical rigour of maximum entropy methods and the relative richness of agent-based modelling.

One of the key developments in this area is Quantal Response Statistical Equilibrium (QRSE) proposed by Scharfenaker and Foley [5]. This approach enabled applications of the maximum entropy method [6,19,20] to a broad class of economic decision-making. The QRSE model was further explored in [21], arguing that "any system constrained by negative feedbacks and boundedly rational individuals will tend to generate outcomes of the QRSE form". The QRSE approach is detailed in Section 3.1.

Ömer [22–24] applies QRSE to housing markets (which we also use as a validating example), modelling the change in the U.S. house price indices over several distinct periods, and explaining dynamics of growth and dips. Yang [25] applies QRSE to a technological change, modelling the adoption of new technology for various countries over multiple years and successfully recovering the macroeconomic distribution of rates of cost reduction. Wiener [26–28] applies QRSE to labour markets, modelling the competition between groups of workers (such as native and foreign-born workers in the U.S.), and capturing the distribution of weekly wages. Blackwell [29] provides a simplified QRSE for understanding the behavioural foundations. Blackwell further extends this in [30], introducing an alternate explanation for skew, which arises due to the agents having different buy (enter) and sell (exit) preferences. Scharfenaker [31] introduces Log-QRSE for income distribution, and importantly, (briefly) mentions informational costs as a possible cause for asymmetries in QRSE. This is captured by measuring utility U as a sum $U[a,x] + C(a|x)$, allowing for higher costs (C) of entrance or exit into a market, where a is an action and x is a rate. Such a separation allows for an "alternative interpretation of unfulfilled expectations".

These developments show the usefulness of maximum entropy methods, where we have placed particular focus on QRSE, for inferring decisions from only macro-level economic data. However, these approaches do not consider the contribution of a priori knowledge to the resulting decision-making process. The key objective of our study is to generalise the QRSE framework by the introduction of the prior beliefs, as well as the information acquisition costs as a measure of deviation from such priors.

3. Underlying Concepts

Two main concepts form the basis for the proposed model. The first is the QRSE approach developed by [5], and the second is a thermodynamics-based concept of decision-making derived from minimising negative free energy, proposed by [4].

3.1. QRSE

The QRSE framework aims to explain macroeconomic regularities as arising from social interactions between agents. There are two key assumptions stemming from the idea of Smithian competition: Agents observe and respond to macroeconomic outcomes, and agent actions affect the macroeconomic outcome, i.e., a feedback loop is assumed. It is this feedback that is deemed to cause the macroeconomic outcome to have a distribution that stabilises around an average value. Given only the macroeconomic outcome, QRSE infers the least biased distribution of decisions, which result in the observed macroeconomic distribution using the principle of maximum entropy. This makes QRSE particularly useful for inferring decisions when the individual decision level data is unobserved. In the following section, we outline the key notions behind QRSE [5].

3.1.1. Deriving Decisions

Agents are assumed to respond (i.e., make decisions) based on the macroeconomic outcome, for example, based on profit rates x. This is captured by the agents' utility U. However, agents are assumed to act in a boundedly rational way, such that they may not always choose the option with the highest U, for example, if it becomes impractical to consider all outcomes. That is, agents are attempting to maximise their expected utility, subject to an entropy constraint capturing the uncertainty:

$$\max \sum_{a \in A} f[a|x] U[a,x] \tag{1}$$

$$
\begin{aligned}
\text{subject to} \sum_{a \in A} f[a|x] &= 1 \\
-\sum_{a \in A} f[a|x] \log f[a|x] &\geq H_{min}
\end{aligned}
\tag{2}
$$

where $f[a|x]$ represents the probability of an agent choosing action a if rate x is observed. The first constraint ensures the probabilities sum to 1, while the second is a constraint on the minimum entropy. The minimum entropy constraint implies a level of boundedness such that there is some limit to the agents' processing abilities, which allows QRSE to deviate from perfect rationality.

Lagrange multipliers can be used to turn the constrained optimization problem of Equation (2) into an unconstrained one, which forms the following Lagrangian function:

$$
\mathcal{L} = -\sum_{a \in A} f[a|x]U[a,x] - \lambda\left(\sum_{a \in A} f[a|x] - 1\right) + T\left(-\sum_{a \in A} f[a|x] \log f[a|x] - H_{min}\right) \tag{3}
$$

taking the first order conditions of Equation (3), and solving for $f[a|x]$ yields:

$$
f[a|x] = \frac{1}{Z} e^{\frac{U[a,x]}{T}} \tag{4}
$$

representing a choice of a mixed strategy by maximising the expected utility subject to an entropy constraint. This problem is dual to maximising entropy of the mixed strategy, subject to a constraint on the expected utility as detailed in Appendix A.1.

3.1.2. Deriving Statistical Equilibrium

From Section 3.1.1 we have a derivation for a decision function, where agents maximise expected utility subject to an entropy constraint introducing bounds in the agents processing abilities. In order to infer the statistical equilibrium based on observed macroeconomic outcomes, the joint probability $f[a,x]$ must be computed.

The joint distribution captures the resulting statistical equilibrium which arises from the individual agent decisions. While there are many potential joint distributions, using the principle of maximum entropy allows for inference of the least biased distribution. From an observer perspective, maximising the entropy of the model accounts for model uncertainty, by providing the maximally noncommittal joint distribution. To compute this, Scharfenaker and Foley [5] maximise the joint entropy with respect to the marginal probabilities (since individual action data is not available), by decomposing the joint entropy into a sum of the marginal entropy and the (average) conditional entropy.

The solution for $f[a|x]$, given by Equation (4), can be used to compute the joint probability $f[a,x]$, as long as marginal $f[x]$ is determined (since $f[a,x] = f[a|x]f[x]$). In order to derive $f[x]$, the approach considers the state dependant conditional entropy, represented as

$$
H[A|x] = -\sum_{a \in A} f[a|x] \log f[a|x] \tag{5}
$$

Scharfenaker and Foley [5] then use the principle of maximum entropy to find the distribution of $f[x]$ which maximises

$$
\max_{f[x] \geq 0} H = -\int_x f[x] \log f[x] dx + \int_x f[x] H[A|x] dx \tag{6}
$$

$$
\begin{aligned}
\text{subject to} \int_x f[x] dx &= 1 \\
\int_x f[x] x dx &= \xi
\end{aligned}
\tag{7}
$$

The first constraint ensures the probabilities sum to 1, and the second constraint applies to the mean outcome (with ξ being the mean from the actual observed data $\bar{f}[x]$). Importantly, there is also an additional constraint which models Smithian competition [32] in the market. Smithian competition models the feedback structure for competitive markets, for example, entrance into a market tends to lower the profit rates, and exit tends to raise the profit rates. This is captured as the difference between the expected returns conditioned on entrance, and the expected returns conditioned on exiting. This competition constraint can be represented as

$$\text{subject to} \int_x f[x](f[a|x] - f[\bar{a}|x])x dx = \delta \tag{8}$$

The combination of the conditional probabilities of Equation (4), which stipulate that the agents enter and exit based on profit rates, and the competition constraint of Equation (8) models a negative feedback loop that results in a distribution of the profit rates around an average (ξ).

Again, using the method of Lagrange multipliers, the associated Lagrangian becomes

$$\begin{aligned} \mathcal{L} = &- \int_x f[x] \log f[x] dx + \int_x f[x] H[A|x] dx - \\ \lambda\Big(\int_x f[x]dx - 1\Big) - \gamma\Big(&\int_x f[x]x dx - \xi\Big) - \rho\Big(\int_x f[x](f[a|x] - f[\bar{a}|x])x dx - \delta\Big) \end{aligned} \tag{9}$$

where taking the first order conditions of Equation (9), and solving for $f[x]$ yields

$$f[x] = \frac{1}{Z_A} e^{H[A|x] - \gamma x - \rho x(f[a|x] - f[\bar{a}|x])} \tag{10}$$

where Z_A is the partition function $Z_A = \int_x e^{H[A|x] - \gamma x - \rho x(f[a|x] - f[\bar{a}|x])} dx$. Note that in Equation (9) we use ρ as the Lagrangian multiplier for the competition constraint. Parameter ρ is referred to as β in [5], we have avoided this notation to avoid confusion with the thermodynamic β (inverse temperature) discussed in later sections.

Equations (4) and (10) comprise a fully defined joint probability. Crucially, QRSE allows for modelling the resultant statistical equilibrium even when the individual actions are unobserved—by inferring these decisions based on the principle of maximum entropy.

3.1.3. Limitations of Logit Response

In Section 3.1.1 we have seen how the logit response function used for decision-making in QRSE is derived from entropy maximisation. Following the Boltzmann distribution well known in thermodynamics, this logit response has seen extensive use throughout the literature arising in a variety of domains. For example, the logit function is used as sigmoid or softmax in neural networks, logistic regression, and in many applications in economics and game theory [33,34]. However, one important development not yet discussed is the incorporation of prior knowledge into the formation of beliefs. Up until now, we have considered a choice to be the result of expected utility maximisation based on entropy constraints from which the logit models have arisen. However, from psychology [35], behavioural economics [36,37], and Bayesian methods [38,39] we know that the incorporation of a priori information is often an important factor in decision-making. Thus, we explore the incorporation of prior beliefs into agent decisions in more detail in the following section (and the remainder of the paper).

Furthermore, one criticism of the logit response arises from the independence of irrelevant alternatives (IIA) property of multinomial logit models (which would extend to the conditional function used in QRSE in a multi-action case), which states that the ratio between two choice probabilities should not change based on a third irrelevant alternative. Initially, this may seem desirable, however, this can become problematic for correlated outcomes (of which many real examples possess). This criticism has been proved correct in several thought experiment studies, showing violations of the IIA assumption [40]. The classical example is the Red Bus/Blue Bus problem [41,42].

Consider a decision-maker who must choose between a car and a (blue) bus, $A = \{\text{car}, \text{blue bus}\}$. The agent is indifferent to taking the car or bus, i.e., $p(\text{car}) = p(\text{blue bus}) = 0.5$. However, suppose a third option is added, a red bus which is equivalent to the blue bus (in all but colour). The agent is indifferent to the colour of the bus, so when faced with $A_1 = \{\text{blue bus}, \text{red bus}\}$ the agent would choose $p(\text{red bus}) = p(\text{blue bus}) = 0.5$. Now suppose the agent is faced with a choice between $A_2 = \{\text{car}, \text{blue bus}, \text{red bus}\}$. As per the IIA property, the ratio $\frac{p(\text{blue bus})}{p(\text{car})}$ (from A, $\frac{0.5}{0.5}$) must remain constant. So adding in a third option, the probability of taking any a becomes $p(a) = \frac{1}{3}$ (for all a), maintaining $\frac{p(\text{blue bus})}{p(\text{car})} = 1$. However, this has reduced the odds of taking the car from 0.5 to 0.33 based on the addition of an irrelevant alternative (i.e., the red bus in which the agent does not care about colour of the bus). In reality, the probability for taking the car should have stayed fixed at $p(\text{car}) = 0.5$, and the probability of taking a bus reduced to 0.25 each. This reduction in the probability of $p(\text{car})$ does not make sense for a decision-maker who is indifferent to the colour of the bus and is the basis for the criticism. This may not be immediately relevant for current QRSE models (especially binary ones), but with potential future applications, for example, in portfolio allocation, this could become an important consideration. For example, if adding an additional stock to a portfolio which is similar to an existing stock, it may not be desriable to reduce the likelihood of selecting other (unrelated) stocks.

3.2. Thermodynamics of Decision-Making

A thermodynamically inspired model of decision-making which explicitly considers information costs, as well as the incorporation of prior knowledge, is proposed by [4]. The proposed approach can be seen as a generalisation of the logit function, where the typical logit function can be recovered as a special case, but in the more general case manages to avoid the IIA property.

Ortega and Braun [4] represent changing probabilistic states as isothermal transformations. Given some initial state $x \in X$ with initial energy potential $\phi_0[x]$, the probability of being in state x is $p[x] = \frac{e^{-\beta\phi_0[x]}}{\sum_{x' \in X} e^{-\beta\phi_0[x']}}$ (from the Boltzmann distribution). Updating state to $f[x]$ corresponds to adding new potential $\Delta\phi_0[x]$. The transformation requires physical work, given by the free-energy difference $\Delta F[f]$. The free energy difference between the initial and resulting state is then

$$\begin{aligned} \Delta F[f] &= F[f] - F[p] \\ &= \sum_{x \in X} f[x]\Delta\phi(x) + \frac{1}{\beta}\sum_{x \in X} f[x]\log\left(\frac{f[x]}{p[x]}\right) \end{aligned} \tag{11}$$

which allows the separation of the prior $p[x]$ and the new potential $\Delta\phi_0[x]$. In economic sense, representing the negative of the new potential as the utility gain, i.e., $U(x) = -\Delta\phi_0[x]$, allows for reasoning about utility maximisation subject to an informational constraint, given here as the Kullback-Leibler (KL) divergence from the prior distribution [4]. Golan [43] shows how the KL-divergence naturally arises as a generalisation of Shannon entropy (of Equation (2)) when considering prior information, and Hafner et al. [44] show how various objective functions can be seen as functionally equivalent to minimising a (joint) KL-divergence, even those not directly motivated by the free energy principle. Such analysis makes the KL-divergence a logical and fundamentally grounded measure of information acquisition costs, captured as the divergence from a prior distribution.

Ortega and Stocker [45] then apply this formulation to discrete choice by introducing a choice set A (space of actions), which leads to the following negative free energy difference, for a given observation x:

$$-\Delta F[f[a|x]] = \sum_{a \in A} f[a|x]U[a,x] - \frac{1}{\beta}\sum_{a \in A} f[a|x]\log\left(\frac{f[a|x]}{p[a]}\right) \quad (12)$$

where again a represents a choice (or action), and U the utility for the agent. The first term of Equation (12) is maximising the expected utility, and the second term is a regularisation on the cost of information acquisition. Again, in this representation, information cost is measured as the KL-divergence from the prior distribution.

Taking the first order conditions of Equation (12) and solving for $f[a|x]$ yields

$$f[a|x] = \frac{p[a]e^{\frac{U[x,a]}{T}}}{\sum_{a' \in A} p[a']e^{\frac{U[a',x]}{T}}} \quad (13)$$

where we have moved from inverse temperature β to temperature T for notational convenience, i.e., $T = \frac{1}{\beta}$. The key formulation here is the separation of the prior probability p from the utility gain (or the new potential from the initial potential). T then arises as the Lagrange multiplier for the cost of information acquisition (as opposed to the entropy constraint of QRSE, described in Section 3.1). We emphasise this aspect in later sections.

Revisiting the IIA property, the incorporation of the prior probabilities in Equation (A7) can adjust the choices away from the logit equation, and thus managing to avoid IIA. However, if desired, the free energy model reverts to the typical logit function in the case of uniform priors, and so this property can be recovered. In economic literature, a similar model is given by Rational Inattention (R.I.) by [2]. The relationship between R.I. and the free energy approach of [4,45] is detailed in Appendix C.

4. Model

In this section, we propose an information-theoretic model of decision-making with prior beliefs in the presence of Smithian competition and market feedback. Given an agent's prior beliefs and an observed macroeconomic outcome (such as the distribution of returns), the model can infer the least biased decisions that would result in such returns. Importantly, the incorporation of prior beliefs allows for reasoning about the decision-making of the agent based upon both their prior beliefs and their utility maximisation behaviour.

We develop upon the maximum-entropy model of inference from [5], and the thermodynamic treatment of prior beliefs formalised by [4], as outlined in Section 3.

4.1. Maximum Entropy Component

The proposed approach can be seen as a generalisation of QRSE, allowing for the incorporation of heterogeneous prior beliefs based on the free-energy principle. The key element is the information acquisition cost, measured as the KL-divergence which arises from the free-energy principle and has been shown to provide a fundamentally grounded application of Bayesian inference [46]. In order to derive decisions $f[a|x]$ for an action or choice a (e.g., buy, hold or sell) given an observed return x (e.g., a return on investment), we maximise the expected utility U subject to a constraint on the acquisition of information measured as the maximal divergence d between the posterior decisions and prior beliefs

$p[a]$. As mentioned, d is measured as the KL-divergence, which is the generalised extension of the original (Shannon) entropy constraint [43] introduced in Equation (2)):

$$\begin{aligned} \max \sum_{a\in A} f[a|x]U[a,x] \\ \text{subject to} \sum_{a\in A} f[a|x]\log\left(\frac{f[a|x]}{p[a]}\right) \leq d \\ \sum_{a\in A} f[a|x] = 1 \end{aligned} \tag{14}$$

The Lagrangian for Equation (14) then becomes

$$\mathcal{L} = \sum_{a\in A} f[a|x]U[a,x] - \lambda\left(\sum_{a\in A} f[a|x] - 1\right) - T\left(\sum_{a\in A} f[a|x]\log\left(\frac{f[a|x]}{p[a]}\right) - d\right) \tag{15}$$

There are two distinct modelling views on such a formulation [47–50]. The first assumes that specific constraints are known from the data, for example, a maximal divergence d may be specified based on actual observations of agent behaviour. The second view, instead, would consider the Lagrange multiplier T to be a free parameter of the model, with the constraint d representing an arbitrary maximum value: Thus, this approach would optimise T in finding the best fit. In this work, we take the second perspective since underlying decision data is unavailable, and a specific restriction on divergent information costs should not be enforced. In other words, T is considered to be a free model parameter corresponding to different information acquisition costs, mapping to different (unknown) cognitive and information-processing limits d.

Looking at the final term in Equation (15), in the case of homogeneous priors, $\log p[a]$ is a constant which drops out of the solution, which is equivalent to the optimisation problem of Equation (3), and thus, recovers the original QRSE model. In the general case, the dependence on $\log(p[a])$ means that T instead serves as the Lagrange multiplier for the cost of information acquisition. Taking the first order conditions of Equation (15) and solving for $f[a|x]$ (as shown in Appendix A.2) yields

$$f[a|x] = \frac{1}{Z_{A|x}} p[a] e^{\frac{U[a,x]}{T}} \tag{16}$$

we see this as a generalisation of the logit function, which allows for the separation of the prior beliefs and the agent's utility function.

In the more general case, $p[a]$ can be heterogeneous for all a. Parameter T therefore controls the deviations from the prior (rather than from the base case of uniformity), that is, it controls the cost of information acquisition. Following [4], we observe the following limits

$$\begin{aligned} \lim_{T\to\infty} f[a|x] &= p[a] \\ \lim_{T\to 0, T\geq 0} f[a|x] &= e^{\frac{U[x,a]}{T}} = \max U[x,a] \\ \lim_{T\to 0, T<0} f[a|x] &= e^{\frac{U[x,a]}{T}} = \min U[x,a] \end{aligned} \tag{17}$$

In the limit $T \to \infty$ (i.e., infinite information acquisition costs), the agent just falls back to their prior beliefs as it becomes impossible to obtain new information. In the limit $T \to 0$, the agent becomes a perfect utility maximiser (i.e., if information is free to obtain, the agent could obtain it all and choose the option that best maximises payoff with probability 1). In the $T < 0$ case, we see this corresponds to anti-rationality. For economic decision-making, we can limit temperatures to be non-negative, $T \geq 0$, although there are specific cases where such anti-rationality may be useful (e.g., modelling a pessimistic observer or adversarial environments [4]). The relationship between temperature and utility is visualised in Figure 1.

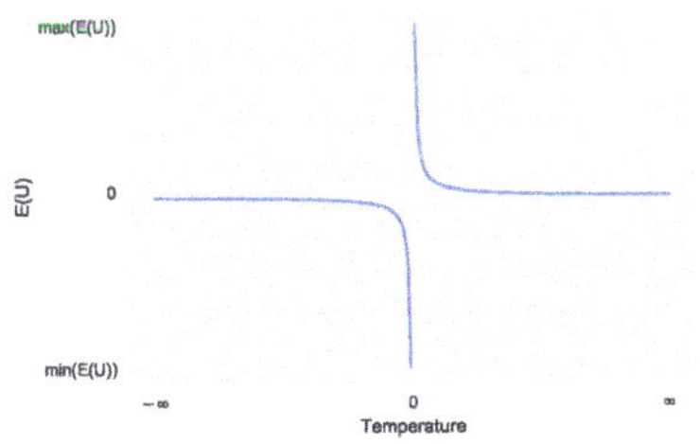

(**a**) Temperate vs. Utility

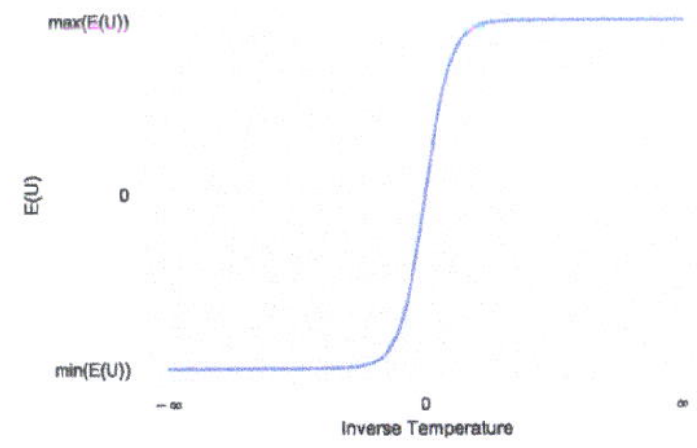

(**b**) Inverse Temperate ($\frac{1}{T}$) vs. Utility

Figure 1. The effect of decision temperature T on the resulting expected payoffs (**a**). for the limits given by Equation (17). The inverse temperature $\frac{1}{T}$ (**b**) conveys the same information but may offer a more useful visualisation due to the continuity.

Crucially, large temperatures (costly acquisition) do not revert to the uniform distribution (as in the typical QRSE case, unless the prior is uniform), instead reverting to prior beliefs. This is visualised in Figure 2, and discussed in more detail in Section 4.3.

4.2. Feedback Between Observed Outcomes and Actions

Following [5], we use a joint distribution to model the interaction between the economic outcome x, and the action of agents a.

To recover a joint probability, we need to determine $f[x]$ (since $f[a,x] = f[a|x]f[x]$) which we do with the maximum entropy principle, as shown in Section 3.1. To do this, we maximise the joint entropy with respect to the marginal probabilities. That is,

$$\begin{aligned}\mathcal{L} = &-\int_x f[x]\log f[x]dx + \int_x f[x]H[A|x]dx - \lambda\left(\int_x f[x]dx - 1\right)\\ &-\gamma\left(\int_x f[x]xdx - \xi\right) - \rho\left(\int_x f[x]\ \frac{p[a]e^{\frac{U[a,x]}{T}} - p[\bar{a}]e^{\frac{U[\bar{a},x]}{T}}}{Z_{A|x}}xdx - \delta\right)\end{aligned} \tag{18}$$

with

$$\begin{aligned}H[A|x] &= -\sum_{a\in A} f[a|x]\log f[a|x]\\ &= -\frac{1}{Z_{A|x}}\sum_{a\in A} p[a]e^{\frac{U[a,x]}{T}}\left(\log p[a] + \frac{U[a,x]}{T} - \log Z_{A|x}\right)\end{aligned} \tag{19}$$

An important point to be made here is that $H[A|x]$ still measures (Shannon) entropy. We have seen above how the new definition for $f[a|x]$ uses the KL-divergence as a generalised extension of entropy when incorporating prior information. In Equation (19), we do not use this divergence for an important reason. In Equation (14) we are measuring divergence from known prior beliefs, however, now when optimising Equation (18) we wish to infer decisions from unobserved decision data. This is where the principle of maximum entropy comes into play, i.e., we wish to maximise the entropy of our new choice data (which was derived from KL-divergence of prior beliefs), but we do not wish to perform cross-entropy minimisation as we do not have the true decisions $\bar{f}[a|x]$. With this in mind, we still utilise the principle of maximum entropy as is done in QRSE for inference to obtain the least biased resulting decisions. This keeps the proposed extensions in the realm of QRSE, but comparisons to the principle of minimum cross-entropy [51,52] could be considered in future work particularly when some target distributions are known directly.

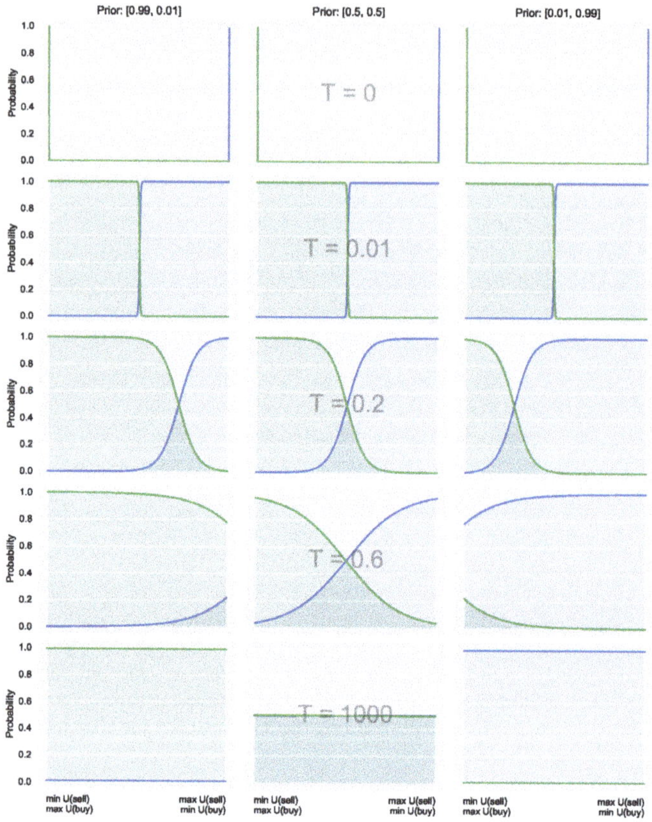

Figure 2. Decision Functions. All cases have equivalent utility functions. Each row has equivalent temperatures, showing how with matched parameters and utility, having an alternate prior can shift the decision-makers preference. Each column has different priors, given along the top of the first row to show how decision-makers decisions change based on their prior beliefs. On the left-hand side, preference is shifted towards the buying case. Likewise, on the right-hand side, preference is given to the selling case. The uniform case with equal preference is shown in the middle.

In Equation (18), ξ is known from the mean of the observed macroeconomic outcome, and so this constraint is used explicitly. This is in contrast to d (and δ) which are unknown as outlined in Section 4.1. The important distinction with Equation (18) is that the $f[a|x]$ functions (and $H[A|x]$) now use the updated expressions for $f[a|x]$, which incorporate the prior beliefs. Taking the partial derivative of $\mathcal{L}$ with respect to $f[x]$, and solving for $f[x]$ gives

$$f[x] = \frac{1}{Z_A} e^{H[A|x] - \gamma x - \rho x \left(\frac{p[a] e^{\frac{U[a,x]}{T}} - p[\bar{a}] e^{\frac{U[\bar{a},x]}{T}}}{Z_{A|x}} \right)} \tag{20}$$

Equation (20) expresses the information acquisition cost in the form of the Lagrange multiplier T (from Equation (15)), and a competition cost in the form of the multiplier ρ.

As we have a solution for $f[a|x]$ (Equation (16)) and $f[x]$ (Equation (20)) in terms of prior beliefs and information acquisition costs, we can then derive all other probability functions using the Bayes rule. That is, we can obtain $f[a,x]$, $f[x|a]$ and $f[a]$ which in turn incorporate these prior beliefs/acquisition costs:

$$
\begin{aligned}
f[a,x] &= f[a|x]f[x] \\
&= \frac{p[a]e^{\frac{U[a,x]}{T}+H[A|x]-\gamma x-\rho x\left(\frac{p[a]e^{\frac{U[a,x]}{T}}-p[\bar{a}]e^{\frac{U[\bar{a},x]}{T}}}{Z_{A|x}}\right)}}{Z_{A|x}Z_A}
\end{aligned}
\tag{21}
$$

We can obtain $f[a]$ by marginalising out x from the joint distribution:

$$
\begin{aligned}
f[a] &= \int_x f[a,x] \\
&= \frac{1}{Z_A}\int_x \frac{1}{Z_{A|x}} p[a]e^{\frac{U[a,x]}{T}+H[A|x]-\gamma x-\rho x\left(\frac{p[a]e^{\frac{U[a,x]}{T}}-p[\bar{a}]e^{\frac{U[\bar{a},x]}{T}}}{Z_{A|x}}\right)}
\end{aligned}
\tag{22}
$$

Finally, $f[x|a]$ can then be computed by a direct application of the Bayes rule: $f[x|a] = f[a,x]/f[a]$.

Given only an expected average value ξ (and the usual normalisation constraints), we have derived a joint probability distribution, which maximises the entropy subject to some information acquisition cost d, along with a competition cost δ. The resulting distribution free parameters (the Lagrange multipliers) are those which fit most closely to the true underlying distribution of returns. Thus, we have provided a generalisation of QRSE, which is fully compatible with the incorporation of prior beliefs.

4.3. Priors and Decisions

The introduced priors affect the conditional probabilities of agent decisions by shifting focus towards these preferred choices. The introduced priors allow the decision-maker to place more focus on particular actions if they have been deemed important a priori.

In Section 3.2 we showed how to separate the initial energy potential and new energy potential for distinguishing prior beliefs and utility functions. It is instructive to interpret these again as potentials, by setting $\alpha_a = T\log p[a]$, which allows us to represent the choice probability as

$$
f[a|x] = \frac{1}{Z_{A|x}}e^{\frac{U[x,a]+\alpha_a}{T}}. \tag{23}
$$

Equation (23) shows how α shifts the likelihood based on the prior preferences. An example of these shifts is visualised in Figure 2. This can be interpreted as placing more emphasis on actions deemed useful a priori as T increases. The information acquisition cost component T then controls the sensitivity between the utility and a priori knowledge, with a high T meaning higher dependence on prior information, and low T indicating a stronger focus on the utility alone.

The majority of binary QRSE models use a simple linear payoff definition for utility:

$$
U[x,a] = x-\mu,\;\; U[x,\bar{a}] = -(x-\mu)\,.
$$

With this definition, a tunable shift parameter μ serves as the expected fundamental rate of return. The relationship between μ and the real markets returns ξ (which was used as a constraint in Equation (7)), serves then as a measure of fulfilled expectations (i.e., if $\mu = \xi$) or unfulfilled expectations ($\mu \neq \xi$). This implies a symmetric shift parameter μ. As a specific example, if $a = \text{sell}$ and $\bar{a} = \text{buy}$, $\mu = 0.25$ means that at $x = 0.25$, buyers and sellers will be equally likely to participate in the market, i.e., $f[\text{sell}|\mu] = f[\text{buy}|\mu] = 0.5$. In this sense, μ can be seen as the indifference point. The symmetry arises from the fact that $f[\text{buy}|x] + f[\text{sell}|x] = 1$. Therefore, in the binary action case, it is possible to find a μ^* with the uniform priors $p = [0.5, 0.5]$ such that the decision functions will be equivalent to μ with any arbitrary priors $p = [c, 1-c]$, with $c \in [0,1]$. In this sense, μ can be seen as encapsulating a prior belief.

However, explicit incorporation of prior beliefs on actions is useful here as it helps to separate the agents' expectations in relation to their prior belief (e.g., a higher μ resulted from needing to change from their past behaviour) and choose the actions for which an agent should emphasise acquiring more information. The introduced prior beliefs are strictly known before any inference is performed, whereas μ is the result of the inference process. The separation of prior beliefs and current expectations is important, as with μ alone this can not capture an agent's predisposition prior to performing any information processing. In addition, this applies more generally to any arbitrary utility functions (as QRSE is, of course, not limited to the linear shift utility function with μ outlined above), or when any preference is known about decisions a priori.

Consider also the three action case, $A = \{\text{buy}, \text{hold}, \text{sell}\}$, with the same utility functions as above but with the extra utility for holding being $U[x, \text{hold}] = 0$. We can see that it would be desirable if buying and selling no longer required this symmetry. The use of priors can introduce this asymmetry, by providing separate indifference points for buy/hold and sell/hold. Such asymmetry alters the resulting frequency distribution of transactions, and may help to explain various trading patterns [16]. The difference of symmetric and asymmetric buy and sell curves is shown in Figure 3. Figure 3 shows that such functions could be recovered by introducing a secondary shift parameter μ_2. Parameter μ_1 (the original μ) then becomes the indifference point for buy and hold, and μ_2 for sell and hold. This is the method proposed in [30]. Introducing priors into this case again allows for separation of expectation μ, from prior belief and follows the same methodology as outlined above for the binary case. Furthermore, if we set $p[\text{hold}] = 0$, we recover the binary case. This highlights that the standard QRSE with binary actions and uniform priors is a special case of the ternary action case with heterogeneous priors.

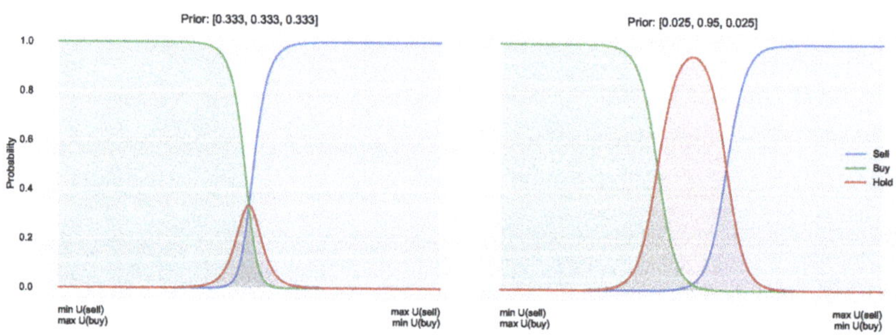

Figure 3. In the three-action case, the priors can introduce asymmetries by biasing the decision functions. This allows for separate indifferent points (**right**) vs. the uniform priors implying a single intersect (**left**).

From this, we can see how introducing priors alters the decision functions by allowing agents to focus on suitable a priori candidate actions. We have also shown how the binary case of a utility function with a shift parameter can be formalised to achieve equivalent results with a uniform prior and altered shift parameter. However, in the multi-action case, the priors allow for asymmetry, and in general, the priors may help with the optimisation process (by providing an alternate initial configuration). This approach also allows for the explicit separation of the two factors affecting an agent's choice, by distinguishing the contributions of prior beliefs and the utility maximisation.

4.4. Rolling Prior Beliefs

The proposed extension is general and allows for the incorporation of any form of prior beliefs, and in this section, we illustrate an example where the priors at time t are set as the resulting marginal probabilities from the previous time $t-1$:

$$p_t[a] = f_{t-1}[a]$$

i.e., the prior belief $p_t[a]$ is set as the previous marginal probability $f_{t-1}[a]$ for taking action a (at $t=0$, we use a uniform prior). Using the previous marginal probability as a prior introduces an "information-switching" cost, where T relates to the divergence from the previous actions, resulting in the following decision function:

$$f_t[a|x] = \frac{1}{Z_{A|x}} f_{t-1}[a] e^{\frac{U[x,a]}{T}}$$

That is, acquiring information on top of the previous knowledge comes at a cost (controlled by T). When the cost of information acquisition is high (large T), the agent reverts to the previously learnt knowledge (i.e., the marginal probabilities from $t-1$). In contrast, when T is extremely small, the agent is able to acquire new information allowing deviation from their prior knowledge at $t-1$. In the special case of $T=0$, information is free, and the agent can become a perfect utility maximiser.

Given the expression for $f_t[a|x]$, we obtain the following solution for $f_t[x]$:

$$f_t[x] = \frac{1}{Z_A} e^{H[A|x] - \gamma x - \rho x \left(\frac{f_{t-1}[a] e^{\frac{U[a,x]}{T}} - f_{t-1}[\bar{a}] e^{\frac{U[\bar{a},x]}{T}}}{Z_{A|x}} \right)}$$

from which we can derive the joint and other probabilities, as shown in Section 4.1. This is exemplified in Section 5, in which we examine various priors for time-dependent applications.

5. Australian Housing Market

To exemplify the model, we use the Greater Sydney house price dataset provided by SIRCA-CoreLogic and utilised in [53,54]. This dataset is outlined in Appendix B. In [54], an agent-based model is used to explain and forecast house price trends and movement patterns as arising from the individual agent's buy and sell decisions. Furthermore, the ABM implemented bounded rational agents driven by social influences (e.g., fear of missing out) and partial information about submarkets. While the resulting dynamics produced by the ABM accurately match the actual price trends, the decision-making mechanism and the bounded rationality of the agents were not theoretically grounded. In the following section, we aim to explain how the bounded rational behaviour of the agents operating in the housing market can be aligned with the model proposed in this study based on prior beliefs of agents and Smithian competition within the market. With this example, Smithian competition can be seen as agent decisions (buying or selling) affecting returns for an area, and agents decisions also being made based on returns for particular areas, i.e., a feedback loop is assumed in the market.

In particular, we want to explore what role an agent's prior beliefs play in their resulting decisions. For example, given equivalent configurations (e.g., utility and returns) and different prior knowledge, how would the agent's behaviour differ? Furthermore, we would like to explore the rationality of the agents, measured in terms of the cost of information acquisition, in order to see how the agents behave. For example, are agents predominantly reliant on past knowledge in times of market growth, resulting in unexpected downturns from mismanaged agent expectations? Alternatively, in deciding if it is a good time to buy or sell, the agents may balance their past knowledge with utility and current returns (i.e., the past knowledge would not be a predominant factor). The proposed model is particularly suited for answering such questions due to the low number of free (and microeconomically) interpretable parameters, as well as the explicit separation of prior beliefs (as opposed to previous QRSE approaches). Our goal is not to infer the "best" prior, but rather to explore and compare dynamics resulting from various priors. In addition, we aim to verify the conjecture that during crises, and periods exhibiting non-linear market dynamics, macroeconomic conditions may become more heterogeneous, and thus, non-uniform priors may outperform uniform ones in such times.

5.1. Model

We use our model of binary actions with prior beliefs introduced in Section 4.1, with actions $A = \{\text{buy}, \text{sell}\}$. The decision functions are then given by

$$\begin{aligned} f_t[\text{buy}|x] &= \frac{1}{Z_{t,x}} p_t[\text{buy}] e^{\frac{U[x,\text{buy}]}{T}} \\ f_t[\text{sell}|x] &= \frac{1}{Z_{t,x}} p_t[\text{sell}] e^{\frac{U[x,\text{sell}]}{T}} \\ Z_{t,x} &= f_t[\text{buy}|x] + f_t[\text{sell}|x] \end{aligned} \tag{24}$$

where we explore a range of p_t (prior at time t) functions, discussing their effects on decision-making and resulting probability distributions.

5.1.1. Priors

While the proposed approach is capable of incorporating any form of prior belief on the choice set A, below we outline several example priors which we explore. In exploring these priors, we highlight differences in resulting agent posterior decisions based on various prior beliefs.

Uniform

We begin with a uniform prior. The uniform probability represents the default case of QRSE, where each action has an equally weighted prior. In the binary case, this corresponds to $p_t[a] = 0.5$ for all t and a. This corresponds to an agent who is agnostic to the available actions before observing U.

Previous

Next we look at a "previous" prior. The previous prior uses the marginal action probabilities from the previous time step as priors to the current timestep. This means at time t, $f_t[a]$ plays the role of a posterior probability of making a decision, however, at time $t+1$ $f_t[a]$ now serves as the empirical prior. This is the example introduced in Section 4.4. This corresponds to $p_t[a] = f_{t-1}[a]$ for $t > 0$, and $p_t[a] = 0.5$ for $t = 0$. The previous prior represents an empirical prior where the decision is conditioned on previous market information, where T controls the level of influence from the previous market stage (in our case, each year). A high T means high influence from the past market state, whereas low T means focusing on current market conditions alone (as measured by U). In the extreme case of $T = \infty$, a backward looking expectations [55] approach is recovered where decisions are assumed to be a function purely of past decisions, however, in the more general case with $T < \infty$, U adjusts the decisions based on the current market state.

Mean

We also consider a mean prior. The mean prior uses the average marginal action probability from all previous timesteps. This corresponds to $p_t[a] = \frac{\sum_{t'=0}^{t-1} f_{t'}[a]}{t}$, for $t > 0$, and $p_t[a] = 0.5$ for $t = 0$. This can be seen as belief evolution, where over time, the previous decisions help build the current prior (modulated by T) at each stage.

Extreme Priors

As two further examples, we introduce extreme priors (more for visualisation/discussion sake as opposed to being particularly useful). The extreme buy prior corresponds to a strong prior preference for the buy action, $p_t[\text{buy}] = 0.99, p_t[\text{sell}] = 0.01$, for all t. Likewise, the extreme sell case is simply the inverse of the buy case, a strong prior preference for selling, i.e., $p_t[\text{sell}] = 0.99, p_t[\text{buy}] = 0.01$, for all t.

However, the formulations provided above by no means represent an exhaustive set of possible priors. For example, Genewein et al. [56] discuss "optimal" priors, which draws parallels with rate-distortion theory and can be seen as building abstractions of decisions (see Appendix C). Adaptive expectations [57] are discussed in [58–60], where priors could be partially adjusted based on some strength term (λ_E), where the strength term adjusts the contribution from some error. For example, an adaptive prior could be represented as $p_t = p_{t-1} + \lambda(p_{t-1} - \hat{p}_{t-1})$, where $\hat{p}_{t-1}$ is the actual known likelihood of actions from the previous time period. With our specific housing market data, we do not have $\hat{p}$, i.e., we do not have the true buying and selling likelihoods, but if known, such information could be used to adjust future beliefs, i.e., over time the adaptive priors would adjust decisions based on the previously observed likelihoods (controlled by λ). The proposed approach makes no assumption about the forms of prior beliefs, so the ideas outlined above can be incorporated into the method outlined here by adjusting the definition of p_t.

5.2. Results

We fit the distributions with the various priors outlined in Section 5.1.1 to the actual underlying return data, to estimate how well we are able to capture this distribution and explore the effects that these priors have on the resulting distribution. The results are presented in Table 1, which summarises the likelihood and the percentage of the explained variability (measured as Information Distinguishability (I.D.) [61]) compared to the underlying distribution. We see that there are no large differences in general between the priors in terms of the explained variability. However, the goal here is not to argue for the "best" prior fitting the dataset in terms of the explained variability, but rather to explore differences in the agent behaviour based on the prior knowledge (using the housing dataset as an example). Thus, the resulting fitted distributions $f[x]$, which are visualised in Figure A5, are more interesting. We observe how altering prior beliefs result in different resulting distributions and discuss how the incorporation of prior beliefs allows for a separation of the agents' utility maximisation behaviour from their previous knowledge. From Figure A5 we can also see how the priors can alter the optimisation process, for example, a good (bad) prior may help (harm) the optimisation by providing alternate initial configurations. The extreme priors can be seen as harmful, for example, in 2012 where the resulting distributions are unable to capture the true underlying distribution. The reason for this is being unable to find suitable T to enable appropriate divergence from the extreme prior beliefs. In contrast, well selected priors can help the optimisation process and result in better fitting distributions, such as in 2016 where the decisions resulting from the mean and previous prior fit the true data significantly better than the uniform prior.

The agents' decision functions $f[a|x]$ are visualised in Figure A7 which makes it clear how each prior adjusts the resulting probability of taking an action (and thus, alters the decisions). From this, we can see different probabilistic behaviours despite having equivalent utility functions and optimisation processes due to varying prior beliefs. For example, with the extreme priors, we observe a clear shift towards the strongly preferred action.

Figure A6 shows the resulting joint distributions $f[a,x]$, combining the results of Figures A5 and A7, since $f[a,x] = f[a|x]f[x]$. Looking at the second row of each plot in Figure A6, we can see a visual representation of how the joint probabilities adjust over time when using the previous year as the prior belief.

Table 1. Resuling likelihood and percentage of variability explained for each year, when compared to the actual underlying distribution (i.e., those given in Figure A2). Optimisation is done by minimising the negative log-likelihood between the resulting distributions and the actual distribution of returns.

	Uniform	Previous	Mean	Extreme Buy	Extreme Sell
2006	1082 (93%)	1082 (93%)	1082 (93%)	885 (59%)	1005 (74%)
2007	1089 (92%)	1089 (92%)	1090 (90%)	939 (68%)	1042 (83%)
2008	998 (95%)	905 (78%)	998 (95%)	998 (95%)	998 (95%)
2009	918 (96%)	918 (96%)	866 (88%)	880 (85%)	875 (85%)
2010	857 (95%)	857 (95%)	857 (95%)	740 (62%)	857 (95%)
2011	1045 (92%)	1044 (91%)	1047 (92%)	1045 (91%)	873 (62%)
2012	1067 (96%)	1067 (96%)	1067 (96%)	162 (6%)	142 (8%)
2013	1080 (90%)	1076 (90%)	1083 (90%)	983 (77%)	1075 (91%)
2014	938 (98%)	851 (74%)	938 (98%)	875 (71%)	938 (98%)
2015	860 (96%)	860 (96%)	860 (96%)	33 (10%)	808 (71%)
2016	873 (84%)	932 (95%)	908 (86%)	817 (70%)	932 (95%)
2017	916 (97%)	916 (97%)	916 (97%)	812 (76%)	916 (97%)
2018	989 (88%)	932 (85%)	933 (85%)	955 (82%)	998 (91%)
2019	1101 (92%)	1103 (92%)	1067 (94%)	1101 (92%)	952 (76%)

The resulting marginal action probabilities are visualised in Figure 4, where we observe clear market peaks and dips which match the actual returns of Figure 5, aligning with the general trends observed in Figure A1. The priors work on either increasing or decreasing the resulting marginal probabilities. For example, in the extreme sell case we see much higher resulting probabilities for $f[\text{sell}]$, likewise in the extreme buying case, we see much higher probabilities for $f[\text{buy}]$. The general peaks/dips remain in both cases. Overall, this shows how the prior belief can influence the resulting marginal probabilities.

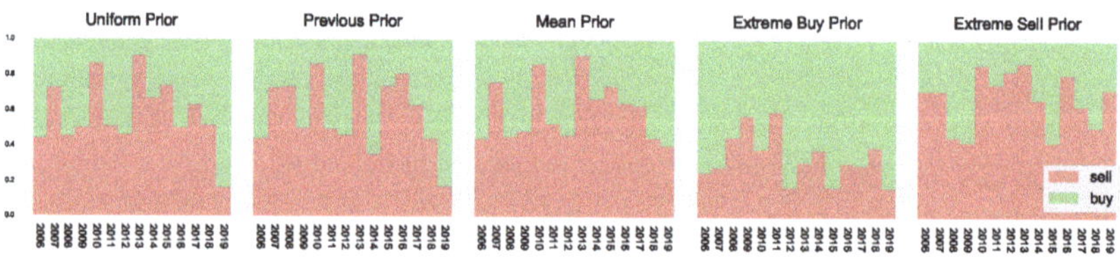

Figure 4. Resulting marginal probabilities $f[a]$ for varying priors. Green represents $f[\text{buy}]$, and red represents $f[\text{sell}]$.

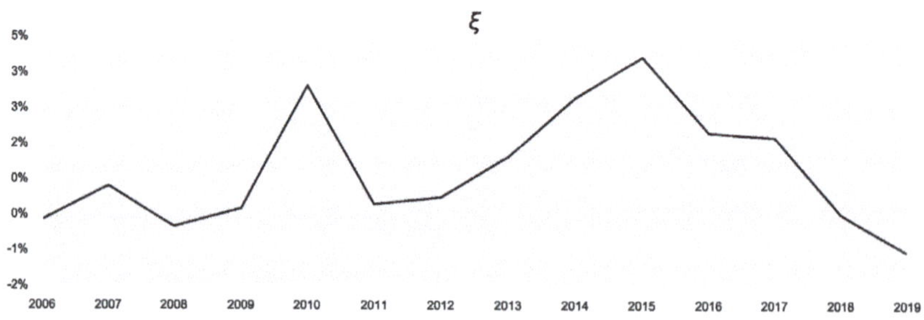

Figure 5. Real Average Returns.

Using the previous year's marginal probability as a prior for the current year has a smoothing effect on the resulting year-to-year marginal probabilities. Comparing the previous prior with the uniform prior in Figure 4, we observe, particularly during 2015–2018, a more defined/well-behaved step-off in $f[\text{sell}]$. This indicates the slowing of returns during these years. At the same time, the uniform priors are more affected by local noise, potentially overfitting to only the current time period, since no consideration can be given to the past behaviour of the market. This results in larger fluctuations in the agent behaviour as they have no concept of market history.

5.3. Role of Parameters

One of the benefits of QRSE is the low number of free parameters which results in a relatively interpretable model. There are four free parameters in the typical QRSE distribution: T, μ, ρ and γ, each with a corresponding microeconomic foundation. In this section, we discuss the two main parameters of interest in this work: The decision temperature T and agent expectations μ, and the effect that prior beliefs have on the resulting values (and interpretation) of these parameters. We also include discussion on the impact of decisions on resulting outcomes ρ and skewness of the resulting distributions γ in Appendix D, since ρ and γ were less affected by the introduced extensions. There is an additional parameter ξ (shown in Figure 5), which is not a free parameter, representing the mean of the actual returns and serving as a constraint on the mean outcome in Equation (7).

5.3.1. Decision Temperature

The decision temperature T controls the level of rationality and deviations from an agent's prior beliefs. An extremely high temperature corresponds to high information acquisition cost and results in choosing actions simply based on the prior belief. In contrast, an extremely low temperature corresponds to utility maximisation, and in the case of free information ($T = 0$) a perfect utility maximiser is recovered (i.e., homo economicus). In the housing example used here, T relates to the ability of an agent to learn all the required knowledge of the market, i.e. the actual profit rates for various areas. With $T = 0$, the agent has perfect knowledge of the current market profitability. With $T > 0$, this represents some friction with acquiring such information, e.g., it can be difficult to gather all the required information to make an informed choice due to, for example, search costs. From a psychological perspective, T can be a measure of the "just-noticeable difference" [62], meaning microeconomically, T is related to the ability of an agent to observe quantitative differences in resulting choices. High T means the agent is unable to distinguish choices based on U, due to high information-processing costs, so instead acts according to their previously learnt knowledge.

Since T is related to the prior, we see differences in the resulting values visualised in Figure 6. What can be observed from looking at the general trends of T is that it peaks in the years with high average growth (large ξ), such as 2015, as these years correspond to a growing market, and agents require less attention to market conditions, although this depends on the prior used.

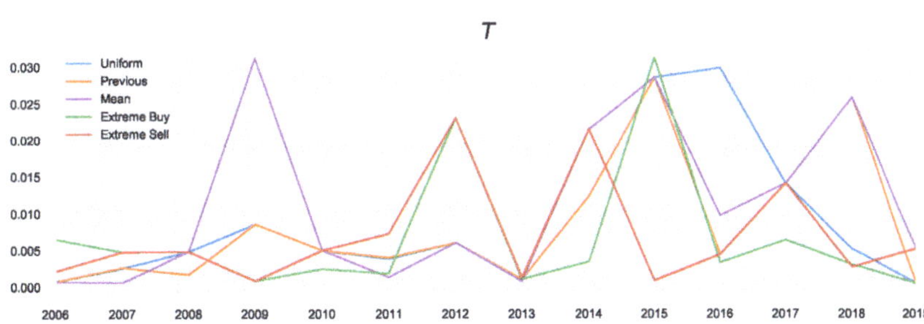

Figure 6. Decision Temperature.

Looking at the previous marginal probability as the prior (the orange profile), we observe in the build-up phase to 2015 increasing decision temperatures corresponding to agents acting on these previous beliefs. As these beliefs were also positive (i.e., agents expected favourable returns), these large returns can be explained by the agents continuously expecting this growth. This pattern changed in 2016, when the market "reverses": Now the agents must focus instead on their current utility since their prior beliefs no longer reflect the current market state. Such market reversals are categorised by low decision temperatures, since using the previous action probabilities now becomes misinformative (in contrast to the "building"/trend-following stages). This indicates an increased focus on agent rationality in times of market reversals. The incorporation of prior beliefs (particularly using the previous priors) is useful as it allows for the discussion to be extended in the temporal sense (as is done here). In other words, we can consider "building" the agent's beliefs as possible underlying causes for market collapses and relating the rationality of agents to the relative state of the market.

5.3.2. Agent Expectations

In microeconomic terms, parameter μ captures the agent's expectations. A large μ corresponds to an optimistic agent, who is expecting high returns from the market. In contrast, a low μ corresponds to a pessimistic agent, who is expecting poor returns from the market. As this works to shift the decision functions, there is a relation between the prior and parameter μ, since the prior also works as shifting preferences towards a priori preferred actions as shown in Section 4.3. There is also a relationship between μ and γ (outlined in Appendix D.2), since γ can help to account for unfulfilled agent expectations by adjusting the skew of the resulting distributions.

Generally, the agent's beliefs are within the $\pm 2.5\%$ range (expecting between a 2.5% quarterly growth or 2.5% dip), which corresponds to the bulk of the area under the curve in Figure A2. This means that the agent's expectations develop in accordance with actual market conditions, as can be seen in Figure 7.

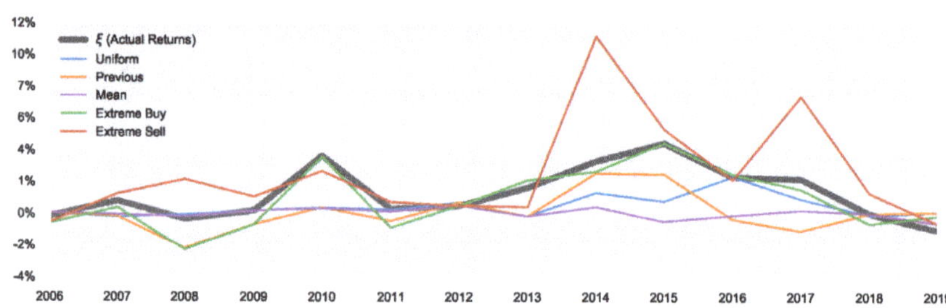

Figure 7. Agent Expectations vs. Actual Returns (in black).

The extreme priors result in larger absolute values of μ since larger shifts are needed to offset the (perhaps) poor prior beliefs. This can be seen in 2014 particularly, where the extreme sell prior has $\mu = 10\%$.

The values of previous prior μ tend to have a larger magnitude than the uniform priors, since as mentioned, these priors can capture build-up of beliefs (and as such some "trend-following" can be captured). For example, the year 2008 saw the lowest average returns ξ, as shown in Figure 5. Using the previous prior, the agents' expectations correctly match the sign of the actual returns in 2008 (i.e., agents correctly expected a decline in house prices). This results in more pessimistic agents than those using the uniform prior since they can reflect on the market performance from 2007. Likewise, during 2013–2015, the values of previous prior μ become larger than those for the uniform prior, since they are building on the previous years expectations which were all positive. In contrast, the period 2015–2017 saw a steady decline in agents expectations of returns with previous priors, reflecting the overall market state which appeared to be in a downward trend. The previous priors were able to capture this trend. Using the uniform priors, the year 2016 had a higher μ than the market peak of 2015. The reason is that uniform priors are unable to capture the fact that the previous timestep had higher (or lower) returns than the current timestep. In this case, the discussion can not be extended in the temporal sense of "building" on beliefs, and agents may miss such crucial temporal information without the incorporation of prior beliefs. This is evidenced by the significantly lower performance of the uniform prior in 2016 in comparison to the previous prior, as shown in Table 1, highlighting the usefulness of non-uniform (and temporal-based) priors in times of market crises and reversals.

5.4. Temporal Effects of Data Granularity on Decisions

In Section 5.2, we have analysed agent decisions over the previous 15 years, where decisions were grouped annually. This level of granularity was chosen to examine different agent behaviour from year to year. However, other levels of grouping can also be explored to give an insight into the impact of noise on the inference process. For example, an extremely granular grouping will likely result in additional noise in the decision-making process, which may or may not be impacted by the incorporation of prior beliefs. Likewise, a low granular grouping can be seen as "pre-smoothed", which may work in a similar fashion to the incorporation of prior temporal-based beliefs at a higher granularity, which we have seen can smooth the resulting decisions. In this section, we examine the usefulness of prior beliefs in such situations, providing comparisons with alternate data representations.

Two additional levels of granularity are considered, one more granular and one less granular than the annual groupings introduced in Section 5.2. We look at quarterly data, as well as aggregate groupings based on market state. In doing so, we have three levels for categorising agent behaviour: Quarterly, annually, and aggregated market state. This allows us to compare resulting agent decisions across different temporal scales, comparing the differences generated by the incorporation of prior beliefs and various data-level modifications.

The aggregate market state data groups years into "terms", which correspond with various "stages" of the market. These are growth and crash phases, highlighted as "Pre Crash" (Mid 2006–2007), "Crash" (2008), "Recovery 1" (2009–Mid 2011), "Small Crash" (Mid 2011–Mid 2012), "Recovery 2" (Mid 2012–Mid 2018) and "Recent Crash" (Mid 2018 to 2020). The overall market trends can be visualised in Figure A1 to see market returns for each corresponding "term".

The resulting decision likelihoods $f[A]$ are presented in Figure 8. In analysing the differences in resulting marginal probabilities between the various granularities, we can observe the impact from data-level modifications, i.e., performing inference on a larger time scale for macroeconomic observations, and how the incorporation of prior information affects such results. In Section 5.2 we have mentioned the previous and mean priors can have a smoothing effect on resulting decisions, in this sense, the lower granularity groupings (the market state based grouping) can also be seen as a smoothed version of the macroeconomic outcomes, i.e. pre-smoothing the data by considering a much larger interval composed of several years for groupings. We see that the incorporation of prior information helps preserve some important information in such settings. Looking at the left-most column of Figure 8 (the uniform priors), we can see the overall "shape" of the peaks and dips in preferences $f[a]$ is lost with aggregate groupings. For example, in the quarterly breakdown, there is a clear preference for selling in the later region in the range 2014–2017, corresponding to the highest growing market, which is labelled as "Recovery 2" in the aggregated version. When considering the "Recovery 2" with uniform priors, such a clear preference is lost, and the "Pre Crash" and "Initial Recovery" have a higher corresponding preference. This is because the agents can not separate past market information from the current market state and act purely based on the current utility. In contrast, with both the mean and the previous prior, such overall trends are preserved across the various granularities since agents can distinguish favourable environments when compared with previous market states (as captured by their prior beliefs). This additional temporal insight provides an important consideration and shows that even with various data-level smoothing or preprocessing (i.e., considering alternate data groupings) the prior information remains useful and highlights various market states and corresponding agent preferences.

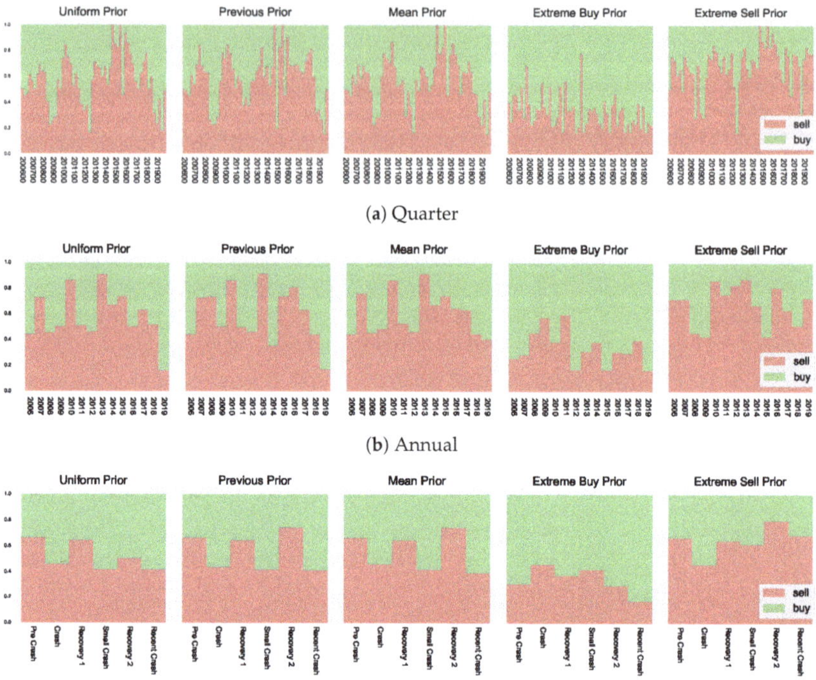

(a) Quarter

(b) Annual

(c) Terms

Figure 8. $f[a]$ for varying granularities.

A key takeaway from this exploration is that the potential for temporal analysis introduced by the prior beliefs provides additional insights into decision-making. These insights can not be generated by simple data-level modifications. Furthermore, the decision temperature T provides a way to modulate market state changes when considering agent decision-making.

6. Discussion and Conclusions

Despite many well-founded doubts of perfect rationality in decision-making, agents are often still modelled as perfect utility maximisers. In this paper, we proposed an approach for inference of agent choice based on prior beliefs and market feedback, in which agents may deviate from the assumption of perfect rationality.

The main contribution of this work is a theoretically grounded method for the incorporation of an agent's prior knowledge in the inference of agent decisions. This is achieved by extending a maximum entropy model of statistical equilibrium (specifically, Quantal Response Statistical Equilibrium, QRSE), and introducing bounds on the agent processing abilities, measured as the KL-divergence from their prior beliefs. The proposed model can be seen as a generalization of QRSE, where prior preferences across an action set do not necessarily have to be uniform. However, when uniform prior preferences are assumed, the typical QRSE model is recovered. The result is an approach that can successfully infer least biased agent choices, and produce a distribution of outcomes matching that of the actual observed macroeconomic outcomes when individual choice level data is unobserved.

In the proposed approach, the agent rationality can vary from acting purely on prior beliefs, to perfect utility maximisation behaviour, by altering the decision temperature. Low decision temperatures correspond to rational actors, while high decision temperatures represent a high cost of information acquisition and, thus, revert to prior beliefs. We showed how varying an agent's prior belief altered the resulting decisions and behaviour of agents, even those with equivalent utility functions. Importantly, the incorporation of prior beliefs into the decision-making framework allowed the separation of two key elements: The agent's utility maximisation, and the contribution of the agent's past beliefs. This separation allowed for a discussion on the decision-making process in a temporal sense, being able to refer to the previous decisions. This allows for investigation into the building of beliefs over time, elucidating resulting microeconomic foundations in terms of the underlying parameters.

It is worth pointing out some parallels with, and differences from, the frameworks of embodied intelligence and information-driven (guided) self-organisation, in which embodiment is seen as a fundamental principle for the organisation of biological and cognitive systems [63–66]. Similar to these approaches, we consider information-processing as a dynamic phenomenon and treat information as a quantity that flows between the agent and its environment. As a result, an adaptive decision-making behaviour emerges from these interactions under some constraints. Maximisation of potential information flows is often proposed as a universal utility for such emergent agent behaviour, guiding and shaping relevant decisions and actions within the perception-action loops [67–70]. Importantly, these studies incorporate a trade-off between minimising generic and task-independent information-processing costs and maximising expected utility, following the tradition of information bottleneck [71].

In our approach, we instead consider specific information acquisition costs incurred when the agents need to update their relevant beliefs in the presence of (Smithian) competition and market feedback. The adopted thermodynamic treatment of decision-making allows us to interpret relevant economic parameters in physical terms, e.g., agent's decision temperature T, the strength of negative feedback ρ, and skewness of the resulting energy distribution γ. Interestingly, the decision temperature appears in our formalism as the Lagrange multiplier of the information cost incurred when switching posterior and prior beliefs (KL-divergence). The KL-divergence can be interpreted as the expected excess code-length that is needed if a non-optimal code that was optimal for the prior (outdated) belief is used instead of an optimal code based on the posterior (correct) belief. Thus, the decision temperature modulates the inference problem of determining the true distribution given new evidence, in a forward time direction [72]. Moreover, the thermodynamic time arrow (asymmetry) is maintained only when decision temperatures are non-zero.

We demonstrated the applicability of the method using actual Australian housing data, showing how the incorporation of prior knowledge can result in agents building on past beliefs. In particular, the agent focus can be shown to shift from utility maximisation to acting on previous knowledge. In other words, during the periods when the market has been performing well, the agents were shown to become overly optimistic based on the past performance.

The generality of the proposed approach makes it useful for incorporating any form of prior information on the agent's choice set. Moreover, we have shown that the default QRSE is a special case of the proposed extension with uniform (i.e., uninformative) priors. Therefore, the proposed approach can be seen as an extension of QRSE, which accounts for prior agent beliefs based on information acquisition costs. As the QRSE framework continues to be expanded, the generalised model proposed here could become an important approach. Particularly, this would be useful whenever prior knowledge on agent decisions is known, as well as in multi-action cases when the IIA property of the general logit function is undesirable. Other relevant applications include scenarios with multiple time periods, allowing for a detailed temporal analysis and exploration of the cost of switching between equilibria (measured as an information acquisition cost from prior beliefs).

Author Contributions: B.P.E. and M.P.; Funding acquisition, M.P.; Software, B.P.E.; Supervision, M.P.; Writing—original draft, B.P.E. and M.P. All authors have read and agreed to the published version of the manuscript.

Funding: This research was funded by the Australian Research Council Discovery Project DP170102927.

Data Availability Statement: The real-estate pricing data used in this work were made available under license for this study by SIRCA-CoreLogic (https://www.corelogic.com.au/industries/residential-real-estate).

Acknowledgments: The authors would like to thank Kirill Glavatskiy and Michael S. Harré for many helpful discussions regarding the Australian housing market, as well as Adrián Carro, Jangho Yang and anonymous reviewers for various comments. The authors would also like to acknowledge the Securities Industry Research Centre of Asia-Pacific (SIRCA) and CoreLogic, Inc. (Sydney, Australia) for their data on Greater Sydney housing prices.

Conflicts of Interest: The authors declare no conflict of interest. The funders had no role in the design of the study; in the collection, analyses, or interpretation of data; in the writing of the manuscript, or in the decision to publish the results.

Appendix A. Derivations

Appendix A.1. Decision Duality

There are two main perspectives, the first is of the agent performing actions within the system, and the second is of the system observer [29].

Each of the two perspectives allows to capture the uncertainty faced by either the actor or the observer, by imposing a constraint on entropy. In this section, we outline the duality that arises from these perspectives, showing that a duality exists between maximum entropy models, and entropy constrained models [7]. Additional discussion on such perspectives is given in [21].

Modelling the actor corresponds to maximising the expected utility subject to a fixed entropy constraint. This is the method outlined in Section 3.1.1. In this case, the agent can be seen as a boundedly rational decision-maker, in that they might not have all of the information required to make a perfectly rational choice.

The alternate perspective, modelling an observer, corresponds to maximising the entropy of the decisions subject to a fixed expected utility. With this perspective, we capture modelling uncertainty from the observer. The observers problem is formulated as follows

$$\begin{aligned} \max -\sum_{a\in A} f[a|x]\log f[a|x] \\ \text{subject to} \sum_{a\in A} f[a|x] = 1 \\ \sum_{a\in A} f[a|x]U[a,x] \geq U_{min} \end{aligned} \tag{A1}$$

where U_{min} represents the minimum expected utility. In order to see the duality of Equations (A1) and (1), we formulate the following Lagrangian for converting Equation (A1) into an unconstrained optimization problem.

$$\mathcal{L} = -\sum_{a\in A} f[a|x]\log f[a|x] - \lambda\left(\sum_{a\in A} f[a|x] - 1\right) + \beta\left(\sum_{a\in A} f[a|x]U[a,x] - U_{min}\right) \tag{A2}$$

where again, taking the first order conditions and solving for $f[a|x]$ yields

$$f[a|x] = \frac{1}{Z}e^{\beta U[a,x]} \tag{A3}$$

We can see Equation (A3) is equivalent with Equation (4) with $\beta = \frac{1}{T}$, which highlights an important dualism between the two perspectives.

Appendix A.2. Decision Function

By setting the partial derivative of the unconstrained optimisation problem given in Equation (15) with respect to $f[a|x]$ to 0, we can obtain the following definition for $f[a|x]$:

$$\begin{aligned} \frac{d\mathcal{L}}{f[a|x]} &= U[a,x] - \lambda - T\log\left(\frac{f[a|x]}{p[a]}\right) = 0 \\ f[a|x] &= e^{\frac{U[a,x]}{T} - \lambda + \log p[a]} \end{aligned} \tag{A4}$$

and, using the normalisation constraint $\sum_{a \in A} f[a|x] = 1$, we obtain the following decision function

$$\begin{aligned} f[a|x] &= \frac{1}{Z_{A|x}} e^{\frac{U[a,x]}{T} + \log p[a]} \\ &= \frac{1}{Z_{A|x}} p[a] e^{\frac{U[a,x]}{T}} \end{aligned} \tag{A5}$$

with the partition function $Z_{A|x} = \sum_{a' \in A} p[a'] e^{\frac{U[a',x]}{T}}$.

Appendix B. Australian Housing Market Data

Data from 2006–2020 is used. Data is split into individual years. We use the rolling median price for each area and then measure the quarterly percentage growth rate for the areas. The month-to-month percentage changes are visualised in Figure A1. The distributions of the returns are visualised in Figure A2.

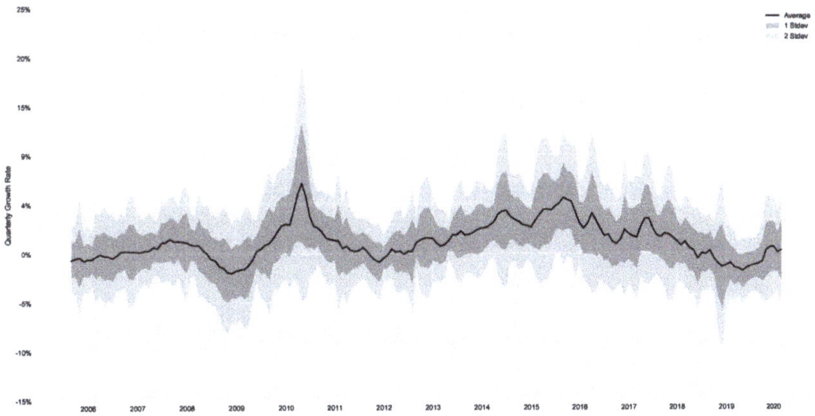

Figure A1. Quarterly returns in the Sydney housing market.

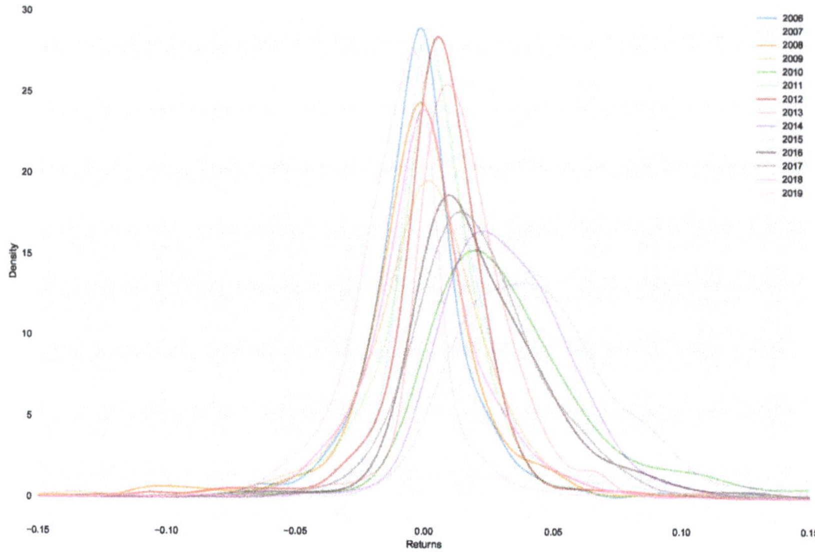

Figure A2. Density plots of returns grouped by year. We can see each year follows a different shape, but shows some striking regularities representing a statistical equilibrium.

Appendix C. Relation to Rational Inattention

In his seminal work, [2] outlined rational inattention "based on the idea that individual people have limited capacity for processing information". This work introduced information-processing constraints into the macroeconomic literature, using mutual information as a measure of such information costs.

Of particular interest are the developments of [3] who showed how to apply rational inattention (RI) to discrete decision-making. The key contribution was the modification to the logit function that arises from considering a cost to decision-makers from deviating from prior knowledge. In this section, we highlight the similarities of R.I. with the thermodynamic approach of [4] and the work proposed here.

The problem to be solved is formulated as follows. A utility-maximising agent must make a discrete choice, while it is costly to acquire information about the options A available:

$$\max f[a,x] \sum_{a\in A} \int_x f[a,x]U[a,x]dx - T\left(- \sum_{a\in A} f[a,x] \log(\frac{f[a,x]}{p[x]f[a]}) \right) \quad \text{subject to} \sum_{a\in A} f[a|x] = 1 \tag{A6}$$

where the first term is the expected utility, and the second a cost of information (following Sims [2], the mutual information). We see this as a similar setup to that of [4], which also corresponds to maximising the expected utility subject to an information cost, however, the information cost in [4] is instead measured as the KL-divergence. A key difference between the two is that Equation (A6) adds a dependence on $f[a]$ into the denominator of the information cost term. We can take the first order conditions of the resulting Lagrangian for (A6) and solve for $f[a|x]$, yielding:

$$f[a|x] = \frac{e^{\frac{U(a,x)}{T} + \log(f[a])}}{\sum_{a'\in A} e^{\frac{U(a',x)}{T} + \log(f[a'])}} = \frac{f[a]e^{\frac{U(a,x)}{T}}}{\sum_{a'\in A} f[a']e^{\frac{U(a',x)}{T}}} \tag{A7}$$

which is not yet fully solved, as there is a dependence on the unconditional probability $f[a]$. Since $f[a] = \int_x f[a|x]p[x]dx$, $f[a]$ depends on $f[a|x]$, and $f[a|x]$ depends on $f[a]$, this must (generally) be solved numerically, for example, with the Blahut–Arimoto algorithm by first making a guess for $f[a]$ and then iterating from there (see Caplin et al. [73] or Matějka and McKay [3] for solutions). It is for this reason, we utilise the configuration of [4] for the decision-making component, which depends only on the prior probabilities, and not the unconditional action probabilities $f[a]$ meaning an analytical solution can be obtained. However, the R.I. framework can be seen as equivalent to choosing an "optimal" prior in the free energy framework of [4], as both can be seen as applications of rate-distortion theory [56].

Further discussion on the relationship between R.I. and QRSE is given in [30].

Appendix D. Additional Parameters

While μ and T are the main parameters of interest in this work, since they have a direct contribution to the modified decision function introduced, ρ and γ are still important, although to a lesser extent as they are indirectly impacted. ρ is the Lagrange multiplier for the competition constraint, and γ controls the skewness of the resulting distribution.

Appendix D.1. Impact of Decisions on Outcomes

Parameter ρ measures the impact of individual decisions on housing prices. A large ρ corresponds to a highly effective market (high impact of actions on the response). In contrast, a low ρ corresponds to a weaker market response, and thus, lower market effectiveness. Parameter ρ, therefore, corresponds to the strength of the negative feedback mechanism, with the case of $\rho = 0$ implying no market feedback (i.e., no impact on the outcome based on the actions). In all cases, we see relatively large ρ's, peaking in 2013 and 2019, indicating the presence of a well-functioning feedback loop across the years. We see little variation between the uniform, previous, and mean prior in Figure A3, perhaps drawn from the fact the priors work as linear weightings in the difference between the conditional action probabilities, as shown in Equation (20).

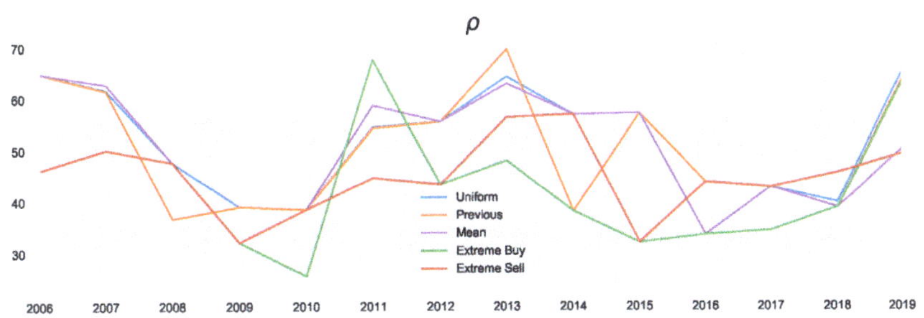

Figure A3. Competition.

Appendix D.2. Skewness

The parameter γ affects the skew of the resulting exponential distribution. This skew arises from (potentially) unfulfilled agent expectations, i.e., where $\mu \neq \xi$ [21]. Parameter γ, therefore, is a measure of skewness in the binary action case. In the asymmetric multi-action QRSE case, γ is replaced by alternate μ's explaining such skew. As mentioned, the priors can also introduce such a skew (without the need for a γ). This is shown in the extreme buy γ in Figure A4 which was almost always near zero, as the buying preference already creates the skew needed to describe the underlying distribution (i.e., the skewness was already explained by p). In contrast, extreme sell needs small γ's to switch their (incorrect) skew.

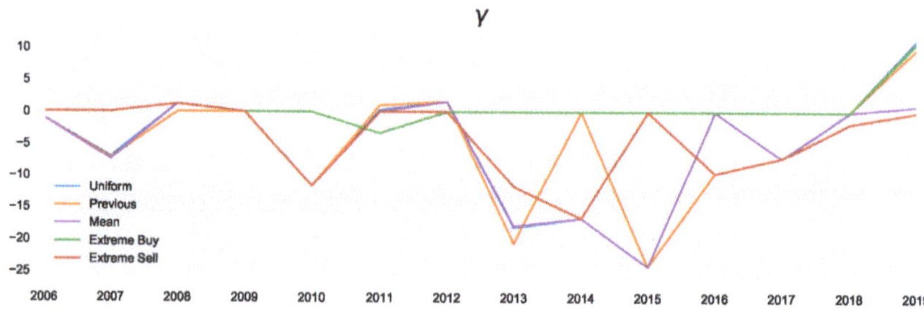

Figure A4. Skewness.

Negative γ corresponds to positive skewness, and positive γ corresponds to negative skewness. In most cases here, we see (at least slightly) positively skewed distributions (resulting in negative γ's), with the exception of 2019, which is negatively skewed, as can be verified in Figure A5.

Generally, γ's for the mean, previous, and uniform priors follow similar paths, except for the 2013–2016 years. In 2014 and 2016, γ's for the previous priors differs from the other priors. This can be explained by the fact that in both cases, the prior had a strong sell preference (shown in Figure 4), meaning an adjusted γ was needed to capture the current distributions shift correctly (and offset the influence of the prior).

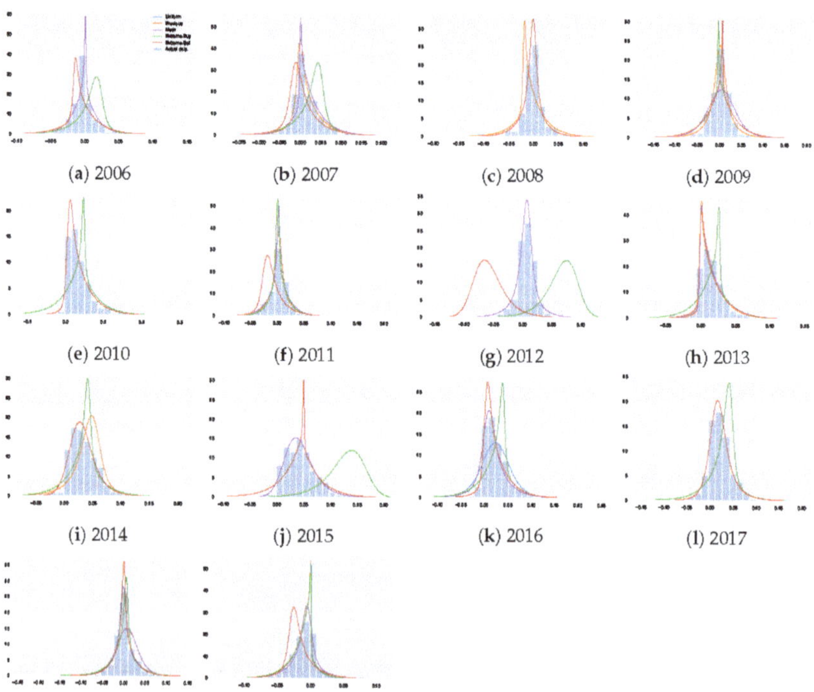

(**a**) 2006 (**b**) 2007 (**c**) 2008 (**d**) 2009
(**e**) 2010 (**f**) 2011 (**g**) 2012 (**h**) 2013
(**i**) 2014 (**j**) 2015 (**k**) 2016 (**l**) 2017
(**m**) 2018 (**n**) 2019

Figure A5. Resulting fitted marginals distributions $f[x]$ for each year. Each coloured line represents a different prior (with the legend given in the top left). The blue bars show the (discretized) actual return distribution.

Appendix E. Probability Plots

In this section, we provide the resulting probability plots for $f[x]$ (Figure A5), $f[a, x]$ (Figure A6), and $f[a|x]$ (Figure A7) across all years analysed.

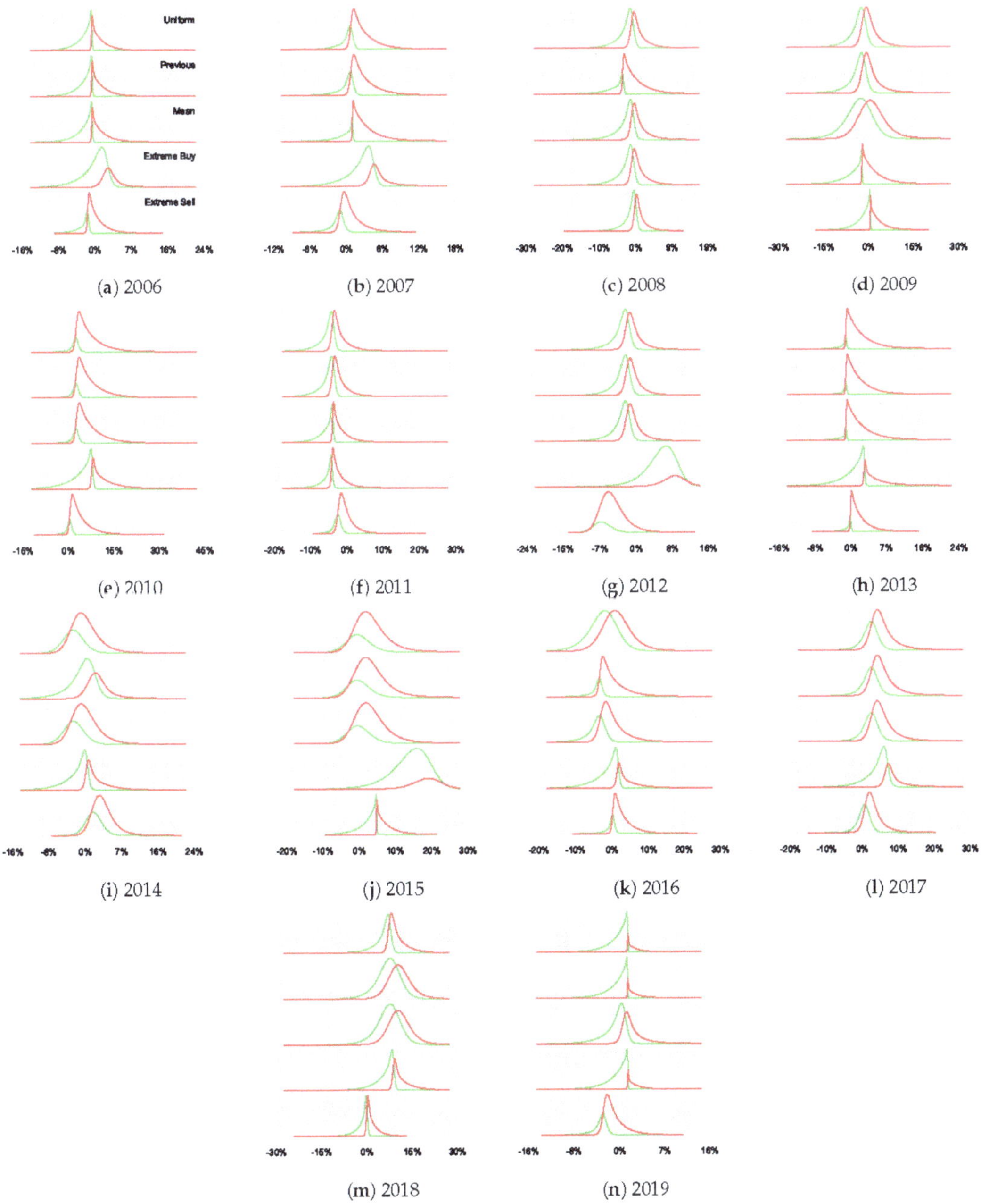

Figure A6. Resulting Joint Distributions. Red lines represent $f[\text{sell}, x]$, and green lines represent $f[\text{buy}, x]$. Each plot from top to bottom shows: Uniform, previous, mean and extreme buy and extreme sell priors (in that order).

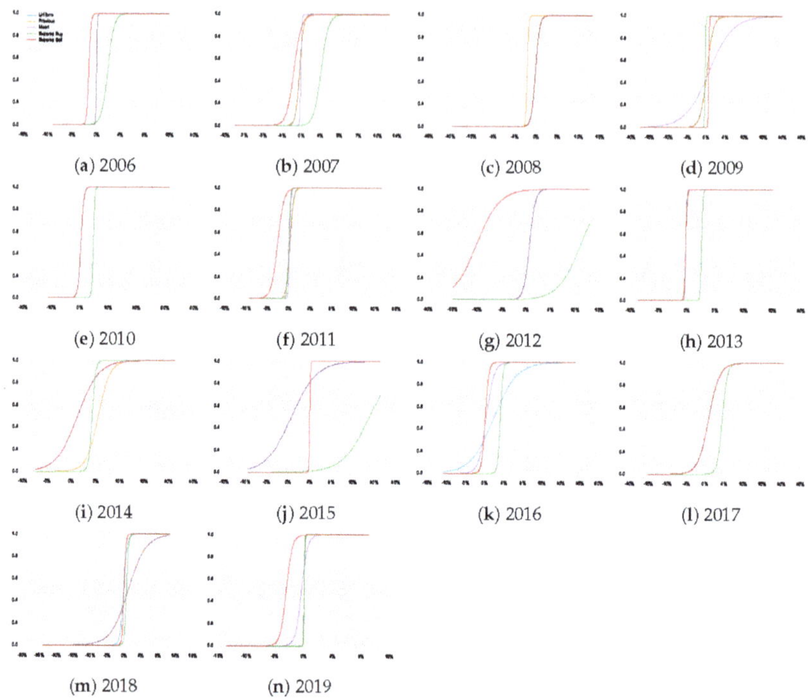

(**a**) 2006 (**b**) 2007 (**c**) 2008 (**d**) 2009 (**e**) 2010 (**f**) 2011 (**g**) 2012 (**h**) 2013 (**i**) 2014 (**j**) 2015 (**k**) 2016 (**l**) 2017 (**m**) 2018 (**n**) 2019

Figure A7. Decision functions for selling. Buying curves are excluded as they are simply the complement (1 − sell). The green lines represent the extreme buy a priori preference, which means the resulting probabilities of selling are shifted far to the right, i.e., the majority of the area comprises buying actions, and only the extreme positive growth rates for sell. In contrast, the red lines represent the sell preference, which "pulls" the area to the left, resulting in a strong resulting conditional preference for selling.

References

1. Simon, H.A. *Models of Man; Social And Rational.*; Wiley: Hoboken, NJ, USA, 1957.
2. Sims, C.A. Implications of rational inattention. *J. Monet. Econ.* **2003**, *50*, 665–690. [CrossRef]
3. Matějka, F.; McKay, A. Rational inattention to discrete choices: A new foundation for the multinomial logit model. *Am. Econ. Rev.* **2015**, *105*, 272–298. [CrossRef]
4. Ortega, P.A.; Braun, D.A. Thermodynamics as a theory of decision-making with information-processing costs. *Proc. R. Soc. A Math. Phys. Eng. Sci.* **2013**, *469*, 20120683. [CrossRef]
5. Scharfenaker, E.; Foley, D.K. Quantal response statistical equilibrium in economic interactions: Theory and estimation. *Entropy* **2017**, *19*, 444. [CrossRef]
6. Jaynes, E.T. Information theory and statistical mechanics. *Phys. Rev.* **1957**, *106*, 620. [CrossRef]
7. Yang, J. Information theoretic approaches in economics. *J. Econ. Surv.* **2018**, *32*, 940–960. [CrossRef]
8. Yakovenko, V.M. Econophysics, Statistical Mechanics Approach to. In *Encyclopedia of Complexity and Systems Science*; Meyers, R.A., Ed.; Springer: New York, NY, USA, 2009; pp. 2800–2826.
9. Scharfenaker, E.; Semieniuk, G. A statistical equilibrium approach to the distribution of profit rates. *Metroeconomica* **2017**, *68*, 465–499. [CrossRef]
10. Scharfenaker, E.; Yang, J. Maximum entropy economics. *Eur. Phys. J. Spec. Top.* **2020**, *229*, 1577–1590. [CrossRef]
11. Wolpert, D.H.; Harré, M.; Olbrich, E.; Bertschinger, N.; Jost, J. Hysteresis effects of changing the parameters of noncooperative games. *Phys. Rev. E* **2012**, *85*, 036102. [CrossRef]
12. Dragulescu, A.; Yakovenko, V.M. Statistical mechanics of money. *Eur. Phys. J. Condens. Matter Complex Syst.* **2000**, *17*, 723–729. [CrossRef]
13. Yakovenko, V.M.; Rosser Jr, J.B. Colloquium: Statistical mechanics of money, wealth, and income. *Rev. Mod. Phys.* **2009**, *81*, 1703. [CrossRef]

14. Foley, D.K. Unfulfilled Expectations: One Economist's History. In *Expectations*; Springer: Berlin/Heidelberg, Germany, 2020; pp. 3–17.
15. Harré, M.S. Information Theory for Agents in Artificial Intelligence, Psychology, and Economics. *Entropy* **2021**, *23*, 310. [CrossRef] [PubMed]
16. Foley, D.K. Information theory and behavior. *Eur. Phys. J. Spec. Top.* **2020**, *229*, 1591–1602. [CrossRef]
17. Ömer, Ö. Maximum entropy approach to market fluctuations as a promising alternative. *Eur. Phys. J. Spec. Top.* **2020**, *229*, 1715–1733. [CrossRef]
18. Yang, J.; Carro, A. Two tales of complex system analysis: MaxEnt and agent-based modeling. *Eur. Phys. J. Spec. Top.* **2020**, *229*, 1623–1643. [CrossRef]
19. Jaynes, E.T. Information theory and statistical mechanics. II. *Phys. Rev.* **1957**, *108*, 171. [CrossRef]
20. Jaynes, E.T. *Probability Theory: The Logic of Science*; Cambridge University Press: Cambridge, UK, 2003.
21. Scharfenaker, E. Implications of quantal response statistical equilibrium. *J. Econ. Dyn. Control.* **2020**, *119*, 103990. [CrossRef]
22. Ömer, Ö. Dynamics of the US Housing Market: A Quantal Response Statistical Equilibrium Approach. *Entropy* **2018**, *20*, 831. [CrossRef]
23. Ömer, Ö. Essays on Modeling Housing Markets, Income Distribution, and Wealth Concentration. Ph.D. Thesis, The New School, New York, NY, USA, 2018.
24. Ömer, Ö. Equilibrium-Disequilibrium Dynamics of the US Housing Market, 2000–2015: A Quantal Response Statistical Equilibrium Approach. Working Papers 1809, New School for Social Research, Department of Economics, 2018. Available online: https://econpapers.repec.org/paper/newwpaper/1809.htm (accessed on 30 September 2020)
25. Yang, J. A quantal response statistical equilibrium model of induced technical change in an interactive factor market: Firm-level evidence in the EU economies. *Entropy* **2018**, *20*, 156. [CrossRef]
26. Wiener, N. Measuring Labor Market Segmentation from Incomplete Data. Working Paper 2018-01, Amherst, MA, 2018. Available online: https://scholarworks.umass.edu/econ_workingpaper/238/ (accessed on 3 October 2020)
27. Wiener, N. Essays on Labor Mobility and Segmentation. Ph.D. Thesis, The New School, New York, NY, USA, 2019.
28. Wiener, N.M. Labor market segmentation and immigrant competition: A quantal response statistical equilibrium analysis. *Entropy* **2020**, *22*, 742. [CrossRef]
29. Blackwell, K. A Behavioral Foundation for Commonly Observed Distributions of Financial and Economic Data. Working Papers 1912, New School for Social Research, Department of Economics, 2019. Available online: https://ideas.repec.org/p/new/wpaper/1912.html (accessed on 8 October 2020)
30. Blackwell, K. Entropy Constrained Behavior in Financial Markets A Quantal Response Statistical Equilibrium Approach to Financial Modeling. Ph.D. Thesis, The New School, New York, NY, USA, 2018.
31. Scharfenaker, E. Statistical Equilibrium Methods in Analytical Political Economy. *J. Econ. Surv.* **2020**. [CrossRef]
32. Smith, A. *The Wealth of Nations: An inquiry into the nature and causes of the Wealth of Nations*; Harriman House Limited: Petersfield, UK, 2010.
33. McKelvey, R.D.; Palfrey, T.R. Quantal response equilibria for normal form games. *Games Econ. Behav.* **1995**, *10*, 6–38. [CrossRef]
34. McKelvey, R.D.; Palfrey, T.R. Quantal response equilibria for extensive form games. *Exp. Econ.* **1998**, *1*, 9–41. [CrossRef]
35. Lord, C.G.; Ross, L.; Lepper, M.R. Biased assimilation and attitude polarization: The effects of prior theories on subsequently considered evidence. *J. Personal. Soc. Psychol.* **1979**, *37*, 2098. [CrossRef]
36. K Levine, D. *Is Behavioral Economics Doomed?: The Ordinary Versus the Extraordinary*; Open Book Publishers: Cambridge, UK, 2012.
37. DellaVigna, S. Psychology and economics: Evidence from the field. *J. Econ. Lit.* **2009**, *47*, 315–372. [CrossRef]
38. Daunizeau, J.; Den Ouden, H.E.; Pessiglione, M.; Kiebel, S.J.; Stephan, K.E.; Friston, K.J. Observing the observer (I): Meta-bayesian models of learning and decision-making. *PLoS ONE* **2010**, *5*, e15554. [CrossRef] [PubMed]
39. Khalvati, K.; Park, S.A.; Mirbagheri, S.; Philippe, R.; Sestito, M.; Dreher, J.C.; Rao, R.P. Modeling other minds: Bayesian inference explains human choices in group decision-making. *Sci. Adv.* **2019**, *5*, eaax8783. [CrossRef] [PubMed]
40. Kruis, J.; Maris, G.; Marsman, M.; Bolsinova, M.; van der Maas, H.L. Deviations of rational choice: An integrative explanation of the endowment and several context effects. *Sci. Rep.* **2020**, *10*, 1–16. [CrossRef]
41. Debreu, G. Review of individual choice behavior by RD Luce. *Am. Econ. Rev.* **1960**, *50*, 186–188.
42. McFadden, D. Conditional logit analysis of qualitative choice behavior. In *Frontiers in Econometrics*, Zarembka, P., Ed.; Academic Press: Cambridge, MA, USA, 1973; pp. 105–142.
43. Golan, A. *Foundations of Info-Metrics: Modeling, Inference, and Imperfect Information*; Oxford University Press: Oxford, UK, 2018.
44. Hafner, D.; Ortega, P.A.; Ba, J.; Parr, T.; Friston, K.; Heess, N. Action and perception as divergence minimization. *arXiv* **2020**, arXiv:2009.01791.
45. Ortega, P.A.; Stocker, A.A. Human decision-making under limited time. In Proceedings of the Advances in Neural Information Processing Systems, Barcelona, Spain, 5–10 December 2016; pp. 100–108.
46. Gottwald, S.; Braun, D.A. The two kinds of free energy and the Bayesian revolution. *PLoS Comput. Biol.* **2020**, *16*, e1008420. [CrossRef] [PubMed]
47. Wilson, A. Boltzmann, Lotka and Volterra and spatial structural evolution: An integrated methodology for some dynamical systems. *J. R. Soc. Interface* **2008**, *5*, 865–871. [CrossRef]

48. Crosato, E.; Nigmatullin, R.; Prokopenko, M. On critical dynamics and thermodynamic efficiency of urban transformations. *R. Soc. Open Sci.* **2018**, *5*, 180863. [CrossRef]
49. Slavko, B.; Glavatskiy, K.; Prokopenko, M. Dynamic resettlement as a mechanism of phase transitions in urban configurations. *Phys. Rev. E* **2019**, *99*, 042143. [CrossRef]
50. Harding, N.; Spinney, R.E.; Prokopenko, M. Population mobility induced phase separation in SIS epidemic and social dynamics. *Sci. Rep.* **2020**, *10*, 1–11. [CrossRef]
51. Shore, J.; Johnson, R. Axiomatic derivation of the principle of maximum entropy and the principle of minimum cross-entropy. *IEEE Trans. Inf. Theory* **1980**, *26*, 26–37. [CrossRef]
52. Kesavan, H.; Kapur, J. Maximum Entropy and Minimum Cross-Entropy Principles: Need for a Broader Perspective. In *Maximum Entropy and Bayesian Methods*; Springer: Berlin/Heidelberg, Germany, 1990; pp. 419–432.
53. Glavatskiy, K.S.; Prokopenko, M.; Carro, A.; Ormerod, P.; Harre, M. Explaining herding and volatility in the cyclical price dynamics of urban housing markets using a large-scale agent-based model. *SN Bus. Econ.* **2021**, *1*, 1–21. [CrossRef]
54. Evans, B.P.; Glavatskiy, K.; Harré, M.S.; Prokopenko, M. The impact of social influence in Australian real estate: Market forecasting with a spatial agent-based model. *J. Econ. Interact. Coord.* **2021**, 1–53.
55. Hommes, C.H. On the consistency of backward-looking expectations: The case of the cobweb. *J. Econ. Behav. Organ.* **1998**, *33*, 333–362. [CrossRef]
56. Genewein, T.; Leibfried, F.; Grau-Moya, J.; Braun, D.A. Bounded rationality, abstraction, and hierarchical decision-making: An information-theoretic optimality principle. *Front. Robot. AI* **2015**, *2*, 27. [CrossRef]
57. Friedman, M. *Theory of the Consumption Function*; Princeton University Press: Princeton, NJ, USA, 2018.
58. Hommes, C.; Wagener, F. Complex evolutionary systems in behavioral finance. In *Handbook of Financial Markets: Dynamics and Evolution*; Elsevier: Amsterdam, The Netherlands, 2009; pp. 217–276.
59. Evans, G.W.; Honkapohja, S. *Learning and Expectations in Macroeconomics*; Princeton University Press: Princeton, NJ, USA, 2012.
60. Chow, G.C. *Usefulness of Adaptive and Rational Expectations in Economics*; Center for Economic Policy Studies, Princeton University: Princeton, NJ, USA, 2011.
61. Soofi, E.S.; Retzer, J.J. Information indices: Unification and applications. *J. Econom.* **2002**, *107*, 17–40. [CrossRef]
62. Dziewulski, P. Just-noticeable difference as a behavioural foundation of the critical cost-efficiency index. *J. Econ. Theory* **2020**, *188*, 105071. [CrossRef]
63. Pfeifer, R.; Bongard, J. *How the Body Shapes the Way We Think: A New View of Intelligence*; MIT Press: Cambridge, MA, USA, 2006.
64. Polani, D.; Sporns, O.; Lungarella, M. How information and embodiment shape intelligent information processing. In *50 Years of Artificial Intelligence*; Springer: Berlin/Heidelberg, Germany, 2007; pp. 99–111.
65. Ay, N.; Bernigau, H.; Der, R.; Prokopenko, M. Information-driven self-organization: The dynamical system approach to autonomous robot behavior. *Theory Biosci.* **2012**, *131*, 161–179. [CrossRef]
66. Montúfar, G.; Ghazi-Zahedi, K.; Ay, N. A theory of cheap control in embodied systems. *PLoS Comput. Biol.* **2015**, *11*, e1004427. [CrossRef]
67. Polani, D.; Nehaniv, C.L.; Martinetz, T.; Kim, J.T. Relevant information in optimized persistence vs. progeny strategies. In Proceedings of the Artificial Life X: Proceedings of the Tenth International Conference on the Simulation and Synthesis of Living Systems, Bloomington, IN, USA, 3–6 June 2006; MIT Press: Cambridge, MA, USA, 2006.
68. Prokopenko, M.; Gerasimov, V.; Tanev, I. Measuring spatiotemporal coordination in a modular robotic system. In Proceedings of the Artificial Life X: Proceedings of the 10th International Conference on the Simulation and Synthesis of Living Systems, Bloomington, IN, USA, 3–6 June 2006; pp. 185–191.
69. Capdepuy, P.; Polani, D.; Nehaniv, C.L. Maximization of potential information flow as a universal utility for collective behaviour. In Proceedings of the 2007 IEEE Symposium on Artificial Life, Honolulu, HI, USA, 1–5 April 2007; pp. 207–213.
70. Tishby, N.; Polani, D. Information theory of decisions and actions. In *Perception-Action Cycle*; Springer: Berlin/Heidelberg, Germany, 2011; pp. 601–636.
71. Tishby, N.; Pereira, F.C.; Bialek, W. The information bottleneck method. *arXiv* **2000**, arXiv:physics/0004057.
72. Spinney, R.E.; Lizier, J.T.; Prokopenko, M. Transfer entropy in physical systems and the arrow of time. *Phys. Rev. E* **2016**, *94*, 022135. [CrossRef] [PubMed]
73. Caplin, A.; Dean, M.; Leahy, J. Rational inattention, optimal consideration sets, and stochastic choice. *Rev. Econ. Stud.* **2019**, *86*, 1061–1094. [CrossRef]

entropy

MDPI

Article

Heterogeneous Criticality in High Frequency Finance: A Phase Transition in Flash Crashes

Jeremy D. Turiel [1,*] and Tomaso Aste [2]

1 Department of Computer Science, University College London, Gower Street, London WC1E 6BT, UK
2 Systemic Risk Centre, London School of Economics and Political Sciences, Houghton Street, London WC2A 2AE, UK; t.aste@ucl.ac.uk
* Correspondence: jeremy.turiel@gmail.com

Abstract: Flash crashes in financial markets have become increasingly important, attracting attention from financial regulators, market makers as well as from the media and the broader audience. Systemic risk and the propagation of shocks in financial markets is also a topic of great relevance that has attracted increasing attention in recent years. In the present work, we bridge the gap between these two topics with an in-depth investigation of the systemic risk structure of co-crashes in high frequency trading. We find that large co-crashes are systemic in their nature and differ from small ones. We demonstrate that there is a phase transition between co-crashes of small and large sizes, where the former involves mostly illiquid stocks, while large and liquid stocks are the most represented and central in the latter. This suggests that systemic effects and shock propagation might be triggered by simultaneous withdrawals or movement of liquidity by HFTs, arbitrageurs and market makers with cross-asset exposures.

Keywords: flash crash; systemic risk; financial networks; high frequency trading; market microstructure; phase transition; criticality

Citation: Turiel, J.D.; Aste, T. Heterogeneous Criticality in High Frequency Finance: A Phase Transition in Flash Crashes. *Entropy* **2022**, *24*, 257. https://doi.org/10.3390/e24020257

Academic Editors: H. Eugene Stanley, Ryszard Kutner and Christophe Schinckus

Received: 4 January 2022
Accepted: 4 February 2022
Published: 10 February 2022

1. Introduction

Flash crashes in financial markets can be defined as extreme changes in the price of one or multiple assets within a short interval of time. These have become increasingly relevant for practitioners and, in particular, market makers whilst being increasingly studied and reported in the quantitative finance literature.

The most notorious flash crash is likely that of 6 May 2010, which involved the major U.S. stock indices (S&P, DJIA, and NASDAQ composite) and caused a ≈9% drop in the DJIA in the 36 min it lasted for. This event led to a variety of empirical and theoretical papers trying to understand the event and its causes, with the aim to shed light on other black swan events too (up/down crashes). High frequency traders are at the center of interest in a large portion of this literature; hence, we report a brief summary of their role in markets and its regulatory concerns.

It has been shown that HFT market players contribute to price efficiency and tighter spreads, thereby improving the price discovery process. These players and electronic trading as a whole have become increasingly dominant in recent years to the point of constituting a large portion of the traded volume in financial markets. On the other hand, some characteristics of HFT players have caused other market players to raise concerns, as the run to incredibly fast execution leaves many behind and allows HFTs to front run other players [1]. The ability of HFTs to process information faster than other players leads to adverse selection and its fixed cost to a size advantage for larger players, which might hurt the overall welfare of market participants [2]. It can now perhaps be argued that the run to faster execution is going beyond price efficiency, which benefits investors and toward an unstable price process driven by competition between large firms. This is supported by a large body of literature on flash crashes, which places HFTs at the center of some disruptive systemic events, as discussed below.

The SEC's report on the flash crash of May 6th [3] finds that most market participants automatically halted their trading due to hard risk constraints triggered by the sudden price change, while some HFT firms kept trading, as it was deemed still profitable by their algorithms. These absorbed most of the original large sell order, but once they reached inventory or loss constraints, they started selling too. This increased the selling pressure in the market, and some works hold that it caused HFTs to trade with each other repeatedly ("hot potato phenomenon"), thereby increasing the traded volume (but not the real liquidity). This apparent increase in liquidity in the form of high trading volume caused large sell orders to get executed faster [4]. This chain of events highlights that the phenomenon has a dangerous positive feedback loop.

The results in [5] show through simulations how reducing either (or both) the number of HFT players or the size of the large sell order greatly reduced the size of the drawdown. Further, other works find that black swan phenomena of duration <1.5 s are about ten times more frequent than longer ones, and their return distribution deviates from the canonical power law distribution of returns. The authors suggest a phase transition to an all-machine environment at $\sim$1 s, as human reaction time is in the order of seconds. The authors also investigate the time scales via additional simulations to show the rise in extreme events and their magnitude around $\sim$1 s in what they define as the all-machine phase [6,7]. Findings along those lines, on the distribution of high frequency black swan events deviating from the canonical return distributions, were also recently published by the authors of this work [8].

From the review above, we see that crashes of different sizes seem to involve a self-perpetuating cycle [5] with positive feedback loops.

This type of self-excited process is also investigated in [9] for the liquidity and information dependence between two sample assets, showing how liquidity shocks to an asset can propagate to related ones (and by extension to the wider market).

The frequency and size (in terms of number of securities involved) of simultaneous-like crashes in HFT is also investigated in the literature. For instance, the works by Lillo and co-authors [10,11] investigate the dynamics of simultaneous flash crashes, and motivate their importance by showing the growth in the number of mini crashes in recent years. Further, they show how the number of simultaneously crashing securities has grown over the last 10 years, thereby highlighting the increasing systemic relevance of this phenomenon.

We recognize that systemic risk is traditionally defined as "the risk of a cascading failure in the financial sector" [12]. In this work, we do not investigate interbank connectivity, but rather the connectivity of trading patterns across financial assets which can lead to breakdowns or temporary dysfunctions in financial markets, as per the definition of systemic risk in [13]. We phrase the concept in a slightly different manner in the context of our work as follows. In this paper, we define systemic risk as the risk component of an event (say a flash crash) that is given by the interconnectedness of assets, likely as a result of correlated actions and arbitrage between market participants. This causes isolated events to spread in the market and affect more assets, thereby increasing their impact and relevance for all market participants. A related concept is that of "synchronization" which is the systemic and concentration aspect that arises from the alignment and interdependence of actions between market players (on a single asset) rather than across assets.

Our phrasing of the concept of systemic risk from [13] highlights our microstructural investigation of the trading dynamics which lead to dysfunctions and disruptions in the orderly functioning of financial markets. Indeed, crashes can be just due to microstructural dynamics, but as price efficiency deteriorates and volatility spikes, investors shy away from financial markets. Financial markets allow investors to provide companies in the real economy with capital, and their dysfunction can turn mere trading issues into real economic panic and crisis. Therefore, even high frequency black swan events can have dramatic effects on the real economy, as proven multiple times in recent history, which ties our interpretation of systemic risk back to Ref. [12] as well.

The systemic risk posed by HFTs was investigated in the literature in the last decade. The work by Paulin et al. [14] simulates flash crashes through agent-based modeling and highlights the importance of market structures in the systemic propagation of extreme events. The works by Abreu and Brunnermeier [15] and Bhojraj et al. [16] investigate the risks of synchronization between arbitrageurs in financial markets and acknowledge its existence. Other works investigate the systemic risk of HFT dynamics. Jain et al. [17] investigate how low-latency HFT trading can worsen extreme systemic events in financial markets and argue for the need to incorporate correlation and market structure in regulating these risks. The work by Harris [18] discusses many mechanisms, among which systemic risks originating from order routing and self-reinforcing mechanisms which cause crashes. The review by De Gruyter [19] summarizes the systemic aspects of HFTs and market structure, such as position correlation and herd behavior, adverse selection in orders and crowding, as well as negative contribution to price discovery at times.

Co-crashes are becoming more frequent and systemic. It is, therefore, important to investigate their structure. In particular, it is relevant to understand which stocks are central to larger systemic events as well as the contagion structure between stocks in the market. This is a central theme in market stability for regulators as well as in risk management for market makers.

The present work joins the two themes of flash crashes and systemic risk by delving deeper into the dynamics of simultaneous flash crashes of different sizes throughout 300 liquid stocks traded on the NASDAQ. We investigate the empirical distribution of crash sizes and the structure of these events in the market. We also investigate whether larger systemic events involve highly unstable stocks (which crash often) or stocks that are more stable in their price dynamics, yet more influential to trigger larger systemic events when subject to liquidity shocks. We apply tools from statistical physics to show the difference between crashes which involve a small or large number of assets. We uncover a phase transition occurring when the crash size exceeds five stocks. Implications for systemic risk in high frequency markets are discussed from both a trading and regulatory perspective.

2. Data

In the present work, we consider a universe of 300 liquid stocks from the NASDAQ exchange between 3 January 2017 and 25 September 2020. High frequency price data are obtained from LOBSTER [20] and sampled to obtain non-overlapping one-minute returns. This frequency was also adopted in [10] and other works in the literature for the detection of price jumps, as it is understood that below this limit, microstructural noise becomes relevant and can impact the validity of the method.

3. Method

3.1. Jump Detection

In the present work, we focus on anomalous movements in the mid-price p_t and their co-occurrence structure. To do so, we detect price jumps (up and down crashes) similarly to [10], at least in principle, in 1 min non-overlapping returns.

Specifically, we apply the basic jump detection method from [21] and detect jumps at the 5% significance level. The intuition behind this method is simple: we consider changes in p_t in the form of log-returns $r_t = \log \frac{p_t}{p_{t-1}}$. Those are normalized so that, in the absence of jumps, their distribution is close to being normal and stationary. The method then exploits extreme value theory to obtain thresholds, above or below which, r_t can be classified as anomalous (i.e., a jump), with a given confidence level.

To achieve a distribution of log-returns close to normal and stationary in time, we must normalize returns locally to account for two known regularities: daily seasonalities and long memory effects [22–25]. Mid-price returns have been shown to have approximate zero mean but a non-stationary variance due to the above [26]. Hence, the method empirically measures and discounts daily seasonality patterns and autocorrelations in return variance from the data. This yields a time series of almost normally distributed returns with stationary variance. Extreme value theory can then be applied as described above.

In addition to the basic features of the method for robust volatility estimation in intraday patterns, we obtain a robust estimate of intraweek periodicity and adjust the return series and jump detection according to [27].

As per the description above, null models are calibrated, and price jumps detected individually for each stock. As we consider 1 min non-overlapping returns, our sampling allows for aligned timestamps. We then consider contemporaneous price jump detection across assets in the universe as simultaneous jumps (a single systemic event).

It is important to highlight in the context of risk that crashes are normally associated with negative price returns of anomalous magnitude. The method used here detects both positive and negative anomalous price movements and we consider both as they are "jumps". In our related work [8], we have shown how both up and down jumps are relevant for risk, as market makers can hold inventory and be exposed in either direction. Further, a short squeeze can potentially be more dangerous, as it is often associated with high levels of leverage. Still, we recognize the importance of investigating down jumps (traditionally termed "crashes") and are looking to include a comparison between down and up jump structures in follow-up works.

3.2. *Crash Size Distribution and Firm Persistence*

We investigate whether co-jumps which involve different numbers of stocks originate from the same dynamic process and present the same distribution. We also consider whether individual stocks are involved to the same extent across co-crashes of different sizes or if a pattern emerges.

We define the unnormalized crash frequency for stock x, in co-crashes with m stocks and time range $t \in [0, T]$ as

$$f_{x,m} = \sum_{t=0}^{T} c_{x,t,m}$$

with

$$c_{x,t,m} = \begin{cases} 1, & \text{if stock } x \text{ is involved in a crash of size } m \text{ at time } t \\ 0, & \text{otherwise} \end{cases}$$

By marginalizing over the ensemble of stocks x, we obtain the frequency distribution across co-crash sizes

$$f_m = \sum_x f_{x,m}$$

The changes in the composition of the crashes are investigated by computing the correlation between the involvement of firms across crashes of different sizes. Namely, for each crash size m, we assign to each firm x a rank in decreasing order by $f_{x,m}$. We then compute the Spearman correlations between these ranks.

3.3. *Statistical Testing*

To support the visual intuition of our results, we apply statistical testing in the form of null models. We applied the Spearman correlation to test for rank similarity between the crash frequency distributions across stocks at different crash sizes m. As the frequency distributions are noisy and fat-tailed, the correlation p-value seems hard to justify as a valid test. Hence, we follow the idea of Mantegna et al. [28] to create a simple null model of correlation significance.

To do so, we sample without replacement the whole list of stocks S_m according to $\propto f_m$ from Section 3.2 to obtain a biased reshuffling $G_{i,m}$ of the stocks according to their crash frequency.

For each shuffled list, we calculate the Spearman correlation coefficient between the sample and the original list to form the empirical null distribution as

$$D_m = Spearman(G_{i,m}, S_m)_{i=1}^{10^5}$$

We then define the significance of the correlation between sizes $m, m+\tau$ as the quantile of $Spearman(S_{m+\tau}, S_m)$ in D_m.

3.4. Crash-Weighted Trading Volume

To investigate the relationship between the crash size and the involvement of highly traded stocks, we define a weighted average daily dollar traded volume for each crash size, where the weighting is given by the normalized crash frequency of each stock.

For crash size m and crash frequency distribution $f_{x,m}$, as per Section 3.2, we define the crash-weighted dollar traded volume DTV_m as

$$\text{DTV}_m = \frac{\sum_x f_{x,m}\,\text{DTV}_{x,m}}{f_m}$$

This measure aims to represent how more highly traded stocks are involved at different crash sizes.

4. Results and Discussion

The plot in Figure 1a shows the frequency distribution f_m of the number of stocks involved in each flash crash. Figure 1b plots the cumulative frequency $f(M \geq m)$. It is evident from both figures that they are heavy-tailed, and there is a change in the slope around $m \approx 5$ and a finite size effect at $\approx 10^2$, which is when the crash involves a large portion of the system (system size is $3 \cdot 10^2$) [29]. This kind of distribution was already reported in [10], where the authors investigated and modeled flash crash sizes and frequency as a single Hawkes process. The authors there suggest that each security's crash dynamics should be modeled as a self-excitation process, but they point out that this would involve tuning a large number of parameters on very noisy data. They therefore decided to model the collective self-excitation process of securities as the frequency of crashes (or co-crashes) and their size. Hence, all crash sizes are treated as instances of a multi-asset Hawkes process in [10], with no distinction between the assets involved in each crash or their co-occurrence structure.

In the present work, we take a more granular approach and move to investigate the structure of co-crashes and the individual susceptibility of each stock.

To further investigate the difference between small and large crash sizes, we report in Figure 2a the Spearman correlation between the ranks of crash frequency for all stocks. Specifically, each line reports the correlation between the rank of the companies in the initial crash size m (correlation 1) and all other crashes of higher sizes $m+\tau$. We indeed observe how crashes of smaller sizes ($m < 5$) have a substantially different composition to crashes of larger sizes. We instead observe that for sizes $m > 5$, a steady state is reached, with a large component of the population having similar ranks in frequency across all crash sizes. These steady states for $m > 5$ are significantly higher than the ones of smaller sizes, as the structure no longer evolves significantly between higher size crashes. The plot in Figure 2b provides a clearer visualization of this. We highlight that already at size 5, the correlation transitions directly to the steady state, albeit a lower one with respect to the ones for crash size 6 and above.

To validate the visual results from Figure 2, we apply the null model of correlation significance between crash frequency distributions.

Figure 3 shows the correlation significance between the starting point m on the horizontal axis and its steady state distribution $\sim[m+2, m+10]$. We observe the first significant value at 1% around $m = 4$, which confirms the intuition from Figure 1a,b that crash sizes up to $\approx$4 belong to a different process than larger crashes. Indeed smaller crashes are dominated by less stable stocks and larger ones by very liquid stocks with high market capitalization. This suggests that more influential and systemic stocks are involved in larger crashes and perhaps even trigger those. A reason for why this is not the case in small crashes can be that these stocks are systemic enough to mostly be involved in (or perhaps even cause) crashes of a larger size. These are then even more relevant for systemic risk. Alternatively, only larger crashes involve enough activity to influence highly traded stocks.

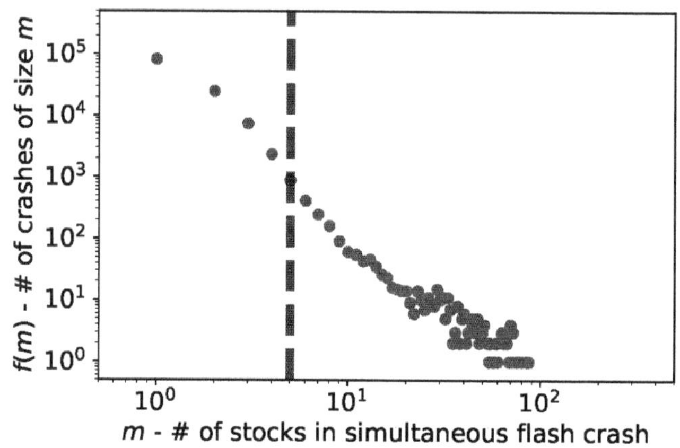

(a)

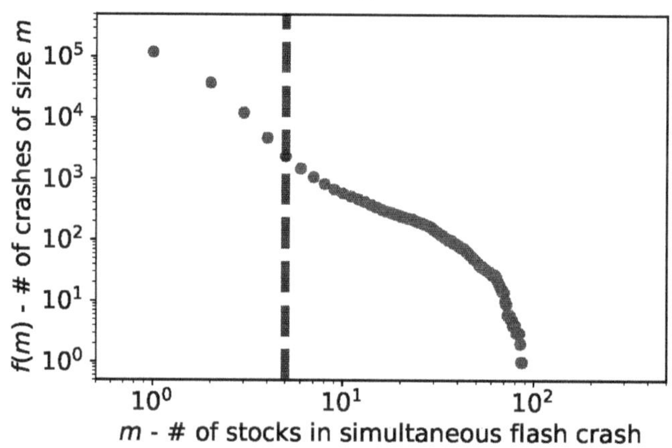

(b)

Figure 1. Heterogeneous crash distribution. Log-log plot of the flash crash size distribution. We observe that sizes lesser than 4 follow a different trend, with lower than expected frequency. This suggests that crashes of this size and onwards do not belong to the same self-organized process, but that this is rather a heterogeneous distribution. (**a**) $f(m)$; (**b**) $f(M \geq m)$.

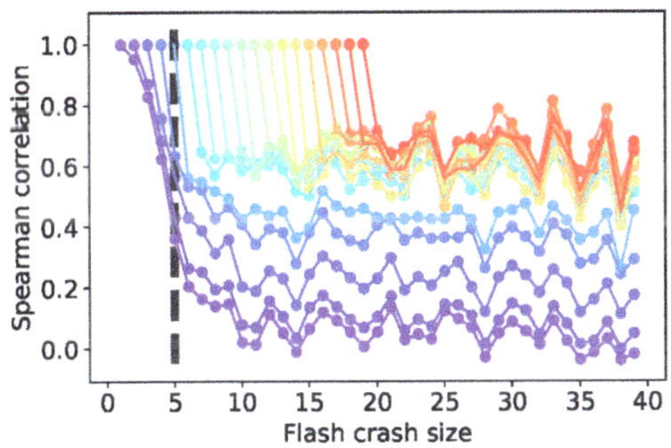

(a)

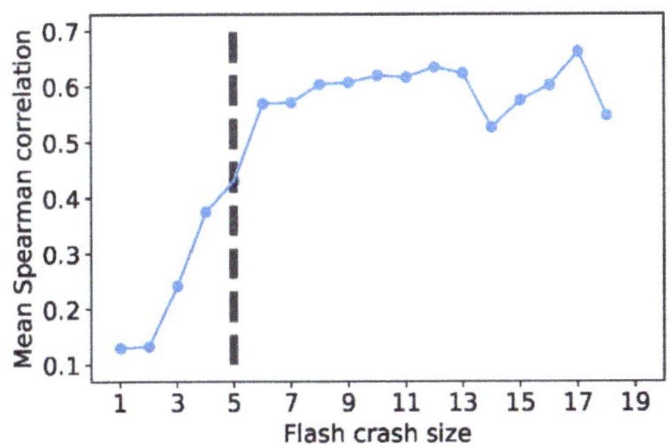

(b)

Figure 2. Crash component rank correlation. Evidence that there is a transition around $m = 5$ with crashes involving a small number of companies ($m < 5$) being substantially differently populated with respect to crashes involving a larger number of companies ($m > 5$). The plot in Figure (**a**) reports the Spearman correlations of ranks in frequency between each starting crash size and higher crash sizes. The plot in Figure (**b**) looks at the average correlation in the range $[m + 2, m + 20]$ for each value of m from Figure (**a**), which offers better visual intuition. (**a**) Spearman correlation between all consecutive crash sizes; (**b**) Spearman steady-state correlation mean in $[m + 2, m + 20]$, $\forall m$.

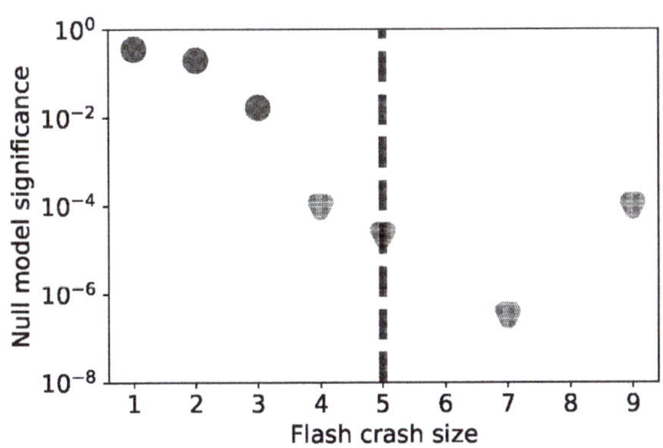

Figure 3. Crash component significance phase transition Evidence of a transition in the dynamics of crashes composition occurring around $m = 5$. The plot reports the steady-state statistical significance of the base crash size's frequency distribution.

This is therefore further evidence of the occurrence of a transition in the process between smaller and larger crashes. The slow decay of smaller crash sizes indicates how these belong to similar distributions of non-systemic events, but as the crash size grows, the steady state gets closer to the large crash level. This suggests that larger crashes have some systemic characteristics.

If we take a closer look at the top ranked stocks at each size, we observe that smaller crash sizes are dominated by very volatile and illiquid stocks, which are subject to large jumps perhaps due to the lack of a smooth price process in their trading. We would expect this though to make them susceptible to larger systemic events as well and, hence, stably ranked. Yet, we observe very low to null rank correlation between individual (and small) crash frequencies and the large crash size steady state. It seems as if these crashes are not only non-indicative, but also, as indicated by the phase transition in Figure 3, they belong to an unrelated ranking and distribution. We highlight that we considered rankings and ranking correlation in order to avoid any sensitivity to large values or outliers at smaller frequencies.

Large crash sizes involve stocks such as Microsoft (MSFT) and Apple (AAPL) as being consistently high ranked. We highlight that these stocks are highly liquid and characterized by a stable price process with very few price jumps. Indeed, the few times they get involved into jumps, they are often part of larger simultaneous crashes, which involve more stable and systemic stocks. Further, when analyzing the co-crash relations between pairs of stocks, we observed a heavy-tailed distribution of centrality for these large systemic stocks, which suggests a community and core-periphery-like structure of the contagion network of co-crashes [30–34].

The above observations prompted us to conduct further analyses on the relation between stock liquidity (where average daily dollar traded volume is used as a proxy) and crash frequency at different crash sizes.

To validate visually and numerically our observation that highly traded stocks are more present in large crashes, we present the plots in Figure 4. The plot in Figure 4a shows the average daily traded volume of a stock per crash size, weighted by its crash frequency, as per the definition in Section 3.4. This is plotted against the crash size to show a clearly increasing trend in crash-weighted traded volume with crash size. This shows how larger crashes see stocks with higher traded volumes being more frequently involved.

This could, however, be the consequence of a subset of crashes which involved highly traded stocks. We therefore test this with the results in Figure 4c, which show how not only the average crashing stock is more "liquid" in larger crashes, but also that the fraction of crashes, which involve at least one of the top 20 stocks by traded volume in our universe, increases with crash size.

In line with this, we test how the traded volume of each stock correlates with its crash frequency for each crash size. We report results for the Spearman correlation coefficient in Figure 4b, where dots are used for correlations significant to the 5% confidence level and crosses otherwise. We see that co-crashes of size 1 and 2 seem to have an inverse or no relation between the volume traded and crash size. At our previously identified phase transition point $m \approx 5$, we see the first significant positive correlation between the volume traded and crash frequency, which stays somewhat stable or is slightly increasing with crash size.

This last result is less clear than the previous one, but still shows a positive correlation between the volume traded (a proxy for liquidity) and crash frequency at crash sizes $m > 5$.

The presence of liquid stocks in most large crashes observed in Figure 4c prompts questions around the periphery structure of the different liquid stocks and implications for systemic risk. Further work in this direction is already underway with promising results and will be the topic of a follow-up work. The causality of such co-crash structures is also a very important topic, albeit harder to investigate rigorously, and should be the subject of future work.

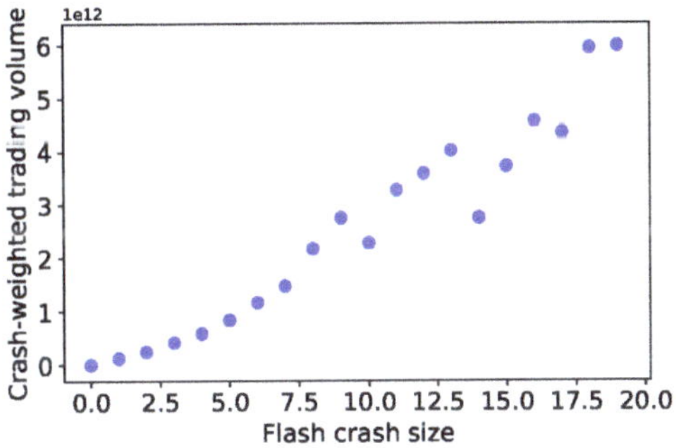

(**a**)

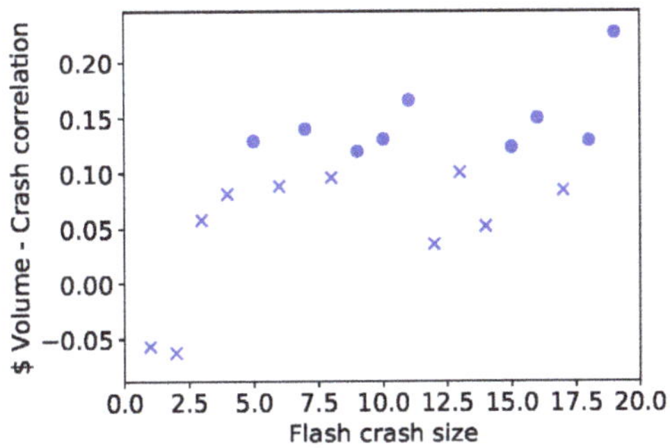

(**b**)

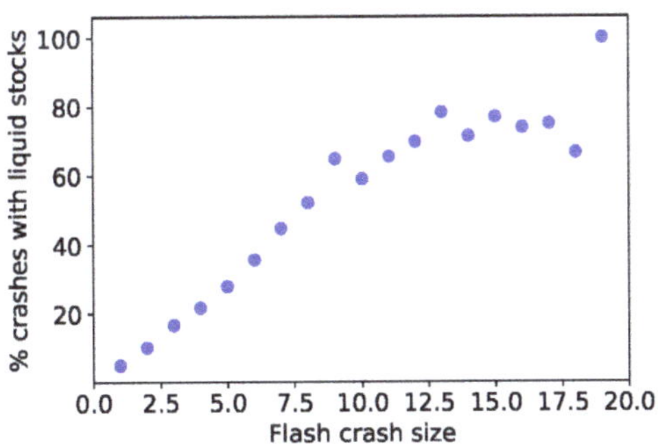

(**c**)

Figure 4. Relation of Traded Volume to crash size. The figures above show evidence of a relationship between the traded volume of stocks and their involvement in crashes of different sizes. (**a**) shows the general positive relation between crash size and the involvement of highly traded stocks. (**b**,**c**) show how the relationship exists not only on average, but also how "liquid" stocks are more involved throughout crashes at higher crash sizes. (**a**) Positive relation between crash-weighted average daily dollar traded volume and crash size m.; (**b**) Spearman correlation between traded volume and crash frequency across crash sizes m; (**c**) positive relation between fraction of crashes involving liquid stocks and crash size $m/$.

5. Conclusions

The present work analyses co-jump structures in high frequency markets. We investigate the distribution of co-jump sizes for 300 stocks on 1 min returns. We highlight features of this distribution, such as the finite size effect in the tail and the divergence of small crash frequencies from the distribution. We show how the ranking and structure of crash frequency throughout stocks changes drastically through a phase transition between small and large crash sizes at size 5. We quantify this with the Spearman correlation between crash frequency ranks at different crash sizes. We then apply a null model of crash frequency at each crash size to test the hypothesis of a phase transition. Finally, we highlight how larger crashes are dominated not by the less liquid stocks present in small crashes, but rather by highly liquid stocks which are present in most flash crashes as the crash size grows. Preliminary results, which we leave for future work, find these stocks to be systemic in communities and core-periphery like structures of co-crashes. We suggest that these systemic events can be viewed as communities centered around these most influential stocks.

We know from the literature that these structures can be indeed vulnerable and highly unstable, as well as fragmented if characterized by multiple cores. One of the possible reasons for this can be inferred from the interviews with different market players following the crash of May 6th [3]. Many HFTs highlight the centralized risk constraints for volatility and P&L, which cause them to withdraw from the market in the case of extreme conditions or losses. As they constitute much of the liquidity in the market in particular for smaller stocks, withdrawing from those causes liquidity droughts. These are often systemic, as players have central risk constraints and withdraw from the entire market as those are triggered. Further, as systemic stocks crash, arbitrageurs come into play to level prices across the market, thus making the isolated event a systemic one. In this view, well-known stocks are not systemic per se, but rather as a result of non-siloed trading by HFTs and ETFs.

In light of the present results, future works shall investigate the asynchronous price changes of securities and model spreading dynamics of flash crashes and their directed structure. Lead–lag investigations of causality of these larger crashes are also suggested for future work. Already from our results, one can monitor, in particular, the most systemic stocks from larger flash crashes for co-jumps of size 5 and higher and induce trading halts or limitations to avoid further spreading of these systemic events. This is crucial, as our results combined with those of [10] suggest a systemic self-excited process in both frequency and magnitude of those crashes.

We leave the investigation of this structure for future work and highlight that this is of high importance for practitioners and regulators when dealing with market efficiency and stability, particularly as trading frequencies rise and electronic trading becomes widespread across securities.

We conclude by observing that volatility and P&L-based trading breaks used by market players may worsen these events and their systemic characteristics since they cause liquidity withdrawals throughout stocks and market players. This introduces systemic synchronization throughout the market and makes individual assets more susceptible to small trading volumes. Further, we suggest to monitor the stocks we find to be systemic throughout larger crashes to model the contagion of liquidity crises and halt trading before these spread and distort a larger number of assets. This should also be topic of future work aimed at smart and efficient regulation in high frequency markets.

Author Contributions: J.D.T. conducted the computational analysis and drafted the manuscript. T.A. guided the work and interpretation of results and reviewed and edited the manuscript. All authors have read and agreed to the published version of the manuscript.

Funding: J.D.T. acknowledges support from EPSRC (EP/L015129/1). TA acknowledges support from ESRC (ES/K002309/1), EPSRC (EP/P031730/1) and EC (H2020-ICT-2018-2 825215).

Data Availability Statement: The data used in this work was obtained from the LOBSTER dataset (https://lobsterdata.com (accessed on 2 January 2022)) under academic license to the Financial Computing and Analytics group at University College London. We are therefore unable to publish the raw data.

Acknowledgments: J.D.T. acknowledges Riccardo Marcaccioli for useful discussions and support with the jump detection method. J.D.T. acknowledges Charles-Albert Lehalle for useful feedback which motivated us to investigate the role of liquid stocks in co-crash structures and prompted further works on the topic.

Conflicts of Interest: The authors declare no conflict of interest.

References

1. Chlistalla, M. *High-Frequency Trading: Better than Its Reputation? Research Briefing, Deutsche Bank Research*; Deutsche Bank: Berlin, Germany, 2011.
2. Biais, B.; Foucault, T.; Moinas, S. Equilibrium high frequency trading. In Proceedings of the Fifth Annual Paul Woolley Centre Conference, London, UK, 7–8 June 2012.
3. CFTCS; SE U. Findings Regarding the Market Events of 6 May 2010. In *Report of the Staffs of the CFTC and SEC to the Joint Advisory Committee on Emerging Regulatory Issues*; CFTC: Washington, DC, USA, 2010; p. 104.
4. Golub, A.; Keane, J.; Poon, S.H. High Frequency Trading and Mini Flash Crashes. Available online: https://papers.ssrn.com/sol3/papers.cfm?abstract_id=2182097 (accessed on 2 January 2022).
5. Paddrik, M. An Agent Based Model of the e-mini s&p 500 applied to Flash Crash Analysis. In Proceedings of the 2012 IEEE Conference on Computational Intelligence for Financial Engineering & Economics (CIFEr), New York City, NY, USA, 29–30 March 2012; pp. 1–8.
6. Johnson, N.Financial black swans driven by ultrafast machine ecology. *arXiv* **2012**, arXiv:1202.1448.
7. Nanex, L.L.C. Nanex Flash Crash Summary Report. 2010. Available online: http://www.nanex.net/20100506/FlashCrashAnalysis_Intro.html (accessed on 9 February 2022).
8. Turiel, J.D.; Aste, T. Self-organised criticality in high frequency finance: The case of flash crashes. *arXiv* **2021**, arXiv:2110.13718.
9. Cespa, G.; Foucault, T. Illiquidity contagion and liquidity crashes. *Rev. Financ. Stud.* **2014**, *27*, 1615–1660. [CrossRef]
10. Calcagnile, L.M.; Bormetti, G.; Treccani, M.; Marmi, S.; Lillo, F. Collective synchronization and high frequency systemic instabilities in financial markets. *Quant. Financ.* **2018**, *18*, 237–247. [CrossRef]
11. Bormetti, G. Modelling systemic price cojumps with hawkes factor models. *Quant. Financ.* **2015**, *15*, 1137–1156. [CrossRef]
12. Huang, X.; Vodenska, I.; Havlin, S.; Stanley, H.E. Cascading failures in bi-partite graphs: Model for systemic risk propagation. *Sci. Rep.* **2013**, *3*, 1–9.
13. Hansen, L.P. *1. Challenges in Identifying and Measuring Systemic Risk*; University of Chicago Press: Chicago, IL, USA, 2014
14. Paulin, J.; Calinescu, A.; Wooldridge, M. Understanding flash crash contagion and systemic risk: A micro–macro agent-based approach. *J. Econ. Dynam. Control* **2019**, *100*, 200–229. [CrossRef]
15. Abreu, D.; Brunnermeier, M.K. Synchronization risk and delayed arbitrage. *J. Financ. Econ.* **2002**, *66*, 341–360. [CrossRef]
16. Bhojraj, S.; Bloomfield, R.J.; Tayler, W.B. Margin trading, overpricing, and synchronization risk. *Rev. Financ. Stud.* **2009**, *22*, 2059–2085. [CrossRef]
17. Jain, P.K.; Jain, P.; McInish, T.H. Does high-frequency trading increase systemic risk? *J. Financ. Market* **2016**, *31*, 1–24. [CrossRef]
18. Harris, L. What to Do about High-Frequency Trading. *Financ. Anal. J.* **2013**, *69*, 6–9.
19. Serrano, A.S. High-frequency trading and systemic risk: A structured review of findings and policies. *Rev. Econ.* **2020**, *71*, 169–195. [CrossRef]
20. Huang, R.; Polak, T. Lobster: Limit Order Book Reconstruction System. 2011. Available online: https://papers.ssrn.com/sol3/papers.cfm?abstract_id=1977207 (accessed on 9 February 2022).
21. Lee, S.S.; Mykland, P.A. Jumps in financial markets: A new nonparametric test and jump dynamics. *Rev. Financ. Stud.* **2008**, *21*, 2535–2563. [CrossRef]
22. Hardiman, S.J.; Bercot, N.; Bouchaud, J.P. Critical reflexivity in financial markets: A hawkes process analysis. *Eur. Phys. J. B* **2013**, *86*, 1–9. [CrossRef]
23. Blanc, P.; Donier, J.; Bouchaud, J.P. Quadratic hawkes processes for financial prices. *Quant. Financ.* **2017**, *17*, 171–188. [CrossRef]
24. Bacry, E.; Delour, J.; Muzy, J.F. Multifractal random walk. *Phys. Rev. E* **2001**, *64*, 026103. [CrossRef] [PubMed]
25. Chronopoulou, A.; Viens, F.G. Stochastic volatility and option pricing with long-memory in discrete and continuous time. *Quant. Financ.* **2012**, *12*, 635–649. [CrossRef]
26. Cont, R. Empirical properties of asset returns: Stylized facts and statistical issues. *Quant. Financ.* **2001**, *1*, 223. [CrossRef]
27. Boudt, K.; Croux, C.; Laurent, S. Robust estimation of intraweek periodicity in volatility and jump detection. *J. Empirical Finan.* **2011**, *18*, 353–367. [CrossRef]
28. Curme, C.; Tumminello, M.; Mantegna, R.N.; Stanley, H.E.; Kenett, D.Y. Emergence of statistically validated financial intraday lead-lag relationships. *Quant. Financ.* **2015**, *15*, 1375–1386. [CrossRef]

29. Christensen, K.; Moloney, N.R. *Complexity and Criticality*; World Scientific Publishing Company: Hackensack, NJ, USA, 2005; Volume 1.
30. Latora, V.; Nicosia, V.; Russo, G. *Complex Networks: Principles, Methods and Applications*; Cambridge University Press: Cambridge, UK, 2017.
31. Rombach, M.P.; Porter, M.A.; Fowler, J.H.; Mucha, P.J. Core-periphery structure in networks. *SIAM J. Appl. Math.* **2014**, *74*, 167–190. [CrossRef]
32. Everett, M.G.; Borgatti, S.P. Extending centrality. *Models Method. Soc. Netw. Anal.* **2005**, *35*, 57–76.
33. Barucca, P.; Tantari, D.; Lillo, F. Centrality metrics and localization in core-periphery networks. *J. Stat. Mech. Theory Exp.* **2016**, *2016*, 023401. [CrossRef]
34. Da Silva, M.R.; Ma, H.; Zeng, A.P. Centrality, network capacity, and modularity as parameters to analyze the core-periphery structure in metabolic networks. *Proc. IEEE* **2008**, *96*, 1411–1420. [CrossRef]

entropy

Article

A New Look at Calendar Anomalies: Multifractality and Day-of-the-Week Effect

Darko Stosic [1], Dusan Stosic [1], Irena Vodenska [2,*], H. Eugene Stanley [3] and Tatijana Stosic [4]

[1] Centro de Informática, Universidade Federal de Pernambuco, Av. Luiz Freire s/n, Recife 50670-901, PE, Brazil; dd.stosic@gmail.com (D.S.); dbstosic@bu.edu (D.S.)
[2] Department of Administrative Sciences, Metropolitan College, Boston University, 1010 Commonwealth Avenue, Boston, MA 02215, USA
[3] Center for Polymer Studies, Department of Physics, Boston University, 590 Commonwealth Avenue, Boston, MA 02215, USA; hes@bu.edu
[4] Departamento de Estatística e Informática, Universidade Federal Rural de Pernambuco, Rua Dom Manoel de Medeiros s/n, Dois Irmãos, Recife 52171-900, PE, Brazil; tastosic@gmail.com
* Correspondence: vodenska@bu.edu

Abstract: Stock markets can become inefficient due to calendar anomalies known as the day-of-the-week effect. Calendar anomalies are well known in the financial literature, but the phenomena remain to be explored in econophysics. This paper uses multifractal analysis to evaluate if the temporal dynamics of market returns also exhibit calendar anomalies such as day-of-the-week effects. We apply multifractal detrended fluctuation analysis (MF-DFA) to the daily returns of market indices worldwide for each day of the week. Our results indicate that distinct multifractal properties characterize individual days of the week. Monday returns tend to exhibit more persistent behavior and richer multifractal structures than other day-resolved returns. Shuffling the series reveals that multifractality arises from a broad probability density function and long-term correlations. The time-dependent multifractal analysis shows that the Monday returns' multifractal spectra are much wider than those of other days. This behavior is especially persistent during financial crises. The presence of day-of-the-week effects in multifractal dynamics of market returns motivates further research on calendar anomalies for distinct market regimes.

Keywords: calendar anomalies; day-of-the-week effect; market indices; multifractal detrended fluctuation analysis

Citation: Stosic, D.; Stosic, D.; Vodenska, I.; Stanley, H.E.; Stosic, T. A New Look at Calendar Anomalies: Multifractality and Day-of-the-Week Effect. *Entropy* **2022**, *24*, 562. https://doi.org/10.3390/e24040562

Academic Editor: Stanisław Drożdż

Received: 6 January 2022
Accepted: 13 April 2022
Published: 17 April 2022

1. Introduction

Market prices should incorporate and reflect all available information at any point in time, according to the Efficient Market Hypothesis (EMH) [1,2]. Yet, various studies [3–6] show that financial markets often become inefficient, and their behavior no longer follows that of a random walk. Stock markets can instead deviate from the rules of the EMH in the form of anomalies. Anomalies can be broadly categorized into calendar, fundamental and technical anomalies [7]. The most studied set of pricing anomalies is calendar or seasonal anomalies that represent systematic patterns of security returns around certain calendar points. Calendar anomalies include the day-of-the-week effect [8–11], turn-of-the-month effect [12–15], turn-of-the-year effect [16–19] and holiday effect [20–23]. The day-of-the-week effect refers to the tendency of stocks to exhibit significantly higher returns on one particular day compared with other days of the week. Cross [24] first provided evidence of day-of-the-week effects on the Standard and Poor's index, reporting that price returns are significantly negative on Mondays. Since then, this phenomenon has been extensively studied and discovered in other financial markets such as specific equity markets [25–27], exchange rates [28,29], fixed-income securities [30], crude oil [31], gold [32] and cryptocurrencies [33]. For a detailed review of seasonal anomalies, please see [34,35].

Financial markets have attracted much attention from researchers in related fields such as econophysics, paving the road for new perspectives and understanding of financial markets by drawing concepts from statistical physics such as fractals and multifractals [36–39], information theory [40,41] and network structures [42–44] (see [45] and the references therein for a comprehensive review). While many well-known conclusions in the literature on an array of financial markets (including market indices, stocks, exchange rates and commodities) can be attributed to econophysics, there are still a number of important phenomena to be investigated from this perspective. To the best of our knowledge, one such phenomenon that remains to be unearthed is the calendar anomaly, and our study makes a contribution in this direction.

In this paper, we use multifractal analysis to evaluate if the temporal dynamics of market returns exhibit calendar anomalies such as day-of-the-week effects. We apply multifractal detrended fluctuation analysis (MF-DFA) [46] to the daily returns of market indices around the world for each day of the week (Monday returns, Tuesday returns and so on). We then compare the multifractal parameters, the position of maximum width and asymmetry of the multifractal spectrum, which quantify long-term correlations, the degree of multifractality and the dominance of large or small fluctuations in the return series for each day of the week. The economic literature states that market practitioners have been aware of the Monday effect as early as the 1920s [47]. For some markets, this effect disappears as the market becomes more efficient [48,49]. Other studies offer insight into the Monday effect being more prominent toward the end of the month [50] and during periods dominated by bad news [51]. To observe this behavior over time, we perform time-dependent multifractal analysis on the United States (GSPC) market by calculating the multifractal spectra of the return series in a sliding window. This computationally intensive and relatively novel approach, which has been implemented in only a few studies [52–54], permits us to analyze the temporal evolution of multifractal parameters which are related to different properties of market fluctuation, leading to better understanding of the underlying stochastic processes. The rest of this paper is organized as follows. Section 2 introduces the MF-DFA and the time-dependent methods. Section 3 describes the market data. Section 4 presents the results, and Section 5 draws the conclusion.

2. Methods

While fractal processes are characterized by long-term correlations that are described by a single scaling exponent, multifractal time series subsets with small and large fluctuations can scale differently, and the analysis of long-term correlations results in a hierarchy of scaling exponents [46]. Multifractal analysis of temporal series can be performed using different methods, such as the wavelet transform modulus maxima (WTMM) method [55], multifractal detrended fluctuation analysis (MF-DFA) method [46] and multifractal detrending moving average method (MF-DMA) [56]. In this work, we employ MF-DFA, which has been found to produce reliable results [57] and has been widely used to analyze physiological signals [58–60], geophysical data [61], weather data [62], and financial time series [63].

The implementation of the MF-DFA algorithm can be described as follows [46]:

i The first step is the integration of the original series $x(i), i = 1, \dots, N$ to produce

$$X(k) = \sum_{i=1}^{k}[x(i) - \langle x \rangle], \quad k = 1, \dots, N, \tag{1}$$

where $\langle x \rangle = \frac{1}{N} \sum_{i=1}^{k} x(i)$ is the average.

ii Next, the integrated series $X(k)$ is divided into $N_n = int(N/n)$ non-overlapping segments of a length n, and in each segment $\nu = 1, \dots, N_n$, the local trend $X_{n,\nu}(k)$ is estimated as a linear or higher order polynomial least square fit and subtracted from $X(k)$.

iii The detrended variance

$$F^2(n,\nu) = \frac{1}{n} \sum_{k=(\nu-1)n+1}^{\nu n} [X(k) - X_{n,\nu}(k)]^2 \tag{2}$$

is calculated for each segment and then averaged over all segments to obtain the qth order fluctuation function:

$$F_q(n) = \left\{ \frac{1}{N_n} \sum_{\nu=1}^{N_n} [F^2(n,\nu)]^{q/2} \right\}^{1/q}, \tag{3}$$

where, in general, q can take on any real value except zero.

iv Repeating this calculation for all box sizes provides the relationship between the fluctuation function $F_q(n)$ and box size n. $F_q(n)$ increases with n according to a power law $F_q(n) \sim n^{h(q)}$ if long-term correlations are present. The scaling exponent $h(q)$ is obtained as the slope of the linear regression of $\log F_q(n)$ versus $\log n$.

The power law exponent $h(q)$ is called the generalized Hurst exponent, where for stationary time series, $h(2)$ is identical to the well-known Hurst exponent H. For positive q values, $h(q)$ describes the scaling behavior of large fluctuations, while for negative q values, $h(q)$ describes the scaling behavior of small fluctuations, while $h(q)$ is independent of q for monofractal time series and a decreasing function of q for multifractal time series.

The generalized Hurst exponents are related to the Renyi exponents $\tau(q)$ defined by the standard partition function-based multifractal formalism $\tau(q) = qh(q) - 1$. For the monofractal signals, $\tau(q)$ is a linear function of q (as $h(q) = const.$) and for multifractal signals $\tau(q)$ is a nonlinear function of q. A multifractal process can also be characterized by the singularity spectrum $f(\alpha)$, which is related to $\tau(q)$ through the Legendre transform:

$$\alpha(q) = \frac{d\tau(q)}{dq}, \tag{4}$$

$$f(\alpha(q)) = q\alpha(q) - \tau(q), \tag{5}$$

where $f(\alpha)$ is the fractal dimension of the support of singularities in the measure with Lipschitz–Holder exponent α. The singularity spectrum of the monofractal signal is represented by a single point in the $f(\alpha)$ plane, whereas the multifractal process yields a single humped function.

Multifractal spectra reflect the level of complexity of the underlying stochastic process and can be characterized by a set of three parameters, which are determined as follows. The singularity spectra are fitted to a fourth degree polynomial:

$$f(\alpha) = A + B(\alpha - \alpha_0) + C(\alpha - \alpha_0)^2 + D(\alpha - \alpha_0)^3 + E(\alpha - \alpha_0)^4 \tag{6}$$

The multifractal spectrum parameters are found as the position of the maximum $\alpha_0 = \arg\max_\alpha f(\alpha)$, the width of the spectrum $W = \alpha_{max} - \alpha_{min}$ obtained from extrapolating the fitted curve to zero, and the skew parameter $r = (\alpha_{max} - \alpha_0)/(\alpha_0 - \alpha_{min})$, where $r = 1$ for symmetric shapes, $r > 1$ for right-skewed shapes and $r < 1$ for left-skewed shapes. These three parameters can be used to evaluate the complexity of the underlying process. A small value of α_0 means that the process is correlated and more regular in appearance. The width W of the spectrum measures the degree of multifractality of the process, where a wider range of fractal exponents leads to "richer" structures. The skew parameter r indicates which fractal exponents are dominant: the $f(\alpha)$ spectrum is right-skewed ($r > 1$), and the process is characterized by a "fine structure" (small fluctuations) if high fractal exponents are dominant, whereas the process is more regular or smooth, the $f(\alpha)$ spectrum is left-skewed ($r < 1$), and the fractal exponents describe the scaling of large fluctuations if the low fractal exponents are dominant. In summary, a signal with a high value of α_0, a wide range W of fractal exponents (higher degree of multifractality) and a right-skewed shape ($r > 1$) may be considered more complex than one with the opposite characteristics [60].

The two sources of multifratality in a time series are (1) a broad probability density function for the values of the time series and (2) different long-term correlations for small and large fluctuations. The type of multifractal can be found by randomly shuffling the series and analyzing its behavior. For multifractals of the second type, the shuffled series exhibits simple random behavior (since long-term correlations are destroyed), and the width of the $f(\alpha)$ spectrum is reduced to a single point. For multifractals of the first type, the width of the $f(\alpha)$ spectrum remains the same (since the probability density cannot be removed), and for multifractals of types 1 and 2, the shuffled series shows weaker multifractality than the original series [46].

The time-dependent MF-DFA algorithm is based on the sliding window technique and yields a temporal evolution of multifractality in the system. Given a time series $x = x_1, \ldots, x_N$, many sliding windows $z_t = x_{1+t\Delta}, \ldots, x_{w+t\Delta}, t = 0, 1, \ldots, \left[\frac{N-w}{\Delta}\right]$ are constructed, where $w \leq N$ is the window size, $\Delta \leq w$ is the sliding step and the operator [.] denotes taking the integer part of the argument. The values of the time series in each window z_t are then used to calculate the multifractal spectrum at a given time t using the method described above. This allowed us to obtain time evolutions for the three complexity parameters.

3. Data

We analyzed the time series of 19 major stock market indices that appear on the website https://finance.yahoo.com/world-indices/ (accessed on 2 January 2022), which are listed in Table 1. The period under study spanned the earliest recorded date for each index up to the end of 2018. For each of the market indices with consecutive workday closing price values $S(t), t = 1, \ldots, N$, we calculated the daily logarithmic returns:

$$R_t \equiv \ln \frac{S(t)}{S(t-1)} \quad t = 2, \ldots, N, \tag{7}$$

where the returns for Monday were calculated using the closing price of the previous Friday, while for other days of the week, two consecutive workday closing price values were used. Next, we constructed time series from the returns R_t for each day of the week (Monday returns, Tuesday returns and so on):

$$R^i = \{R_{t_i}, R_{t_i+5}, \ldots, R_{t_i+5\left[\frac{N}{5}\right]}\}, \tag{8}$$

where $i = 1, \ldots, 5$ denotes the index of the weekday, R_{t_i} corresponds to the first occurrence of day i in the returns series $R_t, t = 2, \ldots, N$ and the operator [.] denotes taking the integer part of the argument. Figure 1 reveals that the fluctuations in the returns varied between different days. While Monday exhibited the most pronounced negative returns, the fluctuations for other days dominated at specific time intervals. This is a well-known day-of-the-week effect which was found for the US market [8,25].

The MF-DFA method was applied to the day-resolved returns R^i of major stock market indices, where local trends were fitted with a second-degree polynomial $m = 2$. Next, we performed a fourth-order polynomial regression on the singularity spectra $f(\alpha)$ to determine the position of the maximum α_0 and the zeros of the polynomial $\alpha_{\max}$ and $\alpha_{\min}$. From the polynomial fits, we calculated three measures of complexity: the position of the maximum α_0, the width of the spectrum $W = \alpha_{\max} - \alpha_{\min}$ and the skew parameter $r = (\alpha_{\max} - \alpha_0)/(\alpha_0 - \alpha_{\min})$. These parameters were then used to determine the multifractal behavior of the day-resolved price returns.

Table 1. Information on analyzed time series for major market indices.

Market	Country	Index	Period
All Ordinaries	Australia	AORD	3 August 1984–26 December 2018
S&P500/ASX 200	Australia	AXJO	22 November 1992–26 December 2018
BEL 20	Belgium	BFX	9 April 1991–24 December 2018
IBOVESPA	Brazil	BVSP	27 April 1993–21 December 2018
Dow30	United States	DJI	29 January 1985–26 December 2018
CAC 40	France	FCHI	1 March 1990–24 December 2018
DAX Performance	Germany	GDAXI	30 December 1987–27 December 2018
S&P500	United States	GSPC	3 January 1950–24 December 2018
S&P/TSX Composite	Canada	GSPTSE	29 June 1979–24 December 2018
Hang Seng Index	Hong Kong	HIS	31 December 1986–27 December 2018
IPSA Santiago de Chile	Chile	IPSA	2 January 2002–26 December 2018
Nasdaq	United States	IXIC	5 February 1971–26 December 2018
Jakarta Composite	Indonesia	JKSE	1 July 1997–27 December 2018
KOSPI Composite	South Korea	KS11	1 July 1997–26 December 2018
Merval	Argentina	MERV	8 October 1996–26 December 2018
IPC Mexico	Mexico	MXX	8 November 1991–26 December 2018
Nikkei 225	Japan	N225	5 January 1965–27 December 2018
NYSE Composite	United States	NYA	31 December 1965–26 December 2018
TSEC Weighted	Taiwan	TWII	2 July 1997–27 December 2018

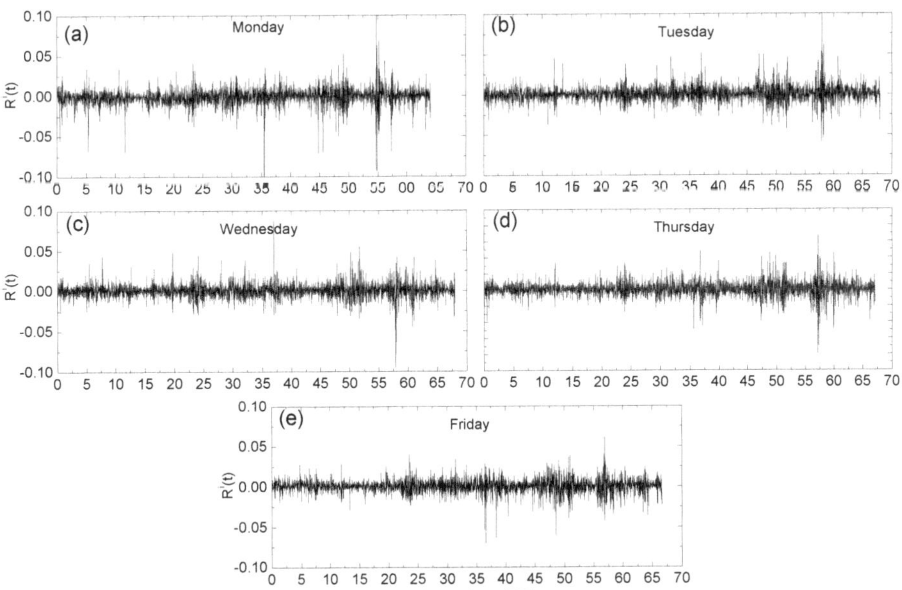

Figure 1. Time series for (**a**) Monday, (**b**) Tuesday, (**c**) Wednesday, (**d**) Thursday and (**e**) Friday day-resolved price returns R^i of the United States (GSPC) market index.

4. Results

4.1. Day-of-the-Week Effect

Complexity measures derived from the singularity spectra were used to study the multifractal behavior of the price returns for every day of the week. We first considered multifractality in the day-resolved price returns from four distinct markets: the United States (GSPC), South Korea (KS11), Chile (IPSA) and France (FCHI). The multifractal spectra for each day using the four markets are illustrated in Figure 2. We observed that the day-of-the-week effects led to significant differences in multifractal behavior: (1) the positions of the maxima α_0 were shifted to the right ($\alpha_0 > 0.5$) for the Monday returns, and (2) the spectrum widths W were wider on Monday than those for returns from other days. There seemed to be no consistent differences in the skew parameter r, which implies that both large and small fluctuations are present for different days of the week (e.g., see Table 2). These results indicate that the Monday returns exhibited more persistent behavior and richer multifractal structures, which led to more complex time series than other day's returns. Our findings are consistent with results obtained from [25], which indicated that Monday had the largest anomalies (day-of-the-week effect) because of the weekend gap in trading hours. Other days of the week did not exhibit any visible patterns in multifractal behavior for either the position or width of the spectrum.

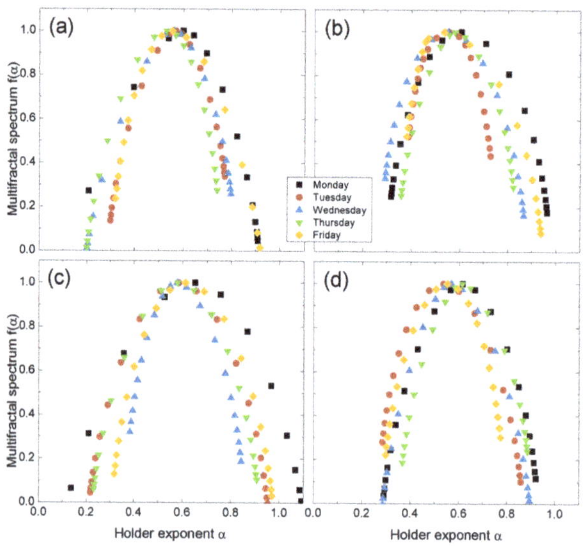

Figure 2. Multifractal spectrum $f(\alpha)$ for day-resolved price returns R^i of (**a**) the United States (GSPC), (**b**) South Korea (KS11), (**c**) Chile (IPSA) and (**d**) France (FCHI) market indices.

Table 2. Multifractal parameters α_0, W and r for day-resolved price returns R^i of major market indices.

Market	Monday			Tuesday			Wednesday			Thursday			Friday			All		
	α_0	W	r	α_0	W	r	α_0	W	r	α_0	W	r	α_0	W	r	α_0	W	r
AORD	0.547	0.570	0.837	0.585	0.684	0.590	0.547	0.628	0.815	0.549	0.549	0.963	0.574	0.603	0.771	0.583	0.579	0.942
AXJO	0.537	0.529	0.990	0.583	0.514	1.201	0.561	0.558	0.897	0.557	0.544	1.180	0.562	0.550	0.866	0.533	0.748	0.913
BFX	0.619	0.633	0.730	0.561	0.662	1.383	0.571	0.541	0.754	0.574	0.553	0.940	0.553	0.556	0.715	0.534	0.676	1.188
BVSP	0.616	0.581	1.562	0.601	0.472	0.932	0.587	0.455	1.492	0.615	0.465	0.939	0.592	0.666	0.943	0.550	0.643	0.917
DJI	0.572	0.826	0.579	0.598	0.643	0.883	0.576	0.586	0.970	0.560	0.661	0.887	0.581	0.669	1.230	0.520	0.690	0.720
FCHI	0.617	0.656	0.969	0.526	0.621	1.257	0.579	0.613	1.087	0.620	0.579	1.090	0.553	0.535	0.894	0.506	0.633	1.174
GDAXI	0.606	0.612	0.682	0.556	0.619	0.886	0.574	0.530	0.986	0.616	0.538	1.034	0.555	0.485	1.397	0.534	0.648	1.176
GSPC	0.590	0.787	0.709	0.565	0.539	0.856	0.557	0.635	0.760	0.528	0.573	0.718	0.551	0.627	1.416	0.528	0.605	0.782
GSPTSE	0.611	0.632	0.683	0.587	0.647	0.841	0.581	0.618	0.956	0.587	0.552	0.733	0.554	0.681	0.775	0.585	0.613	0.928
HIS	0.582	0.823	0.828	0.562	0.669	0.639	0.514	0.730	0.864	0.592	0.509	1.083	0.576	0.749	0.878	0.557	0.609	0.805
IPSA	0.654	0.969	0.832	0.584	0.747	0.984	0.580	0.519	1.250	0.582	0.705	0.938	0.611	0.677	1.174	0.601	0.825	0.801
IXIC	0.641	0.707	0.764	0.585	0.644	0.941	0.615	0.702	0.781	0.563	0.671	1.425	0.587	0.680	1.134	0.591	0.624	0.901
JKSE	0.598	0.848	1.352	0.539	0.674	1.335	0.582	0.725	0.877	0.560	0.566	1.802	0.500	0.907	0.881	0.570	0.518	0.769
KS11	0.607	0.707	1.190	0.539	0.421	1.140	0.540	0.637	1.195	0.590	0.535	1.026	0.526	0.616	2.180	0.530	0.633	0.945
MERV	0.651	0.520	0.927	0.537	0.625	1.265	0.537	0.681	1.163	0.611	0.647	0.652	0.540	0.602	1.135	0.574	0.534	0.985
MXX	0.580	0.805	0.890	0.542	0.666	1.088	0.548	0.577	1.039	0.606	0.690	1.150	0.552	0.557	0.967	0.548	0.617	0.951
N225	0.584	0.472	1.041	0.573	0.745	0.714	0.550	0.639	0.901	0.614	0.505	0.732	0.553	0.530	0.804	0.539	0.406	0.559
NYA	0.593	0.685	0.466	0.579	0.648	0.790	0.550	0.588	0.615	0.559	0.691	0.827	0.526	0.573	0.954	0.522	0.583	0.772
TWII	0.659	0.474	1.661	0.594	0.564	1.453	0.519	0.453	1.069	0.540	0.494	1.584	0.503	0.764	1.303	0.539	0.491	1.053

We expanded our investigation to other markets listed in Table 1. Figure 3 reveals that the multifractal spectra of the Monday returns were dominantly right-shifted ($\alpha_0 > 0.5$) compared with other days for most analyzed markets. Notable exceptions included the United States (DJI), Australia (AORD, AXJO), where the Tuesday returns were more persistent, and Japan (N225), where the Thursday returns exhibited stronger persistency. The width of the multifractal spectrum displayed similar tendencies to its position, where the Monday returns possessed broader multifractal widths. Yet, we found that more markets tended to have other days with richer multifractal structures; the multifractal spectra were the widest for the Friday returns in Taiwan (TWII) and the Tuesday returns in Japan (N225) and Australia (AORD), as opposed to the markets with dominant Monday returns considered so far. It has been noted that the day-of-the-week effect occurs on different distinct days of the week for different markets [25]. Considering both parameters α_0 and W, we observed that the North American, European and some Asian (South Korea, Indonesia and Hong Kong) and Latin American markets (Chile and Mexico) tended to show both stronger persistency and stronger multifractality for the Monday returns, while for Australia, Indonesia and Taiwan, this tendency was found for the Tuesday returns. This is also in agreement with the literature, where it was found that some Asian markets displayed a Tuesday anomaly, which is one day out of phase with North American markets due to different time zones [64]. Patterns in the skew of multifractal spectra for a given day of the week are again hard to discern across distinct markets, where both small and large fluctuations exist. Values of the multifractal complexity parameter are listed in Table 2. Our results indicate that while most markets exhibit more complex behavior for Monday returns, some markets have other days with largest anomalies (day-of-the-week effect) such as Tuesday, Thursday and Friday returns. This is expected from literature where it was found that different day-of-the-week effects exist for different markets [25].

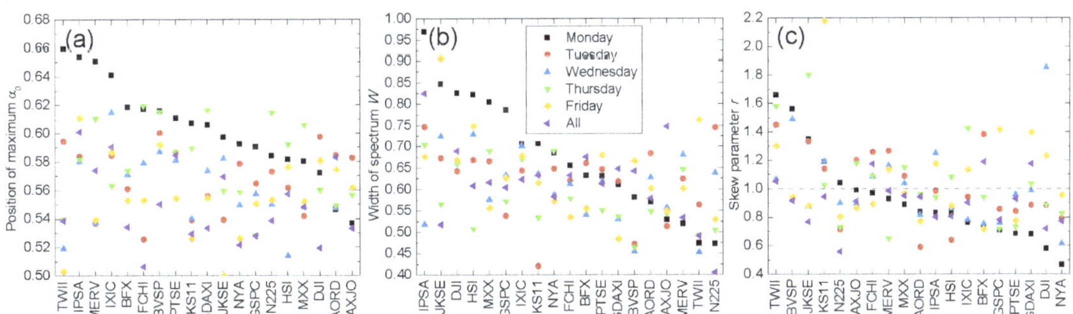

Figure 3. Complexity parameters (**a**) position of maximum α_0, (**b**) spectrum width W, and (**c**) skew parameter r, for day-resolved price returns of the market indices listed in Table 1, sorted from largest to smallest.

4.2. Comparison to Bulk Behavior

The day-resolved multifractal spectra could also be compared to those for the whole time series. The motivation for such a comparison is to provide more insight on the relation between multifractality and the day-of-the-week effect. From Figure 3, we found that many markets (IPSA, KS11, GSPTSE and MMX) exhibited distinct multifractal properties for a particular day (e.g., Monday returns), while the whole series displayed similar multifractal behavior to the bulk, or the remaining days of the week. For other markets (DJI, AXJI and N225), the overall multifractality of the series differed widely from the multifractal spectrum for each day of the week. This suggests that the day-of-the-week effects resulted in different multifractalities for these markets. We could further classify the markets into one of two multifractal behaviors: (1) bulk multifractality, which only differs for one particular day of the week, and (2) day-of-the-week multifractality, which is unique to every day and differs from the bulk behavior.

4.3. Source of Multifractality

We shuffled the time series of the day-resolved returns for the four markets and then applied MF-DFA to determine the source of multifractality. The shuffling procedure performed $1000 \times N$ transpositions on each series and was repeated 100 times with different random number generator seeds in order to obtain statistics such as the mean and standard deviation. Figure 4 reveals that for the United States (GSPC), the right-hand side of the spectrum (effect of small fluctuations) was mildly affected by shuffling on Mondays and Fridays, while the left side of the spectrum (effect of large fluctuations) was affected primarily on Thursdays (and less so on Wednesdays), and the position remained the same for all of the day-resolved returns. This indicates that multifractality arose primarily from a broad probability density function [65], and the long-term correlations had only a minor impact on some days of the week.

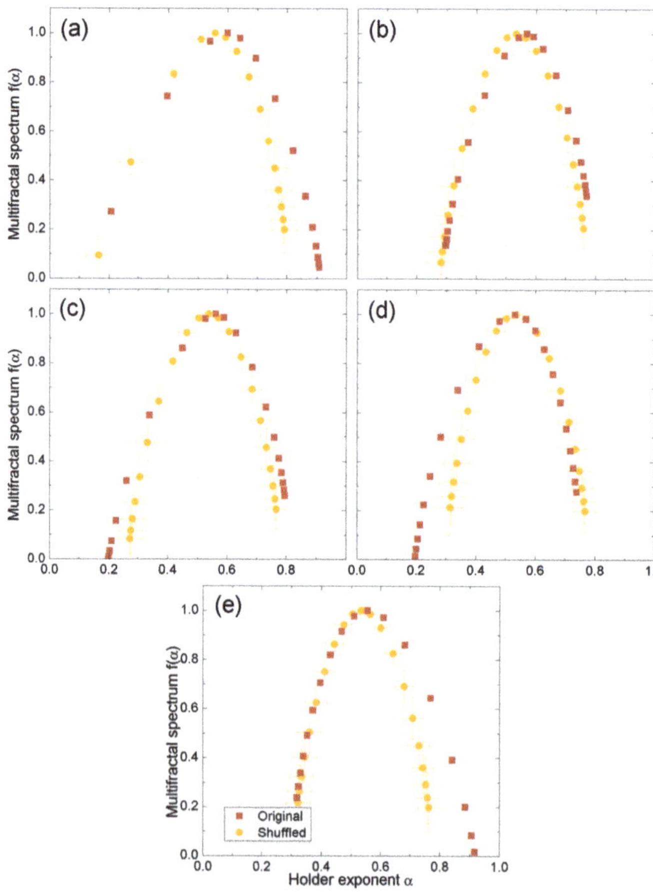

Figure 4. Original and shuffled multifractal spectra $f(\alpha)$ for (**a**) Monday, (**b**) Tuesday, (**c**) Wednesday, (**d**) Thursday and (**e**) Friday day-resolved price returns of the United States (GSPC) market.

While it may be argued that destroying correlations by shuffling leads to strictly monofractal behavior and leaving only finite size effects, as shown for the qGaussian distributions using MFDFA [66] and market volatility data using partition function formalism [67], in the current case, shuffling left the spectrum width only slightly narrowed down, in agreement with previous MFDFA studies of market returns [65]. Even if upon shuffling only a finite size effect remained, different effects on different days of the week on small and large fluctuations provided novel insight into the market behavior.

Table 3 lists the changes in spectra position ($\Delta\alpha_0$) and width (ΔW) after shuffling the day-resolved returns for GSPC, KS11, IPSA and FCHI. We found that the Monday returns tended to exhibit the strongest effect from shuffling, where aside from the probability density function, long-term correlations also contributed to multifractality.

Table 3. Differences in multifractal parameters between original and shuffled day-resolved price returns.

Market	Monday		Tuesday		Wednesday		Thursday		Friday	
	$\Delta\alpha_0$	ΔW	$\Delta\alpha_0$	ΔW	$\Delta\alpha_0$	ΔW	$\Delta\alpha_0$	ΔW	$\Delta\alpha_0$	ΔW
GSPC	0.049	0.115	0.030	0.019	0.021	0.094	0.008	0.058	0.014	0.126
KS11	0.033	0.010	0.034	0.219	0.028	0.052	0.012	0.138	0.044	0.052
IPSA	0.073	0.200	0.022	0.119	0.028	0.069	0.023	0.064	0.051	0.098
FCHI	0.064	0.035	0.023	0.078	0.031	0.054	0.066	0.033	0.002	0.025

4.4. Time Evolution

For intuition on how the multifractal day-of-the-week effects change over time, we could analyze the time evolutions of the multifractal spectra. We considered the United States (GSPC) market, since the day-of-the-week effects over time here are well known [48]. For each day-of-the-week return, we constructed a sliding window of a size $w = 730$ days with a sliding step $\Delta = 5$ days, meaning that we applied the MF-DFA method over a 14-year period in monthly intervals. Figure 5 illustrates the time evolutions of the multifractal spectra for different day-resolved returns. We observed that the spectrum evolved differently for each day of the week. For the Monday returns, the spectrum shifted to the left, which means that the fluctuations became less persistent over time. Other day-of-the-week returns either exhibited small movements in the multifractal spectra or moved back to the same position after some time. For a more quantitative analysis, we calculated the differences over time in the complexity parameters, namely $\Delta\alpha_0$ and ΔW, between Monday and other day-resolved returns. Figure 6a reveals that the spectra position of the Monday returns differed considerably from α_0 of the other day returns in the first 15 years of the recorded period, but their differences dropped to zero in the subsequent years. This indicates the presence of strong day-of-the-week effects between 1950 and 1980 ($\Delta\alpha_0 \to 0$ after 1965, where 1980 is already included because of the 14-year long sliding window), which is consistent with the literature, where it was found that the day-of-the-week effects diminished around 1980 [48].

Fluctuations around $\Delta\alpha_0 = 0$ after 1980 can be attributed to large financial crises that affected the entire market, such as Black Monday in 1987 and the global financial crisis in 2008. Figure 6b illustrates the time evolutions of the differences in the spectra width ΔW between Monday and other day-resolved returns. We observed that the Monday returns exhibited much wider multifractal spectra than other day's returns during either of the two financial crises in 1987 and 2008. The Monday returns were characterized by more complex structures and had significant day-of-the-week effects during the financial crises even after 1980, when the effects from the calendar anomalies should have vanished. A possible explanation for this phenomena is the weekend gap in trading hours, which leads to even more speculative behavior from investors during a crisis.

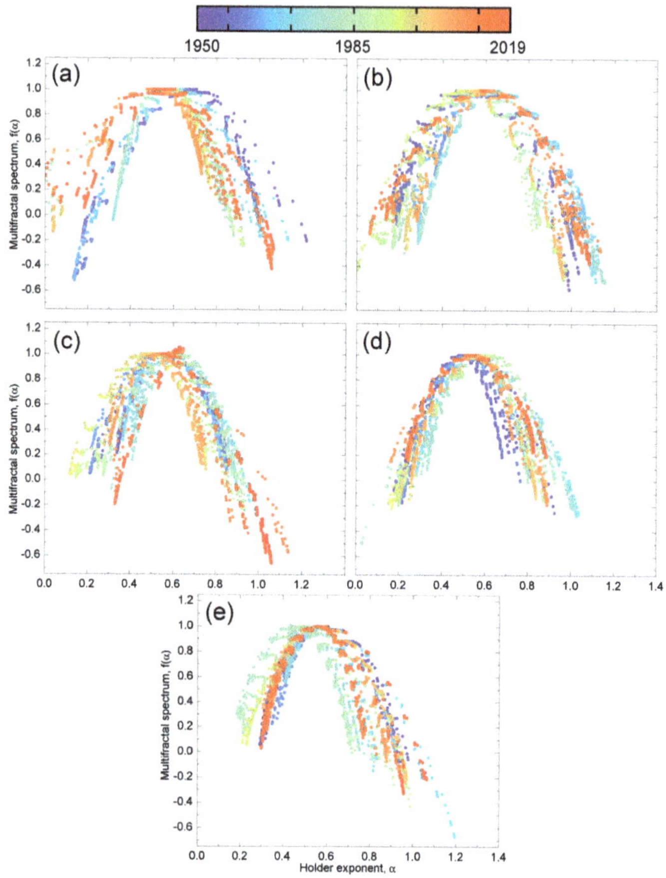

Figure 5. Time evolution of the multifractal spectrum $f(\alpha)$ for (**a**) Monday, (**b**) Tuesday, (**c**) Wednesday, (**d**) Thursday and (**e**) Friday day-resolved price returns of the United States (GSPC) market. A sliding window of 14 years and monthly intervals were used for the period spanning from 1950 to 2019.

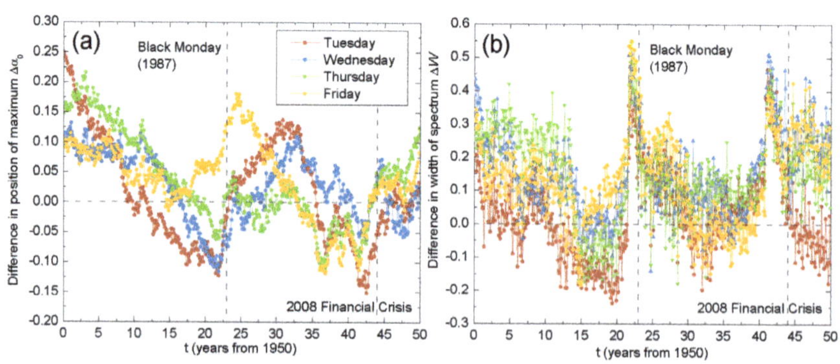

Figure 6. Time evolution of differences in complexity parameters (**a**) α_0 and (**b**) W derived from the multifractal spectra $f(\alpha)$ between Monday and other day-resolved price returns for the United States (GSPC) market. A sliding window of 14 years and monthly intervals were used for the period spanning from 1950 to 2019.

5. Conclusions

This paper investigated the multifractal behavior of the day-of-the-week returns for market indices worldwide. We applied the MF-DFA method to daily returns for each day of the week (Monday returns, Tuesday returns and so on) and calculated the multifractal spectra as well as their complexity parameters. Considering the multifractal parameters' positions of the maximum α_0 and width W of an $f(\alpha)$ spectrum, we observed that distinct multifractal properties were found for the different days of the week, where North American, European and some Asian (South Korea, Indonesia and Hong Kong) and Latin American markets (Chile and Mexico) tended to show both stronger persistency ($\alpha_0 > 0.5$) and stronger multifractality (larger W) for the Monday returns, while for Australia, Indonesia and Taiwan, this tendency was found for the Tuesday returns. This finding agrees with the literature in that different day-of-the-week effects exist for different markets [25]. Some Asian markets displayed the Tuesday anomaly, being one day out of phase with the North American markets due to different time zones [64]. We found that multifractality arose from a broad probability density function and long-term correlations by analyzing shuffled series. The time-dependent multifractal analysis of the United States (GSPC) market revealed that the multifractal spectra for the Monday returns shifted to the left, or the fluctuations became less persistent over time. Other day-of-the-week returns exhibited small movements in the multifractal spectra. While the authors of [48] found that the effects from calendar anomalies vanished after 1980, in our study, we observed that the day-of-the-week effects persisted after the 1980s. Notably, the Monday returns exhibited much broader multifractal spectra compared with other days of the week. This behavior was especially pronounced around Black Monday on 19 October 1987 and the global financial crisis in 2008. A possible explanation for this phenomenon is the weekend gap in trading hours, leading to even more speculative behavior from investors during a crisis. Monday returns in general in the US tend to be different compared with those of other days of the week. This anomaly has been attributed to companies' release of news after the financial markets close on Friday, and hence, the Monday prices reflect the accumulated reaction of investors over the weekend. This unique behavior of financial asset prices on Monday can be informative and useful for investment decision making and can inform policymakers to possibly limit important news releases on Friday afternoon. The Monday effect may be reduced by current tendencies of after-hours trading. However, since the after-hours trading volumes are much lower than the regular trading hours, the Monday effect is still present. Future studies should further investigate the multifractal dynamics and day-of-the-week effects for other financial markets and extend the current analysis to other calendar anomalies.

Author Contributions: Data curation, D.S. (Darko Stosic); Formal analysis, D.S. (Dusan Stosic); Methodology, T.S.; Supervision, H.E.S.; Writing—review & editing, I.V. All authors have read and agreed to the published version of the manuscript.

Funding: This research received no external funding.

Conflicts of Interest: The authors declare no conflict of interest.

References

1. Fama, E.F. Random Walks in Stock Market Prices. *Financ. Anal. J.* **1995**, *51*, 75–80. [CrossRef]
2. Malkiel, B.G.; Fama, E.F. Efficient capital markets: A review of theory and empirical work. *J. Financ.* **1970**, *25*, 383–417. [CrossRef]
3. Lo, A.W.; MacKinlay, A.C. Stock market prices do not follow random walks: Evidence from a simple specification test. *Rev. Financ. Stud.* **1988**, *1*, 41–66. [CrossRef]
4. Lim, K.P.; Brooks, R. The evolution of stock market efficiency over time: A survey of the empirical literature. *J. Econ. Surv.* **2011**, *25*, 69–108. [CrossRef]
5. Hamid, K.; Suleman, M.T.; Ali Shah, S.Z.; Akash, I.; Shahid, R. Testing the weak form of efficient market hypothesis: Empirical evidence from Asia-Pacific markets. *Int. Res. J. Financ. Econ.* **2017**, *58*, 121–133. [CrossRef]
6. Ito, M.; Sugiyama, S. Measuring the degree of time varying market inefficiency. *Econ. Lett.* **2009**, *103*, 62–64. [CrossRef]

7. Latif, M.; Arshad, S.; Fatima, M.; Farooq, S. Market efficiency, market anomalies, causes, evidences, and some behavioral aspects of market anomalies. *Res. J. Financ. Account.* **2011**, *2*, 1–13.
8. French, K.R. Stock returns and the weekend effect. *J. Financ. Econ.* **1980**, *8*, 55–69. [CrossRef]
9. Berument, H.; Kiymaz, H. The day of the week effect on stock market volatility. *J. Econ. Financ.* **2001**, *25*, 181–193. [CrossRef]
10. Zhang, J.; Lai, Y.; Lin, J. The day-of-the-week effects of stock markets in different countries. *Financ. Res. Lett.* **2017**, *20*, 47–62. [CrossRef]
11. Kiymaz, H.; Berument, H. The day of the week effect on stock market volatility and volume: International evidence. *Rev. Financ. Econ.* **2003**, *12*, 363–380. [CrossRef]
12. Ariel, R.A. A monthly effect in stock returns. *J. Financ. Econ.* **1987**, *18*, 161–174. [CrossRef]
13. McConnell, J.J.; Xu, W. Equity Returns at the Turn of the Month. *Financ. Anal. J.* **2008**, *64*, 49–64. [CrossRef]
14. Kunkel, R.A.; Compton, W.S.; Beyer, S. The turn-of-the-month effect still lives: The international evidence. *Int. Rev. Financ. Anal.* **2003**, *12*, 207–221. [CrossRef]
15. Sharma, S.S.; Narayan, P.K. New evidence on turn-of-the-month effects. *J. Int. Financ. Mark. Inst. Money* **2014**, *29*, 92–108. [CrossRef]
16. Reinganum, M.R. The anomalous stock market behavior of small firms in January: Empirical tests for tax-loss selling effects. *J. Financ. Econ.* **1983**, *12*, 89–104. [CrossRef]
17. Zhang, C.Y.; Jacobsen, B. Are monthly seasonals real? A three century perspective. *Rev. Financ.* **2013**, *17*, 1743–1785. [CrossRef]
18. Choudhry, T. Month of the year effect and January effect in pre-WWI stock returns: Evidence from a non-linear GARCH model. *Int. J. Financ. Econ.* **2001**, *6*, 1–11. [CrossRef]
19. Haug, M.; Hirschey, M. The january effect. *Financ. Anal. J.* **2006**, *62*, 78–88. [CrossRef]
20. Ariel, R.A. High Stock Returns before Holidays: Existence and Evidence on Possible Causes. *J. Financ.* **1990**, *45*, 1611–1626. [CrossRef]
21. Chong, R.; Hudson, R.; Keasey, K.; Littler, K. Pre-holiday effects: International evidence on the decline and reversal of a stock market anomaly. *J. Int. Money Financ.* **2005**, *24*, 1226–1236. [CrossRef]
22. Białkowski, J.; Etebari, A.; Wisniewski, T.P. Fast profits: Investor sentiment and stock returns during Ramadan. *J. Bank. Financ.* **2012**, *36*, 835–845. [CrossRef]
23. Meneu, V.; Pardo, A. Pre-holiday effect, large trades and small investor behaviour. *J. Empir. Financ.* **2004**, *11*, 231–246. [CrossRef]
24. Cross, F. The Behavior of Stock Prices on Fridays and Mondays. *Financ. Anal. J.* **1973**, *29*, 67–69. [CrossRef]
25. Dubois, M.; Louvet, P. The day-of-the-week effect: The international evidence. *J. Bank. Financ.* **1996**, *20*, 1463–1484. [CrossRef]
26. Seif, M.; Docherty, P.; Shamsuddin, A. Seasonal anomalies in advanced emerging stock markets. *Q. Rev. Econ. Financ.* **2017**, *66*, 169–181. [CrossRef]
27. Apolinario, R.M.C.; Santana, O.M.; Sales, L.J.; Caro, A.R. Day of the week effect on European stock markets. *Int. Res. J. Financ. Econ.* **2006**, *2*, 53–70.
28. Yamori, N.; Kurihara, Y. The day-of-the-week effect in foreign exchange markets: Multi-currency evidence. *Res. Int. Bus. Financ.* **2004**, *18*, 51–71. [CrossRef]
29. Kumar, S. Revisiting calendar anomalies: Three decades of multicurrency evidence. *J. Econ. Bus.* **2016**, *86*, 16–32. [CrossRef]
30. Johnston, E.T.; Kracaw, W.A.; McConnell, J.J. Day-of-the-Week Effects in Financial Futures: An Analysis of GNMA, T-Bond, T-Note, and T-Bill Contracts. *J. Financ. Quant. Anal.* **1991**, *26*, 23–44. [CrossRef]
31. Auer, B.R. Daily seasonality in crude oil returns and volatilities. *Energy Econ.* **2014**, *43*, 82–88. [CrossRef]
32. Blose, L.E.; Gondhalekar, V. Weekend gold returns in bull and bear markets. *Account. Financ.* **2013**, *53*, 609–622. [CrossRef]
33. Caporale, G.M.; Plastun, A. The day of the week effect in the cryptocurrency market. *Financ. Res. Lett.* **2019**, *31*, 258–269. [CrossRef]
34. Philpot, J.; Peterson, C.A. A brief history and recent developments in day-of-the-week effect literature. *Manag. Financ.* **2011**, *37*, 808–816. [CrossRef]
35. Tadepalli, M.S.; Jain, R.K. Persistence of calendar anomalies: Insights and perspectives from literature. *Am. J. Bus.* **2018**, *33*, 18–60. [CrossRef]
36. Ausloos, M. Statistical physics in foreign exchange currency and stock markets. *Phys. A Stat. Mech. Its Appl.* **2000**, *285*, 48–65. [CrossRef]
37. Matteo, T.D.; Aste, T.; Dacorogna, M. Scaling behaviors in differently developed markets. *Phys. A Stat. Mech. Its Appl.* **2003**, *324*, 183–188. [CrossRef]
38. Matia, K.; Ashkenazy, Y.; Stanley, H.E. Multifractal properties of price fluctuations of stocks and commodities. *Europhys. Lett.* **2003**, *61*, 422–428. [CrossRef]
39. Cajueiro, D.O.; Tabak, B.M. Multifractality and herding behavior in the Japanese stock market. *Chaos Solitons Fractals* **2009**, *40*, 497–504. [CrossRef]
40. Zunino, L.; Zanin, M.; Tabak, B.M.; Pérez, D.G.; Rosso, O.A. Complexity-entropy causality plane: A useful approach to quantify the stock market inefficiency. *Phys. A Stat. Mech. Its Appl.* **2010**, *389*, 1891–1901. [CrossRef]
41. Martina, E.; Rodriguez, E.; Escarela-Perez, R.; Alvarez-Ramirez, J. Multiscale entropy analysis of crude oil price dynamics. *Energy Econ.* **2011**, *33*, 936–947. [CrossRef]

42. Zhao, L.; Wang, G.J.; Wang, M.; Bao, W.; Li, W.; Stanley, H.E. Stock market as temporal network. *Phys. A Stat. Mech. Its Appl.* **2018**, *506*, 1104–1112. [CrossRef]
43. Bonanno, G.; Caldarelli, G.; Lillo, F.; Mantegna, R.N. Topology of correlation-based minimal spanning trees in real and model markets. *Phys. Rev. E* **2003**, *68*, 046130. [CrossRef] [PubMed]
44. Stosic, D.; Stosic, D.; Ludermir, T.B.; Stosic, T. Collective behavior of cryptocurrency price changes. *Phys. A Stat. Mech. Its Appl.* **2018**, *507*, 499–509. [CrossRef]
45. Kutner, R.; Ausloos, M.; Grech, D.; Matteo, T.D.; Schinckus, C.; Stanley, H.E. Econophysics and sociophysics: Their milestones & challenges. *Phys. A Stat. Mech. Its Appl.* **2019**, *516*, 240–253.
46. Kantelhardt, J.W.; Zschiegner, S.A.; Koscielny-Bunde, E.; Havlin, S.; Bunde, A.; Stanley, H. Multifractal detrended fluctuation analysis of nonstationary time series. *Phys. A Stat. Mech. Its Appl.* **2002**, *316*, 87–114. [CrossRef]
47. Pettengill, G.N. A survey of the Monday effect literature. *Q. J. Bus. Econ.* **2003**, *42*, 3–27.
48. Kohers, G.; Kohers, N.; Pandey, V.; Kohers, T. The disappearing day-of-the-week effect in the world's largest equity markets. *Appl. Econ. Lett.* **2004**, *11*, 167–171. [CrossRef]
49. Mehdian, S.; Perry, M.J. The Reversal of the Monday Effect: New Evidence from US Equity Markets. *J. Bus. Financ. Account.* **2001**, *28*, 1043–1065. [CrossRef]
50. Wang, K.; Li, Y.; Erickson, J. A new look at the Monday effect. *J. Financ.* **1997**, *52*, 2171–2186. [CrossRef]
51. Fishe, R.P.; Gosnell, T.F.; Lasser, D.J. Good news, bad news, volume, and the Monday effect. *J. Bus. Financ. Account.* **1993**, *20*, 881–892. [CrossRef]
52. Drożdż, S.; Kowalski, R.; Oświecimka, P.; Rak, R.; Gebarowski, R. Dynamical variety of shapes in financial multifractality. *Complexity* **2018**, *2018*, 7015721. [CrossRef]
53. Stosic, D.; Stosic, D.; Ludermir, T.B.; Stosic, T. Multifractal behavior of price and volume changes in the cryptocurrency market. *Phys. A Stat. Mech. Its Appl.* **2019**, *520*, 54–61. [CrossRef]
54. Stosic, T.; Nejad, S.A.; Stosic, B. Multifractal analysis of Brazilian agricultural market. *Fractals* **2020**, *28*, 2050076. [CrossRef]
55. Muzy, J.F.; Bacry, E.; Arneodo, A. Wavelets and multifractal formalism for singular signals: Application to turbulence data. *Phys. Rev. Lett.* **1991**, *67*, 3515–3518. [CrossRef]
56. Gu, G.F.; Zhou, W.X. Detrending moving average algorithm for multifractals. *Phys. Rev. E* **2010**, *82*, 011136. [CrossRef]
57. Oświęcimka, P.; Kwapień, J.; Drożdż, S. Wavelet versus detrended fluctuation analysis of multifractal structures. *Phys. Rev. E* **2006**, *74*, 016103. [CrossRef]
58. Figliola, A.; Serrano, E.; Rosso, O.A. Multifractal detrented fluctuation analysis of tonic-clonic epileptic seizures. *Eur. Phys. J. Spec. Top.* **2007**, *143*, 117–123. [CrossRef]
59. Amor, T.A.; Reis, S.D.; Campos, D.; Herrmann, H.J.; Andrade, J.S., Jr. Persistence in eye movement during visual search. *Sci. Rep.* **2016**, *6*, 20815. [CrossRef]
60. Shimizu, Y.U.; Thurner, S.; Ehrenberger, K. Multifractal spectra as a measure of complexity in human posture. *Fractals* **2002**, *10*, 103–116. [CrossRef]
61. Telesca, L.; Lovallo, M.; Mammadov, S.; Kadirov, F.; Babayev, G. Power spectrum analysis and multifractal detrended fluctuation analysis of Earth's gravity time series. *Phys. A Stat. Mech. Its Appl.* **2015**, *428*, 426–434. [CrossRef]
62. Telesca, L.; Lovallo, M.; Kanevski, M. Power spectrum and multifractal detrended fluctuation analysis of high-frequency wind measurements in mountainous regions. *Appl. Energy* **2016**, *162*, 1052–1061. [CrossRef]
63. Stošić, D.; Stošić, D.; Stošić, T.; Stanley, H.E. Multifractal analysis of managed and independent float exchange rates. *Phys. A Stat. Mech. Its Appl.* **2015**, *428*, 13–18. [CrossRef]
64. Jaffe, J.; Westerfield, R. The Week-End Effect in Common Stock Returns: The International Evidence. *J. Financ.* **1985**, *40*, 433–454. [CrossRef]
65. Zhou, W.X. The components of empirical multifractality in financial returns. *EPL Europhys. Lett.* **2009**, *88*, 28004. [CrossRef]
66. Drożdż, S.; Kwapień, J.; Oświecimka, P.; Rak, R. Quantitative features of multifractal subtleties in time series. *EPL Europhys. Lett.* **2010**, *88*, 60003. [CrossRef]
67. Zhou, W.X. Finite-size effect and the components of multifractality in financial volatility. *Chaos Solitons Fractals* **2012**, *45*, 147–155. [CrossRef]

entropy

Article

Learning Your Options: Option-Based Model of Export Readiness and Optimal Export

Kirill Ilinski [1,2]

[1] National Economics Research Center (NERC), St.-Petersburg University, 7-9-11 Universitetskaya Embankment, 199034 St. Petersburg, Russia; kni@fusionam.com
[2] Fusion Group, 8-10 Great George St., London SW1P 3AE, UK

Abstract: In this short note we offer a novel quantitative approach to modeling of early stages of firm's internalization, namely stages of accumulation of export readiness and their export debut. In particular, we introduce a new model of export readiness and offer an explicit way of how the export readiness can be accounted in the company share price. The model considers export readiness as a non-observable intangible asset that changes a firm's asset dynamics. This, in the framework of an option-based debt-equity Merton model, affects both the equity and debt of the company. The approach also allows one to define the contribution of export readiness to equity price and to find a self-consistent quantitative solution to the problem of optimal export strategy and the corresponding optimal firm's capital allocation.

Keywords: export readiness; internationalization; options pricing

Citation: Ilinski, K. Learning Your Options: Option-Based Model of Export Readiness and Optimal Export. *Entropy* **2022**, *24*, 173. https://doi.org/10.3390/e24020173

Academic Editors: Ryszard Kutner, Christophe Schinckus and H. Eugene Stanley

Received: 23 December 2021
Accepted: 20 January 2022
Published: 24 January 2022

1. Foreword

This article is written specially for the issue of Entropy, to commemorate 30 years of Econophysics, the discipline which appeared at the beginning of the 1990s at the cross-roads of economics, mainly finance, and theoretical physics, in particular many-body systems, thermodynamics, and phase transitions. It was driven primarily by physicists who, being generally curious about nature but not particularly educated about the specific field at the time, did not see any barriers to tackle almost any problem in finance they could think of. These would be, for example, the non-liner modeling of market prices, derivatives pricing, and non-equilibrium market dynamics, to name a few. This intellectual effort, coupled with a cheeky belief in technical superiority and nearly barbaric economic ignorance led to the situation when, in a relatively short period of time, truly cross-disciplinary problems, previously overlooked or simply perceived, hopelessly too complex, were posed and tackled.

To keep up with the tradition and in the general spirit of things, we do not want to present here just another paper on portfolio theory, generalized Sharpe ratios, or a new market forecasting technique—all the boringly routine daily subjects for finance professionals, the things the author actually thinks he now knows about. Instead, we aim to sketch here a new quantitative solution to an old problem in a field that is relatively new to the author. We consider the problem of transition to internationalization through export, export readiness, and forecasting of export success–the questions lying in the overlap of several economic fields, namely Theory of the Firm, International Trade and Finance, and Economics of Government Intervention. The proposed model utilizes objects and methods familiar to us from the field of quantitative finance, in particular capital structure modeling and derivatives pricing, which to our knowledge have not been used in this context before. In this way, we try to keep up with the original Econophysics tradition of solving problems we knew very little about until recently, using methods that earn us our daily keep.

2. Long Introduction—Setting Up the Context

2.1. Internationalization

Accelerated globalization driven by political development, falling trade barriers, development in shipping, and advances in technology, has resulted in nearly 6-times growth in international trade since the 1970s. This growth is not just proportional to GDP growth—international trade accelerated quicker and increased from 10% to 25% of GDP, being one of the drivers of the world's GDP increase.

Studies of how a firm undergoes internationalization—meaning how it expands sales from its own domestic market to some foreign markets, go back nearly 50 years and span a vast field of extensive research effort. We do not attempt here any sort of comprehensive review or introduction in the field and aim only to identify the main branches of the literature, especially in the context of the problem, which we address in this article. We will try to limit the list of citations in this short note to keep the balance but interested readers can find references in the articles for further reading. The most recent review of literature as well as another empirical study of link export readiness and export success can be found in the article [1].

Johanson and Vahlne in their seminal paper "The Internationalisation process of the firm: A model of knowledge development and increasing foreign commitments" (1977) [2] introduced what is now known as the original Uppsala model. The model views internationalization as a sequential process of firm development, from a pre-internationalization phase, to trial export, export through a partner, establishing a foreign subsidiary, and foreign manufacturing. In modern formulation, the model distinguishes a pre-export stage, "experimental" export with accumulation of experience, a committed exporter stage, and a full-integration (multi-national) stage. An updated Uppsala model (2009) [3] moved from individual interactions to interactions of economic agents inside a business network of contacts and a recursive learning process in this network. These are not the only model, of course—alternatives would include internationalization through business networks [4], "Born Globals" [5,6], model of "cultural distance" [7], among others. We, however, wanted to start with the Uppsala model because it clearly identified the pre-internationalization stage with its accumulation of key resources needed for a trial export as a battle ground to understand the process of crossing from non-exporting to an exporting firm.

The pre-internationalization stage can be viewed as a stage when a firm accumulates knowledge and resources to start an initial export [8]. The firm readies itself for export, i.e., it accumulates export readiness ([9] defines export readiness as preparedness and propensity to commence export). Multiple papers try to assess export readiness (as early as 1990, [10]) and to construct quantitative algorithms to estimate—find a number, an index—which would characterize export readiness of the firm, in hope that this number would define the corresponding success of future export.

It became clear quite early in the study that the firm's readiness to export should be assessed from two angles: Operational readiness and product readiness [11]. This is reflected now in some modern two-stage export readiness tests, which first analyzes general organizational readiness and then overlays it with a "particular product for particular market" analysis [12,13]. Popular research directions in this area include studies of key factors for export readiness, construction of qualitative and quantitative questionnaires, and an evaluation of different methodologies of digitalization of particular qualitative characteristics/answers. All these become inputs into the construction of multiple-item indices to measure export readiness. The indices then tested through logistic regression in a large sample of companies to separate exporters from non-exporters. The Holy Grail of export readiness would be to find an export readiness model that is quantitatively built from both objective and subjective information about the company, such that it would be able to predict future out-of-sample success of the export activity, as well as being able to identify particular problem areas that need to be addressed to increase the chance of this success. It would also be good if the model can time the crossing to export, adding a time dimension to the problem, and explaining how the threshold is reached and when. Indeed, the firm can be "accumulating the knowledge" but never actually becomes involved in the export. How does this jump actually happen? Only an export readiness model with elements of the dynamics can fully and self-consistently answer the question. This article is a step in the quest for the Holy Grail.

2.2. Million Dollar Question—Export Readiness

The problem seems very academic and almost artificial. However, there are a couple of very practical angles to it.

2.2.1. Government Support

First of all, governments, central and regional, try to support export. The importance of this is not purely economical but a socio-political one as well. Typically, it is done by creating specialized agencies whose main role is to educate the firms, promote international trade, and provide specific measures to stimulate firms to explore export sales. In Scotland, for example, the role is played by Scottish Development International and in South Africa it would fall to the Trade and Investment South Africa agency. Firms in Russia's Moscow region (actually comparable to a not-so-small country) can find support from ANO Mosprom, a specialized government agency whose role is to increase the export of Moscow region enterprises. The support has its cost, paid by taxpayers money, which has to be spent in the most efficient way. One of ways to define the measure of this efficiency, one of possible Key Performance Indicators for the government agencies, in this case would be the amount of additional export generated by the firms per one dollar spent by the agencies. Adding to this, a limit of maximal annual aid per company and requirement on the minimal number of the supported firms per year, we come to problems regarding the efficient selection of candidates for export stimuli. The solution has to be objective, transparent, and sufficiently simple to be explainable because nothing is more damaging to budget spending than implications of cronyism and corruption.

Finding a quantitative solution is not a trivial problem because it requires prediction of future export success, which is particularly difficult for companies that have not entered the export market yet and are only planning to do so. Unfortunately, this is also the most practically interesting case for the agencies since the highest marginal effect of government support measures comes from these companies, which usually belong to the Small and Medium (SME) sector of the economy. There are literally millions of companies. One has to have a quantitative screening process to limit this number of hopefuls to a manageable quantity, so that a handful of "expensive" experts can look further. In short, to efficiently distribute government resources allocated to the support, we need to estimate how ready a non-exporting company is for export, what the corresponding probability of export success is, and what the monetary consequences of the company's export would be. The latter question also requires an assumption about the internal optimal allocation of the firm's resources to domestic and export markets, which in turn, again requires probabilistic assumptions about export success.

2.2.2. Corporate Robo-Advice

Government agencies cannot help every company because of the associated costs involved in working with the candidates. Thus, they need to identify the candidates better (this is already covered above), but also to provide everybody else with a proxy service of export consulting, which does not require the same level of personal involvement from the agencies and can be done at virtually zero cost using online advice portals. This is what we call Corporate Robo-Advice, to underline the similarity with Financial Robo-Advice already widely available to individual retail customers whose assets are not sufficient to justify full face-to-face financial advice from specialized regulated financial advisory firms, or who are not prepared to pay full price for the advice. Instead, customers opt to receive proxy advice, generated online using risk profiling and asset allocation algorithms. Numerically, in developed economies, the numbers of potential clients of both Robo-advisories are actually quite similar and are sufficiently large. For example, in the UK number of people currently not receiving financial advice but wanting to receive it and one prepared to pay for it is close to 6 mln [14]. The number of active (employer) Small and Medium (SME) firms in the country is circa 1.4 mln, with the total number of private firms at about 5.5 mln [15]. The number of employer firms in the US would be close to 6 mln. Only the use of algorithms to provide tech-driven advice can help to close both "advice gaps".

In the case of corporate clients, many agencies are keen to develop online portals that can efficiently estimate the export potential of firms. These portals qualitatively and quantitatively define firms' export readiness and estimate the probability of export success, while at the same time provide concrete advice on improving areas of operations linked to all these quantities. Most export agencies, as well as some export-oriented banks (see for example, HSBC [16]), already have online resources with educational literature and simplified questionnaire-based models to estimate export readiness and, often, to highlight problem areas. These models are mostly qualitative, lack predictive power or statistical grounds, and play a primarily marketing role. The development of a full quantitative model of export readiness, forecasting of success probability, and the optimal export firm's exposure is a necessary step in providing a quantitative advisory to SME firms in the context of export. This includes identifying the most important factors, statistically testable functional dependencies on these factors, particular firm's shortfall in areas affecting these factors, and the most cost-efficient way to improve the outcome (so-called goal-oriented advice). This part of consulting should be an integral part of wider Corporate Robo-Advice including treasury advice, financial planning advice, budgeting, and hedging advice.

2.2.3. Probability of Successful Export as Export Readiness Index

A historically popular approach to evaluating export readiness was a construction of numerical index (ex. [9,17]) based on the digitization of answers to a particular export readiness questionnaire and complementing this index with a threshold number—if an index for a particular company was above the threshold, the company was defined as export ready. The selection of questions was driven by a selection of factors that were assumed to be particularly relevant for the internalization initiation. The index is usually calibrated on a mixed set of exporting and non-exporting companies. Parametric ansatz for the probability of future export success can be seen therefore as a particular choice, probably most logical, of an export readiness index. The model described below for the increasing export-preparedness of a company and the corresponding dynamics of its export debut can be seen as an alternative approach to the construction of export-readiness indices, being structural rather than phenomenological.

3. Challenges in Formulating the Model

Hopefully, by now the reader is convinced that the problem at hand is an interesting one to study. So why has it not been already solved, if so much research effort and practical investment has been already dedicated to it? This is because it is notoriously complex, partially due to its ill-definition, multiple possible scenarios of internationalization as well as internal firm's dynamics, and a multitude of factors affecting both the route and dynamics. On top of that, one of the most important factors in export readiness is the internal motivation of management and internal (read—cultural) specifics of the firm. A host of little issues can decide when, if at all, the firm decides to export. We argue here that in the context of multiple random or unknown factors, the dynamics of an export debut should be described by a stochastic process.

Before moving further, let us list common challenges which are facing every model of transition to export.

3.1. Challenge 1—What Is the Event?

Before we calculate a probability, we need first to define an event. We need to define the export success or a particular level of export efficiency which could be seen as a "success". Is it a single, first export transaction? Is it a particular percentage of total revenues of a company coming from export activities? Is it reaching a particular level of export intensity (say, 10%, commonly taken as a border between "export experiment" and "active exporters"; or may be 40% to count the firm as a "committed exporter")? Perhaps it is achieving a particular level of export efficiency? Or may be it is not quantitative at all and is defined by the perception of the firm's management of satisfactorily achieving their export goals (which also may not be purely quantitative, such as reputation of an international firm, personal ambition, protection against political prosecution, etc.). This is not an idle question—all that we have just brought up as examples are, in fact, actual measures. Ref. [18] documents 45 measures of export efficiency. It is complemented by another paper [19], 4 years later, bringing the total listed number of different measures to 50. A comprehensive literature review [20] lists 9 main categories of determinants of export performance and 36 main export performance measures (referencing literally hundreds of scientific publications). The criteria to select a particular measure of export efficiency and, therefore, a definition of export success, is dictated by a wider context of the problem for which one has to find the probability (for example, specific target set of Key Performance Indicators, or a target function to be optimized for a particular agency's development program). In practical terms, different definitions of export success will cause different calibrations of the same model on different information sets.

3.2. *Challenge 2—What Is the Time Horizon?*

We are looking for a probability of the event happening. Strictly speaking, this requires us to define a particular time window in which we observe firms to define whether the positive outcome has happened. What is this time window? Popular choices include 2 years and 5 years. Intuitively it seems that the dynamics of 2-year and 5-year windows are different, economically and functionally, and is led by potentially different factors. A reasonable model must describe this shift in relative importance of the factors, as well as potentially different functional dependencies on them. Basically, to have a self-consistent model for the export transition, we ideally need a model that would describe all windows, the whole term structure curve of probabilities of export success. Current Bayesian logit-linear models of construction of export readiness indices do not address this.

3.3. *Challenge 3—Why Do Different Firms with the Same Parameters Behave So Differently?*

Every firm is different—different corporate cultures, different styles of management, and a different speed of making corporate decisions (the corporate time). We can name so many various idiosyncratic factors that it is impossible, and also not actually desirable to account for them all. We are going to account for one of them—the corporate time, but will treat the rest in a reduced description approach, changing to a stochastic picture of internal firm's dynamics and response to external macro stimuli. In this approach, similar initial conditions will define similar statistical behavior rather than exact matched outcomes. In short, we aim to build a stochastic model of the first (successful) export event and will calculate the probability of a successful export as a result of this model.

3.4. *Challenge 4—What Is "Physical Meaning" of Export Readiness?*

Firm needs to become "export ready" before considering physical investment into resources to access export markets. What is this "export readiness"? Increasing export-readiness, simply according to its definition, makes a company prepared to export successfully. Successful export changes the dynamics of assets of the company, adding new channel for assets growth, which comes with its own associated risks. Therefore, export readiness can be defined as a characteristic that becomes a signal variable for the change of the asset growth process. It is an intangible asset of the company which is, mostly, not reflected in the balance sheet of the company but is vital to defining company dynamics and valuation.

Thinking of intangibles in the context of change in the parameters, or even nature, of company growth is not a new concept. Intangible assets such as skilled workforce, patents and know-how, unique organizational design and processes, even corporate culture represent valuable investments. Export readiness can be seen as a particular type of intangible capital which is required in a necessary quantity to initiate export activity. This is the approach we take in this paper.

Intangible assets are generally divided as intangible capital and intangible effort. Intangible capital is the stock of capital a company possesses, while intangible effort is the expenses spent on developing and maintaining intangible capital. Different accounting treatment and, as a consequence, different tax treatment dictates a recorded split of intangible assets and, for our purposes, obscures the economic picture that could be tested. If the intangible assets are estimated from the split of associated costs analyzing financial information of a company, it is easier to test a positive relationship between an investment in intangibles and export intensity [21–23], but it ignores the fact that a lot of things cannot be priced and are not charged for. Management motivation would be one example. Therefore, here we opted not to consider export-readiness in the resource-based view and firm-specific asset theory [24] and model it within the assets of the firm. Instead,

Proposition 1. *We see export-readiness as a stochastic variable that defines the asset process rather than a component of the assets.*

This is the main difference between our approach and the existing literature on the subject.

Thus, there is no standard way to measure a company's intangible capital because there is no a single accepted definition of intangibles. There are many ways to measure it (paper [25] found nearly 700 papers related to measurements of intangible capital). In general, they are split into cost-based and value-based concepts. However, even in the cost-based approach, there is no single agreed method to define intangible expenses and no standardized accounting method to account for them in financial reporting. In simple terms—it is not clear what you need to add to the assets in the balance sheet, so that you can use structural model for the firm valuation, based on the same model for asset dynamics but with re-defined assets. Therefore, we here take a view that export-readiness, R, is a special type of intangibles for which we define a process which, in turn, will affect the dynamics of the standard (accounting) assets A of the firm. The value of the firm will then be calculated, as in the Merton structural model, as a price of a call option on the assets with the firm's debt as a strike.

4. Quick Primer on the Structured Merton Model

Capital structure arbitrage models are a way to think about the relative pricing of debt and equity of a particular company. Everybody nowadays begin their introduction to the field with the Merton Model, for it is the simplest and most intuitive way to look at the matter. While the model, or rather the whole framework, is referred to as the Merton Model and his paper [26] is mostly cited in this regard, it is fair to add that Black and Sholes in their original paper [27] already considered corporate debt in the context of derivatives pricing. While there is an extensive body of literature on the capital structure models and various extensions and generalization of the Merton Model, we need here only a basic framework. Therefore, we will use its minimal set up, ignoring multiple complications and extensions (another 50 years of research).

Let us consider a company, ABC Limited. The company's balance sheet will show the balance (the clue is in the word) of assets of the company and its liabilities, i.e., means of how these assets are funded. On the one hand side there, are assets, everything which ABC possesses. This might include machinery, stock, patents, leases, furniture, cash, etc. The total value of these assets at time t we denote as $A(t)$.

On the other side, we have sources of capital that were used to finance these assets. These sources usually include debt D maturing, say, at time T, and the Book Equity Capital (Shareholder Funds). Book Equity balances the equation:

$$A(t) = D + BookEquity(t) \quad (1)$$

at any moment of time and contains the initial equity investment from shareholders, subsequent equity placing proceeds, and importantly, the accumulated Profit and Loss of previous periods. As a little clue where it is all going, if assets do not contain some intangibles, like export readiness, Book Equity will not reflect this either. The Market Price of Equity, or offered share price, on the other hand will. Therefore, we need to go from Book Equity to Market Price of Equity.

Since Debt is maturing only at time T in future, its current value is not D but less, and it depends on the probability of the company being able to repay the debt. The insight of the original authors cited above was that both the current value of the debt and current value of equity do not coincide with Book (balance sheet) values but are both derivatives of the current asset value. Indeed, if at time T when Debt matures, the value $A(T)$ is less than D then equity will be worthless and all assets will be sold to re-pay, as much as possible, the Debt:

$$D(T) = D - (D - A(T)) * \theta(D - A(T)) \equiv D - (D - A(T))^{+}.$$

Here $\theta(.)$ is the Heaviside function. In option pricing, this "payout" corresponds to cash D and a short put option on the firm's assets with strike D and maturity T.

At the same time, Equity would be equivalent to a call option:

$$E(T) = (A(T) - D) * \theta(A(T) - D) \equiv (A(T) - D)^{+}$$

with strike D and maturity T. To find values of both debt and equity for the company one has to use option pricing techniques which depend on the complexity of the asset dynamical process. If the process is a simple log-Brownian motion:

$$dA = A\sigma_A dW_A \quad (2)$$

then one can quickly get simple analytical formulae for prices of both corporate debt and equity. In a more general case, the prices are values of the payout functions averaged with the transition probability of the asset values (assuming for simplicity zero interest rates):

$$E(t) = \int_0^{\infty} dA(A - D)^{+} * P(A(t), t, A, T)$$

$$D(t) = \int_0^{\infty} dA(D - (D - A)^{+}) * P(A(t), t, A, T)$$

where $P(A(t), t, A, T)dA$ is the probability of the Asset value finishing in interval dA around A at time T conditional on the value of the asset being $A(t)$ at time t.

These prices will still satisfy the balance condition of equity plus debt to be equal to the value of assets, which in option world is known as Call-Put Parity. If, for some reason, prices of equity and debt change so that the Call-Put Parity brakes, it causes a "risk-less" profitable (arbitrage) trading opportunity, exactly as it happens in option trading. This arbitrage is called Capital Structure Arbitrage to reflect that it is caused by dis-balance between different parts of the company's capital structure.

Options that are used above are European vanilla options, meaning that their payouts are defined only by the value of the Asset at maturity. Black and Cox [28] removed this assumption by stating that for the company to avoid default, the barrier D should always remain un-breached, not only at maturity but also prior to maturity. This condition is called the American barrier, and it models an existence of loan covenants which, if breached, accelerate the debt repayments, thus bringing maturity forward. For our main purpose here, we will stick with the simplest Merton framework and will consider its basic European formulation, adding that all the usual refinements to the model (volatility skew, random barrier, complex debt profiles, etc.—see, for example [29]) can be added at a later stage.

5. Formulating the Model

5.1. Defining the Export Readiness Process

All studies of export-readiness first examine different factors affecting the export readiness. There are different taxonomies of the factors. The factors can be classified as intrinsic or external. They can be defined according to the mechanism of their action—for example, existing contacts with foreign partners, motivation of the management, sufficiency of financial resources, ability to manage risk, ability to modify its product, etc. These factors are typically selected by experts for a particular export-readiness model and are reflected in the corresponding questionnaire. Answers to the questions need to be digitized and the rule to combine the digital answers have to be defined. Examples of these workings can be found in [9,12,17]. The process is as much an art as it is a science and multiple trial-and-error iterations of the models have to appear before the model becomes operational. For our goals here we, however, will use a different classification—we will split factors between static (necessary to begin exporting) and dynamic (able to affect (increase or decrease) export readiness). For the combination of all static factors we will call export barriers, while the dynamic factors will be called export stimuli. Examples of components of export barriers would be: export licenses, knowledge of expected product support in export countries, ability to manage foreign exchange and interest rate risks etc. Examples of components of export stimuli would be company-sponsored foreign language lessons, government support, and education programs to increase awareness of foreign markets, management participation in industry networking events, etc. This classification does not remove the problem of building corresponding questionnaires and digitalization of qualitative answers but how it is done is not critically important for our subject here. It is enough for us to assume that all export barriers answers are digitized and combined in a total Export Barrier B. At the same time, the firm undertakes activities to increase/support/maintain export readiness while also fighting export readiness decay (example, people leave which reduces the expertise). The firm fights "Lateral rigidity" which (see [30]) is seen as one of the most important factors in export commencing. This results in change in export readiness per unit of time. The activity is reflected in the answers to the questionnaire. All Export Stimuli answers are digitized and combined, scaling for a unit of time, to obtain export readiness drift μ_R. Even if the definition is somewhat arbitrary, it has to be consistent across all the companies to allow for effective model calibration. Now, let us assume that, according to a particular questionnaire, the firm is distance B far from the export barrier and has export drift μ_R. The stochastic model for the export readiness R then takes the form: Initial $R(0) = 0$, export indicator $\xi = 0$:

$$dR = \mu_R dt + \sigma_R dW_R$$

and the export event ($\xi = 1$) is defined as R breaching the barrier B for the first time. Export-readiness volatility, the measure of uncertainty of the stochastic process, is a new parameter. This parameter characterizes measure of internal company dynamics—parameter $\frac{1}{\sigma^2}$ can be seen as a measure of internal company-specific time (different firms can have different speed of taking decisions, for example) as well as a measure of firm's susceptibility to external noise. Export readiness R is therefore affected by random noise and by the drift which results from combined Export Stimuli.

In this formulation, the export readiness is an unobservable, hidden variable which reflects an increase of expertise and other resources required to begin export. It does, however, have two important derivative quantities that depend on it and can be estimated directly: Probability of export success and equity value of export readiness.

5.2. Probability of Export Success

The model allows analytical expression for probability of successful export on different time horizons, being simply the probability of breaching the barrier B for different time windows. From these probabilities we can build a curve of export exits that is similar to the CDS curve in credit derivatives. We can also explicitly calculate a probability distribution of time of successful export entry as a probability distribution of first passage time in the described above barrier problem. Explicit formulae for both quantities can be found in any textbook on probability theory. In particular, the probability distribution of the first passage time, the function which we will use below, for fixed barrier B and drift μ_R is given by the following expression [31,32]:

$$\Pi(\tau) = \Pi_\tau = \frac{B}{\sigma_R \sqrt{2\pi}\tau^{3/2}} \exp[-\frac{(B - \mu_R \tau)^2}{2\sigma_R^2 \tau}] \tag{3}$$

while the corresponding survival probability can be written as:

$$\Pi_T = 1 - \int_0^T \Pi(\tau) d\tau = \Phi(\frac{B - \mu_R T}{\sigma_R \sqrt{T}}) - \exp(\frac{2B\mu_R}{\sigma_R^2})\Phi(\frac{-B - \mu_R T}{\sigma_R \sqrt{T}}) \tag{4}$$

where $\Phi(.)$ is Cummulative Error Function. Both types of the quantities can be used to calibrate the model parameters to the existing information set of exporting companies, particular questionnaires, and particular selection of measurements of export success. In this form, the model is also able to explain the relative importance of different factors on different time horizons, since the effects of volatility dominate on shorter time horizons while the drift is the defining factor in the long run.

We end this sub-section with a note on the further use of the model framework rather than the simplified model for R itself. As in the Merton model for credit default, it was long argued that, while the hindsight of the model is definitely valuable, the log-Brownian asset dynamics are too restrictive. It forces us, through model calibration, to use "wrong parameters in the wrong model". One of the approaches to estimate the probabilities of default was suggested by Vasicek and co-authors in the form of the KMV model [33], which, together with KMV Corporation, was acquired in 2002 and is included in services provided by Moody's analytics. The main role in this approach was played by the distance to default which in the option picture corresponds to the moneyness. Using the analogy here we can introduce Distance to Export as:

$$DE = \frac{B - \mu T}{\sigma_R^2 T}.$$

One can group companies according to the value of DE and plot probabilities of successful export as functions of DE. This functional form then substitutes of the Cumulative Error Function appearing in our simplified model and effectively corrects simplified the log-Brownian dynamic assumption. The model can be further expanded for the practical use applying the same technique as in the KMV model in the context of Export Readiness and substituting Distance to Default with Distance to Export.

5.3. Equity Value of Export Readiness

The model allows one to find "observable" equity value of un-observable export readiness. To this end we are to use the Merton model and see how the price of equity changes due to a possible change of asset dynamics if there is a possibility of a new export channel.

We saw above that equity price E of the company can be calculated as a price of call option on the assets A of the company. However, now, rather than to follow asset process (2) assets A of the company, ABC Limited will follow a modified Merton stochastic process with a switch from pure domestic to domestic+export dynamics triggered by export readiness variable R reaching the export barrier B.

Simplified Toy Model

In our toy model, we substitute a simplified assumption for the asset process (2):

$$dA = A\sigma_A dW_A$$

with a more complicated asset process:

$$dA = A\mu_A dt + A\sigma_A dW_A$$

where the parameters are defined as:

$$\mu_A = \mu_0(1-\xi) + \xi\mu_1,$$

$$\sigma_A = \sigma_0(1-\xi) + \xi\sigma_1.$$

Here μ_0 and σ_0 correspond to the company's evolution in a pure domestic market, and μ_1 and σ_1 correspond to evolution of the assets of the company if both domestic and export channels are used. The variable $\xi = \{0,1\}$ is the same signal variable already introduced in Section 5.1 in the context of dynamics of export readiness R. We also assume that processes W_A and W_R are independent, in particular that:

$$< dW_A, dW_R >= 0.$$

In this case, equity price can be calculated as:

$$E(t)_{\mu_R,B} = \int_0^{\infty} dA(A-D)^+ * P_R(A(t),t,A,T)$$

where the transition probability $P_R(A(t),t,A,T)$ accounts now also for the switch to export. Introducing τ as a first passage time (to export barrier B) one can see that $P_R(A(t),t,A,T)$ can be calculated as:

$$P_R(A(t),t,A,T) = \Pi_T P_0(A(t),t,A,T) + \int_0^T d\tau \Pi_\tau * \tilde{P}_0(A(t),t,A,T)$$

where Π_T is the probability of not touching the barrier from time t to time T (survival probability (4) with T substituted by $T-t$), Π_τ is the probability of first passage time being τ ((3) with T substituted by $T-t$), $P_0(A(t),t,A,T)$ is the log-normal transition probability distribution of $\frac{A}{A(t)}$ with parameters μ_0 and σ_0 and, finally, $\tilde{P}_0(A(t),t,A,T)$ is the log-normal transition probability distribution of $\frac{A}{A(t)}$ with parameters $\tilde{\mu}$ and $\tilde{\sigma}$:

$$\tilde{\mu} = \frac{\tau}{T}\mu_0 + \frac{T-\tau}{T}\mu_1,$$

$$\tilde{\sigma}^2 = \frac{\tau}{T}\sigma_0^2 + \frac{T-\tau}{T}\sigma_1^2.$$

These formulae give a semi-analytical solution for the equity price in the case of possible future exports. They also allow us to define the export readiness benefit to the shareholders, which is not reflected in the balance sheet of the company. The quantity, which we call the Export Readiness Benefit (ERB):

$$ERB = E(t)_{\mu_R,B} - E(t)_{\mu_R,B=\infty}$$

defines the monetary contribution of non-observable export readiness into the price of company equity.

5.4. Extended Model

A more realistic but unfortunately more complicated model can be built if we explicitly describe uncertainty in the domestic and in the export channels. In this case the asset process will take the form:

$$dA = A((1-\xi)[\mu_d dt + \sigma_{A,d} dW_{A,d}] +$$
$$\xi[(\omega_d(t,A)\mu_d + \omega_e(t,A)\mu_e)dt + \omega_d(t,A)\sigma_d dW_{A,d} + \omega_e(t,A)\sigma_e dW_{A,e}]).$$

Here we introduced two sets of parameters, with subscripts i and e, which correspond to domestic and export markets, together with a new element—weightings of capital allocations towards the domestic and export markets, ω_d and ω_e. Two Brownian motions, $W_{A,d}$ and $W_{A,e}$ describe the corresponding uncertainties in return from domestic and export markets. In general setup, all three Brownian motions, $W_{A,d}$, $W_{A,e}$, and W_R are mutually correlated. The complexity of the model is not only due to the increased dimensionality of the problem. It is also due to the dynamical nature of changes in the optimal capital allocation between foreign and domestic markets. The weights ω_d and ω_e have to be found self-consistently from the problem of optimization of a particularly selected utility function from the firm's equity value. This is a highly non-linear stochastic problem.

5.5. Self-Consistent Model for Optimal Export Strategy

This is still not the end of the whole story yet. The model for export-readiness process:

$$dR = \mu_R dt + \sigma_R dW_R$$

contains parameters that we have so far held constant. The company can decide to invest more (or less) into export readiness, spending some of the cash accounted in assets (thus adding negative drift into the asset process to account for spent cash) for change in the parameters of export readiness process—bringing more qualified staff, engaging a consulting company, and so forth. Luckily, the problem of the first passage time with time-dependent parameters has been solved by physicists [34] and some explicit formulae exist instead of the simplest expressions (3) and (4), bringing back physicists into the picture. Exchanging cash for change in the values of export barrier B and export stimuli μ_R is a management decision. This decision, once again, is driven by the same utility optimization problem. This makes the optimization problem even more complicated but now complete. The solution of the problem, which would give optimal spending on export readiness as well as optimal capital allocation weights, constitutes a self-consistent solution of an optimal export problem.

Let us pose the problem more formally. One has to choose the control functions to maximize the sharefolder value:

$$max_{(\omega_d(.,.),\omega_d(.,.),f(.,.))} E_T(A_0, B_0, \mu_{R,0}, \sigma_{R,0})$$

where

$$E(t) = \int_0^\infty dA(A-D)^+ * P_R(A(0),0,A,T)$$

and $P_R(A(0),0,A,T)$ is the transition probability for the asset process with the explicit "cash drain" term $f(t)$:

$$dA = A((1-\xi)[\mu_d dt + \sigma_{A,d} dW_{A,d} - f(t,A)dt] +$$
$$\xi[(\omega_d(t,A)\mu_d + \omega_e(t,A)\mu_e)dt + \omega_d(t,A)\sigma_d dW_{A,d} + \omega_e(t,A)\sigma_e dW_{A,e}]).$$

The signal variable ξ is defined, as before, by the export readiness process: Initial $\xi(0) = 0$ and becomes $\xi(\tau) = 1$ when the export readiness process $R(\tau)$ $(R(0) = 0)$:

$$dR = \mu_R(t)dt + \sigma_R(t)dW_R$$

is breaching the barrier $B(\tau)$ for the first time at the first passage time τ. Time-dependent parameters of the export readiness process then are functions of the "cash drain", which we take for simplicity to be linear:

$$d\mu_R(t) = m * f(t, A)Adt\,,\; d\sigma_R(t) = s * f(t, A)Adt\,,\; dB_R(t) = b * f(t, A)Adt$$

with some company-specific efficiencies constants m,s,b.

The solution to the combined problem is not "one fits all" as internal company specifics, the internal cost of changing export readiness parameters, and internal return profiles from domestic and export activity depends on a particular company. Solving this problem opens the way to a quantitative selection criteria for government agency early export support, which we highlighted in the Introduction.

6. Conclusions

In this short note we sketched a new approach to modeling export readiness dynamics and posed the problem of finding an optimal firm's strategy of export debut. While the model is quite complex and requires a combination of analytical and numerical studies, it builds a qualitative and intuitive picture of the transition to export dynamics. The most labor-intensive component of further work is to construct a qualitative questionnaire and the corresponding quantitative digitalization algorithms to estimate the model parameters Export Barrier B, Export Stimuli μ_R, and internal volatility σ_R for different types of companies. As the size of the company can be one of quantitative factors affecting export readiness, it is possible that B, μ_R, and σ_R will be asset-dependent, which will further increase the non-linearity of the problem, causing multiple equilibria and conditional instability typical for this type of complex systems, bringing it even closer to problems studied by Econophysics.

Funding: This research received no external funding.

Data Availability Statement: Not applicable.

Acknowledgments: The author would like to thank their colleagues from AO National Economics Research Center of St Petersburg State University for introducing the author to the subject. The author would like to express their gratitude to Maxim Bouev, Alexander Gurevich, and Nadezhda Ivannik for their many fruitful discussions, as well as to Igor Makarov for his help in preparing this article.

Conflicts of Interest: The author declares no conflict of interest.

References

1. Gerschewski, S.; Scott-Kennel, J.; Rose, E.L. Ready to export? The role of export readiness for superior export performance of small and medium-sized enterprises. *World Econ.* **2020**, *43*, 1253–1276. [CrossRef]
2. Johanson, J.; Vahlne, J. The Internationalization process of the firm: A model of knowledge development and increasing foreign commitments. *J. Int. Bus. Stud.* **1977**, *8*, 23–32. [CrossRef]
3. Johanson, J.; Vahlne, J. The Uppsala internationalization process model revisited: From liability of foreignness to liability of outsidership. *J. Int. Stud.* **2009**, *40*, 1411–1431. [CrossRef]
4. Lee, H.; Kelley, D.; Lee, J.; Lee, S. SME survival: The impact of internationalization, technology resources, and alliances. *J. Small Bus. Manag.* **2012**, *50*, 1–19. [CrossRef]
5. Johanson, J.; Mattson, L.-G. Internationalisation in industrial systems—A network approach. In *Strategies in Global Competition*; Hood, N., Vahlne, J.-E., Eds.; Croom Helm: New York, NY, USA, 1988; pp. 287–314.
6. Weerawardena, J.; Mort, G.S.; Liesch, P.W.; Knight, G. Conceptualizing accelerated internationalization in the born global firm: A dynamic capabilities perspective. *J. World Bus.* **2007**, *42*, 294–306. [CrossRef]
7. Nes, E.B.; Solberg, S.A.; Silkoset, R. The impact of national culture and communication on exporter-distributor relations and on export performance. *Int. Bus. Rev.* **2007**, *16*, 405–424. [CrossRef]
8. Tan, A.; Brewer, P.; Liesch, P. Before the First Export Decision: Internationalisation Readiness in the Pre-Export Phase. *Int. Bus. Rev.* **2007**, *16*, 294–309. [CrossRef]
9. Tan, A.; Brewer, P.; Liesch, P. Measuring export readiness using a multiple-item index. In *Proceedings of the 2010 European International Business Academy Conference*; Tavares-Lehmann, A., Ed.; European International Business Academy: Brussels, Belgium, 2010; pp. 1–33.

10. Cavusgil, S.T.; Nason, R.W. Assessment of company readiness to export. In *International Marketing Strategy*; Thorelli, H.B., Cavusgil, S., Eds.; Pergamon Press: Oxford, UK, 1990; pp. 129–139.
11. Cavusgil, S.T. On the internationalisation process of firms. *Eur. Res.* **1980**, *8*, 273–281.
12. Bouev, M.; Gurevich, A.; Ivannik, N.; Ilinski, K. *Export Readiness Model "ADEPT 7"*, Technical Report; NERC: St. Petersburg, Russia, 2020.
13. David, J.P.; Cariou, G. Evaluating the Firm's Readiness for Internalization: From the Design to the Application of an International Qualification Framework. *Int. J. Bus. Manag.* **2014**, *9*, 1–9. [CrossRef]
14. Open Money Report. Available online: https://www.open-money.co.uk/advice-gap-2021 (accessed on 19 December 2021).
15. National Federation of Self-Employed and Small Businesses. Available online: https://www.fsb.org.uk/uk-small-business-statistics.html (accessed on 19 December 2021).
16. HSBC Export Readiness Tool. Available online: https://www.business.hsbc.uk/en-gb/campaigns/export-resource-centre/tools-and-resources (accessed on 19 December 2021).
17. Van Eldik, S.; Viviers, W. The measurement of export readiness of companies in South Africa. *S. Afr. Bus. Rev.* **2005**, *9*, 1–11.
18. Katsikeas, C.S.; Leonidou, L.C.; Morgan, N.A. Firm-level export performance assessment: Review, evaluation and development. *J. Acad. Mark. Sci.* **2000**, *28*, 493–511. [CrossRef]
19. Sousa, C.M.P. Export performance measurement: An evaluation of the empirical research in the literature. *Acad. Mark. Sci. Rev.* **2004**, *9*, 90–95.
20. Beleska-Spasova, E. Determinants and measures of export performance: Comprehensive literature review. *J. Contemp. Econ. Bus. Issues* **2014**, *1*, 63–74.
21. Amadieu, P.; Maurel, C.; Viviani, J.-L. Intangibles, Export Intensity, and Company Performance in French Wine Industry. *J. Wine Econ.* **2013**, *8*, 198–224. [CrossRef]
22. Kotha, S.; Rindova, V.P.; Rothaermel, F. Assets and action: Firm-specific factors in the internationalization of U.S. Internet firms. *J. Int. Bus. Stud.* **2001**, *32*, 769–791. [CrossRef]
23. López Rodríguez, J.; García Rodríguez, R.M. Technology and export behaviour: A resource-based view approach. *Int. Bus. Rev.* **2005**, *14*, 539–557. [CrossRef]
24. Ruppenthal, T.; Bausch, A. Research on export performance over the past 10 years: A narrative review. *Eur. J. Int. Manag.* **2009**, *3*, 328–364. [CrossRef]
25. Marr, B.; Gray, D.; Neely, A. Why do firms measure their intellectual capital? *J. Intellect. Cap.* **2003**, *4*, 441–464. [CrossRef]
26. Merton, R.C. On the Pricing of Corporate Debt: The Risk Structure of Interest Rates. *J. Financ.* 1974, *29*, 449–470.
27. Black, F.; Scholes, M. Pricing of Options and Corporate Liabilities. *J. Political Econ.* **1976**, *81*, 637–654. [CrossRef]
28. Black, F.; Cox, J.C. Valuing Corporate Securities: Some Effects of Bond Indenture Provisions. *J. Financ.* **1976**, *31*, 351–367. [CrossRef]
29. Schonbucher, P.J. *Credit Derivatives Pricing Models: Models, Pricing and Implementation*; Wiley: Hoboken, NJ, USA, 2003.
30. Tan, A.; Brewer, P.; Liesch, P. Rigidity in SME export commencement decisions. *Int. Bus. Rev.* **2018**, *27*, 46–55. [CrossRef]
31. Cox, D.R.; Miller, H.D. *The Theory of Stochastic Processes*; Chapman & Hall, CRC: Boca Raton, FL, USA, 1965.
32. Feller, W. *An Introduction to Probability Theory and Its Applications*, 3rd ed.; Wiley: Hoboken, NJ, USA, 1971; Volume 2.
33. Vasicek, O.A. Credit Valuation. KMV Corporation. 1984. Available online: http://www.ressources-actuarielles.net/EXT/ISFA/1226.nsf/0/c181fb77ee99d464c125757a00505078/%24FILE/Credit_Valuation.pdf (accessed on 19 December 2021).
34. Molini, A.; Talkner, P.; Katul, G.G.; Porporatoa, A. First passage time statistics of Brownian motion with purely time dependent drift and diffusion. *Physica A* **2011**, *390*, 1841–1852. [CrossRef]

entropy

MDPI

Article

Multifractal Company Market: An Application to the Stock Market Indices

Michał Chorowski * and Ryszard Kutner

Faculty of Physics, University of Warsaw, Pasteur Str. 5, PL-02093 Warsaw, Poland; ryszard.kutner@fuw.edu.pl
* Correspondence: ma.chorowski@student.uw.edu.pl

Abstract: Using the multiscale normalized partition function, we exploit the multifractal analysis based on directly measurable shares of companies in the market. We present evidence that markets of competing firms are multifractal/multiscale. We verified this by (i) using our model that described the critical properties of the company market and (ii) analyzing a real company market defined by the *S*&*P* 500 index. As the valuable reference case, we considered a four-group market model that skillfully reconstructs this index's empirical data. We point out that a four-group company market organization is universal because it can perfectly describe the essential features of the spectrum of dimensions, regardless of the analyzed series of shares. The apparent differences from the empirical data appear only at the level of subtle effects.

Keywords: multiscale partition function; multifractal analysis; company market

PACS: 89.65 Gh; 05.40.-a; 89.75.Da

Citation: Chorowski, M.; Kutner, R. Multifractal Company Market: An Application to the Stock Market Indices. *Entropy* **2022**, *24*, 130. https://doi.org/10.3390/e24010130

Academic Editor: Stanisław Drożdż

Received: 30 December 2021
Accepted: 12 January 2022
Published: 16 January 2022

1. Introduction

In the last two decades, multifractal properties have been the subject of intense research in very different areas of science [1–13]. The fashion for searching for new areas of multifractality is still ongoing. The shape, location, and spread of the spectrum of dimensions (singularities)—the leading multifractality indicator—provide invaluable information about the layout. We use the formalism [14] that describes not only systems in the state of statistical equilibrium but also stationary states. Furthermore, we indicate that formalism can easily be extended to transient states.

Our approach is complementary to the commonly used multifractal detrended fluctuation analysis (MF-DFA) [1,2]. More precisely, in the presence of state intervention, our concept of using (normalized) market shares for multifractal analysis of the market of competing firms is new. It starts with a partition function expressed directly by shares. Thanks to this, it bypasses the onerous preparation of traditional MF-DFA, based on a fluctuation function built with the help of time series.

We demonstrate how our method works with the example of a competing company market model published previously [15]. In this model, we assume that companies can merge, create spin-offs, and go bankrupt in the presence of state intervention. This tendency for firms to disappear from the market can counterbalance the tendency to design firms, leading to critical phenomena. We examined these phenomena in our previous work [15]. In this work, we explore a different aspect of the market model of competing companies, namely, multifractality.

Moreover, we show that the actual market of *S*&*P* 500 companies is multifractal. Finally, we indicate that this market can be (roughly) described by the multifractal formalism, in which companies are divided into four groups differing significantly in market shares.

The paper consists of two parts. The first part consists of Section 1 (Introduction) together with Section 2 (Theory), which on the example of our critical company market model [15] presents the multifractal approach. The second part presents this multifractal approach to the real market of the $S\&P$ 500 index. Moreover, this part compares the obtained results for the actual market with the four-group market model.

2. Theory

2.1. Definition of Partition Function

The multifractal behavior of the market of competing firms is a new concept. We based this concept on the characteristic for this market, the partition function given by the formula [14]

$$\mathcal{Z}(\beta) = \sum_{n=1}^{N} \omega_n^{\beta}, \tag{1}$$

where ω_n is the (normalized) market share of firm n, while N is the number of firms in the market; both a priori given quantities we can obtain, at a given time, from simulations, empirical data, or from theory.

We characterize the market shares of companies using the Quetelet ranking (see Figure 1), i.e., we build a plot of cumulative distribution function (CDF) versus company share value taken from simulation within our model.

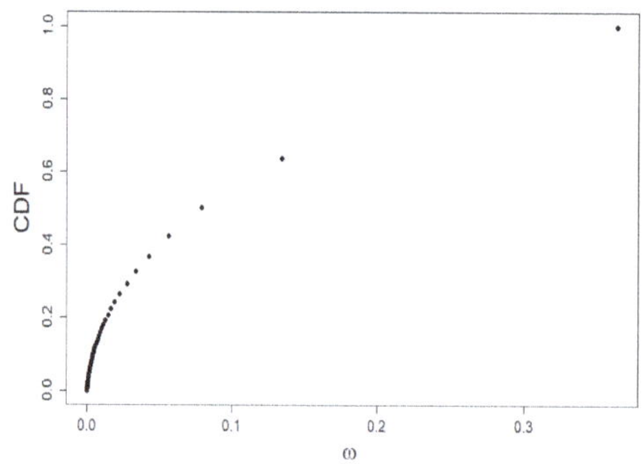

Figure 1. Quetelet curve: the dependence of the standardized rank of companies generated within our model, i.e., CDF, on their shares ω. It is precisely to analyze this simulation data that we use multifractal formalism.

The partition function in the form given by Equation (1) is ready to study the multiscale nature of the ω distribution. This multiscale nature comes from the hierarchical distribution of firms' sizes.

In this section, we limit ourselves to systems in steady states; therefore, we assume that $N = N_{st}$. Recall that in our model N_{st} is clearly related to the level of intervention $0 \leq q \leq 1$, its effectiveness $0 \leq \eta \leq 1$, and the company's activity $0 \leq \lambda \leq 1$ [15]. Figure 2 shows a typical relationship N_{st} vs. q with η (=0.5) and λ (=0.9) fixed. The location of the q_c criticality threshold is clearly visible, signaling a continuous phase transition.

The partition function, $\mathcal{Z}(\beta)$, obeys two basic properties,

$$\mathcal{Z}(\beta = 0) = N, \tag{2}$$

and

$$\mathcal{Z}(\beta = 1) = 1. \tag{3}$$

Of course, Equation (2) describes the size of the multifractal substrate or company market, while Equation (3) comes from the normalization condition of shares.

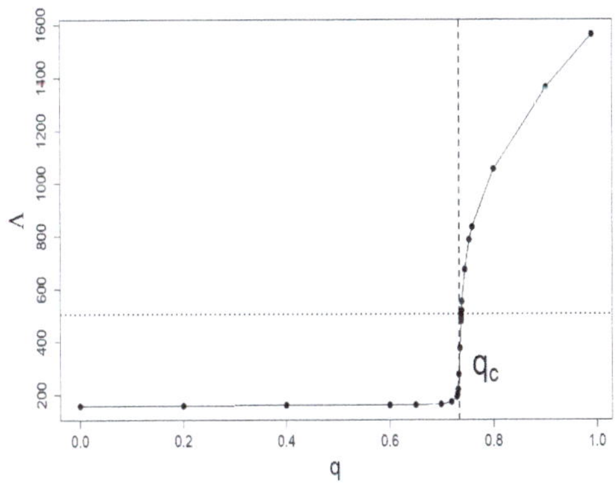

Figure 2. The typical dependence of Λ (=N_{st}) vs. interventionism level q at fixed η (=0.5) and λ (=0.9). It is a flat phase diagram where a continuous phase transition is clearly visible at q_c (=0.734). All other plots in this section have the same η and λ parameters as this plot.

Moreover, using the share limitation from below and above, we get

$$\mathcal{Z}(\beta \to \mp\infty) \approx \begin{cases} (\omega^{\min})^{\beta}, & \beta < 0 \\ (\omega^{\max})^{\beta}, & \beta > 0 \end{cases} \tag{4}$$

where $\omega^{\min}$ and $\omega^{\max}$ determine the marginal values of the companies' market shares.

2.2. Scaling Relations

We continue to show that the partition function $\mathcal{Z}(\beta)$ takes the form of a power law,

$$\mathcal{Z}(\beta) = \Lambda^{-\tau(\beta)} \Leftrightarrow \tau(\beta) = -\frac{\ln \mathcal{Z}(\beta)}{\ln \Lambda}, \tag{5}$$

where $\tau(\beta)$ is the scaling exponent, while the base/scale Λ we define below. Having the partition function at our disposal, we can build a thermodynamic formalism on this basis. We talk more about it in Section 2.5, where we calculate a specific heat.

To prove the correctness of the first equality Equation (5), we use two crucial scaling exponent properties,

$$\tau(\beta) = (\beta - 1)D(\beta) \tag{6}$$

where $D(\beta) \geq 0$ is the Rényi dimensions and

$$\tau(\beta) = \beta h(\beta) - D(\beta = 0), \tag{7}$$

here $h(\beta)$ is a generalized Hurst exponent and $D(\beta = 0)$ is the Hausdorff dimension of the substrate/market, which for our case we can put to 1.

For $\beta \to 1$ the Rényi information approaches the Shannon information that is, it becomes the information dimension,

$$D(\beta = 1) = -\frac{1}{\ln N}\sum_{n=1}^{N} \omega_n \ln \omega_n. \tag{8}$$

For $\beta \to 2$ the partition function (1) reduces to the well-known correlation integral $C(N)$ of Grassbereger and Procaccia [16], i.e.,

$$D(\beta = 2) = -\frac{\ln C(N)}{\ln N}. \tag{9}$$

Furthermore, let us also note that always $D(\beta') \leq D(\beta)$ for $\beta < \beta'$.

Now, we can define basis Λ. We use Equation (5) for this purpose, in which we put $\beta = 0$ followed by Equations (2) and (7). Therefore, we get $\Lambda = N$.

The above result, in combination with the scaling Equation (5), allow us to present the scaling exponent in an explicit asymptotic form,

$$\tau(\beta \to \mp\infty) \approx \begin{cases} -\beta \frac{\ln \omega^{\min}}{\ln N}, & \beta < 0 \\ -\beta \frac{\ln \omega^{\max}}{\ln N}, & \beta > 0. \end{cases} \tag{10}$$

With the above results, we can now present a plot of $\tau(\beta)$ vs. β—this plot and its enlarged version limited to the central values of β (from the range of $[-1.5, 1.5]$), are presented in Figure 3. As one can see, $\tau(\beta)$ is bounded by two diagonal asymptotes defined by Equation (10).

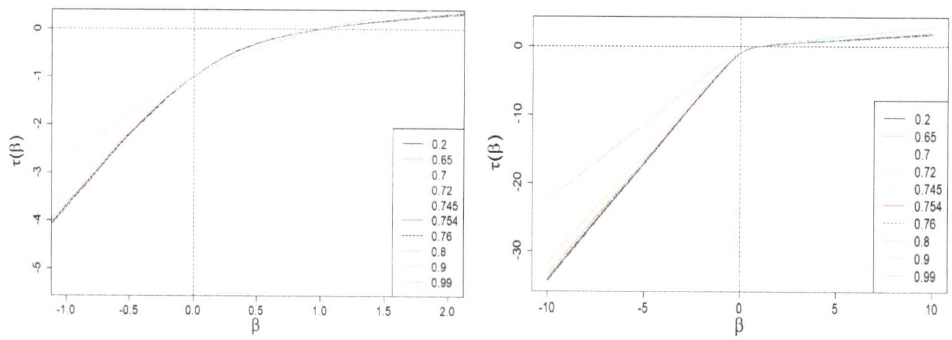

Figure 3. Scaling exponent $\tau(\beta)$ vs. exponent β (the order of scale). Its nonlinear/multifractal behavior in the range of $\beta \in [-1.0, 2.0]$ for interventionism level $0 < q < 1$ is clearly seen (especially on the zoomed plot). On the other hand, the plot on the right shows the existence of oblique asymptotes. Multifractality is present if and only if they are different from each other. For example, we have selected ten characteristic levels of interventionism here (see the legend). The sharp decrease in the slope difference of the asymptotes for $q \approx 1$ (blue dashed curves) is visible. We use the same set of q values in all plots in Section 2.

We consider the next two extreme cases. The first, is when all but one of the company shares disappear (the case of a monopolized market). Then, with Equations (1), (3) and (5), we get immediately that $\tau(\beta)$ is undefined.

The second case is when all shares are equal (the case of the egalitarian market), i.e., $\omega_n = \frac{1}{N}$, $n = 1, 2, \ldots, N$. Then, with Equations (1), (3) and (5), we get

$$\tau(\beta) = \beta - 1, \tag{11}$$

i.e., the scaling exponent is a linear function of β. We continue to deal mainly with cases distant from both of the above extreme cases.

We assume that company shares, ω_n, create the nonuniform/multiscale function ω_n vs. n, a multifractal structure. In other words, we are dealing here with multifractality, the source of which is the heterogeneous distribution of company shares.

2.3. Rényi Dimensions and Generalized Hurst Exponent

In Figures 4 and 5, we present the Rényi dimensions, $D(\beta)$, generalized Hurst exponent, $h(\beta)$, and their spans $\Delta D(\beta) = D(-\beta) - D(\beta)$ and $\Delta h(\beta) = h(-\beta) - h(\beta)$, respectively. The former two quantities are limited by identical horizontal asymptotes:

$$\begin{aligned} & D(\beta \to \mp\infty) = h(\beta \to \mp\infty) \\ = & \begin{cases} D^{\max} = h^{\max} = -\frac{\ln \omega^{\min}}{\ln N}, \beta < 0, \\ D^{\min} = h^{\min} = -\frac{\ln \omega^{\max}}{\ln N}, \beta > 0, \end{cases} \end{aligned} \tag{12}$$

while

$$\begin{aligned} & \Delta D(\beta \to \infty) = D^{\max} - D^{\min} \\ = & \ \Delta h(\beta \to \infty) = h^{\max} - h^{\min} \\ = & \ \ln\left(\frac{\omega^{\max}}{\omega^{\min}}\right). \end{aligned} \tag{13}$$

Equations (12) and (13) are a direct result of the asymptotic scaling exponent properties given by Equation (10) and by Equations (6) and (7), respectively.

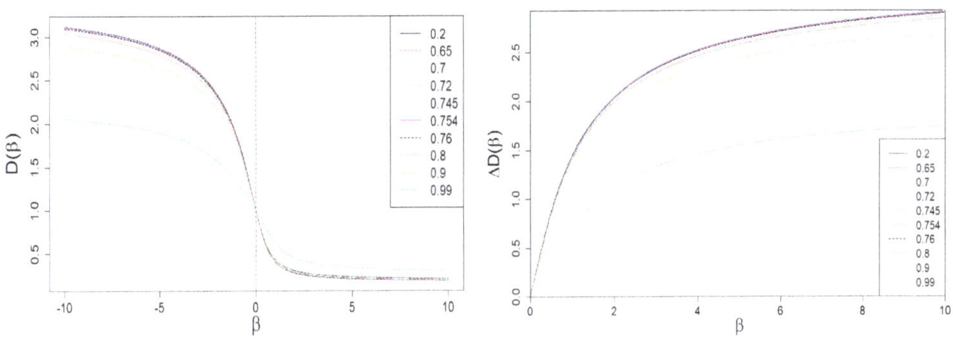

Figure 4. Dependence of Rényi dimensions D on β. A sharp drop in the $\Delta D(\beta)$ span is clearly visible on the right plot for large values of $|\beta|$ and $q \approx 1$ (blue dashed curve). This is the result of the behavior of the $\tau(\beta)$ vs. β curve shown in Figure 3.

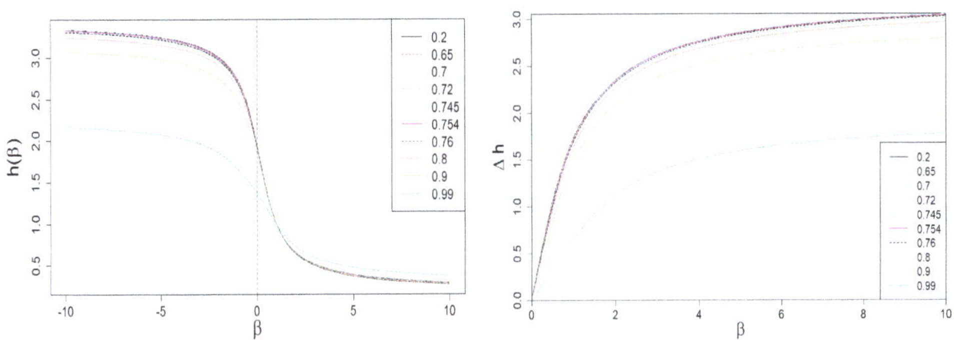

Figure 5. The dependence of the generalized Hurst exponent h and its span Δh on β. A sharp drop in the $\Delta h(\beta)$ span is clearly visible for large values of $|\beta|$ and $q \approx 1$ (blue dashed curve). It is the result of the behavior of the $\tau(\beta)$ vs. β curve shown in Figure 3.

2.4. Spectrum of Dimensions

We now designate the most crucial multifractality signature, i.e., the spectrum of dimensions (singularities), f, given by the Legendre transformation,

$$f(\alpha) = \beta(\alpha)\alpha - \tau(\beta(\alpha)), \tag{14}$$

where the local dimension (singularity or Hölder exponent) is

$$\alpha(\beta) = \frac{d\tau(\beta)}{d\beta} = -\frac{1}{\ln N}\frac{\sum_n \omega_n^{\beta} \ln \omega_n}{\sum_n \omega_n^{\beta}}. \tag{15}$$

Therefore, we obtain a helpful equality locating the maximum spectrum of dimensions $f(\alpha(\beta = 0))$,

$$\alpha(\beta = 0) = -\frac{1}{N \ln N}\sum_n \ln \omega_n. \tag{16}$$

and we get, analogously as in Equation (12),

$$\alpha(\beta \to \mp\infty) \approx \begin{cases} \alpha^{\max} = -\frac{\ln \omega^{\min}}{\ln N}, \\ \alpha^{\min} = -\frac{\ln \omega^{\max}}{\ln N}. \end{cases} \tag{17}$$

As one can see from Equation (12), the quantities D, h, and α have the same lower and upper bounds.

Furthermore, from Equations (14) and (15) we get

$$\beta = \frac{df(\alpha)}{d\alpha}. \tag{18}$$

In Figure 6, we present the dependence of local exponent α and its span $\Delta\alpha$ on β.

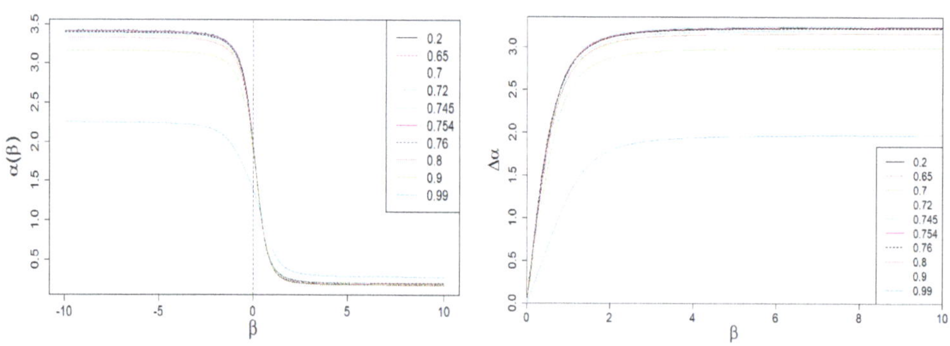

Figure 6. Dependence of the local singularity α on β. A sharp drop in the $\Delta\alpha(\beta)$ span is clearly visible on the right plot for large values of $|\beta|$ and $q \approx 1$ (blue dashed curve). This is the result of the behavior of the $\tau(\beta)$ vs. β curve shown in Figure 3.

In Figure 7, we present the dependence of the local singularity span $\Delta\alpha$ on q at fixed $\beta = 5.0$.

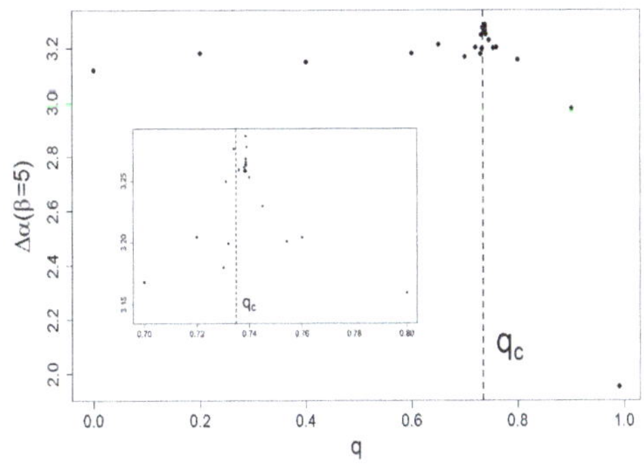

Figure 7. Dependence of the local singularity span $\Delta\alpha$ on q at fixed $\beta = 5$. A slight but distinct peak locates near $q_c = 0.735$, which defines the criticality threshold used by us at earlier work [15]. We also included a magnification of this peak.

Figure 8 shows he dependencies of f on α and on β. The $\alpha(\beta)$ vs. β plot (like $D(\beta)$ and $h(\beta)$ vs. β ones) is limited by two horizontal asymptotes given by Equation (17). This is a direct result of the asymptotic properties of Equation (10).

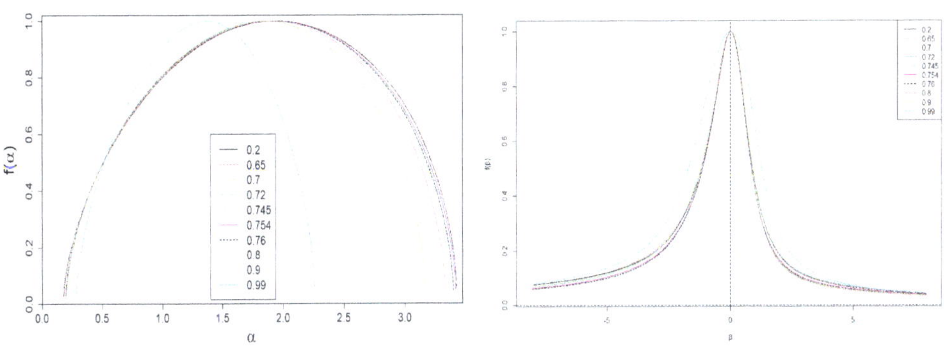

Figure 8. Dependence of spectrum of dimensions, f, from α (left plot) and β (right plot). There is a visible nonlinear dependence of the shape f on the level of interventionism q. Moreover, there is a wide spread in the spectrum of singularities $\Delta\alpha$. As expected, the same applies to the dependence of f on β. In addition, there is a slight asymmetry of f, i.e., $\gamma > 0$, herein.

We present below useful quantities, which characterize the spectrum of singularities:

(i) $f^0 = f(\alpha(\beta = 0) = \alpha^0) = D^0 = D(\beta = 0)$, which results from Equations (7), (14) and (18), and moreover we get $\frac{df(\alpha)}{d\alpha}|_{\beta=0} = 0$;

(ii) for $\beta = 1$ we immediately get from Equation (18) $\frac{df(\alpha)}{d\alpha}|_{\beta=1} = 1$, therefore $f^1 = f(\alpha(\beta = 1)) = \alpha(\beta = 1) = \alpha^1$;

(iii) for $\beta \to -\infty$ we get from Equations (14) and (15), that $f^{\max} = f\left(\alpha = \alpha^{\max} = -\frac{\ln \omega^{\min}}{\ln N}\right) = 0$ and $\frac{df(\alpha)}{d\alpha}|_{\beta\to-\infty} = -\infty$; similarly for $\beta \to \infty$ we get $f^{\min} = f\left(\alpha = \alpha^{\min} = -\frac{\ln \omega^{\max}}{\ln N}\right) = 0$ and $\frac{df(\alpha)}{d\alpha}|_{\beta\to\infty} = \infty$;

(iv) the maximum span of f we determine as follows, $\Delta\alpha|_{|\beta|\to\infty} = \alpha^{\max} - \alpha^{\min} = \frac{1}{\ln N} \ln\left(\frac{\omega^{\max}}{\omega^{\min}}\right)$. We continue to use the simplified designation $\Delta\alpha = \Delta\alpha|_{|\beta|\to\infty}$;

(v) the following asymmetry factor can be used to determine the degree of asymmetry f, $\gamma|_{|\beta|\to\infty} = \frac{\alpha(\beta=0)-\alpha^{min}}{\alpha^{max}-\alpha(\beta=0)}$, where $\alpha(\beta = 0)$ is given by Equation (16). We continue to use the simplified designation $\gamma = \gamma|_{|\beta|\to\infty}$.

It should be emphasized that in general $f(\alpha = \alpha^{min}, \alpha^{max}) \neq 0$. This happens when at least one of the boundary values $\omega^{min}, \omega^{max}$ is degenerated. This is discussed in Section 3.2 for a four-group company market model.

The large span $\Delta\alpha$ visible in Figure 8 indicates a great volatility of competing firms on the market. At the same time, we deal with a wide variety of companies only when it also occurs that $N \gg 1$. However, the shift of the spectrum of dimensions to higher values of α signals the dominance of smaller companies on the market. Let us note that we would deal with a weak multifractality if and only if the span $\Delta\alpha \ll 1$.

One can also analyze asymmetry of f using the coefficient γ. If $\gamma > 1$, then we are talking about the advantage on the market of large companies, as opposed to the situation of $\gamma < 1$. The marginal case $\gamma = 1$ corresponds to the balanced situation.

2.5. Specific Heat

We can now define the specific heat c of the system/market on the reciprocal of the temperature β, as follows [4,14,17]:

$$\begin{aligned} c(\beta) &= -\beta^2\left(\frac{\partial^2(\beta F/V)}{\partial\beta^2}\right)_V \\ &= \frac{1}{\ln N}\beta^2\left(\frac{\partial^2 \ln \mathcal{Z}}{\partial\beta^2}\right)_N, \end{aligned} \tag{19}$$

where $\frac{1}{V}\beta F = -\frac{1}{\ln N}\ln \mathcal{Z}$, while F is the free energy of a company market, and $V = \ln N$ here.

The dependence of $c(\beta)$ on β is presented in Figure 9. Apparently, this dependence is anomalous (both for positive and negative values of β) because it has a local peak, analogous to the Schottky peak for the specific heat of the solid [18,19] related to its internal degrees of freedom. Let us add that the disappearance of $c(\beta = 0)$ in $\beta = 0$ results directly from the second formula (19). Such clear peaks are the result of highly differentiated values of the shares, ω_i, that define partition function $\mathcal{Z}$. They play the role of internal degrees of freedom here. We proof that $\mathcal{Z}$ composed of only two different shares ω^{max} and ω^{min} already leads to the anomalous peaks of specific heat.

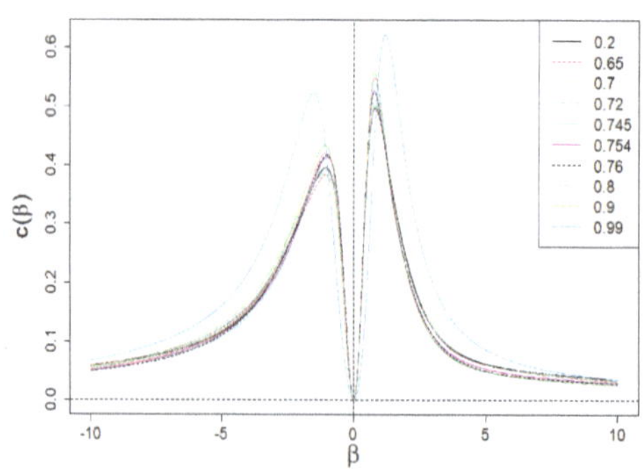

Figure 9. Dependence of specific heat, c, for a constant volume ($V = \ln N_{st}$) on β. The anomalous behavior of c is apparent due to the presence of Schottky peaks for both the positive and negative values of β.

3. Discussion and Concluding Remarks

3.1. Multifractality of Real Company Market

As an example of the method's application, we exploit the '$S\&P$ 500 Companies by Weight' page (from the day 12 November 2021). (The data was taken from the page https://www.slickcharts.com/sp500. Accessing to this page is common and unlimited all the time). The available empirical data covers approximately 70–80% of the total US stock market capitalization. These empirical data directly provide the market daily share values of individual companies, i.e., the data we need.

Let us characterize the market shares of companies using the Quetelet ranking (see Figure 10), i.e., we build a cumulative distribution function (CDF) versus company share value plot. The market structure is visible:

- the market segmentation into the overwhelming majority of companies with a small market share (around 0.01 or less)
- five companies with a market share between 0.02 and 0.03
- three companies with the highest market share between 0.04 and 0.065.

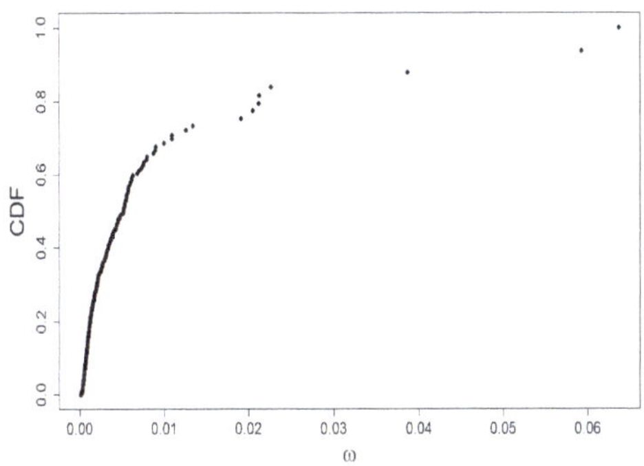

Figure 10. Quetelet curve: the empirical dependence of the standardized rank of companies, belonging to the $S\&P$ 500 index, i.e., CDF, on their shares ω. It is precisely to analyze this data that we use multifractal formalism.

In such a situation, the question of the actual dominance of companies on the market is justified: will small companies dominate large ones, or is the opposite case. For this purpose, we use the multifractal analysis described in Section 2.

It is worth realizing that if the CDF was built on a power, exponential, or Gaussian distribution, we would not be dealing with multifractality. In the first case, the scaling exponent $\tau(\beta)$ would be a linear function of β, in the second case it would be logarithmic, and in the third case, it would be a linear combination of logarithmic and linear functions.

We continue to investigate the empirical relationship shown in Figure 10 with the multifractality approach shown in Section 2. When using Equations (1) and (5), we find the relationship $\tau(\beta)$ vs. β, but we do not go into whether the market is in a steady-state or not, i.e., the number of firms in the index $N = N(t) \neq N_{st}$ may fluctuate around 500 and shares may depend on time. We can use it here because the above considered method applies to both stationary and non-stationary states.

The above-mentioned relationship, $\tau(\beta)$ vs. β, is shown in Figure 11. The presented dependence is a nonlinear function of β, which allows us to carry out the next steps of the method.

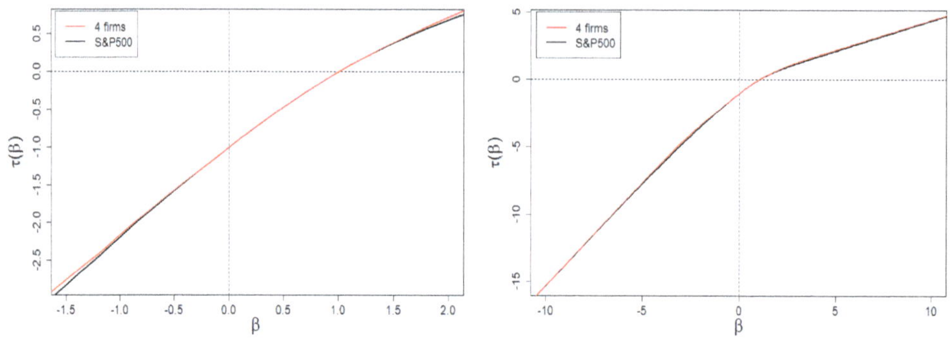

Figure 11. Dependence of $\tau(\beta)$ vs. β for the company market from the $S\&P$ 500 index. The left plot is a magnification of the β range belonging to the $[-1.5, 2.0]$ interval. The right plot shows the one in the full β range, i.e., belonging to the $[-10, 10]$ interval. In the assumed plot's resolution of the whole (right) graph, it is impossible to distinguish the results of the four-group company market model (red curve) from the empirical (black) curve.

In Figure 12, we presented the dependence of the generalized Hurst exponent on the β exponent. Its span is sufficient for the one of the spectra of singularities presented in Figure 13 (cf. the black curve) to define a solid multifractality.

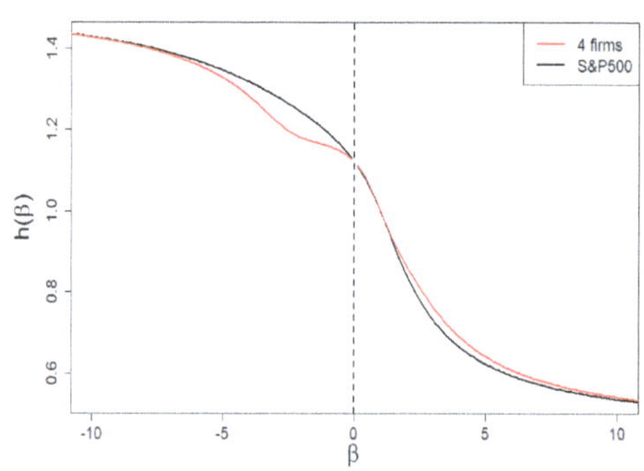

Figure 12. Dependence of the generalized Hurst exponent $h(\beta)$ on the β exponent. Its span is sufficient for one of the spectra of dimensions presented in Figure 13 (both curves have there a common span) to define a solid multifractality. There are slight/subtle local differences between the two curves in both figures (black: the empirical one; red: the four-group company market).

In Figure 14, we show the specific heat $c(\beta)$ vs. β. As in Section 2.5, we see peaks analogous to the Schottky peak—for both positive and negative values of β. There are differences in the predictions of the approach described below in Section 3.2 (in red) from the empirical curve (in black). These are hyper-fine deviations, as they appear at the level of the second order derivative of the scaling exponent τ.

We remind that subtle deviations (of the first order, i.e., at the level of the first derivative) are observed for the Hurst exponent as well as spectral dimension f (Figures 12 and 13, respectively). Deviations regarding the τ curve itself are imperceptible (on the scale of the right plot in Figure 11).

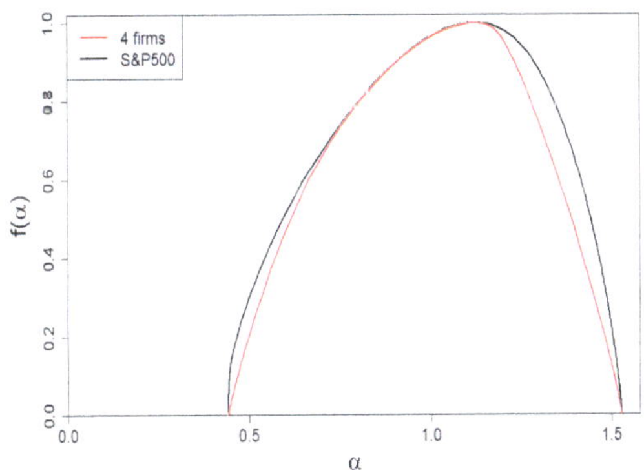

Figure 13. Dependence of the spectrum of dimensions $f(\alpha)$ vs. α for the company market from the *S&P* 500 index (black curve). The f asymmetry favoring large firms is visible. For comparison, we have included the spectra of dimensions for the four-group company market represented by the red curve.

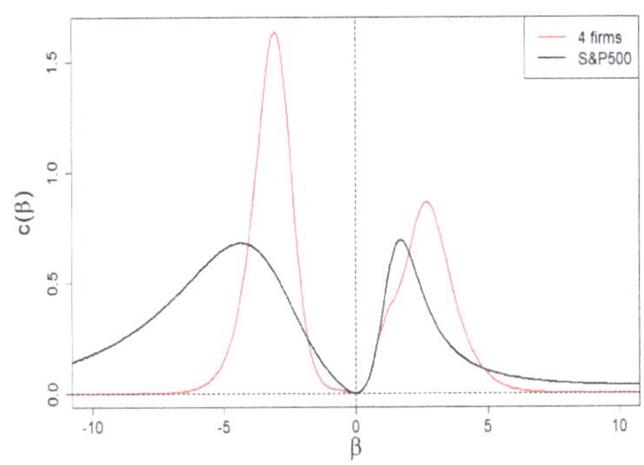

Figure 14. Anomalous dependence of specific heat $c(\beta)$ vs. β for the company market, for example, from *S&P* 500 index. As can be seen, the model of four-group company market shows apparent differences from the empirical data only at the level of the second τ derivative, i.e., at the level of hyper-fine effects.

3.2. Real Market vs. Four-Group Company Market

Now, we answer the question: how should the market of companies be grouped/ organized in order not to violate its diversity, i.e., to recreate its empirical spectrum of dimensions presented in Figure 13 (black curve). It is about its location and the basic shape defined by $(\alpha^{\min}, f^{\min}), (\alpha^1, f^1), (\alpha^0, D^0)$, and $(\alpha^{\max}, f^{\max})$ (see Figure 15 for details).

We use for this purpose the following expression for the scaling exponent (based on the multifractal formalism presented in Section 2),

$$\tau(\beta) = -\frac{\ln \mathcal{Z}_4(\beta)}{\ln N} = -\frac{1}{\ln N} \times \ln\left(M\left(\omega^{\min}\right)^{\beta} + K_1\omega_1^{\beta} + K_2\omega_2^{\beta} + L(\omega^{\max})^{\beta}\right), \quad (20)$$

where $\mathcal{Z}_4(\beta)$ means the partition function obtained from Equation (1) for the four-group company market. This section shows that such a division is enough to recreate the localization and shape of the spectrum of dimensions and other multifractality characteristics such as the scaling exponent, Hurst exponent, local exponent, and specific heat. We can show that two- and three-group company markets are not suitable for describing the multifractality of real company markets. For example, they cannot reproduce a location or a span of the spectrum of dimensions correctly.

Our specific goal is to clearly determine eight unknowns: the size of each of the four groups of companies M, K_1, K_2, L and their shares $\omega^{min}, \omega_1, \omega_2, \omega^{max}$. At least for the four-group company market, we can unambiguously determine the eight wanted unknowns.

Figure 15 shows an example schematic image of spectrum of dimensions—reading the coordinates of some of these points from this spectrum of dimensions allows us to determine the variables we are looking for. We show how to practically do this below.

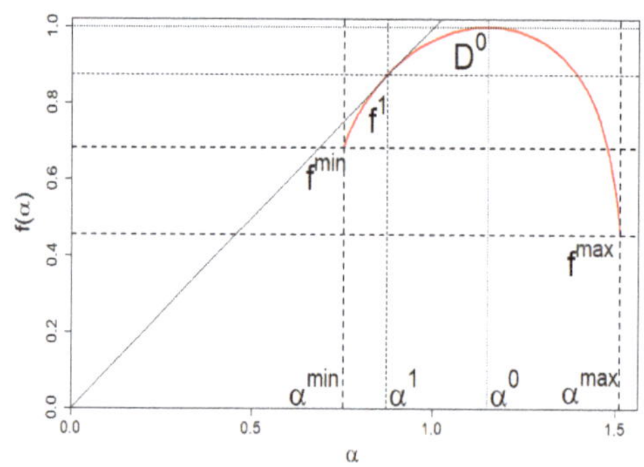

Figure 15. An example plot of the spectrum of dimensions f vs. α for the company market consisting of the four groups. Characteristic coordinates that we read from the graph, define the conditions (considered in the main text), which help us to determine the unknowns M, K_1, K_2, L and $\omega^{min}, \omega_1, \omega_2, \omega^{max}$.

The normalization condition takes the form

$$\mathcal{Z}_4(\beta = 1) = M\omega^{min} + K_1\omega_1 + K_2\omega_2 + L\omega^{max} = 1, \tag{21}$$

while the size of the market is fixed,

$$\mathcal{Z}_4(\beta = 0) = M + K_1 + K_2 + L = N. \tag{22}$$

The point is that N is fixed either as a stationary value or an instantaneous value of the number of firms in the market. Therefore, we take it from empirical data.

We emphasize that Equations (21) and (22) are the first two equations from the system of equations that allow us to find the above-mentioned unknowns we are looking for. Because the shares of ω^{min} and ω^{max} are read directly from the empirical data, in order to find the remaining unknowns, we need four more equations, which we consider below.

From Equation (20), and Definitions (5) and (15), we get

$$\begin{aligned} \alpha(\beta) &= \frac{d\tau(\beta)}{d\beta} = -\frac{1}{\ln N}\frac{1}{\mathcal{Z}_4(\beta)} \\ &\times [M\left(\omega^{min}\right)^{\beta}\ln\omega^{min} + K_1\omega_1^{\beta}\ln\omega_1 \\ &+ K_2\omega_2^{\beta}\ln\omega_2 + L(\omega^{max})^{\beta}\ln\omega^{max}] \end{aligned} \tag{23}$$

From the definition of the spectrum of dimensions (14), we obtain its boundary values for our case,

$$f^{\min} = f(\alpha^{\min} = \alpha(\beta \to \infty)) = \frac{\ln L}{\ln N},$$
$$f^{\max} = f(\alpha^{\max} = \alpha(\beta \to -\infty)) = \frac{\ln M}{\ln N}, \tag{24}$$

which can also be read (to good approximation) from the empirical f shown in Figure 13 (black curve). Thus, the number of unknowns is reduced to two, namely, to ω_1 and ω_2. It should be emphasized that only in the special case, when M or L are equal to 1, i.e., when the marginal values of companies' market shares are non-degenerate, do the boundary values of the spectrum of dimensions (24) disappear. It happens precisely in the case of the empirical data we use here.

Another needed quantity, which we read from the empirical f shown in Figure 13 (black curve), is the location of the center of the peak f given by the formula,

$$\alpha(\beta = 0) = -\frac{1}{N \ln N}$$
$$\times \left(M \ln \omega^{\min} + K_1 \ln \omega_1 + K_2 \ln \omega_2 + L \ln \omega^{\max}\right). \tag{25}$$

The same applies to the point of contact $f(\alpha(\beta = 1)) = \alpha(\beta = 1)$. Therefore,

$$\alpha(\beta = 1) = -\frac{1}{\ln N}$$
$$\times \; [M\omega^{\min} \ln \omega^{\min} + K_1\omega_1 \ln \omega_1$$
$$+ \; K_2\omega_2 \ln \omega_2 + L\omega^{\max} \ln \omega^{\max}]. \tag{26}$$

Both of the above equations have been obtained from Equation (20) and definition (15).

Now we calculate unknowns K_1 and K_2 from Equations (21) and (22) as the function of ω_1 and ω_2. We substitute the obtained quantities into Equations (25) and (26). Thus, we reduce our problem to two transcendental equations. For our case, $M = L = 1$, these equations can be converted to the form

$$\alpha(\beta = 0) N \ln N + \ln\left(\omega^{\min}\omega^{\max}\right)$$
$$= (N-2)\frac{\omega_1 \ln \omega_2 - \omega_2 \ln \omega_1}{\omega_2 - \omega_1} + \Omega\frac{\ln\left(\frac{\omega_1}{\omega_2}\right)}{\omega_2 - \omega_1}, \tag{27}$$

and

$$\alpha(\beta = 1) \ln N + \omega^{\min} \ln \omega^{\min} + \omega^{\max} \ln \omega^{\max}$$
$$= (N-2)\frac{\omega_1\omega_2}{\omega_2 - \omega_1} \ln\left(\frac{\omega_2}{\omega_1}\right) + \Omega\frac{\omega_1 \ln \omega_1 - \omega_2 \ln \omega_2}{\omega_2 - \omega_1}, \tag{28}$$

(where $\Omega = 1 - \omega^{\min} - \omega^{\max}$), which are more convenient for a numerical solution. Thus we have reduced our problem to the above two transcendental equations.

Table 1 presents the empirical data needed here regarding the first and last components of the $S\&P$ 500 index of 12 November 2021, consisting (on this day) of $N = 505$ companies.

Based on these empirical data, we solve numerically Equations (27) and (28) and obtain $\omega_1 = 0.00065$ and $\omega_2 = 0.0101$. Therefore, we have $K_1 = 439$ and $K_2 = 64$. Thus, in our case, we obtain non-degenerate share margins and strongly degenerate (though very different) intrinsic share values. The resulting spectrum of dimensions we presented in Figure 13 by means of a red curve. Likewise, we have presented the remaining results in Figures 11, 12 and 14 by means of red curves.

Table 1. Empirical data on the first and last components of the *S&P* 500 index as of 12 November 2021.

No.	Company	$\omega^{\min}$	$\omega^{\max}$	M	L
1	AAPL (Apple Inc., Cupertino, CA, USA)	–	0.06866056	–	1
505	NWS (New Corporation Class B, New York, NY, USA)	0.00006948	–	1	–

We emphasize that the obtained result is universal in the sense that, starting from the four-group market of companies, we obtain enough equations to describe the location and shape of the multifractality characteristics.

3.3. Conclusions

It is worth realizing how distributions induce common multifractal structures. Therefore, it is not so much about searching for such structures, but about the possibility of comparing them with each other, i.e., answering the question of which structures are more multifractal and which are less. For this, they must first be classified according to their symmetry and degeneration. The larger the logarithm of these steps, the higher these elevations are.

The degree of asymmetry in the multifractal structure is determined by the γ asymmetry coefficient. If $\gamma = 1$, we have a symmetric multifractal structure. If $\gamma > 1$, we have left asymmetry, while for $\gamma < 1$, we have right asymmetry.

The degree of degeneration of the marginal shares determines the elevation of the edges of the spectral dimensions: the left one depends on the degree of degeneration of the maximum share, and the right one depends on the degree of degeneration of the minimum share.

In this way, we have divided multifractal structures into nine groups, where both asymmetries and degenerations match themselves like the symmetry of the left and right hands (see Figure 16 for illustration, there, for example, the first plot in the first column and the last plot in the third column). Only within each group can we introduce a measure that allows us to organize the multifractal structure. The above classification is possible due to the fact that asymmetry and degeneration are independent of each other.

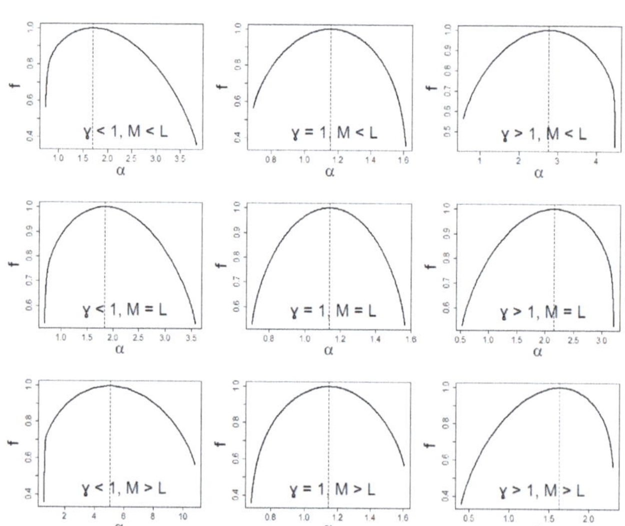

Figure 16. Schematic classification of spectrum of dimensions due to asymmetry γ and degeneration (M, L).

Suppose two multifractal structures have the same span of the spectrum of dimensions and location. One is more multifractal than the other if its degeneration levels are less than the corresponding other.

Another special case is when both multifractal structures' degeneracy levels are equal, while the structures differ in span. Then the more multifractal structure is for, the larger span structure plus f^1.

We introduce a precise definition of the linear multifractal capacity, $\mathcal{M}$, utilizing a definition based on Figure 15 and Equation (24),

$$\mathcal{M} = \Delta\alpha + f^1 + M^{-1} + L^{-1}. \quad (29)$$

Notably, there is no differentiation of multifractality due to location α^0. The proposed phenomenological measure of multifractal capacity, $\mathcal{M}$, is a partial in the sense that it does not take into account the entire fine structure of the spectrum of dimension f.

In conclusion, in this paper, we examine the multifractality/multiscaling coming from shares and not from correlations. In this sense, this work is complementary to our previous one [15]. As a reference case, we have discussed the instructive example of the four-group company market. We have shown that (within the zero-order approximation) each market can be reduced to a four-group company market, which should facilitate market analysis.

Finally, we can say that this is the first time such a multifractal analysis of the market of competing companies has been performed.

Notably, we can apply the approach to any series of shares, e.g., shares of turnover volumes on the stock exchange and shares of companies' quotations on the stock exchange. In short, the approach can be applied to any normalized series of positively defined elements. Moreover, our approach makes it possible to examine the evolution of multifractality of company market especially in the vicinity of crash regions. That is why it is so important to study in the near future the relationship between multifractality and criticality suggested by Figure 7.

Author Contributions: Conceptualization, R.K.; data curation, M.C.; formal analysis, M.C.; methodology, R.K.; resources, M.C.; software, M.C.; supervision, R.K.; validation, M.C.; visualization, M.C.; writing—original draft, R.K.; writing—review & editing, R.K. All authors have read and agreed to the published version of the manuscript.

Funding: This research received no external funding.

Institutional Review Board Statement: Not applicable.

Informed Consent Statement: Not applicable.

Data Availability Statement: Data available in a publicly accessible repository. The data presented in this study are openly available in "*S&P* 500 Companies by Weight" at https://www.slickcharts.com/sp500.

Conflicts of Interest: The authors declare no conflict of interest.

References

1. Kantelhardt, J.W.; Koscielny-Bunde, E.; Rego, H.H.; Havlin, S.; Bunde, A. Detecting long-range correlations with detrended fluctuation analysis. *Phys. A Stat. Mech. Its Appl.* **2001**, *295*, 441–454. [CrossRef]
2. Kantelhardt, J.W.; Zschiegner, S.A.; Koscielny-Bunde, E.; Havlin, S.; Bunde, A.; Stanley, H.E. Multifractal detrended fluctuation analysis of nonstationary time series. *Phys. A Stat. Mech. Its Appl.* **2002**, *316*, 87–114. [CrossRef]
3. Kwapień, J.; Drożdż, S. Physical approach to complex systems. *Phys. Rep.* **2012**, *515*, 115–226. [CrossRef]
4. Perelló, J.; Masoliver, J.; Kasprzak, A.; Kutner, R. Model for interevent times with long tails and multifractality in human communications: An application to financial trading. *Phys. Rev. E* **2008**, *78*, 036108. [CrossRef] [PubMed]
5. Kasprzak, A.; Kutner, R.; Perelló, J.; Masoliver, J. Higher-order phase transitions on financial markets. *Eur. Phys. J. B* **2010**, *76*, 513–527. [CrossRef]
6. Oswiecimka, P.; Kwapien, J.; Drożdż, S. Wavelet versus detrended fluctuation analysis of multifractal structures. *Phys. Rev. E* **2006**, *74*, 016103. [CrossRef] [PubMed]

7. Grech, D.; Pamuła, G. On the multifractal effects generated by monofractal signals. *Phys. A Stat. Mech. Its Appl.* **2013**, *392*, 5845–5864. [CrossRef]
8. Grech, D.; Mazur, Z. On the scaling ranges of detrended fluctuation analysis for long-term memory correlated short series of data. *Phys. A Stat. Mech. Its Appl.* **2013**, *392*, 2384–2397. [CrossRef]
9. Dariusz, G.; Zygmunt, M. On the scaling range of power-laws originated from fluctuation analysis. *Phys. Rev. E* **2013**, *87*, 052809. [CrossRef]
10. Oswiecimka, P.; Drożdż, S.; Forczek, M.; Jadach, S.; Kwapień, J. Detrended cross-correlation analysis consistently extended to multifractality. *Phys. Rev. E* **2014**, *89*, 023305. [CrossRef] [PubMed]
11. Drożdż, S.; Oswiecimka, P. Detecting and interpreting distortions in hierarchical organization of complex time series. *Phys. Rev. E* **2015**, *91*, 030902(R). [CrossRef] [PubMed]
12. Jiang, Z.Q.; Xie, W.J.; Zhou, W.X.; Sornette, D. Multifractal analysis of financial markets: A review. *Rep. Prog. Phys.* **2019**, *82*, 125901. [CrossRef] [PubMed]
13. Klamut, J.; Kutner, R.; Gubiec, T.; Struzik, Z.R. Multibranch multifractality and the phase transitions in time series of mean interevent times. *Phys. Rev. E* **2020**, *101*, 063303. [CrossRef] [PubMed]
14. Stanley, H.E. Fractals and Multifactals: The interplay of Physics and Geometry. In *Fractals and Disordered Systems*, 2nd ed.; Bunde, A., Havlin, S., Eds.; Springer: Berlin/Heidelberg, Germany, 1996.
15. Chorowski, M.; Kutner, R. Critical phenomena in the market of competing firms induced by state interventionism. *Phys. A Stat. Mech. Its Appl.* **2021**, *578*, 126102. [CrossRef]
16. Grassberger, P.; Procaccia, I. On the Characterization of Strange Attractors. *Phys. Rev. Lett.* **1983**, *50*, 346. [CrossRef]
17. Beck, C.; Schögl, F. *Thermodynamics of Chaotic Systems*; An Introduction; Cambridge Nonlinear Science Series 4; Cambridge University Press: Cambridge, UK, 1995.
18. Ferdinand, A.E.; Fisher, M.E. Bounded and Inhomogenous Ising Models. I. Specific Heat anomaly of a finite lattice. *Phys. Rev.* **1969**, *185*, 832–846. [CrossRef]
19. Tari, A. (Ed.) *The Specific Heat of Matter at Low Temperatures*; World Scientific Pub. Co., Imperial College Press: London, UK, 2003; p. 250.

entropy

MDPI

Article

On the Mortality of Companies

Peter Richmond [1,*] and Bertrand M. Roehner [2]

[1] School of Physics, Trinity College Dublin, D02 PN40 Dublin 2, Ireland
[2] Institute for Theoretical and High Energy Physics (LPTHE), Pierre and Marie Curie Campus, Sorbonne University, National Center for Scientific Research (CNRS), 75016 Paris, France; roehner@lpthe.jussieu.fr
* Correspondence: peter_richmond@ymail.com

Abstract: Using data from both the US and UK we examine the survival and mortality of companies in both the early stage or start-up and mature phases. The shape of the mortality curve is broadly similar to that of humans. Even small single cellular organisms such as rotifers have a similar shape. The mortality falls in the early stages in a hyperbolic manner until around 20–30 years when it begins to rise broadly according to the Gompertz exponential law. To explain in simple terms these features we adapt the MinMax model introduced by the authors elsewhere to explain the shape of the human mortality curve.

Keywords: mortality; companies; start-up; FTSE100; Gompertz; MinMax; survival probability distribution

1. Introduction

In 1999, one of the authors of this paper (PR) arrived in Ireland to spend what became a decade in Trinity College. During that year he had the opportunity to attend the first ever European Physical Society sponsored conference on econophysics in Dublin. During the meeting he obtained a copy of the book 'An Introduction to Econophysics' by Rosario N Mantegna and H Eugene Stanley. As for many other physicists, that meeting and the book inspired new research directions. This paper is the latest in a series that have emerged from that initial revelation over two decades ago.

In a series of recent papers [1–3] the present authors have studied human mortality demonstrating how the shape of the mortality function has a bathtub type of shape where the infant mortality decreases with age whereas in old age it increases (Figure 1). In medical terminology infancy refers to new born under one year of age. However, in reality the decrease of the death rate continues until the age of 10, For humans, the increase of the death rate is described by the well-known law of Gompertz [4]. This law can be summarized for by saying that the death rate doubles approximately every 10 years of age. Even the mortality of small animals such as rotifers [3] exhibit similar behaviour as is shown within the inset in Figure 1.

It has been suggested in the literature that non-biological systems obey a similar law however evidence of such behaviour in non-biological systems is not easy to find. Very recently Richmond et al. [5] studied the mortality of systems consisting of soap films and confirmed the bathtub nature of such systems. However, the systems were relatively small and towards the end of life, whilst the mortality increased there was no clear evidence of Gompertz behaviour. In this paper we present evidence for company mortality which mirrors the behaviour shown in Figure 1. The mortality of start-up companies decreases according to a hyperbolic law whereas the mortality of mature companies increases and the long-term trend is in accordance with the Gompertz law. This is shown in the next section. In Section 3, we present a simple model with offers and explanation as to why such behaviour can be expected for complex systems. We close with comments and thoughts for further studies.

Citation: Richmond, P.; Roehner, B.M. On the Mortality of Companies. *Entropy* **2022**, *24*, 208. https://doi.org/10.3390/e24020208

Academic Editors: Ryszard Kutner, Christophe Schinckus and H. Eugene Stanley

Received: 21 December 2021
Accepted: 26 January 2022
Published: 28 January 2022

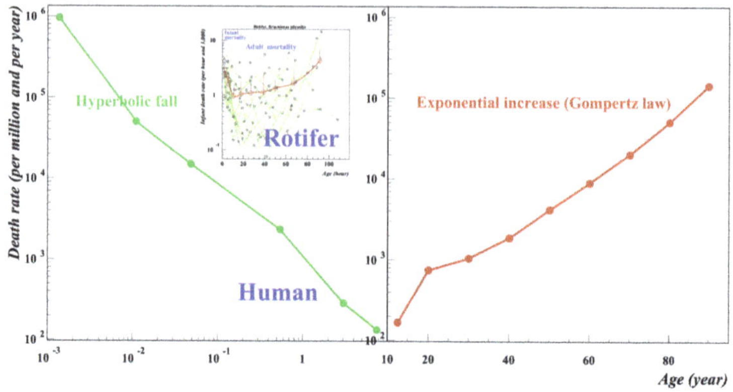

Figure 1. Infant versus old age human mortality. The data are for the US over the period 1999–2016. Between birth and the age of 10 (note the log-log scale) the infant mortality rate falls off as a power law: $\mu(x) = A/x^{\gamma}$ where the exponent γ is 0.99 and usually of the order of 1. After the infant phase comes the aging phase (note the linear-log scale) during which the death rate increases exponentially: $\mu(x) = \mu(0)\exp(\alpha x)$ in agreement with Gompertz's law and for humans $\alpha = 0.079$. Source: Wonder-CDC data base for detailed mortality data.

2. The Mortality of Companies

2.1. Start-Up

Much has been written about the survival of start-up companies. Usually this is directed to reasons why such companies fail and do not manage to survive the so-called valley of death in which companies fail due to inadequate working capital. Many other reasons can lead to failure, poor management, marketing, etc. Here, we are not concerned with these micro details rather we shall explore the mortality from a physics perspective looking for general features which characterize the mortality of all companies. For our purposes, a useful dataset is provided at LinkedIn in a paper by McIntyre [6]. Here, can be found survival data for cohorts of companies from their start-up year of 1994 through to 2021. More data is provided for similar cohorts beginning in 1995 and all years through to 2020. Each dataset consisted of over 500,000 companies ensuring good statistics. Earlier data for the period 1947–1954 is given by Steidl [7]. Steidl differentiates between manufacturing, retail and service industries. We show in Figure 2, survival probabilities for both data sets. The broad trend is similar but clearly the data for 1947–1954 falls more steeply than that for more recent years.

From this data for the survival probability, $\sigma(t)$, we can compute numerically the 'force of mortality', or more simply the mortality, $\mu(t)$. By definition this is the conditional probability that given a person is alive at time t, they will die in within the time interval $[t, t + \Delta t]$. It is equivalent to the rate of death conditional on life at time t. It follows from this formal definition that it is equal to the ratio of the unconditional survival probability density and the survival probability at time t:

$$\mu(t) = -\frac{1}{\sigma(t)}\frac{\Delta\sigma(\mathrm{t})}{\Delta t}$$

The Steindl data is fitted extremely well by a hyperbolic function as can be seen on Figure 3. For the average values we find $\sigma(t) = 0.60t^{-0.48}$. This allows us to compute the mortality directly by simple differentiation. Thus $\mu = 0.48t^{-1.0}$ However, from the figure it is readily seen that neither an exponential nor a hyperbolic function fits the McIntyre data and a simple first order difference procedure was used to compute annual values for the mortality which does however follow a hyperbolic function over much of the timescale. Thus, both McIntyre and Steindl data sets decay in a similar way following closely the hyperbolic trend $y = A/x^{\gamma}$. Decaying with a power law of −1, the Steindl data follow the value observed for human mortality. The more recent data decays more slowly.

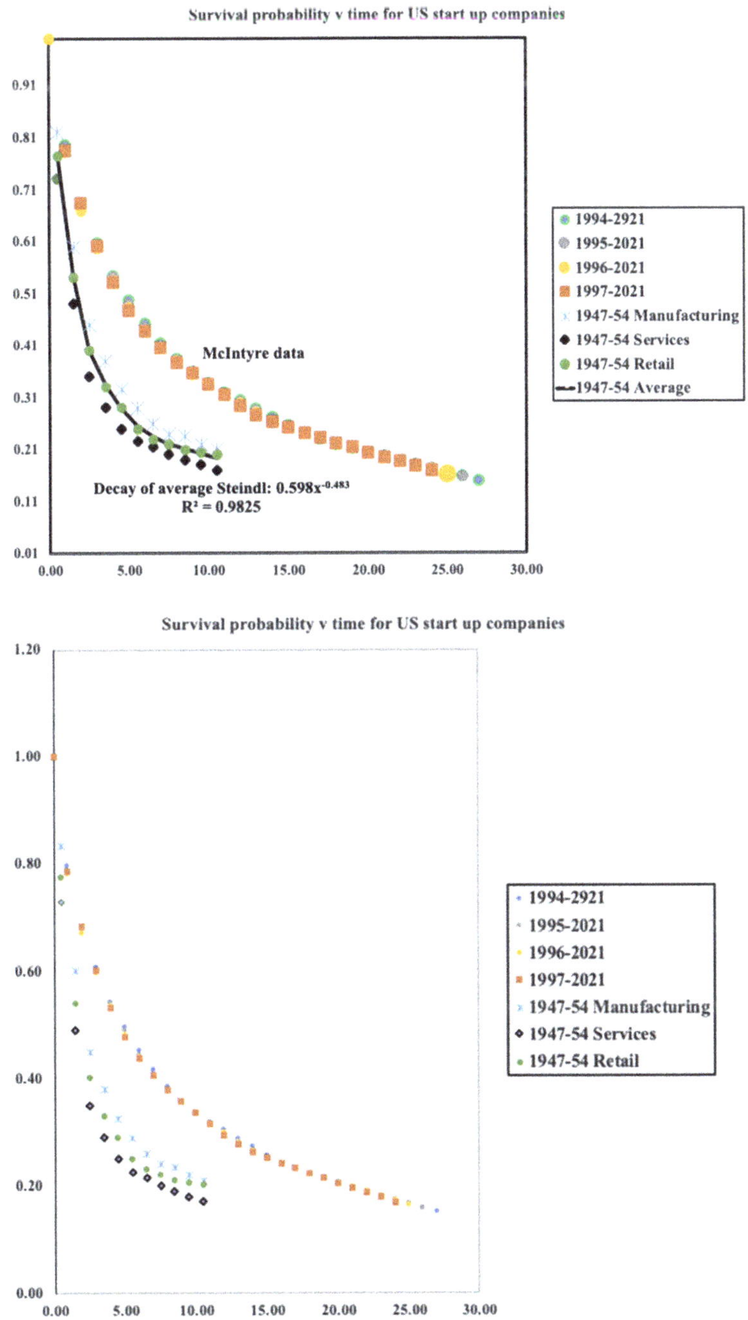

Figure 2. Survival probabilities for US start-up companies over the period 1947–1954 and 1994–2021. Data sources: McIntyre [2] and Steindl [3].

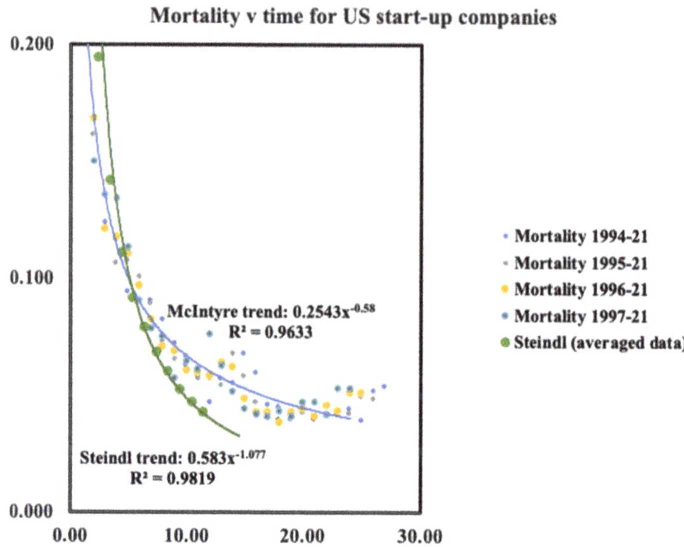

Figure 3. Age specific mortality for US start-up companies using data shown in Figure 2.

However, it may be seen in Figure 3 that for the recent McIntyre data there is hint of a minimum in the mortality versus time after approximately 20 years which is roughly the same as the minimum observed for humans. Such a minimum is not yet evident in the data of Steindl which only available for up to 10 years.

2.2. *Mortality of Mature Companies*

Having easy access to the UK FTSE100 index we chose to begin here. Comparing the composition for the FTSE 100 when it was first established in 1984 with that in 2021 we can establish 53 companies missing from the 2021 index. The company pages on Wikipedia then provides dates for both birth and death.

At this point a word of caution is in order. Within this list some companies did die in the sense of going bankrupt. However, others were taken over or merged into another company. Here, we did not differentiate between these different modes of 'death'. Takeovers and mergers were simply regarded as a point of death. Clearly a takeover or merger 'deaths is different in nature to a simple bankruptcy. In a sense such a death may not be dissimilar to deaths which occur in some biological systems such as that of a caterpillar as it becomes a butterfly. However, our dataset here is small and we leave further investigation of this point for another study for which a larger index or examination of multiple indices is required.

The lifetime of our 53 companies varies from 13 to 259 years. The one with the shortest lifetime is an oil company; the longest is a brewery. In between we see many types of company. For example: food production, electronics and telephone companies, banks and investment trusts. Figure 4 shows the survival probability of the 53 companies. This was computed simply using the data sets.

Unlike the data for early-stage companies, the survival probability shown in Figure 4 for the set of mature companies is clearly not smooth. Applying the route used previously to compute the mortality leads to a result which exhibits a number of anomalous sharp peaks which remain despite extensive smoothing of the data. One of us (PR) is grateful to an anonymous referee for pointing out the folly of this procedure. To work around this problem we followed a different procedure. From the data for the survival probability, we first computed the negative of the logarithm. Numerical derivatives were then computed from the resulting data using the central difference approximation yielding the mortality.

This procedure has the added benefit of avoiding the numerical division by the survival probability, thus:

$$\mu^i = -\frac{Ln\sigma^{i+1} - Ln\sigma^{i-1}}{t^{i+1} - t^{i-1}}$$

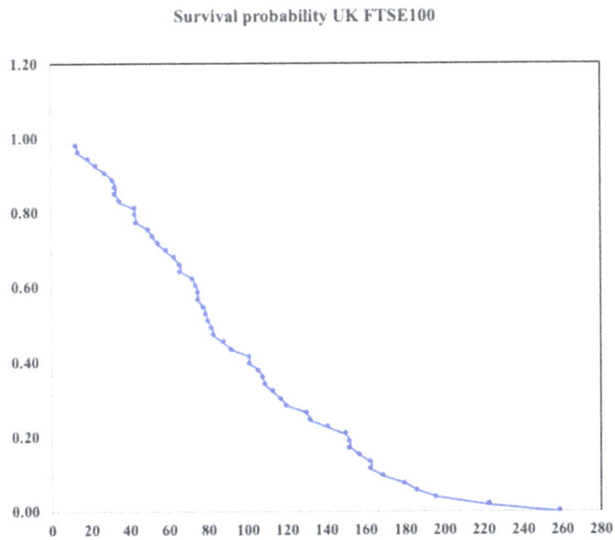

Figure 4. Survival probability for 53 FTSE100 companies which ceased trading between 1921 and 1984, the date the FTSE100 began being compiled.

The result is shown in Figure 5. The dots are computed data points and whilst there is scatter, the solid curve which an exponential fit corresponding to Gompertz behaviour fits reasonably well.

Finally, Figure 6 shows both sets of data (start-up and mature companies together in the manner of Figure 1 for human mortality.) The left-hand curve for small companies is plotted on a log-log scale; the mature company data is shown on a log-linear scale. The ordinate scales are identical.

The general trend follows that for human mortality shown in Figure 1. Start-up or early life mortality falls in a hyperbolic fashion; mature mortality trends upwards in the manner of Gompertz. From Figure 3, we have noted that the minimum for the more recent start-up company data seems to be around 25 years. However, from the figure it is clear there is a gap between where the deathrate appears to rise (~20–30 years) for the early-stage companies and the level at which mature companies has reached at the same age. However, the earlier data from Steindl falls more steeply and assuming no change in the trend the gap could be better closed with a minimum between the mortalities of early stage and mature companies of around 30–40 years. Why might the McIntyre data be so different? We know from studies of human mortality that data from different time eras can behave in this way. For example, modern medicine reduced substantially the mortality for babies with congenital defects. Here, we have two data sets for small companies taken from quite different time eras. The period 1947–1954 was a period of reconstruction after World War 2 and the nature of small companies then depended on large amounts of capital investment as indeed had been the case since the industrial revolution. However, with the advent of modern computers, the situation changed. Since the 1990s it has been to start up a company with little capital being dependent more on knowledge and computers than intensive amounts of capital. Microsoft for example was set up by Bill Gates in his garage and Google began as an undergraduate project. Using a biological term, we might say we are comparing two different species of company before and after the 1990. We would see similar discrepancies comparing say, human infant mortality with the mortality of adult elephants. Ideally, we should have data for a cohort of similar companies which have evolved in similar environments. Our FTSE100 data is taken over an era extending from the late 20th century back to the 17th century. Therefore, it seems not unreasonable to compare against this mature data with the earlier Steindl data than the McIntyre data. Perhaps even earlier data for the start-up companies might trend even more steeply downwards. A much larger group of similar mature companies might be collected from US data. The S&P 500 perhaps although the time period will be more limited going back perhaps only to the middle of the 19th century. More time needs to elapse before we shall see sufficient data for mature companies to compare with the McIntyre data.

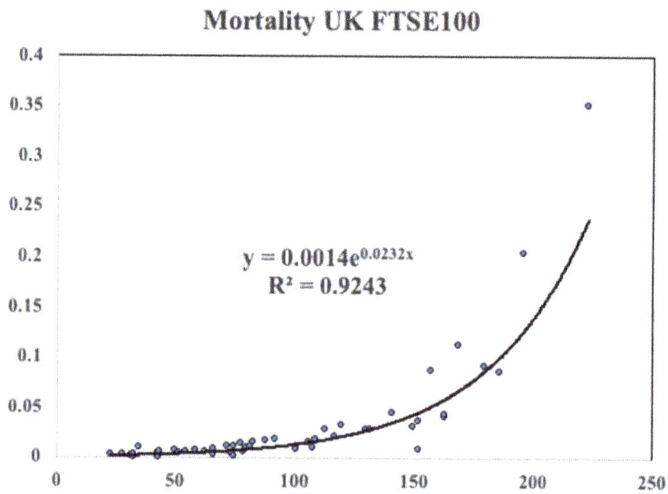

Figure 5. The mortality of the FTSE100 data on a linear plot. The data was computed using the method outlined in the text. The solid line is an exponential fit corresponding to Gompertz like behaviour.

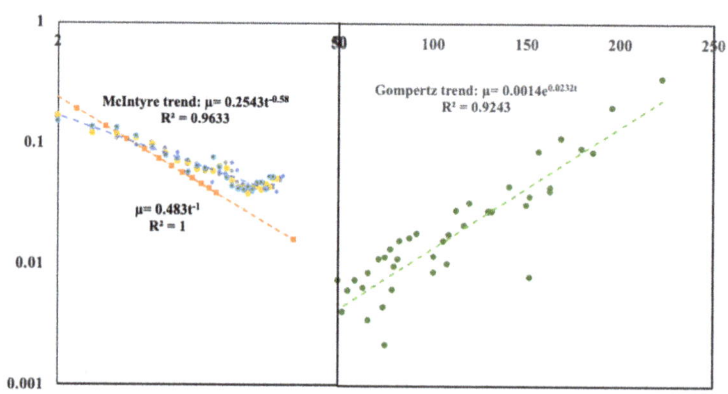

Figure 6. The left-hand graph is the small company mortality data plotted on a log-log plot; the right-hand graph is the mortality data for mature companies plotted on a log-linear plot. The ordinate scales for the mortality or annual deathrate are identical for both data sets. The abscissa for the early-stage companies is a logarithmic and extends to 50 years of age; the abscissa for the mature companies is a linear scale extending from 50 years to 250 years. The solid lines are the same regression fits shown in the earlier figures with details within the insets. However, here we have extended the early stage data trend line further out to around 30 years.

An interesting point is that, whereas for humans medical advances have led to a decrease in mortality, for companies, it seem over time the mortality of early stage companies has increased. The opportunity to set up a company with little capital makes it easier to begin a business, but then perhaps it is also easier too to stop trading. Finally, we note in passing that based on the data we have and extrapolating the trend beyond the maximum data point to where it reaches a value of unity, the results predict a maximum company life time of 283 years. This assumes no takeovers or mergers—which we have seen is not the case. Nevertheless, it will be interesting—for others!—to see if this outcome holds in the modern world.

3. The Minmax Model of Mortality

To explain the different forms of early and mature life mortality Richmond and Roehner offered a 'MinMax' model where the system was decomposed into elements each of which could function correctly or fail in a random way. For full details we refer the reader to the publication [1]. Here, for completeness we summarize the idea and results. In the case of the human these various elements could be thought of as the different organs (for example: heart lung, brain, etc.). For companies we might think of various departments or functions of the company such as marketing, finance, production etc.) such as shown schematically in Figure 7.

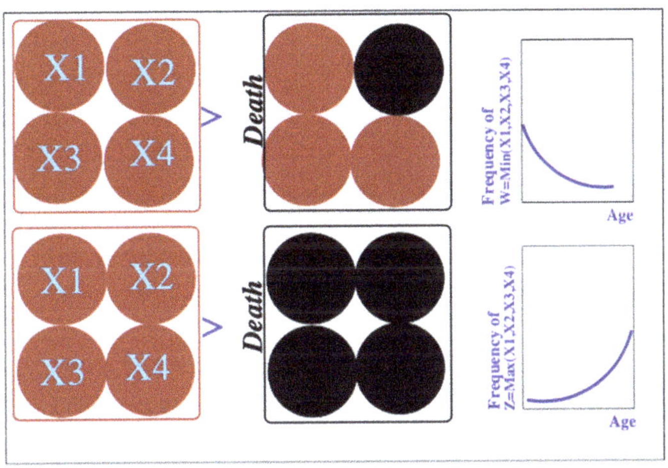

Figure 7. Illustration of the decomposition of an organism into vital organs and the difference between early life and mature life mortality mechanisms. The upper diagrams illustrate early life death. It is the consequence of the failure of a single vital organ. The lower diagrams show a mature death which is a consequence of uniform deterioration of all the vital organs. The graphs on the right-hand side show the implications of these mechanisms in terms of age-specific death rates: decreasing for early life as observed in infant death rates, increasing for mature death as seen in old-age.

We need a way to describe mathematically whether each element as well as the whole is functioning effectively or not. For simplicity we normalize the life span, Xi, of the elements to [0, 1] where 1 represents the maximum life span of an element. Moreover, all elements are supposed identical.

3.1. Mature Mortality

In the model we define the ultimate death of the company to have occurred when all elements of the organization have failed. For the simple 4 element system shown in Figure 8 this may be expressed by saying that if X1 = 0.5, X2 = 0.3, X3 = 0.7, X4 = 0.1 then the age of death is represented by the random variable Z = 0.7, in other words, Z = Max(X1, X2, X3, X4). In [1] we give the complete derivation of the density function $f_p(x)$ for p elements in terms of the density $f(x)$ and cumulative distributions $F(x)$ for the single elements:

$$f_{M,p}(x) = pf(x)F^p(x)$$

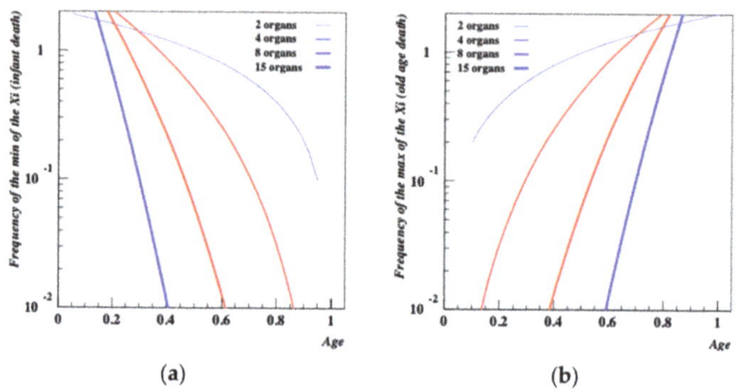

Figure 8. (**a**,**b**) MinMax model density functions for a set of random variables which represent age specific deathrates in early stage and mature companies. When the elemental structure of the company becomes large, deathrates for mature companies become exponential.

For the simple case of a random variable with a uniform density over the interval (0, 1); in this case: for x ∈ (0,1): $f(x) = 1$, $F(x) = x$. Thus for $x \in (0,1)$: $fZ(x)$ = pxp − 1. This function, shown in Figure 8b for p = 2, 4, 8, 15, is a power law function that *increases* fast with age. This is consistent with a Weibull distribution but when p becomes large it has the shape of an exponential which is qualitatively consistent with Gompertz's law according to which the probability of death increases exponentially with age.

3.2. Early-Stage Company Death

Again using the same ideas, early stage death would mean that the age of death is: W = 0.1, the is W = Min(X1, X2, X3, X4). Again we refer the reader to [1] where it is shown that for p elements in terms of the density $f(x)$ and cumulative distributions $F(x)$ for the single elements the density function for early death is

$$f_{W,p} = pf(x)[1 - F(x)]^{p-1}$$

For the simple model above we see that $f_{W,p}(x) = p[1 - x]^{p-1}$. Consequently, the probability of early-stage death, illustrated in Figure 8a is a *decreasing* function of age, consistent with what is expected for infant mortality.

4. Discussion and Conclusions

The interesting conclusion is that the broad trend of company mortality mimics that of humans. Perhaps this is not surprising since companies reflect human behaviour and ingenuity. As for human mortality, the minmax model gives some insight into the behaviour of company mortality. Small companies are known to fail as a result of a particular problem: a new product fails to succeed in a market, new finance is not forthcoming or production is found to be problematic. However, large companies that have evolved beyond the early stage can usually compensate for single department problems. Moreover, it would seem from the data that the trend in early-stage mortality is for it to have risen over the years. Could this be due to it being easier for an entrepreneur to set up a business? Moreover, with the need for limited capital investment relative to earlier times, could it be easier to close down a failing business?

From the minmax idea then we can understand the general shape of the mortality curve and as far as we are aware, this work is the first to show this behaviour for a non-biological system. However, why should the minimum occur around the age of 20–30 years in the manner of human mortality? Could this be linked to the complete passing of the first generation of employees over to a new group who are fully able to grapple with management of complexity as opposed to the skills offered by the initial entrepreneurs? More studies with new data sources are needed to explore this in more detail.

Author Contributions: Conceptualization, P.R.; Investigation, P.R.; Methodology, B.M.R.; Writing—original draft, P.R. All authors have read and agreed to the published version of the manuscript.

Funding: This research received no external funding.

Institutional Review Board Statement: Not applicable.

Informed Consent Statement: Not applicable.

Data Availability Statement: Not applicable.

Conflicts of Interest: The authors declare no conflict of interest.

References

1. Richmond, P.; Roehner, B.M. A joint explanation of infant and old age mortality. *J. Biol. Phys.* **2021**, *47*, 131–141. [CrossRef] [PubMed]
2. Berrut, S.; Pouillard, V.; Richmond, P.; Roehner, B.M. Deciphering infant mortality. *Phys. A: Stat. Mech. Appl.* **2016**, *463*, 400–426. [CrossRef]
3. Bois, A.; Garcia-Roger, E.M.; Hong, E.; Hutzler, S.; Irannezhad, A.; Mannioui, A.; Richmond, P.; Roehner, B.M.; Tronche, S. Infant mortality across species. A global probe of congenital abnormalities. *Phys. A: Stat. Mech. Appl.* **2019**, *535*, 122308. [CrossRef]
4. Gompertz, B. On the nature of the function expressive of the law of human mortality, and on the mode of determining the value of life contingencies. *Philos. Trans. R. Soc.* **1825**, *115*, 513–585.
5. Richmond, P.; Roehner, B.M. Mortality: A physics perspective. *Phys. A* **2021**, *566*, 125660. [CrossRef]
6. McIntyre, G. What Percentage of Small Businesses Fail? (And Other Need-to-Know Stats). Available online: https://www.linkedin.com/pulse/20140915223641-170128193-what-are-the-real-small-business-survival-rates (accessed on 21 December 2021).
7. Steindl, J. *Random Processes and the Growth of Firms*; Charles Griffin: London, UK, 1965.

entropy

Review

Econophysics and the Entropic Foundations of Economics

J. Barkley Rosser, Jr.

Department of Economics, James Madison University, Harrisonburg, VA 22807, USA; rosserjb@jmu.edu

Abstract: This paper examines relations between econophysics and the law of entropy as foundations of economic phenomena. Ontological entropy, where actual thermodynamic processes are involved in the flow of energy from the Sun through the biosphere and economy, is distinguished from metaphorical entropy, where similar mathematics used for modeling entropy is employed to model economic phenomena. Areas considered include general equilibrium theory, growth theory, business cycles, ecological economics, urban–regional economics, income and wealth distribution, and financial market dynamics. The power-law distributions studied by econophysicists can reflect anti-entropic forces is emphasized to show how entropic and anti-entropic forces can interact to drive economic dynamics, such as in the interaction between business cycles, financial markets, and income distributions.

Keywords: econophysics; entropy; complex systems; ecological economics; urban–regional economics; income distribution; financial market dynamics

Citation: Rosser, J.B., Jr. Econophysics and the Entropic Foundations of Economics. *Entropy* **2021**, *23*, 1286. https://doi.org/10.3390/e23101286

Academic Editors: Ryszard Kutner, Christophe Schinckus and H. Eugene Stanley

Received: 5 June 2021
Accepted: 16 September 2021
Published: 30 September 2021

1. Where Econophysics Came From

It has long been argued as for example by Mirowski [1] that economic theorists have drawn on ideas from physics, with an especially dramatic and influential example being Paul Samuelson's Foundations of Economic Analysis [2] from 1947. However, while the influence of physics concepts in Samuelson, as well as many economists much earlier, was enormous and openly acknowledged, it was only much later that the term econophysics would be coined, reportedly at a conference in 1995 Kolkata, India [3] by H. Eugene Stanley, who as a longtime editor of Physica A has played a crucial role in publishing many papers that have been identified as representing and advancing this approach, with the term first appearing in print in 1996 [4]. Curiously when it came to define this multidisciplinary neologism, the emphasis given by Mantegna and Stanley [5] was not upon the ideas or specific theoretical methods involved, but rather on the people doing it: "the activities of physicists who are working on economics problems to test a variety of new conceptual approaches deriving from the physical sciences".

This freshly defined approach involving physicists in particular, sometimes in conjunction with economists, quickly became a self-conscious cottage industry, even though arguably similar efforts had been going on for a long time, if not specifically by self-identified physicists, although some econophysicists have argued that an early inspiration for their work was Ettore Majorana in 1942 [6], whose untimely death gave him dramatic attention as he argued for the profound identity of statistical methods used in social sciences and physics. Important influences on the self-identified econophysicists included statistical mechanics [7,8] and also self-organized criticality models derived from models of avalanches [9] and earthquakes [10]. These approaches led to studies of many subjects in the early days, generally finding distributions that did not follow Gaussian patterns characterizable solely by mean and variance. These subjects included financial market returns [11–18], economic shocks and growth rate variations [19,20], city size distributions [21,22], firms size and growth rate patterns [4,23,24], scientific discovery patterns [25,26], and the distribution of income and wealth [27–29].

While the emerging econophysicists identified themselves as being physicists, an important impetus to their activities came from the intense discussions between economists and physicists at the Santa Fe Institute starting in the late 1980s [30,31]. While some of the economists defended existing economic theory, these discussions often emphasized dissatisfaction with its ability to explain empirical phenomena exhibiting non-Gaussian distributions with skewness and "fat tails" leptokurtosis [32–34]. While indeed most of the economists in these discussions disavowed some of the models developed by the econophysicists, the irony is that some of these models introduced by the physicists that could generate such higher moments as well as scaling properties were originally developed by economists, with the most important example of this being the Pareto distribution [35].

2. The Important Role of the Pareto Distribution

This important distribution that shows so many characteristics interesting to econophysicists was initially developed by the socio-economist Vilfredo Pareto in 1897 [35]. If N is the number of observations of a variable exceeding x, and A and α are positive constants, then

$$N = Ax^{-\alpha}. \tag{1}$$

Scaling can be seen as:

$$\ln(N) = \ln A - \alpha \ln(x), \tag{2}$$

with it possible to stochastically generalize this by replacing N with the probability an observation exceeds x. The log–log form of this is conveniently linear.

Much like the more recent econophysicists, Pareto's original focus was on income distribution, and he believed (inaccurately) that he had found the universally true value of 1.5 for a. In 1931, Gibrat [36] countered Pareto's argument with the idea that instead income distribution followed the lognormal form of the Gaussian distribution that can arise from a random walk, first studied by Bachelier in 1900 [37], with Einstein adopting it to model Brownian motion [38]. However, further studies suggest that combining these two provides a better description of income distribution, with the upper end of the distribution showing a Pareto pattern and lower portions showing lognormal Gaussian forms [39–42].

As it was, the Gaussian random walk would come to dominate a great deal of the modeling of price dynamics and financial market dynamics, including the widely used Black–Scholes formula [43]. Ironically, this triumph of what became the standard economic approach was engineered by the physicist M.F.M. Osborne in 1959 [44]. His model of dynamic prices, with p as the price, R the price increase return, B as the debt, and σ as the Gaussian standard deviation, is given by:

$$\mathrm{d}p = Rp\mathrm{d}t + \sigma p\mathrm{d}B. \tag{3}$$

Nevertheless, parallel developments inspired by Pareto went on through the twentieth century, with some using the stable Lévy distribution developed in 1925 [45] as a generalization of Pareto's distribution. Applications included looking at scientific discovery patterns [46] and city sizes [47]. A singular figure later in the century would be the father of fractal geometry, Benoit Mandelbrot [48,49], who directly posed the rival Pareto distribution as being able to model price dynamics [50] in 1963, in contrast to Osborne's argument. In 1977, Iriji and Simon [51] applied this to firm size distributions, a finding generally ignored until verified by Rob Axtell in 2001 [52].

3. The Influence of Statistical Mechanics

Arguably, the earliest influence of physics on economics was due to Canard in 1801 [53], who posed supply and demand as being "forces" opposing each other in a physics sense. However, a more specific influence on conventional economics would be statistical mechanics, developed by J. Willard Gibbs in 1902 [7]. As noted earlier, Samuelson in 1947 [2], who drew the influence from Irving Fisher [54], drew on Gibbs's approach for his reformulation of standard economic theory, a development much criticized by Mirowski [1], who derided all as economists exhibiting "physics envy".

More recently, there have been a variety of economists using statistical mechanics to develop stochastic models of various economic dynamics, including work by Hans Föllmer in 1974 [55], and then in the 1990s, just as the econophysicists were getting going by Blume [56], Durlauf [31] (pp. 83–104) and [57], Brock [58], Foley [59], and Stutzer [60]. Stutzer applied the maximum entropy formulation of Gibbs with the conventional Black–Scholes model [43], drawing on Arrow–Debreu contingent claims theory [61]. Brock and Durlauf [62] would formalize the general approach within the context of socially interacting heterogeneous agents maximizing utility in a discrete choice setting.

To a substantial degree, most econophysicists were not aware of either the more recent work along these lines, much less the deeper work further in the past, with this leading to some of them making unfortunately exaggerated claims about the originality and transformative nature of what they were doing. These problems were discussed in a critical essay called "Worrying trends in econophysics" by Gallegati et al. in 2006 in *Physica A* [63]. They identified the following as problematic trends: missing knowledge of the existing economics literature, a readiness to believe there may be universal empirical regularities in economics not really there unlike in physics, much use of unrigorous statistical methods sometimes just looking at figures, and relying on inappropriate theoretical foundations such as invalid conservation principles. McCauley responded [64], taking a hard line, that economic theory is so worthless that it should be totally replaced by ideas coming from physics. Reviewing these arguments, Rosser [65,66] agreed that economists often make vacuous assumptions, despite excessively unreal assumptions damaging usefulness of models. One way to deal with this is to have more joint research between economists and physicists.

4. Forms of Entropy

In the Gibbsian statistical mechanics, the question of maximizing entropy is a crucial element, which leads us to the question of what entropy is. Its original formulation came from Ludwig Boltzmann [67], although it was not as many thought the form that appeared on his grave [68] that has long received a great deal of attention. The statistical mechanics problems involve aggregating out of individual molecular interactions to observe systemic averages, such as temperature out of such a motion in a space. Letting S be entropy, k_B be the Boltzmann constant, and W be the statistical weight of the system macroscopic state (also known as the "thermodynamic probability"), then the following equation can be written as:

$$S = k_B \ln W, \tag{4}$$

where the configurational statistical weight of the macrostate of the system, W, defines the number of ways (configurations) of the arrangement of N of the identicalideal classical gas molecules in the microstates of the system (constituting a given macrostate), where N_i is the number of the identical molecules in the microstate i. The author uses this physical interpretation later in the work, given N is the sum of over the n available microstates of the system each given by N_i. Then according to Chakrabarti and Chakraborty [69], this implies that one is dealing with factorials multiplying each other as:

$$W = N!/\Pi N_i!. \tag{5}$$

From this, Boltzmann entropy can be rewritten as:

$$S = k_B \ln (N!/\Pi N_i!). \tag{6}$$

Moreover, the transition to the thermodynamics of an ideal classical gas at a temperature of $T > 0$ requires additional conditions to be taken into account, concerning the consistency of the total number of molecules of the gas, N, and the total energy, E, of all molecules.

Gibbs [7] famously declared that "mathematics is a language", which indeed he viewed as applying to his analysis of entropy within statistical mechanics. However, while we can view the mathematical formulation of Boltzmann entropy as a linguistic matter, it describes the real physical phenomenon of thermodynamics. Thus, it can be viewed as being ontological entropy [70], as it can be applied to more abstract phenomena with less linkage to definite physical processes, thus allowing them to be labeled metaphorical entropy. The first application beyond thermodynamics was information patterns in the form of Shannon entropy [71]. This describes H—the probability distribution of informational uncertainty states for message i that reflects the whole set of information concerning the relevant microstate, $H(p_1 \ldots p_n)$. Therefore, informational entropy involves adding up the individual log probabilities times their probabilities to give [71–73]:

$$H(p_1 \ldots p_n) = -k_B \Sigma p_i \ln p_i. \quad (7)$$

An obvious question arises as to how this widely used and influential metaphorical entropy measure relates to the ontological one of Boltzmann. In fact, they are proportional to each other as the number of possible states, N, approaches infinity, because $p_i = N_i/N$, resulting in [74,75]:

$$S = k_{BN} \Sigma p_i \ln p_i. \quad (8)$$

5. Ontological Entropy, Econophysics, and the Foundations of Growth

Ontological entropy lies at the heart of the econophysics foundation of economic growth due to the profound importance of energy both through the role of steam engines in industrial production and electricity and in agriculture through the thermodynamic transmission of solar energy through the larger global biosphere. The origin of understanding thermodynamics came from Sadi Carnot [76] in 1828 and later more fully Rudolf Clausius [77]. In 1971, Nicholas Georgescu-Roegen, [78] argued that the openness of the global biosphere to the sun allows temporarily overcoming the law of entropy [79]. Even so, there is a limit to solar energy, which implies limits for economic activity. However in an open system, anti-entropic forces can operate to develop order in local areas, drawing on the argument of Schrödinger [80] that life is ultimately an anti-entropic process involving the drawing of energy and matter from outside the living organism until it dies. Georgescu-Roegen also argued for this to extend to broader material resource inputs, subject to a form of the law of entropy. More broadly for Georgescu-Roegen [78] (p. 281), "the economic process consists of a continuous transformation of low entropy into high entropy, that is, into *irrevocable waste*, or, with a topical term, into pollution".

Many ecological economists [81,82] have supported the idea of entropy as an ontological limit to growth. However, while this is clearly true, others have noted that the limit is many orders of magnitude above other limits that are more immediate [83–85]. Drawing down stored fossil fuel energy sources generates climate-changing pollution by releasing CO_2 and thus further limiting growth. Others note the unlimited ingenuity of the human mind, with Julian Simon [86] (p. 347) arguing that "those who view the relevant universe as unbounded view the second law of thermodynamics as irrelevant to the discussion".

6. Ontological Entropy and Economic Value

Another argument has seen ontological entropy as the fundamental source of economic value in a parallel to the labor theory of value. The earliest version of this dates to the turn of the twentieth century in arguments by "energeticist" physicists [87–89]. Julius Davidson [90] saw the economics law of diminishing returns based on the law of entropy, with the law of diminishing marginal returns, probably the only "economic law" that has no exception to it. Davis [91] claimed "economic entropy" underlies the utility of money, but Lisman [92] argued this is not how thermodynamics operates in physics. Samuelson [93] ridiculed such arguments as a "crackpot", even as he drew on entropic ideas of Gibbs [7] and Lotka [81].

Lotka [81] (p. 355) himself noted limits to this argument: "The physical process is a typical case of 'trigger action' in which the ratio of energy set free to energy applied is subject to no restricting general law whatsoever (e.g., a touch of the finger upon a switch may set off tons of dynamite). In contrast with the case of thermodynamics conversion factors, the proportionality factor is here determined by the particular mechanism employed". Georgescu-Roegen [78] saw value as ultimately coming from utility rather than entropy. Thus, most people value the high-entropy beaten egg more highly than the low-entropy raw egg, and nobody valuing low-entropy poisonous mushrooms, due to utility rather than entropy.

7. Thermodynamic Sustainability of Urban–Regional Systems

The ontological entropic analysis of urban and regional systems sees them driven by the second law of thermodynamics based on actual energy transfers as argued by Rees [94], Balocco et al. [95], Zhang et al. [96], Marchinetti et al. [97], and Purvis et al. [98]. Alan Wilson [99] reviews both ontological and metaphorical approaches to the entropic analysis of urban and regional systems.

Considering urban–regional systems as open and dissipative systems, experiences allows the analysis of sustainability, depending on their energy and material flows [81,100]. In open systems, entropy can rise or fall, as energy and materials flow into them, in contrast to closed systems where entropy can only rise. This is the key to Schrödinger's [80] that life is an anti-entropic process with organisms drawing in energy-generating structure and order while life lasts. Anti-entropy is also known *exergy* [101] and also negentropy or "negative entropy".

Three concepts to distinguish are S_{total} as total entropy, S_i as inside entropy, and S_o as outside entropy. Assuming the statistical independence between both the internal states and the external states, then their dynamic relationship can be written as:

$$dS_{total}/dt = dS_i/dt + dS_o/dt, \text{ with } dS_i/dt > 0. \quad (9)$$

Given that dS_o/dt can be either sign, when negative with an absolute value greater than that of S_i, then total entropy may fall as the system absorbs energy and materials creating order, with entropy increasing outside as waste and disorder leave the system. Wackernagel and Rees [102] state, "Cities are entropic black holes" implying, as they produce large ecological footprints, their sustainability becomes impaired.

The maximum amount of the useful work possible to reach a maximum entropy condition of zero has been called *exergy* by Rant [101] initially for chemical engineering. This term is essentially identical to the term "chemical potential" and also "Gibbs-free energy". Rant's original formulation holds, when B is the exergy, U is the internal energy, P is the pressure, V is the volume, T is the temperature, S is the entropy, μ_i is the chemical potential of component i, and N_i is the moles of component i, implying:

$$B = U + PV - TS + \Sigma\mu_i N_i. \quad (10)$$

Recognizing that this is an isolated system implies:

$$dB/dt \leq 0 \boxed{\leftrightarrow} dS/dt \geq 0. \quad (11)$$

The right-hand side of Equation (11) simply holds for an isolated system, from which we see the anti-entropic nature of exergy, determining the irreversible spontaneous time evolution (or "time arrow").

Balocco et al. [95] consider exergy in construction and building depreciation in Castelnuovo Beardenga near Siena, Italy, relying on an adaptation by Moran and Sciubba [103] of Rant's model. Studying particularly the input–output of the construction industry, it is seen that those built in 1946–1960 provide higher sustainability than newer ones.

Zhang et al. [96] use entropy concepts to study sustainable development in Ningbo, China, a city near Shanghai, relying on ideas in [95,102,104,105]. They examine both ontological and metaphoric information entropy measures, as they consider four distinct aspects. The first two are sustaining input entropy and imposed output energy, arising from production. The second two constitute the urban system's metabolic functions, regenerative metabolism and destructive metabolism, which linked to pollution and its cleanup, a measure of environmental harmony. These contrast developmental degree and harmony degree, with the finding during the 1996–2003 period that these two went in opposite directions, with the developmental degree rising (associated with declining entropy) and the harmony degree falling (associated with rising entropy). Thus, we see Chinese urban development sustainability issues clearly.

The dependence versus autonomy of systems on their environment, derived from dissipative structures of open systems considered by Prigogine [100], was formulated by Morin [106] and then used by Marchinetti et al. [97]. This finds urban systems development between autarchy and globalization, either extreme unsustainable, advocating a balanced path they see urban–regional systems as ecosystems operating on energy flows [107] based on a complex wholes emerging out of interacting micro-level components [108].

8. An Anti-Entropic Econophysics Alternative in Urban–Regional Systems

Opposing this entropic version urban and regional systems structure is a power law version. In higher-level distributional systems, entropy ceases to operate and become irrelevant. This reflects a balance of entropic and exergetic forces operating in the relations and distributions within urban–regional systems [109].

Power-law distributions of econophysics reflect dominant anti-entropic forces [70], and urban size distributions seem to show these [22]. For the Pareto power-law distribution of city sizes [35], P is the population, r is the rank, with A and α are constants, implying:

$$rPr^{\alpha} = P_1. \tag{12}$$

For $\alpha = 1$, the population of rank r is written as:

$$P_r = P_1/r. \tag{13}$$

This is the rank-size rule of Auerbach [110] from 1913 and generalized in 1941 as Zipf's law, claimed to be applied to many distributions [47]. Since Auerbach [110] proposed it and Lotka [81] challenged it, there has been much debate regarding the matter. Many urban geographers [111] claim it is a universal law. Many economists have doubted this, saying there is no reason for it, even as urban sizes may show power-law distributions [112,113]. However, Gabaix [22] says Zipf's law holds in the limit if Gibrat's law is true with growth rates, independent of city sizes.

US city size distributions seem to have shown power-law distributions from 1790 to the present, although not precisely following the rank-size rule (the size of Los Angeles is now larger than half the size of New York), according to Batten [112]. A meta-study of many empirical studies by Nitsch [114] finds widely varying estimates over these studies, although showing an aggregate mean of $\alpha = 1.08$, near Zipf's value. Berry and Okulicz-Kozaryn [111] say Zipf's law strongly holds if one uses consistent measures for urban regions across studies, especially the largest ones for megalopolises. Anyway, city size distributions seem to be power-law-distributed, suggesting dominance by anti-entropic econophysics forces in this matter.

Long viewed as foundational for economic complexity, increasing returns may provide a basis for power-law distributional outcomes [115]. Three different kinds of these have been identified for urban systems: firm-level internal economies [116], external agglomeration between firms in a single industry providing localization economies [117], and external agglomeration economies across industries generating yet larger-scale urbanization economies [118].

Papageorgiou and Smith [119] and Weidlich and Haag [120] have shown that rising agglomeration economies can overcome congestion costs to manifest urban concentration. However, such models have been partially replaced by "new economic geography" ones emphasizing economies of scale appearing in monopolistic competition studied by Dixit and Stiglitz [121]. Fujita [122] first applied this approach to urban–regional systems, although Krugman [123] received much more attention for his version [124].

9. General Equilibrium Value and Metaphorical Entropy

Metaphorical Shannon entropy offers a different approach than Arrow–Debreu general equilibrium theory of value. Arrow and Debreu views equilibrium as a fixed point set of steady prices. However, in the reality of a stochastic world, equilibrium may be a probability distribution of prices that are constantly varying everywhere at any point in time for any commodity that can be modeled entropically. The Arrow–Debreu solution is a special case of Lebesgue measure in the space of outcomes. Initially conceived by Föllmer [55], Foley [59] developed it, followed by Foley and Smith [125].

Foley [59] assumes all possible transactions within an economy have equal probability, implying a statistical distribution of behaviors in the economy where a particular transaction is inversely proportional to the exponential of its equilibrium entropy price. This is a shadow price derived from a Boltzmann–Gibbs maximum entropy set. The special case when "temperature" is zero implies Walrasian general equilibrium. The solution is not necessarily Pareto optimal, and it allows for possible negative prices as Herodotus observed in ancient Babylonian bridal auctions, where they sold brides in descending prices that started out positive but then would go negative [126]. Foley emphasizes the crucial importance of constraints in this approach, as one finds in the Arrow–Debreu model.

If there are m commodities, n agents of type k who make a transaction x of which there is $h^k[x]$ proportion of agents type k out of r, which make transaction x out of an offer set A, of which there are mn, then *multiplicity* W of an assignment for n agents assigned to S actions, each of them s, which gives the probabilistic states across these possible transactions as:

$$W[n_{sJ}] = n!/(n_1! \ldots\ n_s! \ldots n_S!). \tag{14}$$

Shannon entropy of this multiplicity involves summing over these proportions similarly to Equation (7) and is written as:

$$H\{h^k[x]\} = -\Sigma_{k=1}{}^{r} W^k \Sigma_{x\epsilon A} h^k[x]. \tag{15}$$

This formulation maximizes entropy subject to certain non-empty feasibility constraints, thus giving the Gibbs solution:

$$H^k[x] = exp[-\Pi x]/\Sigma_x exp[-\Pi x], \tag{16}$$

with Π is the entropy shadow price vectors.

10. Metaphorical Entropic Financial Modeling

Schinkus [127] points out that econophysicists are more willing than most economists to approach data open to more possible distributions or parameter values, while favoring ideas from physics, including entropy for financial modeling. According to Dionisio et al. [128] (p. 161):

> "Entropy is a measure of dispersion, uncertainty, disorder and diversification used in dynamic process, in statistics and information theory, and has been increasingly adopted in financial theory".

Using the entropy law with Shannon or Boltzmann–Gibbs distributions can model distributions involving lognormality, both exhibiting normal Gaussian characteristics, Michael J. Stutzer [60,129] has drawn on both types of entropy to model Black–Scholes [43] formuli. In [129], he uses Shannon entropy, like Cozzolino and Zahner [130], allowing them to derive lognormal stock price distributions at the same time, similar to what Black and Scholes [43] did in deriving their options formuli without using entropy measures. Stutzer [129] considered a discrete form version modeling a stock market price dynamic by:

$$\Delta p/p = \mu\Delta t + \sigma\sqrt{\Delta t}\Delta z, \tag{17}$$

with p is the price, Δp is the random shock, Δt is the time interval, and the second term on the right hand side is the random shock, distributed ~ $N(0, \sigma^2\Delta t)$.

The order-maximizing solution for the neutral density of relative entropy-minimizing conditional risk given by the integral is written as:

$$arg\ min_{\mathrm{d}Q/\mathrm{d}P} \int log\ \mathrm{d}q/\mathrm{d}p\ \mathrm{d}q, \tag{18}$$

which satisfies a martingale restriction with q as a quantity:

$$r\Delta t - E[(\Delta p/p)(\mathrm{d}q/\mathrm{d}p)] = 0. \tag{19}$$

Thus, the Black–Scholes option-pricing formula can be derived from a martingale product density arising from relative entropy minimizing conditional risk for an asset subject to IID normally distributed shocks. Stutzer understood this does not generate non-Gaussian distributions such as econophysics power law ones. He poses using Generalized Auto Regressive Conditional Heteroskedastic (GARCH) processes as an alternative.

More recent studies have expanded the forms of entropy used in studying financial market dynamics. Thus, transfer entropy has been used by Jizba et al. [131] to study differences in related financial times series focusing on spike events by Dimpli and Peter [132] to study cryptocurrency dynamics and by Kim et al. [133] for directional stock market forecasting. In addition, permutation entropy has been used in a variety of financial market econophysics applications [134].

11. Using Statistical Mechanics to Model Income and Wealth Distributions

Income and wealth dynamical systems can be driven by interactions between power-law distributions and non-power-law ones. Wealth dynamics apparently exhibit power-law distributions, while income distribution dynamics look to consist of entropy-related Boltzmann–Gibbs distributions. The former seem to drive the top 2–3 percent of income distributions, while the latter seem to drive income distributions below that level in the UK and US [28,40].

Entropy came to be used in generalizations of various income distribution measures as early as 1981, when Cowell and Kuga [135] presented a generalized axiomatic formulation for additive measures of income distribution. Adding two axioms to the standard model allowed a generalized entropy approach to subsume the well-known Atkinson [136] and Theil measures [137]. The former can distinguish the skewness of tails, while latter has more generality, with Bourgignon [137] showing the Theil to be the only zero-homogeneous decomposable "income-weighted" inequality measure. Adding a sensitivity axiom to others, Cowell and Kuga [135] argued a generalized entropy concept implies the Theil index, even as some argued that this linking was a challenge, with Montroll and Schlesinger [138] (p. 209) declaring

"The derivation of distributions with inverse power tails from maximum entropy formalism would be a consequence only of an unconventional auxiliary condition that involves the specification of the average of a complicated logarithmic function".

It is unsurprising that both wealth and financial market distribution dynamics exhibit power-law distributions taking into account their close link, given Vilfredo Pareto's [35] role in discovering them. Initially trained to be an engineer, Pareto came to study the dynamic social classes relations manifested by income distribution. He claimed a universally true pattern that held throughout "the circulation of elites" he studied, but he was wrong, with ironically his method superior for the study of wealth distributions. He claimed incorrectly that because of the constancy of the income distribution pattern, little can be performed to equalize income, because changes in political leadership simply substitutes one power elite by another with no income distribution change. However, large changes occurred, so his approach went "underground", reappearing for other uses such as for urban metropolitan size distributions [111].

The sociologist, John Angle [139], revived using Pareto's power-law distribution for studying income and wealth distribution dynamics starting in 1986. Then, econophysicists followed up with this, with their finding that wealth distributions follow Pareto's power law view well [27,140,141].

The question arises as to whether we are dealing with ontological or "merely" metaphorical models in studying wealth and income distributional dynamics. Some see the stochastic elements in these distributions associated with thermodynamical processes fundamentally driving the distributional dynamics of income and wealth. However, these do not appear to be direct ontological processes as with Carnot's steam engines. More likely, these reflect dynamics associated with no substantial changes in public distributional policies.

Yakovenko and Rosser [40] show a model with an entropic Boltzmann–Gibbs dynamics for lower-income distribution and a Paretian power-law distributions for higher-level income dynamics. There is an assumption of the conservation of money or income or wealth, which has not held in recent years as top-level incomes have exploded although it did much more so in earlier decades. This is consistent with lognormal entropic dynamics appropriate for the majority of the population below a certain level where wage dynamics predominate, while a Pareto power law is more appropriate for the top level whose income is more determined by wealth dynamics.

Assuming money conservation, m, the Boltzmann–Gibbs entropic equilibrium distribution has probability, P, with m seen as:

$$P(m) = ce^{-m/Tm}, \tag{20}$$

with c is a normalizing constant, and T_m is the "money temperature" thermodynamically, equaling the money supply per capita. The portion of the income distribution below about 97–98 percent seems to be well modeled by this formulation.

If there is a fixed rate of proportional money transfers equaling γ, then the Gamma distribution rather than the Boltzmann–Gibbs distribution better describes the stationary money distribution with a power-law prefactor, m^{β}, such that:

$$\beta = -1 - \ln 2/\ln(1 - \gamma). \tag{21}$$

This Boltzmann–Gibbs version more simply relates to a power law equivalent than that posed by Montrell and Schlesinger [138]. The connection between the models of wealth and income distributions is described as:

$$P(m) = cm^{\beta}e^{-m/T}. \tag{22}$$

Letting m grow stochastically disconnects this outcome from the maximum entropy solution [142], so the stationary distribution becomes Fokker–Planck equation-driven mean field situation, not Boltzmann–Gibbs distribution, although inverse Gamma in [27,142] is a Lotka–Volterra form showing w as the wealth per person and J as the average transfer between agents, with σ being the standard deviation:

$$P(w) = c[(e^{-J/\sigma\sigma w})/(w^{2+J/\sigma\sigma})]. \tag{23}$$

This model provides an empirical explanation of income distribution consistent with Marxist and other classical economic views of socio-economic class dynamics [41,42,143].

Figure 1 exhibits this in the log–log form for the 1997 US income distribution, with the Boltzmann–Gibbs section for the lower 97 percent of the distribution being nonlinear on the left-hand side, while the Pareto section is linear in logs on the right-hand side showing the top 3 percent of the income distribution (Figure 4.5 of [144]).

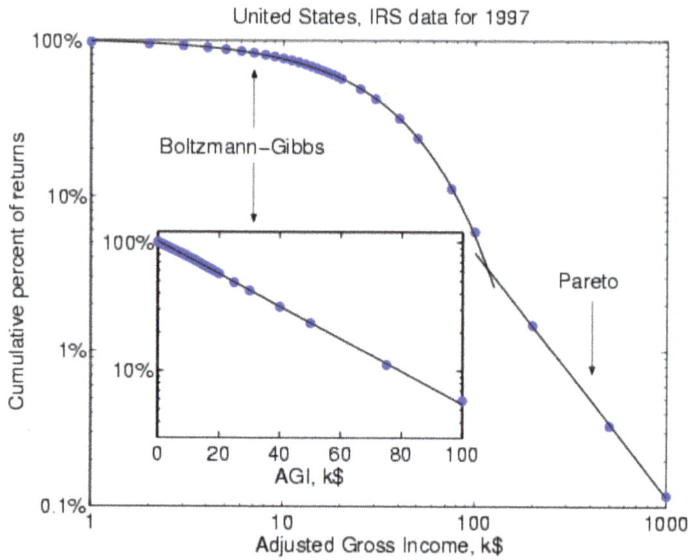

Figure 1. Log–log United States Income Distribution, Boltzmann–Gibbs, and Pareto Sections in 1997 from Yakovenko (Figure 4.6) [144].

There has been further use of variations on the Gamma distribution in studying market dynamics, with Moghaddim et al. [145] using the Beta Prime distribution to study housing market inequality dynamics.

12. The Revenge of Metaphorical Entropy as Bubbles Crash

Financial market dynamics interact with income and wealth distribution dynamics during speculative bubbles following a Minsky process [146–148]. During major bubbles, the top portion of the income and wealth distributions rises noticeably relative to the lower portion. Anti-entropic dynamics drive this process and its reversal, when the bubble crashes, hence the "revenge of entropy". Thus, during a bubble, this upward movement of the Paretian portion also moves its boundary with the Boltzmann–Gibbs portion leftward.

The Great Depression brought the end of the "Gilded Age" after a major financial crash that appears to have lowered the top end of the income distribution, as noted by Smeeding [149]. The 2007–2009 Great Recession had several different bubbles happening, leading to a more complex outcome, with the housing bubble crash badly hurting the middle class, while crashes of the stock market and derivatives markets predominantly hurt the wealthy. The US stock market fell from more than half its value to its bottom in 2009, with total wealth declining by 50 percent. Top 10 percent wealth declined by 13 percent, while top 1 percent wealth declined by 20 percent [149]. However, the stock market quickly turned around, rising more rapidly than in the 1930s or after 2000, while the US housing market grew more slowly. Thus, wealth inequality declined for a while during 2008–2009. It increased again after that as the rising stock market aided those at the top, while the continuing problems of the US housing market held back the middle class. This was the Minsky dynamic at work in a more complex form than seen at other times.

Support for this can be seen looking at the end of the dotcom bubble in 2000, even though somewhat weak, as indicated in Figure 2 (Figure 4.7) [144] showing the log–log relation for the US income distribution for the years 1983–2001, with further discussion in [150] and extension to a sample of 67 nations in [151]. Mostly, the Boltzmann–Gibbs section barely moved, but there were small annual changes in the Paretian part, manifesting gradually increasing inequality over time. However, there is an exception here, the change between 2000 and 2001, with 2000 being the end of the dotcom bubble. This time interval exhibited a reversal, with the 2001 Paretian portion lying below the 2000 portion. This is consistent with a revenge of entropy following the dotcom bubble crash, as the 1990s came to an end.

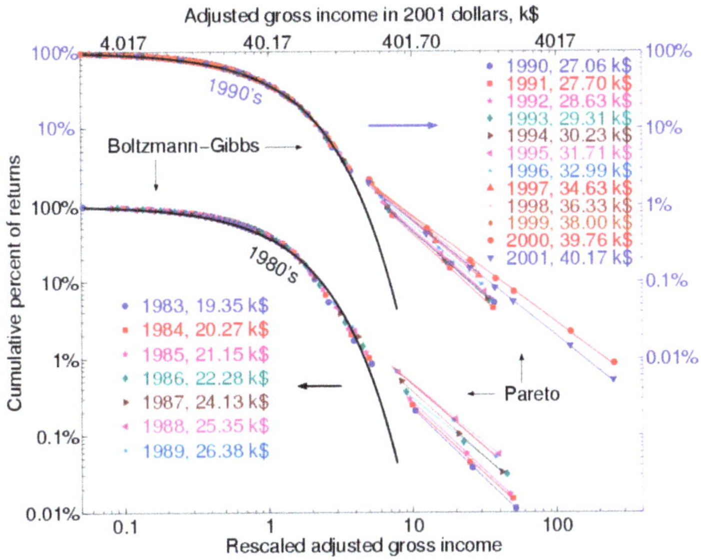

Figure 2. Log–log US Annual Income Distribution during 1983–2001 from Yakovenko (Figure 4.7) [144].

13. Conclusions

The term "econophysics" is of recent vintage, barely a quarter of a century old. However, the idea behind it that ideas and even laws of physics have strongly influenced economicss in a variety of ways is certainly correct. One of such ideas that has deep connections with the newer econophysics is the concept of entropy, which has been applied to many parts of economics, including general equilibrium theory, growth theory, business cycles, ecological economics, urban and regional economics, income and wealth distribution patterns, and financial market dynamics. Some of these applications are ontological in the sense of drawing directly on the second law of thermodynamics as the actual physical driving force involved, such as understanding energy flows through the biosphere and the economy from the Sun. Others are metaphorical, as they draw on models of information theory or other non-specifically physical models using the mathematics of entropy theory. Econophysics has also long emphasized the ubiquity of power-law distributions for many economic phenomena, which in some areas arise from anti-entropic processes that conflict with entropic tendencies. This can generate an underlying dynamic, with an especially dramatic example involving the dynamics of income distribution interacting with business cycles and related financial market dynamics.

Funding: The APC was funded by James Madison University.

Conflicts of Interest: The authors declare no conflict of interest.

References

1. Mirowski, P. *More Heat than Light: Economics as Social Physics: Physics as Nature's Economics*; Cambridge University Press: Cambridge, UK, 1989.
2. Samuelson, P.A. *Foundations of Economic Analysis*; Harvard University Press: Cambridge, MA, USA, 1947.
3. Chakrabarti, B.K. Econophys-Kolkata: A short story. In *Econophysics of Weatlh Distributions*; Chatterjee, A., Yarlagadda, S., Chakrabarti, B.K., Eds.; Springer: Milan, Italy, 2005; pp. 225–228.
4. Stanley, H.E.; Afanasyev, V.; Amaral, I.A.N.; Buldyrev, S.V.; Goldberger, A.I.; Havlin, S.; Leschhorn, H.; Masss, P.; Mantegna, R.N.; Peng, X.-K.; et al. Anomalous fluctuations in the dynamics of complex systems from DNA and physiology to econophysics. *Phys. A* **1996**, *224*, 302–323. [CrossRef]
5. Mantegna, R.N.; Stanley, H.E. *An Introduction to Econophysics: Correlations and Complexity in Finance*; Cambridge University Press: Cambridge, UK, 1999.
6. Majorana, E. Il valore delle leggi statistiche nelle fisica e nelle scienze. *Scientia* **1942**, *36*, 58–66.
7. Gibbs, J.W. *Elementary Principles of Statistical Mechanics*; Dover: New York, NY, USA, 1902.
8. Spitzer, F. *Random Fields and Interacting Particle Systems*; American Mathematical Society: Providence, RI, USA, 1971.
9. Bak, P. *How Nature Works: The Science of Self-Organized Criticality*; Copernicus Press for Springer: New York, NY, USA, 1996.
10. Sornette, D. *Why Stock Markets Crash: Critical Events in Complex Financial Systems*; Princeton University Press: Princeton, NJ, USA, 2003.
11. Mantegna, R.N. Lévy walks and enhanced diffusion in Milan stock exchange. *Phys. A* **1991**, *179*, 232–242. [CrossRef]
12. Levy, M.; Solomon, S. New evidence for the power-law distribution of wealth. *Phys. A* **1997**, *242*, 90–94. [CrossRef]
13. Bouchaud, J.-P.; Cont, R. A Langevin approach to stock market fluctuations and crashes. *Eur. Phys. J. B* **2000**, *6*, 542–550. [CrossRef]
14. Gopakrishnan, P.; Plerou, V.; Amaral, I.A.N.; Meyer, M.; Stanley, H.R. Scaling of the distributions of financial market indices. *Phys. Rev. E* **1999**, *60*, 5305–5316. [CrossRef]
15. Lux, T.; Marchesi, M. Scaling and criticality in a stochastic multi-agent model of a financial market. *Nature* **1999**, *397*, 498–500. [CrossRef]
16. Sornette, D.; Johansen, A. Significance of log-periodic precursors to financial crashes. *Quant. Financ.* **2001**, *1*, 452–471. [CrossRef]
17. Farmer, J.D.; Joshi, S. The price dynamics of common trading strategies. *J. Econ. Behav. Organ.* **2002**, *49*, 149–171. [CrossRef]
18. Li, H.; Rosser, J.B., Jr. Market dynamics and stock price volatility. *Eur. Phys. J. B* **2004**, *39*, 409–413. [CrossRef]
19. Bak, P.; Chen, K.; Scheinkman, J.; Woodford, M. Aggregate fluctuations from independent sectoral shocks: Self-organized criticality in a model of production and inventory dynamics. *Ric. Econ.* **1993**, *47*, 3–30. [CrossRef]
20. Canning, D.; Amaral, I.A.N.; Lee, Y.; Meyer, M.; Stanley, H.E. A power law for scaling the volatility of GDP growth rates with country size. *Econ. Lett.* **1998**, *60*, 335–341. [CrossRef]
21. Rosser, J.B., Jr. Dynamics of emergent urban hierarchy. *Chaos Solitons Fractals* **1994**, *4*, 553–562. [CrossRef]
22. Gabaix, X. Zipf's law for cities. *Q. J. Econ.* **1999**, *114*, 739–767. [CrossRef]
23. Takayasu, H.; Okuyama, K. Country dependence on company size distributions and a numerical model based on competition and cooperation. *Fractals* **1998**, *6*, 67–79. [CrossRef]
24. Botazzi, G.; Secchi, A. A stochastic model of firm growth. *Phys. A* **2003**, *324*, 213–219. [CrossRef]
25. Plerou, V.; Amaral, I.A.N.; Gopakrishnan, P.; Meyer, M.; Stanley, H.E. Similarities between the growth dynamics of university research and competitive economic activities. *Nature* **1999**, *400*, 433–437. [CrossRef]
26. Sornette, D.; Zajdenweber, D. Economic returns of research: The Pareto law and its implications. *Eur. Phys. J. B* **1999**, *8*, 653–664. [CrossRef]
27. Bouchaud, J.-P.; Mézard, M. Wealth condensation in a simple model of economy. *Phys. A* **2000**, *282*, 536–545. [CrossRef]
28. Drăgulescu, A.A.; Yakovenko, V.M. Exponential and power law probability distributions of wealth and income in the United Kingdom and the United States. *Phys. A* **2001**, *299*, 213–221. [CrossRef]
29. Chatterjeee, A.; Yarlagadda, S.; Chakrabarti, B.K. (Eds.) *Econophysics of Wealth Distributions*; Springer: Milan, Italy, 2005.
30. Anderson, P.W.; Arrow, K.J.; Pines, D. (Eds.) *The Economy as a Complex Evolving System*; Addison-Wesley: Redwood City, CA, USA, 1988.
31. Arthur, W.B.; Durlauf, S.N.; Lane, D.A. (Eds.) *The Economy as a Complex Evolving System II*; Addison-Wesley: Reading, PA, USA, 1997.
32. McCauley, J.L. *Dynamics of Markets: Econophysics and Finance*; Cambridge University Press: Cambridge, UK, 2004.
33. Chatterjee, A.; Chakrabarti, B.K. (Eds.) *Econophysics of Stock and other Markets*; Springer: Milan, Italy, 2006.
34. Lux, T. Applications of statistical physics in finance and economics. In *Handbook of Complexity Research*; Rosser, J.B., Jr., Ed.; Edward Elgar: Cheltenham, UK, 2009; pp. 213–258.
35. Pareto, V. *Cours d'Économie Politique*; R. Rouge: Lausanne, Switzerland, 1897.
36. Gibrat, R. *Les Inégalités Économiques*; Sirey: Paris, France, 1931.
37. Bachelier, L. Théeorie de la spéculation. *Ann. Sci. L'école Norm. Supér.* **1900**, *III-17*, 21–86. [CrossRef]
38. Einstein, A. Über die von der molekularkinetischen theorie der warme geforderte bewegung von der ruhenden flüsstigkeiten teichen. *Ann. Phys.* **1905**, *17*, 549–560. [CrossRef]
39. Clementi, F.; Gallegati, M. Power law tails in the Italian personal income distribution. *Phys. A* **2005**, *350*, 427–438. [CrossRef]
40. Yakovenko, V.M.; Rosser, J.B., Jr. Colloquium: Statistical mechanics of money, wealth, and income. *Rev. Mod. Phys.* **2009**, *81*, 1704–1725. [CrossRef]
41. Shaikh, A. *Capitalism: Competition, Conflict, and Crisis*; Oxford University Press: New York, NY, USA, 2016.

42. Shaikh, A.; Jacobo, E.J. Economic arbitrage and the econophysics of income inequality. *Rev. Behav. Econ.* **2020**, *7*, 299–315. [CrossRef]
43. Black, F.; Scholes, M. The pricing of options and corporate liabilities. *J. Political Econ.* **1973**, *81*, 637–654. [CrossRef]
44. Osborne, M.F.M. Brownian motion in stock markets. *Oper. Res.* **1959**, *7*, 134–173. [CrossRef]
45. Lévy, P. *Calcul des Probabilités*; Gauthier-Villars: Paris, France, 1925.
46. Lotka, A.J. The frequency distribution of scientific productivity. *J. Wash. Acad. Sci.* **1926**, *12*, 317–323.
47. Zipf, G.K. *Human Behavior and the Principle of Least Effort*; Addison-Wesley: Cambridge, MA, USA, 1941.
48. Mandelbrot, B.B. *The Fractal Geometry of Nature*; W.H. Freeman: New York, NY, USA, 1982.
49. Mandelbrot, B.B. *Fractals and Scaling in Finance*; Springer: New York, NY, USA, 1997.
50. Mandelbrot, B.B. The variation of certain speculative prices. *J. Bus.* **1963**, *36*, 392–419. [CrossRef]
51. Ijirii, Y.; Simon, H.A. *Skew Distributions and the Size of Business Firms*; North-Holland: Amsterdam, The Netherlands, 1977.
52. Axtell, R.L. Zipf distribution of firm sizes. *Science* **2001**, *293*, 1818–1820. [CrossRef] [PubMed]
53. Canard, N.F. *Principes d'Économie Politique*; 1801. Reprint by Edizioni Bizzari: Rome, Italy, 1969.
54. Fisher, I. *Mathematical Investigations into the Theory of Value and Price*; Yale University Press: New Haven, CT, USA, 1926.
55. Föllmer, H. Random economies with many interacting agents. *J. Math. Econ.* **1974**, *1*, 51–62. [CrossRef]
56. Blume, L.E. The statistical mechanics of strategic interaction. *Games Econ. Behav.* **1993**, *5*, 387–424. [CrossRef]
57. Durlauf, S.N. Nonergodic economic growth. *Rev. Econ. Stud.* **1993**, *60*, 340–366. [CrossRef]
58. Brock, W.A. Pathways to randomness in the economy. *Estud. Econ.* **1993**, *8*, 2–55.
59. Foley, D.K. A statistical equilibrium theory of markets. *J. Econ. Theory* **1994**, *62*, 321–345. [CrossRef]
60. Stutzer, M.J. The statistical mechanics of asset prices. In *Differential Equations, Dynamical Systems, and Control Science: A Festschrift in Honor of Lawrence Markuss*; Elsworthy, K.D., Everett, W.N., Lee, E.B., Eds.; Marcel Dekker: New York, NY, USA, 1994; Volume 152, pp. 321–342.
61. Arrow, K.J. *Essays in the Theory of Risk Bearing*; North-Holland: Amsterdam, The Netherlands, 1974.
62. Brock, W.A.; Durlauf, S.N. Discrete choice with social interactions. *Rev. Econ. Stud.* **2002**, *68*, 235–260. [CrossRef]
63. Gallegati, M.; Keen, S.; Lux, T.; Ormerod, P. Worrying trends in econophysics. *Phys. A* **2006**, *370*, 1–6. [CrossRef]
64. McCauley, J.L. Response to 'Worrying trends in econophysics'. *Phys. A* **2008**, *371*, 601–609. [CrossRef]
65. Rosser, J.B., Jr. Debating the role of econophysics. *Nonlinear Dyn. Psychol. Life Sci.* **2008**, *12*, 311–323.
66. Rosser, J.B., Jr. Econophysics and economic complexity. *Adv. Complex Syst.* **2008**, *11*, 745–761. [CrossRef]
67. Boltzmann, L. Über die eigenschaften monocycklischer und andere damit verwandter systems. *Crelle's J. Reine Angwandte Math.* **1884**, *109*, 201–212.
68. Uffink, J. Boltzmann's work in statistical physics. In *Stanford Encylopedia of Philosophy*; Center for the Study of Language and Information, Stanford University: Stanford, CA, USA, 2014; Available online: https://plato.stanford.edu/entries/statphys-Boltzmann (accessed on 25 May 2021).
69. Chakrabarti, C.G.; Chakraborty, J. Boltzmann-Shannon entropy: Generalization and application. *Mod. Phys. Lett. B* **2006**, *20*, 1471–1479. [CrossRef]
70. Rosser, J.B., Jr. Entropy and econophysics. *Eur. Phys. J. Spec. Top.* **2016**, *225*, 3091–3104. [CrossRef]
71. Jaynes, E.T. Information theory and statistical mechanics II. *Phys. Rev.* **1957**, *108*, 171–180. [CrossRef]
72. Shannon, C.E.; Weaver, W. *Mathematical Theory of Communication*; University of Illinois Press: Urbana, IL, USA, 1949.
73. Rényi, A. On measures of entropy and information. In *Proceedings of the Fourth Berkeley Symposium on Mathematics, Statistics, and Probability, 1960: Contributions to the Theory of Statistics*; University of California Press: Berkeley, CA, USA, 1961; Volume 1, pp. 547–561.
74. Tsallis, C. Possible generalizations of Boltzmann-Gibbs statistics. *J. Stat. Phys.* **1988**, *52*, 479–487. [CrossRef]
75. Thurner, S.; Hanel, R. The entropy of non-ergodic complex systems: A derivation from first principles. *Int. J. Mod. Phys. Conf. Ser.* **2012**, *16*, 105–115. [CrossRef]
76. Carnot, S. *Réflexions sur la Puissance Motrice du Feu etr sur les Machines Propres a Déveloper Cette Puissance*; Vein: Paris, France, 1824.
77. Clausius, R. Über verschiedene für die nverdung bequeme formen du hauptgleichungen du mechanischer warmtheorie. *Ann. Phys.* **1865**, *125*, 353–400. [CrossRef]
78. Georgescu-Roegen, N. *The Entropy Law and the Economic Process*; Harvard University Press: Cambridge, MA, USA, 1971.
79. Rosser, J.B., Jr. *From Catastrophe to Chaos: A General Theory of Economic Discontinuities*; Kluwer: Boston, MA, USA, 1991.
80. Schrödinger, E. *What Is Life? The Physical Aspects of the Living Cell*; Cambridge University Press: London, UK, 1945.
81. Lotka, A.J. *Elements of Physical Biology*; Williams & Wilkens: Baltimore, MD, USA, 1925.
82. Martinez-Allier, J. *Ecological Economics: Energy, Environment and Scarcity*; Blackwell: Oxford, UK, 1987.
83. Gerelli, E. Entropy and the end of the world. *Ric. Econ.* **1985**, *34*, 435–438. [CrossRef]
84. Nordhaus, W.D. Lethal model 2: The limits to growth revisited. *Brook. Pap. Econ. Act.* **1992**, *1992*, 1–59. [CrossRef]
85. Young, J.T. Entropy and natural resource scarcity: A reply to the critics. *J. Environ. Econ. Manag.* **1994**, *26*, 210–213. [CrossRef]
86. Simon, J.L. *The Ultimate Resource*; Princeton University Press: Princeton, NJ, USA, 1981.
87. Helm, G. *Die Lehre von der Energie*; Felix: Leipzig, Germany, 1887.
88. Winiarski, L. Essai sur la mécanique sociale: L'énergie sociale et ses mensurations. *Rev. Philos.* **1900**, *49*, 265–287.
89. Ostwald, W. *Die Energie*; J.A. Barth: Leipzig, Germany, 1908.

90. Davidson, J. One of the physical foundations of economics. *Q. J. Econ.* **1919**, *33*, 717–724. [CrossRef]
91. Davis, H.J. *The Theory of Econometrics*; Indiana University Press: Bloomington, IN, USA, 1941.
92. Lisman, J.H.C. Econometrics and thermodynamics: A remark on Davis's theory of budgets. *Econometrica* **1949**, *17*, 56–62. [CrossRef]
93. Samuelson, P.A. Maximum principles in analytical economics. *Am. Econ. Rev.* **1972**, *62*, 2–17.
94. Rees, W.E. Ecological footprints and appropriated carrying capacity: What urban economics leaves out. *Environ. Urban.* **1992**, *4*, 121–130. [CrossRef]
95. Balocco, C.; Paeschi, S.; Grazzini, G.; Basosi, R. Using exergy to analyze the sustainability of an urban area. *Ecol. Econ.* **2004**, *48*, 211–244. [CrossRef]
96. Zhang, Y.; Yan, Z.; Li, W. Analyses of urban ecosystem based on information entropy. *Ecol. Model.* **2006**, *197*, 1–12. [CrossRef]
97. Marchinetti, N.; Pulselli, F.M.; Tierzi, E. Entropy and the city. *WTI Trans. Ecol. Environ.* **2006**, *93*, 263–272.
98. Purvis, B.; Mao, Y.; Robinson, D. Entropy and its applications to urban systems. *Entropy* **2019**, *21*, 56. [CrossRef]
99. Wilson, A.G. Entropy in Urban and Regional Modelling: Retrospect and Prospect. *Geogr. Anal.* **2010**, *42*, 265–287. [CrossRef]
100. Prigogine, I. *From Being to Becoming*; W.H. Freeman: San Francisco, CA, USA, 1980.
101. Rant, Z. Exergie, ein neues wort fūr "technische arbeitagikeit". *Forsch. Geb. Inginieurwesens* **1956**, *22*, 36–37.
102. Wackernagel, M.; Rees, W.E. *Our Ecological Footprint: Reducing Human Impact on the Earth*; New Society Publishers: Philadelphia, PA, USA, 1996.
103. Moran, M.J.; Sciubba, E. Exergy analysis: Principles and practice. *J. Eng. Gas Turbine Power* **1994**, *116*, 286–290. [CrossRef]
104. Haken, H. *Information and Self Organization*; Springer: New York, NY, USA, 1988.
105. Svirezhev, Y.M. Thermodynamics and ecology. *Ecol. Model.* **2000**, *132*, 11–22. [CrossRef]
106. Morin, E. Le vie della complessita. In *La Sfida della Complessita*; Bocchi, G., Ceruti, M., Eds.; Feltrinelli: Milan, Italy, 1995; pp. 49–60.
107. Odum, E.P. The strategy of ecosystem development. *Science* **1969**, *164*, 262–270. [CrossRef] [PubMed]
108. Ulanowicz, R.E. *Growth and Development: Ecosystems Phenomenology*; Springer: New York, NY, USA, 2012.
109. Rosser, J.B., Jr. *Foundations and Applications of Complexity Economics*; Springer Nature: Cham, Switzerland, 2021.
110. Auerbach, F. Das gesetz der bevölkerungskonzentration. *Peterman's Geogr. Mittelungen* **1913**, *59*, 74–76.
111. Berry, B.J.L.; Okulicz-Kozaryn, A. The city size distribution debate: Resolution for US urban regions and megalopolitan areas. *Cities* **2012**, *48*, 517–523. [CrossRef]
112. Batten, D. Complex landscapes of spatial interaction. *Ann. Reg. Sci.* **2001**, *35*, 81–111. [CrossRef]
113. Fujita, M.; Krugman, P.R.; Venables, A.J. *The Spatial Economy: Cities, Regions, and International Trade*; MIT Press: Cambridge, MA, USA, 1999.
114. Nitsch, V. Zipf zipped. *J. Urban Econ.* **2005**, *57*, 86–100. [CrossRef]
115. Arthur, W.B. *Increasing Returns and Path Dependence in the Economy*; University of Michigan Press: Ann Arbor, MI, USA, 1994.
116. Marshall, A.; Marshall, M.P. *The Economics of Industry*; Macmillan: London, UK, 1879.
117. Marshall, A. *Industry and Trade*; Macmillan: London, UK, 1919.
118. Hoover, E.M.; Vernon, R. *Anatomy of a Metropolis: The Changing Distribution of People and Jobs in the New York Metropolitan Area*; Harvard University Press: Cambridge, MA, USA, 1959.
119. Papageorgiou, Y.Y.; Smith, T.E. Agglomeration as local instability of spatially uniform steady-states. *Econometrica* **1983**, *51*, 1109–1119. [CrossRef]
120. Weidlich, W.; Haag, G. A dynamic phase transition model for spatial agglomeration processes. *J. Reg. Sci.* **1987**, *27*, 529–569. [CrossRef] [PubMed]
121. Dixit, A.; Stiglitz, J.E. Monopolistic competition and optimum product diversity. *Am. Econ. Rev.* **1977**, *67*, 297–308.
122. Fujita, M. A monopolistic competition approach to spatial agglomeration: A differentiated product approach. *Reg. Sci. Urban Econ.* **1988**, *18*, 87–124. [CrossRef]
123. Krugman, P.R. Increasing returns and economic geography. *J. Political Econ.* **1991**, *99*, 483–499. [CrossRef]
124. Rosser, J.B., Jr. *Complex Evolutionary Dynamics in Urban-Regional and Ecologic Systems: From Catastrophe to Chaos and Beyond*; Springer: Heidelberg, Germany, 2011.
125. Foley, D.K.; Smith, E. Classical thermodynamics and general equilibrium theory. *J. Econ. Dyn. Control* **2008**, *32*, 7–65.
126. Baye, M.R.; Kovenock, D.; de Vries, C.G. The Herodotus paradox. *Games Econ. Behav.* **2012**, *74*, 399–406. [CrossRef]
127. Schinkus, C. Economic uncertainty and econophysics. *Phys. A* **2009**, *388*, 4415–4423. [CrossRef]
128. Dionisio, A.; Menezez, R.; Mendes, D. An econophysics approach to analyze uncertainty in financial markets: An application to the Portuguese stock market. *Eur. Phys. J. B* **2009**, *60*, 161–164.
129. Stutzer, M.J. Simple entropic derivation of a generalized Black-Scholes model. *Entropy* **2000**, *2*, 70–77. [CrossRef]
130. Cozzolini, J.M.; Zahner, M.J. The maximum entropy distribution of the future distribution of the future market price of a stock. *Oper. Res.* **1973**, *21*, 1200–1211. [CrossRef]
131. Jizba, P.; Kleinert, H.; Shefaat, M. Rényi's information transfer between financial time series. *Phys. A* **2012**, *391*, 2971–2989. [CrossRef]
132. Dimpli, T.; Peter, F.J. Group transfer entropy with an application to cryptocurrencies. *Phys. A* **2019**, *516*, 534–551.
133. Kim, S.; Ku, S.; Cheng, W.; Song, J.W. Predicting the direction of US stock prices using effective transfer entropy and machine learning technology. *IEEE Access* **2020**, *8*, 111680–111682.

134. Zanin, M.; Zunino, L.; Rosso, O.A.; Papo, D. Permutation entropy and its main biomedical and econophysics applications: A review. *Entropy* **2012**, *14*, 1553–1577. [CrossRef]
135. Cowell, F.A.; Kuga, K. Additivity and the entropy concept: An axiomatic approach to inequality measurement. *J. Econ. Theory* **1981**, *25*, 131–143. [CrossRef]
136. Atkinson, A.B. On the measurement of inequality. *J. Econ. Theory* **1970**, *2*, 244–263. [CrossRef]
137. Bourguignon, F. Decomposition income inequality measures. *Econometrica* **1979**, *47*, 901–920. [CrossRef]
138. Montroll, F.W.; Schlesinger, M.F. Maximum entropy formalism, fractals, scaling phenomena, and 1/f noise: A tale of tails. *J. Stat. Phys.* **1983**, *32*, 209–230. [CrossRef]
139. Angle, J. The surplus theory of social stratification and the distribution of personal wealth. *Soc. Forces* **1986**, *65*, 293–326. [CrossRef]
140. Chakraborti, A.S.; Chakrabarti, B.K. Statistical mechanics of money: How savings propensities affects its distribution. *Eur. Phys. J. B* **2000**, *17*, 167–170. [CrossRef]
141. Solomon, S.; Richmond, P. Stable power laws in variable economics: Lotka-Volterra implies Pareto-Zipf. *Eur. Phys. J. B* **2002**, *27*, 257–261. [CrossRef]
142. Huang, D.W. Wealth accumulation with random redistribution. *Phys. Rev. E* **2004**, *69*, 57–103. [CrossRef] [PubMed]
143. Cockshott, W.P.; Cottrill, A.F.; Michaelson, G.J.; Wright, I.F.; Yakovenko, V.M. *Classical Econophysics*; Routledge: London, UK, 2008.
144. Yakovenko, V.M. Applications of statistical mechanics to economics: Entropic origin of the probability distributions of money, income, and energy consumption. In *Social Fairness and Economics: Economic Essays in the Spirit of Duncan Foley*; Taylor, L., Rezai, A., Michl, T., Eds.; Routledge: London, UK, 2013; pp. 53–82.
145. Moghaddem, M.D.; Miller, J.; Serota, R.A. Generalized prime distributions: Stochastic model of economic exchange and properties of inequality indices. *arXiv* **2019**, arXiv:1906.04833v1.
146. Minsky, H.P. Financial instability revisited: The economics of disaster. *Reapprais. Fed. Reserve Discount Mech.* **1972**, *3*, 97–136.
147. Kindleberger, C.P. *Manias, Panics, and Crashes: A History of Financial Crises*, 4th ed.; Basic Books: New York, NY, USA, 2001.
148. Rosser, J.B., Jr. The Minsky moment and the revenge of entropy. *Macroecon. Dyn.* **2020**, *24*, 7–23. [CrossRef]
149. Smeeding, T.M. *Income, Wealth and Debt, and the Great Recession*; Stanford Center on Poverty and Inequality, Stanford University: Stanford, CA, USA, 2012.
150. Yakovenko, V.M. Monetary economics from econophysics perspective. *Eur. Phys. J. Spec. Top.* **2016**, *225*, 3313–3335. [CrossRef]
151. Tao, Y.; Wu, X.; Zhou, T.; Yan, W.; Huang, Y.; Yu, H.; Mondal, B.; Yakovenko, V.M. Exponential structure of income inequality: Evidence from 67 countries. *J. Econ. Interact. Coord.* **2019**, *14*, 345–376. [CrossRef]

www.ingramcontent.com/pod-product-compliance
Lightning Source LLC
Chambersburg PA
CBHW061446220926
51850CB00004B/24
* 9 7 8 3 0 3 6 5 4 7 4 1 1 *

MDPI
St. Alban-Anlage 66
4052 Basel
Switzerland
Tel. +41 61 683 77 34
Fax +41 61 302 89 18
www.mdpi.com

Entropy Editorial Office
E-mail: entropy@mdpi.com
www.mdpi.com/journal/entropy

entropy

MDPI

Review

Energy, Entropy, Constraints, and Creativity in Economic Growth and Crises

Reiner Kümmel [1,*] and Dietmar Lindenberger [2]

1. Institute for Theoretical Physics und Astrophysics, University of Würzburg, D-97074 Würzburg, Germany
2. Institute of Energy Economics, University of Cologne, D-50827 Cologne, Germany; dietmar.lindenberger@uni-koeln.de

* Correspondence: kuemmel@physik.uni-wuerzburg.de

Received: 29 July 2020; Accepted: 30 September 2020; Published: 14 October 2020

Abstract: The neoclassical mainstream theory of economic growth does not care about the First and the Second Law of Thermodynamics. It usually considers only capital and labor as the factors that produce the wealth of modern industrial economies. If energy is taken into account as a factor of production, its economic weight, that is its output elasticity, is assigned a meager magnitude of roughly 5 percent, according to the neoclassical cost-share theorem. Because of that, neoclassical economics has the problems of the "Solow Residual", which is the big difference between observed and computed economic growth, and of the failure to explain the economic recessions since World War 2 by the variations of the production factors. Having recalled these problems, we point out that technological constraints on factor combinations have been overlooked in the derivation of the cost-share theorem. Biophysical analyses of economic growth that disregard this theorem and mend the neoclassical deficiencies are sketched. They show that energy's output elasticity is much larger than its cost share and elucidate the existence of bidirectional causality between energy conversion and economic growth. This helps to understand how economic crises have been triggered and overcome by supply-side and demand-side actions. Human creativity changes the state of economic systems. We discuss the challenges to it by the risks from politics and markets in conjunction with energy sources and technologies, and by the constraints that the emissions of particles and heat from entropy production impose on industrial growth in the biosphere.

Keywords: energy; economic growth; output elasticities; entropy production; emissions; optimization

---

1. Introduction

Seventy-five years ago Nazi-Germany collapsed. The allied soldiers who liberated the concentration camps, and the camps where more than two million Soviet prisoners of war were starved to death, shocked the world by the documentations of the atrocities commited by a member of European civilization. After unconditional surrender on 8 May 1945, Germany was left with devastated cities, a shattered economy and moral misery.

The rivalry of economic systems and the fortunes of political change saved Germans from more than the usual revenge by the winners of a war. This was especially true for the ones in the west zones as established by the ruling of the Yalta and Potsdam conferences. The antagonism between capitalist market economics of the western occupying powers, who administered what became the Federal Republic of Germany (FRG), and socialist planned economics of the Soviet Union, who occupied what became the German Democratic Republic (GDR), turned allies into adversaries. Tensions between them were enhanced by the Korean War 1950–1953. To strengthen the western camp the FRG was allowed to benefit from the Marshall Plan [1]. Via this European Recovery Program the USA transferred 13.12 billion dollars (corresponding to 139 billion dollars today) between 1948 and 1952 to war-torn

Europe. In contrast, the industrial capital goods of the GDR were transferred to the Soviet Union as reparations.

The so-called "economic miracle" of the FRG, which started in 1949 with the currency reform that brought the Deutsche Mark (DM), was based on a growing capital stock, rebuilt and modernized by skilled labor, and cheap oil from the newly discovered oil-fields in the Middle East, Indonesia and the Americas: Between 1950 and 1970 the price of 1 barrel of crude oil on the world market had fallen from about 20 to 12 $US\$_{2014}$, and economic growth in the western industrialized democracies was up to 7% annually.

Complementing the retrospect on German crash and recovery by the following tale from a physicist [2] shall indicate the limits-to-growth reason that enticed him and other people outside economics to start thinking about economic growth: "Having experienced how industrializiation improves life while I grew up in postwar Germany and then did physics research at the University of Illinois, I joined a project of scientific cooperation between the FRG and the Republic of Colombia. My task was to participate in the development of a master program in physics at the Universidad del Valle. My excellent Colombian colleagues considered the formation of good physicists and engineers as one prerequisite for progress in the industrialization of a, in many aspects, still agrarian society. Right in the beginning they asked me to teach thermodynamics. 'Anything but this. Thermodynamics is boring', I objected. 'Read Reif's book *Statistical and Thermal Physics*' [3], they suggested. I did—and for the first time I really understood entropy. When two years later *The Limits to Growth* was published [4], I was deeply shocked. I told my Colombian students that the world would run into trouble because of the Second Law of Thermodynamics, if the developing countries would follow the path of industrialization Europe and the USA have treaded so far—At the celebration of '50 years Physics Department of the Universidad del Valle' in 2013, some of my former students, now physics professors, told me that they well remember how much I had been shocked.—After three unforgettable years in Colombia I returned to Germany. After having settled at the Julius-Maximilians-Universität I got in touch with economists, and in addition to teaching theoretical physics and continuing research in solid state theory I offered courses on thermodynamics and economics. Good students joined research in this field, and experienced economists helped."

Mostly in plain language, the present article presents a synopsis of the resulting studies on energy and entropy in economic growth. It includes an outlook on options of how to deal with the crises ahead. The review is limited in the extent it covers the literature on energy economics. More on that can be learned from, e.g., Eichhorn et al. [5], Ayres and Warr [6], Hall and Klitgaard [7], Herrmann-Pillat [8], and Ayres [9]. The mathematics, on which our principal findings are based, is packed in an appendix.

2. Basic Physics

Whenever something happens, energy is converted and entropy is produced. This summarizes the First and the Second Law of Thermodynamics. More precisely, the First Law on the conservation of energy says that *energy* consists of the never changing sum of *exergy* and *anergy*. Exergy (with x) is the valuable part of energy, which can be converted into any form of work that is needed to cause a change, and anergy is the useless part of energy, e.g., heat dumped into the environment. Primary energy such as solar radiation, water power, and—in principle, at sufficiently high process temperatures—fossil and nuclear fuels as well, are 100 percent exergy. The Second Law on the increase of entropy—which is the physical measure of disorder—states that irreversible processes produce entropy. All processes that are not infinitely slow are irreversible. They are triggered by removals of constraints.

Energy-converting activities in natural and economic systems are irreversible. Their entropy production involves heat and particle emissions and destroys exergy. Furthermore, if their impact on the biosphere cannot be balanced by thermal radiation into space and processes activated by the exergy radiated from the Sun to Earth, the living species and their societies face problems of adaptation to environmental changes. In principle, pollution by particles such as SO_2, NO_x, dust, CO_2, and by radioactive waste as well, can be mitigated by appropriate removal techniques and sufficient exergy

inputs [10] (Section 3.6). However, even if the emissions of carbon dioxide and other infrared-active trace gases can be curbed so drastically that the anthropogenic greenhouse effect need not worry us any longer, an increasing use of energy from earth-internal sources will cause considerable climate changes, once the heat barrier at about 3×10^{14} Watts (W) of anthropogenic waste-heat emissions will be surpassed. In 2018, global primary energy consumption was 1.75×10^{13} W, and the power of solar radiation received by Earth is 1.2×10^{17} W. [11]

Nicholas Georgescu-Roegen was the first economist to point out the importance of entropy for economic and social evolution in his seminal book *The Entropy Law and the Economic Process* [12]. It stimulated new research on thermodynamics and economics [13–16]. However, claiming to have discovered a "fourth law of thermodynamics" on the dissipation of matter [17,18] he had created some confusion. This was resolved, when it became clear that the dissipation of matter is included in the Second Law of Thermodynamics [19] via the particle-current-density terms, which are one component of the non-negative density of entropy production derived in non-equilibrium thermodynamics [20]; see also [10] (p. 154ff) and [21].

The *empirical* laws on energy conservation and entropy production are the most powerful laws of nature. Any theory that is against them is doomed to failure.

3. Wealth Production and Growth in Conventional Economics

3.1. Concepts of Agrarian Society

In 1776, Adam Smith's "The Wealth of Nations" was published, James Watt's first steam engines were installed in commercial enterprises, and the "Declaration of Independence" was approved by the Second Continental Congress in Philadelphia. "The Wealth of Nations" founded market economics, the steam engine triggered the industrial revolution, and the "Declaration of Independence" proclaimed the human rights, among them "life, liberty, and the pursuit of happiness." The human rights and market economics would not have become ruling principles of free societies had not steam engines and more advanced heat engines provided the energy services that liberate humans from drudgery.

The 18th century had only the Aristotelian notion of *energeia* as a philosophical concept for action or force; entropy was unknown. Adam Smith's economic world was that of the agrarian society, in which the wealth of nations had been produced for about 10,000 years by the factors capital, labor, and land [22]. Nobody saw that energy is present in so many forms such as light, fire, flowing water, wind, wood, wheat, meat, gun powder, and coal.

Only in the 19th century, when investigating the processes of industrial production, people in the natural sciences and engineering discovered energy and entropy and their pivotal role in these processes. In addition, today we know that our universe started about 14 billion years ago, when all its energy, concentrated in a "point", exploded in the Big Bang. Since then all entities of the physical world have evolved from energy, while entropy increases.

In the tradition of Adam Smith, conventional neoclassical textbook economics has worked with the production factors capital, labor, and land until these days. The modern concept of capital includes all energy-conversion devices and information processors, and all buildings and installations necessary for their protection and operation. Energy activates the capital stock and labor handles it. Nevertheless, energy is usually not considered to be a factor of production, despite Tryon's early observation: "Anything as important in industrial life as power deserves more attention than it has yet received from economists ... A theory of production that will really explain how wealth is produced must analyze the contribution of the element energy." [23] Rather, energy has been and still is considered as just one of the many elements in the basket of natural resources, about which the Nobel laureate in economics R.M. Solow [24] stated: "The world can, in effect get along without natural resources", adding, however, that "if real output per unit of resource is effectively bounded—cannot exceed some upper limit of productivity which in turn is not far from where we are now—then catastrophe is unavoidable." Since the useful component exergy of the "natural resource" energy is

unavoidably diminished by entropy production in every economic process, real output per unit of energy *is* effectively bounded. Are we, therefore, heading for catastrophe?

3.2. Economic Growth, Its Actual Importance, and Neoclassical Theory

Obviously, people fear that industrial free-market economies cannot evolve in stability without the economic growth we have known so far. The growth of gross domestic product (GDP) is considered to be vital for the following reasons. The GDP sums up all salaried economic activities that produce the output of value added within a country. It is measured in monetary units [25]. It includes services that mitigate the damages from accidents, crime, pollution, and other harmful occurrences, such as the abuse of drugs and alcohol, and it excludes the domestic care of people for their children and parents, housekeeping by family members, and community services. Thus, it does not measure the overall well being of a country's population. This is common knowledge. Nevertheless, the growth of GDP and the growth of the output of economic sectors such as agriculture, industry and services, are of eminent political and social importance, because GDP measures economic activities. People appreciate these activities, notwithstanding their negative side effects, and go where the action is; this drives the rural exodus to the urban centers. One important reason is that economic activities provide jobs, especially when economic growth opens up new fields whose jobs make up for the traditional jobs that are lost to progress in automation. Thus, voters tend to reelect governments that rule in times of growth, and oust the ones they hold responsible for economic recessions. Migrants from less industrialized parts of the world with low GDP/capita risk their lives to get into highly industrialized countries with high GDP/capita. When in 2020 the Covid-19 pandemic drove the world into the deepest recession since the turn of the century, many billions of US Dollars, Yuans, Yens, and Euros were spent by governments, indebting their countries heavily, in order to reestablish economic growth.

The mainstream neoclassical economic theory of production and growth describes the output Y of goods and services, which is the gross domestic product or parts thereof, by a function of the inputs of capital K and labor L [26]. One special type of such a macroeconomic production function, the Cobb-Douglas function of K and L, had been used by Solow [27,28] in his ground-breaking contribution to the theory of economic growth. He discovered what is called the "Solow residual". This residual is the big difference between the observed economic growth and the much smaller theoretical growth computed with the empirical data of capital and labor. Solow proposed that "technological progress" is responsible for that part of growth that capital and labor cannot explain. Since then, neoclassical growth theory has been based on production functions $Y_{nc}(K, L; t)$ with the factor inputs K and L and a "technological progress" component that depends on time t and is determined by minimizing the Solow residual.

3.3. Oil-Price Shocks

Between 1973–1975 the oil price on the world market nearly tripled when OPEC "punished the West" for supporting Israel in the Yom-Kippur war. The resulting first oil-price shock interrupted the strong economic growth enjoyed after World War 2 especially by the G7 countries Canada, France, the FRG, Italy, Japan, the United Kingdom, and the USA [29,30]. For instance, within these two years the output slumped by more than 5 percent and by nearly 6 percent in the industrial sectors of the USA and the FRG, respectively; simultaneously, these sectors' energy use dropped by more than 7 percent in the USA and more than 8 percent in the FRG [31] (p. 200). Another recession was caused by the second oil-price shock between 1979–1981, when the inflation-corrected market price of oil doubled, shooting up to its 20th century maximum, as a consequence of Iraq's attack on revolutionary Iran and the curb of oil supply from these two major exporters.

The drastic downturns and upswings of economic output and energy use, induced by the oil-price shocks, led economists, in studies such as that by [32–36], to treat energy E as a third factor of production on an equal footing with capital K and labor L, and describe output and its growth by different types of production functions $Y_{nc}(K, L, E; t)$. In a controversial discussion on whether the

first oil-price shock could have been related to the 1973–1975 recession in the USA, the econometrician Denison [37] argued: "Energy gets about 5 percent of the total input weight in the business sector . . . the value of primary energy used by nonresidential business can be put at $42 billion in 1975, which was 4.6 percent of a $ 916 nonresidential business national income. . . . If . . . the weight of energy is 5 percent, a 1 percent reduction in energy consumption with no changes in capital and labor would reduce output by 0.05 percent."

Denison's argument is based on the cost-share theorem, one of the pillars of neoclassical growth theory. The cost-share theorem says that a production factor's economic weight—more precisely: its *output elasticity*, see below—must be equal to the factor's share in total factor cost. In the G7 countries the cost shares have been roughly 25 percent for capital, 70 percent for labor, and 5 percent for energy. Thus, a 7 percent reduction of energy input, as it was observed for the industrial sector of the USA between 1973 and 1975, should have resulted in a (5 percent)×(7 percent) = 0.35 percent reduction of output. As mentioned above, the actually observed output reduction was more than 5 percent.

Consequently, neoclassical production functions $Y_{nc}(K, L, E; t)$ with cost-share weighting of K, L, E neither reproduce the recessions and recoveries spurred by the oil-price explosions, nor can they get rid of Solow residuals without neoclassical "technological progress" functions. From the perspective of orthodox economics energy, even if taken into account as a production factor, matters little in economic growth.

This may lead to illusions about easy paths to sustainability: W. Nordhaus received the 2018 Nobel Price in Economics for his research on climate economics. In his book "A Question of Balance. Weighing the Options on Global Warming Policies" [38] (p. 34) he weighs energy's contribution to production and growth by its cost share [39–43]. Neoclassical growth models are used in integrated assessment models of climate change. Climate activists invoke "the results of science" and demand a rapid and "courageous" exit from the use of oil, gas and coal, which presently satisfy more than 83% of world energy demand. If energy really had an economic weight of only a few percent, a precipitous ban of fossil energy technologies would not cause major economic problems, even if investments in renewables, which are to substitute fossil fuels, should fall way behind. Sufficient were "to wake up politicians" so that they promote the appropriate "technological progress"— whatever that may be.

The dominating role of technological progress "has led to a criticism of the neoclassical model: it is a theory of growth that leaves the main factor in economic growth unexplained", as the founder of neoclassical growth theory, Robert M. Solow, stated himself [44]. Endogenizing technological progress [45–47] does not change the disdain of energy.

The cost-share theorem, which assigns the few-percent weight to energy, results from the conditions for the equilibrium in which an economy is supposed to evolve. These conditions fix the output elasticities of capital, labor and energy in mainstream economics. Roughly speaking, the output elasticity of a production factor gives the percentage of output change when the factor changes by 1 percent [48]. It indicates the economic weight, or productive power, of a production factor.

4. Economic Equilibrium and Technological Constraints

Economic growth depends on the preferences of people and technical possibilities. Aspects that matter are:

1. The economic actors choose the *quantities* of factor inputs at time t according to the expected demand for output.
2. Neoclassical economics assumes:
    (a) Entrepreneurs select the factor *combinations* that maximize profit or overall welfare; the latter is represented by time-integrated utility. (Preferences that may result from drives for power and grandeur are not considered.) The optimized factor combinations define the equilibrium in which the economy is supposed to evolve.
    (b) All combinations of K, L, E are possible.

3. Engineering experience, however, is that not all factor combinations are possible:
   (a) One cannot feed more energy into the machines of the capital stock than they are designed for. If one would try, the machines would break down. Thus, the degree $\eta(K, L, E)$ of capital's capacity utilization cannot exceed 100%.
   (b) The possibility of substituting capital and energy for labor by increasing automation increases with the decreasing mass and volume of information processors. Where the transistor replaces the vacuum tube, it is the density of transistors on a microchip that matters. This density, however, is limited by Joule heating and heat conductivity [49]. Thus, the degree of automation at a given time t, $\rho(K, L, E)$, cannot exceed some technological limit $\rho_T(t)$, which trivially, cannot exceed 100%.

The cost-share theorem is invalid, if one or more of the underlying assumptions 1, 2(a), or (2b) are invalid. For the sake of the argument, we do not question 1 and (2a), but focus only on (2b). It turns out to be sufficient to refute the assumption of the general validity of the cost-share theorem by including the constraints 3(a), 3(b) in the optimization of profit/cost, or overall welfare [10,50]. For this, the constraints $\eta(K, L, E) \leq 1$ and $\rho(K, L, E) \leq \rho_T(t)$ are written in the form of equalities $f_\eta(K, L, E; t) = 0$, $f_\rho(K, L, E; t) = 0$ with the help of slack variables K_ρ, L_η, E_η, which are added to K, L, E in the explicit equations for $\eta(K, L, E)$ and $\rho(K, L, E)$. Optimization subject to the technological constraints in the form of equalities is done by adding these constraints, multiplied by the Lagrange multipliers λ_η and λ_ρ, to the objective function. In the case of profit optimization the objective function is output $Y(K, L, E; t)$ minus total factor cost $p_K K + p_L L + p_E E$, where p_K, p_L, p_E are the prices per unit of K, L, E. Defining $(K, L, E) \equiv (X_1, X_2, X_3)$ and $(p_K, p_L, p_E) \equiv (p_1, p_2, p_3)$, and doing the optimization one obtains the equilibrium conditions, which say: The output elasticities of capital ϵ_1, labor ϵ_2, and energy ϵ_3 must be

$$\epsilon_i = \frac{X_i\,[p_i + s_i]}{\sum_{i=1}^{3} X_i\,[p_i + s_i]}, \quad i = 1, 2, 3\,. \tag{1}$$

Here $s_i \equiv -\lambda_\eta \frac{\partial f_\eta}{\partial X_i} - \lambda_\rho \frac{\partial f_\rho}{\partial X_i}$ are (generalized) shadow prices, which map the technological constraints into monetary terms. "Generalized" indicates that there are additional "soft" constraints that prevent entrepreneurs from managing the economy in the state where a technological constraint is exactly binding. In such a state, there would be only two instead of three independent variables (K, L, E) and, thus, less freedom to adjust production to changes of demand or factor availability. Between 1960 and 1990 the industrial sector of the FRG evolved on a path in the cost mountain that is high above the neoclassical cost minimum and more or less parallel to the barrier from the binding constraint $\eta(K, L, E) = 1$ [50]. From experience, entrepreneurs are aware of the technological constraints and steer clear of the barriers formed by them. Only by calling upon "soft constraints" their behavior agrees with the assumption 2(a) of textbook economics. Anyway, decisive is that entrepreneurs know that the assumption 2(b) is wrong. At the energy prices we have known so far, the cost-share theorem is invalid. Optimization of time-integrated utility yields equilibrium conditions such as Equation (1) with somewhat modified s_i [50].

If there were no technological constraints, the Lagrange multipliers λ_η and λ_ρ would be zero, so would be the s_i, and Equation (1) would be reduced to the cost-share theorem that fixes the output elasticities of neoclassical production functions $Y_{nc}(K, L, E; t)$: The numerator is the cost of the production factor X_i, the denominator is the cost of all factors, and the quotient is the cost share.

The technological constraints on factor combinations, ignored in the derivation of the cost-share theorem, drive the wedge between neoclassical growth theory and what really happens in modern economies [51,52].

5. Wealth Production and Growth: A Biophysical Analysis

5.1. General Outline

The cost-share theorem misleads investigations of economic growth. An alternative biophysical analysis disregards this generally invalid theorem. From neoclassical economics it only adopts the concept of the macroeconomic production function [53–55].

Biophysical production functions $Y(K, L, E; t)$ have the independent variables $K(t), L(t)$ and $E(t)$ [56], which the economic actors choose within given technical and legal constraints according to the expected demand for goods and services and the ends they pursue by their economic activities. The Mathematical Appendix, Section 8, presents the basic equations for computing non-neoclassical output elasticities (compatible with (3a) and (3b) of Section 4 above) and the corresponding production functions. The following summarizes that.

$Y(K, L, E; t)$ is a state function of the economic system—just as internal energy and entropy are state functions of thermodynamic systems in (local) equilibrium. As such $Y(K, L, E; t)$ depends only on the actual magnitudes of the variables $K(t), L(t), E(t)$ and not on the path in (KLE)-space along which the system has arrived at them. Consequently, at any fixed time t, the growth rate of output, dY/Y, is unequivocally determined by the growth rates of capital dK/K, labor dL/L, and energy dE/E, and the respective output elasticities. In total, the *growth equation* is $dY/Y = \alpha \cdot dK/K + \beta \cdot dL/L + \gamma \cdot dE/E + \delta \cdot dt/\Delta t$, where the last term takes into account a possible explicit time dependenc of Y. The second-order mixed derivatives of Y with respect to K, L, E must be equal. The resulting three partial differential equations for the output elasticities of capital, α, labor, β, and energy, γ, are coupled by the requirement of "constant returns to scale", which means that $\alpha + \beta + \gamma = 1$ at any fixed time t [57]. They have innumerable solutions. The trivial solutions are the constants $\alpha_0, \beta_0, \gamma_0 = 1 - \alpha_0 - \beta_0$. Non-trivial, i.e., factor-dependent output elasticities are obtained from (asymptotic) boundary conditions that incorporate economic developments such as the one described by the law of diminishing returns. This law, one of the most famous laws of economics [58], says: "At a given state of technology the additional input of a factor, at constant inputs of the other factors, results in an increase of output. Beyond a certain point, however, the additional return from an additional unit of the variable factor will decrease. This decrease is due to the fact that one unit of the increasing factor is combined with less and less quantities of the fixed factors."

$Y(K, L, E; t)$ abstains from the neoclassical "technological progress function". It depends explicitly on time, if the technology parameters, which result as integration constants of the differential equations, do so. The parameters are determined by minimizing the deviations of theoretical from empirical growth, subject to the conditions that output elasticities must be non-negative. They change in time when human ideas, inventions and value decisions, which summarily are called "creativity", change the state of economic systems; δ in the growth equation is the output elasticity of creativity. Creativity, in this context, has positive and negative components such as human rights, the transistor, and to foster agreement, on the one hand, and racism, cheating software in the exhaust control of Diesel cars, and to obstruct cooperation, on the other hand.

5.2. Observed and Computed Economic Growth

Biophysical production functions have been applied to economic growth in highly industrialized countries since 1982 [31]. Recent results for the USA and the FRG from 1960-2013 are reported by Lindenberger et al. [59]. Figure 1 is an example from the sector "Industries" (I) of the FRG. There, the strongest variations of empirical output and inputs occurred. Since 1990 these variations have been influenced by the only territorial enlargement of a major industrial country after World War 2. They test the sensitivity of production functions to technological and structural changes, and political and psychological perturbations as well. Two production functions were utilized for the reproduction of the observed growth: On the one hand the energy-dependent Cobb-Douglas function Y_{CDE}, Equation (7), whose constant output elasticities turn out to be $\alpha_0 = 0.41, \beta_0 = 0.06, \gamma_0 = 0.53$,

and on the other hand the LinEx function Y_{L1}, Equation (9), with *factor-dependent* output elasticities, whose *time-averages* result to be $\bar{\alpha} = 0.28$, $\bar{\beta} = 0.08$, $\bar{\gamma} = 0.64$, and $\bar{\delta} = 0.13$.

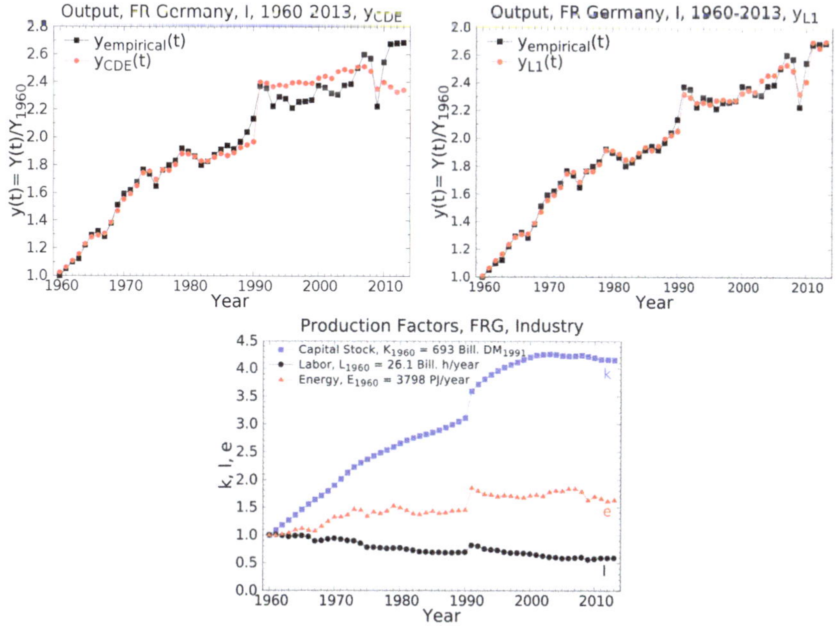

Figure 1. Growth from 1960 to 2013 of the empirical output $y = Y/Y_{1960}$ in the industrial sector of the Federal Republic of Germany (FRG), black squares, and theoretical growth computed with the energy-dependent Cobb-Douglas function, red circles (**left**), and the LinEx function, red circles (**right**). Empirical growth of capital $k = K/K_0$, labor $l = L/L_0$, and energy $e = E/E_0$ (**bottom**). The base year t_0, to which output and inputs are normalized, is 1960. $Y_{1960} = 453.5 \times 10^9$ DM_{1991} [59].

Y_{L1} is the simplest production function of the LinEx-function family, whose members depend *lin*early on one factor, here E, and *ex*ponentially on the quotients of the other factors. More complicated LinEx functions are given in [10,59]. They are all special forms of the general linearly homogeneous, twice differentiable, energy-dependent production functions that solve the growth Equation (2). The latter are shown by Equations (10)–(12) of the Mathematical Appendix, Section 8.

Noteworthy features of empirical and theoretical growth in Figure 1 are:

1. Between 1960 and 1990 the energy-dependent Cobb-Douglas function with its constant output elasticities reproduces observed growth nearly as well as the LinEx function with its factor-dependent output elasticities. After 1990 LinEx is much better. (Its adjusted coefficient of determination is $\bar{R}^2 = 0.99$ and the Durbin-Watson coefficient is $d_w = 1.75$; the statistically best values are 1 and 2, respectively.) Both the time-averaged LinEx and the constant Cobb-Douglas output elasticities are for energy much larger and for labor much smaller than these factors' cost shares. Please note that also the sum of the time-averaged LinEx output elasticities that are related to routine and "creative" activities of humans, $\bar{\beta} + \bar{\delta}$, stays well below energy's output elasticity $\bar{\gamma}$.
2. Creativity's component "value decisions" was activated, when, unexpectedly, the winners of World War 2 agreed to let divided Germany reunite in 1990: Factor inputs and output increase abruptly in 1990. (The LinEx technology parameter "energy demand of the capital stock" does the same [59].)

3. The bidirectional causality that rules the coupling of energy and economic growth shows in the four economic recessions and recoveries and the simultaneous downs and ups of the energy input. Two of them were caused by supply and two by demand, and three were enhanced by feedbacks between the two. The supply side triggered the first and the second oil-price shock 1973–1975 and 1979–1981: The oil-price explosions, caused by OPEC, made investors worry about shortages of power fuel for their machines so that they substantially reduced investments. A demand-side element amplified the shocks: Part of the consumers' buying power had been skimmed by the oil producers. Thus, consumers demanded less goods and services. To satisfy the reduced demand from investors and consumers less energy was needed for production. When the oil-price stopped shooting up, the shocks subsided, and growth of output and energy consumption restarted. Demand-side triggering occurred, when between 1965 and 1966 the ruling conservative-liberal coalition of the FRG became unstable. The resulting economic uncertainties led to reductions of investment, consumption and energy use. Then, for the first time after WW 2, the social democrats became part of the federal government. The new coalition restored confidence in the country, ended the economic crisis, and with increasing demand for goods and services energy consumption rose again. Similarly, the *global* financial crisis 2007–09 was due to a demand-side trigger: After the global breakdown of stock markets, demand for goods and services slumped, machines went idle and did not need energy, until banks were saved by the taxpayers' money so that confidence in the economy came back, and demand for output and energy rose. On the other hand, the burst of the US mortgage bubble, which caused the initial crash of the US stock market, is related by Murray and King [60] to a supply-side effect: Before 2007, the oil price had risen to more than 100 $US\$_{2014}$/barrel. The higly indebted homeowners in the American suburbs were confronted with exploding costs for commuting to their jobs and could not pay their mortgage interests any more.
4. The overall growth of output follows the empirical growth of the capital stock. The latter's flattening and even decrease reflects outsourcing in German industry. The share of the industrial sector in the GDP of the FRG decreased from 51.7% in 1970 to 39.6% in 1992 to 27.1% in 2009 [10] (p. 193) Especially, energy-intensive and polluting industries have been shifted to developing countries and emerging economies. This has stopped the growth of the industrial capital stock and contributes substantially to the reduction of German energy consumption and CO_2-emissions. The decrease of labor input, which is also observed in the total economy of the FRG [10,59], is due to outsourcing and increasing automation.

Growth of output with its ups and downs in the total economies of the FRG and the USA from 1960 to 2013 is also well reproduced by the LinEx function and its factor-dependent output elasticities [59] . Again, the time-averaged output elasticities turn out to be for energy much larger and for labor much smaller than those factors' cost shares. Manrique-Dias and Lemus-Polonia [61] computed economic growth in Colombia from 1925 to 1997. The LinEx function, with "electricity consumption" as the energy variable E, reproduces the empirical growth of Colombian GDP satisfactorily. The output elasticities have time averages similar to the ones of the total economy of the FRG and patterns of temporal variations that somehow resemble those of the total US economy.

Using "useful work" instead of primary energy in a formally modified LinEx function Ayres and Warr [6] computed economic growth in the USA and Japan from 1900 to 2005 (excluding 1941–1948) in good agreement with observed growth. Useful work is the exergy that works directly from the machines on materials plus the physical work performed by animals. The data on it in [62] incorporate efficiency improvements of the energy-converting systems. The magnitudes of the output elasticities that result from this analysis contradict the cost-share theorem, too. This analysis stimulated more research on "exergy economics", such as [63]. Earlier studies on the pivotal role of energy in economic growth led Hall et al. to emphasize "the need to reintegrate the natural laws with economics" [64] .

Computation of future economic growth could be done via scenarios concerning entrepreneurial choices of capital, labor, and energy, in which the crises ahead will challenge creativity. For this,

models such as the HARMONEY model [65], a long-term dynamic growth model that endogenously links biophysical and economic variables in a stock-flow consistent manner, may be useful. Furthermore, production functions with output elasticities that take into account the impact of emission mitigation [21], may also serve as analytical tools. Consistent data on capital, labor, and energy in different sectors of the economy will be important. Studies on past growth have shown that inconsistent data lead to breakdowns of production-function estimations. The sources and structures of the data used in our most recent study on energy and economic growth are documented in [59] (Appendix 3).

6. Crises and Creativity

The strong coupling between energy and economic growth via bidirectional causality has shown especially in times of crises. There have been and will be crises related to politics and markets, and crises involving natural challenges and human responses.

6.1. Politics and Markets

Initially, the two economic recessions in 1973–1975 and 1979–1981 were called "energy crises". However, "oil-price shocks" better indicates the psychology involved. After the oil-price had settled on its 1975 level, the shock wore off, and output resumed growth despite the tripled oil price. The cost share of all energy carriers in total factor cost was still much lower than energy's productive power. Even the next oil-price explosion in 1979 did not change this. However, it caused the second shock and the resulting recession. After the Iraq-Iran war the oil-price collapsed [66], the economic actors in the market economies relaxed, and growth restarted from about the 1978 level. To the recovery also contributed the development of nuclear energy, the discovery of new, non-OPEC oil fields, and the reinvestment of petro dollars in the G7 countries. Here, the solutions to the crises came from the easing of tensions in international politics and markets, the opening up of new energy sources, and the self-interest of the owners of surplus petro dollars.

The 1965–1967 crisis in the FRG ended with the recovery of political stability. The 2007–2009 financial and economic crisis was overcome when central banks, especially the FED and the ECB, did "Whatever it may take" to help tattered firms with direct or indirect subsidies and battered states with bond purchases and cuts of interest rates. This contributed to the mounting public debt and losses on bank deposits.

On May 5, 2020, the Federal Constitutional Court of the FRG, after several years of legal deliberations, ruled that the Public Sector Purchase Program (PSPP) of the ECB has violated the principle of comparativeness insofar as government bonds were also purchased with the aim to keep the inflation rate close to 2%. According to the estimation of the ECB, if inflation were less, deflation would hamper economic growth. Actual inflation had been below the 2% level, because the price of a barrel of crude oil had dropped from nearly 120 US$$_{2014}$ in 2012 to less than 40 US$$_{2014}$ in 2014. Since then it had been fluctuating somewhat until the end of the decade. The prices of most other consumption goods, however, had risen so much that consumers did *not* delay spending in expectation of deflation. However, obviously, the ECB considers energy as just another commodity. A better understanding of the impact of energy and its price on economic growth by decision makers would have avoided that, in the worst case, the Central Bank of Germany will be forced to withdraw from the ECB.

Eichhorn and Solte analyzed the global financial system. They point out that in 2008, new indebtedness of public sector entities world wide was higher than global savings performance, and that global securitized assets exceeded the global stock of central bank money—the only legal tender—by a factor of 50. In the 40 years before, global financial and tangible assets grew more rapidly than global value added (GDP). If the past trends of interest and return on investment (ROI) were to continue in the future, by the year 2030 all of global GDP would be necessary to service the accumulated debts. Nothing would be left to pay employees. [67] (pp. 190–193).

In the long run the most dangerous crises in the field of politics and markets may originate from the inequalities of wealth distribution on national and international scales and their consequences of civic unrest and international conflicts. The inequality of income distribution *within* several OECD countries has been measured by the Luxembourg Income Study [68] by means of the Gini coefficients $G, 0 \leq G \leq 1$, which result from those countries' Lorentz curves [10] (p. 185). The larger G the higher the inequality. According to the study, in the mid-1980s G was close to 20% for Finland, Sweden and Norway, and it exceeded 30% for Switzerland, Ireland, and the USA. The *global* inequality of *wealth* distribution in 2005 is indicated by the shares of the rich and the poor in world's private consumption of *goods and services* per wealth/poverty level [10] (p. 232f), [69]. The wealthiest 10 percent of world's population had a share of 59% of world's private consumption, whereas the share of the world's poorest 50 percent was just 7.2%. By 2005 approximately half the world's population lived in cities and towns, where one out of three urban dwellers (approximately 1 billion people) was living in slum conditions. In developing countries some 2.5 billion humans were forced to rely on biomass—fuelwood, charcoal and animal dung—to meet their energy needs for cooking; this sort of biomass is usually not included in the international energy statistics.

Lawrence, Liu, and Yakovenko [70] analyze the global probability distribution of *energy* consumption per capita around the world from 1980–2010. This impressively complements the statistics on global wealth distribution. Their Lorentz curves "Fraction of World Energy Consumption" vs. "Fraction of World Population" involve the USA, USSR/Russia, France, the UK, China, Brasil, and India, and correspond to Gini coefficients G of 0.66 in 1980, 0.64 in 1990, 0.62 in 2000, and 0.55 in 2010. Thus, within 30 years the global inequality of *energy* consumption per capita has decreased [71]. However, still 70 percent of the world's population in developing and emerging economies had a fraction of less than 40 percent of world energy consumption in 2010. The remaining more than 60 percent of energy consumption went to the 30 percent of world population in the industrialized countries. Many of the latter belong to the wealthiest ones, with high shares of private consumption and small inequalities of income distribution, i.e., Gini coefficients not much above 30%, as mentioned above.

The statistical findings on the distributions of wealth and energy consumption support the econometric findings that energy is an important factor in the production of wealth.

Since the 1960s, the programs of development assistance have aimed at fostering the well being of the people in the developing countries by (a) increasing their countries' GDP and (b) by reducing the inequalities of internal wealth distribution. Aim (a) has been reached to some extent by promoting industrialization and energy consumption world wide. Progress in reaching aim (b) has been slow. It may be advanced by appropriate energy taxation and/or an international agreement on preventing the flight of capital from the developing countries to the highly industrialized countries. However, the threats from emissions and climate change because of entropy production may endanger even further progress towards aim (a). In addition, even more disquieting, Lawrence, Liu and Yakovenko deduce from the principle of maximum entropy production that one may never achieve a less unequal distribution of global energy consumption than the one represented by the Lorentz curve with a Gini coefficient of 0.5 in [70] (Figure 3). The expectation that this may also lead to a corresponding stable global inequality in the distribution of CO_2-emissions has been recently confirmed [72]. Are we approaching a stagnation in which "the world is likely to stay put in the present state of global inequality", because "human development for centuries was driven by geographic expansion, but this era is over" [70] (p. 5573)?

Space industrialization with solar power satellites, discussed below, may provide a way out of stagnation. It may also provide the last resort (for some), if outbreaks of supervolcanoes with high extinction potential that lurk below the Yellowstone Park and the Phlegraean Fields materialize.

6.2. Natural Challenges and Human Responses

1. Risk assessments of energy resources and technologies

On March 11, 2011, one of the worst earthquakes in the history of Japan, and the tsunami it caused,

destroyed the Fukushima 1 nuclear power facility erected right on the Pacific ring of fire on the east coast of Japan. The earthquake severed the connection to the electricity grid and the Tsunami inundated the emergency generators of four reactor blocks, built just 10 m above sea level. The emergency shutdown of three reactors worked well. A fourth reactor had been deactivated, and its nuclear fuel rods were cooled in the fuel pit. Because of the lack of cooling, the nuclear waste heat from β-decay could not be removed, three reactors suffered core meltdowns, and the fourth exploded, most likely because of oxygenhydrogen formation in the hall containing the fuel pit [73,74]. On the whole, the radioactive emissions caused by the Fukushima accident were 10 to 20% of the catastrophe in Chernobyl, where a graphite-moderated reactor blew up in a failed safety experiment. Prior incidents in Japanese nuclear power stations in 2005 and 2007 had already shown that their design, adopted from reactors in the USA, had not been modified properly to meet the known risks that exist in Japan. One had decided to accept them.

In the 2009 electoral campaign for the German Bundestag, the ruling coalition under chancellor Dr. Merkel promised that it would extend the legal operation time for the German water-moderated nuclear reactors by up to 14 years. Otherwise, it was said, Germany would not be able to meet her aims of reducing CO_2-emissions. The coalition was reelected with a comfortable majority, and the parliament passed the law on the operation-time extension. Right after the Fukushima catastrophe, in a U-turn of German energy policy called "Energiewende", the government of Dr. Merkel proposed the total exit from nuclear power, and the parliament decided it. Eight reactors were shut off right away, and of the remaining nine the last one is scheduled to cease operation in 2022. In a mix-up of "known risk" and "residual risk" Dr. Merkel told the public that the reason for the U-turn was the underestimation of the residual risk of German nuclear reactors. Actually, the probability that an accident as in Fukushima would occur in Germany is equal to the probability of a heavy earthquake in Germany *and* that a tsunami destroys the emergency generators of four nuclear power plants in the country.

Germany claims a cutting edge in climate protection [75,76]. Experience will show, how she lives up to that claim. After the banning of nuclear power without changing the German road map for reducing CO_2-emissions, renewable energies must fill the gap in electricity generation that would open up, if coal and lignite power plants would be abolished as planned originally. Success or failure of renewable energies will decide, whether, in the end, the "Energiewende" will turn out as either a positive or a negative element of creativity. The uncertainty results from the phenomenon of *size-dependent risk perception*, which is a fundamental problem faced by energy policy everywhere: When an energy source contributes noticeably to the energy supply of an economy, its inevitable side effects will affect the environment. If people notice them, there will be protests, often pursuant to the NIMBY (Not In My BackYard) principle. Side effects that go unnoticed for some time, may become big problems in the future.

Renewables are an example. In 2018 they contributed just 4% to global primary energy consumption [77]. In Germany, their total share in primary energy was about 13%, with the shares of biomass, wind, and photovoltaics being 7.1%, 2.8%, and 1.1%, respectively [78].

(a) Biomass dominates. It is a storage of solar energy and well accepted by the population. However, the National Academy of Sciences (Leopoldina) points out that biomass has a bad *Energy Return on (Energy) Investment* (EROI) [79], mostly below 3, that its production threatens biodiversity, damages soil quality, pollutes ground water, rivers and lakes, and that financially it has the highest price per saved ton of CO_2 [80].

(b) Wind power is heavily attacked by civic movements. The given reasons are: Onshore wind turbines make noise, cast whirling shadows, kill birds, and spoil the landscape. The high-voltage transmission lines that shall carry electric power from offshore wind parks in the wind-rich north of Germany to southern Germany are rejected for esthetic

reasons and their land requirements. The protesters ignore that the specific total life-cycle CO_2-emissions of wind parks are only 10–20 g CO_2 per kilowatt-hour of electric energy—similar to those of nuclear power plants—and the lowest of all renewables.

(c) Photovoltaics (PV), whose specific total life-cycle CO_2-emissions range from 70 to 150 g CO_2 per kilowatt-hour, is still well accepted. To keep it that way the government has tried to limit the payments of the electricity consumers to the providers of PV power to 10–11 billion Euros annually [21]. Looking into the future, GreenMatch, "a comprehensive guide designed to help you navigate the transition to renewable energy" [81] points out the need to recycle PV panels when their life cycle ends: "If recycling processes were not put in place, there would be 60 million tons of PV panels waste lying in landfills by the year 2050; since all PV cells contain certain amounts of toxic substances that would truly become a not-so-sustainable way of sourcing energy." GreenMatch estimates the amount of solar panel waste (in tons) to be for (a) the USA in 2016: 6500 t, 2030: 400,000 t, 2050: 7500,000 t, (b) Germany in 2016: 3500 t, 2030: 400,000 t, 2050: 4300,000 t, (c) Saudi Arabia in 2016: 200 t, 2030: 3500 t, 2050: 450,000 t. The energy requirements for recycling these quantities of PV waste, and the associated emissions and cost, remain to be estimated.

2. Pandemics
The economic instruments to fight the 2007-09 financial and economic crisis have been reactivated in the Corona crisis that started with the outbreak of the Covid-19 pandemic in Wuhan, China, by the end of 2019. Since then, severe constraints on the interaction between people have been imposed by governments all over the world and successively strangulated commercial, artistic and educational activities. Employment slumped. This has dwarfed the demand for many goods and services, their production ceased, and so did the demand for energy. Occasionally, the oil price even became negative, when the producers of conventional oil and the US-producers of oil from fracking would not or could not reduce oil production, while all the oil-storage facilities were filled up. As in the 2007-09 crisis, the actions of governments and central banks to stabilize economies—and this time also public health—boost public debt. To complicate things, health and environmental protection must be balanced with economic and social losses. The G7-countries are especially vulnerable to the constraints imposed on personal interactions in times of pandemics such as Corona, because the share of their service sectors in both employment and GDP has been roughly 70% since the turn of the century [10] (p. 193)

3. Limits to growth in the biosphere
Two ways of dealing with the thermodynamic limits to industrial growth in the biosphere are (a) to adapt to them via transition to a post-growth economy, and (b) to surmount them via space industrialization.

(a) Niko Paech [82] proposes that the highly industrialized societies adapt to the ecological constraints that exist on the surface of Earth, by changing lifestyles and patterns of supply. This implies a cultural change to sufficiency, and it involves three levels: local subsistence, a regional economy, and a significantly shriveled residual industry. To cushion the reductive transition socially, especially to achieve full employment, a reallocation of the reduced time for gainful occupation will be necessary. 20 h of conventional labor, which are the basis for a reduced monetary income, can be complemented by another 20 h of working for self-sufficiency. Indigenous production, extension of service life, collective use of capital goods etc. will help to continue the use of modern consumption functions and simultaneously realize a higher degree of economic autonomy. Firms can support this development by contributing in many ways to satisfying needs without actually producing new goods.

Contrary to happy "green" utopias, Niko Paech's transition scheme to a post-growth economy is sober and realistic. Sober, because it clearly tells people what drastic changes of personal behavior will be necessary. Realistic, because it combines well-known elements of the stationary societies, in which human civilizations have evolved during the last 10,000 years, with the production facilities of the industrial age, whose growth dynamics now threatens the stability of the biosphere. The problem is that the stationary societies of the past had rigid social structures with little social mobility. Traveling for pleasure was unusual.

Nieto et al. [83] applied an ecological macroeconomics model to the Energy Roadmap 2050 (ER2050) of the European Union; this roadmap has ambitious emission-mitigation targetes, to be achieved by reducing energy use and a transition to renewables. Their "results show that GDP growth and employment creation may be halted due to energy scarcity if the ER2050 targets are met even considering great energy efficiency gains. In addition, the renewables share would increase enough to reduce the energy imports dependency, but not sufficiently to meet the emission targets. Only a Post-Growth scenario would be able to meet the climate goals and maintain the level of employment."

In the present Covid-19 pandemic, people suffer from and complain about constraints on professional and leisure activities, many of which are linked to industrialization. Perhaps we can learn from the pandemic how well modern humans will accept the changes of lifestyle, and of the production and distribution of wealth, which may be necessary for adaptation to the stationary society of a Post-Growth age.

(b) Ancient and modern history tell tales of expansion, when resources become scarce and pioneers, full of vigor and zest for action, set out for new territories with wide-stretching frontiers. The scarce resource of the past was fertile land, whose plants capture the solar energy needed by humans and animals.

Presently, scarce is the space that, without harmful side effects, can absorb the emissions of industrial energy conversion. However, vast is the space beyond the biosphere. For more than four billion years it has absorbed all heat and particle emissions that accompany the production of life-giving sunlight by nuclear fusion in the core of the Sun. Being aware of this, since the early 1970s, and for about two decades, young, middle-aged, and old scientists from many disciplines had tried to promote a grand design of using extraterrestrial resources to surmount the limits to growth. It implies delivery of clean electric energy to Earth via solar power satellites (SPS) and the production of them in space-manufacturing facilities by people who live in large habitats that orbit around the Lagrange libration point L5. The sources of most of the required energy and materials would be the Sun and the Moon.

Peter E. Glaser from Arthur D. Little, Inc., proposed and patented solar power satellites [84–86]. They are to be stationed in geosynchronous Earth orbit, always above the same point on the equator at a maximum distance of 35,785 km. They convert sunlight into electric energy, either by photovoltaic cells or by solar thermal dynamic systems. Klystrons convert the electric energy into microwaves of about 3-GHz frequency, which are beamed from a transmitting antenna, diameter 1km, of the satellite to a receiving antenna on Earth, diameter 10 km. There, the microwave energy is reconverted into electricity, which is fed into the public grid. Typical generating capacities of SPS are 5000–10,000 MW at bus bar on Earth. The total mass of a SPS is between 34,000 and 86,000 t. This and more, e.g., Boeing's SPS design and NASA's system studies, is documented in [87–89].

> The big problem is transportation of people and initially required materials to low Earth orbit via chemical rockets. After the catastrophes of the Challenger and Columbia space shuttles in 1986 and 2003, the USA terminated the Space Shuttle Program in 2011. Since then, for the transportation of US astronauts to the International Space Station (ISS), the USA have bought seats in the Russian Sojus rockets. Finally, US billionaires are coming to the rescue of the US space program. For instance, Elon Musk's commercial "Space X" enterprise builds reusable rockets and space capsules for the transportation of freight and astronauts. There are plans to return to the Moon and go to Mars [90]. China is vigorously pursuing such plans, too. Once on the Moon, one could resuscitate the grand scheme of Princeton physics professor Gerard K. O'Neill to catapult Moon-material via electromagnetic mass drivers [10] (p. 88f) to catchers in the libration point L2 and transfer it to space-manufacturing facilities. There, SPS and habitats for the people who construct and maintain them, would be built [91–94]. Outside the gravitational abysses of planets, traveling large distances requires little energy. O'Neill's scheme to open up "The High Frontier" of space for humanity led Representative Olin Teague to present the "House Concurrent Resolution 451" [93] to the 95 Congress of the USA on 15 December 1977. It was referred to the Committee on Science and Technology and closes with the words: "Whereas the 'High Frontier' of Space does provide valid opportunities whereby we can conserve and enhance humanity's existence on Earth, including but not limited to such social and economic benefits as greater employment, a cleaner environment, new energy sources, new knowledge and understanding. ..: Now, therefore be it *Resolved by the House of Representatives (the Senate concurring)* ...: the Office for Technology Assessment specifically is requested to organize and manage a thorough study and analysis to determine the feasibility, potential consequences, advantages and disadvantages of developing as a national goal for the year 2000 the first manned structures in space for the conversion of solar energy and other extraterrestrial resources to the peaceable and practical use of human beings everywhere."

On November 9, 1989 the Berlin Wall came down. Thereafter, the Iron Curtain dissolved, and the Cold War with its threat of humankind's self-destruction ended. However, the competitive pursuit of power, ingrained in human nature, continues. In the 20th century, those who ruled the seas and the air dominated the world. In the 21st century, the powers in space will become the masters of Earth. If the colonization of space is forgone, humans must tame their competitive drives and dedicate their resources and creativity to dealing with the thermodyamic limits to growth. In either case, cooperation between individuals and nations in strict observation of the constraints from human and natural laws will be needed more than ever.

7. Summary and Conclusions

The laws of physics on energy conversion and entropy production have stimulated economic growth analyses via biophysical production functions of capital, labor, and energy. They are solutions of a set of differential equations and their asymptotic boundary conditions. Three efficiency-related integration constants may become time dependent when human ideas, inventions and value decisions, in short: "creativity", change the state of the economy. The biophysical production functions and their estimation disregard the cost-share theorem of neoclassical economics, because it is flawed: When optimizing profit or overall welfare, one must take into account the technological constraints on factor combinations; these, however, were ignored in the neoclassical derivation of the cost-share theorem. This theorem, which assigns only a small economic weight to energy, is invalid at the low energy prices we have known so far.

The biophysical analyses well reproduce the observed economic growth and its crises in major industrial countries during more than 50 years. The resulting economic weights (output elasticities)

are for energy much larger and for labor much smaller than these factors' shares in total factor cost. While creativity is qualitatively decisive in the long run, its quantitative contribution to growth is much smaller than the one that neoclassical growth theory assigns to "technological progress".

In highly industrialized countries the growth of gross domestic product, and parts thereof, follows the growth of the capital stock. Despite the outsourcing of energy-intensive industries and the shifting of production to the service sector, in times of economic recessions and recoveries economic output and energy consumption decrease and increase simultaneously. This shows the bidirectional causality between energy and economic growth, which follows from energy's economic role of activating the capital stock.

Since energy conversion is a powerful driver of industrial growth, and since it is inevitably coupled to emissions of particles and heat via the entropy law, the stability of the biosphere is threatened. Understanding the production and growth of wealth, and careful assessments of the risks and opportunities involved with energy sources and the technologies of their use, are necessary for successful adaptation to the ecological constraints on growth. Experiences from past crises should be remembered. Once the feasible options for adequate technological and social changes are identified, people will hopefully follow creative leadership on the most promising path of future economic evolution.

8. Mathematical Appendix

The total differential of the production function $Y(K, L, E; t)$, divided by the production function itself, yields the growth equation:

$$\frac{dY}{Y} = \alpha\frac{dK}{K} + \beta\frac{dL}{L} + \gamma\frac{dE}{E} + \delta\frac{dt}{\Delta t}, \qquad \delta \equiv \frac{\Delta t}{Y}\frac{\partial Y}{\partial t}, \tag{2}$$

where

$$\alpha \equiv \frac{K}{Y}\frac{\partial Y}{\partial K}, \; \beta \equiv \frac{L}{Y}\frac{\partial Y}{\partial L}, \; \gamma \equiv \frac{E}{Y}\frac{\partial Y}{\partial E} \tag{3}$$

are the *output elasticities* (productive powers) of capital, labor, and energy, respectively. δ in Equation (2) results formally from the explicit time dependence of the production function via time-dependent technology parameters and economically from the influences of human ideas, inventions and value decisions on economic evolution. These influences are summarized by the concept of *creativity*; $\Delta t = t - t_0$, where t_0 is an arbitrary base year with the factor inputs K_0, L_0, E_0.

Since $Y(K, L, E; t)$ is a state function, its second-order mixed derivatives with respect to K, L, E must be equal. Calculating these derivatives from the growth Equation (2) one obtains the integrability conditions

$$L\frac{\partial\alpha}{\partial L} = K\frac{\partial\beta}{\partial K}, \quad E\frac{\partial\beta}{\partial E} = L\frac{\partial\gamma}{\partial L}, \quad K\frac{\partial\gamma}{\partial K} = E\frac{\partial\alpha}{\partial E}. \tag{4}$$

The growth equation is integrated at a fixed time t, when the production factors are $K = K(t), L = L(t), E = E(t)$. The integral of the left-hand side from $Y_0(t)$ to $Y(K, L, E; t)$ is $\ln \frac{Y(K,L,E;t)}{Y_0(t)}$. It is equal to the integral of the right-hand side:

$$F(K, L, E)_t \equiv \int_{P_0}^{P} \left[\alpha\frac{dK}{K} + \beta\frac{dL}{L} + \gamma\frac{dE}{E}\right] ds \,. \tag{5}$$

This integral can be evaluated along any convenient path s in factor space from an initial point P_0 at (K_0, L_0, E_0) to the final point P at $(K(t), L(t), E(t))$. With $\ln \frac{Y(K,L,E;t)}{Y_0(t)} = F(K, L, E)_t$ the production function becomes

$$Y(K, L, E; t) = Y_0(t) \exp\{F(K, L, E)_t\} \,. \tag{6}$$

The integration constant $Y_0(t)$ is the monetary value of the basket of goods services at time t, if it were produced by the factors K_0, L_0, and E_0. If creativity were dormant during the time interval $t - t_0$, $Y_0(t)$ would also be equal to the output at time t_0.

The partial differential Equations (4) turn into three coupled partial differential equations for α and β, if one uses $\gamma = 1 - \alpha - \beta$ according to "constant returns to scale", as substantiated in Section 5.1.

The trivial solutions of these differential equations are the *constant* output elasticities α_0, β_0 and $\gamma_0 = 1 - \alpha_0 - \beta_0$. With them, and Equations (5) and (6), one obtains

$$Y_{CDE}(K, L, E; t) = Y_0(t) \left(\frac{K}{K_0}\right)^{\alpha_0} \left(\frac{L}{L_0}\right)^{\beta_0} \left(\frac{E}{E_0}\right)^{1-\alpha_0-\beta_0}. \tag{7}$$

This is the simplest energy-dependent production function. It bears the names of Cobb and Douglas, who had constructed a function of such structure, but *without* energy, in the 1920s. The Cobb-Douglas function of capital and labor has been and still is frequently used in neoclassical economics. The simplest non-trivial solutions are the *factor dependent* output elasticities

$$\alpha = a\frac{(L/L_0 + E/E_0)}{K/K_0}, \quad \beta = a\left(c\frac{L/L_0}{E/E_0} - \frac{L/L_0}{K/K_0}\right), \quad \gamma = 1 - a\frac{E/E_0}{K/K_0} - ac\frac{L/L_0}{E/E_0}. \tag{8}$$

The output elasticity of capital, α, satisfies in the simplest way the law of diminishing returns, β is the simplest solution of the partial differential equation that couples α and β, and γ results from constant returns to scale. (More details on the factor dependencies of α and β in view of the capital stock's degrees of utilization and automation are given in [10,31].) With them and Equations (5) and (6) one obtains the (first) LinEx function

$$Y_{L1}(K, L, E; t) = Y_0(t)\frac{E}{E_0} \exp\left[a\left(2 - \frac{L/L_0 + E/E_0}{K/K_0}\right) + ac\left(\frac{L/L_0}{E/E_0} - 1\right)\right]. \tag{9}$$

The parameter c measures the energy demand of the fully utilized capital stock, and the parameter a is a measure of capital's effectiveness in producing output when activated by energy and handled by labor. The technology parameters a and c become time dependent, when creativity is active. They, and $Y_0(t)$, are determined by minimizing the sum of squared errors over all observation times t_i, i.e., $SSE = \Sigma_i |Y_{empirical}(t_i) - Y_{theoretical}(t_i)|^2$, subject to the constraints $\alpha \geq 0, \beta \geq 0, \gamma \geq 0$; the Levenberg-Marquardt algorithm in combination with the Ceres Solver statistics program was applied to this problem of non-linear optimization in [59] (p. 9).

The most general production function, in which the output elasticity of energy is known from $\gamma = 1 - \alpha - \beta$, and α and β have to be determined from their three coupled partial differential equations and appropriate asymptotic boundary conditions, is

$$Y = E\mathcal{F}\left(\frac{L}{K}, \frac{E}{K}\right). \tag{10}$$

Production functions of the general type (10), especially the LinEx function (9), have been used to analyze economic growth in [6,10,31,50,59,61,64], and references therein [95].

The most general production function, in which the output elasticity of labor is known from $\beta = 1 - \alpha - \gamma$, whereas α and γ have to be determined from their three coupled partial differential equations and appropriate asymptotic boundary conditions, is

$$Y = L\mathcal{G}\left(\frac{L}{K}, \frac{E}{K}\right). \tag{11}$$

A special, LinEx-type function of this form has been used to describe the growth of service industries, which also include increasingly digitized processes, e.g., in banking, insurance, and public

administration [96]. Another type may be interpreted as describing the evolution of economies in an early state of industrialization.

Finally, the most general production function, in which the output elasticity of capital is known from $\alpha = 1 - \beta - \gamma$, and β and γ must be determined from their three coupled partial differential equations and appropriate asymptotic boundary conditions, is

$$Y = K\mathcal{H}\left(\frac{L}{K}, \frac{E}{K}\right). \tag{12}$$

The simplest LinEx-type production function of this form may describe a future state of total digitization.

$\mathcal{F}, \mathcal{G}, \mathcal{H}$ are twice differentiable with respect to L/K and E/K.

Author Contributions: This review article is based on the research of its two authors and their collaborators. R.K. wrote the draft of the manuscript, and D.L. improved it. All authors have read and agreed to the published version of the manuscript.

Funding: This research received no external funding.

Acknowledgments: We thank three anonymous reviewers for stimulating comments.

Conflicts of Interest: The authors declare no conflict of interest.

References and Notes

1. That was a much better treatment than the one proposed by the abolished Morgenthau Plan: To transform Germany into an agrarian state.
2. R.K., see also [10] (p. XI).
3. Reif, F. *Fundamentals of Statistical and Thermal Physics*; McGraw-Hill: New York, NY, USA, 1965.
4. Meadows, D.H.; Meadows, D.L.; Randers, J.; Behrens, W.W., III. *The Limits to Growth*; Universe Books: New York, NY, USA, 1972.
5. Eichhorn, W.; Henn, K.; Neumann, K.; Shephard, R.W. (Eds.) *Economic Theory of Natural Resources*; Physica Verlag: Würzburg-Wien, Germany, 1982.
6. Ayres, R.U.; Warr, B. *The Economic Growth Engine*; Edward Elgar: Cheltenham, UK, 2009.
7. Hall, C.S.; Klitgaard, K. *Energy and the Wealth of Nations*; Springer: New York, NY, USA, 2012.
8. Herrmann-Pillath, C. *Foundation of Economic Evolution*; Edward Elgar: Cheltenham, UK, 2013.
9. Ayres, R. *Energy, Complexity, and Wealth Maximization*; Springer: Cham, Switzerland, 2016.
10. Kümmel, R. *The Second Law of Economics: Energy, Entropy, and the Origins of Wealth*; Springer: New York, NY, USA, 2011.
11. Watts (W) measure energy *per unit time*. The BP Statistical Review 2019, p. 8, shows that 13,864.9 MtOE were consumed in the 8760 h of 2018. 1tOE = 11,630 kWh. According to p. 9 of this statistic the shares of the different primary energy carriers in 2018 world energy consumption are: Oil 33.6%, Natural gas 23.8%, Coal 27.2%, Nuclear energy 4.4%, Hydro electricity 6.8%, Renewables 4.0%.
12. Georgescu-Roegen, N. *The Entropy Law and the Economic Process*; Harvard University Press: Cambridge, MA, USA, 1971.
13. van Gool, W.; Bruggink J.J.C. (Eds.) *Energy and Time in the Economic and Physical Sciences*; North-Holland: Amsterdam, The Netherlands, 1985.
14. Faber, M.; Niemes, H.; Stephan, G. *Entropy, Environment, and Resources*; Springer: Berlin, Germany, 1987.
15. Faber, M.; Proops, J. *Evolution, Time, Production, and the Environment*, 2nd ed.; Springer: Berlin, Germany, 1994.
16. Daly, H.E. On Nicholas Georgescu-Roegen's contributions to economics: An obituary essay. *Ecol. Econ.* **1995**, *13*, 149–154. [CrossRef]
17. Georgescu-Roegen, N. *Energy and Economic Myths*; Pergamon: New York, NY, USA, 1976.
18. Georgescu-Roegen, N. The entropy law and the economic process in retrospect. *East. Econ. J.* **1986**, *12*, 3–23.
19. Letters to the Editor: Recycling of Matter. *Ecol. Econ.* **1994**, *9*, 191–196. [CrossRef]
20. Kluge, G.; Neugebauer, G. *Grundlagen der Thermodynamik*; Spektrum Fachverlag: Heidelberg, Germany, 1993.
21. Kümmel, R. The impact of entropy production and emission mitigation on economic growth. *Entropy* **2016**, *18*, 75. [CrossRef]

22. In agrarian societies the factor capital includes hand tools, means of transportation moved by muscle power, water and wind, and buildings to accommodate people, animals, vehicles and tools; the factor land sustains the plants, which by photosynthesis absorb solar energy and convert it into food for humans and animals and wood for construction and fire.
23. Tryon, F.G. An index of consumption of fuel and water power. *J. Am. Stat. Assoc.* **1927**, *22*, 271–282. [CrossRef]
24. Solow, R.M. The economics of resources or the resources of economics. *Am. Econ. Rev.* **1974**, *64*, 1–14.
25. For critique of this measurement see Section 5. Some people even say that the consideration of economic growth does not make any sense at all. Our response to that in this paragraph has partly been taken from Biophys.Econ.Sust. (2020) 5.5, https://doi.org/10.1007/s41247-020-00068-1.
26. Aggrandizement of land is no growth option in times of peace with fixed borders.
27. Solow, R.M. A contribution to the theory of economic growth. *Q. J. Econ. Perspect.* **1956**, *70*, 65–94. [CrossRef]
28. Solow, R.M. Technical change and the aggregate production function. *Rev. Econ. Stat.* **1957**, *39*, 312–320. [CrossRef]
29. In 2013 the G7 countries had a share of 11 percent of global population and of 33 percent of global domestic product (adjusted for purchasing power) [30].
30. O'Donnel, D. *G7 in Figures*; Statistisches Bundesamt: Wiesbaden, Germany, 2015.
31. Kümmel, R. The impact of energy on industrial growth. *Energy* **1982**, *7*, 189–203. [CrossRef]
32. Hudson, E.H.; Jorgenson, D.W. U.S. energy policy and economic growth, 1975–2000. *Bell J. Econ. Manag. Sci.* **1974**, *5*, 461–514. [CrossRef]
33. Griffin, J.M.; Gregory, P.R. An intercountry translog model of energy substitution responses. *Am. Econ. Rev.* **1975**, *66*, 845–857.
34. Berndt, E.R.; Jorgenson, D.W. How energy and its cost enter the productivity equation. *IEEE Spectr.* **1978**, *15*, 50–52. [CrossRef]
35. Berndt, E.R.; Wood, D.O. Engineering and econometric interpretations of energy–capital complementarity. *Am. Econ. Rev.* **1979**, *69*, 342–354.
36. Jorgenson, D.W. The role of energy in productivity growth. *Am. Econ. Rev.* **1984**, *74*, 26–30. [CrossRef]
37. Denison, E.F. Explanation of declining productivity growth. *Surv. Curr. Bus.* **1979**, *59 Pt II*, 1–24.
38. Nordhaus, W. *A Question of Balance: Weighing the Options on Global Warming Policies*; Yale University Press: London, UK, 2008.
39. Herman Daly [40] has summarized the studies of the renowned economists Nordhaus [41], Beckermann [42] and Schelling [43] on the economic impact of climate change, assuming that climate change only concerns agriculture. Since that sector contributed less than 3% to the GDP of the USA in 1992, they concluded that even a drastic decline of agricultural production should only result in small losses of welfare.
40. Daly, H. When smart people make dumb mistakes. *Ecol. Econ.* **2000**, *334*, 1–3. [CrossRef]
41. Nordhaus, W. *Science* **1991**, *1206*. Available online: http://www.paecon.net/PAEReview/issue78/Daly78.pdf (accessed on 6 October 2020)
42. Beckermann, W. *Small is Stupid*; Duckworth: London, UK, 1997.
43. Schelling, T.C. The Cost of Combating Global Warming. *Foreign Aff.* **1997**, *9*, 8–14. [CrossRef]
44. Solow, R.M. Perspectives on growth theory. *J. Econ. Perspect.* **1994**, *8*, 45–54. [CrossRef]
45. Romer, P.M. Increasing returns and long-run growth. *J. Polit. Econ.* **1986**, *94*, 1002–1037. [CrossRef]
46. Lukas, R.E. On the mechanics of economic development. *J. Monet. Econ.* **1988**, *22*, 3–42. [CrossRef]
47. Rebelo, S. Long-run policy analysis and long-run growth. *J. Polit. Econ.* **1991**, *99*, 500. [CrossRef]
48. The precise definition of output elasticities is in the Mathematical Appendix, Section 8.
49. Quantum computers still have a long way to go.
50. Kümmel, R.; Lindenberger, D. How energy conversion drives economic growth far from the equilibrium of neoclassical economics. *New J. Phys.* **2014**, *16*, 125008. [CrossRef]
51. D.I. Stern [52] modified Solow's growth model by adding an energy input. He described *gross output* by a function that "embeds a Cobb-Douglas function of capital (K) and labor (L) in a CES function of value added and energy (E)." The embedded Cobb-Douglas function weighs capital and labor with their cost shares. Time-dependent augmentation indices of labor and energy take care of changes in factor quality and technology. But *gross output* is the sum of GDP *plus* intermediate consumption. As a rule, the theory of economic growth considers GDP.

52. Stern, D.I. The role of energy in economic growth. *Ann. N. Y. Acad. Sci.* **2011**, *1219*, 26–51. [CrossRef] [PubMed]
53. Conceptual objections against macroeconomic production functions such as the ones raised by Joan Robinson [54,55] in the "Cambridge controversy" have been dealt with in [10,31] by defining output and capital in terms of work performance and information processing and relating this to monetary units. In biophysical energy-dependent production functions the production factors are the ones actually used by the economic actors. These inputs are *not* those of neoclassical *production possibility frontiers*, where the degree of factor use is per definition 100 percent.
54. Robinson, J. The production function and the theory of capital. *Rev. Econ. Stud.* **1954**, *21*, 81–106. [CrossRef]
55. Robinson, J. The measure of capital: The end of the controversy. *Econ. J.* **1971**, *81*, 597–602. [CrossRef]
56. The production function's implicit time dependence via the time dependence of $K(t), L(t), E(t)$ is not indicated in $Y(K, L, E; t)$ for notational simplicity.
57. Constant returns to scale implies that the output of two identical production systems is twice the output of one of these systems.
58. Samuelson, P.A. *Economics*, 10th ed.; International Student Edition; MacGraw-Hill Kogagusha LTD.: Tokyo Japan, 1976.
59. Lindenberger, D.; Weiser, F.; Winkler, T.; Kümmel, R. Economic growth in the USA and Germany 1960–2013: The underestimated role of energy. *Biophys. Econ. Resour. Qual.* **2017**, *2*, 10. [CrossRef]
60. Murray, J.; King, D. Oil's tipping point has passed. *Nature* **2012**, *481*, 433–435. [CrossRef]
61. Manrique-Diaz, O.G.; Lemus-Polania, D.F. Procedimiento de optimizacion no lineal para la cuantificacion del aporte de la energia electrica en el crecimiento economico colombiano, 1925–1997. *Lect. Econ.* **2020**, *93*, 65–100. [CrossRef]
62. Ayres, R.U.; Ayres, L.W.; Warr, B. Exergy, power and work in the US economy, 1900–1998. *Energy* **2003**, *28*, 219–273. [CrossRef]
63. Brockway, P.E.; Saunders, H.; Heun, M.K.; Foxon, T.J.; Steinberger, J.K.; Barrett, J.R.; Sorrell, S. Energy rebound as a potential threat to a low-carbon future: Findings from a new exergy-based national-level rebound approach. *Energies* **2017**, *10*, 51. [CrossRef]
64. Hall, C.; Lindenberger, D.; Kümmel, R.; Kroeger, T.; Eichhorn, W. The need to reintegrate the natural sciences with economics. *Bioscience* **2001**, *51*, 663–673. [CrossRef]
65. King, C.W. An integrated biophysical and economic modeling framework for long-term sustainability analysis: The HARMONEY model. *Ecol. Econ.* **2020**, *169*, 106464. [CrossRef]
66. This was the first *negative* oil-price shock for the oil-producing countries. The Soviet Union never recovered from it.
67. Eichhorn, W.; Solte, D. *Das Kartenhaus Weltfinanzsystem*; Fischer Taschenbuch Verlag: Frankfurt, Germany, 2009.
68. Atkinson, A.B.; Rainwater, L.; Smeeding, T.M. *Income Distribition in OECD Countries—Evidence from the Luxembourg Income Study*; OECD: Paris, France, 1995.
69. Shah, A. Poverty Facts and Stats. Global Issues. Available online: http://www.globalissues.org/article/26/poverty-facts-and-stats (accessed on 2 October 2020)
70. Lawrence, S.; Liu, Q.; Yakovenko, V.M. Global inequality in energy consumption from 1980 to 2010. *Entropy* **2013**, *15*, 5565–5579. [CrossRef]
71. The decrease of inequality in energy consumption during the first decade of the 21st century also shows in the increase of the global average of energy consumption from 2 kW/capita in 2004 [10] (p. 61) to 2.5 kW/capita in 2010 [70] (Figure 3, left panel).
72. Semieniuk, G.; Yakovenko, V. Historical evolution of global inequality in carbon emissions and footprints versus redistributive scenarios. *J. Clean. Prod.* **2020**, *264*, 121420. [CrossRef]
73. NISA, JNES. The 2011 off the Pacific Coast of Tohoku Pacific Earthquake and the Seismic Damage to the NPPs, 2011. Available online: www.webcitation.org/5xuhLD1j7 (accessed on 2 October 2020).
74. Gesellschaft für Reaktorsicherheit (GRS). Fukushima Daiichi 11. März 2011: Unfallablauf, Radiologische Folgen, 2. Aufl. GRS, Köln. 2013. Available online: https://www.grs.de/sites/default/files/pdf/GRS-S-53op1.pdf (accessed on 6 October 2020)

75. When assessing German CO_2-emissions one should keep in mind that the official numbers, e.g., 13.2 tons per person and year in 2011, are calculated from the emissions *produced* within the country. If one adds the emission caused by the production of the goods and services imported by and *consumed* in Germany, the number increases to 18.3 t. For comparison, the corresponding emissions in the USA are 23.5 t as per production principle, and 27.9 t as per consumption principle [76].
76. Steininger, K.W.; Lininger, C.; Meyer, L.H.; Muñoz, P.; Schinko, T. Multiple carbon accounting to support just and effective climate policies. *Nat. Clim. Chang.* **2016**, *6*, 35–41. [CrossRef]
77. Section 2, ref. 11.
78. Arbeitsgemeinschaft Energiebilanzen, Stand August 2018, AGEE-Stat.; Stand Ende 2018: AGEB, Energieverbrauch in Deutschland, Tabelle 14, AGEE-Stat. Available online: https://www.ag-energiebilanzen.de/ (6 October 2020)
79. Murphy, D.; Hall, C. Year in review—EROI or energy return on (energy) invested. *Ann. N. Y. Acad. Sci.* **2010**, *1185*, 102–118. [CrossRef]
80. Nationale Akademie der Wissenschaften Leopoldina. *Bioenergie—Möglichkeiten und Grenzen*; Kurzfassung und Empfehlungen; Deutsche Akademie der Naturforscher Leopoldina: Halle, Germany, 2010; pp. 8, 11f.
81. www.greenmatch.co.uk/blog/2017/10/the-opportunities-of-solar-panel-recycling.
82. Paech, N. Liberation from Excess —The Road to a Post-Growth Economy. oekom verlag München, 2. Auflage, 2016. Available online: https://www.oekom.de/buch/liberation-from-excess-9783865813244 (accessed on 6 October 2020)
83. Nieto, J.; Carpintero, O.; Lobejon, L.F.; Miguel, L.J. An ecological macroeconomics model: The energy transition in the EU. *Energy Policy* **2020**, *145*, 111726. [CrossRef]
84. Glaser, P.E. Power from the Sun; its future. *Science* **1968**, *162*, 857–861. [CrossRef]
85. Glaser, P.E. Method and Apparatus for Converting Solar Radiation to Electrical Power. U.S. Patent 3,781,647, 23 December 1973.
86. Glaser, P.E. Solar power from satellites. *Phys. Today*, February 1977; pp. 30–38.
87. Boeing Aerospace Co. *System's Definition—Space Based Power Conversion Systems*; NASA, MSFC, Contract NAS8-31628, Fourth Performance Briefing; National Aeronautics and Space Administration: Washington, DC, USA, 1976.
88. US Department of Energy and the National Aeronautics and Space Administration. *Satellite Power System*; Reference System Report, October 1978, DOE/ER-0023; National Technical Information Service, U.S. Department of Commerce: Springfield, VA, USA, 1979.
89. Lior, N. Power from space. *Energy Convers. Manag.* **2001**, 42, 1769–1805. [CrossRef]
90. Diaz, F.C.; Seedhouse, E. *To Mars and Beyond, Fast*! Springer Praxis Books; Springer International Publishing: Cham, Switzerland, 2017; doi:10.1007/978-3-319-22918-8_1. [CrossRef]
91. O'Neill, G.K. The colonization of space. *Phys. Today* **1974**, 32–40.10.2514/6.1975-2041. [CrossRef]
92. O'Neill, G.K. *The High Frontier—Human Colonies in Space*; William Morrow & Co.: New York, NY, USA, 1977.
93. O'Neill, G.K. The low (profile) road to space manufacturing. *Astronaut. Aeronaut.* **1978**, *16*, 18–32.
94. Grey, J.; Hamdan, L.A. (Eds.) Space Manufacturing 4. In *Proceedings of the Fifth Princeton/AIAA Conference, Princeton, NJ, USA, 18–21 May 1981*; American Institute of Aeronautics and Astronautics: New York, NY, USA, 1981.
95. A publication of scholars, who work in a program of investigating the role of energy and exergy in economic growth, interpreted the LinEx function's linear dependence on E and exponential dependence on quotients of K, L, E as meaning that energy E is the only real factor of production. The exponential was understood as being just a function of time t, because the factors vary in time. In our resulting discussion with the colleagues via e-mail, in which the difference between explicit and implicit time dependence was explained, they raised the question under which conditions energy-dependent production functions will depend linearly on L, or on K, and exponentially on the factor quotients. The two paragraphs that contain Equations (11) and (12) give the general answer to this question. Special, explicit answers are presented elsewhere (in "Energie, Entropie, Kreativität", Springer Spektrum, 2018).
96. Lindenberger, D. Service Production Functions. *J. Econ. (Z. Nationalökon.)* **2003**, *80*, 127–142. [CrossRef]

entropy

MDPI

Article

Aspects of a Phase Transition in High-Dimensional Random Geometry

Axel Prüser [1,*], Imre Kondor [2,3,4] and Andreas Engel [1]

1 Institute of Physics, Carl von Ossietzky University of Oldenburg, D-26111 Oldenburg, Germany; andreas.engel@uol.de
2 Parmenides Foundation, 82049 Pullach, Germany; kondor.imre@gmail.com
3 London Mathematical Laboratory, London W6 8RH, UK
4 Complexity Science Hub, 1080 Vienna, Austria
* Correspondence: axel.prueser@uol.de

Abstract: A phase transition in high-dimensional random geometry is analyzed as it arises in a variety of problems. A prominent example is the feasibility of a minimax problem that represents the extremal case of a class of financial risk measures, among them the current regulatory market risk measure Expected Shortfall. Others include portfolio optimization with a ban on short-selling, the storage capacity of the perceptron, the solvability of a set of linear equations with random coefficients, and competition for resources in an ecological system. These examples shed light on various aspects of the underlying geometric phase transition, create links between problems belonging to seemingly distant fields, and offer the possibility for further ramifications.

Keywords: random geometry; portfolio optimization; risk measurement; disordered systems; replica theory

PACS: 05.20.-y; 05.40.-a; 05.70.Fh; 87.23.Ge

Citation: Prüser, A.; Kondor, I.; Engel, A. Aspects of a Phase Transition in High-Dimensional Random Geometry. *Entropy* **2021**, *23*, 805. https://doi.org/10.3390/e23070805

Academic Editors: Ryszard Kutner, H. Eugene Stanley and Christophe Schinckus

Received: 10 May 2021
Accepted: 17 June 2021
Published: 24 June 2021

1. Introduction

A large class of problems in random geometry is concerned with the collocation of points in high-dimensional space. Applications range from optimization of financial portfolios [1], binary classifications of data strings [2] and optimal stategies in game theory [3] to the existence of non-negative solutions to systems of linear equations [4,5], the emergence of cooperation in competitive ecosystems [6,7], and linear programming with random parameters [8]. It is frequently relevant to consider the case where both the number of points T and the dimension of space N tend to infinity. This limit is often characterized by abrupt qualitative changes reminiscent of phase transitions when an external parameter or the ratio T/N vary and cross a critical value. At the same time, this high-dimensional case is amenable to methods from the statistical mechanics of disordered systems offering additional insight.

Some results obtained in different disciplines are closely related to each other without the connection always being appreciated. In the present paper, we discuss some particular cases. We will show that the boundedness of the expected maximal loss, as well as the possibility of zero variance of a random financial portfolio is closely related to the existence of a linear separable binary coloring of random points called a dichotomy. Moreover, we point out the connection with the existence of non-negative solutions to systems of linear equations and with mixed strategies in zero-sum games. On a more technical level and for the above-mentioned limit of large instances in high-dimensional spaces, we also make contact between replica calculations performed for different problems in different fields.

In addition to uncovering the common random geometrical background of seemingly very different problems, our comparative analysis sheds light on each of them from various angles and points to ramifications in their respective fields.

2. Dichotomies of Random Points

Consider an N-dimensional Euclidean space with a fixed coordinate system. Choose T points in this space and color them either black or white. The coloring is called a dichotomy if a hyperplane through the origin of the coordinate system exists that separates black points from white ones, see Figure 1.

To avoid special arrangements like all points falling on one line, the points are required to be in what is called a general position: the position vectors of any subset of N points should be linearly independent. Under this rather mild prerequisite, the number $C(T,N)$ of dichotomies of T points in N dimensions only depends on T and N and not on the particular location of the points. This remarkable result was proven in several works, among them a classical paper by Cover [2]. Establishing a recursion relation for $C(T,N)$, the explicit result was derived:

$$C(T,N) = 2\sum_{i=0}^{N-1}\binom{T-1}{i}. \tag{1}$$

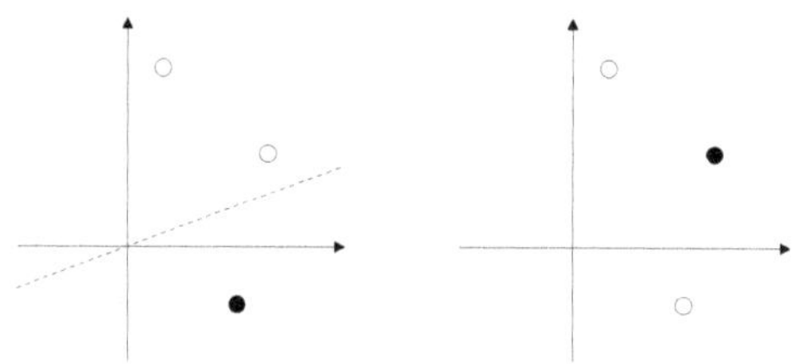

Figure 1. Two colorings of three points in two dimensions. In the **left** one, black and white points can be separated by a line through the origin; this coloring therefore represents a dichotomy. For the **right** one, no such separating line exists.

If the coordinates of the points are chosen at random from a continuous distribution, the points are in a general position with the probability one. Since there are in total 2^T different binary colorings of these points and only $C(T,N)$ of them are dichotomies, we find for the probability that T random points in N dimensions with random coloring form a dichotomy with the cumulative binomial distribution:

$$P_{\rm d}(T,N) = \frac{C(T,N)}{2^T} = \frac{1}{2^{T-1}}\sum_{i=0}^{N-1}\binom{T-1}{i}. \tag{2}$$

Hence, $P_{\rm d}(T,N) = 1$ for $T \leq N$, $P_{\rm d}(T,N) = 1/2$ for $T = 2N$ and $P_{\rm d}(T,N) \to 0$ for $T \to \infty$. The transition from $P \simeq 1$ at $T = N$ to $P \simeq 0$ at large T becomes sharper with increasing N. This is clearly seen when considering the case of constant ratio

$$\alpha := \frac{T}{N} \tag{3}$$

between the number of points and the dimension of space for different values of N, which shows an abrupt transition at $\alpha_c = 2$ for $N \to \infty$, cf. Figure 2.

For later convenience, it is useful to reformulate the condition for a certain coloring to be a dichotomy in different ways. Let us denote the position vector of point $t, t = 1,\ldots,T$, by $\boldsymbol{\xi}^t \in \mathbb{R}^N$ and its coloring by the binary variable $\zeta^t = \pm 1$. If a separating hyperplane exists, it has a normal vector $\mathbf{w} \in \mathbb{R}^N$ that fulfills

$$\zeta^t = \operatorname{sign}(\mathbf{w}\cdot\boldsymbol{\xi}^t), \qquad t = 1,\ldots,T, \tag{4}$$

where we define $\operatorname{sign}(x) = 1$ for $x \geq 0$ and $\operatorname{sign}(x) = -1$ otherwise. With the abbreviation

$$\mathbf{r}^t := \zeta^t\boldsymbol{\xi}^t, \tag{5}$$

Equation (4) translates into $\mathbf{w} \cdot \mathbf{r}^t \geq 0$ for all $t = 1, \ldots, T$ which for points in a general position, is equivalent to the somewhat stronger condition

$$\mathbf{w} \cdot \mathbf{r}^t > 0, \qquad t = 1, \ldots, T. \tag{6}$$

A certain coloring ζ^t of points $\boldsymbol{\xi}^t$ is hence a dichotomy if a vector $\mathbf{w}$ exists such that (6) is fulfilled, that is, if its scalar product with all vectors $\mathbf{r}^t$ is positive. This is quite intuitive, since by going from the vectors $\boldsymbol{\xi}^t$ to $\mathbf{r}^t$ according to the (5), we replace all points colored black by their white-colored mirror images (or vice versa). If we started out with a dichotomy, after the transformation, all points will lie on the same side of the separating hyperplane. The meaning of Equation (6) is clear: For T random points in N dimensions with coordinates chosen independently from a symmetric distribution, there exists with probability $P_d(T, N)$ a hyperplane such that all these points lie on the same side of the hyperplane. This formulation will be crucial in Section 3 to relate dichotomies to bounded cones characterizing financial portfolios.

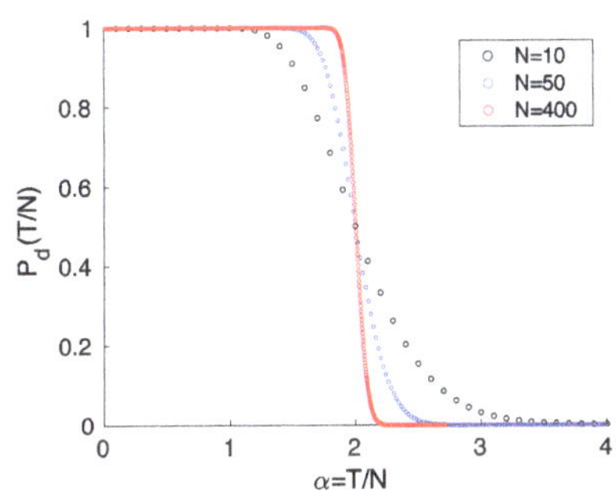

Figure 2. Probability $P_{\mathrm{d}}(T, N)$ that T randomly colored points in a general position in N dimensional space form a dichotomy as a function of the ratio α between T and N for different values of N. The transition between the limiting values $P = 1$ at $\alpha = 1$ and $P = 0$ at large α becomes increasingly sharp when N grows.

Singling out one particular point $s = 1, \ldots, T$, this in turn implies that there is, for any choice of s, a vector $\mathbf{w}$ with

$$\mathbf{w} \cdot \mathbf{r}^t > 0, \quad t = 1, \ldots, T, t \neq s \qquad \text{and} \qquad \mathbf{w} \cdot (-\mathbf{r}^s) < 0. \tag{7}$$

Consider now all vectors $\bar{\mathbf{r}}$ of the form

$$\bar{\mathbf{r}} = \sum_{t \neq s} c^t \mathbf{r}^t, \qquad \text{with} \qquad c^t \geq 0, \; t = 1, \ldots, T, t \neq s, \tag{8}$$

that is, all vectors that may be written as a linear combination of the $\mathbf{r}^t$ with $t \neq s$ and all expansion parameters c^t being non-negative. The set of these vectors $\bar{\mathbf{r}}$ is called the *non-negative cone* of the $\mathbf{r}^t, t \neq s$. Equation (7) then means that $-\mathbf{r}^s$ cannot be an element of this non-negative cone. This is clear since the hyperplane perpendicular to $\mathbf{w}$ separates $-\mathbf{r}^s$ from this very cone, an observation that is known as Farkas' lemma [9]. Therefore, if a set of vectors $\mathbf{r}^t$ forms a dichotomy no mirror image $-\mathbf{r}^s$ of any of them may be written as a linear combination of the remaining ones with non-negative expansion coefficients

$$\sum_{t \neq s} c^t \mathbf{r}^t \neq -\mathbf{r}^s, \qquad \forall c^t \geq 0. \tag{9}$$

Finally, adding $\mathbf{r}^s$ to both sides of (9), we find

$$\sum_t c^t \mathbf{r}^t \neq \mathbf{o}, \qquad \text{with} \qquad c^t \geq 0, t = 1, \ldots, T, \quad \text{and} \quad \sum_t c^t > 0, \tag{10}$$

where $\mathbf{o}$ denotes the null vector in N dimensions. Given T points $\mathbf{r}^t$ in N dimensions forming a dichotomy, it is therefore impossible to find a nontrivial linear combination of these vectors with non-negative coefficients that equals the null vector.

Additionally, this corollary to the Cover result is easily intuitively understood. Assume there were some coefficients $c^t \geq 0$ that were not all zero at the same time, and that realize

$$\sum_t c^t \mathbf{r}^t = \mathbf{o}. \tag{11}$$

If the points $\mathbf{r}^t$ form a dichotomy, then according to (6), there is a vector $\mathbf{w}$ that makes a positive scalar product with all of them. Multiplying (11) with this vector, we immediately arrive at a contradiction, since the l.h.s. of this equation is positive and the r.h.s. is zero.

Note that the inverse of (10) is also true: if the points do not form a dichotomy, a decomposition of the null vector of the type (11) can always be found. This is related to the fact that the non-negative cone of the corresponding position vectors is the complete $\mathbb{R}^N$. For if there were a vector $\mathbf{b} \in \mathbb{R}^N$ that lies not in this cone by Farkas' lemma, there would be a hyperplane separating the cone from $\mathbf{b}$. However, the very existence of this hyperplane would qualify the points $\mathbf{r}^t$ to be a dichotomy in contradiction to what was assumed.

In the limit $N \to \infty$, $T \to \infty$ with $\alpha = T/N$, keeping the problem of random dichotomies constant can be investigated within statistical mechanics. To make this connection explicit, we first note that no inequality in (6) is altered if $\mathbf{w}$ is multiplied by a positive constant. To decide whether an appropriate vector $\mathbf{w}$ fulfilling (6) may be found or not, it is hence sufficient to study vectors of a given length. It is convenient to choose this length as $\sqrt{N}$, requiring

$$\sum_{i=1}^{N} w_i^2 = N. \tag{12}$$

Next, we introduce for each realization of the random vectors $\mathbf{r}^t$ an energy function

$$E(\mathbf{w}) := \sum_{t=1}^{T} \Theta\Big(-\sum_i w_i r_i^t\Big), \tag{13}$$

where $\Theta(x) = 1$ if $x > 0$, and $\Theta(x) = 0$; otherwise it is the Heaviside step function. This energy is nothing but the number of points violating (6) for a given vector $\mathbf{w}$. Our central quantity of interest is the entropy of the groundstate of the system, that is, the logarithm of the fraction of points on the sphere defined by (12) that realize zero energy:

$$S(\kappa,\alpha) := \lim_{N\to\infty} \frac{1}{N} \ln \frac{\int \prod_{i=1}^{N} dw_i\, \delta(\sum_i w_i^2 - N) \prod_{t=1}^{\alpha N} \Theta(\sum_i w_i r_i^t - \kappa)}{\int \prod_{i=1}^{N} dw_i\, \delta(\sum_i w_i^2 - N)}. \tag{14}$$

Here, $\delta(x)$ denotes the Dirac δ-function, and we have introduced the positive stability parameter κ to additionally sharpen the inequalities (6).

The main problem in the explicit determination of $S(\kappa,\alpha)$ is its dependence on the many random parameters r_i^t. Luckily, for large values of N deviations of S from its typical value, S_{typ} becomes extremely rare and, moreover, this typical value is given by the average over the realizations of the r_i^t:

$$S_{\text{typ}}(\kappa,\alpha) = \langle\langle S(\kappa,\alpha)\rangle\rangle. \tag{15}$$

The calculation of this average was performed by a classical calculation [10] which gave rise to the result:

$$S_{\text{typ}}(\kappa,\alpha) = \underset{q}{extr}\left[\frac{1}{2}\ln(1-q) + \frac{q}{2(1-q)} + \alpha \int Dt \ln H\left(\frac{\kappa - \sqrt{q}t}{\sqrt{1-q}}\right)\right], \tag{16}$$

where the extremum is over the auxiliary quantity q, and we have used the shorthand notations

$$Dt := \frac{dt}{\sqrt{2\pi}} e^{-\frac{t^2}{2}} \quad \text{and} \quad H(x) := \int_x^\infty Dt. \tag{17}$$

More details of the calculation may be found in the original reference, and in chapter 6 of [11]. Appendix A contains some intermediate steps for a closely related analysis.

Studying the limit $q \to 1$ of (16) reveals

$$S_{\text{typ}}(\kappa, \alpha) \begin{cases} > -\infty & \text{if} \quad \alpha < \alpha_c(\kappa) \\ \to -\infty & \text{if} \quad \alpha > \alpha_c(\kappa), \end{cases} \tag{18}$$

corresponding to a sharp transition from solvability to non-solvability at a critical value $\alpha_c(\kappa)$. This is because $\kappa = 0$ finds $\alpha_c = 2$ in agreement with (2), cf. Figure 2.

Note that Cover's result (2) holds for all values of T and N, whereas the statistical mechanics analysis is restricted to the thermodynamic limit $N \to \infty$. On the other hand, the latter can deal with all values of the stability parameter κ, whereas no generalization of Cover's approach to the case $\kappa \neq 0$ is known.

3. Phase Transitions in Portfolio Optimization under the Variance and the Maximal Loss Risk Measure

3.1. Risk Measures

The purpose of this subsection is to indicate the financial context, in which the geometric problem discussed in this paper appears. A portfolio is the weighted sum of financial assets. The weights represent the parts of the total wealth invested in the various assets. Some of the weights are allowed to be negative (short positions), but the weights sum to 1; this is called the budget constraint. Investment carries risk, and higher returns usually carry higher risk. Portfolio optimization seeks a trade-off between risk and return by the appropriate choice of the portfolio weights. Markowitz was the first to formulate the portfolio choice as a risk-reward problem [12]. Reward is normally regarded as the expected return on the portfolio. Assuming return fluctuations to be Gaussian-distributed random variables, portfolio variance offered itself as the natural risk measure. This setup made the optimization of portfolios a quadratic programming problem, which, especially in the case of large institutional portfolios, posed a serious numerical difficulty in its time. Another critical point concerning variance as a risk measure was that variance is symmetric in gains and losses, whereas investors are believed not to be afraid of big gains, only big losses. This consideration led to the introduction of downside risk measures, starting already with the semivariance [13]. Later it was recognized that the Gaussian assumption was not realistic, and alternative risk measures were sought to grasp the risk of rare but large events, and also to allow risk to be aggregated across the ever-increasing and increasingly heterogeneous institutional portfolios. Around the end of the 1980s, Value at Risk (VaR) was introduced by JP Morgan [14], and subsequently it was widely spread over the industry by their RiskMetrics methodology [15]. VaR is a high quantile, a downside risk measure (note that in the literature, the profit and loss axis is often reflected, so that losses are assigned a positive sign. It is under this convention that VaR is a high quantile, rather than a low one). It soon came under academic criticism for its insensitivity to the details of the distribution beyond the quantile, and for its lack of sub-additivity. Expected Shortfall (ES), the average loss above the VaR quantile, appeared around the turn of the century [16]. An axiomatic approach to risk measures was proposed by Artzner et al. [17] who introduced a set of postulates which any coherent risk measure was required to satisfy. ES turned out to be coherent [18,19] and was strongly advocated by academics. After a long debate, international regulation embraced it as the official risk measure in 2016 [20].

The various risk measures discussed all involved averages. Since the distributions of financial data are not known, the relative price movements of assets are observed at a number T of time points, and the true averages are replaced by empirical averages from these data. This works well if T is sufficiently large; however, in addition to all the aforementioned problems, a general difficulty of portfolio optimization lies in the fact that

the dimension N of institutional portfolios (the number of different assets) is large, but the number T of observed data per asset is never large enough, due to lack of stationarity of the time series and the natural limits (transaction costs, technical difficulties of rebalancing) on the sampling frequency. Therefore, portfolio optimization in large dimensions suffers from a high degree of estimation error, which renders the exercise more or less illusory (see e.g., [21]). Estimation of returns is even more error-prone than the risk part, so several authors disregard the return completely, and seek the minimum risk portfolio (e.g., [22–24]). We follow the same approach here.

In the two subsections that follow, we also assume that the returns are independent, symmetrically distributed random variables. This is, of course, not meant to be a realistic market model, but it allows us to make an explicit connection between the optimization of the portfolio variance under a constraint excluding short positions and the geometric problem of dichotomies discussed in Section 2. This is all the more noteworthy because analytic results are notoriously scarce for portfolio optimization with no short positions. We note that similar simplifying assumptions (Gaussian fluctuations, independence) were built into the original JP Morgan methodology, which was industry standard in its time, and influences the thinking of practitioners even today.

3.2. Vanishing of the Estimated Variance

We consider a portfolio of N assets with weights w_i, $i = 1, \dots, N$. The observations r_i^t of the corresponding returns at various times $t = 1, \dots, T$ are assumed to be independent, symmetrically distributed random variables. Correspondingly, the average value of the portfolio is zero. Its variance is given by

$$\sigma_p^2 = \frac{1}{T}\sum_t \left(\sum_i w_i r_i^t\right)^2 = \sum_{i,j} w_i w_j \, \frac{1}{T}\sum_t r_i^t r_j^t =: \sum_{i,j} w_i w_j C_{ij}, \tag{19}$$

where C_{ij} denotes the covariance matrix of the observations. Note that the variance of a portfolio optimized in a given sample depends on the sample, so it is itself a random variable.

The variance of a portfolio obviously vanishes if the returns are fixed quantities that do not fluctuate. This subsection is not about such a trivial case. We shall see, however, that the variance optimized *under a no-short constraint* can vanish with a certain probability if the dimension N is larger than the number of observations T.

The rank of the covariance matrix is the smaller of N and T, and for $N \leq T$ the estimated variance is positive with the probability one. Thus, the optimization of variance can always be carried out as long as the number of observations T is larger than the dimension N, albeit with an increasingly larger error as T/N decreases. For large N and T and fixed $\alpha = T/N$, the estimation error increases as $\alpha/(\alpha - 1)$ with decreasing α and diverges at $\alpha \downarrow 1$ [25,26]. The divergence of the estimation error can be regarded as a phase transition. Below the critical value $\alpha_d := 1$, the optimization of variance becomes impossible. Of course, in practice, one never has such an optimization task without some additional constraints. Note that because of the possibility of short-selling (negative portfolio weights), the budget constraint (a hyperplane) in itself is not sufficient to forbid the appearance of large positive and negative positions, which then destabilize the optimization. In contrast, any constraint that makes the allowed weights finite can act as a regularizer. The usual regularizers are constraints on the norm of the portfolio vector. It was shown in [27,28] how liquidity considerations naturally lead to regularization. Ridge regression (a constraint on the ℓ_2 norm of the portfolio vector) prevents the covariance matrix from developing zero eigenvalues, and, especially in its nonlinear form [29], results in very satisfactory out-of-sample performance.

An alternative is the ℓ_1 regularizer, of which the exclusion of short positions is a special case. Together with the budget constraint, it prevents large sample fluctuations of the weights. Let us then impose the no-short ban, as it is indeed imposed in practice on a number of special portfolios (e.g., on pension funds), or, in episodes of crisis, on the

whole industry. The ban on short-selling extends the region where the variance can be optimized, but below $\alpha = 1$ the optimization acquires a probabilistic character in that the regularized variance vanishes with a certain probability, and the optimization can only be carried out when it is positive. (Otherwise, there is a continuum of solutions, namely any combination of the eigenvectors belonging to zero eigenvalues, which makes the optimized variance zero).

Interestingly, the probability of the variance vanishing is related to the problem of random dichotomies in the following way. For the portfolio variance (19) to become zero, we need to have

$$\sum_i w_i r_i^t = 0 \tag{20}$$

for all t. If we interchange t and i, we see that according to (11), this is possible as long as the N points in $\mathbb{R}^T$ with position vectors $\vec{r}_i := \{r_i^t\}$ do not form a dichotomy. Hence, the probability for zero variance is from (2)

$$P_{\text{zv}}(T,N) = 1 - P_{\text{d}}(N,T) = 1 - \frac{1}{2^{N-1}} \sum_{i=0}^{T-1} \binom{N-1}{i} = \frac{1}{2^{N-1}} \sum_{i=T}^{N-1} \binom{N-1}{i}. \tag{21}$$

Therefore, the probability of the variance vanishing is almost 1 for small α, decreases to the value $1/2$ at $\alpha = 1/2$, decreases further to 0 as α increases to 1, and remains identically zero for $\alpha > 1$ [30,31]. This is similar but also somewhat complementary to the curve shown in Figure 2. Equation (21) for the vanishing of the variance was first written up in [30,31] on the basis of analogy with the minimax problem to be considered below, and it was also verified by extended numerical simulations. The above link to the Cover problem is a new result, and it is rewarding to see how a geometric proof establishes a bridge between the two problems.

In [30,31], an intriguing analogy with, for example, the condensed phase of an ideal Bose gas was pointed out. The analogous features are the vanishing of the chemical potential in the Bose gas, resp. the vanishing of the Lagrange multiplier enforcing the budget constraint in the portfolio problem; the onset of Bose condensation, resp. the appearance of zero weights ("condensation" of the solutions on the coordinate planes) due to the no-short constraint; the divergence of the transverse susceptibility, and the emergence of zero modes in both models.

3.3. The Maximal Loss

The introduction of the Maximal Loss (ML) or minimax risk measure by Young [32] in 1998 was motivated by numerical expediency. In contrast to the variance whose optimization demands a quadratic program, ML is constructed such that it can be optimized by linear programming, which could be performed very efficiently even on large datasets already at the end of the last century. Maximal Loss combines the worst outcomes of each asset and seeks the best combination of them. This may seem to be an over-pessimistic risk measure, but there are occasions when considering the worst outcomes is justifiable (think of an insurance portfolio in the time of climate change), and, as will be seen, the present regulatory market risk measure is not very far from ML.

Omitting the portfolio's return again and focusing on the risk part, the maximal loss of a portfolio is given by

$$\text{ML} := \min_{\mathbf{w}} \max_{1 \le t \le T} \left(-\sum_i w_i r_i^t \right) \tag{22}$$

with the constraint

$$\sum_i w_i = N. \tag{23}$$

We are interested in the probability $P_{\text{ML}}(T,N)$ that this minimax problem is feasible, that is, ML does not diverge to $-\infty$. To this end, we first eliminate the constraint (23) by putting

$$w_N = N - \sum_{i=1}^{N-1} w_i. \tag{24}$$

This results in

$$\mathrm{ML} := \min_{\tilde{\mathbf{w}}} \max_{1 \le t \le T} \left(-\sum_{i=1}^{N-1} w_i (r_i^t - r_N^t) - N r_N^t \right) =: \min_{\tilde{\mathbf{w}}} \max_{1 \le t \le T} \left(-\sum_{i=1}^{N-1} w_i \tilde{r}_i^t - N r_N^t \right) \tag{25}$$

with $\tilde{\mathbf{w}} := \{w_1, \ldots, w_{N-1}\} \in \mathbb{R}^{N-1}$ and $\tilde{\mathbf{r}}^t := \{r_1^t - r_N^t, \ldots, r_{N-1}^t - r_N^t\} \in \mathbb{R}^{N-1}$. For ML to stay finite for all choices of $\tilde{\mathbf{w}}$, the T random hyperplanes with normal vectors $\tilde{\mathbf{r}}^t$ have to form a bounded cone. If the points $\tilde{\mathbf{r}}^t$ form a dichotomy, then according to (6), there is a vector $\mathbf{W} \in \mathbb{R}^{N-1}$ with $\mathbf{W} \cdot \tilde{\mathbf{r}}^t > 0$ for all t. Since there is no constraint on the norm of $\tilde{\mathbf{w}}$, the maximal loss (25) can become arbitrarily small for $\tilde{\mathbf{w}} = \lambda \mathbf{W}$ and $\lambda \to \infty$. The cone then is not bounded. We therefore find

$$P_{\mathrm{ML}}(T, N) = P_{\mathrm{d}}(T, N-1) = \frac{1}{2^{T-1}} \sum_{i=0}^{N-2} \binom{T-1}{i} \tag{26}$$

for the probability that ML cannot be optimized.

In the limit $N, T \to \infty$ with $\alpha = T/N$ kept finite, (25) displays the same abrupt change as in the problem of dichotomies, a phase transition at $\alpha_c = 2$. Note that this is larger than the critical point $\alpha_d = 1$ of the unregularized variance, which is quite natural, since the ML uses only the extremal values in the data set. The probability for the feasibility of ML was first written up without proof in [1], where a comparative study of the noise sensitivity of four risk measures, including ML, was performed. There are two important remarks we can make at this point. First, the geometric consideration above does not require any assumption about the data generating process; as long as the the returns are independent, they can be drawn from any symmetric distribution without changing the value of the critical point. This is a special case of the universality of critical points discovered by Donoho and Tanner [33].

The second remark is that the problem of bounded cones is closely related to that of bounded polytopes [34]. The difference is just the additional dimension of the ML itself. If the random hyperplanes perpendicular to the vectors $\tilde{\mathbf{r}}^t$ form a bounded cone for ML according to (25), then they will trace out a bounded polytope on hyperplanes perpendicular to the ML axis at sufficiently high values of ML. In fact, after the replacement $N - 1 \to N$ Equation (26) coincides with the result in Theorem 4 of [34] for the probability of T random hyperplanes forming a bounded polytope in N dimensions (there is a typo in Theorem 4 in [34]; the summation has to start at $i = 0$). The close relationship between the ML problem and the bounded polytope problem, on the one hand, and the Cover problem on the other hand, was apparently not clarified before.

If we spell out the financial meaning of the above result, we are led to interesting ramifications. To gain an intuition, let us consider just two assets, $N = 2$. If asset 1 produces a return sometimes above, sometimes below that of asset 2, then the minimax problem will have a finite solution. If, however, asset 1 dominates asset 2 (i.e., yields a return which is at least as large, and, at least at one time point, larger, than the return on asset 2 in a given sample), then, with unlimited short positions allowed, the investor will be induced to take an arbitrarily large long position in asset 1 and go correspondingly short in asset 2. This means that the solution of the minimax problem will run away to infinity, and the risk of ML will be equal to minus infinity [1]. The generalization to N assets is immediate: if among the assets there is one that dominates the rest, or there is a combination of assets that dominates some of the rest, the solution will run away to infinity, and ML will take the value of $-\infty$. This scenario corresponds to an arbitrage, and the investor gains an arbitrarily large profit without risk [35]. Of course, if such a dominance is realized in one

given sample, it may disappear in the next time interval, or the dominance relations can rearrange to display another mirage of an arbitrage.

Clearly, the ML risk measure is unstable against these fluctuations. In practice, such a brutal instability can never be observed, because there are always some constraints on the short positions, or groups of assets corresponding to branches of industries, geographic regions, and so forth. These constraints will prevent instabilities from taking place, and the solution cannot run away to infinity, but will go as far as allowed by the constraints and then stick to the boundary of the allowed region. Note, however, that in such a case, the solution will be determined more by the constraints (and ultimately by the risk manager imposing the constraints) rather than by the structure of the market. In addition, in the next period, a different configuration can be realized, so the solution will jump around on the boundary defined by the constraints.

We may illustrate the role of short positions for the instability of ML further by investigating the case of portfolio weights w_i that have to be larger than a threshold $\gamma \leq 0$. For $\gamma \to -\infty$, there are no restrictions on short positions, whereas $\gamma = 0$ corresponds to a complete ban on them. For $N, T \to \infty$ with fixed $\alpha = T/N$, the problem may be solved within the framework of statistical mechanics. The minimax problem for ML is equivalent to the following problem in linear programming: minimize the threshold variable κ under the constraints (23), $w_i \geq \gamma$, and

$$-\sum_i w_i r_i^t \leq \kappa \quad \forall t = 1, \ldots, T. \tag{27}$$

Similarly to (14), the central quantity of interest is

$$\Omega(\kappa, \gamma, \alpha) = \frac{\int_\gamma^\infty \prod_{i=1}^N dw_i \, \delta(\sum_i w_i - N) \prod_{t=1}^{\alpha N} \Theta(\sum_i w_i r_i^t + \kappa)}{\int_\gamma^\infty \prod_{i=1}^N dw_i \, \delta(\sum_i w_i - N)}, \tag{28}$$

giving the fractional volume of points on the simplex defined by (23) that fulfill all constraints (27). For given α and γ, we decrease κ down to the point κ_c, where the typical value of this fractional volume vanishes. The ML is then given by $\kappa_c(\alpha, \gamma)$.

Some details of the corresponding calculations are given in the Appendix A. In Figure 3, we show some results. As discussed above, the divergence of ML for $\alpha < 2$ is indeed formally eliminated for all $\gamma > -\infty$, and the functions $\text{ML}(\alpha; \gamma)$ smoothly interpolate between the cases $\gamma = 0$ and $\gamma \to -\infty$. However, the situation is now even more dangerous, since the unreliability of ML as a risk measure for small α remains without being deducible from its divergence.

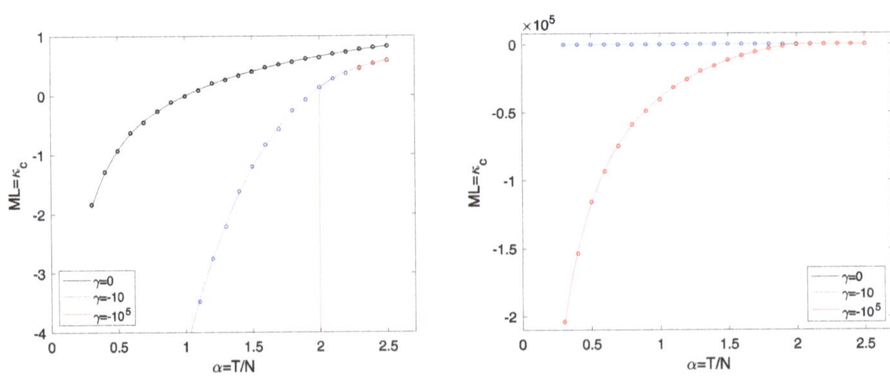

Figure 3. ***Left***: The Maximal Loss ML $= \kappa_c$ as a function of α. The analytical results (solid line) are compared to simulation results (circles) with $N = 200$ averaged over 100 samples. The symbol size corresponds to the statistical error. ***Right***: Same as left with largely extended axis of ML.

The recognition of the instability of ML as a dominance problem has proved very fruitful and led to a series of generalizations. First, it was realized [1] that the instability of the expected shortfall, of which ML is an extreme special case, has a very similar geometric origin. (The current regulatory ES is the expected loss above a 97.5% quantile, whereas ML corresponds to 100%.) Both ES and ML are so-called coherent risk measures [17], and it was proved [35] that the root of this instability lies in the coherence axioms themselves, so every coherent risk measure suffers from a similar instability. Furthermore, it was proved [35] that the existence of a dominant/dominated pair of assets in the portfolio was a necessary and sufficient condition for the instability of ML, whereas it was only sufficient for other coherent risk measures. It follows that in terms of the variable α used in this paper (which is the reciprocal of the aspect ratio N/T used in some earlier works, such as [35–37]), the critical point of ML is a lower bound for the critical points of other coherent measures. Indeed, the critical line of ES was found to lie above the ML critical value of $\alpha_c = 2$ [36]. Value at Risk is not a coherent measure and can violate convexity, so it is not amenable to a similar study of its critical point. However, parametric VaR (that is, the quantile where the underlying distribution is given, only its expectation value and variance is determined from empirical data) *is* convex, and it was shown to possess a critical line that runs above that of ES [37]. The investigation of the semi-variance yielded similar results [37]. It seems, then, that the geometrical analysis of ML provides important information for a variety of risk measures, including some of the most widely used measures in the industry (VaR and ES), and also other downside risk measures.

4. Related Problems

In this section, we list a few problems from different fields of mathematics and physics that are linked to the random coloring of points in high-dimensional space and point out their connection with the questions discussed above.

4.1. Binary Classifications with a Perceptron

Feed-forward networks of formal neurons perform binary classifications of input data [38]. The simplest conceivable network of this type—the perceptron—consists of just an input layer of N units ξ_i and a single output bit $\zeta = \pm 1$ [39]. Each input ξ_i is directly connected to the output by a real valued coupling w_i. The output is computed as the sign of the weighted inputs

$$\zeta = \operatorname{sign}\left(\sum_{i=1}^{N} w_i\, \xi_i\right). \tag{29}$$

Consider now a family of random inputs $\{\xi_i^t\}$, $t = 1, \ldots, T$ and ask for the probability $P_\mathrm{p}(T,N)$ that the perceptron is able to implement a randomly chosen binary classification $\{\zeta^t\}$ of these inputs. Interpreting the vectors $\boldsymbol{\xi}^t := \{\xi_i^t\}$ as position vectors of T points in N dimensions and the required classifications ζ^t as a black/white coloring, we hence need to know the probability that this particular coloring is a dichotomy. Indeed, if a hyperplane exists that separates black points from white ones, it has a normal vector $\mathbf{w}$ that gives a suitable choice for the perceptron weights to get all classifications right. Therefore, we have

$$P_\mathrm{p}(T,N) = P_\mathrm{d}(T,N) = \frac{1}{2^{T-1}} \sum_{i=0}^{N-1} \binom{T-1}{i}. \tag{30}$$

In the thermodynamic limit $N, T \to \infty$, this problem, together with a variety of modifications, can be analyzed using methods from the statistical mechanics of disordered systems along the lines of Equations (14)–(16), see [11].

4.2. Zero-Sum Games with Random Pay-Off Matrices

In game theory, two or more players choose among different strategies at their disposal and receive a pay-off (that may be negative) depending on the choices of all participating

players. A particularly simple situation is given by a zero-sum game between two players, where one player's profit is the other player's loss. If the first player may choose among N strategies and the second among T, the setup is defined by an $N \times T$ pay-off matrix r_i^t, giving the reward for the first player if he plays strategy i and his opponent strategy t. Barring rare situations in which it is advantageous for one or both players to always choose one and the same strategy, it is known from the classical work of Morgenstern and von Neumann [40] that the best the players can do is to choose at random with different probabilities among their available strategies. The set of these probabilities p_i and q^t, respectively, is called a mixed strategy.

For large numbers of available strategies, it is sensible to investigate typical properties of such mixed strategies for random pay-off matrices. This can be done in a rather similar way to the calculation of ML presented in the Appendix A of the present paper [3]. One interesting result is that an extensive part of the probabilities p_i and q^t forming the optimal respective mixed strategies have to be identically zero: for both players, there are strategies they should never touch.

4.3. Non-Negative Solutions to Large Systems of Linear Equations

Consider a random $N \times T$ matrix r_i^t and a random vector $\mathbf{b} \in \mathbb{R}^N$. When will the system of linear equations

$$\sum_t r_i^t x^t = b_i, \qquad i = 1, \ldots, N \tag{31}$$

typically have a solution with all x^t being non-negative? This question is related to the optimization of financial portfolios under a ban of short-selling as discussed above, and also occurs when investigating the stability of chemical or ecological problems [6,41]. Here, the x^t denotes concentrations of chemical or biological species, and hence has to be non-negative. Similar to optimal mixed strategies considered in the previous subsection, the solution typically has a number of entries x^t that are strictly zero (species that died out), the remaining ones being positive (surviving species). Again for $T = \alpha N$ and $N \to \infty$, a sharp transition at a critical value α_c separates situations with typically no non-negative solution from those in which typically such a solution can be found [4].

To make contact with the cases discussed before, it is useful to map the problem to a dual one by again using Farkas' lemma. Let us denote by

$$\bar{\mathbf{r}} = \sum_t c^t \mathbf{r}^t, \qquad c^t \geq 0,\ t = 1, \ldots, T \tag{32}$$

the vectors in the non-negative cone of the column vectors $\mathbf{r}^t$ of matrix r_i^t. It is clear that (31) has a non-negative solution $\mathbf{x}$ if $\mathbf{b}$ belongs to this cone, and that no such solution exists if $\mathbf{b}$ lies outside the cone. In the latter case, however, there must be a hyperplane separating $\mathbf{b}$ from the cone. Denoting the normal of this hyperplane by $\mathbf{w}$, we hence have the following duality: either the system (31) has a non-negative solution $\mathbf{x}$, or there exists a vector $\mathbf{w}$ with

$$\mathbf{w} \cdot \mathbf{r}^t \geq 0 \quad t = 1, \ldots, T \qquad \text{and} \qquad \mathbf{w} \cdot \mathbf{b} < 0. \tag{33}$$

If the r_i^t is drawn independently from a distribution with finite first and second cumulant R and σ_r^2, respectively, and the components b_i are independent random numbers with average B and variance σ_b^2/N, the dual problem (33) may be analyzed along the lines of (14)–(16). The result for the typical entropy of solution vectors $\mathbf{w}$ reads [4]

$$S_{\text{typ}}(\gamma, \alpha) = \underset{q,\kappa}{extr} \left[\frac{1}{2} \ln(1-q) + \frac{q}{2(1-q)} - \frac{\kappa^2 \gamma}{2(1-q)} + \alpha \int Dt \ln H\left(\frac{\kappa - \sqrt{q}t}{\sqrt{1-q}} \right) \right], \tag{34}$$

where the parameter

$$\gamma := \left(\frac{B\,\sigma_r}{R\,\sigma_b}\right)^2 \tag{35}$$

characterizes the distributions of r_i^t and b_i. The main difference to (16) is the additional extremum over κ regularized by the penalty term proportional to κ^2. Considering the limit $q \to 1$ in (34), it is possible to determine the critical value $\alpha_c(\gamma)$ bounding the region where typically no solution $\mathbf{w}$ may be found. For nonrandom $\mathbf{b}$, that is, $\sigma_b \to 0$ implying $\gamma \to \infty$, we find back the Cover result $\alpha_c = 2$.

The problem is closely related to a phase transition found recently in MacArthur's resource competition model [4,6,7], in which a community of purely competing species builds up a collective cooperative phase above a critical threshold of the biodiversity.

5. Discussion

In this paper, we have reviewed various problems from different disciplines, including high-dimensional random geometry, finance, binary classification with a perceptron, game theory, and random linear algebra, which all have at their root the problem of dichotomies, that is, the linear separability of points carrying a binary label and scattered randomly over a high-dimensional space. No doubt there are several further problems belonging to this class; those that spring to mind are theoretical ecology alluded to at the end of the previous Section, or linear programming with random parameters [8]. Some of these conceptual links are obvious, and have been known for decades (for example, the link between dichotomies and the perceptron), and others are far less clear at first sight, such as the relationship with the two finance problems discussed in Section 3. We regard as one of the merits of this paper the establishment of this network of conceptual connections between seemingly faraway areas of study. Apart from the occasional use of the heavy machinery of the replica theory, in most of the paper we offered transparent geometric arguments, where our only tool was basically the Farkas' lemma.

The phase transitions we encountered in all of the problems discussed here are similar in spirit to the geometric transitions discovered by Donoho and Tanner [33] and interpreted at a very high level of abstraction by [42]. One of the central features of these transitions is the universality of the critical point. This universality is different from the one observed in the vicinity of continuous phase transitions in physics, where the value of the critical point can vary widely, even between transitions belonging to the same universality class. The universality in physical phase transitions is a property of the critical indices and other critical parameters. Critical indices also appear in our abstract geometric problems, and they are universal, but we omitted their discussion which might have led far from the main theme.

At the bottom of our geometric problems, there is the optimization of a convex objective function (which is, by the way, the key to the replica symmetric solutions we found). The recent evolution of neural networks, machine learning, and artificial intelligence is mainly concerned with a radical lack of convexity, which points to the direction in which we may try to extend our studies. Another simplifying feature we exploited was the independence of the random variables. The moment that correlations appear, these problems become hugely more complicated. We left this direction for future exploration. However, it is evident that progress in any of these problems will induce progress in the other fields, and we feel that revealing their fundamental unity may help the transfer of methods and ideas between these fields. This may be the most important achievement of this analysis.

Author Contributions: Conceptualization, I.K. and A.E.; formal analysis, A.P., I.K. and A.E.; software, A.P.; writing—original draft, A.P., I.K. and A.E. All authors have read and agreed to the published version of the manuscript.

Funding: This research received no external funding.

Data Availability Statement: The data presented in this study are available on request from the corresponding author.

Acknowledgments: A.P. and A.E. are grateful to Stefan Landmann for many interesting discussions.

Conflicts of Interest: The authors declare no conflict of interest.

Appendix A. Replica Calculation of Maximal Loss

In this appendix, we provide some details for the determination of the maximal loss of a random portfolio using the replica trick. The calculation is a generalization of the one presented in [3] for random zero-sum games. A presentation at full length can be found in [43]. As we pointed out in the main text, maximal loss is a special limit of the Expected Shortfall risk measure, corresponding to the so-called confidence level going to 100%. In [44] a detailed study of the behavior of ES was carried out, including the limiting case of maximal loss. That treatment is completely different from the one in here, so the present calculation can be regarded as complementary to that in [44].

The central quantity of interest is the fractional volume

$$\Omega(\kappa,\gamma,\alpha) = \frac{\int_\gamma^\infty \prod_{i=1}^N dw_i\, \delta(\sum_i w_i - N)\, \prod_{t=1}^{\alpha N} \Theta(\sum_i w_i r_i^t + \kappa)}{\int_\gamma^\infty \prod_{i=1}^N dw_i\, \delta(\sum_i w_i - N)} \tag{A1}$$

defined in (28). Although not explicitly indicated, $\Omega(\kappa,\gamma,\alpha)$ depends on all the random parameters r_i^t and is therefore by itself a random quantity. The calculation of its complete probability density $P(\Omega)$ is hopeless but for large N this distribution gets concentrated around the typical value $\Omega_{\text{typ}}(\kappa,\gamma,\alpha)$. Because Ω involves a product of many independent random factors this typical value is given by

$$\Omega_{\text{typ}}(\kappa,\gamma,\alpha) = e^{\langle\langle \ln\Omega(\kappa,\gamma,\alpha)\rangle\rangle} \tag{A2}$$

rather than by $\langle\langle\Omega(\kappa,\gamma,\alpha)\rangle\rangle$. Here $\langle\langle\ldots\rangle\rangle$ denotes the average over the r_i^t. A direct calculation of $\langle\langle\ln\Omega\rangle\rangle$ is hardly possible. It may be circumvented by exploiting the identity

$$\langle\langle \ln(\Omega(\kappa,\gamma,\alpha))\rangle\rangle = \lim_{n\to 0}\frac{1}{n}[\langle\langle\Omega^n(\kappa,\gamma,\alpha)\rangle\rangle - 1] \tag{A3}$$

For natural n the determination of $\langle\langle\Omega^n\rangle\rangle$ is feasible. The main problem then is to continue the result to real n in order to perform the limit $n \to 0$.

The explicit calculation starts with

$$\langle\langle\Omega(\kappa,\gamma,\alpha)^n\rangle\rangle = \left\langle\!\!\left\langle \frac{\int_\gamma^\infty \prod_{i=1}^N \prod_{a=1}^n dw_i^a\, \prod_{a=1}^n \delta(\sum_i w_i^a - N)\, \prod_{t=1}^{\alpha N}\prod_{a=1}^n \Theta(\sum_i w_i^a r_i^t + \kappa)}{\int_\gamma^\infty \prod_{i=1}^N \prod_{a=1}^n dw_i^a\, \prod_{a=1}^n \delta(\sum_i w_i^a - N)} \right\rangle\!\!\right\rangle. \tag{A4}$$

Using

$$\int_\gamma^\infty \prod_{i=1}^N dw_i\, \delta(\sum_i w_i - N) \sim \exp\{N[1+\ln(1-\gamma)]\} \tag{A5}$$

for large N and representing the δ-functions and Θ-functions by integrals over auxiliary variables E_a, λ_t^a, and y_t^a we arrive at

$$
\begin{aligned}
\langle\langle \Omega(\kappa,\gamma,\alpha)^n \rangle\rangle =& \exp\{-nN[1+\ln(1-\gamma)]\} \\
&\times \int_\gamma^\infty \prod_{i,a} dw_i^a \int \prod_a \frac{dE_a}{2\pi} \exp\left[iN \sum_a E_a \left(\frac{1}{N} \sum_i w_i^a - 1 \right) \right] \\
&\times \int_{-\kappa}^\infty \prod_{t,a} d\lambda_t^a \int \prod_{t,a} \frac{dy_t^a}{2\pi} \exp\left(i \sum_{t,a} y_t^a \lambda_t^a \right) \left\langle\left\langle \exp(-i \sum_{i,t,a} y_t^a w_i^a r_i^t) \right\rangle\right\rangle.
\end{aligned} \tag{A6}
$$

The average over the r_i^t may now be performed for independent Gaussian r_i^t with average zero and variance $\sigma^2 = 1/N$. The result is valid also for more general distributions. First, multiplying the variance by a constant just rescales the maximal loss but does not influence the optimal $\mathbf{w}$. Second, for $N \to \infty$ only the first two cumulants of the distribution matter due to the central limit theorem. Crucial is, however, the assumption of the r_i^t being independent.

Performing the average we find

$$
\begin{aligned}
\left\langle\left\langle \exp\left(-i \sum_{i,t,a} y_t^a w_i^a r_i^t \right) \right\rangle\right\rangle =& \prod_{i,t} \left[\int \frac{dr_i^t}{\sqrt{2\pi\sigma^2}} \exp\left(-\frac{(r_i^t)^2}{2\sigma^2} - ir_i^t \sum_a y_t^a w_i^a \right) \right] \\
=& \exp\left(-\frac{1}{2N} \sum_{i,t} \sum_{a,b} w_i^a w_i^b y_t^a y_t^b \right).
\end{aligned} \tag{A7}
$$

To disentangle in (A6) the w-integrals from those over λ and y we introduce the order parameters

$$
q_{ab} = \frac{1}{N} \sum_i w_i^a w_i^b, \quad a \geq b \tag{A8}
$$

together with the conjugate ones $\hat{q}_{ab}$. Using standard techniques [11] we end up with

$$
\begin{aligned}
\langle\langle \Omega(\kappa,\gamma,\alpha)^n \rangle\rangle =& \int \prod_{a\geq b} \frac{dq_{ab} d\hat{q}_{ab}}{2\pi/N} \int \prod_a \frac{dE_a}{2\pi} \\
&\times \exp\left\{ -iN \sum_{a\geq b} q_{ab}\hat{q}_{ab} - iN \sum_a E_a - nN[1+\ln(1-\gamma)] + NG_s + \alpha N G_E \right\},
\end{aligned} \tag{A9}
$$

where

$$
G_S = \ln\left[\int_\gamma^\infty \prod_a dw^a \exp\left(i \sum_{a\geq b} \hat{q}_{ab} w^a w^b + i \sum_a E_a w^a \right) \right] \tag{A10}
$$

and

$$
G_E = \ln\left[\int_{-\kappa}^\infty \prod_a d\lambda^a \int \prod_a \frac{dy^a}{2\pi} \exp\left(-\frac{1}{2} \sum_{a,b} q_{ab} y^a y^b + i \sum_a y^a \lambda^a \right) \right]. \tag{A11}
$$

For $N \to \infty$ the integrals over the order parameters in (A9) may be calculated using the saddle-point method. The essence of the so-called replica-symmetric ansatz is the assumption that the values of the order parameters at the saddle-point are invariant under permutation of the replica indices a and b. In [43] arguments are given why the replica-symmetric saddle-point should yield correct results in the present context. We therefore assume for the saddle-point values of the order parameters

$$
\begin{aligned}
q_{aa} &= q_1 \qquad i\hat{q}_{aa} = -\frac{1}{2}\hat{q}_1 \qquad iE_a = E \qquad \forall a \\
q_{ab} &= q_0 \qquad i\hat{q}_{ab} = \hat{q}_0 \qquad\qquad\qquad\quad \forall a > b.
\end{aligned} \tag{A12}
$$

which implies various simplifications in (A9)–(A11). Employing standard manipulations [11] we arrive at

$$\langle\langle\Omega(\kappa,\gamma,\alpha)^n\rangle\rangle \sim \exp\left\{N \underset{q_0,\hat{q}_0,q_1,\hat{q}_1,E}{\text{extr}}\left[-\frac{n(n-1)}{2}q_0\hat{q}_0 + \frac{n}{2}q_1\hat{q}_1 - nE - n(1+\ln(1-\gamma)) + G_S + \alpha G_E\right]\right\}. \tag{A13}$$

Using the shorthand notations (17) the functions G_S and G_E are now given by

$$G_S = \ln\int Dl\left[\exp\left(\frac{(\sqrt{\hat{q}_0}l+E)^2}{2(\hat{q}_0+\hat{q}_1)}\right)\sqrt{\frac{2\pi}{\hat{q}_0+\hat{q}_1}}H\left(-\frac{\sqrt{\hat{q}_0}l+E-\gamma(\hat{q}_0+\hat{q}_1)}{\sqrt{\hat{q}_0+\hat{q}_1}}\right)\right]^n \tag{A14}$$

and

$$G_E = \ln\int Dm\, H\left(\frac{\sqrt{q_0}m-\kappa}{\sqrt{q_1-q_0}}\right)^n. \tag{A15}$$

We may now treat n as a real number and perform the limit $n \to 0$. In this way we find for the averaged entropy

$$S(\kappa,\gamma,\alpha) := \lim_{N\to\infty}\frac{1}{N}\langle\langle\ln[\Omega(\kappa,\gamma,\alpha)]\rangle\rangle = \lim_{N\to\infty}\frac{1}{N}\lim_{n\to 0}\frac{1}{n}[\langle\langle\Omega(\kappa,\gamma,\alpha)^n\rangle\rangle - 1] \tag{A16}$$

the expression

$$\begin{aligned} S(\kappa,\gamma,\alpha) = \underset{q_0,\hat{q}_0,q_1,\hat{q}_1,E}{\text{extr}}&\left[\frac{q_0\hat{q}_0}{2} + \frac{q_1\hat{q}_1}{2} - E - 1 - \ln(1-\gamma) + \frac{1}{2}\ln(2\pi) - \frac{1}{2}\ln(\hat{q}_0+\hat{q}_1)\right. \\ &+\frac{\hat{q}_0+E^2}{2(\hat{q}_0+\hat{q}_1)} + \int Dl\ln H\left(-\frac{\sqrt{\hat{q}_0}l+E-\gamma(\hat{q}_0+\hat{q}_1)}{\sqrt{\hat{q}_0+\hat{q}_1}}\right) \\ &\left.+\alpha\int Dm\ln H\left(\frac{\sqrt{q_0}m-\kappa}{\sqrt{q_1-q_0}}\right)\right]. \end{aligned} \tag{A17}$$

The remaining extremization has to be done numerically. Before embarking on this task it is useful to remember that Ω and S are only instrumental in determining the maximal loss which in turn is given by the value κ_c of κ for which Ω tends to zero. At the same time the typical overlap q_0 between two different vectors in Ω has to tend to the self-overlap q_1. To investigate this limit we replace the order parameter q_1 by

$$v := q_1 - q_0 \tag{A18}$$

and study the saddle-point equations for $v \to 0$. In this limit it turns out that the remaining order parameters may either also tend to zero or diverge. It is therefore convenient to make the replacements

$$\hat{q}_0 \to \frac{\hat{q}_0}{v^2}, \quad \hat{q}_1 \to \hat{w} := \frac{\hat{q}_1+\hat{q}_0}{v}, \quad E \to \frac{E}{v}. \tag{A19}$$

Rescaled in this way the saddle-point values of the order parameters remain $\mathcal{O}(1)$ for $v \to 0$. After some tedious calculations the saddle-point equations acquire the form

$$\begin{aligned}
0 &= \hat{w} - \alpha H\left(\frac{\kappa_c}{\sqrt{q_0}}\right) \\
0 &= -\hat{q}_0 + \hat{w}(q_0 + \kappa_c^2) - \alpha\sqrt{q_0}\kappa_c\, G\left(\frac{\kappa_c}{\sqrt{q_0}}\right) \\
0 &= E(1-\gamma) - \hat{w}(q_0 - \gamma) + \hat{q}_0 \\
0 &= \hat{w} - H\left(-\frac{E - \gamma\hat{w}}{\sqrt{\hat{q}_0}}\right) \\
0 &= \hat{w}(E-1) + \sqrt{\hat{q}_0}\, G\left(\frac{E-\gamma\hat{w}}{\sqrt{\hat{q}_0}}\right) + \gamma\hat{w}(1-\hat{w})
\end{aligned} \tag{A20}$$

where

$$G(x) := \frac{1}{\sqrt{2\pi}} e^{-\frac{x^2}{2}}. \tag{A21}$$

From the numerical solution of the system (A20) we determine $\kappa_c(\alpha, \gamma)$ as shown in Figure 3.

References

1. Kondor, I.; Pafka, S.; Nagy, G. Noise sensitivity of portfolio selection under various risk measures. *J. Bank. Financ.* **2007**, *31*, 1545–1573. [CrossRef]
2. Cover, T.M. Geometrical and Statistical Properties of Systems of Linear Inequalities with Applications in Pattern Recognition. *IEEE Trans. Electron. Comput.* **1965**, *EC-14*, 326–334. [CrossRef]
3. Berg, J.; Engel, A. Matrix Games, Mixed Strategies, and Statistical Mechanics. *Phys. Rev. Lett.* **1998**, *81*, 4999–5002. [CrossRef]
4. Landmann, S.; Engel, A. On non-negative solutions to large systems of random linear equations. *Physica* **2020**, *A552*, 122544. [CrossRef]
5. Garnier-Brun, J.; Benzaquen, M.; Ciliberti, S.; Bouchaud, J.P. A New Spin on Optimal Portfolios and Ecological Equilibria. 2021. Available online: https://arxiv.org/abs/2104.00668 (accessed on 17 June 2021).
6. MacArthur, R. Species packing and competitive equilibrium for many species. *Theor. Popul. Biol.* **1970**, *1*, 1–11. [CrossRef]
7. Tikhonov, M.; Monasson, R. Collective Phase in Resource Competition in a Highly Diverse Ecosystem. *Phys. Rev. Lett.* **2017**, *118*, 048103. [CrossRef] [PubMed]
8. Todd, M. Probabilistic models for linear programming. *Math. Oper. Res.* **1991**, *16*, 671–693. [CrossRef]
9. Farkas, J. Theorie der einfachen Ungleichungen. *J. Reine Angew. Math. (Crelles J.)* **1902**, *1902*, 1–27.
10. Gardner, E. The space of interactions in neural network models. *J. Phys. A Math. Gen.* **1988**, *21*, 257–270. [CrossRef]
11. Engel, A.; Van den Broeck, C. *Statistical Mechanics of Learning*; Cambridge University Press: Cambridge, UK, 2001.
12. Markowitz, H. Portfolio selection. *J. Financ.* **1952**, *7*, 77–91.
13. Markowitz, H. *Portfolio Selection: Efficient Diversification of Investments*; J. Wiley and Sons: New York, NY, USA, 1959.
14. JP Morgan. *Riskmetrics Technical Manual*; JP Morgan: New York, NY, USA, 1995.
15. JP Morgan and Reuters. Riskmetrics. In *Technical Document*; JP Morgan: New York, NY, USA, 1996.
16. Acerbi, C.; Nordio, C.; Sirtori, C. Expected Shortfall as a Tool for Financial Risk Management. 2001. Available online: https://arxiv.org/abs/cond-mat/0102304 (accessed on 17 June 2021).
17. Artzner, P.; Delbaen, F.; Eber, J.M.; Heath, D. Coherent Measures of Risk. *Math. Financ.* **1999**, *9*, 203–228. [CrossRef]
18. Acerbi, C.; Tasche, D. Expected Shortfall: A Natural Coherent Alternative to Value at Risk. *Econ. Notes* **2002**, *31*, 379–388. [CrossRef]
19. Pflug, G.C. Some remarks on the value-at-risk and the conditional value-at-risk. In *Probabilistic Constrained Optimization*; Uryasev, S., Ed.; Springer: Berlin/Heidelberg, Germany, 2000; pp. 272–281.
20. Basel Committee on Banking Supervision. *Minimum Capital Requirements for Market Risk*; Basel Committee on Banking Supervision: Basel, Switzerland, 2016.
21. Michaud, R.O. The Markowitz optimization enigma: Is 'optimized' optimal? *Financ. Anal. J.* **1989**, *45*, 31–42. [CrossRef]
22. Kempf, A.; Memmel, C. Estimating the global minimum variance portfolio. *Schmalenbach Bus. Rev.* **2006**, *58*, 332–348. [CrossRef]
23. Basak, G.K.; Jagannathan, R.; Ma, T. A jackknife estimator for tracking error variance of optimal portfolios constructed using estimated inputs. *Manag. Sci.* **2009**, *55*, 990–1002. [CrossRef]
24. Frahm, G.; Memmel, C. Dominating estimators for minimum-variance portfolios. *J. Econom.* **2010**, *159*, 289–302. [CrossRef]

25. Pafka, S.; Kondor, I. Noisy Covariance Matrices and Portfolio Optimization II. *Physica* **2003**, *A 319*, 487–494. [CrossRef]
26. Burda, Z.; Jurkiewicz, J.; Nowak, M.A. Is Econophysics a Solid Science? *Acta Phys. Pol.* **2003**, *B 34*, 87–132.
27. Caccioli, F.; Still, S.; Marsili, M.; Kondor, I. Optimal liquidation strategies regularize portfolio selection. *Eur. J. Financ.* **2013**, *19*, 554–571. [CrossRef]
28. Caccioli, F.; Kondor, I.; Marsili, M.; Still, S. Liquidity Risk And Instabilities In Portfolio Optimization. *Int. J. Theor. Appl. Financ.* **2016**, *19*, 1650035. [CrossRef]
29. Ledoit, O.; Wolf, M. Nonlinear shrinkage estimation of large-dimensional covariance matrices. *Ann. Stat.* **2012**, *40*, 1024–1060. [CrossRef]
30. Kondor, I.; Papp, G.; Caccioli, F. Analytic solution to variance optimization with no short positions. *J. Statitical Mech. Theory Exp.* **2017**, *2017*, 123402. [CrossRef]
31. Kondor, I.; Papp, G.; Caccioli, F. Analytic approach to variance optimization under an ℓ_1 constraint. *Eur. Phys. J. B* **2019**, *92*, 8. [CrossRef]
32. Young, M.R. A minimax portfolio selection rule with linear programming solution. *Manag. Sci.* **1998**, *44*, 673–683. [CrossRef]
33. Donoho, D.; Tanner, J. Observed universality of phase transitions in high-dimensional geometry, with implications for modern data analysis and signal processing. *Philos. Trans. R. Soc. Math. Phys. Eng. Sci.* **2009**, *367*, 4273–4293. [CrossRef]
34. Schmidt, B.K.; Mattheiss, T. The probability that a random polytope is bounded. *Math. Oper. Res.* **1977**, *2*, 292–296. [CrossRef]
35. Kondor, I.; Varga-Haszonits, I. Instability of portfolio optimization under coherent risk measures. *Adv. Complex Syst.* **2010**, *13*, 425–437. [CrossRef]
36. Ciliberti, S.; Kondor, I.; Mézard, M. On the Feasibility of Portfolio Optimization under Expected Shortfall. *Quant. Financ.* **2007**, *7*, 389–396. [CrossRef]
37. Varga-Haszonits, I.; Kondor, I. The instability of downside risk measures. *J. Stat. Mech. Theory Exp.* **2008**, *2008*, P12007. [CrossRef]
38. Hertz, J.; Krogh, A.; Palmer, R.G. *Introduction to the Theory of Neural Computation*; Addison-Wesley: Redwood City, CA, USA, 1991.
39. Rosenblatt, F. *Principles of Neurodynamics: Perceptions and the Theory of Brain Mechanisms*; Spartan: Washington, DC, USA, 1962.
40. Von Neumann, J.; Morgenstern, O. *Theory of Games and Economic Behavior*; Princeton University Press: Princeton, NJ, USA, 1953.
41. May, R. Will a large complex system be stable? *Nature* **1972**, *238*, 413–414. [CrossRef]
42. Amelunxen, D.; Lotz, M.; McCoy, M.B.; Tropp, J.A. Living on the edge: A geometric theory of phase transitions in convex optimization. *Inform. Inference* **2013**, *3*, 224–294. [CrossRef]
43. Prüser, A. Phasenübergänge in Zufälligen Geometrischen Problemen. Master's Thesis, University of Oldenburg, Oldenburg, Germany, 2020.
44. Caccioli, F.; Kondor, I.; Papp, G. Portfolio optimization under expected shortfall: Contour maps of estimation error. *Quant. Financ.* **2018**, *18*, 1295–1313. [CrossRef]

entropy

MDPI

Article

Optimizing Expected Shortfall under an ℓ_1 Constraint—An Analytic Approach

Gábor Papp [1], Imre Kondor [2,3,4,*] and Fabio Caccioli [3,5,6]

1. Institute for Physics, Eötvös Loránd University, 1117 Budapest, Hungary; pg@ludens.elte.hu
2. Parmenides Foundation, 82049 Pullach, Germany
3. London Mathematical Laboratory, London W6 8RH, UK; f.caccioli@ucl.ac.uk
4. Complexity Science Hub, Vienna 1080, Austria
5. Department of Computer Science, University College London, London WC1E 6BT, UK
6. Systemic Risk Centre, London School of Economics and Political Sciences, London WC2A 2AE, UK

* Correspondence: i.kondor@lml.org.uk

Citation: Papp, G.; Kondor, I.; Caccioli, F. Optimizing Expected Shortfall under an ℓ_1 Constraint—An Analytic Approach. *Entropy* **2021**, *23*, 523. https://doi.org/10.3390/e23050523

Academic Editors: Ryszard Kutner and Geert Verdoolaege

Received: 9 March 2021
Accepted: 22 April 2021
Published: 24 April 2021

Publisher's Note: MDPI stays neutral with regard to jurisdictional claims in published maps and institutional affiliations.

Abstract: Expected Shortfall (ES), the average loss above a high quantile, is the current financial regulatory market risk measure. Its estimation and optimization are highly unstable against sample fluctuations and become impossible above a critical ratio $r = N/T$, where N is the number of different assets in the portfolio, and T is the length of the available time series. The critical ratio depends on the confidence level α, which means we have a line of critical points on the $\alpha - r$ plane. The large fluctuations in the estimation of ES can be attenuated by the application of regularizers. In this paper, we calculate ES analytically under an ℓ_1 regularizer by the method of replicas borrowed from the statistical physics of random systems. The ban on short selling, i.e., a constraint rendering all the portfolio weights non-negative, is a special case of an asymmetric ℓ_1 regularizer. Results are presented for the out-of-sample and the in-sample estimator of the regularized ES, the estimation error, the distribution of the optimal portfolio weights, and the density of the assets eliminated from the portfolio by the regularizer. It is shown that the no-short constraint acts as a high volatility cutoff, in the sense that it sets the weights of the high volatility elements to zero with higher probability than those of the low volatility items. This cutoff renormalizes the aspect ratio $r = N/T$, thereby extending the range of the feasibility of optimization. We find that there is a nontrivial mapping between the regularized and unregularized problems, corresponding to a renormalization of the order parameters.

Keywords: portfolio optimization; regularization; renormalization

1. Introduction

A risk measure is a functional on the probability distribution of the fluctuating returns of a security or a portfolio. Since it is impossible to condense all the information in a probability distribution into a single number, there is no unique way to choose the "best" risk measure. In Markowitz's ground breaking portfolio selection theory [1], with the assumption of Gaussian distributed returns, variance offered itself as the natural risk measure. The crises of the late eighties and early nineties led both the industry and regulators to realize that the most dangerous risk lurked in the asymptotically far tail of the return distribution. To grasp this risk, a high quantile of the profit and loss distribution called Value at Risk (VaR) was introduced by J.P. Morgan [2]. For a certain period, VaR became a kind of industry standard, and it was embraced by international financial regulation as the official risk measure in 1996 [3]. Value at Risk is a threshold which losses only exceed with a small probability (such as, e.g., 0.05 or 0.01), corresponding to a confidence level of $\alpha = 0.95$, resp. 0.99. (In this context, it is

customary to regard losses as positive and profits as negative). As a quantile, VaR is not sensitive to the distribution of losses above the confidence level and is not subadditive when two portfolios are combined. This triggered a search for alternatives and led Artzner et al. [4] to formulate a set of axioms that any coherent risk measure should satisfy. The simplest and most intuitive of these coherent measures is the Expected Shortfall (ES) [5,6]. ES is essentially the expected loss above a high quantile that can be chosen to be the VaR itself. After a long debate about the relative merits and drawbacks of ES, whose details are not pertinent to our present study, regulators adopted ES as the current official market risk measure to be used to assess the financial health of banks and determine the capital charge they are required to hold against their risks. The regulators and the industry settled on a confidence level of $\alpha = 0.975$ [7].

ES is mainly designed to be a diagnostic tool. At the same time, it is also a constraint that banks have to respect when considering the composition of their portfolios. It is then in their best interest to optimize ES, in order to keep their capital charge as low as possible. However, the optimization of ES is fraught with problems of estimation error, which is quite natural if one considers that the number of different items N in a bank's portfolio can be very large, whereas the number of observations (the length of the available time series T) is always limited. In addition, at the regulatory confidence level, one has to throw away 97.5% of the data. Moreover, the estimation error increases with the ratio $r = N/T$ and at a critical value of r, it actually diverges, growing beyond any limit. As shown in [8], the instability of the optimization of ES (as well as all the coherent risk measures) follows directly from the coherence axioms [4].

The divergence of ES is the signature of a phase transition. The critical r for ES is smaller or equal to $1/2$, its value depending on the confidence level α. For ES, there is then a line of critical points, a phase diagram, on the $r - \alpha$ plane. A part of this phase diagram has been traced out by numerical simulations in [9], while the full phase diagram has been determined by analytical calculations by Ciliberti et al. [10]. Going beyond merely determining the phase diagram, a detailed study of the estimation error and other relevant quantities has been performed inside the whole feasibility region in [11,12], and it was shown that, due to the nontrivial behavior of the contour lines of constant estimation error, especially in the vicinity of $\alpha = 1$, the number of data necessary to have a reasonably low estimation error was way above any T available in practice.

Because of the large sample fluctuations of ES, its optimization constitutes a problem in high dimensional statistics [13]. A standard tool to tame these large fluctuations is to introduce regularizers, which penalize large excursions. Although the introduction of these penalties may seem an arbitrary statistical trick coming from outside of finance, it was shown in [14] that these regularizers express liquidity considerations, and take into account, already at the construction of the portfolio, the expected market impact of a future liquidation. The regularizers are usually chosen to be some constraints on the norm of the portfolio weights. In [15], we studied the effect of an ℓ_2 regularizer on ES and found that ℓ_2 obviously suppresses the instability and, for sufficiently small r and with a strong enough regularizer, it extends the range where the estimation error is reasonably small by a factor of about 4.

It is interesting to see how an ℓ_1 regularizer works with ES. (The importance of studying the effect of various regularizers in combination with the different risk measures was emphasized by [16]). The regularizer ℓ_1 is known to produce sparse solutions, which means that in order to rein in large fluctuations, it eliminates some of the securities from the portfolio. This obviously contradicts the principle of diversification, but considerations of transaction costs or the technical difficulties of managing large portfolios may make it desirable to remove the most volatile items from the portfolio, and this is precisely what a no-short constraint tends to do.

It has been known for 20 years now that the optimization of ES can be translated into a linear programming problem [17]. Accordingly, as it has been realized in [18], the piece-wise linear ℓ_1 with an infinite slope corresponding to an infinite penalty on short selling can prevent the instability of ES. The purpose of this paper is to determine the effect of ℓ_1-regularization on the phase diagram and also on the behavior of the various quantities of interest inside the region where the optimization of ES is feasible and meaningful. (We will see that as a result of regularization new characteristic lines appear on the $r - \alpha$ plane, beyond which the optimization of ES is still mathematically feasible, but the results become meaningless, as they correspond to negative risk.) In [12], a detailed analytical investigation of the behavior of the estimation error, the in-sample cost, the sensitivity to small changes in the composition of the portfolio, and the distribution of optimal weights were carried out in the non-regularized case. Here, we derive the same quantities for an ℓ_1-regularized ES, including the special case where short selling is banned, that is when the portfolio weights are constrained to be non-negative. The density of the items eliminated from the portfolio, to be referred to as the "condensate" in the following, is also determined. The most striking result of the present study is that the regularized solution can be mapped back onto the unregularized one. We are not aware of a similarly tight relationship between a regularized and an unregularized problem, not only in a finance context, but neither in the general context of machine learning.

2. Method and Preliminaries

If the true probability distribution of returns were known, it would be easy to calculate the true value of Expected Shortfall and the optimal portfolio weights. However, the true distribution of returns is unknown, therefore one has to rely on finite samples of empirical data. This means one observes N time series of length T and estimates the optimal weights and ES on the basis of this information. It is clear that the weights and ES so obtained will deviate from their "true" values. (The latter would be obtained in an infinitely long stationary sample.) The deviation of the estimated values will be the stronger the shorter the length T and the larger the dimension N. Performing this measurement on different samples one would obtain different estimates: there is a distribution of ES and of the optimal weights over the samples. In a real market, one cannot repeat such an experiment multiple times. Instead, one has to squeeze out as much information as possible from a single sample of limited size. There are well-known numerical methods for this, like cross-validation or bootstrap [19]. In contrast, in the present work we aim to obtain *analytic* results. In order to mimic empirical sampling, we choose a simple data generating process, such as a multivariate Gaussian. The true value of ES is easy to obtain for this case, which provides a standard to measure finite sample deviations from. Then we determine ES for a large number of random samples of length T drawn from this underlying distribution, average it over the random samples and finally compare this average to its true value. This procedure will give us an idea about how large the estimation error is for a given dimension N, sample size T, and confidence level α, under the idealized conditions of stationarity and Gaussian fluctuations, and how much it will be reduced when we apply an ℓ_1 regularizer of a given strength. It is reasonable to assume that the estimation error obtained under these idealized circumstances will be a lower bound to the estimation error for real-life processes.

Now we wish to implement this program via analytic calculations. The averaging over the random samples just described is analogous to the averaging over the random realization of disorder in the statistical physics of random systems, which enables us to borrow methods from that field, in particular the replica method [20]. It assumes that both N and T are large, with their ratio $r = N/T$ kept finite (thermodynamic or Kolmogorov limit). A small value of r corresponds to the classical setup in statistics where one has a large number of observations relative to the dimension. Estimates in this case are sharp and close to their true values.

In contrast, when r is of order unity, or larger, we are in the high dimensional limit where fluctuations are large. It is here that the regularizer becomes important.

In the usual application of ℓ_1 in finite dimensional numerical studies, the regularizer eliminates the dimensions one by one, in a stepwise manner, as the strength of the regularizer is increasing. In our present work, the large N, T limit and the averaging over infinitely many samples result in a continuous dependence of the "condensate" density (the relative number N_0/N of the dimensions eliminated by ℓ_1) on the aspect ratio r, the confidence level α, and the strength of ℓ_1. In a study of ℓ_1-regularized variance [21], we found that the stepwise increase of the density of eliminated weights in a numerical experiment nicely follows the continuous curve obtained analytically. It is obvious that the situation is similar in the case of ES, but we have also confirmed this by numerical simulations.

For the sake of simplicity, we will also assume that the returns are independent, that is the true covariance matrix is diagonal. This is not an innocent assumption: it will be seen, for example, that the maximum degree of sparsity that ℓ_1 can achieve in this scheme is one half of the total number of dimensions, whereas for correlated returns the maximum sparsity can be either larger or smaller than $1/2$, according to whether correlations are predominantly positive or negative. Combining ℓ_1 with a non-diagonal covariance matrix poses additional technical difficulties that we wish to avoid in the present account. However, we do allow the diagonal elements σ_i of the covariance matrix to be different from each other.

As a further simplification, we do not impose any other constraint on the optimization of ES beside the budget constraint and the ℓ_1 regularizer. In particular, we do not set a constraint on the expected return, and seek the global minimum of the regularized ES. This is in line with a number of studies, [22–24] among others, which focus on the global minimum in the problem of variance optimization, because of the extremely noisy estimates of the expected return. Furthermore, the global minimum is precisely what one needs in minimizing tracking-errors, that is, when trying to follow, say, a market index as closely as possible [23].

The replica method used below have already been applied with minor variations to various portfolio optimization problems in a number of papers [10–12,14,18,21,25–28], where the replica derivation of the main formulae were repeatedly explained, so we do not need to go through that exercise again here. Then the natural starting point for our present work is the detailed study of the behavior of ES *without* regularization in [12]. The argument there leads to a relationship between ES and an effective cost or free energy per asset f as follows:

$$\mathrm{ES} = \frac{fr}{1-\alpha}. \tag{1}$$

The free energy f itself is given by the minimum of a functional depending on six order parameters

$$\begin{aligned} f(\lambda, \epsilon, q_0, \Delta, \hat{q}_0, \hat{\Delta}) &= \lambda + \frac{1}{r}(1-\alpha)\epsilon - \Delta\hat{q}_0 - \hat{\Delta} q_0 \\ &+ \langle \min_w [V(w,z,\sigma)] \rangle_{\sigma,z} + \frac{\Delta}{2r\sqrt{\pi}} \int_{-\infty}^{\infty} \mathrm{d}s\, e^{-s^2} g\left(\frac{\epsilon}{\Delta} + s\sqrt{\frac{2q_0}{\Delta^2}}\right), \end{aligned} \tag{2}$$

where

$$V(w,z,\sigma) = \hat{\Delta}\sigma^2 w^2 - \lambda w - zw\sigma\sqrt{-2\hat{q}_0} + \eta^+ \theta(w) w - \eta^- \theta(-w) w \tag{3}$$

and the double average $\langle \dots \rangle_{\sigma,z}$ means

$$\int_0^{\infty} \mathrm{d}\sigma\, \frac{1}{N} \sum_i \delta(\sigma - \sigma_i) \int_{-\infty}^{\infty} \frac{\mathrm{d}z}{\sqrt{2\pi}} e^{-z^2/2} \dots \tag{4}$$

Finally, the function g in the integral in (2) is defined as

$$g(x) = \begin{cases} 0, & x \geq 0 \\ x^2, & -1 \leq x \leq 0 \\ -2x-1, & x < -1 \end{cases}. \quad (5)$$

The differences with respect to the setup in [12] are the following: a trivial change of notation (τ there is $1/r$ here); the variable σ has been introduced in (3), which together with the recipe (4) allows us to consider assets with different volatilities σ_i; and the regularizer has been built into the effective potential (3). Note that the ℓ_1 in (3) is asymmetric in order to allow us to penalize long and short positions separately. The usual ℓ_1 corresponds to $\eta^+ = \eta^-$, the ban on short selling to $\eta^- \to \infty$. We will also use the arrangement where there is a finite penalty η^- on short positions and none on long ones $\eta^+ = 0$.

A note on signs: for consistency, the order parameters λ, Δ, q_0, and $\hat{\Delta}$ must be positive, $\hat{q}_0$ negative, and ϵ can be of either sign. Furthermore, λ must be larger or equal to the right slope of the regularizer: $\lambda \geq \eta^+$.

Before setting out to derive the stationarity conditions that determine the optimal value of the free energy and thence of ES, we spell out the meaning of the order parameters. The first of these is the Lagrange multiplier λ that enforces the budget constraint:

$$\sum_{i=1}^{N} w_i = N. \quad (6)$$

Note that the sum of portfolio weights is set to N here, instead of the usual 1. This is to keep the weights of order unity in the large N limit.

Because of the relationship between λ and the budget constraint, λ can be thought of as a kind of chemical potential. It is an important quantity, because, as we shall see later, its value at the stationary point is equal to the free energy, hence directly related to the optimal value of ES. In [12], we argued that this optimal value of ES is, in fact, the in-sample estimate of Expected Shortfall. According to (1), ES is proportional to the product fr, which means f, and hence λ too, must be inversely proportional to r when $r = N/T \to 0$, because ES is certainly finite in this limit: a finite N and $T \to \infty$ corresponds to the case of having complete information. This spurious divergence of f and λ is an artifact, due to our having absorbed a factor $1/r$ in their definition. This is explained purely by convenience: we wish to keep as close to the convention in [12] as possible. The opposite limit, when $\lambda - \eta^+$ vanishes, is another important point: it signals the instability of the portfolio, and the onset of the phase transition.

The next order parameter, ϵ, was suggested by [17] as a proxy for Value at Risk. Indeed, in the limit $r \to 0$ where we know the true distribution of returns, ϵ will be seen to be equal to the known value of VaR for a Gaussian.

The third order parameter, q_0, is of central importance: According to [12], the ratio of the out-of-sample estimate ES_{out} and its true value $\text{ES}^{(0)}$ is given by the square root of q_0. For the case of different σ_is considered here, q_0 has to be amended by a factor depending on the structure of the portfolio [21] as

$$\tilde{q}_0 = q_0 \frac{1}{N} \sum_i \frac{1}{\sigma_i^2}. \quad (7)$$

Then the ratio of the estimated and true ES will be

$$\frac{\text{ES}_{out}}{\text{ES}^{(0)}} = \sqrt{\tilde{q}_0} \quad (8)$$

that is the relative estimation error is $\sqrt{\tilde{q}_0} - 1$.

The fourth order parameter, Δ, measures the sensitivity to a small shift in the returns.

The remaining two order parameters, $\hat{q}_0$ and $\hat{\Delta}$, are auxiliary variables that do not have an obvious meaning, they enter the picture through the replica formalism, and can be eliminated once the stationarity conditions have been established. The stationarity or saddle point conditions are derived by taking the derivative of the free energy with respect to the order parameters and setting them to zero. They will be written up in the next Section.

3. Results

First, we are going to spell out the saddle point conditions in full detail and reduce them to special cases later.

Let us bring the integral in (2) to a more convenient form by integrating by parts:

$$I = \frac{1}{\sqrt{\pi}} \int_{-\infty}^{\infty} \mathrm{d}s\, e^{-s^2} g\left(\frac{\epsilon}{\Delta} + s\sqrt{\frac{2q_0}{\Delta^2}}\right) = \frac{2q_0}{\Delta^2}\left[W\left(\frac{\Delta+\epsilon}{\sqrt{q_0}}\right) - W\left(\frac{\epsilon}{\sqrt{q_0}}\right)\right] - 1 - 2\frac{\epsilon}{\Delta}. \quad (9)$$

With this identity, the free energy becomes

$$f = \lambda - \frac{\alpha\epsilon}{r} - \Delta\hat{q}_0 - \hat{\Delta}q_0 - \frac{\Delta}{2r} + \frac{q_0}{r\Delta}\left[W\left(\frac{\Delta+\epsilon}{\sqrt{q_0}}\right) - W\left(\frac{\epsilon}{\sqrt{q_0}}\right)\right] + \langle \min V \rangle_{\sigma,z}. \quad (10)$$

The function W in the above formulae, together with two related functions Φ and Ψ, will frequently appear in the following; they are integrals of the Gaussian $\frac{1}{\sqrt{2\pi}}e^{-x^2/2}$:

$$\Phi(x) = \int_{\infty}^{x} \mathrm{d}t\, \frac{1}{\sqrt{2\pi}} e^{-t^2/2} \quad (11)$$

$$\Psi(x) = \int_{-\infty}^{x} \mathrm{d}t\, \Phi(t) \quad (12)$$

$$W(x) = \int_{-\infty}^{x} \mathrm{d}t\, \Psi(t). \quad (13)$$

Now we evaluate the minimum of V in (3) and denote the "representative weight" where this minimum is located by w^*. It works out to be

$$w^* = \frac{\lambda + \sigma z\sqrt{-2\hat{q}_0} - \eta^+\Theta(w^*) + \eta^-\Theta(-w^*)}{2\sigma^2\hat{\Delta}}, \quad (14)$$

or

$$w^* = \begin{cases} \frac{\lambda+\sigma z\sqrt{-2\hat{q}_0}-\eta^+}{2\sigma^2\hat{\Delta}}, \text{ if } z \geq \frac{\eta^+-\lambda}{\sigma\sqrt{-2\hat{q}_0}} \\ 0, \text{ if } -\frac{\lambda+\eta^-}{\sigma\sqrt{-2\hat{q}_0}} < z < \frac{\eta^+-\lambda}{\sigma\sqrt{-2\hat{q}_0}} \\ \frac{\lambda+\sigma z\sqrt{-2\hat{q}_0}+\eta^-}{2\sigma^2\hat{\Delta}}, \text{ if } z \leq -\frac{\lambda+\eta^-}{\sigma\sqrt{-2\hat{q}_0}}. \end{cases} \quad (15)$$

With this and (4), one can calculate V^*, the value of V at the minimum, and perform the double averaging to obtain

$$\langle V^* \rangle_{\sigma,z} = \frac{\hat{q}_0}{\hat{\Delta}} \frac{1}{N} \sum_i \left[W\left(\frac{\lambda - \eta^+}{\sigma_i\sqrt{-2\hat{q}_0}}\right) + W\left(-\frac{\lambda+\eta^-}{\sigma_i\sqrt{-2\hat{q}_0}}\right)\right]. \quad (16)$$

Then, the fully explicit form of the free energy becomes

$$\begin{aligned} f &= \lambda - \frac{\alpha\epsilon}{r} - \Delta\hat{q}_0 - \hat{\Delta}q_0 - \frac{\Delta}{2r} + \frac{q_0}{r\Delta}\left[W\left(\frac{\Delta+\epsilon}{\sqrt{q_0}}\right) - W\left(\frac{\epsilon}{\sqrt{q_0}}\right)\right] \\ &+ \frac{\hat{q}_0}{\hat{\Delta}}\frac{1}{N}\sum_i\left[W\left(\frac{\lambda-\eta^+}{\sigma_i\sqrt{-2\hat{q}_0}}\right) + W\left(-\frac{\lambda+\eta^-}{\sigma_i\sqrt{-2\hat{q}_0}}\right)\right]. \end{aligned} \tag{17}$$

It is now straightforward to take the derivatives of f with respect to the order parameters and derive the stationary conditions.

From $\partial f/\partial\lambda = 0$, it follows that

$$1 = \frac{\sqrt{-2\hat{q}_0}}{2\hat{\Delta}}\frac{1}{N}\sum_i\frac{1}{\sigma_i}\left[\Psi\left(\frac{\lambda-\eta^+}{\sigma_i\sqrt{-2\hat{q}_0}}\right) - \Psi\left(-\frac{\lambda+\eta^-}{\sigma_i\sqrt{-2\hat{q}_0}}\right)\right]. \tag{18}$$

The derivative with respect to $\hat{q}_0$ yields

$$2\Delta\hat{\Delta} = \frac{1}{N}\sum_i\left[\Phi\left(\frac{\lambda-\eta^+}{\sigma_i\sqrt{-2\hat{q}_0}}\right) + \Phi\left(-\frac{\lambda+\eta^-}{\sigma_i\sqrt{-2\hat{q}_0}}\right)\right]. \tag{19}$$

From the derivative with respect to $\hat{\Delta}$, we get

$$q_0 = -\frac{\hat{q}_0}{\hat{\Delta}^2}\frac{1}{N}\sum_i\left[W\left(\frac{\lambda-\eta^+}{\sigma_i\sqrt{-2\hat{q}_0}}\right) + W\left(-\frac{\lambda+\eta^-}{\sigma_i\sqrt{-2\hat{q}_0}}\right)\right]. \tag{20}$$

As mentioned before, q_0 determines the out-of-sample estimate for ES and the estimation error.

The derivative with respect to q_0 leads to

$$2r\Delta\hat{\Delta} = \Phi\left(\frac{\Delta+\epsilon}{\sqrt{q_0}}\right) - \Phi\left(\frac{\epsilon}{\sqrt{q_0}}\right), \tag{21}$$

where use has been made of the identity

$$W(x) = \frac{1}{2}x\Psi(x) + \frac{1}{2}\Phi(x). \tag{22}$$

The condition for the derivative with respect to ϵ to vanish is

$$\alpha = \frac{\sqrt{q_0}}{\Delta}\left[\Psi\left(\frac{\Delta+\epsilon}{\sqrt{q_0}}\right) - \Psi\left(\frac{\epsilon}{\sqrt{q_0}}\right)\right]. \tag{23}$$

The derivation of the last equation takes a little more effort. Let us go back to the free energy in (2) and take the derivative with respect to Δ. Noticing that $\langle V\rangle_{\sigma,z}$ does not depend on Δ, and using the integral given in (9), we have

$$\frac{\partial f}{\partial\Delta} = -\hat{q}_0 + \frac{1}{2r}I + \frac{\Delta}{2r}\frac{\partial I}{\partial\Delta} = 0 \tag{24}$$

valid at the stationary point. From here we find

$$\frac{1}{2r}I_{st} = \hat{q}_0 + \frac{2q_0}{r\Delta^2}\left[W\left(\frac{\Delta+\epsilon}{\sqrt{q_0}}\right) - W\left(\frac{\epsilon}{\sqrt{q_0}}\right)\right] - \frac{\epsilon}{r\Delta} - \frac{\sqrt{q_0}}{r\Delta}\Psi\left(\frac{\Delta+\epsilon}{\sqrt{q_0}}\right), \tag{25}$$

where (9) was used again and we denoted by I_{st} the integral I evaluated at the stationary point. Now we apply the identity (22) and the stationary conditions (23), (21) to arrive at

$$\frac{1}{2r}I_{st} = \hat{q}_0 + \frac{2q_0\hat{\Delta}}{\Delta} - (1-\alpha)\frac{\epsilon}{r\Delta}, \tag{26}$$

which, combined with (9), finally leads to

$$\hat{q}_0 + \frac{2q_0\hat{\Delta}}{\Delta} + \alpha\frac{\epsilon}{r\Delta} + \frac{1}{2r} - \frac{q_0}{r\Delta^2}\left[W\left(\frac{\Delta+\epsilon}{\sqrt{q_0}}\right) - W\left(\frac{\epsilon}{\sqrt{q_0}}\right)\right] = 0. \tag{27}$$

The Equations (18)–(23) and (27) constitute the system of equations for the six order parameters. These equations are valid both for the regularized and (setting $\eta^+ = \eta^- = 0$) for the unregularized cases.

Let us now work out the relationship between the free energy and the chemical potential. Comparing (16) and (20), we see that $\langle V^*\rangle_{\sigma,z} = -q_0\hat{\Delta}$, which with (10) and (27), results in the simple formula

$$f = \lambda \tag{28}$$

at the stationary point, as we anticipated before. In [12], we argued that the stationary value of f determines the in-sample estimate of ES through (1).

The last object to determine is the distribution of weights:

$$p(w) = \langle \delta(w - w^*)\rangle_{\sigma,z}. \tag{29}$$

With (14), we find

$$p(w) = n_0\delta(w) + \frac{1}{N}\sum_i \frac{1}{\sigma_w^{(i)}\sqrt{2\pi}}\exp\left(-\frac{1}{2}\left(\frac{w - w_i^+}{\sigma_w^{(i)}}\right)^2\right)\theta(w) \tag{30}$$

$$+ \frac{1}{N}\sum_i \frac{1}{\sigma_w^{(i)}\sqrt{2\pi}}\exp\left(-\frac{1}{2}\left(\frac{w - w_i^-}{\sigma_w^{(i)}}\right)^2\right)\theta(-w), \tag{31}$$

where $\delta(w)$ is the Dirac delta,

$$\sigma_w^i = \frac{\sqrt{-2\hat{q}_0}}{2\hat{\Delta}\sigma_i} \tag{32}$$

is the (estimated) variance of the ith return,

$$w_i^+ = \frac{\lambda - \eta^+}{2\sigma_i^2\hat{\Delta}} \tag{33}$$

is the center of the Gaussian distribution of the (estimated) positive weight i,

$$w_i^- = \frac{\lambda + \eta^-}{2\sigma_i^2\hat{\Delta}} \tag{34}$$

is the same for negative weight i, and finally,

$$n_0 = \frac{1}{N}\sum_i\left[\Phi\left(\frac{\lambda+\eta^-}{\sigma_i\sqrt{-2\hat{q}_0}}\right) - \Phi\left(\frac{\lambda-\eta^+}{\sigma_i\sqrt{-2\hat{q}_0}}\right)\right] \tag{35}$$

is the density of the assets whose weights are set to zero by the regularizer.

We wish to make an important remark here: the right hand side of (19) is just $1 - n_0$. This will prove to be the key to the mapping between the regularized and unregularized cases.

Let us record the condensate density n_0 also for the special case when short positions are excluded ($\eta^- \to \infty$), but long positions are not penalized ($\eta^+ = 0$):

$$n_0 = \frac{1}{N}\sum_i \left[1 - \Phi\left(\frac{\lambda}{\sigma_i\sqrt{-2\hat{q}_0}}\right)\right]. \tag{36}$$

From (36), we can see that, since $\Phi(x)$ is monotonic increasing and, for $x \geq 0$, concave, the contribution to n_0 from assets with larger σ_is is larger than that from smaller σ_is. This means that in the no-short limit, the regularizer ℓ_1 eliminates more volatile assets with larger probability than the less volatile ones. Thus, we can think of the no-short constraint as a smooth upper cutoff in volatility. This is not true in the generic case (35), where the contributions of the small and large volatility items depend on the order parameters and the regularizer's slopes η^+ and η^- in a complicated manner: the probability of an asset with volatility σ_i to be removed is given by the difference of the two term in (35) under the sum. We do not wish to analyze this situation in detail, apart from the remark that a sufficiently large η^- generally favors the elimination of large volatility items.

The integral of $p(w)$ is, of course, 1. Its first moment, $\langle w^* \rangle_{\sigma,z}$, works out to be the same as (18):

$$\langle w^* \rangle_{\sigma,z} = 1\,. \tag{37}$$

The second moment of the weight distribution is readily obtained as

$$\langle (w^*)^2 \rangle_{\sigma,z} = -\frac{\hat{q}_0}{\hat{\Delta}^2}\frac{1}{N}\sum_i \frac{1}{\sigma_i^2}\left[\mathrm{W}\left(\frac{\lambda - \eta^+}{\sigma_i\sqrt{-2\hat{q}_0}}\right) + \mathrm{W}\left(-\frac{\lambda + \eta^-}{\sigma_i\sqrt{-2\hat{q}_0}}\right)\right]. \tag{38}$$

The variance of the weight distribution is then

$$\langle (w^*)^2 \rangle_{\sigma,z} - (\langle w^* \rangle_{\sigma,z})^2\,, \tag{39}$$

which is equal to $q_0 - 1$, when the variances of the assets are all equal to 1. For a portfolio with different σ_i's, however, the relevant quantity that determines the out-of-sample estimate of ES is not the second moment of the weight distribution, but the true variance of the ith asset multiplied by the estimated portfolio weights squared and summed over the different assets, that is

$$\langle \sigma^2 (w^*)^2 \rangle_{\sigma,z}\,, \tag{40}$$

which is precisely q_0 as given in (20), and this is the quantity (multiplied by the correction as in (7)) that enters the formula for the out-of-sample estimate of ES in (8). For a not too inhomogeneous portfolio, the difference between the second moment of the weight distribution and q_0 is not significant, so we can think of q_0 as a measure of the variance of the portfolio.

Now we are ready to consider various special cases.

3.1. The Limit of Complete Information

When we have many observations (very long time series, $T \to \infty$) relative to the dimension N of the portfolio, we are in the $r = N/T \to 0$ limit. As we have already mentioned, this also corresponds to the "chemical potential" λ going to infinity. Obviously, in this limit, the regularizer plays no role.

We need the asymptotic behavior of the functions appearing in our stationary conditions: for $x \to \infty$, $\Phi(x) \to 1$, $\Psi(x) \sim x$, and $W(x) \sim x^2/2$, while for $x \to -\infty$, all three vanish exponentially.

Then from (18) we have

$$1 = \frac{\lambda}{2\hat{\Delta}} \frac{1}{N} \sum_i \frac{1}{\sigma_i^2}. \tag{41}$$

From (19)

$$2\Delta\hat{\Delta} = 1. \tag{42}$$

Combining the two:

$$1 = \lambda\Delta \frac{1}{N} \sum_i \frac{1}{\sigma_i^2}. \tag{43}$$

We know from (1) and (28) that λ must be inversely proportional to r when $r \to 0$. It follows that $\Delta \sim r$ for small r.

Then, from (20) we find

$$q_0 = \Delta^2 \lambda^2 \frac{1}{N} \sum_i \frac{1}{\sigma_i^2}. \tag{44}$$

Combined with the previous equation, this gives

$$q_0 = \frac{1}{\frac{1}{N}\sum_i \frac{1}{\sigma_i^2}}. \tag{45}$$

The "true" ($r \to 0$) value of the order parameter q_0 is thus determined by the structural constant $\frac{1}{N}\sum_i \frac{1}{\sigma_i^2}$, which is given by the variances of the returns σ_i^2. This is in accord with the corresponding result found in the case of the ℓ_1-regularized variance risk measure [21,29]. The above result for q_0 also means that the quantity $\tilde{q}_0$ introduced in (7) is equal to 1, and according to (8) the out-of-sample estimate of ES is equal to its true value $\mathrm{ES}^{(0)}$, the estimation error is zero—an obvious result for the case of complete information.

From (23) with $\Delta \to 0$ we obtain $\alpha = \Phi(\epsilon/\sqrt{q_0})$, or

$$\epsilon = \Phi^{-1}(\alpha)\sqrt{q_0}. \tag{46}$$

Now from (21) we get $r = \Phi'\left(\frac{\epsilon}{\sqrt{q_0}}\right)\frac{\Delta}{\sqrt{q_0}}$, or

$$\Delta = r\sqrt{q_0} \frac{1}{\frac{1}{\sqrt{2\pi}} e^{-\epsilon^2/2q_0}}. \tag{47}$$

However, then we have found

$$\lambda = \frac{q_0}{\Delta} = \frac{1}{r} \frac{1}{\sqrt{2\pi}} e^{-\epsilon^2/2q_0} \sqrt{q_0} = \frac{1}{r} \frac{1}{\sqrt{2\pi}} e^{-\left(\Phi^{-1}(\alpha)\right)^2/2} \sqrt{q_0}. \tag{48}$$

Since $\lambda = f$ and $\mathrm{ES} = fr/(1-\alpha)$, we have the $r \to 0$ limit (the true value) of ES:

$$\mathrm{ES}^{(0)} = \frac{1}{1-\alpha} \frac{1}{\sqrt{2\pi}} e^{-\left(\Phi^{-1}(\alpha)\right)^2/2} \sqrt{q_0}. \tag{49}$$

We record the $r \to 0$ limits of the two auxiliary variables, $\hat{\Delta}$ and $\hat{q}$, for completeness:

$$\hat{\Delta} = \frac{1}{2r\sqrt{q_0}} \frac{1}{\sqrt{2\pi}} e^{-\epsilon^2/2q_0} \tag{50}$$

and

$$\hat{q}_0 \sim -\frac{1}{r}, \tag{51}$$

with a coefficient that will not be needed in the following.

Let us turn to the distribution of weights now.

In the $r \to 0$ limit, the widths of the Gaussians in (30) all vanish, so the Gaussians become delta functions:

$$p = \frac{1}{N}\sum_i \delta(w - w_i^+)\theta(w) + \frac{1}{N}\sum_i \delta(w - w_i^-)\theta(-w). \tag{52}$$

In the $r \to 0$ limit, the weights are all positive, so the second sum disappears.

For the weights, w_i^+ we find

$$w_i^+ \simeq \frac{\lambda}{2\sigma_i^2\hat{\Delta}} = \frac{\lambda\Delta}{\sigma_i^2} = \frac{1}{\sigma_i^2} \frac{1}{\frac{1}{N}\sum_k \frac{1}{\sigma_k^2}}. \tag{53}$$

They sum to N, as stipulated.

The variance of a linear combination of independent random variables with averages w_i^+ and variances σ_i^2 is

$$\sigma_p^2 = \sum_i (w_i^+)^2 \sigma_i^2 = \frac{N}{\frac{1}{N}\sum_k \frac{1}{\sigma_k^2}}. \tag{54}$$

Now we recognize the meaning of the (true value of the) order parameter q_0: it is the normalized (to $\mathcal{O}(1)$) variance of the portfolio. This also explains the correction factor appearing in (7). We also see that (46) and (49) are the standard expressions for Value at Risk and Expected Shortfall indeed.

We emphasize again that all the results presented in this subsection are only valid in the $r \to 0$ limit when we are dealing with a finite dimension N and infinitely long time series T.

For finite r, the sample fluctuations start to broaden the delta spikes in the distribution of weights, the condensation of zero weights begins, λ decreases, and all the formulae above become considerably more complicated. We turn to this situation in the next subsections.

By now, we have learned everything that was to be learned from keeping the variances σ_i different, in particular the tendency of the elimination of the most volatile assets by the regularizer in the case of restriction of short selling. In order to simplify the presentation and avoid the appearance of very large and hardly transparent formulae, henceforth we set all the σ_i's equal to 1. We stress, however, that the main message of this paper, namely the existence of a mapping between the regularized and unregularized cases, depends only on the structure of the equations, and works also with different σ's.

3.2. Without Regularization

In this subsection, we set $\eta^+ = \eta^- = 0$, that is we consider our problem without regularization, and according to what has just been said, put $\sigma_i = 1$. We will make use of the identities

$$\begin{aligned}\Phi(x)+\Phi(-x) &= 1 &(55)\\ \Psi(x)+\Psi(-x) &= x &(56)\\ W(x)+W(-x) &= \frac{1}{2}(x^2+1)\,. &(57)\end{aligned}$$

The free energy (17) becomes

$$f = \lambda - \frac{\alpha\epsilon}{r} - \Delta\hat{q}_0 - \hat{\Delta}q_0 - \frac{\Delta}{2r} + \frac{q_0}{r\Delta}\left[W\left(\frac{\Delta+\epsilon}{\sqrt{q_0}}\right) - W\left(\frac{\epsilon}{\sqrt{q_0}}\right)\right] - \frac{\lambda^2}{4\hat{\Delta}} + \frac{\hat{q}_0}{2\hat{\Delta}}. \tag{58}$$

For the saddle point equations, we find:

$$1 = \frac{\lambda}{2\hat{\Delta}}\,, \tag{59}$$

$$2\Delta\hat{\Delta} = 1\,, \tag{60}$$

$$q_0 = \frac{\lambda^2}{4\hat{\Delta}^2} - \frac{\hat{q}_0}{2\hat{\Delta}^2}\,, \tag{61}$$

$$2r\Delta\hat{\Delta} = r = \Phi\left(\frac{\Delta+\epsilon}{\sqrt{q_0}}\right) - \Phi\left(\frac{\epsilon}{\sqrt{q_0}}\right), \tag{62}$$

$$\alpha = \frac{\sqrt{q_0}}{\Delta}\left[\Psi\left(\frac{\Delta+\epsilon}{\sqrt{q_0}}\right) - \Psi\left(\frac{\epsilon}{\sqrt{q_0}}\right)\right], \tag{63}$$

$$\hat{q}_0 + \frac{2q_0\hat{\Delta}}{\Delta} + \frac{\alpha\epsilon}{r\Delta} + \frac{1}{2r} - \frac{q_0}{r\Delta^2}\left[W\left(\frac{\Delta+\epsilon}{\sqrt{q_0}}\right) - W\left(\frac{\epsilon}{\sqrt{q_0}}\right)\right] = 0\,. \tag{64}$$

These equations are rather similar to their counterparts in the previous subsection, but of course $r \to 0$ is not assumed here. As for their solutions, they were discussed and illustrated in several figures in [12], therefore we will not dwell upon them here. (Some results will be given in Section 3.6.) Instead, we write up the corresponding equations in the case where no short positions are allowed and make a term-by-term comparison between the two sets of equations.

3.3. No Short Selling

Short positions will be excluded by imposing infinite penalty on them by letting η^- go to infinity. The functions $\Phi(x)$, $\Psi(x)$, and $W(x)$ all vanish when $x \to -\infty$. Long positions will not be penalized, so we set $\eta^+ = 0$.

The free energy becomes

$$\begin{aligned} f &= \lambda - \frac{\alpha\epsilon}{r} - \Delta\hat{q}_0 - \hat{\Delta}q_0 - \frac{\Delta}{2r} + \frac{q_0}{r\Delta}\left[W\left(\frac{\Delta+\epsilon}{\sqrt{q_0}}\right) - W\left(\frac{\epsilon}{\sqrt{q_0}}\right)\right] &(65)\\ &+ \frac{\hat{q}_0}{\hat{\Delta}}W\left(\frac{\lambda}{\sqrt{-2\hat{q}_0}}\right). &(66)\end{aligned}$$

The stationary conditions now read as:

$$1 = \frac{\sqrt{-2\hat{q}_0}}{2\hat{\Delta}} \Psi\left(\frac{\lambda}{\sqrt{-2\hat{q}_0}}\right), \tag{67}$$

$$2\Delta\hat{\Delta} = \Phi\left(\frac{\lambda}{\sqrt{-2\hat{q}_0}}\right), \tag{68}$$

$$q_0 = -\frac{\hat{q}_0}{\hat{\Delta}^2} W\left(\frac{\lambda}{\sqrt{-2\hat{q}_0}}\right), \tag{69}$$

$$2r\Delta\hat{\Delta} = \Phi\left(\frac{\Delta+\epsilon}{\sqrt{q_0}}\right) - \Phi\left(\frac{\epsilon}{\sqrt{q_0}}\right), \tag{70}$$

$$\alpha = \frac{\sqrt{q_0}}{\Delta}\left[\Psi\left(\frac{\Delta+\epsilon}{\sqrt{q_0}}\right) - \Psi\left(\frac{\epsilon}{\sqrt{q_0}}\right)\right], \tag{71}$$

$$r\left(\hat{q}_0 + \frac{2q_0\hat{\Delta}}{\Delta}\right) + \frac{\alpha\epsilon}{\Delta} + \frac{1}{2} - \frac{q_0}{\Delta^2}\left[W\left(\frac{\Delta+\epsilon}{\sqrt{q_0}}\right) - W\left(\frac{\epsilon}{\sqrt{q_0}}\right)\right] = 0, \tag{72}$$

the last equation being the same as (64), just multiplied by r.

In the distribution of weights in (30), the second sum of Gaussians will disappear, because for $\eta^- \to \infty$, all the weights (34) go to infinity. The weights (33) become

$$w_i^+ = \frac{\lambda}{2\hat{\Delta}}, \tag{73}$$

while the density of zero weights is now

$$n_0 = 1 - \Phi\left(\frac{\lambda}{\sqrt{-2\hat{q}_0}}\right), \tag{74}$$

which with (68) leads to

$$1 - n_0 = 2\Delta\hat{\Delta}. \tag{75}$$

From (74), we see that $n_0 = 0$ for $r = 0$ and increases as λ decreases, until it reaches its maximal value $1/2$ when λ vanishes. Mathematically, there is nothing to prevent us from continuing to increase r and driving λ to negative values, which would allow n_0 to grow beyond $1/2$, up to $n_0 = 1$, but a negative λ would cause the free energy and thus also ES to change sign—an extreme case of "in-sample optimism", entirely due to the lack of sufficient information. We consider such a situation "unphysical", and never go beyond the point where λ (or $\lambda - \eta^+$ if $\eta^+ > 0$) vanishes anywhere in this paper.

3.4. No-Short Mapping

We are now ready to spell out the mapping between the no-short case and the unregularized one.

The first point to notice is that the only difference between Equation (62) valid in the unregularized case and its counterpart (70) in the no-short case (combined with (75)) appears on their left hand side: the terms r and $(1 - n_0)r$, respectively. This suggests to introduce an effective r:

$$r_{\text{eff}} = (1 - n_0)r. \tag{76}$$

Now $r = N/T$, and n_0 is the density of the assets removed by the regularizer, thus $(1-n_0)r = \frac{N-N_0}{T}$ is the number of surviving assets divided by the length of the time series. As r_{eff} increases from zero to $1/2$, r will increase between zero and 1.

Inspired by the connection between r and r_{eff}, we compare the two sets of equations and recognize that, in fact, the whole system of saddle point equations can be mapped from the regularized case to the unregularized one. A variable that appears in all the subsequent equations is

$$z = \frac{\lambda}{\sqrt{-2\hat{q}_0}}, \tag{77}$$

where the variables λ and $\hat{q}_0$ are those that appear in the no-short equations.

Then the connection between the order parameters belonging to the two cases is the following:

$$q_0 = q_0^{\text{eff}} \frac{z}{\Psi(z)}, \tag{78}$$

$$\Delta = \Delta_{\text{eff}} \sqrt{\frac{z}{\Psi(z)}}, \tag{79}$$

$$\epsilon = \epsilon_{\text{eff}} \sqrt{\frac{z}{\Psi(z)}}, \tag{80}$$

$$\lambda = \lambda_{\text{eff}} \sqrt{\frac{z}{\Psi(z)}} \Phi(z), \tag{81}$$

$$\hat{q}_0 = \hat{q}_0^{\text{eff}} \Phi(z), \tag{82}$$

$$\hat{\Delta} = \hat{\Delta}_{\text{eff}} \sqrt{\frac{\Psi(z)}{z}} \Phi(z). \tag{83}$$

A direct substitution shows that if the order parameters on the left hand sides of the above equations satisfy the no-short equations, then the effective variables satisfy the unregularized ones, provided we also replace r with r_{eff}. In particular, the contour maps of the unregularized order parameters presented in [12] can be taken over and simply blown up by a factor $\frac{1}{1-n_0}$ to obtain the contour maps of the no-short variables. Given the relation between q_0 and the estimation error, we see that the mapping also means that a given error belongs to a larger r in the no-short case than in the unregularized one, in other words, the no-short constrained problem demands $(1-n_0)$ times less data (shorter time series) than the unregularized one.

One may wonder whether this mapping expresses some symmetry of the problem, that is whether the free energy functional is invariant under this mapping. The answer is no: the mapping works only in the saddle point equations, it is a property of the stationary point.

It is important to learn the range of this transformation. In the limit $r \to 0$, the transformation is the identity, but this is trivial: when we have complete information, the regularizer does not play any role. It is more interesting to consider the vicinity of the phase transition in the unregularized case, where q_0^{eff} and Δ_{eff} diverge. These divergences are removed by the mapping, no singularity is found in the no-short case. This is in accord with [18]: the infinite penalty on short positions precludes the phase transition and no singularity shows up in q_0, Δ, or ϵ. Mathematically, we can continue the unregularized solutions into the non-feasible region beyond the phase boundary, but they make no sense there (for example, q_0 changes sign, Δ and ϵ become imaginary, etc.), while their mapped counterparts continue to behave reasonably. According to (76), when r_{eff} reaches the critical point $r_c(\alpha)$, the corresponding value of r in the no-short problem will be twice as large, so the whole phase diagram is multiplied by a

factor 2. Beyond the mapped phase boundary the regularized solutions still survive, but their meaning becomes questionable, because the free energy, hence also ES change sign. As noted in the previous Subsection, we refrain from the discussion of this unphysical region.

3.5. Mapping for Generic ℓ_1 Constraint

The mapping between the generic ℓ_1-constrained ES optimization and the unregularized one is a straightforward generalization of the results in the previous Subsection. The mapping is made more complicated because of the sums and differences of the Ψ, Φ, and W functions appearing on the right hand side of Equations (18)–(20). We introduce the following notation for these combinations:

$$A_\Psi = \Psi\left(\frac{\lambda - \eta^+}{\sqrt{-2\hat{q}_0}}\right) - \Psi\left(-\frac{\lambda + \eta^-}{\sqrt{-2\hat{q}_0}}\right), \tag{84}$$

$$A_\Phi = \Phi\left(\frac{\lambda - \eta^+}{\sqrt{-2\hat{q}_0}}\right) + \Phi\left(-\frac{\lambda + \eta^-}{\sqrt{-2\hat{q}_0}}\right), \tag{85}$$

and

$$A_W = W\left(\frac{\lambda - \eta^+}{\sqrt{-2\hat{q}_0}}\right) + W\left(-\frac{\lambda + \eta^-}{\sqrt{-2\hat{q}_0}}\right), \tag{86}$$

where we have set all the $\sigma_i = 1$.

In terms of these quantities the generic map reads as

$$q_0 = q_0^{\text{eff}} \frac{2A_W - A_\Phi}{(A_\Psi)^2}, \tag{87}$$

$$\Delta = \Delta_{\text{eff}} \frac{\sqrt{2A_W - A_\Phi}}{A_\Psi}, \tag{88}$$

$$\epsilon = \epsilon_{\text{eff}} \frac{\sqrt{2A_W - A_\Phi}}{A_\Psi}, \tag{89}$$

$$\lambda = \lambda_{\text{eff}} \frac{z A_\Phi}{\sqrt{2A_W - A_\Phi}}, \tag{90}$$

$$\hat{q}_0 = \hat{q}_0^{\text{eff}} A_\Phi, \tag{91}$$

$$\hat{\Delta} = \hat{\Delta}_{\text{eff}} \frac{A_\Phi A_\Psi}{\sqrt{2A_W - A_\Phi}}. \tag{92}$$

For the condensate density n_0, we have

$$1 - n_0 = A_\Phi, \tag{93}$$

and for the effective aspect ratio

$$r_{\text{eff}} = 2r\Delta\hat{\Delta} = rA_\Phi = (1 - n_0)r. \tag{94}$$

As before, if the order parameters satisfy the regularized stationarity conditions (18)–(27) (with $\sigma_i = 1$), then the effective parameters will satisfy the unregularized Equations (59)–(64), and vice versa.

Note that the above equations remain invariant if we redefine λ as $\lambda - \eta^+$ and η^- as $\eta^- + \eta^+$. So we can set $\eta^+ = 0$ and $\eta^- + \eta^+ = \eta$ without loss of generality. We will use this setup in the following, in order to reduce the number of parameters when solving the stationarity equations.

3.6. Solutions for the Order Parameters

Except for a few exceptional points, it is impossible to obtain the solutions of the stationarity equations in closed, analytical form, but it is perfectly possible to get them numerically, by a computer. (The case of $\alpha = 1$ is exceptional in several respects and will not be considered here.) In the following, the solutions will be presented in graphical form.

Figure 1 exhibits three special lines, belonging to three different cases: the unregularized case, the one with a finite regularizer, and the one with a no-short constraint.

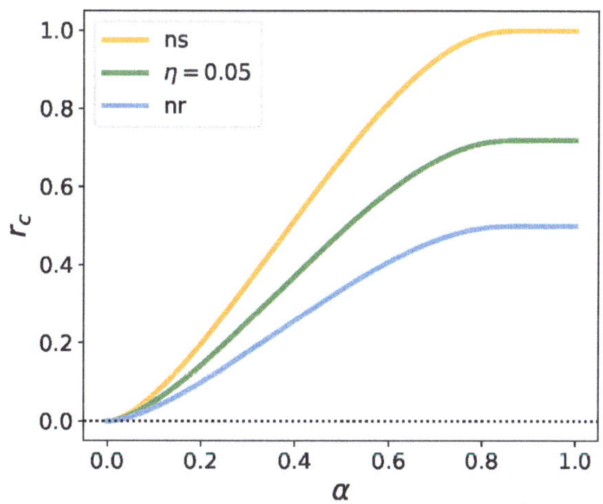

Figure 1. The boundary of the region where the optimization of ES is feasible in the unregularized case (nr); its image under the map for a finite $\eta^- = 0.05$, $\eta + = 0$ regularizer; and the same under the no-short map (ns).

The blue line is the upper boundary of the region where the optimization of unregularized ES is feasible. This line was first determined in [10]. It is a phase boundary, along which a phase transition takes place: q_0, Δ, and ϵ diverge here, while λ becomes zero. The unregularized equations can be solved also above this line, up to the horizontal line at $r = 1$ (not shown in the Figure), but the solutions are meaningless: q_0 is negative, while λ, Δ, and ϵ become imaginary. The unregularized equations do not have any solution above $r = 1$.

The green line is the image of the unregularized phase boundary under the mapping described in the previous Subsection, and corresponds to a one-sided regularizer with $\eta^- = 0.05$, $\eta + = 0$. There is no phase transition when we cross this line, the order parameters remain smooth, finite quantities, but λ (along with the free energy and the in-sample estimate of ES) changes sign, rendering the solution in the region above the green line "unphysical". Nevertheless, if we keep following the solutions beyond the green line we can go up to the image of the $r = 1$ line (mapped into $r \to \infty$), where q_0 and Δ will ultimately diverge. The region between the green line and the image of the $r = 1$ line has an intricate structure, but because it corresponds to negative risk, it is of no interest for us in the present context.

In the no-short case, there is always a solution with the order parameters remaining finite all the way up to infinity, which is the image of the $r = 1$ line under the no-short map. However, as we cross the orange line, λ changes sign, and the region beyond it is meaningless again. The orange line is the unregularized phase boundary (blue line) blown up by a factor $\frac{1}{1-n_0} = 2$. All this is in accord with the picture described in [18] in that the no-short constraint eliminates the critical line. The solutions becoming unphysical beyond a certain r-range could not be foreseen on the basis of the analysis in [18].

Figure 2 shows the η-dependence of q_0 and the density of the zero weights n_0 at criticality, and that of the value of the critical r. In the unregularized case ($\eta \to 0$), $q_0 \to \infty$, while in the no-short case ($\eta \to \infty$) $q_0 \to \pi$. At $\alpha = 0.975$, the value of the critical r_c increases from $r_c \approx 1/2$ in the unregularized case to $\approx$1 for the no-short case. The proportion of the assets eliminated from the portfolio (the condensate density) goes from zero for $\eta = 0$ to $1/2$ for large η.

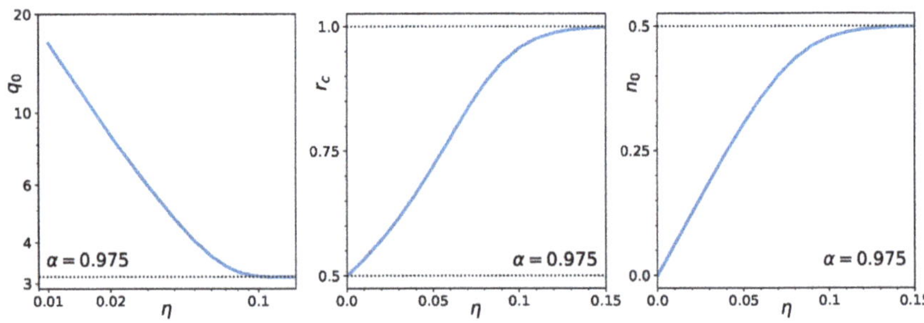

Figure 2. Dependence of q_0 at r_c (**left**), critical point (**middle**), and proportion of zero weights at r_c (**right**) as a function of the regularization strength, $\eta^- = \eta$ ($\eta^+ = 0$). Note the logarithmic scale in the left panel.

In Figure 3, we display the r-dependence of q_0, Δ, and λ for the three cases: unregularized, regularized, and no-short. Without regularization, q_0 and Δ increase with r and diverge at an r_c slightly less than $\frac{1}{2}$; while λ decreases from infinity at $r = 0$ to zero at r_c. (The confidence limit α is set at its regulatory value 0.975 in these figures.) Under the regularizer $\eta^- = 0.05$, $\eta^+ = 0$, q_0, and Δ increases up to the r where λ vanishes. The situation is similar for an infinitely strong (no-short) regularizer, with the limiting value of $q_0 = \pi$ and $\lambda = 0$ at $r \approx 1$.

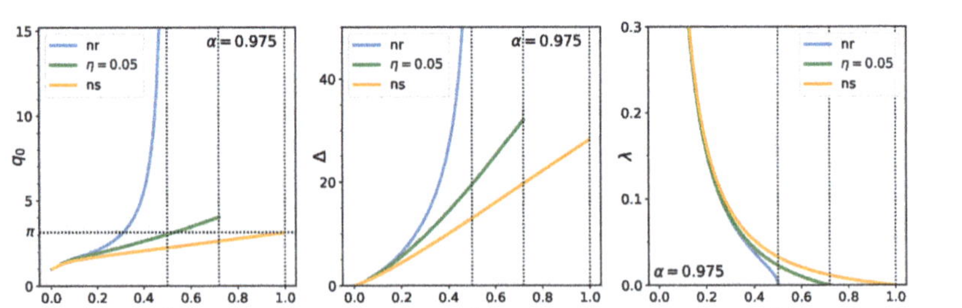

Figure 3. Dependence of q_0(**left**), Δ (**middle**) and "chemical potential" λ (**right**) on $r = N/T$, for the unregularized (blue), $\eta^- = 0.05, \eta^+ = 0$ regularized (green), and no-short (yellow) cases.

The left panel in Figure 4 shows the relative out-of-sample estimation error, which is related to the out-of-sample estimate of ES by (8) ($\tilde{q}_0 = q_0$ now, as we have set all the σ_i =1).

These curves are similar to the curves of q_0 in the previous Figure. It can be seen that the curves of the relative estimation error run very close to each other for small values of r: there is no substantial reduction of the error in this range. Where they fan out and the effect of regularization starts to be felt (say around $r = 0.1$), the relative error is already about 20%.

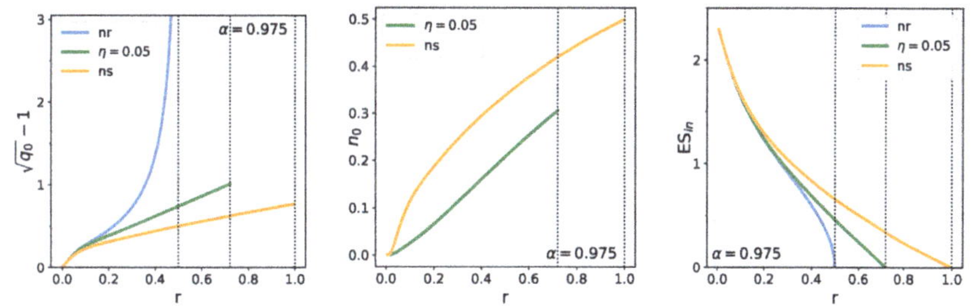

Figure 4. Dependence of the out-of-sample estimation error (**left**), proportion of zero weights (**center**), and in-sample ES (**right**) on $r = N/T$, for the non-regularized (blue), $\eta^- = \eta$ ($\eta^+ = 0$) regularized (green), and no-short (orange) cases.

The middle panel in Figure 4 shows the behavior of the density of zero weights as function of r for the finite η-regularized and the no-short cases. In the no-short case, n_0 reaches its maximal value $\frac{1}{2}$ at $r \approx 1$ (for $\alpha = 0.975$) where λ vanishes. For a regularizer of finite strength, it always remains below $\frac{1}{2}$.

The right panel in Figure 4 displays the behavior of the in-sample estimate of ES for the three cases. This quantity is directly related to λ through (1) and (28). The monotonic and fast decay of these curves demonstrates what is called in-sample optimism, a strong underestimation of risk.

4. Discussion

In the preceding Section we compared the behavior of the order parameters in the three instances considered in this paper: the case of the unregularized, the ℓ_1-regularized, and the no-short constrained Expected Shortfall optimization. We have seen that without regularization, there is a phase transition as we cross the phase boundary $r_c(\alpha)$ shown in Figure 1 with Δ, q_0, and ϵ diverging here, as known since the paper [10]. In contrast, the infinite penalty on short positions suppresses this phase transition, while an ℓ_1 regularizer with finite slopes only shifts the phase boundary. These facts were also known from earlier work [14,18]. However, the picture has turned out to be more complicated than envisaged in [18]. The numerical solution for the order parameters performed in this paper has revealed that new characteristic lines emerge both in the case of finite regularization and the no-short constraint, along which the order parameter λ and, consequently, the free energy and the in-sample estimate of Expected Shortfall change sign. We have determined the position of these new characteristic lines: in the no-short case the new line is the curve $2r_c(\alpha)$, for a finite regularizer it is $\frac{r_c(\alpha)}{1-n_0}$, where $n_0 \leq \frac{1}{2}$. We have omitted the detailed analysis of the regions above these lines, where the estimated risk becomes negative. Instead, we confined ourselves to merely pointing out that the critical line for the no-short constraint is projected out to infinity, so the phase transition is removed indeed, while for a finite slope regularizer the critical line is shifted into the unphysical, negative risk region, where for some values of the regularizer's strength η, it even develops two branches.

We have also found the behavior of the various order parameters, most notably that of q_0 that determines the out-of-sample estimation error of ES, the free energy that gives the

in-sample estimator, and the susceptibility-like quantity Δ, and displayed their behavior for the three cases studied here. It is satisfactory to see that q_0 and Δ remain finite up to the new characteristic lines, that is, the regularizer acts as expected: it suppresses the divergent sample fluctuations in the optimization of ES. Unfortunately, this suppression is not strong enough to bring down the estimation error to acceptable values, except for the range of small $r = \frac{N}{T}$ ratios where it demands far too long time series for any realistic N, and where r is small already without any regularization.

What is the meaning of this phase transition? As analyzed in [8,26] it follows from the coherence axioms that coherent risk measures, including ES, are unstable in the sense that whenever an asset or a combination of assets in the portfolio stochastically dominates the others in a given sample, the investor can take an extremely large long position in the dominant asset and compensate this with an appropriately large short position, without violating the budget constraint. This means that the weight of the dominant asset runs away practically to infinity, resulting in an arbitrarily large negative value of the risk measure. This is a mirage of an arbitrage, which can disappear in the next sample, or change into another arbitrage with a different weight running away to infinity. In practice, there are always constraints that prevent such a divergence from taking place. The ban on short selling is just this sort of constraint. The runaway solutions try to escape, but get arrested at the walls constituted by the constraint, in the case of a no-short ban, at the coordinate planes. This is how the condensate of zero weights builds up. This mechanism is the stronger the larger the ratio $r = N/T$.

There is nothing surprising about solutions sitting on the constraint-walls or at corners in a linearly programmable problem, such as the optimization of ES. In the usual applications of linear programming, the constraints typically express some physical limitation like a finite amount of resources, material or labor, etc. In the present finance problem, such a finite resource would be the limited budget, but if short selling is not constrained, the budget in itself cannot prevent runaway solutions. The ban on short positions corresponds to an infinitely strong ℓ_1 regularizer, which, combined with the budget constraint, is already sufficient to take care of the runaway solutions. So, with a no-short ban on, we can increase r (that is the dimension, or decrease the amount of data) without any mathematical contradiction showing up; neither q_0 nor Δ will diverge. It is clear, however, that the solution based on less and less information becomes increasingly meaningless. In these circumstances, the optimization will not tell us anything useful about the structure of the market, it will be determined more and more by the constraint.

What we regard as the most intriguing result of this paper is the existence of a mapping between the regularized and the unregularized problems.

Author Contributions: Conceptualization, G.P., I.K. and F.C.; formal analysis, G.P. and I.K.; funding acquisition, F.C.; investigation, G.P., I.K. and F.C.; writing—original draft preparation, I.K.; writing—review and editing, G.P. and F.C.; visualization, G.P. All authors have read and agreed to the published version of the manuscript.

Funding: This research received no external funding.

Institutional Review Board Statement: Not applicable.

Informed Consent Statement: Not applicable.

Acknowledgments: We are indebted to Susanne Still and Matteo Marsili for collaboration and useful discussions years ago on joint works preceding the present one. Although they did not participate in this work, their ideas have remained a source of inspiration for us. I.K. is obliged to Risi Kondor for several enlightening discussions.

Conflicts of Interest: The authors declare no conflict of interest.

References

1. Markowitz, H. Portfolio selection. *J. Financ.* **1952**, *7*, 77–91.
2. Morgan, J. *Riskmetrics Technical Manual;* JP Morgan: New York, NY, USA, 1995.
3. Basel Committee on Banking Supervision. *Overview of the Amendment to the Capital Accord to Incorporate Market Risks;* Bank for International Settlements: Basel, Switzerland, 1996.
4. Artzner, P.; Delbaen, F.; Eber, J.M.; Heath, D. Coherent Measures of Risk. *Math. Financ.* **1999**, *9*, 203–228. [CrossRef]
5. Acerbi, C.; Tasche, D. Expected Shortfall: A Natural Coherent Alternative to Value at Risk. *Econ. Notes* **2002**, *31*, 379–388. [CrossRef]
6. Pflug, G.C. Some remarks on the value-at-risk and the conditional value-at-risk. In *Probabilistic Constrained Optimization;* Uryasev, S., Ed.; Springer: Boston, MA, USA, 2000; pp. 272–281.
7. Basel Committee on Banking Supervision. Minimum Capital Requirements for Market Risk. 2016. Available online: https://www.bis.org/bcbs/publ/d352.htm (accessed on 23 April 2021).
8. Kondor, I.; Varga-Haszonits, I. Instability of portfolio optimization under coherent risk measures. *Adv. Complex Syst.* **2010**, *13*, 425–437. [CrossRef]
9. Kondor, I.; Pafka, S.; Nagy, G. Noise sensitivity of portfolio selection under various risk measures. *J. Bank. Financ.* **2007**, *31*, 1545–1573. [CrossRef]
10. Ciliberti, S.; Kondor, I.; Mézard, M. On the Feasibility of Portfolio Optimization under Expected Shortfall. *Quant. Financ.* **2007**, *7*, 389–396. [CrossRef]
11. Kondor, I.; Caccioli, F.; Papp, G.; Marsili, M. Contour Map of Estimation Error for Expected Shortfall. 2015. Available online: http://ssrn.com/abstract=2567876 and http://arxiv.org/abs/1502.0621 (accessed on 23 April 2021).
12. Caccioli, F.; Kondor, I.; Papp, G. Portfolio optimization under expected shortfall: Contour maps of estimation error. *Quant. Financ.* **2018**, *18*, 1295–1313. [CrossRef]
13. Bühlmann, P.; Van De Geer, S. *Statistics for High-Dimensional Data: Methods, Theory and Applications;* Springer Science & Business Media: Berlin/Heidelberg, Germany, 2011.
14. Caccioli, F.; Still, S.; Marsili, M.; Kondor, I. Optimal liquidation strategies regularize portfolio selection. *Eur. J. Financ.* **2013**, *19*, 554–571. [CrossRef]
15. Papp, G.; Caccioli, F.; Kondor, I. Variance-bias trade-off in portfolio optimization under Expected Shortfall with ℓ_2 regularization. *J. Stat. Mech. Theory Exp.* **2019**, *2019*, 013402. [CrossRef]
16. Still, S.; Kondor, I. Regularizing portfolio optimization. *New J. Phys.* **2010**, *12*, 075034. [CrossRef]
17. Rockafellar, R.T.; Uryasev, S. Optimization of Conditional Value-at-Risk. *J. Risk* **2000**, *2*, 21–41. [CrossRef]
18. Caccioli, F.; Kondor, I.; Marsili, M.; Still, S. Liquidity Risk and Instabilities In Portfolio Optimization. *Int. J. Theor. Appl. Financ.* **2016**, *19*, 1650035. [CrossRef]
19. Hastie, T.; Tibshirani, R.; Friedman, J. *The Elements of Statistical Learning, Data Mining, Inference, and Prediction*, 2nd ed.; Springer Series in Statistics; Springer: Berlin/Heidelberg, Germany, 2008.
20. Mézard, M.; Parisi, G.; Virasoro, M.A. *Spin Glass Theory and Beyond;* World Scientific Lecture Notes in Physics Volume 9; World Scientific: Singapore, 1987.
21. Kondor, I.; Papp, G.; Caccioli, F. Analytic approach to variance optimization under an ℓ_1 constraint. *Eur. Phys. J.* **2019**, *92*, 8. [CrossRef]
22. Kempf, A.; Memmel, C. Estimating the global minimum variance portfolio. *Schmalenbach Bus. Rev.* **2006**, *58*, 332–348. [CrossRef]
23. Basak, G.K.; Jagannathan, R.; Ma, T. A jackknife estimator for tracking error variance of optimal portfolios constructed using estimated inputs. *Manag. Sci.* **2009**, *55*, 990–1002. [CrossRef]
24. Frahm, G.; Memmel, C. Dominating estimators for minimum-variance portfolios. *J. Econom.* **2010**, *159*, 289–302. [CrossRef]
25. Ciliberti, S.; Mézard, M. Risk minimization through portfolio replication. *Eur. Phys. J.* **2007**, *B 57*, 175–180. [CrossRef]
26. Varga-Haszonits, I.; Kondor, I. The instability of downside risk measures. *J. Stat. Mech. Theory Exp.* **2008**, *2008*, P12007. [CrossRef]
27. Shinzato, T. Minimal investment risk of portfolio optimization problem with budget and investment concentration constraints. *J. Stat. Mech. Theory Exp.* **2017**, *2017*, 023301. [CrossRef]
28. Kondor, I.; Papp, G.; Caccioli, F. Analytic solution to variance optimization with no short positions. *J. Stat. Mech. Theory Exp.* **2017**, *2017*, 123402. Available online: https://iopscience.iop.org/article/10.1088/1742-5468/aa9684 (accessed on 23 April 2021). [CrossRef]
29. Varga-Haszonits, I.; Caccioli, F.; Kondor, I. Replica approach to mean-variance portfolio optimization. *J. Stat. Mech. Theory Exp.* **2016**, *2016*, 123404. [CrossRef]

entropy

MDPI

Article

Victory Tax: A Holistic Income Tax System

Donald J. Jacobs †

Department of Physics and Optical Science, University of North Carolina at Charlotte, Charlotte, NC 28223, USA; djacobs1@uncc.edu

† Affiliate faculty of the UNC Charlotte School of Data Science.

Abstract: How can an income tax system be designed to exploit human nature and a free market to create a poverty free society, while balancing budgets without disproportional tax burdens? Such a tax system, with universal character, is deduced from the following guiding principles: (1) a single tax rate applies to all income types and levels; (2) the tax rate adjusts to satisfy budget projections; (3) government transfer only supplements the income of households with self-generated income below the poverty line; (4) deductions for basic living expenses, itemized investments and capital losses are allowed; (5) deductions cannot be applied to government transfer. A general framework emerges with three parameters that determine a minimum allowed tax deduction, a maximum allowed itemized deduction, and a maximum deduction defined by income percentage. An income distribution that mimics the United States, and a series of log-normal distributions are considered to quantitatively compare detailed characteristics of this tax system to progressive and flat tax systems. To minimize government dependency while maximizing after-tax income, the effective tax rate (*ETR*) as a function of income percentile takes the shape of the letter, V, inspiring the name victory tax, where the middle class has the lowest *ETR*.

Keywords: income tax; tax deduction; income redistribution; government transfer; government dependency; poverty line; basic income guarantee; effective tax rate; balanced budget; elastic tax

Citation: Jacobs, D.J. Victory Tax: A Holistic Income Tax System. *Entropy* **2021**, *23*, 1492. https://doi.org/10.3390/e23111492

Academic Editors: Ryszard Kutner, H. Eugene Stanley and Christophe Schinckus

Received: 17 October 2021
Accepted: 8 November 2021
Published: 11 November 2021

1. Introduction

Many forms of taxation evolved organically in different political and economic systems over human history [1]. A tax system deals with the flow of money from society to government, as tax revenues are collected and given back to society as government spends tax revenue, including government transfers in relation to wealth redistribution. Historically, the extent and purpose of taxation generally was not to benefit society as a whole [1]. Today, opinions on the extent and purpose of taxation are often politically charged, making it impossible to design a tax system that is acceptable to all positions. Arguments from different philosophical perspectives on relationships between society, economy, and public policy identify complex issues that need to be reconciled to achieve a rational tax system. The impetus of this work comes from many laudable debates in the United States (US) about the federal tax system regarding how to set tax rates, and on related issues that affect the nature of taxation that include government spending, short-term deficits, long-term debt, government dependency, poverty, income inequality, disproportionate tax burdens, and the diminishing wealth of the middle class. These ongoing debates indicate that the structure of a tax system has a significant impact on the economy and well-being of society. The aim of this work is to construct the foundation for a holistic tax system with universal appeal divorced from ideology, by taking a pragmatic approach to solving a wide range of problems faced by modern society.

1.1. Motivations for a Holistic Tax System with Desirable Characteristics

Based on the Gini index [2] from 1980 to the present, the middle class of the US is steadily shrinking, suggesting that the federal individual income tax could be changed to

strengthen the middle class. Although a strong middle class fuels a consumer economy, the distribution of income (and wealth) in free markets is empirically found to be highly skewed toward a tiny percent of its population [3–5]. Remarkably, there is universality in heavy tailed distributions for income and wealth across a population in free markets, regardless of government form and tax law. This work accepts the inevitability of highly skewed income dispersion in free-market economies, and then uses the free market and human nature to its advantage.

It is prudent to create incentives for economic growth, while discouraging government dependency, which can lead to complacency and a less productive society. In this context, regressive and progressive taxation affects the economy and redistribution of wealth in different ways. When government redistributes wealth, the effective tax rate (*ETR*) will vary between households. A (higher, lower) *ETR* makes it (more, less) difficult for a household to generate wealth due to a (higher, lower) tax burden. A pragmatic reason for the government to redistribute income is that this practice creates an environment in which all segments of society can accumulate wealth.

A widely employed definition of a flat tax involves taxing labor income at a single marginal tax rate with an allowed deduction [6]. In this work, a flat tax is defined as having the *ETR* independent of income type and level. In practice, a flat tax is achieved by not allowing tax deductions or government transfers, so that no redistribution of income is made. In the deductive approach taken here, the *ETR* dependence on income level is not a priori assumed. Rather, the *ETR* will be a result of a set of guiding principles concerned with fairness, maintaining long-term stability for society, and creating an incentive for personal economic growth at all income levels. There should be a simple way for the government to collect tax revenues without runaway deficits and not impose excessive tax burdens on the population. If a tax system achieves these goals using a strange looking *ETR*, so be it.

A holistic tax system should create a net benefit to society while promoting individual interests, and therefore can be implemented by any political system provided the government is sincere about respecting human dignity and wants to maintain a stable free-market economy. Minimally, all persons in society should have equal opportunity to generate personal wealth from a market economy. The tax system should create incentives for individuals to generate wealth. The mathematical framework of the tax system should not be tied to specific policies. The tax system should have a structure independent of a population's income distribution. Moreover, it should be easy for a household to determine its tax burden, pay owed taxes, and receive government assistance when needed. Likewise, it should be easy for the government to manage administratively, deter runaway deficits, and adapt to society's needs. Applied across the income spectrum, the tax system should capitalize on the free market to create income growth opportunities, encourage self-reliance, and minimize government dependence.

1.2. Contributions from a Scientific Approach

To my knowledge, the tax system developed herein is novel, although there are similarities with the concepts of a negative income tax [7] and basic income guarantee [8]. Specifically, the principles for a holistic tax system developed in Section 2.1 based on pragmatic considerations identify negative income tax and basic income guarantee as inadmissible. Positive and negative aspects of these other proposed tax systems have been extensively discussed [9–16]. The tax system developed here escapes the pitfalls of negative income tax [17]. The main criticism of a negative income tax is that it removes an incentive for people to work in low-wage jobs when it is more lucrative to obtain greater income when not working. This problem is generally referred to as a welfare trap [18]. Common criticisms against a basic income guarantee are that marginal tax rates become very high [19–22], it is too risky [19], and it is cost prohibitive on large scales [20,21]. Following five principles that are conceptually justified below, a simple holistic tax system emerges, and it is shown in Section 3 to be feasible and cost-effective.

The concept of a basic income guarantee was discovered independently many times under different names [16]. In this paper, the similar concept of a need-based income guarantee is the result of the development of a pragmatic tax system based on a deductive approach. The critical difference to previously proposed schema [13] is in the way government transfer is handled. After constructing five guiding principles, which include a single tax rate with three types of tax deductions, the derived tax system provides a surprisingly low population average *ETR*, where the most tax relief goes to the middle class. The *ETR* for the ultra-rich is no higher than for the extreme poor. Moreover, low-income households are subjected to a regressive *ETR* that eliminates the welfare trap and encourages the poor to gain financial independence and move into the middle class. The tax system is unified with the welfare system, so that poverty and deficits can be virtually eliminated without imposing disproportionate tax burdens.

For the rest of this paper, Section 2 develops the tax system first conceptually and then mathematically. Next, several model economies are constructed to quantify the characteristics of the tax system. In Section 3, the parameters of the tax system are explored, leading to a set of parameters that maximize after-tax income for the vast majority of the population with minimal government dependency. In this scenario, it turns out that the ETR takes the shape of a "V", where the middle class enjoys near zero effective tax rates. This V-signature inspires the name victory tax, which has been adopted because of its general applicability to widely varying economies. The victory tax system is then compared with a flat and linear progressive tax system. In Section 4, the benefits of a simple tax structure are discussed, as well as possible public policy decisions and future work. The conclusions of this paper are given in Section 5. The main conclusion is that the victory tax as a holistic income tax system is constrained in such a way that, with minimal government involvement, households at all levels of income can reap benefits by using this tax system selfishly to gain wealth, which broadly helps society achieve a higher standard of living.

2. Model and Methods

The proposed tax system is deduced from five guiding principles that form the basis of a mathematical framework. This section starts with the conceptual framework, in which the rationale for the guiding principles is discussed. Recognizing that there are different perspectives, each principle is rationalized by questioning whether it has universal appeal. The goal is to transcend political biases as much as possible by rejecting what is not universal. An effort is made to distill an income tax system into essential elements. A mathematical framework is then developed with parameters that encapsulate a family of income tax systems that differ by the parameter values set by the state of the economy. This allows governments to adapt to society's needs over time.

2.1. Conceptual Framework

In a bottom-up approach, five guiding principles for a holistic tax system are first listed. The rationale for each principle is then discussed as subsections. These principles shape the tax system by imposing constraints, which leads to a tax system that is easily parameterized within a mathematical framework. The principles are listed as:

1. Income of all types is taxed at the same rate, independent of the income level.
2. A single tax rate adjusts to ensure government fiscal stability.
3. Government transfer is only used to establish a minimum standard of living.
4. Three types of tax deductions incentivize wealth accumulation.
   (a) A basic deduction to offset living expenses.
   (b) Itemized deductions that promote better standard of living.
   (c) Capital loss deductions that promote economic growth.
5. Tax deductions cannot be applied on government transfer.

2.1.1. Income of All Types Is Taxed at the Same Rate, Independent of the Income Level

Taxing one type of income differently from another creates social discrimination. For example, it could be argued that income from work done by a teacher should be taxed twice as high as income from work done by a welder. Intrinsic to this argument is that the value of the work of a welder is more important than a teacher, or perhaps simply to offset the risks inherent in welding. However, an objective truth of this differentiation is not self-evident, as many teachers would argue otherwise. Many of these comparisons can be made, all of which end with arbitrary conclusions. Logic suggests that it is not the place of government to judge the intrinsic value of income beyond the definition of legal and illegal activities. In practice, assigning different tax rates to different types of income creates a complicated system that leads to endless debate, because there is no universal truth for all cultures, all types of economy, all types of government, and certainly not a constant in time. The same logic is true when it comes to distinguishing income from labor versus investments or other forms of income, such as gifts, winnings, insurance or inheritance.

For free-market economies, the amount of income a person generates in terms of salary or return on investments determines the value society attaches to occupation and investment. Income is determined by tangible factors, such as supply and demand, investment decisions, the wealth potential of occupations, and the desire of an individual to be wealthy. For example, the income of a surgeon can be higher or lower than a professional athlete, depending on various factors. Therefore, the allocation of different tax rates on different types or income levels must be rejected. A household with orders of magnitude more income than another will pay proportionally that much more tax, which does not discriminate on income levels. Although sale taxes can coexist with this principle, no other form of taxation of a person's income is allowed. For example, this means that in the US, the separate payroll tax must be eliminated, as there can only be one tax rate on income, which is comprehensively taken care of by the income tax system within a holistic approach.

2.1.2. A Single Tax Rate Adjusts to Ensure Government Fiscal Stability

A pertinent question is: What should the single tax rate be? A constant value (say 9%) could be argued as optimal, but this value is not self-evident. Indeed, any specific value would not be universally optimal for all cultures, economies, governments, and for all time, as the state of the economy fluctuates over time. Therefore, a variable tax rate that adapts to the revenue needed to cover projected government spending is required. A dynamic tax rate allows a government to control fiscal stability while adapting to short- and long-term economic conditions. Cycles of high and low tax rates induce an elastic response to balancing budgets (with limited liability), which ensures reliability in public services and mitigates the accumulation of long-term debt. Both of these attributes are necessary for long-term stability of society.

It is worth pointing out that policy makers have responsibility for developing debt accumulation or reduction plans. Regardless of the directions that policy makers decide, the adjustable tax rate makes tax revenue collecting responsive to government policies. For example, if more funding is appropriated toward infrastructure or defense, the single tax rate will increase, and society can monitor and judge the benefits for increased taxes. In summary, a single tax rate offers transparency in government spending, and in combination with other social-economic measures, the value that the government attaches to society's well-being becomes transparent.

2.1.3. Government Transfer Is Only Used to Establish a Minimum Standard of Living

For what reason, if any, should government transfer be used to supplement household income? Arguably, government transfers should not be used for anything other than to help a household achieve a minimal standard of living. This principle does not exclude government support in other forms, such as tax deductions and public services. For example, government spending on entitled health care, education, or other infrastructure

does not constitute government transfer. However, public services must be independent of the income level, void of any income qualification.

Providing public services to poor subpopulations is unnecessary, because the poorest households will live at the poverty line, which sets a minimum standard of living. Rather than designing social programs to help the poor, public services should be designed to help society. This paradigm shift of shared interest will ensure that social programs are of high quality. It is worth noting that public services will reduce poverty by reducing basic living costs. Conversely, the poverty line rises as free public services decrease. Importantly, this principle prohibits the use of government transfers for unemployment or retirement compensation. Consequently, policy decisions will involve common interests in diverse and large segments of the population.

It is obvious that if government transfers are used to subsidize households for anything other than supporting a minimum standard of living, an arbitrary number of good reasons to redistribute income will lead to a complex tax system that is not universal. Why, however, should government redistribute income to set a minimum standard of living? Elimination of poverty is a singular case that appeals universally because of the innate human desire to live healthy and securely with dignity. This institutes the responsibility of the government to provide the means for all individuals in society to live securely with dignity over countless generations.

In practice, the poverty line must be set to balance competing factors. The poverty line should not be set too low, because more productivity in the entire population will result if society as a whole has a functional standard of living. Conversely, if the poverty line is set too high, the tax rate will rise too high, stifling economic growth. As such, the minimum standard of living that society can tolerate sets the poverty line for households. Although the way policy makers define this poverty line is left open, it must be based on income (not savings or wealth). Government transfers supplement income to establish a minimum standard of living as a safety net. If a household starts with considerable assets and then unexpectedly finds itself without income, this household can survive at the poverty line, with basic needs fulfilled. Moreover, this household can use its savings, albeit a finite resource, to live a higher standard of living.

A consequence of having one specific reason for government transfer is that the government has minimal involvement in a free market. As another example, if a household with a large accumulation of debt suddenly loses its income, it is likely to lose its possessions if an agreement with its lenders cannot be reached. Responsibility and risk tolerances exist between lenders and households taking loans. Government transfer is used only to maintain a minimum standard of living, and this results in keeping the net amount of transfer to a minimum, and hence keeps the tax rate to a minimum.

The COVID-19 pandemic is an unfortunate example of a situation in which the victory tax system maintains a stable economy during a crisis. Households automatically receive government transfers when their income falls below the poverty line due to job loss. Because of guaranteed basic income, the debate in the US on the scope of COVID-19 relief packages would be unnecessary since government transfer creates a safety net of security. Nevertheless, financial losses from businesses and households would be expected. While many lenders would be eager to force foreclosure, other lenders would use unfortunate events as a growth opportunity to attract new (sound) customers by covering businesses from bankruptcy and households from personal losses. The free market would solve the vast majority of the problems, with government regulations perhaps requiring debt collectors to exercise patience. As jobs reemerge, low-income households quickly increase their after-tax income, avoiding long-term economic stagnation.

2.1.4. Three Types of Tax Deductions Incentivize Wealth Accumulation

Tax deductions are used to reduce the tax burden on households for various reasons. For a certain amount of revenue to be collected, reducing the tax burden on a subset of households requires other households to pay disproportionately higher taxes. Of course,

different tax rates applied to different income levels or types cause disproportionate tax burdens. However, even if a single tax rate is applied to all income levels and types (e.g., nominally a flat tax), tax deductions create a non-flat ETR dependent on household income. Importantly, different types of tax deductions produce different relative benefits for households at different income levels. For example, the basic tax deduction that offsets minimum living expenses provides proportionately more benefit to low-income households. Itemized deductions predominately help households with middle-range income. Capital loss deductions on investment losses primarily benefit high-income households. In fact, low-income households cannot capitalize on capital loss tax deductions.

Among the different types of tax deductions, a balance should be sought between benefits versus the disproportionate tax burdens created across the income spectrum of households. The structure of tax deductions should benefit society, as taxpayers seek to obtain the maximum after-tax income possible from their own interests. Tax deductions therefore offer specific redistribution mechanisms for government support to motivate households to accumulate wealth, which in turn maintains a stable and growing economy. The rationale for the basic, itemized, and capital-loss tax deductions is discussed next, while key variables for the victory tax system are introduced.

Basic tax deduction: A minimum income is needed to live functionally in modern society. In the past, most people could live on natural resources or farm land. Unless free public services take the place of natural resources, job loss literally becomes life threatening. Hence, a basic deduction, BD, is incorporated to cover minimum living costs for a household. Although BD is a free parameter, it is appropriately related to the poverty line. Furthermore, BD should only depend on the number of dependents in a household and the cost for necessities (which is location dependent), as its sole purpose is to offset minimal living costs in the context of a social norm.

Itemized tax deduction: To incentivize financial independence, optional itemized deductions are allowed. Itemized deductions offer the government flexibility in the tax code to encourage certain measures, such as buying a house, accumulating a retirement portfolio, compensating costs for professional training, education or medical needs, or making donations to charities. As such, itemized tax deductions create self-interest incentives for households to take measures that also benefit society as a whole. The net itemized deduction, ID, is incorporated into the general framework of the victory tax. The total income that can be deducted is capped at a maximum. Setting a maximum deduction prevents all households from not paying tax. Two methods are used to set the maximum total deduction. A maximum deduction, MD, and a maximum percentage, MP, of net income, NI. The total deduction, TD, allowed by a household is given by:

$$TD = \min(BD + ID, MD, MP \times NI) \tag{1}$$

For a household with a net income above the poverty line with no itemized deductions, its taxable income, TI is given as:

$$TI = \max(0, NI - TD) \quad \forall \ NI > BD \tag{2}$$

The equation for taxable income developed thus far is easy to understand. The combined total of basic and itemized deductions cannot exceed the maximum allowed deduction, nor a maximum percentage of income. Once the total deduction, TD, is determined, it will be used to reduce net income in order to achieve the taxable income. However, if the deduction is greater than the net income, NI, then the taxable income, TI, is set to zero, as it cannot be negative. The net income will be precisely defined after capital loss deductions are considered.

Capital loss deduction: To encourage households to increase their wealth through investments, capital loss deductions are used to mitigate risk. It is self-evident that government cannot rescue all households from financial loss. Hence, under what circumstance, if any, should government aid households to recover lost wealth? Imagine an individual

that invests \$100,000 in a company, and subsequently loses this investment due to the bankruptcy of that company. Another individual buys a \$100,000 painting, which is inadvertently destroyed in a fire. Should both scenarios be treated equally, or should a distinction between these two losses be made? The answer rests with public policy makers who create tax law, where the answer can range from no capital loss deduction for anything to almost everything. The concern addressed next is that if the tax burden for the wealthy is disproportionately reduced, the middle class has a much greater tax burden, as the poor contribute little to tax revenue. Therefore, to justify a capital loss deduction, it is prudent to make an analogy with government-run health care, which shares an equivalent concept of large-scale group insurance.

Capital loss tax deduction is similar to an insurance program managed by the government. Specifically, all household incomes are being taxed, but the government only aids households suffering capital losses. A greater capital loss begets more government aid. The practice of spreading investment risk across the entire population to cover only households that made poor investments is like a health insurance program. That is, sick and healthy individuals are taxed, but the government only aids those suffering sickness. The greater the sickness begets more government aid. Again, aid only goes to a subpopulation to keep people functional and productive, which is beneficial to society as a whole. Although only a subpopulation will benefit from the insurance, a priori it is unknown who will use it.

The arguments against universal health care (lack of resources, highly skewed redistribution of income and poor government management) amplify against the rationale for capital loss tax deductions. The most troubling aspect is that only households with the highest income are predisposed to benefit. Thus, it is not self-evident that capital loss deductions should be included in a tax system. Nevertheless, creating opportunities that ensure the well-being of society must be the responsibility of government. When viewed as insurance, policy makers need only debate the scope of coverage. From this point of view, there is no universal answer for the scope of capital loss tax deduction or free health care services, since a weak economy cannot support the same level of coverage as a strong economy. As such, public policy debates will ultimately affect government budgets, tax rates and poverty line. The proposed tax system is designed to support the outcome of these debates within the constraints inherent in the tax system.

Since the capital loss deduction benefits society as a growth mechanism, it is included in the victory tax system to ensure generality. Nevertheless, since only a subset of households reaps the benefits, it is prudent to limit this redistribution of wealth to prevent a higher tax burden on households with much lower income. This limitation is similar to a maximum coverage limit in an insurance policy. Note that government transfer to the poor is limited by the poverty line, and a limit to the maximum itemized deduction was also introduced. In the same spirit, a cap on capital loss deductions is set by not counting capital losses that exceed capital gains within a given year. From a consistency point of view, paying tax on income must be over the same time period regardless of the income type or income level of a household. To roll over capital losses to future years, the same time period must apply to all income types. A tax system in which taxes are due annually seems reasonable, compared to alternatives such as every four months or every four years. For the prototype tax system constructed and demonstrated in this paper, capital losses are limited to one-year windows, which correspond to taxes collected annually. Although there is a maximum deduction for capital losses per year, no lifetime limit for capital losses is set. Likewise, there are annual limits, but no lifetime limit to government transfer or itemize deductions.

The definition of net income within the victory tax system is given as:

$$NI = E + \max(0, CIG - CIL) \tag{3}$$

where E defines earnings from employment, CIG defines capital income gained, and CIL defines capital income lost. Note that CIG includes all types of income that are not earnings from employment or government transfers. Upon inspection of (3) within a

given year, the government will maximally allow a household to deduct as much capital loss as gained. The amount of risk assumed is therefore determined by the skill of the investor. If an investor incurs more capital loss in a given year than capital gains, the government will not allow this excess loss to be deducted on the grounds that the investor creates too much risk for society to absorb. This restriction strengthens a free market: investors will exercise prudent judgments to ensure that gains are greater than losses when government assistance is limited. In practice, the tax system encourages investors to focus on fundamentals and long-term investments, spreading losses over multiple years. Placing annual limits on capital loss deductions (in fact all types of tax deductions) minimizes government dependency.

2.1.5. Tax Deductions Cannot Be Applied on Government Transfer

This principle is self-evident, because income from government transfers is at the expense of all taxpayers who make a productive contribution to society through earnings and/or capital gains. Note that allowing deductions on government transfer would amplify government assistance. This guideline minimizes government dependency.

2.1.6. Unique Property of the Victory Tax System

To emphasize the unique properties of the victory tax system compared to other basic income guarantee tax systems, an important consequence follows when the first, third and fifth principles are combined. Since government transfer income is taxed, but deductions cannot be applied to this part of household income, a regressive ETR emerges as a function of income percentile for the poor. The regressive ETR enables low-income households receiving government aid to become self-reliant and achieve higher income levels without a welfare trap (the analog to a nucleation barrier). The formation of a low-income regressive ETR will become clear in the results section.

2.2. *Mathematical Framework*

The five principles examined above are now considered axioms to construct the general mathematical framework of the victory tax system. Relevant variables for a household to calculate tax liability are described in Table 1 for convenient reference.

Table 1. Alphabetically ordered list of variables for the victory tax system and their description.

Variable	Variable Description
ATI	After-tax income.
BD	Basic deduction for minimum living expenses.
CIG	Capital income gain.
CIL	Capital income loss.
E	Earnings through employment.
ETR	Effective tax rate.
GTI	Government transfer income to a household.
ID	Itemized deduction taken by a household.
MD	Maximum deduction allowed for a household.
MP	Maximum percent of household income that can be deducted.
NI	Net income of a household after capital loss deductions.
PL	Poverty line.
TAX	Tax liability.
TD	Tax deduction.
TI	Taxable income.
TTI	Total taxable income over the entire population.
TTR	Total tax revenue from the entire population.
VTR	Victory tax rate applied to all households and income levels.

In addition to (3) defining net income, the victory tax formulas are given as:

$$GTI = \max(0, BD - NI) \tag{4}$$

$$TD = \min(BD + ID, \max(BD, MD), MP \times NI) \quad (5)$$

$$TI = GTI + \max(0, NI - TD) \quad (6)$$

$$TAX = VTR \times TI \quad (7)$$

$$ATI = GTI + NI - TAX \quad (8)$$

$$ETR = \frac{TAX}{GTI + NI} \quad (9)$$

The variables $\{E, CIG, CIL\}$ quantify income characteristics of a household. Notice that the net income of a household can never be negative. The maximum income for a household eligible for government transfer is determined by the basic deduction, BD. The word "eligible" underscores the restriction that households with incomes above BD cannot receive government transfer. Otherwise, GTI defined in (4) covers the income deficiency of a low-income household in order to put it on the poverty line. The parameters $\{BD, MD, MP\}$ define the maximum limits on allowed deductions. After the total tax deduction is determined from (5), the taxable income is calculated by (6). The victory tax rate, VTR, multiplies the taxable income of a household to obtain tax liability (7). The after-tax income is calculated from (8), which adds government transfer to net income minus paid tax. The ETR is defined in (9) as tax paid divided by the total income. Since no deductions can be applied to government transfer, it works out that $ETR \rightarrow VTR$ when $NI \rightarrow 0$. Although counterintuitive, the poorest households pay the highest effective tax rate among the entire population.

Within the victory tax system, VTR, depends on the target tax revenue, TTR, deduced from projected budget needs for the next year set by policy makers. In addition, BD, should be proportional to the poverty line, PL which is the income required to maintain a minimum living cost. Reflecting the state of the economy, BD will change annually to ensure that the least possible after-tax income, ATI, corresponds to PL. From Equations (3) and (4) a household with $NI = 0$ will have a taxable income equal to BD from government transfer, and after-tax income will be given by $ATI = BD - VTR \times BD$. This shows the insightful relationship that $BD = PL/(1 - VTR)$, which indicates that as VTR increases, the basic tax deduction increases at a higher rate. When public policy makers propose to increase VTR for building infrastructure or defense, BD will automatically increase even if the poverty line remains constant.

The two parameters $\{MD, MP\}$ determine the shape of the ETR. In practice, MD and MP will be functions of the number of dependents in a household, denoted as d. For simplicity, MP is considered independent of d. Although MD can be set with considerable latitude through tabulation, a sound approach is to set MD proportional to PL. By assuming $MD = k \times PL(d)$, the parameterization details for MD as a function of dependents are inherited from $PL(d)$. For the US, the poverty line for a household with a certain number of dependents is publicly available in tables [23]. More generally, an objective economic measure will be used to define $PL(d)$. Although not considered here to keep the analyses clear, the poverty line will generally depend on location (region within a country), since all regions do not have the same cost of living.

Parameters for the victory tax system are explored in Section 3. It is found that when $MP = 0$ and $ID = 0$, a V-shape signature emerges for the ETR as a function of income percentile, with $ETR = 0$ marking the bottom of the V. Importantly, (5) determines the allowed tax deduction after taking into account basic and itemized deductions and maximum percent of NI. Since itemized plus basic deductions increase total deduction, the term $\max(BD, MB)$ that appears in (5) is needed to enforce consistency. In particular, when MD is set below BD as an independent variable, the basic deduction is still offered, but itemized deductions are no longer allowed. However, MP can reduce the maximum allowed deduction below BD without inconsistency, because $MP \times NI$ is a competing restriction on tax deductions. Note that a flat tax corresponds to the limit $MP \rightarrow 0$, where ETR is constant, independent of income percentile.

As MP gradually changes from 1 to 0, the V-shape morphs into a U-shape as the "U" becomes shallower, with the minimum ETR increasing as VTR decreases. These shape changes create an elastic tax system [24]. At $MP = 0$ a flat tax emerges as a special case of a victory tax system, where BD sets the threshold income in which government transfer is no longer received. A prototypical victory tax system considered in this work is defined by: $MD = kPL$ and $MP = \min(r, 1)$, where r is the coefficient of variation in net income over the population. Specifically, r is the ratio of the standard deviation in NI to the mean NI, which serves as a convenient objective measure of the economy. The application of objective measures updates PL and r each year, allowing the victory tax system to respond dynamically to changes in the free market and public policy.

The basic deduction is determined in the prototypical victory tax once parameter k is specified together with the poverty line, $PL(d)$. By combining equations of the victory tax system, the tax liability of a household with d dependents is given as $TAX = VTR \times f_d(x|ID, VTR)$ from Equation (7), and its taxable income is expressed as:

$$f_d(x|ID, VTR) = \max\left[0, \frac{PL(d)}{1-VTR} - x\right] + \max\left\{0, x - \min\left[\left(\frac{PL(d)}{1-VTR} + ID\right), \max\left(kPL(d), \frac{PL(d)}{1-VTR}\right), rx\right]\right\} \tag{10}$$

where x is the net income (replacing NI for simpler mathematical notation) and the subscript d denotes the fact that the poverty line depends on the number of dependents in a household. Equation (10) is the result of substituting all the relevant variables described above in the arguments of Equation (6). Note that $f_d(x|ID, VTR)$ depends on VTR, which is the dependent variable to be determined. In addition, $VTR = TTI/TTR$ where TTI is the total taxable income over the population, and TTR is the total tax revenue to be collected over the population. To calculate $VTR = TTI/TTR$, we must have TTI, which is given by the net sum of $f_d(x|ID, VTR)$ over all households in the population. However, to calculate TTI, we must have VTR because $f_d(x|ID, VTR)$ depends on VTR. Despite this circular dependence, it is straightforward to numerically solve for VTR iteratively. Uncertainties in VTR will primarily arise from estimates in TTR that will lead to surpluses or deficits at the end of a tax year when government spending deviates from budgeted allocations. These calculations are not technically difficult. The tax agency will have all income data from previous years, and TTI, based on the latest tax records, can be calculated with all details from the tax code. However, the aim of this paper is to analyze the general characteristics of the victory tax system, rather than focus on nuanced details.

For clarity and without loss of generality, the characteristics of the victory tax system are analyzed using an average household size and with the subscript d suppressed. This allows TTI to be expressed through the simple function $N(x)$, which gives the income distribution over households of the average size. Defining N_o to be the number of such households, and $p(x)$ the probability density function quantifying how income is distributed over these households, $N(x) = N_o p(x)$. Assuming all households will take the maximum itemized deduction based on (Equation (10)), a lower-bound estimate for TTI is obtained. With $f_d(x|ID_{max}, VTR) \to f(x|VTR)$ from the simplifying assumptions, the total taxable income of the entire population is given by:

$$TTI = N_o \int_0^\infty f(x|VTR)\, p(x)\, dx \ . \tag{11}$$

2.3. Test Economies

The victory tax system will be characterized by a series of test economies. A test economy is modeled by the income distribution of the population and poverty line. In this paper, the characteristics of the US economy are used as a starting template and for comparison in discussing the importance of the results. However, the main interest is on investigating general trends that show how the victory tax system responds to dramatic

changes in income distribution. Therefore, several test economies are considered that systematically deviate from the US economy, where the size of the middle class gradually shifts from the largest segment of society to the smallest. In particular, the standard deviation in income distribution is used to expand and shrink the middle class, which respectively decreases and increases the income gap between the low middle class and the wealthiest households. Quantitative comparisons are made between a series of test economies in which the variance in household income is systematically varied as the average household income is fixed.

2.3.1. Income Distribution

Accurate modeling of income distribution over a population has received much attention [3–5]. The income distribution of a population is represented as a probability density function (PDF) denoted as $p(x)$, where x is net income. Income distributions are modeled by the κ-generalized statistics [3–5] and log-normal statistics. Although the log-normal PDF is a qualitatively adequate model for free markets, it underestimates population density with very high or very low incomes. The κ-generalized PDF provides a more accurate model description. In particular, the empirically observed Pareto power law tail [25] is recovered for high incomes (ultra-rich) and also more statistical weight is given to the extreme poor (both effects take away statistical weight from the middle class). The κ-generalized PDF is defined as:

$$p(x) = \frac{\alpha\beta\left(\frac{x}{\mu}\right)^{\alpha-1}\exp_\kappa\left[-\beta\left(\frac{x}{\mu}\right)^{\alpha}\right]}{\sqrt{1+\kappa^2\beta^2\left(\frac{x}{\mu}\right)^{2\alpha}}} \tag{12}$$

$$\beta = \frac{1}{2\kappa}\left[\frac{\Gamma\left(\frac{1}{\alpha}\right)\Gamma\left(\frac{1}{2\kappa}-\frac{1}{2\alpha}\right)}{\kappa+\alpha\Gamma\left(\frac{1}{2\kappa}+\frac{1}{2\alpha}\right)}\right]^{\alpha} \tag{13}$$

$$\Gamma(y) = \int_0^\infty t^{y-1}\, e^{-t}\, dt \tag{14}$$

$$\exp_\kappa(y) = \left(\sqrt{1+\kappa^2 y^2}+\kappa y\right)^{\frac{1}{\kappa}} \tag{15}$$

The two parameters $\{\alpha,\kappa\}$ are adjusted to fit to empirical data. Although it is not prohibitively difficult to evaluate the κ-generalized PDF and other properties from κ-generalized statistics [5], the form of this distribution is not convenient to create a systematic series of test economies. Log-normal statistics are therefore used to quantitatively analyze systematic trends in a number of economies that range from a strong to a weak middle class.

The simpler log-normal PDF is defined as:

$$p(x) = \frac{1}{x\gamma\sqrt{2\pi}}\exp\left[\frac{-(\ln x-\lambda)^2}{2\gamma^2}\right] \tag{16}$$

$$\gamma = \sqrt{\ln\left[1+\left(\frac{\sigma}{\mu}\right)^2\right]} \tag{17}$$

$$\lambda = \ln(\mu) - \frac{\sigma^2}{2} \tag{18}$$

Again, only two parameters $\{\lambda,\gamma\}$ characterize the PDF. However, they easily relate to the mean, μ, and standard deviation, σ, of income of the population. The mean household income is defined by the total net income from all households, divided by the total number of households. Both log-normal and κ-generalized distributions use the same mean

household income, but they require different standard deviations to best fit the empirical adjusted gross income data for the US, as explained below.

2.3.2. Test Economy Parameterization

The IRS data [26] from 1979 to 2009 was invoked to mimic the income distribution of the US. The IRS method of statistical weighting normalized the data to produce an effective number of dependents per household, and the reported income was based on 2009 dollars after adjusting for inflation. Year 2003 is chosen as an illustration. The US economy had just recovered after a market correction and began to grow over the next four years. The snapshot of the 2003 economy (in 2009 dollars) reflects a time of stability and the beginning of economic growth. The reported IRS data deals with adjusted gross income (AGI), which is analogous to net income (NI). At the level of analysis presented in this work, IRS data [26] (and NIH data [23]) are invoked to obtain realistic parameters while providing context for discussions. However, because AGI and NI are not the same, the test economies constructed in this work are best viewed as hypothetical examples.

The best fit for the IRS 2003 AGI data using the κ-generalized distribution yields $\alpha = 1.50$ and $\kappa = 0.56$ with a relative fit error of $\pm 5\%$ across all income brackets. This result gives a market income distribution (synonymous with AGI) compared to total income, which includes government transfer and other forms of US subsidies, such as food stamps, etc. Henceforth, this market income distribution will be associated with the economy A, shown in Figure 1. At the far right end of the tail, 10 households with an average income of approximately $113,000,000 per year are captured. The mean household income over the entire population is $75,300 with a standard deviation of $171,712. The median income is $44,612 indicating that ultra high-income households skew the distribution, causing the mean to be considerably larger than the median. As Figure 1 shows, the most probable market income (the mode) is approximately $10,000 per year.

The log-normal distribution, which fits well to the 2003 IRS income data, has the same mean income and a smaller standard deviation of $114,475. This log-normal distribution is employed as another test economy, henceforth referred to as economy B. A smaller standard deviation in market income indicates a larger middle class, since more households picked at random are likely to be closer to the mean income. Differences in market income between economy A and B are shown in Figure 2 on a log–log plot. The κ-generalized and log-normal distributions describe similar market incomes between $4000 and $2,000,000, but the peak in the log-normal distribution shifts upward near $12,000 to compensate for depleting probability from extreme income ranges. The Lorenz curves for these two test economies, shown in Figure S1, appear virtually identical.

A series of six log-normal economies, all with the same mean income of $\mu =$ $75,000 is also considered, with standard deviations spaced by approximately powers of 2, such that $\sigma =$ $7000, $14,000, $28,600, $57,240, $114,475 and $229,000 where the second largest standard deviation is economy B. In Figure S2, the six distributions are compared on a log–log plot, and their Lorenz curves are compared in Figure S3. All six log-normal economies have equal total adjustable income generated by the population, enabling a systematic method to study the dependence of a tax system on the strength of the middle class.

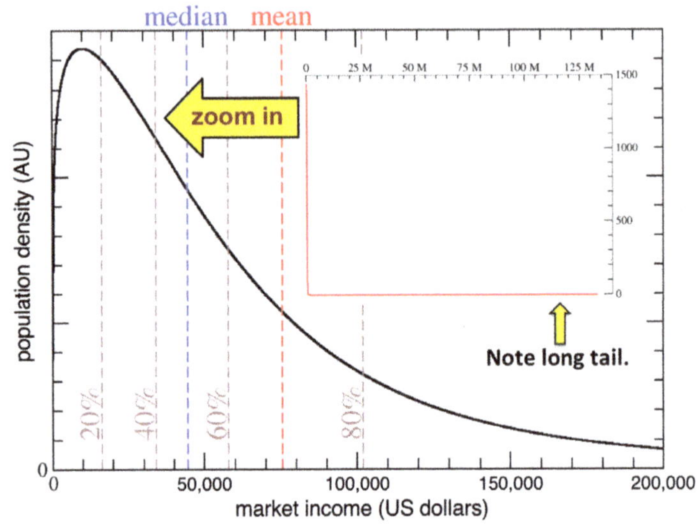

Figure 1. Market income distribution for economy A: The household AGI probability density ranging from \$0 to \$200,000 is shown. Population quintiles are marked with light brown dashed vertical lines at 20%, 40% 60%, and 80%. Median and mean incomes are respectively marked as blue and red dashed vertical lines. (**inset**) The same distribution is shown without cutting out data from high income households. The horizontal line at the bottom of the graph highlights the heavy tailed distribution, indicating only a tiny number of households reach this level of income.

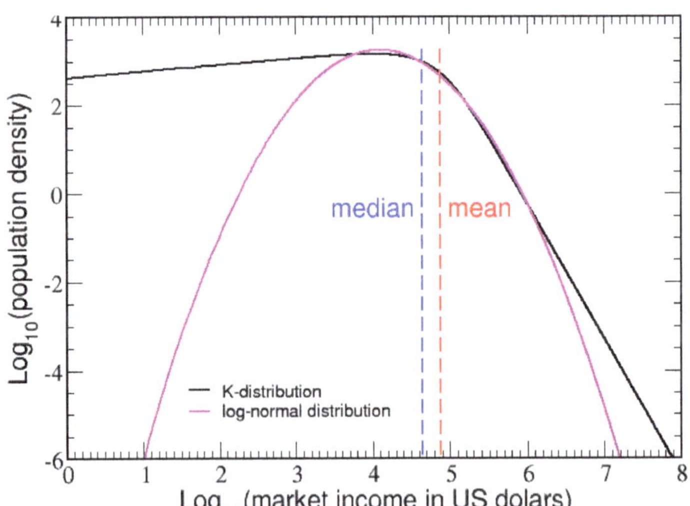

Figure 2. Market income distribution comparison: On a log–log scale the κ-distribution defining economy A and log-normal distribution defining economy B are compared. The κ-distribution puts more statistical weight for the ultra-rich and extreme-poor subpopulations as reflected in the wings of the distribution. The empirical median and mean incomes for the US 2003 economy are shown as blue and red vertical dashed lines.

2.3.3. Poverty Line, Tax Revenue and Government Transfer

Other information characterizes an economy, beyond income distribution, such as the number of people in the population and the poverty line. For the US economy in 2003, the number of tax-paying households was approximately $112,100,000 and the population among those households was approximately 290 million, yielding an effective household size of approximately 2.6 people. For illustration, and without altering the conclusions of the analysis, the poverty line for a mean household size of 2.6 is estimated by interpolating data from the 2009 Department of Health and Human Services data [23], yielding $16,814.

Unlike the victory tax system, which is the only source of government transfer to a household, there are many state and federal aid programs for the poor in the US to offset housing, energy, food, education, health care, and so on. Once other forms of government assistance are eliminated, a better estimate for the poverty line is $22,306. Examples are juxtaposed with both estimates to quantify how much tax burden increases when the poverty line is raised, which takes into account removal of public services for low-income households. In the victory tax system, there can be public services, but without income qualifications. As another point of reference, the poverty line in 2021 is listed as $21,960 for a household of 3 [23].

The target net tax revenue is set at $854,182,445,000 for all test economies, which corresponds to tax revenue collected by the IRS in 2003 after all government transfers were distributed. To compare all tax systems and economies, the total government transfer is calculated and added to the target net tax revenue of this hypothetical budget. When the income distribution has (more, less) households below the poverty line, the total tax revenue, TTR, to be collected will (increase, decrease). It is worth noting that approximately 35 percent of US federal tax revenue was derived from payroll tax, and 19 percent came from other sources [27]. Payroll tax and all other forms of individual taxation at the federal level outside of individual income tax are eliminated in the victory tax system. Although corporate tax can coexist with a victory tax system, no corporate tax is considered for simplicity of analysis in this paper. Here, all tax revenues come from individual income tax, making the VTR of the test economies an upper bound.

For comparison, it is insightful to examine how US tax revenues were redistributed in 2003. Total tax revenues collected were $1,952,929,045,000, of which $1,098,746,600,000 was then redistributed to households through government transfers. As such, more than 56% of the tax revenue collected was redistributed to tax payers, as captured in Figure S4. Despite the enormous amount of tax revenue redistributed to households, unfortunately approximately 15% of households live in poverty in the US [23]. In the US, tax revenues are redistributed to households at all income levels, including high-income brackets, creating both complexity and inefficiency.

3. Results

The adaptive nature of the victory tax requires VTR to be calculated each year. The procedure to calculate VTR from (10) favors tax revenue surpluses by assuming that all households take the maximum itemized deduction allowed. This situation yields the maximum VTR. Note that VTR must increase as itemized deductions increase to generate the same target revenue. The source of surplus is households that do not utilize all their itemized deductions. For analysis purposes, comparisons are made for a household that takes all or no itemized deductions to establish bounds for relevant quantities, such as ETR and ATI. Quantities are usually expressed as a function of the percentile of households. Percentile of households is calculated by ranking all households by net income, and then counting numbers of households at or below a certain income level to obtain a normalized scale from 0% to 100%.

3.1. Parameter Exploration

The shape of ETR as a function of the percentile of households depends on parameters $\{BD, MD, MP\}$, which are first explored as independent variables to elucidate how they

affect the tax structure. Note that MP defines a percentage of total income, while BD and MD are in dollar amounts. However, when convenient, BD and MD will be specified in terms of the percentile of households. When stated $BD = 20\%$ and $MD = 50\%$, this means BD and MD are respectively set to the income level at the 20 and 50 percentile of households. For example, for economy A, these percentiles translate to $BD = \$16{,}331$ and $MD = \$44{,}612$, which are the dollar amounts used in Figure 3.

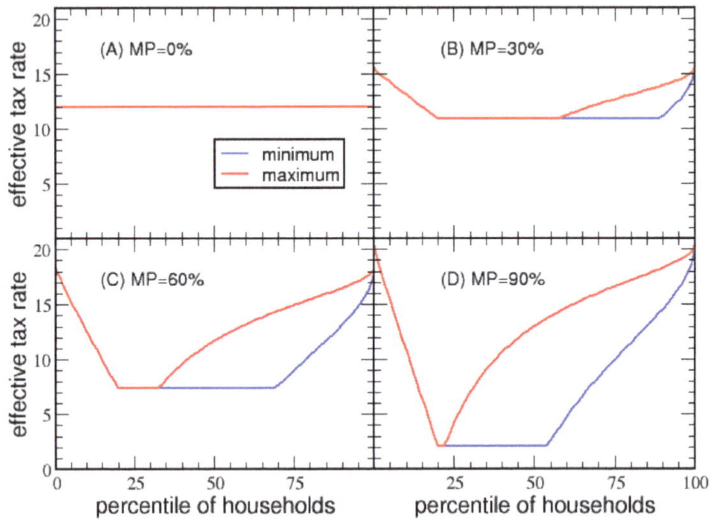

Figure 3. Effective tax rate comparisons: For economy A the minimum ETR (blue) and maximum ETR (red) are shown for $BD = 20\%$ and $MD = 50\%$ in each of the panels with different MP given by: (**A**) 0% (red line covering blue line); (**B**) 30%; (**C**) 60%; (**D**) 90%.

Illustrated in Figure 3, as MP increases from 0 to 100% more deductions are allowed, leading to a gradual change in shape from a flat horizontal line to a "U"-shape. In Figure 3A, a flat tax appears when $MP = 0$, causing $ETR = VTR$. In general, VTR is the y-intercept of the ETR plots, and $ETR < VTR$ for households with $NI > 0$. For MP values of 0%, 30%, 60%, 90%, the respective VTR are 11.97%, 15.49%, 18.35% and 20.37%. The maximum deduction implies $ETR \rightarrow VTR$ from below for high-income households. Note that ETR is regressive for low-income households until government transfer is discontinued. This point occurs at an income level equal to the basic deduction, which creates a kink in the ETR. Thereafter, a flat ETR applies to all households until the maximum income a household can deduct is equal to the basic deduction. At this point, a bifurcation can occur where households can use itemized deductions. The red and blue lines correspond to taking only the basic deduction versus the maximum deduction. The last segment of percentile of households has a progressive ETR. As MP increases, more deductions are possible, causing VTR to increase and the range for a flat ETR to decrease. Only at $MP = 1$ will $ETR = 0$ at the bottom of the dip. The same analysis for economy B results in the same qualitative behavior as shown in Figure S5.

As shown in Figure 4, when $MP = 0$ the flat segment of ETR goes to 0%, starting at BD and ending at MD for households that take the maximum deduction. The red line tracks the maximum ETR for households without itemized deductions. As the basic deduction rises from 10% to 40% in steps of 10%, the VTR is 18.72%, 20.90%, 24.47% and 29.44%, respectively. The same analysis for economy B results in the same qualitative behavior as shown in Figure S6.

In Figure 5, VTR and average ETR are plotted for five different BD values, as a function of MP and MD. Panel A shows that VTR increases as more tax deductions are allowed.

Panel C, on the other hand, shows that as more deductions become available, the average ETR decreases, where the average ETR reaches a minimum when $MP = 100\%$. The flat plateaus observed in panels B and D, which extend longer for greater BD, appear because the maximum deduction cannot be less than the basic deduction. Generally, a low average ETR is the result of the greatest tax relief for the middle class. This result suggests that MP should be set to 100 percent for economies with large net income variations. The same qualitative behavior is shown in Figure S7 for economy B.

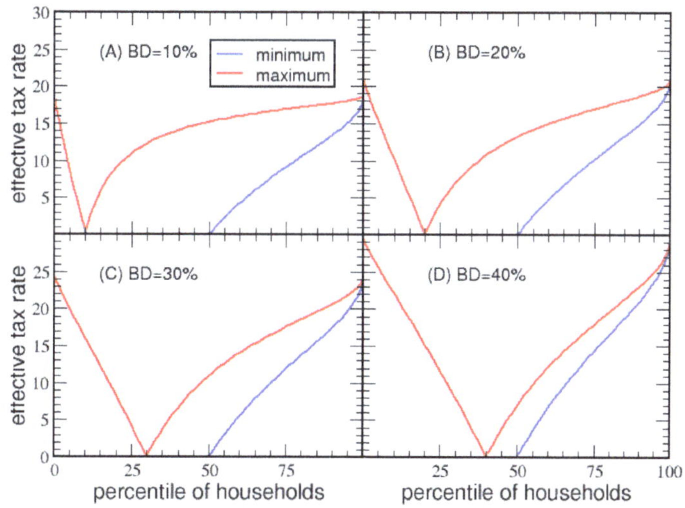

Figure 4. Effective tax rate comparisons: For economy A, the lower- and upper-bound ETRs are shown for $MD = 50\%$ and $MP = 100\%$ in each of the panels with different BD given by: (**A**) 10%; (**B**) 20%; (**C**) 30%; (**D**) 40%. At large BD, the signature "V" shape appears for the maximum ETR.

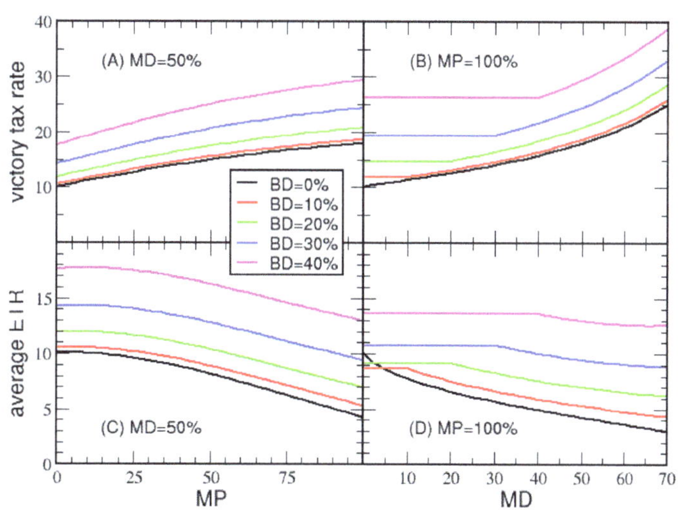

Figure 5. Tax rate comparisons: Trends in tax rates for economy A are explored. The legend applies to all panels, where different color lines represent a BD of 0%; 10%; 20%; 30%; 40%. The victory tax rate is shown as a function of (**A**) MP; (**B**) $MD\%$. The average ETR is shown as a function of: (**C**) MP; (**D**) $MD\%$. In panels A and C, $MD = 50\%$, and in panels B and D $MP = 100\%$.

In Figure 6, panels A and B, respectively, plot the minimum and maximum *ATI* as a function of percentile of households for different values of *BD*. Figure 6C compares the minimum and maximum *ATI* to the *ATI* from a flat tax. Since more tax revenue is needed for government transfer when *BD* increases, *VTR* increases too. The advantage of increasing *BD* is that lower-income households receive considerably more ATI as *BD* increases. However, the gains in *ATI* decrease as household income increases until a point where *ATI* decreases for high-income households. As shown in Figure 6D, households that take the maximum deduction beyond the 80 percentile have less *ATI* compared to a flat tax. Clearly, tax deductions are paid for by the progressive nature of the victory tax on high-income households (especially ultra-high-income households). Figure S8 shows the same qualitative behavior in economy B.

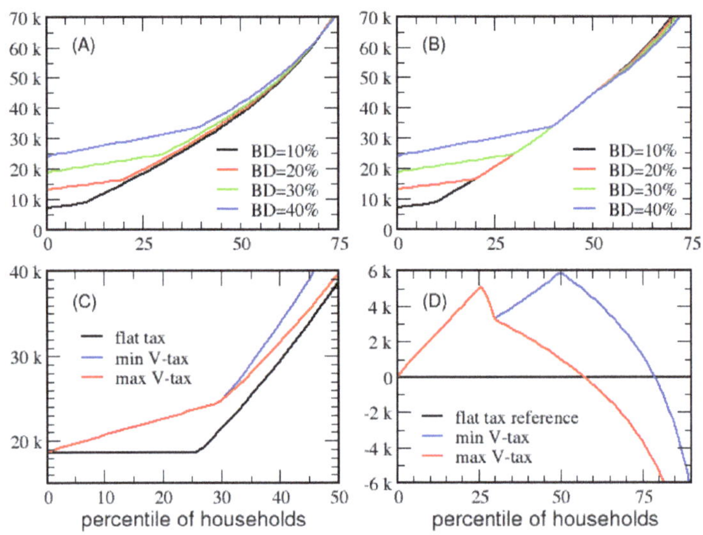

Figure 6. After-tax income comparisons: Trends in *ATI* are explored for economy A. With $MD = 50\%$; $MP = 100\%$, and *BD* ranging from 10% to 40% the *ATI* as a function of household percentile is shown for the case: (**A**) minimum *ATI*; (**B**) maximum *ATI*. (**C**) Comparing a flax tax to the victory tax (V-tax), the minimum *ATI* (max V-tax), maximum *ATI* (min V-tax) and flat tax *ATI* are shown with $BD = 30\%$, $MD = 50\%$, $MP = 100\%$. (**D**) For the same parameters used in panel C, the difference in min/max V-tax *ATI* relative to the *ATI* for a flax tax is shown.

Since $BD = PL/(1 - VTR)$, and *PL* can be quantitatively measured, *MD* is the only parameter left to determine. A balanced approach must compromise the desire for a generous maximum itemized deduction compared to the desire for a low *VTR*. As living costs decrease, *PL* will decrease, suggesting that the maximum deduction should not be large, as purchasing power is strong. Conversely, the maximum deduction should increase with an increase in living costs to ensure that itemized deductions have a positive impact on households. This leads to a prototypical victory tax system in which the maximum deduction is proportional to the basic deduction, where $MD = kBD$. The choice of an effective value of *k* is examined in Figure 7. The minimum ETR is shown in Figure 7A,C when the poverty line is at $16,814 and $22,306 while considering three values for *k*. Summarizing many of these calculations, Figure 7B,D show the average *ETR* and *VTR* as a function of *k*. Taken together, $k = 2$ is a good compromise in tax structure. Moreover, for $k < 2.5$ the results are insensitive to *PL*. The same qualitative behavior is observed in Figure S9 for economy B.

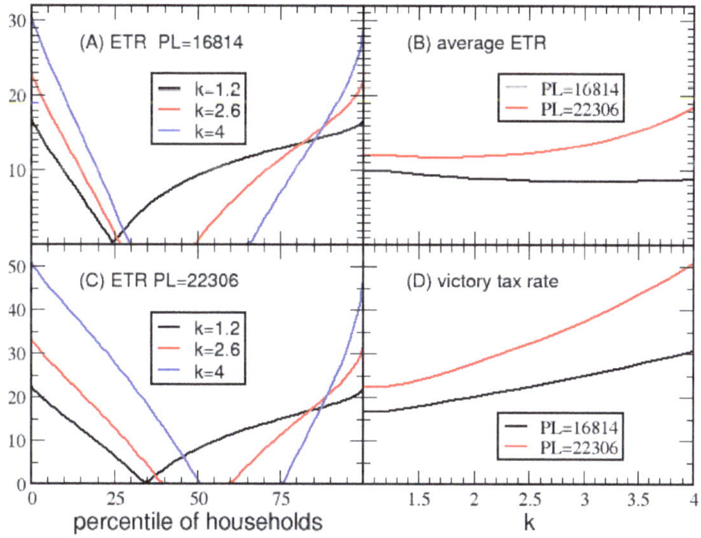

Figure 7. Maximum deduction exploration: For economy A, the *ETR* as a function of percentile of households is shown in panels (**A**) and (**C**) for a poverty line of $16,814 and $22,306, respectively. The maximum deduction is set to be proportional to the poverty line, where different color lines show different proportionality constants set at 1.2, 2.6 and 4. As a function of *k* and for two different poverty levels, panel (**B**) shows the average *ETR* and panel (**D**) shows the victory tax rate.

The question of how the prototypical victory tax system responds to extreme variation in the economy is addressed by considering a series of six log-normal test economies, characterized by a Gini index ranging from 0.05 to 0.72 (see Figures S1–S3). Figure 8A,C show the minimum *ETR* for these test economies, with poverty lines of $16,418 and $22,306, respectively. As the middle class expands, the *ETR* flattens, resulting in the lowest possible *VTR*. This flattening occurs because $MP = \min(r, 1)$. For the 6 test economies from highest to lowest Gini index, the coefficient of variations (i.e., $r = \sigma/\mu$) are, respectively, 304%, 152%, 76%, 38%, 19% and 9%. As income dispersion increases, *VTR* increases, making *ETR* very low for middle-class households. Next, Figure 8B plots *VTR* and the average *ETR* for the test economies as a function of Gini index. In Figure 8D the ratio defined by the total government transfer for eradicating poverty to the total tax revenue collected is plotted against the Gini index. For a Gini index of 0.56 (modeling the 2003 US economy), the total government transfer amounts to 23.50% or 42.30% of the total tax revenue collected when the poverty line is $16,418 or $22,306, respectively (recall 56% was used in 2003 from IRS data).

Based on the above exploration of parameters {*BD*,*MD*,*MP*}, henceforth the prototypical victory tax system will have: $BD = PL/(1 - VTR)$, $MD = 2PL$ and $MP = \min(\sigma/\mu, 1)$. The objective measures of the economy dynamically alter the victory tax structure, where it becomes flatter as the middle class becomes stronger. When the middle class shrinks as dispersion of net income increases, the victory tax increases *VTR* and lowers *ETR* for the middle class as it morphs into the V-signature. The steepness of V increases as the dispersion in net income increases. A steep regressive tax at low incomes provides the poor with the means to move upwards into the middle class. The progressive segment of the victory tax on high-income households provides the additional tax revenue necessary to form the V. These results show that the victory tax system is a type of governor to maintain a strong middle class.

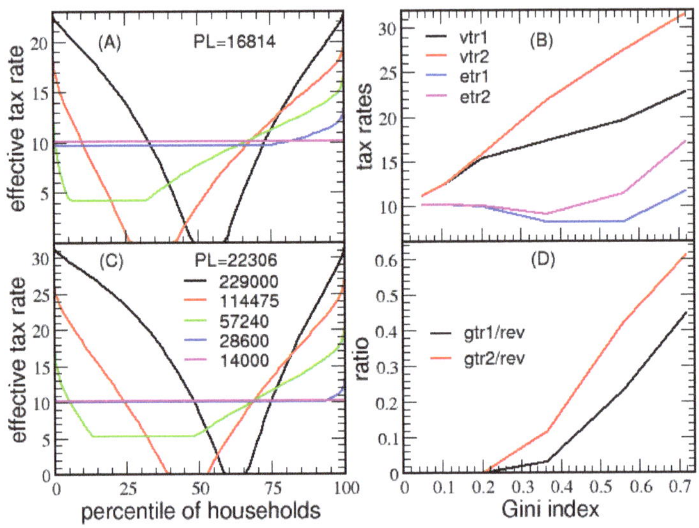

Figure 8. Systematic variation in net income dispersion: Trends in *ETR* with respect to income dispersion are shown in panels (**A**) and (**C**) for six economies described by a log-normal distribution with standard deviations ranging from \$7000 to \$229,000 about a mean income of \$75,300. The panel (**C**) legend also applies to panel (**A**). The *ETR* for the \$7000 standard deviation is not shown because if plotted, it is flat and hidden under the magenta line. As a function of Gini index, panel (**B**) plots the victory tax rate (vtr1 or vtr2) and average *ETR* (etr1 or etr2) for cases 1 and 2 corresponding to a poverty line of \$16,814 and \$22,306. For the same two cases, panel (**D**) plots total government transfer divided by total tax collected (gtr1/rev or gtr2/rev).

3.2. Flat, Linear Progressive and Victory Tax Comparisons

Here, the victory tax is compared with a flat and linear progressive tax under identical conditions (e.g., the same income distribution, total tax revenue and poverty line). The flat tax is a victory tax with $MP = 0$. A linear progressive tax system is defined when ETR is a linear function of the percentile of households with $ETR = 0$ for a household with no self-generated income. The progressive tax rate, PTR, sets a maximum tax rate adjusted to collect the desired amount of tax revenue. For example, households at 20 and 50 percentiles will have $ETR = 0.2PTR$ and $ETR = 0.5PTR$, respectively. Note that the linear progressive tax system does not satisfy the guiding principles 1 and 4. The ETR for each tax system is shown in Figures S10 and S11 for poverty lines \$16,814 and \$22,306, respectively. For a poverty line of \$22,306, the flat tax rate, FTR, is 14.80%, while PTR is 18.60% and *VTR* is 27.74%, whereas the population average *ETR* is 14.80%, 9.30% and 11.76%, respectively.

Comparisons of *ATI* in Figure 9A show that each tax system ensures the lowest possible *ATI* is at *PL*, but this occurs at different percentiles and with different trends. The linear progressive tax creates a welfare trap. That is, starting with no employment and full dependency on government transfer, the initial *ATI* is above *PL*, and then *ATI* decreases as employment increases until $ATI = PL$, and thereafter *ATI* starts increasing. Clearly, 30% of the lowest income households are better off not working than to work for any amount of time in a low-wage job. Although the flat tax does not penalize part-time employment and/or working low-wage jobs, it does not offer any advantage for individuals to work in a low-wage job. Quite clearly, the victory tax provides an incentive for low-income households to seek employment, where they will gain significant wealth as they move into the middle class. Consider, for example, two households living on the poverty line, the first fully dependent on government transfers and the second fully self-supporting. The second household enjoys \$8565 more *ATI* (a 38% increase) due to the basic tax deduction.

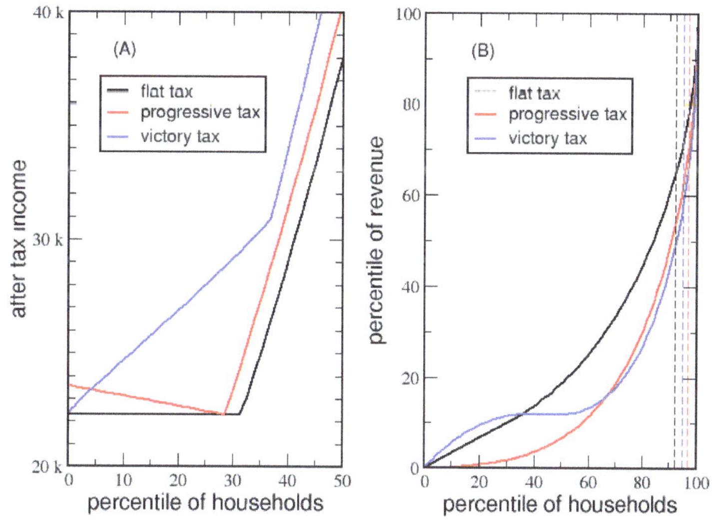

Figure 9. Tax system comparisons for economy A with a $22,306 poverty line: (**A**) After-tax income for flat, linear progressive and victory tax systems are shown in different colors; (**B**) The Lorenz curves for tax revenue are shown. The dashed vertical lines of corresponding color to the tax system indicate the percentile of households from which point onward the collected tax revenue is sufficient to pay for all government transfer needed to eliminate poverty.

For each tax system, a Lorenz curve is shown in Figure 9B for tax revenues. The Gini index for these Lorenz curves quantifies how uniform the tax burden is across income levels. A proportionate tax burden will show a similar Gini index for tax revenue as that for income received. For economy A, the Gini index for income received is 56% (see Figure S1). A flat tax without taxing government transfers yields an identical Gini index of 56% for tax revenues. However, with government transfers taxed, the flat tax gives a Gini index of 49% on tax revenues, with a higher tax burden on low-income households. The linear progressive tax system has a Gini index of 68% for tax revenues, which puts the greatest tax burden on high-income households. The victory tax reduces the tax burden on high-income households with a Gini index of 62% on tax revenues. Thus, a victory tax maintains a reasonably balanced tax burden over all income levels, where the flat tax is a special limit of the victory tax. The Lorenz curve for the victory tax is generally not monotonic, showing that low and high-income households inherit the greatest tax burden, which is why a low *ETR* is possible for the middle class.

For a (flat, progressive, victory) tax system, revenue from (7.7%, 3.3%, 5.2%) of households with the highest income corresponds to (35.2%, 30.7%, 42.3%) of the total tax revenue redistributed as government transfer. Clearly the victory tax reinvests the most back into society, which is responsible for removing the welfare trap [18] through its regressive *ETR* for low-income households. In general, as government transfers decrease due to an expanded middle class, the degree of disproportionate tax burden decreases. Figure S12 shows the same qualitative behaviors with a lower poverty line.

Now the question of how these three tax systems respond to the poverty line is addressed. Tax rates as a function of PL for the flat, linear progressive and victory tax systems are shown in Figure 10A–C, respectively. This analysis is applied to the series of six log-normal economies described in Section 2.3.2 to elucidate the effect of income dispersion for a fixed mean income. Here, the characteristic tax rate of a tax system is tracked as the poverty line is varied for different income dispersion scenarios. The lowest dispersion of $7000 represents the situation in which the vast majority of households fall into the middle class, with only rare cases of poor or rich households. As income dispersion increases,

the middle class shrinks, and the percentage of extreme poor significantly increases. In contrast, only a small increase in ultra-rich households occurs. At high income dispersion, wealthy households pay much more tax revenue, because most of the taxable income generated across the population comes from wealthy households. Moreover, as income dispersion increases, more households fall below the poverty line, which increases tax revenues that must be collected to cover the increase in government transfer. As a reference point, the final value of the considered standard deviation ($229,000) is approximately twice the dispersion found in the 2003 US economy. Although income dispersion in the US economy has steadily increased since 2003, it is still lower than the largest dispersion considered here.

In the victory tax system (including the special case of flat tax), it is seen that *VTR* increases as *PL* increases. In particular, as the poverty line is increased, *VTR* must increase more rapidly, as population dispersion is larger. Consequently, a large middle class keeps the *VTR* low, and the progressive part of the victory tax on high-income households will be most shallow.

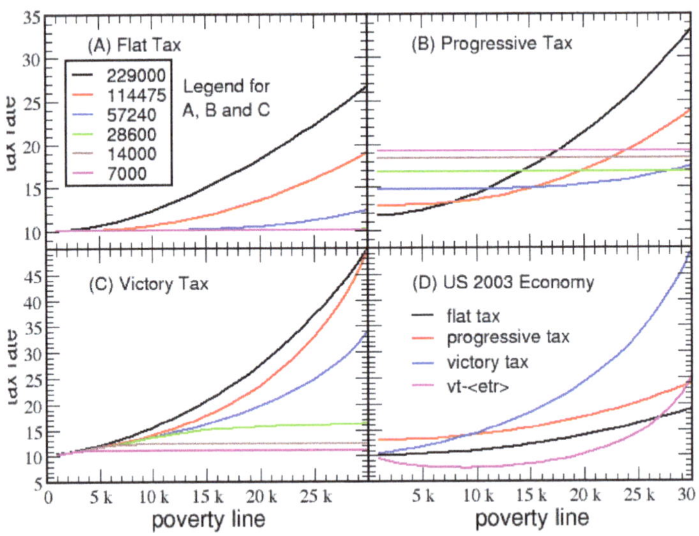

Figure 10. Tax rate dependence on poverty line: A series of six log-normal economies with income dispersion characterized by standard deviation ranging from $7000 to $229,000 about a mean income of $75,300. (**A**) flat; (**B**) linear progressive; (**C**) victory tax. (**D**) The tax rate for a flat, linear progressive and victory tax are shown together for economy A, which is used to mimic the 2003 US economy. The average effective tax rate over the population for the victory tax (denoted as vt-etr) is also shown as a function of poverty line.

Figure 10B shows that a linear progressive tax system requires an increase in PTR as the middle class expands. Interestingly, PTR is generally insensitive to the poverty line. Only when there is a wide income gap will the dependence on the poverty line become substantial. As the middle class shrinks, the tax rate rises quickly, but unfortunately there is no mechanism for the middle class to reexpand. Contrary to the linear progressive tax, as shown in Figure 10A,C, the flat and victory tax have the intuitive rank ordering that *VTR* decreases as the strength of the middle class increases.

For economy A shown in Figure 10D, *VTR* rapidly increases as a function of *PL*, while the average *ETR* remains markedly insensitive to *PL*. A low average *ETR* can be obtained even when *VTR* is high, because most of the population substantially benefits from itemized tax deductions, except for households for which the maximum itemized deduction is dwarfed relative to their net income. As the middle class shrinks, the tax

base also shrinks, and the VTR increases. This creates an elastic response to income gaps, which means the tax structure makes it easier for the middle class to expand when market pressures act to shrink it. Conversely, when the middle class expands, the V flattens, encouraging accumulation of wealth in households above the middle class level. In summary, the victory tax has desirable elasticity [24] to create stability in the economy, where the sharpness of the V-shape increases as income dispersion becomes extreme, and flattens as the middle class expands, fueling a consumer-based economy.

3.3. Distribution of Government Assistance

Government transfers and tax deductions are two forms of government assistance. However, the deductions on capital losses were not reflected in the above analyses, as net income distributions account for capital gains and losses. To quantify the degree of government assistance in terms of optional deductions, estimates of the average amount of itemized and capital loss deductions must first be modeled. Simple models are presented to quantify the total itemized and capital deductions in the population as a function of the percentile of households. These models are not part of the victory tax system. However, they are useful in characterizing how government assistance is distributed across households. The qualitative estimates made in this subsection are only intended for illustration and discussion purposes.

The net itemized deduction, NID, is modeled as: $NID = \frac{f^2}{2}\max(0, NI - BD)$ where f is the percentile of households. For economy A with a poverty line of \$22,306, the basic deduction is \$30,871 and the maximum deduction is \$44,612. The net itemized deduction for a household at the (20, 40, 60, 80) percentile, with corresponding NI of (\$16,331, \$33,878, \$57,787, \$101,844), produce a mean itemized deduction of (\$0, \$240, \$4844, \$22,711). The factor of f^2 models the qualitative trend that the more income a household has, the more likely it will utilize itemized deductions up to MD.

The total capital loss, TCL, is modeled as: $TCL = NIf^4CL/(1 - CL)$. The factor of NIf^4 models household capital gains, where it rapidly tends to zero as $f \to 0$, and tends to NI as $f \to 1$ to reflect the empirical observation that high-income households have larger portions of their income from investments. The parameter CL is a ratio of capital losses to capital gains for the tax year. Since CL reflects an average over the population, the range from 0.1 to 0.7 suffices to quantify the fraction of government assistance applied to capital loss deductions. Illustrating this qualitative model: For $CL = 20\%$ and for households at the (20, 40, 60, 80) percentiles, TCL is modeled as (\$7, \$216, \$1877, \$10,429). For example, a household with gross income of \$112,273 and capital loss of \$10,429 has a net income of \$101,844.

Employing the models for NID and TCL, government assistance, GA, is given as:

$$GA = GTR \times (1 - VTR) + VTR \times (TD + TCL) \quad (19)$$

where TD is total allowed deduction from basic and itemized deductions. In addition to government transfer, reducing the tax burden on households through tax deductions contributes to government assistance. The average tax deductions made by households at the (20, 40, 60, 80) percentiles are estimated to be (\$16,338, \$31,329, \$37,592, \$55,042). The average of $(TD + TCL)$ over the population gives the average tax deduction, which can exceed MD because TCL is included. Due to TCL, a plot of government assistance versus percentile is not informative, because government assistance to the top 0.1% of households dwarfs all other forms of assistance on an absolute scale. For example, with $CL = 20\%$, the average tax deduction at the 99.9 percentile is \$436,583. However, relative comparisons in government assistance can be made. Here, GA is divided by $GTR + NI$ to define a fraction of assistance that shows how government assistance is distributed over the percentile of households. Note that the assistance fraction for the extreme poor is not 100% for the victory tax. For a household with $NI = 0$, which receives all its income from government transfers, the assistance fraction is equal to $GA/GTR = 1 - VTR$. For economy A, and PL

of \$22,306, recall *VTR* is 27.74%, meaning 0.7226 is the fraction of government assistance when $NI = 0$. The relative government assistance decreases if only basic and itemized deductions are considered.

The percentage of government aid as a function of percentile of households for different *CL* ranging from 10% to 70% is shown in Figure 11A,B. The deviation of CL considered causes "fanning" in the tail of the assistance fraction. Qualitative characteristics and general trends are not sensitive to the models used for tax deductions. High-income households with high profit-to-loss ratios have the smallest assistance fractions, but they receive much more absolute assistance than poor households. Middle-class households have too much income to benefit from government transfers, and too little income to get much government assistance from tax deductions. However, in the victory tax system, middle-class households have the lowest *ETR*. As the poverty line increases, the assistance fraction increases for households at all income levels, not just low-income households. Together, these trends allow the victory tax system to be relatively balanced, without creating serious disproportionate tax burdens. In short, the extreme poor and ultra-wealthy pay the highest tax rates, but also receive the greatest government assistance through various mechanisms.

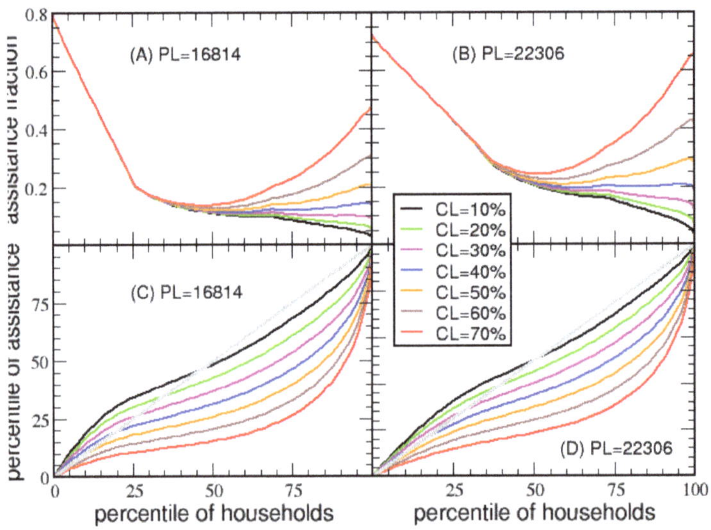

Figure 11. Government Assistance Distribution: Analysis of tax deductions for economy A: (**A**,**C**) consider a \$16,814 poverty line and (**B**,**D**) consider a \$22,306 poverty line. Panels (**A**,**B**) show the assistance fraction as a function of percentile of households. Panels (**C**,**D**) show the Lorenz curves for government assistance from (**A**,**B**), respectively. The diagonal line colored light grey in panels (**C**,**D**) is drawn to guide the eye.

Figure 11C,D show Lorenz curves for government assistance. The associated Gini index of these Lorenz curves will be 0 for proportionate government assistance at all income levels. It is noticed that the Gini index depends on the average ratio of population loss to profit. For (10%, 30%, 50%, 70%), the Gini index for government assistance is (0.00, 0.17, 0.34, 0.53). The relatively low Gini index values indicate an adequate balance in the way government assistance is distributed across income levels. Since the Lorenz curve is mostly below the diagonal line (when $CL > 10\%$), ultra-high-income households receive the most government assistance on an absolute scale. This results suggest the ultra-rich will significantly increase government dependency in economic downturns, where massive losses occur (i.e., CL large). This is another sign of elasticity to maintain social-economic stability, because government aid to the wealthy is strongest at the time when the risk of investing is highest.

4. Discussion

In the US, there is a bad connotation when discussing the poor taking advantage of the welfare system, or the rich taking advantage of tax breaks. With the five guiding principles satisfied, all taxpayers are encouraged to take advantage of the victory tax system for personal benefit, regardless of their station in life. As taxpayers take advantage of the victory tax system based on rational self-interest, a synergic benefit for society as a whole arises from the constraints on the tax structure. The mathematical framework enables a pragmatic approach to taxation, because competing interests within the population are represented within a small parameter space, which helps funnel opposing public policy options into transparent objectives. For illustration, several issues that are important to the US are discussed herein.

4.1. Consolidation of Diverse Interest Groups

A victory tax system necessarily eliminates myriad assistance programs in the US that require a need-based qualification, such as food stamps, housing/energy assistance, unemployment, social security and medicare. All types of income based welfare programs are consolidated into a single mechanism for distributing government transfer based on need, without judgment qualifications. Social security is an excellent topic of discussion. Unfortunately, the US social security program often fails to meet the basic needs of the elderly, and its solvency is questionable because new revenue from the workforce is not synchronized with recipient needs. Moreover, social security benefits have expanded beyond a retirement fund tied to age.

In the victory tax system, the unemployed and retirees belong to the same interest group. A large diverse subpopulation of recipients of government transfer will lobby for public policy to overestimate the poverty line (requiring a higher tax rate), which will counteract other large diverse subpopulations that will lobby for a lower tax rate to stimulate economic growth. In addition, as the middle class lobbies for greater itemized deductions, this also requires increasing the poverty line. Notice that the vast majority of advocates for increasing the poverty line will not belong to the marginalized subpopulation living in poverty. Competing interests from large influential groups will encourage an objective measure of the poverty line to be set within public policy. It is critical for large powerful and diverse groups to argue for specific changes within a small parameter space to produce effective outcomes.

4.2. Legal Requirements

If proof of citizenship or legal residency is required, illegal immigrants seeking public welfare are discouraged from living in the country because they will be identified. Furthermore, it would be disadvantageous for households with incomes below the poverty line not to report income for work.

4.3. Tax Form Simplicity

It is envisioned that a simple form determines tax liability. The main page of the tax form will be less than 1 page long, with optional worksheets for specifying itemized deductions and capital gain/loss information. A look-up table could be invoked to determine the basic and maximum deductions depending on the number of dependents in a household. Although public policy determines the tax code, the presence of maximum deductions does not warrant complexity in the code. For illustrative purposes, an example of the first page of a victory tax form is given in Table 2.

Table 2. Calculation of tax liabilities for twelve exemplar households. The first column gives the line numbers on the tax form. The second column gives the instructions. The six columns afterward represent example answers for households at different percentiles, f. The first 11 rows of the table correspond to the first 11 line numbers: #1 government transfer; #2 earnings income; #3 other income; #4 deductible income; #5 basic deduction; #6 itemized deductions; #7 total deductions; #8 reduced income; #9 net income; #10 taxable income; #11 tax owed. The *ETR* is given in the bottom row. All examples are based on economy A with a poverty line of $22,306 for a household of 3 with 1 dependent. The models in Section 3.3 for itemized and capital loss deductions are used to fill table entries in lines #3 and #6.

Line #	Instructions	f = 0%	f = 10%	f = 20%	f = 30%	f = 40%	f = 50%
1	direct data entry	$30,871	$22,383	$14,540	$6228	$0	$0
2	direct data entry	$0	$8488	$16,326	$24,603	$33,705	$44,064
3	data from worksheet	$0	$0	$5	$40	$173	$558
4	add line 2 and line 3	$0	$8488	$16,331	$24,643	$33,878	$44,622
5	data from lookup table	$30,871	$30,871	$30,871	$30,871	$30,871	$30,871
6	data from worksheet	$0	$0	$0	$0	$241	$1719
7	add line 5 and line 6	$30,871	$30,871	$30,871	$30,871	$31,112	$32,590
8	subtract line 7 from line 4	−$30,871	−$22,383	−$14,540	−$6228	$2766	$12,032
9	greater of line 8 or $0	$0	$0	$0	$0	$2766	$12,032
10	add line 1 and line 9	$30,871	$22,383	$14,540	$6228	$2766	$12,032
11	multiply line 10 by 0.277443	$8565	$6210	$4034	$1728	$768	$3338
	effective tax rate =	27.7%	20.1%	13.1%	5.6%	2.3%	7.5%
Line #	**Instructions**	**f = 60%**	**f = 70%**	**f = 80%**	**f = 90%**	**f = 95%**	**f = 99%**
1	direct data entry	$0	$0	$0	$0	$0	$0
2	direct data entry	$56,289	$71,677	$93,501	$135,429	$189,913	$411,893
3	data from worksheet	$1498	$3616	$8343	$20,455	$36,958	$97,951
4	add line 2 and line 3	$57,787	$75,293	$101,844	$155,884	$226,871	$509,844
5	data from lookup table	$30,871	$30,871	$30,871	$30,871	$30,871	$30,871
6	data from worksheet	$4845	$10,883	$13,741	$13,741	$13,741	$13,741
7	add line 5 and line 6	$35,716	$41,754	$44,612	$44,612	$44,612	$44,612
8	subtract line 7 from line 4	$22,071	$33,539	$57,232	$111,272	$182,259	$465,232
9	greater of line 8 or $0	$22,071	$33,539	$57,232	$111,272	$182,259	$465,232
10	add line 1 and line 9	$22,071	$33,539	$57,232	$111,272	$182,259	$465,232
11	multiply line 10 by 0.277443	$6123	$9305	$15,879	$30,872	$50,566	$129,075
	effective tax rate =	10.6%	12.4%	15.6%	19.8%	22.3%	25.3%

To cover a diverse range of possibilities, Table 2 compares 12 examples of filled tax forms for a range of household percentiles from 0 to 90 in steps of 10, as well as 95 and 99. A maximum deduction of $44,612 (being twice the poverty line) together with a $30,871 basic deduction create a cap of $13,174 on itemized deductions. From line #6 on the tax form, households at and above the 80 percentile request the maximum deduction possible. Households within the percentile range from 40 to 70 pay more taxes than they need, because they are likely not to have enough surplus income. Allowed itemized deductions help society and the household. For example, an important itemized deduction should be to invest in a retirement fund. As a household achieves greater net income and/or reduce expenses, more itemized deductions for a retirement fund are possible, among other allowed reasons.

4.4. Financial Security from Job Loss

Regardless of a household's savings or previous income levels, a household automatically gains a minimum guaranteed income at the poverty line if job losses lead to no income. A household with a living standard far above the poverty line would inevitably exhaust their savings due to an extended period of job loss. However, it is not the role of government to maintain differences in wealth in households, even for a short time.

4.5. *Right to Work*

The basic guarantee of income in the victory tax system eliminates the need for a minimum wage. Without a minimum wage, many companies are likely to lower wages below a living wage. Nevertheless, a low-wage job will increase *ATI* above the poverty line, as the regressive *ETR* leads to significant increases in *ATI* from modest income increases. This increase in net income was not the case with the negative income tax, which was tested in the late 1960s to early 1970s in North America [9]. Within a victory tax system, low-wage entry-level jobs can benefit society. For example, a young person with no previous work experience earning low wages increases the collective *ATI* of a household. This is a win–win–win situation: A company acquires cheap labor, a new worker gains valuable training, and the household increases its net income. The steep regressive tax encourages workers to accept low paying jobs in exchange for building skills. After gaining experience, workers should expect to move into higher-paying positions, creating rapid turnover in entry-level jobs. Companies will have to adjust their wages to balance the turnover rate with the costs of training new workers.

4.6. *Right Not to Work*

As companies become dependent on the government to pay low-wage workers, if left unchecked, this practice will inevitably develop into a modern form of slavery [28], when work is required in exchange for government transfer. Enforcing work for government transfer at the poverty level is analogous to forcing companies to pay workers a high minimum wage. To prevent exploitation of workers, work requirements cannot be applied to the basic income guarantee. In particular, the right not to work is a necessary balance in a free market in which individuals are free agents who promote their own agenda for wealth accumulation. Guaranteed basic income subsidizes both workers and owners. When workers are independent agents, the main reasons for unions become unnecessary. For example, a person can refuse to work in conditions they consider inappropriate, unworthy of their talents, too little pay, or because the job is uninteresting. This gives workers free time to develop new skills and seek better-paying jobs based on their merits. Whether a person receives government transfer because of retirement from the workforce, does not find work or decides not to work is irrelevant to the victory tax system.

4.7. *Productivity in Society*

The victory tax system promotes a productive society, not by providing comfortable financial security to people, but rather by providing incentives for people across the income spectrum to take jobs, become more demanding for higher salaries and better benefits, and to regularly make financial investments. Of course, there will be a subpopulation of people that will take guaranteed basic income and never work. A victory tax system allows the free market to determine workforce equilibrium, and it embraces the income distribution from that free market that includes non-working households. The victory tax system makes no judgments about why people work or not.

It is important to stress that productivity cannot be quantified in monetary terms, because not generating taxable income is not the same as unproductive. For example, a basic income guarantee can help low-income parents meet household needs to raise children, which is a productive activity for society. People in low-income households often have health problems that prevent them from working [15]. Some people will choose to live a life near the poverty level while providing community benefit through good deeds. Allowing people to retire at the age of their choice also eliminates the arbitrary mandatory retirement age set by government. In short, the victory tax system offers each individual the opportunity to achieve success in their own terms, which can evolve over time.

Consider the case in which a person (young or old) wishes to pursue creative interests. Such passions can be pursued productively, rather than inhibited by the need to work for mere survival without dignity. Observing peers accumulating personal wealth creates a powerful incentive for the vast majority of people not to indefinitely pursue personal

interests strictly at the poverty line. Generally, wealth accumulation occurs over one's lifetime. Consider the most likely scenario when young adults leave their parents' home. For an inexperienced worker, it is generally difficult to find a good-paying job that meets basic needs. The victory tax system allows young adults to become independent and productive taxpayers sooner due to the regressive nature of the ETR. Starting with government transfer, as an individual's economic status increases government assistance changes to itemized and capital loss tax deductions. Thus, all forms of government assistance help create a productive society as individuals capitalize on their rational self-interest.

4.8. Security in Personal Wealth

The victory tax system offers short- and long-term opportunities for prosperity through itemized deductions. For example, the cost of investing in stocks and bonds can be an itemized deduction. In the short term, this itemized deduction contributes to wealth accumulation by encouraging households to reduce their *ETR* and increase their *ATI* by investing in the economy. In the long run, the accumulated wealth of a household can be exploited during retirement as other income, which can substantially increase *ATI* above the poverty line for the middle class. For lower-income households, modest but significant gains in *ATI* will result from supplementing government transfer. Furthermore, the wealth generated by a taxpayer stays with the taxpayer at all times. Households can sell assets for income at any time without imposing penalties for early withdrawal or waiting until a certain retirement age.

Policy makers should allow a wide range of itemized deductions to offer opportunities for prosperity. For example, allowing itemized deductions on accumulating assets, such as a house, or to offset the cost of higher education or for training on workforce skills. Other forms of security could include itemized deductions on health insurance or health care costs. In this way, households can pay less taxes by taking measures to strengthen their financial independence and well-being. In summary, many allowed itemized tax deductions in the tax code will give households the opportunity to use surplus income for personal gains that create benefits for society.

4.9. Catalyst for Micro-Businesses

The victory tax system creates a supportive environment for micro-enterprises to form. As low-wage jobs are subsidized, new businesses can rely on this to reduce startup costs. For example, new businesses can form a mission to attract low-skilled workers to help the local community while building skills for their recruited workers. This paradigm replaces working in low-wage jobs without growth opportunities. With a foundation for social security, the private sector has the means to solve local community problems without high barriers. The safety net of guaranteed basic income enables low- to middle-income households to take risks in entrepreneurial endeavors that would otherwise be prohibitive. Over time, micro-enterprises can grow into larger businesses.

4.10. Responsible Government

The floating tax rate helps avoid runaway deficits, as it can adapt to government budgets that take into account debt and repayment plans to control the ratio of debt to gross domestic product (GDP) through public policy. It is worth noting that public policy may promote large debt accumulation to keep the victory tax rate lower, but this jeopardizes long-term stability. For the US, an open budget to the public provides transparency to determine if taxation has representation. With only a few key tax parameters, debates will focus on why tax rates, poverty lines, or deductions should be changed. Furthermore, any proposed changes to the parameters can be modeled and tested with the consequences predicted. This clarity will make public policy debates more substantive, such as evaluating the effectiveness of expanding free public services compared to increasing the poverty line.

Government transfers should be distributed continuously, as they are part of a steady income of a household. An efficient and convenient method of government transf

collection of taxes owed is consistent with Adam Smith's four principles [29], the third of which states: "Every tax ought to be levied at the time, or in the manner, in which it is likely to be convenient for the contributor to pay". This logic also applies to government transfers in the victory tax system, which eliminates the need for government to administer complex assistance programs. However, there will be subpopulations in society that will refuse government support on the grounds of principle or incompetence. In the latter case, the government should support institutions that care for people who are dependent on others, such as nursing homes, institutions for mentally ill or homeless, etc. If a person must be put in a public or private institution because they cannot live independently, the institution becomes their household, and will receive government transfer in their behalf. Financial administrators of these institutions will be obliged to pay income tax on the government transfer received in aggregate form. In this way, the victory tax system supports the entire population, including the "forgotten" people in society who must live as dependents.

4.11. Role of Corporate and Other Taxes

In the analysis of the victory tax system, all other sources of tax revenue, such as sales and corporate tax, were dismissed. As the example economy of Table 2 shows, *VTR* approaches 30% when there is a weak middle class and highly skewed net income distribution, as is currently the case in the US. An argument often made is that low corporate taxes create GDP growth. For the analyses given here, corporate taxes are 0% across the board, regardless of the size of the company/business. Remember that in 2003, approximately 19 percent of tax revenues came from other admissible sources. Therefore, it is feasible to shift *VTR* down by approximately 5% when other tax revenues are taken into account. Ignoring all other admissible tax revenue sources, the 27.7% *VTR* is not prohibitively high, demonstrating that the victory tax system is cost effective. Discussion of an appropriate corporate tax system goes beyond the scope of this paper.

4.12. Future Work

This work can be expanded in several ways. The consequences of a victory tax system should be quantified by large-scale agent-based modeling to simulate an evolving economy [30]. Different initial economic conditions under different public policy constraints should be systematically investigated and compared with other tax systems. A few key questions should be addressed: Does the victory tax system stabilize the middle class? What is the impact on market income distribution? How will GDP be affected? Other features to be explored include how to deal with spatial (regional) inhomogeneity, how to synergistically combine this system with corporate taxation, and how to transition from an existing tax system into a victory tax system. The victory tax system, along with other tax systems, should be tested in economies that will be largely automated. For those interested in further testing or developing the victory tax system, a C++ program that is used to generate all the results presented here (including other types of Lorentz curves that are not shown) is available in supplementary materials. Although many aspects and consequences for universal basic income have recently been addressed [8], these findings, albeit insightful, are not directly applicable to the victory tax system.

5. Conclusions

Five guiding principles have been developed for a pragmatic income tax system that generates revenue for the government while providing basic welfare, and encourages households across the income spectrum to accumulate wealth. These principles limit the way in which government redistributes wealth, but the details of implementation are left to public policy. These principles lead to a mathematical framework describing a family of tax systems based on a small parameter space where the parameters are informed by the state of the economy. A prototypical victory tax system was created that shows the feasibility of a holistic paradigm for a poverty-free and productive society without

disproportional or excessive tax burdens. To highlight the effect of the constraints imposed by the mathematical framework, four key features of the victory tax are summarized.

1. Every year, a victory tax rate (*VTR*) is set to cover projected government spending based on public policy, which plays a role in incurring or reducing debt. The *VTR* increases when government spending increases, and vice versa.
2. As the maximum allowed deduction increases, the *VTR* increases, but the effective tax rate (*ETR*) for the middle class remains markedly low, and possibly zero, due to its V-shape as a function of household income percentile. Conversely, as the maximum allowed deduction decreases, the V-shape of the *ETR* becomes shallower and gradually flattens until a flat tax appears when no deductions are allowed, resulting in the lowest possible *VTR*.
3. The combination of government transfer and a basic deduction creates a regressive tax for low-income households and guarantees a basic income for all households, which is set at the poverty line after taxes are paid. By taxing government transfers at the *VTR*, no welfare trap is formed, creating a substantial incentive for households to generate income. The poverty line sets the minimum standard of living that society can tolerate, which depends on availability of public services. An increase in the poverty line will increase the *VTR*, and vice versa. Each year, the poverty line and income dispersion are objectively measured and updated to keep the victory tax responsive to changes in the economy.
4. Government assistance in the form of government transfer, basic deductions, itemized deductions and capital loss deductions create opportunity for wealth accumulation in households across the income spectrum. The *ETR* on taxable income for the ultra-rich is no higher than for the extreme poor. As the middle class shrinks, the *VTR* increases, and the ETR for low and high-income households become more regressive and progressive, which stabilizes the middle class.

The victory tax makes it easy for taxpayers to calculate tax liability, collect government transfer when needed, and for government to set tax rates that will generate the projected revenue to cover its expenditures. While the victory tax cannot prevent runaway deficits, the adjustable tax rate provides the government with a means to control debt, which is an important factor in setting the single tax rate. The constraints within the tax system make public policy objectives transparent, suggesting that policy debates will become meaningful for the typical taxpayer. This is because government assistance, either through direct transfers or deductions, offer incentives for households to take advantage of the victory tax system out of rational self-interest. As low-income households accumulate wealth, society's standard of living can rise significantly, suggesting that the middle class will be the dominant segment of the population.

Supplementary Materials: The following are available online at https://www.mdpi.com/1099-4300/23/11/1492/s1, Figure S1: Lorenz curve for economies A and B; Figure S2: Systematic series of log-normal distributions as middle class size increases; Figure S3: Lorenz curve for the systematic series of log-normal distributions; Figure S4: Average tax rate from the US congressional budget office; Figure S5: Effective tax rates for economy B for illustrative set of parameters; Figure S6: Effective tax rates for economy B for additional parameter combinations; Figure S7: victory tax rate and average effective tax rate for economy B; Figure S8: After-tax income for economy B; Figure S9: Effective and victory tax rates for economy B; Figure S10: effective tax rate for low poverty line; Figure S11: effective tax rate for high poverty line; Figure S12: After-tax income comparisons for different types of tax systems. A separate zipped file contains a C++ program to generate the results, and data files from the US government used to obtain tax and poverty information.

Funding: This research received no external funding.

Acknowledgments: I thank Yihuan Song, a Charlotte Research Scholar from the Belk College of Business in assisting me to obtain the publicly available information on US federal taxes and poverty data.

Conflicts of Interest: The author declares no conflict of interest.

References

1. Tiley, J. *Studies in the History of Tax Law*; Hart Publishing: Oxford, UK; Portland, OR, USA, 2002.
2. Gini, C. Measurement of Inequality of Incomes. *Econ. J.* **1921**, *31*, 124–126.
3. Clementi, F.; Gallegati, M.; Kaniadakis, G. κ Generalized Statistics in Personal Income Distributions. *Eur. Phys. J. B* **2007**, *57*, 187–193.
4. Clementi, F.; Gallegati, M.; Kaniadakis, G. κ-Generalized Statistical Mechanics Approach to Income Analysis. *J. Stat. Mech. Theory Exp.* **2009**, *2*, P02037.
5. Clementi, F.; Gallegati, M. *The Distribution of Income and Wealth: Parametric Modeling with the κ-Generalized Family*, 1st ed.; New Economic Windows Series 2039–2411x; Springer International Publishing: Berlin/Heidelberg, Germany, 2016.
6. Keen, M.; Kim, Y.; Varsano, R. The "Flat Tax(es)": Principles and Evidence. *IMF Work. Pap.* **2006**, 15, 712–751.
7. Constantine, A.; Thompson, B.S. Negative income taxes, inequality and poverty. *Can. J. Econ.* **2016**, *49*, 1016–1034.
8. Gentilini, U.; Grosh, M.; Rigolini, J.; Yemtsov, R. *Exploring Universal Basic Income: A Guide to Navigating Concepts*, 1st ed.; World Bank Publications: Washington, DC, USA, 2019.
9. Widerquist, K. A failure to communicate: What (if anything) can we learn from the negative income tax experiments? *J. Socio-Econ.* **2005**, *34*, 49–81.
10. Lopez-Daneri, M. NIT picking: The macroeconomic effects of a Negative Income Tax. *J. Econ. Dyn. Control* **2016**, *68*, 1–16.
11. Bryan, J.B. Targeted programs v the basic income guarantee: An examination of the efficiency costs of different forms of redistribution. *J. Socio-Econ.* **2005**, *34*, 39–47.
12. Gamel, C.; Balsan, D.; Vero, J. The impact of basic income on the propensity to work. *J. Socio-Econ.* **2006**, *35*, 476–497.
13. Tondani, D. Universal Basic Income and Negative Income Tax: Two different ways of thinking redistribution. *J. Socio-Econ.* **2009**, *38*, 246–255.
14. Forget, E.L. The Town with No Poverty: The Health Effects of a Canadian Guaranteed Annual Income Field Experiment. *Can. Public Policy-Anal. Polit.* **2011**, *37*, 283–305.
15. Bryan, J.B. Have the 1996 welfare reforms and expansion of the earned income tax credit eliminated the need for a basic income guarantee in the US? *Rev. Soc. Econ.* **2005**, *63*, 595–611.
16. Van Parijs, P. Basic Income: A Simple and Powerful Idea for the Twenty-First Century. *Politics Soc.* **2004**, *32*, 7–39.
17. Allen, J.T. Negative Income Tax. In *The Concise Encyclopedia of Economics*; Library of Economics and Liberty: Carmel, IN, USA, 2008. Available online: http://www.econlib.org/library/Enc1/NegativeIncomeTax.html (accessed on 10 October 2021).
18. Maag, E.; Steuerle, C.E.; Chakravarti, R.; Quakenbush, C. How Marginal Tax Rates Affect Families at Various Levels of Poverty. *Natl. Tax J.* **2012**, *65*, 759–782.
19. Ingles, D.; Oliver, K. Options for Assisting Low Wage Earners. *Econ. Labour Relat. Rev.* **2000**, *11*, 76–107.
20. Henderson, D.R. A Philosophical Economist's Case against a Government-Guaranteed Basic Income. *Indep. Rev.* **2015**, *19*, 489–502.
21. Freedman, D.H. Basic Income: A Sellout of the American Dream. *Technol. Rev.* **2016**, *119*, 48–53.
22. Monitor, C. Against a basic income guarantee. *New Econ. Dir.* **2016**, *23*, 23–26.
23. Office of the Assistant Secretary for Planning and Evaluation (ASPE). Available online: https://aspe.hhs.gov/topics/poverty-economic-mobility/poverty-guidelines (accessed on 10 October 2021).
24. Saez, E. Using elasticities to derive optimal income tax rates. *Rev. Econ. Stud.* **2001**, *68*, 205–229.
25. Pareto, V. La legge della domanda. Giornale degli Economisti. *Riv. Politica Econ.* **1997**, *87*, 645. (In English)
26. Internal Revenue Service (IRS) Historical Data Tables. Available online: https://www.irs.gov/uac/soi-tax-stats-historical-data-tables (accessed on 10 October 2021).
27. Tax Policy Center (TPC) Urban Institute and Brookings Institution. Available online: http://www.taxpolicycenter.org (accessed on 10 October 2021).
28. Zelleke, A. Distributive justice and the argument for an unconditional basic income. *J. Socio-Econ.* **2005**, *34*, 3–15.
29. Brunori, D. Principles of Tax Policy and Targeted Tax Incentives. *State Local Gov. Rev.* **1997**, *29*, 50–61.
30. Banzhaf, W. The effects of taxes on wealth inequality in artificial chemistry models of economic activity. *PLoS ONE* **2021**, *16*, e0255719.

entropy

MDPI

Article

Highway Freight Transportation Diversity of Cities Based on Radiation Models

Li Wang [1], Jun-Chao Ma [1,2], Zhi-Qiang Jiang [1,2,3], Wanfeng Yan [2,4] and Wei-Xing Zhou [1,3,4,*]

[1] School of Business, East China University of Science and Technology, Shanghai 200237, China; shellyly@ecust.edu.cn (L.W.); jcma@mail.ecust.edu.cn (J.-C.M.); zqjiang@ecust.edu.cn (Z.-Q.J.)
[2] Zhicang Technologies, Beijing 100016, China; wanfeng.yan@google.com
[3] Research Center for Econophysics, East China University of Science and Technology, Shanghai 200237, China
[4] Department of Mathematics, East China University of Science and Technology, Shanghai 200237, China
* Correspondence: wxzhou@ecust.edu.cn

Abstract: Using a unique data set containing about 15.06 million truck transportation records in five months, we investigate the highway freight transportation diversity of 338 Chinese cities based on the truck transportation probability p_{ij} from one city to another. The transportation probabilities are calculated from the radiation model based on the geographic distance and its cost-based version based on the driving distance as the proxy of cost. For each model, we consider both the population and the gross domestic product (GDP), and find quantitatively very similar results. We find that the transportation probabilities have nice power-law tails with the tail exponents close to 0.5 for all the models. The two transportation probabilities in each model fall around the diagonal $p_{ij} = p_{ji}$ but are often not the same. In addition, the corresponding transportation probabilities calculated from the raw radiation model and the cost-based radiation model also fluctuate around the diagonal $p_{ij}^{\rm geo} = p_{ij}^{\rm cost}$. We calculate four sets of highway truck transportation diversity according to the four sets of transportation probabilities that are found to be close to each other for each city pair. It is found that the population, the gross domestic product, the in-flux, and the out-flux scale as power laws with respect to the transportation diversity in the raw and cost-based radiation models. It implies that a more developed city usually has higher diversity in highway truck transportation, which reflects the fact that a more developed city usually has a more diverse economic structure.

Keywords: econophysics; highway freight transportation; radiation model; transportation network; network diversity; power law; economic development

Citation: Wang, L.; Ma, J.-C.; Jiang, Z.-Q.; Yan, W.; Zhou, W.-X. Highway Freight Transportation Diversity of Cities Based on Radiation Models. *Entropy* **2021**, *23*, 637. https://doi.org/10.3390/e23050637

Academic Editor: Ryszard Kutner

Received: 19 April 2021
Accepted: 13 May 2021
Published: 20 May 2021

1. Introduction

The growing volumes of passenger and freight transport around regionally and globally witness their important role for economic development of different countries [1–5]. Aviation, railway, highway and shipping are four main transportation methods in modern societies. Unlike other three ones, information about highway transportation is less publicly available. In mainland China, the highway system has experienced a very rapid development since the Reform and Opening-up of China, forming a rapidly expanding multiplex network which contains national highways, provincial highways, county highways and countryside highways [6]. China has the longest expressway network in the world, which includes about 0.143 million kilometers expressways.

In the past decades, the gravity law is the most adopted in understanding transportation networks and predicting transportation fluxes [7–11], which reads

$$W_{ij} \sim \frac{M_i^{\alpha} M_j^{\beta}}{d_{ij}^{\gamma}}, \tag{1}$$

where W_{ij} is the flow between locations i and j, M_i (or M_j) is usually the population or gross domestic product (GDP) of location i (or j), d_{ij} is the distance between i and j, and α, β and γ are the model parameters. Very relevantly, the gravity law has been investigated and confirmed in the Korean highway network between 30 largest cities [7], the express bus flow in Korea consisting of 74 cities and 170 bus routes with 6692 operating buses per day [12], and the urban bus networks of Korean cities [13], and the highway freight transportation networks of 338 Chinese cities [6].

However, the gravity model has several limitations, especially the requirement of previous traffic data to fit the parameters [14]. To overcome those limitations, the radiation model has been proposed [14], in which the predicted flux $\tilde{F}_{ij}$ from city i to city j is obtained as follows

$$\tilde{F}_{ij} = F_i^{\text{out}} \frac{M_i M_j}{(M_i + S_{ij})(M_i + M_j + S_{ij})}, \tag{2}$$

where S_{ij} is the total "mass" (population or GDP) in the circle of radius d_{ij} centered at i but excluding the source and destination population, and F_i^{out} is total out-flux departing from city i

$$F_i^{\text{out}} = \sum_{j \neq i} F_{ij}, \tag{3}$$

where F_{ij} is the real flux from i to j. Obviously, the data of F_i^{out} are much easier to collect than F_{ij}.

In the raw radiation model, d_{ij} is the geographic distance between i and j. The cost-based radiation model has been soon proposed based on the intuition that an individual will choose the site that has the lowest travel cost on the network, where the travel cost can be measured by the path length or travel time from i to j [15]. In this work, d_{ij} is measure by the path length or driving distance from i to j. Later, to better estimate the fluxes at different spatial scales, a scaling parameter is introduced into the radiation model [16]. By combining memory effect and population-induced competition, a general model has been developed to enable accurate prediction of human mobility based on population distribution only, which also has a parameter qualifying the memory effect [17].

Although the radiation model has been adopted in the study of trip distributions [9,18–21], applications to freight transportation are rare. In this work, using a unique data set about the highway freight transportation by trucks between 338 cities in mainland China, we investigate the transportation probability p_{ij} between two cities i and j and the transportation diversity of a city calculated from p_{ij}. Although most studies dealt with undirected transportation networks [6,22,23], radiation models enable us to consider directed transportation networks due to the availability of data [24]. The raw radiation model and the cost-based radiation model are adopted because they are parameter free.

It has been reported that higher social network diversity provides greater access to social and economic opportunities and has a strong correlation with the economic development [25]. With the highway freight transportation data between Chinese cities available, we aim to investigate the relationship between highway freight transportation network diversity and economic development of cities. Such an analysis has not been conducted due to the difficulty in obtaining the highway freight transportation data. Our analysis shows that the population, the gross domestic product, the in-flux, and the out-flux scale as power laws with respect to the transportation diversity in the raw and cost-based radiation models, which implies that a more developed city usually has higher diversity in highway truck transportation. This finding reflects the fact that a more developed city usually has a more diverse economic structure.

The remainder of this work is organized as follows. Section 2 describes the data sets we analyze. Section 3 studies the basic properties of transportation probability. Section 4 deals with the transportation diversity of cities and their relationship with population and GDP. We discuss and summarize in Section 5.

2. Data Sets

The data set we analyze was provided by a leading truck logistics company in China, which records the highway truck freight transportation between 338 cities in mainland China over the period from 1 January 2019 to 31 May 2019 [6]. The data cleaning was done by the company, who used the data set in their truck scheduling and route planning. There are about 15.06 million truck freight transportation records in total, each entry containing the origin and destination cities and the starting date of the transportation. We can construct the flux matrix $\mathbf{F} = [F_{ij}]_{338\times338}$, where F_{ij} stands for the number of trucks with freights driven from city i to city j. Unloaded trucks are not counted in. Because radiation models do not consider intra-city transportation, we set that

$$F_{ii} = 0. \tag{4}$$

It is obvious that F_{ij} is not necessary to be equal to F_{ji} for $i \neq j$.

The GDP and population data for the 338 Chinese cities in 2017 were retrieved online from the Complete Collection of World Population (http://www.chamiji.com, accessed on 18 May 2021), which are publicly available except for a few cities. We supplemented the missing data by searching Baidu Encyclopedias (https://baike.baidu.com, accessed on 18 May 2021).

The geographic distance $d_{ij}^{\rm geo}$ is the shortest surface distance between two cities located by the longitude and latitude, which is the length of the great circle arc connecting two points on the surface of the earth. The longitude and latitude of each city can be easily obtained online for free. The data set of the driving distances $d_{ij}^{\rm cost}$ between pairs of cities was provided by the same truck logistics company, which were collected by their truck drivers. The driving distance between two cities are usually "optimized" by the truck drivers because they always have the motivation to find a path connecting the two cities with the least cost (time and money). Such an optimization is achieved either by their own experience or by information from buddy truck drivers they trust. It is obvious that

$$d_{ij}^{\rm geo} < d_{ij}^{\rm cost} \tag{5}$$

for all pairs of cities. The difference between these two distances increases when the two cities are farther away to each other. By definition, the geographic distance matrix is symmetric, that is,

$$d_{ij}^{\rm geo} = d_{ji}^{\rm geo}. \tag{6}$$

In contrast, the driving distance matrix is asymmetric, i.e.,

$$d_{ij}^{\rm cost} \neq d_{ji}^{\rm cost}, \tag{7}$$

which is mainly due to the fact that, besides highways, there are often local roads that a truck driver has to take from one city to the other.

3. Transportation Probability

3.1. Formulae

According to the radiation models (2) we adopt, the transportation probability p_{ij} from city i to city j is

$$p_{ij} = \frac{M_i M_j}{(M_i + S_{ij})(M_i + M_j + S_{ij})}. \tag{8}$$

When we choose population P for M, the transportation probability becomes

$$p_{ij} = \frac{P_i P_j}{(P_i + S_{ij})(P_i + P_j + S_{ij})}, \tag{9}$$

where S_{ij} is the total population in the circle of radius d_{ij} centered at i but excluding the source and destination population. Alternatively, when we use GDP as the proxy, we have

$$p_{ij} = \frac{G_i G_j}{(G_i + S_{ij})(G_i + G_j + S_{ij})}, \tag{10}$$

where S_{ij} is the total GDP in the circle of radius d_{ij} centered at i but excluding the source and destination population.

The transportation probabilities p_{ij} of the raw radiation model using geographic distance and the cost-based radiation model using driving distance are calculated with respect to population P in Equation (9) and gross domestic product G in Equation (10).

3.2. Power-Law Distribution of p_{ij}

Figure 1 illustrates the four empirical distributions of the transportation probability p_{ij} between two cities for the two radiation models with $M = P$ and $M = G$, respectively. We observe a nice power-law tail in each case and the exponents are the same for the four cases:

$$f(p_{ij}) \sim p_{ij}^{-\alpha-1}, \tag{11}$$

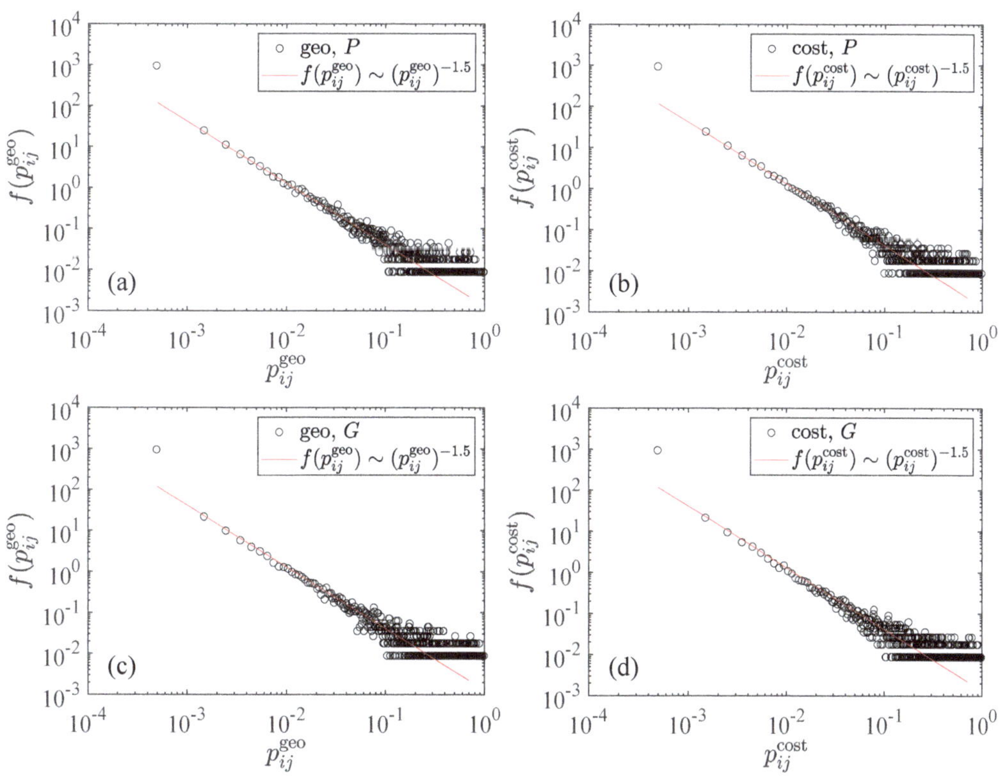

Figure 1. Power-law tailed distribution of the transportation probability between two cities. The solid lines are power laws with the same exponent of −1.5. (**a**) Population P is used in the raw radiation model with the geographic distance. (**b**) Population P is used in the cost-based radiation model with the driving distance. (**c**) Gross domestic product (GDP) G is used in the raw radiation model with the geographic distance. (**d**) Gross domestic product G is used in the cost-based radiation model with the driving distance.

where the tail exponents $\alpha \approx 0.5$ and the intercepts are almost the same. The power-law relationship holds over three orders of magnitude. The smallest transportation probabilities deviate from the power-law distributions with higher probability density. Theoretically, we know that two cities with longer distance usually have a smaller transportation probability. Indeed, it we plot p_{ij} with respect to d_{ij}, we find that the points fluctuate around a power-law scaling with an exponent of -4:

$$p_{ij} \sim d_{ij}^{-4}, \tag{12}$$

which corresponds to the case of uniform population (or GDP) density [14]. The standard deviation of the data points from this reference power law quantifies the strength of heterogeneity of the spatial distribution of population and GDP in mainland China.

3.3. *Asymmetric Relationship between p_{ij} and p_{ji}*

We illustrate in Figure 2 the asymmetric relationship between p_{ij} and p_{ji} for the two radiation models using population. The results for GDP is very similar for each model. It is striking that the predicted values of transportation probability span nine orders of magnitude. We also find that the scatter points lies close to the diagonal $p_{ij} = p_{ji}$. The points from the cost-based model in Figure 2b concentrate more to the diagonal than the points in Figure 2a and thus the transportation probability matrix $\{p_{ij}\}$ is less asymmetric. The two dashed lines impose a restriction on the transportation probability values, requiring that

$$p_{ij} + p_{ji} = 1, \tag{13}$$

which is more visible if we use linear coordinates. This restriction can be derived as follows.

According to Equation (9), the probability of transportation from city j to city i is

$$p_{ji} = \frac{P_i P_j}{(P_j + S_{ji})(P_i + P_j + S_{ji})}. \tag{14}$$

For two given cities i and j, it is easy to notice that p_{ij} and p_{ji} reach their maxima when the two cities are adjacent, that is

$$S_{ij} = S_{ji} = 0. \tag{15}$$

In this case, we have

$$p_{ij} = \frac{P_j}{P_i + P_j} \tag{16}$$

and

$$p_{ji} = \frac{P_i}{P_i + P_j}. \tag{17}$$

The restriction shown in Equation (13) is thus obtained. This argument holds for both of the radiation models, because the derivation is independent of the definition of the distance between two cities. It also applies to the two models based on GDP, as expressed in Equation (10).

3.4. *Comparison between p_{ij}^{geo} and p_{ij}^{cost}*

We compare the predicted transportation probabilities from the two models. The results are shown in Figure 3. We find that the points fluctuate around the diagonal line

$$p_{ij}^{\text{cost}} = p_{ij}^{\text{geo}}. \tag{18}$$

The insets show that there are many points that fall exactly on the diagonal. These points correspond to the situations when

$$S_{ij}^{\text{geo}} = S_{ij}^{\text{cost}}. \tag{19}$$

Usually, this condition (19) is more likely to be fulfilled when the two cities i and j are close. As a special case, when city j is the closest city of city i, we have $S_{ij}^{geo} = S_{ij}^{cost} = 0$. In this case, the two transportation probabilities p_{ij}^{geo} and p_{ij}^{cost} are identical.

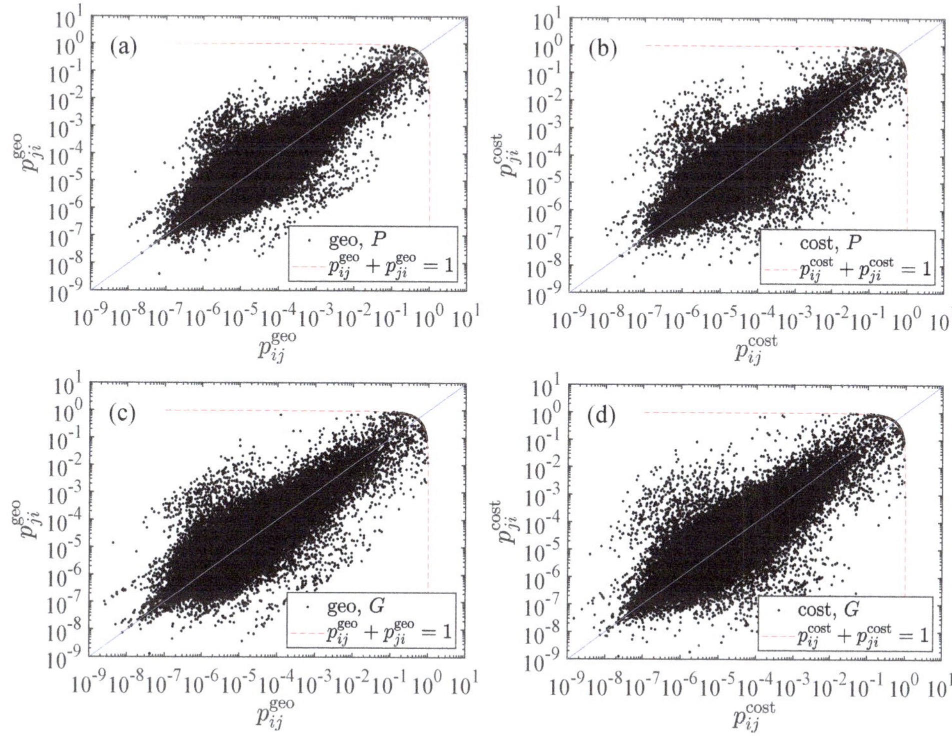

Figure 2. Asymmetric relationship between p_{ij} and p_{ji}. (**a**) Population P is used in the raw radiation model with the geographic distance. (**b**) Population P is used in the cost-based radiation model with the driving distance. (**c**) Gross domestic product G is used in the raw radiation model with the geographic distance. (**d**) Gross domestic product G is used in the cost-based radiation model with the driving distance.

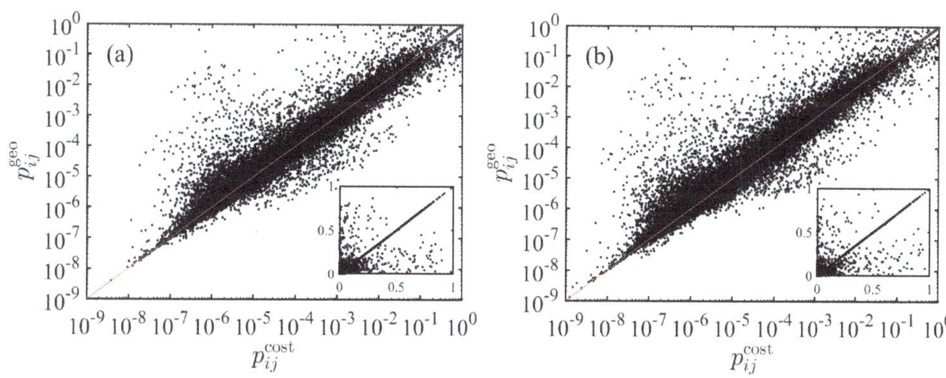

Figure 3. Comparison of the transportation probabilities p_{ij} from the two models based on geographic distance and driving distance. The insets are the same data in linear coordinates. (**a**) The radiation models are based on population. (**b**) The radiation models are based on GDP.

4. Transportation Diversity

We now define the transportation diversity of a city i based on its transportation probability p_{ij} as follows

$$D_i = -\sum_{i \neq j} p_{ij} \ln p_{ij}, \tag{20}$$

where p_{ij} can be calculated from the two radiation models using either population P or gross domestic product G. We calculate four sets of diversity $D_i^{M,d}$, where $M = P$ or $M = G$ and $d = d^{\text{geo}}$ or $d = d^{\text{cost}}$. Indeed, human mobility or communication diversity has been proposed and studied [25–27].

4.1. Comparison of Diversity Based on Population and Gross Domestic Product

In Figure 4, we compare six pairs of any two diversity sets obtained. The two plots in the top row show the influence of distance on diversity for fixed choice of M, while the two plots in the bottom row illustrate the influence of the choice of M on diversity in a given model. We find that, in each plot, there is a nice linear relationship:

$$D_i^{M^{(1)},d^{(1)}} = D_i^{M^{(2)},d^{(2)}}. \tag{21}$$

It is found that the influence is weaker for the choice of model than for the choice of M.

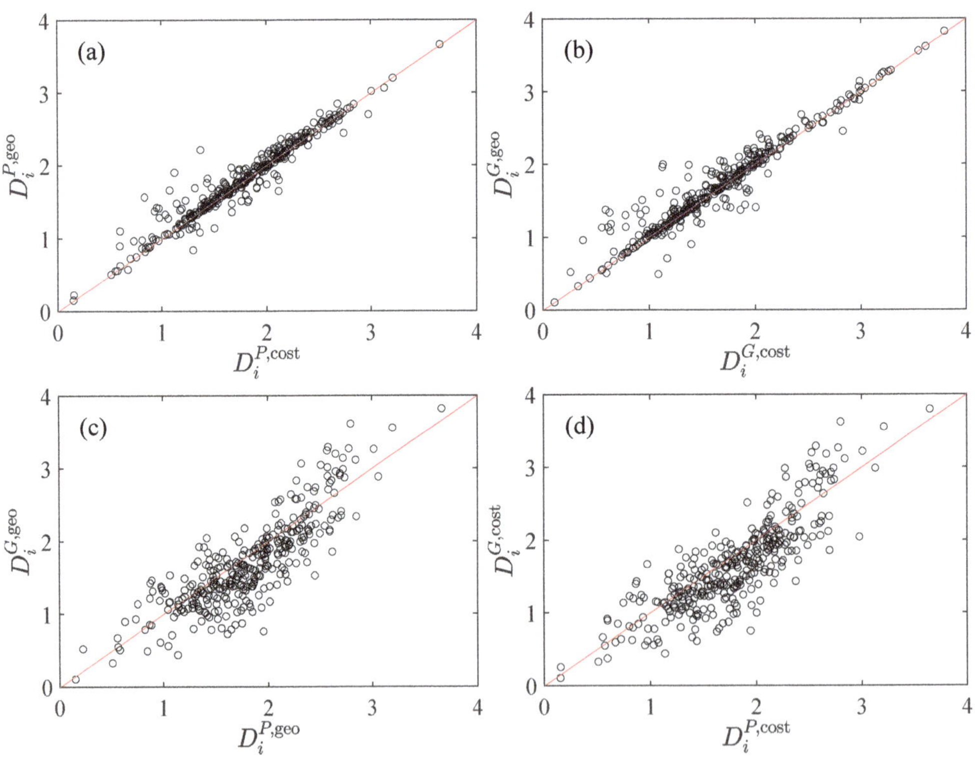

Figure 4. Comparison of the two transportation diversity measures $D_i^{M^{(1)},d^{(1)}}$ and $D_i^{M^{(2)},d^{(2)}}$ calculated using population P and gross domestic product G for the raw radiation model and the cost-based radiation model. (**a**) $M^{(1)} = M^{(2)} = P$, $d^{(1)} = d^{\text{geo}}$ and $d^{(2)} = d^{\text{cost}}$. (**b**) $M^{(1)} = M^{(2)} = G$, $d^{(1)} = d^{\text{geo}}$ and $d^{(2)} = d^{\text{cost}}$. (**c**) $d^{(1)} = d^{(2)} = d^{\text{geo}}$, $M^{(1)} = G$, and $M^{(2)} = P$. (**d**) $d^{(1)} = d^{(2)} = d^{\text{cost}}$, $M^{(1)} = G$, and $M^{(2)} = P$. The solid lines are the diagonal lines.

4.2. Dependence of City Traits on Diversity

We further check the dependence of city traits (P, G, F^{out}, or F^{in}) on the truck transportation diversity D_i, where F_i^{in} is total in-flux arriving at city i

$$F_i^{\text{in}} = \sum_{j \neq i} F_{ji}. \tag{22}$$

The results are depicted in Figure 5. In the four plots of Figure 5e–h for $D_i^{P,\text{cost}}$, we observe two outliers that seem isolated from other points. These outliers correspond to two same cities, Shennongjia Forestry District and Ali District. The diversities of these two cities are respectively 0.1496 and 0.1529.

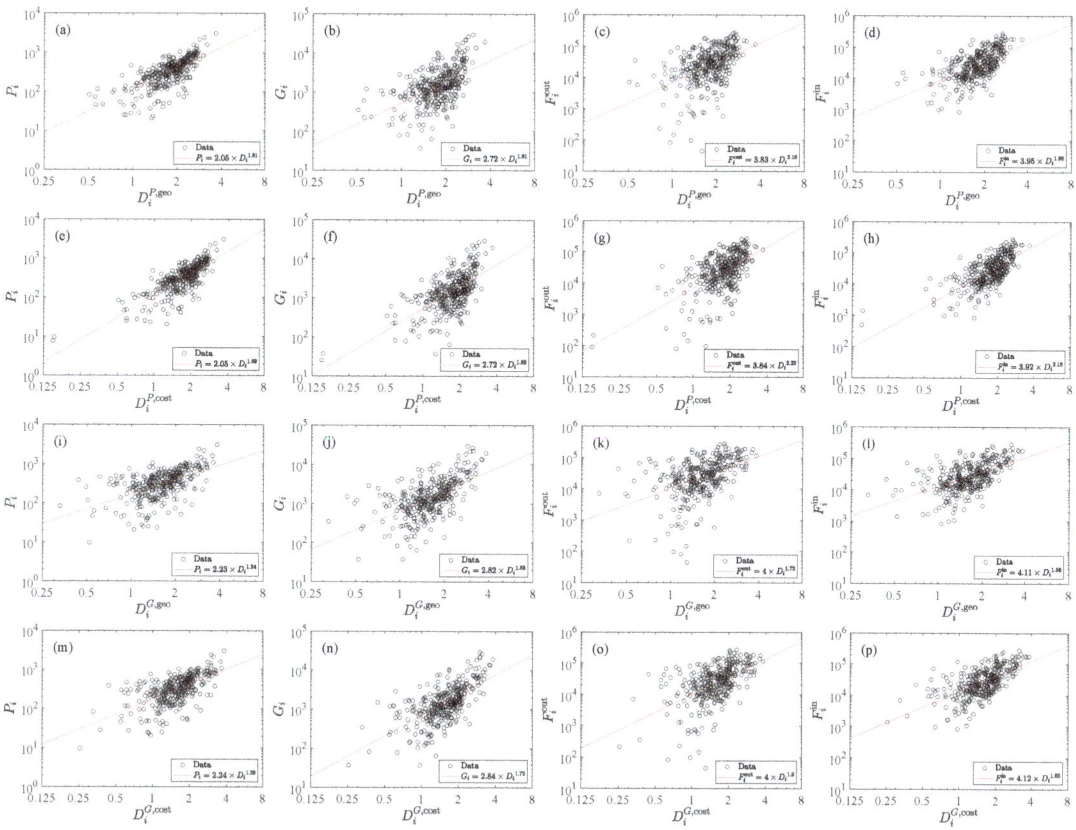

Figure 5. Dependence of city traits (P, G, F^{out}, and F^{in}) on truck transportation diversity ($D^{P,\text{geo}}$). The diversity is calculated from the raw radiation model based on population and geographic distance. The solid lines are power-law fits.

We observe power-law dependence in each plot. We can write that

$$Y_i \sim (D_i^{M,d})^{\beta(Y,M,d)}, \tag{23}$$

where Y represents P, G, F^{out} or F^{in}, M stands for population P or gross domestic product G in the radiation model, and d determines the geographic or driving distance. The power-law exponents $\beta(Y, M, d)$ are estimated with the ordinary least-squares regression, which are presented in Table 1. For a given city trait and the chosen M, the two power-law exponents are similar in the raw radiation model and the cost-based radiation model.

In contrast, the power-law exponent is larger when we use population P as M in the radiation models.

Table 1. Power-law exponents $\beta(Y, M, d)$ for the cost-based radiation model.

Model	$Y = P$	$Y = G$	$Y = F^{\text{out}}$	$Y = F^{\text{in}}$
d^{geo}, P	1.8111	1.8063	2.1558	1.9829
d^{cost}, P	1.8863	1.8890	2.2277	2.1775
d^{geo}, G	1.2384	1.6523	1.7299	1.5613
d^{cost}, G	1.2838	1.7246	1.8990	1.6471

5. Discussion and Conclusions

In this work, we investigated the highway freight transportation diversity of 338 Chinese cities based on the transportation probability p_{ij} from one city to the other. The transportation probabilities are calculated from the raw radiation model based on geographic distance and the cost-based radiation model based on driving distance as the proxy of cost.

We found that, in either the raw radiation model or the cost-based radiation model, the results obtained with the population and the gross domestic product are quantitatively similar. It is mainly due to the nice power-law scaling between population and GDP of Chinese cities, where the power-law scaling exponent is estimated to be 1.15 ± 0.08 [6,28].

We investigated several important properties of the truck transportation probability p_{ij}. It is found that the transportation probabilities are distributed broadly with a nice power-law tail and the tail exponents are close to 0.5 for the four models. It is also found that the transportation probability matrix in each model is asymmetric such that p_{ij} does not necessary equal to p_{ji}, which is consistent with our intuition.

We also found that the population, the gross domestic product, the in-flux, and the out-flux scale as power laws with respect to the transportation diversity in the raw radiation model and the cost-based radiation model. It is intuitive that a city with higher GDP (often with larger population) usually has higher diversity in its industrial structure. These cities usually have higher diversity in highway freight transportation.

The strong correlation between transportation diversity and economic development implies a strong association between industry diversity and economic development. Although a causal direction of this relationship cannot be established through our analysis, transportation diversity at least provides a structural signal for the economic development of a city, highlighting the potential benefit of industry-targeted policies for economic development. Further research is required to obtain reliable policy implications. In particular, longitudinal data sets for transportation networks and economic development are required to establish a possible causal relationship.

Author Contributions: Funding acquisition, W.-X.Z.; Investigation, L.W., J.-C.M., Z.-Q.J. and W.Y.; Methodology, L.W. and W.-X.Z.; Supervision, W.-X.Z.; Writing—original draft, L.W. and W.-X.Z.; Writing—review & editing, L.W., J.-C.M., Z.-Q.J., W.Y. and W.-X.Z. All authors have read and agreed to the published version of the manuscript.

Funding: This work was partly supported by the Fundamental Research Funds for the Central Universities.

Institutional Review Board Statement: Not applicable.

Informed Consent Statement: Not applicable.

Data Availability Statement: We signed a confidentiality agreement with the transportation company who provided us the data used in this work. Hence the data will not be shared.

Conflicts of Interest: The authors declare no conflict of interest.

References

1. Belokurov, V.; Spodarev, R.; Belokurov, S. Determining passenger traffic as important factor in urban public transport system. *Transp. Res. Proc.* **2020**, *50*, 52–58. [CrossRef]
2. Albayrak, M.B.K.; Ozcan, I.C.; Can, R.; Dobruszkes, F. The determinants of air passenger traffic at Turkish airports. *J. Air Transp. Manag.* **2020**, *86*, 101818. [CrossRef]
3. Pasha, J.; Dulebenets, M.A.; Kavoosi, M.; Abioye, O.F.; Theophilus, O.; Wang, H.; Kampmann, R.; Guo, W. Holistic tactical-level planning in liner shipping: An exact optimization approach. *J. Ship. Trade* **2020**, *5*, 8. [CrossRef]
4. Enoch, M.P.; Cross, R.; Potter, N.; Davidson, C.; Taylor, S.; Brown, R.; Huang, H.; Parsons, J.; Tucker, S.; Wynne, E.; et al. Future local passenger transport system scenarios and implications for policy and practice. *Transp. Policy* **2020**, *90*, 52–67. [CrossRef]
5. Dulebenets, M.A. An adaptive island evolutionary algorithm for the berth scheduling problem. *Memet. Comput.* **2020**, *12*, 51–72. [CrossRef]
6. Wang, L.; Ma, J.C.; Jiang, Z.Q.; Yan, W.; Zhou, W.X. Gravity law in the Chinese highway freight transportation networks. *EPJ Data Sci.* **2019**, *8*, 37. [CrossRef]
7. Jung, W.S.; Wang, F.Z.; Stanley, H.E. Gravity model in the Korean highway. *EPL (Europhys. Lett.)* **2008**, *81*, 48005. [CrossRef]
8. Masucci, A.P.; Serras, J.; Johansson, A.; Batty, M. Gravity versus radiation models: On the importance of scale and heterogeneity in commuting flows. *Phys. Rev. E* **2013**, *88*, 022812. [CrossRef]
9. Lenormand, M.; Bassolas, A.; Ramasco, J.J. Systematic comparison of trip distribution laws and models. *J. Transp. Geogr.* **2016**, *51*, 158–169. [CrossRef]
10. Barbosa, H.; Barthelemy, M.; Ghoshal, G.; James, C.R.; Lenormand, M.; Louail, T.; Menezes, R.; Ramasco, J.J.; Simini, F.; Tomasini, M. Human mobility: Models and applications. *Phys. Rep.* **2018**, *734*, 1–74. [CrossRef]
11. Piovani, D.; Arcaute, E.; Uchoa, G.; Wilson, A.; Batty, M. Measuring accessibility using gravity and radiation models. *R. Soc. Open Sci.* **2018**, *5*, 171668. [CrossRef] [PubMed]
12. Kwon, O.; Jung, W.S. Intercity express bus flow in Korea and its network analysis. *Phys. A* **2012**, *391*, 4261–4265. [CrossRef]
13. Hong, I.; Jung, W.S. Application of gravity model on the Korean urban bus network. *Phys. A* **2016**, *462*, 48–55. [CrossRef]
14. Simini, F.; González, M.C.; Maritan, A.; Barabási, A.L. A universal model for mobility and migration patterns. *Nature* **2012**, *484*, 96–100. [CrossRef] [PubMed]
15. Ren, Y.; Ercsey-Ravasz, M.; Wang, P.; Gonzalez, M.C.; Toroczkai, Z. Predicting commuter flows in spatial networks using a radiation model based on temporal ranges. *Nat. Commun.* **2014**, *5*, 5347. [CrossRef] [PubMed]
16. Yang, Y.; Herrera, C.; Eagle, N.; Gonzalez, M.C. Limits of predictability in commuting flows in the absence of data for calibration. *Sci. Rep.* **2014**, *4*, 5662. [CrossRef]
17. Yan, X.Y.; Wang, W.X.; Gao, Z.Y.; Lai, Y.C. Universal model of individual and population mobility on diverse spatial scales. *Nat. Commun.* **2017**, *8*, 1639. [CrossRef]
18. Lenormand, M.; Huet, S.; Gargiulo, F.; Deffuant, G. A universal model of commuting networks. *PLoS ONE* **2012**, *7*, e45985. [CrossRef]
19. Gargiulo, F.; Lenormand, M.; Huet, S.; Espinosa, O.B. Commuting network models: Getting the essentials. *J. Artif. Soc. Soc. Simul.* **2012**, *15*, 6. [CrossRef]
20. Lenormand, M.; Huet, S.; Gargiulo, F. Generating French virtual commuting networks at the municipality level. *J. Transp. Land Use* **2014**, *7*, 43–55. [CrossRef]
21. Xia, N.; Cheng, L.; Chen, S.; Wei, X.; Zong, W.; Li, M. Accessibility based on gravity-radiation model and Google Maps API: A case study in Australia. *J. Transp. Geogr.* **2018**, *72*, 178–190. [CrossRef]
22. Serrano, M.A.; Boguñá, M. Topology of the world trade web. *Phys. Rev. E* **2003**, *68*, 015101(R). [CrossRef] [PubMed]
23. Garlaschelli, D.; Loffredo, M. Patterns of link reciprocity in directed networks. *Phys. Rev. Lett.* **2004**, *93*, 268701. [CrossRef]
24. Zhou, W.X.; Wang, L.; Xie, W.J.; Yan, W. Predicting highway freight transportation networks using radiation models. *Phys. Rev. E* **2020**, *102*, 052314. [CrossRef]
25. Eagle, N.; Macy, M.; Claxton, R. Network diversity and economic development. *Science* **2010**, *328*, 1029–1031. [CrossRef]
26. Yan, X.Y.; Han, X.P.; Wang, B.H.; Zhou, T. Diversity of individual mobility patterns and emergence of aggregated scaling laws. *Sci. Rep.* **2013**, *3*, 2678. [CrossRef] [PubMed]
27. Liu, E.J.; Yan, X.Y. A universal opportunity model for human mobility. *Sci. Rep.* **2020**, *10*, 4657. [CrossRef] [PubMed]
28. Bettencourt, L.M.A.; Lobo, J.; Helbing, D.; Kuhnert, C.; West, G.B. Growth, innovation, scaling, and the pace of life in cities. *Proc. Natl. Acad. Sci. USA* **2007**, *104*, 7301–7306. [CrossRef] [PubMed]